Transgenic Crops of the World

Transgenic Crops of the World

Essential Protocols

Edited by

Ian S. Curtis

National Institute of Agrobiological Sciences (NIAS),
Department of Biotechnology, Ibaraki, Japan

KLUWER ACADEMIC PUBLISHERS

DORDRECHT / BOSTON / LONDON

A C.I.P. Catalogue record for this book is available from the Library of Congress.

ISBN 1-4020-2332-4 (HB)
ISBN 1-4020-2333-2 (e-book)

Published by Kluwer Academic Publishers,
P.O. Box 17, 3300 AA Dordrecht, The Netherlands.

Sold and distributed in North, Central and South America
by Kluwer Academic Publishers,
101 Philip Drive, Norwell, MA 02061, U.S.A.

In all other countries, sold and distributed
by Kluwer Academic Publishers,
P.O. Box 322, 3300 AH Dordrecht, The Netherlands.

Dedicated to my parents

Gordon and Brenda Curtis

For bringing me into a world of many challenges

If the trees flower, let us hope they bear fruit...

IAN S. CURTIS

B.Sc. Hons. (Wales), P.G.Dip. (Leic.), M.Sc. (Newc.), Ph.D. (Notts.).
National Institute of Agrobiological Sciences (NIAS), Department of Biotechnology,
Kannondai 2-1-2, Tsukuba, Ibaraki 305-8602, Japan

Dr. Ian Scott Curtis is a Visiting Associate Professor at the National Institute of Agrobiological Sciences (NIAS), Japan. In the early 1990s, he developed the genotype-independent transformation system for lettuce under the supervision of Drs. MR Davey and JB Power (University of Nottingham, UK). In 2001, he successfully modified the floral-dip method towards the production of agronomically useful transgenic radish in the laboratory of Dr HG Nam (POSTECH, South Korea).

TABLE OF CONTENTS

viii

Preface

Since the first transgenic plants were produced back in the early 1980s, there have been substantial developments towards the genetic engineering of most crops of our world. Initial studies using isolated plant cells and removing their cell walls to form protoplasts, offered the possibility of transferring genetic material by *Agrobacterium*-mediated gene transfer, chemical agents or electrical charges. However, in those cases were isolated protoplasts could be transformed, often, a shoot regeneration system was not available to induce the production of transgenic plants and any such regenerated plants were subject to mutation or chromosomal abnormalities. By the mid-1980s, the use of cultured plant organs, such as leaf disks, offered the convenience of combining gene transfer, plant regeneration and selection of transformants in a single system. This approach, enabled the production of stable, phenotypically-normal, transgenic potato and tomato plants in culture. By the late 1980s, the use of biolistics offered a means of inserting foreign genes into plant cells which where inaccessible to *Agrobacterium* infection. Even today, this technology is now standard practice for the production of some transgenic plants. Extensive research in the improvement of tissue culture conditions, development of new *Agrobacterium* strains and vectors, novel selectable marker genes and evaluations of various plant organs, tissues and cells towards shoot regeneration capability, in association with gene transfer technology enabled the more recalcitrant plants to be transformed. For example, the identification that the scutellum tissue of rice was both amenable to *Agrobacterium* infection and regenerable in culture from embryogenic calli, enabled the production of stably, transgenic rice plants by the mid-1990s. This discovery formed a platform towards the acceleration in gene transfer towards several major cereals. By the turn of the 21st Century, the use of *in planta* transformation strategies, such as floral-dipping, allowed the first transgenic pakchoi and radish plants to be produced without the necessity of tissue culture. To date, there remains considerable research to be undertaken to be able to routinely transform all the major crops. Indeed, many of the procedures described in the literature are often genotype-dependent. For this reason, a manual has been constructed which offers ideas and sometimes solutions towards extending the possibility of genetically modifying our desired germplasms. unhealth

The need to produce crops with improved agronomic traits is paramount in a world where more than 700 million people are undernourished and within the next 20 years the world's population will probably double. Transgenic plant technology offers one potential opportunity to improve crop production to help circumvent the need in the Third World. Recent developments in gene discovery and functional genomics should allow 'designer' crops to be grown in the field. In addition, such technology has great potential in facilitating novel approaches in genetics, biochemistry and physiology. Our food is produced from many countries of the globe. For this reason, *Transgenic Crops of the World – Essential Protocols*, contains contributions from many experts in the field of plant transformation across the world, describing a vast array of state-of-the-art methods in the production of genetically modified plants. Some chapters are review-orientated, the majority, gives detailed current and most versatile procedures known in the literature. Due to the lack of a single, genotype-independent system for most crops, in some chapters, we offer possibilities on how specified protocols could be modified to circumvent the transformation of specific genotypes. Such a designed manual should facilitate the production of transgenics in many laboratories involved in fundamental and applied biology. Finally, in the wake of environmental concerns over transgene dispersal via pollen, the use of antibiotic-free markers and chloroplast transformation systems are also described.

I would like to thank all contributing authors in preparing this excellent book and the people of Kluwer Academic Publishers, especially Noeline Gibson, for helping to assemble this essential practical manual.

Ian S. Curtis

Part I: Cereals and Grasses

Chapter 1

TRANSGENIC RICE PLANTS

M. ASHIKARI, M. MATSUOKA AND S.K. DATTA[1]

Bioscience and Biotechnology Center, Nagoya University, Japan
[1]International Rice Research Institute, Philippines
Corresponding author: S. Datta. E-mail: S.Datta@CGIAR.ORG

1. INTRODUCTION

1.1. Importance of Rice

Rice (*Oryza sativa* L.) is a hugely important food crop. It has been estimated that 50% of the human population depends on rice as its main source of nutrition. It is the most important crop for people living in the tropical areas and to some extent in the subtropical areas of Asia, where it has a long history of cultivation. Rice is deeply ingrained in the daily lives of Asians.

Several important crops worldwide such as rice, wheat, barley, and maize belong to the cereals (*Gramineae*), and they are derived from the same ancestral plant. Therefore, the rice genome shows apparent synteny (a close genome relationship) with many other important cereals (*1, 2*). Rice has the smallest genome (430 Mbp) relative to other important cereal plants; for instance, its genome size is estimated to be about 1/10 of that of maize and 1/40 of that of wheat, and it is just 3 times that of *Arabidopsis*, a model plant of dicotyledons (dicots). The compact nature of the rice genome thus provides a distinct advantage in gene isolation and genomic sequencing as opposed to other cereal crops. Because of its small genome size and genome synteny with other cereals, rice is established as a model plant of monocotyledons (monocots).

During the last decade, technological innovations in science enabled dramatic advances in the field of plant genomics (genome research). The Rice Genome Research Program (RGP) in Japan was launched in 1991 and it provides very useful information for plant biology and plant breeding (*3*). Subsequently, in 1997, the International Rice Genome Sequencing Project (IRGSP) started (*4*) its work on complete gene sequencing of rice. In 2002, the IRGSP announced the partial draft sequence and released the data (*5, 6*). In the same year, the Beijing Genomics Institute (BGI) and Syngenta also published the draft rice genome sequence (*7, 8*). Rice became the first cereal being successfully sequenced. These efforts contribute not only to basic biology research but also to applied molecular rice breeding.

1.2. Need for Genetic Improvement

The world population is expanding rapidly and it is predicted that it will surpass 8 billion people by 2025, with the greatest increase occurring in developing countries. In addition, the availability of land for cultivation has declined dramatically as a result

3

I.S. Curtis (ed.), Transgenic Crops of the World - Essential Protocols, 3-18.
© 2004 Kluwer Academic Publishers. Printed in the Netherlands.

of desertification caused by reckless deforestation and construction. For these reasons, the rate of world population growth has exceeded the rate of growth in food-grain production. We cannot now avert an impending food crisis in the near future. Indeed, many people living in Asia and Africa already suffer from malnutrition.

Providing food for the ever-growing population now requires an expansion in world grain production by 26 million tonnes per year. In total, world food-grain production must increase by another 40% (9). To feed a world population of 8 billion by 2025, we need to act now and develop scientific breeding to meet this challenge.

Genetic engineering and biotechnology hold great potential for plant breeding as they can expedite the time required to produce new crop varieties with desirable characteristics. Because rice breeding by traditional selection methods may not be able to keep up with the increasing global food demand, other non-conventional strategies are required. One such strategy is genetic engineering. Among the cereals, rice can be used efficiently for genetic transformation. Improving rice cultivars through genetic engineering/transformation is one of the most important strategies for overcoming food shortages.

The global climate is changing and possible global warming is of particular concern since it may have a negative effect on world food security. Because rice plants can be cultivated under a wide range of environments, from arid highlands to flooded lowlands and under high humidity or high temperature, rice may be suitable for producing reasonable yield under predicted warmer conditions. Rice genomics research provides applications for plant biology and breeding, and its possible role in overcoming problems associated with climatic changes and food production should be increased (*10*).

1.3. Genetic Improvement of Rice

To date, a large number of rice mutants have been reported and most of them have been produced artificially by using physical mutagens such as irradiation, chemical compounds, and DNA insertion, including T-DNA (*11*), *Ac/Ds* transposons (*12*), and retrotransposons, such as *Tos17* (*13*). These mutants are significant materials for functional genomics, and can be used for rice molecular breeding.

The Rice Genome Research Program has been very helpful since a large number of DNA markers are already developed and a high-resolution linkage map and physical map are being constructed. Finally, the draft genome sequences of rice became open to the public. These genomic tools, materials, and sequencing data will accelerate the cloning of genes, related to important agricultural traits. Recently, besides the major genes, QTLs have also been recognized to be associated with important agricultural traits, such as photoperiod (*14-17*). Once such genes become available, they will provide us with opportunities to improve rice production quantitatively and qualitatively by genetic engineering.

Genetic engineering gives us possibilities to improve our well-being and overcome the impending food crisis. Transgenic plants may have improved traits such as increased grain production, better nutrition in the grain, and tolerance of different stresses, including drought, cold, and disease. Therefore, rice could be planted on adverse land, which is not possible using current germplasms.

Now, 124 million children worldwide suffer from vitamin-A deficiency (VAD) which causes malfunctioning of the immune system. Night blindness is a major symptom of VAD. Through genetic engineering, a model japonica rice cultivar T309, was developed with biosynthesis of beta-carotene in transgenic rice seeds (*18*). Indica rice feeds more than 2 billion people and a recent report highlighted the development of a golden indica rice. Because of its success in producing provitamin A, it could produce a greater impact in rice breeding and in farmers' fields (*19*). Bioengineered high-iron rice has also been reported, and it could help in reducing problems of anemia in Asia and Africa (*20*).

1.4. Genetic Transformation of Rice

Rice transformation can be achieved by several methods depending on the tissue culture ability of the genotypes and suitability of the method for a particular genotype. Transformation methods that have been used include microinjection, macroinjection, laser beam techniques, pollen-tube pathway, dry seed imbibition, and cell and tissue electroporation. However, the most reproducible and unambiguous results have been obtained from protoplast, biolistic, and *Agrobacterium* methods (*21*). In terms of the genotypes used, japonica rice cultivars are generally responsive to tissue culture and work very well with known transformation systems. Taipei 309 was initially used as a model for rice transformation, followed by Nipponbare, Yamabiko, Yamahoushi, Norin 14, Zhounghus 6, etc., and they all belong to japonica rice (*22-25*). Indica-type rice is more difficult to transform; thus, it took a little longer to establish a routine transformation system. The first report of genetic engineering of indica rice came in 1990 from Datta's group using anther culture–derived embryogenic suspension culture, and homozygous plants from transformed protoplasts were recovered (*26*). It is now possible to transform many indica cultivars, such as IR64, IR72, IR58, IR54, IR50, IR68144, IR68899B, IR58025B, CBII, Dinorado, Limpopo, Basmati 370, BPT5204, and Mediterranean rice, using protoplast, *Agrobacterium,* and biolistic transformation systems (*27*). Many agronomically important genes have now been transferred into indica rice and some products are being evaluated in the field (*28, 29*).

For the protoplast system, the single-cell origin, non-chimeric nature, and genetic fidelity of plants derived from somatic embryogenesis are very attractive features of this kind of transformation method. Early success was based on the following (*26*):

- a) Establishment of embryogenic cell suspension, the key source of regenerable totipotent protoplasts.
- b) The use of a suitable plant transformation vector with a selectable marker gene(s).
- c) Continuous efforts to improve tissue culture protocols by different laboratories, for example, using maltose instead of sucrose, 2,4-D, nurse culture, and osmotic adjustments, etc.

The biolistic system is now routinely used for several plants, including rice. Microprojectile bombardment employs high-velocity metal particles to deliver biologically active DNA into plant cells and whole plants can be recovered from transformed cells through selection. Genotype-independent transformation can be carried out with this system, as there is no biological limitation to DNA delivery. Japonica and indica rice cultivars have been transformed with this system (*25, 30*).

Agrobactrium-mediated-transformation has been extensively used for both dicot and monocot species, including indica rice (*21, 27, 31*). In this report, a detailed protocol using this method carried out in our laboratory is introduced. All 3 systems of transformation described above work well for both japonica and indica rice (*21*).

2. MATERIALS

2.1. Target Gene Preparation and Bacterial Strain

The target gene is first introduced into the binary vector using standard molecular biology techniques. Plasmid DNA is introduced into *Agrobacterium* strain EHA101 or LBA4404 (*21, 31*) by electroporation (*see* Note 1). The *A. tumefaciens* is grown on AB medium (Table 1) at 28°C. Several new binary vectors suitable for rice transformation have been reported (*21*).

2.2. Plant Materials

Immature embryos and immature embryo-derived primary and secondary calli have been reported as suitable explants for DNA delivery. Immature embryos are suitable for callus induction, particularly if the materials are available throughout the year, as in the case at IRRI. Donor plants grown under a light intensity of 185 µmol/m^2/sec at 27-31°C (day) and 21-24°C (night) are most suitable for obtaining reproducible results.

2.3. Culture Media

The culture media and stock solutions are stored at 4°C for at least 3-12 months. Vitamins/hormone solutions are stored for 1-2 months at the same temperature. Most of the chemicals are purchased from Sigma or mentioned otherwise.

Table 1. AB medium

Component	Amount (g/500 ml)
Stock I	
K_2HPO_4	30
NaH_2PO_4	10
Stock II	
NH_4Cl	10
$MgSO_4 \bullet 7H_2O$	3
KCl	1.5
$CaCl_2$	0.1
$FeSO_4 \bullet 7H_2O$	0.025
Stock III	900/ml
Glucose	5
Agar	15

Autoclave the 3 stock solutions separately (*see* Note 2).
Add 25 ml each of stocks I and II to 450 ml of stock III.
Add appropriate antibiotics as required.

Table 2. AAM medium

Component	Amount (g/500 ml)
Macronutrients (AA medium) I	
$CaCl_2 \cdot 2H_2O$	150
$MgSO_4 \cdot 7H_2O$	250
$NaH_2PO_4 \cdot H_2O$	150
KCl	2950
Micronutrients	
KI	0.75
H_3BO_3	3
$MnSO_4 \cdot H_2O$	10
$ZnSO_4 \cdot 7H_2O$	2
$Na_2MoO_4 \cdot 2H_2O$	0.25
$CuSO_4 \cdot 5H_2O$	0.025
$CoCl_2 \cdot 6H_2O$	0.025
Iron composition (MS medium)	
Na_2 EDTA	37.3
$FeSO_4 \cdot 7H_2O$	27.8
Vitamins (MS medium)	
Nicotinic acid	0.5
Pyridoxine-HCl	0.5
Thiamine-HCl	1
Glycine	2
Myo-inositol	100
Others	
L-glutamine	876
Aspartic acid	266
Arginine	174
Casamino acid	500
Sucrose	34.3
Glucose	18
Acetosyringone	200 μM

pH 5.2
Mix components separately in water before adjusting pH. Filter-sterilize.

Table 3. Modified MS medium for callus induction and proliferation

Component	Quantity (mg/l)
NH_4NO_3	1650
KNO_3	1900
$CaCl_2 \bullet 2H_2O$	440
$MgSO_4 \bullet 7H_2O$	370
KH_2PO_4	170
KI	0.83
H_3BO_3	6.3
$MnSO_4 \bullet 4H_2O$	22.3
$ZnSO_4 \bullet 7H_2O$	8.6
$Na_2MoO_4 \bullet 2H_2O$	0.25
$CuSO_4 \bullet 5H_2O$	0.025
$CoCl_2 \bullet 6H_2O$	0.025
$FeSO_4 \bullet 7H_2O$	27.8
Na_2 EDTA	37.3
Nicotinic acid	0.5
Pyridoxine-HCl	0.5
Thiamine-HCl	1
Glycine	2
Casein hydrolysate	300
Myo-inositol	100
2,4-D	2
Sucrose or maltose	30 g/l
Agar	8 g/l

pH 5.8; sterilize by autoclaving.

Table 4. Composition of media for growth of transformants

Component	[a]N6 medium (mg/l) (*32*)	Hormone-free MS medium (mg/l) (*33*)	Regeneration medium MSNK (mg/l)
$(NH_4)_2SO_4$	463	-	-
KH_2PO_4	400	170	170
KNO_3	2830	1900	1900
NH_4NO_3	-	1650	1650
$CaCl_2 \cdot 2H_2O$	166	440	440
$MgSO_4 \cdot 7H_2O$	185	370	370
Na_2EDTA	37.3	37.3	37.3
$FeSO_4 \cdot 7H_2O$	27.8	27.8	27.8
$MnSO_4 \cdot 4H_2O$	4.4	22.3	22.3
H_3BO_3	1.6	6.3	6.3
$ZnSO_4 \cdot H_2O$	1.5	8.6	8.6
KI	0.8	0.83	0.83
$CoCl_2 \cdot 6H_2O$	-	0.025	0.025
$CuSO_4 \cdot 5H_2O$	-	0.025	0.025
$Na_2MoO_4 \cdot 2H_2O$	-	0.25	0.25
Thiamine-HCl	1	1	1
Nicotinic acid	0.5	0.5	0.5
Pyridoxine-HCl	0.5	0.5	0.5
Glycine	2	2	2
Myo-inositol	100	100	100
Kinetin	-	-	2
NAA	-	-	1
2,4-D	2	-	-
Casamino acids	300	300	300
Cefotaxime	-	-	-
Carbenicillin	-	-	250
Hygromycin	-	-	50
Glucose	10	-	-
Maltose (g/l)	-	30	-
Sucrose (g/l)	30	-	30
Sorbitol (g/l)	-	-	10
Agar (g/l)	8	7	-
Agarose(g/l) (Fischer)	-	-	6[b]
Gelrite (g/l)	-	-	2.3[b]
Acetosyringone	200 μM	-	-
	pH 5.2	pH 5.8	pH 5.8

[a]Modified N6 medium is supplemented with 100 mg/l cefotaxime, 250 mg/l carbenicillin and 50 mg/l hygromycin.
[b]Either gelling agent can be used.

3. METHODS

Here, we report the *Agrobacterium*-mediated transformation and plant regeneration of rice (*see* Note 3). *Agrobacterium* is grown in liquid AAM medium (Table 2) and maintained on semi-solidified AB medium (Table 1). A detailed report has been published elsewhere (*21*).

3.1. Callus Induction

1. Remove the hull of rice seeds with a Satake machine (SATAKE Corporation, Japan; Fig. 1A).
2. Place approx. 100 seeds into a 50 ml disposable tube containing sterile water and shake briefly (Fig. 1B). Remove the water, then immerse seeds in 70% ethanol and shake manually for 2-3 min. Remove the ethanol, and add 1.8% sodium hypochlorite solution and sterilize the seeds on a shaker (60 rpm) for 30 min. Discard the bleach solution and wash the seeds with sterile water 3 times.
3. Transfer the seeds to MS modified medium (Table 3) or N6 medium (Table 4; 100 seeds/9 cm diam. Petri dish; 20 ml medium/dish) (Fig. 1C). In the case of immature seeds, the embryos are isolated and cultured on the same medium. Incubate at 30°C in the dark for 3 weeks (Fig. 1D).
4. After 2-3 weeks, callus formation is visible on the surface of the scutellum (Fig. 1E).

3.2. Pre-culture of Calli and Pre-treatment of **Agrobacterium** *Strain*

1. Transfer 3 week-old calli (size approx. 1.5-3 mm in diam.) to fresh N6 medium (approx. 200 calli/dish) and incubate at 30°C in the dark for 3 days (Fig. 1F).
2. Inoculate *Agrobacterium* strain EHA101 on AB medium (Table 1) and grow for 3 days at 28°C under dark conditions (Fig. 1G).

3.3. Infection

1. Add 30 ml of AAM medium (Table 2) with 30 µl of 200 µM acetosyringone in a 50 ml disposable tube.
2. Take one loopful of *Agrobacterium* culture from AB medium and place into AAM medium (Fig. 1H) and mix well (Fig. 1I).
3. Transfer calli to a nylon mesh (500 µm) (Fig. 1J).
4. Transfer buffer onto the calli (Fig. 1K).
5. Submerge calli in the infection medium and mix well for 2 min by hand (Fig. 1L).
6. Absorb the infection medium from the callus using a sterile paper towel (Fig. 1M).
7. Transfer calli to N6 medium (Table 4) and incubate the plates for 3 days at 28°C under dark conditions (Fig. 1N).

3.4. *Removing the* Agrobacterium *from Calli*

1. Add sterile water to a 50 ml disposable tube.
2. Transfer the calli from the medium to the above tube (Fig. 1O). Shake well by hand and change water 5 times.
3. Wash calli with carbenicillin buffer (30 ml sterile water with 60 µl carbenicillin (250 mg/ml stock solution) to remove *Agrobacterium.*
4. Transfer the calli to modified N6D medium (Table 4) containing hygromycin and carbenicillin (N6 medium with 1ml hygromycin (50 mg/ml stock) and 1 ml carbenicillin (250 mg/ml) and 1 ml cefotaxime (100 mg/ml) and incubate at 30°C in dark conditions (Fig. 1P).

3.5. *Plant Regeneration*

1. Transfer the hygromycin resistant calli to the MS-NK medium (Table 4) containing hygromycin and carbenicillin as before (*see* Section 3.4.4.) and 1 ml of NAA (0.2 mg/ml stock) and 20 ml kinetin (0.1 mg/ml stock) per litre of medium. The efficiency of shoot regeneration varies considerably between genotypes (*see* Note 4).
2. Incubate cultures at 30°C under continuous light conditions for japonica cultivars and a 16 h photoperiod for indica cultivars at 74 µmol/m^2/sec (*see* Note 5).
3. Regenerated shoots are usually observed within 3-4 weeks (Fig. 1Q).
4. Transfer shoots to hormone-free MS medium (Table 4) and culture for 2-3 weeks with/without the presence of hygromycin. Shoots start to develop roots between 7-10 days (Fig. 1R).

3.6. *Transfer of Plants to Soil in the Glasshouse*

1. Well rooted plants are transferred to normal paddy soil in a pot (2 litre-capacity, plastic) and watered every morning and with usual fertilizer applied in rice field (*21*).

4. NOTES

1. *A. tumefaciens* strains EHA101 and LBA4404 have been extensively used in our laboratory and elsewhere for both japonica and indica rice transformation. Other strains also work with relatively less efficiency.
2. Stock solutions are made separately to improve solubility and to avoid precipitation.
3. Plant regeneration varies considerably because indica cultivars are more recalcitrant in tissue culture response than japonica rice. A few japonica cultivars (e.g. T309, Nipponbare) and indica cultivars (Tetep, CBII, IR68899B) respond well for plant transformation production with a frequency between 10-80% for japonica and 0.1-10% for indica cultivars. Calculations are based on generation of transgenic lines from primary explants.
4. Presence of hygromycin and carbenicillin in MSNK medium are preferable but not essential.
5. Light intensity at 74 µmol/m^2/sec is suitable for both types of cultivars. However, most japonica type rice cultivars respond more favourably under continuous light conditions.

6. The hpt gene is reliable for rice transformation along with the pmi gene (19). However, marker-free transgenic rice is now achievable along with a gene of interest and perhaps will be used for future plant breeding (34).

EXAMPLES OF MOLECULAR BREEDING

Controlling Plant Height

Plant hormones are associated with several plant growth and development processes. The gibberellin hormones are involved in many aspects of development throughout the life-cycle of plants, one of which is stem elongation. We have attempted to change plant height by controlling the expression of a GA-biosynthetic gene, GA3 oxidase. In the GA biosynthetic pathway, GA3 oxidase catalyzes the reactions from inactive GA_9 to bioactive GA_4 and inactive GA_{20} to bioactive GA_1. For this, we employed reducing GA3 oxidase gene expression by constructing an antisense GA3 oxidase gene to decrease the production of active GAs. Transgenic plants carrying an antisense construct of the GA3 oxidase gene gave a semi-dwarf phenotype and we have successfully produced dominant semi-dwarf rice plants (Fig. 2A, *35*). Because GAs are usually present in plants and are used to control plant growth and development, these data suggest that it may be possible to control plant height of other crops by modifying the genes controlling GA synthesis.

Improving Rice Plant Architecture

The erect-leaf phenotype is a desirable agricultural trait for high yield in rice because plants with erect leaves can be densely planted in the field and can also receive sunlight more efficiently. Brassinosteroids (BRs) are a group of plant hormones that are found at low levels in pollen, seeds, and young vegetative tissues throughout the plant kingdom. BRs share structural similarity with animal steroid hormones, and have been shown to regulate gene expression and stimulate cell division and differentiation. They also control leaf angle in rice. To produce an erect-leaf rice plant, we employed the dominant negative strategy to reduce the function of the BR receptor gene. The transgenic plants showed erect leaves without dwarfism (Fig. 2B).

Developing of Indica Golden Rice containing Beta-carotene in Rice Endosperm

Vitamin-A deficiency is a major malnutrition problem in South Asia and developing countries worldwide. Genes for beta-carotene biosynthesis (*psy, crtI*) driven by endosperm-specific promoters (glutelin promoter, Gt-1) were introduced into indica cultivars (IR64, BR29, Mot Bui, NHCD, etc.) with *hpt* and *pmi* selection system that were well adapted to different agroclimatic zones of Asia (*see* Note 6; Fig. 3A). The yellow colour of the polished rice grain evidenced the carotenoid accumulation in the rice grain (Fig. 3B). HPLC analysis further confirmed the levels of beta-carotene biosynthesis in uncooked as well as cooked polished rice seeds (*19*). Marker free Golden indica rice has also been developed.

CONCLUDING REMARKS

Today, gene discovery and the functionality of gene(s) through the production of transgenic plants are commonly used in plant science. Such technologies are very powerful and efficient for producing "ideal" crop plants. Transgenic crop plants, commonly referred to as genetically modified organisms (GMOs), with traits such as insect resistance and herbicide tolerance, have already been commercialized after passing stringent field trials and food-safety analysis. We believe that, in the near future, the use of GMOs in everyday life will be a common feature and this technology can help to attain food security as a complementary tool of plant breeding.

REFERENCES

1.   Moore G, Devos KM, Wang Z and Gale MD (1995). Grasses, line up and form a circle. *Current Biology,* **5**: 737-739.
2.   Gale MD and Devos KM (1998). Comparative genetics in the grasses. *Proceedings of the National Academy of Sciences USA,* **95**: 1971-1974.
3.   Sasaki T (1998). The rice genome project in Japan. *Proceedings of the National Academy of Sciences USA,* **95**: 2027-2028.
4.   Sasaki T and Burr B (2000). International Rice Genome Sequencing Project: the effort to completely sequence the rice genome. *Current Opinion of Plant Biology,* **3**: 138-141.
5.   Sasaki T, Matsumoto T, Yamamoto K, Sakata K, Baba T, Katayose Y, Wu J, Niimura Y *et al.* (2002). The genome sequence and structure of rice chromosome 1. *Nature,* **420**: 312-316.
6.   Feng Q, Zhang Y, Hao P, Wang S, Fu G, Huang Y, Li Y, Zhu J *et al.* (2002). Sequence and analysis of rice chromosome 4. *Nature,* **420**: 316-320.
7.   Yu J, Hu S, Wang J, Wong GK, Li S, Liu B, Deng Y, Dai L *et al.* (2002). A draft sequence of the rice genome (*Oryza sativa* L. ssp. *indica*). *Science,* **296**: 79-92.
8.   Goff SA, Ricke D, Lan TH, Presting G, Wang R, Dunn M, Glazebrook J, Sessions A *et al.* (2002). A draft sequence of the rice genome (*Oryza sativa* L. ssp. *japonica*). *Science,* **296**: 92-100.
9.   Khush G S (1999). Green revolution: preparing for the 21st century. *Genome,* **42**: 646-655.
10.   Ashikari M and Matsuoka M (2002). Application of rice genomics to plant biology and breeding. *Botanical Bulletin of Academia Sinica,* **43**:1-11.
11.   Jeon JS, Lee S, Jung KH, Jun SH, Jeong DH, Lee J, Kim C, Jang S *et al.* (2000). T-DNA insertional mutagenesis for functional genomics in rice. *The Plant Journal,* **22**: 561-570.

12. Greco R, Ouwerkerk PB, Taal AJ, Favalli C, Beguiristain T, Puigdomenech P, Colombo L, Hoge JH et al. (2001). Early and multiple Ac transpositions in rice suitable for efficient insert ional mutagenesis. Plant Molecular Biology, 46: 215-27.
13. Hirochika H (2001). Contribution of the Tos17 retrotransposon to rice functional genomics. Current Opinion of Plant Biology, 4: 118-22.
14. Yano M, Harushima Y, Nagamura Y, Kurata N, Minobe Y and Sasaki T (1997). Identification of quantitative trait loci controlling heading date in rice, using a high-density linkage map. Theoretical and Applied Genetics, 95: 1025-1032.
15. Yano M, Katayose Y, Ashikari M, Yamanouchi U, Monna L, Fuse T, Baba T, Yamamoto K et al. (2000). Hd1, a major photosensitivity quantitative trait locus in rice, is closely related to the Arabidopsis flowering time gene CONSTANS. The Plant Cell, 12: 2473-2483.
16. Takahashi Y, Shomura A, Sasaki T, and Yano M (2001). Hd6, A rice quantitative trait locus involved in photoperiod sensitivity, encodes the α-subunit of protein kinase CK2. Proceedings of the National Academy of Sciences USA, 98: 7922-7927.
17. Yano M (2001). Genetic and molecular dissection of naturally occurring variation. Current Opinion of Plant Biology, 4: 130-135.
18. Ye X, Al-Babli S, Kloti A, Zhang J, Lucca P, Beyer P and Potrykus I (2000). Engineering the provitaminA (β-carotene) biosynthetic pathway into (carotenoid free) rice endosperm. Science, 287: 303-305.
19. Datta K, Baisakh N, Oliva N, Torrizo L, Abrigo E, Tan J, Rai M, Rehana S et al. (2003). Bioengineered golden indica rice cultivars with β-carotene metabolism in the endosperm with hygromycin and mannose selection systems. Plant Biotechnology Journal, 1: 81-90.
20. Vasconcelos M, Datta K, Oliva N, Khalekuzzaman M, Torrizo L, Krishnan S, Oliveira M, Goto F et al. (2003). Enhanced iron and zinc accumulation in transgenic rice with the ferritin gene. Plant Science, 164: 371-378.
21. Datta K and Datta SK (2000). Plant Transformation. In: Gilmartin PM, Bowler C (eds.), Molecular Plant Biology, Vol. 1. A Practical Approach, Oxford University Press, UK, pp.13-32.
22. Toriyama K, Arimoto Y, Uchimiya H and Hinata K (1986). Transgenic rice plants after direct gene transfer into protoplasts. Bio/Technology, 6: 881-888.
23. Shimamoto K, Terada R, Izawa T and Fujimoto H (1989). Fertile transgenic rice plants regenerated from transformed protoplast. Nature, 337: 274-276
24. Peterhans A, Datta SK, Datta K, Godall GJ, Potrykus I and Paszkowski J (1990). Recognition efficiency of Dicotyledoneae-specific promoter and RNA processing signals in rice. Molecular and General Genetics, 22: 361-368.
25. Christou P, Ford TF and Kofron M (1991). Production of transgenic rice (Oryza sativa) plants from agronomically important indica and japonica varieties via electric discharge particle acceleration of exogenous DNA into immature embryos. Bio/Technology, 9: 957-962.
26. Datta SK, Peterhans A, Datta K and Potrykus I (1990). Genetically engineered fertile indica rice plants recovered from protoplasts. Bio/Technology, 8: 736-740.
27. Datta SK (2000). Potential benefit of genetic engineering in plant breeding: rice, a case study. Agricultural Chemistry and Biotechnology, 43: 197-206.
28. Tu J, Datta G, Datta K, Xu C, He Y, Zhang Q, Khush GS and Datta SK (2000). Field performance of transgenic elite commercial hybrid rice expressing Bacillus thuringiensis β-endotoxin. Nature Biotechnology, 18: 1101-1104.
29. Tu J, Datta K, Khush GS, Zhang Q and Datta SK (2002). Field performance of Xa21 transgenic indica rice (Oryza sativa L.), IR72. Theoretical and Applied Genetics, 101: 15-20.
30. Datta K, Vasquez A, Tu J, Torrizo L, Alam MF, Oliva N, Abrigo E, Khush GS et al. (1998). Constitutive and tissue-specific differential expression of cryIA(b) gene in transgenic rice plants conferring enhanced resistance to rice insect pests. Theoretical and Applied Genetics, 97: 20-30.
31. Hei Y, Ohta S, Komari T and Kumashiro T (1994). Efficient transformation of rice (Oryza sativa) mediated by Agrobacterium and sequence analysis of the boundaries of the T-DNA. The Plant Journal, 6: 271-282.
32. Chu CC, Wang CC, Sun SS, Hsu C, Yin KC, Chu CY anf Bi YF (1975). Establishment of an efficient medium for anther culture of rice through comparative experiments on the nitrogen sources. Scientific Sinica, 18: 659-668.
33. Murashige T and Skoog F (1962). A revised medium for a rapid growth and bioassay with tobacco tissue culture. Physiologia Plantarum, 15: 473-497.

34.   Tu J, Datta K, Oliva N, Zhang G, Xu C, Khush G S, Zhang Q and Datta SK (2003). Site independent integrated transgenes in the elite restorer rice line Minghui 63 allow removal of a selectable marker from the gene of interest by self-segregation. *Plant Biotechnolgy Journal,* **1**: 155-165.

35.   Itoh H, Ueguchi-Tanaka M, Sakamoto T, Kayano T, Tanaka H, Ashikari M and Matsuoka M (2002). Modification of rice plant height by suppressing the height-controlling gene, *D18*, in rice. *Breeding Science,* **52**: 215-218.

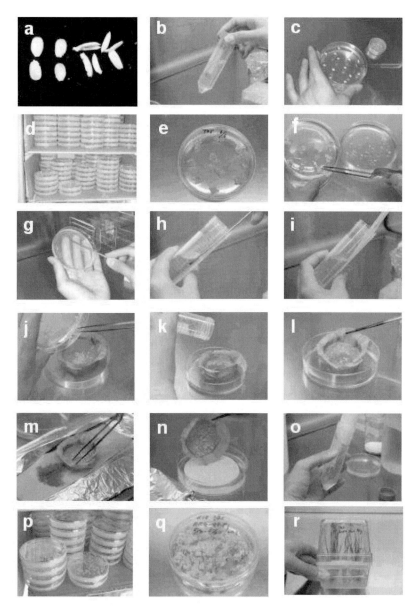

Figure 1. A step-wise procedure of rice transformation based on Agrobacterium tumefaciens (details has been described in the text and methodology).

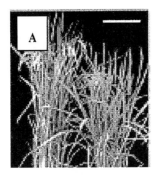

Figure 2. Modification of plant morphology by the down-regulation of plant hormones in japonica rice. A. Comparison of wildtype and antisense GA3 oxidase transformant; B. wildtype and a transformant with reduced function of the BR receptor gene (bar = 8.5 (left) and 9.0 cm (right), respectively.

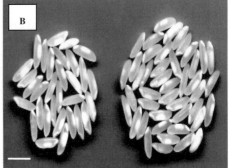

Figure 3. Golden indica rice. A. IR64 Golden indica Rice at IRRI grown under glasshouse conditions (bar = 0.6 cm). B. Transgenic seeds showing beta-carotene expression (right-side) compared with wildtype (left; bar = 2.6 cm).

Chapter 2

TRANSFORMATION OF WHEAT BY BIOLISTICS

C.A. SPARKS AND H.D. JONES

CPI Division, Rothamsted Research, Harpenden, Hertfordshire AL5 2JQ, UK.
Corresponding author: H. D. Jones. E-mail: huw.jones@bbsrc.ac.uk.

1. INTRODUCTION

1.1. Importance and Distribution of Wheat

Food products derived from wheat are one of the most important sources of calorific intake worldwide and have formed an important part of man's diet since Neolithic times. In 2002, 573 million tonnes of wheat grain were produced worldwide, of which approx. three quarters was eaten by humans (*1*). Wheat grain is rich in carbohydrates, proteins and essential vitamins and minerals such as vitamins B and E, magnesium and phosphorous, as well as fibre. It is the only cereal with enough gluten to make leavened bread and is a major constituent of many other foods including biscuits, cakes, breakfast cereal and pasta. Low-grade wheat and industrial wheat by-products are used for animal feed. Wheat is highly adaptable and is grown throughout the world, from the Arctic Circle to south of the Tropic of Capricorn, although it is most suited to more temperate latitudes between 30°N–50°N and 25°S–30°S. The global area of land under wheat cultivation has fluctuated greatly over recent decades. In the early 1960s, approx. 210 million hectares were grown. By the early 1980s, this had increased to approx. 240 million hectares but by the year 2000 the area had fallen again to 213 million hectares. However, in the same time frame, the world wheat production has steadily increased from 200 million tonnes to nearly 600 million tonnes per year today. It is projected to increase further to 860 million tonnes by 2030 (*2*).

1.2. Genetic Origins of Modern Wheats

The precise genetic origin of modern cultivated wheats is uncertain, but the first step probably began some ten thousand years ago in the Middle East with the spontaneous crossing of the diploid species *Triticum uratu* and *Aegilops speltoides* to produce tetraploid wild emmer wheat (*T. turgidum* ssp. *dicoccoides*). Further hybridizations and domestication gave rise to cultivated emmer (*T. turgidum* ssp. *dicoccon)* and early forms of durum wheat (*T. turgidum* ssp. *durum*). Hexaploid bread wheat finally evolved when cultivated emmer crossed with *Ae. Tauschii (*also known as *Ae. Squarrosa*) in the southern Caspian plains. This evolution was accelerated by an expanding geographical range of cultivation and by human selection, and played an important part in displacing hunter-gathering as a way of human existence. Systematic breeding based on Mendel's laws of genetics, to improve productivity and consistency, began at the turn of the 20th Century. A quantum leap in productivity occurred in the so called 'green revolution' of the 1960s with the development of semi-dwarf varieties created by the introgression of

19

I.S. Curtis (ed.), Transgenic Crops of the World - Essential Protocols, 19-34.
© 2004 Kluwer Academic Publishers. Printed in the Netherlands.

dwarfing alleles originating from Japan. Modern breeding methods are now moving from 'Mendelian to molecular', and future varietal improvement increasingly utilizes biotechnological tools such as marker-assisted selection, genome mapping and genetic transformation.

1.3. Development of Transformation Methods

Wheat was the last of the major cereals to be transformed, mainly due to difficulties of regenerating adult plants via tissue culture. The first successful production of adult wheat plants was reported by Vasil *et al.* (*3*) who used biolistics to introduce a construct containing the marker genes *bar* and *uid*A (GUS) into regenerable embryogenic callus. This method took over one year to produce transgenic plants, with efficiencies of approx. 0.2%. Improvements to DNA delivery protocols and the use of tissue culture based on isolated immature scutella have now reduced this to three and a half months, with efficiencies of more than 5% (*4-9*). Throughout the 1990s, DNA delivery based on biolistics became the established method of choice for wheat species. Biolistics tends to generate transgene events with variable copy number (5-15 copies) and expression levels, however lines giving the appropriate expression levels can be identified by screening independent transgenic plants. Recent reports indicate that *Agrobacterium*-mediated transformation is a viable additional method for wheat transformation giving efficiencies similar to biolistic approaches (*10-13*). This chapter describes an established method for the introduction of plasmid vectors into immature wheat scutella by biolistics and a tissue culture/selection regime to recover fertile plants via somatic embryogenesis.

1.4. Targets for Genetic Improvement

Conventional wheat breeding has been successful in exploiting variation in available germplasm, however the great potential of genetic transformation lies in its ability to manipulate traits in a way not possible by conventional means. Genetic manipulation allows the insertion of novel genes from non-sexually compatible plants, or enables the expression of native genes at different levels, in specified tissues or under novel developmental patterns of expression. Specific targets for wheat genetic engineering fall into two broad areas. Firstly, "input traits" that improve crop performance and productivity such as pest/disease resistance, tolerance to abiotic stresses, herbicide tolerance, and alterations of plant architecture. Secondly, "output traits" that modify quality, composition and use, such as the production of novel compounds, alterations to grain hardness, starch quality, and the high-molecular-weight glutenin subunit composition of seed endosperm for improved bread-making. With the exception of herbicide tolerance, for which commercial wheat varieties have been produced, input traits are generally under complex genetic control. Some grain quality traits such as hardness, storage protein composition and starch quality are becoming better understood and are now clear targets for genetic improvement (*14*).

2. MATERIALS

2.1. Growth of Donor Plants

Wheat *(Triticum aestivum* L.) plants (*see* Note 1), are grown 5 plants per 21 cm diam. plastic pot (Nursery Trades (Lea Valley) Ltd., Hertfordshire, UK) filled with soil (75% fine-grade peat, 12% screened sterilized loam, 10% 6 mm screened lime-free grit, 3% medium vermiculite, 2 kg osmocote plus/m^3 (slow-release fertilizer, 15N/11P/13K plus micronutrients), 0.5 kg PG mix/m^3 (14N/16P/18K granular fertilizer plus micronutrients (Petersfield Products, Leicestershire, UK)). Winter wheat varieties are vernalized from seed for 8 weeks at 4-5°C. Plants are maintained at 18-20°C day and 14-15°C night temperatures under a 16 h photoperiod provided by banks of HQI lamps 400W (Osram Ltd., Berkshire, UK) in growth rooms (intensity ~700 µmol/m^2/sec photosynthetically active radiation (PAR)) (*see* Note 2). Initially all plants are top watered in order to monitor water requirements and thereby provide sufficient water without water-logging. Once the root system reaches the base of the pot, the plants are watered using an automated flooding system. Pests and disease are kept to a minimum by good housekeeping practices but *Amblyseius caliginosus* (Nursery Trades (Lea Valley) Ltd.) is used as a biological control agent to manage thrips, and the fungicide Fortress (DOW Agrosciences Ltd., Hertfordshire, UK) is applied as a preventative spray to avoid mildew. If plants appear diseased, they are discarded immediately.

2.2. Sterilization Materials for Wheat Caryopses

1.   70% (v/v) aqueous ethanol.
2.   10% (v/v) aqueous Domestos (Lever Fabergé Ltd., Surrey, UK)
3.   Sterile water (*see* Note 3).

2.3. Stock Solutions and Tissue Culture Media

Detailed below are the recipes for stock solutions of basal culture media components (*see* Section 2.3.1.), hormones (*see* Section 2.3.2.), AgNO$_3$ (*see* Section 2.3.2.), selection agents (*see* Section 2.3.3.) and agargel (*see* Section 2.3.4.), from which the final tissue culture media (*see* Section 2.3.5.) are prepared (*see* Notes 3 and 4).

2.3.1. Stock Solutions of Basal Culture Media Components

1.   MS macrosalts (x10): 16.5 g/l NH$_4$NO$_3$ (Fisher Scientific UK, Leicestershire, UK), 19.0 g/l KNO$_3$ (Sigma-Aldrich, Dorset, UK), 1.7 g/l KH$_2$PO$_4$ (Fisher Scientific UK), 3.7 g/l MgSO$_4$.7H$_2$O (Fisher Scientific UK), 4.4 g/l CaCl$_2$.2H$_2$O (Fisher Scientific UK) (*see* Note 5). Autoclave at 121°C for 20 min and store at 4°C (*see* Note 6).
2.   MS vitamins (-glycine) (x1000): 0.1 g/l thiamine-HCl (Sigma-Aldrich), 0.5 g/l pyridoxine-HCl (Sigma-Aldrich), 0.5 g/l nicotinic acid (Sigma-Aldrich). Prepare 100 ml at a time. Filter-sterilize (*see* Note 7) and store at 4°C (*see* Note 6).

3. L7 macrosalts (x10): 2.5 g/l NH_4NO_3, 15 g/l KNO_3, 2.0 g/l KH_2PO_4, 3.5 g/l $MgSO_4.7H_2O$, 4.5 g/l $CaCl_2.2H_2O$ (*see* Note 5). Autoclave at $121^{\circ}C$ for 20 min and store at $4^{\circ}C$ (*see* Note 6).

4. L7 microsalts (x1000): 15 g/l $MnSO_4$ (Fisher Scientific UK) (*see* Note 8), 5 g/l H_3BO_3 (Fisher Scientific UK), 7.5 g/l $ZnSO_4.7H_2O$ (Fisher Scientific UK), 0.75 g/l KI (Fisher Scientific UK), 0.25 g/l $Na_2MoO_4.2H_2O$ (VWR International Ltd., Leicestershire, UK), 0.025 g/l $CuSO_4.5H_2O$ (Fisher Scientific UK), 0.025 g/l $CoCl_2.6H_2O$ (Sigma-Aldrich). Prepare 100 ml at a time. Filter-sterilize (*see* Note 7) and store at $4^{\circ}C$ (*see* Note 6).

5. L7 vitamins/inositol (x200): 40 g/l myo-inositol (Sigma-Aldrich), 2.0 g/l thiamine-HCl, 0.2 g/l pyridoxine-HCl, 0.2 g/l nicotinic acid, 0.2 g/l Ca-Pantothenate (Sigma-Aldrich), 0.2 g/l ascorbic acid (Sigma-Aldrich). Store at $-20^{\circ}C$ in 10 ml aliquots (*see* Note 6).

6. 3AA amino acids (x25): 18.75 g/l L-glutamine (Sigma-Aldrich), 3.75 g/l L-proline (Sigma-Aldrich), 2.5 g/l L-asparagine (Sigma-Aldrich). Store solution at $-20^{\circ}C$ in 40 ml aliquots (*see* Note 6).

2.3.2. Hormones and AgNO₃

1. 2,4-D (Sigma-Aldrich): 1 mg/ml in ethanol/water (dissolve powder in ethanol then add water to volume). Mix well. Filter-sterilize (*see* Note 7) and store at $-20°C$ in 1 ml aliquots (*see* Note 6).

2. Zeatin mixed isomers (Sigma-Aldrich): 10 mg/ml in HCl/water (dissolve powder in small volume 1M HCl and make up to volume with water). Mix well/vortex. Filter-sterilize (*see* Note 7) and store at $-20°C$ in 1 ml aliquots (*see* Note 6).

3. Silver nitrate ($AgNO_3$) solution (Sigma-Aldrich): 20 mg/ml in water. Mix well. Filter-sterilize (*see* Note 7) and aliquot into 1 ml volumes. Store at $-20°C$ in the dark (*see* Notes 6 and 9).

2.3.3. Selection Agents

1. Glufosinate ammonium (Greyhound Chromatography and Allied Chemicals, Cheshire, UK) (synthetic PPT – *see* Note 10): 10 mg/ml in water. Mix well/vortex. Filter-sterilize (see Note 7) and store at $-20°C$ in 1 ml aliquots (*see* Note 6).

2. Geneticin disulphate (G418) (Melford Laboratories Ltd., Suffolk, UK) (*see* Note 11): 50 mg/ml in water. Mix well/vortex. Filter-sterilize (*see* Note 7) and store at $-20°C$ in 1 ml aliquots (*see* Note 6).

2.3.4. Agargel

1. Agargel (x2) (Sigma-Aldrich): Prepare in 400 ml volumes at 10 g/l and sterilize by autoclaving at $121^{\circ}C$ for 20 min. Store at room temperature and melt in microwave before use (*see* Note 12).

2.3.5. Tissue Culture Media for Wheat Immature Scutella

Induction media:
1. MSS 3AA/2 9%S (x2): 200 ml/l MS macrosalts, 2 ml/l L7 microsalts, 20 ml/l Ferrous sulphate chelate solution (x100) (Sigma-Aldrich), 2 ml/l MS (-glycine) vitamins, 200 mg/l *myo*-inositol (Sigma-Aldrich), 40 ml/l 3AA amino acids (*see* Note 13), 180 g/l (9% final concentration) sucrose (Sigma-Aldrich) (*see* Note 14). Adjust pH to 5.7 with 5M NaOH or KOH. Osmolarity should be within the range 800-1100 mOsM. Filter-sterilize (*see* Note 7) and store at $4^{°}$C (*see* Note 6).
2. MS9%0.5DAg: Mix an equal volume of MSS 3AA/2 9%S (x2) with sterilized, melted agargel. Add 0.5 mg/l 2,4-D (*see* Note 15) and 10 mg/l $AgNO_3$ and pour into 9 cm diam. Petri dishes (Bibby Sterilin Ltd., Staffordshire, UK) (~28 ml per dish). Store at $4^{°}$C in the dark (*see* Notes 9 and 16).

Regeneration media:
1. R (x2): 200 ml/l L7 macrosalts, 2 ml/l L7 microsalts, 20 ml/l Ferrous sulphate chelate solution (x100), 10 ml/l L7 vitamins/inositol, 60 g/l maltose (Melford Laboratories Ltd.). Adjust pH to 5.7 with 5M NaOH or KOH. Osmolarity should be within the range 269-298 mOsM. Filter-sterilize (*see* Note 7) and store at $4^{°}$C (*see* Note 6).
2. RZDAg: Mix an equal volume R (x2) with sterilized, melted agargel. Add 5 mg/l zeatin, 0.1 mg/l 2,4-D and 10 mg/l $AgNO_3$ and pour into 9 cm Petri dishes (~28 ml per dish). Store at $4^{°}$C in the dark (*see* Notes 9 and 16).

Selection media:
1. RZDPPT4 or RZDG50: Mix an equal volume of R (x2) with sterilized, melted agargel and add 5 mg/l zeatin, 0.1 mg/l 2,4-D and 4 mg/l glufosinate ammonium (PPT4) (*see* Note 10) or 50 mg/l G418 (G50). Pour into 9 cm Petri dishes (~28 ml per dish) or Magenta GA-7 vessels (Sigma-Aldrich) (~60 ml per vessel). Store at $4^{°}$C (*see* Note 16).
2. RPPT4 or RG50: Mix an equal volume of R (x2) with sterilized, melted agargel and add 4 mg/l glufosinate ammonium (PPT4) (*see* Note 10) or 50 mg/l G418 (G50). Pour into 9 cm Petri dishes (~28 ml per dish) or Magenta GA-7 vessels (~60 ml per vessel). Store at $4^{°}$C (*see* Note 16).

2.4. Bombardment Materials/Consumables

1. 2.5M Calcium chloride (Fisher Scientific UK): Dissolve 3.67 g $CaCl_2.2H_2O$ in 10 ml water. Mix well/vortex. Filter-sterilize (*see* Note 7) and store at -20°C in 50 μl aliquots (*see* Note 6).

2. 0.1M Spermidine free-base (Sigma-Aldrich): Prepare 1M stock from powder in sterile water and maintain at -80°C in 20 μl aliquots. Prepare the 0.1M working solution by making a 1:10 dilution of 1M stock in sterile water under sterile conditions. Mix well, aliquot in 10 μl volumes and store immediately at -20°C (*see* Note 17).

3. Gold particles: 0.6 μm (sub-micron) gold particles (BIO-RAD Laboratories, Hertfordshire, UK) (*see* Note 18). (For preparation, *see* Section 3.2.1.).

4. Macro-carriers, stopping screens, 650 or 900 psi rupture discs (all BIO-RAD Laboratories) (*see* Note 19).

5. Plasmid DNA: Prepare using Qiagen Maxi–prep kit (Qiagen Ltd., West Sussex, UK) and resuspend at 1 mg/ml in sterile TE buffer or sterile water. Store in 20 μl aliquots at -20°C (*see* Note 20).

3. METHODS

3.1. Preparation of Donor Material

The method described below is optimized for transformation of immature scutella but immature inflorescences can be used as alternative explants. These may be more responsive than immature scutella for certain varieties e.g. *T. aestivum* vars. Baldus and Brigadier, and other *Triticeae* e.g. *T. turgidum* ssp. *durum* and tritordeum (s*ee* Note 21).

3.1.1. Collection and Sterilization of Wheat Caryopses

1. Collect spikes from growth room-grown plants at ~10-12 weeks after sowing. Embryos at the correct stage are usually found ~12-16 days post anthesis (*see* Section 3.1.2.). A few caryopses can be opened at the time of collection to determine the size of the embryos within.

2. Remove the caryopses from the panicles (*see* Notes 22 and 23).

3. Sterilize the caryopses by rinsing in 70% (v/v) aqueous ethanol for 5 min then soak for 15-20 min in 10% (v/v) Domestos with gentle shaking on a platform shaker (~ 60 rpm).

4. Rinse copiously with at least 3 changes of sterile water. Maintain the sterilized caryopses in moist conditions but do not keep immersed in water.

3.1.2. Isolation and Pre-culture of Immature Scutella

1. Isolate the embryos microscopically in a sterile environment (Fig. 2A) and remove the embryo axis to prevent precocious germination. Embryos of size approximately 0.5–1.5 mm are generally most responsive but there is genotypic variation (*see* Note 24).
2. Place 25-30 scutella per 9 cm Petri dish containing induction medium (MS9%0.5Dag), locating them within a central target area (*see* Note 25) and orientating them with the uncut scutellum uppermost i.e. the uncut side is bombarded (Fig. 2B).
3. Pre-culture the prepared donor material in the dark at 26°C for 1-2 days prior to bombardment (*see* Note 26).

3.2. Gold Particles

3.2.1. Preparation of Gold Particles
1. In a 1.5 ml Eppendorf, place 20 mg BIO-RAD sub-micron gold particles (0.6 μm), and add 1ml 100% ethanol. Sonicate for 2 min, pulse spin in a microfuge for 3 sec and remove the supernatant. Repeat this ethanol wash twice more.
2. Add 1 ml sterile water and sonicate for 2 min. Pulse spin in a microfuge for 3 sec and remove the supernatant. Repeat this step.
3. Resuspend fully by vortexing in 1 ml sterile water. Aliquot 50 μl amounts into sterile 1.5 ml eppendorf tubes, vortexing between taking each aliquot to ensure an equal distribution of particles. Store at -20°C.

3.2.2. Coating of Gold Particles with DNA for Bombardment

The following procedure should be carried out on ice, in a sterile environment.
1. Allow a 50 μl aliquot of prepared gold (*see* Section 3.2.1.) to thaw at room temperature then sonicate for 1-2 min (*see* Note 27). The tubes can be vortexed following sonication to ensure total re-suspension, particularly if the aliquots are to be sub-divided for smaller preparations (*see* Note 28).
2. Add 5 μl DNA (1 mg/ml in TE *see* Note 29) or water (*see* Note 30) and vortex briefly to ensure good contact of DNA with the particles (*see* Note 31).
3. Place 50 μl 2.5 M $CaCl_2$ and 20 μl 0.1 M spermidine into the lid of the Eppendorf tube and mix together. Briefly vortex into the gold and DNA solution (*see* Note 32). Centrifuge 13,000 rpm for 3-5 sec in a microfuge to pellet the DNA-coated particles. Discard the supernatant.
4. Add 150 μl 100% ethanol to wash the particles, re-suspending them as fully as possible (*see* Notes 33 and 34).
5. Centrifuge at 13,000 rpm for 3-5 sec in a microfuge to pellet the particles and discard the supernatant.
6. Resuspend fully in 85 μl 100% ethanol and maintain on ice (*see* Note 35).

3.3. Particle Bombardment using the PDS-1000/He Particle Gun [BIO-RAD] (see Note 36)

Note: The delivery system involves the use of high pressure to accelerate particles to high velocity. Appropriate safety precautions should be taken and safety spectacles should be worn when operating the gun.

3.3.1. Delivery of DNA-coated Gold Particles

In any bombardment experiment, controls should be included to monitor regeneration and selection efficiencies (*see* Note 37).

1. DNA-coated gold particles (*see* Section 3.2.2.) are delivered using the PDS-1000/He particle gun [BIO-RAD] (Fig. 1) according to the manufacturer's instructions. The following settings are maintained as standard for this procedure (*see* Note 38): gap 2.5 cm (distance between rupture disc and macro-carrier), stopping plate aperture 0.8 cm (distance between macro-carrier and stopping screen), target distance 5.5 cm (distance between stopping screen and target plate), vacuum 91.4 – 94.8 kPa, vacuum flow rate 5.0, vent flow rate 4.5 (7).
2. Sterilize the gun chamber and component parts by spraying with 90% (v/v) ethanol which should be allowed to evaporate completely (~5 min).
3. Sterilize macro-carrier holders, macro-carriers, stopping screens and rupture discs by dipping in 100% ethanol and allow the alcohol to evaporate completely on a mesh rack in a flow hood (*see* Note 39). Place the macro-carrier holders into sterile 6 cm Petri dishes and introduce one macro-carrier into each holder.
4. Briefly vortex the coated gold particles (*see* Section 3.2.2.), take 5 µl and drop centrally onto the macro-carrier membrane. Allow to dry naturally, not in the air-flow (*see* Note 40).
5. Load a rupture disc (650 or 950 psi) (*see* Note 19) into the rupture disc retaining cap (Fig. 1) and screw into place on the gas acceleration tube, tightening firmly using the mini torque wrench (*see* Note 41).
6. Place a stopping screen into the fixed nest. Invert the macro-carrier holder containing macro-carrier and gold particles/DNA and place over the stopping screen in the nest and maintain its position using the retaining ring. Mount the fixed nest assembly onto the second shelf from the top to give a gap of 2.5 cm (Fig. 1).
7. Place a sample on the target stage on a shelf to give the desired distance; fourth shelf from the top gives a target distance of 5.5 cm.
8. Draw a vacuum of 91.4 – 94.8 kPa and fire the gun (*see* Note 42).
9. Once the vacuum has been released, remove the sample and disassemble the component parts, discarding the ruptured disc and macro-carrier (*see* Note 43).
10. Place the macro-carrier holder and stopping screen in 100% ethanol to re-sterilize if they are to be re-used for further shots, otherwise place in 1:10 dilution Savlon (Novartis Consumer Health, West Sussex, UK) to soak. Sonicate for 10 min prior to re-use.

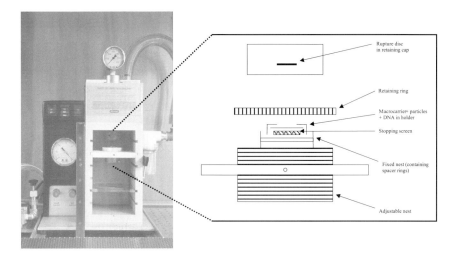

Figure 1. The PDS1000/He particle gun [BIO-RAD] (left) and enlargement of component parts described in Section 3.3.1. (right).

3.4. Culture of Immature Scutella Following Bombardment

3.4.1. Induction of Embryogenic Callus

1. Following bombardment, spread the scutella more evenly across the medium, dividing each replicate between 2-3 plates of induction medium (MS9%0.5Dag) in 9 cm Petri dishes i.e.~10 scutella per plate (*see* Note 44).
2. Seal the plates with Nescofilm® (Fisher Scientific UK) and incubate at 26°C in the dark for induction of embryogenic callus (*see* Notes 45 and 46; Fig. 2C).

3.4.2. Regeneration and Selection

1. After 3-5 weeks on induction medium, transfer any callus bearing somatic embryos to regeneration medium (RZDAg) in 9 cm Petri dishes. Whole calli should be transferred without division, placing ~10 calli per plate. Incubate at 26°C in the light (*see* Note 47) for 3-4 weeks.
2. After the first 3-4 week round of regeneration, transfer calli without division to RZD with selection in 9 cm Petri dishes (the medium is dependent on the selectable marker gene in the transforming plasmid: RZDPPT4 for *bar* gene or RZDG50 for *npt*II gene (*see* Note 48). Incubate at 26°C in the light (*see* Note 47).

3. After a further 3-4 weeks, and every subsequent 3-4 weeks, transfer surviving calli to regeneration medium with selection but without hormones (RPPT4 or RG50) in 9 cm Petri dishes (*see* Notes 48 and 49; Fig. 2D). Incubate at 26°C in the light (*see* Note 47).

4. Once regenerating shoots are clearly defined and can be separated easily from the callus, transfer these to regeneration medium with selection but without hormones (RPPT4 or RG50) in Magenta GA-7 vessels, placing no more than 4 shoots per Magenta. Incubate at 26°C in the light (*see* Note 47).

5. Transfer plantlets to soil once the leaves have out-grown the Magenta vessel (typically >10 cm) and a reasonable root system has been established. This usually takes at least 3 months from bombardment. Carefully remove plantlets from the agargel-solidified medium (rinsing the roots of excess agargel with water if necessary) and pot into soil in 8 cm square plastic pots (Nursery Trades, (Lea Valley) Ltd.). Plantlets are grown in a GM containment glasshouse (initially within a propagator to provide a high humidity for 1-2 weeks to acclimatize them from tissue culture) (*see* Notes 50 and 51).

6. Once established (3-4 leaves), plants are re-potted to 13 cm diam. pots (Nursery Trades (Lea Valley) Ltd.) and grown on under the same glasshouse conditions. Plants should reach maturity in 3-4 months (Fig. 2G).

7. Transgenic plants can be analysed in a number of ways: Marker gene expression can be assessed using for example, UV visualization of GFP (Fig. 2E), the histochemical GUS test (*15*) for *uid*A (*see* Fig. 2F), and herbicide leaf paint assay (*13*) and/or the ammonium test (*16*) for *bar*. Gene integrations can be determined using PCR, Southern analysis and fluorescent *in situ* hybridisation (FISH).

4. NOTES

1. The transformation method detailed in this chapter has been successfully used for a number of commercial wheat varieties but with a range of efficiencies; Cadenza, Canon and Florida have given the highest efficiencies (up to 13%) (*7-9*). It is also possible to transform *T. turgidum* ssp. *durum* (e.g. cvs. Ofanto and Venusia) based on this method (*17, 18*) but for these and alternative wheat varieties, modifications may be required (*19, 20*).

2. Glasshouse-grown plants can be used but these are generally less consistent due to seasonal changes.

3. The water used for all solutions is always reverse osmosis, polished water with a purity of 18.2 MΩ/cm.

4. For alternative varieties or wheat species, modifications to the media detailed here may be required. For example, the choice of basal salts (MS or L7), the concentration of sugars (sucrose or maltose), the level of hormones etc. need to be empirically determined.

5. Dissolve $CaCl_2.2H_2O$ in water before mixing with other components.

6. Sterile stock solutions can be stored at 4°C for 1-2 months. Some settling of salts may occur during storage, so the medium should be shaken well prior to use. Stock solutions stored at -20°C should remain effective for at least a year, provided no freeze/thawing has occurred.

7. Filter-sterilization is carried out using 0.2 μm filter size. For large volumes use MediaKap® (NBS Biologicals Ltd., Cambridgeshire, UK), for smaller volumes use a Nalgene syringe filter (Fisher Scientific UK).

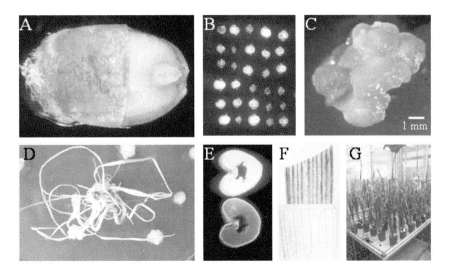

Figure 2. A . Dissected wheat caryopsis showing immature embryo. B. Isolated immature scutella plated for bombardment. C. Induction of embryogenic callus. D. Regenerating plantlet surviving selection. E. Transverse section of wheat seed expressing GFP (top), control seed (bottom). F. Leaf section showing GUS expression (top), control leaf (bottom). G. Regenerated transformed plants in GM containment glasshouse.

8. $MnSO_4$ may have various hydrated states which will alter the required weight. For $MnSO_4.H_2O$, add 17.05 g/l, $MnSO_4.4H_2O$, add 23.22 g/l, or $MnSO_4.7H_2O$, add 27.95 g/l.

9. $AgNO_3$ is used to promote embryogenesis; Silver thiosulphate (a mix of silver nitrate and sodium thiosulphate) at 10 mg/l can be used as an alternative. Both are photosensitive so the stock solutions and any media plates containing them should be kept in the dark.

10. Glufosinate ammonium is synthetically produced PPT bound to ammonium, and is the active component in herbicides such as Basta[TM]. Bialaphos (Melford Laboratories Ltd.) at 3-5 mg/l is a successful alternative selection agent.

11. Kanamycin, paromomycin and neomycin are alternative aminoglycoside antibiotics that can be used for selection with the *npt*II gene. Although they may be successful for some plant species, they are not recommended for wheat as untransformed tissues exhibit natural resistance.

12. The agargel solution should be shaken well both before and after autoclaving to avoid non-uniform solidification which leads to difficulties when re-melting.

13. Instead of using the 3AA stock solution, 0.75 g/l L-Glutamine, 0.15 g/l L-Proline, and 0.1 g/l L-Asparagine can be added individually.

14. 9% sucrose partially plasmolyses the cells during pre-culture which may allow them to withstand bombardment. However, this is variety and species dependent and 3% sucrose is often suitable e.g. for *T. turgidum* ssp. *durum* scutella. The osmolarity for 3% sucrose medium should be within the range 355-398 mOsM.

15. Picloram (Sigma-Aldrich) can be used as an alternative auxin at 2-6 mg/l (*20, 21*).

16. Tissue culture media should be prepared as fresh as possible and not to be stored for more than 2-3 weeks in Petri dishes and Magenta vessels. However, they should be prepared a few days in advance to allow any contamination to be detected before use. To minimize condensation in the plates, allow the agargel to cool once melted, and pour the final medium at ~50°C.

17. Spermidine deaminates with time and solutions are hygroscopic and oxidizable, therefore they should be maintained below –20°C, preferably at –80°C, and any unused aliquots once thawed, should be discarded.

18. Heraeus gold particles of 0.4-1.2 μm diam. (W. C. Heraeus GmbH & Co. KG, Hanau, Germany) have also been used successfully for transformation, however the submicron BIO-RAD particles give more consistent results for wheat. The smaller, more uniform size of the latter particles is preferable for small wheat cells but for other species, larger particles may be suitable.

19. Rupture pressures of 650 or 900 psi have been found to be optimal for the wheat varieties reported here; 450 or 1100 psi pressures will result in successful transformation but with lower efficiency. Rupture discs of 450, 650, 900, 1100, 1350, 1550, 1800, 2000, 2200 psi are available, and a range should be tested if attempting transformation of any new variety or species.

20. Plasmids tend to be pUC-based and contain one or more gene cassettes. To allow selection of transformed tissues, a selectable marker gene is required, for example the *bar* or *npt*II gene under the control of a constitutive promoter (e.g. Maize Ubiquitin or CaMV35S) and a suitable terminator (e.g. *nos*). The *bar* gene confers resistance to the herbicides Basta™ (Glufosinate ammonium/PPT) and Bialaphos and the *npt*II gene confers resistance to the antibiotics geneticin disulphate (G418), kanamycin, neomycin, paromomycin etc. (*see* Note 11). A reporter gene (e.g. *uid*A, *luc* or *GFP*) is also a useful tool to monitor both transient and stable transformation (Figs. 2E and F). These marker genes can be located in the same plasmid or on separate plasmids co-precipitated onto the gold particles.

21. Immature inflorescences are an alternative explant for transformation as they can have high regeneration potential. Some modifications to the described method may be required. For references on the use of immature inflorescences with *T. aestivum* varieties, see *7, 22*; for *T. turgidum* ssp. *durum*, see *17, 18*; for tritiordeum (a fertile cereal amphidiploid obtained from crosses between *Hordeum chilense* and durum wheat cultivars, and containing the genome H^{CH}H^{CH}AABB)) (*19, 23, 24*; and for barley, *20*).

22. Avoid using the inner caryopses of the spikelet which generally contain smaller embryos due to asynchronous development.

23. Although it is not encouraged, if the caryopses are not used on the same day, it is possible to store the spikes intact at 4°C, with stems in water.

24. 0.5-1.5 mm is the most responsive size range for the varieties reported here. Smaller and larger embryos may respond but with much lower efficiencies.

25. Typically the gun shot fires most gold particles within a central circular area of a Petri dish of ~ 2 cm diam.. Arranging scutella within this area maximizes particle delivery (as shown by transient expression studies (*7*)).

26. The pre-culture allows the tissues to recover from the isolation procedure before being subjected to bombardment. It also allows any contamination to be detected prior to bombardment. If donor material is difficult to sterilize leading to contaminated explants, Plant Preservative Mixture (PPM™) (Plant Cell Technology Inc., Washington DC, USA) can be included in tissue culture media at 1 ml/l. This is a non-toxic broad-spectrum preservative and biocide which does not interfere with callus proliferation or regeneration.

27. If the sonication has worked efficiently the gold particles should be resuspended in the liquid rather than be present as a pellet at the bottom of the tube. The particles should not be sonicated for too long however, as there is evidence that over-sonication can cause aggregation.

28. If fewer shots are required or a variety of DNAs are to be compared, the gold preparation can be sub-divided and volumes scaled down accordingly.

29. If plasmids are not at a concentration of 1 mg/ml, re-calculate the volume to give 5 μg and add to the gold.

30. In a bombardment experiment, control plates are required to monitor regeneration and selection efficiencies (*see* Note 37). Some particles should therefore be prepared by this process but without DNA; replacing the DNA solution with sterile water.

31. The standard amount of DNA is 5 μg/50 μl gold suspension. If using more than one plasmid i.e. for co-bombardment, the amounts of DNA added should be calculated so that equimolar quantities are used, with 5 μl of each DNA. Alternatively, as the DNA is already in excess, a total of 5 μg DNA for the two plasmids can be added, which may reduce clumping of gold particles.

32. The $CaCl_2$ and spermidine act to bind, stabilize and precipitate the DNA. Precipitation onto the gold particles is very rapid so it is better that the $CaCl_2$ and spermidine are mixed first so that coating is as even as possible.

33. Resuspend the particles as well as possible by scraping the side of the tube with the pipette tip to remove clumps, and drawing up and expelling the solution repeatedly. Make sure the gold is fully resuspended at this stage as clumps cannot be removed during later resuspension steps. Vortexing will not aid resuspension.

34. Although the coated particles should be used as soon as possible, they can be kept on ice at this stage (but for no longer than an hour), continuing with the rest of the protocol just prior to use.

35. Do not aspirate too much at this stage otherwise the ethanol evaporates and so increases the final concentration of particles. Although 85 μl should be sufficient for 16-17 shots (5 μl/shot), evaporation already means there is generally enough for only 10-12 shots. Once resuspended, to reduce evaporation of the ethanol, the Eppendorf lids can be sealed with Nescofilm® (Fisher Scientific UK) until the particles are required. However, it is advisable to use coated gold particles as soon as possible.

36. Recent advances in *A. tumefaciens*-mediated transformation of wheat makes this a viable alternative DNA delivery system. Various factors important for successful transformation have been reported (*13, 25*).

37. Various control plates should be included within each experiment: unbombarded – to monitor the development/regeneration of donor tissue, bombarded with gold (no DNA) and unselected – to monitor tissue culture response following bombardment, and bombarded with gold (no DNA) and no selective agent – to monitor the effects of the selection on regeneration.

38. Although these distances were found to be optimal for some wheat varieties, they can be altered as desired for different species or varieties.

39. The rupture discs should be sterilized for not more than 10 min or the laminate layers may become separated.

40. Once the coated particles have been dispensed onto the macro-carriers, the ethanol should be allowed to evaporate slowly. In order to create an even spread of dried particles on the macro-carrier, place macro-carriers within their Petri dishes outside of the flow hood on a non-vibrating surface in order to avoid particle agglomeration caused by vibration from the air flow. Only a few macro-carriers should be loaded with gold at any one time such that they are used when recently dried. Macro-carriers can be examined microscopically prior to bombardment to determine the uniformity and spread of particles, discarding any that have agglomerated clumps of gold which will reduce transformation efficiency.

41. The helium pressure on the cylinder should be set to approx. 200 psi more than the intended rupture pressure.

42. The helium pressure builds up and breaks the rupture disc, sending the macro-carrier onto the stopping plate, thus releasing and dispersing the particles. The actual pressure at which the rupture disc bursts should be monitored to ensure a successful shot, otherwise transformation efficiencies may be affected.

43. The macro-carrier can be observed microscopically after the shot to visualize the mesh pattern left by the stopping screen. This will reveal how much gold has been released/retained.

44. The scutella are spread more evenly in order to reduce the culture density and prevent competition for nutrients.

45. The induction period for somatic embryogenesis is usually 3-5 weeks, however the explants should be observed regularly to check for contamination. Judgement and experience is required to monitor development in order to determine the ideal time for transfer to regeneration medium; transfer is carried out when the embryogenic callus has mature somatic embryos some of which may just be forming small shoots.

46. Transient assays e.g. histochemical GUS assay, can be carried out after 2-3 days depending on the construct i.e. strength of the promoter.

47. Incubation is carried out in a controlled environment room with lighting levels ~250 μmol/m^2/sec PAR from cool white fluorescent tubes for a 12 h photoperiod.

48. Selection is generally applied at the second and subsequent transfers, until all control plantlets have been killed (*see* Note 37). The selection agent should be used at a concentration which is known to fully inhibit the growth of non-transformed explants, however, the concentration should be gauged

according to the development of the cultures at each transfer stage. Generally use within the range 2-4 mg/l glufosinate ammonium (PPT) and 25-50 mg/l G418.

49.   At this stage, any dead or unresponsive callus can be removed from the regenerating areas to leave only healthy tissue. If the regenerating calli to be transferred are large, the number of calli per 9 cm Petri dish should be reduced to prevent overcrowding.

50.   Tissue cultured plantlets have little or no waxy cuticle and so are particularly prone to desiccation after transfer to soil.

51.   Glasshouse conditions are 18-20°C day and 14-16°C night temperatures with a 16 h photoperiod provided by natural light supplemented with banks of Son.T 400W sodium lamps (Osram Ltd.) giving 400-1000 μmol/m^2/sec PAR.

CONCLUSIONS

Though initially one of the more recalcitrant species, wheat has, in the last decade, become readily transformable by biolistics. There are variations in reported transformation efficiencies between different laboratories. These probably result from slight differences in protocols used, the wheat varieties chosen and quality of donor plants. In our experience, specific factors that have a marked influence on the success of wheat transformation include; the age of donor plants, size/age of immature embryos, DNA-delivery parameters, media constituents, timing and strength of selection, and the responsiveness of particular wheat varieties. In addition to the regeneration of fertile transgenic wheat plants, the DNA-delivery methods described in this chapter form a powerful tool to study transient gene expression in a variety of tissues.

ACKNOWLEDGEMENTS

Rothamsted receives grant-aided support from the Biotechnological and Biological Sciences Research Council, UK. We acknowledge other members of the Rothamsted Cereal Transformation Group, past and present, for their significant contribution to the protocols described here.

REFERENCES

1.   Food and Agriculture Organization of the United Nations. www.fao.org.

2.   Marathee, J-P and Gomez-MacPherson H (2001). Future world supply and demand. In: Bonjean AP, WJ Angus (eds.), *The World Wheat Book.*, Intercept, Hampshire, UK & Lavoisier, Paris. pp. 1107-1116.

3.   Vasil V, Castillo AM, Fromm ME and Vasil IK (1992). Herbicide Resistant Fertile Transgenic Wheat Plants Obtained by Microprojectile Bombardment of Regenerable Embryogenic Callus. *Bio/Technology,* **10**: 667-674.

4.   Weeks JT, Anderson OD and Blechl AE (1993). Rapid production of multiple independent lines of fertile transgenic wheat (*Triticum aestivum*). *Plant Physiology, 102*: 1077-1084.

5.   Altpeter F, Vasil V, Srivastava V, Stöger E and Vasil IK (1996). Accelerated production of transgenic wheat (*Triticum aestivum* L.) plants. *Plant Cell Reports, 16*: 12-17.

6.   Zhang L, Rybczynski JJ, Langenberg WG, Mitra A and French R (2000). An efficient wheat transformation procedure: transformed calli with long-term morphogenic potential for plant regeneration. *Plant Cell Reports, 19*: 241-250.

7.   Rasco-Gaunt S, Riley A, Barcelo P and Lazzeri PA (1999). Analysis of particle bombardment parameters to optimise DNA delivery into wheat tissues. *Plant Cell Reports, 19*: 118-127.

8.   Pastori GM, Wilkinson MD, Steele SH, Sparks CA, Jones HD and Parry MAJ (2001). Age-dependent transformation frequency in elite wheat varieties. *Journal of Experimental Botany, 52*: 857-863.

9.   Rasco-Gaunt S, Riley A, Cannell M, Barcelo P and Lazzeri PA (2001). Procedures allowing the transformation of a range of European elite wheat (*Triticum aestivum* L.) varieties *via* particle bombardment. *Journal of Experimental Botany, 52*: 865-874.

10.   Cheng M, Fry JE, Pang SZ, Zhou HP, Hironaka CM, Duncan DR, Conner TW and Wan YC (1997). Genetic transformation of wheat mediated by *Agrobacterium tumefaciens*. *Plant Physiology, 115*: 971-980.

11.   Weir B, Gu X, Wang MB, Upadhyaya N, Elliott AR and Brettell RIS (2001). *Agrobacterium tumefaciens*-mediated transformation of wheat using suspension cells as a model system and green fluorescent protein as a visual marker. *Australian Journal of Plant Physiology, 28*: 807-818.

12.   Khanna HK and Daggard GE (2003). *Agrobacterium tumefaciens*-mediated transformation of wheat using a superbinary vector and a polyamine-supplemented regeneration medium. *Plant Cell Reports, 21*: 429-436.

13.   Wu H, Sparks C, Amoah B and Jones HD (2003). Factors influencing successful *Agrobacterium*-mediated genetic transformation of wheat. *Plant Cell Reports, 21*: 659-668.

14.   Cannell ME and Jones HD (2001). Transgenic wheat. In: Jaiwal PK, Singh RP (eds.), *Plant Genetic Engineering: Improvement of Food Crops.*, Sci-Tech Pub. Co., Houston, Texas. Vol. 2, (in press).

15.   Jefferson RA, Kavanagh TA and Bevan MW (1987). GUS fusions – Beta-glucuronidase as a sensitive and versatile fusion marker in higher plants. *The EMBO Journal, 6*: 3901-3907.

16.   Rasco-Gaunt S, Riley A, Lazzeri PA and Barcelo P (1999). A facile method for screening for phosphinothricin (PPT)-resistant transgenic wheats. *Molecular Breeding, 5*: 255-262.

17.   He GY and Lazzeri PA (2001). Improvement of somatic embryogenesis and plant regeneration from durum wheat (*Triticum turgidum* var. *durum* Desf.) scutellum and inflorescence cultures. *Euphytica, 119*: 369-376.

18.   Lamacchia C, Shewry PR, Di Fonzo N, Forsyth JL, Harris N, Lazzeri PA, Napier JA, Halford NG *et al.* (2001). Endosperm-specific activity of a storage protein gene promoter in transgenic wheat seed. *Journal of Experimental Botany, 52*: 243-250.

19.   Barcelo P and Lazzeri PA (1995). Transformation of cereals by microprojectile bombardment of immature inflorescence and scutellum tissues. In: Jones H (ed.), *Methods in Molecular Biology: Plant Gene Transfer and Expression Protocols*, Humana Press, Totowa. Vol. 49, pp. 113-123.

20.   Barro F, Martin A, Lazzeri PA and Barcelo P (1999). Medium optimisation for efficient somatic embryogenesis and plant regeneration from immature inflorescences and immature scutella of elite cultivars of wheat, barley and tritordeum. *Euphytica, 108*: 161-167.

21.   Barro F, Cannell ME, Lazzeri PA and Barcelo P (1998). The influence of auxins on transformation of wheat and tritordeum and analysis of transgene integration patterns in transformants. *Theoretical and Applied Genetics, 97*: 684-695.

22.   Rasco-Gaunt S and Barcelo P (1999). Immature inflorescence culture of cereals: a highly responsive system for regeneration and transformation. In: Hall RD (ed.), *Methods in Molecular Biology: Plant Cell Culture Protocols*, Humana Press, Totowa. Vol. 111, pp. 71-81.

23.   Barcelo P, Hagel C, Becker D, Martin A and Lörz H (1994). Transgenic cereal (tritordeum) plants obtained at high efficiency by microprojectile bombardment of inflorescence tissue. *The Plant Journal, 5*: 583-592.

24.   Barcelo P, Vazquez A and Martin A (1989). Somatic embryogenesis and plant regeneration from tritordeum. *Plant Breeding*, **103**: 235-240.
25.   Amoah BK, Wu H, Sparks C and Jones HD (2001). Factors influencing *Agrobacterium*-mediated transient expression of *uid*A in wheat inflorescence tissue. *Journal of Experimental Botany,* **52**: 1135-1142.

Chapter 3

GENETIC TRANSFORMATION OF BARLEY (*HORDEUM VULGARE* L.) BY CO-CULTURE OF IMMATURE EMBRYOS WITH *AGROBACTERIUM*

G. HENSEL AND J. KUMLEHN

Institute of Plant Genetics and Crop Plant Research (IPK), Plant Reproductive Biology Group, Corrensstr. 3, D-06466 Gatersleben, Germany. Corresponding author: G. Hensel, E-mail: hensel@ipk-gatersleben.de

1. INTRODUCTION

1.1. Importance and Distribution of Barley

Barley is one of the major and most widely distributed crops worldwide. It has already been used intensively as a model species for cereals in the area of classical genetics and is still today of extraordinary importance as an experimental object for fundamental and applied research. In past years, a tremendous amount of genetic resources have been generated which include genomic DNA sequences, full-length cDNAs and expressed sequence tags. In barley, sequences from more than twenty thousand different genes are available (*1, 2*). For a comprehensive functional analysis of an expressed sequence, a known, reliable transformation technology is required. By standard approaches such as overexpression, knock out, translational reporter gene-fusions or promoter-reporter gene combination, biological functions and expression patterns can be assigned to a given gene. In 2002, the world production of barley was 132,215,617 metric tonnes (*3*). Main producers are Australia, Europe, Canada, Russian Federation and Ukraine. Barley constitutes a key input for the malting industry and breweries as well as for livestock production. Recently, it has been shown that barley can be successfully employed as a bioreactor to produce large quantities of valuable protein (*4*).

1.2. Need for Genetic Improvement

With regard to the limitations of classical breeding, genetic engineering has increasingly been considered to have a huge potential to continuously generate new variation as a basis for ongoing crop improvement. Genetic transformation of barley will contribute to improve yield, malting or feed quality, resistance towards pathogens as well as tolerance to harsh conditions (*5-13*). Genetic engineering approaches may also be pursued to produce valuable products, e.g. technical enzymes, pharmaceuticals, recombinant antibodies or neutriceuticals.

35

I.S. Curtis (ed.), Transgenic Crops of the World - Essential Protocols, 35-44.
© 2004 Kluwer Academic Publishers. Printed in the Netherlands.

1.3. Genetic Transformation of Barley

Stable transformation of barley was first reported in 1994 (*14*), and since then there have been many reports of successful barley transformation (*15*). Most studies have used microparticle bombardment for transgene insertion, but the use of *Agrobacterium tumefaciens* for transformation of barley (*16*) offers an alternative methodology. When conducting tissue culture of immature scutella as was developed for biolistic transformation, an *Agrobacterium*-based method can be considerably more efficient. Furthermore, *Agrobacterium*-mediated transformation was shown to result in higher quality transgenic plants regarding the intactness of the sequences to be transferred, the integration of backbone DNA, the transcriptional activity of the integration sites within the target genome and the transgene copy number. The method described here was developed for the model cultivar Golden Promise and is based upon protocols elaborated by Tingay *et al.*, 1994 (*16*), Trifonova *et al.*, 2001 (*17*) and Kumlehn (unpublish.). Unfortunately, *Agrobacterium*-mediated transformation of barley is not genotype-independent. *Agrobacterium*-mediated transformation of immature barley embryos of lines other than Golden Promise has not been reported so far.

2. MATERIALS

2.1. A. tumefaciens *Strain*

Agrobacterium tumefaciens strain AGL1 (*18*) carrying the binary vector pYF133 (*19*) was used for the transformation studies. The binary vector includes an *hpt* selectable marker gene driven by the CaMV35S-promoter, an *sgfp* (S65T) reporter gene (*20*) driven by the maize ubiquitin promoter with first intron (*21*) and the vector backbone from pCAMBIA1300 (*22*) with its borders derived from a nopaline Ti plasmid. The vector plasmid was introduced into AGL1 by electroporation. The primers used for PCR analysis of the transgenic plants are shown (Table 1).

Table 1. PCR-Primer used for the analysis of transgenic plants

Primer	Sequence 5' – 3'
35S-F3	TTC GCA AGA CCT TCC TCT A
Hyg-R1	GTG CCG ATA AAC ATA ACG ATC
GH-GFP-F1	GGT CAC GAA CTC CAG CAG GA
GH-GFP-R2	TAC GGC AAG CTG ACC CTG AA

2.2. Donor Plants

Seeds of barley (*Hordeum vulgare* L.) cultivar Golden Promise were grown in pots in a growth chamber at 14/12°C day/night temperature with a 12 h photoperiod and 370 µmol/m^2/sec for up to 3 months. After this time, plants were transferred to controlled glasshouse conditions with 18/16°C day/night temperature and a 16 h photoperiod at 463 $\mu mol/m^2/sec$.

2.3. Growth of Agrobacteria

Agrobacteria were grown overnight at 28°C in 10 ml of MG/L medium (*24*, Table *2*) in 100 ml Erlenmeyer flasks with shaking at 180 rpm. For culture initiation, a glycerol stock (200 µl from a growing culture with an OD_{600} of 0.2 and 200 µl of 15% glycerol) from −80°C were thawed and added to the medium.

Table 2. Medium composition for Agrobacteria

	MG/L
Macroelements (mg/l)	
KH$_2$PO$_4$	250
NaCl	100
MgSO$_4$·7H$_2$O	100
Vitamins (mg/l)	
Biotin	0.001
Amino acids (g/l)	
L-Glutamic acid	1.0
Sugars (g/l)	
Mannitol	5.0
Miscellaneous (g/l)	
Tryptone	5.0
Yeast extract	2.5

2.4. Plant Tissue Culture Media

The culture media used are summarised (Table 3). The protocol included 3 different media, a liquid co-culture medium (CCM) as well as semi-solid media for callus induction (CIM) and regeneration (RM). CCM and CIM are based on MS salts (*23*) supplemented with additional components as shown (Table 3). The pH was adjusted to 5.8 prior to filter-sterilization of the solutions. For semi-solid media, one part of concentrated solution were appropriately mixed with 3 parts of either GelRite (Duchefa, The Netherlands) or Phytagel (Sigma) which had been autoclaved with distilled water.

Table 3. Media composition for plant material

	CCM	CIM	RM
Macroelements (mg/l)			
NH_4NO_3	1650	1650	320
KNO_3	1900	1900	3640
KH_2PO_4	170	170	340
$CaCl_2 \cdot 2H_2O$	441	441	441
$MgSO_4 \cdot 7H_2O$	331	331	246
Microelements (mg/l)			
H_3BO_3	6.20	6.20	3.10
$MnSO_4 \cdot 4H_2O$	22.40	22.40	11.20
$ZnSO_4 \cdot 7H_2O$	8.60	8.60	7.20
KI	0.83	0.83	0.17
$Na_2MoO_4 \cdot 2H_2O$	0.25	0.25	0.12
$CuSO_4 \cdot 5H_2O$	0.025	1.275	0.1275
$CoCl_2 \cdot 6H_2O$	0.025	0.025	0.024
$Na_2FeEDTA$	36.70	36.70	36.70
Vitamins (mg/l)			
Nicotinic acid	0.50	0.50	1.00
Pyridoxine HCl	0.50	0.50	1.00
Thiamine HCl	1.10	1.10	10.00
Amino acids (mg/l)			
L-Cysteine	800		
L-Glutamine			146
L-Proline	690	690	
Sugars (g/l)			
Maltose monohydrate	30	30	36
Growth regulators (mg/l)			
DICAMBA	2.50	2.50	
6-BAP			0.225
Miscellaneous (g/l)			
Acetosyringone	0.098		
Casein hydrolysate	1	1	
Gelrite		35	
Myo-inositol	350.00	350.00	100.00
Phytagel			30

2.5. Isolation of Immature Embryos and Co-culture with **Agrobacterium**

For the isolation of immature embryos and their subsequent co-culture with *A. tumefaciens*, the following materials are needed:

1. Forceps and needles
2. Preparation microscope
3. 6-well Petri dishes (Greiner, Germany)

4. Vacuum pump
5. Pipettes and disposable tips (200-1000 µl and 1000-5000 µl, autoclaved)
6. Exsiccator and magnetic stirrer

3. METHODS

3.1. Growth of Donor Plants

1. Germinate seeds in a substrate mix ("Spezialmischung Petuniensubstrat", Klasmann, Germany) in a growth chamber with controlled conditions (14/12°C day/night, 12 h photoperiod, 370 µmol/m^2/sec) for 10-12 weeks (*see* Note 1).
2. Fertilize the plants at the beginning of tillering with Osmocote (40 g/7.5 l) (*see* Note 2).
3. At 2 week intervals, plants need to be watered with 0.3% Hakaphos Blau (Compo, Germany) until stems start to elongate (*see* Note 3).
4. Transfer plants to a temperature-controlled glasshouse (18/16°C day/night, 16 h photoperiod, 463 µmol/m^2/sec).
5. Plants are to be watered once with 0.3% Hakaphos Grün (Compo, Germany) when heading stage commences (*see* Note 3).

3.2. Isolation of Immature Embryos

1. Harvest developing caryopses at around 12 days post pollination.
2. Immerse caryopses for 60 sec in 70% ethanol.
3. Incubate the material in 5% sodium hypochlorite for 20 min.
4. Rinse the material 4-5 times in sterile, distilled water.
5. Prepare a 6-well plate with 2.5 ml CCM per well.
6. Excise immature embryos (1-1.5 mm in size) from the caryopses by using forceps and a lanzette needle (*see* Note 4).
7. Remove the embryonic axes of the embryos by use of a lanzette needle.
8. Transfer the immature embryos to liquid CCM (*see* Notes 5-7).

3.3. Growth of Bacteria

1. Add a glycerol stock of bacterial cells to 10 ml of mg/l medium without antibiotics (*see* Note 8).
2. Incubate bacterial cells at 28°C with shaking at 180 rpm for around 24 h until an OD$_{600}$ of 0.2-0.25 is obtained.

3.4. Co-culture of Immature Embryos and Agrobacteria

1. Assess the optical density of the Agrobacteria.
2. Remove the CCM completely from the 6-well plate by using a 1000-5000 µl pipette and add 600 µl of bacterial culture per well.

3. Place the plate in an exsiccator and vacuum infiltrate for 60 sec at 500 mbar.
4. Keep the plate for 10 min inside the laminar hood without agitation.
5. Remove the bacteria and wash the embryos twice using 2.5 ml of CCM per well with a 5 min period between the washing steps.
6. Add 2.5 ml of CCM per well and incubate the plate at 21°C in the dark for 2-3 days without agitation.

3.5. Callus development

1. After co-culture, the embryos are transferred to CIM supplemented with 50 mg/l hygromycin B (Boehringer, Germany) (*see* Note 9).
2. Place 10 embryos per 10 cm Petri dish, the scutellum side facing the medium and incubate the sealed dishes at 24°C in the dark.
3. Replace the medium after 2 weeks (Figs. 1A and B).

3.6. Regeneration and Rooting

1. Four weeks after transformation, the calli obtained need to be transferred to RM supplemented with 25 mg/l hygromycin B (*see* Note 10). Transfer the dishes to the light at 24°C, 16 h photoperiod, 370 μmol/m^2/sec.
2. Replace the RM every 2 weeks until the emergence of new regenerants has ceased (Figs. 1C and D).
3. Transfer the regenerants obtained (leaf length 2-3 cm) to glass tubes (height 100 mm and outer diam. 25 mm; Schütt, Germany) containing 4.5 ml of RM supplemented with 25 mg/l hygromycin B until the plants form roots (Figs. 1E and F).
4. Transfer the plants to the glasshouse and allow to grow to maturity under the same conditions as was described for the donor plants (*see* Note 11).

3.7. Analysis of Transgenic Plants

After successful transfer of the plants to soil, genomic DNA of the plants can be analysed for integration of the transgenes. The presence of the transgene can be tested by PCR. The transgene copy number should be analysed by Southern blotting.

3.7.1. PCR Analysis

For PCR analysis, genomic DNA can be isolated by use of commercially available extraction kits (e.g. DNAzol, Invitrogen, Germany) according to the manufacturer's instructions.

1. Harvest approx. 100 mg leaf material and maintain the sample in liquid nitrogen until DNA extraction.
2. Isolate genomic DNA and estimate the DNA content.

3. Perform standard PCR reactions with the primers specified (Table 1) using 100 ng genomic DNA per plant.
4. Confirm the PCR products by gel electrophoresis (Fig. 1G).

3.7.2. Southern Blot Analysis

Plants which are positive by PCR should be further analysed for their copy number. For this purpose larger amounts of high quality DNA is required. The protocol described by Sambrook *et al.* (*25*) is recommended.
1. Prepare genomic DNA of leaf material and estimate the DNA content.
2. Digest 25 µg genomic DNA with the appropriate restriction enzyme.
3. Run a gel for separation of the fragments and transfer to a membrane.
4. Hybridize the genomic DNA with a gene-specific probe (labelled either with radioactive dCTP or DIG).
5. Place a film or screen on the membrane and incubate as long as needed.
6. Develop the film or screen and analyze the resulting signal pattern (Fig. 1H).

4. NOTES

1. The substrate mix is a special white peat substrate plus clay for an improved buffering effect. Seeds were germinated individually in 18 cm pots with a volume of 2.5 litres.
2. Osmocote is a general long-term fertilizer that contains 19 % N, 6 % P and 12 % K. Hakaphos Blau is a general fertilizer that contains 15 % N, 10 % P and 15 % K. Hakaphos Grün is a general fertilizer that contains 20 % N, 5 % P and 10 % K.
3. The developmental stage of the embryos is more crucial than their size. For the protocol described here, embryos at the transition stage from translucent to white are ideal for transformation.
4. Contradictory results have been published regarding the effect of acetosyringone on *Agrobacterium*-mediated transformation of immature barley embryos (*7, 16, 19*). The addition of acetosyringone results in higher transformation efficiency under the conditions described here.
5. The addition of L-cysteine was reported to prevent browning of the embryos and to increase the transformation efficiency in soybean (*26*).
6. Thirty to forty embryos can be cultured per well.
7. In general, there is a risk to drop a plasmid when bacteria are grown in the absence of antibiotics. In the protocol described, a substantial loss of the vectors could not be detected (checked several times via plasmid-preparation from the *Agrobacteria* used for transformation).
8. The increased $CuSO_4$ concentration (*17*) results in formation of more green plants compared with the medium described by Tingay *et al.* (*16*).
9. FHG medium (*14-16, 19, 27*) was successfully used for regeneration of shoots in a number of published experiments. Yet, a direct comparison of the two media revealed that RM is superior to FHG.
10. The transformation efficiency obtained by the method described is between 15-60 % without escapes.

CONCLUSIONS

We present here a contemporary technique to genetically transform barley. This protocol will be suitable for comprehensive functional analyses of gene and promoter sequences on a large scale. Furthermore, the protocol constitutes a promising basis for applied research aiming to improve e.g. disease resistance, tolerance towards abiotic stresses as well as product quality of barley. For crop improvement approaches it will be necessary to couple the method with techniques, which enable the removal of selectable markers such as co-transformation or the use of recombination sites. Another important challenge is to make genetic transformation of barley more genotype-independent. The high efficiency of the protocol presented here promises successful screening of current elite lines of both spring and winter barley, yet some modifications might be indispensible to adapt the method to another genetic background.

ACKNOWLEDGEMENTS

The authors thank C. Marthe and I. Otto for their excellent technical assistance. This work was performed in the scope of the BMBF project PTJ-BIO/0312281. Thanks also to H.-H. Steinbiss (Max-Planck-Institute for Plant Breeding Research, Cologne) for his constructive advice and discussions.

REFERENCES

1.   http://pgrc.ipk-gatersleben.de/research.php.
2.   http://ukcrop.net/barley.html.
3.   FAO Statistics (2002). http://apps.fao.org/page/collections?subset=agriculture.
4.   Schünmann PHD, Coia G and Waterhouse PM (2002). Biopharming the SimpliRED™ HIV diagnostic reagent in barley, potato and tobacco. *Molecular Breeding*, **9**: 113-121.
5.   Matthews PR and Jacobsen J (2001). *Proceedings of the 10th Australian Barley Technical Symposium*.
6.   McGrath PF, Vincent JR, Lei CH, Pawlowski WP, Torbert KA, Gu W, Kaeppler HF, Wan Y *et al.* (1997). Coat protein-mediated resistance to isolates of barley yellow dwarf in oats and barley. *European Journal of Plant Pathology*, **103**: 695-710.
7.   Patel M, Johnson JS, Brettell RIS, Jacobsen J and Xue GP (2000). Transgenic barley expressing a fungal xylanase gene in the endosperm of the developing grains. *Molecular Breeding*, **6**: 113-123
8.   Matthews PR, Wang M-B, Waterhouse PM, Thornton S, Fieg SJ, Gubler F and Jacobsen JV (2001). Marker gene elimination from transgenic barley, using co-transformation with adjacent 'twin T-DNAs' on a standard *Agrobacterium* transformation. *Molecular Breeding*, **7**: 195-202.
9.   Dahleen LS, Okubara A and Blechl AE (2001). Transgenic approaches to combat fusarium head blight in wheat and barley. *Crop Science*, **41**: 628-637.
10.   Hayes PM, Castro A, Marquez-Cedillo L, Corey A, Henson C, Jones BL, Kling J, Mather D *et al.* (2003) Genetic diversity for quantitatively inherited agronomic and malting quality traits. In: von Bothmer R, Knupffer H, van Hintum T, Sato K (eds.), *Diversity in Barley*. Elsevier Scientific Publishers, Amsterdam, The Netherlands (in press).
11.   Horvath H, Huang J, Wong O, Kohl E, Okita T, Kannangara CG and Wettstein D von (2000). The production of recombinant proteins in transgenic barley grains. *Proceedings of the National Academy of Sciences USA*, **97**: 1914-1919.
12.   Nuutila AM, Ritala A, Skadsen, RW, Mannonen L and Kauppinen V (1999). Expression of fungal thermotolerant endo 1,4-β-glucanase in transgenic barley seeds during germination. *Plant Molecular Biology*, **41**: 777-783.
13.   Xue GP, Patel M, Johnson JS, Smyth DJ and Vickers CE (2003). Selectable marker-free transgenic barley producing a high level of cellulase (1,4-β-glucanase) in developing grains. *Plant Cell Reports*, **21**: 1088-1094.
14.   Wan Y and Lemaux PG (1994). Generation of large numbers of independently transformed fertile barley plants. *Plant Physiology*, **104**: 37-48.

15. Lemaux PG, Cho MJ, Zhang S and Bregitzer P (1998). Transgenic Cereals: *Hordeum vulgare* L. In: Vasil IK (ed.), *Molecular Improvement of Cereal Crops*. Kluwer Academic Publishers, pp. 255-316.

16. Tingay S, McElroy D, Kalla R, Feig S, Wang M, Thornton S and Brettell R (1997). *Agrobacterium tumefaciens*-mediated barley transformation. *The Plant Journal*, **11**: 1369-1376.

17. Trifonova A, Madsen S and Olesen A (2001). *Agrobacterium*-mediated transgene delivery and integration into barley under a range of *in vitro* culture conditions. *Plant Science*, **162**: 871-880.

18. Lazo GR, Stein PA and Ludwig RA (1991). A DNA transformation-competent *Arabidopsis* genomic library in *Agrobacterium*. *Bio/Technology*, **9**: 963-967.

19. Fang Y-D, Akula C and Altpeter F (2002). *Agrobacterium*-mediated barley (*Hordeum vulgare* L.) transformation using green fluorescent protein as a visual marker and sequence analysis of the T-DNA::barley genomic DNA junctions. *Journal of Plant Physiology*, **159**: 1131-1138.

20. Chiu W-L, Niwa Y, Zeng W, Hirano T, Kobayashi H and Sheen J (1996). Engineered GFP as a vital reporter in plants. *Current Biology*, **6**: 325-330.

21. Christensen AH and Quail PH (1996). Ubiquitin promoter-based vectors for high-level expression of selectable and/or screenable marker genes in monocotyledonous plants. *Transgenic Research*, **5**: 213-218.

22. Roberts CS, Rajagopal S, Smith LA, Nguyen TA, Yang W, Nugroho S, Ravi KS, Vijayachandra K *et al.* (1996) A comprehensive set of modular vectors for advanced manipulations and efficient transformation of plants by both *Agrobacterium* and direct DNA uptake methods. *5th Annual Meeting National Rice Biotechnology Network Proceedings*. IARI. New Delhi. November 13-16, 1996.

23. Murashige T and Skoog F (1962). A revised medium for rapid growth and bioassays with tobacco tissue cultures. *Physiologia Plantarum*, **15**: 473-497.

24. Garfinkel, DJ, and Nester EW (1980). *Agrobacterium tumefaciens* mutants affected in crown gall tumorigenesis and octopine catabolism. *Journal of Bacteriology*, **144:** 732–743.

25. Sambrook J, Fritsch EF and Maniatis T (1989). *Molecular Cloning: A Laboratory Manual* 2nd edn., Cold Spring Harbor Laboratory Press, New York, USA.

26. Olhoft PM and Somers DA (2001). L-Cysteine increases *Agrobacterium*-mediated T-DNA delivery into soybean cotelydonary-node cells. *Plant Cell Reports*, **20**: 706-711.

27. Hiei Y, Ohta S, Komari T and Kumashiro T (1994). Efficient transformation of rice (*Oryza sativa* L.) mediated by *Agrobacterium* and sequence analysis of the boundaries of the T-DNA. *The Plant Journal*, **6**: 271-282.

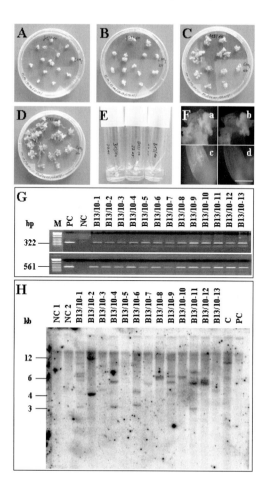

Figure 1. Procedure for the genetic transformation of barley by co-culture with A. tumefaciens. Barley immature embryo-derived callus culture 2 (A) and 4 (B) weeks after co-culture with Agrobacteria on CIM medium supplemented with 50 mg/l hygromycin B. Regenerating shoots after 2 (C) and 3 (D) weeks on RM containing 25 mg/l hygromycin B (scale bar = 2 cm). Shoots were separated from calli and transferred to tubes with the same medium (E) (scale bar = 2 cm). GFP-fluorescence analysis of callus (a, b) and an emerging shoot (c, d) (scale bar = 2 mm) (F). Gel analysis of PCR products (G) by using hpt- *(upper) or* gfp- *(lower) specific primers and genomic DNA of barley plants obtained after transformation with a binary plasmid harbouring the* gfp *and* hpt *genes within the same T-DNA. PC: plasmid control, NC: negative control (wt), M: 1 kb ladder. Southern blot analysis (H) of the same barley plants used for PCR (G). 25 µg genomic DNA samples were digested with* BamHI, *separated on a 0.8% agarose gel, transferred to a nylon membrane and hybridized against a* gfp-*specific probe. PC: plasmid control, NC: negative control (wt), C: DNA of a* gfp-*positive barley plant previously confirmed by Southern blot analysis.*

Chapter 4

MAIZE TRANSFORMATION

K. WANG AND B. FRAME

Plant Transformation Facility, Department of Agronomy, Iowa State University, Ames, IA 50011-1010, USA. Corresponding author: K. Wang. E-mail: kanwang@iastate.edu

1. INTRODUCTION

Maize (*Zea mays* L.) is the world's leading staple grain crop for food and feed. It is grown on over 140 million hectares worldwide with the USA accounting for 40.3% of global production in 2001/2002, followed by China (19.1%), EU-15 (6.6%), Brazil (5.9%), Mexico (3.4%), and Argentina (2.4%). According to the Economic Research Service of USDA (*1*), about 75 million acres of land in USA were planted to corn in 2002, with an average yield of approx. 130 bushels per acre (8.2 tonnes per hectare). While the majority of this crop (58%) is used as the main energy ingredient in livestock feed, maize is also processed into a multitude of food and industrial products including fuel ethanol (9%), high fructose corn syrup (6%), starch (3%), sweeteners (2%), corn oil, beverage and industrial alcohol (*2*).

Originating in Mexico, maize has evolved from a wild grass (a strain of teosinte, *Zea mays* ssp. *parviglumis*) to a highly productive domesticated crop possessing the largest kernel size of all cereal seeds. Its ranking as the most important industrial crop today is attributed to successful human invention and intervention (*3*). USA maize yields tripled between 1920 and 1990 due to rigorous breeding programs and improved cultural practices. While agricultural practices such as higher planting density and use of nitrogen fertilizers have accounted for 40-50% of USA maize yield gain since the 1930s, the remaining 50-60% is attributed to the genetic yield potential of US corn belt hybrids possessing improved tolerance to abiotic and biotic stresses (*4*). Because modern maize does not exist in the wild, its very survival and status depend on continued human intervention.

While conventional breeding programs continue to play a major role in maize improvement, recent technologies such as molecular marker assisted breeding and genetic transformation are showing great potential. Genetic transformation, a technique for inserting genetic materials from an unrelated species into maize, allows breeders to improve corn yield or enhance its value. The USA was the first country to commercialize transgenic maize in 1996 (*5*) with the introduction of *Bt* (*Bacillus thuringiensis*) maize to control European corn borer. Herbicide tolerant maize was commercialized in 1997 followed by the stacked genes of *Bt* and herbicide tolerance in 1998 (*6*). In 2002, genetically engineered corn occupied 12.4 million hectares, or 21% of the global area of transgenic crops (*7*).

45

I.S. Curtis (ed.), Transgenic Crops of the World - Essential Protocols, 45-62.
© 2004 Kluwer Academic Publishers. Printed in the Netherlands.

The first report of an attempt to genetically transform maize was published some 35 years ago, when Coe and Sarkar (*8*) injected total maize seedlings DNA from a purple-anthered variety to apical meristems of developing maize seedling of a regular variety. No positively transformed plants were obtained. The authors attributed the lack of any phenotypic change during the development of their recipient plants to the incompetence of the targeted cells and an ineffective delivery mechanism.

During the formative years of plant transformation in the 1980s, *Agrobacterium*-mediated transformation – a biological gene delivery system – was considered an effective way to transform dicotyledonous, but not monocotyledonous plants (*9*). Efforts in maize transformation were therefore focused primarily on using physical delivery systems to target cell cultures or protoplasts. Black Mexican Sweet (BMS) maize suspension cultures were widely used in these early optimization experiments (*10, 11*), but plants could not be produced from this cell line.

The first transgenic maize plants were reported by Rhodes *et al.* (*12*) and were obtained by electroporating protoplasts isolated from a regenerable cell culture of the maize inbred line A188 with a selectable *npt*II gene. Unfortunately, these transgenic plants were sterile. It was not until 1990 that Gordon-Kamm *et al.* (*13*) succeeded in producing fertile, transgenic maize plants by using the particle gun to deliver the *bar* gene to embryogenic suspension culture cells and selecting this material on medium containing the herbicide bialaphos to recover transgenic events. This work represents a milestone in the history of maize transformation and now, a decade later, the biolistic gun mediated transformation method is routine in most maize transformation laboratories.

While it is considered a reliable method for obtaining transgenic maize, the frequency of high copy number events recovered using the biolistic delivery method has been a challenging issue for researchers. Multiple transgene insertion often leads to transgene silencing in subsequent generations (*14*). Conversely, the *Agrobacterium*-mediated transformation method theoretically inserts fewer copies of a more intact piece of DNA into the plant genome.

Early efforts to transform maize using the *Agrobacterium*-mediated method include that of Grimsley *et al.* (*15*) in which the authors were the first to demonstrate that *Agrobacterium* was capable of infecting maize. This was followed by observations of *Agrobacterium* attachment to the surfaces of maize cells (*16*), identification of *vir*-inducing substances by some monocots, including maize (*17*), and expression of the *gus*A gene in *Agrobacterium*-infected maize tissues (*18-21*). In 1996, Ishida *et al.* (*22*) reported obtaining fertile maize plants from a large number of transgenic events recovered after infecting immature embryos of the maize inbred line A188 with an *Agrobacterium* strain carrying a "super-binary" vector system in which additional copies of *virB, C* and *G* genes are present (*23*). This seminal publication represents the first well-documented evidence for reproducibly transforming maize using the *Agrobacterium*-mediated delivery method. Subsequent work has demonstrated the efficacy of using a standard *Agrobacterium* vector system (*24*) for maize transformation as well.

Fertile transgenic maize plants can be produced using a number of other delivery systems, including electroporation of immature embryos or embryonic callus cultures (*25*), silicon carbide whisker-mediated embryogenic cell suspension culture (*26*) or callus cultures (*27*), and PEG-mediated protoplast transformation (*28*). There are several review articles addressing each system in detail (*29-31*).

In this chapter, we will focus on the two most efficient maize transformation protocols that are also the standard systems used in our laboratory, the biolistic gun and the *Agrobacterium*-mediated (standard binary vector) systems. Briefly, we use the Hi II hybrid (*32*) and the herbicide bialaphos to select transgenic events expressing the *bar* gene. Using the biolistic gun we target either immature zygotic embryos or embryogenic callus cultures derived from the same (*33*) and have reported transformation efficiencies ranging from 4-18% for the embryos and 10% for callus. Using the *Agrobacterium*-mediated transformation method we target immature zygotic embryos and a standard binary vector for an average efficiency of 5% (*24*).

2. MATERIALS

2.1. Plasmids and A. tumefaciens *Strains*

Plasmid DNA used for biolistic gun-mediated maize transformation usually contains a selectable marker gene cassette and a gene of interest (GOI) cassette. These two gene cassettes can be on one plasmid or two separate plasmids. Bombardment of a two-plasmid DNA mixture is called co-bombardment. One of the most frequently used selectable marker genes for maize transformation is the *bar* gene (*34*), the phosphinothricin acetyl transferase gene from *Streptomyces hygroscopicus* that confers resistance to the herbicide phosphinothricin and its derivatives (*35*).

Agrobacterium tumefaciens strain EHA101 (*36*) carries the 11.6 kb standard binary vector pTF102 (*24*). pTF102 contains a broad host range origin of replication (pVS1, *37*) and a spectinomycin-resistant marker gene (*aad*A) for bacterial selection. The CaMV 35S promoter is used to drive both the *bar* selectable marker gene and the *gus* reporter gene. A tobacco etch virus (TEV) translational enhancer (*38*) was inserted at the 5' end of the *bar* gene and the soybean vegetative storage protein terminator (*39*) was cloned to the 3' end of the *bar* gene. The *gus*A gene contains a portable intron in its codon region (*40*) to prevent GUS activity in *Agrobacterium* cells.

2.2. Stock Solutions and Culture Media for A. tumefaciens

1. Kanamycin sulphate (Sigma): 10 mg/ml stock in ddH$_2$O. Sterilize by filtration through a 0.2 μm membrane (Fisher Scientific Inc, Pittsburgh, PA, USA). Dispense in 0.25 ml aliquots in Eppendorf tubes and store at – 20°C for up to 6 months.
2. Spectinomycin sulphate (Sigma): 100 mg/ml stock in ddH$_2$O. Sterilize by filtration, aliquot (0.05 ml) and store at – 20°C for up to 6 months.

3. Chloramphenicol (Sigma): 25 mg/ml stock in ddH$_2$O. Sterilize by filtration, aliquot (0.05 ml) and store at – 20°C for up to 6 months.
4. YEP medium (*41*): 5 g/l yeast extract (Fisher), 10 g/l peptone (Fisher), 5 g/l NaCl$_2$ (Fisher), pH 6.8. For solid medium, add 15 g/l Bacto-agar (Fisher). Appropriate antibiotics are added to autoclaved medium after it cools to 50°C. For the strain EHA101 containing pTF102, the final antibiotic concentrations in YEP are: 50 mg/l kanamycin (for maintaining of the disarmed Ti plasmid pEHA101), 100 mg/l spectinomycin (for maintaining the binary vector plasmid pTF102) and 25 mg/l chloramphenicol (for the *Agrobacterium* chromosome).

2.3. Plant Materials

Collect ears of the maize Hi II genotype (A188 x B73 origin, *32*) 9-13 days after pollination from glasshouse-grown plants. Use immature zygotic embryos as target tissue for bombardment (*33, 42*) or *Agrobacterium* infection (*24, 43*). Alternatively, dissect immature zygotic embryos and culture on N6E medium for initiation of Type II callus that can also be targeted for transformation using the biolistic gun (*33*). Because Type II callus is friable and fast growing, post transformation selection pressure can be effectively applied to these cells with minimal manipulation and no escapes (*33*). A major disadvantage of using Type II callus for maize transformation, however, is that it is not readily induced in most inbred germplasms.

2.4. Culture Media for Maize Transformation

2.4.1 Stock Solutions for Transformation

1. Bialaphos: 100 mg of Bialaphos (Shinyo Sangyo, Japan) is dissolved in 100 ml of ddH$_2$O. Stock solution (1 mg/ml) is filter-sterilized and stored at 4°C for up to 6 months.
2. Glufosinate: 100 mg of glufosinate ammonia (Sigma) is dissolved in 100 ml of ddH$_2$O. Stock solution (1 mg/ml) is filter-sterilized and stored at 4°C for up to 6 months.
3. Acetosyringone: 0.392 g of acetosyringone (Sigma) is dissolved in 10 ml of DMSO. This solution is diluted 1:1 with ddH$_2$O and filter-sterilized. Aliquots (0.5 ml) of stock solution (100 mM) are stored at –20°C for up to 6 months.
4. Cefotaxime: 1.0 g of cefotaxime (Phytotechnology Laboratories, USA) is dissolved in 5 ml ddH$_2$O. The stock solution (200 mg/ml) is filter-sterilized, aliquoted (0.625 ml) and stored at –20°C for up to 4 weeks.
5. Cysteine: 500 mg of L-cysteine (Gibco, Rockville, MD, USA) is dissolved in 5 ml of ddH$_2$O. The stock solution (100 mg/ml) is filter-sterilized and used the same day for making co-cultivation medium. Any extra, unused stock solution is discarded.
6. Silver Nitrate: 0.85 g of silver nitrate (Fisher) is dissolved in 100 ml of ddH$_2$O. The stock solution (50 mM) is filter-sterilized and stored at 4°C in a foil-wrapped container to avoid exposure to the light for up to 1 year.

2.4.2 Media for Biolistic-gun Transformation

All media described below use 100 x 15 mm Petri dish (Fisher). Chemical source is either from Sigma or Fisher unless stated otherwise.

1. N6E (callus initiation, *44*): 4 g/l N6 salts (*45*), 1 ml/l (1000X) N6 vitamin stock (*see* Note 1), 2 mg/l 2,4-D, 100 mg/l *myo*-inositol, 2.76 g/l L-proline, 30 g/l sucrose, 100 mg/l casein hydrolysate, 2.5 g/l Gelrite, pH 5.8. Filter-sterilized silver nitrate (25 μM) is added after autoclaving.
2. N6osm (pre and post bombardment osmotic treatment, *46*): 4 g/l N6 salts, 1 ml/l N6 vitamin stock, 2 mg/l 2,4-D, 100 mg/l *myo*-inositol, 0.69 g/l proline, 30 g/l sucrose, 100 mg/l casein hydrolysate, 36.4 g/l sorbitol, 36.4 g/l mannitol, 2.5 g/l Gelrite, pH 5.8. Filter-sterilized silver nitrate (25 μM) is added after autoclaving.
3. N6S (selection): 4 g/l N6 salts, 1 ml/l N6 vitamin stock, 2 mg/l 2,4-D, 100 mg/l myo-inositol, 30 g/l sucrose, 2.5 g/l Gelrite, pH 5.8. Filter-sterilized silver nitrate (5 μM) and bialaphos (2 mg/l) are added after autoclaving.

2.4.3 Media for Agrobacterium-mediated Transformation

All media described below use 100 x 25 mm Petri dishes (Fisher). Media is after Zhao *et al.* (*43*) with the addition of cysteine (300 mg/l) to co-cultivation medium and the use of cefotaxime instead of carbenicillin for counter-selection of *Agrobacterium* after co-cultivation.

1. Infection (liquid): 4 g/l N6 salts, 1 ml/l N6 vitamin stock, 1.5 mg/l 2,4-D, 0.7 g/l L-proline, 68.4 g/l sucrose, and 36 g/l glucose, pH 5.2. This medium is filter- sterilized and stored at 4°C. Acetosyringone is added immediately prior to use (100 μM).
2. Co-cultivation (make fresh and use within 8 days): 4 g/l N6 salts, 1.5 mg/l 2,4-D, 0.7 g/l L-proline, 30 g/l sucrose, and 3 g/l Gelrite, pH 5.8. Filter-sterilized N6 vitamins (1 ml/l), silver nitrate (5 μM), acetosyringone (100 μM), and L-cysteine (300 mg/l) are added after autoclaving.
3. Resting: 4 g/l N6 salts, 1.5 mg/l 2,4-D, 0.7 g/l L-proline, 30 g/l sucrose, 0.5 g/l MES, and 8 g/l purified agar, pH 5.8. Filter-sterilized N6 vitamins (1 ml/l), cefotaxime (250 mg/l), and silver nitrate (5 μM) are added after autoclaving.
4. Selection I: 4 g/l N6 salts, 1.5 mg/l 2,4-D, 0.7 g/l L-proline, 30 g/l sucrose, 0.5 g/l MES, and 8 g/l purified agar, pH 5.8. Filter-sterilized N6 vitamins (1 ml/l), cefotaxime (250 mg/l), silver nitrate (5 μM), and bialaphos (1.5 mg/l) are added after autoclaving.
5. Selection II: The same as Selection I except that bialaphos concentration is increased to 3 mg/l. This prevents excessive proliferation of non-transformed callus during this later phase of selection.

2.5. Culture Media for Regeneration

All media described below use 100 x 25 mm Petri plates. Media is after Armstrong and Green (*47*) and McCain *et al.* (*48*).

1. Regeneration I: 4.3 g/l MS Salts (*49*), 1 ml/l (1000X) MS vitamin stock (*see* Note 2), 100 mg/l *myo*-inositol, 60 g/l sucrose, 3 g/l Gelrite, pH 5.8. Filter-sterilized glufosinate ammonia (4 mg/l) is added after autoclaving. For regeneration of *Agrobacterium*-derived transgenic events, cefotaxime (250 mg/l) is also added after autoclaving.
2. Regeneration II: The same as Regeneration I except that sucrose concentration is reduced to 30 g/l.

3. METHODS

3.1. Biolistic Gun-mediated Transformation

3.1.1. Embryo Dissection

1. Dehusk ear and break off the top 1 cm of the cob. Into the tip of each ear, insert a pair of numbered forceps to help identify each ear and to serve as a "handle" for aseptic manipulation during dissection. In a laminar flow bench, place 2 or 3 ears in a single sterile mason jar, with forceps protruding.
2. Add ~600 ml of sterilizing solution [50% commercial bleach (5.25% hypochlorite) in water and 1 drop of surfactant Tween 20] to cover ears. During the 20 min disinfection, occasionally tap the mason jar on the bench surface to dislodge air bubbles. Pour off bleach solution and rinse the ears 3 times in generous amounts of sterilized water. The final rinse is drained off and the ears are ready for embryo dissection.
3. In a large (150 x 15mm) sterile Petri dish, cut off the kernel crowns (the top 1-2 mm) with a sharp scalpel blade. Steriguard 350 bead sterilizers (Inotech Biosystems International, Rockville, MD, USA) are used for intermittent re-sterilization of utensils throughout this protocol.
4. Excise the embryos by inserting the narrow end of a sharpened spatula between the endosperm and pericarp at the basipetal side of the kernel (towards the bottom of the cob) popping the endosperm out of the seed coat. This exposes the untouched embryo which is then gently coaxed onto the spatula tip and plated with the embryo-axis side down (scutellum side up), as follows:
    a. For immature zygotic embryo bombardment: onto a filter paper (Whatman No. 4, 5.5 cm, Fisher) laying on the surface of a plate of N6E medium. Embryos are arranged in a 2 cm^2 grid (30 embryos/plate) and bombarded 3 days after dissection.

b. For callus initiation: to N6E medium directly (30 embryos/plate). Type II callus developed from the immature embryo explant can then be used for bombardment.

5. Wrap plates with vent tape (Vallen Safety Supply, USA) and incubate at 28°C (dark).

3.1.2. Initiation of Type II Callus Lines for Bombardment

1. Incubate embryos dissected for callus initiation for 2 weeks. Friable, rapidly growing embryogenic callus will proliferate from the scutellum surface of ~100% of the immature zygotic embryo explants. Subculture callus to fresh N6E medium. Wrap plates with parafilm (28°C, dark).

2. To develop callus lines, each originating from an independent immature zygotic embryo, carry out weekly subcultures of this callus to fresh N6E medium using a 40X dissecting scope. Lines are used for bombardment beginning 4-6 weeks after initiation and are discarded within 4 months.

3.1.3. Pre-bombardment treatment

3.1.3.1. Immature Embryos

1. After 3 days incubation at 28°C, a raised ridge is visible at the base of the swollen immature zygotic embryos indicating that Type II callus initiation is underway and embryo scutellar tissue is ready for bombardment.

2. Four hours prior to bombardment, use two sterile forceps to gently transfer the embryos and filter paper onto osmotic medium (N6osm), centring the embryo grid, not the filter paper, on the 3.5 cm diam. circle drawn on the bottom-centre of each osmotic medium plate. Embryos should be facing scutellum side up at bombardment since it is from this surface that callus initiation continues and from which transformed cells are subsequently selected.

3.1.3.2. Callus

1. Draw a 3.5 cm diam. circle on the bottom, centre of a plate of osmotic medium (N6osm). This defines the target area to which callus pieces are loaded for bombardment.

2. Four hours prior to the bombardment, use a microscope to transfer 30 pieces (4 mm) of friable, rapidly growing callus directly to the medium surface within the target area.

3.1.4. Micro-projectile Bombardment

1. Use the PDS 1000/He biolistic gun (Bio-Rad, USA).
2. Gold particle preparation (*see* Note 3).
3. Load 650 psi rupture disk, and tighten the barrel on the helium chamber nozzle.

4. Load the launch assembly by first laying in place a stopping screen followed by an inverted, pre-loaded macro-carrier holder (*see* Note 3). Screw on the launch assembly lid to hold these parts in place.
5. Slide the launch assembly into place immediately below the helium nozzle, and set the gap distance (6 mm).
6. Slide the opened Petri dish containing the target tissue onto the shelf at a selected distance from the stopping screen (6 cm).
7. Close the vacuum chamber, pull a vacuum, and fire the gun in time for the rupture disk to break as soon as the vacuum reaches 28 inches of Hg.
8. Vent the chamber, remove the plate containing the bombarded tissue, and prepare the gun for the next bombardment by replacing the spent rupture disk, macro-carrier and stopping screen (disposables). Dispose of all plasmid waste in biohazard bags for autoclaving.
9. Repeat steps 3-8 for the next shot.

3.1.5. Post-bombardment Treatment

3.1.5.1. Immature Embryos

Gently wrap bombarded embryos (still on filter paper on N6osm) with vent tape and incubate at 28°C (dark). The next day (16-20 h after bombardment), transfer embryos individually off the filter paper on N6osm and onto the surface of N6E medium to continue callus initiation. Orient embryos scutellum side up.

3.1.5.2. Callus

Leave bombarded callus pieces on N6osm medium for 1 h and then transfer to N6E. The autonomy of each piece of callus is maintained for all subsequent transfers. Wrap plates with vent tape (28°C, dark).

3.2. Agrobacterium-*mediated Transformation*

3.2.1. Agrobacterium *Preparation*

The vector system, pTF102 in EHA101, is maintained on antibiotic-containing YEP plate. For weekly experiments, initiate bacteria cultures from stock plates stored for up to 1 month at 4°C. Refresh this "mother" plate from long-term, -80°C glycerol stocks every month.

3.2.2. Embryo Dissection

Embryos are dissected directly in the liquid infection medium therefore no initial orientation is necessary (*see* section 3.1.1.).

3.2.3. Agrobacterium *infection*

1. Grow *Agrobacterium* cultures for 3 days at 19°C (or 2 days at 28°C) on semi-solid YEP medium amended with antibiotics as described.
2. Scrape one full loop (3 mm) of bacteria culture from the plate and suspend in 5 ml infection medium supplemented with 100 μM acetosyringone in a 50 ml Falcon tube (Fisher). Fix the tube horizontally to a bench-top shaker or a Vortex Genie (Fisher) platform head and shake on low speed (~75 rpm) for 2-5 h at room temperature.
3. For infection, dissect up to 100 immature zygotic embryos (1.5-2.0 mm) directly to a 2 ml Eppendorf tube filled with bacteria-free infection medium (with 100 μM acetosyringone). Remove this wash. Wash embryos a second time with 1 ml of the same medium. Remove the final wash and add 1 ml of *Agrobacterium* suspension (adjusted to OD_{550} = 0.35-0.45 using a spectrophotometer). To infect the embryos, gently invert the tube 20 times before resting it upright for 5 min with embryos submerged in the *Agrobacterium* suspension. Embryos are not vortexed at any time during this procedure.

3.2.4. Co-cultivation

1. After infection, transfer embryos to the surface of co-cultivation medium containing 300 mg/l cysteine (*see* Note 4). Pipet off any excess *Agrobacterium* suspension. Orient embryos scutellum side up.
2. Wrap plates with vent tape and incubate at 20°C (dark) for 3 days.

3.2.5. Resting

After 3 days co-cultivation, transfer all embryos to resting medium at 28°C (dark) for 7 days. After 3 days on freshly prepared co-cultivation medium containing 300 mg/l cysteine, 1.5 mm embryos are flaccid and less swollen than embryos co-cultivated on non-cysteine medium. Although the cysteine treated embryos produce callus more slowly than non-cysteine treated embryos, do not discard them. Instead, transfer all embryos from co-cultivation medium to resting and from resting to selection media, regardless of their appearance (*see* Note 4).

3.3. Selection for Stable Transformation Events

3.3.1. Biolistic Gun Events

1. After 10-14 days on initiation medium (N6E), transfer the bombarded embryos to N6S selection medium (2 mg/l bialaphos) to begin the recovery of transformed cells.
2. Three weeks later, transfer embryos to fresh N6S. Within 6-8 weeks of bombardment, newly emerged, embryogenic callus sectors growing rapidly on bialaphos containing N6S are visible on a subset of embryos. Subculture each of these putative, stable transformation events to its own plate of N6S for

maintenance and labelling. Efficiency of recovery is 10%, or 10 independent, bialaphos resistant calli per 100 embryos bombarded.

3.3.2. Agrobacterium *Events*

1. After 7 days on resting medium, transfer all embryos to selection medium (30/plate) containing 1.5 mg/l bialaphos for 2 weeks followed by two more 2 week passages on 3 mg/l bialaphos.
2. As early as 5 weeks after infection, putative, stable transformation events are visible on a subset of embryos as described above. Transfer each putative transformation event to its own plate for maintenance and naming. Efficiency of recovery is 5%, or 5 independent, bialaphos resistant calli per 100 infected embryos.

3.4. Regeneration of Transgenic Plants

1. Use a 40X dissecting microscope to transfer 12-15 small pieces of bialaphos resistant, embryogenic callus (stocked embryos suspended in a friable callus matrix) from an independent transformation event to a plate of Regeneration Medium I. Wrap plates with vent tape and incubate at 25°C (dark).
2. After 2-3 weeks on this maturation medium, somatic embryos appear swollen, opaque and white with the coleoptile already emerging from some pieces. Use a dissecting microscope to transfer these mature somatic embryos (~12 pieces) to the surface of Regeneration Medium II for germination in the light (25°C, 80-100 $\mu E/m^2$/sec light intensity, 16 h photoperiod). Wrap plates with vent tape. Plantlets sprout leaves and roots on this medium within 7 days and are ready for transfer to soil within 7-10 days.

3.5. Plant Acclimatization

3.5.1. Moving Plants from Petri dish to Soil

Transplanting is done on a per plant basis when plantlet leaves are ~5 cm long and a good-sized root is developed. Smaller plantlets in the same Petri dish are returned to the light until they are large enough for transplant to soil.

3.5.2. Plant Hardening in the Growth Chamber and Glasshouse

1. Fill small plastic pots (6.4 cm^2) with Redi-Earth (Hummert International, USA) and place into 25 x 51 cm flat with drainage holes. Wet the soil (until it is sticky but not soaked).
2. In a laminar flow bench, use sterile forceps to transfer the plantlets from Petri dish to the soil surface. Remove any media still clinging to the roots. Handle plants by the callus ball at the base of the stem to avoid breaking off the fragile leaf.

3. Gently press the plantlet roots into the soil covering them completely and lightly press the soil surrounding the base of the stem. After transplanting, thoroughly soak the flat – be careful not to dislodge the transplants during this step.

4. Cover plants with a plastic humi-dome (Hummert International) in which one ventilation hole has been cut; place the flat in a growth chamber (CMP 3244, Conviron, Canada) with a 16 h photoperiod, at 350 $\mu E/m^2/sec$ light intensity (plant height of 30 cm). Day and night temperatures are 26°C and 22°C, respectively.

5. If soaked thoroughly after transplanting, transplants should not need water for 48 h. Thereafter, check plants daily and water individual plants as required. Prevent overwatering that may lead to fungal growth at the stem base of transplants and to poorly developed roots.

6. Just before plants touch the top of the humi-dome, remove it. One week later, move the flat to the glasshouse (*see* Note 5).

7. In the glasshouse, check plantlet soil moisture daily but water only those that are dry. Fertilize once before transplanting to larger pots using liquid Miracle Gro Excel 15-5-15 water-soluble fertilizer supplemented with calcium and magnesium (0.15 g/l, Hummert International).

8. When root density is large enough that the soil adheres to the root ball if the plantlet is lifted out of the small pot, transfer the plantlets (about 16 cm leaf height) to a larger pot as described (*see* Section 3.5.3.).

3.5.3. Transplant to Larger Pots

1. Half-fill a 7.6 litre nursery pot (4 drainage holes) with Sunshine Universal Mix SB300 (Consumer Supply, IA) soil mix. Mix-in ~15 g) of Sierra 16-8-12 controlled release fertilizer with trace elements (Hummert International). Fill remainder of pot to within 5 cm of the top edge.

2. Lay plantlet on soil surface in middle of the large pot and gently press the roots into the soil. Cover all roots, ensuring that the plant is transplanted deeply enough that it doesn't topple over as it grows.

3. Drench by bringing water to top edge of the pot and draining completely. Repeat.

4. In general, new transplants will not need watering for the next 7-10 days.

3.5.4. Watering

1. Resume watering when the pot can be lifted easily due to low soil moisture. At this time, water thoroughly and then wait to water again until soil moisture is depleted. For their first 5 waterings after transplant, use liquid fertilizer as described (*see* Section 3.5.5.1.).

2. Frequency of watering will depend on plant size and season. Avoid over-watering smaller plants. During the summer, large mature plants may need to be watered twice a day.

3.5.5. Fertilization

1. Water plants with 50 ppm Miracle Gro Excel 15-5-15 (Hummert International) supplemented with calcium and magnesium for the first 5 times after transfer to larger pots.
2. Just prior to internode elongation, top dress plants with an additional teaspoon of Sierra 16-8-12 slow release fertilizer.
3. Treat plants showing calcium deficiency symptoms (rippled leaf margins, unpigmented lesions on leaf, stunting) with mineral solution (*see* Note 6).

3.5.6. Pollination

1. Just prior to silk emergence, carefully cover ears with wax-treated paper shoot bags (Lawson Pollinating Bags, USA; Cat #217) to ensure controlled pollination of transgenic ears. When silks have emerged for 2-3 days, remove the shoot bag long enough to cut off the top 1.5 inches of the ear, and immediately replace and date it. This provides a uniform lawn of silks for pollination the following day, and is a particularly helpful pollination method if donor pollen is limited.
2. Use fresh pollen from a non-transgenic donor plant tassel to pollinate transgenic silks (*see* Note 7). Lift the shoot bag, and sprinkle pollen on the emerged tuft of silks. Immediately cover the pollinated ear with a labelled pollination bag.
3. Striped pollination bags (Lawson Pollinating Bags, USA; Cat #404) can be used to differentiate transgenic and non-transgenic glasshouse pollinations. Use thick, black permanent markers to label bags with the cross ID (female plant ID x male plant ID) and date.

3.5.7. Dry Down, Seed Harvest and Storage

1. Continue to water plants as needed 3 weeks after pollination, but do not overwater during this period or seeds may subsequently sprout on the cob. At 21-25 days after pollination, watering should be stopped altogether, and plants moved to a dry-down area.
2. At 15 days post-pollination, lift the pollination bag off the ear to facilitate air-drying of the cob. Ten to fifteen days later pull down the husks to allow the kernels to dry further.
3. At 40 days post-pollination, check the seed on the ear. If the seed shows no milk line and a black layer at the seed base, harvest the seed.
4. Seed is inventoried and securely stored in the cold, in a dark, humidity-controlled environment.

3.6. Selection for T1 Transgenic Plants

1. Fill a 25 x 51 cm flat (with drainage holes) with Sunshine Universal Mix SB300. Wet the soil and sow 30-40 transgenic seeds, 2.5 cm apart. Include 2 transgenic maize seeds (bialaphos resistant) as a positive control and 2 non-transgenic Hi II

seeds as a negative control in each flat. Cover the flat with a humi-dome and water as needed.

2. Because T0 seeds are produced by pollinating the silks of a regenerated, transgenic plant with pollen from a non-transgenic donor plant (genotype B73) we expect T1 plants to segregate 1:1 for the *bar* gene if the transgene is integrated into the chromosome as a single dominant locus, and is being expressed. We can test this hypothesis, and identify T1 progeny transgenic for the *bar* gene, by spraying the population of seedlings with the herbicide Liberty (AgrEvo, USA). Liberty contains 200 mg/ml glufosinate, a less expensive alternative to bialaphos with the same mode of action. Segregants expressing the *bar* gene will detoxify the active ingredient in the herbicide (phosphinothricin) and survive the spray, while those not expressing the *bar* gene are killed by the herbicide.

3. A solution of 250 mg/l glufosinate is prepared from the concentrated Liberty stock by mixing 1.25 ml Liberty in 1 litre of water. The surfactant Tween 20 is also added to this solution at a rate of 1 ml/l to favour better contact of the herbicide with the leaf surfaces. This mixture is made fresh for each spraying.

4. Seven to nine days after sowing, emerged plantlets are sprayed with this diluted Liberty solution using a hand held spray bottle. Care is taken to thoroughly spray each plantlet in the flat. Plantlets are sprayed a second time 2 days later. When the negative control is dead (within 5 days of the first spray), the population of plants can be scored for dead or living plants.

5. The surviving, T1 transgenic plants, are then transplanted to large pots and grown to maturity (*see* Section 3.5.3.).

4. NOTES

1. N6 Vitamin Stock (1000X): 1.0 g/l thiamine HCl, 0.5 g/l pyridoxine-HCl, 0.5 g/l nicotinic acid, 2.0 g/l glycine.

2. MS Vitamin Stock (1000X): 0.5 g/l thiamine-HCl, 0.5 g/l pyridoxine-HCl, 0.05 g/l nicotinic acid, 2.0 g/l glycine

3. Gold Coating Procedure (*50*):

Washing Gold
a) Weigh 15 mg of 0.6 μm gold particles (Bio-Rad) and transfer to a sterile, 1.5 ml Eppendorf tube. This is considered a 5X quantity of gold.
b) In the laminar flow hood, add 500 μl 100% ethanol (from freezer) and sonicate in an ultrasonic water bath (Fisher) for 15 sec. Tap the tube to gather all droplets to the tube bottom and allow to stand for 30 min.
c) Spin in a tabletop centrifuge for 60 sec at 3000 rpm, and remove the ethanol supernatant (keep the droplet-shaped pellet facing down or it will fall into the pipet). To rinse, add 1 ml cold, sterile ddH$_2$O. Slightly disturb the pellet by finger tapping the tube and allow the gold to settle out again. Spin at 3000 rpm for 60 sec. Repeat this rinse-step two more times, the third time centrifuging at 5000 rpm for 15 sec. After removing the final wash, suspend the pellet in 250 μl sterile water. Ultra-sonicate this suspension for 15 sec, then immediately place the tube on a multi-head vortex to keep it shaking (vortex setting of 3).

Aliquoting Gold to 1X tubes
While the 5X tube is shaking, aliquot 25 µl of the gold suspension to each of 5 Eppendorf tubes in sequence. Beginning with the last tube, aliquot another 25 µl of the gold suspension to each tube. Each "1X" tube of gold should contain 3 mg gold in 50 µl water. Store at -20°C.

Coating the gold with DNA
a) On the day of bombardment, thaw one, 1X tube of gold for each 8-10 plates to be bombarded. Ultra-sonicate the tube for 15 sec. In laminar flow bench, add in sequence, the appropriate amount of selectable marker construct and gene of interest (GOI) construct to each tube. We use a ratio of 1:3 (molar ratio) for selectable marker construct vs GOI construct. For a 1X gold tube, we use 0.3 µg of selectable marker construct. The amount of GOI is calculated as following:

$$\text{GOI to be added to 1 X tube (µg)} \quad = \quad 0.9 \ \times \ \frac{\text{Size of GOI plasmid or fragment (kb)}}{\text{Size of marker plasmid or fragment (kb)}}$$

For example, the size of the *bar* gene construct is 7 kb and the GOI construct is 14 kb. For a 1X tube, we add 0.3 µg of the *bar* construct and 1.8 µg of the GOI construct. Briefly vortex, then tap the tube on the bench to gather all the droplets to the bottom of the tube. Add 50 µl $CaCl_2$ (Fisher, 2.5 M sterile stock stored at 4°C). Using the same pipet tip, gently suck the suspension up and down once, then place it on the multi-head vortex at setting 2-3. To this shaking tube, add 20 µl spermidine free-base (Sigma, 0.1 M sterile stock stored in small aliquots at –20°C). Finger vortex well, return tube to the vortex, and let shake for 10 min.

b) Remove from vortex and let the gold settle out for several minutes. Centrifuge for 15 sec at 5000 rpm (just enough to pellet the gold), and remove the supernatant. Add 250 µl cold, 100% ethanol to the 1X pellet. Finger vortex to dislodge the pellet, then rock the tube back and forth until the gold achieves a very "silty" smooth consistency dispersed on the base of the tube. Finger vortex several times, then allow to settle until the gold settles out (3-5 min). Centrifuge for 15 sec at 5000 rpm. Remove the supernatant, add 120 µl 100% ethanol and close tube immediately. Finger vortex well to ensure complete suspension of the gold pellet and place on the vortex (setting 2-3) for loading the macro-carriers.

Loading the macro-carriers
Fit pre-sterilized (10 min soak in 70% ethanol then air-dry overnight) macro-carriers (Bio-Rad) into their stainless steel holders in a sterile dish resting on a bed of indicating drierite. While the 1X tube of DNA-coated gold particles is shaking, aliquot 10 µl of the suspension onto the centre of each macro-carrier. Draw a spiral with the pipet tip from the centre of each macro-carrier outwards to cover an area of 1 cm^2. Work quickly to avoid evaporation of the remaining suspension. After loading the macros, let them sit for 5-10 min to be sure they are dry before bombardment.

4. The anti-oxidant cysteine is attributed with moderating the hypersensitive response of maize cells to infection with *Agrobacterium*. This is thought to increase transformation efficiency by favouring the survival of infected cells (*51*). While we have shown that the inclusion of cysteine during co-cultivation increases the rate of stable clone recovery using the described standard binary vector (*24*), the effect of its inclusion in co-cultivation medium on embryo response in the resting phase can be detrimental, and is dependent on embryo size and freshness of the co-cultivation medium. Response of smaller embryos (<1.5 mm) co-cultivated on 1 day-old medium containing 300 or 400 mg/l cysteine can be reduced to half that of non-cysteine co-cultivated control embryos after 7 days on resting medium. However, all embryos on resting medium, responding or not, are transferred to selection medium 10 days after infection (3 days co-cultivation and 7 days resting).
5. When the outdoor temperature is very cold or below freezing, and if the glasshouse is not in the same building as the laboratory, transplants should be re-covered by the humi-dome and wrapped with a large plastic bag then transferred to the glasshouse in a preheated vehicle.
6. Calcium/Magnesium Treatment (Dr. Charles Block, Iowa State University, pers. comm.)
    a) Stock #1: Dissolve 720 g calcium nitrate·4 H_2O (Fisher) in 1 litre of water.
    b) Stock #2: Dissolve 370 g magnesium sulphate·7 H_2O (Fisher) in 1 litre of water.
    c) Store the two stock solutions separately.

d) To 3.8 litres of water, add 5 ml Stock #1. Mix well. Add 5 ml Stock #2. Do not premix the two stocks before adding to water.

e) Water plants directly with this solution (1000 ppm calcium and 500 ppm magnesium).

7. To provide an uninterrupted source of non-transgenic pollen for crossing to T0 transgenic ears, seed donor plants are sown on a weekly basis beginning when transgenic material is transferred to the light on Regeneration II medium. This is generally about 2 weeks before regenerated transgenic plants are transplanted to soil.

CONCLUSION

Here, we describe the biolistic gun-mediated and the *Agrobacterium*-mediated transformation methods, both of which are routinely used for maize Hi II genotype transformation in our laboratory. Immature zygotic embryos or embryo-derived callus cultures can be transformed using the biolistic gun method while to date we have transformed only immature embryos using the *Agrobacterium*-mediated method. Transformation efficiencies (measured as the number of bialaphos resistant callus lines produced per hundred bombarded or infected embryos or callus pieces) in our laboratory are 10-15% using the biolistic method and 5% using the *Agrobacterium* method. Importantly, our *Agrobacterium* method uses a standard binary vector system (*24*), thereby avoiding the additional construct assembly step in *Agrobacterium* required for the super-binary vector system (*22*).

Although the efficiency of *Agrobacterium*-mediated transformation using a standard binary vector is one-half that of particle bombardment in our laboratory, the resulting transformants have lower transgene copies, and higher and more stable gene expression than their bombardment-derived counterparts. Our recent study (*51*) showed that the majority of the *Agrobacterium*-derived transgenic events (>90%) contained 1-3 copies of the transgene, whereas most of the bombardment-derived events (~30%) contained more than 10, and as high as 200 copies of the transgene. In addition, the *Agrobacterium*-derived events generally showed higher transgene expression than the bombardment-derived events. The average amount of transgene transcript in *Agrobacterium*-derived events was 4 times more than that of bombardment-derived events. Transgene expression in the T2 generation was greater than the T1 progenitor for 77% of the *Agrobacterium*-derived events. In contrast, 89% of the bombardment-derived events had lower transgene expression in the T2 than in the T1 generation.

One interesting observation with the *Agrobacterium* method, however, was that more than 70% of these events contained various lengths of the bacterial plasmid backbone DNA sequence, indicating that the *Agrobacterium*-mediated transformation were not as precise as previously perceived, using the current binary vector system (*52*). It is possible that vector backbone contamination may be decreased by further improvement of the binary vector system using current genomic information about *A. tumefaciens*.

On the other hand, it is also possible to achieve low- and single- copy transgenic plants by precipitating low concentrations of DNA onto gold particles for bombardment (*53;* L. Marcell, unpublish.).

In conclusion, while these two maize transformation methods are effective and offer distinct advantages, they also have limitations. When choosing a maize transformation method, it is important to examine the overall goals and resources available for the project, the space and conditions in your plant growth facility, the research experience of your personnel, and the long range plan for your transgenic plants and their progeny.

ACKNOWLEDGEMENTS

The authors wish to thank the maize team members, Lise Marcell, Jennifer McMurray and Tina Fonger, for their assistance. This research is supported by the National Science Foundation (DBI #0110023), the Iowa Corn Promotion Board, the Agricultural Experiment Station, the Office of Biotechnology, the Plant Science Institute and the Baker Endowment Advisory Council for Excellence in Agronomy at Iowa State University.

REFERENCES

1. USDA (2003). http://www.ers.usda.gov/Briefing/Corn/.
2. USA National Corn Growers Association (2003). http://www.ncga.com/03world/main/consumption.htm.
3. Salvador RJ (1997). Maize. In: Werner MS (ed.), *The Encyclopedia of Mexico: History, Culture and Society*, Vol II.. Fitzroy Dearborn Publishers, Chicago and London. pp. 769–775.
4. Duvick (1999). Heterosis: Feeding people and protecting natural resources. In: Coors JG, Pandey S (eds.), *The Genetics and Exploitation of Heterosis in Crops*. American Society of Agronomy, Inc., Crop Science Society of America, Inc., Soil Science Society of America, Inc., Madison, WI, USA. pp. 19-29.
5. James C and Krattiger AF (1996). Global review of the field testing and commercialization of transgenic plants: 1986-1995 – the first decade of crop biotechnology. *ISAAA Brief no. 1*. The International Service for the Acquisition of Agri-biotech Applications. Ithaca, NY, USA.
6. James C (1998). Global review of commercialized transformation crops: 1998. *ISAAA Brief No. 8*. The International Service for the Acquisition of Agri-biotech Applications. Ithaca, NY, USA.
7. James C (2002). Global status of commercialized transgenic crops 2002. *ISAAA Brief No. 27*. The International Service for the Acquisition of Agri-biotech Applications. Ithaca, NY, USA.
8. Coe EH and Sarkar KR (1966). Preparation of nucleic acids and a genetic transformation attempt in maize. *Crop Science*, **6**: 432-435.
9. De Cleene M (1985). The susceptibility of monocotyledons to *Agrobacterium tumefaciens*. *Phytopathologie Zeitschrift*, **113**: 81-89.
10. Fromm ME, Taylor LP and Walbot V (1986). Stable transformation of maize after gene transfer by electroporation. *Nature*, **319**: 791-793.
11. Klein TM, Fromm M, Weissinger A, Tomes D, Schaaf S, Slettern M, and Sanford JC (1988). Transfer of foreign genes into intact maize cells using high velocity microprojectiles. *Proceedings of the National Academy of Sciences USA*, **85**: 4305-4309.
12. Rhodes CA, Pierce DA, Mettler IJ, Mascarenhas D and Detmer JJ (1988). Genetically transformed maize plants from protoplasts. *Science*, **240**: 204-207.
13. Gordon-Kamm WJ, Spencer TM, Mangano ML, Adams TR, Daines RJ, Start WG, O'Brien JV, Chambers SA *et al.* (1990). Transformation of maize cells and regeneration of fertile transgenic plants. *The Plant Cell*, **2**: 603-618.
14. Vaucheret H and Fagard M (2001). Transcriptional gene silencing in plants: targets, inducers and regulators. *Trends in Genetics*, **17**: 29-35.

15. Grimsley NH, Hohn B, Hohn T and Walden R (1986). "Agroinfection", an alternative route for viral infection of plants by using the Ti plasmid. *Proceedings of the National Academy of Sciences USA*, **83**: 3282-3286.

16. Graves AE, Goldman SL, Banks SW and Graves AC (1988). Scanning electron microscope studies of *Agrobacterium tumefaciens* attachment to *Zea mays, Gladiolus* sp., and *Triticum aestivum*. *Journal of Bacteriol*ogy, **170**: 2395-2400.

17. Primich-Zachwieja S and Minocha SC (1991). Induction of virulence response in *Agrobacterium tumefaciens* by tissue explants of various plant species. *Plant Cell Reports*, **10**: 545-549.

18. Gould J, Devey M, Hasegawa O, Ulian EC, Peterson G and Smith RH (1991). Transformation of *Zea mays* L. using *Agrobacterium tumefaciens* and the shoot apex. *Plant Physiology*, **95**: 426-434.

19. Ritchie SW, Lui C-N, Sellmer JC, Kononowicz TK, Hodges TK and Gelvin SB (1993). *Agrobacterium tumefaciens*-mediated expression of *gus*A in maize tissue. *Transgenic Research*, **2**: 252-265.

20. Schläppi M and Horn B (1992). Competence of immature maize embryos for *Agrobacterium*-mediated gene transfer. *The Plant Cell*, **4**: 7-16.

21. Shen W-H, Escudero J, Schläppi M, Ramos C, Hohn B and Koukolíková-Nicola Z (1993). T-DNA transfer to maize cells: histochemical investigation of β-glucuronidase activity in maize tissues. *Proceedings of the National Academy of Sciences USA*, **90**: 1488-1492.

22. Ishida Y, Saito H, Ohta S, Hiei Y, Komari T and Kumashiro T (1996). High efficiency transformation of maize (*Zea mays* L.) mediated by *Agrobacterium tumefaciens*. *Nature Biotechnology*, **14**: 745-750.

23. Komari T and Kubo T (1999). Methods of genetic transformation: *Agrobacterium tumefaciens*. In: Vasil I (ed.), *Molecular Improvement of Cereal Crops*. Kluwer Academic Publishers, Dordrecht/Boston/London pp. 43-82.

24. Frame BR, Shou HX, Chikwamba RK, Zhang ZY, Xiang CB, Fonger TM, Pegg SE, Li BC *et al.* (2002). *Agrobacterium tumefaciens*-mediated transformation of maize embryos using a standard binary vector system. *Plant Physiology*, **129**: 13-22.

25. D'Halluin K, Bonne E, Bossut M, De Beuckeleer M and Leemans J (1992). Transgenic maize plants by tissue electroporation. *The Plant Cell*, **4**: 1495-1505.

26. Frame BR, Drayton PR, Bagnall SV, Lewnau CJ, Bullock WP, Wilson HM, Dunwell JM, Thompson JA *et al.* (1994). Production of fertile transgenic maize plants by silicon carbide whisker-mediated transformation. *The Plant Journal*, **6**: 941-948.

27. Petolino JF, Hopkins NL, Kosegi BD and Skokut M (2000). Whisker-mediated transformation of embryogenic callus of maize. *Molecular and General Genetics*, **207**: 245-250.

28. Golovkin VM, Abraham M, Morocz S, Bottka S, Fefer A and Dudits D (1993). Production of transgenic maize plants by direct DNA uptake into embryogenic protoplasts. *Plant Science*, **90**: 41-52.

29. Armstrong CL (1999). The first decade of maize transformation: a review and future perspective. *Maydica*, **44**: 101-109.

30. Armstrong CL, Spencer TM, Stephens MA and Brown SM (2000). Transgenic maize. In: O'Brien L, Henry RJ (eds.), *Transgenic cereals*. American Association of Cereal Chemists, St. Paul, Minnesota, USA. pp. 115-152.

31. Wang K, Frame BR and Marcell L (2003). Maize Genetic Transformation. In: Jaiwal PK, Singh RP (eds.), *Plant Genetic Engineering: improvement of food crops*. Sci-Tech Publication, Houston, Texas, USA. pp. 175-217.

32. Armstrong CL, Green CE and Phillips RL (1991). Development and availability of germplasm with high Type II culture formation response. *Maize Genetic Cooperation Newsletter*, **65**: 92-93.

33. Frame B, Zhang H, Cocciolone S, Sidorenko L, Dietrich C, Pegg S, Zhen S, Schnable P *et al.* (2000). Production of transgenic maize from bombarded Type II callus: effect of gold particle size and callus morphology on transformation efficiency. *In Vitro Cellular and Developmental Biology-Plant*, **36**: 21-29.

34. Gordon-Kamm WJ, Baszczynski CL, Bruce WB and Tomes DT (1999). Transgenic cereals – *Zea mays* (maize). In: Vasil I (ed.), *Molecular Improvement of Cereal Crops*. Kluwer Academic Publishers, Dordrecht/Boston/London, pp. 43-82.

35.   White J, Chang S-YP, Bibb MJ and Bibb MJ (1990). A cassette containing the *bar* gene of *Streptomyces hygroscopicus*: a selectable marker for plant transformation. *Nucleic Acids Research*, **18**: 1062.

36.   Hood EE, Helmer GL, Fraley RT and Chilton MD (1986). The hypervirulence of *Agrobacterium tumefaciens* A281 is encoded in a region of pTiBo542 outside of T-DNA. *Journal of Bacteriology*, **168**: 1291-1301.

37.   Hajdukiewicz P, Svab Z and Maliga P (1994). The small, versatile pPZP family of *Agrobacterium* binary vectors for plant transformation. *Plant Molecular Biology*, **25**: 989-994.

38.   Carrington JC and Freed DD (1990). Cap-independent enhancement of translation by a plant potyvirus 5' nontranslated region. *Journal of Virology*, **64**: 1590-1597.

39.   Mason HS, DeWald DB and Mullet JE (1993). Identification of a methyl jasmonate-responsive domain in the soybean vspB promoter. *The Plant Cell*, **5**: 241-51.

40.   Vancanneyt G, Schmidt R, O'Connor-Sanchez A, Willmitzer L and Rocha-Sosa M (1990). Construction of an intron-containing marker gene: splicing of the intron in transgenic plants and its use in monitoring early events in *Agrobacterium*-mediated plant transformation. *Molecular and General Genetics*, **220**: 245-50.

41.   An G, Ebert P, Mitra A and Ha SB (1988). Binary vectors. In: Gelvin SB, Schilperoort RA (eds.), *Plant Molecular Biology Manual*, pp. 1-19. Kluwer Academic Publishers, Dordrecht, Boston.

42.   Songstad DD, Armstrong CL, Petersen WL, Hairston B and Hinchee MAW (1996). Production of transgenic maize plants and progeny by bombardment of Hi-II immature embryos. *In Vitro Cellular and Developmental Biology-Plant*, **32**: 179-183.

43.   Zhao Z-Y, Gu W, Cai T, Tagliani LA, Hondred D, Bond D, Schroeder S, Rudert M *et al.* (2001). High throughput genetic transformation mediated by *Agrobacterium tumefaciens* in maize. *Molecular Breeding*, **8**: 323-333.

44.   Songstad DD, Armstrong CL and Petersen WL (1991). AgNO$_3$ increases type II callus production from immature embryos of maize inbred B73 and its derivatives. *Plant Cell Reports*, **6**: 699-702.

45.   Chu CC, Wang CC, Sun CS, Hsu C, Yin KC, Chu CY and Bi FY (1975). Establishment of an efficient medium for anther culture of rice through comparative experiments on the nitrogen source. *Scientific Sinica*, **18**: 659-668.

46.   Vain P, McMullen MD and Finer JJ (1993). Osmotic treatment enhances particle bombardment-mediated transient and stable transformation of maize. *Plant Cell Reports*, **12**: 84-88.

47.   Armstrong CL and Green CE (1985). Establishment of friable, embryogenic maize callus and the involvement of L-proline. *Planta*, **164**: 207-214.

48.   McCain JW, Kamo KK and Hodges TK (1988). Characterization of somatic embryo development and plant regeneration from friable maize callus cultures. *The Botanical Gazette*, **149**: 16-20.

49.   Murashige T and Skoog F (1962). A revised medium for rapid growth and bioassays with tobacco tissue cultures. *Physiologia Plantarum*, **15**: 473-497.

50.   Sanford JC, Smith FD and Russell JA (1993). Optimizing the biolistic process for different biological applications. *Methods in Enzymology*, **217**: 483-509.

51.   Olhoft PM and Somers DA (2001). L-cysteine increases *Agrobacterium*-mediated T-DNA delivery into soybean cotyledoanry-node cells. *Plant Cell Reports*, **20**: 706-711.

52.   Shou H, Frame B, Whitham S and Wang K (2003). Assessment of transgenic maize events produced by particle bombardment or *Agrobacterium*-mediated transformation. *Molecular Breeding*, (in press).

53.   Duncan DR and Spencer M (2003). An efficient system for rapid generation of low copy transgenic plants using the maize inbred H99. *Proceeding of the Annual Conference of American Society of Plant Biologists*, poster #959.

Chapter 5

GENETIC ENGINEERING OF OAT (*AVENA SATIVA* L.) VIA THE BIOLISTIC BOMBARDMENT OF SHOOT APICAL MERISTEMS

S.B. MAQBOOL[1], H. ZHONG[2] AND M.B. STICKLEN

Department of Crop and Soil Sciences, Michigan State University, East Lansing, MI 48824, USA ; [1]Department of Biology, Syracuse University, 130 College Place, Syracuse, NY 13244, USA; [2]SABRI, 3054 Cornwallis Road, RTP, NC27709-2257, USA. Corresponding author: M.B. Sticklen. E-mail: stickle1@msu.edu

1. INTRODUCTION

1.1. Importance and Distribution of Oats

Oat (*Avena sativa* L.) is a temperate cereal crop which ranks after wheat, maize, rice, barley and sorghum in world cereal production (*1*). Oat originated in North Africa, the Near East, and temperate Russia (*2*). These areas are well suited for oat production because of the cool and moist climate (*3*), with few oat varieties being adopted coast to coast (*4*). The common oat varieties used in temperate and mountain areas were derived from wild oat (*Avena fatua* L.), whereas Algerian and red oat varieties were derived from *Avena byzantina* K. Koch (*5, 6*). Oats are adapted to different soil types, thus can be grown regardless of soil type but limited to temperature and moisture conditions (*7, 8*). Oat requires sufficient water for growth and grain production and the major production areas include Russia, Canada, USA, Finland and Poland (*9*).

Oats have been mainly used for livestock and human foods for several centuries (*10*). The use of oat in human food has increased recently because of its characteristic to reduce blood cholesterol and in regulating gastro-intestinal function (*9*). Oat is also used for hay, pasture, green manure or as a cover crop. In the latter case, it enhances soil life, suppress weeds, provides erosion control and increases the organic component of soils (*2, 8*).

1.2. Need for Genetic Engineering

Oat is sensitive to hot dry weather. Yield of oat grain increases linearly with the availability of sufficient water and it decreases as the soil water-deficiency is experienced during growth season (*11*). There is very little information available on salt tolerance of oat cultivars. However, it is considered to be a less salt tolerant

63

I.S. Curtis (ed.), Transgenic Crops of the World - Essential Protocols, 63-78.
© 2004 Kluwer Academic Publishers. Printed in the Netherlands.

species compared with other cereal or forage crops (*12*). Adverse affects of soil salinity condition have been observed on seed germination and subsequent development in different oat cultivars (*12, 13*). Oat is very sensitive to lodging, which may cause severe losses in grain yield and quality. Oat is also limited in natural resistance to barley yellow dwarf viruses (*14, 15*).

Moreover, the sensitivity of oat towards insect attack has also been observed. Oats are often infected by several grain aphids such as bird cherry-oat aphid *Rhopalosiphum padi* and English grain aphid *Sitobion avenae* that attract lady beetles. Oat varieties have adapted rapidly to resist its pathogens, which have shown only minor improvements (*4*). Therefore, wise selection of varieties with genetic potential or the development of resistant cultivars via genetic engineering may help to improve oat production under stress conditions.

1.3. Desired Traits to be Integrated into Oats

Altered traits that result in the induction of stress resistance, disease resistance, insect resistance and herbicide resistance mechanisms in crop plants, as well as seed traits that result in grain with modified oil, protein, carbohydrate content and composition would be of great value to farmers, producers and consumers. The nutritional values of cereal crops as feed for livestock and food for man depends on the amount of nutritional components present in the grain. It is difficult to control these components genetically and to breed successfully in combinations in cereals like oat (*16*). However, the gene transfer technology with the ability to engineer multiple genes simultaneously increases the possibilities for producing novel products in plants (*17*).

1.4. General Account of Tissue Culture and Transformation Technologies Applied to Oats

During the last decade, major achievements in the improvement of cereal crops have been made via tissue culture and genetic engineering of useful gene traits (*17-19*). However, the information about genotype independency for *in vitro* regeneration and genetic engineering systems of oat is limited. In the beginning, to establish a highly regenerable callus cultures, different explants of oat such as immature embryos, mature embryos, intact mature caryopses, young seedling mesocotyl and apical meristems have been used (Table 1). Among them, immature embryos appeared to produce the highest percentage of regenerable callus cultures (*20*). During these experiments, utilizing different cultivars, significant differences have been observed in the percentage induction, characteristic of callus and in the ability of regeneration (*20-22*). In 1985, Rinse and Luke (*23*) showed that a selected line GAF (agronomic less important variety) produced regenerable calli at the highest frequency among all other tested cultivars. Therefore, Somers *et al.* (*24*) attempted for the first time to establish a transformation system in oat utilizing a GAF-derived specific genotype (GAF-30/Park). They successfully produced fertile transgenic oat plants by particle bombardment of friable embryogenic callus. However, the induction of a friable embryogenic callus of oat was found limited to a few genotypes (*19*).

Therefore, attempts were made to develop a genotype-independent and efficient regeneration system from shoot apical meristems for the genetic transformation of commercial oat cultivars (*1*). Zhang *et al.* (*1*) tested four oat cultivars Prairie, Porter, Ogle and Pacer for the development of differentiated multiple shoots from shoot apical meristems. All these oat cultivars generated a genotype-independent *in vitro* differentiated multiple shoots from shoot apical meristems at a high frequency. This system was suggested as an alternative genotype-independent regenerable target tissue for genetic transformation of commercial cultivars of oat (*1*).

Later studies describe the use of Biolistc ^{TM} mediated-transformation system for genetic transformation of oat. For example, Torbert *et al.* (*25*) described the transformation efficiency of selectable marker *npt*II gene using mature embryo-derived callus as the target tissue. Gless *et al.* (*26*) described the transformation of oat plants utilizing leaf base segments of young seedlings with *uid*A gene. Cho *et al.* (*27*) described the transformation of mature oat embryo-derived callus using *hpt* and *gus*A genes. Zhang *et al.* (*28*) described the transformation of commercial oat cultivars using shoot meristematic cultures with *bar* and *uid*A genes. Kaeppler *et al.* (*29*) described the transformation of oat utilizing *gfp* to establish a visual selection system for transformation. Kuai *et al.* (*30*) described the transformation of primary embryogenic calli derived from immature embryos with two plasmids containing *bar* or *uid*A genes.

However, it was suggested that the application of the genotype-dependent embryo-derived callus system to routine oat transformation may not be feasible and the prolonged tissue culture may generate undesirable somaclonal variations in the callus cultures (*31, 32*).

1.5. Genetic Transformation via Shoot Apical Meristems

An efficient and reproducible *in vitro* culture system to produce multiple shoot primordia from cereal shoot apical meristems was developed in our laboratory (*1,33-36*). The advantages of adopting this system for oat transformation may include the production of multiple shoot meristem cultures from mature seeds with highest frequency of regeneration and reproducibility as well as the highest level of fertility and genomic stability in regenerated transgenic plants.

Using this system, 3 oat cultivars have been successfully transformed using microprojectile bombardment with two plasmids BY520 (containing *hva1* and *bar*) and Act1-D (containing *gus*A gene) (*37*). Putative transgenic oat plants as well as T0, T1, and T2 progenies were analysed at the molecular and biochemical levels for transgene integration, expression stability and inheritance. All transgenic plants that showed the integration of the *bar* gene also showed the cointegration of *hva1*. The frequency of cointegration of all 3 transgenes (*hva1, bar* and *gus*A) was 61.6%. The histochemical localization of the GUS in T0 and T1 plants revealed that the vascular tissues and the pollen grains of mature flowers expressed GUS at the highest levels. All developmental stages of transgenic plants showed the expression of HVA1, however, particularly higher during the early seedling stages. Transgenic plants (T2 progeny of 5 independent transgenic lines) were tested *in vitro* for tolerance to osmotic (salt and mannitol) and water-deficit stresses. Transgenic plants maintained a higher growth and showed significantly (P<0.05) increased tolerance to stress conditions than non-transgenic control plants. This study suggests that HVA1 protein may aid developing tolerance in oats to salinity and possible water-deficient or drought stress conditions.

Here, we describe the method of production of shoot apical meristems and biolistic-mediated genetic transformation. We also describe the biochemical analysis of transgenic oat plants and the experiments to test the salinity and water-deficit tolerance in HVA1 expressing transgenic plants.

2. MATERIALS

2.1. Plant Cultivars

The following 3 oat cultivars were used for transformation studies:

1. Ogle: Brave/2/Tyler/Egdoion 23 released by University of Illinois.
2. Pacer: Coachman x CI 1382 released by the Michigan Agricultural Experiment Station.
3. Prairie: IL73-5743 X Ogle released by the University of Wisconsin-Madison.

2.2. Plasmids

Two different plasmids BY520 and Act1-D were used for genetic engineering of oat cultivars (*37*). The plasmid BY520 contained two genes, a selectable marker and herbicide resistant phosphinothricin acetyl transferase the *bar* gene from *Streptomyces hygroscopicus* and the stress-resistance *hva1* gene from barley. The selectable marker *bar* is under the control of CaMV35S promoter and the 3' non-coding region from NOS. The stress-resistance *hva1* gene is under the control of rice actin 1 promoter (Act1) and 3' non-coding region (pin II) of Potato protease inhibitor II gene. The plasmid Act1-D contained the *gus*A gene under the control of the Act1 promoter and the NOS terminator.

2.3. Culture Media

1. Germination medium MS1: 4.3 g/l Murashige and Skoog (MS, *38*) basal salts and vitamins (GIBCO BRL Rockville, Maryland) supplemented with 30 g/l sucrose (Sigma) and 3 g/l Phytagel (Sigma), pH 5.6.
2. Shoot multiplication medium MS2: MS medium supplemented with 30 g/l sucrose, 500 mg/l casein enzymatic hydrolysate (Sigma), 0.5 mg/l 2,4-D (Sigma), 2 mg/l BAP (Sigma) and 3 g/l Phytagel, pH 5.6.

3. Shoot multiplication medium with low selection pressure MS3: MS2 medium containing 5 mg/l of glufosinate ammonium (Sigma) or 2 mg/l of bialaphos (Meiji Seika Kaisha, Japan).

4. Shoot multiplication medium with high selection pressure MS4: MS2 medium containing 10 mg/l of glufosinate ammonium or 3 mg/l of bialaphos.

5. Shoot regeneration medium MS5: MS medium supplemented with 20 g/l sucrose, 0.5 mg/l BAP, 0.5 mg/l IBA (Sigma) 15 mg/l of glufosinate ammonium or 5 mg/l of bialaphos and 3-5 g/l Phytagel, pH 5.6

6. Rooting medium MS6: 2.15 g/l MS medium, 10 g/l sucrose, and 15 mg/l of glufosinate ammonium or 5 mg/l of bialaphos and 3-5 g/l Phytagel, pH 5.8.

7. Salt-stress medium MS7: 2.15 g/l MS medium, 10 g/l sucrose, 100 mM NaCl and 3-5 g/l Phytagel, pH 5.8.

8. Osmotic-stress medium MS8: 2.15 g/l MS medium, 10 g/l sucrose, 200 mM mannitol (Sigma) and 3-5 g/l Phytagel, pH 5.8.

All media were sterilized by autoclaving at 121°C for 20 min. Antibiotics were added to the medium after autoclaving (40-50°C).

3. METHODS

3.1. In Vitro *Morphogenesis of Shoot Tip Cultures*

1. Use mature seeds of oat cultivars (e.g., Ogle, Pacer and Prairie) for shoot tip cultures.

2. Remove the lemmas and the paleas from the grains (caryopses) using fingernails.

3. Surface-sterilize the grains first in 70% ethanol (Sigma) for 5 min, wash with sterile (autoclaved) distilled water once, and then soak in 20% Clorox bleach (Clorox Professional Products Company, Oakland, CA, USA) for 30 min with constant shaking at 120 rpm in a rotary shaker (VWR International, CT, USA) (*see* Note 1).

4. Wash the surface-sterilized seeds 3 times with sterile distilled water.

5. Germinate 10-12 surface-sterilized seeds aseptically in each Petri dish (10 cm diam.) containing germination medium (MS1; 25 ml/dish) in the growth room chamber or incubator set at 25°C under dark and allow to grow for 7 days (*see* Note 2).

6. Seven days after seed germination, excise the small sections of shoot apices (3-5 mm in length) containing apical meristems, 2-3 leaf primordia and leaf bases aseptically using sterilized scalper blades (Sigma) (*see* Note 3).

7. Culture 5-7 excised shoot apices on shoot multiplication medium (MS2; 25 ml/dish) horizontally in a Petri dish (10 cm diam.), with 7-8 dishes for each cultivar and incubate under constant light (60 μmol/m^2/sec from cool-white 40W Econ-o-watt fluorescent lamp; Philips Westinghouse, USA) at 25°C for 4 weeks (*see* Note 4).

8. At weekly intervals, remove physically the elongating leaves, coleoptiles and stems of cultured shoot apices and cut to 3-5 mm in length during each subculture.
9. Maintain multiple shoot cultures by subculturing on MS2 medium at 2 week intervals.
10. Calculate the relative frequency of differentiation of multiple shoots after 8 weeks of shoot apex cultures as the percentage of multiplied shoot apices that produce multiple shoots in total cultured shoot apices for each cultivar (*see* Note 5).

3.2. Preparation of Particles for Microprojectile Bombardment

1. For bombardment, weigh 30 mg of Gold particles (1 μm diam.; Bio-Rad, Richmond, CA, USA) or Tungsten particles (0.9 μm diam.; GTE Sylvania, Towanda, PA, USA) in a 1.5 ml microcentrifuge tube (VWR) and sterilize using 1 ml of 100% ethanol with continuous vortexing at high speed for 30 min.
2. Pipette an aliquot of 50 μl of the particle-ethanol suspension into a 1.5 ml microcentrifuge tube while vortexing continuously, centrifuge at 13,000 rpm using a microcentrifuge (Brinkman, Westbury, NY 11590) for 30 sec and wash with 1 ml sterile distilled water, vortex and centrifuge again. After washing twice, resuspend the particles in 332 μl of sterile distilled water.
3. To the resuspended particles, add successively a total of 15 μl of plasmid DNA (15 μg of pBY520 and pAct1-D at a 1:1 molar ratio), 225 μl of 2.5 M CaCl$_2$ (Sigma) and 50 μl of 0.1 M spermidine (Sigma) and vortex for 5 min at room temperature (*see* Note 6).
4. Incubate the particle DNA on ice for 10 min and centrifuge at 13,000 rpm for 1 min.
5. Wash the particles with 500 μl of absolute ethanol, vortex for 30 sec and centrifuge for 60 sec and finally resuspend in 100 μl of absolute ethanol.

3.3. Microprojectile Bombardment of Multiple Shoot Tips

1. Use 4 week-old multiple shoot cultures, differentiated from the shoot meristems of multiple shoots for microprojectile bombardment (Fig. 1a).
2. Prior to bombardment, physically expose multiple shoot cultures by removal of the coleoptiles and leaves, if necessary.
3. Position the shoot tips (2-3 shoot clumps depending upon the size) in the area of 1.5 cm diam. on shoot multiplication medium (MS2; 25 ml/dish containing 5 g/l Phytagel) in a Petri dish below the Microprojectile-stopping screen.
4. For each bombardment, pipette 10 μl of the particle suspension onto the centre of the macrocarriers and use for bombardment as soon as the ethanol evaporates.
5. Use a biolistic particle acceleration device (PDS 1000/He, Bio-Rad) for bombardment under a chamber pressure of 26 mm of Mercury at a distance of

1.5, 2 and 6.5 cm from the rupture disc to the macrocarriers to the stopping screen to the target, respectively, with a Helium pressure of 1550 psi.

6. Bombard the multiple shoot culture twice with an interval of 2 h (*see* Note 7).
7. Transfer the bombarded multiple shoot cultures to fresh shoot multiplication medium (MS2; 25 ml/dish) without any selection pressure for 4 weeks in continous light (60 μmol/m^2/sec) at 25°C with one subculture (Fig. 1b).

3.4. Selection of Transformants

1. After 4-weeks of bombardment and subculture, divide the shoot clumps into small clumps of 5-10 mm in diam. carefully (avoid injury to shoot meristems) and transfer (6-8 clumps/dish) to shoot multiplication medium with low selection pressure MS3 (25 ml/dish) to select the transformants.
2. Subculture the green shoot clumps onto the same medium after 2 weeks and discard the yellow or brown shoot tip cultures.
3. After 4 weeks of selection on MS3 medium, further divide the green shoot clumps into 5-10 mm in diam. and subculture on selection medium with high selection pressure MS4 (25 ml/dish) for 4-6 weeks.
4. Transfer the fast growing multiple shoot clumps after a total of 4 months of selection and multiplication to Magenta GA-7 boxes (Sigma; 2-3 clumps/Magenta) containing regeneration medium MS5 (50-70 ml/Magenta) for vegetative growth and then to MS6 (50-70 ml/Magenta) for root development (Fig. 1c).
5. Transplant regenerated putative transgenic plantlets (5-10 cm in height, 2-3 leaves) to plastic pots (8 cm square, Griffin Green House and Nursery supplies, Tewksbury, MA, USA; 1 plant/pot) containing a soil mix composed of 1:1 (v:v) peat and perlite (Griffin Green House and Nursery supplies). Fertilize the plants weekly if necessary with Peters 20:20:20 fertilizer (Griffin Green House and Nursery supplies) and grow in a glasshouse under 16 h photoperiod (150 μmol/m^2/sec) and 70-80% humidity until maturity (Fig. 1d) (*see* Note 8).

3.5. Production and Analysis of Progeny

1. Allow the transgenic plants to self-pollinate and harvest the seeds and store individually at 4°C and 70% humidity.
2. To screen transgenic T0, T1, and T2 progenies for herbicide resistance, germinate the mature caryopses (10-12 seeds/dish) on MS6 medium (25 ml/dish) for 7 days (*see* Note 9). Alternatively, germinate the seeds in multi-cell plastic trays (Griffin Green House and Nursery supplies) using a soil mixture (360 Metro Mix; Griffin Green House and Nursery supplies) and allow to grow in the glasshouse for 2 weeks and then treat with 1% Ignite® (Hoechst-Roussel Agri-Vet Company, NJ, USA) (*see* Note 10).
3. After 7 days, count the number of germinated seeds on selection medium or surviving seedlings after herbicide treatment and the non-germinated seeds and the dead seedlings.

4. Perform the χ^2 test (*39*) for segregation data analysis in T0, T1 and T2 progenies.

3.6. Histochemical Analysis of GUS

1. For histochemical analysis of GUS expression, analyze the whole plant tissues or different organs from both transgenic and non-transformed plants (Figs. 1e-g). Also *see* Maqbool *et al.* (37).
2. Remove the chlorophyll from green tissues by incubating tissues first in 70% ethanol for 2 h and then in 100% ethanol for overnight.
3. Provide a little vacuum to the tissues to remove the trapped air.
4. Immerse the samples in GUS substrate mixture (10 mM EDTA (pH 7.0), 0.1M NaPO$_4$ (pH 7.0) and 1-5 mM X-gluc; X-gluc can be dissolved in DMSO (Sigma) immediately followed by vacuum treatment and incubate at 37°C as described (*40*).
5. For localization of GUS activity, prepare the tissues using free-hand cross sections and examine under a Zeiss SV8 stereomicroscope or a Zeiss Axioskop routine microscope.

3.7. Response of Transgenic Plants under Stress-Conditions

Evaluate the performance of the transgenic plants expressing the *hva1* gene (the stress-resistance gene) under stress-conditions as follows:

3.7.1. In Vitro Salt-Stress Condition

The following is the test used to evaluate the response of transgenic plants expressing the *hva1* gene to salt-stress conditions.

1. Surface-sterilize 50-60 progeny seeds (T0, T1 or T2; *see* Note 11) from each transgenic and non-transgenic lines. After acclimatization, germinate (10-12 seeds/dish) in the dark at 25°C on MS6 medium (25 ml/dish) for 4 days (*see* Note 12). Use MS6 medium without glufosinate ammonium or bialaphos for non-transgenic seeds.
2. Divide the germinated seedlings in 2 sets. One set to grow without stress and the other to grow under stress conditions.
3. Transfer 4-5 growing seedlings from each independent transgenic and non-transgenic line to Magenta GA-7 boxes containing MS7 medium (50-70 ml/Magenta) with and without NaCl and grow under light (60 μm/m^2/sec) at 25°C in the form of 4 replicates.
4. Analyze the growth of 10 day-old seedlings on NaCl containing medium (Fig. 1h).
5. Collect the data by taking measurements such as plant fresh weight (*see* Note 13), plant height, root length and plant dry weight (*see* Note 14; Table 2).
6. Analyze the data by analysis of variance (*45*).

7. Separate means by using Tukey's Studentized range test at 95% confidence level.
8. Transfer the plants to a soil mixture composed of 1:1 (v:v) peat:perlite for further growth and development in the glasshouse.

3.7.2. Water-Deficit Stress **In Vitro**

1. Use mannitol in the culture medium to create a water-deficit condition for *in vitro* grown plants.
2. Germinate seeds from T2 progenies on MS6 medium (*see* Section 3.7.1.).
3. Divide the germinated seedlings in two sets. Grow one set without stress and the other on stress-inducing medium.
4. Transfer 4-5 growing seedlings from each independent transgenic and non-transgenic line to Magenta GA-7 boxes containing MS8 medium (50-70 ml/Magenta) with and without mannitol and grow under light at 25°C, 4 replicates per treatment.
5. Analyze the growth of young seedlings (10 day-old) on mannitol containing medium.
6. Record and analyze the data as mentioned for salt-stress test.
7. Transfer the plants to a soil mixture composed of 1:1 (v:v) peat:perlite for further growth and development in the glasshouse.

3.7.3. Water-Deficit Stress **In Vivo**

Perform the following test to analyze the response of HVA1 activity in transgenic lines growing in the glasshouse according to the protocol described by Xu *et al.* (*42*) with some modifications:

1. Germinate 30-40 seeds from T2 transgenic plants on MS6 medium and select transformants.
2. Transfer 7 day-old growing seedlings on selection medium into small pots (8 x 8 cm) containing a soil mix composed of 1:1 (v:v) peat and perlite (1 plant per pot).
3. Maintain the pots in water filled trays and allow the seedlings to grow in the glasshouse for 14 days before initiating the water-deficit treatment.
4. Divide the growing plants into two sets of experiments (a) watered (non-stressed) and (b) water-deficit (stressed) and use 8-10 plants for each transgenic line and non-transformed plants for each treatment. Measure the initial plant height and leaf numbers of each plant before initiating the stressed treatment.
5. Water the first set of plants (non-stressed) continuously from the trays for the whole week.
6. Water the second set of plants (water-stressed) only for 2 days.
7. After 2 days of watering, remove the water completely from the trays of the second set of plants to create a water-deficit condition for the rest of the week.
8. Repeat this treatment for 5 weeks.

9. Measure plant height and leaf number of each plant at the end of the experiment. Compare the difference in growth of stressed and non-stressed of transgenic and non-transgenic plants (Table 3).

10. Analyze the data by analysis of variance (*41*).

11. Separate means by using Tukey's Studentized range test at 95% confidence level.

12. *See* Note 14 for salinity test *in vivo*.

4. NOTES

1. Seeds can also be sterilized at first in 70% ethanol for 2 min, wash with sterile water once, and then soak in 50% Clorox bleach for 15 min with constant shaking at 120 rpm.

2. Before sowing sterilized seeds on germination medium MS1, remove excess water from the seeds using sterilized filter paper sheets (3 mm; VWR).

3. Apical meristems usually can be distinguished as a swollen part close to the leaf base.

4. For optimal results, use various combinations of 2, 4-D (0 and 0.5 mg/l) and BAP (0, 0.5, 1.0, 2.0, 4.0 and 8.0 mg/l) when testing several cultivars.

5. To calculate the relative frequency of differentiation of multiple shoots *see* Maqbool, *et al.* (*37*).

6. Plasmid concentrations can also be used as 3:1, 2:1 (for 2 plasmids) and 1:1:1 or 3:2:1 (for 3 plasmids). The ratio 1 should be used for the plasmid that contains the selection-marker.

7. After first bombardment, incubate the tissues in the dark at 25°C before the second bombardment.

8. Transfer the plants to larger pots (12 x 12 cm) after 1 month of growth in the glasshouse to enhance the vegetative growth.

9. For acclimatisation, place the seeds into a Petri dish containing moist filter paper and cover with a wet filter paper and incubate at 3-5°C for 2-3 days. Supplement water regularly to keep the seeds moist.

10. Herbicide application: Non-selective 1% Ignite herbicide contains 200 g/l (16.222%) of glufosinate as the active ingredient. The herbicide was initially painted on the youngest leaf of the 3-leaved plants and at the 6-leaf stage the whole plant is sprayed.

11. Use progeny (T0, T1 or T2) seeds from different independent transgenic lines. We used the T2 progeny from 5 independent transgenic lines Ogle BRA-82, Ogle BRA-17, Ogle BRA-8, Ogle BRA-19 and Ogle BRA-41 in this experiment.

12. Seeds from transgenic T2 lines were grown on MS6 medium with selection to screen transgenic and non-transgenic plants. Only those seedlings that grow on selection were further tested for stress-resistance.

13. Remove the whole plant from the Magenta carefully without damaging the plant, wash the medium away from roots, blot dry the roots using paper towels and then weigh using a balance (Precision Weighing Balances, Brad ford, MA 01835).

14. Place the whole plant on weighed 3 mm filter papers (already dried to constant weight) and then wrap them inside aluminum foil individually and place in an oven (VWR) set at 110°C for 24 h. Remove the plants from the oven and keep inside a desiccator (VWR) to bring at room temperature and avoid moisture absorption before weighing. Weigh each plant with filter papers and calculate individual plant weight after subtracting the filter paper weight.

15. The performance of *in vivo* grown transgenic plants expressing *hval* can also be evaluated for salinity tolerance using the protocol described by Xu *et al.* (*42*).

CONCLUSION

Oat plants have been genetically engineered using the immature embryo system. We established an efficient and reproducible *in vitro* culture as well as a genetic engineering system for oat using shoot apical meristems derived from mature seeds. The use of multiple meristematic tissues, highly capable of foreign DNA integration and plant regeneration, as targets for genetic transformation via the Biolistic™ bombardment provides a high rate of independent transformation events with high plant fertility. This report also demonstrates that the oat shoot apical meristem system may be genotype-independent. Furthermore, the production of salt- and water deficit-tolerant transgenic oat plants expressing the *hva1* gene may prove significant in the genetic improvement of oat and for producing salinity and/or drought tolerant varieties.

REFERENCES

1.　Zhang S, Zhong H and Sticklen MB (1996). Production of multiple shoots from apical meristems of oat (*Avena sativa* L.). *Journal of Plant Physiology*, **148**: 667-671.

2.　McLeod E (1982). *Feed the soil*. Organic Agriculture Research Institute, P.O. Box 475, Graton, CA., USA, pp. 95444.

3.　Coffman FA (1977). Oat History, Identification and Classification. Technical Bulletin No. 1516, United States Department of Agriculture, Agricultural Research service, Washington, D. C., USA.

4.　Stoskopf NC (1985). Barley and Oat. In: Stoskopf, NC (ed.), *Cereal Grain Crops*, pp. 444-458. Reston Publishing Company, Inc., Reston, Virginia, USA.

5.　Hitchcock AS (1971). *Manual of the grasses of the United States*. Originally published in 1950 as U.S.D.A. Miscellaneous publication No.200. Dover, New York. 2nd edn. revised by Agnes Chase.

6.　Munz PA and Keck DD (1973). *A California Flora* (with Supplement by P.A. Munz). University of California Press. Berkeley, California, USA.

7.　Madson BA (1951). Winter Covercrops. Circular 174, California Agricultural Extension Service, College of Agriculture, University of California, USA, June 1951.

8.　Johnny's selected seeds (1983). *Green Manures-A Mini Manual*. Johnny's selected seeds, Box 2580 Albion, Maine 04910, USA.

9.　Gibson L and Benson G (2002). Origin, history, and uses of oat (*Avena sativa*) and wheat (*Triticum aestivum*). Course Agronomy 212, Iowa State University, Department of Agronomy, Iowa, USA.

10.　Forsberg RA and Shands HL (1989). Oat breeding. In: Janick J (ed.), *Plant Breeding Reviews*. Vol. 6. Timber Press, Portland, OR, USA pp.167-207.

11.　Martin RJ, Jamieson PD, Gillespie RN and Maley S (2001). Effect of timing and intensity of drought on the yield of oats (*Avena sativa* L.). *Proceedings of the 10ᵗʰ Australian Agronomy Conference*, Hobart, Australia.

12.　Murty AS, Misra PN and Haider MM (1984). Effect of different salt concentrations on seed germination and seedling development in few oat cultivars. *Indian Journal of Agricultural Research*, **18**: 129-132.

13.　Verma OPS and Yadava RBR (1986). Salt tolerance of some oats (*Avena sativa* L.) varieties at germination and seedling stage. *Journal of Agronomy Crop Sciences*, **156:** 123-127.

14.　Schonbeck MW (1988). Cover Cropping and Green Manuring on Small Farms in New England and New York: An Informal Survey. Research Report 10, New Alchemy Institute, 237 Hatchville Road, East Falmouth, MA 02536.

15.　Koev G, Mohan BR, Dinesh-Kumar SP, Torbert KA, Somers DA and Miller WA (1998). Extreme reduction of disease in oats transformed with the 5' half of the barley yellow dwarf virus PAV genome. *Phytopatholgy*, **88**: 1013-1019.

16.   Susanne G, Shahryar K, Ronald PL, Howard RW, Deon SD, Darrell W and Fulcher RG (2000). Associations between grain morphology and grain quality traits in hexaploid oat revealed by QTL analysis. TEKTRAN, United States Department of Agriculture, Agricultural Research Service, St. Paul, Minnesota, USA.

17.   Mazur B, Krebbers E and Tingey S (1999). Gene discovery and product development for grain quality traits. *Science*, **285:** 372-375.

18.   Rines HW, Phillips RL and Somers DA (1992). Application of tissue culture to oat improvement. In: Marshall HG, Sorrels ME (eds.), *Oat Science and Technology*. American Society of Agronomy and Crop science Society of America, 677s. Segoe Rd. Madison, WI 53711, USA. pp. 777-791.

19.   Somers DA, Torbert KA, Pawlowski WP and Rines HW (1994). Genetic engineering of oat. In: Henry RJ, Ronalds JA (eds.), *Improvement of Cereal quality by Genetic Engineering*. Plenum Press, New York, USA, pp. 37-46.

20.   Cummings DP, Green CE and Stuthman DD (1976). Callus induction and plant regeneration in oats. *Crop Science*, **16**: 465-470.

21.   Rines HW and McCoy TJ (1981). Tissue culture initiation and plant regeneration in hexaploid species of oats. *Crop Science*, **21**: 837-842.

22.   Bregitzer P, Bushnell WR, Somers DA and Rines HW (1989). Development and characterization of friable, embryogenic oat callus. *Crop Science*, **29**: 798-803.

23.   Rines HW and Luke HH (1985). Selection and regeneration of toxin insensitive plants from tissue cultures of oat (*Avena sativa*) susceptible to *Helminthosporium victoriae*. *Theoretical and Applied Genetics*, **71**: 16-21.

24.   Somers DA, Rines HW, Gu W, Kaeppler HF and Bush-Nell WR (1992). Fertile transgenic oat plants. *Bio/Technolgy*, **10**: 1589-1594.

25.   Torbert KA, Rines HW and Somers DA (1998). Transformation of oat using mature embryo-derived tissue cultures. *Crop Science*, **38**: 226-231.

26.   Gless C, Lorz H and Jahne-Gartner A (1998). Establishment of a highly efficient regeneration system from leaf base segments of oat (*Avena sativa* L.). *Plant Cell Reports*, **17**: 441-445.

27.   Cho MJ, Jiang W and Lemaux PG (1999). High frequency transformation of oat via microprojectile bombardment of seed-derived highly regenerative cultures. *Plant Science*, **148**: 9-17.

28.   Zhang S, Cho MJ, Koprek T, Yun R, Bregitzer P and Lemaux PG (1999). Genetic transformation of commercial cultivars of oat (*Avena sativa* L.) and barley (*Hordeum vulgare* L.) using *in vitro* shoot meristematic cultures derived from germinated seedlings. *Plant Cell Reports*, **18**: 959-966.

29.   Kaeppler HF, Menon GK, Skadsen RW, Nuutila AM and Carlson AR (2000). Transgenic oat plants via visual selection of cells expressing green fluorescent protein. *Plant Cell Reports*, **19**: 661-666.

30.   Kuai B, Perret S, Wan SM, Dalton SJ, Bettany AJE and Morris P (2001). Transformation of oat and inheritance of *bar* gene expression. *Plant Cell, Tissue and Organ Culture*, **66**: 79-88.

31.   Somers DA (1999). Genetic engineering of oat. In: Vasil I, Phillipes R (eds.), *Molecular Improvement of Cereal Crops*. Kluwer Academic Publishers, Dordrecht, The Netherlands.

32.   Choi HW, Lemaux PG and Cho MJ (2001). High frequency of cytogenetic aberration in transgenic oat (*Avena sativa* L.) plants. *Plant Science*, **160**: 761-762.

33.   Zhong H, Srinivasan C and Sticklen MB (1992). *In vitro* morphogenesis of corn (*Zea mays* L.). I. Differentiation of ear and tassel clusters from cultured shoot apices and immature inflorescences. *Planta*, **187:** 483-489.

34.   Zhong H, Wang W and Sticklen MB (1998). *In vitro* morphogenesis of *Sorghum bicolor* (L.) Moench: efficient plant regeneration from shoot apices. *Journal of Plant Physiology*, **153**: 719-726.

35.   Devi P, Zhong H and Sticklen MB (2000). *In vitro* morphogenesis of pearl millet (*Pennisetum glaucum* (L.) R.Br.): efficient production of multiple shoots and inflorescences from shoot apices. *Plant Cell Reports*, **19**: 546-550.

36.   Ahmad A, Zhong H, Wang W and Sticklen MB (2001). Shoot apical meristem: *In vitro* plant regeneration and morphogenesis in wheat (*Triticum aestivum* L.). *In Vitro Cellular and Developmental Biology-Plant*, **38**: 163-167.

37.   Maqbool SB, Zhong H, El-Maghraby Y, Ahmad A, Chai B, Wang W, Sabzikar R and Sticklen MB (2002). Competence of oat (*Avena sativa* L.) shoot apical meristems on integrative

transformation, inherited expression, and osmotic tolerance of transgenic lines containing the *hval*. *Theoretical and Applied Genetics*, **105**: 201-208.

38.   Murashige T and Skoog F (1962). A revised medium for rapid growth and bioassays with tobacco tissue cultures. *Physiologia Plantarum,* **15**: 473-497.
39.   Strickberger MW (1985). *Genetics*, 3rd edn. Macmillan, New York, USA, pp. 126-146.
40.   Jefferson RA, Kavanagh TA and Bevan MW (1987). GUS fusions: β-glucuronidase as a sensitive and versatile gene fusion marker in higher plants. *The EMBO Journal*, **6**: 3901-3907.
41.   SAS Institute (1985). *SAS User's Guide: Statistics*, 5th edn. SAS Institute, Cary, NC, USA.
42.   Xu D, Duan X, Wang B, Hong B, Ho TD and Wu R (1996). Expression of a late embryogenesis abundant protein gene *hval*, from barley confers tolerance to water deficit and salt stress in transgenic rice. *Plant Physiology*, **110**: 249-257.
43.   Heyser JW and Nabors MW (1982). Long-term plant regeneration, somatic embryogenesis and green spot formation in secondary oat (*Avena sativa* L.) callus. *Zeitschrift für Pflanzenphysiologie*, **107**: 153-160.
44.   Lorz H, Harms CT and Potrykus I (1976). Regeneration of plants from callus in *Avena sativa* L. *Zeitschrift für Pflanzenzuchtung,* **77**: 257-259.
45.   Lamb CRC, Milach SCK, Pasquali G and Barro RS (2002). Somatic embryogenesis and plant regeneration derived from mature embryos of oat. *Pesquisa Agropecuaria Brasileira,* **37**: 123-130.
46.   Hassan G, Zipf A, Sharma GC and Wesenberg D (1999). Plant regeneration from mature *Avena* tissue explants. *Cereal Research Communication*, **27**: 25-32.

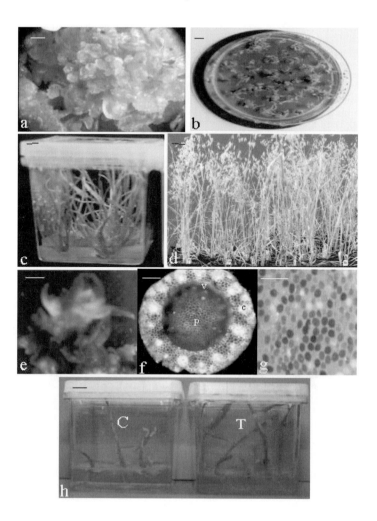

*Figure 1. Genetic engineering of oat (*Avena sativa *L.). (a) Shoot apical meristems after 4 weeks of culture; bar = 0.5 mm. (b) Multiplication of multiple shoot cultures 2 weeks after bombardment without selection; bar = 5 mm. (c) Regeneration of transgenic shoots from multiple-shoot cultures on selection medium; bar = 10 mm. (d) Mature fertile transgenic oat plants in glasshouse; bar = 100 mm. (e) GUS expression in multiple shoot clump; bar = 2 mm (f) Cross-section of stem showing GUS expression, p: pith, v: vascular bundles, c: cortex; bar = 0.5 mm. (g) GUS expression in mature pollen; bar = 0.2 mm. (H) T2 transgenic (T) and non-transgenic (C) oat plants grown on salt-stress medium; bar = 10 mm.*

Table 1. Plant regeneration in oat (Avena sativa L.) from in vitro cultures

Origin of Explant	Regeneration Studies	Reference
Immature embryos	Callus induction and plant regeneration	19, 20, 21, 43
Mature embryos	Callus induction, somatic embryogenesis and plant regeneration	19, 20, 25, 44, 45
Intact mature caryopses	Callus induction and plant regeneration	21, 43
Young seedling mesocotyl	Callus induction, somatic embryogenesis and plant regeneration	21, 43
Apical meristems	Callus induction and plant regeneration	19
Mature tissue explant	Callus induction, somatic embryogenesis and plant regeneration	46
Leaf base segment	Callus induction and plant regeneration	26

Table 2. Effect of salt-stress on the growth of in vitro *grown young seedlings of T2 transgenic oat cultivars*

Lines	Shoot length		Root length		Fresh weight	
	MS cm	MS + NaCl cm	MS Cm	MS + NaCl cm	MS cm	MS + NaCl cm
Non-transgenic	10.67 ± 0.45	6.26 ± 1.13	4.02 ± 0.45	2.08± 0.58	0.20 ± 0.01	0.16 ± 0.02
Ogle BRA-82	12.23 ± 0.52*	9.05 ± 0.50*	5.81± 0.88*	3.3 ± 0.48*	0.31 ± 0.02*	0.17 ± 0.13
Ogle BRA-17	10.83 ± 0.77	6.98 ± 0.64	5.50 ± 0.69*	2.55 ± 0.37	0.25 ± 0.03	0.17 ± 0.17
Ogle BRA-8	10.78 ± 0.67	9.30 ±0.40*	3.64 ± 0.61	2.86 ± 0.33	0.22 ± 0.02	0.20 ± 0.01*
Ogle BRA-19	10.92 ± 0.83	8.48 ± 0.94	3.85 ± 0.43	3.80 ± 0.58*	0.16 ± 0.02	0.22 ± 0.02*
Ogle BRA-41	12.27 ± 0.86*	8.48 ± 0.84	4.33 ± 0.94	2.68 ± 0.40	0.28 ± 0.04*	0.22 ± 0.03

* $P < 0.05$

GENERATION OF TRANSGENIC RYE (*Secale cereale L.*) PLANTS WITH SINGLE AND DEFINED T-DNA INSERTS, FOLLOWING *AGROBACTERIUM*-MEDIATED GENE TRANSFER

F. ALTPETER AND J.C. POPELKA[1]

University of Florida - IFAS, Agronomy Department,
Laboratory of Molecular Plant Physiology, 2191 McCarty Hall, P.O. Box 110300,
Gainesville FL 32611-0300, USA;
[1]CSIRO Plant Industry, GPO Box 1600, Canberra, ACT, 2601, Australia.
Corresponding author: F. Altpeter. E-mail: faltpeter@mail.ifas.ufl.edu

1. INTRODUCTION

1.1. Importance and Distribution of Rye

Rye (*Secale cereale* L.) is one of the most recently domesticated cereals. Its likely origin is the Caucasus region from where it spread as a weed in wheat, representing a so called "secondary crop" (*1*). The cool temperate zones of Europe are the major growing areas for rye since the early Middle Ages due to its high winter hardiness, high tolerance to diseases, drought, and nutrient stress, and, therefore, high yielding stability. More than 90% of the world production is harvested in Russia, Belarus, Ukraine, Poland and Germany (*2*). Rye is second to wheat in bread-making; about 50-75% of the yearly rye harvest is used for this purpose. The remaining portion is used for feeding, the production of alcohol, like for example rye whiskey, pasture and green manure (*3*).

1.2. Need for Genetic Improvement

Rye is widely grown on poor, sandy soils where it has a higher relative performance than wheat and triticale. Improvement of the nutrient uptake efficiency and drought tolerance of rye will enhance its competitiveness on marginal land. The most important diseases in rye are snow mold (*Microdochium nivale*), leaf rust (*Puccinia recondita*), powdery mildew (*Erysiphe graminis*), foot rot (*Pseudocercosporella herpotrichoides, Fusarium* spp.), head blight (*Fusarium graminearum, F. culmorum, M. nivale*), and ergot (*Claviceps purpurea*). The production of hybrid cultivars of rye has improved yield significantly compared to traditional population varieties. However, hybrids are genetically more uniform than population varieties and breeding for disease resistance is therefore urgently needed, especially for those diseases that cannot be prevented by chemical means and those that are associated with mycotoxin accumulation, like Fusarium head blight and ergot. In

I.S. Curtis (ed.), Transgenic Crops of the World - Essential Protocols, 79-88.

order to improve seed quality, a large kernel weight and resistance to pre-harvest sprouting are important goals. Rye bread has a high nutritional value and long fresh-keeping properties compared to wheat bread. However, the pure rye dough has low technical properties and needs to be mixed with gluten rich wheat flours for bread preparation. If rye is used as feed, the high pentosan content compared to other cereals, has been associated with lower digestibility of rye in pigs and chicken (4, 5). Genetic improvement is an important task but conventional breeding of rye is limited due to its cross-pollinating nature.

1.3. Genetic Improvement of Rye

Fifty years ago, rye was the most cultivated cereal in middle and northern Europe. With a reported decrease of more than 50% during the last 40 years, in 2002 the cultivated area was estimated to be only 9,519,040 hectares (2). The main reason for this drastic decrease is the low breeding advance of rye, being the only cross-pollinated species among the small grain cereals. Selfing is naturally prevented by an effective gametophytic self-incompatibility mechanism. Self-fertile forms have been found in several populations and are routineously used for developing inbred lines. Selfing results in strong inbreeding depression and hybrids display a high heterosis. In order to exploit heterosis, pre-selected inbred lines of different gene pools need to be crossed. Classical breeding methods have been successfully employed for rye improvement but the breakthrough came late, with the discovery of the male sterility inducing cytoplasm by Geiger and Schnell (6). Male sterility is caused by introgression of the 'Pampa' (P) cytoplasm of Argentinean origin that is environmentally highly stable. Pollen fertility in the hybrids is restored by the use of dominant nuclear-coded restorer genes from European or exotic populations. Systematic search for gene pools with maximal heterosis soon revealed that the two German populations Petkus and Carstens were particularly well matched. Inbred lines from the two pools are used successfully for the development of hybrids. With the release of the first hybrid varieties in 1984 (1), the productivity was increased and the original importance of rye may be recovered partially. Molecular markers have also been identified in rye and will support marker assisted breeding and genotyping (7, 8). Rye was also used to develop the most important artificial amphydiploid cereal crop, *Triticale*, a hybrid between *Triticum aestivum* and *Secale cereale*. Genome analysis of rye has considerable potential. Rye has a diploid genome, its outbreeding nature results in a high degree of polymorphism and rye has a superior biotic and abiotic stress tolerances compared to related monocotyledonous model organisms. Furthermore, rye represents the only hybrid crop among the small grain cereals and could serve as a model for breeding hybrids in other small grains in the future. The genetic transformation of rye has a high potential to complement traditional breeding programmes.

1.4. Genetic Transformation of Rye

Genetic transformation provides a powerful method to introduce novel and useful genes into crops, even from unrelated species and organisms. The use of such technologies requires the development of reliable and efficient methods for DNA delivery into plant cells and the regeneration of normal and fertile plants from transgenic *in-vitro* cultures. Rye is known as one of the most recalcitrant species for regeneration from tissue cultures and genetic transformation.

An early attempt to genetically transform rye was the injection of plasmid DNA directly into floral tillers (*9*). The first convincing molecular evidence for the generation of a few transgenic rye plants has been reported after biolistic gene transfer into embryogenic callus (*10*). This milestone in rye biotechnology encouraged us for the development of a highly reproducible rye transformation system. Rye is a strictly cross-pollinating species and the identification of inbred lines displaying a good regeneration response from tissue cultures (*11*) and the optimization of gene transfer and selection parameters significantly increased the reproducibility and efficiency of biolistic transformation experiments in rye (*12*).

Agrobacterium-mediated gene transfer offers potential advantages over biolistic gene transfer, including preferential integration of defined T-DNAs into transcriptionally active regions of the chromosome (*13, 14*) with exclusion of vector DNA (*15-17*). Transfer of large DNA fragments enabled by *Agrobacterium* mediated gene transfer (*18*) will support positional cloning and pathway engineering. Unlinked integration of co-transformed T-DNAs after *Agrobacterium*-mediated gene transfer allows elimination of selectable marker genes in subsequent generations by segregation analysis (*19, 20*).

Here, we describe a detailed protocol for the efficient stable genetic transformation of rye by co-cultivation of pre-cultured immature embryos with *A. tumefaciens*. This genetic transformation protocol will support the development of genetically improved cultivars of rye in the near future. The over-expression of high molecular weight glutenin subunits of wheat in transgenic rye resulted in a reversion of the prolamin to glutelin ratio of the rye endosperm, associated with improved functional properties of the rye dough (*21*).

2. MATERIALS

2.1. Plasmid and A. tumefaciens *Strain*

The binary plasmid pJF*npt*II (*22*) containing the selectable marker gene *npt*II under control of the maize ubiquitin promoter with first intron and the CaMV35S terminator inserted in the pPZP111 vector-backbone was introduced into the *A. tumefaciens* strain AGLO (*23*) by electroporation (*24*) (*see* Note 1).

2.2. Culture Media for A. tumefaciens

1. Luria Broth (LB): 10 g/l tryptone, 5 g/l yeast extract (Merck), 10 g/l NaCl, pH 7.5.
2. Rifampicin (Duchefa, Haarlem, The Netherlands): 100 mg/ml stock in DMSO. Dispense 1 ml aliquots into Eppendorf tubes and store at –20°C.

3. Kanamycin sulphate (Duchefa): 100 mg/ml stock in water. Sterilize by filtration through a 0.2 µm membrane. Dispense 1 ml aliquots into Eppendorf tubes and store at –20ºC.

4. Liquid culture medium: LB medium containing 50 mg/l kanamycin sulphate and 50 mg/l rifampicin.

5. Agar culture medium: LB medium, plus 10 g/l agar (Bacto-agar, Difco, Detroit, MI, USA), pH 7.5, with 50 mg/l kanamycin sulphate, and 50 mg/l rifampicin.

2.3. Plant Materials

Donor plants of the rye (*Secale cereale* L.) spring inbred line L22 (provided by the Lochow-Petkus GmbH in Bergen/Germany) were grown in the glasshouse at 16-20°C during the day and 12-16°C during night with 12 h photoperiod until end of tillering; 16/8 h light/dark cycle after tillering (*see* Notes 2 and 3). The soil mixture consisted of a 1:3:1 mixture of sand:topsoil:peat, and plants were fertilized biweekly with Peters-fertilizer (St. Louis, MO, USA), following manufacturer's recomendations. A minimum of 360 µE/m²/sec of light intensity was provided by sodium vapour lights (Sonagro, Phillips). Immature caryopses were harvested, when immature embryos were approx. 2 mm in size (Fig. 1a), corresponding to developmental stage 3 (*25*) (*see* Note 4).

2.4. Tissue Culture

1. Surface sterilants: 70% (v/v) ethanol, sodium hypochlorite solution (Roth, Karlsruhe, Germany; 2.4% active Chlorine) containing 0.1% (w/v) of Tween 20 (Merck) followed by 5 washes with previously autoclaved (121°C, 1.5 bar for 15 min), distilled water.

2. Vitamin stock: Murashige and Skoog (MS, *26*) Vitamin Mixture 1000x (Duchefa). Dissolve 10.31 g of the premixed vitamins in 100 ml water, previously purified with a Milli-Q water purification system (Millipore, Schwalbach, Germany) Sterilize by filtration through a 0.2 µm membrane (Minisart, Sartorius, Göttingen, Germany). Dispense 1 ml aliquots into Eppendorf tubes and store at –20°C.

3. 2,4-D (Duchefa): 2 mg/ml stock solution. Prepare by dissolving the powder in a minimum amount of 1 M KOH and dissolve by heating to 60°C. Make to volume with Milli-Q purified water. Store at 4ºC.

4. Acetosyringone (Sigma) 100 mM: Dissolve 196.2 mg of acetosyringone in 10 ml DMSO.

5. Paromomycin sulphate 50 mg/ml (Duchefa) aqueous stock solution. Sterilize by filtration through a 0.2 µm membrane. Dispense 1 ml aliquots into Eppendorf tubes and store at –20°C.

6. Timentin (Duchefa): 150 mg/ml aqueous stock solution. Sterilize by filtration through a 0.2 µm membrane and use immediately.

7. Basic medium (BM): 4.3 g MS salts (*26*), 1 ml/l vitamin stock, 100 mg/l casein hydrolysate (Duchefa), 500 mg/l glutamine (Duchefa).

8. Callus induction medium (CIM): BM plus 30 g/l sucrose, 2.5 mg/l 2,4-D, pH 5.8, 3.0 g/l Phytagel (Sigma).
9. Osmotic treatment medium (OTM) BM plus 30 g/l sucrose, 6.0 mg/l 2,4-D, 72.9 g/l mannitol, pH 5.8.
10. Co-cultivation medium liquid (CCML): BM plus 15 g/l sucrose, 15 g/l glucose, 6.0 mg/l 2,4-D, pH 5.2, 200 µM acetosyringone (*see* Notes 5 and 6).
11. Co-cultivation medium solid (CCMS): CCML with 3 g/l Phytagel (Sigma).
12. Sub-culture medium (SCM): CIM with 150 mg/l timentin.
13. Shoot regeneration medium I (SRM I): BM plus 30 g/l sucrose, pH 5.8, 5 g/l Agarose type I (Sigma), 150 mg/l timentin and 30 mg/l paromomycin sulphate (*see* Note 7).
14. Shoot regeneration medium II (SRM II): BM plus 30 g/l sucrose, pH 5.8; 5.0 g/l Agarose type I, 150 mg/l timentin and 50 mg/l paromomycin sulphate (*see* Note 7).

All media are sterilized by autoclaving at 121°C, 1.5 bar for 15 min. Antibiotics, and vitamins are added to the medium as concentrated, filter-sterilized solutions after autoclaving. Acetosyringone is added after autoclaving.

Media containing Timentin are used immediately after preparation; others are stored at room temperature for up to 2 weeks.

3. METHODS

3.1. Explants and Agrobacterium Preparation

1. Surface-sterilization: Rinse immature caryopses for 3 min in 70% (v/v) ethanol and for 20 min in sodium hypochlorite solution (Roth, Karlsruhe, Germany; 2.4% active Chlorine) containing approximately 0.1% (v/v) Tween 20 (Merck) while shaking at 50 rpm, followed by 5 washes with previously autoclaved (121°C, 1.5 bar for 15 min), distilled water.
2. Excise immature embryos, in developmental stage 3 (*25*) and approx. 2 mm in size from the caryopses and place on CIM with the coleoptile in contact with the medium. Approx. 50 explants per 9 cm Petri dish are equally distributed on the surface of the medium and cultured in the dark at 25° C for 5 days.
3. Grow *A. tumefaciens* on LB agar culture medium (with antibiotics) at 28°C for 2 days. Pick one colony of bacteria, transfer to 2 ml of LB liquid culture medium (with antibiotics) and grow overnight at 28°C on an orbital shaker at 230 rpm (*see* Notes 8 and 9).
4. Measure absorbance at 660 nm of a 1 ml aliquot of the bacterial overnight culture in a spectrophotometer (expected OD_{660} value is 2.0-2.5).
5. Centrifuge 1ml of the bacterial culture at 5000 rpm for 5 min in a sterile Eppendorf centrifuge tube, discard the supernatant, resuspend and dilute the pellet to a concentration equivalent to an OD_{660} value of (1.5–2.0) in 1/1 (v/v) LB medium/CCML (without antibiotics) and incubate at 28°C on an orbital shaker at 230 rpm for 2 h before co-cultivation.

3.2. Inoculation, Co-cultivation, Selection and Recovery of Whole Transgenic Plants

1. Before co-cultivation with *Agrobacterium*, place 25-50 pre-cultured immature embryos per well of a 6×macroplate (Greiner, Cellstar, Frickenhausen, Germany) and suspend in 5 ml OTM medium per well for 4-6 h (Fig. 1b).
2. Remove OTM medium with a pipet and inoculate 25-50 pre-cultured immature embryos per 39 mm well with approx. 300 µl *Agrobacterium* suspension (*see* Section 3.1.5.), vacuum treat at 500-800 mbar for 1 min and finally keep for 10 min in the laminar hood before removing excess *Agrobacterium* suspension.
3. Remove excess *Agrobacterium* suspension with a pipet and rinse explants gently twice with 5 ml CCML medium to remove excess *Agrobacterium*, followed by a co-culture in 5 ml CCML medium overnight at 22°C at 80 rpm in the dark (*see* Note 10).
4. After 14-16 h co-cultivation in CCML medium, rinse explants thoroughly (at least 5 times) in CCML, blot dry the explants with sterile filter paper, followed by an immediate transfer to CCMS medium (25 explants/9 cm dish). With the coleoptile in contact with the medium, culture for 2 days at 22°C in the dark (*see* Note 10).
5. Transfer explants to CIM medium to promote callus and suppress *Agrobacterium* growth after the co-cultivation. Maintain cultures in the dark for 2 weeks at 25°C (*see* Notes 10 and 11).
6. Transfer calli (Fig. 1c) to 9 cm Petri dish with SRM I medium and culture for 3 weeks at a 16 h photoperiod, 60 µE/m^2/sec illumination at 25°C for shoot regeneration.
7. Transfer 5 regenerating calli to a 60 x 100 x 100 mm (L x W x H) container (Green Box, Duchefa) with 60 ml SRM II medium and culture at a 16 h photoperiod, 60 µE/m^2/sec illumination at 25°C for shoot elongation and root formation of transgenic shoots within 3 weeks (*see* Note 12; Fig. 1e).
8. Transfer rooted transgenic plantlets (Fig. 1f) to soil (*see* Section 2.3.).
9. Regenerated plants and their progeny (Fig. 1g) can be assessed for expression of the *npt*II gene by a commercially available ELISA-kit (Agdia, Elkhart, IN) (*see* Note 13) and by leaf painting with 1% (w/v) paromomycin sulphate (Fig. 1h). Southern blot analyzes must be performed to confirm the stable integration of the transgene/s, and Northern blot and Western blot analyzes are used to confirm the expression of the transgenes (*see* Notes 14 and 15).

4. NOTES

1. We have demonstrated that *A. tumefaciens* strain AGL0 is able not only to efficiently transform rye but also barley (*16*) and ryegrass (unpublish.).
2. Identification of genotypes with the potential to initiate embryogenic callus cultures and their plant regeneration potential is the main limitation in establishing a rye transformation system (*12*). Immature embryos are the superior explant to initiate tissue cultures in rye. In cross-pollinating species like rye, every embryo represents a different genotype. Genotypic differences in tissue culture response reduce reproducibility. To overcome this problem we prefer working with inbred lines. Screening of tissue culture response of 21 different inbred lines and their crosses revealed that tissue

culture response was highly variable from genotype to genotype in rye and was not affected by inbred depression (*12*).

3. Tissue culture response is highly affected by donor plant quality. In this respect it is important to avoid, if at all possible, the use of systemic pesticides and maintain the temperature below 20°C.

4. Developmental stage 3 is normally reached 10 days after pollination. However, differences in genotypes, and glasshouse conditions suggest to closely monitor the developmental stage rather than sticking to a rigid schedule.

5. Co-cultivation in a medium rich in auxins enhances dedifferentiation, induction of cell division and callus proliferation of plant cells, which might enhance their transformation competence.

6. We have not investigated if the supplementation of CCML and/or CCMS medium with acetosyringone, used prior and during co-cultivation has a significant effect on transformation efficiencies in rye. In *Agrobacterium*-mediated barley transformation, the use of acetosyringone was not required for stable transformation and did not significantly effect transient expression (*16*).

7. Many gelling agents cause precipitation of paromomycin. In order to avoid precipitation of paromomycin it is important to use agarose as a gelling agent.

8. The binary plasmid used has incorporated bacterial resistance to kanamycin. *A. tumefaciens* strain AGL0 has chromosomal resistance to rifampicin.

9. If a series of transformation experiments with the same plasmid is planned, reproducibility of results can be improved by using 20 μl of an *Agrobacterium* stock stored in glycerol as an inoculum instead of plate colonies (*see* Section 3.1.3.). For the preparation of glycerol stocks a 100 ml *Agrobacterium* suspension can be grown overnight (as a subculture of the culture initiated in 100 ml LB medium). After mixing 1:1 (vol:vol) with an autoclaved glycerol solution (30%) the aliquoted suspension can be stored for several months at –80°C.

10. Rye tissue cultures are very sensitive to *Agrobacterium* overgrowth. The described liquid co-cultivation protocol and period followed by thorough rinsing of the explants and the use of timentin reduces the potential of *Agrobacterium* overgrowth.

11. Rye tissue cultures lose their plant regeneration capacity quickly during extended tissue culture and in the presence of selective agents (Fig. 1d). Therefore this optimized protocol avoids selection during callus culture and uses the plant regeneration period for selection and has a very short tissue culture period.

12. Selection is applied only during the regeneration phase of tissue cultures. Paromomycin, compared to other selective agents, is superior in suppressing elongation of non-transgenic shoots (Fig. 1). Due to the short selection period, the escape rate is usually high and has ranged from 30-80% of the regenerated plants being non-transgenic plants.

13. NPT II expression can be rapidly monitored by a commercially available ELISA kit or by leaf painting with paromomycin (Fig. 1). However, paromomycin is highly toxic and should not be applied to plants without appropriate safety equipment. The removal of the selectable marker genes from transgenic elite events can be achieved by *Agrobacterium*-mediated co-transformation of un-linked T-DNAs followed by segregation analysis in sexual progenies (*19, 20*). We have also demonstrated that selectable marker free transgenic rye plants can be generated without the use of selectable marker genes by PCR screening. For this approach, the timing between gene transfer and initiation of regeneration is most critical. (*27*).

14. Transformation frequencies obtained in the described protocol are usually in the range of 1-4% (inoculated pre-cultured embryos regenerating independent transgenic plants).

15. The majority of plants generated with this protocol showed a simple transgene integration pattern with a single transgene copy, defined T-DNA borders and exclusion of vector DNA, which supports the stability of transgene expression in subsequent generations (*17*).

CONCLUSIONS

Monocots are not among the natural hosts of *Agrobacterium* and only in the last decade convincing molecular evidence of stable *A. tumefaciens* mediated gene transfer into monocotyledonous crops including rice (*28*), wheat (*29*) and barley (*30*) was presented. The transfer of T-DNA and its integration into the plant genome is influenced by several *Agrobacterium* and plant tissue specific factors. These include the vector-plasmid (*31*) bacterial strain (*20, 32*), the addition of *vir*-gene inducing synthetic phenolic compounds (*28*), culture media composition and culture conditions (*33*) and osmotic stress treatments (*34*) during and before *Agrobacterium* infection. The selected plant genotype and the tissue to be transformed, as well as the suppression and elimination of *Agrobacterium* after co-cultivation (*35*) are further influencing factors for the efficient production of stable transformed plants. We describe a detailed protocol for the stable genetic transformation of rye by co-cultivation of pre-cultured immature embryos with *A. tumefaciens*. Rye is one of the most recalcitrant plants in tissue culture. Most important factors in establishing a transformation protocol for rye include the identification of inbred lines with a good regeneration response, a co-cultivation procedure that reduces the risk for *Agrobacterium* overgrowth, a short callus culture period without using a selective agent, selection of the transgenic events during regeneration and elongation of shoots. The majority of plants generated with this protocol showed a simple transgene integration pattern with a single transgene copy, defined T-DNA borders and exclusion of vector DNA, which supports the stability of transgene expression in subsequent generations (*17*) and will support genetic engineering of improved bread-making quality (*21*), feed quality or stress tolerance in rye in the near future.

ACKNOWLEDGEMENTS

The authors wish to thank Erika Gruetzemann for her excellent technical assistance and Peer Wilde, Brigitta Schmiedchen and Jutta Menzel, Lochow-Petkus GmbH, Bergen, Germany for helpful discussions. This research was supported by the German Bundesministerium fuer Bildung, Wissenschaft, Forschung und Technologie (BMBF) and Lochow-Petkus GmbH, Bergen, Germany. Fig. 1 and partial contents from Introduction and Material and Methods sections were reproduced from *Molecular Breeding*, **11**: 203-211, 2003, *Agrobacterium*-mediated genetic transformation of rye (*Secale cereale* L.). J.C. Popelka and F. Altpeter (*17*), copyright Kluwer Academic Publishers, with kind permission of Kluwer Academic Publishers.

REFERENCES

1.   Miedaner T (1997). Roggen: vom Unkraut zur Volksnahrung. DLG-Verlag: Frankfurt/Main.

2.   FAO (2002). http://apps.fao.org/page/collections.

3.   Bushuk W (2001). *RYE: Production, Chemistry, and Technology*, 2nd edn., AACC, St Paul, MN, USA.

4.   Rakowska M, Zebrowska T, Neumann M, Medynska K and Raczynska-Bojanowska K (1990). The apparent ileal and faecal digestibilities of amino acids and carbohydrates of rye, barley and triticum grains in pigs. *Archiv fur Tierernahrung*, **40**: 695-701.

5.   Boros D, Marquardt RR, Guenter W and Brufau J (2002). Chick adaptation to diets based on milling fractions of rye varying in arabinoxylans content. *Animal Feed Science and Technology*, **101**: 135-149.

6.   Geiger HH and Schnell FW (1970). Die Züchtung von Roggensorten aus Inzuchtlinien. *Theoretical and Applied Genetics*, **40**: 305-311.

7. Saal B and Wricke G (1999). Development of simple sequence repeat markers in rye (*Secale cereale* L.). *Genome*, **42**: 964-972.
8. Matos M, Pinto-Carnide O and Benito C (2001). Phylogenetic relationships among Portuguese rye gased on isozyme, RAPD and ISSR markers. *Hereditas*, **134**: 229-236.
9. De la Pena A, Lorz H and Schell J (1987). Transgenic rye plants obtained by injecting DNA into young floral tillers. *Nature* **325**: 274-276.
10. Castillo AM, Vasil V and Vasil IK (1994). Rapid production of fertile transgenic plants of rye (*Secale cereale* L.). *Bio/Technology*, **12**:1366-1371.
11. Popelka JC and Altpeter F (2001). Interactions between genotypes and culture media components for improved *in vitro* response of rye (*Secale cereale* L.) inbred lines. *Plant Cell Reports*, **20**: 575-582.
12. Popelka JC and Altpeter F (2003a). Evaluation of rye (*Secale cereale* L.) inbred lines and their crosses for tissue culture response and stable genetic transformation of homozygous rye inbred line L22 by biolistic gene transfer. *Theoretical and Applied Genetics*, **107**: 583-590.
13. Czernilofsky AP, Hain R, Baker B and Wirtz U (1986). Studies of the structure and functional organization of foreign DNA integrated into the genome of *Nicotiana tabacum*. *DNA*, **5**: 473-482.
14. Koncz C, Martini N, Mayerhofer R, Koncz-Kalman Z, Körber H, Redei GP and Schell J (1989). High-frequency T-DNA-mediated gene tagging in plants. *Proceedings of the National Academy of Sciences USA*, **86**: 8467-8471.
15. Hiei Y, Komari T and Kubo T (1997). Transformation of rice mediated by *Agrobacterium tumefaciens*. *Plant Molecular Biology*, **35**: 205-218.
16. Fang Y-D, Akula C and Altpeter F (2002). *Agrobacterium*-mediated barley (*Hordeum vulgare* L.) transformation using green fluorescent protein as a visual marker and sequence analysis of the T-DNA::genomic DNA junctions. *Journal of Plant Physiology*, **159**: 1131-1138.
17. Popelka JC and Altpeter F (2003b). *Agrobacterium tumefaciens*-mediated genetic transformation of rye (*Secale cereale* L.). *Molecular Breeding*, **11**: 203-211.
18. Hamilton CM, Frary A, Lewis C and Tanksley SD (1996). Stable transfer of intact high molecular weight DNA into plant chromosomes. *Proceedings of the National Academy of Sciences USA*, **93**: 9975-9979.
19. McKnight TD, Lillis MT and Simpson RB (1987). Segregation of genes transferred to one plant cell from two separate *Agrobacterium* strains. *Plant Molecular Biology*, **8**: 439-445.
20. Komari T, Hiei Y, Saito Y, Murai N and Kumashiro T (1996). Vector carrying two separate T-DNAs for co-transformation of higher plants mediated by *Agrobacterium tumefaciens* and segregation of transformants free from selection markers. *The Plant Journal*, **10**: 165-174.
21. Altpeter F, Popelka JC, Kiefer R and Wieser H (2003). Stable expression of 1Dx5 and 1Dy10 high molecular weight glutenin subunit genes in transgenic rye, drastically increases the polymeric glutelin fraction in rye flour (in prep.).
22. Altpeter F and Xu J (2000). Rapid production of transgenic turfgrass (*Festuca rubra* L.) plants. *Journal of Plant Physiology*, **157**: 441-448.
23. Lazo GR, Stein PA and Ludwig RA (1991). A DNA transformation-competent *Arabidopsis* genomic library in *Agrobacterium*. *Bio/Technology*, **9**: 963-967.
24. Mersereau M, Pazour GJ and Das A (1990). Efficient transformation of *Agrobacterium tumefaciens* by electroporation. *Gene*, **90**: 149-151.
25. Zimny J and Loerz H (1989). High frequency of somatic embryogenesis and plant regeneration of rye (*Secale cereale* L.). *Plant Breeding*, **102**: 89-100.
26. Murashige T and Skoog F (1962). A revised medium for rapid growth and bioassays with tobacco tissue cultures. *Physiologia Plantarum*, **15**: 473-497.
27. Popelka JC, Xu J and Altpeter F (2003). Generation of rye (*Secale cereale* L.) plants with low transgene copy number after biolistic gene transfer and production of instantly marker-free transgenic rye. *Transgenic Research*, (in press).
28. Hiei Y, Ohta S, Komari T and Kumashiro T (1994). Efficient transformation of rice (*Oriza sativa* L.) mediated by *Agrobacterium* and sequence analysis of the boundaries of the T-DNA. *The Plant Journal*, **6**:271-282.
29. Cheng M, Fry JE, Pang S, Zhou H, Hironaka CM, Duncan DR, Conner TW and Wan Y (1997). Genetic transformation of wheat mediated by *Agrobacterium tumefaciens*. *Plant Physiology*, **115**: 971-980.

30.   Tingay S, McElroy D, Kalla R, Fieg S, Wang M, Thornton S and Brettell R (1997). *Agrobacterium tumefaciens*-mediated barley transformation. *The Plant Journal*, **11**: 1369-1376.
31.   Klee H (2000). A guide to *Agrobacterium* binary Ti vectors. *Trends in Plant Science*, **5**: 446-451.
32.   Hoekema A, Hirsch PR, Hooykaas PJJ and Schilperpoort RA (1983). A binary plant vector strategy based on separation of *vir*- and T-region of the *Agrobacterium tumefaciens* Ti-plasmid. *Nature* **303**: 179-180.
33.   Alt-Mörbe J, Kühmann H and Schröder J (1989). Differences in induction of Ti-plasmid virulence genes *virG* and *virD* and continued control of *virD* expression by four external factors. *Molecular and Plant-Microbe Interactions*, **2**: 301-308.
34.   Usami S, Okamoto S, Takebe I and Machida Y (1988). Factor inducing *Agrobacterium tumefaciens vir* gene expression is present in monocotyledonous plants. *Proceedings of the National Academy of Sciences USA*, **85**: 3748-3752.
35.   Nauerby B, Billing K and Wyndaele R (1997). Influence of the antibiotic timentin on plant regeneration compared to carbenicillin and cefotaxime in concentrations suitable for elimination of *Agrobacterium tumefaciens*. *Plant Science*, **123**: 169-177.

Figure 1. Transformation and selection of rye. (a-b) Immature embryo explants were (a) pre-cultured for 5 days (bar = 2 mm) and (b) co-cultured on liquid medium with Agrobacterium *suspension (bar = 3 cm). (c-d) Induced rye callus 3 weeks after subculture on (c) non-selective (bar = 1.5 mm) and (d) selective callus induction medium containing 100 mg/l paromomycin sulphate (bar = 1.5 mm). (e-f) In-vitro regeneration in (e) non-selective treatment (bar = 2 cm) and (f) on selective regeneration medium (50 mg/l paromomycin sulphate) with one regenerated, transgenic rye plant (arrow) (bar = 2 cm). (g-h) Screening of segregating progeny by (g) leaf painting assay for NPTII expression analysis (bar = 10 cm) showing (h) symptoms after application of 1% (w/v) paromomycin sulphate solution on transgenic control (C), wildtype plants (Wt) and transgenic progeny (T1) (bar = 1.5 cm).*

Chapter 7

PARTICLE INFLOW GUN-MEDIATED TRANSFORMATION OF *SORGHUM BICOLOR*

S.B. WILLIAMS, S.J. GRAY[1], H.K.C. LAIDLAW AND I.D. GODWIN

School of Land and Food Sciences, The University of Queensland, Brisbane, Queensland, Australia, 4072; [1]Bureau of Sugar Experiment Stations, Indooroopilly, Queensland, Australia, 4068. Corresponding author: I. Godwin. Email: i.godwin@uq.edu.au

1. INTRODUCTION

Sorghum (*Sorghum bicolor* L.) is an important cereal throughout the semi-arid tropical and sub-tropical regions of the world. Sorghum biotechnology research is not as advanced as other important cereals. Much of the genomics research that has been undertaken with sorghum is a reflection of its close co-linearity with the maize genome, rather than the interest in sorghum *per se*. It is in advances in genetic engineering research that sorghum particularly lags behind the progress made with other cereals.

Sorghum is commonly considered a "poor man's crop". This perception stems predominantly from the fact that sorghum is mainly cultivated in agriculturally marginal areas, the areas of least economic development. Even in developed countries such as Australia and the USA, sorghum is mostly cultivated in areas too hot and/or dry for reliable yields of maize or other higher value crops. Indeed, it is the adaptation of sorghum that makes it a preferable proposition in more marginal areas of production. Sorghum crops can yield under low water conditions and will produce viable pollen under higher temperatures than alternative crops, such as maize, which tends to suffer from tissue damage and poor pollination under high temperatures. Sorghum is remarkably tolerant of high temperatures, with the rate of metabolic processes unaffected up to 38°C (*1*), and reversible high temperature damage does not occur until leaf temperature exceeds 46°C (*2*).

In semi-arid and arid tropics, sorghum and millet are the major cereals for human consumption. Sorghum is commonly, erroneously, grouped with the millets as broom millet. The true millets are actually of the genera *Panicum* and *Pennisetum*, and finger millet is *Eleusine*. Approx. 300 million people in sub-Saharan Africa and India rely on sorghum as a major staple food. In many African and Asian developing countries, sorghum is a multi-purpose crop, with the stover widely used as forage for cattle, goats and sheep, and as a building material and firewood. In the Americas and Australia, sorghum is important as an animal feed grain and a forage crop.

89

I.S. Curtis (ed.), Transgenic Crops of the World - Essential Protocols, 89-102.

Sorghum's lower value as a crop is a result of two factors:

1. The major grains (wheat, rice and maize) are preferred as food by many people, either through taste preference or the variety of utilization
2. Sorghum is less digestible to both humans and livestock than wheat, barley, maize and oats (Table 1).

Table 1. Total grain digestibility (%) of the major feed grains in cattle (6)

Digestibility	maize	sorghum	barley	wheat
Rumen	76	64	87	89
Post-rumen	66	63	73	85
whole tract	93	87	93	98
% energy excreted in faeces	5.7	9.7	4.0	1.4

1.1. Crop Utilization

In tropical and sub-tropical areas of Africa and Asia, sorghum grain is grown predominantly for human consumption. These sorghums are almost exclusively of the white pericarp and white or yellow endosperm types, as the red or brown sorghums have high tannin levels which adversely affect palatability. The grain is usually milled for flour and used for a variety of unleavened breads and couscous, or more coarsely ground for use in porridges. Sorghum grain is also used for beer-making, commonly the opaque beers of southern Africa, or malted dark beers such as those found in West Africa. In China, sorghum grain is commonly used to produce wine and spirit.

In Australia and the Americas, the grain is used almost exclusively for stock-feed. Grain is fed-milled or steam-flaked to beef and dairy cattle, and to poultry and pigs as a major energy source. Forage is sometimes grazed *in situ*, harvested and fed directly or used to produce silage. In some parts of the southern USA, some of the sweet sorghum types have minor importance as a source of syrup for human consumption. Godwin and Gray (*3*) have extensively reviewed the utilization constraints for sorghum, and how these may be improved by molecular genetic means.

While sorghum is a useful feed grain, it is commonly discounted against the other major feed grains. This reflects its significantly lower digestibility in comparison with maize, barley and wheat (Table 1). Almost 10% of the total energy in sorghum grain remains undigested when fed to ruminant animals such as cattle. The comparison becomes more unfavourable when fed to monogastric mammals, including humans. Various researchers have demonstrated that it is common for more than 20% of the sorghum grain carbohydrate to remain undigested in human diets (*4*). The world's most important food crop, rice, has 96-97% digestible energy (*5*). Hence, improving the nutritional quality is a major goal of many sorghum improvement programmes. There is considerable scope for the transgenic approach to make important contributions to this aim.

Improving the quality of sorghum grain for human and animal nutrition, and the development of sorghum-based cereal products (food and non-food) is essential for maintaining sorghum's place in the top cereals grown internationally.

1.2. Constraints to Production and Quality

As sorghum is predominantly grown in the tropical developing world, or the more marginal cropping areas of the developed nations, heat and drought remain the major abiotic stresses to production (7).   Much of the plant breeding focus is on maintaining production under the typical conditions of high temperature and low water availability at some stage of the summer production cycle. As with other cereals, nitrogen use efficiency is also a major issue in crop improvement and management.   In tropical areas of Africa and South America, acid soils are a limitation to cereal and other grain crop cultivation, with associated problems of aluminium toxicity and phosphorus deficiency.

Worldwide, insect pests are the major biotic stress faced by sorghum producers, with an estimated grain loss well in excess of US$1 billion annually (8). In terms of economic damage, the most destructive pests are the lepidopterans, heliothis (*Helicoverpa armigera*) and stem borer (*Chilo partellus*, yellow stem borer) and the dipterans, midge (*Stenodiplosis sorghicola*) and shootfly (*Atherigona soccata*). Genetic sources of resistance have been combined to produce midge resistant hybrids in Australia (9) and India (10), although this appears to have come at the expense of genetic variability (11).   There are no sources of true resistance to heliothis, stem borer or shootfly.

Diseases are a constraint to both production and quality of grain and forage sorghums. Downy mildew is a major limitation to grain yield in the USA and other parts of the world (12).   Other fungal leaf diseases usually do not impact greatly on grain yield, but some such as rust and leaf blight are important in reducing forage quality. Viral diseases such as Johnson Grass Mosaic Virus (JGMV) and Maize Dwarf Mosaic Virus (MDMV) can be particularly damaging to yield and forage quality, particularly when infection occurs early in the crop cycle (13).

Sorghum ergot, or sugary disease, is caused by 2 different fungi (*Claviceps sorghi*, *C. africana*).   The disease has been important in Asia and Africa, and in the late 1990s, spread throughout the Americas and Australia.   The disease can lead to total crop loss, and recent reports have demonstrated that the alkaloids produced by the fungus are toxic to livestock (14, 15).   Ergot is particularly devastating to susceptible cytoplasmic male sterile parents in hybrid seed production (16, 17).

Genetic transformation techniques can be used to transfer agronomically useful genes from any source into the crop of interest.   For example, chitinase genes produce pathogenesis-related proteins with the potential to protect plants against fungal pathogens and insect pests that contain chitin as structural components. Researchers from Kansas State University have transformed sorghum with a rice chitinase gene (18). Of the six T0 plants produced, five expressed both the *bar* and chitinase genes and one plant expressed only the *bar* gene even though the intact chitinase gene was detected by Southern blot analysis. Zhu *et al.* (*18*) believe this to be caused by a position effect and possibly other mechanisms such as methylation, reverse methylation and antisense suppression. They found the chitinase gene to be inherited in a Mendelian manner among the T1 progeny, although some results indicated gene silencing at different stages of plant growth and instability of the silenced transgene.   Analysis of T2 and T3 lines with constitutive chitinase expression showed that they were significantly more resistant to stalk rot than control plants and transgenic plants with no detectable expression of chitinase (*19*).

At the University of Queensland, we have produced transgenic sorghum families expressing either a synthetic *cry1Ab* or *cry1B* gene driven by the maize ubiquitin promoter and a *bar* gene for selection (*20, 21*). These synthetic *cry* genes confer proteins with *in vitro* efficacy against the sorghum stem borer, *Chilo partellus*.

With the availability of functional genomics programmes to elucidate the gene networks involved in plant development and response to biotic and abiotic stresses, transgenics are a powerful tool.   The ability to test the operation and interactions of particular genes, both under homologous and heterologous control of transcription, will require a high throughput transformation system. Provision of a high throughput transformation system is an integral part of functional genomics capability.   This will require not only a rapid, high frequency, transformation system, but also a low level of somaclonal variation, and techniques for controlling gene expression levels (*22*).

This chapter will focus on the enabling technologies required to successfully genetically engineer sorghum.

2. MATERIALS

2.1. Tissue Culture, Selection and Regeneration

Successful sorghum transformation technology has relied on the use of microprojectile bombardment of embryogenic callus or immature explants, with regeneration via somatic embryogenesis (*23-25*).   The genotypic specificity of plant regeneration has meant that reported transformation frequencies are low, and restricted to a few amenable genotypes.   A more efficient regeneration system via organogenesis has been developed (*20, 21*).   The advantages of the organogenic pathway, which is described here, include the use of mature seed-derived meristems as explants, limited callus proliferation and broader genotype amenability.   The tissue culture and transformation process is illustrated (Fig. 1).

1.   Mature seeds of *Sorghum bicolour.*   The organogenic tissue culture system works well for all genotypes tested to date which includes 296B, 90562-1-1, M35-1, Tx430, Tx623, QL41, QL27, QL12, TAM422, SA281, Ks48, Ks51, N38, P898012, Ajebsido and SPV462. Of these 296-B, 90562-1-1, M35-1 and Tx623 have been used to generate transgenic plants.

2. 2,4-D (Sigma): 1 mg/ml stock solution. Prepare by dissolving the powder in absolute ethanol and make up to volume. pH to between 6 and 7 to prevent precipitation. Store at 4°C. 2,4-D has a storage life of approx. 2 months.

3. BAP (Sigma): 1 mg/ml stock solution. Prepare by dissolving in 1 N NaOH and make up to volume with deionised water. Store at 4°C. BAP has an approx. storage life of 2 months.

4. Germination Medium (GM): Murashige and Skoog (MS) macro and micro-nutrients (26) with Gamborg's B5 vitamins (27) (Sigma), 20 g/l sucrose, 8 g/l agar (Sigma), pH 5.8.

5. Organogenic Medium (OM): MS nutrients with B5 vitamins (Sigma), 30 g/l sucrose, 2 g/l casein hydrolysate, 0.5 mg/l 2,4-D, 2 mg/l BAP, 4.9 μM CuSO$_4$, 3 g/l Phytagel (Sigma), pH 5.8.

6. Osmoticum Medium (OsM): OM with 0.2 M Mannitol and 0.2 M Sorbitol.

7. Shoot Elongation Medium (SEM): OM without 2,4-D.

8. Rooting Medium (RM): OM with 20 g/l sucrose and without 2,4-D and BAP.

9. Selection medium (OMB): organogenic medium supplemented with 2 mg/l bialaphos (bialaphos sodium: working standard grade. Meiji Seika Kaisha Ltd, Japan).

10. 0.1% Mercuric chloride.

11. Petri dishes: 90 x 15 mm and 90 x 20 mm UV sterilized (Labtek, Australia).

12. Tissue culture vessels (approx. 200 ml capacity) (Nalgene Nunc International).

2.2. Transformation

Sorghum transformation has been reported using microprojectile (28-30, 20, 21) and *Agrobacterium tumefaciens* methods (31). The sorghum transformation protocol described here utilizes a Particle Inflow Gun (PIG). The PIG is a simple and inexpensive particle bombardment device for the delivery of DNA into plant cells. DNA-coated tungsten particles are accelerated using pressurized helium in combination with a partial vacuum (32, 33). Transient or stable transformation have been reported for many species including soybean (32) maize (33) tobacco, alfalfa, barley, apple, rice, wheat (34) and sorghum (28) using a PIG.

1. Particle inflow gun and high pressure, high purity helium tank
2. Autoclaved baffles for each genotype of sorghum to be shot
3. Autoclaved filters for each plasmid used
4. 2.5 M Calcium chloride
5. Tungsten M-10 powder
6. Spermidine (Sigma)
7. Plasmids at 1 μg/μl
8. Deconex detergent (Comet, Australia).

2.3. Potting Plantlets

1. Sterile pots: for initial planting into pots use 8 cm diam. x 12 cm tall pots. For transplanting of transgenics into larger pots use at least 20 cm diam. x 25cm tall sized pots.
2. Stock fertilizer mix (17.8% w/w Blood 'n' Bone (7% nitrogen, 5% phosphorus), 2.9% w/w potassium nitrate, 1.5% w/w potassium sulphate, 17.8% w/w superphosphate, 29.7% w/w dolomite, 17.8% w/w hydrated lime, 8.9% w/w gypsum, 3.6% slow release trace element mix).
3. Steam-sterilized UC mix soil (1/3 cu. M. sand, 1/6 cu. M. peat (compressed 2:1), 4 kg of stock fertilizer mix (based on University of California potting mix B, fertilizer II, with readily available nitrogen plus moderate reserve of nitrogen, pH 6.5).
4. Commercial seed raising mix (Yates, Australia).
5. Plastic bags or mini-glasshouses to maintain high humidity.

3. METHODS

3.1. Tissue Culture and Regeneration

1. Soak seeds in 70% (v/v) ethanol for 5 min followed by one rinse in sterile water, and then a 20 min soak under vacuum in 0.1% (w/v) mercuric chloride. Rinse in sterile water, 3 times, leaving the seeds to soak for 30 min in the final rinse. Drain off water and rinse in fresh sterile water (*see* Note 1).
2. Place seeds onto Petri dishes containing GM medium (15-20 seeds/dish) and incubate in the dark for 4-5 days at $26 \pm 2°C$.
3. Excise a 10-15 mm section of the germinated seedling containing the meristem and lie horizontally on the surface of Organogenic Medium (OM) (Fig. 1A), sinking the meristem slightly into the medium (20-30 explants/dish). Incubate at $26 \pm 2°C$ under 35-40 $\mu mol/m^2/sec$ NEC 37W fluorescent lights with a 16 h daylength (*see* Note 2).
4. Shoot elongation will occur from the original explant so it is essential to trim the explant back to the original size to prevent apical dominance (Figs. 1B and C). Trim explants and transfer onto fresh OM medium every 2 weeks.
5. After approx. 4 weeks of culture, axillary and adventitious buds are visible and after a further month explants develop multiple buds and/or shoots (Fig. 1D).
6. Explants can be divided into smaller pieces (no smaller than 1 cm^2) (*see* Notes 3 and 4).
7. Four to sixteen hours before shooting, trim off shoots from explants so that new axillary and adventitious buds can develop. Subculture trimmed explants onto OsM medium to induce osmotic stress on the explants in preparation for shooting.
8. Shoot tissue on OsM medium as described (*see* Section 3.2.).

3.2. Transformation

3.2.1. Preparation of sterile tungsten

1. Perform all of the following steps in a laminar flow cabinet.
2. Weigh out 25 mg of tungsten powder (*see* Note 5).
3. Add x10 the weight in microlitres of absolute ethanol (for example, if weighed out 25 mg of tungsten, add 250 µl of ethanol).
4. Vortex the suspension for 10 sec and then allow to settle for 5 min. Repeat 3 times.
5. Spin to pellet the tungsten.
6. Remove the ethanol and add the same volume of sterile water to the tube.
7. Vortex and spin. Remove the water.
8. Repeat steps 6 and 7 three times.
9. Add 20 µl less water than volume previously added (i.e. if adding 250 µl of water in step 6, add 230 µl) since some water remains in tungsten suspension. Vortex and immediately dispense 50 µl into individual tubes.

3.2.2. Preparation of Plasmid – Tungsten Mix and use of the PIG

1. Set up the PIG in a laminar flow cabinet (Fig. 1E), clean with ethanol, activate vacuum pump.
2. Set helium tank to a pressure of 2200 kPa.
3. Open aperture valve on PIG by one full turn and ensure that the timer for each shot is set to 0.05 sec.
4. Prepare fresh spermidine to a final concentration of 100 mM (*see* Note 6).
5. Combine 50 µl sterile tungsten, 10 µl plasmid (from stock 1 µg/µl), 50 µl $CaCl_2$ and 20 µl diluted spermidine (*see* Notes 7 and 8).
6. Vortex the solution and sit on ice for 5 min. Tungsten particles will settle during this time. Discard 100 µl of the supernatant. Keep remaining plasmid-tungsten mix on ice.
7. Rasp the microcentrifuge tube vigorously across rack to mix well. Take 4 µl from bottom of tube and transfer onto filter (*see* Note 9). Place the filter into the chamber.
8. Place tissue for transformation 15 cm from the filter inside the chamber.
9. Draw vacuum to –90 psi and shoot tissue.
10. Release the vacuum and remove the tissue.
11. Repeat steps 7-10 for each tissue sample to be transformed.
12. When all shooting is completed, all baffles and filters should be soaked in Deconex (or equivalent detergent) solution for 24 h, before washing, rinsing and autoclaving for the next experiment.
13. Leave tissue on OsM media for 4-16 h in the dark. Subculture onto fresh OM medium and leave for 7 days before subculturing onto selection media.

3.3. Selection Strategy

1. Transfer 5-10 explants/dish onto OMB medium (*see* Note 10) and subculture every 2–4 weeks (*see* Note 11; Fig. 1F).
2. After 3 months, transfer vigorously growing shoots to culture vessels (approx. 200 ml capacity) containing SEM (*see* Notes 12 and 13; Fig. 1G).
3. When shoots reach 3 cm in height, transfer to equivalent vessels containing RM medium. It can take up to 4 months to induce shoots, and an additional 1-2 months to develop roots that are suitable for transferring to soil (*see* Note 14).

3.4. Potting Putative Transgenic Plants

1. After the establishment of a good root system, carefully remove the plant from its container and gently wash away any gelling medium.
2. Pot individual plantlets into commercial seed raising mix (*see* Notes 15 and 16). Cover plantlets with sealed plastic bags, or mini-glasshouse chambers to maintain high humidity (Fig. 1H).
3. Grow plants under lights at $26°C \pm 2°C$ under 35-40 $\mu mol/m^2/sec$ NEC 37W fluorescent lights with a 16 h daylength.
4. After 2-4 weeks of acclimatization, transfer plants to the glasshouse.
5. Remove or cut open bags around plants after 7 days in the glasshouse. After plants begin to grow transplant into larger pots containing sterilized UC mix to allow plants to mature.
6. Fertilize plants weekly with a complete liquid fertilizer and water plants gently 2-3 times per week. Plants will require an application of urea every 4 weeks until panicle emergence (approx. 1g per pot).

3.5. Frequency of Transformation and Tissue Culture

1. Transformation efficiency of sorghum is very low, with frequencies of 1-5% generally reported (*18, 21*). However, PCR analysis of the most recently generated transgenic plants has shown the efficiency to be as high as 14%. These plants were bombarded with two genes of interest and a selectable marker gene. Of those plants which were transgenic, 52% contained both of the genes of interest whilst 48% contained only one of the transgenes in their genomes. These plants are currently being tested using Southern blot analysis to confirm transgenicity.

Figure 1. Procedure for the genetic transformation of Sorghum bicolour. A. *After seed sterilization and germination on GM the shoot apices are excised and placed on organogenic medium to induce axillary and adventitious bud formation (bar = 1 cm).* B. *Sorghum apical meristem swelling whilst on OM (bar = 3 cm).* C. *Consistent trimming is required to break apical dominance and induce bud development (bar = 1 cm).* D. *and* F. *Axillary and adventitious bud formation on the organogenic explant (D. bar = 2 cm, F. bar = 2.3 cm). These explants are trimmed 4-16 h prior to particle bombardment with* E. *the PIG (bar = 10 cm). After particle bombardment, explants are placed on selection medium to select for putative transgenic plantlets.* G. *Putative transgenic plantlets are subcultured into larger culture vessels as they grow (bar = 2.5 cm), prior to* H. *potting into soil (bar = 3.5 cm). During acclimatization plantlets are initially covered by plastic bags to maintain high humidity before transfer to larger pots in the glasshouse.*

2. Transgenicity is confirmed through both PCR and Southern blot analysis, and
    expression of the transgene is confirmed through RT-PCR and Northern blot
    analysis. Transgenic progeny are grown from the seed of T0 plants and tested
    via PCR and Southern analysis.

3.6. Additional Considerations

1. ### Gene Silencing

    Transgenic plants generated via particle bombardment techniques often contain
    multiple insertions of the transgene (*35-37, 29*). These plants have an increased
    likelihood of gene silencing, as demonstrated in sorghum where complete
    silencing of the *gus* reporter gene was found in some T0 and T1 progeny (*29*).
    Emani *et al.* showed that silencing of *gus* was methylation–mediated, and the
    addition of 5-azacytidine reversed the silencing in 80% of the plants. Similarly,
    insertions of a transgene into a non-transcribed region of the genome can occur,
    and result in null expression.

2. ### Somaclonal Variation

    Sorghum is difficult to transform and has a long duration in tissue culture. One
    of the major benefits of the organogenic tissue culture system is the ability to
    maintain cultures almost indefinitely in culture. However, as reported in other
    crops such as maize (*38*) and barley (*39*) long durations in tissue culture can
    increase the frequency of somaclonal variation.

3. ### Chimerism

    This organogenic tissue culture method utilizes the shoot apices of germinated
    seedlings of sorghum which produce multiple axillary and adventitious buds.
    Transformation of these buds via particle bombardment can result in transgenic
    chimeric plants (e.g. only part of the meristem is transformed at bombardment,
    and therefore integration of the transgene is not facilitated in all cells of the
    plant). This was reported in transgenic wheat to occur at low frequencies (*40*).
    Consequently, it is important to check for stable transformation in second
    generation plants.

4. ### Promoters

    The best constitutive promoter for transient expression in sorghum is the maize
    Ubiquitin promoter compared with Actin1 and CaMV35S promoters (*30*).

Other promoters may be used for transformation, but preliminary transient experiments are recommended to test efficacy of the promoter in sorghum.

4. NOTES

1. Mercuric chloride is an environmental pollutant and its use is restricted in some countries. As an alternative, sodium hypochlorite can be used for surface-sterilization of the seeds. It is recommended to use sodium hypochlorite only if the seeds are clean and fresh. Soak the seeds in 70% ethanol for 5 min, surface-sterilize in 4% (v/v) sodium hypochlorite for 20-30 min and rinse 3 times with sterile water.

2. It is important to give the explants sufficient space to grow, so it is recommended to use deeper Petri dishes (90 x 25 mm) for culture of the meristem explants.

3. One of the advantages of the organogenic tissue culture system is that explants at this stage can be kept almost indefinitely. However, as somaclonal variation typically increases with time in tissue culture it is recommended that culture times be kept to a minimum. As this system starts with mature seed, cultures can be set up at any time of the year, unlike embryogenic systems that require a supply of immature embryos.

4. It is recommended that explants should not be divided into small pieces. Many sorghum genotypes produce phenolics in culture and when the explants are wounded by cutting into smaller pieces they can be killed by oxidized phenolics.

5. Five mgs of tungsten makes 1 mix for shooting (1 mg/μl). Make up sufficient tungsten for the number of shots required (4 shots/mix).

6. Defrost spermidine on ice, and keep cold.

7. Vortex each reagent before adding. Work quickly and use each mix immediately after preparation.

8. If co-bombarding with 2 plasmids, add 5 μl of each plasmid at 1 μg/μl.

9. Tungsten-plasmid mix should be uniform in colour. Excessive aggregation of the tungsten particles is detrimental as large clumps of tungsten damage the tissue and inhibit transformation. Do not use this mix, and prepare a new one.

10. This selection system relies on the use of the *bar* gene. The concentration of bialaphos may be increased if the genotype is particularly vigorous in tissue culture. An alternative selection system involves the use of the *hpt* gene with 50 mg/l hygromycin B (*30*) in the selection medium. Positive selection systems, such as the *man*A gene and mannose as the selective agent (*41*), are being investigated.

11. Subculturing should occur more frequently if the genotype produces substantial levels of oxidative phenolic compounds (these are observed as regions of blackening in the growth medium).

12. A suitable plantlet for transfer into the larger culture vessels is an explant which has a green shoot of approx. 2 cm in height. As individual plantlets become evident from the explant mass it is useful to place only 1 plantlet per culture vessel to reduce losses from contamination spread. It can be difficult to distinguish and separate individual plantlets from the explant clump so care must be taken here.

13. Extra care must be taken to ensure that aseptic technique is strictly adhered to as moving plantlets into bigger containers can increase contamination frequency.

14. A suitable plantlet for transfer from the larger culture vessels into soil is one that has a green shoot of approx. 10 cm in height and root tissue of at least 4 cm in length.

15. Great care must be taken to ensure a clean environment is provided for the sorghum plantlets during potting as plantlets are susceptible to fungal contaminants.

16. Different soil mixes have been used for potting plants. Commercial seed raising mix, sterilized UC mix with or without vermiculite (ratio 2:1 or 3:1) have been utilized. The potting of plants from the sterile culture vessels into soil is a very sensitive part of the whole transformation procedure. Many plants die at this phase so great care must be taken. Further work is required to reduce losses during this stage.

CONCLUSION

Genetic transformation of sorghum is now routine and reliable. Transgenic plants may be regenerated at an acceptable frequency using the methodology described. The organogenic methodology has overcome many of the problems associated with genotypic limitations of somatic embryogenic techniques. However, there are problems which require additional research to further advance sorghum transformation to become high-throughput. A major shortcoming of the methodology is the selection process. The development of better *in vitro* selection strategies, particularly using positive selection strategies, to avoid problems of high escape frequencies is a high priority.

REFERENCES

1.   Ludlow MM and Wilson GL (1971). Photosynthesis of tropical pasture plants 1. Illuminance, carbon dioxide concentration, leaf temperature and leaf-air vapour pressure difference. *Australian Journal of Biological Science*, **24**: 449-470.
2.   Ludlow MM (1987). Light stress at high temperature. In: Kyle D, Arntzen C, Osmond CB (eds.), *Photoinhibition. Topics in Photosynthesis.*, Elsevier, Amsterdam, The Netherlands, Vol 9, pp. 89-110.
3.   Godwin ID and Gray SJ (2000). Overcoming productivity and quality constraints in sorghum: the role for genetic engineering. In: O'Brien L, Henry RJ (eds.), *Transgenic Cereals.*, AACC Publishing, St Paul, MN, USA, pp. 153-177.
4.   Hamaker BR (1997). Chemical and physical aspects of food and nutritional quality of sorghum and millet. *INTSORMIL Annual Report*, pp. 113-119.
5.   Chang TT (1987). The impact of rice on human civilization and population expansion. *Interdisciplinary Science Reviews*, **12**: 63-69.
6.   Rowe JB and Pethick DW (1994). Starch digestion in ruminants – problems, solutions and opportunities. *Proceedings of the Nutrition Society in Australia*, **18**: 40-52.
7.   Doggett H (ed.) (1988). *Sorghum.* Longmans Group, London, UK, 2nd edn.
8.   Nwanze KF, Seetharama N, Sharma HC, and Stenhouse JW (1995). Biotechnology in pest management improving resistance in sorghum to insect pests. *African Crop Sciences Journal*, **3**: 209-215.
9.   Henzell RG and Hare BW (1996). Sorghum breeding in Australia – public and private breeding endeavours. In: Foale MA, Henzell RG, Kniepp JF (eds.), *Proceedings of the 3rd Australian Sorghum Conference*, Tamworth, February. 1996. Australian Institute of Agricultural Science, Melbourne, Occasional Publication No. 93, pp. 159-172.
10.   Sharma HC, Abraham CV, Vidyasagar P and Stenhouse JW (1996). Gene action for resistance in sorghum to midge, *Contarinia sorghicola. Crop Science*, **32**: 1091-1098.
11.   Jordan DR, Tao Y, Godwin ID, Henzell RG, Cooper M and McIntyre CL (1998). Loss of genetic diversity associated with selection for resistance to sorghum midge in Australian sorghum. *Euphytica*, **107**: 1-7.
12.   Frederikson RA (1980). Sorghum downy mildew in the United States: Overview and outlook. *Plant Disease*, **67**: 903-908.
13.   Franzmann BA, Persley DM and Murray DAH (1996). Protecting sorghum against pests and diseases. In: Foale MA, Henzell RG, Kniepp JF (eds.), *Proceedings of the 3rd Australian Sorghum Conference,*. Tamworth, February. 1996. Australian Institute of Agricultural Science, Melbourne, Occassional Publication No. 93: pp. 103-112.
14.   Moss RJ, Blaney BJ, Casey ND, Gobius NR, Jonsson NN and Corbett JL (1999). Ergot (*Claviceps africana*) contamination of sorghum grain reduces milk production. *Recent Advances in Animal Nutrition Australia*, **12**: 21A.

15. Bailey CA, Fazzino JJ, Ziehr MS, Sattar M, Haq AU, Odvody G and Porter JK (1999). Evaluation of sorghum ergot toxicity in boilers. *Poultry Science,* **78**: 1391-1397.

16. Meinke H and Ryley M (1997). Effects of sorghum ergot on grain sorghum production: a preliminary climate analysis. *Australian Journal of Agricultural Research*, **48**: 1241-1247.

17. Isakeit T, Odvody GN and Shelby RA (1998). First report of sorghum ergot caused by *Claviceps africana* in the United States. *Plant Disease, 82*: 591.

18. Zhu H, Muthukrishnan S, Krishnaveni S, Wilde G, Jeoung J-M and Liang GH (1998). Biolistic transformation of sorghum using a rice chitinase gene. *Journal of Genetics and Breeding*, **52**: 243-252.

19. Krishnaveni S, Jeoung JM, Muthukrishnan S and Liang GH (2001). Transgenic sorghum plants constitutively expressing a rice chitinase gene show improved resistance to stalk rot. *Journal of Genetics and Breeding,* **55**: 151-158.

20. Gray SJ, Zhang S, Rathus C, Lemaux P and Godwin ID (2003). Development of sorghum transformation: organogenic regeneration and gene transfer methods. In: Seetharama N, Godwin ID (eds.), *Sorghum Tissue Culture and Transformation.* Science Publishers Inc New Hampshire, USA (in press).

21. Gray SJ, Nguyen T-V, Rathus C. and Godwin ID (2003). Development of regeneration and transformation systems for sorghum (*Sorghum bicolor* L.). *Field and Crops Research* (in press).

22. Able JA and Godwin ID (2000). Enhancing transgene expression in plants: a role for matrix attachment regions (MARs). *Current Topics in Plant Biology*, **2**: 117-124.

23. Gamborg OL, Shyluk JP, Brar DS and Constabel F (1977). Morphogenesis and plant regeneration from callus of immature embryos of sorghum. *Plant Science Letters*, **10**: 67-74.

24. Ma H, Gu M and Liang GH (1987). Plant regeneration from cultured immature embryos of *Sorghum bicolor* (L.) Moench. *Theoretical and Applied Genetics*, **73**: 389-394.

25. Rathus C and Godwin ID (2000). Transgenic Sorghum (*Sorghum bicolor*). In: Bajaj YPS (ed.), *Biotechnology in Agriculture and Forestry, Transgenic Crops I.* Springer-Verlag, Berlin Heidelberg, Germany, Vol 46, pp. 76-83.

26. Murashige T and Skoog F (1962). A revised medium for rapid growth and bioassays with tobacco tissue cultures. *Physiologia Plantarum*, **15**: 473-497.

27. Gamborg OL, Muller RA and Ojima K (1968). Nutrient requirements of suspension cultures of soybean root cells. *Experimental Cell Research*, **50**: 151-158.

28. Able JA, Rathus C and Godwin ID (2001). The investigation of optimal bombardment parameters for transient and stable transgene expression in sorghum. *In Vitro Cellular and Developmental Biology – Plant*, **37**: 341-348.

29. Emani C, Sunilkumar G and Rathore KS (2002). Transgene silencing and reactivation in sorghum. *Plant Science*, **162**: 181-192.

30. Hagio T, Blowers AD and Earle ED (1991). Stable transformation of sorghum cell cultures after bombardment with DNA coated microprojectiles. *Plant Cell Reports,* **10**: 260-264.

31. Zhao Z, Tishu C, Tagliani L, Miller M, Wang N, Pang H, Rudert M, Schroeder S *et al.* (2000). *Agrobacterium*-mediated sorghum transformation. *Plant Molecular Biology*, **44**: 789-798.

32. Finer JJ, Vain P, Jones MW and McMullen MD (1992). Development of the particle inflow gun for DNA delivery to plant cells. *Plant Cell Reports*, **11**: 232-238.

33. Vain P, Keen N, Murillo J, Rathus C, Nemes C and Finer JJ (1993). Development of the Particle Inflow Gun. *Plant Cell, Tissue and Organ Culture*, **33**: 237-246.

34. Abumhadi N, Trifonova A, Takumi S, Nakamura C, Todorovska E, Getov L, Christov N and Atanassov A (2001). Development of the particle inflow gun and optimizing the particle bombardment method for efficient genetic transformation in mature embryos of cereals. *Biotechnology and Biotechnological Equipment*, **15**: 87-96 Suppl. S 2001.

35. Matzke MA and Matzke AJM (1995). How and why do plants inactivate homologous (trans) genes? *Plant Physiology,* **107**: 679-685.

36. Kohli A, Gahakwa D, Vain P, Laurie DA and Christou P (1999). Transgene expression in rice engineered through particle bombardment: molecular factors controlling stable expression and transgene silencing. *Planta, 208*: 88-97.

37. Demeke T, Huci P, Baga M, Caswell K, Leung N and Chibbar RN (1999). Transgene inheritance and silencing in hexaploid spring wheat. *Theoretical and Applied Genetics*, **99**: 106-118.

38. Armstrong CL and Phillips RL (1988). Genetic and cytogenetic variation in plants regenerated from organogenic and friable, embryogenic tissue cultures of maize. *Crop Science*, **28**: 363-369.

39. Hang A and Bregitzer P (1993). Chromosomal variations in immature embryo-derived calli from six barley cultivars. *Journal of Heredity*, **84**: 105-108.

40. Stoger E, Williams S, Keen D and Christou P (1998). Molecular characteristics of transgenic wheat and the effect on transgene expression. *Transgenic Research*, **7**: 463-471.

41. Joersbo M, Donaldson I, Kreiberg J, Petersen SG, Brunstedt J and Okkels FT (1998). Analysis of mannose selection system used for transformation of suger beet. *Molecular Breeding*, **4**: 111-117.

Chapter 8

SUGARCANE TRANSFORMATION

S.J. SNYMAN

South African Sugar Association Experiment Station, Biotechnology Department, Private Bag X02, Mount Edgecombe, KwaZulu Natal, 4300, South Africa. E-Mail: snyman@sugar.org.za

1. INTRODUCTION

1.1. Sugarcane as a Crop

Sugarcane is a hybrid (*Saccharum* species hybrids) belonging to the grass family, *Gramineae*. It is best grown commercially in tropical and sub-tropical areas that are characterized by warm temperatures, high incident solar radiation and annual rainfall, and deep fertile soils (*1*). Sugarcane is cultivated in over 100 countries and is the source of approx. 70% of the world's sugar (*2*). In addition, it is used as a raw material for ethanol production in Brazil, which is the largest producer of cane sugar in the world (*3*).

Sugar industries are driven to increase output but are limited by a number of factors. These include sucrose storage limits of 14% (g sucrose/100 g fresh mass plant) reached by the genotypes currently in use (*4*) and increasing world sugar consumption from 118 to 130 million Mg during the period 1996 to 2001. In addition, loss of productive capacity of soils caused by declining soil health under sugarcane monoculture has resulted in only marginal increases in sugar yield over the last 20 years (*5*). Another major economic factor worldwide affecting yield is crop damage caused by insects. Large stands of single crop species have long been recognized as favouring dramatic increases in pest species, despite the use of chemical pesticides, insect parasitoids and modified cropping practices (*6, 7*). Although conventional breeding may result in the production of pest and disease resistant cultivars, there are limitations to this process in sugarcane.

1.2. Limitations of Conventional Sugarcane Breeding

Sugarcane breeding as we know it today began in the late 1800s in Java, in response to the need for resistance to sereh disease. One of a few locally available *Saccharum spontaneum* clones, Glagah (2n = 112), was used in an interspecific hybridization with *S. officinarum. S. officinarum* generally has the qualities of high sucrose content and low fiber, which led to it being referred to as a 'noble cane', while *S. spontaneum* has the attributes of resistance to many pests and diseases, tolerance to environmental stress and its adaptability (*8*). A selected hybrid was backcrossed to noble types, and the process was repeated in order to dilute the negative effects of the wild germplasm in a process that became known as nobilization (*8, 9*). As a

I.S. Curtis (ed.), Transgenic Crops of the World - Essential Protocols, 103-114.
© 2004 Kluwer Academic Publishers. printed in the Netherlands.

result of a limited number of parent genotypes used during nobilization, current sugarcane breeding populations show a relatively high degree of co-ancestry (*10, 11*). Consequently, commercial sugarcane is derived from a narrow genetic base, consisting of the germplasm of about twenty *S. officinarum* clones and less than ten *S. spontaneum* derivatives (*12*).

Cytological studies by Bremer (*13*) showed that nobilization is characterized by asymmetric chromosome transmission. In a cross between *S. officinarum* (2n = 80) as the female parent and *S. spontaneum* (2n = 40-128) as the male parent, *S. officinarum* generally transmits two haploid chromosome sets, while *S. spontaneum* transmits one. This 2n + n chromosome transmission that occurred in early nobilizations is due to a phenomenon known as 'female restitution', and the number of chromosomes contributed by the female parent, *S. officinarum,* increased with respect to the parent. Due to their high polyploidy and interspecific origin, modern sugarcane cultivars are genetically very complex (*11, 14*), and as a result, the outcome of crosses in breeding programmes is unpredictable (*15*).

Sugarcane is therefore a prime candidate for the application of genetic engineering, as single characters can be manipulated in a complex genetic background, a) to allow the introduction of novel traits and b) to salvage or correct germplasm that may have commercial potential except for one negative factor, such as disease susceptibility.

1.3. Advances in Sugarcane Improvement shown by Genetic Engineering

The genetic engineering of plants has allowed genes to be transferred to the target species from a wide variety of sources, including higher and lower plants, bacteria and animals. In many cases there was no source of the desired trait in the gene pool for conventional breeding. Since genes from nearly any source can be engineered to express in plants, the range of potential products that can be produced by transgenic plants is extensive (*16*). Below are some of the traits that have been introduced into sugarcane:

- insect resistance via the δ-endotoxin gene from *Bacillus thuringiensis* (*17*), proteinase inhibitor genes (*18, 19*) and mannose-binding lectins (*20*),
- an array of disease resistance genes including antimicrobial peptides (*20, 21*),
- resistance to sugarcane mosaic virus (SCMV) (*22, 23*),
- altered sucrose content via down-regulation of pyrophosphate-dependent phosphofructokinase (*24*), and
- herbicide resistance via the *pat* gene (*25-28*).

1.4. An Overview of in vitro Culture Systems and Transformation Technologies in Sugarcane

Although sugarcane is vegetatively propagated commercially, a substantial amount of work has been carried out on *in vitro* culture systems for the purposes of somatic cell improvement through culture-induced mutations (*29-31*), the production of disease-free plants (*32*), *in vitro* micropropagation (*33-35*) and more recently, genetic transformation (*17, 25, 27, 36-38*). It is evident from these reports that sugarcane can be regenerated via several pathways, and these are summarized (Fig. 1).

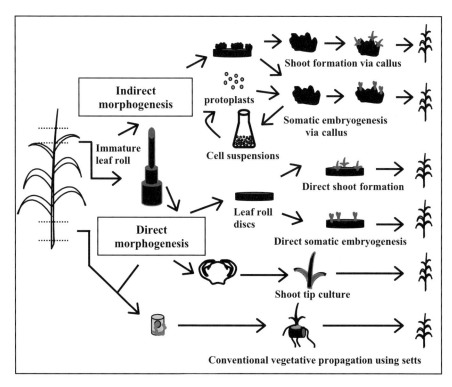

Figure 1. Diagrammatic representation of the different routes by which sugarcane plants can be regenerated in vitro *and* in vivo.

Regeneration of sugarcane can occur through two main routes, direct and indirect morphogenesis. With direct morphogenesis, plants are regenerated directly from tissue such as immature leaf roll discs. Indirect morphogenesis involves the initial culturing of leaf roll sections or inflorescences on an auxin-containing medium to produce an undifferentiated mass of cells, or callus. The source of auxin used in sugarcane tissue culture is normally 2,4-D at a concentration of 3 mg/l (*33, 39-41*), but other synthetic auxins such as picloram (*42*) and dicamba have been used (*43*).

Direct organogenesis (or shoot morphogenesis) is the way in which sugarcane is propagated vegetatively on a commercial basis. Sections of stalk or setts containing a bud and root primordia give rise to plantlets following the inhibition of apical dominance (*35*). Direct organogenesis *in vitro* can be achieved by shoot tip culture (*33, 34*) or shoot multiplication from leaf discs (*32, 34, 44, 45*).

Direct somatic embryo formation on sugarcane leaf discs (i.e. without callus formation) *in vitro* was first reported by Irvine and Benda (*32*) during a process of rapid regeneration of plantlets in an attempt to eliminate sugarcane of SCMV. It was recently reported again during studies on suspension cultures for protoplast isolation (*46*), and in a transformation protocol using leaf discs as target material for bombardment, with plantlet regeneration occurring via embryo formation directly on leaf discs (*47-49*). Although this approach in a transformation programme could substantially reduce the length of time in culture and avoid prolonged exposure to high levels of 2,4-D, the technique is currently unsuitable for widespread adoption. Existing limitations include the production of chimaeric plants (*50*) and possible intellectual property restrictions.

The delivery of foreign DNA to sugarcane has been achieved via a range of procedures such as electroporation of suspension culture cells (*38*) and *Agrobacterium*-mediated transformation (*26, 37, 51*). However, the most widespread approach is that of microprojectile bombardment of embryogenic calli (*23, 25, 27, 36, 52*), although transgenic plant production ranges widely from 1-20 plants per bombardment (*27* and *36*, respectively). Factors such as the choice of cultivar, genotypic responses *in vitro* and selection regime may account for the differences in efficiency reported.

This chapter outlines the production of transgenic sugarcane plants by microprojectile bombardment of embryogenic callus, which, with a few minor modifications, is based on the published protocol of Bower *et al.* (*36*). A co-bombardment strategy is used, where the *npt*II selectable marker gene is driven by the Emu monocot-specific promoter (*53*) and selection is carried out on medium containing the antibiotic geneticin. For technical simplicity, transformation efficiency is assessed using herbicide resistance as a model trait. The *pat* gene confers resistance to the herbicidal compound glufosinate ammonium and the phenotypic expression of the gene in transgenic plants can be facilitated by herbicide application to plants in the glasshouse.

2. MATERIALS

2.1. Plasmid Constructs

1. Plasmids are maintained and generated in *E.coli* JM109. Plasmid EmuKN (ES Dennis, CSIRO, Plant Industry, GPO Box1600, Canberra ACT2601, Australia) is used for selection and contains the *npt*II selectable marker gene under the Emu monocot specific promoter. Plasmid UbiAck (Aventis, France) contains the synthetic herbicide resistance gene *pat* (which confers resistance to glufosinate ammonium) under the control of the maize ubiquitin promoter.
2. Luria broth (LB): 10 g/l bacto-tryptone, 5 g/l yeast extract, 10 g/l NaCl, pH 7.5.
3. Liquid culture medium: LB medium containing 50 mg/l ampicillin (Sigma).
4. Plasmid DNA is purified using the Nucleobond AX100 kit (Macherey-Nagel, Germany).
5. TE buffer: 10 mM Tris-HCl, pH 7.5 and 1 mM EDTA.

2.2. Tissue Culture

1. Sugarcane cultivars NCo310, N27 and the non-commercial cultivar 88H0019.
2. MS3 medium: 4.3 g/l Murashige and Skoog (MS, *54*) salts and vitamins (Highveld Biological, South Africa), 20 g/l sucrose, 0.5 g/l casein (Sigma), 3 mg/l 2,4-D (Sigma), 5 g/l agargel (Sigma), pH 5.8 (with KOH).
3. Osmoticum medium: MS3 medium with 0.2 M mannitol, 0.2 M sorbitol, and 2.75 g/l Gelrite (Polychem, SA).
4. Selection medium: MS3 medium with 45 mg/l geneticin (Roche), filter-sterilize through a 0.2 μm membrane (Cameo 25AS; Osmonics Inc.) to autoclaved medium.
5. Regeneration medium: 4.3 g/l MS salts and vitamins, 20 g/l sucrose, 0.5 g/l casein, 0.5 mg/l kinetin (Sigma), 2 g/l Gelrite, pH 5.8, with 45 mg/l geneticin.
6. Magenta GA-7 vessel medium: regeneration medium made with half-strength MS salts and vitamins.
7. 2,4-D (Sigma): solubilize in a small volume of 100% (v/v) ethanol, make up to volume with water and store as a 100 x stock solution at 4°C.
8. Kinetin: stock solution of 20 mg/ml, solubilize in dilute HCl. Sterilize solution by filtration, aliquot in sterile microfuge tubes and store at –20°C.
9. Surface sterilant: 70% (v/v) ethanol.
10. Sterile Petri dishes: 10 x 2 cm (Corning, UK) and 9 x 1 cm (Carbi, LASEC, South Africa).
11. Magenta GA-7 vessels (Sigma).
12. Parafilm (Pechiney Plastic Packaging, Chicago, IL, USA) and 3M medical micropore tape.
13. All media are sterilized by autoclaving at 121°C for 20 min. Antibiotics are added to the medium after autoclaving (40°C for agar medium).

2.3. Microprojectile Bombardment

1. Particle Inflow Gun (PIG): based on helium propulsion of particles, as designed by Finer *et al.* (*55*).
2. Coating of microprojectiles: size M10 tungsten particles (Bio-Rad Laboratories, CA), 0.1 M spermidine free base (Sigma), 2.5 M $CaCl_2$ (Saarchem, South Africa).
3. Filter holder: 13 mm Swinney filter unit (Millipore, Germany).
4. Nylon mesh screen: 200 $μm^2$ nylon mesh is fitted to the top of a Magenta vessel (Sigma) that has had part of the base cut away so that the final height of the jar is 4 cm.

2.4. Glasshouse

1. Potting soil (90 % composed bark, 10 % river sand; Grovida, South Africa).
2. Nutrient solution (seedling hydroponic nutrient solution (Hygrotech Seed, South Africa)): 1 g/l irrigation water, applied weekly to plants in pots.

3. Herbicide: Buster (active ingredient: 200 mg/l glufosinate ammonium; Aventis, France).
4. Knapsack sprayer: hydraulically operated Matabi sprayer (Goizper, Spain).

3. METHODS

3.1. Plasmid Preparation

1. Plasmids are purified using a Nucleobond AX100 kit (Macherey-Nagel, Germany).
2. Plasmid integrity is verified by restriction analysis followed by agarose gel electrophoresis.
3. Plasmid concentration is assessed on the fluorometer (Hoefer DyNA Quant 200) and is adjusted to a final concentration of 500 ng/µl with TE. Plasmid is aliquoted in sterile microfuge tubes and stored at -20°C.

3.2. Generation of Embryogenic Callus

1. The apical region of the stem of field-grown plants is collected. The older, loosely attached leaves are removed and a 20-30 cm-long leaf roll containing the apical meristem is briefly swabbed with 70% (v/v) ethanol to remove microbial contaminants. The remaining outermost mature leaves are aseptically removed and the young, inner, tightly furled leaves are sliced into transverse sections approx. 3 mm thick.
2. Leaf roll discs (21 per stalk, taken from 10-15 cm directly above the uppermost node) are placed on MS3 callus induction medium. A total of 7 discs are placed per Petri dish (9 x 1 cm in size), containing 25 ml medium. Plates are sealed with parafilm and incubated in the dark at 26 ± 1 °C for 8 weeks. Explants are transferred to fresh MS3 medium every 2 weeks.
3. Type 3 embryogenic callus (*see* Note 1) is removed and subcultured 2-4 days prior to bombardment. Only the white, nodular embryogenic pieces (*see* Note 2) are used, and this is cut away from other callus pieces on the original explant (*see* Note 3).
4. On the day that the callus is to be bombarded, the osmoticum medium (*see* Note 4) surface is scored with sterile forceps in a 2.5 cm diam. circle in the centre of a Petri dish (9 x 1 cm in size). Callus pieces are placed in the scored area and pushed down slightly in to the medium. The callus is incubated on osmoticum for 4 h prior to and after bombardment (*56*).

3.3. Microprojectile Bombardment

1. Tungsten particles (100 mg/ml) are sterilized in absolute ethanol for 5-10 min. The particles are pelleted by microfuging for 60 sec. The supernatant is removed, and the particles are washed twice with sterile distilled water,

centrifuge (benchtop microfuge at room temperature) for 5 min each time to pellet the particles.

2. In non-stick microfuge tubes (Merck, SA), a total of 10 μg of plasmid DNA (or a 1:1 mixture of two plasmids) in a volume of 20 μl is precipitated on to 50 μl of the tungsten suspension by the sequential addition of 50 μl $CaCl_2$ and 20 μl spermidine. The components are mixed by continuous vortexing on a moderate setting. The mixture is vortexed for a further 30 sec and placed on ice for 5 min. A volume of 110 μl is removed from the supernatant, leaving 30 μl behind. The suspension is maintained on ice for the duration of the bombardments.

3. The helium line of the PIG is cleared of air by passing helium through the line prior to the bombardment session.

4. The Petri dish containing the callus on osmoticum medium is placed on a shelf in the PIG vacuum chamber, 6 cm below the unit containing the DNA-coated tungsten particles. A nylon mesh screen is placed over the calli to reduce calli movement during microprojectile delivery.

5. The microfuge tube containing the DNA is mixed prior to each bombardment by running the base of the tube over a microfuge rack several times. A volume of 4 μl of the particle suspension is placed in the centre of a 1 mm^2 metal grid in a disassembled Swinney filter holder. The filter unit is reassembled and screwed into an adapter above the calli. The chamber is evacuated to a pressure of 10 kPa and the particles are discharged when the helium (1000 kPa) is released by a solenoid linked to a timer relay (0.05 sec).

3.4. Selection and Regeneration of Transformants

1. Bombarded calli recover on MS3 medium for 2 days, after which they are transferred to the selection medium containing geneticin. Media (50 ml per Petri dish) is dispensed in 10 x 2 cm Petri dishes (see Note 5). Callus is subcultured fortnightly for a total of 10-12 weeks, and maintained in the dark at 26 °C (see Note 6).

2. Transgenic calli are transferred to a 16 h photoperiod (45 $\mu mol/m^2/sec$ photosynthetic active radiation) on regeneration medium with 45 mg/l geneticin 10-12 weeks after bombardment or when they are approx. 0.5 cm^2 in size (see Note 7).

3. After 6-8 weeks or when plantlets are approx. 2 cm in height with visible roots, they are transferred to Magenta vessels (see Note 8).

4. When plantlets are approx. 10 cm tall and have a well-established root system, transfer to the glasshouse in moist potting soil either in small pots covered with a clear plastic bag, or in 72-well polystyrene seedling trays covered with plastic wrap. The plants are kept covered for 4 days, after which the plastic covering is removed for 1-4 h on 4 consecutive days. Once the plastic covering has been removed, the plants are watered for 60 sec every 4 h with a fine mist during natural daylight hours for a period of 2 weeks. Thereafter, the plants are transferred to plastic pots (first to 3 litre volume, and then to 20 litre pots), watered for 10 min twice a day and fertilized once a week.

3.5. Phenotypic Analysis by Glasshouse Spraying

1. Herbicide is applied to sugarcane plants that are approx. 4 months old. The commercial herbicide Buster is applied at a rate of 4 litres/hectare using a knapsack fitted with a Teejet nozzle (110°C) spraying at a pressure of 100 kPa.
2. Plants are rated for their susceptibility 3 weeks after herbicide application (*see* Note 9).

4. NOTES

1. Recognition of different types of callus is crucial to the success of transformation. Although most authors state the use of 'embryogenic callus' as target material for microprojectile delivery, the callus morphology is important, with Falco *et al.* (*27*) stating specifically that they use 'nodular, compact and white calli' as target material, and this is identified microscopically. Bower and Birch (*52*) observed that this type of callus showed the highest levels of transient expression. Later studies (*36*) utilizing Type 3 callus, as classified by Taylor *et al.* (*40*) (Table 1), recorded transformation efficiencies of up to 20 plants per bombardment, which are the highest reported for sugarcane.

Table 1. A description of three sugarcane callus morphologies observed in all cultivars in this study according to the classification proposed by Taylor et al. *(40).*

Parameter	Type 2	Type 3	Type 4
Colour	Yellow-grey	White	Cream or yellow
Growth rate	Slow(er)	Fast	Fast
Texture	Wet/shiny, sticky	Dry, nodular, compact	Wet, grainy, friable
Overall characteristics	A shiny, wet matrix, no aggregates visible	Homogenous mixture of early somatic embryos, giving a compact, nodular appearance	Homogenous mixture of smooth-surfaced pre-embryogenic aggregates

2. In cultivars where Type 3 callus is not produced in copious amount, it is beneficial to cycle the callus between MS3 medium and MS medium containing 1 mg/l of 2,4-D. Cycling callus between high and low concentrations of 2,4-D has been shown to maintain the morphogenic competence of cells (*57*) and fortnightly transfer between the two media is beneficial in optimizing the proportion of Type 3 callus for bombardment (*58*).
3. When cutting Type 3 callus away from other callus types, where possible, try to keep callus attached to original explant, as this minimizes the occurrence of small pieces during multiple transfers on the day of bombardment.
4. The osmoticum medium needs to be poured while it is quite hot (55-60°C), and a volume of 50 ml per Petri dish needs to be dispensed, otherwise the media tends to splatter out of the Petri dish during bombardment.
5. Double the amount of selection medium (50 ml) can be dispensed in the deeper Petri dishes and there is more 'head space', so visually, calli appear to retain embryogenic characteristics when compared to culture in regular 9 cm dishes. In addition, the 10 cm Petri dishes are not sealed with parafilm, but with two strips of tape (we use medical micropore tape, but any adhesive tape would be suffice) on opposite sides of the dish. It is speculated that this allows for increased aeration of the cultures.
6. Small, white outgrowths are visible on bombarded callus 4-6 weeks after placement on selection medium. Initially, the newly formed callus has a mucilaginous appearance and by 8-10 weeks the

embryogenic in nature can be determined. By comparison, calli that are not transformed either: a) show some signs of necrosis 2-4 weeks after placement on selection medium, and by 10 weeks are black, or b) do not turn black on selection medium, but do not produce any visible new growth, take on a shiny appearance and are not regenerable. The use of 0.8% tetrazolium red (2,3,5 triphenyltetrazolium chloride (Sigma) in 0.05 M sodium phosphate buffer) was valuable initially in identifying that the pieces of callus described (*see* Note 2) were not viable, and therefore should not be transferred to regeneration medium. Callus pieces are transferred to 0.5 ml of tetrazolium red in a microfuge tube and incubated in the dark overnight. Viable callus stains red.

7. Frequently, when calli are transferred to the light, weekly subculturing is required, as the production of phenolic compounds seems to increase directly prior to regeneration. The production of phenolics may also be cultivar-dependent. The time taken for callus to turn green on regeneration medium is approx. 2-3 weeks.

8. If young plantlets root poorly, they are transferred to MS medium containing 0.5 mg/l IBA to stimulate root production before being placed in Magenta vessels. If in spite of the IBA treatment rooting is not vigorous, plants do not acclimatize to glasshouse conditions.

9. Although transformation efficiency varies according to genotype, in this study 1 transgenic plant per bombardment was recorded for all 3 genotypes utilized.

CONCLUSIONS

The establishment of reliable techniques for *in vitro* culture and transformation has enabled the production and field assessment of transgenic sugarcane containing a range of cloned genes. These include the *Bt* endotoxin gene (*17*), herbicide resistance (*25, 28*) and SCMV resistance (*23*). Despite these successes, intellectual property restrictions may limit commercial application of transgenic sugarcane. Public resistance to foods from transgenic sources and resultant marketing constraints are further negative factors at present. However, novel work by commercial partners in developing sugarcane for the production of alternative commodities such as collagen and biodegradable plastics (*59, 60*) may overcome these barriers. Nonetheless, there are still some constraints in the production of transgenic plants (*61*). Future sugarcane transformation research could be directed to the following areas: expanding the genotype range, dealing with variable transformation efficiencies, overcoming the effects of somaclonal variation that have frequently resulted in undesirable phenotypes in the field, including yield loss (*62-64*), and addressing the complicated issue of gene silencing that has made interpretation of genetic manipulations difficult.

ACKNOWLEDGEMENTS

I thank Gwethlyn Meyer and Zama Mtshali for technical assistance and Barbara Huckett for critical review of the manuscript.

REFERENCES

1.   Barnes AC (1974). *The sugarcane,* 2nd edn. Leonard Hill Books, London UK. pp. 572.
2.   Anonymous (2000). South African Sugar Association industry directory. Published by SASA Public Affairs Division.
3.   http://apps.fao.org.
4.   Heinz DJ (1987). *Sugarcane Improvement Through Breeding.* Elsevier, The Netherlands, pp.1-6.
5.   Garside AL (1997). Yield decline research in the Australian sugar industry. *Proceedings of the South African Sugar Technology Association, 71*: 3-8.
6.   Thomas JC, Adams DG, Keppenne VD, Wasmann CC, Brown JK, Kanost MR and Bohnert HJ (1995a). *Manduca sexta* protease inhibitors expressed in *Nicotiana tabacum* provide protection against insects. *Plant Physiology and Biochemistry, 33*: 611-614.
7.   Thomas JC, Adams DG, Keppenne VD, Wasmann CC, Brown JK, Kanost MR and Bohnert HJ (1995b). Protease inhibitors of *Manduca sexta* expressed in transgenic cotton. *Plant Cell Reports, 14*: 758-762.
8.   Daniels J and Roach BT (1987). Taxonomy and Evolution. In: Heinz DJ (ed.), *Developments in Crop Science II. Sugarcane improvement through breeding.* Elsevier, Amsterdam, pp. 7-84.
9.   Stevenson GC (1965). *Genetics and breeding of sugarcane.* Longmans, London, pp. 284.
10. Berding N and Roach RT (1987). Germplasm collection, maintenance and use. In: Heinz D (ed.), *Developments in Crop Science III. Sugarcane Improvement Through Breeding.* Elsevier, The Netherlands, pp.143-210.
11. Butterfield MK, D'Hont AD and Berding N (2001). The sugarcane genome: a synthesis of current understanding, and lessons for breeding and biotechnology. *Proceedings of the South African Sugar Technology Association, 75*: 1-5.
12. Arceneaux G (1965). Cultivated sugarcanes of the world and their botanical derivatives. *Proceedings of the International Society of Sugar Cane Technology, 12*: 844-855.
13. Bremer G (1961). Problems in breeding and cytology of sugarcane. IV The origin of the increase of chromosome number in species hybrids *Saccharum. Euphytica, 10*: 325-342.
14. Bennet MD and Smith JD (1976). Nuclear DNA amounts in Angiosperms. *Philosophical Transactions of the Royal Society of London Series B, 274*: 227-274.
15. Price S (1967). Interspecific hybridization in sugarcane breeding. *Proceedings of the International Society of Sugar Cane Technology, 12*: 1021-1026.
16. Dunwell JM (1999). Transgenic crops: the next generation, or an example of 2020 vision. *Annals of Botany, 84*: 269-277.
17. Arencibia A, Vazquez RI, Prieto D, Tellez P, Carmona ER, Coego A, Hernandez L, de la Riva GA *et al.* (1997). Transgenic sugarcane plants resistant to stem borer attack. *Molecular Breeding, 3:* 247-255.
18. Allsopp PG, McGhie TK, Hickman KA and Smith GR (1996). Increasing the resistance of sugarcane to attack from whitegrubs by introducing novel insecticidal genes. In: JR Wilson, DM Hogarth, JA Campbell, AL Garside (eds.), *Sugarcane: Research Towards Efficient and Sustainable Production.* Sugar 2000 Symposium, CSIRO Division of Tropical Crops and Pastures, Brisbane, Australia, pp. 141-143.
19. Nutt KA, Allsopp PG, McGhie TK, Shepard KM, Joyce PA, Taylor GO, McQualter RB, Smith GR *et al.* (1999). Transgenic sugarcane with increased resistance to canegrubs. *Proceedings of the Australian Society of Sugar Cane Technology,* Townsville, Queensland, Australia, 27-30 April 1999, pp. 171-176.
20. Irvine JE and Mirkov TE (1997). The development of genetic transformation of sugarcane in Texas. *Sugar Journal June 1997,* pp. 25-29.
21. Zhang L and Birch RG (1996). Biocontrol of sugarcane leaf scald disease by an isolate of *Pantoea dispersa* which detoxifies albicidin phytotoxins. *Letters in Applied Microbiology, 22*: 132-136.
22. Joyce PA, McQualter RB, Handley JA, Dale JL, Harding RM and Smith GR (1998). Transgenic sugarcane resistant to sugarcane mosaic virus. *Proceedings of the Australian Society Sugar Cane Technology, 20*: 204-210.

23.   Ingelbrecht IL, Irvine JE and Mirkov TE (1999). Postranscriptional gene silencing in transgenic sugarcane. Dissection of homology-dependent virus resistance in a monocot that has a complex polyploid genome. *Plant Physiology*, **119**: 1187-1197.

24.   Groenewald J-H and Botha FC (2001). Manipulating sucrose metabolism with a single enzyme: pyrophosphate-dependent phosphofructokinase (PFP). *Proceedings of the South African Sugar Technology Association*, **75**: 101-103.

25.   Gallo-Meagher M and Irvine JE (1996). Herbicide resistant sugarcane containing the *bar* gene. *Crop Science*, **36**: 1367-1374.

26.   Enriquez-Obregon GA, Vazquez-Padron RI, Prieto-Samsonov DL, de la Riva GA and Selman-Housein G (1998). Herbicide resistant sugarcane plants by *Agrobacterium*-mediated transformation. *Planta* **206**: 20-27.

27.   Falco MC, Tulmann Neto A and Ulian EC (2000). Transformation and expression of a gene for herbicide resistance in Brazilian sugarcane. *Plant Cell Reports*, **19**: 1188-1194.

28.   Leibbrandt, NB and Snyman SJ (2003). Stability of gene expression and agronomic performance of a transgenic herbicide-resistant sugarcane line in South Africa. *Crop Science*, **43**: 671-677.

29.   Heinz DJ and Mee GWP (1969). Plant differentiation from callus tissue of *Saccharum* species. *Crop Science*, **9**: 346-348.

30.   Heinz DJ, Krishnamurthi M, Nickell LG and Maretzki A (1977). Cell, tissue and organ culture in sugarcane improvement. In: Reinert J, Bajaj YPS (eds.), *Applied and Fundamental Aspects of Plant Cell, Tissue and Organ Culture.* Springer, Berlin, Germany, pp. 3-17.

31.   Larkin PJ and Scowcroft WR (1981). Somaclonal variation- a novel source of variability from cell cultures for plant improvement. *Theoretical and Applied Genetics*, **60**: 197-214.

32.   Irvine JE and Benda GTA (1985). Sugarcane mosaic virus in plantlets regenerated from diseased leaf tissue. *Plant Cell, Tissue and Organ Culture*, **5**: 101-106.

33.   Lee TSG (1987). Micropropagation of sugarcane (*Saccharum* spp.). *Plant Cell, Tissue and Organ Culture*, **10**: 47-55.

34.   Grisham MP and Bourg D (1989). Efficiency *of in vitro* propagation of sugarcane plants by direct regeneration from leaf tissue and by shoot-tip culture. *Journal of the American Society of Sugarcane Technology*, **9**: 97-102.

35.   Moutia M and Dookun A (1999). Evaluation of surface sterilization and hot water treatments on bacterial contaminants in bud culture of sugarcane. *Experimental Agriculture*, **35**: 265-274.

36.   Bower R, Elliott AR, Potier BAM and Birch RG (1996). High-efficiency, microprojectile-mediated cotransformation of sugarcane, using visible or selectable markers. *Molecular Breeding*, **2**: 239-249.

37.   Arencibia AD, Carmona ER, Tellez P, Cahn M-T, Yu S-M, Trujilo LE and Oramas P (1998). An efficient protocol for sugarcane (*Saccharum spp.* L.) transformation mediated by *Agrobacterium tumefaciens. Transgenic Research*, **7**: 213-222.

38.   Arencibia AD, Carmona E, Cornide MT, Menendez E and Molina P (2000). Transgenic sugarcane (*Saccharum* spp.). In: Bajaj SS (ed.), *Biotechnology in Agriculture and Forestry 46. Transgenic crops 1.* Springer, Heidelberg, Germany, pp. 188-206.

39.   Nadar HM, Soepraptopo S, Heinz DJ and Ladd SL (1978). Fine structure of sugarcane (*Saccharum* spp.) callus and the role of auxin in embryogenesis. *Crop Science*, **18**: 210-216.

40.   Taylor PWJ, Ko H-L, Adkins SW, Rathus C and Birch RG (1992). Establishment of embryogenic callus and high protoplast yielding suspension cultures of sugarcane (*Saccharum* spp. hybrids). *Plant Cell, Tissue and Organ Culture*, **28**: 69-78.

41.   Fitch MM and Moore PH (1993). Long-term culture of embryogenic sugarcane callus. *Plant Cell, Tissue and Organ Culture*, **32**: 335-343.

42.   Fitch MM and Moore PH (1990). Comparison of 2,4-D and picloram for selection of long-term totipotent green callus cultures of sugarcane. *Plant Cell, Tissue and Organ Culture*, **20**: 157-163.

43.   Brisibe EA, Miyake H, Taniguchi T and Maeda E (1994). Regulation of somatic embryogenesis in long-term callus cultures of sugarcane (*Saccharum officinarum* L.). *New Phytologist*, **126**: 301-307.

44.   Gambley RL, Ford R and Smith GR (1993). Microprojectile transformation of sugarcane meristems and regeneration of shoots expressing β-glucuronidase. *Plant Cell Reports*, **12**: 343-346.

45.   Gambley RL, Bryant JD, Masel NP and Smith GR (1994). Cytokinin-enhanced regeneration of plants from microprojectile bombarded sugarcane meristematic tissue. *Australian Journal of Plant Physiology,* **21**: 603-612.

46.   Aftab F and Iqbal J (1999). Plant regeneration from protoplasts derived from cell suspension of adventive somatic embryos in sugarcane (*Saccharum* spp. hybrid cv. CoL-54 and cv. CP-43/33). *Plant Cell, Tissue and Organ Culture,* **56**: 155-162.

47.   Snyman SJ, Watt MP, Huckett BI and Botha FC (2000). Direct somatic embryogenesis for rapid, cost effective production of transgenic sugarcane (*Saccharum* spp. hybrids). *Proceedings of the South African Sugar Technology Association,* **74**: 186-187.

48.   Snyman SJ, Huckett BI, Botha FC and Watt MP (2001). A comparison between direct and indirect somatic morphogenesis for the production of transgenic sugarcane (*Saccharum* spp. hybrids). *Acta Horticulturae,* **560**: 105-108.

49.   Elliot AR, Geijskes RJ, Lakshmanan P, McKeon MG, Wang LF, Berding N, Grof CPL and Smith GR (2002). Direct regeneration of transgenic sugarcane following microprojectile transformation of regenerable cells in thin transverse section explants. *10th International Association for Plant Tissue Culture and Biotechnology Congress*, 23-28 June, Orlando, USA, poster#1376.

50.   Snyman SJ, Huckett BI, Botha FC and Watt MP (2002). The use of sugarcane leaf roll discs as target material and regeneration of transgenic plants via direct embryogenesis: problems and potential. 10th International Association for Plant Tissue Culture and Biotechnology Congress, 23-28 June, Orlando, USA, poster#1460.

51.   Elliot AR, Campbell JA, Brettell RIS and Grof CPL (1998). *Agrobacterium*-mediated transformation of sugarcane using GFP as a screenable marker. *Australian Journal of Plant Physiology,* **25**: 739-743.

52.   Bower R and Birch RG (1992). Transgenic sugarcane plants via microprojectile bombardment. *The Plant Journal,* **2**: 409-416.

53.   Last DI, Brettell RIS, Chamberlain DA, Chaudhury AM, Larkin PJ, Marsh EL, Peacock WJ and Dennis ES (1991). pEmu: an improved promoter for gene expression in cereal cells. *Theoretical and Applied Genetics,* **81**: 581-588.

54.   Murashige T and Skoog F (1962). A revised medium for rapid growth and bioassays with tobacco tissue cultures. *Physiologia Plantarum,* **15**: 473-497.

55.   Finer JJ, Vain P, Jones MW and McMullen MD (1992). Development of the particle inflow gun for DNA delivery to plant cells. *Plant Cell Reports,* **11**: 323-328.

56.   Vain P, McMullen MD and Finer JJ (1993). Osmotic treatment enhances particle bombardment-mediated transient and stable transformation of maize. *Plant Cell Reports,* **12**: 84-88.

57.   Chen WH, Davey MR, Power JB and Cocking EC (1988). Control and maintenance of plant regeneration in sugarcane callus cultures. *Journal of Experimental Botany,* **39**: 251-261.

58.   Snyman SJ, Meyer GM, Carson D and Botha FC (1996). Establishment of embryogenic callus and transient gene expression in selected sugarcane varieties. *South African Journal of Botany,* **62**: 151-154.

59.   http://agnews.tamu.edu/dailynews/stories/SOIL/Apr2103a.htm.

60.   http://www.bses.org.au/crcbid/Biofactory/default.htm.

61.   Birch RG (1997). Plant transformation: problems and strategies for practical application. *Annual Review of Plant Physiology and Plant Molecular Biology,* **48**: 297-326.

62.   Arencibia AD, Carmona ER, Cornide MT, Castiglione S, O'Relly J, Chinea A, Oramas P and Sala F (1999). Somaclonal variation in insect-resistant transgenic sugarcane (*Saccharum* hybrid) plants produced by cell electroporation. *Transgenic Research,* **8**: 349-360.

63.   Grof C (2001). Molecular manipulation of sucrose metabolism. *Proceedings of the International Society of Sugar Cane Technology,* **24**: 586-587.

64.   Harrison D, Waldron J, Ramage C, Peace C, Cox M, Birch RG and Carroll BJ (2001). New DNA technologies provide insights into the molecular basis of somaclonal variation in sugarcane. *Proceedings of the International Society of Sugar Cane Technology,* **24**: 580-582.

Chapter 9

BIOLISTIC TRANSFORMATION OF FESCUES AND RYEGRASSES

G. SPANGENBERG AND Z.Y. WANG[1]

Plant Biotechnology Centre, Department of Primary Industries, La Trobe University,Bundoora, 3086 Victoria, Australia; [1]Samuel Roberts Noble Foundation, Ardmore,Oklahoma 73401, USA
Corresponding author: G. Spangenberg
E-mail: German.Spangenberg@dpi.vic.gov.au

1. INTRODUCTION

1.1. Importance and Distribution of Ryegrasses and Fescues

Worldwide permanent pasture is estimated to cover 70% of agriculturally cultivated area. Ryegrasses (*Lolium* spp.) together with the closely related fescues (*Festuca* spp.) are of significant value in temperate grasslands (*1*). The commercially most important ryegrasses are Italian or annual ryegrass (*L. multiflorum* Lam.) and perennial ryegrass (*L. perenne* L.). They are the key forage species in countries where livestock production is an intensive enterprise, such as the Netherlands, UK and New Zealand (*2*). *L. multiflorum* is a biennial (var. *italicum* Beck) or annual (var. *westerwoldicum* Mansh.) highly palatable, nutritious grass which shows a rapid establishment from seed, a good production in the seeding year and a rapid recovery after defoliation (*2, 1*).

L. perenne is also a palatable but persistent grass of high tillering density that shows resistance to treading and good response to high nitrogen application. It was probably the first herbage grass to become a crop plant (*3, 1*). In New Zealand, over 7 million hectares are grown to perennial ryegrass providing high quality forage to support 60 million sheep and cattle (*4*). In the US, perennial ryegrass is grown in permanent pastures, particularly in the Pacific Coast and the Southern States; it is also a common but minor component of lawn grass mixtures (*1*).

Commercially important fescues include tall fescue (*Festuca arundinacea* Schreb.) and meadow fescue (*F. pratensis* Huds.). *F. arundinacea* is a wind-pollinated, highly self-infertile polyploid perennial cool-season forage, turf and conservation grass. It is indigenous to Europe, also naturally occurring on the Baltic coasts throughout the Caucasus, in western Siberia and extending into China. Introductions have been made into North and South America, Australia, New Zealand, Japan, and South and East Asia (*5*).

Tall fescue has become the predominant cool-season perennial grass species in the USA where it is grown on approx. 14 million hectares (*4*). Some of the acreage has resulted from natural seeding, but much of it is due to introduced seedings. Tall

115

I.S. Curtis (ed.), Transgenic Crops of the World - Essential Protocols, 115-128.
© *2004 Kluwer Academic Publishers. Printed in the Netherlands.*

fescue is used in pastures, lawns, parks, golf courses and football fields, highway medians and roadsides. It serves as perennial ground cover for millions of acres of erodible land (5). Tall fescue also provides forage to millions of sheep and cattle in different grassland countries. Thus, tall fescue is important for grazing, stabilizing soil for agriculture, and enhancing the environment through multiple uses (6).

F. pratensis is a major cool-season high-yielding forage grass of agricultural importance in the temperate region. It has a wide range of distribution on the northern hemisphere, mainly in Europe (7). Since meadow fescue has good digestibility, good winter hardiness and longevity under a system with frequent cutting or grazing, it has become an increasingly important crop species in leys and pastures (8).

1.2. Need for Biotechnological Approaches in Genetic Improvement of Ryegrasses and Fescues

Genetic improvement of the highly self-incompatible ryegrasses and fescues by conventional plant breeding is difficult and selection schemes are complex, which results in slow breeding progress. Biotechnological approaches such as genetic transformation for the direct introduction of agronomically useful genes show promise - when considered as part of ryegrass and fescue improvement programmes - to complement and accelerate conventional breeding efforts (9).

Gene technology and the production of transgenic plants offers the opportunity to generate unique genetic variation, when the required variation is either absent or has very low heritability. In recent years, the first transgenic ryegrasses and fescues with simple 'engineered' traits have reached the stage of field-evaluation. While gaps in our understanding of the underlying genetics, physiology and biochemistry of many complex plant processes are likely to delay progress in many applications of transgenesis in forage plant improvement, gene technology is a powerful tool for the generation of the required molecular genetic knowledge. Consequently, applications of transgenesis to temperate grass improvement are focused on the development of transformation events with unique genetic variation and in studies on the molecular genetic dissection of plant biosynthetic pathways and developmental processes of high relevance for forage production (10).

1.3. Genetic Transformation of Fescues and Ryegrasses

The potential of biotechnology in the development of improved forage grass cultivars has been recognized in recent years (9, 11). Efficient plant regeneration systems have been established for fescues (*Festuca* spp.) and ryegrasses (*Lolium* spp.) based on embryogenic cell cultures (12-17). Transgenic forage and turf grass plants have been produced in tall fescue (10, 18-27), meadow fescue (23), red fescue (28, 13, 24), perennial ryegrass (21, 29-31), Italian ryegrass (18, 21, 30, 32, 33), creeping bentgrass (34-37), orchardgrass (38-40), switchgrass (41), blue grama grass (42) and bahiagrass (43).

Efforts have been made to generate transgenic forage and turf grass plants with potentially useful agronomic genes or to evaluate novel strategies for grass improvement. Available reports in this aspect are: herbicide resistant transgenic tall and red fescues (*13, 26*), herbicide resistant creeping bentgrass (*34-37*), transgenic Italian ryegrass with altered fructan accumulation (*44*), transgenic perennial ryegrass for down-regulation of pollen allergens (*10, 45*), transgenic tall fescue expressing sulphur-rich sunflower albumin genes for improved nutritional quality (*27*) and virus resistant transgenic perennial ryegrass (*46*).

Gene transfer to fescues and ryegrasses has been reported over the last decade through a variety of transformation methods: direct gene transfer to protoplasts, surrogate transformation with transformed endophytes, biolistic- and whiskers-mediated transformation of embryogenic and meristematic cells and *Agrobacterium*-mediated transformation (Table 1). The efficient genetic transformation procedure based on biolistic gene transfer of embryogenic suspension cells in tall fescue using the liquid selection scheme, resulting in 1.2 Hm resistant calli per bombarded dish is described.

2. MATERIALS

2.1. Culture Media and Equipment

A source for the standard components of the culture media required is given (Table 2). Gelrite (Duchefa) is used as gelling agent. Other Chemicals used were $Ca(OCl)_2$ (Fluka), Tween 80 (Merk), LiCl (Fluka), ethanol, spermidine (Sigma), glycerol, hygromycin (CalBiochem) and gold particles (gold powder, spherical, 1.5-3.0 µm, 99.9%, Aldrich).

Parafilm was used as sealing tape. Sterile disposable bottle-top filters (Zapcap S, 0.2 µm) and disposable filter holders (FP 030/3, 0.2 µm) were filter-sterilization units from Schleicher & Schuell. Also used were 10 ml sterile disposable pipettes, 9 cm Petri dishes (Bibby Sterilin Ltd), 5 cm sterile plastic culture dishes (Bibby Sterilin Ltd.), sterile plastic culture vessels (6 cm diam., 305 ml, Greiner), filter paper disks (Schleicher & Schüll, LS14, 4 cm diam.), 0.5 ml Eppendorf tubes and 10 ml, 100 ml and 1000 ml automatic pipettes with corresponding tips. Culture glass jars (400 ml glass jars, ABS) with twist-off metalic caps perforated with one 8 mm diam. hole, and closed with a ceapren plug (22 mm, Greiner), provided optimal gas exchange conditions for growth of plantlets.

A laminar flow hood with horizontal flow (type CHR400, Concept GmbH), a rotary shaker (Infors type RC-406, Infors AG), UV sterilization apparatus (e.g. UV Stratalinker 2400, Stratagene), long forceps (16 cm, bent tip, Gerald, Medicon) and Eppendorf microfuge were used.

A particle inflow gun (PIG), constructed according to Finer *et al.* (*47*), or a Biorad PDS-1000/He was used as the biolistic device to deliver DNA into embryogenic suspension cells.

Table 1. Genetic transformation and production of transgenic fescues and ryegrasses

Plant species	Transgenes	Method	Outcome	References
Festuca arundinacea (tall fescue)	hpt, bar	Protoplasts	Transgenic plants	Wang et al. (1992), (26)
	hpt, gusA	Protoplasts	Putative transgenic plants	Ha et al. (1992), (55)
	hpt	Biolistics	Transgenic plants	Spangenberg et al. (1995c), (24)
	hpt	Protoplasts	Transgenic plants	Dalton et al. (1995), (20)
	gusA	Protoplasts	Transformed calli	Kuai and Morris (1996), (56)
	hpt	Whiskers	Transgenic plants	Dalton et al. (1998), (21)
	hpt, bar, gusA	Biolistics	Transgenic plants	Cho et al. (2000), (57)
	sfa8, hpt	Biolistics	Transgenic plants	Wang et al. (2001b), (27)
	hpt, gusA	Biolistics	Transgenic plants	Wang et al. (2003), (25)
	hpt, gusA	Agrobacterium	Transgenic plants	Bettany et al. (2003), (18)
Festuca pratensis (meadow fescue)	bar	Protoplasts	Transgenic plantlets	Spangenberg et al. (1995b), (23)
Festuca rubra (red fescue)	bar	Protoplasts	Transgenic plants	Spangenberg et al. (1994b), (13)
	hpt	Biolistics	Transgenic plants	Spangenberg et al. (1995c), (24)
	nptII	Biolistics	Transgenic plants	Spangenberg et al. (1998), (9)
	hpt, gusA, gfp	Biolistics	Transgenic plants	Cho et al. (2000), (57)
	nptII	Biolistics	Transgenic plants	Altpeter et al. (2000), (28)
Lolium multiflorum (Italian ryegrass)	nptII	Protoplasts	Transformed calli	Potrykus et al. (1985), (58)
	hpt, gusA	Biolistics	Transgenic plants	Ye et al. (1997), (33)
	nptII, gusA	Protoplasts	Transgenic plants	Wang et al. (1997), (32)
	hpt, gusA	Whiskers	Transgenic plants	Dalton et al. (1998), (21)
	hpt, gusA	Biolistics	Transgenic plants	Dalton et al. (1999), (30)
	sacB, hpt	Biolistics	Transgenic plants	Ye et al. (2001), (44)
	hpt, gusA	Agrobacterium	Transgenic plants	Bettany et al. (2003), (18)
Lolium perenne (perennial ryegrass)	hpt	Biolistics	Transformed calli	Hensgens et al. (1993), (59)
	hpt, gusA	Biolistics	Transformed calli	Van der Maas et al. (1994), (60)
	hpt, gusA	Biolistics	Transgenic plants	Spangenberg et al. (1995d), (31)
	nptII, gusA	Protoplasts	Transgenic plants	Wang et al. (1997), (32)
	Lol p1, Lol p2	Biolistics	Transgenic plants	Wu et al. (1997), (45)
	hpt	Whiskers	Transgenic plant	Dalton et al. (1998), (9)
	hpt, gusA	Biolistics	Transgenic plants	Dalton et al. (1999), (30)
	nptII	Biolistics	Transgenic plants	Altpeter et al. (2000), (29)
	RgMV-CP	Biolistics	Transgenic plants	Xu et al. (2001), (46)
Lolium rigidum (wimmera ryegrass)	Lol p5, nptII	Biolistics	Transgenic plants	Bhalla et al. (1999), (61)

Lol p: ryegrass pollen allergen; *RgMV-CP*: ryegrass mosaic virus coat protein; *sacB*: levansucrase; *sfa8*: sunflower albumin8.

2.2. Plasmids

Plasmid AcH1 (5.6 kb) bearing a chimeric 3' modified *hpt* gene (*48*) under control of the rice actin 1 5' regulatory signals (*49*) was used for stable transformation experiments.

2.3. Plant Material

Mature seeds (>1,000) each from different cultivars of tall fescue were used for callus induction and establishment of cell suspensions.

3. METHODS

3.1. Establishment of Embryogenic Cell Suspensions

This procedure is modified from that of Wang *et al.* (*50, 51*).

3.1.1. Solutions and Culture Media

The composition of the media used at the final concentrations of their individual ingredients is given (Table 2). Gelrite is used as gelling agents (2.5 g/l) for the semi-solidified media M1, M5 and MSK.

a) Media M1, M5 and MSK: Modified from Murashige and Skoog (52)
b) Medium AAF: Modified from Müller and Grafe (53)
c) Media Stocks: For all culture media used, prepare macro-elements as 10-fold concentrated stocks. Micro-elements and vitamins stock solutions are 100-fold concentrated in all cases. For preparation of 1000 ml micro-elements stock, dissolve first Na_2EDTA (3.73 g) and $FeSO_4.7H_2O$ (2.78 g) separately in 400 ml distilled water each. Mix dissolved solutions and heat at ca. 60°C. After cooling down, add remaining stock ingredients dissolved in distilled water and complete to 1 litre.
d) Preparation of Culture Media:
   Prepare all stock solutions in glass distilled water and keep at -20°C. Stock solutions can be stored at -20°C indefinetly. Filter-sterilize stock solutions of vitamins and keep at 4°C for up to 6 months. Dissolve hormones 2,4-D and kinetin in 1 M KOH and dilute with distilled water to prepare 100 mg/l concentrated stocks. Keep at 4°C for up to 6 months.
   Prepare the culture media using stock solutions (as indicated above) and dilute with distilled water. Adjust to pH 5.8 with KOH for media M1, M5, M1-5 and MSK, add 0.25% (w/v) Gelrite, and autoclave (121°C, 1 bar, 15 min.). Filter-sterilize medium AAF (0.2 µm pore size).

3.1.2. Steps in Procedure

1. Surface-sterilize 500 seeds per cultivar of tall fescue by incubating them with agitation in hypochlorite solution [containing 3% (w/v) $Ca(ClO)_2$ and 0.05% (v/v) Tween 80] for 90 min, rinse the seeds by washing 3 times in sterile distilled water, keep moist at room temperature overnight.
2. Sterilize the seeds again for 30 min in hypochlorite solution (see Section 3.1.2.1.) and rinse the seeds 3 times in sterile distilled water.
3. Place 10-15 seeds per 9 cm Petri dish containing M5 medium. Keep dishes, sealed with Parafilm, in the dark at 25°C for 6 weeks for the establishment of proliferating calli.
4. Subculture embryogenic friable yellowish calli (Fig. 1A), developing from mature seed-derived embryos onto new dishes containing M5 medium. Keep dishes under the same culture conditions for 3-4 weeks.
5. Select and transfer single genotype-derived friable, yellowish, embryogenic callus (Fig. 1B) separately into 6 cm culture vessels containing 14 ml AAF medium. Keep culture vessels at 25°C in the dark on a rotary shaker at 60 rpm. Subculture weekly by replacing 2/3 of the culture medium with fresh one. Two months later, embryogenic suspensions can be established (Fig. 1C).

6. To test the embryogenic potential of established suspension cultures, plate cell suspension aggregates onto M1 medium in 9 cm Petri dishes. Keep culture dishes in the dark at 25°C for 3 weeks to allow for proliferation of plated cells (Fig. 1D).
7. Transfer the cultures onto solid MSK medium in 9 cm Petri dishes for differentiation of preformed somatic embryos on the surface of the compact calli. Keep the culture dishes at 25°C under fluorescent light conditions (16 h photoperiod; 55 μmol/m^2/sec) for 4 weeks until regenerating shoots are obtained (Fig. 1E).

3.2. Microprojectile Bombardment and the Generation of Transgenic Plants

This procedure is modified from those of Finer *et al*. (*47*) and Spangenberg *et al*. (*23, 54*).

3.2.1. Preparation of Embryogenic Suspension Cells for Microprojectile Bombardment

3.2.1.1. Solutions and Culture Media

The composition of the medium M1-5 used at the final concentrations of their individual ingredients is given (Table 2). Gelrite is used as gelling agents (2.5 g/l) for the semi-solidified M1-5 medium.

3.2.1.2. Steps in Procedure

1. Remove the liquid culture medium AAF from the cell suspensions with 10 ml sterile pipettes.
2. Add 14 ml of liquid M1-5 medium to the suspension cells for a osmoticum treatment, shake the culture vessels at 60 rpm in dimlight on a rotary shaker for 30 min.
3. Remove the liquid M1-5 medium from the culture vessels with 10 ml sterile pipettes.

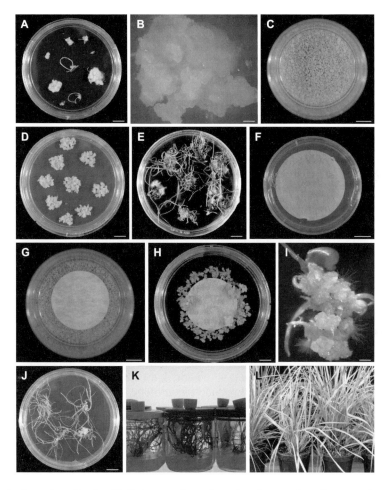

Figure 1. Procedure for the biolistic transformation of embryogenic cell suspensions and recovery of transgenic tall fescue (F. arundinacea) plants. A, Embryogenic callus induced from seeds/caryopses. B, Detailed view of an embryogenic callus used to establish suspension cultures. C, Embryogenic suspension cultures established from single genotype-derived embryogenic callus. D, Embryogenic cell clusters plated on proliferation medium. E, In vitro regeneration of green shoots from embryogenic suspension cultures. F, Suspension cells plated on filter paper disk prior to microprojectile bombardment. G, Selection of bombaraded suspension cells in liquid AAF medium containing hygromycin. H, Resistant calli obtained after continued selection in solid medium containing 250 mg/l hygromycin. I,J, Shoot differentiation of hygromycin resistant calli after transfer onto regeneration medium. K, In vitro transgenic plantlets recovered after microprojectile bombardment of embryogenic cells. L, Transgenic tall fescue plants growing under glasshouse conditions. Size bars: image A,C,D,E,F,H and J, 1 cm; image B and I, 1 mm.

4. Distribute the osmotically pre-treated suspension cells with forceps as a 2 cm diam. monolayer on a sterile filter paper disk placed onto semi-solidified M1-5 medium in 5 cm Petri dishes (Fig. 1F).
5. Keep the dishes opened in the laminar flow hood for approx. 30 min to partially dry remaining liquid medium.

3.2.2. Preparation of Gold Particles

3.2.2.1. Solutions and Culture Media

1. Sterilized DNA solution: Precipitate 500 µg DNA (enough to bombard 50 dishes of suspension cells) overnight in a 1.5 ml Eppendorf tube by adding 1/10 vol. of 4 M LiCl solution and 3 vol. of absolute ethanol. Centrifuge for 10 min at room temperature in a Eppendorf microfuge at full speed (14,000 rpm), wash the DNA pellet by adding 1 ml filter-sterilized 70% ethanol and repeat centrifugation for 5 min. Let the pellet dry in a laminar flow hood and dissolve in sterile distilled water at a final concentration of 1 µg/µl.
2. $CaCl_2$ solution (2.5 M): Dissolve 36.75 g $CaCl_2.2H_2O$ in 100 ml distilled water and filter-sterilize the solution.
3. Spermidine solution (100 mM): Dissolve 1.27 g spermidine in 100 ml distilled water and filter-sterilize the solution, store at -20°C.
4. Gold particles: Autoclave gold particles in punctured 1.5 ml Eppendorf tubes (50 mg/tube) overlayed by 1 ml of glycerol. After autoclaving, transfer the gold particles together with glycerol to new autoclaved Eppendorf tubes (unpunctured) and store at room temperature. Before use, remove the glycerol and wash the particles twice by adding 1 ml filter-sterilized absolute ethanol and spin down briefly in Eppendorf microfuge (approx. 6 sec). Fill up with filter-sterilized absolute ethanol to 500 µl to obtain a final concentration of 100 µg/µl.

3.2.2.2. Steps in Procedure

1. Fill a 0.5 ml autoclaved Eppendorf tube with 20 µl of sterilized DNA solution (1 µg/µl).
2. Add 10 µl of gold particles (100 µg/µl) in filter-sterilized absolute ethanol.
3. Add 25 µl of 2.5 M $CaCl_2$ solution.
4. Add 10 µl of 100 mM spermidine solution.
5. Mix with pipette (100 µl).
6. Incubate for 10 min at 4°C.
7. Spin the gold particles down briefly in Eppendorf microfuge, and remove supernatant.
8. Wash the particles with ethanol by adding 100 µl filter-sterilized absolute ethanol and spin down briefly. Repeat this washing procedure.
9. Let the pellet dry in the laminar flow hood for 10 min.

10. Add 20 µl of sterile distilled water to the pellet. The amount of particles in one tube is sufficient for 2 shots (500 µg of gold particles coated with 10 µg of DNA per shot).

3.2.3. *Microprojectile Bombardment of Embrygenic Cell Suspensions and the Regeneration of Transgenic Plants*

1. UV sterilize baffle meshes (500 µm) each side for 15 min and autoclave the Gelman filters.
2. Adjust bombardment parameters of the PIG as follows: 6 bar bombardment pressure, 12 cm baffle and 15 cm target-bombardment distances.

Table 2. Culture media for the genetic transformation and production of transgenic tall fescue plants

Media component		M1	M5	M1-5	AAF	MSK
Macroelements (mg/l final concentration)						
KNO₃	[Merck]	1900	1900	1900		1900
NH₄NO₃	[Merck]	1650	1650	1650		1650
CaCl₂ x 2H₂O	[Merck]	440	440	440	440	440
MgSO₄ x 2H₂O	[Merck]	370	370	370	370	370
KH₂PO₄	[Merck]	170	170	170	170	170
KCl	[Merck]				2950	
Microelements (mg/l final concentration)						
Na₂EDTA	[Fluka]	37.3	37.3	37.3	37.3	37.3
FeSO₄ x 7H₂O	[Merck]	27.8	27.8	27.8	27.8	27.8
H₃BO₃	[Merck]	6.2	6.2	6.2	6.2	6.2
KI	[Merck]	0.83	0.83	0.83	0.83	0.83
MnSO₄ x H₂O	[Merck]	16.9	16.9	16.9	16.9	16.9
ZnSO₄ x 7H₂O	[Merck]	8.6	8.6	8.6	8.6	8.6
CuSO₄ x 5H₂O	[Merck]	0.025	0.025	0.025	0.025	0.025
Na₂MoO₄ x 2H₂O	[Merck]	0.25	0.25	0.25	0.25	0.25
CoCl₂ x 6H₂O	[Merck]	0.025	0.025	0.025	0.025	0.025
Carbohydrates (g/l final concentration)						
D(+) Sucrose	[Fluka]	30	30	30	20	30
D-Sorbitol	[Sigma]			45.5	20	
D-Mannitol	[Sigma]			45.5		
Hormones (mg/l final concentration)						
2,4-D	[Serva]	1.5	5.0	1.5	1.5	
Kinetin	[Serva]					0.2
Vitamins (mg/l final concentration)						
Pyridoxine HCl	[Merck]	0.5	0.5	0.5	1.0	0.5
Thiamine HCl	[Merck]	0.1	0.1	0.1	10.0	0.1
Nicotinic acid	[Merck]	0.5	0.5	0.5	1.0	0.5
Inositol	[Merck]	100	100	100	100	100
Other organics (mg/l final concentration)						
Glycine	[Serva]	2	2	2	7.5	2
L-Glutamine	[Fluka]				877	
L-Asparagine	[Fluka]				266	
L-Arginine	[Fluka]				174	
Casein hydrolysate	[Fluka]	400	400	400		

3. Mix particles by pipetting up and down, load 9 µl of DNA particles suspension into a Gelman filter unit and fix the filter unit to the PIG. Immediately place the target 5 cm dish with suspension cells on the shelf, and bombard the cells when the vacuum pressure reaches 0.8 bar.
4. Release vacuum, remove the dish and seal with Parafilm.
5. Incubate the dishes with bombarded cells for 3-4 days at 25°C in the dark.
6. Transfer the bombarded cells from two dishes into a 6 cm culture vessel containing 14 ml liquid culture medium AAF with 50 mg/l hygromycin. Keep the culture vessels at 25°C in the dark on a rotary shaker at 60 rpm (Fig. 1G).
7. After 2 weeks, replace the culture medium completely by fresh one containing 100 mg/l hygromycin, and maintain for 2-3 weeks at 25°C in the dark on a rotary shaker at 60 rpm.
8. Remove the liquid medium in the culture vessels and transfer the cells with forceps to 9 cm Petri dishes with M1 medium containing 250 mg/l hygromycin. Keep the dishes for 4 weeks at 25°C in the dark (Fig. 1H).
9. Transfer 4-6 resistant calli onto dishes containing medium MSK without hygromycin. Keep the culture dishes at 25°C under fluorescent light conditions (16 h photoperiod; 55 µmol/m^2/sec) for 4 weeks (Figs. 1I and J).
10. Transfer 3-4 regenerated shoots per callus to glass jars containing 50 ml hormone-free half strength MSK medium. Keep cultures at 25°C under fluorescent light conditions (16 h photoperiod; 55 µmol/m^2/sec) for 3-4 weeks (Fig. 1K).
11. Transfer rooted plantlets to soil and grow to maturity (Fig. 1L) under glasshouse conditions (16 h photoperiod; 23°C/18°C; fluorescent light 145 µmol/m^2/sec). Maintain plants in 20 cm pots containing standard composed pinebark soiless-mix (Propine Nursery Supplies Pty. Ltd.).

4. NOTES

1. Using the protocol described, it is possible to obtain transgenic plants from different *Festuca* (*F. arundinacea* and *F. rubra*) and *Lolium* (*L. perenne* and *L. multiflorum*) species, as described by Spangenberg *et al.* (*23, 54*) and Ye *et al.* (*33*).
2. The transgenic nature of regenerated plants should be confirmed by standard molecular analysis (Southern and Northern hybridization analysis, enzyme assays etc.).
3. All manipulations with suspension cells are to be performed under axenic conditions (in a horizontal laminar flow hood) and sterilized materials are to be used.
4. For the establishment of embryogenic suspensions the correct choice of single genotype-derived friable yellowish embryogenic callus is crucial.

CONCLUSIONS

This protocol describes a robust plant transformation system for the reproducible generation of fertile transgenic forage and turf grass fescues and ryegrasses using a chimeric *hpt* gene as selectable marker and embryogenic suspension cells as direct targets for microprojectile bombardment.

ACKNOWLEDGEMENTS

The authors wish to thank XD Ye, XL Wu and J Nagel for their contribution to the development of biolistic-mediated transformation methodology in fescues and ryegrasses. Funding from the Swiss Federal Institute of Technology, the Swiss National Science Foundation, the Australian Dairy Research and Development Corporation, the Victorian Department of Primary Industries and the Samuel Roberts Noble Foundation is gratefully acknowledged.

REFERENCES

1.   Jauhar PP (1993). Cytogenetics of the *Festuca-Lolium* complex. Relevance to breeding. In: Frankel R, Grossman M, Linskens HF, Maliga P, Riley R (eds.), *Monographs on Theoretical and Applied Genetics*, vol 18. Springer, Berlin Heidelberg New York, 243 pp.

2.   Buckner RC, Todd JR, Burrus PB and Barnes RF (1967). Chemical composition, palatability, and digestibility of ryegrass-tall fescue hybrids, 'Kenwell', and 'Kentucky 31' tall fescue varieties. *Agronomy Journal*, **59**: 345-349.

3.   Borrill M (1976). Temperate grasses. *Lolium, Festuca, Dactylis, Phleum, Bromus* (Gramineae). In: Simmonds NW (ed.), *Evolution of Crop Plants*. Longman, London, UK, pp 137-142.

4.   Siegel MR, Latch GCM and Johnson MC (1985). *Acremonium* fungal endophytes of tall fescue and perennial ryegrass: significance and control. *Plant Disease*, **69**: 179-183.

5.   Barnes RF (1990). Importance and problems of tall fescue. In: Kasperbauer MJ (ed.), *Biotechnology in Tall Fescue Improvement*; 2-12, Boca Raton, Ann Arbor, Boston, CRC Press Inc.

6.   Buckner RC and Bush LP (1979). *Tall Fescue*. American Society of Agronomy, Madison, WI., USA.

7.   Hulten E (1971). *Atlas of the Distribution of Vascular Plants in Northwestern Europe*. Generalstabens Litografiska Anstaltsförlag, Stockholm.

8.   Aastveit AH and Aastveit K (1989). Genetic variations and inheritance of quantitative characters in two populations of meadow fescue (*Festuca pratensis*, Huds.) and their hybrid. *Hereditas*, **111**: 103–114.

9.   Spangenberg G, Wang ZY and Potrykus I (1998). Biotechnology in forage and turf grass improvement. In: Frankel R, Grossman M, Linskens HF, Maliga P, Riley R (eds.), *Monographs on Theoretical and Applied Genetics*, vol 23. Springer, Berlin Heidelberg New York.

10. Spangenberg G, Kalla R, Lidgett A, Sawbridge T, Ong EK and John U (2001). Breeding forage plants in the genome era. In: *Molecular Breeding of Forage Crops. Proceedings of the 2nd International Symposium, Molecular Breeding of Forage Crops*, Lorne and Hamilton, Victoria, Australia, 19 24 November, 2000. 2001, 1 39; 154 ref. Kluwer Academic Publishers; Dordrecht; The Netherlands.

11. Wang ZY, Hopkins A and Mian R (2001a). Forage and turf grass biotechnology. *Critical Reviews in Plant Science*, **20**: 573-619.

12. Altpeter F and Posselt UK (2000). Improved plant regeneration from cell suspensions of commercial cultivars, breeding- and inbred lines of perennial ryegrass (*Lolium perenne* L.). *Journal of Plant Physiology*, **156**: 790-796.

13. Spangenberg G, Wang ZY, Nagel J and Potrykus I (1994b). Protoplast culture and generation of transgenic plants in red fescue (*Festuca rubra* L.). *Plant Science*, **97**: 83-94.

14. Wang ZY, Legris G, Nagel J, Potrykus I and Spangenberg G (1994). Cryopreservation of embryogenic cell suspensions in *Festuca* and *Lolium* species. *Plant Science*, **103**: 93-106.

15. Wang ZY, Legris G and Spangenberg G (1995a). Establishment of and plant regeneration from embryogenic cell suspensions and their protoplasts in forage grasses. In: Potrykus I, Spangenberg G (eds.), *Gene Transfer to Plants*, pp. 295-304. Springer, Berlin, Germany.

16. Wang ZY, Legris G, Valles MP, Potrykus I and Spangenberg G (1995b). Plant regeneration from suspension and protoplast cultures in the temperate grasses *Festuca* and *Lolium*. In: Terzi M,

Cella R, Falavigna A (eds.), *Current Issues in Plant Molecular and Cellular Biology*, pp. 81-86. Kluwer Academic Publishers; Dordrecht; The Netherlands.

17. Wang ZY, Nagel J, Potrykus I and Spangenberg G (1993). Plants from cell suspension-derived protoplasts in *Lolium* species. *Plant Science*, **94**: 179-193.

18. Bettany AJE, Dalton SJ, Timms E, Manderyck B, Dhanoa MS and Morris P (2003). *Agrobacterium tumefaciens*-mediated transformation of *Festuca arundinacea* (Schreb.) and *Lolium multiflorum* (Lam.). *Plant Cell Reports*, **21**: 437-444.

19. Bettany AJE, Dalton SJ, Timms E and Morris P (1998). Stability of transgene expression during vegetative propagation of protoplast-derived tall fescue (*Festuca arundinacea* Schreb.) plants. *Journal of Experimental Botany*, **49**: 1797-1804.

20. Dalton SJ, Bettany AJE, Timms E and Morris P (1995). The effect of selection pressure on transformation frequency and copy number in transgenic plants of tall fescue (*Festuca arundinacea* Schreb.). *Plant Science*, **108**: 63-70.

21. Dalton SJ, Bettany AJE, Timms E and Morris P (1998). Transgenic plants of *Lolium multiflorum*, *Lolium perenne*, *Festuca arundinacea* and *Agrostis stolonifera* by silicon carbide fibre-mediated transformation of cell suspension cultures. *Plant Science*, **132**: 31-43.

22. Kuai B, Dalton SJ, Bettany AJE and Morris P (1999). Regeneration of fertile transgenic tall fescue plants with a stable highly expressed foreign gene. *Plant Cell, Tissue and Organ Culture*, **58**: 149-154.

23. Spangenberg G, Wang ZY, Valles MP and Potrykus I (1995b). Genetic transformation in *Festuca arundinacea* Schreb.(tall fescue) and *Festuca pratensis* Huds.(meadow fescue). In: Bajaj YPS (ed.), *Biotechnology in Agriculture and Forestry*, pp.183-203. Springer, Berlin, Heidelberg, Germany.

24. Spangenberg G, Wang ZY, Wu XL, Nagel J, Iglesias VA and Potrykus I (1995c). Transgenic tall fescue (*Festuca arundinacea*) and red fescue (*F. rubra*) plants from microprojectile bombardment of embryogenic suspension cells. *Journal of Plant Physiology*, **145**: 693-701.

25. Wang ZY, Bell J, Ge YX and Lehmann D (2003). Inheritance of transgenes in transgenic tall fescue (*Festuca arundinacea*). *In Vitro Cellular and Developmental Biology-Plant*, **39**: 277-282.

26. Wang ZY, Takamizo T, Iglesias VA, Osusky M, Nagel J, Potrykus I and Spangenberg G (1992). Transgenic plants of tall fescue (*Festuca arundinacea* Schreb.) obtained by direct gene transfer to protoplasts. *Bio/Technology*, **10**: 691-696.

27. Wang ZY, Ye XD, Nagel J, Potrykus I and Spangenberg G (2001b). Expression of a sulphur-rich sunflower albumin gene in transgenic tall fescue (*Festuca arundinacea* Schreb.) plants. *Plant Cell Reports*, **20**: 213-219.

28. Altpeter F and Xu JP (2000). Rapid production of transgenic turfgrass (*Festuca rubra* L.) plants. *Journal of Plant Physiology*, **157**: 441-448.

29. Altpeter F, Xu JP and Ahmed S (2000). Generation of large numbers of independently transformed fertile perennial ryegrass (*Lolium perenne* L.) plants of forage- and turf-type cultivars. *Molecular Breeding*, **6**: 519-528.

30. Dalton SJ, Bettany AJE, Timms E and Morris P (1999). Co-transformed, diploid *Lolium perenne* (Perennial ryegrass), *Lolium multiflorum* (Italian ryegrass) and *Lolium temulentum* (Darnel) plants produced by microprojectile bombardment. *Plant Cell Reports*, **18**: 721-726.

31. Spangenberg G, Wang ZY, Wu XL, Nagel J and Potrykus I (1995d). Transgenic perennial ryegrass (*Lolium perenne*) plants from microprojectile bombardment of embryogenic suspension cells. *Plant Science*, **108**: 209-217.

32. Wang GR, Binding H and Posselt UK (1997). Fertile transgenic plants from direct gene transfer to protoplasts of *Lolium perenne* L. and *Lolium multiflorum* Lam. *Journal of Plant Physiology*, **151**: 83-90.

33. Ye XD, Wang ZY, Wu XL, Potrykus I and Spangenberg G (1997). Transgenic Italian ryegrass (*Lolium multiflorum*) plants from microprojectile bombardment of embryogenic suspension cells. *Plant Cell Reports*, **16**: 379-384.

34. Asano Y, Ito Y, Fukami M, Sugiura K and Fujiie A (1998). Herbicide-resistant transgenic creeping bentgrass plants obtained by electroporation using an altered buffer. *Plant Cell Reports*, **17**: 963-967.

35. Lee L (1996). Turfgrass biotechnology. *Plant Science*, **115**: 1-8.

36.   Liu CA, Zhong H, Vargas J, Penner D and Sticklen M (1998). Prevention of fungal diseases in transgenic, bialaphos- and glufosinate-resistant creeping bentgrass (*Agrostis palustris*). *Weed Science*, **46**: 139-146.

37.   Yu TT, Skinner DZ, Liang GH, Trick HN, Huang B and Muthukrishnan S (2000). *Agrobacterium*-mediated transformation of creeping bentgrass using GFP as a reporter gene. *Hereditas*, **133**: 229-233.

38.   Cho MJ, Choi HW and Lemaux PG (2001). Transformed T0 orchardgrass (*Dactylis glomerata* L.) plants produced from highly regenerative tissues derived from mature seeds. *Plant Cell Reports*, **20**: 318-324.

39.   Denchev PD, Songstad DD, McDaniel JK and Conger BV (1997). Transgenic orchardgrass (*Dactylis glomerata*) plants by direct embryogenesis from microprojectile bombarded leaf cells. *Plant Cell Reports*, **16**: 813-819.

40.   Horn ME, Shillito RD, Conger BV and Harms CT (1988). Transgenic plants of orchardgrass (*Dactylis glomerata* L.) from protoplasts. *Plant Cell Reports*, **7**: 469-472.

41.   Richards HA, Rudas VA, Sun H, McDaniel JK, Tomaszewski Z and Conger BV (2001). Construction of a GFP-BAR plasmid and its use for switchgrass transformation. *Plant Cell Reports*, **20**: 48-54.

42.   Aguado SGA, Rascon CQ, Cabrera PJL, Martinez HA, Olalde PV and Herrera EL (2002). Transgenic plants of blue grama grass, *Bouteloua gracilis* (H.B.K.) Lag. ex Steud., from microprojectile bombardment of highly chlorophyllous embryogenic cells. *Theoretical and Applied Genetics*, **104**: 763-771.

43.   Smith RL, Grando MF, Li YY, Seib JC and Shatters RG (2002). Transformation of bahiagrass (*Paspalum notatum* Flugge). *Plant Cell Reports*, **20**: 1017-1021.

44.   Ye XD, Wu XL, Zhao H, Frehner M, Nosberger J, Potrykus I and Spangenberg G (2001). Altered fructan accumulation in transgenic *Lolium multiflorum* plants expressing a *Bacillus subtilis sacB* gene. *Plant Cell Reports*, **20**: 205-212.

45.   Wu XL, Ye XD, Wang ZY, Potrykus I and Spangenberg G (1997). Gene transfer to ryegrasses: Down-regulation of major pollen allergens in transgenic plants. In: *Proceedings of the XVIII International Grassland Congress*, pp. 35-36.

46.   Xu J, Schubert J, Altpeter F and Xu JP (2001). Dissection of RNA-mediated ryegrass mosaic virus resistance in fertile transgenic perennial ryegrass (*Lolium perenne* L.). *The Plant Journal*, **26**: 265-274.

47.   Finer JJ, Vain P, Jones MW and McMullen MD (1992). Development of the particle inflow gun for DNA delivery to plant cells. *Plant Cell Reports*, **11**: 323-328.

48.   Bilang R, Iida S, Peterhans A, Potrykus I and Paszkowski J (1991). The 3'-terminal region of the hygromycin-B-resistance gene is important for its activity in *Escherichia coli* and *Nicotiana tabacum*. *Gene*, **100**: 247–250.

49.   McElroy D, Blowers AD, Jenes B and Wu R (1991). Construction of expression vectors based on the rice actin 1 (Act1) 5' region for use in monocot transformation. *Molecular and General Genetics*, **231**: 150–160.

50.   Wang ZY, Valles MP, Montavon P, Potrykus I and Spangenberg G (1993). Fertile plant regeneration from protoplasts of meadow fescue (*Festuca pratensis* Huds.). *Plant Cell Reports*, **12**: 95-100.

51.   Wang ZY, Legris G, Valles MP, Potrykus I and Spangenberg G (1995). Plant regeneration from suspension and protoplast cultures in the temperate grasses *Festuca* and *Lolium*. In: Terzi M, Cella R, Falavigna A (eds.), *Current Issues in Plant Molecular and Cellular Biology*. Kluwer Academia Publishers, Dordrecht, The Netherlands, pp. 81-86.

52.   Murashige T and Skoog F (1962). A revised medium for rapid growth and bioassays with tobacco tissue culture. *Physiologia Plantarum*, **15**: 473-497.

53.   Müller AJ and Grafe R (1978). Isolation and characterization of cell lines of *Nicotiana tabacum* lacking nitrate reductase. *Molecular and General Genetics*, **161**: 67–76.

54.   Spangenberg G, Wang ZY, Legris G, Montavon P, Takamizo T, Perez Vicente R, Valles MP, Nagel J *et al.* (1995a). Intergeneric symmetric and asymmetric somatic hybridization in *Festuca* and *Lolium*. *Euphytica*, **85**: 235-245.

55.   Ha SB, Wu FS and Thorne TK (1992). Transgenic turf-type tall fescue (*Festuca arundinacea* Schreb.) plants regenerated from protoplasts. *Plant Cell Reports*, **11**: 601–604.

56. Kuai B and Morris F (1996). Screening for stable transformants and stability of beta-glucuronidase gene expression in suspension cultured cells of tall fescue (*Festuca arundinacea*). *Plant Cell Reports*, **15**: 804-808

57. Cho MJ, Ha CD and Lemaux PG (2000). Production of transgenic tall fescue and red fescue plants by particle bombardment of mature seed-derived highly regenerative tissues. *Plant Cell Reports*, **19**: 1084-1089.

58. Potrykus I, Saul MW, Petruska J, Paszkowski J and Shillito R (1985). Direct gene transfer to cells of a graminaceous monocot. *Molecular and General Genetics*, **199**: 183–188.

59. Hensgens LAM, de Bakker EPHM, van Os-Ruygrok EP, Rueb S, van de Mark F, van der Maas HM, van der Veen S, Kooman-Gersmann M, Hart L *et al.* (1993). Transient and stable expression of *gus*A fusions with rice genes in rice, barley and perennial ryegrass. *Plant Molecular Biology*, **22**: 1101-1127.

60. Van der Maas HM, de Jong ER, Rueb S, Hensgens LAM and Krens FA (1994). Stable transformation and long-term expression of the *gus*A reporter gene in callus lines of perennial ryegrass (*Lolium perenne* L.). *Plant Molecular Biology*, **24**: 401-405.

61. Bhalla PL, Swoboda I and Singh MB (1999). Antisense-mediated silencing of a gene encoding a major ryegrass pollen allergen. *Proceedings of the National Academy of Sciences USA*, **96**: 11676-11680.

Part II: Woody Plants

Chapter 10

TRANSFORMATION OF BANANA USING MICROPROJECTILE BOMBARDMENT

D.K. BECKER AND J.L. DALE[1]

School of Life Sciences, Faculty of Science, Queensland University of Technology,
[1]Science Research Centre, Faculty of Science, Queensland University of Technology,
GPO Box 2434 Brisbane, QLD 4001, Australia. Corresponding author: D.K. Becker.
E-mail: d.becker@qut.edu.au

1. INTRODUCTION

1.1. Importance and Origin of Bananas and Plantains

Bananas and plantains (*Musa* spp.) are grown throughout the humid tropics and subtropics where they are of great importance both as a subsistence crop and as a source of domestic and international trade. World production is estimated at more than 100 million tonnes annually (*1*), the majority of which is grown by smallholders for their own consumption and/or traded locally. The total area under production is estimated at over 5.5 million hectares (*1*) with the major growing regions being Asia, Africa and Latin America.

The natural range of wild *Musa* species is from the northern end of Australia to southern China and west to India. Edible, seedless banana and plantain cultivars are derived from intra- and interspecific hybridization of the two seeded, wild diploid species, *Musa acuminata* (A genome) and *M. balbisiana* (B genome). These hybrids are believed to be the result of naturally occurring crosses which were then selected by early farmers in the Asian region. The haploid genome of both *M. acuminata* and *M. balbisiana* consist of 11 chromosomes, however, the *acuminata* genome has been estimated as being slightly larger, 610 Mbp cf. 560 Mbp for *balbisiana* (*2*). Many edible hybrids are parthenocarpic, female sterile and triploid with the relative contribution of each species to the genome being annotated by either A or B. Dessert bananas are usually AAA, plantains AAB and cooking bananas ABB. In recent times, breeding programs have also generated tetraploid hybrids. For brevity, all groups will be referred to collectively as bananas.

1.2. Need for Genetic Improvement and Limitations of Conventional Breeding

Banana breeding programs have largely focussed on generating pest and disease resistant cultivars while at the same time retaining acceptable yield and fruit quality. The two major fungal diseases of banana are Sigatoka leaf spots (*Mycosphaerella*

131

I.S. Curtis (ed.), Transgenic Crops of the World - Essential Protocols, 131-143.

spp.) and Fusarium wilt (*Fusarium oxysporum* f. sp. *cubense*). Viruses have a significant impact on banana production and include banana bunchy top virus (BBTV), banana bract mosaic virus (BBrMV) and banana streak virus (BSV). Among the many pests affecting banana, nematodes, particularly the burrowing nematode (*Radopholus similis*), are the most serious. Various strains of the bacteria, *Pseudomonas solanacearum* infect the vascular tissue (Moko and blood disease) and fruit (Bugtok) of banana. The significance of these pests and diseases in particular banana-growing regions varies depending on banana cultivar, distribution of the pathogen and cultural practices. Due to infertility, triploidy and long generation time, very few conventionally bred cultivars have reached commercial release (*3*). There have been significant successes in breeding disease resistant plantains for use by subsistence growers (*4, 5*) however, problems have been encountered with BSV. Many cases of BSV infection in these new hybrids are believed to be the result of a viral sequence being integrated into the genome of a commonly used breeding parent which is reactivated by the hybridization and/or tissue culture process (*6*). As a result of the aforementioned difficulties encountered with conventional breeding, a great amount of effort has been directed towards improving banana through manipulation of somatic cells and more recently genetic engineering (*3*).

1.3. Genetic Transformation of Banana

Around the world, banana biotechnology programmes are mainly focused on genetically engineered disease resistance although other applications such as delayed ripening (*7*) and edible vaccines (*8*) have also been investigated. Transgenic fungal resistance strategies include the use of antifungal proteins (*9*) and resistance genes (R-genes) derived from wild banana (*10*). Transgenic virus resistance strategies include posttranscriptional gene silencing (*11*), use of mutated viral replication proteins (*12*) and a novel approach utilizing viral activation of suicide genes (*13*).

Although there has been one published report of banana transformation targeting meristematic tissue rather than a regenerable cell culture (*14*), most banana transformations systems involve the use of embryogenic cell suspensions (ECSs). This is due to the low efficiency and chimerism experienced when transforming meristems.

ECSs have been generated using a range of explants including corm tissue (*15*), leaf bases (*15*), immature zygotic embryos (*16*), meristems (*17*) and immature male (*18*) and female (*19*) flowers. Currently, ECSs are most commonly derived from immature flowers or from 'scalps', the term used for meristems derived from a particular form of proliferating shoot culture (*16*). Both these techniques appear to be applicable to a wide range of genotypes including seeded diploids, edible triploids and also recently developed tetraploids derived from breeding programmes (*20, 21*). If there is convenient access to field-grown bananas, immature flowers are the preferred explant for initiation of ECSs because (a) embryogenic callus can be induced directly from floral explants rather than requiring periods of up to 9 months to generate the 'scalps' from which embryogenic cultures are then initiated and (b) some varieties require very high concentrations of growth regulators to form 'scalps' which, in combination with the long culture periods, can add to the likelihood of somaclonal variation.

Both biolistics and *Agrobacterium tumefaciens* have been used to generate transgenic banana from ECSs (*11, 22*). *Agrobacterium*-mediated transformation has the advantage of few transgene rearrangements and low transgene copy number however it is a more difficult system to establish due to complex interactions between the host and vector. Biolistic transformation is a relatively easy system to establish and the construction of expression vectors used for this method is less complex, especially as various transgenes can be mixed and matched as required through co-transformation (*23*). Due to its simplicity and flexibility, most banana transformation in our laboratory is via biolistics. However, we have recently established an *Agrobacterium*-mediated transformation (unpublish.) and use it for applications where it is critical to avoid transgene rearrangements.

Agents used to select transgenic banana include kanamycin and geneticin (resistance provided by *npt*II gene) (*11, 14, 20*), hygromycin (resistance provided by *hpt* gene) (*24*) and chlorosulfuron (resistance provided by *als* gene) (*22*). Reporter genes used in banana include *uid*A gene or GUS (*14, 20, 24-30*) and GFP (*20, 25, 31*). GUS has been most commonly used because quantification is relatively simple but GFP has proven to be a useful tool. We have found GFP to be very useful for monitoring transformation efficiency (*20*) and routinely use it as a control in transformation experiments. Because the assay is non-destructive, GFP can be visualized in transient assays as a measure of cell line transformation competence and during later stages as a measure of selection and regeneration efficiency. Many versions of GFP are now available with some expressed more efficiently in certain plant species than others. We have found sGFP(S65T) (*32*) under the control of the CaMV 35S promoter to be expressed at a level where fluorescence is easily discernable above the red autofluorescence of chlorophyll without the need for additional filters (*20*).

A range of promoters have been shown to be active in banana including those derived from viruses (*29-31, 33*), plant genes (*27*) and *Agrobacterium* (*14, 22*). Techniques to enhance expression such as the use of leader sequences (*24*) and introns (*25, 26*) have also been employed. Promoters shown to have high activity in banana include those derived from badnaviruses (banana streak virus (*30*), sugarcane bacilliform virus (*29*), taro bacilliform virus (*33*)) and the maize polyubiquitin gene (*20, 27-30*). The CaMV 35S promoter provides sufficient activity for detection of reporter genes and, when fused to selectable marker genes, resistance to selective agents.

The following protocol describes biolistic transformation from immature flower-derived ECSs which has been established in our laboratory since 1997, and has progressively been modified and improved since that time. It should be noted that the major hurdle to banana transformation is not the transformation procedure itself but rather the generation of competent ECSs. Therefore, a considerable proportion of this protocol is devoted the generation of such cultures.

2. MATERIALS

2.1. Source of Plant Material

Bananas are monoecious and parthenocarpic, bearing female flowers in the basal hands and male flowers on the distal hands of the inflorescence. Therefore, in most cases, the immature male flowers still encased in their bracts (the bell) can be removed from the end of the inflorescence without damaging the fruit. The bells are harvested when 10 cm or less below the point of attachment of the lowest hand of fruit. Some cultivars, such as False Horn Plantains, do not produce male flowers. In this instance, immature female flowers can be used by harvesting the inflorescence before emergence from the pseudostem (*19*). This ensures sufficiently immature flowers are obtained. Cultures can be successfully initiated up to one week after harvest and inflorescences are tolerant of shipping including air travel. It is important to note, that seasonal effects on the embryogenic response from immature banana flowers have been observed (*see* Note 1).

2.2. Tissue Culture

All "M#" mediums are based on those described by Côte *et al.* (*18*). BL medium is based on the suspension maintenance medium described by Dhed'a *et al.* (*17*). All media components are prepared from Plant Cell Culture Tested products from Sigma (Sigma-Aldrich) unless otherwise stated. Solutions stored at −20°C may be kept for up to 6 months without any apparent loss of activity. Solutions stored at 4°C should be replaced after 4 weeks. Once prepared, media can be safely stored one month if kept cool (< 25°C) and in the dark. The above information also applies to solutions and media as described (*see* Section 2.3.).

1. Surface sterilant: 70% ethanol (v/v).
2. Murashige and Skoog (MS, *34*) vitamin 100 x stock solution: 10g/l *myo*-inositol, 50 mg/l nicotinic acid, 50 mg/l pyridoxine-HCl, 10 mg/l thiamine-HCl, 200 mg/l glycine. Measure into 10 ml aliquots and store at −20°C.
3. BL vitamin 100 x stock solution (*17*): 50 mg/l nicotinic acid, 50 mg/l pyridoxine-HCl, 40 mg/l thiamine-HCl, 200 mg/l glycine. Measure into 10 ml aliquots and store at −20°C.
4. Morel and Wetmore (MW) vitamin 100 x stock solution (*35*): 10 g/l *myo*-inositol, 100 mg/l thiamine-HCl, 100 mg/l nicotinic acid, 100 mg/l pyridoxine-HCl, 100 mg/l calcium panthothenate, 1 mg/l biotin, 1000 mg/l folic acid. Measure into 10 ml aliquots and store at −20°C.
5. 2,4-D: 1 mg/ml stock solution. Prepare by dissolving in absolute ethanol and make up to volume with double distilled water. Store at 4°C.
6. IAA: 1 mg/ml stock solution. Prepare by dissolving in absolute ethanol and make up to volume with double distilled water. Filter-sterilize and store at −20°C.

7. NAA: 1 mg/ml stock solution. Prepare by dissolving in 1N NaOH and make up to volume with double distilled water. Store at 4°C.

8. Zeatin: 0.5 mg/ml stock solution. Prepare by dissolving in 1N NaOH and make up to volume with double distilled water. Filter-sterilize and store at −20°C.

9. BAP: 0.1mg/ml stock solution. Prepare by dissolving in 1N NaOH and make up to volume with double distilled water. Store at 4°C.

10. 2iP: 0.1 mg/l stock solution. Prepare by dissolving in 1N NaOH and make up to volume with double distilled water. Store at 4°C.

11. Kinetin: 0.1 mg/l stock solution. Prepare by dissolving in 1N NaOH and make up to volume with double distilled water. Store at 4°C.

12. M1 medium (initiation medium): Murashige and Skoog (MS) salts (*34*), 10 ml/l MS vitamin stock, 1 mg/l biotin, 4 ml/l 2,4-D stock, 1 ml/l IAA stock, 1 ml/l NAA stock, 30 g/l sucrose, 7 g/l SeaKem LE agarose (FMC BioProducts, Rockland, ME, USA), pH 5.7 (*see* Note 2).

13. BL medium (suspension culture medium): MS salts with half strength macronutrients, 10 ml/l BL vitamins, 10 mg/l sodium ascorbate, 1.1 ml/l 2,4-D stock, 0.5 ml/l zeatin stock, 20 g/l sucrose, pH 5.7 (*see* Note 3).

14. M3 medium (embryo formation/maturation medium): Schenk and Hildebrandt (SH) salts (*36*), 10 ml/l MS vitamin stock, 1 mg/l biotin, 100 mg/l L-glutamate, 230 mg/l proline, 100 mg/l malt extract, 0.2 ml/l NAA stock, 0.1 ml/l zeatin stock, 0.1 ml/l kinetin stock, 1.4 ml/l 2iP stock, 10 g/l lactose, 45 g/l sucrose, 3 g/l Phytagel, pH 5.7.

15. M4 medium (germination medium): MS salts, 10 ml/l MW vitamin stock, 0.45 ml/l BAP stock, 0.2 ml/l IAA stock, 30 g/l sucrose, 2 g/l Phytagel, pH 5.7.

16. M5 medium (rooting medium): M4 medium without the growth regulators.

For all steps of culture initiation, transformation and regeneration, cultures are maintained in the light (16 h photoperiod) at 27°C. The light regime does appear critical for banana culture although there are some reports of improved embryogenic response in immature male flowers when cultured in the dark. We maintain all cultures in the light using cool white fluorescent lamps providing 50-100 μmol/m^2/sec.

2.3. Transformation

1. Gold microcarrier (1.0 μm, Bio-Rad): To prepare, weigh 120 mg of gold in a 1.5 ml-capacity Eppendorf tube. Add 1ml absolute ethanol and vortex well (~1 min). Centrifuge for 5 min in a microfuge, remove supernatant and discard. Repeat ethanol wash twice more. Wash 3 times in 1 ml sterile distilled water as per ethanol wash. Add 1 ml of sterile 50% (v/v) glycerol and resuspend gold by vortexing well. Place 25 μl aliquots in 1.5 ml microfuge tubes and store at −20°C.

2. Calcium chloride (CaCl$_2$): 2.5 M solution in double distilled water. Filter-sterilize and store at −20°C.

3. Spermidine: 0.1 M solution in double distilled water. Filter-sterilize and store at
    −20°C.
4. Plasmids: Use clean plasmid prepared using maxiprep kits (eg BRESApure
    Plasmid Maxi Kit, GeneWorks Pty Ltd, Thebarton, SA, Australia) or caesium
    chloride gradient. Both the CaMV 35S and maize polyubiquitin-1 promoters
    provide sufficient activity to visualize GUS and GFP reporter gene expression.
    Promoters with relatively low activity such as the nopaline synthase gene of
    Agrobacterium (NOS) provide sufficient NPTII activity to confer antibiotic
    resistance to transformed cells.
5. Kanamycin sulphate: 100 mg/ml stock solution. Prepare by dissolving in double
    distilled water. Filter-sterilize and store at −20°C.
6. BL, M3, M4 and M5 selection media: 1 ml/l of kanamycin stock (*see* Note 4).

All media are sterilized by autoclaving at 121°C for 15 min. Antibiotics, zeatin and
IAA are added to the medium after autoclaving.

3. METHODS

3.1. Initiation of Flower Cultures

The unopened inflorescence consists of layers of bract-covered flowers in a spiral
arrangement. Under each bract is a tightly packed cluster of flowers. Each cluster is
referred to as a "hand". Removing layers of bracts and flower hands will reveal
flowers of decreasing maturity until the inflorescence apical meristem is reached.
Only a small percentage of flowers will give an embryogenic response. The flowers
from at least 10 inflorescences should be cultured.

1. Progressively remove bracts and flower hands from the inflorescence until it is
    5-8 cm long. This can be done by cutting the inflorescence transversely at the
    stem end and peeling away the detached bracts and flowers. Repeat until desired
    size is reached. Place in sterilant (70% ethanol) for 10 min and move to a
    laminar flow cabinet. Transfer to individual containers of sterile distilled water
    (*see* Note 5).
2. Remove bracts and flower hands by cutting their point of attachment to the stem
    of the inflorescence until the region of flower hands still enclosed in the bracts
    is 2 cm long. This will leave a bare stem that is convenient for handling.
3. It is recommended that a dissection microscope be used at this stage. From this
    point on, keep all flower hands in the order in which they are excised. Keep in a
    closed Petri dish to prevent desiccation. Continue to excise decreasingly mature
    flower hands until the apical meristem of the inflorescence is exposed. Select
    the most immature hand and the previous 14 hands for culture on M1 medium.
    As a guide, individual flowers in the most immature hand selected will be
    approximately 0.5 mm in length (*see* Note 6).

4. Place individual flower hands in 30 ml-capacity polycarbonate culture tubes (Sarstedt Australia Pty Ltd, Adelaide, SA) containing 8 ml of M1 medium (*see* Note 7). Place the cut edge of the flower hand in firm contact with the media. Be careful not to push the small flower hands below the surface of the media and that they are not immersed in condensation.
5. Observe cultures under a microscope at fortnightly intervals. Explants should not be subcultured as this inhibits embryo formation.

3.2. Initiation of Suspension Cultures

Somatic embryos will start to appear from between 5 weeks and 5 months of culture. They are translucent, elongated and have a distinct epidermis (Fig. 1A). In general, somatic embryos will form on yellow nodular callus. There may be an intervening layer of white friable callus visible between the somatic embryos and the yellow callus.

1. Embryogenic cultures should be transferred to liquid media as soon as they are identified. Remove the whole embryogenic complex, including embryos and yellow callus, from the original explant and place in 10 ml of BL medium in a 100 ml Erlenmeyer flask. Maintain liquid cultures on an orbital shaker at 70–100 rpm.
2. After 10 days, add 10 ml of fresh BL medium.
3. Subsequent subculturing should be performed every 10 days. Do this by allowing cells to settle in the flask, removing 10 ml of old medium and adding 10 ml of fresh BL medium.
4. Once cell growth from the original inoculant is obvious, transfer to a 250 ml Erlenmeyer flask and make up to 50 ml with fresh BL medium. Cell clump size is highly heterogeneous (*37*), some being too large to enter a standard sized pipette. Transfer all cells using a wide bore pipette.
5. Again, subculture every 10 days, this time by removing 25 ml of old medium and replacing with 25 ml of fresh medium. If cell growth is sufficiently dense the cell line can be multiplied by splitting into two flasks (*see* Note 8).

3.3. Screening Cell Lines

Cell lines vary markedly in their regeneration potential. In some lines, most cells become necrotic when plated on M3 medium and very few somatic embryos are produced. In other lines, most cells retain their cream colour and somatic embryo production is prolific (Fig. 1B). Four months after initiating the suspension cultures, cell lines should be screened to allow selection of those with high regeneration potential.

1. Pass cells through a 500 μm filter and place in a tall narrow graduated vessel. We use 50 ml graduated plastic Falcon tubes (Becton Dickinson Biosciences Discovery Labware, USA).
2. Allow the cells to settle for 10 min and then remove liquid medium to leave a settled cell volume:media ratio of 1:4 (*see* Note 9).
3. Using a 1 ml micropipette with the end of the pipette tips cut to increase the bore width, plate 250 μl aliquots of cells on sterile filter paper discs (7 cm diam. Whatman #1, Whatman International Ltd, Maidstone, Kent, UK) overlaid on 25 ml M3 in a Petri dish (90 x 14 mm, Sarstedt Australia Pty Ltd, SA), using one filter paper per plate (*see* Note 10). As cells settle quickly, mix gently before taking each aliquot.
4. Observe cultures for 1 month for somatic embryo formation. Somatic embryos derived from ECSs are spherical, translucent and have a distinct epidermal layer. Translucent somatic embryos are not mature and have a poor germination rate.
5. As somatic embryos mature they increase in size and change from translucent to opaque white (*see* Note 11). To germinate, remove embryos from filter paper using forceps and place in direct contact with 40 ml of M4 medium in a deep Petri dish (100 x 20 mm, Sarstedt Australia Pty Ltd, SA).
6. Once germinated embryo shoots are 2 cm in length, transfer to M5 medium for root formation and shoot elongation.

3.4. Transformation

1. Four to six days after subculturing suspensions, plate cells as described previously. Plate on BL medium solidified with 7 g/l agar (Research Organics Inc., USA). Plate cells at least 24 h before bombardment to allow time to dry onto the filter paper. This will prevent dispersal of cells during bombardment.
2. Vortex a 25 μl aliquot of gold particles thoroughly (approx. 60 sec) then mix with 2 μg of plasmid DNA (equimolar concentrations if co-transforming plasmids), 25 μl CaCl$_2$ solution, and 5 μl spermidine solution. Keep all solutions on ice.
3. Keep gold in suspension for 5 min by vortexing briefly every 30 sec.
4. Allow gold particles to precipitate for 10 min and remove 22 μl of supernatant.
5. Vortex the remaining suspension and use 5 μl aliquots of the wet suspension for bombardment in a PIG (*38*). Vortex between taking aliquots. Bombard with cells placed 7.5 cm from the point of particle discharge and covered by a 210 μm stainless steel mesh baffle. Use a helium pressure of 550 kPa and chamber vacuum of –84 kPa (*see* Note 12).
6. Culture cells for 3 days before transferring to 25 ml semi-solid BL selection medium in a Petri dish (90 x 14 mm, Sarstedt Australia Pty Ltd, SA). Move cells by transferring the whole filter paper disc. Include a plate of non-bombarded cells as a control. Keeps cells on BL selection medium for a total 3 months, subculturing onto fresh medium at monthly intervals.
7. Transfer cells (still on filter papers) to 25 ml M3 selection in a Petri dish (90 x 14 mm). Subculture monthly for 3 months.

8. At this point, clumps of transgenic somatic embryos should be visible on the surface of the filter paper. As cells on the filter paper have not been disturbed after bombardment, individual clumps of somatic embryos represent different transformation events (Fig. 1C). Keeping these clumps separate, remove from the filter paper and place directly into contact with to 40 ml of M4 selection medium in a deep Petri dish (100 x 20 mm) (Fig. 1D). As cells have been on selection medium for 6 months, any escaped non-transgenic embryos will fail to germinate on M4 selection medium.

9. Some embryos may be slow to germinate. Transfer to fresh M4 selection medium at monthly intervals.

10. After germination, transfer plantlets to M5 selection medium for root formation and shoot elongation (Fig. 1E).

11. Acclimatize rooted plants by transferring to propagation tubes (50 mm diam.) using any good quality seed raising mix available from a retail nursery. We use Searles® Seed Raising Mixture (J.C & A.T. Searle Pty Ltd, Kilcoy, Qld, Australia). Place in a high-sided tray (125 mm) and cover with clear plastic film. Remove plastic after 2 weeks. Alternatively, a propagation mister may be used instead of plastic film. If acclimatizing under natural light rather than in a phytotron, light intensity should be reduced using shade cloth (50-70%) for the first 2 weeks. After the first 2 weeks, remove from the high humidity environment and re-pot to larger containers (150 mm diam. pots) using potting mix (Searles® Premium Potting Mix).

4. NOTES

1. We have found that the frequency of embryogenesis is greatest in flowers harvested during June/July in both tropical and subtropical Australia (*20*). Escalant *et al.* (*39*) reported that in Costa Rica, embryogenic frequency had two peaks in March/April and in September/October. Unfortunately there does not appear to be a correlation between these two countries in respect to environmental conditions at these peak times.

2. M1 medium contains 4 mg/l 2,4-D. In some cultivars, we have found that reducing 2,4-D levels markedly increases the frequency of embryogenesis (*20*). Recommended level for Cavendish, (AAA) Ladyfinger (AAB) and FHIA-1 (Goldfinger; AAAB) is 2 mg/l and for Bluggoe (ABB) is 1 mg/l. When initiating cultures from a previously untested cultivar, we examine the response using 1, 2 and 4 mg/l 2,4-D.

3. The original protocol as described by Côte *et al.* (*18*) uses a suspension culture medium designated as M2. We have initiated ECSs using both M3 and BL. There is a substantial amount of variability between cell lines, however, in general, those initiated and maintained in BL appear to have a higher population of embryogenic cells (*20*).

4. We routinely use kanamycin as the selection agent but have also used 10 mg/l geneticin. Other groups have also used 50 mg/l hygromycin (*24*) and 0.01 mg/l chlorosulphuron (Glean®) (*22*).

5. After becoming proficient, a single technician can initiate cultures from 10 inflorescences in 1 day. All inflorescences to be dissected for the day can be disinfested and transferred to sterile distilled water where, prior to dissection, they can be left for several hours with no ill effect.

6. Once the inflorescence apical meristem is exposed, layers of bracts and flowers are still present behind the meristem. These flowers are too small to survive *in vitro* and are not used. Flowers larger than the previous 15 excised are too mature and will continue to develop into flowers rather than form regenerable callus.

7. We culture flower hands in individual culture tubes. Other groups culture multiple hands within one Petri dish. If a Petri dish is used, use deep dishes (20 mm) and ensure good spacing as explants expand greatly during culture.

8. Cell density should be kept low, between 1-2% settled cell volume. High cell density encourages the growth of non-regenerable cells.
9. Embryogenic cells have a dense cytoplasm and settle quickly. After cells have been allowed to settle for 10 min, some may remain in suspension. These cells are not embryogenic and can be discarded.
10. Filter paper discs are satisfactory however there appears to be better cell survival and more somatic embryos produced when suspension cells are plated on glass microfibre filters (Whatman International Ltd, UK).
11. Embryos will be mature enough to germinate 3 months after the initial plating on M3 medium. Longer periods of maturation (up to 6 months) will increase the germination rate. During this time, subculturing is not required but does increase the maturation rate. Subculture monthly by moving the whole filter paper with embryos attached to fresh M3 medium.
12. We use a PIG as they are less expensive to purchase, cheaper to operate and require shorter preparation time in comparison to the Bio-Rad PDS 1000/He System. However, if a PIG is not available, the Bio-Rad system may be used. Because of the different mechanisms involved in particle acceleration, the required helium pressure and target distance would need to be determined empirically.

CONCLUSION

As previously discussed, the major limitation to banana transformation is not the transformation system itself but rather the reliable generation of competent ECSs. Cell lines do not remain regenerable indefinitely and longer periods of *in vitro* culture may contribute to somaclonal variation. Two areas of research which may contribute to alleviating this problem is cryopreservation of highly competent ECSs and further studies of the complete embryogenic process from the influence of field conditions on the explant through to initiation and maintenance of the suspension cultures.

There is a strong case for applying transgenic technologies to banana, and indeed many vegetative crops. Pests and diseases are a major limitation to banana production and although resistant wild seeded banana have been identified incorporating these traits into edible banana has proven problematic. The identification of resistance genes in seeded banana and their transfer to edible banana via transformation would both increase production and reduce reliance and agricultural chemicals. This is particularly true for Sigatoka leaf spots and nematodes, which are controlled by intensive chemical usage, and Fusarium wilt for which there is no control the only solution being plant resistance. The use of "natural" resistance genes may also be more acceptable to the public. However, for this to become a reality, a better understanding of resistance mechanisms at the molecular level is required.

ACKNOWLEDGEMENTS

The authors wish to thank Dr Mike Smith (Department of Primary Industries, Queensland) for his contribution to the development of the banana biotechnology program at QUT. This research was supported by Horticulture Australia Limited and Queensland Fruit and Vegetable Growers Limited.

REFERENCES

1. FAO (2002). FAOSTAT Database. http://apps.fao.org.
2. Kumate K, Brown S, Durand P, Bureau JM, De Nay D and Trinh TH (2001). Nuclear DNA content and base composition of 28 taxa of *Musa*. *Genome*, **44**: 622-627.
3. Smith MK, Hamill SD, Becker DK and Dale JL (2003). Musaceae. In: Litz RE (ed.), *Biotechnology of Fruit and Nut Crops*;; CAB International: Wallingford, UK (in press).
4. Ortiz R and Vuylsteke D (1998). 'PITA-14': A black sigatoka-resistant tetraploid plantain hybrid with virus tolerance. *HortScience*, **33**: 360-361.
5. Ortiz R, Vuylsteke D, Crouch H and Crouch J (1998). TM3x: Triploid black sigatoka-resistant *Musa* hybrid germplasm. *HortScience*, **33**: 362-365.
6. Dallot S, Acuna P, Rivera C, Ramirez P, Côte F, Lockhart BEL and Caruana ML (2001). Evidence that the proliferation stage of micropropagation procedure is determinant in the expression of banana streak virus integrated into the genome of the FHIA 21 hybrid (*Musa* AAAB). *Archives Virology*, **146**: 2179-2190.
7. Balint-Kurti P, Firoozabady E, Moy Y and Gutterson N (2000). Transgenic banana research at DNAP. In: *Proceedings of the 2nd International Symposium of the Molecular and Cellular Biology of Banana*. Byron Bay, New South Wales, Australia, p. 17.
8. Walmsley AM and Arntzen CJ (2000). Plants for delivery of edible vaccines. *Current Opinion in Biotechnology*, **11**: 126-129.
9. Chakrabarti A, Ganapathi TR, Mukherjee PK and Bapat VA (2003). MSI-99, a magainin analogue, imparts enhanced disease resistance in transgenic tobacco and banana. *Planta*, **216**: 587-596.
10. Taylor K, McMahon JA, Becker DK, Harding RM and Dale JL (2000). Isolation of potential disease resistance genes from banana. In: *Proceedings of the Keystone Symposia, Signal and Signal Perception in Biotic Interactions in Plants*. Taos, New Mexico, USA, p. 53.
11. Becker DK, Dugdale B, Smith MK, Harding RM and Dale JL (2000). Genetic transformation of Cavendish banana (*Musa* spp. AAA group) cv. "Grand Nain" via microprojectile bombardment. *Plant Cell Reports*, **19**: 229-234.
12. Webb M, Tsao T, Hastie M, Harding R and Dale J (2002). Competitive inhibition of banana bunchy top nanovirus (BBTV) replication. In: *Proceedings of the XIIth International Congress of Virology*, Paris, France, p. 413.
13. Dale JL, Dugdale B, Hafner G, Hermann SR, Becker DK, Harding RM and Chowpongpang S (2001). A construct capable of release in closed circular form from a larger nucleotide sequence permitting site specific expression and/or developmentally regulated expression of selected genetic sequences. Patent Number WO0172996A1, published 4th October 2001.
14. May GD, Afza R, Mason HS, Wiecko A, Novak FJ and Arntzen CJ (1995). Generation of transgenic banana (*Musa acuminata*) plants via *Agrobacterium*-mediated transformation. *Bio/Technology*, **13**: 486-492.
15. Novak FJ, Afza R, Van Duren M, Perea-Dallos M, Conger BV and Xialolang T (1989). Somatic embryogenesis and plant regeneration in suspension cultures of dessert (AA and AAA) and cooking (ABB) bananas (*Musa* spp.). *Bio/Technology*, **7**: 154-159.
16. Escalant JV and Teisson C (1989). Somatic embryogenesis and plants from immature zygotic embryos of the species *Musa acuminata* and *Musa balbisiana*. *Plant Cell Reports*, **7**: 665-668.
17. Dhed'a D, Dumortier F, Panis B, Vuylsteke D and De Lange E (1991). Plant regeneration in cell suspension cultures of the cooking banana cv. 'Bluggoe' (*Musa* spp. ABB group). *Fruits*, **46**: 125-135.
18. Côte FX, Domergue R, Monmarson S, Schwendiman J, Teisson C and Escalant JV (1996). Embryogenic cell suspensions from the male flower of *Musa* AAA cv. Grand Nain. *Physiologia Plantarum*, **97**: 285-290.
19. Grapin A, Ortiz JL, Lescot T, Ferriere N and Côte FX (2000). Recovery and regeneration of embryogenic cultures from female flowers of False Horn Plantain. *Plant Cell, Tissue and Organ Culture*, **61**: 237-244.
20. Becker DK (1999). The transformation of banana with potential virus resistance genes. Doctoral dissertation, Queensland University of Technology, Brisbane, Australia.
21. Schoofs H (1997). The origin of embryogenic cells in *Musa*. Doctoral dissertation, Katholic University of Leuven, Leuven, Belgium.

22.   Ganapathi TR, Higgs NS, Balint-Kurti PJ, Arntzen CJ, May GD and Van Eck JM (2001). *Agrobacterium*-mediated transformation of embryogenic cell suspensions of the banana cultivar Rasthali (AAB). *Plant Cell Reports,* **20**: 157-162.

23.   Remy S, Francois I, Cammue BPA, Swennen R and Sagi L (1998). Co-transformation as a potential tool to create multiple and durable disease resistance in banana (*Musa* spp.). *Acta Horticulturae,* **461**: 361-365.

24.   Sagi L, Panis B, Remy S, Schoofs H, Desmet K, Swennen R and Cammue BPA (1995). Genetic transformation of banana and plantain (*Musa* spp) via particle bombardment. *Bio/Technology,* **13**: 481-485.

25.   Dugdale B, Becker DK, Beetham PR, Harding RM and Dale JL (2000). Promoters derived from banana bunchy top virus DNA-1 to -5 direct vascular-associated expression in transgenic banana (*Musa* spp.). *Plant Cell Reports,* **19**: 810-814.

26.   Dugdale B, Becker DK, Harding RM and Dale JL (2001). Intron-mediated enhancement of the banana bunchy top virus DNA-6 promoter in banana (*Musa* spp.) embryogenic cells and plants. *Plant Cell Reports,* **20**: 220-226.

27.   Hermann SR, Harding RM and Dale JL (2001). The banana actin 1 promoter drives near-constitutive transgene expression in vegetative tissues of banana (*Musa* spp.). *Plant Cell Reports,* **20**: 525-530.

28.   Hermann SR, Becker DK, Harding RM and Dale JL (2001). Promoters derived from banana bunchy top virus associated components S1 and S2 drive transgene expression in both tobacco and banana. *Plant Cell Reports,* **20**: 642-646.

29.   Schenk PM, Sagi L, Remans T, Dietzgen RG, Bernard MJ, Graham MW and Manners JM (1999). A promoter from sugarcane bacilliform badnavirus drives transgene expression in banana and other monocot and dicot plants. *Plant Molecular Biology,* **39**: 1221-1230.

30.   Schenk PM, Remans T, Sagi L, Elliot AR, Dietzgen RG, Swennen R, Ebert PR, Grof CPL *et al.* (2002). Promoters for pregenomic RNA of banana streak badnavirus are active for transgene expression in monocot and dicot plants. *Plant Molecular Biology,* **47**: 399-412.

31.   Dugdale B, Beetham PR, Becker DK, Harding RM and Dale JL (1998). Promoter activity associated with the intergenic regions of banana bunchy top virus DNA-1 to -6 in transgenic tobacco and banana cells. *Journal of General Virology,* **79**: 2301-2311.

32.   Chiu WL, Niwa Y, Zeng W, Hirano T, Kobayashi H and Sheen J (1996). Engineered GFP as a vital reporter in plants. *Current Biology,* **6**: 325-330.

33.   Yang IC, Iommarini JP, Becker DK, Hafner GJ, Dale JL and Harding RM (2003). A promoter derived from taro bacilliform badnavirus drives strong expression in transgenic banana and tobacco plants. *Plant Cell Reports,* **21**: 1199-1206.

34.   Murashige T and Skoog F (1962). A revised medium for rapid growth and bio-assays with tobacco tissue cultures. *Physiologia Plantarum,* **15**: 473-497.

35.   Morel G and Wetmore RH (1951). Tissue culture of monocotyledons. *American Journal of Botany,* **38**: 138-140.

36.   Schenk RU and Hildebrandt AC (1972). Medium and techniques for induction and growth of monocotyledonous plant cell cultures. *Canadian Journal of Botany,* **50**: 199-204.

37.   Georget F, Domergue R, Ferriere N and Côte FX (2000). Morphohistological study of the different constituents of a banana (*Musa* AAA, cv. Grande Naine) embryogenic cell suspension. *Plant Cell Reports,* **19**: 748-754.

38.   Finer JJ, Vain P, Jones MW and McMullen MD (1992). Development of a particle inflow gun for DNA delivery into plant cells. *Plant Cell Reports,* **11**: 323-328.

39.   Escalant JV, Teisson C and Côte F (1994). Amplified somatic embryogenesis from male flowers of triploid banana and plantain cultivars (*Musa* spp.). *In Vitro Cellular and Developmental Biology-Plant,* **30**: 181-186.

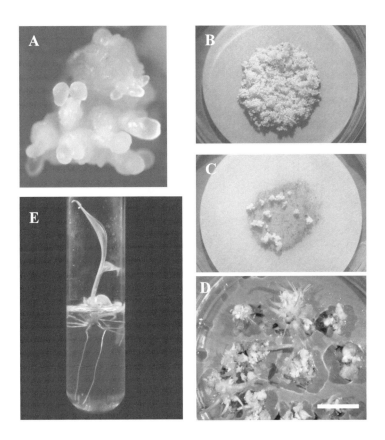

Figure 1. (A) Embryogenic callus and somatic embryos formed on immature male flowers 3 months after initiation. (B) Somatic embryos derived from ECS by culturing on M3 (embryo formation) medium. Cells were plated on filter paper overlaid on culture medium. (C) Somatic embryos formed following microprojectile bombardment and selection on kanamycin. Clumps of embryos represent individual transformation events. (D) Transgenic somatic embryos following transfer to M4 medium for germination. Note embryo clumps were kept separate. (E) Germinated somatic embryo trasferred to M5 (root formation) medium. Scale bar = (A) 1 mm;(B) and (C) 19 mm; (D) 20 mm; (E)18 mm.

Chapter 11

AGROBACTERIUM-MEDIATED TRANSFORMATION OF CITRUS

L. PEÑA, M. CERVERA, C. FAGOAGA, R. PÉREZ, J. ROMERO, J. JUÁREZ, J.A. PINA AND L. NAVARRO

Departamento Protección Vegetal y Biotecnología. Instituto Valenciano de Investigaciones Agrarias (IVIA). Apartado Oficial. 46113-Moncada. Valencia. Spain. Corresponding author: L. Peña. E-mail: lpenya@ivia.es

1. INTRODUCTION

1.1. Importance and Distribution of Citrus

Citrus is the most important fruit crop in the world, with a production of almost 100 million tonnes and acreage of 7.2 million hectares in 2001 (*1*). It is grown in more than 100 countries all over the world, mainly in tropical and subtropical areas (approximately 40° latitude in each side of the equator) where favourable soil and climatic conditions occur. Citrus fruits are marketed fresh or as processed juice and canned segments.

The general area of origin of citrus is believed to be south-eastern Asia, including south China, north-eastern India and Burma, though its introduction into cultivation probably started in China. Commercial citrus species and related genera belong to the order *Geraniales*, family *Rutaceae*, subfamily *Aurantoidea*. All used rootstocks and varieties are included in the genus *Citrus*, except for kumquats (*Fortunella* spp.), and trifoliate orange (*Poncirus trifoliata* L. Raf.), which are used exclusively as a rootstock. Commercial citrus fruits fall into several main groups: sweet oranges (*C. sinensis* (L.) Osb.), mandarins (*C. reticulata* Blanco, *C. deliciosa* Ten.), satsumas (*C. unshiu* (Mak.) Marc.), clementines (*C. clementina* Hort. ex Tan.), grapefruits (*C. paradisi* Macf.), pummelos (*C. grandis* (L.) Osb.), lemons (*C. limon* (L.) Burm. f.), and limes (*C. aurantifolia* (Christm.) Swing.). There are other species of relative importance in certain areas, such as sour oranges (*C. aurantium* L.), citrons (*C. medica* L.), and bergamots (*C. bergamia* Risso & Poit.). In addition, there are some commercial hybrids which are used as rootstocks include citranges (sweet orange x trifoliate orange) and citrumelos (grapefruit x trifoliate orange), and also important variety hybrids such as tangelos (mandarin x grapefruit or pummelo), tangors (mandarin x sweet orange) and mandarin hybrids.

1.2. Need for Genetic Improvement

Many different citrus genotypes are commercially grown in a wide diversity of soil and climatic conditions, implicating that trees are subjected to abiotic and biotic

I.S. Curtis (ed.), Transgenic Crops of the World-Essential Protocols, 145-157.

stresses that limit crop production. The main abiotic stresses are acid, alkaline, and salty soils, flooding and drought, freezing and high temperatures. Citrus trees are also affected by many pests and diseases caused by nematodes, fungi, oomycetes, bacteria, spiroplasmas, phytoplasmas, viruses and viroids. Some diseases are spread throughout the world, as those produced by the oomycete *Phytophthora* sp., or by the *Citrus tristeza virus* (CTV), that preclude the use of certain excellent rootstocks, and severely restrict fruit production and quality of important varieties in some countries. Other diseases are restricted to specific geographic areas, as those produced by the bacteria *Xylella fastidiosa* in Brazil and *Liberobacter asiaticum* in most countries of south-east Asia. In both cases, these bacteria are currently devastating millions of trees and there are no means for efficient control.

At the same time that citrus industry is threatened by important biotic and abiotic stresses, markets of developed countries demand fruit of increasing quality. In this situation, genetic improvement of citrus is a major priority but conventional breeding of citrus has important limitations.

1.3. Genetic Improvement of Citrus

Citrus species have a complex reproductive biology. Some important genotypes have total or partial pollen and/or ovule sterility that cannot be used as parents in breeding programmes. There are many cases of cross- and self-incompatibility. Most species are apomictic, which means that adventitious embryos initiate directly from maternal nucellar cells precluding the development of zygotic embryos, and thus the recovery of sexual progeny populations. Citrus species also have a long juvenile period and most need at least 5 years to initiate flowering in subtropical areas. All these features together with their large plant size, high heterozigosity, lack of basic knowledge about how the most important horticultural traits are inherited, and quantitative inheritance of most characters have greatly impeded the genetic improvement of citrus through conventional breeding methods.

Although some breeding programmes for citrus improvement started more than 100 years ago (*2*), nowadays most rootstocks are citrus species without any improvement, and most important varieties originated by budsport mutations and chance seedlings. Only a few hybrid varieties are economically relevant in certain local markets. Probably, the most significant results from any improvement programme performed in the world are the hybrids Carrizo and Troyer citranges, originated in a cross made in 1909 that was originally oriented to introduce cold tolerance into edible fruits (*3*). To date, citranges are widely used as rootstocks in countries such as Spain and the USA.

Recently, the development of genetic markers is providing a new potential tool for citrus breeding. Linkage maps have been performed using isozymes, RFLPs, RAPDs, sequence-characterized amplified regions (SCARs), AFLPs, microsatellites (single sequence repeat; SSR), and cleaved amplified polymorphic sequences (CAPS). Although these studies have served to determine the mode of inheritance of these traits and could be useful for breeding purposes, map-based cloning of the corresponding genes remains unsolved. The only exception could be the CTV-resistance gene from *Poncirus trifoliata*. Cloning of this gene is under way in several laboratories (*4, 5*). Other promising biotechnologies for citrus improvement include somatic hybridization (*6*) and embryo rescue mainly for obtaining triploid (seedless) varieties (*7*).

Genetic transformation may provide an efficient alternative for citrus improvement opening the way for the introduction of specific traits into known genotypes without altering their genetic elite background.

1.4. Genetic Transformation of Citrus

Until recently, citrus species have been recalcitrant to transformation. Kobayashi and Uchimiya (*8*) obtained transgenic calli from Trovita sweet orange by PEG treatment of cell suspensions with a plasmid containing the marker gene *npt*II, but regeneration of transgenic plants from such calli was unsuccessful. Vardi *et al.* (*9*) produced transgenic calli from Rough lemon (*C. jambhiri* Lush) by PEG treatment of protoplasts with a plasmid containing the marker genes *cat* and *npt*II, and obtained several stably transgenic embryogenic lines, but only 2 lines regenerated whole plants. Hidaka *et al.* (*10*) produced transformed calli of Washington navel and Trovita oranges by co-cultivation of embryogenic cell suspension lines with *Agrobacterium tumefaciens*, but only 1 transgenic plantlet of Washington navel was regenerated. Moore *et al.* (*11*) produced 2 transgenic plantlets of Carrizo citrange by co-cultivation of internodal stem segments from *in vitro* grown seedlings with *A. tumefaciens*. More recently, this group reported a slight increase in transformation efficiency using basically the same procedure (*12*). Hidaka and Omura (*13*) obtained transformed Ponkan mandarin (*C. reticulata* Blanco) calli by electroporation of protoplasts, but no plants were regenerated. Yao *et al.* (*14*) reported transformation of Page tangelo embryogenic cells using particle bombardment, producing 15 transgenic embryo lines, but no further progress was reported.

The first report of efficient and reliable production of citrus transgenic plants was published by Kaneyoshi *et al.* in 1994 (*15*). A transformation efficiency of 25% was obtained by co-cultivating etiolated epicotyl segments of trifoliate orange with *A. tumefaciens*. This study allowed the efficient incorporation of the human epidermal growth factor (*hEGF*) (*16*) and *rol*C gene from *A. rhizogenes* (*17*) into this species. The same procedure with slight modifications has been used to transform Washington navel orange (*18*) and grapefruit (*19, 20*). Pérez-Molphe and Ochoa-Alejo (*21*) have reported the efficient production of transgenic lime plants by co-cultivation of internodal stem segments from *in vitro* grown seedlings with *A. rhizogenes*. Yang *et al.* (*22*) used *A. tumefaciens* to transform epicotyl segments of Rio Red grapefruit, and reported production of transgenic plants containing an untranslatable version of the coat protein gene from CTV and the *Galanthus nivalis* agglutinin gene. A different approach was followed by Fleming *et al.* (*23*) who used the *gfp* gene both as a selectable and reporter marker to transform Itaborai sweet orange protoplasts using PEG to produce embryogenic calli to recover whole transgenic plants through somatic embryogenesis. Li *et al.* (*24*), recently reported the generation of Ponkan transgenic plants through *Agrobacterium*-mediated transformation of embryogenic calli, using Basta® at 50 mg/l. Here, a ribonuclease gene under the control of a tapetum-specific promoter was introduced to produce pollen sterile transformants.

The recalcitrance of citrus to genetic transformation is mainly due to several factors: citrus species are not natural hosts of *Agrobacterium*, problems of regenerating shoots from transformed cells and difficulties in rooting transgenic shoots. Our group has been able to develop efficient and reliable procedures to produce transgenic plants from Carrizo citrange and *P. trifoliata* (*25-27*), using epicotyl segments from *in vitro* grown seedlings. In addition, transgenic plants have also been produced from Pineapple sweet orange (*28, 29*), lime (*30, 31*), sour orange (*32*), alemow (*C. macrophylla* Wester), lemon, and Cleopatra mandarin (*C. reshni* Hort. ex. Tan.) (*33*), using internodal stem segments from glasshouse-grown seedlings. The use of an *Agrobacterium* strain carrying a supervirulent vector, the optimization of infection and co-cultivation conditions and the establishment of appropriate culture media have enabled high output of transgenic citrus plants. In addition, the use of good physiological state source plant material, identification of competent cells for transformation in explants, the use of appropriate marker genes, and the rapid production of whole transgenic plants through grafting of regenerating transgenic shoots into vigorous rootstocks first *in vitro* and later in the glasshouse have also been crucial towards the production of transgenic plants. Furthermore, we have been able to transform mature plant material and recover transgenic plants that flowered and set fruits 1-2 years after *Agrobacterium* infection (*29*). This has allowed us to incorporate transgenes of potential agricultural interest into citrus species which could be used as citrus rootstocks and cultivars with improved characters such as resistance to CTV by introducing viral genes and sequences into Mexican lime (*31, 34-36*) and sour orange plants (*32*). Citrus has also been demonstrated by higher tolerance to *Phytophthora* by introducing a pathogenesis-related (PR) antifungal protein from tomato into Pineapple sweet orange plants (*37*), higher tolerance to salinity by introducing the *HAL2* gene from *Saccharomyces cerevisiae* into Carrizo citrange plants (*38*), and also shortening of the long juvenile period of citrus by introducing the meristem-identity *LFY* or *AP1* genes from *Arabidopsis thaliana* into Carrizo citrange plants (*39*). Our genetic transformation procedures are the subjects of this chapter.

2. MATERIALS

2.1. A. tumefaciens *Strain*

A. tumefaciens strain EHA105 (*40*) carrying a binary plasmid is used as a vector system for transformation (*see* Note 1). The T-DNA of the binary plasmid must contain, apart from the expression cassette/s of interest, a selectable marker gene, such as *npt*II, and a reporter marker gene, *uid*A or *gfp*, both under the control of constitutive promoter and terminator sequences (*see* Note 2). The binary plasmid is introduced into *Agrobacterium* by electroporation.

2.2. Culture Media for A. tumefaciens

1. Luria broth (LB) medium: 10 g/l tryptone (Merck, Darmstadt, Germany), 5 g/l yeast extract (Merck), 10 g/l NaCl, pH 7.5.
2. Kanamycin sulphate (Sigma): 100 mg/ml stock in water. Sterilize by filtration through a 0.2 μm membrane (Minisart, Sartorius, Göttingen, Germany). Dispense 1 ml aliquots into Eppendorf tubes and store at –20°C.
3. Nalidixic acid (Sigma): 25 mg/ml stock in NaOH/water (dissolve the powder in a few drops of 1N NaOH and then add water to volume). Sterilize by filtration and store at –20°C.
4. Liquid culture medium: LB medium containing 25 mg/l kanamycin sulphate and 25 mg/l nalidixic acid.
5. Agar culture medium: LB medium, plus 10 g/l agar (Bacto-agar, Difco, Detroit, MI, USA), pH 7.5, with 25 mg/l kanamycin sulphate, and 25 mg/l nalidixic acid.

2.3. Plant Materials

1. For Carrizo citrange and *P. trifoliata*, stored seeds coming from the same tree stock are used as the source of tissue for transformation.
2. For sweet orange, sour orange, lime, alemow, lemon, and mandarin, vigorous shoots from 4-12 month-old glasshouse-grown (18-27°C) seedlings are used as the source of tissue for transformation.
3. For the transformation of sweet orange mature tissues, buds are collected from trees maintained in a screenhouse (pathogen-free Germplasm Bank Collection of the IVIA) and are grafted onto seedlings of a vigorous rootstock under glasshouse conditions (18-27°C), and newly elongated shoots are used as starting material (*see* Note 3).

2.4. Tissue Culture

1. Seed germination medium: 4.3 g/l Murashige and Skoog (MS, *42*) salts, 10 g/l agar (Difco), pH 5.7.
2. Surface sterilant: 2% (v/v; stems) or 0.5% (v/v; seeds) sodium hypochlorite solution containing 0.1% (v/v) Tween 20 (Merck).
3. Vitamin stock: 100 mg/l *myo*-inositol (Duchefa, Haarlem, The Netherlands), 0.2 mg/l thiamine-HCl (Duchefa), 1 mg/l pyridoxine-HCl (Duchefa), 1mg/l nicotinic acid (Duchefa).
4. Inoculation medium (IM): 4.3 g/l MS salts, 10 ml/l vitamin stock solution, 30 g/l sucrose, pH 5.7.
5. 2,4-D (Sigma): 5 mg/100 ml stock solution. Prepare by dissolving the powder in a few drops of DMSO. Make to volume with double distilled water. Store at 4°C.
6. IAA (Sigma): 5 mg/100 ml stock solution. Prepare as for 2,4-D and store at 4°C.
7. 2,i-P (Sigma): 5 mg/100 ml stock solution. Prepare as for 2,4-D and store at 4°C.
8. BAP (Sigma): 5 mg/100 ml stock solution. Prepare by dissolving the powder in a few drops of 1N NaOH. Complete final volume with double distilled water. Store at 4°C.

9. NAA (Sigma): 5 mg/100 ml stock solution. Prepare as for BAP and store at 4°C.
10. Co-cultivation medium (CM): IM plus 2 mg/l 2,4-D, 2 mg/l IAA, 1 mg/l 2,i-P, 8 g/l agar, pH 5.7.
11. Shoot regeneration medium (SRM): IM plus: a) 3 mg/l BAP for sweet orange, *P. trifoliate*, and citrange; b) 1 mg/l BAP for lime, lemon, alemow and mandarin; c) 1 mg/l BAP and 0.3 mg/l NAA for sour orange. Medium is semi-solidified with 10 g/l agar, pH 5.7, supplemented with 100 mg/l kanamycin sulphate, 500 mg/l cefotaxime and 250 mg/l vancomycin.
12. Kanamycin sulphate: (*see* Section 2.2.2.).
13. Cefotaxime (Aventis Pharma, Madrid, Spain): 250 mg/ml aqueous stock solution. Sterilize by filtration through a 0.2 μm membrane and store at –20°C.
14. Vancomycin (Abbott Laboratories, Madrid, Spain): 250 mg/ml aqueous stock solution. Sterilize by filtration through a 0.2 μm membrane and store at –20°C.
15. Shoot-tip grafting medium: 4.3 g/l MS salts, 10 ml/l vitamin stock, 75 g/l sucrose, pH 5.7.
16. Rooting medium I (RM I): SRM medium supplemented with 0.3 mg/l BAP.
17. Rooting medium II (RM II): SRM medium with 5 mg/l IBA (Sigma).
18. IBA (Sigma): 5 mg/100 ml. Prepare as for BAP and store at 4°C.
    All media are sterilized by autoclaving at 121°C for 20 min. Antibiotics are added to the medium after autoclaving.

3. METHODS

3.1. Explants and Agrobacterium Preparation

1. For Carrizo citrange and *P. trifoliata*, peel seeds, removing both seed coats, disinfect for 10 min in surface-sterilant, and rinse 3 times with sterile distilled water.
2. Sow individual seeds onto 25 ml aliquots of germination medium contained in 25 x 150 mm glass tubes (Bellco glass, Vineland, NJ, USA). Incubate at 26°C in darkness for 2 weeks. Transfer cultures to conditions of 16 h photoperiod and illumination of 45 μE/m^2/sec for 3 additional weeks.
3. For sweet orange (juvenile and mature tissues), sour orange, lime, alemow, lemon, and mandarin, strip stem pieces (20 cm in length) of their leaves and thorns, disinfect for 10 min in surface sterilant, and rinse 3 times with sterile distilled water (*see* Note 4).
4. Cut transversely 1 cm long epicotyl (citrange and *P. trifoliata*) or internodal stem (sweet orange, sour orange, lime, alemow, lemon, and mandarin) segments with forceps and sterile scalpel.
5. Grow *A. tumefaciens* on LB agar culture medium (with antibiotics) at 28°C for 2 days. Take 1 loopful of bacteria and transfer to 100 ml of LB liquid culture medium (with antibiotics) and grow overnight at 28°C on an orbital shaker at 200 rpm.
6. Measure absorbance at 600 nm of a 1 ml aliquot of the bacterial overnight culture in an spectrophotometer, and then calculate bacterial concentration

according to the formula based on the standard curve from the linear regression between optical density and the log of cells/ml of strain EHA 105 (*43*):

$$Bacterial\ concentration = 1.3716\ x\ 10^9\ x\ A_{600} - 5.5376\ x\ 10^7$$

7. Centrifuge the bacterial culture at 3500 rpm for 10 min in 40 ml sterile centrifuge tubes with cap (Beckman Instruments, Palo Alto, CA, USA), discard the supernatant, resuspend and dilute the pellet with IM medium to a concentration of approx. 4 x 10^7 cells/ml (*see* Note 5). Pour 15 ml aliquots of the diluted culture into sterile glass 10 cm diam. Petri dishes.

3.2. Inoculation, Co-cultivation, Selection and Recovery of Whole Transgenic Plants

1. Immerse the explants into the *A. tumefaciens* culture for 15 min and incubate by gentle shaking, blot-dry on sterile filter paper, and place horizontally on CM (30 explants/Petri dish) for a 3 day co-cultivation period at 26°C at a low light intensity (10 µEm^2/sec, 16 h photoperiod) (*see* Notes 6 and 7).
2. After co-cultivation, blot-dry the explants with sterile filter paper, and transfer to SRM medium (10 explants/dish; *see* Note 8). Maintain cultures in the dark for 2-4 weeks at 26°C and transfer to a 16 h photoperiod, 45 µE/m^2/sec illumination at 26°C (*see* Note 9).
3. Shoots should develop from the cut ends 3-5 weeks after co-cultivation (Fig. 1B). Check the transgenic nature of the regenerated shoots by performing a histochemical GUS assay (*44, 25*) or by testing GFP expression (*45, 46*) (*see* Note 10). GUS- or GFP-negative shoots are considered as non-transformed, and commonly called escape shoots (*see* Note 11).
4. Graft *in vitro* apical portions of the GUS- or GFP-positive shoots onto decapitated seedlings (*47, 25*; Fig. 1C). Rootstock preparation is as follows: peel seeds, remove both seed coats, disinfect for 10 min in surface-sterilant, and rinse 3 times with sterile distilled water. Sow individual seeds onto 25 ml aliquots of germination medium contained in 25 x 150 mm glass tubes and incubate at 27°C in the dark for 2 weeks. Decapitate seedlings leaving 1-1.5 cm of the epicotyls. Shorten the roots to 4-6 cm and remove the cotyledons and their axillary buds. Place the regenerated shoot onto the apical end of the cut surface of the decapitated epicotyl, in contact with the vascular ring (*see* Note 12).
5. Culture grafted plants in a shoot-tip grafting medium and maintain at 25°C, 16 h of photoperiod and 45 µE/m^2/sec of illumination (*see* Note 13). Scions develop 2-4 expanded leaves 3-4 weeks after grafting.
6. Grafting of *in vitro*-grown plants onto vigorous rootstocks in the glasshouse allows the rapid acclimatization and development of the plants (Fig. 1D).
7. Alternatively, the rooting of transgenic shoots can also be performed (*see* Note 14). Cut 0.5-1 cm regenerated shoots from the explants and transfer to RM I medium for 7-10 days, and then to RM II medium (25°C, 16 h of photoperiod, 45 µE/m^2/sec).

8. Regenerated plants can be assessed by PCR to detect the presence of the transgene/s in leaf samples. Southern blot analyses must be performed to confirm the stable integration of the transgene/s, and Northern blot and Western blot analyses would confirm their expression in transgenic plants (*see* Note 15).

9. Transgenic plants flower and set fruits 12 months after *Agrobacterium*-inoculation (Fig. 1E). Further confirmation of the transgenic nature of harvested fruits can also be monitored using whole fruits (Figs. 1F and H) and embryos from seeds (Fig. 1G).

4. NOTES

1. We have investigated the use of appropriate *Agrobacterium* strains to efficiently transform citrus genotypes. We have demonstrated that *A. tumefaciens* strain A281 is able not only to efficiently transform citrus but is super-virulent towards many citrus genotypes (*41*). In addition, we have found a correlation between super-virulence and the super-transformation ability of its disarmed derivative EHA105 (*33*). Super-virulence of A281 in many plants has been attributed to higher expression of *vir* genes after induction. We have demonstrated that the *vir* region of Ti plasmid pTiBo542 provides this ability in citrus (*33*).

2. The binary plasmid used has incorporated bacterial resistance to kanamycin, as it indeed occurs for most binary plasmids. *A. tumefaciens* strain EHA105 has chromosomal resistance to nalidixic acid.

3. Alternatively, new shoots from the Pineapple sweet orange tree maintained in a screenhouse (at the Germplasm Bank Collection) can be directly used as starting material for transformation.

4. The use of vigorous shoots from plants growing in the glasshouse as source of material for transformation in the case of citrus varieties provides us with the possibility of using the same procedure first established to transform juvenile tissues of a given species to also transform adult materials from the same species. Transformation of mature tissues allows overcoming the long juvenile period of most citrus genotypes, and thus opening the possibility to analyze fruit features very early in new genetically modified genotypes.

5. A higher bacterial concentration (~10^8) in the IM medium results in a low transformation frequency because plant cells become stressed; a low bacterial concentration (~10^6) results in a low transformation frequency but in this case fewer cells at the cut ends become transformed (*25*).

6. Co-cultivation in a medium rich in auxins provides to the wounded plant cells an appropriate treatment to shift them to a competent state for transformation, involving dedifferentiation, induction of cell division and callus proliferation (*30, 26, 31*; Fig. 1A).

7. Prolonging or shortening co-cultivation period does not increase transformation frequency. In Carrizo citrange, when the explants were transferred to selective medium just after inoculation with *Agrobacterium* (no co-cultivation), no transformation was observed. Transformation frequency was very low after 1 day co-cultivation, but increased rapidly when the co-culture was prolonged to 3 days, reaching a maximum on day 5. Although prolonged co-cultivation periods of more than 3 days have been successfully used for certain plants, 2-3 day co-cultivation has been usually used in most reported transformation protocols since longer co-cultivation periods frequently result in *Agrobacterium* overgrowth. In citrange, a 5 day co-cultivation period also resulted in abundant proliferation of the bacteria and subsequent decrease in regeneration frequency of transformed shoots. Therefore, a 3-day co-cultivation is routinely used (*26*).

8. For sour orange, the combination BAP/NAA in the SRM medium is more favourable than BAP alone in stimulating cell divisions and re-differentiation from the transgenic competent cells to form transformation events (*32*).

9. Culture of explants in the dark improves callus formation and the progress of transformation events to regenerate transgenic shoots and, so avoiding the regeneration of escape shoots that could be stimulated by the exposure of explants directly to the light (*30, 26*). Culturing in the dark varies between 2-4 weeks depending on the genotype and the physiological and ontological state of the plant material. Indeed, the explants should be kept in darkness until they develop a prominent visible callus formed at the cambial ring.

10. Since some citrus genotypes are particularly recalcitrant to transformation, we have investigated the use of a range of marker genes with the aim to increase the efficiency of generating transgenic citrus plants. To develop transformation procedures in these cases, it is important to localize the sites of transgene expression in order to favour the regeneration of whole plants from such competent cells. The use of reporter genes such as *uid*A or *gfp* allowed the localization of competent cells for transformation in de-differentiated callus from the cambium tissue of citrus explants (*46*). Therefore, treatments favouring the development of such callus tissue are necessary to increase transformation frequencies and to enhance a more efficient recovery of citrus transgenic plants.

11. GFP provides the possibility to perform *in vivo* monitoring of *Agrobacterium*-inoculated plant tissues and thus enabling the efficient selection of transgenic shoots and so avoiding escapes and chimeras (shoots with some transgenic cells or sectors) which is a common problem in citrus. GFP expression permits a rapid and easy discrimination of transgenic and escape shoots in citrus. Competition for growth between transformed and non-transformed shoots could be avoided by eliminating the escapes soon after their emergence (*46*). In addition, the GFP-positive shoots can be regenerated from explants inoculated with *Agrobacterium* and cultured in a medium without kanamycin and so opening the possibility of producing transgenic plants without using selective agents (*46*).

12. When larger than 0.4 cm, shoots can be inserted into a lateral incision or in a vertical incision along the length of the epicotyl, at the point of decapitation.

13. Shoots of 0.1 cm in length can be used to regenerate transgenic plants following this protocol. A frequency of 100% successful grafts is usually obtained.

14. For certain purposes, rooting of transgenic shoots, allow the transgene to be assessed in terms of its affect on aerial and root development (this is important in terms of determining tolerance to salinity and plant size, etc.). Although development of whole plants is slower and less efficient than performing *in vitro* grafting, rooted plants can be obtained. In the case for lime and lemon, roots can develop within 3-6 weeks with an efficiency of 90-100% efficiency.

15. Transformation efficiency is the frequency of whole transgenic regenerated plants established in a glasshouse per *Agrobacterium*-inoculated explant. To date, we have obtained the following efficiencies: more than 40 % for citrange and *P. trifoliata*, approx. 20% for sweet orange and lime, between 5-10% for alemow and sour orange and less than 5% for lemon and mandarin.

CONCLUSIONS

We have described a genetic transformation system for citrus based on the co-cultivation of epicotyl or internodal stem segments with an *Agrobacterium tumefaciens* strain carrying a supervirulent Ti plasmid, the regeneration of plants via organogenesis from actively dividing callus cells derived from the cambium which are competent for transformation, and the production of whole plants by grafting *in vitro* of regenerating transgenic shoots into vigorous rootstocks. This general procedure has allowed: a) the efficient and reliable recovery of transgenic plants from many economically important citrus genotypes; b) the generation of transgenic adult plants, overcoming the long juvenile periods of these species; and c) the successful incorporation of transgenes of interest into citrus with the aim of improving rootstock and variety performance.

ACKNOWLEDGEMENTS

The authors wish to thank C. Ortega and A. Navarro for their excellent technical assistance and personnel from the glasshouses. This research was supported by grants from the INIA (RTA01-120) and from MCYT (AGL2003-01644).

REFERENCES

1.   FAO (2001). http.//apps.fao.org/lim500/nph-wrap.pl.
2.   Soost RK and Cameron JW (1975). Citrus. In: Janick J, Moore JN (eds.), *Advances in Fruit Breeding*, Purdue University Press, West Lafayette, USA, pp. 507-540.
3.   Savage EM and Gardner FE (1965). The Troyer and Carrizo citranges. *California Citrograph*, **50**: 112-116.
4.   Deng Z, Huang S, Ling P, Yu C, Tao Q, Chen C, Wendell MK, Zhang H-B *et al.* (2001). Fine genetic mapping and BAC contig development for the citrus tristeza virus resistance gene locus in *Poncirus trifoliata* (Raf.). *Molecular and General Genomics*, **265**: 739-747.
5.   Yang Z-N, Ye X-R, Choi S, Molina J, Moonan F, Wing RA, Roose ML and Mirkov TE (2001). Construction of a 1.2-Mb contig including the citrus tristeza virus resistance gene locus using a bacterial artificial chromosome library of *Poncirus trifoliata* (L.) Raf. *Genome*, **44**: 382-393.
6.   Grosser JW, Ollitrault P and Olivares-Fuster O (2000). Somatic hybridization in Citrus: an effective tool to facilitate variety improvement. *In Vitro Plant*, **36**: 439-449.
7.   Ollitrault P, Dambier D, Sudahono S, Mademba-Sy F, Vanel F, Luro F and Aubert B (1998). Biotechnology for triploid mandarin breeding. *Fruits*, **53**: 307-317.
8.   Kobayashi S and Uchimiya H (1989). Expression and integration of a foreign gene in orange (*Citrus sinensis* Osb.) protoplasts by direct DNA transfer. *Japanese Journal of Genetics*, **64**: 91-97.
9.   Vardi A, Bleichman S and Aviv D (1990). Genetic transformation of citrus protoplasts and regeneration of transgenic plants. *Plant Science*, **69**: 199-206.
10.   Hidaka T, Omura M, Ugaki M, Tomiyama M, Kato A, Ohshima M and Motoyoshi F (1990). *Agrobacterium*-mediated transformation and regeneration of Citrus spp. from suspension cells. *Japanese Journal of Breeding*, **40**: 199-207.
11.   Moore GA, Jacono CC, Neidigh JL, Lawrence SD and Cline K (1992). *Agrobacterium*-mediated transformation of citrus stem segments and regeneration of transgenic plants. *Plant Cell Reports*, **11**: 238-242.
12.   Gutiérrez MA, Luth DE and Moore GA (1997). Factors affecting *Agrobacterium*-mediated transformation in *Citrus* and production of sour orange (*Citrus aurantium* L.) plants expressing the coat protein gene of citrus tristeza virus. *Plant Cell Reports*, **16**: 745-753.
13.   Hidaka T and Omura M (1993). Transformation of citrus protoplasts by electroporation. *Journal of the Japanese Society for Horticultural Science*, **62**: 371-376.
14.   Yao J-L, Wu J-H, Gleave AP and Morris BAM (1996). Transformation of citrus embryogenic cells using particle bombardment and production of transgenic embryos. *Plant Science*, **113**: 175-183.
15.   Kaneyoshi J, Kobayashi S, Nakamura Y, Shigemoto N and Doi Y (1994). A simple and efficient gene transfer system of trifoliate orange. *Plant Cell Reports*, **13**: 541-545.
16.   Kobayashi S, Nakamura Y, Kaneyoshi J, Higo H and Higo K (1996). Transformation of kiwifruit (*Actinidia chinensis*) and trifoliate orange (*Poncirus trifoliata*) with a synthetic gene encoding the human epidermal growth factor (*hEGF*). *Journal of the Japanese Society for Horticultural Science*, **64**: 763-769.
17.   Kaneyoshi J and Kobayashi S (1999). Characteristics of transgenic trifoliate orange (*Poncirus trifoliata* Raf.) possessing the *rolC* gene of *Agrobacterium rhizogenes* Ri plasmid. *Journal of the Japanese Society for Horticultural Science*, **68**: 734-738.
18.   Bond JE and Roose ML (1998). *Agrobacterium*-mediated transformation of the commercially important citrus cultivar Washington navel orange. *Plant Cell Reports*, **18**: 229-234.
19.   Luth D and Moore G (2000). Transgenic grapefruit plants obtained by *Agrobacterium tumefaciens*-mediated transformation. *Plant Cell, Tissue and Organ Culture*, **57**: 219-222.
20.   Costa MGC, Otoni WC and Moore GC (2002). An evaluation of factors affecting the efficiency of *Agrobacterium*-mediated transformation of *Citrus paradisi* (Macf.) and production of transgenic plants containing carotenoid biosynthetic genes. *Plant Cell Reports*, **21**: 365-373.
21.   Pérez-Molphe E and Ochoa-Alejo N (1998). Regeneration of transgenic plants of Mexican lime from *Agrobacterium rhizogenes*-transformed tissues. *Plant Cell Reports*, **17**: 591-596.
22.   Yang Z-N, Ingelbrecht IL, Louzada E, Skaria M and Mirkov TE (2000). *Agrobacterium*-mediated transformation of the commercially important grapefruit cultivar Rio Red (*Citrus paradidi* Macf.). *Plant Cell Reports*, **19**: 1203-1211.

23. Fleming GH, Olivares-Fuster O, Fatta Del-Bosco S and Grosser JW (2000). An alternative method for the genetic transformation of sweet orange. *In Vitro Cellular and Developmental Biology-Plant*, **36**: 450-455.

24. Li DD, Shi W and Deng XX (2002). *Agrobacterium*-mediated transformation of embryogenic calluses of Ponkan mandarin and the regeneration of plants containing the chimeric ribonuclease gene. *Plant Cell Reports*, **21**: 153-156.

25. Peña L, Cervera M, Juárez J, Ortega C, Pina JA, Durán-Vila N and Navarro L (1995). High efficiency *Agrobacterium*-mediated transformation and regeneration of citrus. *Plant Science*, **104**: 183-191.

26. Cervera M, Pina JA, Juárez J, Navarro L and Peña L (1998). *Agrobacterium*-mediated transformation of citrange: factors affecting transformation and regeneration. *Plant Cell Reports*, **16**: 271-278.

27. Cervera M, Pina JA, Juárez J, Navarro L and Peña L (2000). A broad exploration of a transgenic population of citrus: stability on gene expression and phenotype. *Theoretical and Applied Genetics*, **100**: 670-677.

28. Peña L, Cervera M, Juárez J, Navarro A, Pina JA, Durán-Vila N and Navarro L (1995). *Agrobacterium*-mediated transformation of sweet orange and regeneration of transgenic plants. *Plant Cell Reports*, **14**: 616-619.

29. Cervera M, Juárez J, Navarro A, Pina JA, Durán-Vila N, Navarro L and Peña L (1998). Genetic transformation and regeneration of mature tissues of woody fruit plants bypassing the juvenile stage. *Transgenic Research*, **7**: 51-59.

30. Peña L, Cervera M, Juárez J, Navarro A, Pina JA and Navarro L (1997). Genetic transformation of lime (*Citrus aurantifolia* Swing.): factors affecting transformation and regeneration. *Plant Cell Reports*, **16**: 731-737.

31. Domínguez A, Guerri J, Cambra M, Navarro L, Moreno P and Peña L (2000). Efficient production of transgenic citrus plants expressing the coat protein gene of citrus tristeza virus. *Plant Cell Reports*, **19**: 427-433.

32. Ghorbel R, Domínguez A, Navarro L and Peña L (2000). High efficiency genetic transformation of sour orange (*Citrus aurantium* L.) and production of transgenic trees containing the coat protein gene of citrus tristeza virus. *Tree Physiology*, **20**: 1183-1189.

33. Ghorbel R, LaMalfa S, López MM, Petit A, Navarro L and Peña L (2001). Additional copies of *virG* from pTiBo542 provide a super-transformation ability to *Agrobacterium tumefaciens* in citrus. *Physiology and Molecular Plant Pathology*, **58**: 103-110.

34. Domínguez A, Hermoso de Mendoza A, Guerri J, Cambra M, Navarro L, Moreno P and Peña L (2002). Pathogen-derived resistance to citrus tristeza virus (CTV) in transgenic Mexican lime (*Citrus aurantifolia* (Christ.) Swing.) plants expressing its *p25* coat protein gene. *Molecular Breeding*, **10**: 1-10.

35. Domínguez A, Fagoaga C, Navarro L, Moreno P and Peña L (2002). Regeneration of transgenic citrus plants under non selective conditions results in high-frequency recovery of plants with silenced transgenes. *Molecular and General Genomics*, **267**: 544-556.

36. Ghorbel R, López C, Moreno P, Navarro L, Flores R and Peña L (2001). Transgenic citrus plants expressing the citrus tristeza virus p23 protein exhibit viral-like symptoms. *Molecular Plant Pathology*, **2**: 27-36.

37. Fagoaga C, Rodrigo I, Conejero V, Hinarejos C, Tuset JJ, Arnau J, Pina JA, Navarro L *et al.* (2001). Increased tolerance to *Phytophthora citrophthora* in transgenic orange plants overexpressing a tomato pathogenesis related protein PR-5. *Molecular Breeding*, **7**: 175-181.

38. Cervera M, Ortega C, Navarro A, Navarro L and Peña L (2000). Generation of transgenic citrus plants with the tolerance-to-salinity gene HAL2 from yeast. *Journal of Horticultural Science & Biotechnology*, **75**: 26-30.

39. Peña L, Martín-Trillo M, Juárez J, Pina JA, Navarro L and Martínez-Zapater JM (2001). Constitutive expression of Arabidopsis LEAFY and APETALA1 genes in citrus reduces their generation time. *Nature Biotechnology*, **19**: 263-267.

40. Hood EE, Gelvin SB, Melchers LS and Hoekema A (1993). New *Agrobacterium* helper plasmids for gene transfer to plants. *Transgenic Research*, **2**: 208-218.

41. Cervera M, López MM, Navarro L and Peña L (1998). Virulence and supervirulence of *Agrobacterium tumefaciens* in woody fruit plants. *Physiology and Molecular Plant Pathology*, **52**: 67-78.

42.   Murashige T and Skoog F (1962). A revised medium for rapid growth and bioassays with tobacco tissue cultures. *Physiologia Plantarum*, **15**: 473-497.

43.   Koch AL (1994). Growth measurements. In: Gerhardt P, Murray RGE, Wood WA, Krieg NR (eds.), *Methods for General and Molecular Bacteriology*. American Society for Microbiology, Washington, D.C., USA, pp. 248-277.

44.   Jefferson RA, Kavanagh TA and Bevan MW (1987). GUS fusions: β-glucuronidase as a sensitive and versatile gene fusion marker in higher plants. *The EMBO Journal*, **6**: 3901-3907.

45.   Chalfie M, Tu Y, Euskirchen G, Ward WW and Prasher DC (1994). Green fluorescent protein as a marker for gene expression. *Science*, **263**: 663-664.

46.   Ghorbel R, Juárez J, Navarro L and Peña L (1999). Green fluorescent protein as a screenable marker to increase the efficiency of generating transgenic woody fruit plants. *Theoretical and Applied Genetics*, **99**: 350-358.

47.   Navarro L, Roistacher CN and Murashige T (1975). Improvement of shoot-tip grafting *in vitro* for virus-free citrus. *Journal of American Society for Horticultural Science*, **100**: 471-479.

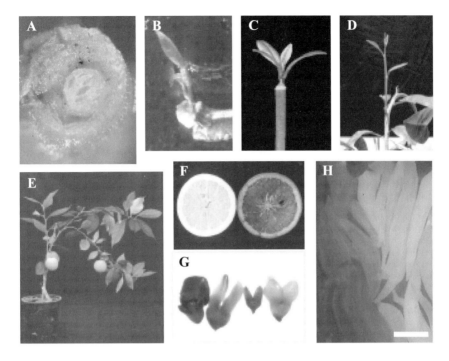

Figure 1. Procedure for the genetic transformation of citrus plants. After co-cultivation with A. tumefaciens, explants are transferred to a regeneration-selection medium. A. Co-cultivation in a medium rich in auxins provides wounded plant cells an appropriate treatment to shift them to a competent state for transformation; competent cells are localized in de-differentiated callus from the cambium tissue (bar = 1 mm). B. Cultivation of explants in selection medium promotes the regeneration of transgenic shoots (bar = 3 mm). C. Apical portions of shoots are grafted in vitro *on decapitated citrange seedlings (bar = 1 cm). D. Several weeks after grafting, developed scions are grafted again onto vigorous rootstocks in the glasshouse (bar = 12.5 cm). E. After mature tissues are transformed, transgenic plants start to flower and set fruits 12 months after* Agrobacterium *inoculation and so confirming their mature nature. Transformed plants undergo normal development and are fertile (bar = 16.5 cm). F. GUS expression of a transgenic sweet orange fruit (right) and a non-transgenic control fruit (left) (bar = 5 cm). G. GUS expression of cultured embryos recovered from seeds of an adult transgenic plant (bar = 2 mm). H. GFP expression of juice vesicles from a transgenic (right) and a non-transgenic (left) lime fruit (bar = 3 mm).*

Chapter 12

COFFEA SPP. GENETIC TRANSFORMATION

T. LEROY AND M. DUFOUR

Centre de Coopération Internationale en Recherche Agronomique pour le Développement, Département des Cultures Pérennes (CIRAD-CP). TA 80/02. 34398 Montpellier Cedex 5. France. Corresponding author : T. Leroy. E-mail : leroy.t@cirad.fr

1. INTRODUCTION

1.1. Importance of Coffee and Breeding Tasks

Coffee is an extremely important agricultural crop with more than 7 millions tonnes of green beans produced every year on about 11 millions hectares. In terms of economic importance on the international markets, it is second only to oil and contributes to more than US$ 9,000 million.

Traditional breeding is aimed at improving the income of the planters, who are mainly small farmers. It is time consuming, due to the biological cycle of the coffee tree. It takes at least 3 years to harvest the first crop of fruits from one progeny and 5 years are necessary for yield evaluation. Two major species, *Coffea arabica* (self-pollinated and allotetraploid: 2n = 44, 68% of the global production) and *Coffea canephora* (self-sterile and diploid: 2n = 22) are cultivated all over the tropics. Arabica breeding is traditionally based on pure line selection, but recently, a F1 hybrid selection strategy has been developed (*1*). Desired traits are principally pest and disease resistances to introduce into elite varieties. On the opposite, Canephora breeding is more oriented towards improving yield, technological and organoleptic quality through creation of hybrids between genotypes of different genetic groups (*2*), or selection of improved clones (*3*).

As for other perennial crops, coffee has a long juvenile period and conventional breeding for the introduction of new traits, mainly related to resistance or quality, can take between 25-35 years. It is a major drawback for coffee improvement; therefore genetic engineering could potentially shorten this time by allowing the incorporation of known genes into elite genetic backgrounds.

Reviews on biotechnology techniques applied to coffee have been published recently (*4, 5*). Most studies deal with large scale propagation techniques through micro-cuttings or somatic embryogenesis, somaclonal variation, *in vitro* preservation of coffee germplasm, and genetic transformation.

I.S. Curtis (ed.), Transgenic Crops of the World - Essential Protocols, 159-170.
© 2004 Kluwer Academic Publishers. Printed in the Netherlands.

1.2. Breeding for Insect Resistance

Together with leaf rust, coffee berry disease, and nematodes, one of the major pests threatening this production is the leaf miner *Leucoptera* spp., which has an economically important impact in East Africa and Brazil (*6*). Since the caterpillar develops inside coffee leaves, insecticidal sprays cannot come into direct contact with the pest. For a treatment to be efficient, the miner should ingest the insecticide or insecticidal protein. The only strategy available would therefore be the transgenic approach. The use of *Bacillus thuringiensis* genes to transform plants for protection to insect is currently the most reliable strategy (*7, 8*) and preliminary investigations were therefore conducted to determine the susceptibility of *Leucoptera* spp. to *B. thuringiensis* insecticidal proteins and identify candidate genes for transformation of coffee (*9*). The toxin expressed by the *cry1Ac* gene, widely used to confer resistance to lepidopterae (*10*), has been demonstrated to be the most effective. Current studies are in progress to select other *B. thuringiensis* strains active against Coffee Berry Borer and White Stem Borer (*11*), two coleopteran pests of economic importance.

1.3. Genetic Transformation of Coffee

The first reported genetic transformation of coffee cells (*12*) was achieved using protoplast electroporation. Genetic engineering using *Agrobacterium* sp. has also been reported (*13, 14*). Regeneration of transgenic coffee trees was obtained after transformation of somatic embryos *via Agrobacterium rhizogenes* (*15-17*), or *via A. tumefaciens* (*18-20*). Although somatic embryogenesis is still a tedious process for some coffee species (*21, 22*) regeneration was easily obtained.

Recent research has been focused on the transfer of pest and disease resistance genes (*23*), cup quality improvement genes (*24*) and specific promoters (*25*). Genetic tools like linkage mapping (*26, 27*) and BAC DNA libraries (*28*) can be useful for identifying the location of such genes.

The need for a suitable transformation protocol is of utmost importance, either to evaluate gene functionality, or to introduce new genes of interest into coffee elite genotypes. Starting with an existing protocol (*15*), we developed our own, higher yielding procedure (*20*), which we describe here, as well as modifications for use with different coffee genotypes.

2. MATERIALS

Three genotypes were used for the first experiments, one *Coffea canephora* and two *Coffea arabica*. The *C. canephora* genotype (126) was a selected clone of good agronomic value; a Catimor (8661-4) and a F1 hybrid (Et29 X Ca5) were used as *C. arabica* genotypes.

In a second step, 10 genotypes from each species were used, in order to have a suitable general protocol for coffee transformation.

2.1. Transformation Vectors

Three *A. tumefaciens* disarmed strains were used for transformation: LBA4404, C58 and EHA105. Constructs have been integrated in pBin19 (*29*) and pCAMBIA A1 plasmids. Three genes were introduced: (a) the *uid*A bacterial gene isolated from *E. coli* coding for β-glucuronidase, with an additional intron for specific expression in plants (*30*). The gene was controlled by the CaMV 35S promoter and terminator; (b) the *csr1-1* gene isolated from *Arabidopsis thaliana* (*31*) conferring resistance to the herbicide chlorsulfuron. This herbicide was used for selection of transformed cells. The gene was controlled by the CaMV 35S promoter with a duplicated enhancer sequence (*32*), and by the *csr1-1* terminator; (c) the *cry1Ac* gene from *B. thuringiensis* in a modified form synthesized at the University of Ottawa (*33*). After preliminary studies on coffee cells demonstrating its efficiency in transient expression (*34*), the EF1α promoter from *A. thaliana* (*35*) was chosen, together with a promoter enhancer sequence Ω' derived from tobacco mosaic virus (*36*) and the NOS terminator.

2.2. Culture Media for Agrobacterium Strains

1. YEP medium : 10 g/l bacto-peptone (Difco), yeast extract (Merck), 10 g/l, NaCl 5g/l, pH 7.0.
2. Kanamycin sulphate (Sigma): 50 mg/ml stock solution in water. Sterilize by filtration through a 0.2 μm membrane. Dispense 1 ml aliquots in Eppendorf tubes and store at –20°C (used for the 3 strains).
3. Streptomycin sulphate (Sigma): 100 mg/ml stock solution in water. Sterilize by filtration and store at –20°C (used for LBA strain).
4. Rifampicin (Sigma): 25 mg/ml stock solution in water, use a few drops of HCl 1N for correct dilution. Sterilize by filtration and store at –20°C (used for the 3 strains).
5. Gentamycin (Sigma): 10 mg/ml stock solution in water. Sterilize by filtration and store at –20°C (used for C58 strain).
6. YEP liquid medium containing 50 mg/l kanamycin, 50 mg/l rifampicin, 100 mg/l streptomycin (LBA) or 20 mg/l gentamycin (C58), pH 7.5.

2.3. Tissue Culture

2.3.1. Coffea canephora Genotypes

1. BAP (Sigma): 10 mg/100 ml stock solution. Prepare by dissolving in 1 ml of 1N HCl before making up to 100 ml with RO (Reverse Osmosis) water. Store at 4°C.
2. Yasuda medium (*21*): modified Murashige and Skoog (MS, *37*) salts, Gamborg B5 vitamins (*38*: 10 mg/l thiamine-HCl (Sigma), 1 mg/l pyridoxine-HCl

(Sigma), 1 mg/l nicotinic acid (Sigma), 100 mg/l *myo*-inositol (Sigma), 1.125 mg/l BAP, 30 g/l sucrose, 6 g/l Phytagel, pH 5.6.

2.3.2. Coffea arabica *Genotypes*

1. 2,4-D (Sigma): 10 mg/100 ml stock solution. Prepare by dissolving in 1 ml of 1N NaOH before making up to 100 ml with RO water. Store at 4°C.
2. 2-iP (Sigma): 10 mg/100 ml stock solution. Prepare by dissolving in 1 ml of 1N NaOH before making up to 100 ml with RO water. Store at –20°C.
3. IBA (Sigma): 10 mg/100 ml stock solution. Prepare by dissolving in 1 ml of 1N NaOH before making up to 100 ml with RO water. Store at 4°C.
4. Kinetin (Sigma): 10 mg/100 ml stock solution. Prepare by dissolving in 1 ml of 1N NaOH before making up to 100 ml with RO water. Store at 4°C.
5. T1B medium (*22*): Half-strength MS salts, 10 mg/l thiamine-HCl, 1 mg/l pyridoxine-HCl, 1 mg/l nicotinic acid, 1 mg/l glycine (Sigma), 100 mg/l *myo*-inositol, 100 mg/l casein hydrolysate (Sigma), 400 mg/l malt extract (Sigma), 0.5 mg/l 2,4-D, 1 mg/l IBA, 2 mg/l 2-iP, 30 g/l sucrose, 2 g/l Phytagel, pH 5.6.
6. T2B medium (*22*): Half-strength MS salts, 20 mg/l thiamine-HCl, 20 mg/l glycine, 40 mg/l L-cystein (Sigma), 200 mg/l *myo*-inositol, 60 mg/l adenine sulphate (Sigma), 200 mg/l casein hydrolysate, 800 mg/l malt extract, 1 mg/l 2,4-D, 4 mg/l BAP, 30 g/l sucrose, 2 g/l Phytagel, pH 5.6.
7. T3B medium (*22*): Half-strength MS salts, 5 mg/l thiamine-HCl, 0.5 mg/l pyridoxine-HCl, 0.5 mg/l nicotinic acid, 10 mg/l L-cystein, 50 mg/l *myo*-inositol, 100 mg/l casein hydrolysate, 200 mg/l malt extract, 1 mg/l 2,4-D, 1 mg/l kinetin, 15 g/l sucrose, pH 5.6.
8. Surface sterilant : 8% HClO (w/v).
9. Cefotaxime (Claforan, Roussel): 200 mg/ml in water. Sterilize by filtration and store at –20°C.
10. Chlorsulphuron (Kalys, Roubaix, France): 200 µg/ml, use 3N KOH for correct dilution, sterilize by filtration and store at –20°C.
11. Acetosyringone (Fluka Chemica): 1 mM in ethanol, store at –20°C. Shake well before use.
12. Germination medium: semi-solid MS germination medium with half-strength salts, Morel vitamins (*39*: 1 mg/l thiamine-HCl, 1 mg/l pyridoxine-HCl, 1 mg/l nicotinic acid, 100 mg/l *myo*-inositol, 1 mg/l calcium pantothenate and 0.01 mg/l biotine), 0.225 mg/l BAP, 15 g/l sucrose and 3 g/l Phytagel, pH 5.6.
13. Rooting medium: identical to germination medium, without BAP.
14. Media sterilized by autoclaving for 20 min at 121°C; antibiotics and chlorsulphuron added to cooled sterile media.

3. METHODS

Cut leaves from mother trees in the glasshouse. Optimal stage is fully expanded leaves, but "immature" (not as glossy as older leaves). Sterilize for 20 min with 8% HClO (w/v) and rinse 3 times with sterile RO water. Culture small pieces (0.25 cm²) in 5.5 cm diam. Petri dishes (12.5 ml medium and 4 explants per dish).

3.1. Coffea canephora *genotypes*

1. Culture leaf explants on semi-solid medium, at low light (10 µE/m²/sec), 16 h photoperiod (*see* Section 2.3.1.). Sub-culture every 5 weeks until embryogenic callus appears on the cut edges of the explants.
2. Six weeks before transformation, sub-culture embryogenic calli (4 weeks after appearance) onto fresh medium at low cell density (1 g per dish) on a 50 µm nylon mesh (Ref. 87 mn PA 50/27, SAATI, Sailly, France; same container, same light conditions).

3.2. Coffea arabica *genotypes*

1. Leaf explants are cultured successively on the two media defined for the F1 hybrid genotypes: T1B, in 5.5 cm diam. Petri dishes (12.5 ml medium per plate) 4 weeks in the dark; T2B, in 10 cm diam. Petri dishes (25 ml medium per plate) 3-8 months in low-light conditions (10 µE/m²/sec, 16 h per day; *see* Section 2.3.2.). Sub-culture every 5 weeks on 10 cm diam. Petri dishes for 3-5 months until embryogenic callus appears on the cut edges of the explants.
2. Six weeks before transformation, sub-culture embryogenic calli (1 month after appearance) onto fresh medium at low cell density (1 g per dish) on a 50 µm nylon mesh under the same light conditions.

3.3. General Procedure

1. Three weeks before transformation, transfer coffee cells from *C. canephora* or *C. arabica* genotypes to T3B medium with appropriate density (1 g/50 ml of semi-solidified medium contained in a 10 cm diam. Petri dish with a 50 µm nylon mesh). Keep at low light (10 µE/m²/sec), for a 16 h photoperiod.
2. Prepare the *A. tumefaciens* strain containing the appropriate construct as a 20% glycerol solution from a culture of approx. 1.5 O.D.$_{600nm}$; deep-freeze (-80°C) (*see* Note 1).
3. Grow bacteria from a 20% glycerol solution in YEP medium with appropriate antibiotics: 50 mg/l kanamycin, 50 mg/l rifampicin and eventually 100 mg/l streptomycin (for LBA strains) or 20 mg/l gentamycin (C58 strains), and supplemented with acetosyringone at a final concentration of 100 µM. The amount of glycerol solution inoculated into YEP medium depends on bacterium strain and time of growing. The bacteria cultures need to be at 0.3-0.5 O.D.$_{600nm}$

for direct use. This O.D. is usually obtained after a 16 h or overnight culture (*see* Notes 2-5).

4. Centrifuge the bacteria (4000 rpm, 10 min), resuspend in a 0.9% NaCl solution supplemented with 200 µM acetosyringone.

5. Soak coffee cells in 0.9% NaCl solution together with bacteria for 2 h in 250 ml flasks, shaking constantly (100 rpm) (*see* Notes 6 and 9).

6. Gently eliminate the excess of bacterium culture and co-cultivate coffee cells with bacterium, in the dark, for 2 days on semi-solidified Yasuda medium in 10 cm diam. Petri dish containing 25 ml medium with 1.125 mg/l BAP and 30 g/l sucrose, on a 50 µm nylon mesh, with 200 µM acetosyringone.

7. After co-cultivation, rinse cells in liquid Yasuda medium supplemented with 1.125 mg/l BAP and 1 g/l cefotaxime in 500 ml Erlenmeyer flasks for 4-5 h, shaking constantly (100 rpm). Use 50 ml medium and 0.1 g coffee cells per flask.

8. Gently blot-dry coffee cells on sterile Whatman paper, and culture on semi-solidified Yasuda medium with 1.125 mg/l BAP and 30 g/l sucrose, supplemented with 400 mg/l cefotaxime, on a 10 cm diam. Petri dish with a 50 µm nylon mesh. Keep at low light (10 $\mu E/m^2/sec$), for a 16 h photoperiod.

9. After a period of 28 days, transfer cells to the same medium supplemented with selective agent (80-200 µg/l chlorsulphuron), under the same light conditions (*see* Notes 7, 10 and 11).

10. Subculture cells every 4 weeks by simply transferring the nylon mesh containing the cells from the old medium to a fresh medium. This subculture is necessary due to the decreasing efficiency of the herbicide under culture conditions. Incubate the cultures under low-light conditions (10 $\mu E/m^2/sec$), with a 16 h photoperiod (*see* Note 8).

11. Three months after co-culture, the medium is still supplemented with selective agent but cefotaxime is removed.

12. After 3-8 months on the selective medium, small calli or embryos appear (potentially transgenic). Subculture these calli onto fresh medium without mesh, but containing selective agent to aid regeneration. Incubate the cultures under direct light (45 $\mu E/m^2/sec$) at a 16 h photoperiod (*see* Note 12).

13. Cultivate young (torpedo to cotyledonary) embryos on MS germination medium with Morel vitamins, 0.225 mg/l BAP and 15 g/l sucrose under direct light.

14. After development, transfer the embryos to rooting medium (hormone-free germination medium).

15. Two to three months later, acclimatize small plantlets in the glasshouse. Dip basal end of the stem into a commercial rooting powder (containing IBA) and transfer to a sand-soil (75:25) mixture in small (50 x 30 x 30 cm) propagators (Puteaux S.A., Le Chesnay, France) under very high humidity. Start reducing humidity after 5 days by opening the vents progressively (over a 2 week period). Finally remove the covers after weaning.

16. After 1-2 months, transfer plantlets to small pots containing a standard horticultural soil mixture and maintain in under glasshouse conditions.

17. GUS histochemical assay: Incubate calli, shoots, leaves or roots overnight at 37°C in the classical medium defined by Jefferson (*40*), modified by the use of

phosphate buffer (0.2 M at pH 7.0). Add methanol (20% v/v) to eliminate any eventual endogenous or parasitic expression of the non-integrated GUS gene.

18. PCR analysis: Extract total DNA from plants that present a positive reaction to the GUS histochemical test. Choose several pairs of primers according to constructs used, in order to analyze the quality of DNA and the structure of the integrated T-DNA. Perform amplifications using 25 ng of genomic DNA in a final volume of 50 μl containing 1 X PCR buffer, dNTP, primers and *Taq* DNA polymerase as recommended by the manufacturer (Stratagene). Run 40 cycles of amplification (94°C, 1 min; Thyb, 1 min; 72°C, 2 min), where Thyb is a specific temperature for each pair of primers. Use a pre-denaturation step of 3 min and a final elongation step of 7 min. Visualize the amplification products after electrophoresis in 1 % agarose gels by staining with ethidium bromide and exposing to UV light.

19. Southern blotting analysis: Extract DNA from 5 g of fresh leaves. Digest total DNA (5 μg) with adapted enzymes according to constructs used. Separate DNA on a 0.7% agarose gel in TRIS-borate-EDTA buffer, then transfer to a Hybond N+ membrane (Appligene) and hybridize with α^{32}P-labelled DNA probes.

20. Western blot hybridization and bioassays: Extract proteins from approx. 1 g of fresh leaves, and perform the detection of the insecticidal protein by Western blotting using a rabbit polyclonal antiserum raised against the previously purified Cry1Ac protein (dilution of 1/1200 v/v). Use a secondary goat anti-rabbit antiserum alkaline phosphatase conjugate (Sigma) for final detection, at a dilution of 1/1000 (v/v).

21. Perform bioassays (*see* Note 13) (*41*).

22. Transfer a sample of plants to the field for agronomic and pest resistance evaluation. A 5 year experimental trial is necessary to obtain reliable data and confirm the expression of the introduced gene (*42, 43*).

4. NOTES

1. Several *A. tumefaciens* strains can be used for coffee transformation. Based on previous results on other plants, LBA4404 was first used and transformed plantlets were obtained. Later, two C58 strains (GV2260 and PMP90) have also been used with similar results. Recently, the EHA105 strain has also been used, producing better results on the transformation of coffee cells.

2. Growth of bacteria is slightly slower in LB medium than in YEP medium.

3. Some of the first experiments were conducted with the armed *A. rhizogenes* strain A4 using the same protocols as for *A. tumefaciens* strains. A few plants have been regenerated from explants inoculated with *A. rhizogenes*, but never developed in the field.

4. Several binary plasmids have been used for coffee transformation: pBin19 was the first used, and allowed regeneration of transformed plantlets for all genotypes investigated. Nevertheless, pCAMBIAA1, recently used with new constructs, appears to be more efficient for cell transformation and improving the rate of transformation.

5. A synthetic *cry1Ac* gene, modified for its G+C content for suitable expression in plants (*44*), was used in this study. However, in previous experiments, we used a native *cry1Ac* gene with limited success.

6. As described (*20*), somatic embryos could also be transformed by *Agrobacterium*. Such explants need to be wounded with a scalpel before soaking in the bacterium suspension. Subsequently, the embryos are cultivated like cells, and transformed calli or embryos appear on these primary

explants. It is difficult to obtain a large number of embryos at the correct stage (late torpedo), and the rate of transformation always remain low (under 10%).

7. Chlorsulfuron has been the main herbicide used for the selection of transformed events. Bialaphos has also been used successfully in coffee transformation where the *bar* gene has been used as a selectable marker. However, the *bar* gene used in our constructs was not a synthetic gene, and worked poorly in our experiments. Hygromycin has also been used for coffee transformation (*18*), and kanamycin should be used at high concentrations (more than 400 mg/l).

8. Instead of using a semi-solid medium for cell culture after co-culture, a liquid medium in flasks could also be used. Some transformation events have been obtained with this method, but it remains difficult to control the growth of transformed calli, and to avoid mixtures of different transformation events.

9. Soaking of cells with bacteria can be achieved in liquid MS medium without hormones or directly in the bacterial medium.

10. In the case of herbicide selection, coffee cells can be cultured on Yasuda (MS/4) medium. However, for some *C. arabica* genotypes, like fixed lines, it could be better to grow cells on a slightly richer medium such as MS/2.

11. The working dose for selection on chlorsulfuron is very variable, depending on genotype used. For *C. canephora* and Catimor genotypes, the classical dose of 80 µg/l is ideal. For some *C. arabica* genotypes, like hybrid varieties, a higher dose of up to 200 µg/l is necessary.

12. Transformation efficiency is very variable from one genotype to another. For *C. canephora* genotypes, the efficiency can be 50% of transformed calli growing from transformed cells. For *C. arabica* genotypes, the efficiency is lower at approx. 30%. The second critical step in the production of transformed plants is the regeneration of transformed embryos from callus. For this to occur, cefotaxime needs to be removed as soon as possible, and transformed calli need to be cultured on the selective medium at a very low density.

13. We have performed bioassays using a leaf miner species from Tanzania (*Leucoptera caffeina*). Plants to be tested were transferred in cages for 24 h to allow adult insects to lay their eggs on the leaves. After 15 days, an overall score was attributed to the plants, according the following scale: 0, death of larvae after hatching, no galleries observed; 1, galleries shorter than 3 cm, death of the larvae usually within 7 days; 2, galleries larger than 3 cm before the death of the larvae; 3, living larvae on the leaves, possible development of pupae; 4, complete development of the larvae with formation of pupae (*41*).

CONCLUSIONS

We have developed our own methodology for the genetic transformation of coffee using somatic embryos, immature embryogenic cultures or suspensions. We were able to apply this methodology to 20 different genotypes among the two cultivated coffee species *C. arabica* and *C. canephora*. Although we used specific constructs designed for providing insect resistance, our technique can be for the transfer of other agronomic traits. Field study is an important step in genetic engineering studies and our preliminary results show consistency with laboratory bioassays as well as good agronomic behaviour.

ACKNOWLDGEMENTS

We wish to acknowledge the efficient technical assistance of C. Carasco-Lacombe, I. Jourdan and C. Fenouillet, as well as help from B. Perthuis and J.-L. Pradon in monitoring the coffee plants in the field.

REFERENCES

1.	Bertrand B, Aguilar G, Santacreo R, Anthony F, Etienne H, Eskes AB and Charrier A (1997). Comportement d'hybrides F1 de *Coffea arabica* pour la vigueur, la production et la fertilité en Amérique centrale. In: ASIC (eds.), *17th International colloquium on coffee science*, Nairobi (Kenya), 20-25 July 1997, ASIC, Paris (France). pp. 415-423.

2.	Leroy T, Montagnon C, Cilas C, Yapo A, Charmetant P and Eskes AB (1997). Reciprocal recurrent selection applied to *Coffea canephora* Pierre. III. Genetic gains and results of first cycle intergroup crosses. *Euphytica*, **95**: 347-354.

3.	Capot J (1977). L'amélioration du caféier Robusta en Côte d'Ivoire. *Café Cacao Thé*, **21**: 233-244.

4.	Spiral J, Leroy T, Paillard M and Petiard V (1999). Transgenic Coffee (*Coffea* species). In: Bajaj YPS (ed.), *Biotechnology in Agriculture and Forestry, Vol. 44 Transgenic Trees.* Springer-Verlag Berlin Heidelberg (Germany) pp. 55-76.

5.	Etienne H, Anthony F, Dussert S, Fernandez D, Lashermes P and Bertrand B (2002). Biotechnological applications for the improvement of coffee (*Coffea arabica* L.). *In Vitro Cellular and Developmental Biology-Plant*, **38**: 129-138.

6.	Guerreiro Filho O, Penna Medina FH, Gonçalves W and Carvalho A (1990). Melhoramento do cafeeiro: XLIII. Selecao do cafeeiros resistentes ao bicho-mineiro. *Bragantia*, **49**: 291-304.

7.	Estruch JJ, Carozzi NB, Desai N, Duck NB, Warren GW and Koziel MG (1997). Transgenic plants: an emerging approach to pest control. *Nature Biotechnology*, **15**: 137-141.

8.	Schuler TH, Poppy GM, Kerry BR and Denholm I (1998). Insect-resistant transgenic plants. *Trends in Biotechnology*, **16**: 168-175.

9.	Guerreiro O, Denolf P, Peferoen M, Decazy B, Eskes AB and Frutos R (1998). Susceptibility of the coffee leaf miner (*Perileucoptera* spp.) to *Bacillus thuringiensis* δ-endotoxins: a model for transgenic perennial crops resistant to endocarpic pests. *Current Microbiology*, **36**: 175-179.

10.	Dandekar AM, McGranahan GH, Vail PV, Uratsu SL, Leslie CA and Tebbets JS (1998). High levels of expression of full-length *cryIA(c)* gene from *Bacillus thuringiensis* in transgenic somatic walnut embryos. *Plant Science*, **131**: 181-193.

11.	Surekha K, Royer M, Naidu R, Vassal JM, Philippe R, Jourdan I, Fenouillet C, Leroy T and Dufour M (2002). Bioassay of *Bacillus thuringiensis* toxins against two major coffee pests, i.e. coffee berry borer (*Hypothenemus hampei*) and coffee white stem borer (*Xylotrechus quadripes*). *SIP, Annual Meeting of the Society for Invertebrate Pathology*. 35, 2002/08/18-23, Foz de Iguassu, Brésil, Program and Abstracts, p. 85.

12.	Barton CR, Adams TL and Zarowitz MA (1991). Stable transformation of foreign DNA into *Coffea arabica* plants. In: ASIC (eds.), *14th International Colloquium on Coffee Science*, San Francisco (USA), 14-19 July 1991, ASIC, Paris (France). pp. 853-859.

13.	Feng Q, Yang MZ, Zheng XQ, Zhen XS, Pan NS and Chen ZL (1992). *Agrobacterium*-mediated transformation of coffee (*Coffea arabica* L.). *Chinese Journal of Biotechnology*, **8**: 255-260.

14.	Freire AV, Lightfoot DA and Preece JE (1994). Genetic transformation of coffee *(Coffea arabica* L.) by *Agrobacterium* spp. *HortScience*, **29**: 454.

15.	Spiral J and Pétiard V (1993). Développement d'une méthode de transformation appliquée à différentes espèces de caféiers et régénération de plantules transgéniques. In: ASIC (eds.), *15th International Colloquium on Coffee Science,* Montpellier, France, 6-11th June 1993, ASIC, Paris (France). pp. 115-122.

16.	Spiral J, Thierry C, Paillard M and Pétiard V (1993). Obtention de plantules de *Coffea canephora* Pierre (Robusta) transformées par *Agrobacterium rhizogenes*. *Compte Rendu le l'Academie des Sciences de Paris*, **316**: 1-6.

17.	Sugiyama M, Matsuoka C and Takagi T (1995). Transformation of coffee with *Agrobacterium rhizogenes*. In: ASIC (eds.), *16th International Colloquium on Coffee Science,* Kyoto, (Japan), 9-14th April 1995, ASIC, Paris (France), pp. 853-859.

18.	Hatanaka T, Choi YE, Kusano T and Sano H (1999). Transgenic plants of coffee *Coffea canephora* from embryogenic callus via *Agrobacterium tumefaciens*-mediated transformation. *Plant Cell Reports*, **19**: 106-110.

19.	Leroy T, Paillard M, Royer M, Spiral J, Berthouly M, Tessereau S, Legavre T and Altoosaar I (1998). Introduction de gènes d'intérêt agronomique dans l'espèce *Coffea canephora* Pierre par transformation avec *Agrobacterium* sp. In: ASIC (eds.), *17th International Colloqium on Coffee Science*, Nairobi (Kenya), 20-25 July 1997, ASIC, Paris (France). pp. 439-445.

20. Leroy T, Henry AM, Royer M, Altosaar I, Frutos R, Duris D and Philippe R (2000). Genetically modified coffee plants expressing the *Bacillus thuringiensis cry1Ac* gene for resistance to leaf miner. *Plant Cell Reports*, **19**: 382-389.

21. Yasuda T, Fujii Y and Yamaguchi T (1985). Embryogenic callus induction from *Coffea arabica* leaf explants by benzyladenine. *Plant Cell Physiology*, **26**: 595-597.

22. Berthouly M and Michaux-Ferrière N (1996). High frequency somatic embryogenesis in *Coffea canephora*. *Plant Cell, Tissue and Organ Culture*, **44**: 169-176.

23. Noir S, Anthony F, Bertrand B, Combes MC and Lashermes P (2003). Identification of a major gene (Mex-1) from *Coffea canephora* conferring resistance to *Meloidogyne exigua* in *Coffea arabica*. *Plant Pathology*, **52**: 97-103.

24. Marraccini P, Pereira LPP, Ferreira LP, Vieira LGE, Cavalari AA, Geromel C and Mazzafera P (2003). Biochemical and molecular characterization of enzyme controlling sugar metabolism during coffee bean development. *ISPMB Conference*, Barcelona (Spain), 23-28 June 2003, poster#S19-14.

25. Marraccini P, Deshayes A, Pétiard V and Rogers WJ (1999). Molecular cloning of the complete 11S seed storage protein gene of *Coffea arabica* and promoter analysis in transgenic tobacco plants. *Plant Physiology and Biochemistry*, **37**: 273-282.

26. Paillard M, Lashermes P and Pétiard V (1996). Construction of a molecular linkage map in coffee. *Theoretical and Applied Genetics*, **93**: 41-47.

27. Lashermes P, Combes MC, Prakash NS, Trouslot P, Lorieux M and Charrier A (2001). A genetic linkage map of *Coffea canephora*: effect of segregation distorsion and analysis of recombination rate in male and female meiosis. *Genome*, **44**: 589-595.

28. Leroy T, Dufour M, Montagnon C, Lashermes P, Marraccini P, Sabau X, Glaszmann JC and Piffanelli P (2003). Characterisation of the first *Coffea canephora* coffee-tree BAC library. *ISPMB Conference*, Barcelona (Spain), 23-28 June 2003, poster#S04-17.

29. Bevan M (1984). Binary *Agrobacterium* vectors for plant transformation. *Nucleic Acids Research*, **12**: 8711-8721.

30. Vancanneyt G, Schmidt R, O'Connor-Sanchez A, Willmitzer L and Rocha-Sosa M (1990). Construction of an intron-containing marker gene. Splicing of the intron in transgenic plants and its issue in monitoring early events in *Agrobacterium*-mediated plant transformation. *Molecular and General Genetics*, **220**: 245-250.

31. Brasileiro ACM, Tourneur C, Leplé JC, Combes V and Jouanin L (1992). Expression of the mutant *Arabidopsis thaliana* acetolactate synthase gene confers chlorsulfuron resistance to transgenic poplar plants. *Transgenic Research*, **1**: 133-141.

32. Kay R, Chan A, Daly M and McPherson J (1987). Duplication of CaMV 35S promoter sequences creates a strong enhancer for plant genes. *Nature*, **236**: 1299-1302.

33. Sardana R, Dukiandjiev S, Giband M, Cheng XY, Cowan K, Sauder C and Altosaar I (1996). Construction and rapid testing of synthetic and modified toxin gene sequences CryIA (*b&c*) by expression in maize endosperm culture. *Plant Cell Reports*, **15**: 677-681.

34. Van Boxtel J, Berthouly M, Carasco C and Eskes AB (1995). Transient expression of β glucuronidase following biolistic delivery of foreign DNA into coffee tissues. *Plant Cell Reports*, **14**: 748-752.

35. Curie C, Liboz T, Bardet C, Gander E, Médale C, Axelos M and Lescure B (1991). *Cis* and *trans*-acting elements involved in the activation of *Arabidopsis thaliana A1* gene encoding the translation elongation factor EF-1α. *Nucleic Acids Research*, **19**: 1305-1310.

36. Gallie DR and Kado CI (1989). A translational enhancer derived from tobacco mosaic virus is functionally equivalent to a Shine-Dalgarno sequence. *Proceedings of the National Academy of Sciences USA*, **86**: 129-132.

37. Murashige T and Skoog F (1962). A revised medium for rapid growth and bioassays with tobacco tissue cultures. *Physiologia Plantarum*, **15**: 473-497.

38. Gamborg OL, Miller RA and Ojima K (1968). Nutrient requirements of suspension cultures of soybean root cells. *Experimental Cell Research*, **50**: 151- 158.

39. Morel G and Wetmore R (1951). Tissue culture of monocotyledons. *American Journal of Botany*, **38**: 138-140.

40. Jefferson R (1987). Assaying chimeric genes in plants: the GUS gene fusion system. *Plant Molecular Biology Reporter*, **5**: 387-405.
41. Dufour M, Philippe R, Fenouillet C, Carasco-Lacombe C, Gruchy D, Jourdan I and Leroy T (2002). Analysis of genetically transformed coffee plants (*Coffea canephora* Pierre) for resistance to coffee leaf miner: bioassays, molecular and immunological analyses. [CD-ROM]. In: ASIC (eds.), *19th International Colloquium on Coffee Science,* Trieste (Italy), 14-18th May 2001. 1 disque optique numérique (CD-ROM). ASIC, Paris (France).
42. Leroy T, Philippe R, Royer M, Frutos R, Duris D, Dufour M, Jourdan I, Lacombe C and Fenouillet C (2000). Genetically modified coffee-trees for resistance to leaf miner. Analysis of gene expression, resistance to insects and agronomic value. In: ASIC (eds.), *18th International Colloquium on Coffee Science,* Helsinki (Finland), 2-6 August 1999, ASIC, Paris (France). pp. 332-338.
43. Perthuis B, Philippe R, Pradon J-L, Dufour M and Leroy T (2002). Premières observations sur la résistance au champ de plantes de *Coffea canephora* génétiquement modifiées contre la mineuse des feuilles *Perileucoptera coffeella* Guérin-Méneville. [CD-ROM]. In: ASIC (eds.), *19th International Colloquium on Coffee Science,* Trieste (Italy), 14-18th May 2001. 1 disque optique numérique (CD-ROM). ASIC, Paris (France).
44. Adang MJ, Straver MJ, Rocheleau TA, Leighton J, Barker RF and Thomson V (1985). Characterised full-length and truncated plasmid clones of the crystal protein of *Bacillus thuringiensis* subsp. Kurstaki HD-73 and their toxicity in *Manduca sexta*. *Gene*, **36**: 289-300.

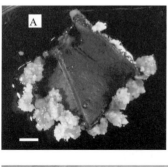

Figure 1. Genetic transformation of Coffea *spp. A. Embryogenic callus growing at the edge of the foliar explants is used for co-cultivation with* A. tumefaciens *strains (bar = 0.2 cm). B. Young transformed plantlets of* C. arabica *on rooting medium 3 months after selection of embryos growing from the co-cultivated callus on selective medium (bar = 1 cm). C. Transformed plants from Catimor genotype compared with a non transformed in* vitro *control (on the right), 6 months after transfer to the glasshouse (bar = 5cm). D. Twelve month- old transformed plants of* C. canephora *during bioassays. Both plants were transformed for resistance to coffee leaf miner with the* cry1Ac *gene. Plants were exposed to the insects at the same time. Plant 1 (upper) shows leaves susceptible to the pest with many galleries. Plant 2 is resistant to the pest (bar = 5 cm).*

Chapter 13

GENETIC TRANSFORMATION OF TEA

A. BHATTACHARYA, T.K. MONDAL, I. SANDAL, O. PRAKASH, S. KUMAR, AND P.S. AHUJA

Division of Biotechnology, Institute of Himalayan Bioresource Technology, Palampur-176061, H. P., India. Corresponding author: P. S. Ahuja. E-mail: dihbt@csir.res.in

1. INTRODUCTION

1.1. Importance and Distribution of Tea

Tea is an economically important plantation crop of many countries of the globe in terms of foreign exchange earnings, employment and revenue generation (*1, 2*). Since its leaves are rich in polysaccharides, essential oils, vitamins, minerals, purines, flavonoids and alkaloids like caffeine and polyphenols like catechins, it is not surprising that tea is also a popular caffeine containing beverage with anticancer, anti-ageing and medicinal properties (*3*). Tea leaves are processed in 3 different manners and each yield a distinct beverage that have antibacterial and free radical capturing (anti-oxidizing) activities (*4*). While the flavonoids have antioxidant, anti-inflammatory, anti-allergic, antibacterial and antiviral effects with an ability to strengthen veins and decrease permeability, the tannins or catechins are detoxifying agents. Tea is also considered to be a potent therapeutic plant drink with antiseptic and antioxidant properties (*4, 5*). Moreover, the high contents of vitamin C in the Black, Green or Oolong tea makes it equivalent to 3 glasses of orange juice or 2 capsules (200 mg) of vitamin C when consumed as 1 or 2 cups/day.

Tea is one of the 200 species that belong to the genus *Camellia* and have originated from the regions that lie between the longitudes 85° and $150^{\circ}E$ and latitudes $37^{\circ}N$ and $10^{\circ}S$. Although tea originated in South and South-eastern regions of China (*6*), it spread to most parts of the globe through monks, preachers and traders. Today, tea is cultivated in more than 20 countries of Asia, Africa and South America and amongst them India continues to maintain its leadership in the international market supporting about 2% of the Indian populace. The tea industry with its associated and ancillary activities in distribution, retailing, warehousing, tea shops etc. has a turnover of US$ 1.5 billion and supports more than 50,000 small tea growers, 1.1 million workers and more than 4 million dependants of these workers (*7*). While India, Sri Lanka, Indonesia, China and Japan are the main black tea producing countries, China, Japan and parts of former USSR are the primary producers of green tea (*8*).

I.S. Curtis (ed.), Transgenic Crops of the World-Essential Protocols, 171-185.
© 2004 Kluwer Academic Publishers. Printed in the Netherlands.

1.2. Improvement of the Tea Crop

Despite the fact that tea is gaining immense popularity as a 'health drink' all throughout the world, total production is not sufficient to meet the demands of the world market (*9*). While the yield of tea is further reduced by several biotic and abiotic stresses, the scope for extending tea cultivation at the expense of important food crops is also greatly limited even in countries where tea is significantly important.   Unlike other crops where higher yield per unit area is of primary importance, tea not only demands higher yield but also better adaptability and cup characters (*10*). Crop improvement combining both yield and quality is of utmost importance because the world market has critical standards for tea from different parts of the globe and products generally need to conform to these in order to attain significant commercial value (*11*).

Generally, biclonal or polyclonal seeds are produced through natural hybridization between parents with previous performances of yield, quality or disease resistance (*12*). However, their reproducibility is very low because dependence on natural crossing is sure to bring about genetic recombination and further variability. Moreover, the pedigree of the selected elites is largely unknown (*13*). Apart from being time consuming and labour intensive, the other important constraints include, (a) perennial nature of the plant (b) long gestation period (c) high inbreeding depression (d) self-incompatibility (e) unavailability of distinct mutants of different biotic and abiotic stresses (f) lack of distinct selection criteria (g) low success rate of hand pollination (h) short flowering time of only 2-3 months (i) long duration of 12-18 months from anthesis to seed maturation and (j) the differences in the time of flowering and fruit bearing capability.

Although genetic manipulation can overcome most of these constraints, however, it has failed to yield any success in tea. The different methods of genetic manipulation that have been attempted include production of cold tolerant somaclonal variants (*14*), aneuploids (2n = 18-22) from anther cultures (*15-17*) and protoplast culture (*18, 19*). Although a large number of transgenics have been produced in a number of genera, production of transgenic tea plants remains a difficult problem mainly due to low competence for transformation as well as regeneration (*13*). The presence of high levels of polyphenols with germicidal property has been considered to be the main obstacle to plant transformation (*20*). Nevertheless, *Agrobacterium tumefaciens*-mediated genetic transformation of tea has been attempted by several workers using explants such as leaves and somatic embryos (*21-24*). Leaf explants have shown to be difficult to transform and have been reported to yield only stably transformed callus which failed to differentiate further (*20-22*). Hairy root formation through the use of *A. rhizogenes* was also attempted in tea (*25*). Leaves from 35 days-old *in vitro*-grown plants were infected with *A. rhizogenes* strain A4 at a cell density of 10^8/ml for 2 min followed by blotting on sterile filter papers and co-cultivation in dark for 2 days. Hairy roots were induced 35 days after inoculation and such isolated roots were successfully grown in liquid culture medium. Mannopine was extracted from these roots when analyzed by paper electrophoresis and thus confirming the stable integration of this gene. Konwar *et al.* (*26*) transformed 4-6 months old *in vitro*-grown tea shoots by infecting the basal portion with *A. rhizogenes*, followed by co-cultivation in liquid basal Murashige and Skoog (MS, *27*) medium supplemented with IBA (5 mg/l) and rifampicin (100 mg/l). Root initiation from the basal portion in 66% explants after 32-45 days enabled convenient hardening of the shoots in pots or nursery beds.

Circumvention of the developmental blocks in tea somatic embryogenesis and morphogenesis led to the successful production of transgenic tea plants for the first time in our laboratory. The procedures for production of transgenic tea are discussed in this chapter.

2. MATERIALS

2.1. Agrobacterium-*mediated Transformation of Somatic Embryos*

2.1.1. *Plant Materials*

Seeds of mature green fruits collected from tagged plants of Chinary type tea (*Camellia sinensis* L. O. Kuntze cv. Kangra jat) were used to obtain de-embryonated cotyledon explants (*see* Note 3).

2.1.2. A. tumefaciens *Strain*

A. tumefaciens strain EHA105, harbouring a binary vector (*28*) was used throughout this study. The p35SGUS-INT vector is a pBin19 derivative, containing the *gus*A gene, truncated by the potato LS-1 intron, under the control of CaMV35S promoter, and the *npt*II gene conferring kanamycin resistance under the control of the NOS promoter.

2.1.3. *Media for* A. tumefaciens

1. Liquid Yeast Mannitol Broth (YMB) medium (*29*): Di-potassium phosphate (0.5 g/l), sodium chloride (0.5 g/l), 0.8 ml of 1 M solution of magnesium sulphate, mannitol (10 g/l) and yeast extract (1 g/l). Dispense into 20 ml aliquots of the medium prepared in distilled water into 100 ml flasks. Adjust pH to 7.5 and sterilize by autoclaving at 121°C for 20 min. Add sterile kanamycin and ampicillin (*see* Section 2.1.4.) to a final concentration of 50 mg/l.
2. Agar culture medium: Add 1.5% agar (Difco agar, Qualigens, India) to YMB medium and autoclave at 121°C for 20 min. After cooling to 45°C, add sterile kanamycin and ampicillin (*see* Section 2.1.4.) to a final concentration of 50 mg/l.

2.1.4. *Antibiotics*

1. Kanamycin sulphate (Sigma): 100 mg/ml stock in water. Filter-sterilized through a 0.22 μm membrane (Stericup[TM,] Millipore Express[TM], Bangalore, India) and stored at –20°C for 2-3 months (*see* Note 2).

2. Ampicillin (Hi Media, India): 50 mg/ml stock solution dissolved in purified water. Filter-sterilize as above and store as 1 ml aliquots at –20°C for 2-3 months.
3. Carbenicillin (Hi Media) and cephalexin (Sporidex; Ranbaxy, India): 1 g/10 ml aqueous stock solution of each is prepared. Sterilize by filtration through a 0.22 μm membrane and store at –20°C for 2-3 months. Antibiotics are added to the autoclaved medium after it is cooled to about 45°C.

2.1.5. Growth Regulators

1. BAP (Hi Media): 56.3 mg/100 ml stock solution. Dissolve in a few drops of 1N NaOH, then make to final volume with water. Store at 4°C for 2-3 months.
2. IBA (Hi Media): 101.6 mg/100 ml stock solution dissolved in water. IBA is heat labile and should be filter-sterilized and added to autoclaved media. Store at 4°C for 2-3 months.
3. NAA (Hi Media): 93.1 mg/100 ml stock solution. Dissolve in a few drops of alcohol and then make to final volume with water. Store at 4°C for 2-3 months.
4. Gibberellic acid (GA$_3$, Hi Media): 0.56 g/100ml stock solution by dissolving in 70% alcohol and store at 4 °C for 2-3 months.

2.1.6. Tissue Culture

Tissue culture media are based on MS medium. Adjust the pH of the media to 5.6-5.7 and sterilize by autoclaving at 121°C and 15 psi.

1. Inoculation or Seed germination medium (SGM): Half strength basal MS salts (Difco, Qualigens, India), 3% sucrose and 0.7% agar dissolved in 1 litre medium. Dispense 20 ml medium into 80 ml test tubes after sterilization.
2. Somatic embryo induction medium (SIM): Half strength basal MS salts, 20 g/l sucrose, 2.5 mg/l NAA, 0.2 mg/l BAP and 0.7% agar. Dispense 100 ml medium into 250 ml flasks after sterilization.
3. Co-cultivation or Reccurrent somatic embryogenesis medium (REM): Full strength MS basal salts modified with half strength nitrates of potassium (950 mg/l) and ammonium (825 mg/l), potassium sulphate (300 mg/l), BAP (2.0 mg/l), IBA (0.2 mg/l), L-glutamine (1.0 g/l), 30 g/l sucrose, 50 mg/l kanamycin and 0.7% agar. Adjust pH to 5.6 and dispense 20 ml medium into 90 mm Petri dishes.
4. Selection or Somatic embryo germination medium (SGMa): Full strength basal MS medium, maltose (40 g/l), *trans*-cinnamic acid (3.0 g/l), 0.7% agar. Sterilize by autoclaving after pH adjustment to 5.7 and cool the medium to about 45°C. Add filter-sterilized 500 mg/l carbenicillin, 500 mg/l sporidex and 200 mg/l kanamycin and dispense 20 ml into 90 mm Petri dishes.
5. Selection or Somatic embryo germination medium (SGMb): Full strength basal MS medium, sucrose (30 g/l), GA$_3$ (1.5 g/l) and 0.7% agar. Sterilize by autoclaving after pH adjustment to 5.7 and cool the medium to about 45°C. Add filter-sterilized 500 mg/l carbenicillin, 500 mg/l sporidex and 200 mg/l kanamycin, and dispense 20 ml into 90 mm Petri dishes.

2.2. Transformation of Somatic Embryos by Particle Gun Bombardment

2.2.1. Plant Materials

In vitro raised shoot cultures of cv. Kangra jat were used as the source of leaf explants (*see* Note 4).

2.2.2. Plasmid

Plasmid RT99GUS is a 6.71 kb pUC 18 derivative containing the *neo* gene coding for NPTII and *gus*A gene coding for β-glucuronidase (GUS) both of which are driven by the CaMV35S promoter and flanked by the NOS terminator. The plasmid is introduced into *E. coli* cells through electroporation.

2.2.3. Media for Transformed E. coli

1. Liquid culture medium: 10 g/l Luria Broth (LB) (Hi Media, India) dissolved in purified distilled water. Dispense into 20 ml aliquots in 100 ml flasks, sterilize by autoclaving at 121°C for 20 min and store at room temperature.
2. Agar culture medium: 40 g/l LB Agar (Hi Media). Suspend 40 g of the powder in 1 litre of purified distilled water. Autoclave at 121°C for 20 min. After cooling to 45°C, add sterile kanamycin and ampicillin (*see* Section 2.2.4.) to a final concentration of 50 mg/l.

2.2.4. Antibiotics (see Section 2.1.4.).

2.2.5. Growth Regulators

1. 2, 4-D (Hi Media): 110.5 mg /100 ml stock solution. First dissolve in 1N NaOH, then make to final volume with water. Store at 4°C.
2. TDZ (Hi Media): 40.1 mg/40 ml stock solution. First dissolve in 1N NaOH, then make to final volume with water and store at 4°C.

2.2.6. Tissue Culture

All media used for particle gun bombardment are based on MS medium (*29*).

1. Culture initiation medium (CIM): Full strength MS medium, 30 g/l sucrose, 0.7% agar, 2.0 mg/l BAP, 0.2 mg/l IBA. Adjust to pH 5.6 and autoclave at 121°C for 20 min.
2. Leaf callus induction medium (LCM): Full strength basal MS salts, 30 g/l sucrose, 5.0 mg/l 2,4-D, 0.7% agar and 200 μg/ml kanamycin sulphate. Adjust pH to 5.7, sterilize and dispense 20 ml medium into 90 mm Petri dishes.

3. Adventitious bud regeneration medium (SRM): Full strength basal MS salts, 30 g/l sucrose, 1000 µg/ml kanamycin sulphate, 0.7% agar, 2.0 mg/l BAP and 0.2 mg/l IBA for rhizogenesis and removal of IBA from the above for caulogenesis. Dispense 20 ml medium into 90 mm Petri dishes.

4. Selection or Shoot multiplication medium (SMM): Full strength basal MS salts, 30 g/l sucrose, 5 µM TDZ, 1000 mg/l kanamycin in 1 litre sterile distilled water. Dispense 20 ml liquid medium into 250 ml flask after sterilization.

All media are sterilized by autoclaving at 121°C and 15 psi for 20 min. Antibiotics and IBA are added to the medium after autoclaving.

2.2.7. Particle Gun Bombardment

1. BioRad accessories for the PDS-1000/He particle gun are as follows: 1100 psi rupture disks, macrocarriers, stopping screens, 1.0 µm gold microcarriers.

2. $CaCl_2$ 2.5 M stock solution dissolved in water, filter-sterilize through a 0.22 µm Minisart (Sartorius) membrane, store at –20°C as 1 ml aliquots.

3. Spermidine (Sigma) 0.1 mM stock solution dissolved in water and filter-sterilize through a 0.22 µm Minisart (Sartorius) membrane, store at –20°C as 1 ml aliquots.

4. Hormona-free basal MS medium or 0.25 M sorbitol for osmotic pre-treatment after sterilization by autoclaving.

5. Absolute ethanol.

6. Sterile 3MM Whatman filter paper.

7. Re-suspension buffer: 50 mM Tris-HCl pH 8.0, 10 mM EDTA, 100 µg/ml RNAse A. Store at 4°C.

8. Lysis buffer: 200 mM NaOH, 1 % SDS. Store at room temperature.

9. Neutralization buffer: 3.0 M potassium acetate pH 5.5. Store at room temperature or 4°C.

10. Equilibration buffer: 750 mM NaCl, 50 mM MOPS, pH 7.0, 15% ethanol, 0.15% Triton X -100. Store at room temperature.

11. Wash buffer: 1.0 mM NaCl, 50 mM MOPS, pH 7.0, 15% ethanol. Store at room temperature.

12. Elution buffer: 1.25 mM NaCl, 50 mM Tris-HCl, pH 8.5, 15% ethanol. Store at room temperature.

13. GUS solution: 100 mM sodium phosphate pH 7.0, 10 mM EDTA pH 8.0, 1 mM potassium ferrocyanide, 1 mM potassium ferricyanide, 0.1% TritonX-100 and 1 mM of X-Gluc (Sigma).

3. METHODS

3.1. Agrobacterium-*mediated Transformation of Somatic Embryos*

3.1.1. Sterilization of Cotyledon Explants and Induction of Somatic Embryogenesis

Disinfect approx. 50 naked embryos comprising of cotyledons and embryonic axes in Tween 80 (Merck) for 2 min followed by 4% (w/v) calcium hypochlorite solution (100 ml) for 7-8 min under the laminar hood. After 10 min, discard the hypochlorite solution and wash the embryos 5-6 times with 100 ml sterile distilled water and culture on 100 ml SGM medium in 250 ml flask for 15 days ($26 \pm 2°C$, 52 $\mu mol/m^2/sec$, 12 h photoperiod). The embryonic axes from these embryos are removed when the cotyledons turn green and the de-embryonated cotyledons are cut into 0.5 cm slices and then cultured on SIM medium in 250 ml flasks for induction of somatic embryos on the cut surfaces. About 10-15 slices are cultured per flask. After 6-8 weeks, the low frequency of primary embryos that are formed are cultured on REM medium for secondary embryogenesis (Fig. 1B) and increasing the number of somatic embryo production.

3.1.2. Preparation of Agrobacteria Inoculum

1. Grow a colony of *A. tumefaciens* strain EHA105 on YMB agar semi-solidified medium (with 50 $\mu g/ml$ kanamycin) at 28°C in dark for 48 h. Take 1 loopful of bacteria and transfer to 20 ml aliquots of YMB liquid culture medium (with antibiotics) in 100 ml narrow mouth flasks and grow overnight at 28°C on an orbital shaker at 150-200 rpm. An overnight bacterial culture corresponding to an $OD_{600} = 0.6$ is used.
2. Centrifuge the bacterial culture at 6000 rpm for 10 min in 50 ml sterile tubes with screw caps and discard the supernatant. Resuspend and dilute the bacterial pellet with 20 ml YMB liquid culture medium.
3. Measure the optical density (absorbance) of 1 ml aliquot of the above bacterial culture in a spectrophotometer (Hitachi, Japan) at 600 nm and calculate the bacterial concentration according to the formula: x X OD X 3 X 1 X 10^9 cells/ml = y X 1 X 10^9 cells/ml (where x = the amount bacterial culture taken from the 10 ml stock culture prepared as above and y = the total amount of culture required for transformation in order to obtain an optimal cell density of 1 X 10^9 cells/ml where 1.0 OD $_{600\ nm}$ = 3 X 10^9 cells is the standard conversion number).
4. Take the calculated amount of bacterial culture (such that the bacterial cell density is about 10^9 cells/ml) in a 30 mm Petri dish in order to make up a volume of 20 ml bacterial culture.

3.1.3. Step by Step Methodology for Agrobacterium-*mediated Transformation*

1. Immerse the somatic embryos (100 per dish) into the *A. tumefaciens* culture prepared as above and vacuum infiltrate for 20 min under sterile conditions using

400 mm/17 inches of mercury. Remove excess *A. tumefaciens* by lightly blotting the embryos on sterile filter paper and incubate on antibiotic free REM medium for 5 days at 22°C in the dark.

2. After co-cultivation, wash off the *A. tumefaciens* growing on the embryo surface with liquid antibiotic free REM and randomly select a few embryos in order to check for transient GUS expression (*30, see* Note 4). Then, transfer the remaining somatic embryos on solid kanamycin free REM medium for 2 weeks at a low light intensity (25 ± 2°C and 52 μ mol/m^2/sec, 12 h photoperiod). After 2 weeks, transfer the somatic embryos (free of *A. tumefaciens*) to REM medium supplemented with kanamycin and allow the putative transformants to grow without subculturing for 14-16 weeks until profuse secondary embryogenesis occurs (Figs. 1A and B). Separate each embryo and culture on SGMa medium for 7 days followed by a transfer to SGMb medium for another 7 days. Remove the germinated embryos (Figs. 1E and F) into Hikko trays (634.25cm^3 i.e. 29.5 cm X 21.5 cm X 10.0 cm in size and with a capacity to hold 20 plants in cells each of 5 cm diam.; Ajay Kumar and Co., India) containing a potting mixture of 2 parts of garden soil, 1 part river bed sand (approx. < 2mm) and 1 part farmyard manure (well rotted autoclaved cow dung). Maintain the plantlets in a containment facility (Fig. 1G).

3. Confirm the integration of transgenes in the genome of tea plants derived from somatic embryos by PCR amplification followed by Southern hybridization. Confirm the expression of the transgene by Northern blot analyses.

3.2. Transformation of Somatic Embryos by Particle Gun Bombardment

3.2.1. Sterilization of Nodal Segments and Establishment of Aseptic Cultures

1. Collect tender shoots from tagged bushes and cut stem pieces of about 2.5-3.0 cm, wash thoroughly for 15 min in a beaker containing 0.5% Tween 20 with the aid of a sable hair brush. Disinfect for 5 min in 0.1% mercuric chloride solution in a beaker and rinse 3 times with sterile distilled water. Transversely cut 0.5-1.0 cm long nodal segments and remove 1.5-2.0 mm slices from the exposed surfaces.

2. Inoculate in 20-25 ml aliquots of SMM medium contained in 25 X 150 mm glass tubes and maintain under culture laboratory conditions of 26 ± 2°C at an illumination of 52 μmol/m^2/sec of 12 h photoperiod for 4 weeks and then for additional 4 weeks on fresh SMM medium until the shoots attain a height of 3 cm.

3. Excise first leaves from these shoot cultures and use for bombardment (*see* Note 4).

3.2.2. Preparation of Bacterium

1. Grow *E. coli* cells carrying the engineered plasmid RTGUS99 on LB agar culture medium with 50 μg/ml kanamycin sulphate using the spread plate

method at 37°C for 12 h. Take a single colony and inoculate in 10 ml of LB (Hi Media) containing 50 µg/ml kanamycin sulphate and grow overnight.

2. Transfer the 10 ml bacterial culture (grown for 12 h to mid log phase) to a 3 litre flask containing 1500 ml of LB medium with 50 µg/ml kanamycin and incubate overnight at 37°C in dark at 200 rpm.

3. Pellet the overnight grown bacterial cells by centrifugation at 6000 rpm for 30 min at 25°C. Dissolve the pellet in 10 ml of re-suspension buffer and then add 10 ml of lysis buffer. After 5 min, add 10 ml of chilled neutralization buffer.

4. Centrifuge at 10,000 rpm for 30 min at 4°C and then remove the supernatant promptly. Repeat the centrifugation at 10,000 rpm for 15 min at 4°C and again remove the supernatant promptly. Alternatively, filter the supernatant over a pre-wetted folded, sterile Whatman filter paper.

5. Equilibrate a Qiagen tip 500 by applying 10 ml of equilibration buffer QBT and allow the column to empty by gravity flow and then apply the supernatant to the Qiagen tip and allow to enter the resin by gravity flow. Wash the Qiagen-tip with 2 X 30 ml 'Wash buffer' and elute the DNA with 5-15 ml elution buffer 'Buffer QF'.

6. Precipitate the DNA with 0.7 volumes of iso-propanol at room temperature, centrifuge immediately at 13,000 rpm for 30 min at 4°C and remove the supernatant carefully. Wash the DNA with 5 ml of 70% ethanol, air dry for 5 min and re-dissolve in a suitable volume of buffer (approx. 500 µl). Quantify the DNA (see Note 6) and prepare a working DNA solution of 10 ng/µl by diluting in sterile distilled water and store at 4°C before use.

7. For bombardment, take the first leaves from in vitro-grown plants (see Note 4) and treat with liquid 0.25 M sorbitol or alternatively with hormone free liquid basal MS medium on a shaker for 4 h at 100 rpm. Blot the leaves slightly on sterile filter paper and arrange with their abaxial surfaces in full contact with LCM medium so as to form a circle (3 cm diam.) in 9 cm Petri dishes.

8. For bombardment of leaves, first coat the micro-projectiles. For this, take 60 µg of gold particles and sterilize by washing with 70% alcohol followed by sterile water 3 times each and finally suspend in about 50 µl of sterile distilled water. Dispense the suspension into 1.5 ml Eppendorf tubes for each bombardment. Mix 50 µl of the gold suspension with 10 µl of pRT99GUS DNA (1.0 µg/µl), 50 µl of 2.5M $CaCl_2$ and 10 µl of phosphate free spermidine (Sigma). Vortex the suspension from time to time followed by spinning for 10 sec at 10,000 rpm. Remove the supernatant, wash the DNA coated micro-projectiles with 70% ethanol and finally suspend the gold particles or micro-projectiles in 60 µl of 100% ethanol (Fluka). Finally, suspend 10 µl of the plasmid DNA coated gold particles on sterile macro-carriers (BioRad) by immediate vortexing.

9. Bombard the leaves arranged as above using rupture discs of 1100 psi burst pressure, 6 cm target distance (TD), 16 mm macro-carrier flight distance (MFD) and gap distance of 3/8+1/4+ ¼ inches (see Note 7). Helium powered Particle Delivery system, PDS-1000/He (Bio-Rad) is used under a chamber pressure of 25 inches mercury and 1.0 µm gold particles coated with 1 µg/µl plasmid DNA are used for gene transfer. Bombard only the adaxial surfaces of the leaf

explants of each plate twice after changing the direction of the Petri plate and then turn the bombarded leaves upside down on the regeneration medium with their abaxial surfaces up so that the bombarded surface touches the regeneration medium (*see* Note 8). Culture the explants in the dark for 2 days at 25 ± 2°C.

10. Assay some randomly selected bombarded leaves for GUS expression (Fig. 2A) following the method of Jefferson, (*30, see* Note 5). Transfer the rest of the explants to LCM medium for 4-6 weeks for callusing (Fig. 2B) and then to SRM medium for another 10 weeks for adventitious shoot regeneration (Figs. 2C and D). Excise the shoot buds (about 1.5-2.0 cm in length) and grow in SMM medium (*31, see* Note 9). Finally, select the vigorously growing micro-shoots and micro-graft them onto recipient plants (Figs. 2E-G) using the method of Prakash *et al.* (*32*). This grafting procedure has a success rate of about 70-75%. Excise leaves of healthy plants after 16 weeks and confirm the integration of the transgenes in the genome of tea plants by subjecting them to PCR amplification (*see* Note 10) followed by Southern Hybridization. Confirm the expression of the transgenes by Northern blot analyses.

4. NOTES

1. Plasmid 35SGUSINT (*33*) is a derivative of the binary vector pBin19 (*34*) and has a NOS promoter-*npt*II-NOS terminator (polyA) cassette and the *gus*A reporter gene with a plant intron *gus*-int under the regulatory control of the CaMV 35S promoter and a CaMV-pA sequence at the left border. For routine use, the engineered *Agrobacterium* strain EHA105 is grown in the dark at 28°C in agar semi-solidified YMB medium (*28*) supplemented with kanamycin sulphate.

2. All antibiotics are filter-sterilized and are added to the autoclaved medium (pH 7.0) after the latter is cooled to 45°C prior to solidification.

3. There are 3 major types of tea: the Chinary, Assam and the Cambod. The Kangra jat is a cultivar of the Chinary type and is extensively propagated in the Kangra valley located at the Himalayan foothills of Himachal Pradesh, India to yield the highly valued Kangra tea. Although the quality is similar to the Darjeeling tea, Kangra tea has a special flavour that makes it highly popular in India. Moreover, the plants can tolerate the adverse conditions more than the other cultivars. Thus, certain plants of this cultivar that are growing in our Tea Experimental Farm, Banuri, Palampur, India (32°N and 76°E and 1292 m above sea level) were tagged for the cotyledon explants. For this, only fruits at full maturity are collected during November. The fruit coat is first removed by hand and, viable seeds are selected by soaking them overnight in water and selecting only the 'sinkers'. The seed coats of the sinkers are removed and embryos comprising of both cotyledons and embryonic axes are used as explants for *Agrobacterium*-mediated transformation.

4. For biolistic mediated transformations, the 'first leaves' or young leaves (10 x 0.8 mm) which are closely attached to the apical buds from the *in vitro* grown cultures serve as explants. The *in vitro* cultures are established from new shoots that sprout during the first, second and third flushes of the growing season after the intermittent periods of bud dormancy. All explants are collected from tea bushes growing in the Tea Experimental Farm, Banuri, Palampur, India (32°N and 76°E and 1292 metres above sea level). Use of leaf explants instead of seed-derived cotyledons maintains the true to type nature of elite characteristics and prevents the segregation of genetic traits as observed in seed derived explants.

5. During the assay of GUS transient expression, the β-D-glucuronidase enzyme produced by the transformed leaf explants and substrate X-gluc react to yield a blue coloured 'reaction product' This indicates the transfer of the transgene into the host genome and its transcription.

6. Quantify the DNA by taking a 5 µl aliquot into a 1 ml quartz cuvette and make up the volume to 1 ml with sterile distilled water and measure the absorbance first at 260 and then at 280 nm. A 1.0 OD at 260 nm = 50 µg/ml of double stranded DNA. When the ratio of A_{260}/A_{280} = 1.8, the DNA is

considered to be pure. If the ratio is > 0.8, then the DNA sample is considered to be contaminated by protein and phenols.

7. While MFD is the distance between macro-carrier and stopping screen, target distance is the distance between the micro-projectile stopping screen and target tissue. The burst pressure ranges between 650-1350 psi and is the pressure that can burst the rupture disc (BioRad). For increasing the surface area for maximum particle penetration, the minimum cell damage/injury and maximum regeneration efficiency parameters requires optimization. The tissue damage due to gas shock and high particle dispersion is circumvented by increasing the target distance for optimal particle dispersion and simultaneously the tissue damage due to off-centred flight of microprojectile flight distance is overcome by decreasing the gap distance

8. When the bombarded leaves are turned upside down on the regeneration medium with their abaxial surfaces upwards and cultured in dark for 2 days, the injured leaf surface in contact with the regeneration medium are healed quickly and regeneration response is observed.

9. Healthy explants and micro-shoots growing vigorously on medium supplemented with lethal doses of kanamycin (1000 μg/ml) indicates that the transformed leaves are putative transformants.

10. For PCR analysis, DNA is prepared from 200-300 mg callus tissue using a CTAB buffer at 65°C (*35*). Primers that are used for these analyses include:

*npt*II
5'-CCA-TCG-GCT-GCT-CTG-ATG-CCG-CCG-T-3' (25 mer)
5'- AAG-CGA-TAG-AAG-GCG-ATG-CGC-TGC-3' (24 mer)
*gus*A
5'-GGT-GGG-AAA- GCG-CGT-TAC-AAG-3' (21 mer)
5'-TGG-ATC-CCG-GCA-TAG-TTA-AA-3'   (20 mer)

Primers are so designed as to give amplification products of the internal sequence of the *npt*II and *gus*A genes of 693 bp and 650 bp, respectively. Approx. 500 ng DNA is used for each reaction. While both transformed and untransformed DNA are used, plasmid DNA is used as the positive control. Samples are subjected to 32 cycles following the method of Mondal *et al*. (24). Stratagene Robocycler Gradient 40 with the fastest available transitions between temperatures is used for amplification and the products are separated by 1.4% agarose gel electrophoresis and photographed under a UV trans-illuminator (Fotodyne MP-St) equipped with MP4 Polaroid Instant Camera System.

The amplification cycle comprises of an initial cycle of 94°C for 4 min, 55°C for 1 min and 72°C for 4 min followed by 30 cycles consisting of (a) denaturing at 94°C – 1 min, (b) annealing at 55°C – 1 min, (c) extension at 72°C – 1 min and a final additional 7 min at 72°C before its rapid cooling to room temperature.

CONCLUSIONS

Genetic transformation protocols both via *A. tumefaciens* and also by micro-projectile bombardment have been described. The somatic embryos are co-cultivated with *A. tumefaciens* for 5 days at 22°C in the dark and made to undergo recurrent embryogenesis. The secondary embryos germinate to yield stably transformed transgenic tea plants with a transformation efficiency of approx. 0.5%. The biolistic-mediated genetic transformation of tea is important when leaves are used as explants as the presence of high contents of polyphenols are reported to inhibit *Agro*-infection. Leaves bombarded using PDS-1000/He (Bio-Rad) at a chamber pressure of 25 inches mercury and 1.0 μm gold particles coated with 1 μg/μl plasmid DNA at 1100 psi burst pressure, 6 cm TD, 16 mm MFD and GD of 3/8+1/4+ ¼ inches yield stable transformations and high frequency of transgenic plant recovery (0.75-1.0 %). The protocol enables the transfer of any useful genes into leaves of elite clones for the first time. Both methods have been found to be reproducible and pave the way for the biotechnological improvement of tea in a time effective manner for the production of agronomically useful germplasms of tea.

REFERENCES

1.   Kabra GD (1999). Tea statistics for 1999. *Tea time*, **8**: 30-31.
2.   Sivaram B (1999). SAARC and Tea. *Tea time,* **8:** 8-10.
3.   Jankun J, Selman SH and Swiercz R (1997). Why drinking green tea could prevent cancer. *Nature*, **5:** 561.
4.   Chu DC (1997). Green Tea- Its cultivation, processing of the leaf leaves for drinking materials, and kinds of green tea. In: Yamamoto T, Juneja LR, Chu DC (eds.), *Chemistry and Applications of Green Tea*. CRC Press, Boca Raton, New York, USA, pp. 1-12.
5.   Chen Z (1999). Pharmacological Functions of Tea. In: Jain NK (ed.), *Global Advances in Tea Science*. Aravali Books International (P) Ltd., New Delhi, India, pp. 333-358.
6.   Smartin MC and Smartin AP (1988). La Camelia. Un Regalo Oriental para Occidente. Editorial Everest, S. A. Leon.
7.   Jhawar MS (2000). Future-Bound with tea (IMT Convention). *Planters' Chronicle*, 203-207.
8.   Vieitez AM (1994). Somatic embryogenesis in *Camellia* spp. In: Jain S, Gupta P, Newton R (eds.), *Somatic Embryogenesis in Woody Plants*. Kluwer Academic Publishers, The Netherlands, pp. 235-276.
9.   Bora PC and Deka A (1999). Tea Industry in India. In: Jain NK (ed.), *Global Advances in Tea Science*. Aravali Books International (P) Ltd., New Delhi, India, pp. 43-64.
10. Bhattacharya A and Ahuja PS (2003). Prospects of transgenics in tea crop improvement. In: Singh RP, Jaiwal PK (eds.), *Plant Genetic Engineering. Improvement of Commercial Plants-I*. Scientific Technology Publishers, LLC, USA, pp. 115-130.
11. Jain NK (1999). Impact of global advances in tea science and technology on economic parameters of tea industry. In: Jain NK (ed.), *Global Advances in Tea Science*. Aravali Books International (P) Ltd., New Delhi, India, pp. 265-296.
12. Barua DN (1989). The tea plant of commerce. In: Barua DN (ed.), *Science and Practice in Tea Culture*. Tea Research Association Calcutta, India, pp. 53-68.
13. Mondal TK, Bhattacharya A and Ahuja PS (2004). Recent advances in Tea Biotechnology. *Plant Cell, Tissue and Organ Culture*, **76**: 195-254.
14. Raj Kumar R and Ayyappan P (1992). Somatic embryogenesis from cotyledonary explants of *Camellia sinensis* (L.) O. Kuntze. *The Planter's Chronicle*, May: 227-229.
15. Saha SK and Bhattacharya NM (1992). Stimulating effects of elevated temperature treatments on production of meristemoids from pollen callus of tea, *Camellia sinensis*) L.) O. Kuntze. *Indian Journal of Experimental Biology*, 30: 83-86.
16. Shimokado T, Murata T and Miyaji Y (1986). Formation of embryoids by anther culture of tea. *Japanese Journal of Breeding*, **36**: 282-283.
17. Chen Z and Liao H (1983). A success in bringing out tea from the anthers. *China Tea*, **5**: 6-7.
18. Nakamura Y (1983). Isolation of protoplasts in from tea plant. *Tea Research Journal* **58:** 36-37.
19. Kuboi T, Suda M and Konishi S (1991). Preparation of protoplasts from tea leaves. *Proceedings of International Symposium. Tea Science,* Shizuoka, Japan, pp. 427-430.
20. Biao X, Toru K, Jian X and Yongyan B (1998). Effect of polyphenol compounds in tea transformations. In: *American Society of Plant Physiologists, Plant Biology*, Abstract No. 314.
21. Matsumoto S and Fukai M (1998). *Agrobacterium tumefaciens* mediated gene transfer in tea plant (*Camellia sinensis*) cells. *Japan Agricultural Research Quarterly*, **32**: 287-291.
22. Matsumoto S and Fukai M (1999). Effect of acetosyringone application on *Agrobacterium* mediated gene transfer in tea plant (*Camellia sinensis*). *Bulletin of the National Research Institute of Vegetables, Ornamental Plants and Tea*, Shizuoka, Japan, **14**: 9-15.

23. Mondal TK, Bhattacharya A, Sood A and Ahuja PS (1999). Production of transgenic tea plants from *Agrobacterium tumefaciens* transformed somatic embryos. In: Altman A, Ziv M, Izhar SE (eds.), *Plant Biotechnology and In Vitro Biology in the 21st century*. Kluwer Academic Publishers, Dordrecht, The Netherlands, pp. 181-184.

24. Mondal TK, Bhattacharya A, Ahuja PS and Chand PK (2001). Transgenic tea (*Camellia sinensis* (L.) O. Kuntze cv. *Kangra Jat*) plants obtained by *Agrobacterium*-mediated transformation of somatic embryos. *Plant Cell Reports*, **20**: 712-720.

25. Biao X, Liu ZS and Liang YR (1997). Genetic transformation of tea plants mediated by *Agrobacterium rhizogenes*. *Journal of Tea Science*, **17**: 155-156.

26. Konwar BK, Das SC, Bordoloi BJ and Dutta RK (1998). Hairy root development in tea through *Agrobacterium rhizogenes* mediated genetic transformation. *Two and A Bud*, **45**: 19-20.

27. Murashige T and Skoog F (1962). A revised medium for rapid growth and bioassays with tobacco tissue cultures. *Physiologia Plantarum*, **15**: 473-497.

28. Hood EE, Gelvin SB, Melchers LS and Hoekema A (1993). New *Agrobacterium* helper plasmids for gene transfer to plants. *Transgenic Research*, **2**: 208-218.

29. Hooykas PJJ, Klapwijk PM, Nuti MP, Schilperoort RA and Rorsch A (1977). Transfer of the *Agrobacterium tumefaciens* Ti plasmid to avirulent *Agrobacteria* and to *rhizobium* explants. *Journal of General Microbiology*, **98**: 477-484.

30. Jefferson RA (1987). Assaying chimeric genes in plants: The GUS gene fusion system. *Plant Molecular Biology Reporter*, **5**: 389-405.

31. Sandal I, Bhattacharya A and Ahuja PS (2001). An efficient liquid culture system for tea shoot proliferation. *Plant Cell, Tissue and Organ Culture*, **65**: 75-80.

32. Prakash O, Sood A, Sharma M and Ahuja PS (1999). Grafting micro-propagated tea (*Camellia sinensis* (L.) O. Kuntze) shoots on tea seedlings - a new approach to tea propagation. *Plant Cell Reports*, **18**: 883-888.

33. Vancanneyt G, Schmidt R, O'Conor-Sanchez A, Willmitzer L and Rocha-Sosa M (1990). Construction of an intron-containing marker gene: splicing of the intron in transgenic plants and its use in monitoring early events in *Agrobacterium*-mediated plant transformation. *Molecular and General Genetics*, **220**: 245-250.

34. Bevan M and Chilton MD (1982). T-DNA of the *Agrobacterium* Ti and Ri plasmids. *Annual Review of Genetics*, **15**: 337-384.

35. Doyle JJ and Doyle JL (1990). Isolation of plant DNA from fresh tissue. *BRL Focus*, **12**: 13-15.

36. Sambrook J, Fritsch EF and Maniatis T (1989). *Molecular Cloning, A Laboratory Manual*. (2nd edn.) Cold Spring Harbor, NY, Cold Spring Harbor Laboratory Press.

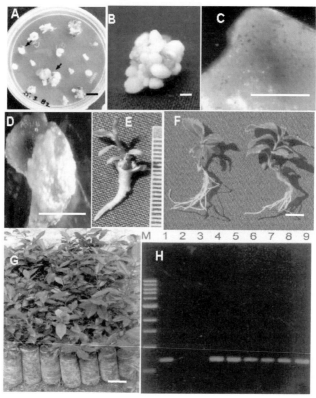

Figure 1. Procedure for the production of A. tumefaciens *mediated transgenic tea plants using somatic embryos. Primary embryos obtained from de-embryonated cotyledons are transformed with* A. tumefaciens *and co-cultivated on medium enriched with L-glutamine and potassium sulphate and plant growth regulators. Co-cultivation on such a medium not only enables rapid proliferation of the few putatively transformed embryos but also enhances the competence for transformation A. Few probable transformants growing on selection medium containing 200 mg/l kanamycin sulphate (see arrows). Proliferation on such a medium allows selection of transformants only (bar = 2.5 mm) B. Recurrent somatic embryos from transformed primary embryos on selection medium (bar = 0. 5 mm); C. GUS transient expression of transformed primary embryos (bar = 0.25 mm); D. Consistent GUS expression of secondary somatic embryos obtained through recurrent embryogenesis (bar = 0.25 mm); E. Germinated embryo, F. Plantlets from germinated somatic embryos (bar = 1 cm) ; G. Healthy plants growing in the field (bar = 1 cm); H. PCR analysis to detect the presence of the* nptII *and* gusA *genes in transgenic tea plants, Lane M is 1 kb ladder; Lanes 2, 3 are DNA from untransformed tea plants; Lanes 4-9 are DNA from transformed tea plants.*

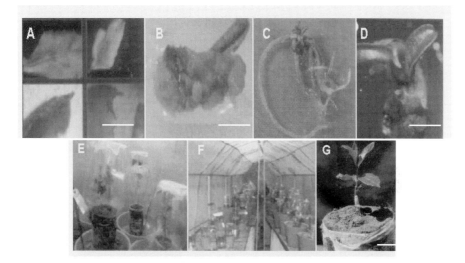

Figure 2. Procedure for the biolistic-mediated production of transgenic tea plants using leaf explants of elites. Use of this procedure allows the addition of useful traits to high yielding and superior tea plants. A. GUS expression of bombarded tea leaf (bar = 5 mm) B. Callusing on bombarded first leaf of selected elite of tea on regeneration medium (LCM) (bar = 5 mm); C and D. Rhizogenesis with shoot bud initiation and caulogenesis (bar = 5 mm) on regeneration medium (SRM1) containing 1000 mg/l kanamycin; E, F and G. Grafted transgenic micro-shoots growing in pots under controlled conditions (bar = 1 cm).

Chapter 14

MICROPROJECTILE-MEDIATED TRANSFORMATION OF PINEAPPLE

M.R. DAVEY, S. SRIPAORAYA[1], P. ANTHONY AND J.B. POWER

Plant Sciences Division, School of Biosciences, University of Nottingham, Sutton Bonington Campus, Loughborough LE12 5RD, UK
[1]Faculty of Agriculture of Nakhonsithammarat, Rajamangala Institute of Technology, Tungyai, Nakhonsithammarat 80240, Thailand
Corresponding author: M.R. Davey. E-mail: mike.davey@nottingham.ac.uk

1. INTRODUCTION

1.1. Importance of Pineapple as a Commercial Crop

Pineapple (*Ananas comosus* L.) is cultivated in all tropical and sub-tropical countries and ranks third in production amongst tropical fruits. In 2002, the world production of pineapple was 14.9 million tonnes, which represents increases of 23% and 43% compared to 1998 and the early 1980s, respectively. Production in Asia is centred in Thailand (*ca.* 14% of world production), The Philippines (10.6%), China (8.6%) and India (7.3%). Africa, Central America and South America contribute approx. equally to the remainder (*1*). About 70% of pineapple is consumed as fresh fruit, with The Philippines, Cote d'Ivoire and Costa Rica predominating in the global production of fresh pineapple. Collectively, these 3 countries account for nearly two-thirds of this world trade, estimated to be about 620,000 tonnes and worth about 200 million $US. The European Union is the largest importer of pineapple and handles in excess of 200,000 tonnes annually, of which France consumes about 30%, followed by Germany, Italy and Spain (*2*). The trade in processed products consists of canned slices (approx. 1.0 million tonnes) and juice (approx. 170,000 tonnes), supplied mainly by Thailand and The Philippines.

The edible portion of the fruit, which constitutes about 60% of the fresh weight, contains about 85% water, 0.4% protein, 14% sugar (sucrose), 0.1% fat and 0.5% fibre (*3*). The fruit is rich in vitamins A and B, while the juice has 75-83% sucrose and 7-9% citric acid on a dry weight basis. The flesh and juice also contain a protein-digesting enzyme, bromelain. The leaves yield a strong, white, silky fibre suitable for production of a fine fabric, pina cloth, which is woven in The Philippines and Taiwan. One reason for the popularity of pineapple is that it is very amenable to large-scale plantation cultivation. Various chimeric forms of pineapple, with green, yellow and pink leaf stripes, are cultivated and marketed as ornamental plants. Discussion of the commercial aspects of pineapple production and

187

I.S. Curtis (ed.), Transgenic Crops of the World-Essential Protocols, 187-197.
© 2004 Kluwer Academic Publishers. Printed in the Netherlands.

processing, the physiology of the crop, breeding programmes and aspects of the application of biotechnological approaches to crop improvement, can be found in an excellent recently published volume (*4*).

1.2. Genetic Improvement of Pineapple

Pineapple breeders aim to generate plants with rapid growth, spineless leaves or leaves having only spiny tips, a response to flower induction to maximize yield, increased vitamin C content and resistance to diseases such as heart and root rots, as well as tolerance to some herbicides. High sucrose content and a pleasant flavour are also essential. Numerous sexual crosses have been made between pineapple cultivars, or between *A. comosus* and related species. Although hybridization programmes have been in progress for many years in several pineapple-producing countries, few of the F1 hybrids produced have been completely satisfactory with respect to introgressed traits, necessitating continued hybridization, selection and back-crossing. Biochemical and molecular markers are being used to confirm genetic relationships in pineapple (*5*, *6*) and these approaches should assist in identifying germplasms most suitable for future breeding programmes.

Pineapple is propagated vegetatively in the field from crowns, slips, suckers or hapas (bases of the fruit slips), with each of the propagules requiring different periods from planting to fruit harvest. This is about 13-15 months for suckers and 2 –3 months longer for crowns. The rate of production of propagules is often slow and the propagules are limited in number. Consequently, micropropagation through tissue culture offers an additional approach to conventional methods of plant multiplication, not only for mass production *per se*, but also to clone elite plants with desirable agronomic characteristics. Virus-free plants can also be generated through meristem culture, while somatic hybridization and transformation are potentially useful technologies for the genetic improvement of pineapple.

1.3. In vitro-based Approaches for Multiplication and Genetic Improvement of Pineapple

During the last three decades, procedures have been published summarizing the composition of culture media and physiological factors for the rapid multiplication of several pineapple cultivars in order to generate large populations of plants for subsequent field cultivation. Earlier reports are discussed in the more recent publications. Whilst shoot regeneration from cultured tissues is predominately by organogenesis, somatic embryogenesis has also been described (*7*). A recent publication describes cost-effective *in vitro* propagation procedures using bioreactors for the pineapple cultivar Smooth Cayenne, since this remains the "standard" cultivar grown most extensively worldwide (*8*).

The ability to micropropagate pineapple from cultured explants also provides an important baseline for genetic improvement of the crop using somatic cell techniques. Indeed, *in vitro* approaches represent an important adjunct to conventional breeding, although some of the methodology is still at the early stages of development. For example, there are no reports, to date, of somatic hybridization to combine the genetic material from different cultivars of pineapple, or to introduce traits into *A. comosus* from related species. Whilst protoplasts can be isolated enzymatically from leaves of cultured shoots, they remain recalcitrant to sustained mitotic division, tissue formation and shoot regeneration (*9*).

More success has been reported for the genetic transformation of pineapple, undoubtedly stimulated by advances made in transforming cereals (*10, 11*). In the first report of *Agrobacterium*-mediated transformation of pineapple, embryogenic tissues of the cultivar Smooth Cayenne were incubated with the disarmed strain, C58C1, of *A. tumefaciens*, the bacterium carrying a binary vector with either an *als* gene, conferring resistance to the selective herbicide chlorsulfuron, or the *npt*II gene, enabling transformed plant cells to grow on medium containing the antibiotics neomycin, kanamycin or geneticin (G418) (*12*). In these experiments, the *gus*A gene was also located on the T-DNA of the binary vector. Following co-cultivation with *Agrobacterium*, transformed tissues were selected on medium containing chlorsulfuron or G418. Selected, transformed embryogenic callus produced somatic embryos, which developed into plants. Transformation was confirmed by a GUS histochemical assay and molecular analyses using PCR and Southern hybridization. A number of plants from several independently transformed lines were transferred to the glasshouse. In the near future, a combination of bioreactor and transformation technologies may form the basis for high-throughput systems to introduce novel characteristics into pineapple (*8*). Other workers have also investigated the possibility of exploiting *Agrobacterium* for gene transfer into pineapple. For example, *A. tumefaciens* strain 1065 carrying a binary vector with the *gus*A and *npt*II genes was used to inoculate embryogenic cell suspensions, this strain of *Agrobacterium* being chosen because of its supervirulent phenotype (*13*). Although GUS-positive cells were observed following inoculation, it was not possible to select transformed tissues and to regenerate transgenic plants. A reliable procedure for pineapple transformation based on *Agrobacterium*-mediated gene delivery is still awaited.

Several workers have evaluated particle bombardment to transform pineapple. In a preliminary report, leaf base-derived tissues (protocorm-like bodies) maintained in liquid medium were used as targets for gene delivery (*14*). Following bombardment with DNA constructs carrying *gus* and *npt*II genes, transformed tissues were selected by their ability to grow in liquid medium supplemented with G418, followed by plant regeneration on agar-solidified medium. Transgenic plants contained the *gus* and *npt*II genes inserted at 1-3 loci, as determined by Southern hybridization. These transgenic plants were multiplied and transferred to the glasshouse for further evaluation. This transformation procedure formed a basis for subsequent studies using agronomically important genes, such as those for disease resistance, by the same authors. Publication of the results of such experiments is expected in due course.

A transformation procedure was also reported for the cultivar Smooth Cayenne using leaf-derived callus as the target for bombardment, with *gus* and *gfp* genes being used to optimize the conditions for transient and stable gene expression (*15*). G418 resistant transgenic callus and plants were obtained following co-transformation with constructs carrying the *npt*II gene. The authors recovered 15 independent transgenic lines that were GUS and GFP positive, with regenerated plants carrying multiple copies of the introduced genes. It was indicated that this transformation system was being exploited to introduce genes into pineapple to control the post-harvest disorder blackheart injury, based on inhibition of polyphenol oxidase expression. Blackheart, resulting from discolouration of the core of the fruit, normally occurs when winter temperatures fall below 20°C, or during cool storage for export. This preliminary report indicated that subsequent field evaluations would determine the merit of Biolistic®-mediated gene delivery for improving fruit quality of the ubiquitous pineapple cultivar Smooth Cayenne.

In more recent investigations, herbicide tolerance was introduced into the Thai pineapple cultivar Phuket by microprojectile-mediated delivery of pAHC25 into the bases of leaves excised from micropropagated shoots (*16*). Plasmid AHC25 carried the *gus* reporter gene and the *bar* gene as a selectable marker, both driven by the maize ubiquitin promoter. The *bar* gene from *Streptomyces hygroscopicus* codes for phosphinothricin acetyltransferase (PAT), which confers resistance to the non-selective herbicide bialaphos by acetylation of the free amide group of the active component, phosphinothricin (PPT). PAT also converts glufosinate ammonium, the active component of the commercially available herbicide Basta™ (Riedel-de-Haën AG, Seeize, Germany), into an inactive form (*17*). Transformed plants were recovered from bombarded leaf bases using a rapid shoot regeneration procedure (*7*), eliminating the need for more complex regeneration pathways involving somatic embryogenesis. Bombarded leaf bases were cultured on Murashige and Skoog (MS) based medium (*18*) containing 0.5 mg/l 2,4-D, 2.0 mg/l BAP and 0.5 mg/l PPT, followed by transfer to medium with 1.0 mg/l BAP and 0.1 mg/l PPT. Subsequently, regenerated plants were micropropagated on MS medium with 1.0 mg/l BAP, but with the PPT concentration increased to 2.0 mg/l. Shoots were rooted on MS medium lacking growth regulators supplemented with 2.0 mg/l PPT. Micropropagated transgenic plants were transferred to the glasshouse. Integration and expression of 4-8 copies of the *bar* gene in the transgenic plants was confirmed by Southern analysis and RT-PCR analyses, respectively.

Regenerated plants were assessed for their herbicide tolerance both *in vitro* and in the glasshouse. Studies *in vitro* involved culture of plants on agar-solidified MS-based medium supplemented with Basta™, the latter being diluted to give concentrations of glufosinate ammonium of 0, 3, 5, 7, 10, 15 or 20 mg/l in the culture medium. The herbicide tolerance of transgenic plants after 28 days of culture was compared to that of non-transformed plants. Concentrations in excess of 3.0 mg/l of glufosinate ammonium in the medium resulted in an inhibition of growth and loss of pigmentation, within 4 days of exposure to the herbicide, in non-transformed plants. In contrast, all PPT-tolerant plants remained green and developed normally in the presence of the herbicide. Transgenic and non-transformed plants, which had been acclimatized to glasshouse conditions for 75 days, were sprayed to run-off with aqueous solutions of Basta™ diluted to contain 100, 200, 400, 600, 800, 1200 or 1400 mg/l of glufosinate ammonium. Plant responses were scored 14 days later. Application of the herbicide to the leaves of non-transformed plants resulted in necrosis after 4 days, followed by browning within a further 14 days and subsequent death of the plants. However, leaves of

PPT-tolerant plants remained green and the plants continued to grow vigorously, even when treated with a herbicide solution containing 1400 mg/l of glufosinate ammonium. There was no apparent link between the number of copies of the *bar* gene integrated into the genome of transgenic plants and the extent of their herbicide tolerance.

Bombardment of totipotent tissues with DNA coated microprojectiles is, at present, the most reliable option for genetic transformation of pineapple. A simple procedure for transferring herbicide tolerance into pineapple by inserting the *bar* gene for bialaphos resistance (*16*), is summarized in this report

2. MATERIALS

2.1. Plant Material and Tissue Culture

1. Target pineapple cultivar(s): Shoots maintained *in vitro* of the Thai cv. Phuket.
2. Reverse osmosis water: Sterile (autoclaved, 120°C for 20 min) and stored in suitable containers (*e.g.* 500 ml screw-capped bottles each containing approx. 300 ml of water).
3. Surface sterilant:"Domestos" bleach (Lever Fabergé, Kinston-upon-Thames, UK) diluted to 20% and 10% (v/v) with sterile tap water.
4. MT medium: Based on the formulation of Murashige and Tucker (*19*) containing 3% (w/v) sucrose, 2.0 mg/l IBA, 2.0 mg/l NAA, 2.0 mg/l kinetin and semi-solidified with 1% (w/v) Bacto agar (Becton Dickinson UK Ltd., Oxford, UK), pH 5.8.
5. MSB2 medium: Based on the formulation of Murashige and Skoog (*18*) with 3% (w/v) sucrose, 2.0 mg/l BAP and semi-solidified with 0.8% (w/v) agar (Sigma), pH 5.8. (*see* Note 1).
6. MSDB medium: Based on the MS formulation (*18*) with 3% (w/v) sucrose, 1.0 mg/l 2,4-D, 1.0 mg/l BAP and 0.8% (w/v) agar (Sigma), pH 5.8.
7. Petri dishes: 9 cm diam. (Bibby-Sterilin, Stone, UK).
8. Screw-capped glass bottles: 175 ml capacity (Beatson Clark, Rotherham, UK) (*see* Note 2).
9. IBA, Sigma: 1 mg/ml stock solution. Prepare by dissolving the powder in a few drops of absolute ethanol. Make up to volume with 70% (v/v) ethanol. Store at 5°C for no more than 6 months.
10. NAA, Sigma: 1 mg/l stock solution. Prepare and store as for IBA.
11. BAP, Sigma: 1 mg/l stock solution. Prepare by dissolving the powder in a few drops of 1 N NaOH. Make up to volume with reverse osmosis water. Store at 5°C for no more than 6 months.
12. 2,4-D, Sigma: 1mg/ml stock solution. Prepare and store as for IBA.
13. Kinetin (Sigma): 1mg/ml stock solution. Prepare by dissolving the powder in reverse osmosis water. Store at 5°C for no more than 6 months.

2.2. Materials for Microprojectile-mediated Gene Delivery and the Selection of Transgenic Plants

1. Plasmid AHC25 containing the ß-glucuronidase (*gus*A; *uid*A) reporter gene and the *bar* gene, both driven by the maize ubiquitin promoter with its first exon and first intron (*20*) (*see* Note 3).
2. Plasmid isolation kit: Wizard Megaprep plasmid isolation kit (Promega, Southampton, UK) (*see* Note 4).
3. TE buffer: 1 mM Tris HCl, pH 7.8 (Sigma), 0.1 mM Na$_2$EDTA (Sigma).
4. CaCl$_2$ (Sigma): 2.5 M solution.
5. Spermidine, free base (Sigma): 0.1 M solution.
6. Gold particles (Heraeus GmbH, Karlsruhe, Germany): 0.4–1.3 µm diam.
7. Absolute ethanol.
8. Vortex mixer.
9. Instrument for microprojectile bombardment: *e.g.* Helium driven PDS-1000/HE device (Bio-Rad Laboratories Ltd., Hemel Hempstead, UK) (*see* Note 5).
10. Pipettes with disposable tips to deliver µl volumes.
11. MSB medium: Based on the MS formulation (*18*) with 3% (w/v) sucrose, 1.0 mg/l BAP and 0.8% (w/v) agar (Sigma), pH 5.8. Supplement some preparations with 1.0 mg/l or 2.0 mg/l PPT (Melford Laboratories Ltd., Chelsworth, Ipswich, UK).
12. PPT: 1 mg/ml stock solution. Prepare by dissolving the powder in water. Store at room temperature.
13. Petri dishes: 9 cm diam.

2.3. Materials for Evaluating Herbicide Tolerance of Selected Plants **in vitro** and in the Glasshouse

1. MS0 medium: MS based medium (*18*) with 3.0% (w/v) sucrose, pH 5.8, lacking growth regulators. Supplement some preparations with 2.0 mg/l PPT or with Basta™ diluted to give glufosinate ammonium concentrations in the medium of 0, 3, 5, 7, 10, 15 or 20 mg/l.
2. Plastic pots (9 cm diam.) with potting mixture and a suitable liquid fertilizer (*see* Note 6).
3. Aqueous solutions of Basta™ diluted to contain 100, 200, 400, 600, 800, 1200 or 1400 mg/l glufosinate ammonium.

3. METHODS

3.1. Introduction of Pineapple into Culture

1. Excise suckers, each 1-2 months old and approx. 30 cm in length with about 20 leaves, from field-grown plants (*see* Note 7). Remove the outer leaves and wash the suckers under running tap water for at least 60 sec.

2. Sterilize the suckers by immersion in 20% (v/v) "Domestos" bleach for 40 min, followed by 10% (v/v) "Domestos" for 10 min. Incubate the explants statically in the sterilant. Wash the explants thoroughly with sterile reverse osmosis water.

3. Excise the axillary and terminal buds from the suckers and insert into agar-solidified MT medium. (*see* Note 8).

4. Incubate for 35 days at $25 \pm 2°C$ with a 16 h photoperiod (47 μmol/m^2/sec Cool White fluorescent tubes; Thorn EMI Ltd., Hayes, UK). If bacterial contamination of cultured explants is a problem, transfer the plant material to new medium with a suitable antibiotic (*see* Note 8).

5. Excise developing shoots from the explants and transfer to MSB2 medium (*see* Note 8). Multiply and maintain the shoots by sub-culture to the same medium every 35-42 days.

3.2. Microprojectile-mediated Transformation of Pineapple

1. Maintain and amplify pAHC25 in *E. coli* strain HB101; isolate and purify the plasmid from the bacteria using a plasmid isolation kit. Resuspend the plasmid in TE buffer; adjust the plasmid concentration to 1 μg/μl. Store at -20°C until required.

2. Coat microprojectile particles with DNA using a CaCl$_2$/spermidine precipitation technique. Vortex 5 μl of plasmid, 50 μl of CaCl$_2$ and 20 μl of spermidine with 50 μl of gold particles in a sterile 1.5 ml capacity microfuge tube. Continue to vortex for 3 min. Centrifuge (10 sec pulse).

3. Remove the supernatant and resuspend the particles in 250 μl absolute ethanol. Repeat the centrifugation as in Step 2.

4. Resuspend the DNA coated particles in 75 μl of absolute ethanol.

5. Pipette 10 μl of ethanol containing the DNA coated particles into the centre of a macroprojectile of the bombardment instrument. Dry for 10-20 min at room temperature in the laminar flow hood (*see* Note 9).

6. Excise 4 leaves (numbers 4-7 from the apex, each approx. 1.5 cm in length) from each cultured pineapple shoot 28-42 days after sub-culture of the shoots to new medium.

7. Incubate the leaves for 2 h before bombardment on mannitol-supplemented MSDB medium (*see* Note 10).

8. Place each Petri dish containing the leaves in the bombardment instrument and bombard the leaves (*see* Note 11).

3.3. Selection of Transformed Plants

1. Transfer the bombarded leaves to the surface of 30 ml aliquots of semi-solid MSDB medium lacking mannitol in 9 cm Petri dishes with 10 leaves/dish. Place the leaves with their lower surfaces in contact with the medium. Maintain for 28 days under the same conditions of temperature and photoperiod (*see* Section 3.1.4.). The same temperature and photoperiod are also used in Steps 2-6 inclusive, below.

2. Transfer the leaves to 30 ml aliquots of semi-solid MSDB medium supplemented with 0.5 mg/L PPT in 9 cm Petri dishes (10 leaves/dish) for 56 days. Sub-culture to new medium every 10-14 days.
3. Transfer the leaves to semi-solid MSB medium with 1.0 mg/l PPT. Incubate on this medium until shoots develop from the leaf bases (*see* Note 12).
4. Excise regenerated shoots from the leaf bases when the shoots are 0.5-1.0 cm in height. Multiply the shoots by transfer to MSB medium containing 2.0 mg/l PPT.
5. Transfer PPT-resistant shoots to semi-solid MS0 medium lacking growth regulators, but supplemented with 2.0 mg/l PPT, to induce the formation of roots on the regenerated shoots.
6. In all experiments, set up suitable controls in which leaves are bombarded with particles which have not been coated with DNA. Subject these leaves to the same media treatments, both with and without PPT, as bombarded leaves.

3.4. Evaluation of the Herbicide Tolerance of Selected Plants in vitro and in the Glasshouse

1. Assess the herbicide tolerance *in vitro* of plants selected on PPT-supplemented medium. Transfer micropropagated shoots to MS0 agar medium lacking growth regulators but supplemented with Basta™ diluted to give a range of concentrations of glufosinate ammonium as described (*see* Section 2.3.1.). Evaluate at least 8 plants per herbicide concentration with 2 plants in each 175 ml capacity jar (50 ml medium/jar).
2. After 28 days, plants with green leaves are considered herbicide tolerant; chlorotic/necrotic and/or dead plants are sensitive to the herbicide (*see* Note 14).
3. Set up suitable controls with micropropagated plants regenerated from leaves bombarded with microprojectiles not coated with DNA (*see* Note 14).
4. Transfer rooted, PPT-resistant shoots to 9 cm diam. plastic pots, each containing a suitable potting mixture (*see* Note 6). Transfer plants regenerated from leaf bases not bombarded with plasmid (controls) to the same potting mixture.
5. Maintain the potted plants in the glasshouse with day/night temperatures of 28 ± 2°C and 24 ± 2°C, respectively, under natural daylight supplemented, as necessary, with fluorescent illumination (*see* Note 13).
6. Spray the plants to run off (approx. 2 ml/plant) with aqueous solutions of Basta™ diluted to contain a range of concentrations of glufosinate ammonium as indicated (*see* Section 2.3.3.). Spray a minimum of 3 plants/concentration.
7. Score the tolerance of plants 14 days after application of the herbicide by observing the same parameters as for plants exposed to herbicide *in vitro* (*see* Note 15).

4. NOTES

1. For convenience, use medium commercially available in powder form and prepare according to the instructions supplied by the manufacturer (*e.g.* Duchefa, Haarlem, The Netherlands). Add any supplements that are required by the experimentation.

2. These jars are relatively inexpensive. Other glass or plastic vessels may also be used. Adjust the volume of the medium in relation to the size of the vessel.

3. The *bar* gene carried on pAHC25 is convenient to transfer herbicide tolerance into pineapple. Workers may have a requirement to use other plasmids carrying different genes for herbicide tolerance.

4. Plasmid isolation kits are available from several commercial suppliers. The kit used will depend upon the plasmid being isolated and the preference of the operator.

5. The PDS-1000/HE device was used to introduce the *bar* gene into the pineapple cultivar Phuket. The advantage of this instrument is that it is employed in many laboratories worldwide, facilitating technology transfer. However, other instruments, the designs of which have been described in the literature, may be used for gene delivery.

6   A suitable potting mixture consists of 6:6:2:2 by volume of Scotts M3 compost (Fisons plc, Ipswich, UK), John Innes No. 3 compost (J. Bentley, Barton-on-Humber, UK) Veriperl perlite (Silvaperl Ltd., Gainsborough, UK) and Veriperl vermiculite. Plants should be given a feed with a liquid fertilizer, such as Maxicrop® Seaweed Extract plus Sequestered Iron [Maxicrop (UK) Ltd., Corby, UK], delivered as a root drench every 4 weeks or a foliar spray every 14 days.

7. The procedure described here has been used for the Thai cultivar Phuket. It should be applicable to other pineapple cultivars such as the widely-grown cultivar Smooth Cayenne, and other Thai cultivars such as Nanglae, Pattavia, Sawee and Tainan.

8. Place 1 explant in 50 ml of medium in a 175 ml capacity screw capped glass bottle. Any bacterial contamination can be eliminated by including cefotaxime (Claforan, Roussel Laboratories, Uxbridge, UK) at 250–500 mg/l in the culture medium.

9. Microprojectile bombardment should be performed in a laminar flow hood to reduce the incidence of contamination of plant samples by micro-organisms. Each macroprojectile for the PDS-1000/He instrument should be placed within its macrocarrier holder prior to drying of the DNA coated gold particles onto the surface of the macroprojectile. Macrocarrier holders can be sterilized by immersion in 70% (v/v) ethanol for 20 min, followed by air drying in the laminar flow hood.

10. Dispense 30 ml aliquots of mannitol-supplemented MSDB medium into 9 cm diam. Petri dishes and place 40-60 leaves on the surface of the semi-solid medium with their lower surfaces in contact with the medium. Overlap the upper portions of the laminae, leaving the basal regions exposed. The basal regions of the leaves are the targets for microprojectile bombardment, since only these regions regenerate shoots.

11. Remove the lids of the Petri dishes immediately prior to bombardment. The bombardment conditions which have been successful for transforming pineapple using the PDS-1000/He instrument are a rupture disc pressure of 9000 KPa (1350 psi), a vacuum of 72.5 cm Hg with a distance of 30 mm between the rupture disc and macrocarrier and 70 mm from the stopping screen to the target leaves, the latter being on mannitol-supplemented MSDB medium.

12. Shoot regeneration in the presence of 1.0 mg/l PPT should occur after approx. 35 days in the cultivar Phuket. The time for regeneration may vary with the cultivar.

13. In the UK, it is beneficial, especially during the winter months, to supplement natural daylight in the glasshouse with artificial illumination. A 16 h photoperiod provided by 180 µmol/m^2/sec Cool White fluorescent illumination has been employed at Nottingham for the vegetative growth of pineapple. Initiation of flowering requires reduction of the photoperiod to 10–11 h until the inflorescence develops, after which plants can be returned to a 16 h photoperiod for fruit production.

14. PPT-resistant plants and their micropropagated clonal material of the cultivar Phuket remain green at herbicide concentrations including 20 mg/l glufosinate ammonium. Such plants develop normally following herbicide treatment. Non-transformed plants of the same cultivar remain green only on medium lacking Basta™. The inhibitory effect of Basta™ at all concentrations of glufosinate ammonium evaluated (3, 5, 7, 10, 15 or 20 mg/l) is normally observed within 4 days of herbicide application.

15. PPT-tolerant plants and their clones of the cultivar Phuket remain green and show vigorous growth even after applications of Basta™ containing 100-1400 mg/l glufosinate ammonium, the higher concentration being in excess of the concentration normally used under field conditions in Thailand (*16*). Non-transformed potted plants undergo extensive necrosis within 4 days of spraying with herbicide, followed by extensive browning within 14 days and death. As in Note 14, these herbicide concentrations should be used as a guide

CONCLUSIONS

Cultured shoots provide a readily available source of leaves whose bases, because of the totipotency of their cells, are a convenient target for gene delivery by microprojectile bombardment. In addition, reproducible shoot regeneration from leaf base tissues eliminates the requirement to establish more complex plant regeneration protocols, such as those involving somatic embryogenesis (*7*), and minimizes the probability of undesirable somaclonal variation. The transgenic status of selected, herbicide tolerant plants should be confirmed at the molecular level using PCR, RT-PCR and Southern analyses. The primers for the *bar* and *gus* genes for PCR and RT-PCR analyses and details of these procedures, together with the method for Southern blot analysis, can be found in the published literature (*16*). Whilst the *gus* gene is not essential for the selection of transgenic plants and, indeed, it is preferable longer-time for this reporter gene to be absent from the DNA used for transformation, it can be exploited to indicate the success and extent of gene delivery into target tissues. The transformation protocol involving the shoot regeneration and selection procedure developed for the cultivar Phuket and described here, should be applicable to a range of commercially grown pineapple cultivars.

REFERENCES

1. FAOSTAT (2002). Agriculture data. www.app.fao.org/page/collections subset = agriculture.
2. FAO (1998). Avocados, Mangoes, Pineapple. *Food and Agricultural Organization Yearbook Production*, **52**: 161-162.
3. Py C, Lacoeuilhe JJ and Tesson C (1987). *The Pineapple: Cultivation and Uses.* GP Maisonneuve *et* Larose, Paris, p. 568.
4. Bartholomew DP, Paull RE and Rohrbach KG (2003). *The Pineapple. Botany, Production and Uses.* CABI Publishing, CAB International, Wallingford, Oxon, UK, p. 301.
5. Duval MF, Noyer JL, Perrioer X, Coppens d'Eeckenbrugge G and Hamon P (2001). Molecular diversity in pineapple assessed by RFLP markers. *Theoretical and Applied Genetics*, **102**: 83-90.
6. Sripaoraya S, Marchant R, Power JB, Lowe KC and Davey MR (2001). Relationships in pineapple by random amplified polymorphic DNA (RAPD) analysis. *Plant Breeding*, **120**: 265-267.
7. Sripaoraya S, Marchant R, Power JB and Davey MR (2003). Plant regeneration by somatic embryogenesis and organogenesis in commercial pineapple (*Ananas comosus* L.). *In Vitro and Cellular Developmental Biology–Plant*, **39**: 450-454.
8. Firoozabady E and Gutterson N (2003). Cost-effective *in vitro* propagation methods for pineapple. *Plant Cell Reports*, **21**: 844-850.
9. Sripaoraya S (2001). Genetic Manipulation of Pineapple (*Ananas comosus*). Ph.D. Thesis, University of Nottingham, UK.
10. Barsby T, Power JB, Freeman J, Ingram HM, Livesey NL, Risacher T and Davey MR (2001). Transformation of wheat. In: Bonjean AP, Angus WJ (eds.), *The World Wheat Book. A History of Wheat Breeding.* Lavoisier Publishing, Paris, pp. 1081-1103.
11. Ke J, Khan R, Johnson T, Somers DA and Das A (2001). High-efficiency gene transfer to recalcitrant plants by *Agrobacterium tumefaciens*. *Plant Cell Reports*, **20**: 150-156.
12. Firoozabady E, Heckert M, Oeller P and Gutterson N (1997). Transformation and regeneration of transgenic pineapple plants. 5th *International Congress of Plant Molecular Biology, Singapore.* Abstract No. 1358.

13.   Curtis IS, Power JB, Blackhall NW, de Laat AMM and Davey MR (1994). Genotype-
       independent transformation of lettuce using *Agrobacterium tumefaciens*.   *Journal of
       Experimental Botany*, **45**: 1441-1449.
14.   Nan G-L and Nagai C (1998). Genetic transformation of pineapple (*Ananas comosus* [L.] Merr.)
       via particle bombardment.   *In Vitro Cellular and Developmental Biology–Plant*, **34**: 55-A,
       Abstract No. 1052.
15.   Ko HL, Graham MW, Hardy VG, Jobin M, O'Hare TJ and Smith MK (2000). Transformation of
       pineapple using Biolistics. *6th International Congress of Plant Molecular Biology, Quebec,
       Canada.* Abstract No. S03-64.
16.   Sripaoraya S, Marchant R, Power JB and Davey MR (2001). Herbicide-tolerant transgenic
       pineapple (*Ananas comosus*) produced by microprojectile bombardment. *Annals of Botany*, **88**:
       597-603.
17.   D'Halluin K, De Block M, Denecke J, Janssens J, Leemans J, Reyaerts A and Botterman J
       (1992). The *bar* gene as a selectable and screenable marker in plant engineering. In: Wu R (ed.),
       Methods in Enzymology Vol. 216, Recombinant DNA, Part C. Academic Press, New York, USA,
       pp. 415-426.
18.   Murashige T and Skoog F (1962). A revised medium for rapid growth and bioassays with
       tobacco tissue cultures. *Physiologia Plantarum*, **15**: 473-497.
19.   Murashige T and Tucker DPH (1969). Growth factor requirements of Citrus tissue culture. In:
       Chapman HD (ed.), *Proceedings of the First International Citrus Symposium, Vol. III*, Riverside
       Color Press, Riverside. pp. 1155-1161.
20.   Christensen AH, Sharrock RA and Quail PH (1992). Maize polyubiquitin genes: structure,
       thermal perturbation of expression and transcript splicing, and promoter activity following
       transfer to protoplasts by electroporation. *Plant Molecular Biology*, **18**: 675-689.

Chapter 15

REGENERATION AND GENETIC TRANSFORMATION OF APPLE (*MALUS* SPP.)

S.M.W. BULLEY AND D.J. JAMES[1]

HortResearch, Mt Albert Research Centre, Private Bag 92169, Auckland, New Zealand; [1]Horticulture Research International, East Malling, Kent, ME196BJ, UK. Corresponding author: S.M.W. Bulley. E-mail: SBulley@hortresearch.co.nz

1. INTRODUCTION

1.1. The Importance of Apple in World Fruit Production

In terms of world production, apple ranks in the top three of major deciduous, citrus and tropical fruits, grapes and berries. In 2001, just over 61 million metric tonnes of apple fruit was produced, accounting for 14% of total world fruit production. Oranges and bananas, the other top three members, account for 14 and 15% of world fruit production, respectively (*1*). Over the past decade production has grown while demand has remained static, so producer returns have been diminishing in real terms for a number of years.

One strategy taken to maintain/increase profitability over the past 20 years has been to introduce novel varieties derived from traditional breeding and mutant 'sports' found in the orchard, such as, Fuji and Gala and its sport, Royal Gala. Worldwide, these varieties are steadily replacing older varieties such as Red Delicious and Granny Smith. However, as volumes increase the premium they return steadily erodes. Conventional breeding will remain a key part of future strategy for profitability but the length of time required to breed and test new cultivars (10-20 years), even with marker assisted selection, as well as the investment required to achieve market penetration/acceptance are major barriers. Such problems are further compounded by changing trends in food consumption towards new exotic fruits and convenience foods eroding demand for traditional foods such as apples (*1*). On the other hand, novelty is not always desirable. Such is the case for bittersweet apple cultivars used for juice/cider production as well as rootstocks. These types of apple tend to be well established and valued for their existing characters, which are often lost/altered through conventional breeding. There are also many old cultivars that might be commercially viable if it were not for a few negative traits e.g. susceptibility to a particular disease, poor storage qualities or skin finish. Therefore conventional breeding is not so attractive when, in many cases, alteration of only a few traits is desired.

1.2. Opportunities for Cultivar Improvement by Genetic Modification

Genetic engineering offers new opportunities to add value and to increase productivity in a vastly shorter time frame (2-5 years) than that required by

199

I.S. Curtis (ed.), Transgenic Crops of the World - Essential Protocols, 199-215.
© *2004 Kluwer Academic Publishers. Printed in the Netherlands*

conventional breeding. Targets on the production side include disease resistance (e.g. fireblight, scab, *Phytophthera* and powdery mildew), post harvest characters, pest resistance (e.g. insect/nematode pests), tree architecture (including control of vigour), flowering time, reducing biennialism, and stress resistance (e.g. to water/salt/heat stress). Most of these requirements are region-specific and so not all need to be deployed in a particular locality. Consumer-orientated traits include improving health characteristics, eliminating allergenicity, altering flavour components and colours, and providing novel benefits such as increased protection from tooth decay or immunisation.

Assuming further advances in sequencing and outputs from the human genome, future possibilities may include matching an individual's genotype with diet. This will create many new opportunities, and possible pitfalls, for apple producers. Current public opinion towards genetic engineering in many countries outside the USA indicates that consumer-benefiting traits may be better received than the genetically engineered crops currently under cultivation, which are seen by many consumers as benefiting producers only. This is an expression of our personal view and experience. We also would like to stress that modification of many production orientated traits result in indirect benefits to the consumer in terms of reduced pesticide inputs, reduced pollution and other effects that reduce impacts on the wider environment.

1.3. Genetic Modification of Apple

Greensleeves was the first apple cultivar to be transformed (*2*) and with slight modifications regularly returns transformation frequencies between 8-16% on a per explant basis (S. Bulley, F. Wilson, A. Passey and S. Vaughan, unpublish.). The method is based on the leaf disk transformation method using disarmed strains of *Agrobacterium tumefaciens*, originally described by Horsch *et al.* (*3*). The transgenes in the transformed apple plants produced this way displayed stable patterns of expression in fruit and Mendelian segregation in the progeny (*4*). To date, variations of this method have been used to transform many different dessert apple varieties (scions) and a number of rootstocks (Table 1).

Factors that are important for successful transformation in Greensleeves include a reliable method for regenerating shoots from leaf tissue, and the use of the hypervirulent *A. tumefaciens* strain EHA101 otherwise known as A281 (*25*). The hypervirulence of EHA101 is encoded in a region of pTiBo542 outside the T-DNA (*26*). Other strains have been used successfully with Greensleeves but do not give such high transformation frequencies in terms of blue stained callus with a construct containing the 35S-*gus*A reporter gene (*25*). In other cultivars, EHA101 has been found to be the most effective strain for transfection and transformation from four strains tested (*7*, *27*).

This chapter updates a previous chapter (*28*) and includes changes in the protocol. It outlines the method of leaf disk transformation of the apple cultivar Greensleeves (*2*) with modifications as described (*5*). This can be used as a starting point when attempting to transform a variety where transformation has not yet been reported. With this in mind, the relative importance of each step is discussed and alternatives for most steps are presented in the Notes Section as well as further suggestions for achieving or improving transformation.

Table 1. Summary of apple transformation using the Agrobacterium-mediated leaf disk transformation method (in order of date published) showing major media components. Reports without Southern blot data are not included.

Cultivar	Agrobacterium tumefaciens strain	Hormones used for regeneration	Nutrients and sugar	Reference
Greensleeves	(LBA4404)[A], (EHA101)[B]	(BAP, NAA)[A], (BAP, NAA, TDZ)[B]	(MS with sorbitol or sucrose)[AB]	[A](2); [B](5 and 6)
M26	(LBA4404)[C], (C58C1)[D]	(BAP, NAA)[CD]	(N$_6$ + MS with sucrose)[C], (MS with sucrose)[D]	[C](7); [D](8)
Delicious	Not stated (GV?)	NAA, TDZ	MS with sucrose	(9)
M7[+]	LBA4404	BAP, NAA	N$_6$ + MS with sucrose	(10)[+]
Royal Gala	LBA4404	BAP, NAA	MS with sucrose	(11)
Braeburn	EHA101	IBA, TDZ, 2iP, GA$_3$	MS with sucrose	(12)
Elstar	(EHA101)[E], (AGL0)[F]	(IBA, TDZ, 2iP, GA$_3$)[E], (NAA, TDZ)[F]	(MS with sucrose)[E], (MS with sorbitol)[F]	[E](12); [F](13)
Fuji	EHA101	IBA, TDZ, 2iP, GA$_3$	MS with sucrose	(12)
Gala	(EHA101)[E], (AGL0)[F]	(IBA, TDZ, 2iP, GA$_3$)[E], (BAP, NAA, TDZ)[F]	(MS with sucrose)[E], (MS with sorbitol)[F]	[E](12); [F](13)
Golden Delicious	(EHA101)[E], (AGL0)[F]	(IBA, TDZ, 2iP, GA$_3$)[E], (BAP, NAA, TDZ)[F]	(MS with sucrose)[E], (MS with sorbitol)[F]	[E](12); [F](13)
Jonagold	EHA101	IBA, TDZ, 2iP, GA$_3$	MS with sucrose	(12)
Jonagold King	EHA101	IBA, TDZ, 2iP, GA$_3$	MS with sucrose	(12)
Jonagold Red	EHA101	IBA, TDZ, 2iP, GA$_3$	MS with sucrose	(12)
Merlijin	EHA101	IBA, TDZ,	MS with sucrose	(12)

Table1(cont.)

		2iP, GA$_3$		
Falstaff	LBA4404	BAP, NAA, TDZ	MS with sorbitol	*(14)*
Pink Lady	Not stated (GV?)	NAA, TDZ	MS with sucrose and glucose	*(15)*
Marshall McIntosh	EHA105	NAA, TDZ	N6 with sucrose	*(16)*
Jork 9	EHA101, C58C1	NAA, TDZ	MS with sucrose	*(17)*
M9/29	C58C1	NAA, TDZ	MS with sorbitol	*(18)*
A2	GV3101	NAA, TDZ	MS with sucrose	*(19)*
Marubakaido	EHA101	TDZ, ABA, IAA	MS with sucrose	*(20)*
Queen Cox	EHA101	NAA, TDZ	DKW with sorbitol	*(21)*
Michelin	EHA101	BAP, NAA, TDZ	MS with glucose	Bulley & James, unpublish.

+Originally reported as transformation of M26 but subsequently found to be M7 (*16*).
Murashige and Skoog(MS, *22*) medium; DKW: Driver and Kuniyuki medium (*23*); N$_6$ macroelements (*24*).
[A-F]Specific references.

2. MATERIALS

2.1. Equipment Required

For micropropagation of apple shoot cultures we use square base cell counting vials (Fig. 1C) by Ratiolab® (Ratiolab GmbH, Am Siebenstein 5, Dreieich-Buchschlag, Germany) containing 8-10 ml of appropriate medium. For bacterial cultivation and regeneration from leaf explants we use 9 cm diam. disposable Petri dishes (Bibby Sterilin Ltd., Stone, UK) containing 25-30 ml of appropriate medium.

2.2. Hormones and Media Preparation

Plant hormones used: BAP (Sigma), IBA (Sigma), gibberellic acid (GA$_3$, Sigma), NAA (Sigma), TDZ (Sigma). Make stock solutions by dissolving 100 mg hormone in either 5 ml ethanol (for GA$_3$, IBA, BAP), or 5 ml 1N NaOH (for NAA) and then make up to 100 ml with double distilled water. Dissolve TDZ in methanol. The 1 mg/ml stocks are dispensed into 5 ml aliquots and stored at –20°C. Keep working stocks at 4°C for no longer than 4 weeks.

2.3. Media Components for Plant Culture

2.3.1. Shoot Multiplication Medium (A17) – 1 litre

1. Sucrose	30.00 g
2. Murashige & Skoog (MS) macroelements (Sigma-Aldrich)	4.40 g
3. Add double distilled water and mix using magnetic stir bar	500 ml
4. Add 1 mg/l BAP stock	1.0 ml
5. Add 1 mg/l IBA stock	0.1 ml
6. Add 1 mg/l GA$_3$ stock	1.0 ml
7. Make volume up to 1 litre with double distilled water	
8. Adjust pH to 5.2	
9. Add gelling agent (Agar No. 3, Oxoid)	7.50 g
10. Autoclave, 20 min, 120°C, 15 psi.	
11. If required, add filter-sterilized selective agent and antibiotics prior to dispensing the medium.	

2.3.2. Root Induction Medium (R13) – 1 litre

1. Sucrose	30.00 g
2. MS macroelements	4.40 g
3. Add double distilled water and mix using a magnetic stir bar	500 ml
4. Add 1 mg/l IBA stock	3.0 ml
5. Make volume up to 1 litre with double distilled water	
6. Adjust pH to 5.2	
7. Add gelling agent (Agar No. 3, Oxoid)	7.50 g
8. Autoclave, 20 min, 120°C, 15 psi.	
9. If required, add filter-sterilized selective agent and antibiotics prior to dispensing the medium.	

2.3.3. Root Elongation Medium (R37) – 1 litre

1. Sucrose	30.00 g
2. MS macroelements	2.20 g
3. Make volume up to 1 litre with double distilled water	
4. Adjust pH to 5.2	
5. Add agar (Agar No. 3, Oxoid)	7.50 g
6. Autoclave, 20 min, 120°C, 15 psi.	
7. If required, add filter-sterilized selective agent and antibiotics prior to dispensing the medium.	

2.4. Media Components for Transformation and Agrobacteria Cultivation

2.4.1. Greensleeves Regeneration Medium (BNZ511) – 1 litre

1. Sorbitol   30.00 g
2. MS macroelements   4.40 g
3. Add double distilled water and mix using magnetic stir bar   500 ml
4. Add 1 mg/l BAP stock   5.0 ml
5. Add 1 mg/l NAA stock   1.0 ml
6. Add 1 mg/l TDZ stock   1.0 ml
7. Make volume up to 1 litre with double distilled water
8. Adjust pH to 5.2
9. Add gelling agent (Phytagel, Sigma-Aldrich)   7.50 g
10. Autoclave, 20 min, 120°C, 15 psi.
11. Add kanamycin (Sigma) for selection of *npt*II marker and cefotaxime (Claforan®, Roussel) by filter-sterilization (0.22 µm pore) to a final concentration of 100 and 200 mg/l, respectively.
12. Dispense into Petri dishes (~30 ml per dish) or Replidishes (~2.5 ml per well). If Replidishes are used dispense a portion into Petri dishes (about 10 dishes) for the co-cultivation step.

2.4.2. Co-cultivation Medium (MS20) – 1 litre (after 29, with modifications)

Prepare this on the day of transformation and store at 4°C until required.
1. Sucrose   20.00 g
2. Murashige and Skoog (MS) macroelements (Sigma-Aldrich)   2.20 g
3. Make volume up to 1 litre with double distilled water
4. Add betaine hydrochloride (Fischer) and acetosyringone (Sigma-Aldrich) to 1 mM and 0.1 mM final concentrations, respectively.
5. Adjust pH to 5.2.
6. Filter sterilise solution (0.22 µm pore).

2.4.3. YEP Medium – 1 litre

1. Yeast extract (Oxoid Ltd.)   10.00 g
2. Peptone (Oxoid Ltd.)   10.00 g
3. NaCl   5.00 g
4. Make volume up to 1 litre with double distilled water
5. Adjust pH to 7.2
6. Autoclave, 20 min, 120°C, 15 psi.
7. Add required antibiotics from filter-sterilized stocks to appropriate concentrations just before use (see methods for concentrations).

2.4.4. LB Medium – 1 litre

1. Yeast extract	5.00 g
2. Tryptone (Oxoid Ltd.)	10.00 g
3. NaCl	10.00 g
4. Make volume up to 1 litre with double distilled water	
5. Adjust pH to 7.2	
6. Add BactoAgar (Difco)	15.00 g
7. Autoclave, 20 min, 120°C, 15 psi.	

8. Add required antibiotics from filter-sterilized stocks to appropriate concentrations just before use (see methods for concentrations).

3. METHODS

3.1. Micropropagation of Greensleeves

The sources of tissue used for transformation are sterile micropropagated shoot cultures. For Greensleeves, shoots (3-4 per pot) are grown in square base cell counting vials (Ratiolab®) pots containing between 8-10 ml A17 media under artificial lighting (70-80 µmol/m^2/sec PAR, 400-700 nm) in a growth room (16 h photoperiod at 24°C). If these are unavailable, any autoclavable glassware or sterile plastic ware can suffice. To achieve optimum results, it is vital the vessels are sealed well so that evaporation is minimized. To maintain vigorous and healthy growth, cut healthy shoot tips and transfer to fresh media every 4 weeks (this is referred to as subculture). To cut down on subculture when tissue is not required it is possible to grow the shoot cultures at 4°C under the same light regime. This way the shoots need only be subcultured every 3-4 months. Before using such cultures for transformation, it is recommended to return the cultures to normal temperature for two subcultures (3 weeks between each) before rooting.

3.2. Introducing a New Cultivar into Sterile Tissue Culture

To introduce a new cultivar into sterile micropropagation it is best to use buds as starting tissue. In spring, harvest between 20-60 breaking buds from the desired tree and surface-sterilize with three 5 min washes in 0.7% bleach (v/v) solution. After washing in bleach solution, rinse 3 times in sterile water and place in A17 media supplemented with PPM™ (Plant Cell Technology). Use the same growth conditions as described for Greensleeves but subculture more frequently, e.g. once every three weeks instead of four. Reducing the temperature (e.g. to 20°C) may improve the overall health of the culture. Depending on the health and rate of growth, different media formulations may be required.

Determining the appropriate medium to use is largely an empirical process. Therefore, we can offer few suggestions other than trying out published formulations. A17 medium is a good starting point for micropropagation media but not all cultivars will grow well on this substrate. We encountered this with some bittersweet apple cultivars that were introduced from the field. Replacing MS macroelements with Driver and Kuniyuki (DKW) medium (*23*) and adding supplementary vitamins gave better growth and healthier tissue. We also saw a pronounced change in growth (for the better) over time as the tissue adapted to sterile culture. However, there were also continual problems with contamination, presumably from two or three different bacterial saprophytes from the surface of the plant due to changes to the plant physiology/morphology induced by tissue culture. This problem was addressed by only using material from containers that appeared clean and performing washes in PPM™.

3.3. Transferring Plants from Sterile Micropropagation to Soil

This method is based (with slight modifications) on that described in (*30*) and is summarized as follows. Induce vigorous healthy shoots to root by excising 2 cm long shoots and placing them in R13 root induction medium for 4 days in darkness. Following this, transfer the shoots to R37 medium, maintain in the dark for 4 more days and then transfer the culture vessels to the light to promote growth and leaf expansion (*see* Section 3.1.). Wait until the roots begin to reach the base of the container (approx. 3-4 weeks) then remove the plants from their containers and rinse in water to remove residual agar. Carefully transfer the plant in autoclaved potting compost, misting each plant well with a fine spray of water to prevent drying out, as the plants are very sensitive at this stage. Maintain the plants in a propagator with vents closed and out of direct light for 3-4 days. After this, place the propagators under direct light (16 h photoperiod at $20 \pm 2°C$) for 5 days. Next, gradually open the vents over a further 5-day period until completely open. After 2 more days, remove the propagator cover and leave for a further 7 days. These plants can now be considered weaned and are ready for transfer to a glasshouse or growth chamber for further growth/analysis.

Suitable glasshouse growth conditions are a 14-18 h photoperiod and temperatures of 20-30°C (day) and 14-20°C (night). For putative transgenic lines, supplement root induction medium with 50 mg/l kanamycin (or the appropriate selective agent for your particular selectable marker). We have found that non-transgenic Greensleeves shoots (escapes) will not form roots under kanamycin selection.

3.4. Preparation of Tissue

Root healthy shoots as described in the section above (*see* Note 1). When the tissue is to be used for transformation, it is important that the material used is no more than 4 weeks-old after being transferred to R37 medium. De Bondt *et al.* (*27*) showed that for a number of cultivars, that the optimum age of harvesting shoots, in terms of the number of GUS zones per leaf, is between 20 and 30 days. Later research showed that the percentage of kanamycin resistant cv. Jonagold shoots arising from 33-day-old explants was approx. 50% that of 27 day old explants (*12*).

3.5. Transformation of Greensleeves

Perform the following under sterile conditions in laminar flow cabinets and if available, Class II downward flow cabinets (for bacterial work). Wear gloves when handling bacterial cultures and follow sterile protocol.

1. Four days before leaf disks are to be cut, streak *A. tumefaciens* strain EHA101, containing the appropriate transformation construct (*see* Note 2), from a frozen glycerol stock onto LB agar plates containing 25 mg/l neomycin sulphate, 50 mg/l chloramphenicol, 25 mg/l gentamycin (the rates and type of antibiotics are those used with pSCV constructs). Incubate at 28°C for 2 days growth then inoculate one colony into 5 ml YEP media containing 25 µg/ml neomycin sulphate, 50 µg/ml chloramphenicol and 15 µg/ml gentamycin in a 50 ml Falcon screw top tube (Sarstedt). Leave the lid loose to allow ventilation. Incubate at 28°C with shaking at 200 rpm for 24 h.
2. Inoculate 4 x 9 ml lots of fresh YEP media containing the appropriate antibiotics with 1 ml of the previous culture and incubate at 28°C with shaking at 200 rpm for 24 h as before. Prepare holding media (BNZ511 regeneration medium without selection), regeneration medium BNZ511 with selection and co-cultivation medium. Dispense the media into sterile Petri dishes.
3. Day of transformation: Check the optical density (OD) of cultures is 1.7-2.0 at 420 nm by reading 1 ml of a 10X dilution in YEP with YEP as a blank (an O.D. of 1.0 at 420 nm is equivalent to approx. 5×10^8 colony forming units). Do not forget to multiply the reading by 10.
4. Centrifuge the remainder at 3500 rpm for 15 min at 4°C. Discard the supernatant and resuspend the pellet immediately in MS20 medium to give equivalent bacterial concentration of 0.5 OD (the pellet may have to be broken up after centrifugation using sterile inoculating loop before total resuspension).

Example: if original reading is 0.15 (x10 = 1.5) then a dilution by a third is needed to give 0.5 OD, thus an original 10 ml culture would need to be resuspended in 30 ml MS20 medium.

5. Aliquot 10 ml of bacterial suspension into sterile 15 ml Falcon tubes (Sarstedt) and incubate at 20°C with gentle shaking for 5 h (*see* Note 3). During this time prepare explant tissue as described in the next step.
6. Excise young healthy leaves from shoots rooted no more than 4 weeks previously and cut leaf disks (approx. 7 mm in diam.) with a sharp sterile number 2 size cork borer. Lightly score the leaf disks with two cuts across the centre of the disk taking care not to cut through the leaf (*see* Note 4). As you cut the leaf disks place them on holding media without selection (regeneration

medium BNZ511 for Greensleeves) to prevent desiccation. Alternatively, cut leaf segments with a scalpel instead of using a cork borer (*see* Note 5).

7. After *Agrobacterium* induction is complete, place 50 leaf disks in each tube containing the bacterial solution (10 ml volumes) and incubate at room temperature with gentle shaking for 20 min.

8. Remove excess liquid from each disk by blotting on sterile filter paper (Whatman) and place abaxial side down onto filter paper (wetted with induction media) overlaid on BNZ511 regeneration medium (without selection) in Petri dishes. Seal the plates and co-cultivate for 3 days at 25°C in darkness (*see* Note 6).

9. After co-cultivation, remove the Agrobacteria by washing in filter-sterilized 500 mg/l augmentin and 200 mg/l cefotaxime solution (pH 5.2) with gentle shaking for 5 h at 20°C. Blot the leaf disks dry as before and then transfer, placing abaxial side in contact (*see* Note 7) with BNZ511 regeneration medium containing kanamycin (or appropriate selective agent) and cefotaxime at 100 mg/l and 200 mg/l, respectively. Incubate at 25°C in darkness (*see* Note 8) and transfer to fresh media every 4 weeks. Shoots generally appear 4 weeks after inoculation (Figs. 1A and B). Once the third leaf appears, excise and transfer to A17 medium containing 50 mg/l kanamycin (or appropriate selective agent) and grow in the light (16 h photoperiod, 24°C day, 16°C at night). Record each shoot as an individual line and keep separate from other lines. Excise shoots (approx. 2 cm in length) initially to R13 medium for root induction and then transfer to R37 medium for root elongation as described (*see* Section 3.3.; Fig. 1C). Maintain the plants in culture then transfer to soil as described previously.

10. Plants are transferred to pots (approx. 18 cm diam.) containing Sinclair's SHL All Purpose Potting Compost (William Sinclair Holding PLC, Lincoln, UK) and grown in propagating boxes enclosed in a growth chamber (16 h photoperiod, 70-80 $\mu mol/m^2/sec$ light intensity at 20ºC). Details of the weaning process of plants are described (*see* Section 3.3.). Once plants are weaned (Fig. 1D), transfer to a designated transgenic glasshouse under ambient light conditions set at 25ºC. Plants are watered every 2-3 days and a liquid fertilizer (Vitafeed from Vitax ltd., Coalville, UK) is applied every third watering. After autumn leaf fall, potted plants are transferred to a cold room at 4°C for 6 weeks and then returned to the glasshouse to flower in the spring.

4. NOTES

1. It has been shown that leaf disks excised from rooted *in vitro* plantlets gave significantly increased levels of regeneration compared to unrooted controls (*31*). Additionally, we feel that the resultant explant material is more uniform in terms of regeneration response than material taken directly from micropropagation. Furthermore, we have found that by inducing the shoots to root in culture causes the leaves to expand more and thus resulting in more leaf disks per leaf.

2. For Greensleeves we use constructs based on the pSCV (Plasmid Shell Clean Vector) series of plasmids (*6*). For initial transformation experiments pSCV1.6 is useful as the T-DNA region contains the selectable marker gene *npt*II for selection with aminoglycoside antibiotics and the *gus*A gene as a visual marker. Moreover the *mob* functions from the backbone of the vector have been deleted to prevent transconjugation with other bacteria and thereby prohibit lateral gene transfer.

The CaMV 35S promoter drives expression of each gene. However, although we use kanamycin as our selectable agent, Norelli & Aldwinckle (*32*) concluded that neomycin and paromomycin are likely to be the more useful aminoglycosides for the selection of *npt*II transgenic plants than either kanamycin or geneticin. There are alternative non-antibiotic based marker systems such as the PMI system (Syngenta). This system has been successfully used for transformation of many different plant species (*33*). Another non-antibiotic marker system using xylose isomerase has been shown to be an effective selectable marker in tomato, potato and tobacco (*34*). Finally, using vectors that incorporate additional *vir* genes has been shown to enhance transformation (*25*).

3. Essential conditions for induction of *vir* genes occur at temperatures below 28°C in the presence of 2-3% sucrose and at acidic pH between 5.0 and 5.5 (*29*). Addition of the phenolic compound acetosyringone is not vital to transformation as chemical elicitors, which include acetosyringone (*35*), arising from wounded leaf tissue is sufficient to induce transfection. However, additional treatment with acetosyringone and betaine synergistically enhanced *Agrobacterium*-mediated transformation in apple (*5*). Incubating bacteria with acetosyringone at pH 5.2 before inoculation induces transcription of virulence genes (*36*), but an acidic pH reduces bacterial growth. This is countered by including betaine (in the form of betaine glycine or betaine HCl), which acts as an osmoprotectant and improves growth and virulence induction by aiding the bacteria to adjust to pH changes (*37*).

4. Further wounding of the explant tissue increases the area able to be infected as well as increasing concentrations of transfection inducing wound chemicals. With Greensleeves we have found scoring to be unnecessary as the rate of transformation was sufficient enough for our purposes. However, wounding in addition to excision of the explant may increase transformation frequencies in other cultivars. Techniques that can be used include vortexing in a solution of silicon carbide whiskers, scoring/cutting with a scalpel (as suggested here), and crushing with forceps (*32, 14*). It should also be noted that Igarashi *et al.* (*20*) obtained transgenic Marubakaido rootstock plants by using fully expanded leaves from 4 week-old micropropagated shoots as explants without any excision/wounding.

5. Whether you use leaf disks or cut leaf segments is a matter of personal preference. The main reason we use leaf disks is that they exhibit less curling and therefore tend to have better contact with the medium than cut leaf segments (*28*). This allows better and more uniform penetration and uptake of media components including selective agents. When testing regeneration media formulations for transformation we recommend the use of leaf disks because it minimizes the between-explant variance contributed by leaf area. This gives greater confidence in statistical comparisons between treatments, especially when variance is normally high and differences between treatments might be small. On the other hand, a benefit of using cut segments compared to cutting leaf disks is that more explants can be prepared in a set time period. In addition, more explants are able to be prepared from the same amount of tissue because no tissue is wasted.

6. The length of co-cultivation is a compromise between allowing transformation to occur while not allowing bacterial overgrowth of tissue that result in tissue necrosis. When transforming Jonagold using EHA101, the number of GUS zones per leaf peaked after 4 days co-cultivation (*27*), therefore increasing the length of co-cultivation may be an option when attempting transformation studies of a different cultivar.

7. Placing the abaxial surface (underside of the leaf) in contact with the media is the most popular way of culturing explants. This was found to be ideal for Greensleeves (*31*), but may not be the case for other cultivars.

8. Different cultivars vary in the optimum temperature required for regeneration. Determination of the correct temperature is largely empirical. It was found that lowering the incubation temperature from 25°C to 20°C increased the number of viable transformed shoots as reported by Falstaff (*14*). Incubation in the dark is most commonly cited in the literature; however, successful regeneration has been achieved using a period of culture in the dark (2-4 weeks) with subsequent incubation in the light (*7, 11*).

FURTHER SUGGESTIONS FOR ATTEMPTING TO TRANSFORM A NEW CULTIVAR

Transformation Steps

As mentioned in the introduction, *A. tumefaciens* strain EHA101 (*25*) gives high rates of transfection so we recommend that this strain to be used. We also recommend that altering induction and co-cultivation steps detailed in the methods should be treated as a low priority when attempting transformation. In our experience, the regeneration response varies greatly between varieties and we recommend this be the main point of focus when trying to transform a cultivar where transformation has not been reported. Using kanamycin as the selective agent and *gfp* gene as a non-destructive reporter, Maxinova *et al.* (*38*) concluded that factors other than *Agrobacterium* interaction and T-DNA transfer are rate-limiting steps in *Agrobacterium*-mediated transformation of apple. They estimated that "the two steps of DNA transformation and shoot regeneration reduced the efficiency of the transformation and regeneration process by about 10,000-fold" (*38*). This loss of efficiency may not be as great if a positive rather than a negative selection method is used because with positive selection, transformed cells do not detoxify the selective agent, but instead, modify a substrate to a utilizable form (e.g. allowing xylose to be used as a carbon source).

Regeneration vs. Transformation and Regeneration

One strategy is to transform and then place subsets of leaf disks on different regeneration media (*11, 13, 17*). Whether you choose to do this is a matter of preference. We prefer to test regeneration media first before attempting transformation. This way we can perform 'kill curves' with the best medium in order to determine the correct amount of selective agent to use. A kill curve is a plot of antibiotic concentration and numbers of shoots that regenerate from explants. The lowest concentration of selective agent that inhibits shoot regeneration is used. When using kanamycin we have found that there is a wide genotypic response of concentrations required that range between 5 µg/ml and 100 µg/ml. Greensleeves requires a high concentration of kanamycin for selecting transformed shoots, and typically for other cultivars one would begin in the range of 30-60 µg/ml.

Conditioning of Explant Tissue

Conditioning of apple shoots for several days in a liquid medium containing MS salts, vitamins, glycine, sucrose and hormones enhanced the regenerative capacity and increased transformation of leaf explants derived from shoots (*15*). The authors demonstrated that this caused increased porosity of cell walls and speculated that this may aid the insertion of T-DNA through the cell wall.

Regeneration

A survey of available literature shows that three phytohormones are commonly used in regeneration media for transformation of a range of cultivars. The auxin, NAA, is used together with either or both the cytokinins BAP and TDZ. For some cultivars, NAA appears to be required for shoot formation because, when omitted, shoot regeneration of apple dwarfing rootstocks was inhibited (*39*). The hormone concentrations used range from 0-22.2 μM for BAP, 0.5-5.4 μM for NAA and 0-10 μM for TDZ. Other forms of auxin and cytokinin have been used for regeneration but in the main BAP, NAA and TDZ have been most effective in terms of transformation of a wide variety of cultivars (Table 1). Therefore, it is recommended that combinations of these hormones provide the basis for initial work. For cultivars where numbers of regenerated shoots are low, it can be useful to begin regeneration on media containing auxin and cytokinin for 1-2 months then to subculture to media without auxin but containing the cytokinin TDZ. Caboni *et al.* (*40*) used this method to regenerate apple shoots from callus of Jork9, M26, Gala and McIntosh shoot apices. This was succesfully employed (with different hormone formulations) to produce transgenic Michelin (S. Bulley, unpublish.) and has also been used successfully for strawberry (S. Vaughan, pers. comm.). Be aware that prolonging the undifferentited callus phase can lead to somaclonal variation and polyploidy of regenerated shoots (*41*).

Gibberellin biosynthesis inhibitors (e.g. paclobutrazol) have been used successfully in other plant species to increase regeneration frequencies (*42, 43*). Gibberellic acid (GA$_3$) treatment inhibited regeneration of apple (S. Bulley, unpublish.) and other species (*42-44*). Therefore, GA biosynthesis inhibitors may be of use in trying to increase shoot regeneration.

Other components that can have an impact are macro-elements, sugar source and gelling agent. Testing different types/concentrations of sugar and gelling agent should take higher precedence over basal macro-elements, as MS gives good results for most of the cultivars studied. BNZ511 media containing 3% (w/v) glucose instead of sorbitol increased shoot number and quality, and almost abolished callusing in untransformed Greensleeves leaf disks (A. Passey, pers. comm.). We have not included this modification as this observation has not been verified by transformation. The type of gelling agent can have a significant effect on regeneration, in that, Phytagel (Sigma-Aldrich) has given optimal results for tested cultivars.

Tissue Type

Reports of apple transformation to date have used leaf tissue for explants, but stem internodes may also be useful. Shoots have been regenerated from stem internodes (*45-47*), but at low frequencies. Apparent transformation in the form of GUS expressing calli using Royal Gala stem internodes was reported by Liu *et al.* (*48*), who also showed that etiolating the explants before *Agrobacterium*-inoculation promoted high frequency of shoot organogenesis.

CONCLUSIONS

A reliable and reproducible model transformation system for apple, based on the cultivar Greensleeves, has been described in detail. Its ease of cultivation, both in tissue culture and *in solium* and the relative ease with which it can be genetically transformed means that we have been able use this system to study gene regulatory function, gene silencing and the role of various promoters in tissue-specific gene expression in fruit trees. The ease of genetic transformation allows the relatively fast generation of large numbers of transgenic tree plants for experimental use. The system is based on the use of leaf explant tissue together with disarmed *A. tumefaciens* strains. This system is not limited to Greensleeves, as it has been successfully applied to a number of different apple cultivars - sometimes with modifications but generally with lower transformation efficiencies. However, it can be used as the basis for attempting to genetically transform a cultivar with no previous reports of transformation.

REFERENCES

1.   World Apple Review (2002). Published by Belrose, Inc. 1045 N.E. Creston Lane, Pullman, WA 99163-3806, USA.
2.   James DJ, Passey AJ, Barbara DJ and Bevan M (1989). Genetic transformation of apple (*Malus pumila* Mill.) using a disarmed Ti-binary vector. *Plant Cell Reports*, **7**: 658-661.
3.   Horsch RB, Fry JE, Hoffman NL, Eicholtz D, Rogers SG and Fraley RT (1985). A simple and general method of transferring genes into plants. *Science*, **227**: 1229-1231.
4.   James DJ, Passey AJ, Baker SA and Wilson FM (1996). Transgenes display stable patterns of expression in apple fruit and Mendelian segregation in the progeny. *Bio/Technology*, **14**: 56-60.
5.   James DJ, Uratsu S, Cheng J, Negri P, Viss P and Dandekar AM (1993). Acetosyringone and osmoprotectants like betaine synergistically enhance *Agrobacterium*-mediated transformation. *Plant Cell Reports*, **12**: 559-563.
6.   Gittins JR, Pellny TK, Hiles ER, Rosa C, Biricolti S and James DJ (2000). Transgene expression driven by heterologous ribulose-1,5-biophosphate carboxylase/oxygense small-subunit gene promoters in the vegetative tissues of apple (*Malus pumila* Mill.). *Planta*, **210**: 232-240.
7.   Maheshwaran G, Welander M, Hutchinson J, Graham M and Richards D (1992). Transformation of apple rootstock M26 with *Agrobacterium tumefaciens*. *Journal of Plant Physiology*, **139**: 560-568.
8.   Welander M, Pawlicki N, Holefors A and Wilson FM (1998). Genetic transformation of the apple rootstock M26 with the *Rol*B gene and its influence on rooting. *Journal of Plant Physiology*, **153**: 371-380.
9.   Sriskandarajah S, Goodwin PB and Speirs J (1994). Genetic transformation of the apple scion cultivar 'Delicious' via *Agrobacterium tumefaciens*. *Plant Cell, Tissue and Organ Culture*, **36**: 317-329.
10.   Norelli JL, Aldwinckle HS, Destéfano-Beltrán L and Jaynes JM (1994). Transgenic 'Malling 26' apple expressing the attacin E gene has increased resistance to *Erwinia amylovora*. *Euphytica*, **77**: 123-128.
11.   Yao J, Cohen D, Atkinson R, Richardson K and Morris B (1995). Regeneration of transgenic plants from the commercial apple cultivar Royal Gala. *Plant Cell Reports*, **14**: 407-412.
12.   De Bondt A, Eggermont K, Penninckx I, Goderis I, and Broekaert WF (1996). *Agrobacterium*-mediated transformation of apple (*Malus* x *domestica* Borkh.): an assessment of factors affecting regeneration of transgenic plants. *Plant Cell Reports*, **15**: 549-554.

13. Puite KJ and Schaart JG (1996). Genetic modification of the commercial apple cultivars Gala, Golden Delicious, and Elstar via an *Agrobacterium tumefaciens*-mediated transformation method. *Plant Science*, **119**: 125-133.

14. Wilson FM and James DJ (1998). Regeneration and transformation of apple cultivar Falstaff. In: Davey MR *et al.* (eds.), *Tree Biotechnology towards the Millenium*, Nottingham: Nottingham University Press, UK, pp. 95-100.

15. Sriskandarajah S and Goodwin PB (1998). Conditioning promotes regeneration and transformation in apple leaf explants. *Plant Cell, Tissue and Organ Culture*, **53**: 1-11.

16. Bolar JP, Brown SK, Norelli JL and Aldwinckle HS (1999). Factors affecting the transformation of 'Marshall McIntosh' apple by *Agrobacterium tumefaciens*. *Plant Cell, Tissue and Organ Culture*, **55**: 31-38.

17. Sedira M, Holefors A and Welander M (2001). Protocol for transformation of the apple rootstock Jork 9 with the *rol*B gene and its influence on rooting. *Plant Cell Reports*, **20**: 517-524.

18. Zhu L, Holefors A, Ahlman A, Xue Z and Welander M (2001). Transformation of the apple rootstock M.9/29 with the *rol*B gene and its influence on rooting and growth. *Plant Science*, **160**: 433-439.

19. Zhu L, Ahlman A, Li X and Welander M (2001). Integration of the *rol*A gene into the genome of the vigorous apple rootstock A2 reduced plant height and shortened internodes. *Journal of Horticultural Science and Biotechnology*, **76**: 758-763.

20. Igarashi M, Ogasawara H, Hatsuyama Y, Saito A and Suzuki M (2002). Introduction of *rol*C into Marubakaidou [*Malus prunifolia* Borkh. var. *ringo* Asami Mo 84-A] apple rootstock via *Agrobacterium tumefaciens*. *Plant Science*, **163**: 463-473.

21. Wilson FM and James DJ (2003). Regeneration and transformation of the premier UK apple (*Malus x pumila* Mill.) cultivar Queen Cox. *Journal of Horticultural Science and Biotechnology*, (in press).

22. Murashige T and Skoog F (1962). A revised medium for rapid growth and bioassays with tobacco tissue cultures. *Physiologia Plantarum*, **15**: 473-497.

23. Driver JA and Kuniyuki AH (1984). *In vitro* propagation of Paradox walnut rootstock. *HortScience*, **19**: 507 - 509.

24. Chu CCC, Wang CC, Sun CS, Hus C, Yin DC and Chu CY (1975). Establishment of an efficient medium for anther culture of rice through comparative experiments on the nitrogen sources. *Scientific Sinica*, **18**: 659-668.

25. Dandekar AM, Uratsu SL and Matsuta N (1990). Factors influencing virulence in *Agrobacterium*-mediated transformation of apple. *Acta Horticulturae*, **280**: 483-494.

26. Hood EE, Helmer GL, Fraley RT and Chilton M (1986). The hypervirulence of *Agrobacterium tumefaciens* A281 is encoded in a region of pTiBo542 outside of T-DNA. *Journal of Bacteriology*, **168**: 1291-1301.

27. De Bondt A, Eggermont K, Druart P, De Vil M, Goderis I, Vanderlayden J and Broekaert WF (1994). *Agrobacterium*-mediated transformation of apple (*Malus x domestica* Borkh.): an assessment of factors affecting gene transfer efficiency during early transformation steps. *Plant Cell Reports*, **13**: 587-593.

28. James DJ and Dandekar AM (1991). Regeneration and Transformation of Apple (*Malus pumila* Mill.). In: Lindsey K (ed.), *Plant Tissue Culture Manual*, B8, 1-18. Kluwer Academic Publishers, The Netherlands.

29. Alt-Moerbe J, Kuhlmann H and Schroder J (1989). Differences in induction of Ti plasmid virulence genes *virG* and *virD* and continued control of *virD* expression by four external factors. *Molecular Plant-Microbe Interactions*, **2**: 301-308.

30. James DJ and Thurbon IJ (1979). Rapid *in vitro* rooting of the apple rootstock M.9. *Journal of Horticultural Science*, **54**: 309-311.

31. James DJ, Passey AJ and Rugini E (1988). Factors affecting high frequency plant regeneration from apple leaf tissues cultured *in vitro*. *Journal of Plant Physiology*, **132**: 148-154.

32. Norelli JL and Aldwinckle HS (1993). The role of aminoglycoside antibiotics in the regeneration and selection of neomycin phosphotransferase-transgenic apple tissue. *Journal of the American Society for Horticultural Science*, **118**: 311-316.

33. Privalle LS, Wright M, Reed J, Hansen G, Dawson J, Dunder EM, Chang Y, Powell ML *et al.* (2000). Phosphomannose Isomerase, A Novel Selectable Plant Selection System: Mode of Action and Safety Assessment. In: Fairbairn C, Scoles G, McHughen A (eds.), *Proceedings of the 6th*

International Symposium on The Biosafety of Genetically Modified Organisms, Saskatoon, Canada. University Extension Press, University of Saskatchewan, pp. 171-178.

34. Haldrup A, Guldager Petersen S and Thyge Okkels F (1998). The xylose isomerase gene from *Thermoanaerobacterium thermosulfurogenes* allows effective selection of transgenic plant cells using D-xylose as the selection agent. *Plant Molecular Biology*, **37**: 287-296.

35. Stachel SE, Messens E, Van Montague M and Zambryski PC (1985). Identification of the signal molecules produced by wounded plant cells that activate T-DNA transfer in *Agrobacterium tumefaciens*. *Nature*, **318**: 624-629.

36. Rogowsky PM, Close TJ, Chimera JA, Shaw JJ and Kado CI (1987). Regulation of *vir* genes and *Agrobacterium tumefaciens* plasmid pTiC58. *Journal of Bacteriology*, **169**: 5101-5112.

37. Vernade D, Herrera-Estrella A, Wang K and Van Montagu M (1988). Glycine betaine allows enhanced induction of the *Agrobacterium tumefaciens vir* genes by acetosyringone at low pH. *Journal of Bacteriology*, **170**: 5822-5829.

38. Maximova SN, Dandekar AM and Guiltinan MJ (1998). Investigation of *Agrobacterium*-mediated transformation of apple using green fluorescent protein: high transient expression and low stable transformation suggest that factors other than T-DNA transfer are rate-limiting. *Plant Molecular Biology*, **37**: 549-559.

39. Welander M and Maheswaran G (1992). Shoot regeneration from leaf explants of dwarfing apple rootstocks. *Journal of Plant Physiology*, **140**: 223-228.

40. Caboni E, Lauri P and D'Angeli S (2000). *In vitro* plant regeneration from callus of shoot apices in apple shoot culture. *Plant Cell Reports*, **19**: 755-760.

41. Lavania UC and Srivastava S (1990). Evolutionary genomic change paralleled by differential responses of *2x* and *4x* calli cultures. *Experientia*, **46**: 322-324.

42. Sankhla D, Davis TD and Sankhla N (1993). Effect of gibberellin biosynthesis inhibitors on shoot regeneration from hypocotyl explants of *Albizzia julibrissin*. *Plant Cell Reports*, **13**: 115-118.

43. Ezura H and Harberd NP (1995). Endogenous gibberellin levels influence *Arabidopsis thaliana* (L.) heynh. *Planta*, **197**: 301-305.

44. Gaba V, Elman C, Watad AA and Gray DJ (1996). Ancymidol hastens *in vitro* bud development in melon. *HortScience*, **31**: 1223-1224.

45. James DJ, Passey AJ and Malhotra SB (1984). Organogenesis in callus derived from stem and leaf tissues of apple and cherry rootstocks. *Plant Cell, Tissue and Organ Culture*, **3**: 333-341.

46. Welander M (1988). Plant regeneration from leaf and stem segments of shoots raised *in vitro* from mature apple trees. *Journal of Plant Physiology*, **132**: 738-744.

47. Belaizi M, Paul H, Sangwan RS and Sangwan-Norreel BS (1991). Direct organogenesis from internodal segments of *in vitro* grown shoots of apple cv. Golden Delicious. *Plant Cell Reports*, **9**: 471-474.

48. Liu Q, Salih S and Hammerschlag F (1998). Etiolation of 'Royal Gala' apple (*Malus* x *domestica* Borkh.) shoots promotes high-frequency shoot organogenesis and enhanced β-glucuronidase expression from stem internodes. *Plant Cell Reports*, **18**: 32-36.

Figure 1. Steps involved in the generation of transgenic Greensleeves apple plants. Depending on the genetic material being inserted, the entire process from co-cultivation to a transgenic plant in the glasshouse can take between 6-12 months to complete. After 3 days co-cultivation with A. tumefaciens, *leaf disks are washed in antibiotic solution (cefotaxime) to remove the bacteria. The disks are then placed on regeneration medium in the dark (with subculture every 4 weeks) and approx. 1-2 months later, putative transgenic shoots begin to arise on medium with selection (A, bar = 10 mm). It is useful to include one plate of disks under no selection in a transformation experiment to act as a control for tissue quality. Without selection, Greensleeves leaf disks show a high regeneration capacity (B, bar = 10 mm). Once the regenerated shoot has formed 3-4 leaves it is transferred to a micropropagation medium and grown under light, and when large enough (having 8 or more leaves), it is induced to root in an auxin (IBA) rich medium (R13 medium) and then transferred to minimal medium (R37) to allow the roots to develop and grow. When the roots reach the base of the container (C) the plant is transferred to soil. The culture vessel shown is a square based cell counting vial. D: A plant successfully weaned from tissue culture to soil.*

Chapter 16

GENETIC TRANSFORMATION OF PEAR VIA *AGROBACTERIUM*-MEDIATED GENE TRANSFER

L.-H. ZHU AND M. WELANDER

Department of Crop Science, Swedish University of Agricultural Sciences, Box 44, SE-230 53 Alnarp, Sweden. Corresponding author: M. Welander. Email: margareta.welander@vv.slu.se

1. INTRODUCTION

1.1. Importance and Distribution of Pear

Pear, a pome fruit, belongs to the family *Rosaceae*, subfamily *Maloideae* and genus *Pyrus*. Over 20 species of pears exist around the world and all originated in regions of temperate Asia, Europe and the mountainous areas of North Africa (*1-3*). These species can be roughly divided into two groups, the European pears and the Asian pears. The most important edible species are *P. communis* (European pear and some of its related species) and *P. pyrifolia (*syn. *serotina*) (Asian pear) *(4, 5)*. The European pears are mainly cultivated in Europe, North America, South America, Africa, and Australia. The Asian pears are mainly grown in China and Japan (*1, 2*). It is believed that the first species to be domesticated was *P. pyrifolia*, originated in China, because the wildtype is edible without selection (*3*). Most of the pear cultivars are diploids and have 34 chromosomes *(4)*, but recently, triploid and tetraploid cultivars have been found in China (*6*). Pear can be consumed fresh or processed as perry (pear cider).

Pear is a major fruit tree of the temperate climates. The world's production of pears lies third after grapes and apple among the temperate fruit tree species with a production of 17 million metric tonnes and area of 1.78 million hectares in 2002 (*7*). The main production areas are Asia and Europe (Table 1).

1.2. Need for Genetic Improvement

Almost all pear scion cultivars in commercial fruit production are propagated on rootstocks because seeds cannot be used for reproducing true-to-type varieties due to their high levels of heterozygosity. Pear rootstocks can be divided into two groups, the *Pyrus* group and quince (*Cydonia oblonga*) group. Seedlings of *P. communis* and

I.S. Curtis (ed.), Transgenic Crops of the World - Essential Protocols, 217-228.

Table 1. Location and production of pear according to FAO (7)

Continent	Area (hectare)	Production(metric tonnes)
Africa	48,322	530,547
North Central America	31,872	956,740
South America	34,597	802,798
Asia	1,375,653	10,996,503
Europe	280,971	3,626,020
Oceania	9,058	202,597
Total	1,780,473	17,115,205

P. pyrifolia are widely used as pear rootstocks. Other *Pyrus* species, such as *P. calleryana, P. betulaefolia, P. pashia* and *P. ussuriensis*, are also used as pear rootstocks where environmental conditions allow their cultivation (*8*). *Pyrus* rootstocks have excellent compatibility with scions, the trees are well anchored and tolerate abiotic stresses, such as, drought, high soil pH and winter cold. Unfortunately, they often result in very vigorous trees which are not precocious and not suitable for high density planting systems. Trees on *Pyrus* rootstocks may sometimes produce smaller fruits than those on quince rootstocks. Clonal quince selections are often used as pear rootstocks, especially for the European pear in many countries in Europe. They result in smaller trees and earlier cropping than *Pyrus* rootstocks, which are excellent for high density planting orchards. However, quince rootstocks are not compatible with many scion cultivars of the European, and almost all of the Asian pear. Besides, quince rootstocks are not winter hardy and intolerant to alkaline soil conditions and highly susceptible to fire blight (*8, 9*). After decades of pear rootstock breeding and selection, the range of rootstocks available for commercial pear production is still very limited and there are no satisfactory rootstocks available. Thus, breeding new pear rootstocks of *Pyrus* type is a high priority.

Like all of the other crops, pears are attacked by a number of economically important pests and diseases that reduce yields and quality of fruits. Fire blight, caused by the bacterium *Erwinia amylovora* (*10*) is a threat to pear production in areas where European pear is grown due to the high susceptibility of the European pear to the disease. This disease has spread to most of the pear production areas. Although studies have been carried out for several decades, there is still no satisfactory remedy to the disease. This is mainly because of inadequate understandings of the host range and genetic diversity of the pathogen and its relationship to epidemiology and to host-pathogen interaction (*10*). Pear scab caused by either *Venturia pirina* or *V. nashicola* is another disease that cause mainly foliar and fruit damage and can bring a serious problem to pear production (*11-13*). Pear psylla (*Psylla pyricola*) is an important pest which can cause a high economic loss if not well-controlled (*14-16*). Pear decline caused by a mycoplasma organism which is transmitted by pear psylla can also be a serious problem in some areas where pear psylla is present (*3*). Certain pear species are highly susceptible to pear decline and when used as a rootstock they cause decline or death of the tree. In addition, serious viruses and virus-like diseases also affect pear production (*17*). Therefore, selection of new cultivars and rootstocks with high disease and pest resistance is of great importance.

Furthermore, the improvement of fruit quality is always a very important goal in pear breeding and many aspects regarding fruit quality (fruit colour, flavour and shelf-life) need to be improved in the future. Besides, early bearing, regular cropping and good storage quality are the other important traits to be improved (*18*).

1.3. The Use of Genetic Engineering in the Improvement of Pear

Traditional breeding by crossing and selection is an inefficient method for producing new varieties and rootstocks of pear because of its heterozygosity, long generation time and self-incompatibility. In this context, gene technology offers a great potential for the improvement of pear, by introducing one or a few desirable genes into the current valuable cultivars or rootstocks. The greatest advantage of this technique is that the recipient plant gains only one or a few foreign genes while the main genetic framework remains unchanged. As a result, only one or a few traits of a plant can be modified. This makes selection of transgenic plants with new traits more straightforward and time-saving. Meanwhile, gene technology has greatly expanded the genetic variation available to plant breeders as it can overcome barriers between species. This is very valuable when it comes to improvement of disease resistance since long-lasting resistance to insects and pests can be derived only from wild pear species. Due to these advantages, gene technology has been extensively applied in improving plant properties of different plant species including fruit trees. Since pear can be readily propagated vegetatively, it is easy to keep the new desirable traits once a rootstock or a cultivar has been improved. Moreover, the breeding process can be greatly accelerated by using the genetic technology.

1.4. Achievements of Genetic Transformation in Pear

Pear is a natural host of *Agrobacterium tumefaciens* (*18*). This makes it possible to use the bacterium with disarmed plasmids to transfer genes of interest into important pear cultivars or rootstocks. However, research on pear genetic transformation has not been so intensive as on other fruit crops like apple. Up to now, *Agrobacterium*-mediated gene transfer has been the main method for pear transformation. The purposes of genetic transformation have been focused on designing the transformation protocol, improving the rooting ability, increasing disease (mainly fire blight) resistance and reducing plant stature.

The first report on pear transformation was published by Mourgues *et al.* in 1996 (*19*). They used the *A. tumefaciens* strain EHA101 with a plasmid carrying the reporter gene *gus*A and the selectable *npt*II gene to transform three pear (*P. communis* L.) cultivars. Transgenic clones were obtained from all three cultivars tested, but a great difference in transformation efficiency was found among the cultivars. Bell *et al.* (*20*) obtained transgenic clones of the pear cultivar Beurre Bosc with the *rol*C gene, originated from *A. rhizogenes*, using the *A. tumefaciens* strain

EHA101. Preliminary results showed that the transgenic clones reduced plant height, fewer nodes and a reduced leaf area compared with untransformed plants (*20*). Lebedev *et al.* (*21*) transformed the pear rootstock GP217 with strain CBE21 harbouring a plasmid containing the bialaphos resistant gene (*bar*) in order to obtain transgenic clones of pear rootstocks resistant to commercial herbicides, such as bialaphos or Basta®. The *bar* gene encodes phosphinotricin acetyltransferase which breaks down the active ingredient phosphinotricin. Some transgenic clones were obtained with resistance to Basta®, but the transformation efficiency was very low since 95% were escapes.

In order to improve the resistance to pear fire blight, Chevreau *et al.* have been working on the genetic transformation of several pear cultivars. They have been using genes encoding for lytic peptides (cecropin B and its synthetic derivatives SB-37 and Shiva-1 as well as attacin E) and lysozymes (T4 lysozyme) as well as lactoferrin. The effect of lytic peptides is to form pores in bacterial membranes. Lysozymes are ubiquitous enzymes with specific hydrolytic activities directed against the bacterial-cell-wall peptidoglycan. Lactoferrin is an iron-binding glycoprotein known to have antibacterial properties (*22*). So far, they have obtained transformed clones of the pear cultivar Passe Crassane (*P. communis*) containing either the attacin E, or SB-37, or cecropin B or T4 lysozyme (*23, 24*). Ten days after the *in vitro* leaf inoculation with a virulent strain of *E. amylovora* CFBP 1430, some transgenic clones showed an enhanced resistance to fire blight. However, variation in resistance was found among the different transgenic clones and the resistance was considerable lower than the natural resistant cultivar Old Home. More recently, Malnoy *et al.* (*25*) transformed the depolymerase gene, encoding an enzyme that degrades the capsular exopolysaccharide necessary for *E. amylovora* to cause normal disease symptoms, into the pear cultivar Passe Crassane and obtained 15 transgenic clones. Two clones showed a slight reduction of fire blight symptoms in an *in vitro* test. Malnoy *et al.* (*26*) examined the activity of three pathogen-inducible promoters of tobacco in transgenic pear after abiotic and biotic elicitation and suggested that two of them could be used to drive the expression of transgenes to promote bacterial disease resistance although their activities were lower than that of CaMV35S.

Dwarfing rootstocks are commonly used for reducing tree size to facilitate high-density planting systems, and thus achieving high production efficiency. However, there are few ideal dwarfing rootstocks for pear cultivars. Breeding new dwarfing rootstocks or improving the current available rootstocks are of great importance. The pear rootstock BP10030, selected in Balsgård, Sweden, is of *Pyrus* type. It is dwarf, cold hardy and compatible with most pear varieties tested, but very difficult to root, thus limiting its commercial use. Improving the rooting ability of this rootstock will improve its commercial value. The *rol*B gene, originated from *A. rhizogenes*, has been reported to increase rooting ability of apple (*27-29*), pear (*30, 31*) and rose (*32*). One of our research goals is to improve the rooting ability of dwarfing fruit tree rootstocks. We have been using the *rol*B gene to transform apple and pear rootstocks and have made substantial progress. So far, we have obtained transgenic clones of the pear rootstock BP10030 and of the two apple rootstocks M9 and M26 (*27, 28, 31*). For BP10030, *in vitro* rooting results showed that rooting ranged from 67-100%

for different transgenic clones and no rooting for untransformed plants on hormone free rooting medium (Fig. 1A). The *ex vitro* rooting by cuttings revealed that rooting ranged from 71-100% for the transgenic clones and only 5% for untransformed controls without auxin treatment (*31*). Moreover, the root number was also greatly increased for the transgenic clones with an even distribution of roots on the cut surface compared to untransformed plants with roots distributed only on one side of the cut surface (Figs. 1B and C) (*31*). Growth analysis showed that some transgenic clones had a shortened stem length (Fig. 1D) (*28, 33*). In this chapter, we describe the pear transformation procedure used in our laboratory.

2. MATERIALS

2.1. A. tumefaciens *Strain*

A. tumefaciens strain C58C1 carrying the helper plasmid pGV3850 and the binary vector pCMB-BGUS were kindly provided by Dr. C. Maurel (Biochimie et Physiologie Moleculaire des Plantes, ENSA-M/INRA/CNRS URA 2133, F-34 060 Montpellier, France), were used in our study. The binary vector contains the *rol*B, *npt*II and *gus* genes. The *rol*B and *gus*A genes are under the *rol*B promoter and the *npt*II gene under the NOS promoter (*34*).

2.2. *Culture Media for* A. tumefaciens

1. LB medium (*35*): 10 g/l tryptone (Duchefa, Haarlem, The Netherlands), 5 g/l yeast extract (Duchefa), 10 g/l NaCl (Duchefa), pH 7.0.
2. Liquid LB medium: LB medium supplemented with 200 mg/l carbenicillin disodium (Duchefa), 100 mg/l kanamycin sulphate (Duchefa), 75 mg/l neomycin sulphate (Sigma).
3. Semi-solid LB medium: liquid LB medium plus 15 g/l agar (Bacto).
4. Carbenicillin aqueous stock (100 mg/ml): Dissolve the powder in sterile double distilled water. Sterilize by filtration through a 0.2 μm membrane filter (Sarstedt, Nümbrecht, Germany). Dispense 5-7 ml into a 8 ml plastic tube (Sarstedt) and store at -20°C.
5. Kanamycin and neomycin aqueous stock (25 mg/ml): Prepare as for carbenicillin, (*see* Section 2.2.4.).
6. Quoirin and Lepoivre (QL) (*36*) macro nutrient stock: 4 g/l NH_4NO_3, 18 g/l KNO_3, 12 g/l Ca $(NO_3)_2.4H_2O$, 3.6 g/l $MgSO_4.7H_2O$, 2.7 g/l KH_2PO_4. Dispense 100 ml into a plastic bottle. Store at -20°C.
7. Murashige and Skoog (MS) (*37*) micro nutrient stock: 4000 mg/l Fe-Na-EDTA, 860 mg/l $ZnSO_4.7H_2O$, 620 mg/l H_3BO_3, 1690 mg/l $MnSO_4.H_2O$, 2.5 mg/l

$CuSO_4.5H_2O$, 83 mg/l KI, 25 mg/l $Na_2MoO_4.2H_2O$, 2.5 mg/l $CoCl_2.6H_2O$. Dispense 10 ml into a plastic tube. Store at -20°C.

8. MS vitamin stock: 10000 mg/l myo-inositol (Merck, Damstadt, Germany), 200 mg/l glycine (Sigma), 50 mg/l nicotinic acid (Sigma), 50 mg/l pyridoxine-HCl (Sigma), 10 mg/l thiamine-HCl (Sigma). Dispense 10 ml into a plastic tube. Store at -20°C.

9. Liquid medium for suspending bacteria (LM): 100 ml QL macro nutrient stock, 10 ml MS micro nutrient stock, 10 ml MS vitamin stock, 30 g/l sucrose, pH 4.8 (*see* Note 1).

2.3. Plant Materials

In vitro-grown shoots of the pear dwarfing rootstock BP10030 (*see* Note 2) were used as plant materials for transformation.

2.4. Tissue Culture

1. QL stock: (*see* Section 2.2.6.).
2. MS micro nutrient stock: (*see* Section 2.2.7.).
3. MS vitamin stock: (*see* Section 2.2.8.).
4. Shoot multiplication medium (SMM): 100 ml QL macro nutrient stock, 10 ml MS micro nutrient stock, 10 ml MS vitamin stock, 0.5 mg/l BAP (Duchefa), 30 g/l sucrose (*see* Note 3), 6 g/l agar, pH 5.5.
5. BAP stock (1 mg/ml): Prepare by dissolving the powder in a few drops of 1N NaOH and complete the final volume with sterile double distilled water. Store at 4°C.
6. Liquid medium used for wetting leaf explants during transformation: (*see* Section 2.2.9.; Note 1).
7. Modified QL macro nutrient stock (MQL) (*38*): 6.07 g/l NH_4NO_3, 11.5 g/l KNO_3, 2.97 g/l Ca $(NO_3)_2.4H_2O$, 1.85 g/l $MgSO_4.7H_2O$, 0.84 g/l KH_2PO_4. Dispense 100 ml into a plastic bottle. Store at -20°C.
8. MS micro nutrient stock: (*see* Section 2.2.7.).
9. Modified MS vitamin stock (*38*): 10000 mg/l *myo*-inositol, 400 mg/l glycine, 100 mg/l nicotinic acid, 100 mg/l pyridoxine-HCl, 80 mg/l thiamine-HCl. Dispense 10 ml into a plastic tube. Store at -20°C.
10. Co-culture medium (CM): 100 ml MQL macro nutrient stock, 5 ml MS micro nutrient stock, 10 ml modified MS vitamin stock, 0.2 mg/l NAA (Duchefa), 3.3 mg/l TDZ (Duchefa), 40 g/l sorbitol (*see* Note 4), 2.5 mg/l Gelrite (Duchefa), pH 5.5.
11. TDZ stock (0.22 mg/ml): Prepare by dissolving the powder in a few drops of 1N NaOH (*see* Note 5). Complete the final volume with double distilled water. Store at 4°C.
12. NAA stock (1 mg/ml): Prepare as for BAP and store at 4°C.

13. Regeneration medium (RM): CM plus 200 mg/l cefotaxime, 50 mg/l kanamycin sulphate (*see* Note 6).

14. Cefotaxime aqueous stock (100 mg/ml): Prepare as for carbenicillin and store at $-20°C$.

15. Kanamycin aqueous stock: (*see* Section 2.2.5.).

16. Rooting medium (RTM): 50 ml QL macro nutrient stock, 10 ml QL micro nutrient stock, 10 ml Walkey (*39*) vitamin stock, 6 mg/l agar, pH 5.5.

17. QL macro nutrient stock: (*see* Section 2.2.6.).

18. QL micro nutrient stock: 4000 mg/l Fe-Na-EDTA, 860 mg/l $ZnSO_4.7H_2O$, 620 mg/l H_3BO_3, 76 mg/l $MnSO_4. H_2 O$, 2.5 mg/l $CuSO_4.5H_2O$, 8 mg/l KI, 25 mg/l $Na_2MoO_4.2H_2O$, 2.5 mg/l $CoCl_2.6H_2O$. Dispense 10 ml into a plastic tube. Store at $-20°C$.

19. Walkey vitamin stock: 10000 mg/l *myo*-inositol, 50 mg/l nicotinic acid, 50 mg/l pyridoxine-HCl, 20 mg/l thiamine-HCl. Dispense 10 ml into a plastic tube. Store at $-20°C$.

3. METHODS

3.1. *Explants and* Agrobacterium *Preparation*

1. Excise 2-3 youngest unfolded leaves from 2 week-old shoot cultures (*see* Note 6). Place the leaves on a wetted sterile filter paper in a 9 cm diam. Petri dish with LM.

2. Wound the leaves on the adaxial surface using the tips of clinical forceps so that tissues become watersoaked, but not macerated (*see* Note 7).

3. Grow *A. tumefaciens* on semi-solid LB medium at $28°C$ for 2 days. Take one loop of bacteria and transfer to 25 ml of the liquid LB medium and grow overnight at $28°C$ on a shaker at 200 rpm.

4. Measure absorbance of a 1 ml aliquot of the overnight bacterial culture at 400 nm in a spectrophotometer.

5. Centrifuge the bacterial culture at 3500 rpm for 20 min in a 40 ml sterile centrifuge tube with cap (Nalgene, Kirkland, WA, USA), discard the supernatant, resuspend and dilute the pellet with the LM medium to a concentration of OD_{400} (optical density) at 0.5-1.0. Pour 25 ml aliquots of the diluted culture into sterile Petri dishes.

3.2. *Inoculation, Co-cultivation, Selection and Recovery of Transgenic Plants*

1. Immerse the wounded leaf explants into the bacterial suspension for 20 min (*see* Note 8) and place on CM with adaxial side in contact with the medium (10 leaves/Petri dish) for a 3 day co-culture at $23°C$ in dark (*see* Note 9).

2. After co-cultivation, rinse the leaf explants in LM medium supplemented with 500 mg/l cefotaxime, blot-dry on sterile filter paper, and place on RM and culture at $23°C$ in the dark.

3. Transfer the leaf explants to fresh RM medium every 2-3 weeks until new shoots develop (*see* Note 10). Shoot regeneration occurs from the mid-vein and basal region of the leaf, after 3-4 weeks on RM.

4. Transfer the leaf explants from which shoots have formed to a 16 h photoperiod with a 33 μmol/m^2/sec light intensity (cool white fluorescent tubes) at 23°C. Cover with a nylon cloth at the beginning. After 1-2 weeks, the shoots (0.5-1 cm in length) can be excised and placed on SMM medium (*see* Note 11).

5. Check putative transformants by performing GUS histochemical staining (*see* Note 12; *27*) or by performing PCR. GUS-negative or PCR-negative shoots are usually considered as non-transformed and considered as escapes (*see* Note 13).

6. The putative transformants can also be tested by rooting on RTM medium containing 50 or 100 mg/l kanamycin. Excise regenerated shoots (approx. 0.5-1.0 cm in length) and transfer to RTM medium and incubate in the dark for 4 days, then transfer to the light (16 h photoperiod, 33 μmol/m^2/sec light intensity) at 23°C. Rooted shoots with green leaves are considered as transformed and non-rooted shoots with yellowish leaves are regarded as untransformed.

7. Southern blot analysis is needed to confirm the copy number of the foreign gene integrated into the plant genome. Transgene expression needs to be analyzed by Northern blot and Western blot analyses (*see* Note 14).

4. NOTES

1. Alternatively, the liquid medium used for suspending the bacteria and wetting leaf explants can be the liquid regeneration medium without NAA and TDZ.

2. Two pear rootstocks were tested in the regeneration experiments, namely, dwarfing rootstock BP10030 and semi-dwarfing rootstock OHF333. The former resulted in a 98% of regeneration rate and 66% for the latter, indicating the difference in regeneration ability between different genotypes under the conditions tested. The transformation work was carried out using BP10030.

3. In order to promote shoot elongation, sucrose can be substituted with sorbitol with the addition of 0.5 mg/l gibberellic acid (GA$_3$) in the shoot multiplication medium (unpublish. results). However, shoot cultures need to be transferred to GA$_3$-free medium for one subculture before rooting to avoid the possible negative effect of GA$_3$ on rooting.

4. We have examined both sorbitol and sucrose as carbon source in the regeneration experiments on the two pear rootstocks. For BP10030, the high regeneration rate of 98 % was achieved with sucrose and 88 % with sorbitol. For OHF333, sorbitol resulted in the highest regeneration rate of 66%, while 22% with sucrose (*30*). However, in the transformation experiments on the rootstock BP10030, sucrose resulted in more calli and less shoots than sorbitol, while the transformation efficiency was not significantly different between the two sugars. If sucrose is used in the medium, 30 mg/l is required.

5. Earlier, we dissolved TDZ in 1 ml 1N KOH for a 100 ml aqueous stock solution at 1 mM as described by Huetteman and Preece (*40*). However, we have found that, for making a 50 ml 1 mM stock solution, TDZ is easily dissolved in a few drops of 1N NaOH, and then finish the final volume with double distilled water. The TDZ aqueous solution can be stored at 4°C for approx. 2-4 weeks.

6. Dark and cold treatments at 4°C of shoot cultures prior to transformation did not influence the transformation result, nor did the age of shoot cultures between 2-4 weeks (*30*). Leaves from apical shoots gave an optimum regeneration rate whereas those from lateral shoots resulted in a lower regeneration rate (unpublish. data). Selection pressure, namely, the kanamycin concentration in the medium, can be crucial for successful transformation. If the concentration is too high, a low recovery of putative transformants can be expected, probably due to the poor survival of supporting tissues. On the other hand, if the concentration is too low, many escapes will be present, which usually result in an inefficient selection.

7. We have tested two different wounding methods for regeneration: wounding with the tips of special clinical forceps with carbide-coated tips or wounding with normal scalpels. The former resulted in a significantly higher regeneration rate and shoot number than the latter (*30*). Therefore, we have been using the forceps wounding method in the transformation experiments. However, when we tested the two methods on apple transformation, no differences in transformation efficiency were found (unpublish. results).

8. Comparison of inoculation time of 10 and 20 min did not result in differences in transformation efficiency (unpublish. data).

9. Alternatively, 30 leaves can be placed in one Petri dish during the co-culture period. Co-culture for 4 or 5 days did not improve the transformation efficiency compared with a 3 day co-culture period (unpublish. results).

10. After 3-4 weeks, the concentration of NAA can be reduced to half and that of TDZ can be increased to 4.4 mg/l to promote shoot formation.

11. It is better to keep newly developed shoots attached to the leaf explant for some time to secure their growth. In order to accelerate growth, shoots can be first grown on SMM medium semi-solidified with 3.5 g/l agar or 1.2 g/l Gelrite.

12. Since the *gus*A gene in the binary vector pCMB-BGUS is under the control of the *rol*B promoter, the GUS activity is mainly found in the meristematic zone of roots and shoot tips as well as in procambial strands of shoots (*27*). In order to observe GUS activity, very young leaves are needed for GUS staining.

13. Sometimes even though the clones are transformed, PCR results can be negative either due to partial deletion of the gene, or incorrect PCR conditions. It is important to further verify the putative transformants by Southern blot analysis (*28*).

14. Northern blot analysis offers information about the transgene expression and Western blot analysis about the protein gene product in transformed plants.

CONCLUSION

Pear belongs to the family *Rosaceae* and genus *Pyrus*. It is a major fruit tree of the temperate climates. The world's production of pears is 17 million metric tonnes from an area of 1.78 million hectares in 2002.

Pears in commercial fruit production are propagated on rootstocks. However, there are no ideal rootstocks available. Meanwhile, pears suffer a number of economically important pests and diseases, such as fire blight, pear scab and psylla. Furthermore, the improvement of fruit quality is also a very important goal in pear breeding. Thus, selection of pear rootstocks with dwarfing characteristics, winter hardiness and good compatibility with scions, and cultivars with better fruit quality, high disease and pest resistance are of great importance. Traditional breeding is an inefficient method for pear because of its heterozygosity, long generation time and self-incompatibility. Gene technology offers great potential for the improvement of pear, by introducing one or a few desirable genes into the current valuable cultivars or rootstocks.

The research on genetic transformation of pear is relatively less intensive compared with apple or citrus. For pear transformation, *Agrobacterium*-mediated gene transfer has been the main method and the purposes have been focused on improving the transformation protocol, enhancing the rooting ability of dwarfing pear rootstocks, increasing disease (mainly fire blight) resistance and reducing plant stature.

In this chapter, we describe a successful transformation protocol for pear by *A. tumefaciens* used in our laboratory. By using this protocol, we have obtained *rol*B transgenic clones of the dwarfing pear rootstock BP10030, which have shown improved rooting ability.

ACKNOWLEDGEMENTS

The financial support from The Swedish Research Council for Environment, Agricultural Sciences and Spatial Planning (FORMAS) is highly acknowledged.

REFERENCES

1. Bell RL, Quamme HA, Layne REC and Skirvin RM (1996). Pears. In: Janick J, Moore JN (eds.), *Fruit Breeding volume I Tree and Tropical Fruits*. John Wiley & Sons, Inc., New York, pp. 441-514.
2. Janick J (1986). *Horticultural Science*, 4th edn., W.H. Freeman and Company, New York, p.628.
3. Lombard PB and Westwood MN (1987). Pear rootstocks. In: Rom RC, Carlson RF (eds.), *Rootstocks for Fruit Crops*. John Wiley & Sons, Inc, Toronto, pp.145-183.
4. Ryugo K (1988). *Fruit Culture-Its Science and Art*. John Wiley & Sons, New York, p.252.
5. Westwood MN (1978). *Temperate-Zone Pomology*. Timber Press, Oregon, pp. 45-46.
6. Cao Y and Huang L (2002). Genetics of ploidy and hybridised combination types for polyploid breeding in pear. *Acta Horticulturae*, **587**: 207-210.
7. FAO (2003). http://appa.fao.org/. FAOSTAT Agriculture DATA.
8. Webster AD (1998). A brief review of pear rootstock development. *Acta Horticulturae*, **475**: 135-141.
9. Jacob HB (1998). Pyrodwarf, a new clonal rootstock for high density pear orchards. *Acta Horticulturae*, **475**: 169-177.
10. Vanneste JL (2000). *Fire Blight-The Disease and its Causative Agent*, Erwinia amylovora. CABI Publishing, Wallingford.
11. Deckers T and Schoofs H (2002). The world pear industry and research: present situation and future development of European pears (*Pyrus communis*). *Acta Horticulturae*, **587**: 37-54.
12. Latorre BA (1998). Phytosanitary status of pear in Chile with special reference to the phytopathological situation. *Acta Horticulturae*, **475**: 439-448.
13. Park P, Ishii H, Adachi Y, Kanematsu S, Ieki H and Umemoto S (2000). Infection behaviour of *Venturia nashicola*, the cause of scab on Asian pear. *Phytopathology*, **90**: 1209-1216.
14. Deckers T and Daemen E (1998). Pear growing in Belgium: production systems and problems. *Acta Horticulturae*, **475**: 49-58.
15. Niederholzer F, Seavert CF and Riedl H (1998). Demonstration and implementation of integrated fruit production (IFP) on pears in northern oregon: introduction. *Acta Horticulturae*, **475**: 59-66.
16. Nikolić M (1998). Pear research and production in Yugoslavia. *Acta Horticulturae*, **475**: 85-89.
17. Hartmann HT, Kester DE, Davies FT, Geneve Jr.RL (1997). *Plant Propagation: Principles and Practices*, 6th edn., Prentice Hall International, Inc. Simon & Schuster/A Viacom Company, New Jersey, USA, pp. 653-655.
18. Chevreau E and Skirvin RM (1992). Pear. In: Hammerschlag FA, Litz RE (eds.), *Biotechnology in Agriculture No.8-Biotechnology of Perennial Fruit Crops*. C.A.B International, Bristol, p.263.
19. Mourgues F, Chevreau E, Lambert C and Bondt An de (1996). Efficient *Agrobacterium*-mediated transformation and recovery of transgenic plants from pear (*Pyrus communis* L.). *Plant Cell Reports*, **16**: 245-249.
20. Bell RL, Scorza R, Srinivasan C and Webb K (1999). Transformation of 'Beurre Bosc' pear with the *rol*C gene. *Journal of American Society for Horticultural Science*, **124**: 570-574.
21. Lebedev VG, Dolgov SV and Skryabin KG (2002). Transgenic pear clonal rootstocks resistant to herbicide "Basta". *Acta Horticulturae*, **569**: 193-197.

22. Mourgues F, Brisset M-N, and Chevreau E (1998). Strategies to improve plant resistance to bacterial diseases through genetic engineering. *Trends in Biotechnology*, **16**: 203-210.

23. Chevreau E, Malnoy M, Mourgues F and Brisset MN (2000). Genetic engineering of pear for increased fire blight resistance. *Acta Horticulturae*, **538**: 639-643.

24. Reynoird JP, Mourgues F, Norelli J, Aldwinckle HS, Brisset MN and Chevreau E (1999). First evidence for improved resistance to fire blight in transgenic pear expression the *attacin E gene* from *Hyalophora cecropia*. *Plant Science*, **149**: 23-31.

25. Malnoy M, Chevreau E and Brisset MN (2002). Expression of a depolymerase gene in transgenic pears increased only slightly their fire blight resistance. *Acta Horticulturae*, **590**: 401-405.

26. Malnoy M, Venisse J-S, Reynoird JP and Chevreau E (2003). Activation of three pathogen-inducible promoters of tobacco in transgenic pear (*Pyrus communis* L.) after abiotic and biotic elicitation. *Planta*, **216**: 802-814.

27. Welander M, Pawlicki N, Holefors A and Wilson F (1998). Genetic transformation of apple rootstock M26 with *rol*B gene and its influence on rooting. *Journal of Plant Physiology*, **53**: 371-380.

28. Zhu LH, Holefors A, Ahlman A, Xue ZT and Welander M (2001). Transformation of the apple rootstock M.9/29 with the *rol*B gene and its influence on rooting and growth. *Plant Science*, **160**: 433-439.

29. Welander M and Zhu LH (2000). The rooting ability of *rol*B transformed clones of the apple rootstock M26 and its relation to gene expression. *Acta Horticulturae*, **521**: 133-138.

30. Zhu, LH and Welander M (2000). Adventitious shoot regeneration of two dwarfing pear rootstocks and the development of a transformation protocol. *Journal of Horticultural Science and Biotechnology*, **75**: 745-752.

31. Zhu LH, Li XY, Ahlman A and Welander M (2003). The rooting ability of the dwarfing pear rootstock BP10030 (*Pyrus communis*) was significantly increased by introduction of the *rol*B gene. *Plant Science*, **165**: 829-835.

32. Van der Salm TPM, van der Toorn CJG, tenCate CHH, van der Krieken WM and Dons HJM (1996). The effects of exogenous auxin and *rol* genes on root formation in *Rosa hybrida* L. 'Moneyway'. *Plant Growth Regulation,* **19**: 123-131.

33. Zhu LH and Welander M (2000). Growth characteristics of the untransformed apple rootstock M26 with the *rol*A and *rol*B genes under steady-state nutrient supply conditions. *Acta Horticulturae*, **520**: 139-146.

34. Maurel C, Leblanc N, Barbier-Brygoo H, Perrot-Rechenmann C, Bouvier-Durand M and Guern J (1994). Alterations of auxin perception in *rol*B-transformed tobacco protoplasts. *Plant Physiology*, **105**: 1209-1215.

35. Sambrook J and Russell DW (2001). *Molecular Cloning- A Laboratory Manual*, 3rd edn., Cold Spring Harbor Laboratory Press, New York. p. A2.2.

36. Quoirin M, Lepoivre P and Boxus P (1977). Un premier bilan de dix années de recherche sur les cultures de méristèmes et la multiplication *in vitro* de fruitiers ligneux (in French), *Compte rendu des recherches*, Station des Cultures Fruitières et Marrîchères de Gembloux 1976-1977: 93-117.

37. Murashige F and Skoog F (1962). A revised medium for rapid growth and bioassys with tobacco tissue cultures. *Physiologia Plantarum*, **15**: 473-492.

38. Chevreau E and Leblay C (1993). The effect of mother plant pretreatment and explant choice on regeneration from *in vitro* pear leaves. *Acta Horticulturae*, **336**: 263-266.

39. Walkey DG (1972). Production of apple plantlets from axillary bud meristems. *Canadian Journal of Plant Science*, **52**: 1085-1087.

40. Huetteman CA and Preece JE (1993). Thidiauron: a potent cytokinin for woody plant tissue culture. *Plant Cell, Tissue and Organ Culture*, **33**: 105-119.

Figure 1. The integration of the rolB gene into the dwarfing pear rootstock BP10030 significantly increased the rooting percentage and reduced plant size in some clones. A: Numbers 1-3 represent three rolB transformed clones and C is untransformed control on the hormone free rooting medium from an in vitro *rooting experiment (bar = 0.8 cm). B: Numbers 1-4 represent four independent rolB transgenic clones and C is the untransformed control from the cutting experiment in the glasshouse (bar = 2 cm). C: Numbers 1-2 represent two rolB transformed clones and C is the untransformed control (bar = 1 cm). D: Pot plants of four rolB transformed clones (1-4) and of the untransformed control (C) (bar = 5 cm). Printed with permission from Elsevier.*

Chapter 17

AGROBACTERIUM-MEDIATED TRANSFORMATION OF GRAPE EMBRYOGENIC CALLI

A. PERL, V. COLOVA-TSOLOVA[1] AND Y. ESHDAT

Department of Fruit Tree Sciences, Institute of Horticulture, Agricultural Research Organization, P.O. Box 6, 50250 Bet-Dagan, Israel; [1]*Center for Viticulture Science, Florida A&M University, Tallahassee, FL 32307, USA. Corresponding author: A. Perl; E-mail: perlx@int.gov.il*

1. INTRODUCTION

The use of breeding and genetics to boost crop productivity and quality, and the use of agricultural chemicals to protect crops and enhance plant growth, has been the two prominent features of agriculture in the 20th Century (*1*). The integration of chemicals and plant breeding resulted in food sufficiency and variety, helping to meet the needs of an ever-increasing population (*2*).

In the 21st Century, crops are also expected to improve consumers' health and provide necessary nutrients. Integrating conventional technologies with those based on molecular biology and genomics could make this objective possible, as genetic resources to improve plants will be derived from the ongoing effort to discover, isolate and characterize genes that encode agriculturally important traits (*3*).

1.1. Importance of Grapes

The Old World species, *Vitis vinifera*, is the grape of antiquity often mentioned in the Bible. Viticulture and wine-making have been part of human culture for thousands of years. Most table-grapes, wine, and raisin grapes are produced from this variety, originated in the regions between the south of the Caspian and the Black Sea in Asia Minor (*4*). It was probably in Northern Iran or Armenia where the grapevine was found growing wild some 8000 years ago. More comprehensive data were lost during generations, but masses of crushed grape pips, stems, and skins were found by paleontologists indicating that wine and table-grapes were widely known and highly popular in distant times.

Grapevine is the most widespread fruit crop in the world with vineyards occupying more than 13 million hectares. World production of grapes is currently in excess of 75 million metric tonnes.

229

I.S. Curtis (ed.), Transgenic Crops of the World - Essential Protocols, 229-242.
© *2004 Kluwer Academic Publishers. Printed in the Netherlands.*

1.2. Need for Genetic Improvement of Grapes

In traditional grape breeding, traits of donor plants are combined by crossing and back-crossing to provide good yield and disease resistant varieties. However, this procedure is very costly and time-consuming. Hybrid seedlings from crosses need to be selected very carefully, as many carry undesired traits. In the cases of annual crops like grapes, the time scale (range) from the beginning of the selection procedure to the approval of a new successful variety takes around 10-15 years. Since all crop plants in viticulture are perennial, thus having a long juvenile period, the breeding process takes even longer as several years are needed to obtain the first yield that would determine whether the cross was successful.

Moreover, grapes are highly heterozygous and the characters which constitute a good cultivar are polygenic in their inheritance. Thus, the probability of recombining in a hybrid the desired set of genes, which determine the essential properties of a given cultivar, is very low.

In light of these limitations, the most economical breeding strategy would be to transfer individual traits as single genes into an already available and desirable genetic background. In such a case, only the desired trait would be transferred, with a minimum disturbance to the original genome.

In the course of the last three decades, a number of biotechnological approaches have been developed to enhance conventional breeding. One method is to introduce a certain gene for a desirable trait into the genome of a variety by the use of genetic engineering techniques. It seems very likely that many, if not most, of the changes grape consumers may expect to achieve, will be through the application of genetic engineering. Although currently controversial, we believe genetic engineering will be invaluable to the future improvement of agricultural systems in general and particularly in viticulture.

1.3. Embryogenesis in Grapes

A prerequisite for achieving an efficient and synchronous transformation system for grape is the establishment and long term maintenance of a highly regenerative embryogenic cell suspension suitable for the currently used *Agrobacterium-* or biolistic-mediated transformation. The type and quality of this cell suspension is a key factor to enable successful transformation. Due to the importance of embryogenic cell lines for grape transformation, this chapter will review some of the recent developments in this field.

Somatic embryogenesis has been previously reviewed in detail (5, 6). The technique was recently refined and extended to other genotypes (7-11). Although anther and ovule tissues have been most often used for the induction of embryogenic cultures, other tissues have also been used successfully. The production of somatic embryos from leaves, petioles, tendrils and protoplasts have been achieved in several cases (12-15). The ability to obtain successful regenerants from sporophytic anther tissues is highly dependent on both genotype and salts in the culture media (16). Perrin *et al.* (17) designated a set of novel media combining significant changes in several salt concentrations (those of KNO_3, NH_4Cl, KH_2PO_4, $CaCl_2$, $MnSO_4$ and

ZnSO$_4$), thus optimizing this media for somatic embryogenesis and plant regeneration in grapevine. Regeneration and germination of somatic embryos may be problematic as embryogenic cultures may age and accumulate mutations during subculturing. Gibberellic acid was found to promote germination of torpedo stage embryos (*13, 18*). Recently, several studies investigated the long term maintenance of the embryogenic potential of grape cultures. Size fractionation by ultrasound during subcultures was found to be a major factor in maintaining actively growing morphogenetic cultures (*19*). Jayasankar and Bondada (*20*), using a 960 mm mesh screens, were able to identify a unique morphotype with enhanced shoot production during embryo germination. Cryopreservation of embryogenic lines was also reported as an efficient tool promoting embryogenesis and subsequent plant germination (*21, 22*). An updated complementary list of the different embryogenic cell lines that are currently utilized for transformation experiments (*23*) is provided (Table 1).

Table 1. Update of additional embryogenic cell suspensions currently utilized for transformation experiments

Cultivar	Genetic origin	Explant origin	Stock location	Reference
Beauty seedless	*V. vinifera*	Leaves	India	*12*
Brachetto	*V. vinifera*	Leaves	Italy	*24*
Bianca	*V. vinifera*	Petioles	Ukraine	*13*
Cabernet sauvignon	*V. vinifera*	Anthers	France	*17*
Cabernet sauvignon	*V. vinifera*	Anthers	Australia	*25*
Cabernet sauvignon	*V. vinifera*	Anthers	Ukrine	*26*
Chardonnay	*V. vinifera*	Anthers	Australia	*25*
Chardonnay (02Ch)	*V. vinifera*	Anthers	USA	*27*
Chardonnay	*V. vinifera*	Anthers	France	*16*
Chenin Blanc	*V. vinifera*	Anthers	Australia	*25*
Danuta	*V. vinifera*	Anthers	France	*16*
Dornfelder	*V. vinifera*	Anthers	Germany	*28*
Fredonia	hybrid	Ovary	USA	*29*
Georgikon 28	hybrid	Anther	Hungary	*30*
Gewurztraminer	*V. vinifera*	Anthers	France	*17*
Intervitis magaracha	hybrid	Petioles	Ukraine	*13*
Kober 55			Italy	*24*

Table 1. (Cont.). Update of additional embryogenic cell suspensions currently utilized for transformation experiments

Cultivar	Genetic origin	Explant origin	Stock location	Reference
Krona 42	*V. vinifera*	Anthers	Ukrine	*26*
Magaracha	*V. vinifera*	Anthers	Ukrine	*26*
Muller Thurgau	*V. vinifera*	Anthers	Germany	*28*
Muscat Ottonel	*V. vinifera*	Anthers	France	*17*
Muscat Gordo Blanco	*V. vinifera*	Anthers	Australia	*25*
Nashik	*V. vinifera*	Leaves	India	*12*
Neo Muscat	*V. vinifera*	Ovules	Japan	*31*
Neo Mat	*V. vinifera*	Anthers	Japan	*32*
Niagra	hybrid	Ovary	USA	*9*
Perlett	*V. vinifera*	Leaves	India	*12*
Pinot Noir	*V. vinifera*	Anthers	France	*17*
Pinot Noir	*V. vinifera*	Anthers	Australia	*25*
Podarok	*V. vinifera*	Anthers	Ukrine	*26*
Portan	*V. vinifera*	Anthers	France	*16*
Portan	*V. vinifera*	Anthers	France	*17*
Regent	*V. vinifera*	Anthers	Germany	*28*
Riesling	*V. vinifera*	Anthers	Australia	*25*
Riesling	*V. vinifera*	Anthers	Germany	*28*
Rubinovyi	*V. vinifera*	Anthers	Ukrine	*26*
Russalka 3	*V. vinifera*	Embryos	Bulgaria	*33*
Sauvignon Blanc	*V. vinifera*	Anthers	Australia	*25*
Semillion	*V. vinifera*	Anthers	Australia	*25*
Seyval Blanc	*V. vinifera*	Leaves	Germany	*34*
Shiraz	*V. vinifera*	Anthers	France	*16*
Tas-e-Ganesh	*V. vinifera*	Tendrils	India	*14*
Podarok Magaracha	hybrid	Petioles	Ukraine	*13*
Pusa seedless	*V. vinifera*	Leaves	India	*12*
Riesling	*V. vinifera*	Anthers	France	*17*
Shiraz	*V. vinifera*	Anthers	USA	*25*
Sonaka	*V. vinifera*	Tendrils	India	*35*
Syrah	*V. vinifera*	Anthers	France	*17*
Thompson Seedless	*V. vinifera*	Leaves	USA	*36*
Thompson Seedless	*V. vinifera*	Anthers	Australia	*37*

1.4. Genetic Transformation in Viticulture

As in traditional wine and grape-growing regions, current genetic developments concentrate on the production of disease and pest resistant varieties (*38, 39*). Many of the varieties cultivated today have been bred for high yields. At the same time, most are highly susceptible to grape diseases which lead to great losses in fruit yield and require the use of large amounts of pesticides. Other approaches aim to accelerate ripening, enhance optical quality and flavour, or induce parthenocarpy and seedlessness (*40, 41*). Different studies focused on optimizing the expression levels in transgenic grapes by screening different constitutive promoters (*36, 42-44*) or selectable markers (*45*).

Since 1999, forty-seven transgenic grape varieties have been released, mostly in the USA, but also in France, Italy, Germany (*46*), Canada and Australia (*25, 37*). Research has also been carried out in South Africa, Japan and Spain. Most of the released grapevines carried resistance genes to bacterial and fungal pathogens. Special attention was given in the USA to the so called Pierce's disease, which is mediated by a bacterium and has become a serious problem recently causing great damage to vineyards.

Besides disease control, biotechnological approaches have focused more on fruit traits, such as sugar content (*47*), colour, seedlessness in table grapes and higher fruit yield (*48*). In Canada and Bulgaria, cold-tolerant grape varieties have been developed (*33*). A few examples of additional recent genetically engineered grape products are as follows:

Viral Resistance. The coat protein of the *Arabis mosaic nepovirus* (ArMV) was expressed in very low levels in leaves of transgenic *Vitis rupestris*. These plants also did not accumulate the ArMV CP at levels detectable by ELISA (*49*).

Bacterial Resistance. The most important pathogen in viticulture in the USA is the bacterium *Xylella fastidiosa*, the causative agent of Pierce's disease. Recent studies further implemented the original findings achieved using Thompson Seedless (*50*). Researchers at the University of Florida transformed the grape varieties Merlot and Chardonnay with the synthetic version of the ceropin antimicrobial gene (*51*). Vidal *et al.* (*52*) reported the production of transgenic Chardonnay plants via bombardment were co-transformed with different antimicrobial genes. Further glasshouse and field trials are still necessary to determine whether these plants are indeed more tolerant towards Pierce's disease.

Fungal Resistance. In many places, but especially in Europe, fungal diseases are the main threat to vineyards. The fungal pathogens downy and powdery mildew and grey mould are the main target in fungal tolerant transgenic grapes. Transgenic grapes contained pathogen-related genes from barley or rice which encode for chitinases, and/or glucanases, both known as cell wall-degrading enzymes. Some transgenic plants showed enhanced resistance against powdery mildew, caused by *Uncinula necator*, and some resistance against *Elisinoe ampelina* (*31*). Resveratrol, a stylbene anti-fungal compound, was shown very elegantly to play a major role in tolerance against *Botrytis* in transgenic plants of the 41B rootstock (*53*). Recently, a eutypine detoxifying gene (*Vr-ERE*), encoding an NADPH-dependent aldehyde reductase, was expressed by transgenic plants from the rootstock 110 Richter. The overexpression of the *Vr-ERE* gene increased their detoxification capacity and enhanced their resistance to *Eutypa lata* (*54-56*). Ribosome-inhibiting proteins were able to block the fungal ribosomes and were also considered capable to block fungal growth. The *Run1* gene, that was isolated from *Muscadinia rotundifolia*, was able to confer total resistance to powdery mildew, and transgenic *Vinifera* plants expressing this gene are under evaluation for fungal disease tolerance (*57*).

Fruit Quality - Seedlessness: Consumers clearly prefer seedless varieties of grapes and this demand dominates the marketplace. Studies were launched to explore novel strategies to convert seeded grapes to seedless using biotechnological approaches. At least several different strategies have been attempted to convert seeded to seedless grapes. Koltunow *et al.* (*58*) was the first to prevent the formation of hard seed-coat layer, yet allowing normal development of fruit and seeds. A later approach focused on directed overexpression of auxins in the fertilized ovule. Overproduction in the ovules prior to fertilization was shown to stimulate the production of parthenocarpic seedless fruits (*40, 41, 59;* Figs. 1H and J).

2. MATERIALS

2.1. A. tumefaciens *Strain*

Agrobacterium tumefaciens strain EHA105 (*60*) carrying a binary plasmid is utilized for transformation. The T-DNA of the binary plasmid contains the gene of interest and a selectable marker such as *npt*II, *hpt* or others. A reporter gene such as *uid*A or *gfp* is not necessary since selection is very efficient for all cultivars transformed according to this protocol. All chemicals of the highest purity were purchased from Sigma unless otherwise mentioned and were stored according to manufacturer's recommendations (*see* Note 1).

2.2. Culture Media for A. tumefaciens

1. YEB medium: 10 g/l yeast extract, 10 g/l peptone, 5 g/l NaCl , pH 7.0.
2. Liquid culture medium (LCM): YEB medium containing 50 mg/l kanamycin sulphate and 30 mg/l rifampicin, pH 7.0.
3. Virulence induction medium (VIM): YEB containing 50 mg/l kanamycin sulphate, 30 mg/l rifampicin, 5 mg/l $MgSO_4$, 200 μM acetocyringone, (Aldrich), pH 5.2.
4. Agar culture medium: YEB medium plus 15 g/l agar (Difco granulated), 50 mg/l kanamycin sulphate, 30 mg/l rifampicin, pH 7.0.
5. Kanamycin sulphate stock solution (100 mg/ml in water). Sterilize by filtration through 0.2 μm membrane (Schleicher & Schuell, Germany). Store in aliquots at -20°C.
6. Rifampicin stock solution (30 mg/ml in methanol). Store in aliquots at -20°C.

2.3. Plant Material and Tissue Culture

1. Flower surface sterilant: 1.3% (v/v) sodium hypochlorite solution containing 0.1% (v/v) Tween 20.
2. Anthers inoculation medium (AIM): 4.4 g/l Murashige and Skoog (MS, *61*) salts and vitamins mixture, Duchefa, Haarlem, The Netherlands), 0.25% Gelrite, 3% sucrose (BDH), 2 mg/l 2,4-D, 0.2 mg/l BAP, 5 mg/l AgNO$_3$ (filter-sterilized), pH. 6.2 (*see* Note 2).
3. Semi-solid maintenance medium (SMM): 4.4 g/l MS, 0.25% Gelrite, 6% sucrose, 0.25% activated charcoal, 5 mg/l IASP (filter-sterilized), 2 mg/l 2,4-D, 0.2 mg/l BAP (*see* Note 3).
4. Liquid maintenance medium (LMM): 2.1 g/l Nitsch and Nitsch salts and vitamins mixture (*62*; Duchefa), 0.25% Gelrite, 3% sucrose, 18 g/l maltose, (Duchefa), 1 g/l casein enzymatic hydrolysate, 4.6 g/l glycerol, 2 mg/l NOA (stock solution: 2 mg/ml in ethanol stored at 4°C), pH 6.0.
5. Co-cultivation medium (CM): 4.4 g/l MS medium, 2% sucrose, 1% glucose, 200 µM acetocyringone, 1 mg/l NOA, 0.26% Gelrite. Place a sterile Whatman paper (No. 41, 7 cm) on top of every 9 cm diam. Petri dish (*see* Note 4).
6. Liquid regeneration medium (LRM): As LMM, but with stepwise addition from 5-30 mg/l paromomycin sulphate, 300 mg/l claforan and lacking NOA (*see* Note 5).
7. Plant rooting medium (PRM): 2.46 g/l Lloyd and McCown woody plant medium (*63*), (Duchefa), 0.26% Gelrite, 3% sucrose, 0.2% activated charcoal, 0.2 mg/l NAA, 0.2 mg/l BAP, 100 mg/l *myo*-inositol, 100 mg/l paromomycin sulphate, 400 mg/l claforan (*see* Note 6).
8. Acetocyringone: 200 mM stock solution in ethanol. Store in aliquots at -20°C.

3. METHODS

3.1. Embryogenic Culture Preparation and Maintenance

1. Collect from the vineyard, small (2-3 mm), unpollinated flower buds, about 12-14 days before anthesis.
2. Store flowers in an empty container with minimal moisture for 3 days at 4°C in the dark, which was found to improve embryogenic calli formation.
3. Immerse flowers for 30 sec in 70% ethanol and further disinfect for 10 min without agitation in surface-sterilant; rinse 3 times with sterile distilled water.
4. Dissect flowers under a stereo-microscope, collect approx. 30-40 intact anthers (without the filament) and transfer to 5 cm diam. Petri dishes containing AIM medium. Incubate at 25°C in darkness for 4 weeks, which was found to improve embryogenic calli formation. Transfer yellow-white emerging calli to fresh medium for 4 additional weeks.
5. After about 2 months, visually select for soft, friable, off-white callus containing proembryogenic clusters. Transfer calli to SMM medium for long-term

maintenance at 25°C in darkness for 4 weeks. Subculture to fresh SMM medium every month.

6. Establish a liquid culture of fine-embryogenic cells by transferring calli from SMM medium to LMM medium. Subculture cells to fresh LMM every week. When subculturing to fresh medium, sieve cells through a metal net (200 μm mesh, Sigma) to maintain only small aggregates (Figs. 1A and B). Incubate at 25°C in darkness on an orbital shaker at 80 rpm (*see* Note 7).

3.2. Agrobacterium *Preparation*

1. Grow *A. tumefaciens* on YEB agar medium (with antibiotics) at 28°C for 2 days. Take one loopful of bacteria and transfer to 50 ml of YEB culture medium (with antibiotics) and grow overnight at 28°C on an orbital shaker at 250 rpm. Measure absorbance at 600 nm to reach an O.D. of 1.0.
2. Centrifuge in 50 ml sterile disposable plastic tubes at 5000 rpm (Hettich Universal 16 table centrifuge) for 10 min. Discard the supernatant and resuspend in 2-3 ml of fresh VIM medium. Add 40 ml of VIM medium and continue to grow for additional 4 h at 28°C on an orbital shaker at 250 rpm.
3. Centrifuge at 5000 rpm for 15 min. Resuspend the bacterial pellet in 10 ml of liquid CM medium.

3.3. *Inoculation, Co-cultivation, Selection and Regeneration of Transgenic Plants*

1. Collect 2-4 g of blotted-dry embryogenic cells and immerse into liquid CM medium containing *A. tumefaciens* in suspension. Use a sterile disposable 10 ml pipette to remove the entire medium from a 250 ml Erlenmeyer flask in which the fine cell suspension was growing. In order to avoid losing cells, empty the media while pressing the mouth of the pipette against the bottom of the Erlenmeyer so the cells will not penetrate the pipette.
2. Fill the Erlenmeyer with 20-30 ml of *Agrobacterium* culture, close the Erlenmeyer opening slightly with a sterile cotton-wool cork covered with an aluminum foil and transfer to a desiccator. Apply vacuum (7.4 psi) for 60 sec using a vacuum pump. Culture for an additional 30 min without shaking.
3. After co-cultivation, blot-dry the cells and transfer to a No. 41 Whatman paper overlying a plate of CM medium. Add few drops of liquid CM medium on top of the cell suspension to prevent the plant material from drying out. Seal with parafilm and culture for 3 days at 25°C in the dark.
4. Collect cells from filter paper and wash 3 times in a sterile 250 ml Erlenmeyer with 50 ml LMM medium supplemented with 1g/l claforan and 100 mg/l DTT to avoid oxidation. Maintain cultures in LMM medium supplemented with 5 mg/l paromomycin, for 4 days in the dark at 25°C on a rotary shaker set at 80 rpm (*see* Note 8).
5. Subculture to fresh LMM medium every 2-3 days. At 7 day intervals, elevate the paromomycin sulphate concentration from 5-10 mg/l, then to 20 mg/l and finally to a concentration of 30 mg/l (*see* Note 9).

6. Embryos are visualized by the naked eye and should be allowed to develop in LMM medium in 4-6 weeks after co-cultivation (Figs. 1C-D and G).

7. Plate regenerated embryos (Fig. 1E) on top of 70 ml PRM medium in Magenta GA-7 boxes. Culture at 25°C in the light (45 µE/m^2/sec). Subculture to fresh PRM every 3-4 weeks until mature rooted grape plantlets (Fig. 1F) are obtained (*see* Note 10).

8. Transfer rooted healthy plants for hardening to a glasshouse. Before transfer to a soil mixture (Peat Moss:Vermiculite; 1:1, vol:vol), trim the *in-vitro* roots leaving approx. 1 cm of root. Dip the roots into rooting powder and transplant to the soil mixture in 10 cm diam. pots. Irrigate once, and then maintain plants in the shade using cheese cloth and a shade net. Gradually decrease moisture while removing both the shade net and the cheese cloth (*see* Note 11).

The different stages leading to generation of transgenic grape plants are illustrated in Figure 1.

4. NOTES

1. In general, the method described in this chapter enabled successful transformation of several grape cultivars and is considered to be general and not specific to a given cultivar. We have successfully applied this procedure to the following cultivars: Prime, Sugarone, Red-Globe, Richter 110, 41B, Thompson Seedless, Italia, Velika, Gamay and others. We have screened different *Agrobacterium* strains for their efficiency to transform embryogenic lines of different *Vitis* species. We have found that most of the stains studied, namely GVE3301, LBA4404 and EHA101, were able to transiently transform embryogenic lines. However, best results were obtained using the EHA strains.

2. The addition of AgNO$_3$ to anthers inoculation medium (AIM) is a major factor in increasing the efficiency of embryogenic callus development from anthers. It is important at this stage to block ethylene synthesis during callus development. Other ethylene inhibitors, such as silver thiosulphate (STS), may also be utilized at 2-5 mg/l.

3. The presence of IASP was found to be very important for the long term maintenance of embryogenic calli on semi-solidified medium as back-up cultures. This conjugated auxin enables a slow release of free auxin during the culture period and may also reduce chromosomal mutation. A constant level of auxin in the media was found to inhibit premature germination of the pro-embryogenic explants which may lead to the loss of the culture.

4. Whatman filter paper were found to reduce the growth of *Agrobacterium* during co-cultivation, a step that may avoid overgrowth of the bacteria at later stages leading subsequently to calli necrosis and death.

5. We have screened different antibiotics, such as kanamycin sulphate, paromomycin and G418, to find the most suitable for the step by step selection procedure. Paromomycin was found to be the most effective selective agent while enabling efficient selection without any non-specific toxic side effects.

6. Rooting and conversion into normal grape plantlets is the major bottle neck before hardening the plants in the glasshouse. Some grape cultivars may prefer to grow on less-rich media at this stage. We have successfully utilized a modified MS medium which contains reduced concentrations of some macro salts: 400 mg/l NH4NO$_3$, 1000 mg/l KNO$_3$, 200 mg/l MgSO$_4$.7H$_2$O, and the addition of 200 mg/l glutamine with 100 mg/l glycine.

7. The shift from semi-solidified cultures to liquid cultures should be performed while keeping the ratio between the volume of cells and the liquid media in the flask at a constant. Select suitable cells and incubate them in 5-8 ml of LMM medium in the dark with an agitation of 80 rpm. After 5-7 days, in case cells multiply, start a step by step increase of media volume in the 250 ml flasks reaching a final volume of 60-70 ml.

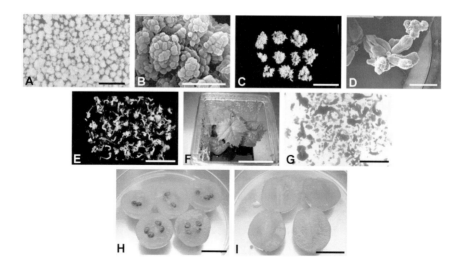

Figure 1. Procedure for the development of transgenic grape plants. A. Yellowish-friable embryogenic cell line of V. vinifera *cv. Red Globe. B. Long term maintenance in liquid culture retains its morphogenetic potential and proembryogenic structure. C-F. Transfer to regeneration medium promotes embryo development (C), germination (D,E) and establishment of mature grape plants (F), respectively. G. After co-cultivation with A.* tumefaciens, *cells are transferred to liquid selection medium in which transformed cells may develop into transgenic embryos. H, I. Transgenic seedlessness: An RNAse gene, expressed under the control of the embryo specific promoter Tob Rb7, enabled the ablation of seeds from a seeded cultivar (H) to promote genetically engineered seedless berries (I). Scale bars = 1 cm (A), 100 μm (B), 3 cm (C, E and F), 1 mm (D), 4 mm (G), 2 cm (H and I).*

8. We have previously recommended the use of different antioxidants during co-cultivation. This step is crucial if the culture contains pro-embryogenic type callus which may necrotize upon exposure to *Agrobacterium*. When true callus is used, DTT may be supplemented to the medium only after the termination of the co-cultivation stage. In a case where hygromycin is preferred for selection, start selection with 5 mg/l and the increase its concentration gradually up to 18 mg/l.

9. Frequent subculturing to fresh LMM medium is an important step enabling efficient selection and regeneration. Several subcultures enable a constant supply of fresh antibiotic for selection and the removal of compounds excreted from the cells that may block regeneration.

10. During subculture to PRM medium it is important to induce the development of a root system. Plants that are poorly rooted will not survive the hardening stage. During transfer, it is important to cut the base of the plantlet before inserting the plantlet into the gelled medium. Upon each subculture, it is important to renew the cutting to further improve rooting.

11. During hardening, it is important not to water the plant directly on the leaves but to ensure high humidity in the glasshouse. Plants should be kept covered with a cheese cloth, and during days of high light intensities one should use a shade net. The appearance of new leaves is an indication of rooting and is the best time to gradually reduce the humidity level. Irrigation with anti-fungal agents, such as Ryzolex, may increase the survival of plantlets.

CONCLUSIONS

Vitis embryogenic calli are the most frequently used explants for genetic transformation in this genus. Co-cultivation of such cells with *A. tumefaciens* should enable the establishment of genetically modified grape plants. Such plants are regenerated through embryogenesis and converted into morphologically normal grape plantlets. We believe that genetic engineering will prove to be invaluable to the future improvement of *Vitis*. The key phenotypes of a grape cultivar (berry, quality, yield and vigour) is a result of complex interactions between the environment, management regime and the genotype of the cultivar. The protocol described in this Chapter provides an average transformation efficiency of 5-10% for all cultivars tested, thus may contribute to the successful incorporation of transgenes of interest into grape, paving the way for improving rootstock and variety performance.

ACKNOWLEDGMENT

This work was supported in part by grant 204-0449 from the Chief Scientist of the Israel Ministry of Agriculture.

REFERENCES

1. Dandekar AM and Gutterson N (2000). Genetic engineering to improve quality, productivity, and value of Crops, *California Agriculture,* **54**: 49- 56.
2. Chrispeels MJ and Sadava DE (1994). Human population growth: Lessons from demography. In: Jones G, Barlett TL (eds.), *Plants, Genes and Agriculture* (pp.13-24). Boston, MA: Jones and Bartlett Publishers.
3. Martin GB (1998). Gene discovery for crop improvement. *Current Opinion in Biotechnology,* **9**: 220-226.
4. Winkler AJ, Cook JA, Kliewer WM and Lider LA (1974). Development and composition of grapes. In: *General Viticulture* (pp. 151-157). Berkeley, Los Angeles, London: University of California Press.
5. Martinelli L and Garibaudo I (2001). Somatic embryogenesis in grapevine (*Vitis* spp.). In: Roubelakis-Angelakis KA (ed.), *Molecular Biology and Biotechnology of Grapevine* (pp. 327-352). The Netherlands: Kluwer Academic Publishers.
6. Martinelli L and Mandolino G (2001). Transgenic transformation in *Vitis*. In: Bajaj YPS (ed.), *Biotechnology in Agriculture and Forestry, Transgenic Crops II,* (Vol. 47, pp. 325-338). Berlin: Germany, Springer-Verlag.
7. Jayasankar S, Gray DJ and Litz RE (1999). High-efficiency somatic embryogenesis and plant regeneration from suspension cultures of grapevine. *Plant Cell Reports,* **18**: 533-537.
8. Jayasankar S, Van Aman M, Li Z and Gray DJ (2001). Direct seedling of grapevine somatic embryos and regeneration of plants. *In Vitro Cellular and Developmental Biology-Plant,* **37**: 476-479.
9. Motoike SY, Skirvin RM, Norton MA and Otterbacher AG (2001). Somatic embryogenesis and long term maintenance of embryogenic lines from fox grapes. *Plant Cell, Tissue and Organ Culture,* **66**: 121-131.
10. Martinelli L, Candioli E, Costa D, Poletti V and Rascio N (2001a). Morphogenic competence of *Vitis rupestris* secondary somatic embryos with a long culture history. *Plant Cell Reports,* **20**: 279-284.

11.   Martinelli L, Garibaudo I, Bertoldi D, Candioli E and Poletti V (2001b). High efficiency embryogenesis and plant germination in grapevine cultivars Chardonnay and Brachetto a grappolo lungo. *Vitis*, **40**: 111-115.

12.   Das DK, Reddy MK, Upadhyaya KC and Sopory SK (2002). An efficient leaf-disc culture method for the regeneration via somatic embryogenesis and transformation of grape (*Vitis vinifera* L.). *Plant Cell Reports*, **20**: 999-1005.

13.   Zlenko VA, Kotikov IK and Troshin LP (2002). Efficient GA3-assisted plant regeneration from cell suspensions of three grape genotypes via somatic embryogenesis. *Plant Cell, Tissue and Organ Culture*, **70**: 295-299.

14.   Salunkhe CK, Rao PS and Mhatre M (1997). Induction of somatic embryogenesis and plantlets in tendrils of *Vitis ninifera* L.. *Plant Cell Reports*, **17**: 65-67.

15.   Zhu YM, Hoshino Y, Nakano M, Takahashi E and Mii M (1997). Highly efficient system of plant regeneration from protoplasts of grapevine (*Vitis vinifera* L.) through somatic embryogenesis by using embryogenic callus cultures and activated charcoal. *Plant Science*, **123**: 151-157.

16.   Torregrosa L, Locco P and Thomas MR (2002b). Influence of *Agrobacterium* strain, culture medium, and cultivar on the transformation efficiency of *Vitis vinifera* L.. *American Journal of Enology and Viticulture*, **53**: 183-190.

17.   Perrin M, Martin D, Joly D, Demangeat G, This P and Masson JE (2001). Medium-dependent response of grapevine somatic embryogenic cells. *Plant Science*, **161**: 107-116.

18.   Takeno K, Koshioka M, Pharis RP, Rajasekaran K and Mullins MG (1983). Endogenous gibberellin-like substances in somatic embryos of grape (*Vitis vinifera* x *Vitis rupestris*) in relation to embryogenesis and the chilling requirement for subsequent development of mature embryos. *Plant Physiology*, **73**: 803-808.

19.   Maitz M (2000). Use of an ultrasound cell retension system for the size fractionation of somatic embryos of woody species. *Plant Cell Reports*, **19**: 1057-1063.

20.   Jayasankar S and Bondada BR (2002). A unique morphotype of grapevine somatic embryogenesis exhibits accelerated germination and early plant development. *Plant Cell Reports*, **20**: 907-911.

21.   Wang Q, Gafny R, Sahar N, Mawassi M, Tanne E and Perl A (2002). Cryopreservation of grapevine (*Vitis vinifera* L.) embryogenic cell suspensions and subsequent plant regeneration by encapsulation-dehydration. *Plant Science*, **162**: 551-558.

22.   Wang Q, Mawassi M, Sahar N, Li P, Colova-Tsolova V, Gafny R, Sela I, Tanne E *et al.* (2003). Cryopreservation of grapevine *(Vitis* spp.) embryogenic cell suspensions by encapsulation-vitrification. *Plant Cell, Tissue and Organ Culture*, (in press).

23.   Perl A and Eshdat Y (1998). DNA transfer and gene expression in transgenic grapes. In: Tombs MP (ed.), *Biotechnology & Genetic Engineering Reviews* (Vol. 15, pp. 365-386). Andover, England: Intercept Ltd.

24.   Semenzato M, Poletti V and Martinelli L (2002). The use of phosphomannose isomerase as a selectable marker to transfer foreign genes in grape (*Vitis* spp.), *Proceedings of the XLVI Italian Society of Agricultural Genetics – SIGA Annual Congress, Giardini Naxos, Italy, 18-21 September*.

25.   Locco P, Franks T and Thomas MR (2001). Genetic transformation of major wine grape cultivars of *Vitis vinifera* L.. *Transgenic Research*, **10**: 105-112.

26.   Rubtsova MA and Levenko BA (1999). Transgenic grapevine plants resistant to the herbicide phosphinothricin and to the crown gall disease. *Fiziologia I Biokhimia Kul'turykh Rastenii*, **31**: 214-219.

27.   Jayasankar S (2000). *In vitro* selection of *Vitis vinifera* Chardonay with *Elsinoe ampelina* culture filtrate is accompanied by fungal resistance and enhanced secretion of chitinase. *Planta*, **211**: 200-208.

28.   Borhoff BA and Harst M (2000). Establishment of embryo suspension of grapevines (*Vitis* L.). *Vitis*, **39**: 27-29.

29.   Motioike SY, Skirvin RM, Norton MA and Otterbacher AG (2002). Development of methods to genetically transform American grape (*Vitis* labrusca L.H. Bailey). *Journal of Horticultural Science & Biotechnology*, **77**: 691-696.

30.   Mozsar L, Viczian O and Sule S (1998). *Agrobacterium*-mediated genetic transformation of a interspecific grapevine. *Vitis*, **37**: 127-130.

31.   Yamamoto T, Iketani H, Leki H, Nishizawa Y, Hibi T, Hayashi T and Matsuta N (2000). Transgenic grapevine plants expressing a rice chitinase with enhanced resistance to fungal pathogens. *Plant Cell Reports*, **19**: 639-646.

32. Nakano M, Watanabe Y and Hoshino Y (2000). Histological examination of callogenesis and adventitious embryogenesis in immature ovary culture of grapevine (*Vitis vinifera* L.). *Journal of Horticultural Science & Biotechnology, 75*: 154-160.

33. Gutoranov GP, Tsvetkov IJ, Colova-Tsolova VM and Atanassov AI (2001). Genetically engineered grapevines carrying GFLV coat protein and antifreeze genes. *Agriculturae Conspectus Scientificus, 66*: 69-74.

34. Buck S (1999). Genetic transformation studies on *Vitis vinifera* cv. Seyval blanc. Doctoral dissertation, University of Hohenheim, Stuttgart, Germany.

35. Salunkhe CK, Rao PS and Mhatre M (1999). Plantlet regeneration via somatic embryogenesis in anther callus of *Vitis latifolia* L.. *Plant Cell Reports, 18*: 670-673.

36. Li Z, Jayasankar S and Gray DJ (2001a). Expression of a bifunctional green fluorescent protein (GFP) fusion marker under the control of three constitutive promoters and enhanced derivatives in transgenic grape (*Vitis vinifera*). *Plant Science, 160*: 877-887.

37. Franks T, Gang HD and Thomas M (1998). Regeneration of transgenic shape *Vitis vinifera* L. Sultana plants: genotypic and phenotypic analysis. *Molecular Breeding, 4*: 321-333.

38. Kikkert JR (2001). Grapevine genetic engineering. In: Roubelakis-Angelakis KA (ed.), *Molecular Biology and Biotechnology of the Grapevine* (pp. 393-463). The Netherlands: Kluwer Academic Publishers.

39. Vivier MA and Pretorius IS (2002). Genetically tailored grapevines for the wine industry. *Trends in Biotechnology, 20*: 472-478.

40. Perl A, Sahar N, Spiegel-Roy P, Gavish S, Elyassi R, Orr E and Bazak H (2000). Conventional and biotechnological approaches in breeding seedless table grapes. *Acta Horticulturae, 528*: 607-612.

41. Mezzetti B, Pandolfini T, Navacchi O and Landi L (2002). Genetic transformation of *Vitis vinifera* via organogenesis. *Bio-Med Central Biotechnology, 2*: 18.

42. Gollop R, Farhi S and Perl A (2001). Regulation of the leucoanthocyanidin dioxygenase gene expression in *Vitis vinifera*. *Plant Science, 161*: 579-588.

43. Gollop R, Even S, Colova-Tsolova V and Perl A (2002). Expression of the grape dihydroflavonol reductase gene and analysis of its promoter region. *Journal of Experimental Botany, 53*: 1397-1409.

44. Torregrosa L, Verrios C and Tesniore C (2002a). Grapevine (*Vitis vinifera* L.) promoter analysis by biolistic-mediated transient transformation of cell suspension. *Vitis, 41*: 27-32.

45. Torregrosa L, Lopez G and Bouquet A (2000). Antibiotic sensitivity of grapevine: A comparison between the effect of hygromycin on shoot development of transgenic 110 Richter rootstock (*Vitis Berlandieri X Vitis rupestris*). *South African Journal of Enology and Viticulture, 21*: 32-39.

46. Harst M, Bornhoff BA, Zyprian E and Topfer R (2000). Influence of culture technique and the genotype on the efficiency of *Agrobacterium*-mediated transformation of somatic embryos (*Vitis vinifera*) and their conversion to transgenic plants. *Vitis, 39*: 99-102.

47. Davis C and Boss PK (2000). The use of molecular biology techniques to study and manipulate the grapevine: Why and how? *Australian Journal of Grape and Wine Research, 6*: 159-167.

48. Vivier MA and Pretorius IS (2000). Genetic improvement of grapevine: Tailoring grape varieties for the third millennium – a review. *South African Journal of Enology and Viticulture, 21*: 5-26.

49. Spielmann A, Krastanova S, Douet-Orhand V and Gugerli P (2000). Analysis of transgenic grapevine (*Vitis vinifera*) and *Nicotiana benthamiana* plants expressing an *Arabis mosaic virus* coat protein gene. *Plant Science, 156*: 235-244.

50. Scorza R, Cordts JM, Gray DJ, Gonsalves D, Emershad RL and Ramming DW (1996). Producing transgenic "Thompson seedless" grape (*Vitis vinifera*) plants. *Journal of American Society for Horticultural Science, 121*: 616-619.

51. Li Z, Jayasankar S and Gray DJ (2001b). An improved enzyme-linked immunoabsorbent assay protocol for the detection of small lytic peptides in transgenic grapevines (*Vitis vinifera*). *Plant Molecular Biology Reporter, 19*: 341-351.

52. Vidal JR, Kikkert JR, Wallace PG and Reisch BI (2003). High-efficiency biolistic co-transformation and regeneration of 'Chardonnay' (*Vitis vinifera* L.) containing *npt*II and antimicrobial peptide genes. *Plant Cell Reports, 22*: 252-260.

53. Coutos-Thevenot P, Poinssot B, Bonomelli A, Year H, Breda C, Buffard D, Esnault R, Hain R *et al.* (2001). *In vitro* tolerance to *Botrytis cinerea* of grapevine 41B rootstock in transgenic plants expressing the stilbene synthase *Vst1* gene under the control of a pathogen-inducible PR 10 promoter. *Journal of Experimental Botany, 52*: 901-910.

54.   Guille'n P, Guis M, Martinez-Reina G, Colrat S, Dalmayrac S, Deswarte C, Bouzayen M, Roustan JP *et al.* (1998). A novel NADPH-dependent aldehyde reductase gene from *Vigna radiata* confers resistance to the grapevine fungal toxin eutypine. *The Plant Journal,* **16**: 335-343.

55.   Amborabe BE, Fleurat-Lessard P, Bonmort J, Roustan JP and Robin G (2000). Effects of eutypine, a toxin from *Eutypa lata*, on the plant cell plasma membrane. *Plant Physiology and Biochemistry,* **38**: 51-58.

56.   Legrand V, Dalmayarc S, Latche A, Pech JC, Bouzayen M, Fallot J, Torregrosa L, Bouquet A *et al.* (2003). Constitutive expression of *Vr-ERE* gene in transformed grapevines confers enhanced resistance to eutypine, a toxin from *Eutypa lata*. *Plant Science,* **164**: 809-814.

57.   Bouquet A, Pauquet J, Adam-Blondon AF, Torregrosa L, Merdinoglu D and Wiedemann-Merdinoglu J (2000). Towards the obtention of grapevine varieties resistant to downy mildews by conventional breeding and biotechnology. *Progros Agricole et Viticole,* **117**: 383-389.

58.   Koltunow AM, Brennan P, Bond JE and Barker SJ (1998). Evaluation of genes to reduce seed size in *Arabidopsis* and tobacco and their application to *Citrus*. *Molecular Breeding,* **4**: 235-251.

59.   Rotino GL, Perri E, Zottini M, Sommer H and Spena A (1997). Genetic engineering of parthenocarpic plants. *Nature Biotechnology,* **15**: 1398-401.

60.   Hood EE, Gelvin SB, Melchers LS and Hoekema A (1993). New *Agrobacterium* helper plasmids for gene transfer to plants. *Transgenic Research,* **2**: 208-218.

61.   Murashige T and Skoog F (1962). A revised medium for the rapid growth and bioassays with tobacco tissue cultures. *Physiologia Plantarum,* **15**: 473-497.

62.   Nitsch JP and Nitsch C (1969). Haploid plants from pollen grains. *Science,* **163**: 85-87.

63.   Lloyd G and McCown B (1980). Commercially feasible micropropagation of mountain laurel, *Kalmia latifolia*, by the use of shoot-tip cultures. *International Plant Propagation Society Proceedings,* **30**: 421-427.

Chapter 18

AGROBACTERIUM-MEDIATED GENETIC TRANSFORMATION OF COTTON

K. RAJASEKARAN
Southern Regional Research Center, USDA-ARS, New Orleans, Louisiana, U.S.A
E-mail: krajah@srrc.ars.usda.gov

1. INTRODUCTION

1.1. Importance and Distribution of Cotton

Cotton is the most important fibre crop of the world with an annual production of about 20 million metric tones from about 33.5 million hectares in 2002 (*1*). Cotton seed is also an important oilseed crop and is the world's third largest in terms of global crushings from an annual production of about 33 million metric tonnes in 2002 and a source of high quality protein meal (*2*). The genus *Gossypium*, a member of the *Malvaceae*, contains 49 species distributed throughout most tropical and subtropical regions of the world (*3*). The most common commercially grown cotton varieties belong to four species of *Gossypium* – *G. arboreum* L., *G. barbadense* L., *G. herbaceum* L. *and G. hirsutum* L. Over 90% of the annual cotton crop in the world is produced from the upland cotton varieties of *G. hirsutum*. This species is generally thought to have a natural origin that involved the combining of genomes from plants related to extant diploid species from the Old World (A genome) and the New World (D genome). Diploid (2n = 2x = 26) species - *G. arboreum* and *G. herbaceum* (AA) are still being grown in the African and Asian continents whereas the allotetraploid (4n = 4x = 52) species - *G. hirsutum* and *G. barbadense* (AADD) are being grown worldwide. The largest cotton producers are China, USA and India.

1.2. Need for Genetic Improvement

Cotton was one of the first crops to be genetically engineered for improved agronomic and fibre traits due to the following reasons:

1. Cotton requires intensive management practices to obtain a healthy crop since it is susceptible to attack by several insect pests, nematodes, fungal and other microbial pathogens. Engineering cotton to express anti-pathogen compounds would make cotton production more cost effective. In addition, it is primarily a fibre crop and does not elicit negative public perception regarding genetic modification as with edible food and feed crops.

I.S. Curtis (ed.), Transgenic Crops of the World - Essential Protocols, 243-254.

2. Improvement of cotton by conventional plant breeding practices has limitations due to time-consuming selection programmes and lack of availability of wild relatives with the desired agronomic and fibre traits. Most of the commercial varieties currently available have been produced by hand-crossing and recurrent selection in progeny rows, which requires approximately 6-7 years for production. Availability of suitable molecular markers could accelerate this selection programme. Hybrid cotton can only be produced currently by hand emasculation and pollination due to the unavailability of methods for inducing male sterility and pollen restoration (*4*). The procedure is laborious and expensive; it is used on a limited scale in India and China.

3. Cotton is primarily cultivated in arid regions with little or no irrigation; thus, it is subject to environmental and other stress factors.

4. Not every useful trait can be found in the cotton germplasm and not all species of *Gossypium* can be domesticated in one location for the purpose of breeding. There are instances where incorporation of resistance genes from exotic germplasm resulted in large-scale yield reductions in commercial lines (*5*). Even in successful introgression of exotic traits, numerous cycles of selection must follow to be of commercial value.

1.3. Genetically Modified Cotton

In terms of genetically engineered crops intended for industrial applications, cotton, is the leading fibre crop of the world. The estimated global area of transgenic cotton for 2002 is 6.8 million hectares, 20% of the total area under cotton cultivation (*6*). In the USA, 77% of cotton acreage was under transgenic cotton in 2002. The first wave of transgenic cotton that became popular since their introduction in 1996 was those with insect resistance and herbicide tolerance. Transgenic cotton with novel agronomic traits (mostly insect resistance and/or herbicide tolerance) have been commercialized in eight countries as of 2003 – Argentina, Australia, China, India, Indonesia, Mexico, South Africa and the USA. The next generation of transgenic cotton will target the most important product from the cotton plant - the fibre, which accounts for about 85% of the growers' income (*7*). The production of superior quality cotton fibre is extremely important to the textile industry worldwide.

1.4. Genetic Transformation Methods for Cotton

The commercially available insect-resistant and herbicide-tolerant cottons are the products of *Agrobacterium*-mediated transformation (*7*). Although particle bombardment has been used to introduce genes into cotton, no transgenic variety has been produced yet using this method (*see* Note 1). Transformation by the *Agrobacterium* method is discussed in detail below.

1.4.1. Agrobacterium-*mediated Transformation of Cotton*

Transformation of cotton using *Agrobacterium* is simple and efficient and this has become the method of choice in many laboratories. Tumor- or root-inducing, wild-type strains of *Agrobacterium tumefaciens* or *A. rhizogenes* infect cotton explants and nearly 100% of the infected explants develop tumors or roots within 7 days of infection (*8*). These observations indicate that transformation of cotton genotypes by *Agrobacterium* is not the limiting factor; however, regeneration of fertile plants through somatic embryogenesis is highly genotype dependent (*8*). The first successful transformation of cotton using non-oncogenic *Agrobacterium* was reported by Firoozabady *et al.* (*9*) and Umbeck *et al.* (*10*). A general protocol for *Agrobacterium*-mediated transformation of seedling explants (cotyledon and hypocotyls) is detailed below that works well with a broad range of genotypes including the obsolete Coker varieties (Coker Pedigreed Seed Co., Hartsville, SC purchased in 1989 by the Stoneville Pedigreed Seed Company, Memphis, TN), the source of all the transgenic cotton varieties available as of 2002. Improved Coker cultivars, once popular during the years 1965-1975, gave way to modern cultivars, with better lint yield and quality, belonging to other cultivar groups such as Acala, Deltapine, Stoneville and Paymaster. Of all these cotton varieties, Coker varieties are among the easiest to transform and regenerate, while elite upland and Pima cotton varieties are among the most difficult. Consequently, transgenic cotton lines are currently produced first with the highly regenerative Coker varieties, and then, through backcrossing, the transgenes are transferred into the desired commercial variety. Regeneration procedures via somatic embryogenesis vary with commercially grown varieties and appropriate procedures are referenced in the text.

2. MATERIALS

2.1. A. tumefaciens *Strain*

A. tumefaciens strain LBA4404 carrying pBI121 (derived from pBIN19; BD Biosciences Clontech, Palo Alto, CA, USA) is used as a model vector system for transformation for the purpose of demonstration in this report (*see* Note 2). The T-DNA of pBI121 contains the selectable marker gene, *npt*II with its own NOS promoter and terminator, and the reporter gene (*see* Note 3), *gus*A under the control of the CaMV35S constitutive promoter and terminator sequences. The plasmid is introduced into competent cells of *A. tumefaciens* strain LBA4404 (Invitrogen Life Technologies, Carlsbad, CA, USA) by tri-parental mating (*11*), electroporation or by heat-thaw method (*12*).

2.2. Stock Solutions

1. Acetosyringone (Sigma-Aldrich) – prepare 200 mM stock in DMSO (Sigma-Aldrich); filter-sterilize using a membrane filter (0.22 μm; Millipore Corp., Bedford, MA, USA). Store at 7°C for up to 3 months.

2. 2iP (Sigma-Aldrich) – 1.0 mg/ml stock - dissolve in a small volume of 0.1N KOH and dilute to volume; add before autoclaving medium. Store in siliconized glassware for up to 3 months at 7°C or longer at -20°C.

3. BAP (Sigma-Aldrich) – 1.0 mg/ml stock – dissolve in a small volume of 0.1N KOH and dilute to volume; add before autoclaving medium. Store in siliconized glassware for up to 3 months at 7°C or longer at -20°C.

4. Carbenicillin (AgriBio, North Miami, FL, USA) – 100 mg/ml in water;   filter-sterilize using a membrane filter (0.22 μm); add to warm medium after autoclaving. Stock solutions preferably stored for not more than a week at 7°C.

5. Cefotaxime (AgriBio) – 100 mg/ml in water; filter-sterilize using a membrane filter (0.22 μm); add to warm medium after autoclaving. Stock solutions preferably stored for not more than a week at 7°C.

6. Geneticin sulphate (G418, Sigma-Aldrich) – 50 mg/ml stock in water; filter-sterilize using a membrane filter (0.22 μm); add to warm medium after autoclaving. Aliquots of stock solutions can be stored at -20°C for 6 months.

7. IAA (Sigma-Aldrich) – 1.0 mg/ml stock in a small volume of 0.1N KOH and dilute to volume; always use a freshly prepared solution;   filter-sterilize using a membrane filter (0.22 micron) before adding to the medium.

8. Kanamycin (Sigma-Aldrich) – 50 mg/ml stock in water; filter-sterilize using a membrane filter (0.22 μm); add to warm medium after autoclaving. Stock solutions preferably stored for not more than a week at 7°C.

9. Kinetin (Sigma-Aldrich) – 1.0 mg/ml stock in a small volume of 0.1N KOH and dilute to volume, add before autoclaving medium. Store in siliconized glassware for up to 3 months at 7°C or longer at -20°C.

10. NAA (Sigma-Aldrich) – 1.0 mg/ml stock in a small volume of 0.1N KOH and dilute to volume; add before autoclaving medium. Store at 7°C for up to 6 months.

11. Streptomycin (Sigma-Aldrich) – 50 mg/ml stock in water; filter-sterilize using a membrane filter (0.22 μm). Stock solutions preferably stored for not more than a week at 7°C.

2.3. *Culture Media for* A. tumefaciens

1. Yeast Extract Broth (YEB) ingredients: 0.1% yeast extract (Difco, Detroit, MI, USA), 0.5% peptone (Difco), 0.5% beef extract (Difco), 0.5% sucrose, 2 mM MgSO$_4$, pH 7.2. YEB can be autoclaved and stored for several months at room temperature. Add MgSO$_4$ and antibiotics just prior to initiating cultures. For culture of LBA4404/pBI121, add to YEB medium 50 mg/l kanamycin (plasmid marker) and 50 mg/l streptomycin (chromosomal marker). If clumping of LBA4404 cells occur, YM medium (Invitrogen Life Technologies) can be used to alleviate the problem.

2. Other media can also be used, for example, Luria Broth (LB) medium – 1% tryptone (Difco), 0.5% yeast extract (Difco), 1.0% NaCl, pH 7.5, or YEP medium – 1% bactopeptone (Difco), 1% yeast extract, 0.5% salt and 20 mM CaCl$_2$, pH 7.0.

3. Single colonies of the vector can also be streaked onto YEB agar plates (semi-solidified with 15g/l Bacto Nutrient Agar, Difco), with antibiotics; *Agrobacterium* cells can be scraped from YEB plates and used for transformation.
4. In our laboratory, we routinely use glycerol stocks of *Agrobacterium* vectors stored at −86°C. To prepare glycerol stocks mix actively growing cells of *Agrobacterium* in YEB liquid medium with equal volume of sterile glycerol; quick-freeze 500 μl aliquots in pre-sterilized microfuge (Eppendorf, Hamburg, Germany) tubes in liquid nitrogen and store at −86°C.

2.4. Plant Materials

Cotton seed freshly harvested or stored for less than 2 years at room temperature is preferable. Delint cotton seed to remove fuzz fibres by carefully stirring the seed with concentrated sulphuric acid for 2-3 min in a fume hood. Decant the acid; carefully neutralise the residual acid with copious amounts of sodium bicarbonate and water. Rinse several times before air-drying and store in a cool (21 ± 3°C) and dry environment.

2.5. Tissue Culture Media

1. Seed Germination medium (SG): Murashige and Skoog (MS, *13*) or White's medium as modified by Singh and Krikorian (*14*). Add BAP or 2iP (5 mg/l) to promote uniform germination and cotyledon unfolding.
2. Surface sterilant: sodium hypochlorite solution (dilute to obtain 1% available chlorine; Clorox Co., Oakland, CA, USA) containing 0.1% (v/v) Tween 20 (Sigma-Aldrich) as a dispersing agent.
3. Callus initiation (CI) medium – MS supplemented with kinetin (1.0 mg/l) with NAA or IAA (2.0 mg/l) and 3% glucose (Sigma-Aldrich) as the carbon source (*see* Note 4). This medium has been successfully used for a broad range of commercial and obsolete varieties including the Coker genotype (*8, 15*). Media formulations for other commercial varieties have been listed in several publications (*see* Note 5).
4. Embryogenic Callus Maintenance medium (ECM) – MS with 2 mg/l NAA and 2% sucrose.
5. Embryo Germination (EG) Medium: MS half strength; add freshly prepared solution of IAA (2 mg/l) if root formation by somatic embryos is a problem.
6. All plant tissue culture media are solidified with 6-8 g/l agar-agar (Caisson Labs, Sugar City, ID, USA) and adjusted to a final pH of 5.8. When gellan gum from several commercial sources (e.g., Phytogel from Sigma-Aldrich; Gel-Gro from ICN Biochemicals, Cleveland, OH, USA; or Gelrite from Merck, Rahway, NJ, Kelco Division, USA) is used as the solidifying agent at 2 g/l, add an additional source of bivalent cation (e.g. 0.75 g/l $MgCl_2$) to aid gelling.

3. METHODS

3.1. Seed Germination

1. Surface-sterilize seeds for 20-25 min, with intermittent shaking, followed by 3-4 washes in sterile water.
2. Soak the seeds in sterile water (add 5 mg/l BAP or 2iP) for 1-2 days until the radicle emerges through the split seed coat. Newer seeds take longer than older seeds. The addition of a cytokinin during soaking helps in the development of uniform emergence of seedlings and unfolding of cotyledons - the source of explants.
3. Remove seed coat by squeezing the chalazal (broad) end of the seed – the folded cotyledon with the embryo usually emerges from the testa.
4. Plant the seed on SG medium contained in Mason jars or Magenta GA-7 boxes (Magenta Corporation, Chicago, IL, USA), 4-5 seeds per jar.
5. Grow the seedlings at 26 $\pm$ 2 $^{\circ}$C in dark or sub-optimal light conditions (10-20 μE/m^2/sec). Hypocotyls elongate in the dark and provide ample material for explant preparation.

3.2. Explant Preparation

1. Prepare cotyledon and hypocotyl explants from young (< 10 day old) seedlings.
2. Cotyledon explants are prepared by cutting 1 x 1 cm segments (6-8 segments can be made from a single cotyledon).
3. Hypocotyl sections (1 cm long) should be split longitudinally.
4. Keep the explants moist (use only few drops of MS liquid medium) for up to 1 h in a deep Petri dish until treatment with *Agrobacterium*.

3.3. Agrobacterium *Preparation*

1. Initiate *Agrobacterium* cultures from glycerol stocks stored at -86°C or from freshly inoculated YEB plates. Inoculate 50 ml YEB liquid medium with 500 µl of the glycerol stock. Use antibiotics in the medium as appropriate for the specific *Agrobacterium* strain in use. Several different host strains, such as LBA4404, MP-90, A136/542, EHA101, EHA105, have been successfully used to transform cotton. Cultures grown for about 16-18 h at 26 $\pm$ 2°C on a rotary shaker (120 rpm) will have the desired optical density (A_{600}) of between 0.6-0.8. Dilute the cultures in liquid MS medium to obtain the desired optical density, if necessary.
2. Pellet the *Agrobacterium* from YEB cultures in 50 ml centrifuge tubes by centrifugation at 3000 rpm for 10 min at 4°C. Discard the supernatant and resuspend the pellet in MS liquid medium to obtain the desired OD. Keep the bacterial suspension at room temperature until ready for use.
3. Add acetosyringone (100–200 µM) to the bacterial suspension to induce *vir* genes for efficient transformation (*see* Note 6). This step is usually not

necessary since transformation of seedling explants is not a problem in cotton
(*29*).

3.4. Transformation, Selection and Regeneration of Transgenic Plants

1. Treat explants with the *Agrobacterium* cell suspension for approx. 15 min with
   intermittent shaking. Drain the bacterial suspension and plate the explants onto a
   sterile 12 cm diam. filter paper (Whatman No.1) overlying freshly made callus
   initiation medium with glucose as the carbon source in large (15 cm diam.) Petri
   dishes. Incubate the plates in an incubator (21 ± 2°C, dark) for 48-72 h
   (cocultivation period) (*see* Note 7).
2. After the cocultivation period, rinse explants once in sterile water to remove
   excess bacteria and then 3 times in MS liquid medium containing cefotaxime
   (400 mg/l) and carbenicillin (200 mg/l). Use enough liquid to immerse all the
   explants in a pre-sterilized glass beaker. After the final rinse, blot the explants
   onto a sterile filter paper.
3. Plate explants on freshly made CI medium with glucose as the carbon source.
   This medium should contain cefotaxime (200 mg/l) and carbenicellin (200 mg/l)
   to prevent *Agrobacterium* regrowth and a selection agent depending on the
   vector used. In the case of LBA4404/pBI121, use either kanamycin 25-50 mg/l
   or its analog G418 at 5-10 mg/l. Plate 5-7 cotyledon segments per Petri dish (9
   cm diam.) and hypocotyls at 7-10 per dish. Split hypocotyl segments are plated
   with the longitudinal cut end in contact with the medium.
4. Place the cultures in incubators (26 ± 2°C, 16 h photoperiod; 40-60 $\mu E/m^2/sec$)
   for 3 weeks. After the initial culture period, subculture the explants onto
   freshly made CI medium (same as before). Remove roots, if present, from
   cotyledon explants and avoid transferring contaminated explants.
5. After the second culture period of 3-4 weeks, isolate callus clumps (>3 mm
   size) that show active growth and transfer to selection medium (10 per dish).
   Plate them onto a fresh medium with the same composition as before, including the selection agent. During this third subculture period (8-12 weeks
   after the initiation of the experiment), check for antibiotic tolerance of the
   putatively transformed callus lines. Transformed lines will show at least 5-20
   times more fresh weight gain compared to non-transformed callus lines. In our
   laboratory, 70-90% of these callus lines will be NPT II-ELISA (Agdia, Elkhart,
   IN, USA)- positive.
6. Transfer the putative resistant callus colonies to fresh medium (change the
   carbon source from 3% glucose to 2% sucrose) and culture further to induce
   embryogenesis. Maintain selective pressure at all times to avoid a high
   frequency of escapes.
7. Identify embryogenic callus colonies from putative antibiotic resistant callus
   (*see* Note 8). Keep individual transformed callus lines separate.
8. Use ECM medium to grow and multiply embryogenic callus from individual
   transformation events.

9. Test a sample of embryogenic callus for the presence of the antibiotic marker gene by PCR or the gene product by NPTII assay. The reporter gene *gus*A which encodes a β-glucuronidase can be assayed by several methods (*16, 17*): a) spectrophotometric assay - catalyze the enzyme substrate, p-nitrophenyl glucuronide (Duchefa, Haarlem, The Netherlands) quantify the reaction product, p-nitrophenol at A_{415}; b) fluorometric assay - incubate with the substrate MUG (CalBiochem, La Jolla, CA, USA) for 1-3 h and measure fluorescence at excitation 365 nm and emission 455 nm; or c) histochemical assay - observe for blue precipitate after staining for 3-24 h at 37°C with the substrate (*see* Note 9) X-gluc (BD BioSciences Clontech).

10. Transfer heart-shaped or mature somatic embryos to EG medium to germinate into plantlets (*see* Note 10). Embryos take about 4-6 weeks to develop into 4-5 leafed plantlets in Magenta boxes. Test leaf segments from each plantlet for the presence of the transgenic character.

11. Transfer transgenic plants to soil contained in 10 cm diam. pots – use a soil mix that provides good drainage (e.g., perlite:peat:sand at 4:4:2; vol:vol). Cover plants with plastic cups or plastic bags (Ziploc bags, SC Johnson Co., Racine, WI, USA) for about 5 days before acclimatization to ambient glasshouse conditions.

12. Keep the top soil in pots dry and water sparingly. Too much soil moisture often results in poor root growth and fungal growth around the stem. Fertilize the plants once a week with nutrient solution (e.g. Hoagland's nutrient solution). Commercially available slow-release nutrient pellets (e.g., Osmocote 14-14-14 from Scotts Co., Marysville, OH, USA) can also be used. Plants flower about 3-4 months after transfer to the glasshouse.

13. On average, *Agrobacterium*-mediated transformation results in the integration of 1-5 copies of the transgene and it is not uncommon to find transgenic cotton plants with only one copy of the gene at a frequency of >50% (*15, 22*). Under glasshouse conditions, in the absence of insect pollinators, cotton flowers are pre-dominantly self-pollinated. Thus, most of the T1 progeny seedlings of transformants with a single copy of the transgene (75%) will be either homozygous or heterozygous with the remainder being double recessives for the transgenic trait. In a large-seeded crop such as cotton, selection for antibiotic resistance or herbicide tolerance at the seed-level is difficult. However, several screening techniques could be employed to identify transgenic traits. In addition to PCR analysis, histochemical staining of leaf disks for GUS provides a quick qualitative screening for the presence of *gus*A. Quantitative expression of the *npt*II gene is possible by NPTII ELISA (*see* Section 3.4.5.) and the GUS protein by fluorometric analysis (*see* Section 3.4.9.). Southern blot analyses of T0 and T1 plants are always essential to confirm the stable integration and transgene copy number followed by Northern and Western blot analyses to confirm their expression at the RNA and protein levels, respectively. When pollen from transgenic T0 Coker plants are used to pollinate a commercial upland variety, progeny seedlings from such outcrosses follow a typical

Mendelian segregation ratio of 1:1 (*15, 18*). Modern commercial transgenic varieties are products of such repeated backcrossing and selection practices.

4. NOTES

1. We have demonstrated in our laboratory that stable transformation of embryogenic cell suspension cultures of cotton can be obtained at a high frequency (over 4% of transiently expressing cells) using multiple biolistic bombardment of cultures during the rapid growth phase and by gradually increasing the selection pressure. However, care should be taken to avoid using long-term cultures (more than 6 months), which are known to accumulate somatic mutations and cytogenetic abnormalities leading to undesirable morphological and fertility-related problems in regenerated cotton plants (*18-20*). We also demonstrated that it is possible to overcome the problems with long-term cell cultures by using freshly initiated or cryopreserved embryogenic cell cultures (*21*). Biolistic transformation of cotton shoot meristems is an extremely tedious, labour-intensive task because it involves careful excision of meristems from imbibed zygotic embryos followed by the surgical removal of leaf primordia to expose the meristems to biolistic bombardment. The transformation efficiency of bombarded shoot meristems is very low (0.001 to 0.01%), and transformation in L2 and L3 layers, which gives rise to stable germline, is very sporadic and requires careful identification and processing by selective pruning (*7, 8*).
2. Different *Agrobacterium* host strains have been used in several published reports. Presence of the *vir* region of the super-virulent Ti plasmid pTiBo542 has been shown to promote transformation efficiency (*15, 22*).
3. *GFP* gene from jellyfish has also been successfully used as a reporter gene in cotton transformation (*23*).
4. Use of simple sugars, for example, glucose (*21*) and maltose (*24*) is recommended to reduce phenolic oxidation and browning, which are detrimental to healthy callus production.
5. Callus initiation medium according to several published reports include a) Firoozabady *et al.* (*25*), 5 mg/l 2iP and 0.1 mg/l NAA; b) Trolinder and Goodin (*20*), 0.5 mg/l kinetin and 0.1 mg/l 2,4-D; c) Davidonis and Hamilton (*26*), 1.0 mg/l kinetin and 2.0 mg/l NAA; d) Gawel *et al.* (*27*), kinetin 1.0 mg/l and 4.0 mg/l NAA.
6. Acetosyringone induces vir genes of *Agrobacterium* strains and the length of time needed for induction is at least 30 min.
7. Co-cultivation at this relatively lower temperature (21°C) compared to 25°C resulted in higher transformation frequencies (*22*). Plating on filter paper helps to limit pathogenic infection of explants by *Agrobacterium* by providing attachment sites for the excess bacterial cells.
8. The success rate of embryogenesis is highly genotype-dependent. Coker varieties are the easiest and most of the commercial varieties (e.g., Acalas, Delta and stripper varieties) are recalcitrant. Coker varieties produce embryogenic callus in approx. 2-5 months; whilst, commercial varieties take an average of 10 months.
9. X-gluc substrate penetration into cotton leaf disks is rather slow and the blue staining occurs only along the cut-ends (*18*). Staining for a longer period is necessary for uniform staining. Fluorometric detection of GUS is more sensitive than histochemical staining.
10. Poor germination of somatic embryos is another bottleneck in cotton regeneration. Germination percentage can vary from 10-70% depending on the genotype and the quality of somatic embryos. Dehydration of somatic embryos and subsequent germination in a low-salt medium in vented (0.22 μm) jars (Sigma-Aldrich) are often recommended and practiced in several laboratories to improve germination and subsequent acclimatization in soil (*24, 28*).

CONCLUSIONS

A step-by-step procedure on *Agrobacterium*-mediated transformation of cotton seedling explants is provided in this chapter. This procedure is equally effective on other explants such as leaf and petiole segments. *A. tumefaciens* infects cotton seedling explants with relative ease and it is not uncommon to observe a transformation efficiency of up to 100% (*29*). However, regeneration of fertile plants through somatic embryogenesis pathway is cumbersome, time-consuming and highly genotype-dependent. In addition, low germination rate of somatic embryos and subsequent survival in soil also limit the success in transformation experiments. In spite of these drawbacks, cotton was one of the first crops to be transformed with improved agronomic traits and marketed successfully. The success is largely due to the use of obsolete, yet highly regenerable Coker varieties in transformation experiments. Attempts are underway in several laboratories around the world to improve the fibre quality of the cotton crop, a polygenic trait that is controlled by the spatial and temporal expression of several thousand genes. Improvement of cotton fibre quality through conventional breeding and biotechnology is vital to the survival of textile industry worldwide.

ACKNOWLEDGEMENTS

The author wishes to thank the current and the former members of his laboratory at the USDA-ARS-Southern Regional Research Center, New Orleans and Phytogen Laboratories, Pasadena, CA, USA, respectively, for their assistance in developing and refining the techniques presented here.

REFERENCES

1. Anon (2003). Cotton: Review of the World Situation. *ICAC (International Cotton Advisory Committee)*, **56**: 1-20.
2. Anon (2000). Oilseeds: World Production. *Oil World Monthly*, **43**: 478.
3. Fryxell PA (1984). Taxonomy and Germplasm Resources. In: Kohel RJ, Lewis CF (eds.), *Cotton* (pp. 27-57). Madison, Wisconsin: American Society of Agronomy.
4. Jenkins JN (1993). Cotton. In: *Traditional Crop Breeding Practices: An Historical Review to Serve as a Baseline for Assessing the Role of Modern Biotechnology* (pp. 61-70). Paris: Organisation for Economic Co-operation and Development.
5. Robinson M, Jenkins JN and McCarty JC Jr (1997). Root-knot nematode resistance of F-2 cotton hybrids from crosses of resistant germplasm and commercial cultivars. *Crop Science*, **37**: 1041-1046.
6. James C (2002). *Global Status of Commercialized Transgenic crops: 2002*. ISAAA Briefs No. 27: Preview edn. Ithaca, NY: ISAAA.
7. Wilkins TA, Rajasekaran K and Anderson DM (2000). Cotton Biotechnology. *Critical Review in Plant Science*, **19**: 511-550.
8. Rajasekaran K, Chlan CA and Cleveland TE (2001). Tissue culture and genetic transformation of cotton. In: Jenkins JN, Saha S (eds.), *Genetic Improvement of Cotton* (pp. 269-290). Enfield, NH: Science Publishers, Inc.

9.   Firoozabady E, DeBoer DL, Merlo DJ, Halk EL, Amerson LN, Rashka KE and Murray EE (1987). Transformation of cotton (*Gossypium hirsutum* L.) by *Agrobacterium tumefaciens* and regeneration of transgenic plants. *Plant Molecular Biology*, **10:** 105-116.

10.   Umbeck P, Johnson G, Barton K and Swain W (1987). Genetically transformed cotton (*Gossypium hirsutum* L.) plants. *Bio/Technology*, **5:** 263-266.

11.   Van Haute E, Joos H, Maes S, Warren G, Van Montagu M and Schell J (1983). Intergeneric transfer and exchange recombination of restriction fragments cloned in pBR322: a novel strategy for reversed genetics of the Ti plasmids of *Agrobacterium tumefaciens*. *The EMBO Journal*, **2:** 411-418.

12.   Hofgen T and Willmitzer L (1988). Storage of competent cells for *Agrobacterium* transformation. *Nucleic Acids Research*, **16:** 9877.

13.   Murashige T and Skoog F (1962). A revised medium for rapid growth and bioassays with tobacco tissue culture. *Physiologia Plantarum*, **15:** 473-497.

14.   Singh M and Krikorian AD (1981). White's standard nutrient solution. *Annals of Botany*, **47:** 133-139.

15.   Rajasekaran K, Grula JW, Hudspeth RL, Pofelis S and Anderson DM (1996). Herbicide-resistant Acala and Coker cottons transformed with a native gene encoding mutant forms of acetohydroxyacid synthase. *Molecular Breeding*, **2:** 307-319.

16.   Jefferson R (1987). Assaying chimeric genes in plants: The GUS gene fusion system. *Plant Molecular Biology Reporter*, **5:** 387-405.

17.   Jefferson RA, Burgess SM and Hirsh D (1986). β-glucuronidase from *Escherichia coli* as a gene-fusion marker. *Proceedings of the National Academy of Sciences USA*, **83:** 8447-8451.

18.   Rajasekaran K, Hudspeth RL, Cary JW, Anderson DM and Cleveland TE (2000). High-frequency stable transformation of cotton (*Gossypium hirsutum* L.) by particle bombardment of embryogenic cell suspension cultures. *Plant Cell Reports*, **19:** 539-545.

19.   Rajasekaran K, Grula JW and Anderson DM (1996). Selection and characterization of mutant cotton (*Gossypium hirsutum* L.) cell lines resistant to sulfonylurea and imidazolinone herbicides. *Plant Science*, **119:** 115-124.

20.   Trolinder NL and Goodin JR (1987). Somatic embryogenesis and plant regeneration in cotton (*Gossypium hirsutum* L.). *Plant Cell Reports*, **6:** 231-234.

21.   Rajasekaran K (1996). Regeneration of plants from cryopreserved embryogenic cell suspension and callus cultures of cotton (*Gossypium hirsutum* L.). *Plant Cell Reports*, **15:** 859-864.

22.   Sunilkumar G and Rathore KS (2001). Transgenic cotton: factors influencing *Agrobacterium*-mediated transformation and regeneration. *Molecular Breeding*, **8:** 37-52.

23.   Sunilkumar G, Mohr L, Lopata-Finch E, Emani C and Rathore KS (2002). Developmental and tissue-specific expression of CaMV 35S promoter in cotton as revealed by GFP. *Plant Molecular Biology*, **50:** 463-474.

24.   Kumria R, Sunnichan VG, Das DK, Gupta SK, Reddy VS, Bhatnagar RK and Leelavathi S (2003). High-frequency somatic embryo production and maturation into normal plants in cotton (*Gossypium hirsutum*) through metabolic stress. *Plant Cell Reports*, **21:** 635-639.

25.   Firoozabady E and DeBoer DL (1993). Plant regeneration via somatic embryogenesis in many cultivars of cotton (*Gossypium hirsutum* L.). *In Vitro Cellular and Developmental Biology-Plant*, **29:** 166-173.

26.   Davidonis GH and Hamilton RH (1983). Plant regeneration from callus tissue of *Gossypium hirsutum* L. *Plant Science Letters*, **32:** 89-93.

27.   Gawel NJ, Rao AP and Robacker CD (1986). Somatic embryogenesis from leaf and petiole callus cultures of *Gossypium hirsutum* L. *Plant Cell Reports*, **5:** 457-459.

28.   Bayley C, Trolinder N, Ray C, Morgan M, Quisenberry JE and Ow DW (1992). Engineering 2,4-D resistance into cotton. *Theoretical and Applied Genetics*, **83:** 645-649.

29.   Rajasekaran K (2003). A rapid assay for gene expression in cotton cells transformed by oncogenic binary *Agrobacterium* strains. *Journal of New Seeds*, **5:** 179-192.

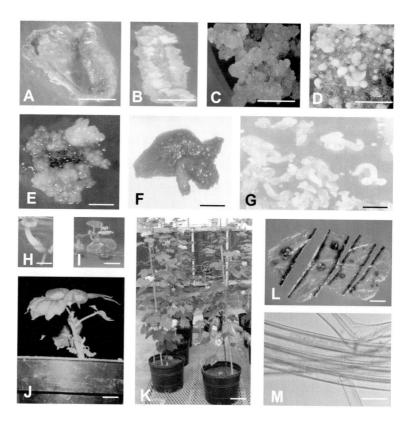

Figure 1. Agrobacterium-*mediated genetic transformation of cotton. After co-cultivation with* A. tumefaciens *carrying a binary vector, explants are transferred to a callus-inducing (CI) medium with a selection agent. Cotyledon (A) and hypocotyl (B) segments producing callus after 2 weeks of growth on CI medium with G418 (10 mg/l) selection. Subculture of actively growing callus lines (C) for 2-4 passages produce embryogenic callus (D) characterized by anthocyanin-rich, cytoplasmic-dense, small, non-vacuolated cells, often with early stages of somatic embryos (globular stage embryos visible in D). Somatic embryos produced on selection medium (E) expressing the* gusA *gene (F) after histochemical assay with the substrate, x-gluc to produce the blue colour. After multiplication of somatic embryos on ECM medium (G), mature embryos (H) are then transferred to EG medium (I) to produce plantlets. Plantlets are then acclimatized in well-drained soil in small pots (J) before transfer to glasshouse where they flower in about 3-4 months (K). GUS expression after incubating with x-gluc in leaf segments (L) of a transgenic plantlet and immature fibres (M) from a morphologically normal, fertile, transgenic plant (K). Scale bars indicate 10 mm in A, B, C, D and I; 2 mm in E, and G; 1 mm in F, and H; 2 cm in J; 30 cm in K; 5 mm in L; and 30 μm in M.*

Part III: Root Crops

Chapter 19

AGROBACTERIUM-MEDIATED TRANSFORMATION OF POTATO

S. MILLAM

Gene Expression, Scottish Crop Research Institute, Invergowrie, Dundee DD2 5DA, UK. E-mail: smilla@scri.sari.ac.uk

1. INTRODUCTION

1.1. Importance and Distribution of Potato

The potato consists of two forms, *Solanum tuberosum* subsp. *tuberosum* (long-day adapted) which is cultivated in temperate and sub-tropical climates such as Europe, USA and Asia, and *Solanum tuberosum* subsp. *andigena* (short-day adapted) which is confined mainly to the Andes. Potato production represents approx. half the world annual output of all root and tuber crops and potato is the fourth most important global food crop, with an annual world production of over 293 million tonnes and covers more than 18 million hectares, grown in over 130 countries (*1*). Although mainly grown for human consumption, in small areas the crop is grown for livestock use and also sugar, starch or alcohol production. The potato sector worldwide is in transition, with some 30% of all production being grown in developing countries, and as such, potato is becoming increasingly important as a source of food, rural employment and income for the growing populations in these regions. Originating in South America, where a major centre of diversity exists in the Andes of southern Peru and northern Bolivia*, Solanum tuberosum* L. is the main, but not only, species grown as a tuber crop today. However, there is a wide geographical distribution for their cultivation and ecological adaptations. In addition, a number of tuber-bearing diploid lines, related to *S.tuberosum* are grown in South America, and for species such as *S. phureja*, can be found in limited trials within Europe.

1.2. The Need for Genetic Improvement

For efficient uptake of potato into new markets and agronomic systems, allied to the increasing diversity of products derived from the crop, there is a requirement for rapid and targeted improvement systems in potato. Potential targets for genetic improvement would include pest and disease resistance, enhanced nutritional qualities of tubers such as modified amino acid components, higher protein content and additional ascorbic acid levels, and industrial targets such as specific starch production. Pests and diseases are major contributors to reduced crop yields and vary according to location. Resistance to Late Blight (*Phytopthora infestans*) remains

257

I.S. Curtis (ed.), Transgenic Crops of th0e World-Essential Protocols, 257-270.

a key global target, and the focus of much research activity, however, the complex resistance system is not yet clearly understood. However, developing resistance to major pests such as Colorado Beetle (*Leptinotarsa decemlineata*) is a more realistic target for transformation technology. The high yields of potato and ease of processing offer scope for using the tubers as efficient plant bio-processors for a range of products including specific starches, antibodies and therapeutics (Table 1) all of which can be mediated by adopting a targeted transformation strategy. In addition, the rapidly-changing consumer trends and short shelf-life of cultivars for processing requirements necessitates a rapid production of new varieties to meet consumer needs.

1.3. Genetic Improvement of Potato

The cultivated potato *Solanum tuberosum* subsp. *tuberosum* functions as an autotetraploid (2n = 4x = 48) with four sets of similar chromosomes, and is highly heterozygous. Many of the breeding problems associated with this species are related to its complex evolutionary history. It was not until the late 1930s that breeders recognised that *Solanum tuberosum* was in fact a tetraploid, which displays tetrasomic inheritance. Even after this discovery, potato breeding remained empirical and unsophisticated, essentially a continuation of the seed selection methods first used in the early 19th Century. Contemporary breeding methods are dominated by the mating systems of the species, causing problems as tetraploid potato has the following characteristics: outbreeding, inbreeding depression, erratic sterility, high heterozygosity and a heavy reliance on phenotypic selection.

Previous conventional breeding strategies involved an initial crossing schedule and the selection of up to 100,000 seedlings followed by up to 15 years of further selection for the creation of one new cultivar. Significant progress has been made in recent years, and accelerated breeding schemes have been devised, often incorporating progeny tests, thus simplifying procedures by merely selecting for single traits. However, even these schemes can take 7-8 years. Inarguably, there are still areas of the breeding process that could be made more efficient and cost-effective.

Potato has been used for a range of biotechnological studies of both a fundamental and applied nature, and is considered a model species for such methods as somatic hybridization and *Agrobacterium*-mediated transformation. Clearly, the problems of tetrasomic inheritance and the lengthy breeding programmes involved in the creation of potato lines with improved characters establish potato as a prime candidate for improvement by *Agrobacterium*-mediated transformation. Potato was indeed one of the first crop plants to be successfully transformed. A large number of single-gene traits have been engineered into potato, these include both input traits e.g. herbicide resistance, pest resistance, virus resistance, fungus/bacteria resistance, and output traits e.g. modified starch, sugars and the production of pharmaceuticals in tubers (Table 1). Potato has also been widely used in molecular mapping studies for such objectives as the identification by molecular markers of resistance genes to pathogens, where RFLP maps have been constructed which are highly saturated with molecular markers (2), the evaluation by molecular markers of introgression and recombination between different genomes and the characterization by molecular markers of potato diploid clones which produce 2n gametes (*3*). Species boundaries have been assessed with three molecular markers; AFLP, RAPD and chloroplast simple sequence repeats (cpSSRs) for all six species of wild potatoes (*Solanum* section Petota) assigned to ser. *Longipedicellata*: *Solanum fendleri, S. hjertingii, S. matehualae, S. papita, S. polytrichon* and *S. stoloniferum* (*4*).

1.4. Genetic Transformation of Potato

Following preliminary reports of infection of potato tissue by various strains of *Agrobacterium rhizogenes* (*5*), the first direct evidence of transformation was obtained by Ooms *et al.* in 1986 (*6*). In this report, clones of the potato cultivar Desirée were regenerated from *A. rhizogenes* infected tissue. This initial work was extended to several other cultivars by Ooms *et al.* (*7*) and to monohaploids and diploids of *S. tuberosum* (*8*). Reports of the transformation of cultivars Desiree and Bintje using *A. tumefaciens* were made by Stiekma *et al.* (*9*), and de Block (*10*) reported a genotype-independent method for transformation using leaf discs as the target tissue. Visser *et al.* (*11*) published a two-stage regeneration and transformation method using stem and leaf explants that is the basis for many protocols used today.

Advancing from reports purely devoted to protocol development, publications arose on the scientific and commercial applications of transformation technology in potato. Visser *et al.* (*12*) reported the inhibition of the expression of the gene for granule-bound starch synthase (GBSS) in potato by antisense constructs with implications for the starch:sugar balance in tubers of transgenic lines. Mackenzie *et al.* (*13*) reported the genetically engineered resistance to Potato Virus-S in the important USA processing cultivar Russet Burbank. One of the first reports of the effect of transgene insertion was made by Brown *et al.* (*14*) citing findings on insert copy number, chromosome number, pollen stainability, and crossability of *Agrobacterium*-transformed diploid potato. Virus resistance was among the first traits investigated for the genetic enhancement of potato and the first report of antisense RNA mediated resistance (to Potato Leafroll Virus, PLRV) was made by Kawchuk *et al.* (*15*) again using the variety Russet Burbank.

Further developments in potato transformation methodology and application were made by Hulme *et al.* (*16*) who reported an efficient genotype-independent method for the regeneration and transformation of potato plants from leaf tissue explants. This was extended to the report of a total of over thirty-five potato varieties which were assessed in a review by Dale and Hampson (*17*) for their morphogenic and transformation efficiency. A range of tissue explants have been successfully used for regeneration and transformation in potato. These include leaf discs (*10, 16*), internodal stem sections (*11*) and microtubers (*18*). Reports of the direct DNA uptake through cut shoots leading to genetic transformation of the related species *Solanum aviculare* Forst. have been made (*19*), citing transformation efficiencies of 5%, considerably less than those made for other systems which are often around 100% of explants forming independent transgenic plants. In the majority of transgenic potato plants developed to date, kanamycin resistance has been used for selection. Other selectable markers which have been successfully used for potato transformation include methotrexate resistance, hygromycin resistance and interestingly, the use of "benign" markers such as galactose mediated by xylose isomerase, (20) or a UDP-glucose: galactose-1-phosphate uridyltransferase gene (21). With regard to the media formulations commonly employed in regeneration and transformation systems, many protocols employ a two-stage regeneration system. The first stage of approximately 2 weeks duration stimulates callus formation, and the second stage is designed to induce de novo shoot outgrowth. In the first media, cytokinin (often in the form of zeatin riboside (ZR), or less commonly BAP) ratio to auxin (often NAA) is between 20 and 200:1. In many protocols, gibberellic acid is used in the second media to enhance shoot outgrowth. Many variations exist in media formulations and if introducing novel germplasm into a transformation system some degree of optimization may be required.

A commonly reported problem in potato transformation is variation in transgene expression among the progeny (9, 22) and has been attributed to the random integration of the transgene into different sites of the plant genome. A detailed analysis of significant populations of transgenic lines of potato has additionally revealed a number of phenotypic changes and substantially reduced tuber yields in field trials. Such changes have been attributed to epigenetic and genetic events occurring during the regeneration phase of transformation. The frequency of these off-types has been recorded as 15-80%, depending on the potato cultivar (23), and often do not become apparent until plants are grown in the field (24). Clearly, data derived from glasshouse trials or first generation tubers derived from microplants needs to be treated with caution. The first report of the field evaluation of transformed potato was made in 1991 by Kuipers et al. (25) who described an evaluation of the antisense RNA mediated inhibition of GBSS gene expression in potato. Conner et al. (24) reported on a field trial of transgenic potatoes undertaken in New Zealand. On an commercial basis, Monsanto's "New Leaf" potato, which was initially the variety Russet Burbank (but later included the varieties Atlantic and Superior) were transformed to contain a Bacillus thuringiensis (Bt) gene conferring resistance to the Colorado Potato Beetle, was first approved by US regulatory agencies in early 1995. This was followed, in late 1998, by a second type of transgenic potatoes: a Russet Burbank marketed as "New Leaf Plus," which combines the Bt resistance trait with resistance to PLRV and a third version, "New Leaf Y," combining the Bt gene with resistance to the Potato Virus Y (PVY). However, plantings of these lines never amounted to more than 2-3% of the total US potato on two sites in Scotland between 1996 and 1998 (27). The objective was to produce genetically modified cultivars of current potato chipping (crisp) varieties, in this case the variety Saturna, with a low level of reducing sugars where sprouting was reduced or eliminated at cold temperatures. In these experiments a double transformant containing two antibiotic resistance genes performed optimally in these trials, but would not be commercially acceptable in that form.

Table 1. Examples of applications of transformation technology in potato

Trait	Approach	Ref
Virus Resistance	Resistance against PVY by transformation with PVYN coat protein gene.	28
	Sense and antisense RNA-mediated resistance to Potato Leaf Roll Virus	15
	PLV expression of PVX coat protein gene plants.	29
	Potato Virus Yp 1 gene sequence.	30
	Reduction of Potato Mop-Top Virus accumulation with a modified triple gene block gene of PMTV.	31
Modified Carbohydrate Pathways	Complementation of the amylose-free starch mutant by the gene encoding granule-bound starch synthase.	32
	Production of cyclodextrins	33
	Expression of *E. coli* glycogen synthase	34
	Modified starch synthesis using wheat genes for ADP glucose pyrophosphorylase	35
Insect Resistance	Transformation with a Delta endotoxin gene from *Bacillus thuringiensis*	36
	Cry1Ac9 gene conferring resistance to potato tuber moth.	37
Herbicide resistance	Evaluation of Phosphinitricin resistance.	38
	Bromoxynil resistance in transgenic potato clones expressing the *bxn* gene.	39
Fungal Resistance	Expression of a fungal glucose oxidase	40
	Gene for the antifungal peptide from *Amaranthus caudatus*	41
Pharmaceutical Production	Cloning and expression of human calcitonin genes	42
	Expression of two subtypes of human IFN-alpha	43
	Expression of human alpha-interferon	44
	Expression and production of recombinant human interleukin-2	45
	Expression of antibodies and Fab fragments	46
Drought Resistance	Introduction of the trehalose-6-phosphate synthase (TPS1) gene from *Saccharomyces cerevisiae*.	47
Increased Nutritional Value	Expression of a nonallergenic seed albumin gene from *Amaranthus hypochondriacus*.	48
	Expression of the Brazil nut methionine-rich protein	49

2. MATERIALS

2.1. Agrobacterium *Vectors*

A wide range of *Agrobacterium* vectors have been used for the transformation of
potato [for an extensive guide to available *Agrobacterium* vectors read Hellens *et
al.* (*50*)], furthermore selection systems based on kanamycin, hygromycin and
glufosinate resistance have all been widely used (*see* Note 1). The protocols
detailed use *A. tumefaciens* strain LBA4404, containing a range of binary vectors
would be broadly applicable. In addition, a number of promoters have been
investigated in this species and should be chosen for the specific purpose of the
experiment e.g. the use of patatin promoters for tuber expression or CaMV35S for
constitutive expression.

2.2. *Medium for Culture of* A.tumefaciens

1. *Agrobacterium* stocks should be maintained as glycerol stocks held at -80°C.
2. *Agrobacterium* should be grown overnight at 28°C on a shaker in 5 ml LB
   medium, 10g/l tryptone (Difco, Detroit, USA), 10 g/l Yeast Extract (Difco)
   10 g/l NaCl, pH 7.5. In the case of an LBA4404 line, the medium should
   contain rifampicin and kanamycin at appropriate concentrations of 100 mg/l.
   (*see* Note 2).
3. Antibiotics should be prepared in advance as stock solutions. For kanamcyin
   sulphate (Melford Laboratories, Chelworth, UK) 100 mg/ml stocks are
   dissolved in water, filter-sterilized through a 0.2 μm filter into a sterile
   disposable container and dispensed aseptically into sterile Eppendorf tubes then
   kept frozen at -20°C. For rifampicin (Sigma), a similar procedure is employed
   except that ethanol is used as solvent.
4. A 5 ml starter culture is initiated from either a glycerol stock or plate culture
   comprising of LB medium, plus 18 g/l Microbiologie Agar (VWR) and 100
   mg/l kanamycin and rifampicin. This starter culture is grown at 28°C on an
   orbital shaker at 200 rpm.
5. After overnight incubation, the 5 ml culture is transferred to a 50 ml of LB with
   appropriate antibiotics in a 250 ml Erlemeyer flask and cultured under the same
   conditions.
6. Following a further overnight incubation under the same conditions as above,
   45 ml of the culture is transferred to a sterile tube and spun at 3000 rpm for 15
   min in a Jouan B3.11 centrifuge (St-Nazaire, France) at room temperature for
   use in transformation experiments.
7. The remaining 5 ml is used to make glycerol stocks. Six hundred and fifty μl of
   culture medium is added to 350 μl of 50% glycerol (v:v) and the samples flash
   frozen and stored at -80°C.

2.3. Plant Materials

1. *Solanum tuberosum* cv. Desiree (obtained from a verified source e.g. the Scottish Agricultural Science Agency (SASA), http://www.sasa.gov.uk/) is the model cultivar for transformation. However, this protocol has been shown to be applicable to a wide range of cultivars, breeding lines and related species (*see* Note 3).

2. Plant material is either obtained as microplants or established *in vitro* from tubers. If establishing from tubers, a four step surface-sterilization protocol is employed, first vigorously washing the excised shoots in running water, soaking the material in 70% ethanol for 60 sec, followed by 15 min in 10% Domestos (Lever Brothers, UK) and 5 subsequent washes in sterile water prior to establishing on 90 mm Petri dishes (Sterilin) containing 25 ml of basal MS20, Murashige and Skoog, (MS, *51*) basal medium plus vitamins (Duchefa, Haarlem, The Netherlands), 20 g/l sucrose, pH 5.8, 8.0 g/l Microagar (Duchefa).

3. Importantly, for successful development of potato *in vitro*, the culture vessels need to enable adequate gaseous exchange. The use of vented Magenta GA-7 vessels, Suncaps (Sigma) or Vitro Vent Vessels (Duchefa) are ideal containers. For culture in Petri dishes, the plates should be sealed with Nescofilm (VWR) and a 30 mm slit made in the seal, which is then covered with micropore tape (VWR).

4. Environmental conditions used are 18-22°C, 16 h photoperiod (80-110 μE/ m^2/sec) which enables a 3 weekly cycling period for nodal sections grown for stock plant material (*see* Note 4).

2.4. Tissue Culture Medium

1. Medium for growth of stock plant material is MS macro-, and micro-elements with vitamins, 20 g/l sucrose, 8 g/l agar (*see* Note 5).

2. MS20 (liquid): MS basal medium (plus vitamins), 20 g/l sucrose, pH 5.8

3. NAA (Duchefa): made up as 1 mg/ml stocks by dissolving 20 mg NAA in 1 ml 1N NaOH and making up to volume with 19 ml distilled water, store at 4°C.

4. Gibberellic acid (GA_3, Duchefa): made up as 1mg/ml stocks by dissolving 20 mg GA_3 in 20 ml 50% v:v ethanol:distilled water, store at 4°C.

5. ZR (Duchefa): made up as 1 mg/ml stocks by dissolving 20 mg in 1 ml 1N NaOH and making up to volume with 19 ml distilled water, store at -20°C.

6. Stock growth regulator solutions are made in advance according to the formulations required, sufficient for 10 x 250 ml complete media. The solutions are filter-sterilized as above, aliquoted into sterile Eppendorf tubes and stored at -20°C (*see* Note 6).

7. Co-cultivation medium (HB1): MS basal medium (plus vitamins), 30 g/l sucrose, 0.2 mg/l NAA, 0.02 mg/l GA_3, 2.5 mg/l ZR, pH 5.8, 8 g/l Microagar.

8. First stage regeneration medium - HB1Cef medium: MS basal medium (with vitamins), 30 g/l sucrose, 0.2 mg/l NAA, 0.02 mg/l GA_3, 2.5 mg/l ZR, 250-500

mg/l cefotaxime (Claforan®, Cefotaxime powder, Roussel, Uxbridge, UK) pH 5.8, 8 g/l agar (*see* Note 7).

9. Cefotaxime is prepared as 125 mg/ml stock solution in water, filter-sterilized and stored at -20°C.

10. Second stage regeneration medium HB2Cef/Kan medium: MS basal medium (plus vitamins), 30 g/l sucrose, 0.02 mg/l NAA, 0.02 mg/l GA_3, 2 mg/l ZR, 250-500 mg/l cefotaxime (filter-sterilized), 50-100 mg/l kanamycin (filter-sterilized), pH 5.8, 8 g/l agar (*see* Note 8).

11. Selection medium, MS20cef/kan medium: MS basal medium (plus vitamins), 20 g/l sucrose, pH 5.8, 8.0 g/l agar, 250-500 mg/l cefotaxime (filter-sterilized), 50-100 mg/l kanamycin (filter-sterilized) (*see* Note 9).

12. Kanamycin sulphate (*see* Section 2.2.2.).

All media is autoclaved for 20 min at 121°C, and antibiotics and growth regulator formulations added to media prior to pouring plates in a laminar flow cabinet.

3. METHODS

3.1. Agrobacterium *and Explant Preparation*

1. Sample preparation (*see* Sections 2.2.1.-2.2.5.).
2. The OD of the suspension at 600 nm is recorded and adjusted to OD = 0.5.
3. The final suspension is dispensed into sterile tubes and centrifuged at 2000 rpm for 10 min (*see* Section 2.2.6.), then resuspended in 15 ml of MS20.
4. Tissue explants of 5-10 mm internodal sections are aseptically excised from 3 week old plant tissue culture stocks in a laminar flow cabinet. Thirty explants are placed in each 90 mm sterile Petri dishes (Sterilin) containing 15 ml liquid MS20 (*see* Note 10).

3.2. *Inoculation ,Co-cultivation, Selection and Regeneration of Independent Transgenic Plants*

1. Explants are inoculated with 1 ml of *Agobacterium* suspension (*see* Section 2.2.4.) per 9 cm Petri dish. The dishes are sealed with Nescofilm and placed on an orbital shaker at 50 rpm for 45 min at 22°C (*see* Note 11).
2. The *Agrobacterium* suspension is poured into a container for inactivation before being disposed according to the Local Rules on handling GM material (*see* Note 12).
3. Explants are gently blotted-dry using sterile filter paper (Whatman) and plated onto HB1 medium (25 ml medium/dish; 30 explants/dish). The plates are sealed (*see* Section 2.3.3.) and co-cultivated at 18-22°C, in low light (20 $\mu E/m^2/sec$) for 2 days (*see* Note 13).

4. Following co-cultivation, the explants are subcultured onto HB1Cef regeneration medium (*see* Note 8).

5. After 14 days on HB1Cef medium, the explants are subcultured onto HB2Cef/Kan medium.

6. Explants are subcultured onto freshly prepared plates of HB1Cef/Kan medium (25 ml/dish) at 14 day intervals (*see* Note 14).

7. Callus and shoot formation occurs after 4 weeks and continues for several weeks. At approx. the third transfer onto fresh HB2Cef/Kan, carefully excise developing shoots (5-10 mm in size) and plate onto selection medium MS20Cef/kan.

8. Continue excising shoots from explants (*see* Note 15) taking care to ensure that the shoots taken are from independent lines.

9. After 14 days on selection medium, remove surviving shoots (those with roots derived from cut ends of shoots only; *see* Note 16) onto MS20Cef/kan.

10. At this stage, a leaf sample can be taken for DNA isolation using the method of Edwards *et al.* (*52*) for PCR analysis to determine preliminary data on transgenic status (*see* Note 17).

11. Following two stages of selection, the surviving material can be subcultured and replicates planted in soil composed of peat, sand, limestone, perlite and Celcote to aid water retention and allow slow release of nutrients (*see* Note 18). Proprietory brands such as Levington compost can be used for growing potato. Plants are grown in an appropriate containment glasshouse for the provision of sufficient leaf material to enable Southern blot analysis to be undertaken to confirm the transgenic status of the material.

12. Transgenic lines can be maintained *in vitro* as catalogued microplants, or as tuber stocks derived from microplants and stored in appropriate low temperature (4°C, dark) containment stores.

4. NOTES

1. Although antibiotic resistance systems work efficiently, attention should be drawn to the use of such systems being phased-out under EU regulations, thus other systems should be investigated.

2. In this example, the selectable marker used is kanamycin where selection levels used are 50-100 mg/l. In the case of hygromycin, the level of selection should be determined empirically per cultivar used, but will be lower than that used for kanamycin often in the range of 16-25 mg/l.

3. Desiree was one of the first varieties to be transformed, and is still widely used as a model variety. The transformation system described has been successfully applied to a wide range of cultivars, including recalcitrant processing lines such as Saturna and wild species.

4. The environmental conditions used of 18-22°C, 16 h photoperiod at 80-110 μE/m^2/sec enables a 3 weekly cycling period for nodal sections grown for stock plant material of most cultivars of potato tested. For other related species such as *S. phureja*, a longer culture period may be necessary.

5. Stock plants can be successfully maintained on half-strength MS medium or slight variants.

6. Example of stock plant growth regulator formulation preparation. HB1 formulation: for 16 x 250 ml aliquots (i.e. 4 litres medium) = 0.8 ml NAA stock growth regulator solution, 0.08 ml/lGA$_3$ stock growth regulator solution, 10 ml ZR stock growth regulator solution (*see* Section 2.4.5.) 12

ml water (total 16 ml i.e. 1 ml/250 ml complete formulation) filter-sterilized, aliquoted into 1 ml sterile Eppendorf tubes and stored at –20°C. These stocks can be kept for up to 6 months.

7. A two-stage regeneration medium has been found to be optimum for many transformation systems. Some authors cite the use of the less expensive BAP as the principal cytokinin source, but this chemical has been found to be markedly less efficient in regeneration of shoots from potato explants under the conditions employed.

8. The system described actually delays applying selection until the second regeneration medium. This differs from some other authors who apply selection directly after the co-cultivation period. In extensive experiments we have found little advantage in applying immediate selection, in fact regeneration rates can be depressed.

9. It is important to maintain cefotaxime levels at this time as *Agrobacterium* overgrowth can be a problem if this antibiotic is omitted at this stage.

10. The regeneration protocols cited are also applicable for 5 mm leaf strip sections and excised 5 mm petiole sections. Regeneration rates vary between cultivars but internodes are often the most responsive and the easiest system for high throughput systems. In addition, the nodes and shoot tips can be "re-cycled" onto basal MS for stock material for subsequent experiments.

11. Co-cultivation methods vary between reports. In some cases a heavy inoculum of *Agrobacterium* is used for a reduced co-cultivation period.

12. For examples of GM legislation in use in the UK refer to http://www.hse.gov.uk/aboutus/hsc/iacs/.

13. Co-cultivation periods also vary between reports. In some cases, 1 day is given, in others 3-4 days is employed. However, using the system described, we find 2 days is optimum. A faint halo of *Agrobacterium* can be seen round the explants at this stage. In the case of overgrowth, washing the explants in 25 ml of 250 mg/l cefotaxime in MS20 for 1 h in sealed 90 mm Petri dishes on an orbital shaker at 50 rpm prior to plating should alleviate this problem.

14. Antibiotics have been previously shown to degrade in the light and thus subculturing at 14 day intervals ensures that the selection pressure is maintained and that levels of antibiotics used are sufficient to control any risk of *Agrobacterium* overgrowth. In addition, there are suggestions that there may be a depletion of growth regulator activity (notably zeatin) and thus regular subculture ensures that the growth regulator component is maintained at a constant level.

15. If shoot regeneration rates are high (over 50 shoots per explant) then the problem of ensuring that shoots taken for selection are truly independent transgenics can be alleviated by only taking one shoot per explant. However, in many cases, the first shoots arising from explants may be escapes, hence a careful cataloguing system must be employed (e.g. by numbering each plate, each explant and sequentially numbering shoots taken from each numbered explant) to ensure that even if several shoots per explant survive selection, only one is taken forward for further analysis.

16. Visual selection of putative transgenics is relatively easy in potato with the transformed status being confirmed by the ability of a shoot to form roots direct from the cut surface. This must not be confused with adventitious rooting from the stem above the cut end. Often the cut end exhibits a curling away from the surface of the medium if it is not transformed. In our hands, two rounds of selection enable a 95% success rate (confirmed by Southern blotting) of transformation.

17. As in many other cases we recommend that PCR is only used as a preliminary indicator of transformation event and that only Southern analysis can constitute complete proof of the transgenic nature of any given plant.

18. Compost mix composition per 1400 litre: mix, 12 parts Irish Moss Peat (bedding grade), 1 part Pavoir sand and 1 part Perlite. To this, mix add 2.5 kg limesone (calcium), 2.5 kg limestone (magnesium), 1.5 kg Celcote wetting agent (LBS Horticulture, Lancashire, UK), 1.5 kg Sincrostart base fertilizer (William Sinclair, Lincoln, UK), 2 kg Osmocote controlled release fertilizer (Scotts, UK) and 390 g Intercept insecticide (Bayer).

CONCLUSION

Potato was one of the first important crop species to be transformed, and the wide range of applications to which transgenic technology can be applied in this species have been described. Potato is relatively easy to transform and thus lends itself well to high throughput experiments of both a commercial and fundamental nature. The system described has been shown to be widely applicable to a range of cultivars, dihaploids and related germplasms. The use of a two-stage regeneration system increases shoot production and the simple selection protocol described has been demonstrated to limit "escapes" to less than 5%. In our hands, the method has been applied to the production of large numbers of independent transformed lines for both field-release and for fundamental high throughput projects on promoter-trapping in potato. Prospects for further work include more applied commercial projects for enhanced nutritional qualities of the tuber such as protein improvement, and further gene discovery studies such as the production of large numbers (>10000) of independent activation-tagged lines, via seed generation from an initial transgenic population.

ACKNOWLEDGEMENTS

SCRI is supported by the Scottish Executive Rural Affairs Department. The technical assistance of Susan Mitchell is acknowledged in developing these protocols.

REFERENCES

1. http://www.china-potato.com/potacenter/cip%20potato%20facts.htm.
2. Gebhardt C, Ritter E, Barone A, Debener T, Walkemeier B, Schachtschabel U, Kaufmann H, Thompson RD *et al.* (1991). RFLP maps of potato and their alignment with the homeologous tomato genome. *Theoretical and Applied Genetics*, **83**: 49-57.
3. Barone A, Carputo D and Frusciante L (1994). Selection of potato diploid hybrids for 2n gamete production. *Journal of Genetics and Breeding*, **47**: 313-318.
4. van den Berg RG, Bryan GJ, del Rio A and Spooner DM (2002). Reduction of species in the wild potato *Solanum* section *Petota* series *Longipedicellata*: AFLP, RAPD and chloroplast SSR data. *Theoretical and Applied Genetics*, **105**: 1109-1114.
5. Ooms G and Lenton JR (1985). T-DNA genes to study plant development - precocious tuberization and enhanced cytokinins in *A. tumefaciens* transformed potato. *Plant Molecular Biology*, **5**: 205-212.
6. Ooms G, Bossen ME, Burrell MM and Karp A (1986). Genetic manipulation in potato with *Agrobacterium rhizogenes*. *Potato Research*, **29**: 367-379.
7. Ooms G, Burrell MM, Karp A, Bevan M and Hille J (1987). Genetic-transformation in 2 potato cultivars with T-DNA from disarmed *Agrobacterium*. *Theoretical and Applied Genetics*, **73**: 744-750.
8. Devries E and Gilissen LJW (1987). Transformation of monohaploid and diploid potato genotypes by hairy root-inducing *Agrobacterium* strains. *Acta Botanica Neerlandica*, **36**: 182-182.
9. Stiekema WJ, Heidekamp F, Louwerse JD, Verhoeven HA and Dijkhuis P (1988). Introduction of foreign genes into potato cultivars Bintje and Desiree using an *Agrobacterium tumefaciens* binary vector. *Plant Cell Reports*, **7**: 47-50.
10. De Block M (1988). Genotype-independent leaf disk transformation of potato (*Solanum tuberosum*) using *Agrobacterium tumefaciens*. *Theoretical and Applied Genetics*, **76**: 767-774.
11. Visser RGF, Jacobsen E, Hesselingmeinders A, Schans MJ, Witholt B and Feenstra WJ (1989). Transformation of homozygous diploid potato with an *Agrobacterium tumefaciens* binary vector

system by adventitious shoot regeneration on leaf and stem segments. *Plant Molecular Biology,* **12**: 329-337.

12. Visser RGF, Somhorst I, Kuipers GI, Ruys NJ, Feenstra WJ and Jacobsen E (1991). Inhibition of the expression of the gene for granule-bound starch synthase in potato by antisense constructs. *Molecular and General Genetics*, **225**: 289-296.

13. Mackenzie DJ, Tremaine JH and McPherson J (1991). Genetically engineered resistance to potato virus-S in potato cultivar Russet Burbank. *Molecular Plant Microbe Interactions*, **4**: 95-102.

14. Brown CR, Yang CP, Kwiatkowski S and Adiwilaga KD (1991). Insert copy number, chromosome-number, pollen stainability, and crossability of *Agrobacterium* transformed diploid potato. *American Potato Journal*, **68**: 317-330.

15. Kawchuk LM, Martin RR and McPherson J (1991). Sense and antisense RNA-mediated resistance to potato leafroll virus in Russet Burbank potato plants. *Molecular Plant Microbe Interactions,* **4**: 247-253.

16. Hulme JS, Higgins ES and Shields R (1992). An efficient genotype-independent method for regeneration of potato plants from leaf tissue. *Plant Cell, Tissue and Organ Culture*, **31**: 161-167.

17. Dale PJ and Hampson KK (1995). An assessment of morphogenic and transformation efficiency in a range of varieties of potato (*Solanum tuberosum*). *Euphytica,* **85**: 101-108.

18. Snyder GW and Belknap WR (1993.) A modified method for routine *Agrobacterium* mediated transformation of *in vitro* grown potato microtubers. *Plant Cell Reports,* **12**: 324-327.

19. Gahan PB, Wyndaele R, Mantell S and Boggetti B (2003). Evidence that direct DNA uptake through cut shoots leads to genetic transformation of *Solanum aviculare* Forst. *Cell Biochemistry and Function,* **21**: 11-17.

20. Haldrup A, Noerremark M and Okkels FT (2001). Plant selection principle based on xylose isomerase. *In Vitro Cellular and Developmental Biology-Plant,* **37**: 114-119.

21. Joersbo M, Jorgensen K and Brunstedt J (2003). A selection system for transgenic plants based on galactose as selective agent and a UDP-glucose:galactose-1-phosphate uridyltransferase gene as selective gene. *Molecular Breeding*, **11**: 315-323.

22. Wenzler HC, Mignery GA, Fisher LM and Park WD (1989). Analysis of a chimeric class-I patatin-*gus* gene in transgenic potato plants - high-level expression in tubers and sucrose-inducible expression in cultured leaf and stem explants. *Plant Molecular Biology,* **12**: 41-50.

23. Jongedijk E, Deschutter AJM, Stolte T, Vandenelzen PJM and Cornelissen BJC (1992). Increased resistance to potato virus-x and preservation of cultivar properties in transgenic potato under field conditions. *Bio/Technology*, **10**: 422-429.

24. Conner AJ, Williams MK, Abernethy DJ, Fletcher PJ and Genet RA (1994). Field performance of transgenic potatoes. *New Zealand Journal of Crop Horticulture*, **22**: 361-371.

25. Kuipers GJ, Vreem JTM, Meyer H, Jacobsen E, Feenstra WJ and Visser RGF (1991). Field-evaluation of antisense RNA mediated inhibition of GBSS gene-expression in potato. *Euphytica*, **59**: 83-91.

26. http://www.geo-pie.cornell.edu/crops/potato.html.

27. http://www.nf-2000.org/secure/Fair/S1140.htm.

28. Okamoto D, Nielsen SVS, Albrechtsen M and Borkhardt B (1986). General resistance against potato virus Y introduced into a commercial potato cultivar by genetic transformation with PVYN coat protein gene. *Potato Research*, **39**: 271-282.

29. Feher A, Skryabin KG, Balazs E, Preiszner J, Shulga OA, Zakharyev VM and Dudits D (1992). Expression of PVX coat protein gene under the control of extensin gene promoter confers virus-resistance on transgenic potato plants. *Plant Cell Reports*, **11**: 48-52.

30. Pehu TM, Makivalkonen TK and Valkonen JPT (1994). Potato plants transformed with a potato virus Y P1 gene sequence are resistant to PVY degrees. *American Potato Journal*, **72**: 523-532.

31. Melander M, Lee M and Sandgren M (2001). Reduction of potato mop-top virus accumulation and incidence in tubers of potato transformed with a modified triple gene block gene of PMTV. *Molecular Breeding*, **8**: 197-206.

32. Vanderleij FR, Visser RGF, Oosterhaven K, Vanderkop DAM, Jacobsen E and Feenstra WJ (1991). Complementation of the amylose-free starch mutant of potato (*Solanum tuberosum*) by the gene encoding granule-bound starch synthase. *Theoretical and Applied Genetics*, **82**: 289-295.

33. Oakes JV, Shewmaker CK and Stalker DM (1991). Production of cyclodextrins, a novel carbohydrate, in the tubers of transgenic potato plants. *Bio/Technology, 9*: 982-986.

34. Shewmaker CK, Boyer CD, Wiesenborn DP, Thompson DB, Boersig MR, Oakes JV and Stalker DM (1994). Expression of *Escherichia coli* glycogen synthase in the tubers of transgenic potatoes (*Solanum tuberosum*) results in a highly branched starch. *Plant Physiology, 104*: 1159-1166.

35. Vardy KA, Emes MJ and Burrell MM (2002). Starch synthesis in potato tubers transformed with wheat genes for ADPglucose pyrophosphorylase. *Functional Plant Biology, 29*: 975-985.

36. Cheng J, Bolyard MG, Saxena RC and Sticklen MB (1992). Production of insect resistant potato by genetic-transformation with a delta-endotoxin gene from *Bacillus thuringiensis* var. Kurstaki. *Plant Science, 81*: 83-91.

37. Davidson MM, Jacobs JME, Reader JK, Butler RC, Frater CM, Markwick NP, Wratten SD and Conner AJ (2002). Development and evaluation of potatoes transgenic for a *cry1Ac9* gene conferring resistance to potato tuber moth. *Journal of American Society for Horticultural Science, 127*: 590-596.

38. Figueira ES, Figueiredo LFA, Monteneshich DC (1994). Transformation of potato (*Solanum tuberosum*) cv Mantiqueira using *Agrobacterium tumefaciens* and evaluation of herbicide resistance. *Plant Cell Reports, 13*: 666-670.

39. Eberlein CV, Guttieri MJ and Steffen-Campbell J (1998). Bromoxynil resistance in transgenic potato clones expressing the *bxn* gene. *Weed Science, 46*: 150-157.

40. Felcher KJ, Douches DS, Kirk WW, Hammerschmidt R and Li W (2003). Expression of a fungal glucose oxidase gene in three potato cultivars with different susceptibility to late blight. *Journal of American Society for Horticultural Science, 28*: 238-245.

41. Lyapkova NS, Loskutova NA, Maisuryan AN, Mazin VV, Korableva NP, Platonova TA, Ladyzhenskaya EP and Evsyunina AS (2001). Transformed potato plants carrying the gene of the antifungal peptide of *Amaranthus caudatus*. *Applied Biochemistry and Microbiology, 37*: 301-305.

42. Ofoghi H, Moazami N, Domonsky NN and Ivanov I (2000). Cloning and expression of human calcitonin genes in transgenic potato plants. *Biotechnology Letters, 22*: 611-615.

43. Ohya K, Matsumura T, Ohashi K, Onuma M, Sugimoto C and Matsumura T (2001). Expression of two subtypes of human IFN-alpha in transgenic potato plants. *Journal of Interferon and Cytokine Research, 21*: 595-602.

44. Sawahel WA (2002). The production of transgenic potato plants expressing human alpha-interferon using lipofectin-mediated transformation. *Cellular and Molecular Biology Letters, 7*: 19-29.

45. Park Y and Cheong H (2002). Expression and production of recombinant human interleukin-2 in potato plants. *Protein Expression and Purification, 25*: 160-165.

46. De Wilde C, Peeters K, Jacobs A, Peck I and Depicker A (2002). Expression of antibodies and Fab fragments in transgenic potato plants: a case study for bulk production in crop plants. *Molecular Breeding, 9*: 271-282.

47. Yeo ET, Kwon HB, Han SE, Lee JT, Ryu JC and Byun MO (2000). Genetic engineering of drought resistant potato plants by introduction of the trehalose-6-phosphate synthase (TPS1) gene from *Saccharomyces cerevisiae*. *Molecules and Cells, 10*: 263-268.

48. Chakraborty S, Chakraborty N and Datta A (2000). Increased nutritive value of transgenic potato by expressing a nonallergenic *seed* albumin gene from *Amaranthus hypochondriacus*. *Proceedings of the National Academy of Sciences USA, 97*: 3724-3729.

49. Tu HM, Godfrey LW and Sun SSM (1998). Expression of the Brazil nut methionine-rich protein and mutants with increased methionine in transgenic potato. *Plant Molecular Biology, 37*: 829-838.

50. Hellens R, Mullineaux P and Klee H (2000). A guide to *Agrobacterium* binary Ti vectors. *Trends in Plant Science, 5*: 446-451.

51. Murashige T and Skoog F (1962). A revised medium for rapid growth and bioassays with tobacco tissue cultures. *Physiologia Plantarum, 15*: 473-497.

52. Edwards K, Johnstone C and Thompson C (1991). A simple and rapid method for the preparation of plant genomic DNA for PCR analysis. *Nucleic Acids Research, 19*: 1349-1349.

Chapter 20

GENETIC TRANSFORMATION OF RADISH (*RAPHANUS SATIVUS* L.) BY FLORAL-DIPPING

I.S. CURTIS

National Institute of Agrobiological Sciences (NIAS), Department of Biotechnology, Kannondai 2-1-2, Tsukuba, Ibaraki 305-8602, Japan. E-mail: curtisis2004@yahoo.co.uk

1. INTRODUCTION

1.1. Importance and Distribution of Radish

Radish, *Raphanus sativus* L., is cultivated globally and exists as two broad categories according to the size of the swollen hypocotyl and taproot. Throughout temperate regions, a small-rooted, short-season radish is cultivated exclusively as a salad crop. Such radishes differ in shape (elongated to flattened spheres) and skin colour (white or red), with the main economic varieties exhibiting a spherical root with red skin and white flesh. The other major group of radish, the large-rooted types, are widely grown in the Far East and are adapted to both temperate and tropical conditions and exhibit a greater diversity of phenotype compared to Western varieties in terms of colour of skin (green, yellow, purple and black) and flesh (white, red, purple and green). Two other minor forms of radish are also cultivated, but not as a root crop. Mougri-radish (rat-tail) is mainly grown in south-east Asia for its edible leaves and very long (80 cm in length) immature seed pods. Fodder radish, is also cultivated for its foliage but is used as fodder or green manure in south-east Asia and parts of Western Europe.

The large-rooted radish is by far the most important grown globally. For example, Japan produces 30 times more radish by weight compared to the whole of Europe (*1*). In the Far East, the radish is the most widely grown root crop in China (1.2 million hectares; *2*) and Korea (35,313 hectares; *3*) and is ranked the fourth-most important vegetable crop in Japan (47,700 hectares; *4*). Nutritionally, the root is rich in vitamins B and C and the seeds and root have important medicinal properties (*5, 6*). Recent studies have revealed two possible chemicals in the roots, which may identify the reason why radish is a highly-valued healthy vegetable. Peroxidase, an oxido-reductase present in rich amounts in radish roots, was shown to contribute to the prevention of hyperlipidemia in mice fed on a diet rich in chloesterol and fat (*7*). With hyperlipidemia being the major cause of coronary heart diseases, consumption of radish may help to eradicate this problem in man. In addition, the root also contains an important group of chemicals known as the isothiocyanates, which can induce antimicrobial, antimutagenic and anticarcinogenic activities. One such member, 4-(methylthio)-3-butenyl isothiocyante (MTBITC), is

271

the dominant form in radish and has shown to have antimutagenic activity against UV-induced *E.coli* B/r WP2 cells (*8*). Although this study utilized only a bacterial antimutagenic assay, if further results demonstrate that MTBITC is potent in mammalian and human cell mutation assays then one of the carcinogenic properties of radish would be known.

1.2. Need for Genetic Improvement

In Europe, the breeding of small-rooted radish has mainly been focused on earliness of bulbing throughout the year. Such a trait, avoids the root becoming 'pithy' or having excessive amounts of nitrate. In addition, the production of edible roots during the winter season (*9*) and the avoidance of bolting under high summer temperatures are also important traits which have been introduced into radish. However, breeding for disease-resistance in the small-rooted types is of low priority due to the crop having a short growing-season (3-4 weeks).

The breeding of the large-rooted, Asiatic radish, however, has targeted the introduction of resistance genes towards fungi and viruses as a priority. Indeed, many Chinese and Japanese ecotypes of radish have shown to contain resistance genes to *Fusarium* (*10*), *Albugo candida* (*11*) and some viruses (*12*). Due to the higher economical value of the large-rooted types, hybrid seed production using the incompatibility system and the Ogura (*13*) cytoplasmic male sterility (CMS) started in the 1960s, leaving the European varieties behind. At this time, Japanese seed companies exploited the incompatibility system in the production of F_1 long white 'mouli' type to western growers. However, recent studies have shown that the Ogura CMS, now identified to being under the control of *orf138* (*14*) from *R. raphanistrum* (*15*), has been transferred into the small-rooted radish and F_1 varieties have started to emerge on the market.

In Korea, the breeding for out-of-season radish is of major importance. Korean varieties of radish are cold-sensitive and so the production of high quality of roots during the autumn has been a huge problem. Attempts of transferring the late-flowering trait from the Japanese variety Tokinashi into existing Korean germplasms produced hybrids with greater tolerance to the cold but their roots were of poor quality (*16*). However, although in later studies, breeders were able to produce autumn cropping varieties, such plants were prone to bolting during the hot summer season. Even today, breeders are continually trying to produce high quality roots of radish which are suited to a wide range of climates.

Although conventional breeding has greatly improved the radish both in terms of developing varieties suited to a wide range of climatic conditions and the transfer of disease resistance genes, such methods are time-consuming and labour-intensive. Over the last 20 years, progress made in plant genetic engineering has opened new opportunities to accelerate the introduction of specific gene traits, which can ultimately improve crop performance. The progress made in both plant tissue culture and *in planta* transformation systems will be discussed in terms of improving radish as a crop.

1.3. Tissue Culture and Plant Regeneration

Although many members of the Cruciferae are amenable to regeneration from cultured cells and tissues, radish remains one of the most recalcitrant amongst this group of crops. Plant regeneration via organogenesis from hypocotyl explants (17) and somatic embryogenesis from hypocotyls (18) and microspores (19, 20) produced shoots at a frequency too low for practical usage. However, the discovery that ethylene produced by cultured explants of mustard impaired shoot regeneration (21) and that the addition of polyamines in culture improved shoot yield in Chinese cabbage (22), prompted researchers to test such treatments in the tissue culture of radish. The supplementation of AgNO₃ (an inhibitor of ethylene action) and AVG (inhibitor of ethylene synthesis) to the regeneration medium N1B2 enabled cultured radish hypocotyl explants to regenerate shoots at a frequency of 40% (23). In the same study, the addition of 10-25 mM of putrescine with 30 μM AgNO₃ or AVG greatly improved the regeneration of hypocotyl explants. In a similar study, the effects of ethylene on the regeneration of cotyledon and hypocotyl explants of the commercially important Korean variety Jin Ju Dae Pyong further confirmed the importance of ethylene as a negative regulator to shoot regeneration (24). Strikingly, it was shown that excised cotyledons could regenerate at a frequency of 60% when cultured on CR medium (Murashige & Skoog (MS, 25) medium, 3% sucrose, 20 μM BAP, pH 5.8) containing 10 μM AVG; hypocotyls regenerated at a frequency of 40% when cultured on N1B2 medium with 20 μM AVG. Despite the recent improvement of radish seedling explants in culture, there remains no report on the production of transgenic radish plants in tissue culture.

1.4. Transgenic Radish by Floral-dipping

To date, most transgenic plants can be produced in culture through the transformation of individual plant cells followed by the regeneration of whole plants from those genetically-modified cells. However, in some crop plants, although single cells can be transformed there remains an inability of such cells to regenerate into transgenic plants, these are commonly termed recalcitrant plants. For example, the cells of explants of radish can be routinely transformed using *Agrobacterium tumefaciens* strain AGL1 (26), but in culture, their ability to regenerate into whole plants is severely impaired in the presence of antibiotic selection (unpublish., Curtis). For this reason, alternative procedures were investigated based on avoiding a tissue culture-based system.

Arabidopsis thaliana, is routinely transformed either by vacuum infiltration or using the floral-dip method. This procedure involves the submergence of a plant in its early stages of flowering into a suspension of Agrobacteria (with or without vacuum) in a sucrose/surfactant medium, re-planting, collection of seeds and screening for transformed seedlings using selective agents (27). This strategy of plant transformation has proved useful in the transformation of other plants namely pakchoi (28) and *Medicago truncatula* (29). With Arabidopsis and radish belonging to the same plant family, the *Brassicacae*, studies were performed on evaluating floral-dipping as a technique of transferring foreign genes into the crop. Below is an account of how the first transgenic radish plants were produced (*30*), which has subsequently been utilized in the production of agronomically useful germplasms (*31*).

2. MATERIALS

2.1. A. tumefaciens *Strain*

A. tumefaciens strain AGL1 (*26*) carrying pCAMBIA3301 is used in transformation studies (*see* Note 1). The binary vector contains the selectable marker *bar* and the reporter *gus*A genes both under the control of the CaMV 35S constitutive promoter (*see* Note 2). The vector is transferred into *Agrobacterium* cells by electroporation.

2.2. Culture Media for A. tumefaciens

1. YEP medium: 10 g/l tryptone (Difco, Becton Dickinson and Company, Sparks, USA), 10 g/l yeast extract (Difco), 5 g/l NaCl.
2. Kanamycin sulphate (Sigma): 50 mg/ml stock in water. Sterilize by filtration through a 0.2 μm membrane (Minisart, Epson, UK). Store at $-20°C$ for 3-4 months.
3. Rifampicin (Sigma): 4 mg/ml stock in methanol. Sterilize by filtration and store at $-20°C$ for 3-4 months.
4. Liquid culture medium: YEP medium supplemented with 50 mg/l kanamycin sulphate and 50 mg/l rifampicin.
5. Agar-solidified medium: YEP medium, plus 14 g/l agar (Bacto-agar, Difco, Detroit, MI, USA), with 50 mg/l kanamycin sulphate and 100 mg/l rifampicin.

2.3. Plant Material

1. Seeds of the Korean variety Jin Ju Dae Pyong (Kyoungshin Seeds Co. Ltd., Nr. Seoul, South Korea) is used due to its economic importance and suscepti-bilty to *A. tumefaciens* strain AGL1 transformation.

2.4. Floral-dip Materials

1. Inoculation medium: 50 g/l sucrose, 0.05% (v/v) Silwet L-77 (Osi Specialities, Inc., Danbury, CT, USA), pH 5.2.
2. Sterile centrifuge tubes.
3. Measuring cylinders and beakers of 1 and 2 litre-capacity.
4. Narrow-pointed scissors.

2.5. GUS Histochemical Solution

1. GUS substrate mixture: 10 mM Na_2EDTA, 0.1 M $NaPO_4$ buffer (pH 7.0), 0.1% Triton X-100, 0.5 M $K_3Fe(CN)_6$, 1-5 mM X-gluc (the latter is dissolved in

ethylene glycol monoethyl ether (EGMGE) as a 1 mg/ml stock prior to use, store for maximum period of 14 days at -20°C).

3. METHODS

3.1. Preparation of Agrobacterium

1. Grow *A. tumefaciens* strain AGL1 carrying pCAMBIA3301 on YEP agar medium with selection at 28°C for 2-3 days in the dark. Transfer one loopful of bacteria to 10 ml of YEP liquid medium with antibiotics and grow overnight, in the dark, on an orbital shaker (180 rpm).
2. Transfer the overnight culture of Agrobacteria to a 2 litre-capacity conical flask containing 1 litre of YEP medium with 50 mg/l kanamycin sulphate and grow overnight on an orbital shaker as described above.
3. Determine the optical density of the bacterial culture by spectrophotometry. A culture with an optical density of 0.7-0.8 at a wavelength of 600 nm is ideal for the transformation of radish by floral-dipping.
4. Transfer the bacterial culture to sterile centrifuge tubes and centrifuge in a Beckman for 20 min at 4,500 rpm (at room temperature).
5. Remove the supernatant by decanting and then resuspend the cells in 1 litre of inoculation medium by gentle agitation.

3.2. Preparation of Plant Material

1. Seeds are sown in seed trays (at a density of 1 seed/3 cm^2) in a mixture of Vermiculite (Samsung, Pusan, Korea), Perlite (Samsung), TKS 2 Instant Soil-based Compost (Floragard, Oldenburg, Germany) and Peat (SunGro Horticulture Inc., Washington, USA) (12:4:3:1 by volume) under natural daylight supplemented with 61 μE/m^2/sec daylight fluorescent illumination at 26°C (day) and 18°C (night).
2. At the 4-leaf stage of development, individual plants are carefully transferred to plastic pots (20 cm diam., 30 cm height) containing the same soil mixture as described (*see* Section 3.2.1.) but using a different ratio of components (2:1:1:2) and then the plants are allowed to grow for a further 10 days (6-leaf stage of growth).
3. Plants are transferred to a cold chamber set at 4 ± 2 °C (16 h photoperiod, 45 μE/m^2/sec, daylight fluorescent tubes) for 10 days to promote bolting.
4. Potted plants are returned to the glasshouse to form primary bolts prior to floral-dipping.

3.3. Floral-dipping

1. Plants exhibiting a primary bolt with many immature floral buds are ideal for transformation studies (*see* Note 3). Shortly before dipping, all siliques, flowers and floral-buds with petal colour are removed using pointed-headed scissors (*see* Note 4). Such wounded areas in the inflorescence can facilitate the Agrobacteria to infect the plant.
2. Transfer the resuspended Agrobacteria (in inoculation medium) to either a 1 or 2 litre beaker or measuring cylinder (*see* Notes 5 and 6).
3. Carefully invert a potted plant into the inoculation medium so that the whole inflorescence is completely submerged in the medium. Gently agitate the plant in the inoculum for 5 sec, then remove the plant from the culture and cover the inflorescence in a polythene bag. Carefully remove as much of the air inside the bag then tie the bag just below the inflorescence (*see* Note 7). Transfer the infected plants to the shade (such as under the glasshouse staging) and then leave overnight.
4. Remove the bag from the inflorescence and then transfer the plants to the light to grow to maturity (*see* Note 8). To facilitate seed set, open flowers are hand-pollinated with a small fine-haired paintbrush (*see* Note 9).
5. Seeds are harvested from dry siliques on the plant (*see* Note 10) and then incubated at 30°C for 10 days to aid seed ripening prior to sowing.

3.4. Screening for Transformed Plants

1. Seeds from floral-dipped plants (T1 generation) are sown in seed trays as described (*see* Section 3.2.1.). In order to screen for herbicide resistance, plants are sprayed at 15, 22 and 29 days post-sowing with 0.03% (v/v) Basta[TM] (*see* Note 11; Fig. 1A).
2. Plants exhibiting resistance to the herbicide after the third spray are further screened for GUS activity to confirm whether both transgenes are active in the putative transformant (Figs. 1B-H). During GUS histochemical staining, plant tissues are immersed in assay buffer overnight at 37°C in the dark. Tissues are then immersed in 95% ethanol (1-2 h) and finally stored in 70% until all the chlorophylls are removed.
3. Herbicide-resistant plants exhibiting GUS activity through histochemical staining are deemed transformed and are individually transferred to pots and grown to the 6-leaf stage of development. Prior to cold-treating the plants to promote bolting, leaves from putative transformed plants are harvested for the extraction of DNA (for PCR and Southern blotting) and RNA (Northern blotting) to further confirm that such plants are transformed (*see* Note 12). Finally all transgenic plants are cold-treated to promote bolting and then grown to maturity as described previously (*see* Sections 3.2. and 3.3.).
4. Following the harvest of seeds from T1 transformed plants, seeds can be germinated and analyzed for GUS activity, 2 days after sowing (Figs. 1I and J). Data acquired from this initial screen can give us information on the segregation pattern of GUS activity within the populations of transgenic plants.

4. NOTES

1. Preliminary studies using seedling explants of radish variety Jin Ju Dae Pyong inoculated with *A. tumefaciens* strain AGL1 carrying pCAMBIA vectors revealed that T-DNA transfer were successful in culture. Due to the susceptibility of this radish genotype to AGL1, all studies focused on this host-*Agrobacterium* combination.

2. Plasmid CAMBIA3301 is an ideal vector for floral-dip transformation. The transfer of the selectable marker *bar* gene into plant cells allows the effective screening for transformed T1 seedlings *in solium* by spraying with the herbicide Basta™. The inclusion of the *gus*A reporter gene along the same T-DNA fragment enables a second screen to be conducted on putative transformants by histochemical staining and so minimizes the chance in selecting non-transformed or 'escape' plants.

3. It has been demonstrated that radish plants exhibiting a single primary bolt with many immature floral buds produce the highest frequency of transformed seeds (1.4% transformation frequency) compared to plants with secondary bolts (0.2%). Tertiary bolted plants usually fail to yield transformed seeds.

4. The site of T-DNA transfer into radish by floral-dip is thought to be the unfertilized ovule. Since the fertilization of the ovule occurs prior to flower opening, all flowers must be removed before the plants are dipped into the inoculation medium to maximize transformation efficiency.

5. The size of container used for the dipping of radish plants depends on the height of the inflorescence. It is important that the depth of the inoculation medium is sufficient to allow the complete submergence of the inflorescence to maximize the contact between the bacteria and host plant and so improve the chance of obtaining transformed seeds.

6. The supplementation of a surfactant to the inoculation medium is critical for the successful transformation of radish. Silwet L-77, a tri-siloxane, is the most effective surfactant compared to non-ionic detergents Pluronic-F68 and Tween 20. The concentration of surfactant used in floral-dipping is an important factor in plant transformation. For example, Silwet L-77 used at 0.05% in the presence of 5% sucrose is a superior treatment compared to Silwet L-77 used at 0.01 and 0.1% (*30*).

7. The polythene bag prevents the inoculated inflorescence from drying out which would be fatal in the transformation process. Maintaining a film of moisture around the inflorescence allows the bacteria to move and so improve the chance of ovules being transformed.

8. After returning the plants to the glasshouse, a general liquid fertilizer is used at 14 day intervals to prevent the plants from suffering from leaf chlorosis and to allow the plants to produce a thick stem from which many floral buds can develop.

9. In the absence of wind and insects in the glasshouse, open-flowers of radish must be hand-pollinated to allow successful seed set. To maximize seed set, flowers should be pollinated on 3 consecutive days when the pollen is loose (usually between 11 am and 2 pm). Plants grown in the absence of pollinating insects fail to set seed due to a sexual incompatibility mechanism.

10. Prior to seed harvest, it is important to note where the seed originated (plant number, location of silique on the plant). Southern blot analysis data has revealed that between 50-60% of all transformed plants exhibit the same T-DNA insertion pattern (commonly referred to as siblings). Our studies have shown that these sibling populations evolved from siliques from the same bolt or inflorescence stem. Hence, to produce a population of independently transformed plants one must select transformants which originated from different bolted plants.

11. Plants which exhibit necrosis 30 days after sowing are regarded as non-transformed or 'escapes' and are discarded.

12. Genomic DNA is extracted from leaves by the method of Dellaporta *et al.* (*32*). Total RNA is extracted from plant tissues using a RNA isolation kit (TRI Reagent™, MRC, Cincinnati, OH, USA) which has been improved for the removal of high concentrations of polysaccharides as described by Chomczynshi and Mackey (*33*).

CONCLUSIONS

Radish has become one of the latest crops to be successfully transformed by *A. tumefaciens*. The delay in producing transgenic plants has been the result of radish seedling explants being recalcitrant in culture and their inability to regenerate shoots.

The application of *in planta* transformation methods, such as floral-dip, enabled transformation studies to be conducted in a non-tissue culture environment, which ultimately provided an 'open window' for the genetic modification of radish.

ACKNOWLEDGEMENTS

I wish to thank Dr. Hong Gil Nam (Pohang University of Science and Technology (POSTECH), South Korea) for allowing me to conduct research on the transformation of radish. Thanks also to Dr. Liwang Liu (Nanjing Agricultural University, PRC) for providing information on radish cultivation in China and Ms. Yosimi Kuroyama (RIKEN, Wako-shi, Japan) for helpful translations. Figure 1. was reproduced from *Transgenic Research,* **10**: 363-371, 2001, Transgenic radish (*Raphanus sativus* L. var. *longipinnatus* Bailey) by floral-dip method – plant development and surfactant are important in optimizing transformation efficiency. I.S. Curtis and H.G. Nam (*30*), copyright Kluwer Academic Publishers, with kind permission Kluwer Academic Publishers.

REFERENCES

1.   Crisp P (1995). Radish. In: Smartt J, Simmonds NW (eds.), *Evolution of Crop Plants*, 2nd edn., Longman Group UK, Harlow, UK pp. 86-89.
2.   Chinese Agricultural Press (2000). (ed.) Ministry of Agriculture of the Peoples Republic of China.
3.   Annual Agricultural Reports of Korea (1998).
4.   Japanese Governmental Figures Report (2000).
5.   Watt JM and Breyer-Brandwijk MG (1962). In: *The Medicinal and Poisonous Plants of Southern and Eastern Africa*, 2nd edn., E. & S. Livingstone, Edinburgh and London.
6.   Curtis IS (2003). The noble radish: past, present and future. *Trends in Plant Science*, **8**: 305-307.
7.   Wang L, Wei L, Wang L and Xu C (2002). Effects of peroxidase on hyperlipidemia in mice. *Journal of Agricultural Food Chemistry*, **50**: 868-870.
8.   Nakamura Y, Iwahashi T, Tanaka A, Koutani J, Matsuo T, Okamoto S, Sato K and Ohtsuki K (2001). 4-(Methylthio)-3-butenyl isothiocyanate, a principal antimutagen in daikon (*Raphanus sativus*; Japanese white radish). *Journal of Agricultural Food Chemistry*, **49**: 5755-5760.
9.   Banga O and Van Bennekom JL (1962). Breeding radish for winter production under glass. *Euphytica*, **11**: 311-326.
10.   Hida K and Ashizawa M (1985). Breeding of radishes for Fusarium resistance. *Japan Agricutural Research Quarterly*, **19**: 190-195.
11.   Williams PH and Pound GS (1963). Nature and inheritance of resistance to *Albugo candida* in radish. *Phytopathology*, **53**: 1150-1154.
12.   Shimizu S, Kanazawa K, Kono H and Yokota Y (1963). Studies on breeding radish for resistance to virus. *Bulletin of the Horticultural Research Station*. Series A, Hiratsuka, **2**: 83-106.
13.   Ogura H (1968). Studies on the new male sterility in Japanese radish with special reference to utilization of this sterility toward the practical raising of hybrid seeds. *Memoirs of the Faculty of Agriculture, Kagoshima University*, **6**: 39-78.
14.   Bonhomme S, Budar F, Lancelin D, Small I, Defrance MC and Pelletier G (1992). Sequence and transcript analysis of the *Nco25* Ogura-specific fragment correlated with cytoplasmic male sterilty in *Brassica* cybrids. *Molecular and General Genetics*, **235**: 340-348.

15. Yamagishi H and Terachi T (1997). Molecular and biological studies on male-sterile cytoplasm in the Cruciferae. IV. Ogura-type cytoplasm found in the wild radish, *Raphanus raphanistrum*. *Plant Breeding*, **116**: 323-329.

16. Lee S-S (1987). Bolting in radish. In: Asian and Pacific Council (eds.), *Improved Vegetable production in Asia*. (pp. 60-70) Food & Fertilizer Technology Center, Book series no. 36, Taipei, Taiwan, Republic of China.

17. Matsubara S and Hegazi HH (1990). Plant regeneration from hypocotyl callus of radish. *HortScience*, **25**: 1286-1288.

18. Jeong WJ, Min SR and Liu JR (1995). Somatic embryogenesis and plant regeneration in tissue cultures of radish (*Raphanus sativus* L.). *Plant Cell Reports*, **14**: 648-651.

19. Lichter R (1989). Efficient yield of embryoids by culture of isolated microspores of different *Brassicaceae* species. *Plant Breeding*, **103**: 119-123.

20. Takahata Y, Komatsu H and Kaizuma N (1996). Microspore culture of radish (*Raphanus sativus* L.): influence of genotype and culture conditions on embryogenesis. *Plant Cell Reports*, **16**: 163-166.

21. Pua E-C and Lee JEE (1995). Enhanced de novo shoot morphogenesis in vitro by expression of antisense 1-aminocyclopropane-1-carboxylate oxidase gene in transgenic mustard plants. *Planta*, **196**: 69-76.

22. Chi G-L, Lin W-S, Lee JEE and Pua E-C (1994). Role of polyamines on *de novo* shoot morphogenesis from cotyledons of *Brassica campestris* ssp. *pekinensis* (Lour) Olsson *in vitro*. *Plant Cell Reports*, **13**: 323-329.

23. Pua E-C, Sim G-E, Chi G-L and Kong L-F (1996). Synergistic effect of ethylene inhibitors and putrescine on shoot regeneration from hypocotyl explants of Chinese radish (*Raphanus sativus* L. var. *longipinnatus* Bailey) *in vitro*. *Plant Cell Reports*, **15**: 685-690.

24. Curtis IS, Nam HG and Sakamoto K (2004). Optimized shoot regeneration system for the commercial Korean radish 'Jin Ju Dae Pyong'. *Plant Cell, Tissue and Organ Culture*, **77**: 81-87.

25. Murashige T and Skoog F (1962). A revised medium for rapid growth and bioassays with tobacco tissue cultures. *Physiologia Plantarum*, **15**: 473-497.

26. Lazo GR, Stein PA and Ludwigb RA (1991). A DNA transformation-competent *Arabidopsis* genomic library in *Agrobacterium*. *Bio/Technology*, **9**: 963-967.

27. Clough SJ and Bent AF (1998). Floral-dip: a simplified method for *Agrobacterium*-mediated transformation of *Arabidopsis thaliana*. *The Plant Journal*, **16**: 735-743.

28. Qing CM, Fan L, Lei Y, Bouchez D, Tourneur C, Yan L and Robaglia C (2000). Transformation of pakchoi (*Brassica rapa* L. ssp. *chinensis*) by *Agrobacterium* infiltration. *Molecular Breeding*, **6**: 67-72.

29. Trieu AT, Burleigh SH, Kardailsky IV, Maldonado-Mendoza IE, Versaw WK, Blaylock LA, Shin H, Chiou T-J *et al*. (2000). Transformation of *Medicago truncatula* via infiltration of seedlings or flowering plants with *Agrobacterium*. *The Plant Journal*, **22**: 531-541.

30. Curtis IS and Nam HG (2001). Transgenic radish (*Raphanus sativus* L. var. *longipinnatus* Bailey) by floral-dip method – plant development and surfactant are important in optimizing transformation efficiency. *Transgenic Research*, **10**: 363-371.

31. Curtis IS, Nam HG, Yun JY and Seo KH (2002). Expression of an antisense *GIGANTEA* (*GI*) gene fragment in transgenic radish causes delayed bolting and flowering. *Transgenic Research*, **11**: 249-256.

32. Dellaporta SL, Wood J and Hicks IB (1983). A plant DNA minipreparation: version II. *Plant Molecular Biology Reporter*, **1**: 19-21.

33. Chomczynski P and Mackey K (1995). Modification of the TRI Reagent™ procedure for isolation of RNA from polysaccharide – and proteoglycan – rich sources. *Biotechniques*, **19**: 942-945.

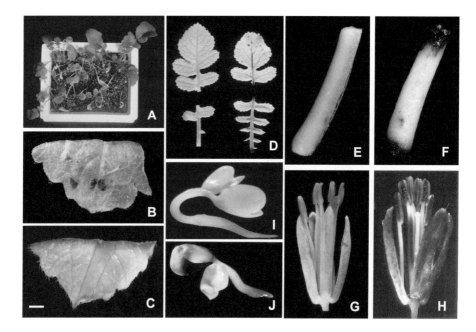

Figure 1. Screening and histochemical GUS staining of transformed radish plants. (A).
Putative transformed plant (top-right) exhibiting herbicide-tolerance after 3 weekly sprayings
with Basta^{TM} (bar = 65 mm). (B). No GUS activity in lamina of a herbicide-sensitive plant;
(C) herbicide-tolerant plant staining positive for GUS (bar = 2 mm). (D). GUS activity
detected in a mature leaf from a Basta-tolerant plant (right), but no GUS detected in wildtype
plant (left; bar = 15 mm). (E). No GUS activity in hypocotyl of wildtype; (F) strong GUS
staining in putative transformant (bar = 4 mm). (G) Flower from wildtype plant showing no
GUS staining; (H) herbicide-tolerant plant shows GUS activity particularly in sepals,
filaments and anthers (bar = 3 mm). Two day-old seedling from a hand-pollinated wildtype
plant (I) and transformed T1 plant (J; bar = 3 mm).

Chapter 21

GENETIC TRANSFORMATION OF *ALLIUM CEPA* MEDIATED BY *AGROBACTERIUM TUMEFACIENS*

S.-J. ZHENG, B. HENKEN, F.A. KRENS AND C. KIK

Plant Research International, Wageningen University and Research Center, P. O. Box 16, 6700 AA Wageningen, The Netherlands. Corresponding author: C. Kik. E-mail: chris.kik@wur.nl

1. INTRODUCTION

1.1. Importance of Onion and Shallot

Onion (*Allium cepa* L. group Common Onion) and shallot (*A. cepa* L. group Aggregatum) are very important vegetable crops worldwide. They are members of the genus *Allium* (Family *Alliaceae*), a genus which comprises more than 600 species (*1*). Onions have a biennial life cycle, however, as they are cultivated for their bulbs, their actual growing period in the field is 4-6 months. Onions are propagated by seeds or bulbs. Onion bulbs are quite variable with respect to shape, size, skin and flesh colour, pungency, skin retention, storage ability, hardiness and dry matter content. Shallot differs from onion primarily in bulb characteristics. The bulbs of onion are large, normally single, and plants are grown from seeds or from seed-grown bulbs. The bulbs of shallot are smaller compared to the bulbs of onion, they form an aggregated cluster of small bulbs, so-called sets, as a result of the rapid formation of lateral bulbs or shoots. Onion and shallot are vulnerable to a number of diseases and pests (*2*). In temperate zones *Botrytis* and *Fusarium* diseases, downy mildew (*Peronospora destructor*), white rot (*Sclerotium cepivorum*), thrips (*Thrips tabaci*) and onion fly (*Delia antiqua*) can cause substantial yield losses. In tropical zones purple blotch (*Alternaria porri*), anthracnose (*Colletotrichum gloeosporioides*) and beet armyworm (*Spodoptera exigua*) are threatening onion and shallot cultivations.

1.2. Need for Genetic Improvement

Due to their biennial life cycle, conventional breeding in onion and shallot is time-consuming. Therefore, the development of an efficient system for genetic transformation is a highly valuable extension of the tools for onion and shallot improvement. Desired traits to be integrated into onion and shallot are firstly herbicide resistance and secondly insect and disease resistance. Herbicide resistance is important because the species has a low relative growth rate due to its biennial life cycle and, is therefore, vulnerable to fast growing weeds. Insect resistance and disease resistance, as mentioned before, are also important to be integrated as the

I.S. Curtis (ed.), Transgenic Crops of the World - Essential Protocols, 281-290.

crop is very vulnerable towards pathogen attack. Thirdly, traits that present a health benefit effect for humans are potentially important to be integrated. In this respect, the ability to modify the organo-sulphur pathway is of significant importance because sulphur is one of the six macronutrients required by plants and incorporated into a large variety of metabolites such as cysteine, methionine, sulpholipids and sulphated glucosides. In order to establish a successful *Allium* genetic transformation system, two key factors should be taken into account. One is the development of sophisticated methods to recover intact plants, either from fully dedifferentiated tissue or from organized tissues that are easy to regenerate. The other is the refinement of methods for the introduction of exogenous DNA into *Allium* germplasm.

1.3. Regeneration Systems of Allium Crops

In *Allium*, various plant regeneration systems have been developed using different starting material. Eady (*3*) reviewed different source materials used for *in vitro* culture of *Allium* species. The most successful regeneration systems in *Allium* are using immature/mature embryos, root tips (segments), flower buds, suspension cultures or protoplasts as starting materials. A key point in *Allium* regeneration protocols is the use of tissues consisting of actively dividing cells, such as calli derived from immature/mature embryos or young calli induced from apical and non-apical root segments of *in vitro* plantlets (*4-6*). Such young and actively dividing callus material is of prime importance in genetic transformation.

1.4. Genetic Transformation of Allium Crops

Until recently, the genetic transformation of *Allium* species was lagging behind other monocot crops (*7-13*). However, during the last few years, reports were published showing that genetic transformation of *Allium* crops has become possible. The progress of transformation research in *Allium* in the last 10-15 years is summarized (Table 1). Klein *et al.* (*14*) first developed a high-velocity microprojectile method and demonstrated that epidermal tissue of onion could incorporate foreign DNA sequences. Wang (*15*) obtained transgenic leek plants by particle bombardment with *barnase* gene and *barstar* gene, and it was shown that the genes were integrated in the leek genome. Transient expression was also shown with particle bombardment in garlic (*16, 17*). Dommisse *et al.* (*18*) demonstrated that onion is a host for *Agrobacterium* because of tumorigenic responses and opine production inside these tumours. Eady *et al.* (*19*) developed a stable *Agrobacterium tumefaciens* transformation protocol using immature embryos of *A. cepa*. Kondo *et al.* (*20*) used highly regenerative calli derived from shoot primordia-like tissues to produce transgenic garlic plants by *Agrobacterium*-mediated gene transfer. Zheng *et al.* (*21, 22*) developed a reproducible and stable transformation protocol using calli derived from mature embryos of *A. cepa* or using calli derived from apical and non-apical root segments of *in vitro* plantlets of garlic (Zheng *et al.*, unpublish.) via

Agrobacterium-mediated transformation. Nowadays, it has become possible to perform onion, shallot and garlic transformations routinely.

Transgenic onions containing a herbicide resistance gene which confers resistance to Basta[TM] have been produced, next to onions which contain an antisense allinase gene for milder taste (*23-25*). Furthermore, shallots and garlic have been successfully transformed with a *cry* gene making them resistant to beet armyworm (*Spodoptera exigua*; Zheng *et al.*, unpublish.). By using particle bombardment, Park *et al.* (*26*) reported that chlorsulfuron-resistant transgenic garlic plants were generated. Overall, the monocotyledonous nature of *Allium* species currently no longer prevents the application of transformation techniques for the transfer of useful genes for crop improvement of these important crop species. Given the fact that the *Agrobacterium*-mediated gene transfer system has proved to be a reliable and efficient genetic modification system in different laboratories worldwide, this method is currently preferred for genetically modifying *Allium* species. Here, we describe a routine procedure developed at PRI Wageningen for the production of transgenic onion and shallot plants.

2. MATERIALS

2.1. A. tumefaciens *Strains*

Agrobacterium strains can be classified as octopine, nopaline and L, L-succinamopine types, depending on which opine synthesis is encoded by the T-DNA. Two strains of *A. tumefaciens,* known as EHA105 and LBA4404, were used in onion and shallot transformation studies. EHA105 is an L, L-succinamopine strain with a C58 chromosome background. It contains as virulence helper pEHA105, originally derived from supervirulent pTiBo542 (*30, 31*). Plasmid CAMBIA1301 is a binary vector originally from the Center for Application of Molecular Biology to International Agriculture, Canberra, Australia (CAMBIA) having *hpt* and intron-*uid*A genes in the T-DNA region. LBA4404 is an octopine strain with Ach5 chromosomal background carrying pAL4404 as virulence plasmid (*32*). Plasmid TOK233 (*13*) is a superbinary vector carrying the *virB*, *virC* and *virG* genes of pTiBo542 and having the *npt*II, *hpt* and intron-*uid*A genes in the T-DNA region. Because both pCAMBIA1301 and pTOK233 have an intron-interrupted *uid*A gene, the expression of *uid*A only occurs in transformed calli and plants rather than in *A. tumefaciens* itself (*33*).

2.2. Culture Media for A. tumefaciens

1. Ultra pure water machine: generate demineralized water (MQ) (Salm and Kipp, Breukelen, The Netherlands).
2. Luria broth (LB) medium: 10 g/l Oxoid Bacto tryptone, 5 g/l Oxoid Bacto-yeast extract, 10 g/l NaCl, 1 g/l glucose, 15 g/l Oxoid Bacto Agar Bacteriological, pH 7.0.

Table 1. Genetic transformation research in different Allium species

Species	Target tissue	Transformation method	Result	Reference
A. cepa	Epidermal tissue	High-velocity microprojectiles	Transient expression of a foreign gene (*cat* gene)	*14*
A. cepa	Epidermal tissue	High-velocity microprojectiles	Transient expression of GFP	*27*
A. cepa	Bulb	*A. tumefaciens, A. rhizogenes, A. rubi*	Tumorigenic response and opine production	*18*
A. cepa	Zygotic mature embryo after *in vitro* culturing for 12 days	*A. tumefaciens*	Transient expression of *gusA*	*28*
A. cepa	Microbulbs from germinating mature seeds, immature embryo after *in vitro* culturing for 14 days	Particle bombardment, *A. tumefaciens*	Transient expression of *gusA*	*29*
A. cepa	Immature embryo	*A. tumefaciens*	Stable expression of *nptII* and *m-gfp5-ER*	*19*
A. cepa	Calli derived from mature embryo	*A. tumefaciens*	Stable expression of *gusA* and *hpt*	*21, 22*
A. cepa	Immature embryo	*A. tumefaciens*	Stable expression of *bar* and antisense versions of alliinase genes	*23, 24*
A. cepa	Calli derived from mature embryo	*A. tumefaciens*	Stable expression of *gusA*, *hpt* and *Bt* genes (*Cry1Ca*)	Zheng *et al.*, unpublish.
A. porrum	Embryogenic callus derived from shoot base	Biolistic system	Stable expression of *gusA* and *bar*	*15*
A. sativum	Leaf, immature bulb, callus from basal plate	Biolistic system	Transient expression of *gusA*	*16*
A. sativum	Embryogenic calli, leaves and basal plate discs	Biolistic system	Transient expression of *gusA*	*17*
A. sativum	Highly regenerative calli derived from shoot primordial-like tissues	*A. tumefaciens*	Stable expression of *gusA* and *hpt*	*20*
A. sativum	Calli derived from apical meristem of cloves	Biolistic system	Stable expression of *gusA*, *hpt* and *als*	*26*
A. sativum	Calli derived from apical and non-apical root segments of *in vitro* plantlets	*A. tumefaciens*	Stable expression of *gusA*, *hpt*, *gfp* or *Bt* genes (*Cry1Ca*)	Zheng *et al.*, unpublish.

3. Rifampicin (Duchefa, The Netherlands): 25 mg/ml stock. Prepare by dissolving 25 mg in 1 ml of 1N HCl and add MQ water to a final volume of 25 ml. Sterilize by filtration through a 0.2 μm Acrodisc®syringe filter membrane and store at -20°C.

4. Kanamycin sulphate (Duchefa): 50 mg/ml stock in water. Sterilize by filtration and store at -20°C.

5. Acetosyringone (Aldrich, Germany): 100 mM stock. Dissolve 196.2 mg first in 96% ethanol, then add MQ water to a final volume of 10 ml, sterilize by filtration and store at -20°C.

6. Liquid culture medium: LB medium containing 50 mg/l kanamycin sulphate, 25 or 50 mg/l rifampicin.

7. Autoclave (Labo autoclave, Sanyo, Japan).

2.3. Tissue Culture

1. Seeds of onion (*Allium cepa* L. group Common Onion) cvs. Sturon and Hyton and shallot (*A. cepa* L. group Aggregatum) cvs. Tropix, Kuning, Bawang Bali and Atlas. Sturon is from Syngenta Seeds, The Netherlands; Hyton, Tropix and Atlas are from Bejo Zaden, The Netherlands; Kuning and Bawang Bali are from Indonesia.

2. MS30 medium: 4.41 g/l Murashige and Skoog (MS, *34*) medium (micro- and macro-elements including vitamins, Duchefa), 30 g/l sucrose, 4 g/l Phytagel (Duchefa).

3. 2,4-D (Duchefa): 1 mg/ml stock solution. Prepare by dissolving 20 mg in 1 ml of 1N KOH and add MQ water to a final volume of 20 ml. Store at -20°C.

4. Casein hydrolysate 0.2 g/l (Duchefa).

5. Cefotaxime sodium (Duchefa): 100 mg/ml stock in water. Sterilize by filtration and store at -20°C.

6. Vancomycin-HCl (Duchefa): 50 mg/ml ml stock in water. Sterilize by filtration and store at -20°C.

7. Hygromycin B: (Duchefa): 50 mg/ml ml stock in water. Sterilize by filtration and store at -20°C.

8. Kinetin (Duchefa): 1 mg/ml stock solution. Prepare by dissolving 20 mg in 1 ml of 1N KOH and add MQ water to a final volume of 20 ml. Store at -20°C.

9. Surface sterilant: 15% (w/v) sodium hypochlorite (Boom, Meppel, The Netherlands).

10. GUS solution: make 0.5 M Na Phoshate buffer ($NaH_2PO_4.2H_2O$ and Na_2HPO_4) pH 7.5. For 400 ml GUS solution, add solution as below: 308 ml MQ water, 8 ml 0.5M EDTA, 80 ml $NaPO_4$ buffer pH 7.5, 4 ml ferri/ferro (0.5 mM potassium hexacyanoferrate III, $K_3[Fe(CN)_6]$); 0.5 mM potassium hexacyanoferrate II, $K_4[Fe(CN)_6]\cdot3H_2O$), Merck, Germany), 200 mg X-gluc (Duchefa), 0.4 ml Triton 100%.

3. METHODS

1. Onion and shallot seeds are rinsed in tap water for 10 min. Discard small and shrunken seeds, then transfer to 70% ethanol for 30 sec. Wash the seeds in sterile MQ water once for 30 sec.
2. Seeds are disinfected in 10% (w/v) sodium hypochlorite (each 100 ml containing 2 drops of Tween 20 (Sigma) for 1 h, continuous agitation.
3. Wash seeds in sterile MQ water 10 times over a 2 h period, continuous agitation.
4. Store seeds in sterile MQ water at 4°C overnight.
5. Seeds are disinfected in 5% (w/v) sodium hypochlorite (containing 2 drops of Tween 20 per 100 ml) for 10 min, continuous agitation.
6. Wash seeds in sterile MQ water 10 times over a 2 h period, continuous agitation (*see* Note 1).
7. Excise part of the embryo containing both the shoot and root apices in 9 cm diam. Petri dishes using fine forceps and scalpel under a dissection microscope in a laminar airflow hood. Remove cotyledon part of embryo with scalpel.
8. Place 20 explants on each 11 cm diam. Petri dish containing 20 ml callus induction medium MS30 with 1 mg/l 2,4-D and 0.2 g/l casein hydrolysate. Incubate for 3 weeks in the dark at 25°C (*see* Note 2, Fig. 1A).
9. A full loop of *Agrobacterium* from a single colony plate was collected and suspended for further culture in 2 ml LB liquid medium (or *Agrobacterium* from −80°C stock directly to LB liquid medium) with an appropriate concentration of antibiotics in an orbital shaker (300 rpm) at 28 °C for 24 h.
10. Culture EHA105::pCAMBIA1301 with 25 mg/l rifampicin and 50 mg/l kanamycin sulphate, or LBA4404::pTOK233 with 50 mg/l rifampicin and 50 mg/l kanamycin sulphate.
11. Take 2 ml of *Agrobacterium* solution from the previous step and add to 20 ml LB liquid medium with an appropriate concentration of antibiotics and 100 μM acetosyringone on an orbital shaker (300 rpm) and grow for 24 h at 28°C.
12. Suspensions are centrifuged at 3,000 rpm for 10 min, then the *Agrobacterium* pellet is resuspended in 20 ml of MS30 with 1 mg/l 2,4-D and 0.2 g/l casein hydrolysate liquid medium to a final concentration of 100 μM acetosyringone and optical density of 0.5-1.0 (OD_{600nm}).
13. Bacterial suspensions are transferred to Greiner bio-one jars (190 ml-capacity, Greiner, The Netherlands). Immerse the 3 week-old calli (500-1000 calli) in the bacterium suspension for at least 10 min (*see* Note 3).
14. After immersing calli in the bacterium suspension, chop individual callus in a Petri dish (*see* Note 4).
15. Transfer 20 individually chopped calluses to a 11 cm diam. Petri dish containing 20 ml of semi-solidified co-cultivation medium: MS30 with 1 mg/l 2,4-D, 0.2 g/l casein hydrolysate, 10 g/l glucose and 100 μM acetosyringone at an ambient temperature of 25°C in the dark.
16. After a co-cultivation period of 4 days, approx. 25 calli derived from various inoculations with different strains and plasmid combinations are examined for GUS transient expression (*see* Notes 5-7; Figs. 1B-D).
17. After callus co-cultivation with *Agrobacterium*, the remaining calli are placed onto selection medium: MS30 with 1 mg/l 2,4-D, 0.2 g/l casein hydrolysate, 50

mg/l hygromycin, 400 mg/l cefotaxime and 100 mg/l vancomycin. Incubate calli at 25°C in the dark.

18. Selection is made for at least 2 months until surviving microcalli appear of an appropriate size (*see* Note 8; Fig. 1E). Every 2-4 weeks, surviving calli are transferred to fresh selection medium.

19. Chop surviving callus into small pieces to obtain homogeneous transgenic calli during each transfer (*see* Notes 9 and 10; Fig. 1F).

20. Actively growing, resistant, and putatively transformed calli are transferred to regeneration medium: MS30 with 1 mg/l kinetin, 50 mg/l hygromycin for 2 months (Fig. 1G). Subculture every month to the same regeneration medium. Incubate calli at 25°C with a 16 h photoperiod (ca. 60 $\mu E/m^2/sec$; lamps used: Philips, TLD 50W/840HF, and Electronic NG).

21. After 2 months on regeneration medium, transgenic shoots are excised from the calli and transferred to MS30 without hormones for further root development (*see* Note 11).

22. Transgenic plants are further grown on MS30 medium and a GUS assay can be performed on the regenerated shoots and roots (Figs. 1H and I).

23. Healthy transgenic plants from different independent transformation events with fully developed roots are transferred to the glasshouse and covered with plastic trays for 7 days to aid acclimation (*see* Note 12).

24. After 7 days in the glasshouse, transgenic plants can be treated normally (*see* Note 13; Fig. 1J) at 16 / 20°C (night / day) with additional light for 16 h (approx. 100 $\mu E/m^2/sec$; lamps used: SON-T 400 Watt).

25. After several months in the glasshouse, bulbs from both onion and shallot are harvested (*see* Notes 14 and 15; Fig. 1K).

4. NOTES

1. Double sterilization of seeds with two concentrations of sodium hypochlorite including an overnight interval in MQ water at 4°C are very important to obtain embryos, clear from seeds. These steps will eliminate bacteria inside the seeds. Only fully developed seeds which are free of pathogen infection can be used.

2. Instead of 3 week-old callus derived from embryos as explants, immature embryos can be used. In addition, 2-6 week-old callus derived from embryos are also useful as explants.

3. Immersing calli into *Agrobacterium* suspensions for 10-120 min does not damage the calli.

4. Chop individual callus in a Petri dish to improve the contact between callus cells and *Agrobacterium*. It is not necessary to remove excessive *Agrobacterium* cells with a dry sterile paper, as there is no problem with later *Agrobacterium* overgrowth. It appears that onion and shallot calli inhibit bacterial growth.

5. A shorter co-cultivation period of 2 days is also efficient for transformation, but a 4 day co-cultivation period is most convenient in practice.

6. After co-cultivation, transfer calli to selective medium immediately. Transient expression of GUS is found to differ between pre-selection for 7 days versus no pre-selection. Without pre-selection gives improved GUS staining.

7. The use of different *Agrobacterium* strains and plasmid combinations (strain LBA4404 with supervirulent pTOK233 or hypervirulent strain EHA105 and binary pCAMBIA1301), does not alter transformation efficiency.

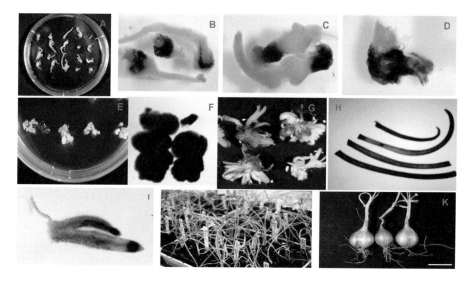

Figure 1. An overview of the transformation procedure developed for onion and shallot (A. cepa L.) at PRI, Wageningen, The Netherlands (from Zheng et al., 2001a, with permission). A. Three week-old callus derived from mature embryos (bar = 2 cm). B. Transient expression of GUS in 2 week-old callus derived from mature embryos of Bawang Bali; infection with LBA4404 (pTOK233) after 4 days of co-cultivation (bar = 0.5 cm). C. Transient expression of GUS in 2 week-old callus derived from immature embryos of Bawang Bali; infection with LBA4404::pTOK233 after 4 days of co-cultivation (bar = 0.5 cm). D. Transient expression of GUS in 3 week-old callus derived from mature embryo of Kuning; infection with EHA105::pCAMBIA1301 after 4 days of co-cultivation. Callus was chopped and the elongated part of the cotyledon was removed (bar = 0.5 cm). E. Hygromycin-resistant callus of Kuning after 2 months growth on selective medium (bar = 1 cm). F. Stable and uniform expression of GUS in hygromycin-resistant callus of Kuning (bar = 0.5 cm). G. Plant regeneration of Kuning. The photograph was taken 4 weeks after the hygromycin-resistant callus had been transferred to regeneration medium with hygromycin (bar = 1 cm). H. Expression of GUS in the leaves of a transformant (bar = 1 cm). I. Expression of GUS in the root of a transformant (bar = 1 cm). J. Transgenic onion and shallot plants in the glasshouse (bar = 10 cm). K. Bulbs from transgenic shallot plants (bar = 4 cm).

8. Most resistant microcalli become visible 4 weeks after transfer to selection medium. Although calli turn brown on selection medium, continue to transfer callus that has produced resistant colonies for 2-3 passages to allow more resistant microcalli to develop.

9. Resistant calli usually form a dry and compact morphology. Reducing the size of the calli during subculture enhances the exposure to the selective agent. This procedure significantly improves the selection efficiency.

10. A more reliable way to check whether transgenic calli are homogeneous is to stain part of the calli in X-gluc solution. Transgenic tissue will turn homogeneous blue within a few hours at 37°C.

11. Plant differentiation occurs 2-3 weeks after transferring resistant calli to regeneration medium. For some callus lines, several subcultures may be required to induce shoot differentiation. Multiple-shoot formation is common in onion and shallot cultures.

12. Southern hybridization shows that most transgenic plants have one copy of T-DNA integrated into their genomes. However, in some cases, callus lines have yielded different plants representing at

least 3 independent transformation events. This suggests that T-DNA integration can occur in separate, individual cells within the same callus line.

13. Avoid damaging the roots of plants during transfer to the glasshouse. Use a plastic tray to cover the plants to avoid desiccation. The first 7 days of acclimation is crucial for the survival of transgenic plants.

14. Transgenic plants appear normal in morphology and produce bulbs (Figs. 1J and K).

15. Subspecies (onion and shallot) and cultivar are important factors for successful transformations: shallot is more efficient than onion with shallot cvs. Kuning and Tropix being the most amenable to *Agrobacterium*-mediated transformation.

CONCLUSION

We have developed a system for the production of stable transformants both for onion and shallot using young callus derived from mature embryos inoculated with A. *tumefaciens*. To date, a large number of transgenic plants have been produced globally. Transgenic onions containing a herbicide resistance gene which confers resistance to Basta™ have been produced. Furthermore, shallots have been successfully transformed with a *cry* gene which are resistant to the beet armyworm. Our transformation method takes approx. 6 months to generate transgenic onion and shallot plants from callus derived from mature zygotic embryos. Together with Eady *et al.* (*19, 23-25*) we have shown that an *Agrobacterium*-mediated gene transfer is now available as a routine method for the future genetic modification of onion and shallot.

REFERENCES

1. Hanelt P (1990). Taxonomy, Evolution and History. In: Rabinowitch HD, Brewster JL (eds.), *Onion and Allied Crops*. CRC Press Inc, Boca Raton, Florida, USA. Vol I. pp.   1-26.

2. Rabinowitch HD (1997). Breeding alliaceous crops for pest resistance. *Acta Horticulturae*, **433**: 223-246.

3. Eady CC (1995). Towards the transformation of onions (*Allium cepa*). *New Zealand Journal of Crop and Horticultural Science,* **23**: 239-250.

4. Zheng SJ, Henken B, Sofiari E, Jacobsen E, Krens FA and Kik C (1998). Factors influencing induction, propagation and regeneration of mature zygotic embryo-derived callus from *Allium cepa* L. *Plant Cell, Tissue and Organ Culture*, **53**: 99-105.

5. Zheng SJ, Henken B, Sofiari E, Keizer P, Jacobsen E, Kik C and Krens FA (1999). The effect of cytokinins and lines on plant regeneration from long-term callus and suspension cultures of *Allium cepa* L.. *Euphytica*, **108**: 83-90.

6. Zheng SJ, Henken B, Krens FA and Kik C (2003). The development of an efficient cultivar independent plant regeneration system from callus derived from both apical and non-apical root segments of garlic (*Allium sativum* L.). *In Vitro Cellular and Developmental Biolog -Plant*, **39**: 288-292.

7. Arencibia AD, Carmona ER, Tellez P, Chan M, Yu S, Trujillo LE and Oramas P (1998). An efficient protocol for sugarcane (*Saccharum spp.* L.) transformation mediated by *Agrobacterium tumefaciens*. *Transgenic Research*, **7**: 213-222.

8. Cheng M, Fry JE, Pang SZ, Zhou HP, Hironaka CM, Duncan DR, Conner TW and Wan YC (1997). Genetic transformation of wheat mediated by *Agrobacterium tumefaciens*. *Plant Physiology*, **115**: 971-980.

9. Hiei Y, Ohta S, Komari T and Kumashiro T (1994). Efficient transformation of rice (*Oryza sativa* L.) mediated by *Agrobacterium* and sequence analysis of the boundaries of the T-DNA. *The Plant Journal*, **6**: 271-282.

10. Hiei Y, Komari T and Kubo T (1997). Transformation of rice mediated by *Agrobacterium tumefaciens*. *Plant Molecular Biology*, **35**: 205-218.

11. Ishida Y, Saito H, Ohta S, Hiei Y, Komari T and Kumashiro T (1996). High efficiency transformation of maize (*Zea mays* L.) mediated by *Agrobacterium tumefaciens*. *Nature Biotechnology*, **14**: 745-750.

12. Rashid H, Yokoi S, Toriyama K and Hinata K (1996). Transgenic plant production mediated by *Agrobacterium* in indica rice. *Plant Cell Reports*, **15**: 727-730.

13. Tingay S, McElroy D, Kalla R, Fieg S, Wang M, Thornton S, Brettell R and Wang MB (1997). *Agrobacterium tumefaciens*-mediated barley transformation. *The Plant Journal*, **11**: 1369-1376.

14. Klein TM, Wolf ED, Wu R and Sanford JC (1987). High-velocity microprojectiles for delivering nucleic acids into living cells. *Nature*, **327**: 70-73.

15. Wang H (1996). Genetic engineering male sterility in leek (*Allium porrum* L.). PhD Thesis, Universiteit Gent, Belgium.

16. Barandiaran X, Pietro AD and Martin J (1998). Biolistic transfer and expression of a *uid*A reporter gene in different tissues of *Allium sativum* L. *Plant Cell Reports*, **17**: 737-741.

17. Ferrer E, Linares C and Gonzalez JM (2000). Efficient transient expression of the beta-glucuronidase reporter gene in garlic (*Allium sativum* L.). *Agronomy*, **20**: 869-874.

18. Dommisse EM, Leung DWM, Shaw ML and Conner AJ (1990). Onion is a monocotyledonous host for *Agrobacterium*. *Plant Science*, **69**: 249-257.

19. Eady CC, Weld RJ and Lister CE (2000). *Agrobacterium tumefaciens*-mediated transformation and transgenic-plant regeneration of onion (*Allium cepa* L.). *Plant Cell Reports*, **19**: 376-381.

20. Kondo T, Hasegawa H and Suzuki M (2000). Transformation and regeneration of garlic (*Allium sativum* L.) by *Agrobacterium*-mediated gene transfer. *Plant Cell Reports*, **19**: 989-993.

21. Zheng SJ, Khrustaleva L, Henken B, Sofiari E, Jacobsen E, Kik C and Krens FA (2001a). *Agrobacterium tumefaciens*-mediated transformation of *Allium cepa* L.: the production of transgenic onions and shallots. *Molecular Breeding*, **7**: 101-115.

22. Zheng SJ, Henken B, Sofiari E, Jacobsen E, Krens FA and Kik C (2001b). Molecular characterization of transgenic shallots (*Allium cepa* L.) by adaptor ligation PCR   (AL-PCR) and sequencing of genomic DNA flanking T-DNA borders. *Transgenic Research*, **10**: 237-245.

23. Eady CC (2002). Genetic transformation of onions. In: Rabinowitch HD, Currah L (eds.), *Allium Crop Science: Recent Advances.*, CAB International, Wallingford, UK. pp. 119-144.

24. Eady CC, Davis S, Farrant J, Reader J and Kenel F (2003a). *Agrobacterium tumefaciens*-mediated transformation and regeneration of herbicide resistant onion (*Allium cepa* L.) plants. *Annals of Applied Biology*, **142**: 213-217.

25. Eady CC, Reader J, Davis S and Dale T (2003b). Inheritance and expression of introduced DNA in transgenic onion plants (*Allium cepa* L.). *Annals of Applied Biology*, **142**: 219-224.

26. Park MY, Yi NR, Lee HY, Kim ST, Kim M, Park JH, Kim JK, Lee JS, Cheong JJ and Choi YD (2002). Generation of chlorsulfuron-resistant transgenic garlic plants (*Allium sativum* L.) by particle bombardment. *Molecular Breeding*, **9**: 171-181.

27. Scott A, Wyatt S, Tsou PL, Robertson D and Allen NS (1999). Model system for plant cell biology: GFP imaging in living onion epidermal cells. *BioTechniques*, **26**: 1125-1132.

28. Joubert P, Sangwan RS, Aouad MEA, Beaupere D and Sangwan-Norreel BS (1995). Influence of phenolic compounds on *Agrobacterium vir* gene induction and onion gene transfer. *Phytochemistry*, **40**: 1623-1628.

29. Eady CC, Lister CE, Suo Y and Schaper D (1996). Transient expression of *uid*A constructs in *in vitro* onion (*Allium cepa* L.) cultures following particle bombardment and *Agrobacterium*-mediated DNA delivery. *Plant Cell Reports*, **15**: 958-962.

30. Hood EE, Helmer GL, Fraley RT and Chilton MD (1986). The hypervirulence of *Agrobacterium tumefaciens* A281 is encoded in a region of pTiBo542 outside of T-DNA. *Journal of Bacteriology*, **168**: 1291-1301.

31. Hood EE, Gelvin SB, Melchers LS and Hoekema A (1993). New *Agrobacterium* helper plasmids for gene transfer to plants. *Transgenic Research*, **2**: 208-218.

32. Hoekema A, Hirsch PR, Hooykaas PJJJ and Schilperoort RA (1983). A binary plant vector strategy based on separation of *vir*- and T-region of the *Agrobacterium tumefaciens* Ti-plasmid. *Nature*, **303**: 179-180.

33. Ohta S, Mita S, Hattori T and Nakamura K (1990). Construction and expression in tobacco of a beta-glucuronidase (GUS) reporter gene containing an intron within the coding sequence. *Plant Cell Physiology*, **31**: 805-813.

34. Murashige T and Skoog F (1962). A revised medium for rapid growth and bioassays with tobacco tissue cultures. *Physiologia Plantarum*, **15**: 473-497.

Chapter 22

TRANSFORMATION OF CARROT

M. HARDEGGER AND R. SHAKYA[1]

Federal Office for Agriculture, Mattenhofstrasse 5, CH-3003 Bern, Switzerland.
[1]*Institute of Plant Biology,University of Zürich, Zollikerstrasse 107, CH-8008
Zürich, Switzerland. Corresponding author: M. Hardegger.
E-mail: markus.hardegger@blw.admin.ch*

1. INTRODUCTION

Carrot (*Daucus carota* L.) is a biennial plant grown for its edible taproot. It has been estimated that some 1 million hectares were grown in 2002, producing 21 million tonnes for human consumption. This represents about 6% of the area devoted to vegetable production worldwide and 9% of the yield. Under optimal conditions, up to 6 kg of roots can be harvested from 1 m^2 of soil after 90-120 days of growth (*1*). Carrot is regarded as a minor crop whose value, either raw or cooked, is essentially the provision of dietary variety in flavour, colour, fibre and provitamin A. Under appropriate conditions of low temperatures and high humidity, carrots can be stored for several months and are therefore available for almost the entire year (*2*).

As for many other plants, genetic modifications of carrot aim to improve agricultural properties, such as resistance to diseases and pests, nutritional quality, or to produce pharmaceutical products. Of high interest, are plants expressing antibodies or synthetic vaccines to be used orally as passive immunization, for which carrots have ideal physiological characteristics. So far, transgenic carrots modified for root knot nematode or fungi resistance (i.e. leaf blight disease caused by *Alternaria daucii*), altered nutritional quality or shelf-life were granted field releases in the USA and the European Union (*3, 4*). The spread of transgenes to the wild carrot populations present in all continents can easily be controlled by removing plants that flower in the first year (bolters).

Skoog and Miller discovered in 1957 that organ formation by plant cells depend on a change in the ratio of auxins and cytokinins (*5*). Subsequently, Stewart and Reinert independently succeeded in regenerating embryos from callus clumps and suspension cell cultures of *Daucus carota* (*6-8*). Now, carrot is often the experimental system of choice for analysis of somatic embryogenesis because of the broad knowledge collected by many subsequent studies and the simple induction of embryogenesis by removing the hormone 2,4-D from cell suspension cultures. Hundreds of transgenic plants can be regenerated from a single cell suspension culture derived from a single transformation event. Artefacts of tissue culture and variations in transgene expression can be minimized by analyzing several clones during each stage of development.

I.S. Curtis (ed.), Transgenic Crops of the World - Essential Protocols, 291-300.

Carrot cultivars used for transformation were selected for characteristics such as shape, colour, taste, nutritional value, yield and resistance towards pests and diseases. A comparison of the published protocols for the generation of transgenic plants of carrot (*9-16*) showed that *Agrobacterium*-mediated DNA transfer, selection of transformed cells with the antibiotic kanamycin, and plant regeneration via somatic embryogenesis were the preferred methods. In each of the protocols, various types of carrot cells and explants were combined with different bacterial strains and different transformation vectors. The results obtained indicated that the identification of `best-matching-partners` and the search for optimal cell cultures conditions are essential aspects of the development of an efficient carrot transformation system (*11*). Therefore, several critical steps for carrot transformation were systematically examined. Different tissues were tested for their susceptibility to two different *Agrobaterium* strains (*17*). The hypocotyl segments from 7 day-old seedlings were chosen because of ease in handling and high efficiency of T-DNA transfer (*18*). With test series of media and combinations of the hormones BAP, NAA and 2,4-D, the conditions for optimal callus formation were found. In addition, various antibiotics were systematically tested for their inhibitory effects on callus formation. The initiation of cell suspension cultures from callus tissue, the maintenance and the embryogenesis from cell suspension cultures were performed as described (*19, 20*). Transgenic plants with taproots of 1 cm in diam. were obtained in approx. 8 months.

Optimization of the conditions for many steps, from DNA-transfer to plant cells up to the regeneration of plants in soil, resulted in a carrot transformation procedure without narrow bottlenecks and a reasonable overall transformation efficiency of about 20%. Variation in transformation efficiency from 5-50% was observed in various experiments with different constructs. This indicates that other factors, which were not tested or optimized, may have influenced transformation efficiencies.

2. MATERIALS

2.1. Plant Material and Bacterial Strains

1. Seeds of carrot (*Daucus carota* L. cv. Nantaise) were obtained from the seed producer Hild (Mambach, Germany). Nantaise, a French carrot cultivar with good agronomic characteristics is well known and widely used in Europe.
2. *Agrobacterium tumefaciens* strain LBA4404 (Clontech, Palo Alto, CA, USA) containing the helper *vir* plasmid pAL4404 or the strain GV3101 (*21*) with the *vir* plasmid pPM6000 in combination with the binary vector pBI101 (Clontech) are used. LBA4404 is derived from Ach5 and produces octopine, whereas GV3101 originated from C58 and produces nopaline.

2.2. Culture Media

1. LB medium: 10 g/l Bacto tryptone (Difco, Detroit, MI, USA), 5 g/l Bacto yeast extract (Difco), 10 g/l NaCl, 18 g/l Bacto agar (Difco), pH 7.2. Autoclave the solution and pour into Petri dishes in a laminar flow hood.
2. YEP medium: 10 g/l Bacto peptone (Difco), 5 g/l Bacto yeast extract (Difco), 5 g/l NaCl, 15 g/l Bacto agar (Difco), 10 mM magnesium sulphate, pH 6.5. Autoclave the solution and pour into Petri dishes in a laminar flow hood.
3. Agar 0.8%: Add 8 g Bacto agar (Difco) per litre water and autoclave the solution. Pour into Petri dishes in a laminar flow hood.
4. B5 medium (Duchefa, Haarlem, Netherlands): 3.29 g/l B5 medium, 30 g/l sucrose and 5 ml/l of 14% MES (Fluka, Buchs, Switzerland) buffer pH 6.0 .Use 1 M KOH to adjust 14% MES stock solution to pH 6.0, store the buffer at 4°C. Add antibiotics and hormones as described (*see* Section 2.3.). Place 50 ml liquid B5 medium into 250 ml Erlenmeyer and autoclave. Store autoclaved flasks in the dark.
5. Semi-solidified B5 medium: B5 medium prepared as described above with 8 g/l Bacto agar (Difco). Pour the autoclaved liquid into Petri dishes in a laminar flow hood. Store the plates in the dark.
6. Soil: Prepare a low nutrient soil such as GS90 or ED73 for carrots. The soil GS90 contains white peat, clay, coconut fibre, pH 5.5-6.5, <1.5 g/l salts (N 50-300 mg/l, P_2O_5 80-400 mg/l, K_2O 400 mg/l) whereas the soil ED73 components are white peat, clay, pumice-stone, pH 6.0, and 3 g/l of salts (N 250 mg/l, P_2O_5 300 mg/l, K_2O 400 mg/l). Both soil types can be purchased from Gebr. Patzer GmbH & Co. (KG, Hameln, Germany).

2.3. Antibiotics and Hormones

Antibiotic and hormone powders are generally weighed in a sterile, soft plastic tube with a volume of 12-15 ml, dissolved in 10 ml sterile water or solvent (as detailed below) and stored at –20°C up to 6 month. If more than 10 ml stock solution is prepared, aliquots of 10 ml or less are prepared and stored at –20°C. If desired, the antibiotic solution can be filtered through a 0.2 μm membrane.

1. Kanamycin sulphate: 10 mg/l stock, dissolve kanamycin sulphate (Serva, Heidelberg, Germany) in sterile water.
2. NAA (Sigma): 200 mg/l stock, dissolve in ethanol.
3. BAP (Sigma): 200 mg/l stock, dissolve in a few drops 1 N KOH and then add sterile water.
4. 2,4-D (Sigma): 200 mg/l stock, dissolve 20 mg powder in 100 ml ethanol.
5. Geneticin (G418, Roche Diagnostics, Rotkreuz, Switzerland): 50 mg/l stock, dissolve in sterile water.
6. Vancomycin (Duchefa): 100 mg/l, dissolve in sterile water.
7. Claforan (Sigma): 200 mg/l stock, dissolve in sterile water.

3. METHODS

1. Surface-sterilize carrot seeds by soaking for 15 min in 70% ethanol and 10% sodium hypochlorite, followed by extensive washing in sterile water. In a laminar flow cabinet, transfer the seeds to Petri dishes containing agar (0.8%) and allow to germinate at 24°C for 7 days in the dark (*see* Note 1).

2. Propagate *A. tumefaciens* strain LBA4404 in 50 ml LB medium or strain GV3101 in 50 ml YEP medium at 200 rpm overnight at 28°C. Overnight cultures of *Agrobacterium* with an optical density (OD) of 0.3-1.5 at 600 nm can be used without any dilution (*see* Note 2).

3. All subsequent manipulations of plant material are performed in a laminar flow cabinet. Cut the hypocotyls of carrot plantlets into pieces of approx. 1 cm in length using sterile forceps and a scalpel inside a Petri dish. Submerge the pieces into 5-15 ml overnight culture of *Agrobacterium* in 3-9 cm diam. Petri dishes, respectively. After a few minutes, dry the hypocotyl pieces using a sterile filter paper, and then place (about 100 pieces per plate) on semi-solidified B5 medium containing 3% sucrose, 1 mg/l NAA and 0.5 mg/l BAP. Perform co-cultivation at 24 °C in the dark (e.g. in a cardboard box) (*see* Note 3).

4. After 2-3 days of co-cultivation, wash the hypocotyl pieces as follows. Transfer the hypocotyl pieces to a 9 cm Petri dish containing 20 ml liquid B5 medium with 0.1 mg/l 2,4-D. Swirl the Petri dish several times. Transfer the hypocotyl pieces (approx. 50 pieces per plate) onto semi-solidified B5 medium containing 3% sucrose, 1 mg/l NAA, 0.5 mg/l BAP, 200 mg/l claforan and 200 mg/l vancomycin in a 9 cm Petri dish. Seal the plates with parafilm, stack them in a cardboard box and store at 24°C for 2 weeks (*see* Note 4).

5. Transfer the explants to fresh semi-solidified B5 medium containing geneticin (G418, 10 mg/l) in addition to the reagents described above. Seal the plates with parafilm, stack them in a paper box and store at 24°C for 4-6 weeks (*see* Note 5).

6. Split the fast-growing calli into small pieces (approx. 5 mm diam.) and transfer to fresh semi-solidified medium of the same composition (*see* Note 6).

7. Cell suspension cultures are initiated from callus tissue derived from a single transformation event. For each new suspension culture, prepare a sterile 250 ml cell suspension flask containing 50 ml liquid B5 medium containing 0.1 mg/l 2,4-D, 10 mg/l geneticin and 200 mg/l claforan. Using a sterile pipette, transfer 10 ml of the medium to a sterile 9 cm diam. Petri dish. Add 1-2 g (fwt.) of callus tissue to the medium and crush into small pieces using sterile forceps. Return the chopped material to the 250 ml cell suspension flask using a wide-bore sterile pipette. Place the flasks on a shaker (120 rpm) at 24°C in the dark (*see* Note 7).

8. Subculture fast-growing cell suspension cultures weekly by transferring 10 ml of the suspension to 50 ml of fresh medium. Slow-growing cultures are subcultured at 2 week intervals (*see* Note 8).

9. After the third subculture, cell suspension cultures can be used for the first time to initiate somatic embryogenesis. Attach a 100 μm mesh sieve on top of a 50 ml sterile plastic tube and pour the contents of the cell suspension flask through the sieve. Close the tube and centrifuge at 800 rpm for 10 min. Gently pour off the supernatant, add 10 ml of hormone-free B5 medium containing 200 mg/l

claforan and then resuspend the cells. Repeat centrifugation, remove the supernatant and finally transfer the cells to the remaining 40 ml of hormone-free B5 medium in the cell suspension flask (*see* Note 9).

10. Somatic embryos, visible 2 weeks after induction, can be isolated manually using a wide-bore pipette and transfer to plates containing semi-solidified B5 medium with 1.5% sucrose, 10 mg/l geneticin and 200 mg/l claforan. Remove as much liquid medium as possible from the plates and then incubate in a 16 h photoperiod at 24°C, with high humidity for 2 weeks. Do not seal the plates with parafilm (*see* Note 10).

11. Transfer the small plantlets with forceps to soil (e.g. GS90) in 20 cm deep pots. Gently press the soil down around the plantlets. Cover the pots with thin light-permeable plastic foil. Make a few thin holes in the foil using a syringe needle or sharp scalpel tip. Place the pots in a phytochamber or growth chamber with a 16 h photoperiod at 22°C (*see* Note 11).

12. Remove the plastic foil after 4 weeks. At this point, most explants will have formed the first true leaves (*see* Note 12).

13. Water the plants once or twice per week. Take appropriate measures if plants are infected by fungi or insects (*see* Note 13).

14. Transgenic calli, embryos, plantlets or plants can be analyzed by PCR, Southern, Northern or Western blotting, or other methods for answering specific questions (*see* Note 14).

4. NOTES

1. Seeds, with their protective hull, can easily be surface-sterilized without affecting the embryos and large amounts of aseptic material can be generated by subsequent germination. The optimum time for which seeds from different varieties should be soaked in each of the sterilization reagents should be determined for best results. In only very rare cases, single carrot plantlets were surrounded by visible fungi after 7 days. About 50-80 seeds are placed in a regular pattern on an agar plate using sterile forceps. This is achieved by laying a chequered paper under the Petri dish and by placing a seed in each second field. Germination in the dark results in long, white and thin hypocotyls which are more susceptible towards *Agrobacterium* infection compared to short, green and thick hypocotyls of seedlings germinated under a 16 h photoperiod.

2. Several strains with different characteristics were described in the literature for the *Agrobacterium*-mediated transformation of carrot (*10-16*). The behaviour of the *A. tumefaciens* strains LBA4404 (*22*) and GV3101 (*21*) were tested using the carrot variety Nantaise. The two strains differed in their efficiency of transferring T-DNA to various carrot tissues (*see* Note 4) as well as in their T-DNA integration pattern (*19*). When the *Agrobacterium* strain LBA4404 was used, the majority of transformants had multiple inserts, predominantly 2-3. Only one single-copy insertion was found among 15 independent transgenic lines. In contrast, all three cell lines obtained by transformation with the *Agrobacterium* strain GV3101 had single copies of T-DNA. The same T-DNA integration pattern was found in transgenic plants and the original callus tissue.

3. The efficiency of T-DNA transfer to various carrot plant tissues has been extensively studied (*19, 20*). Two different *Agrobacterium* strains with different Ti-plasmids, both containing the *gus*A marker gene driven by the CaMV35S promoter, and the *npt*II gene providing kanamycin resistance, were tested. Both strains efficiently transformed hypocotyl segments of 7 day-old seedlings but apparently had different cell-type specificities. Root segments of 7 day-old seedlings were 20 times less efficiently transformed with LBA4404 compared to GV3101. Cells of the intrafascicular cambium of taproot slices were also efficiently transformed, but because of problems with taproot sterilization and handling, such material was not analyzed any further. In general, the addition of acetosyringone

to the *Agrobacterium* suspension did not improve T-DNA transfer efficiency. Hypocotyl pieces are chosen for transformation as large amounts of aseptic explants can be generated, are capable of forming callus, highly susceptible to transformation, do not readily dry out like root pieces of 7 day-old seedlings and are easily manipulated with forceps and scalpels.

4. The bacterial layer around hypocotyl pieces is removed by washing and further bacterial growth is inhibited by claforan and vancomycin, which have no effect on callus formation at the recommended concentration. Optimizing media composition and concentrations of hormones so that tissues can develop under ideal conditions is tedious and appears unnecessary at the outset, but our experience proved the contrary. Approx. 800 hypocotyl pieces were co-cultivated for 2 days with *A. tumefaciens* strain GV3101, washed, and then transferred to selection medium as described in the carrot transformation protocol (*16*) containing 2 mg/l NAA and 0.3 mg/l BAP, 200 mg/l vancomycin, 200 mg/l claforan and 20 mg/l geneticin. Only 3 large calli with GUS activity developed within 9 weeks, whereas numerous explant sections failed to form visible calli but stained blue in the presence of X-gluc. To optimize the formation of transgenic cells, different growth media and different hormone combinations and concentrations were examined. Callus formation was not improved in any of the tested combinations of hormones together with MS medium (*23*). Changing the growth medium composition from MS medium to B5 medium and optimizing the concentration of hormones improved the formation of transgenic calli from hypocotyl and root segments of seedlings by 20-100 fold (Figs. 1A and B). B5 medium contains about half the amount of nitrogen (i.e. in the form of ammonium and nitrate) compared to MS medium. Callus formation on B5 medium occurred along the explant pieces whereas on MS medium this occurred at the segment ends. Efficient formation of calli on hypocotyl or root pieces required 1-2 mg/l NAA and 0.5-0.75 mg/l BAP. In the absence or at low BAP concentrations in the medium, strong root development was observed especially in root derived pieces. The synthetic auxin 2,4-D, frequently used in carrot tissue culture, gave optimal results at low concentrations (0.1 mg/l), but the calli formed within 4 weeks were approx. 4-fold smaller than those formed in the presence of optimal concentrations of NAA and BAP (*17*).

5. Antibiotics and herbicides act as plant hormones at low concentrations, causing mild to severe growth inhibition and toxicity at higher concentrations. Kanamycin, several kanamycin derivatives and a herbicide were tested to evaluate their effect on the growth of callus tissue from the carrot cultivar Nantaise in order to avoid the selection of false positive transformants. Geneticin was the kanamycin derivative that had the strongest inhibitory effect at a concentration 10 times lower than for kanamycin. Direct selection after the co-cultivation period reduced the transformation efficiency. Therefore, selection on medium containing geneticin was applied 2 weeks after callus induction. Because transgenic plants harbouring resistance genes against antibiotics that are applied in human and veterinary medicine will be forbidden in certain countries for cultivation and field releases (*24*), it is recommended to use herbicide or other resistance genes for the selection of transgenic plant cells. Phosphinotricin inhibits carrot callus cell growth at very low concentrations (Fig. 2). In addition, the concentration dependence curve is not as steep as in the case of geneticin.

6. Artificial media has an effect on proliferating plant cells. Effects are exhibited in various forms such as different morphological characteristics and by producing secondary metabolites. Using this protocol, the structure of calli varies from very soft to hard. Soft calli usually grow faster and yield ideal cell suspension cultures.

7. Soft calli produce ideal cell suspension cultures within 7 days, whereas hard calli release cells into the suspension solution very slowly. Some cell lines need a longer adaptation of up to 2-3 months until they proliferate rapidly, and other lines do not grow in liquid culture.

8. Transgenic cell lines that, after the above mentioned adaptation time, do not produce sufficient amounts of suspension cells within 2 weeks after subculture are discarded.

Figure 1: Effect of medium composition on callus formation efficiency. Root segments of 7 day-old seedlings are cultivated on B5 (A) or MS medium (B) for 4 weeks. Somatic embryos at torpedo stage are transferred to a Petri dish containing semi-solidified B5 medium and are incubated in a day/night cycle for 2 weeks (C). At this stage, the plantlets are transferred to soil and about 20% develop leaves (D). A taproot slice of a transgenic carrot (E), harbouring the gusA gene driven by a 323 bp fragment of the promoter from the soluble acid β-fructosidase isoenzyme II from carrot (18) is stained for GUS activity. Bars = 1 cm.

9. Somatic embryogenesis is induced in rapidly growing, transgenic suspensions by omitting hormones and the selection agent geneticin. Embryogenesis is possible 3 weeks after the initiation of suspension cultures and reaches an optimum after 5-12 weeks. The formation of somatic embryos was analyzed in the presence of various concentrations of kanamycin or geneticin. Somatic embryogenesis was not affected in the presence of 1 mg/l kanamycin or 0.05 mg/l geneticin. Only a few globular and heart stage embryos formed when the medium contained 15 mg/l kanamycin or 0.5 mg/l geneticin, and more than 15 mg/l kanamycin or 2 mg/l geneticin inhibited embryogenesis completely. Claforan and vancomycin at concentrations up to 200 mg/l alone or in combination had no effect on somatic embryogenesis. Transgenic carrot cells appeared to grow well in the presence of 10 mg/l geneticin, but as this concentration inhibited somatic embryogenesis, the antibiotic was omitted during this process (*17, 18*).

10. If the plates are sealed with parafilm, the embryos develop into ethylated like plantlets, which generally dry out and die after transfer to soil. Therefore, we do not seal the plates at this step. In contrast, high humidity maintained in a phytochamber prevents the embryos from drying out during development. Within 2 weeks, torpedo stage embryos develop into small plantlets with green cotyledons and roots (Fig. 1C). Further development of the plantlets, especially the root, is inhibited in agar-containing synthetic media and so the plantlets are transferred to soil at this stage.

11. Usually, 20-50% of the explants transferred to soil form leaves within 4 weeks (Fig. 1D). In contrast, on synthetic medium, only 1% of the explants continue to develop. Many plantlets form calli on synthetic medium. After the development of leaves, the growth rate of transgenic carrot plants in soil is indistinguishable from that of non-transformed plants.

12. Remove the foil gradually over a period of about 2 weeks. Remove the foil first at two opposite sides of the pot leaving a gap of approx. 1 cm. Increase the gaps every 2 days.

13. If the potted plants are watered too often, fungi may develop. Carrots with taproots are quite resistant to dryness.

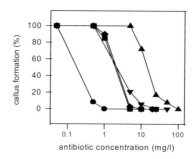

Figure 2: Effect of antibiotics and a herbicide on callus formation from hypocotyl segments of 7 day-old seedlings. Phosphinotricin (PPT, ●), Geneticin (■), Hygromycin (♦), Paromomycin(▼), Kanamycin (▲).

14. With this protocol, transgenic plants for studying tissue- or cell-specific expression of genes were generated. For example, a taproot slice exhibiting the expression of the *gus*A gene driven by a 323 bp fragment of the promoter derived from a soluble acid β-fructosidase isoenzyme II is shown in carrot (*18*; Fig. 1E). In addition, plants expressing genes in antisense orientation were produced to repress the expression of sucrose-cleaving enzymes (*25*). Transgenic callus tissue was screened for reduced expression of specific proteins using Western blotting. Only materials without detectable or reduced Western blot signals were used to generate transgenic plants.

CONCLUSIONS

We have established an efficient *Agrobacterium*-mediated transformation and regeneration protocol for carrot. The efficient transformation of a plant species or even different cultivars of the same species depends on various factors. The presented protocol is the result of the systematic evaluation of factors influencing specific steps during the transformation and regeneration procedures. Therefore, the protocol can be used as a guide for the establishment or adaptation of specific transformation methods. Crucial steps are the identification of plant cells or tissues which are efficiently transformed by the most efficient *A. tumefaciens* strain, the exact composition of salts, hormones and antibiotics in media for the fast proliferation and selection of transformed cells as well as for the establishment of an efficient plant regeneration procedure as presented through the establishment of cell suspension cultures and embryogenesis that allow subsequent plant development.

ACKNOWLEDGEMENTS

The authors wish to thank Margaret Collinge for her critical reading of the manuscript. This research was performed at the Friederich Miescher Institute in Basel, Switzerland.

REFERENCES

1. FAOSTAT, FAO Statistical Database, http://apps.fao.org/default.htm.
2. Gemüsepflanzen (2002). In: Krug H, Liebig H-P, Stüzel H (eds.), *Ein Lehr-und Nachschlagwerk für Studium und Praxis*, Eugen Ulmer Verlag, Stuttgart.
3. United States Department of Agriculture (USDA), Animal and Plant Health Inspection Service (APHIS), Plant Protection and Quarantine, Current Status of Application, http://www.aphis.usda.gov/bbep/bp/status.html.
4. European Comission, Deliberate releases and placing on the EU market of Genetically Modified Organisms (GMOs), http://gmoinfo.jrc.it.
5. Skoog F and Miller C (1957). Chemical regulation of growth and organ formation in plant tissues cultured *in vitro*. *Symposium Society for Experimental Biology*, **11**: 118-131.
6. Reinert, J. (1958): Morphogenese und ihre kontrolle an gewebekulturen aus karotten. *Naturwissenschaften*, **45**: 34-345.
7. Reinert J (1959). Über die Kontrolle der Morphogenese und die Induktion von Adventivembryonen an Gewebekulturen aus Karotten. *Planta*, **53**: 318-333.
8. Steward FC, Mapes MO and Mears K (1958). Growth and organized development of cultured cell. II. Organization in cultures grown from freely suspended cells. *American Journal of Botany*, **45**: 705-708.
9. Scott R and Draper J (1987). Transformation of carrot tissues derived from proembryogenic suspension cells: A useful model system for gene expression studies in plants. *Plant Molecular Biology*, **8**: 265-274.
10. Wurtele E and Bulka K (1989). A simple, efficient method for the *Agrobacterium*-mediated transformation of carrot callus cells. *Plant Science*, **61**: 253-262.
11. Thomas J, Guiltinan M, Bustos S, Thomas T and Nessler C (1989). Carrot (*Daucus carota*) hypocotyl transformation using *Agrobacterium tumefaciens*. *Plant Cell Reports*, **8**: 354-357.
12. Fuji N and Uchimiya H (1991). Conditions favorable for the somatic embryogenesis in carrot cell culture enhance expression of the *rol*C promoter-*GUS* fusion gene. *Plant Physiology*, **95**: 238-241.
13. Balestrazzi A, Carbonera D and Cella R (1991). Transformation of *Daucus carota* hypocotyls mediated by *Agrobacterium tumefaciens*. *Journal of Genetics and Breeding*, **45**,: 135-140.
14. Pawlicki N, Sangwan R and Sangwan-Norreel B (1992). Factors influencing the *Agrobacterium tumefaciens*-mediated transformation of carrot (*Daucus carota* L.). *Plant Cell, Tissue and Organ Culture*, **31**: 129-139.
15. Mattsson J, Borkird C and Engström P (1992). Spatial and temporal expression patterns direct by the *Agrobacterium tumefaciens* T-DNA gene 5 promoter during somatic embryogenesis in carrot. *Plant Molecular Biology*, **18**: 629-637.
16. Gogarten P, Fichmann J, Braun Y, Morgan L, Styles P, Lee Taiz S, DeLapp K and Taiz L (1992). The use of antisense mRNA to inhibit the tonoplast H$^+$ ATPase in carrot. *The Plant Cell*, **4**: 851-864.
17. Hardegger M and Sturm A (1998). Transformation and regeneration of carrot (*Daucus carota* L.). *Molecular Breeding*, **4**: 119-127.
18. Hardegger M (1997). Molecular characterization of sucrose-cleaving enzymes from carrot. *Dissertation submitted to the Swiss federal Institute of Technology Zürich, Naturwissen ETH Zürich, No.* 12338.
19. Guzzo F, Baldan B, Mariani P, Lo Schiavo F and Terzi M (1994). Studies on the origin of totipotent cell in explants of *Daucus carota* L.. *Journal of Experimental Botany*, **45**: 1427-1432.
20. Zimmermann JL (1993). Somatic embryogenesis: a model for early development in higher plants. *The Plant Cell*, **5**: 1411-1423.
21. Rossi L, Escudero J, Hohn B and Tinland B (1993). Efficient and sensitive assay for T-DNA dependent transient gene expression. *Plant Molecular Biology Reporter*, **11**: 220-229.
22. Hoekema A, Hirsch P, Hooykas P and Schilperoort R (1983). A binary plant vector strategy based on separation of *vir*- and T-region of the *Agrobacterium tumefaciens* Ti plasmid. *Nature*, **303**: 179-180.
23. Murashige T and Skoog F (1962). A revised medium for rapid growth and bioassays with tobacco tissue cultures. *Physiologia Plantarum*, **15**: 473-497.

24.   Directive 2001/18/EC of the European Parliament and of the Council of 12 March 2001 on the deliberate release into the environment of genetically modified organisms. *Official Journal* L, **106**: 1-39.

25.   Tang G-Q, Lüscher M and Sturm A (1999). Antisense repression of vacuolar and cell wall invertase in transgenic carrot alters early plant development and sucrose partitioning. *The Plant Cell,* **11**: 177-189.

Chapter 23

PRODUCTION OF TRANSGENIC CASSAVA (*MANIHOT ESCULENTA* CRANTZ)

P. ZHANG AND W. GRUISSEM

Institute of Plant Sciences, ETH-Zentrum/LFW E17, Universtätstrasse 2, CH-8092, Zürich, Switzerland. Corresponding author: P. Zhang. E-mail: zhang.peng@ipw.biol.ethz.ch

1. INTRODUCTION

1.1. Cassava as a Staple Root Crop

Cassava (*Manihot esculenta* Crantz, *Euphorbiaceae*, 2n = 36) is a perennial root crop in the tropics. Its starchy storage roots provide a source of staple food and livelihood for over 600 million people worldwide. Cassava grows in large tropical and subtropical regions of Africa, Asia and Latin America. An average of 10 tonnes per hectare of cassava roots can be produced in a 12 month growing season. The annual world production of cassava in 2002 was 180 million metric tonnes from 16.9 million hectares. Of the total production, Africa, Asia and Latin America account for 54%, 28% and 18% respectively (*1*). In sub-Saharan Africa, cassava provides up to 60% of the daily calorie intake with more than 80% of the harvest used as food (*2, 3*). In certain regions, the leaves, which contain appreciable quantities of protein and vitamins, are also consumed as a major component of the diet to provide supplementary protein, vitamins and minerals. Cassava is a key component of food security in developing countries and especially in Africa, where the food insecurity is the most severe in the world (*4*).

The ability of cassava to tolerate unfavourable ecological conditions, e.g. drought and low fertility soils, makes it suitable to be planted by small-scale and subsistence farmers in marginal areas unable to support any other crop. Also the vegetative propagation mode using stem cuttings reduces input for the farmers. A unique advantage of cassava over e.g. cereals, is its flexible harvesting time that makes it an excellent famine reserve. Cassava is one of the most reliable and cheapest sources of food. However, cassava has its problems. The storage roots are a good source of dietary carbohydrate but deficient in protein. Due to the low protein content (1-2% in dry weight) and limited amounts of sulphur-containing amino acids, additional food sources are required to ensure a diet balanced in protein, vitamins and minerals (*2*). During the long culture period (up to 18 months),

I.S. Curtis (ed.), Transgenic Crops of the World - Essential Protocols, 301-319.

repeated attacks by various insect pests and virus diseases cause 20-50% yield losses worldwide, and locally they can lead to total crop failure (*5, 6*). Furthermore, cassava storage roots suffer from postharvest physiological deterioration during transport, storage and marketing. In addition, all parts of the plants contain toxic levels of cyanogenic glucosides, which need to be removed by laborious processing before cassava can be safely consumed.

The efforts to tackle these problems using conventional breeding are severely constrained by the high heterozygosity of the allopolyploid plants with low natural fertility, as well as by the lack of genes of interest in the sexually compatible germplasm. Although some improved cassava varieties have been introduced by the Consultative Group on International Agricultural Research, some are resistant to pests and diseases (*7*), but traditional breeding is difficult, time consuming and laborious. Biotechnology, identified as a powerful tool to overcome these limitations, not only provides an alternative approach but also complements the efforts in traditional breeding.

1.2. Genetic Engineering of Cassava

1.2.1. Plant Regeneration in vitro

Several regeneration systems of cassava have been developed using a variety of starting materials (Fig. 1). Apical or axillary bud-derived meristems can be used as explants to obtain virus- or disease-free planting material or to multiply shoot cultures of elite cassava varieties. Generally, somatic embryogenesis is the most routinely used method of *de novo* plant regeneration of cassava *in vitro* (Fig. 1). Primary somatic embryos can be induced on cotyledons of zygotic embryos (*8*), immature leaves (*9-13*), shoot apical or axillary meristems (*10, 14*) and floral tissue (*15, 16*) in MS basic medium (*17*) supplemented with picloram (1-12 mg/l) or 2,4-D (1-8 mg/l). Mature embryos with green cotyledons can be developed after transfer of the primary embryos to hormone-free medium or to a medium supplemented with low amounts of cytokinin and/or auxin. Plants regenerate from somatic embryos via germination or shoot organogenesis. The green cotyledons of somatic embryos ('somatic cotyledon' for short) can be used as explants to induce secondary somatic embryos on auxin-containing medium. Continuous culture of somatic embryos on auxin-supplemented medium allows a cyclic system of somatic embryogenesis. Such cultures can be maintained as a continuous explant source for plant regeneration and transformation studies (Fig. 1). The efficiency of regeneration from somatic embryos could be considerably improved by using desiccation treatment of somatic embryos and additional silver nitrate or maltose instead of sucrose in the culture medium (*12, 18, 19*).

In some genotypes, e.g. TMS60444, cycling mature somatic embryos produce friable embryogenic callus (FEC), a kind of less organised embryogenic tissue, on a medium containing GD salts and vitamins (*20*) with 10-12 mg/l picloram (*21*). By the combination of picloram with NAA, FEC was also induced from other cultivars (*22*). Suspension cultures are initiated by culturing FEC in liquid Schenk and

Hildebrandt (*23*) medium with 10-12 mg/l picloram and 60 g/l sucrose (Fig. 1). Suspension cells develop into mature embryos after transfer to a maturation medium and germinate after desiccation treatment (*21, 24*). A recent study has also demonstrated that protoplasts isolated from FEC of cassava genotype TMS60444 divided and developed into plants via the FEC stage and embryogenesis (*25*; Fig. 1). One disadvantage of this system is the long time span (≥20 weeks) from explant to suspension culture and to regenerated plantlets, which may result in somaclonal variation and loss of regeneration capacity, the most common problem associated with suspension cultures. Another limiting factor is the production of FEC, which is strongly genotype-dependent.

Another plant regeneration mode is shoot organogenesis from somatic cotyledons. Adventitious shoots are induced from somatic cotyledons of cassava after culturing somatic cotyledons on a MS medium containing 1 mg/l BAP and 0.5 mg/l IBA for 20 days (*26, 18*). After a passage on elongation medium supplemented with 0.4 mg/l BAP, regenerated shoots are easily rooted in hormone-free medium

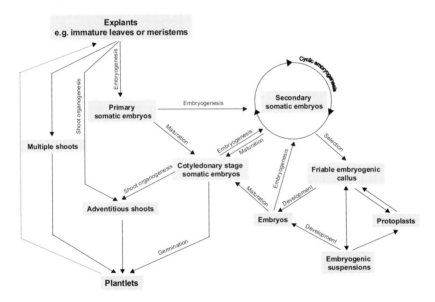

Figure 1. Scheme of plant tissue culture de novo *in cassava*

and can be successfully transplanted to soil (Fig. 1). This system has been successfully adapted to 16 cultivars from Africa, Asia and South America (*27, 28*). During shoot organogenesis from somatic cotyledons, adding $AgNO_3$ to the regeneration medium improved the regeneration frequency and reduced callus formation in all tested cassava cultivars (*27*). Both the extent of the response and the optimum concentration of $AgNO_3$ were cultivar-dependent. Shoots regenerated from explants cultured on a medium containing $AgNO_3$ were more elongated than those cultured on a medium without the salt. The application of $AgNO_3$ did not change the dose response of shoot organogenesis for the selective agents, e.g. hygromycin and mannose (27). Compared to regeneration via germination of embryos derived from suspensions, shoot organogenesis has been proved to be faster, requiring less time in tissue culture, which may reduce the risk of somaclonal variation. The regeneration efficiency of different cultivars varies between 20-95%, which may constrain the use of genetic engineering in cultivars with low regeneration capacity. Much effort is still required in the future to improve the capacity of organogenesis from somatic embryos of various cultivars.

1.2.2. Genetic Transformation

Seven years ago, two independent groups simultaneously published their reports on the successful genetic transformation of cassava (26, 29). So far, there are about a dozen reports on regeneration of transgenic cassava plants (24, 30-37).

After the development of the FEC and suspension system of the genotype TMS60444, a reproducible transformation method has been established by using these tissues in several laboratories. In contrast to the organized primary or secondary somatic embryos, the new embryogenic units in FEC develop from the surface cells of the globular embryogenic clusters, and appear to be of single cell origin, which makes them suitable to *Agrobacterium* and biolistic transformation methods. The suspensions have been transformed using biolistic- or *Agrobacterium*-mediated DNA delivery with different selection systems, such as negative selection using paromomycin and hygromycin (29, 31, 35, 36), positive selection based on mannose system (35, 36), or visual selection using firefly luciferase as a screenable marker gene (24, 32, 30). The method involves a number of different steps, including the selection of antibiotic-resistant/luciferase-positive embryogenic tissue, the formation of somatic embryos and regeneration from these embryos on a medium without a selective agent. Generally, it takes 20-30 weeks to regenerate transgenic shoots. A system allowing plant regeneration from only transgenic lines was designed by combining antibiotic selection with histochemical GUS assays (35). After the primary selection with hygromycin, a part of the hygromycin resistant FEC is used for GUS staining. Only the GUS-positive callus lines showing the characteristics of friable embryogenic callus are transferred to regeneration medium, resulting in a higher efficiency of transformation. By the modification of plant regeneration protocol from suspensions, transgenic cassava plants can be obtained in less than 15 weeks (35).

Li *et al.* (26) obtained transgenic plants via shoot organogenesis from green somatic cotyledons by using *Agrobacterium*-mediated transformation. The shoots regenerated via organogenesis and develop from cells at or close to the cut edge of the somatic cotyledon, which makes them suitable targets for *Agrobacterium* mediated gene transfer. The conditions for optimal transformation were studied, including different *Agrobacterium* strains, binary vectors, an additional phenolic inducer (acetosyringone) and the developmental state of cotyledons. The transformation frequency in this experiment was 0.4%. This transformation system is easy to handle and many explants can be treated within a reasonable time, which should allow regeneration of several transformants from each experiment. Also, as the time required for shoot regeneration is relatively short, the risk of somaclonal variation is less than in long-term cultures. However, as many escapes have been found after selection, the selection system needs further improvement. Based on biolistics and regeneration via organogenesis, transgenic cassava also regenerated from bombarded somatic cotyledons (34). The effects of different parameters for particle bombardment efficiency, including the amount of DNA used, the distance of the projectiles and the pre- and post-plasmolysis time of the target tissue, were evaluated and the conditions were partially optimized. A rooting test was developed for screening the regenerants for antibiotic resistance to reduce the number of escapes obtained after primary selection (34).

Primary and secondary somatic embryos develop from groups of cells, usually located at or near the vascular tissue. The multicellular origin of cassava somatic embryos makes them poorly suited for genetic transformation, and their location under the plant epidermis also limits their accessibility to *Agrobacterium*. However, the green cotyledons of mature somatic embryos can be used as targets for transformation via somatic embryogenesis. Sarria *et al.* (33) showed that a wildtype strain of *A. tumefaciens* (CIAT1182) was able to transfer and insert a T-DNA carrying *npt*II, *bar* and *uid*A genes into cells of the somatic-embryo-derived cotyledons of MPeru183, and transgenic plants could then be regenerated through embryogenesis. However, the long time required for regeneration and very low efficiency makes this method not practicable for routine use.

1.3. Potential Applications of Cassava Biotechnology

Problems that may be efficiently solved by biotechnology include: pest and disease resistance, improved nutritional quality of cassava roots and increased yield. With the progress made in cassava genetic transformation, new improved cultivars have been obtained in several laboratories during the last few years.

1.3.1. Disease Resistance

A number of bacteria and viruses cause severe problems for cassava cultivation. African cassava mosaic virus (ACMV) disease, the most important disease of cassava in Africa, can cause local losses of up to 80%, with losses throughout Africa reaching 36% of total cassava production (6). ACMV belongs to the whitefly-transmitted *Begomovirus* of the family *Geminivirdae* with a genome that consists of two circular single-stranded DNA (ssDNA) molecules denoted DNA A and DNA B (38). Different approaches on the genetic engineering of ACMV resistance have been tested in the past few years. By using the transgenic model plant *Nicotina benthamiana*, a variable increase in ACMV resistance could be observed. Constitutive expressions of defective interfering DNA (39), of a NTP-binding domain mutated Rep, movement protein or coat protein gene (*40-43*) and the virus-induced expression of dianthin (*44*) were tested in stably transformed plants. Transgenic cassava plants expressing antisense for Rep, REn or TrAP and the NTP-binding domain mutated Rep have shown reduced ACMV accumulation in transient assays as well as reduced disease symptoms in mature plants (Zhang P, unpublish. data; Taylor N, pers. comm.). Therefore, these strategies can perhaps be used in transgenic cassava to engineer sustainable resistance to ACMV.

Due to its long growing cycle (8-24 months), cassava can be subject to repeated and prolonged attacks from various insect pests (*5*). There are over 10 major anthropod pests of cassava, including whiteflies, mites, mealybugs and hornworms, which cause yield and planting material losses in Africa, Asia and South America. Transgenic plants expressing the *cry* genes, which encode insect specific δ-endotoxins (Bt toxins), are able to protect themselves against insect pests. Expression of the *cry* genes in transgenic cassava would complement the available methods for pest control in an environmentally and economically sustainable way. Several groups are currently working in this area.

1.3.2. Quality and Yield

Cassava storage roots only contain 1-2% crude protein (dry weight) and have low levels of most essential amino acids. Methionine, tryptophan and cysteine are limiting amino acids in the roots. In the tropics and subtropics, vegetarian populations which strongly depend on cassava root intake may suffer from qualitative malnutrition unless they can supplement their diet with other food sources containing enough essential amino acids. There has been much interest in improving the protein content of cassava, but so far traditional breeding has not been able to improve the protein content of cassava roots. The recent development of cassava transformation methods will allow in the near future the addition of agriculturally valuable traits for improved root quality. A number of storage proteins of seed and root crops, such as maize zeins and HMW subunits of wheat, may contain high contents of certain amino acids and can be expressed in transgenic plants, resulting in improved protein content (*45, 46*). The expression of an amino acid rich protein gene (*asp1*) in cassava has been shown to possibly improve protein content of cassava storage roots via genetic transformation (*37*).

Prolonging the life of cassava leaves could improve the root yield and quality as well as permit more frequent foliage harvesting. Leaf life could be prolonged in transgenic tobacco plants carrying the *ipt* gene encoding cytokinin under the control of a senescence-regulated promoter (SAG12) from *Arabidopsis thaliana* (*47*). A similar approach could be exploited in cassava for prolonged leaf life.

This chapter outlines the detailed methods of cassava transformation using somatic cotyledons and embryogenic suspensions via *Agrobacterium*-mediated DNA delivery. Transgenic cassava plants regenerate from either transformed cotyledons via shoot organogenesis or embryogenic suspensions via embryogenesis under antibiotic selection. Other selection methods, e.g. mannose-based positive selection, are discussed to compare with the antibiotic-based negative selection.

Conditions for cocultivation, selection and regeneration are optimized according to our routine protocols.

2. MATERIALS

2.1. Plant Materials

1. Shoot cultures of cassava cultivars MCol22 and TMS60444, obtained from CIAT (Centro Interancional de Agricultura Tropical, Cali, Columbia) and IITA (International Institute of Tropical Agriculture, Ibadan, Nigeria) and maintained in jars (330 ml, Greiner Bio-One Ltd., UK) containing 30 ml CBM (*see* Section 2.2.2.1.) at 4 week intervals.
2. Four-week-old somatic embryos of cassava cultivar Mcol22 on Petri dishes containing 25 ml CIM medium (*see* Section 2.2.2.3.).
3. Three-month-old friable embryogenic callus of cultivar TMS60444 on Petri dishes containing 25 ml GD medium (*see* Section 2.2.2.5.).

2.2. Cassava Tissue Culture Media

2.2.1. Stock Solutions

1x MS (*17*) basal salt mixture including vitamins (Duchefa Biochemie).
1x SH (*23*) basal salt mixture (Duchefa Biochemie).
1x GD (*20*) basal salt mixture including vitamins (Duchefa Biochemie).
1000× MS (*17*) vitamins (Duchefa Biochemie), dissolve in distilled water and store at -20°C in aliquots.
2 mM $CuSO_4$, store at 4°C.
1 M $MgSO_4$, filter-sterilized and store at 4°C.
1 mg/ml BAP, dissolve in 1 N NaOH and be stable at 4°C for up to 3 months.
1 mg/ml NAA, dissolve in 1 N NaOH and be stable at 4°C for up to 3 months.
12 mg/ml picloram, dissolve in 1 N NaOH and store at -20°C in aliquots.

2.2.2. Media

1. Basic shoot culture medium, CBM: 1x MS salts with vitamins, 2 μM $CuSO_4$, 2% sucrose, 0.3% Gelrite, pH 5.8, autoclave.
2. Axillary bud enlargement medium, CAM: 1x MS salts with vitamins, 2 μM $CuSO_4$, 10 mg/l BAP, 2% sucrose, 0.3% Gelrite, pH 5.8, autoclave.
3. Somatic embryo induction medium, CIM: 1x MS salts with vitamins, 2 μM $CuSO_4$, 12 mg/l picloram, 2% sucrose, 0.3% Gelrite, pH 5.8, autoclave.
4. Embryo maturation medium, CMM: 1x MS salts with vitamins, 2 μM $CuSO_4$, 0.1 mg/l BAP, 2% sucrose, 0.3% Gelrite, pH 5.8, autoclave.
5. Friable embryogenic callus medium, GD: 1x GD salts with vitamins, 12 mg/l picloram, 2% sucrose, pH 5.8, autoclave.

6. Suspension culture medium, SH: 1x SH salts, 1x MS vitamins, 12 mg/l picloram, 6% sucrose, pH 5.8, autoclave.

7. Shoot elongation medium, CEM: 1x MS salts with vitamins, 2 μM $CuSO_4$, 0.4 mg/l BAP, 2% sucrose, 0.3% Gelrite, pH 5.8, autoclave.

8. Somatic embryo emerging medium, MSN: 1x MS salts with vitamins, 1 mg/l NAA, 2% sucrose, 0.3% Gelrite, pH 5.8, autoclave.

9. *Agrobacterium* growth medium, YEB: 1 g/l Bacto[TM] yeast extract, 5 g/l Bacto[TM] beef extract, 5 g/l Bacto[TM] peptone, 5 g/l sucrose, pH 7.2. After autoclaving add 2 ml of filter-sterilized 1M $MgSO_4$.

10. *Agrobacterium* wash medium, MS: 1x MS salts with vitamins, 2% sucrose, pH 5.3, autoclave.

2.3. Other Chemicals

1. Acetosyringone: 200 mM stock solution, prepare by dissolving the powder in DMSO (Sigma) and keep at -20°C in aliquots.

2. Carbenicillin: 250 mg/ml stock solution, prepare by dissolving the powder in water and sterilize by filtration (22 μm pore size). Keep at -20°C in aliquots.

3. Hygromycin: 25 mg/ml, dissolve in water and sterilize by filtration. Keep at -20°C in aliquots.

4. Rifampicin: 25 mg/ml stock solution, prepare by dissolving the powder in 0.1 M HCl and sterilize by filtration. Keep at -20°C in aliquots.

5. Spectinomycin: 50 mg/ml stock solution, prepare by dissolving the powder in water and sterilize by filtration. Keep at -20°C in aliquots.

All the above mentioned chemicals are unstable at high temperature. Thus, they should only be added to medium at 60°C after autoclaving. Swirl to mix before pouring the medium to containers.

2.4. Culture Conditions

All plant materials are cultured at 26°C under a 16 h photoperiod (90-110 μmol/ m^2/sec, TRUE-LITE®, Interlectric Corporation, Warren, PA, USA) in an environmentally controlled growth chamber (Weiss Klimatechnik GmbH, Reiskirchen-Lindenstruth, Germany) unless otherwise stated. Embryogenic suspension cells are cultured in jars (190 ml, Greiner Bio-One Ltd., UK) on a gyratory shaker (Infors type RC-406, Infors AG, Wettingen, Switzerland) at 108 rpm and 28°C under continuous light (~50 μmol/m^2/sec) and subcultured at 3 day intervals.

2.5. Agrobacterium tumefaciens Strain and Example Plasmid Used

Agrobacterium tumefaciens strain LBA4404 (*48*) carrying the binary vector pHMG (Fig. 2) is used for genetic transformation experiments. The plasmid pHMG harbours an intron-interrupted *hpt* gene and an intron-interrupted *uid*A gene. Both genes are driven by the CaMV 35S promoters. Another selectable marker gene, *pmi*, is also present in the T-DNA region (*35*). In the following procedures the *hpt* is used as an example of selectable marker. The bacteria are grown on YEB agar plates with 25 mg/l rifampicin and 100 mg/l spectinomycin.

Figure 2. Schematic representation of the T-DNA region of binary vector pHMG. RB right border, LB left border, p35S CaMV 35S promoter, 35! and Nos! CaMV 35S and Agrobacterium *nopaline gene terminator, respectively.*

2.6. β-glucuronidase Assay Buffer

10 mM $Na_2EDTA \cdot H_2O$, 0.1% Triton X-100, 0.3% X-gluc, 0.1 M NaH_2PO_4 (pH 7.0), 0.5 M $K_3Fe(CN)_6$) (*49*).

3. METHODS

The prerequisite for cassava genetic engineering is to establish a reproducible transformation system consisting of reliable plant regeneration, efficient DNA deliveries and effective selection of transformed plants at a satisfactory frequency. The methods described below outline: (a) induction of primary somatic embryos and cyclic embryogenic cultures; (b) induction of friable embryogenic callus (FEC) and embryogenic suspensions; (c) shoot organogenesis from transformed somatic cotyledons and regeneration of transgenic plants; (d) somatic embryogenesis from transformed somatic cotyledons; (e) FEC transformation and regeneration of transgenic plants. Here we take the antibiotic-based negative selection as an example for selection of transgenic cassava plants (*see* Note 1).

3.1. Induction of Primary Somatic Embryos and Cyclic Embryonic Cultures

3.1.1. Primary Embryogenesis from Axillary Buds, Immature Leaves and Leaf Lobes

1. Cut 1 cm long nodal explants (young stem cuttings with axillary buds) with a sharp scalpel from 4 week-old *in vitro* plants of Mcol22 or TMS60444, grown in jars (330 ml, Greiner Bio-One Ltd.) containing 30 ml of CBM.

2. Place explants horizontally on Petri dishes (ø9 cm, Bibby Sterilin Ltd, UK; 40 explants/Petri-dish) containing 25 ml of CAM for 6 days at 26°C.

3. Remove the enlarged axillary buds (2-4 mm) from the nodal explants with syringe needles (φ 1.20 x 40 mm, Carl Roth GmbH+Co., Germany) and transfer onto plates containing CIM (25 ml/Petri dish).

4. Alternatively, isolate apical meristems (1-2 mm) and 1-6 mm long immature leaf lobes from the shoot tips of Mcol22 or TMS60444 and culture on CIM (*see* Note 2).

5. Incubate the plates at 26°C in the dark or a constant weak light (~10 μmol/ m^2/sec) (*see* Note 3). After 2 weeks, check the emerging embryos on the explants under a stereo microscope. Once new globular and torpedo embryos appear (Fig. 3A), subculture them onto fresh CIM (25 ml/dish) with syringe needles.

3.1.2. Establishment of Cyclic Embryonic Cultures

1. Subculture somatic embryos/embryo clusters on CIM (25 ml/dish) every second week (*see* Note 4). Somatic embryos can be multiplied via secondary embryogenesis by this way.

2. Alternatively, after 2-3 cycles, harvest and then transfer somatic embryos (Fig. 3B) onto CMM (25 ml/dish) at 26°C with a 16 h photoperiod to induce embryo maturation and production of green cotyledons.

3. Collect 2 week-old green cotyledons (Fig. 3C), cut into 0.25 cm^2 pieces with a scalpel and place onto CIM (25 ml/dish) to induce secondary embryogenesis (40 pieces/dish).

4. After 2 weeks, transfer the new somatic embryos/embryo clusters that develop from the cut-ends of the cotyledons onto CIM (25 ml/dish) to establish cyclic embryogenic cultures.

3.2. Induction of FEC and Embryogenic Suspensions (*see* Note 5)

1. Transfer the somatic embryo clusters of TMS60444 (30 clusters/dish) with a pair of sterile fine forceps onto GD (30 ml/dish) and culture at 26°C in the dark.

2. Subculture the cultures at 4 week intervals. After 2-3 cycles, check the embryos under a stereo microscope.

3. Separate the FEC (a light yellow, highly friable callus with numerous spherical embryogenic units and a dense cytoplasm, Fig. 3D) from the surface of embryogenic structures with a syringe needle and place onto GD (25 ml/dish) at 26°C under a 16 h photoperiod.

4. After 2 weeks, select pure FEC under a stereo microscope to eliminate undesirable tissues, such as non-embryogenic friable calli and somatic embryos, and transfer onto fresh GD (25 ml/dish).

5. Maintain the FEC cultures on GD (25 ml/dish) at 3-week intervals.

6. Transfer ~ 0.5 g FEC into 30 ml of SH medium in 190 ml Jars (Greiner Bio-One Ltd.) on a gyratory shaker (108 rpm) at 28°C under continuous light (~50 μmol/m^2/sec).

7. Remove the old medium using a pipette and replace with 30 ml of fresh SH every 3 days.

8. After 3 weeks, sieve the cultures through a ø500 μm metallic net (Saulas, Montreuil, France) to obtain embryogenic units ranging in size from 250-500 μm. Subculture aliquots of 1 ml settled cell volume (SCV) of embryogenic suspension cells (Fig. 3E) into new 190 ml jars with fresh SH (30 ml/jar). Repeat this filtering every 3 weeks.

3.3. Shoot Organogenesis from Transformed Somatic Cotyledons and Regeneration of Transgenic Plants

3.3.1. Transformation of Somatic Cotyledons with Agrobacterium tumefaciens

1. Pick up a single colony of *A. tumefaciens* harbouring pHMG from an agar plate using a toothpick and inoculate into a 14 ml Falcon® round-bottom plastic tube (Becton Dickinson labware, Meylan, France) containing 5 ml liquid YEB with 25 mg/l rifampicin and 100 mg/l spectinomycin. Incubate overnight on a shaker (240 rpm) at 28°C.

2. Transfer aliquots of 25 μl bacterial suspensions to 50 ml fresh medium in 250 ml Duran® flasks (Schott Glas, Mainz, Germany) and culture for 12-20 h to an OD_{600} 0.5-1.0.

3. Centrifuge the bacterial solution at 6000 rpm and 4°C for 10 min in 50 ml Falcon® tubes, wash once with 50 ml liquid MS (pH 5.3) and centrifuge again. Resuspend the bacterial pellets in liquid MS supplemented with 200 μM acetosyringone at an OD_{600} 1.0.

4. Grow the *Agrobacterium* cultures in jars (190 ml, Greiner Bio-One Ltd.) for 2 h at 28°C on an orbital shaker at 80 rpm. Simultaneously, cut the 2 week-old cotyledons into 0.25 cm^2 pieces with a sharp scalpel and place on CBM (25 ml/dish) to prevent the explants from drying.

5. Inoculate approx. 500 cotyledon pieces with *Agrobacterium* in jars (190 ml, Greiner Bio-One Ltd.) containing 20-30 ml of bacterial suspensions and place on a shaker at 80 rpm.

6. After 45 min, remove the bacterial suspension with a pipette and transfer the cotyledon pieces onto plates containing 25 ml of COM supplemented with 100 μM acetosyringone and cocultivate at 25°C for 3 days with a 16 h photoperiod (*see* Note 6).

7. After cocultivation, bacteria can be visible at the edges of the cotyledon explants on the medium. Wash the inoculated explants 3 times with 30 ml of sterile water in jars (190 ml, Greiner Bio-One Ltd.). Then, wash the explants twice with 30 ml of liquid MS medium (pH 5.3) supplemented with 500 mg/l carbenicillin. Place explants onto sterile filter papers (9 cm, Schleicher & Schuell) to draw off excess liquid.

8. Place several cotyledon pieces into GUS buffer for transient assay. Incubate at 37°C for 6 h, wash several times with 96% ethanol and then store in 96%

ethanol. Successful T-DNA transfer is indicated by the dark blue precipitant in the cotyledon pieces (Fig. 3F). Transfer the remaining explants onto regeneration and selection medium (*see* Section 3.3.2.).

3.3.2. Selection and Regeneration of Transgenic Plants via Shoot Organogenesis

1. Transfer the washed cotyledon pieces (40 explants/dish) onto selection medium I (COM supplemented with 4 mg/l AgNO$_3$ (*see* Note 7), 10 mg/l hygromycin and 500 mg/l carbenicillin; 25 ml/dish) in the dark at 26°C for 7 days.
2. Transfer the explants to selection medium II (COM with 20 mg/l hygromycin and 500 mg/l carbenicillin, 25 ml/dish) at 26°C in a 16 h photoperiod for 2 weeks (*see* Note 8).
3. Cut the resistant shoot clusters (3-6 mm) with shoot primordia (Fig. 3G) and place onto CEM in jars (30 ml/jar) under the same culture conditions.
4. After 2-3 weeks, cut off emerging shoots and transfer to CBM for further growth. Once the *in vitro* plants are large enough, multiply them as stem cuttings.
5. Excise one leaf or root from the shoots and test in the GUS assay buffer. GUS positive plant lines are ready for molecular analysis.
6. Alternatively, excise the shoots (~ 1 cm long) developing from axillary buds of stem cuttings and transfer to CBM supplemented with 8 mg/l hygromycin (25 ml/dish) for a rooting screen to eliminate escapes from the selection. Use shoots of wildtype plants as a negative control (*see* Note 9).
7. Check the shoots after 7 days. Transgenic lines can root normally, while non-transformed shoots fail to produce roots.
8. Transgenic plant lines are then ready for molecular analyses, such as PCR, Southern and Northern blotting.

3.4. Embryogensis from Transformed Somatic Cotyledons

1. Place the inoculated and washed somatic cotyledon pieces (*see* Section 3.3.1.7.) onto plates containing CIM supplemented with 10 mg/l hygromycin and 500 mg/l carbenicillin (25 ml/dish) in the dark at 26°C for 7 days.
2. Transfer the explants to fresh CIM with 20 mg/l hygromycin and 500 mg/l carbenicillin (25 ml/dish) at the same culture conditions (*see* Note 8).
3. After 2 weeks, excise the new emergent somatic embryos that appear on the explants (Fig. 3H) and place on CIM supplemented with 10 mg/l hygromycin (25 ml/dish) for a further 2 weeks.
4. Transfer the multiplied somatic embryo lines (Fig. 3I) onto CMM supplemented with 12.5 mg/l hygromycin (25 ml/dish) for embryo maturation. Simultaneously, separate a small part of embryo clusters and transfer to the GUS assay buffer for 3 h. Transgenic embryos will show dark blue staining (Fig. 3J).
5. Somatic embryos with green cotyledons are ready for germination as described (*see* Section 3.5.4.).

3.5. Embryogenesis from Transformed Embryogenic Suspensions and Regeneration of Transgenic Plants (see Note 4)

3.5.1. Transformation of Embryogenic Suspensions

1. Transfer aliquots of 2 ml SCV of embryogenic cassava suspensions by a sterile pipette to jars (190 ml, Greiner Bio-One Ltd.) containing 10 ml of bacterial suspension (see Section 3.3.1.4.).
2. After 45 min, remove excess bacterial liquid with a pipette. Spread the inoculated tissues using a pipette onto sterile filter papers (ϕ 9 cm, Schleicher & Schuell) and transfer to Petri dishes (16 x 90 mm, Bibby Sterilin Ltd., Staffordshire, UK) containing 25 ml of semi-solidified SH supplemented with 100 µm acetosyringone. Cocultivate at 25°C for 3 days.

3.5.2. Selection and Establishment of Transformed Cell Suspensions

1. Collect the transformed tissues (2 ml SCV) from the filter papers with a pair of tweezers and transfer to jars (190 ml, Greiner Bio-One Ltd.) containing 30 ml of SH.
2. Wash the tissues by pipetting up and down several times. Draw off the liquid and repeat the wash twice.
3. Transfer the transformed tissues into new jars containing 30 ml of SH with 12.5 mg/l hygromycin and 500 mg/l carbenicillin. Culture for 3 days on a shaker at 137 rpm under continuous light (~50 µmol/m^2/sec).
4. Replace the culture medium with SH containing 25 mg/l hygromycin and 500 mg/l carbenicillin. Culture on the shaker at 108 rpm. Refresh the medium at 3 day intervals (see Note 8).
5. After 2 weeks cultivation in selective medium, the antibiotic-resistant suspended cells develop into yellowish, friable embryogenic clusters indistinguishable from control FEC grown under non-selective conditions. Stain a small part of cell clusters in GUS assay buffer for 6 h at 37°C. Transgenic cells are easily distinguished by the dark blue precipitate (Fig. 3K) from a background of white and brown-coloured dead tissues.

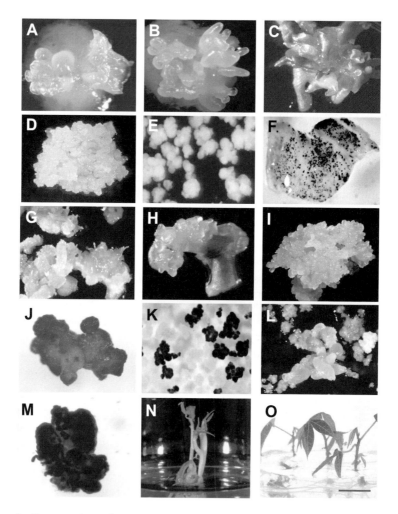

Figure 3. Regeneration of transgenic cassava via shoot organogenesis and somatic embryogenesis. A, Primary embryos. B, secondary embryos. C, cotyledon stage somatic embryos. D, friable embryogenic callus. E, embryogenic suspensions. F, GUS assay of somatic cotyledon after cocultivation. G, shoot organogenesis from somatic cotyledons on selection medium. H, somatic embryogenesis from somatic cotyledons on selection medium. I, Multiplied transgenic somatic embryo cluster. J, GUS assay of transgenic somatic embryos. K, GUS assay of transformed embryogenic suspensions after 3-week selection. L, Formation of somatic embryos from embryogenic suspension cells. M, GUS assay of somatic embryos regenerated from transformed embryogenic suspensions. N, shoots regenerated from transgenic somatic embryos. O, shoot cultures of transgenic cassava. Scale bar = 1 mm in (F, J), 2 mm in (A, B, D, E, H, I, K, L, M), 4 mm in (C), 5 mm in (G), 1.5 cm in (N, O).

3.5.3. Somatic Embryogenesis from Transformed Suspension Cultures

1. Spread the suspensions, which have grown under selection for 2-3 weeks to Petri dishes containing 25 ml of MSN with 10 mg/l hygromycin. Culture at 26°C under a 16 h photoperiod to allow embryo development.
2. After 2-4 weeks, resistant somatic embryos (Fig. 3L) are formed. Use several embryos from each embryo cluster for GUS assay. Blue stain is detectable in transgenic lines (Fig. 3M). Transfer GUS positive embryo lines individually onto CMM containing 12.5 mg/l hygromycin (25 ml/dish) for cotyledon emergence.

3.5.4. Plant Regeneration from Transformed Somatic Embryos (see Note 9)

1. Transfer maturing/mature somatic embryos from either cotyledon transformation experiments or FEC events onto CEM (30 ml/jar) in jars (190 ml, Greiner Bio-One Ltd.) for germination. Generally, it takes 2-4 weeks for new shoots to develop (Fig. 3N) on the cotyledonary embryos.
2. Transfer the shoots onto CBM (30 ml/jar) for further growth (Fig. 3O). After 3 weeks, excise the shoots (~ 1 cm long) developing from axillary buds of the stem cuttings and transfer to CBM medium supplemented with hygromycin 8 mg/l (25 ml/dish) for a rooting screen. Use shoots of wildtype plants as a negative control.
3. Check the shoots after 7 days. Transgenic lines can root normally, while control shoots fail to produce any roots (see Note 10).
4. The plant lines are then ready for molecular analysis, such as PCR.

4. NOTES

1. Both the antibiotic-based negative selection and the mannose-based positive selection have been tested in cassava transformation studies (34-36). Recently, there are public concerns with regards to the use of antibiotic or herbicide resistance genes, (such as nptII, hpt and bar) in transgenic plants as potential risks to the environment, and so, future developments in transformation methods should move towards either using other selection systems or towards methods which allow the elimination of selectable markers from transgenic plants. The mannose-based positive selection is based on the E.coli pmi gene as a selectable marker and mannose as the selective agent. In plant cells, mannose is phosphorylated by a hexokinase to mannose-6-phosphate which, when accumulating in the cells, could lead to energy starvation and severe growth inhibition (50, 51). However, cells transformed with the pmi gene are able to convert mannose-6-phosphate to fructose-6-phosphate which can be utilized as a carbon source, enabling these cells a metabolic advantage in culture. Therefore, transgenic plants could be regenerated using mannose selection. The mannose-based selection system which does not cause any risk to animal, human or to the environment, and has been recently shown to be more efficient for potato, tomato and sugar beet transformations than methods based on antibiotic selection. In cassava transformation, the selection efficiency of mannose is lower than that of hygromycin in shoot organogenesis-based transformation but the same as that of hygromycin in FEC transformation. It is critical to find a ratio of mannose and sucrose in the selection medium which allows developing transformed cells to embryos or shoots with minimized escapes. Paramomycin can also be used for efficient selection of FEC when using the nptII gene as selectable marker (29, 31). Using the luciferase reporter gene (luc), transgenic cassava was also produced by selecting transgenic tissues from bombarded suspensions (24, 32). This method, however, requires access to costly equipment including a coupled device camera for detection and localization of the bioluminescence, which cannot be used routinely in every laboratory.

2. Primary somatic embryogenesis of cassava is an explant-dependent event. Therefore, careful selection of explants should be considered. Meristems from axillary buds or apical shoots are the ideal candidates. Immature leaves smaller than 6 mm can be used, but the compact embryo-like structures that are formed on them rarely develop to maturing somatic embryos. Care should be taken to place the explants with the upper side (meristem tip, adaxial side of the leaf) up on the medium to ensure efficient embryo production.

3. In a recent study, we have shown that weak light (~10-20 μmol/m²/sec) can enhance the frequency of somatic embryogenesis in most tested cultivars.

4. To maintain the vigorous growth of somatic embryos with minimal callus development, somatic embryos and embryo clusters need to be carefully separated from the callus and other non-morphogenic tissues and transferred onto fresh CIM every 2 weeks.

5. The formation of FEC in cassava is cultivar-dependent, but works reliably with e.g. TMS60444. Because the production of FEC and plant regeneration from FEC takes a relatively long time (4-5 months), the risk of mutations should be minimized. We therefore suggest using FEC less than 6 months old for transformation.

6. The use of acetosyringone (100-300 μM) during pre-induction, inoculation and cocultivation could increase the frequencies of *Agrobacterium*-mediated cassava transformation with both somatic cotyledons and FEC (*26, 31*).

7. Silver nitrate (2-8 mg/l) can enhance the efficiency of plant regeneration and reduce callus formation in cassava shoot organogenesis from somatic cotyledons (*27*). Application of silver nitrate in selection and plant regeneration media allows for a higher transformation frequency.

8. Stepwise selection protocols are strongly recommended to allow recovery of transformed cells during earlier stage of selection and the inhibition of non-transformed cells.

9. The sensitivity of root development to different selective agents is dose- and genotype-dependent. The selection pressure used for organogenesis is not very tight. Therefore, in addition to truly transgenic plants, escapes will be also regenerated. Rooting assays allow the elimination of these escapes. Furthermore, shoots emerging from cotyledon explants can also be chimeric. By repeated GUS assays combined with node culture of the originally chimeric transgenics, true transgenic plants can be produced.

10. Transformation efficiency depends on the methods for DNA delivery as well as the explants used. Generally, the frequency of *Agrobacterium*-mediated transformation is higher than that of biolistic-mediated transformation. Many independent transgenic plant lines can be obtained from FEC transformation because a large amount of independent embryogenic cells can be transformed.

CONCLUSION

Production of transgenic plants is becoming routine in cassava biotechnology. Transgenic cassava plants are produced both from somatic cotyledons via shoot organogenesis or somatic embryogenesis and from FEC via embryogenesis using *Agrobacterium*. Improvements on plant regeneration and transformation increase the transformation efficiency. GUS assays and rooting tests can be used to screen for escapes from antibiotic selection, which improves the regeneration rate of truly transgenic plants. The current protocols allow more efficient use and improvement of cassava with traits of interest.

REFERENCES

1. FAO (2002). FAO production Yearbook, FAOSTAT database. http://apps.fao.org/default.htm.
2. Cock JH (1985). *Cassava: new potential for a neglected crop*. Westview, London, UK.
3. Nweke FI, Spencer DSC and Lynam JK (2002). *The cassava transformation: Africa's best-kept secret*. Michigan State University Press, East Lansing, USA.
4. FAO (2000). Cassava can play a key role in reducing hunger and poverty. FAO press release 00/25. http://www.fao.org/waicent/ois/press_ne/presseng/2000/pren0025.htm.

5.   Belloti AC, Smith L and Lapointe SL (1999). Recent advances in cassava pest management. *Annual Review of Entomology*, **44**: 343-370.

6.   Thresh JM, Fargette D and Otim-Nape GW (1994). Effects of African cassava mosaic geminivirus on the yield of cassava. *Tropical Science*, **34**: 26-42.

7.   IITA (2000). Improving cassava-based systems. Annual Report, Project 6.

8.   Konan NK, Sangwan RS and Sangwan-Norreel BS (1994). Somatic embryogenesis from cultured mature cotyledons of cassava (*Manihot esculenta* Crantz). *Plant Cell, Tissue and Organ Culture*, **37**: 91-102.

9.   Stamp JA and Henshaw GG (1987). Somatic embryogenesis from clonal leaf tissue of cassava. *Annals of Botany*, **59**: 445-450.

10.   Szabados L, Hoyos R and Roca W (1987). *In vitro* somatic embryogenesis and plant regeneration of cassava. *Plant Cell Reports*, **6**: 248-251.

11.   Raemakers CJJM, Bessembinder J, Staritsky G, Jacobsen E and Visser RGF (1993). Inducion, germination and shoot development of somatic embryos in cassava. *Plant Cell, Tissue and Organ Culture*, **33**: 151-156.

12.   Mathews H, Schopke C, Carcamo R, Chavarroaga P, Fauquet C and Beachy RN (1993). Improvement of somatic embryogenesis and plant recovery in cassava. *Plant Cell Reports*, **12**: 328-333.

13.   Li HQ, Huang YW, Liang CY and Guo Y (1995). Improvement of plant regeneration from cyclic somatic embryos in cassava. In: Cassava Biotechnology Network (ed.) *Proceedings of the 2nd International Scientific Meeting*, Bogor, Indonesia, CIAT Working Document 150. pp. 289-302.

14.   Puonti-Kaerlas J, Frey P and Potrykus I (1997). Development of meristem gene transfer techniques for cassava. *African Journal of Root Tuber Crops*, **2**: 175-180.

15.   Mukherjee A (1995). Embryogenesis and regeneration from cassava calli of anther and leaf. In: Cassava Biotechnoglogy Network (ed.) *Proceedings of the 2nd International Scientific Meeting*, Bogor, Indonesia, CIAT Working Document 150. pp. 375-381.

16.   Woodward B and Puonti-Kaerlas J (2001). Somatic embryogenesis from floral tissue of cassava (*Manihot esculenta* Crantz). *Euphytica*, **120**: 1-6.

17.   Murashige T and Skoog F (1962). A revised medium for rapid growth and bioassays with tobacco tissue cultures. *Physiologia Plantarum*, **15**: 473-497.

18.   Li HQ, Huang YW, Liang CY, Guo JY, Liu HX, Potrykus I and Puonti-Kaerlas J (1998). Regeneration of cassava plants via shoot organogenesis. *Plant Cell Reports*, **17**: 410– 414.*

19.   Zhu J, Huang YW and Liang CY (1998). Improvement of plant regeneration from cyclic secondary somatic embryos in cassava (*Manihot esculenta* Crantz). *Journal of Tropical and Subtropical Botany*, **6**: 144-151.

20.   Gresshoff P and Doy C (1974). Derivation of a haploid cell line from *Vitis vinifera* and the importance of the stage of meiotic development of the anthers for haploid culture of this and other genera. *Zeitschrift für Pflanzenphysiologie*, **73**, 132-141.

21.   Taylor NJ, Edwards M, Kiernan RJ, Davey CDM, Blakesley D and Henshaw GG (1996). Development of friable embryogenic callus and embryogenic suspension culture systems in cassava (*Manihot esculenta* Crantz). *Nature Biotechnology*, **14**: 726-730.

22.   Taylor NJ, Kiernan RJ, Henshaw GG and Blakesley D (1997). Improved procedures for producing embryogenic tissues of African cassava cultivars: Implications for genetic transformation. *African Journal of Root Tuber Crops*, **2**: 200-204.

23.   Schenk RU and Hildebrandt AC (1972). Medium and techniques for induction and growth of monocotyledonous and dicotyledonous plant cell cultures. *Canada Journal of Botany*, **50**: 199-204.

24.   Raemakers CJJM, Sofiari E, Taylor N, Henshaw G, Jacobsen E and Visser RGF (1996). Production of transgenic cassava (*Manihot esculenta* Crantz) plants by particle bombardment using luciferase activity as selection marker. *Molecular Breeding*, **2**: 339-349.

25.   Sofiari E, Raemakers CJJM, Bergervoet JEM, Jacobsen E and Visser RGF (1998). Plant regeneration from protoplasts isolated from friable embryogenic callus of cassava. *Plant Cell Reports*, **18**: 159-165.

26.   Li HQ, Sautter C, Potrykus I and Puonti-Kaerlas J (1996). Genetic Transformation of Cassava (*Manihot esculenta* Crantz). *Nature Biotechnology*, **14**: 736-740.

27.   Zhang P, Phansiri S and Puonti-Kaerlas J (2001). Improvement of cassava shoot organogenesis by the use of silver nitrate *in vitro*. *Plant Cell, Tissue and Organ Culture*, **67**: 47-54.

28.   Hankoua BB (2003). Regeneration and transformation of African cassava (*Manihot esculenta* Crantz) germplasm. Ph.D. Thesis, University of Ibadan, Ibadan, Nigeria.

29.   Schöpke C, Taylor N, Carcamo R, Konan NK, Marmey P, Henshaw GG, Beachy RN and Fauquet C (1996). Regeneration of transgenic cassava plants (*Manihot esculenta* Crantz) from microbombarded embryogenic suspension cultures. *Nature Biotechnology*, **14**: 731-735.

30.   Snepvangers SCHJ, Raemakers CJJM, Jacobsen E and Visser RGF (1997). Optimization of chemical selection of transgenic friable embryogenic callus of cassava using the luciferase report gene system. *African Journal of Root Tuber Crops*, **2**: 196-200.

31.   González AE, Schöpke C, Taylor NJ, Beachy RN and Fauquet CM (1998). Regeneration of transgenic cassava plants (*Manihot esculenta* Crantz) through *Agrobacterium*-mediated transformation of embryogenic suspension cultures. *Plant Cell Reports*, **17**: 827-831.

32.   Munyikwa TRI, Reamakers CCJM, Schreuder M, Kok R, Schippers M, Jacobsen E and Visser RGF (1998). Pinpointing towards improved transformation and regeneration of cassava (*Manihot esculenta* Crantz). *Plant Science*, **135**: 87-101.

33.   Sarria R, Torres E, Angel F, Chavarriaga P and Roca WM (2000). Transgenic plants of cassava (*Manihot esculenta*) with resistance to Basta obtained by *Agrobacterium*-mediated transformation. *Plant Cell Reports*, **19**: 339-344.

34.   Zhang P, Legris G, Coulin P and Puonti-Kaerlas J (2000). Production of stably transformed cassava plants via particle bombardment. *Plant Cell Reports*, **19**: 939-945.

35.   Zhang P, Potrykus I and Puonti-Kaerlas J (2000). Efficient production of transgenic cassava using negative and positive selection. *Transgenic Research*, **9**: 405-415.

36.   Zhang P and Puonti-Kaerlas J (2000). PIG-mediated cassava transformation using positive and negative selection. *Plant Cell Reports*, **19**: 1041-1048.

37.   Zhang P, Jaynes JM, Potrykus I, Gruissem W and Puonti-Kaerlas J (2003b). Transfer and expression of an artificial storage protein (*ASP1*) gene in cassava (*Manihot esculenta* Crantz). *Transgenic Research*, **12**: 243-250.

38.   Stanley J (1983). Infectivity of cloned geminivirus genome requires sequences from both DNAs. *Nature*, **305**: 643-645.

39.   Stanley J, Frischmuth T and Ellwood S (1990). Defective viral DNA ameliorates symptoms of geminivirus infection in transgenic plants. *Proceedings of the National Academy of Sciences USA*, **87**: 6291-6295.

40.   Hong Y and Stanley J (1996). Virus resistance in *Nicotiana benthamiana* conferred by African cassava mosaic virus replication-associated protein (AC1) transgene. *Molecular Plant-Microbe Interactions*, **9**: 219-225

41.   Von Arnim A and Stanley J (1992). Inhibition of African cassava mosaic virus systemic infection by a movement protein from the related geminivirus tomato golden mosaic virus. *Virology*, **187**: 555-564.

42.   Duan YP, Powell C, Webb S, Purcifull D and Hiebert E (1997). Geminivirus resistance in transgenic tobacco expressing mutated BC1 protein. *Molecular Plant-Microbe Interactions*, **10**: 617-623.

43.   Sangaré A, Deng D, Fauquet CM and Beachy RN (1999). Resistance to African cassava mosaic virus conferred by a mutant of the putative NTP-binding domain of the Rep gene (*AC1*) in *Nicotiana benthanmiana*. *Molecular Breeding*, **5**: 95-102.

44.   Hong Y, Saunders K, Hartley M and Stanley J (1996). Resistance to geminivirus infection by virus-induced expression of dianthin in transgenic plants. *Virology*, **220**: 119-127.

45.   Blechl AE and Anderson OD (1996). Expression of a novel high-molecular-weight glutenin subunit gene in transgenic wheat. *Nature Biotechnology*, **14**: 875-879.

46.   Bagga S, Adams H, Rodriquez JD, Kemp JD and Sengupta-Gopalan C (1997). Co-expression of the maize δ- and β-zein genes results in stable accumulation of δ-zein in ER-derived protein bodies formed by β-zein. *The Plant Cell*, **9**: 1683-1686.

47.   Gan SS and Amasino RM (1995). Inhibition of leaf senescence by autoregulated production of cytokinin. *Science*, **270**: 1986-1988.

48. Hoekema A, Hirsch PR, Hooykaas PJJ and Schilperoort RA (1983). A binary plant vector strategy based on separation of *vir*- and T-region of the *Agrobacterium tumefaciens* Ti-plasmid. *Nature,* **303**: 179-180.

49. Jefferson RA (1987). Assaying chimeric genes in plants: the GUS gene fusion system. *Plant Molecular Biology Reporter,* **5**: 387-405.

50. Ferguson JD, Street HE and David SB (1958). The carbohydrate nutrition of tomato roots. IV. The inhibition of excised root growth by galactose and mannose and its reversal by dextrose and xylose. *Annals of Botany,* **22**: 525-538.

51. Malca I, Endo RM and Long MR (1967). Mechanism of glucose counteraction of inhibition of root elongation by galactose, mannose, and glucosamine. *Phyopathology*, **57**: 272-278.

Part IV: Legumes, Brassicas, Fruits and Oilseed Crops

Chapter 24

SOYBEAN TRANSFORMATION USING THE *AGROBACTERIUM*-MEDIATED COTYLEDONARY-NODE METHOD

P.M. OLHOFT[1] AND D.A. SOMERS

Department of Agronomy and Plant Genetics, University of Minnesota, St. Paul, MN 55108, USA;
[1]Current address: BASF Corporation, 26 Davis Drive, PO Box 13528, Research Triangle Park, NC 27709-3528, USA.
Corresponding author: D.A. Somers. E-mail: somers@biosci.cbs.umn.edu

1. INTRODUCTION

1.1. Importance and Distribution of Soybean

Soybean is grown as a source of vegetable oil and protein meal. Soybean is often referred to as the "miracle crop" because its seed constituents are valuable components of a myriad of foods, animal feeds and industrial products. The crop also plays a dominant role in world agriculture. On a global scale, soybean production was 194 million metric tonnes in 2002 and accounted for 60 % of world oilseed production (www.soystats.com). Major soybean producing countries include the USA, Brazil, Argentina and China. In addition to the traditional uses of soybean seed, there is a resurgence of interest to use soybean oil as a biodiesel fuel. Considering that the human population is projected to increase in the foreseeable future with commensurate increases in demands for meat and energy, soybean production will continue to increase to satisfy the food, feed, industrial and energy needs of a growing world population.

1.2. Genetic Improvement of Soybean

Soybeans were originally domesticated in China and are now produced throughout the world over a wide range of latitudes. Expansion of soybean production over this huge geographical area has been accomplished by conventional plant breeding. Simultaneously, plant breeders have improved soybean yield, disease and pest resistance and seed compositional quality. Genetic engineering also has contributed to soybean improvement especially in the development of herbicide-resistant cultivars for weed management. Soybeans transformed for resistance to the herbicide glyphosate, and referred as Roundup Ready®, are planted on substantial acreages worldwide. For example, in 2003 more than 75% of soybeans planted in the USA were Roundup Ready® soybeans. Beyond herbicide resistance, genetic engineering is being applied to the development of pest and disease resistances and

323

I.S. Curtis (ed.), Transgenic Crops of the World - Essential Protocols, 323-336.
© 2004 Kluwer Academic Publishers. Printed in the Netherlands.

further modification of seed composition. Genetic engineering-based genetic modifications offer the opportunity to augment conventional soybean breeding efforts either by altering the expression of soybean genes or introducing genes from other organisms. Both types of genetic modifications are not achievable by applications of Mendelian genetics, wide crosses or mutation breeding. In the first strategy, soybean genes can be either silenced, overexpressed or ectopically expressed. Examples of modifying soybean gene expression include high oleic acid oil (1) and enhancing or decreasing isoflavone content (2). Genes from any organism may be used to improve or modify soybean as exemplified by the Roundup Ready® trait conferred by a bacterial gene for a resistant form of the herbicide's target enzyme (3). Further applications of these types of genetic modifications are actively being pursued. Clearly, there are numerous opportunities for future development of transgenic traits in soybean.

1.3. Genetic Transformation of Soybean

Transgenic soybean plants are produced using various DNA delivery methods and plant tissues including microprojectile bombardment of shoot meristems (4), embryogenic suspension cultures (5), Agrobacterium tumefaciens-mediated T-DNA delivery into immature cotyledons (6-8), embryogenic suspension cultures (9) and axillary meristematic tissue located in seedling cotyledonary-nodes (10). This diversity of methods for soybean transformation is positive in that it provides researchers with a range of approaches to produce transgenic plants. However, application of these soybean transformation methods to both crop improvement and functional genomics are still somewhat constrained to a limited range of genotypes and by relatively inefficient production of transgenic plants. Thus, continued efforts to improve soybean transformation are warranted (11) to establish a robust method that will be readily adopted by public researchers involved in basic and applied soybean research.

This chapter is focused on the cotyledonary-node transformation method originally reported by Hinchee et al. (10) and recently improved by Zhang et al. (12) and Olhoft et al. (13). In our opinion, this method comes closest to meeting the criterion for an ideal transformation method for soybean. An ideal method should: (a) be simple, inexpensive and rapid; (b) provide efficient production of transgenic plants; (c) be successful with a range of cultivars and A. tumefaciens strains and binary plasmids; and (d) result in the production of simple (non-rearranged) transgene loci that will be most amenable to regulatory characterization and eventual commercial development. The cotyledonary-node method is simple, inexpensive, and rapid because incorporation of genes of interest between the T-DNA borders of binary plasmids and transformation of the finished binary plasmid into the appropriate disarmed Agrobacterium strain is no more complicated than construction of plasmids and their purification for direct DNA delivery. There is no requirement for a DNA delivery device, since Agrobacterium performs this task autonomously. Finally, the explant is readily available and inexpensive to produce because it is derived from germinated seeds, thus requiring little labour and care compared to growing plants for the production of immature embryos or maintaining cultures of somatic embryos. Because the method is based on *Agrobacterium*-mediated T-DNA delivery, it is likely that the frequency of transgenic plants with simple transgene loci will be increased compared to plants transformed using direct DNA delivery methods. In a recent study, we evaluated transgene locus complexity in a population of 95 independent soybean plants. On average, each plant exhibited two T-DNA loci and one of these was a simple locus that exhibited all portions of the T-DNA, no rearranged T-DNA or binary plasmid backbone, and was transmitted in a Mendelian manner (Olhoft *et al.*, in prep.).

We have investigated factors for improving the production of transgenic plants using the cotyledonary-node method. Inclusion of thiol-containing compounds during co-cultivation significantly increased *A. tumefaciens* infection and T-DNA transfer into cotyledonary-node cells *(14, 15)*. These compounds increased transient GUS expression in a range of cultivars using a range of *A. tumefaciens* strains containing binary plasmids. Production of transgenic plants was increased from 0.9% transformation efficiency (independent Southern-confirmed plants produced per cotyledonary-node explant) for non-thiol treated explants followed by PPT selection to a transformation efficiency of 2.1% for explants treated with L-cysteine *(14)*. Thiol compounds appear to improve T-DNA delivery by inhibiting the activity of plant pathogen- and wound-response enzymes, such as peroxidases and polyphenol oxidases *(15)*. In our work with PPT-selection we noted production of a large proportion of non-transgenic selection escapes. Selection efficacy can be improved by adjusting the glufosinate selection regime for the genotype under investigation *(16)*. However, we found that tighter selection of L-cysteine-treated explants afforded by hygromycin B-containing media increased production of transgenic plants to 7.7% *(13)*. We also observed that co-cultivation of explants on medium containing mixtures of L-cysteine, sodium thiosulphate, and DTT combined with hygromycin B selection further increased transformation efficiency to an average of 16.4%. This improved transformation method has been used to produce transgenic plants with multiple *A. tumefaciens* strains and binary plasmids. However, the cotyledonary-node method still does not meet all characteristics of an ideal transformation system. Although we have been able to produce transgenics from different soybean genotypes, doing so requires empirical adjustments of selection regime and other system components. Based on these attributes and remaining limitations of the cotyledonary-node transformation system, we believe that more research to improve the method is warranted. Therefore, our focus on the cotyledonary-node transformation method in this chapter is intended to provide researchers with a foundation for further improving the method.

2. MATERIALS

2.1. A. tumefaciens *Strains*

A variety of *A. tumefaciens* strains have been used for transformation using the cotyledonary-node method including C58, AGL1, EHA101, EHA105, and LBA4404

(12, 13, 15, 17). Transient and stable GUS expression in our laboratory has been the greatest using strains EHA101 and EHA105, however, an equal number of transformants have been recovered using LBA4404. In our experience, C58 was not as good as the other above mentioned strains at T-DNA delivery into the particular soybean cultivars we used. Binary vectors containing a *gus*A gene should be included when establishing the system to monitor both transient and stable GUS expression throughout the regeneration process (*see* Note 1).

2.2. Plant Materials

Mature dried soybean seed that were grown in a disease free environment are used for explant preparation. We use the cultivar Bert, which is a maturity group III genotype and is available from the Minnesota Crop Improvement Association. Maturity groups are assigned according to the latitude that the cultivar is grown *(18)*.

2.3. Media Stocks

The stock solutions used in this protocol should be prepared every month to ensure stability of each solution. All stocks are made with double distilled water (ddH$_2$O) unless noted:

1. B5 major salts (10X stock). Store at 4°C.
a.	0.25 M KNO$_3$	25 g/l
b.	0.01 M CaCl$_2$.2H$_2$O	1.5 g/l
c.	0.01 M MgSO$_4$.7H$_2$O	2.5 g/l
d.	0.01 M (NH$_4$)$_2$SO$_4$	1.34 g/l
e.	0.01 M NaH$_2$PO$_4$.H$_2$O	1.5 g/l

2. B5 minor salts (100X stock). Store at 4°C.
a.	5 mM H$_3$BO$_3$	0.3 g/l
b.	10 mM MnSO$_4$.H$_2$O	0.76 g/l
c.	0.7 mM ZnSO$_4$.7H$_2$O	0.2 g/l
d.	0.45 mM KI	75 mg/l
e.	0.1 mM Na$_2$MoO$_4$.2H$_2$O	25 mg/l
f.	0.01 mM CuSO$_4$.5H$_2$O	2.5 mg/l
g.	0.01 mM CoCl$_2$.6H$_2$O	2.5 mg/l

3. B5 vitamins (100X stock). Store at 4°C.
a.	0.055 M myo-inositol	10 g/l
b.	0.8 mM nicotinic acid	0.1 g/l
c.	0.5 mM pyridoxine-HCl	0.1 g/l
d.	3 mM thiamine-HCl	1 g/l

4. MS major salts (10X stock). Store at 4°C.
   a. 0.2 M NH_4NO_3 16.5 g/l
   b. 0.2 M KNO_3 19 g/l
   c. 30 mM $CaCl_2 . 2H_2O$ 4.4 g/l
   d. 15 mM $MgSO_4 . 7H_2O$ 3.7 g/l
   e. 12.5 mM KH_2PO_4 1.7 g/l

5. MS minor salts (100X stock). Store at 4°C.
   a. 10 mM H_3BO_3 0.62 g/l
   b. 13 mM $MnSO_4 . H_2O$ 2.23 g/l
   c. 3 mM $ZnSO_4 . 7H_2O$ 0.86 g/l
   d. 0.5 mM KI   83 mg/l
   e. 0.1 mM $Na_2MoO_4 . 2H_2O$ 25 mg/l
   f. 0.01 mM $CuSO_4 . 5H_2O$ 2.5 mg/l
   g. 0.01 mM $CoCl_2 . 6H_2O$ 2.5 mg/l

6. MSIII Iron (100X stock). Store at 4°C. (*see* Note 2)
   a. 10 mM $FeSO_4 . 7H_2O$ 2.78 g/l
   b. 10 mM Na_2EDTA 3.72 g/l

7. IAA (Sigma): 1 mg/ml stock solution.
   Dissolve 20 mg in 20 ml 95% ethanol. Make 1 ml aliquots and store at -20°C.
8. Gibberellic acid (GA_3, Sigma): 1 mg/ml stock solution. Dissolve 50 mg GA_3 in 50 ml 70% ethanol. Store at 4°C.
9. BAP (Sigma): 1 mg/ml stock solution. Dissolve 100 mg BAP in 0.2N HCl; bring solution up to 100 ml with ddH_2O. Store at 4°C.
10. *trans*-zeatin riboside (ZR): 1 mg/ml stock solution. Dissolve 50 mg ZR in 0.2N HCl; bring solution up to 50 ml with ddH_2O. Store at -20°C.
11. IBA (Sigma): 1 mg/ml stock solution. Dissolve 20 mg IBA in 20ml 95% ethanol. Filter-sterilize. Store at –20°C.
12. Silver nitrate ($AgNO_3$, Sigma): 5 mg/ml stock solution. Dissolve 200 mg $AgNO_3$ in 40 ml ddH_2O. Filter-sterilize. Store at 4°C.
13. Hygromycin B (Boehringer Mannheim, Mannheim, Germany): 50 mg/ml stock solution. Dissolve 1g hygromycin B in 20 ml ddH_2O. Filter-sterilize. Store at -20°C.

2.4. Culture Media for A. tumefaciens (see Note 3)

1. YEP medium (semi-solid and liquid) per litre:
   a. To ddH_2O: 10 g Bacto-peptone (Difco; Becton, Dickinson, and Co., Cockeysville, MD, USA), 5 g Yeast-extract (Difco), 5 g NaCl, 12 g Granulated agar (Difco) solid only; pH 7.0.
   b. Autoclave.
   c. Add appropriate antibiotics for bacterial selection when cool (e.g. 1 ml of 50 mg/l hygromycin B). Pour medium into 15 x 100 mm Petri dishes (approx. 40 plates per litre medium).

2. Co-cultivation medium CCM (liquid) per litre:
   a. To ddH$_2$O: 10 ml 10X B5 major salts, 1ml 100X B5 minor salts, 1 ml 100X MSIII iron stock, 30 g sucrose, 3.9 g MES, pH 5.4.
   b. Autoclave.
   c. Dissolve 40 mg acetosyringone (AS) in 5 ml 95% ethanol then add: 10 ml 100X B5 vitamins, 0.25 µl GA$_3$, 1.67 ml BAP. Filter-sterilize mixture with 25 mm syringe filter with a 0.2 µM size pore Acrodisc® filter.
   d. Add mixture to cooled medium.

2.5. Soybean Tissue Culture Media

1. Germination medium GM (semi-solid) per litre:
   a. To ddH$_2$O: 100 ml 10X B5 major salts, 10 ml 100X B5 minor salts, 10 ml 100X MSIII iron stock, 20 g sucrose, 10 ml 100X B5 vitamins, 8 g Noble Agar (Difco); pH 5.8.
   b. Autoclave.
   c. Pour medium into 25 x 100 mm Petri dishes (approx. 18 plates per litre medium).

2. Co-cultivation medium CCM (semi-solid) per litre:
   a. To ddH$_2$O: 10 ml 10X B5 major salts, 1 ml 100X B5 minor salts, 1 ml 100X MSIII iron stock, 30 g sucrose, 3.9 g MES, 5 g Noble Agar; pH 5.4.
   b. Autoclave.
   c. Dissolve 40 mg AS in 5 ml 95% ethanol then add: 10 ml 100X B5 vitamins, 0.25 µl GA$_3$, 1.67 ml BAP. Filter-sterilize mixture with a 25 mm syringe filter with a 0.2 µM size pore Acrodisc® filter.
   d. Dissolve thiol compounds (8.8 mM or 1 g L-cysteine, Sigma), 1 mM or 154.2 mg DTT (Fisher Scientific, Fair Lawn, NJ, USA), or 1 mM (245 mg) sodium thiosulphate (Mallinckrodt, Paris, KY, USA) into separate tubes with 5 to 10 ml ddH$_2$O. Filter-sterilize each solution with a 25 mm syringe filter with a 0.2 µM size pore Acrodisc® filter (see Note 4).
   e. Add mixture and thiol compounds to cooled medium. Pour medium into 15 x 100 mm Petri dishes (approx. 40 plates per litre medium).
   f. After medium solidifies, place a single sterile Whatman #1 (70 mm) filter paper (Whatman International Ltd, Maidstone, England) on top of each plate.

3. Shoot induction medium SI (liquid and semi-solid) per litre:
   a. To ddH$_2$O: 100 ml 10X B5 major salts, 10 ml 100X B5 minor salts, 10 ml 100X MSIII iron stock, 30 g sucrose, 0.59 g MES, 8 g Noble Agar semi-solid only; pH 5.6.
   b. Autoclave.
   c. Mix together 10 ml 100X B5 vitamins, 1.67 ml BAP, 500 mg ticarcillin (SmithKline Beecham Pharmaceuticals, Philadelphia, PA, USA), 100 mg cefotaxime (Hoechst-Roussel Pharmaceuticals Incorporated, Somerville,

NJ, USA). Filter-sterilize mixture with a 25 mm syringe filter with a 0.2 μM size pore Acrodisc® filter.

d. Add selective agent when appropriate (e.g. 5 mg/l hygromycin B).

e. Optional: Add 4 ml AgNO₃ (*see* Note 5).

f. Add the mixture to the cooled medium. Pour solid medium into 25 x 100 mm Petri dishes (approx. 16 plates per litre medium).

4. Shoot elongation medium SE (semi-solid) per litre:
   a. To ddH₂O: 100 ml 10X MS major salts, 10 ml 100X MS minor salts, 10 ml 100X MSIII iron stock, 30 g sucrose, 0.59 g MES, 8 g Noble Agar; pH 5.6.
   b. Autoclave.
   c. Mix together 10 ml 100X B5 vitamins, 50 mg L-asparagine, 100 mg L-pyroglutamic acid, 0.1 ml IAA, 0.5 ml GA₃, 1 ml ZR, 500 mg ticarcillin, 100 mg cefotaxime. Filter-sterilize mixture with a 25 mm syringe filter and a 0.2 μM size pore.
   d. Add selective agent when appropriate (e.g. 10 mg/l hygromycin B).
   e. Optional: Add 4 ml AgNO₃.
   f. Add the mixture to the cooled medium. Pour solid medium into 25 x 100 mm Petri dishes (approx. 16 plants per litre medium).

5. Rooting medium RM (semi-solid) per litre:
   a. To ddH₂O: 50 ml 10X B5 major salts, 5 ml 100X B5 minor salts, 10 ml 100X MSIII iron stock, 20 g sucrose, 0.59 g MES, 8 g Noble Agar; pH 5.6.
   b. Autoclave.
   c. Add 1 ml IBA to the cooled medium. Pour semi-solid medium into sterile 25 x 100 mm PYREX culture tubes (Corning Inc., New York, NY, USA) (*see* Note 6).

3. METHODS

3.1. Seed Sterilization and Seedling Growth

1. Seeds are sterilized by exposure to chlorine gas. Layer seeds in a 15 x 100 mm plastic Petri dish and place into a glass desiccator with a tightly fitting lid. Place a 250 ml beaker with 100 ml bleach (5.25% sodium hypochlorite) into the chamber and slowly add 3.5 ml 12N HCl. Place the lid on the desiccator and sterilize the seeds for at least 24 h. For extremely contaminated seeds, repeat this procedure. Seeds can be stored until use.

2. Germinate ca. 16 seedlings on a single plate containing GM under fluorescent lighting (90-150 μmol/m²/sec) in a 18 h photoperiod and grow until the cotyledons turn green but before the first true leaves grow completely out of the cotyledon (*see* Note 7). The seedlings can be used immediately for transformation or stored at 4°C overnight or until the *A. tumefaciens* is ready for

inoculation. We have found that seedlings are susceptible for transformation even after 5 days at 4°C.

3.2. A. tumefaciens *Preparation*

1. Streak *A. tumefaciens* carrying the desired binary vector (from a permanent glycerol stock) onto semi-solid YEP growth medium and incubate for about 2 days at 25°C or until colonies appear. Depending on the selectable marker genes present on the Ti plasmid, the binary vector, and the bacterial chromosomes, different selectable agents will be used for *A. tumefaciens* selection in the YEP semi-solid and liquid media (*see* Note 8). As long as there are no other *hpt* genes residing on the Ti plasmid or bacterial chromosome, 50 mg/l hygromycin can be used to select for *A. tumefaciens* carrying a binary vector containing the *hpt* gene.

2. Pick a single colony with a sterile toothpick and place the toothpick in 50 ml of liquid YEP medium with antibiotics and shake on a rotating table (175 rpm) at 25°C for approx. 2 days. After an OD_{600} between 0.8-1.0 is reached, make a 30 ml 15% glycerol stock with the broth, aliquot 1 ml of stock into 30 x 1.5 ml Eppendorf tubes, and store at -80°C (*see* Note 9).

3. One day before explant inoculation, add 2-3 Eppendorf tubes of *A. tumefaciens* glycerol stocks plus appropriate antibiotics (200 μl Hygromycin B stock) to 200 ml YEP liquid medium in a 500 ml Erlenmeyer flask. Shake the flask overnight at 25°C as stated above until an OD_{600} between 0.8-1.0 is reached. We find that *A. tumefaciens* infection is optimal at these concentrations.

4. Before cutting and inoculating soybean explants, divide the broth into 50 ml aliquots and pellet *A. tumefaciens* by centrifugation for 10 min at 5,500 rpm using a Sorvall JA-12 rotor at 20°C. Resuspend the pellet in 25 ml liquid CCM and place at room temperature at least 30 min before use.

3.3. *Explant Preparation and Inoculation*

1. After the bacterial pellet is resuspended in 25 ml liquid CCM begin cutting the soybean seedlings for transformation. For each seedling, the roots and the majority of the hypocotyls are removed approx. 3-5 mm below the cotyledonary-node by cutting with a number 15 Personna Plus surgeon's blade (American Safety Razor Company, Staunton, Virginia, USA). Two explants are obtained by separating the cotyledons and cutting vertically through the hypocotyl region. This will result in two equal halves each containing a cotyledon, half of the hypocotyl, and half of the epicotyl tissue. Remove the epicotyl at the base of the cotyledonary-node by a single cut and remove all preformed axillary shoots (easily identified as 'hairy' vegetative growth). All preformed shoots must be removed to terminate apical dominance thereby inducing *de novo* growth from the axillary meristems. Finally, wound the axillary meristems and the cotyledonary-node by cutting approx. 10 times with the blade perpendicular to the hypocotyl. When wounding, cut deep enough to

access the meristematic tissue, however, not too much that the tissue is actually removed (*see* Note 10). An example of an explant ready for *A. tumefaciens* inoculation is given (Fig. 1A).

2. Transfer each explant to a Petri dish containing 25 ml of CCM/*A. tumefaciens* suspension for approx. 30 min. Explants can be either moved to the suspension after each wounding or after a group of explants are wounded. After inoculation, randomly place 5 explants with the adaxial or wounded side down on the Whatman paper overlaying semi-solid CCM (Fig. 1B). This filter paper prevents *A. tumefaciens* overgrowth on the soybean explants. Wrap 5 plates with Parafilm "M" (American National Can, Chicago, Illinois, USA) and incubate for 5 days in the dark at 25°C (*see* Note 11).

3.4. Selection and Plant Regeneration

1. After 5 days incubation, remove excess *A. tumefaciens* by briefly immersing the explants into liquid SIM (*see* Note 12). Remove the explants from the liquid SIM and carefully imbed 5 explants into a plate of semi-solid SIM not containing hygromycin B (*see* Note 13). Place the explant into the medium so that it is perpendicular to the surface of the medium with the hypocotyl and wounded cotyledonary-node tissue under the medium and most of the cotyledon out of the medium. Wrap plates with Scotch 394 venting tape (3M, St. Paul, Minnesota, USA) and place in a growth chamber for 2 weeks with a temperature averaging 25°C under 18 h photoperiod at 90-150 $\mu E/m^2/sec$ (*see* Note 14).

2. After 2 weeks, *de novo* shoot growth begins to appear at the cotyledonary-node on the explant. These shoots are new growth from the axillary meristem cells. At this time, sub-culture the explants onto new SIM with 5 mg/l hygromycin B after carefully removing the hypocotyl. Also remove very long shoots at this point, since they are preformed shoots that are not removed during the time of explant preparation. Seal the plates and incubate in the growth chamber for an additional 14 days.

3. After 2 weeks on SIM, there is considerable growth of shoots at the node and an enlargement of the hypocotyls into a hard callus tissue (Fig. 1C). Remove the cotyledon and any dead tissue and place into fresh SEM containing 10 mg/l hygromycin B. Seal the plates and incubate in growth chamber for an additional 14 days (*see* Notes 15 and 16).

4. After every 2-3 weeks, transfer the explants to fresh SEM medium containing 10 mg/l hygromycin B after carefully removing dead tissue (*see* Note 17). The explants should hold together and not fragment into pieces and retain somewhat healthy sectors up until 8 weeks after co-cultivation (4 weeks on SEM; Figs. 1D and E). After this time, it is normal that many shoots are lost and some explants will turn a caramel brown colour, however, there should also be some explants with healthy vigorous shoot growth and shoot elongation at this time (Figs. 1F and G). To increase shoot production in culture, place the developing shoot under the medium to induce axillary shoot growth at the nodes.

5. Once healthy, vigorous shoots have at least 3 sets of leaves and are preferably over 4 cm in length, remove the shoot and transfer to RM (Fig. 1H). In a typical experiment, at least 90% of the shoots form roots at the cut site after 5-14 days on RM. It is common that roots also form while still in SEM. When this happens, transfer to RM for several days before transferring to the glasshouse.

6. Transfer rooted shoots directly to the glasshouse. Plants are potted into 1-2 gallon pots with steamed soil mixed with standard vermiculite and perlite in a 2:1:1 ratio. To reduce moisture loss and to protect the shoot from the surrounding environment, cover with a clear plastic container and remove when the first set of new leaves open. When the first set of trifoliate leaves open, leaf samples are taken, genomic DNA extracted, and Southern blotting performed to determine T-DNA integration and copy number in T0 plants. In addition, if the gusA gene was present on the binary plasmid, GUS staining was also performed on fertile T0 plants (Fig. 1I) and on T1 seeds (Fig. 1J) for segregation analysis.

7. T1 seeds are planted into 48-cell flats or 1-2 gallon pots with the above soil mixture. T1 transgenic plants are analyzed using GUS staining and/or Southern analysis. Soybean T1 genomic DNA is extracted from 200 mg of leaf tissue harvested from the second trifoliate using a 96-well plate genomic DNA extraction protocol developed by Christenson et al. (19). Desirable plants in the flats can be transplanted into 1-2 gallon pots after analysis to increase T2 seed production.

4. NOTES

1. This protocol has been optimized for hygromycin B selection with the hpt gene and the soybean cultivar Bert. For any other selectable marker genes, the levels of selection and strategy of selection pressure can be found in other literature [nptII, (10); bar, (12, 16); cp4, (17)]. Of the four selectable marker genes used in the cotyledonary-node method, selection using the nptII gene (kanamycin selection) is the least stringent. If using other cultivars, the hygromycin B selection may need to be increased or decreased for optimal selection.

2. Prepare the iron stock by first dissolving the iron sulphate in 200 ml ddH$_2$0. In a separate container, boil EDTA for 2 min in 200 ml ddH$_2$0 and then slowly add into the iron sulphate solution. Allow the mixture to cool to room temperature before bringing the final volume to 1 litre.

3. When establishing the cotyledonary-node method, follow the protocol as closely as possible. Substituting components or conditions may drastically alter the Agrobacterium-plant interaction or morphology of the developing shoots and lead to a reduction in transformation efficiency. One such example is substituting agarose for purified agar.

4. In Olhoft et al. (13), we found that the addition of all three compounds in the CCM resulted in the highest transformation efficiency over the addition of only one or two of the thiol compounds using the cotyledonary-node explant. For other explant tissues or plant species, both concentrations and combinations of thiol compounds in the CCM should be tested for optimal transformation efficiencies.

5. The use of AgNO$_3$ in the shoot induction and shoot elongation medium results in a varied morphological response in shoot production from experiment to experiment. Selection for hygromycin-resistant tissues is not as stringent in the first four weeks for explants treated with AgNO$_3$. However, this does not always lead to more transgenic shoots in the long run. However, we recently discovered that silver nitrate may influence the number of T-DNA copies per transformant as well as integration of non-T-DNA binary plasmid sequences (Olhoft et al., in prep.).

6. In this protocol, hygromycin B selection is stringent enough to prevent untransformed shoots elongating at the rooting stage (SE medium). Therefore, we have opted to omit hygromycin B. In addition, the antibiotics have been eliminated from the medium since they can inhibit root formation in soybean (T.E. Clemente, pers. comm.).

7. The actual length of time on GM is dependent on numerous factors including genotype, health of the seed and light intensity. Normally, it takes from 3 days for small seeded cultivars to 5-7 days for large seeded cultivars.

8. Unless a minimal nutrient medium is required (e.g. AB medium) for bacterial growth, YEP is used for both semi-solid and liquid medium.

9. It is common for some A. tumefaciens strains like LBA4404 to grow in clumps in the liquid medium. This will not affect transformation.

10. It is best to dissect the soybean explants under magnification. The target cells are located in the tissue between the hypocotyl and the cotyledon, which can be hard to reach using a scalpel. However, on a good explant, one can see circular tissue growth around the target tissue in some explants.

11. We find that infection is optimal after 4-5 days incubation rather than 1-3 days. A. tumefaciens overgrowth is not a problem with long incubations using this explant tissue.

12. At this time, it is helpful to sample the explants for transient GUS expression if the gusA gene was on the binary vector used for transformation. Although not accurate, this test helps to determine if there were any overall problems with transformation (14, 15). In a good transformation experiment, 80-100% of explants will have transient GUS expression on the hypocotyl, cotyledonary-node region, and/or cotyledons.

13. Explants at this stage in development are too sensitive to hygromycin B and will die if exposed to even low concentrations of hygromycin B. Therefore, the concentration of hygromycin B is gradually increased to reduce initial cell death throughout shoot induction to shoot elongation. For this protocol, the concentrations reported are for the soybean cultivar Bert. If using other cultivars, the concentration of hygromycin B may need to be increased or decreased throughout shoot initiation and elongation. Nevertheless, for the first two weeks on SIM no selection is recommended for all cultivars.

14. Daylength is very important. For Northern cultivars (e.g. Groups 0, 1), including Bert, a long-day length is required to prevent flowering and to induce vegetative growth e.g. 16-20 h. The daylength can be shortened only for those soybean cultivars belonging to higher (Southern) maturity groups.

15. The frequency of regeneration can be calculated by: [(number of explants with shoots/total number of explants inoculated) x 100]. We find a good transformation experiment with the Bert genotype gives us 80-100% regeneration of shoots. If the frequency is lower, there are usually problems with A. tumefaciens overgrowth or poor transformation. The regeneration rate is also cultivar dependent, therefore, one should test the regeneration capacity of a prospective cultivar before beginning a transformation experiment.

16. It is helpful to sample the explants for stable GUS expression after 4 weeks on shoot induction if the gusA gene is on the binary vector during transformation (14). We routinely sample explants from transformation experiments and observe for GUS expressing shoots and callus tissue. In a typical experiment, approx. 20% of the explants will have large sectors of transformed shoots with significant GUS staining in the callus tissue.

17. Using thiol compounds in the co-cultivation medium leads to an increase of transgenic cells on any explant surface that was initially cut before inoculation. When using other methods for selection, such as PPT or glufosinate selection, it may be necessary to increase the concentration of selective agent used over explants not treated with thiol compounds during co-cultivation to effectively kill the non-transgenic cells in the callus/shoot mass (16).

CONCLUSION

The cotyledonary-node transformation method can be used to routinely produce fertile transgenic soybean plants with normal Mendelian inheritance of the integrated T-DNA. On average, two independently segregating loci are recovered per transgenic plant where one locus is a single non-rearranged locus without incorporation of binary backbone sequences (Olhoft *et al.* in prep.). Recent improvements using thiol compounds to increase the susceptibility of the plant to *Agrobacterium* infection *(14, 15)* coupled with stringent selection using hygromycin resistance *(13)* has greatly improved transformation efficiency in our laboratory. The cotyledonary-node method presented in this chapter provides a framework for further improving soybean transformation for applications to basic and applied research goals.

REFERENCES

1.   Buhr T, Sato S, Ebrahim F, Xing A, Zhou Y, Mathiesen M, Schweiger B, Kinney A *et al.* (2002). Ribozyme termination of RNA transcripts down-regulate seed fatty acid genes in transgenic soybean. *The Plant Journal*, **30**: 155-163.

2.   Yu O, Shi J, Hession AO, Maxwell CA, McGonigle B and Odell JT (2003). Metabolic engineering to increase isoflavone biosynthesis in soybean seed. *Phytochemistry*, **63**: 753-763.

3.   Padgette SR, Kolacz KH, Delannay X, Re DB, LaVallee BJ, Tinius CN, Rhodes WK, Otero YI *et al.* (1995). Development, identification, and characterization of a glyphosate-tolerant soybean line. *Crop Science*, **35**:1451-1461.

4.   McCabe DE, Swain WF, Martinell BJ.and Christou P (1988). Stable transformation of soybean (*Glycine max*) by particle acceleration. *Bio/Technology*, **6**: 923-926.

5.   Finer JJ and McMullen MD (1991). Transformation of soybean via particle bombardment of embryogenic suspension culture tissue. *In Vitro Cellular and Developmental Biology-Plant*, **27**: 175-182.

6.   Parrott WA, Hoffman LM, Hildebrand DF, Williams EG and Collins GB (1989). Recovery of primary transformants of soybean. *Plant Cell Reports*, **7**: 615-617.

7.   Yan B, Srinivasa Reddy MS, Collins GB and Dinkins RD (2000). *Agrobacterium tumefaciens*-mediated transformation of soybean [*Glycine max* (L.) Merrill.] using immature zygotic cotyledon explants. *Plant Cell Reports*, **19**: 1090-1097.

8.   Ko T-S, Lee S, Krasnyanski S and Korban S (2003). Two critical factors are required for efficient transformation of multiple soybean cultivars: *Agrobacterium* strain and orientation of immature cotyledonary explant. *Theoretical and Applied Genetics*, **107**: 439-447.

9.   Trick HN and Finer JJ (1998). Sonication-assisted *Agrobacterium*-mediated transformation of soybean [(*Glycine max* (L.) Merrill] embryogenic suspension culture tissue. *Plant Cell Reports*, **17**: 482-488.

10.   Hinchee MAW, Connor-Ward DV, Newel CA, McDonnell RE, Sato SJ, Gasser CS, Fischhoff DA, Re DB *et al.* (1988). Production of transgenic soybean plants using *Agrobacterium*-mediated DNA transfer. *Bio/Technology*, **6**: 915-922.

11.   Trick HN, Dinkins RD, Santarem ER, Di R, Samoylov V, Meurer C, Walker D, Parrott WA *et al.* (1997). Recent advances in soybean transformation. *Plant Tissue Culture and Biotechnology*, **3**: 9-26.

12.   Zhang Z, Xing A, Staswick P and Clemente T (1999). The use of glufosinate as a selective agent in *Agrobacterium*-mediated transformation of soybean. *Plant Cell, Tissue and Organ Culture*, **56**: 37-46.

13.   Olhoft PM, Flagel LE, Donovan CM and Somers DA (2003). Efficient soybean transformation using hygromycin B selection in the cotyledonary-node method. *Planta*, **216**: 723-735.

14.   Olhoft PM and Somers DA (2001). L-Cysteine increases *Agrobacterium*-mediated T-DNA delivery into soybean cotyledonary-node cells. *Plant Cell Reports*, **20**: 706-711.

15.   Olhoft PM, Lin K, Galbraith J, Nielsen NC and Somers DA (2001). The role of thiol compounds in increasing *Agrobacterium*-mediated transformation of soybean cotyledonary-node cells. *Plant Cell Reports*, **20**: 731-737.

16.   Zeng P, Vadnais DA, Zhang Z and Polacco JC (2004). Refined glufosinate selection in *Agrobacterium*-mediated transformation of soybean [*Glycine max* (L.) Merrill]. *Plant Cell Reports*, **22**: 478-482.

17.   Clemente TE, LaVallee BJ, Howe AR, Conner-Ward D, Rozman RJ, Hunter PE, Broyles DL,
       Kasten DS *et al.* (2000). Progeny analysis of glyphosate selected transgenic soybeans derived
       from *Agrobacterium*-mediated transformation. *Crop Science*, **40**: 797-803.
18.   Palmer RG and Kilen TC (1987). Qualitative genetics and cytogenetics. In: Wilcox JR (ed.),
       Soybeans: Improvement, Production and Uses, (2nd edn.). ASA-CSSA-SSSA, Madison, WI,
       USA pp. 135-209.
19.   Christenson JR, Flagel LE, Gustus CD, Olhoft PM, Matthews PD and Somers DA (2003). Cheap
       but not dirty! High throughput, 96-well-uniplate format for extraction of plant genomic   DNA
       - produces DNA suitable for restriction digestion, Southern analysis and PCR.
       http://bioplasticscollaborative.coafes.umn.edu/uniplate_DNA_extraction.htm#.

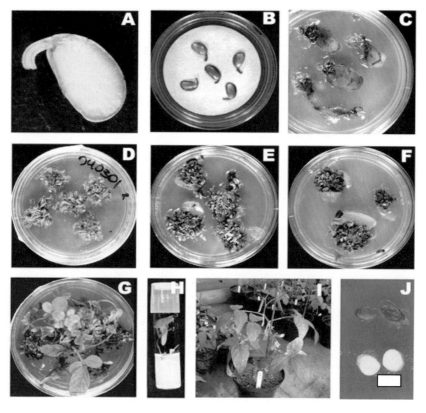

Figure 1. A modified cotyledonary-node method using hygromycin B for selection of transformed cells. Explants were prepared from 5 day-old seedlings by removing the roots and the majority of the hypocotyl from the cotyledons and wounding the axillary meristematic tissue at the cotyledonary-node (A; bar = 0.4 cm). Explants were inoculated with A. tumefaciens *and co-cultivated in co-cultivation medium containing thiol compounds (B; bar = 2.5 cm). After 5 days, the cotyledonary-node and hypocotyl of the explant were embedded into semi-solid shoot induction medium (SIM) to stimulate* de novo *shoot formation from the wounded axillary meristematic tissue (C; bar = 1.8 cm). Explants were cultured on SIM for 28 days, the first 14 days without hygromycin and the second 14 days with 5 mg/l hygromycin (D; bar = 2.5 cm). Significant death of non-transformed shoots and callus tissue was observed after hygromycin levels were raised to 10 mg/l in shoot elongation medium (SEM) for 28 days (E; bar = 2.5 cm). Two months after co-cultivation, explants were maintained on SEM with a maximum of 10 mg/l hygromycin B. Examples of explants undergoing selection of transformed shoots during this time are shown: shoot elongation on an explant 3 months after co-cultivation (F; bar = 2.5 cm) and shoot proliferation 4 months after co-cultivation (G; bar = 2.5 cm). After shoots elongated to at least 4 cm in length, they were placed in rooting medium (H; bar = 2.5 cm) and rooted shoots were directly transferred to a glasshouse and grown to maturity (I). A sample of T1 seeds from each T0 plant was stained for GUS expression; an example (J; bar = 1 cm) of a cross-section of a GUS-positive seed alongside a GUS-negative seed is shown.*

Chapter 25

IN VITRO REGENERATION AND TRANSFORMATION OF VICIA FABA

T. PICKARDT, P. BÖTTINGER AND A. DE KATHEN

Institute for Applied Genetics, Department of Biology/Chemistry/Pharmacy of the Free University of Berlin, 14195 Berlin, Germany. Corresponding author: T. Pickardt. E-mail: pickardt@zedat.fu-berlin.de

1. INTRODUCTION

1.1. Importance of Vicia faba

Vicia faba (faba-, field-, or broad bean) is by far the most economically important species of the genus *Vicia*. With a world production of 3.7 million tonnes in 2002 (*1*) *V. faba* ranks among the most important grain legume crops. The largest producer of faba beans in 2002 was China (1.6 million tonnes, MT), followed by Ethiopia (0.45 MT), Egypt (0.44 MT), France (0.28 MT) and Australia (0.2 MT). Although not comparable with soybean or common bean on a global production scale, this should not ignore its national and regional importance in countries like Egypt or the Magrheb region (Algeria, Morocco, Tunisia), where faba beans account for more than 50% of the area devoted to food legumes.

1.2. Problems in Faba Bean Production and Genetic Improvement

Several biotic and abiotic stressors can be identified, affecting yield and yield stability in faba beans. Chocolate spot (*Botrytis fabae*) is one of the most widespread and devastating diseases in all production areas, followed by Ascochyta blight (*Ascochyta fabae*). Viral diseases are more important in East Africa, whereas nematodes and the parasitic weed *Orobanche crenata* are important production constraints in West Asia and North Africa (*2, 3*). The limited cold tolerance and its susceptibility to drought stress rate first and second amongst the abiotic stresses affecting faba beans. A major bottleneck for the genetic improvement of faba beans is its genetic or reproductive isolation, presumably as a result of its domestication (*4*). The genus *Vicia* comprises of more than 150 species of annual and perennial herbaceous legumes distributed in temperate and subtropical areas of the world (*5*). Wild relatives of *V. faba* or related, compatible species are not yet known (*6*), limiting the accessible secondary and tertiary gene pool. Furthermore, interspecific hybridization within the *Leguminosae* appears to be quite difficult (*7, 8*) and successful generation of interspecific hybrids has only been documented in *Glycine* (*9*) and *Phaseolus* (*10*). Attempts to generate fertile hybrids between *V. faba* and other *Vicia* species, using tissue culture based embryo-rescue techniques, has been

337

I.S. Curtis (ed.), Transgenic Crops of the World - Essential Protocols, 337-350.

unsuccessful (*11*) as well as efforts to create interspecific hybrids between faba bean and *V. narbonensis* (*12, 13*), the latter being accessible to genetic engineering (*14*). Recent phylogenetic analysis based on RFLP, RAPD (*15*) and isozyme patterns (*16*) revealed that *V. faba* and *V. narbonensis* may need to be placed in different monophyletic branches, despite their morphological similarities.

Obviously, present breeding programmes of the partially allogamous *V. faba* could be supplemented by recombinant DNA technology. The incorporation of fungal resistance, nuclear encoded/artificial male sterility, or the generation of glufosinate resistant cultivars to control *Orobanche* infestation could significantly increase yield and yield stability. The modification of amino acid synthesis pathways and/or storage protein gene sequences could improve storage protein composition and nutritional value. In faba bean, techniques for *in vitro* culture and genetic manipulation are still lagging behind those for many other crop plants. However, the recently obtained results encourage to dedicate further research to a plant species which is known to be very difficult to manipulate *in vitro*.

1.3. History of Faba Bean Tissue Culture and Genetic Engineering

Although the generation of transgenic plants bypassing a tissue-culture phase has been described for a few plant species, including *Arabidopsis thaliana* (*17*) and the legume *Medicago truncatula* (*18*), these 'in planta' transformation systems do not appear to be feasible for large seeded legumes. Yet successful tissue culture appears to be the prerequisite for regenerating transgenic *V. faba*.

1.3.1. Callus and Suspension Culture

The first attempts to cultivate *V. faba in vitro* were focused on the optimal growth of callus tissue or suspension cultures rather than the induction of shoot morphogenesis and plant regeneration (*19-21*). Poor growth rates and an increase of necrotic tissue during cultivation were described and Röper (*22*) was the first to mention the low morphogenic potential of *V. faba* cells cultured *in vitro*.

1.3.2. Organogenesis from Meristems

During the last decade a number of reports were published describing the cultivation of tissues containing shoot apical meristems and the subsequent recovery of shoots (*23-30*). Excised apical meristems, nodal buds and cotyledonary nodes were exposed to media containing cytokinins (in most cases BAP) alone or in combination with low amounts of auxin. Under these conditions shoots preferentially develop from pre-existing meristems. Since cytokinins are effective in removing apical dominance (*31*), this treatment continuously promotes the development of young meristems of the apical dome(s) (*32*) to lateral buds and shoots, which are again themselves suppressed during further growth. This cycle of simultaneous induction and suppression finally gives rise to multiple bud/shoot proliferation. Similar protocols have repeatedly been described for soybean, chickpea, cowpea, peanut, common bean, lentil and pea (*33-36*).

Today, several transformation protocols in grain legumes are based on the BAP-induced shoot development from embryo axes and cotyledonary nodes, using *Agrobacterium tumefaciens* (*37, 38*) or the biolistic approach (*39-41*). The recovery of both clonal as well as chimeric primary transformants were repeatedly described in these studies (*42*), indicating the occurrence of single and multiple cell origin of shoots.

1.3.3. Somatic Embryogenesis

Griga *et al.* (*43*) were the first to report somatic embryogenesis in callus and suspension cultures derived from immature cotyledons of *V. faba*. Somatic embryogenesis was also described for *V. narbonensis* in two independent reports. In the study of Albrecht and Kohlenbach (*44*), somatic embryogenesis was induced on leaf-derived callus. Pickardt *et al.* (*45*) described a protocol in which fertile plants were recovered from somatic embryos developing in callus derived from shoot tips. In both protocols, somatic embryogenesis is achieved by the application of chlorinated auxins followed by a subsequent omission of auxins in the culture medium, as it has been described earlier by Ammirato (*46*).

An improved protocol was subsequently combined with the *A. tumefaciens*-mediated gene transfer system (*14*) and is one of the rare examples for a reliable and reproducible transformation protocol in grain legumes (*47-50*).

1.3.4. Protoplast Culture and Regeneration

Protoplast isolation and culture from leaves and shoot apices of *V. faba* was reported by Binding and Nehls (*51, 52*). Röper (*53*) isolated protoplasts from suspension cells. Both groups were able to initiate and maintain cell division and callus growth but attempts to regenerate plants were unsuccessful. In further experiments on somatic cell hybridization between *V. faba* and *Petunia hybrida*, Binding and Nehls (*52*) obtained three hybrid clones; one of them was propagated for at least 9 months. The fusion hybrids contained predominantly nuclei or chromosomes of one or the other species and a few chromosomes of the second parent.

During the next 14 years, no further reports on protoplast regeneration in *Vicia* species were published. Only Tegeder *et al.* (*54*) demonstrated the recovery of fertile, mature plants from protoplasts of *V. faba*. The apparent key step in this study was the application of the phenyl-urea herbicide, thidiazuron (*55*), in a subsequent culture phase on semi-solidified medium. With the exception of Thynn and Werner (*56*) this remained the only report on shoot formation from a tissue lacking axillary shoot meristems in faba bean.

The extension of these studies to *V. narbonensis* (*57*), showed that also in this species, plant regeneration from protoplast-derived calli via shoot morphogenesis could be achieved using the thidiazuron pathway. In addition, the protocol for the induction of somatic embryogenesis (originally developed for shoot tip-derived calli of *V. narbonensis*, *45*) was also successful. It is important to mention, that in *V. narbonensis,* regeneration via somatic embryogenesis occurs with a considerable higher efficiency compared to shoot morphogenesis induced by thidiazuron. Unfortunately, all attempts to induce somatic embryogenesis in *V. faba* were unsuccessful (Pickardt 1988 unpublish., *54*).

Despite the protocols for plant regeneration from protoplasts established for pea (*58, 59*), faba bean (*54*) and soybean (*60*), there remains no single report on the successful production of transgenic grain legumes following direct DNA transfer into protoplasts.

1.4. Genetic Engineering of Vicia faba

Schiemann and Eisenreich (*61*) published the first report on the transfer of chimeric genes into *V. faba.* Seedlings were inoculated with *A. rhizogenes* strains (A4 and 15834) containing the binary vector pGSGluc1 transferring *npt*II and *gus*A genes under the control of the bidirectional TR1/2-promoter. GUS positive roots developing at the inoculation sites were propagated on hormone-free medium. Callus established from these roots maintained GUS activity. Regeneration of shoots was not reported. Ramsay and Kumar (*62*) performed a similar study using *A. rhizogenes*/pBin19 (*63*) for inoculation of *V. faba* cotyledons and stem tissue. The transgenic nature of established root clones were confirmed by hormone autotrophy and NPTII dot blot assays. Attempts to initiate shoot morphogenesis in root cultures were not described.

Jelenic *et al.* (*64*) used three different faba bean cultivars and nine different *A. tumefaciens* and *A. rhizogenes* strains (including shooter and rooter mutants and transconjugants) in stem inoculation experiments in order to determine the ideal cultivar-strain combination. Regeneration was not attempted and *in vitro* transformation of stem segments failed.

In order to develop a transformation system for *V. faba,* we followed a procedure based on a protocol for *de novo* regeneration of shoot initials from dedifferentiated cells that was initially developed for plant regeneration from protoplasts (*54*), combined with the *Agrobacterium*-mediated gene transfer. Shoot morphogenesis was achieved from internodal stem segments after a callus phase and a shoot initiation phase, using thidiazuron as growth regulator. A subsequent transfer of shoot-buds to BAP-containing medium was found necessary for stem-elongation and leaf-development (see below).

The first transgenic line of *V. faba* transmitting *npt*II and *gus*A activity to the progeny in a Mendelian fashion, was presented by Böttinger *et al.* (*65*).

2. MATERIALS

2.1. Agrobacterium tumefaciens *Strains*

Disarmed hypervirulent succinamopine-type *Agrobacterium* strains EHA101 and EHA105 (*66, 67*) have been used harbouring various binary vectors. As plant selectable marker the plasmid vector T-DNAs used so far contain the kanamycin resistance gene *npt*II (driven by the nopaline synthase promotor or the bi-directional TR1/2 promotor, *68*) or the phosphinothricin resistance gene *bar* (*69*) under control of the nopaline synthase promotor (unpublish.). The binary plasmid pGSGluc1 used by Böttinger *et al.* (*65, 68*) contains the *uid*A gene as screenable marker, driven by the bi-directional TR1/2 promotor.

2.2. Culture Media for A. tumefaciens

1. *Agrobacterium* cells are grown in YEB medium: 5 g/l beef extract, 1 g/l yeast extract, 5 g/l peptone (all three ingredients by Difco/USA), 5 g/l sucrose (Merck/Germany), 0.492 g/l MgSO$_4$.7H$_2$O (Merck), pH 7.2, optional 10 g/l Agar (Merck).
2. Depending upon the binary vector used, the strains are selected on medium containing 10 mg/l tetracyclin (Sigma), 50 mg/l kanamycin (Duchefa) or 100 mg/l streptomycin (Sigma). Antibiotics are dissolved in water (10 mg/ml tetracyclin, 50 mg/ml kanamycin, 100 mg/ml streptomycin), sterilized by filtration through 0.2 µm membranes and kept as stock at -20°C.
3. *A. tumefaciens* strains are stored as glycerol stocks: aliquots of an overnight bacterial culture at 28°C were mixed with 1/3 volume sterile glycerol, frozen in liquid N$_2$ and kept at –70°C. Prior to inoculation of stem-explants (*see* Section 3.2.) the bacteria are streaked out on YEB agar plates containing the appropriate antibiotics and kept at 28°C for 48 h.

2.3. Plant Material

1. Transformation experiments are performed with the German cultivars Mythos and Albatross (both ssp. minor) obtained from Norddeutsche Pflanzenzucht/H.Lembke, Hohenlieth, Germany, www.npz.de. In earlier investigations, both cultivars were found amenable for *in vitro* culture (production of moderate amounts of polyphenolic compounds) and capable of *de novo* shoot regeneration from protoplasts (*54*, unpublish. results).
2. Seeds are surface-sterilized by immersion for 60 sec in 70% ethanol and for 5 min in sodium hypochlorite solution (4% active chlorine) containing a few drops of Tween 80, followed by 5 rinses with sterile tap water. After an additional soaking for 6–8 h in sterile tap water, seeds are germinated in the dark at 20°C on half-strength MS basal medium (lacking sucrose).
3. After approx. 10 days, internodal segments of the arising main shoot (Fig. 1a) are used for cocultivation with *A. tumefaciens* (*see* Section 3.2.). The remaining seedlings are kept in the dark, and secondary shoots arising from the cotyledonary bud during subsequent weeks are used as explants for further transformation experiments. Thus, on average, each seed provides 30-40 internodal segments.

2.4. Chemicals, Media, Vessels and Growth Conditions for Plant Tissue Culture

1. MS basal medium (Duchefa, *70*) is generally used, supplemented with 3% sucrose. The plant growth regulators IBA, NAA, 2,4-D and BAP are obtained from Duchefa. TDZ is acquired from Riedel-de-Haen. Plant growth regulators are used from 1 mg/ml stock solutions (0.2 μm filter-sterilized and then stored in 1 ml aliquots at –20°C). Antibiotics are obtained from Duchefa (kanamycin) and Smithkline-Beecham (Betabactyl, a combination of ticarcillin/potassium clavunate, identical to timentin). DL-Phosphinothricin (PPT) was acquired from Hoechst/Germany (now Aventis). Antibiotics and PPT are added post autoclaving at 55°C.

2. Media were made semi-solid by 0.3% Gelrite (Roth) and adjusted to pH 5.7 prior to autoclaving (15 min at 121°C). The following abbreviations apply for the different plant media:
   - TNZ: MS with 3% sucrose, 0.5 mg/ml 2,4-D, 0.5 mg/l NAA, 0.5 mg/l TDZ
   - TNZ1: MS with 3% sucrose, 0.5 mg/ml 2,4-D, 0.5 mg/l NAA, 0.5 mg/l TDZ, 500 mg/l Betabactyl
   - TNZ2K: MS with 3% sucrose, 0.5 mg/ml 2,4-D, 0.5 mg/l NAA, 0.5 mg/l TDZ, 300 g/l Betabactyl, 100 mg/l kanamycin (*see* Note 7)
   - TNZ2P: MS with 3% sucrose, 0.5 mg/ml 2,4-D, 0.5 mg/l NAA, 0.5 mg/l TDZ, 300 g/l Betabactyl, 2 mg/l PPT (*see* Note 8)
   - MTN: MS with 3% sucrose, 7.5 mg/l TDZ, 0.75 mg/l NAA
   - MB1: MS with 3% sucrose, 2 mg/l BAP, optional: + 25 mg/l G418 (*see* Note 7) or 2 mg/l PPT (*see* Note 8)
   - MSR: MS with 3% sucrose, 2 mg/l IBA

3. Culture vessels: plastic Petri dishes (9 cm) and 250 ml glass containers (diam. 8 cm, height 6 cm, covered with glass lids) are used. Petri dishes are generally sealed with Parafilm, while glass containers are not sealed. Petri dishes contain 25 ml medium, glass containers 40 ml.

4. Cultures are kept at 20°C under cool white fluorescent lights (80 μmol/m²/sec) under a 16 h photoperiod unless otherwise stated.

3. METHODS

3.1. General Advice

V. faba tissue is extremely sensitive to manipulation. The use of absolutely clean forceps and fresh blades (only briefly heated, *see* Note 1) is highly recommended for all steps. Blades should be exchanged frequently, e.g. do not cut more than 30 explants (*see* Section 3.2.3.) and do not perform more than 3 graftings (*see* Sections 3.4.1. and 3.4.2.) with a single blade. Blades lose their sharpness quickly. We prefer razor blades (which can be dismounted into 4 pieces simply by buckling), fixed in razor blade-holders, which can be purchased from World Precision Instruments (http://www.wpiinc.com).

3.2. Inoculation Procedure

1. *Agrobacterium* cells are taken from agar plates (*see* Section 2.2.3.) and grown in 50 ml YEB medium with appropriate antibiotics, 12 h at 28°C in 250 ml Erlenmeyer flasks in a rotary shaker at 180 rpm.
2. Bacteria are harvested by centrifugation (3000 rpm, 10 min) and resuspended in an equal volume of liquid TNZ medium. Twenty-five ml aliquots of the bacterial suspension are poured into sterile glass Petri dishes (diam. 10 cm).
3. Epicotyl segments are transversely cut in the bacterial suspension with sharp razor blades (*see* Note 1) into segments of 0.2-0.4 mm in length.
4. After 30 min, the bacterial suspension is removed by the use of a pipette. Inoculated segments are transferred to glass containers (covering the whole medium surface, i.e. approx. 100 explants per container) with semi-solidified TNZ medium and cocultivated in the dark for 48 h at 20°C.
5. After cocultivation, segments are transferred to 500 ml-capacity bottles (Schott) and thoroughly washed by repetitive rinsing with sterile tap water. For this purpose, explants remain in the bottle (only exchange tap water several times) in order to reduce mechanical manipulation.

3.3. Callus Development under Selection Pressure and Regeneration of Shoots

1. Explants are transferred to semi-solidified TNZ1 medium in glass containers (approx. 50 explants per container) and cultured for 4 days in the dark at 20°C.
2. Explants are subsequently transferred to plastic Petri dishes (15 explants per dish) containing TNZ2K or TNZ2P medium (depending on the selection marker). Explants are subcultured to fresh TNZ2K/P medium every 2 weeks for a period of 3-4 months and cultured in the dark at 20°C.
3. Kanamycin or phosphinothricin-resistant calli (*see* Note 3) of approx. 5 mm in diam. (Fig. 1B) are transferred to MTN medium contained in 250 ml glass containers and cultured at 20°C under cool white fluorescent lights (ca. 80 μmol/m^2/sec) under a 16 h photoperiod (*see* Note 4).
4. Every 3-4 weeks, calli are subcultured onto fresh medium for a period of more than 12 months (Fig. 1C; *see* Note 5). The appearance of shoot primordia varies between 4 and 12 months.
5. Shoot primordia (Fig. 1D) are carefully excised and then transferred to MB1 medium and subcultured every 3-4 weeks. To confirm the transgenic state of regenerated shoots, the MB1 medium can be supplemented with 25 mg/l G418 (*see* Note 7) or 2 mg/l PPT (*see* Note 8).
6. These cultures are maintained at 15°C with light conditions unchanged. Culturing explants at a lower temperature does not only reduce the formation of phenolic compounds as described previously for faba bean (*29*), but also enhances internode elongation of developing shoots and leaf development.

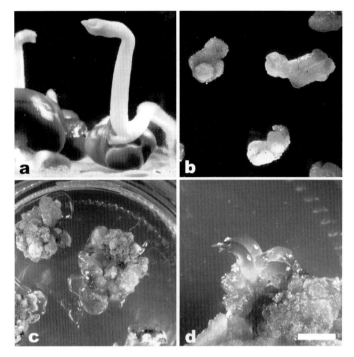

Figure 1. (a) Dark-grown seedlings used as explant source (bar = 1 cm), (b) callus development from stem explants under selection pressure (100 mg/l kanamycin) on TNZ2 medium (bar = 0.5 cm), (c) callus on MTN medium (bar = 1 cm), and (d) shoot-bud development on MTN medium (bar = 0.2 cm).

3.4. Rooting and Grafting

Developing shoots of more than 1 cm in length are rooted on MSR medium. Alternatively, plants can be recovered by grafting developing shoots onto 5-10 day old etiolated sterile seedlings (used also as explant source):

1. The seedlings are decapitated (1-1.5 cm above the cotyledonary node), the remaining stem is split by a longitudinal cut with a razor-blade approx. 0.5 cm deep.
2. The transgenic shoot is 'sharpened' by two minor cuts to form a wedge-shaped base, fitting into the prepared seedling. The shoot is inserted between the split stem of the seedling. The elasticity of the stem-halves provides sufficient support, and so taping is unnecessary. The grafted seedlings are kept on half-strength MS medium (lacking sucrose) in 250 ml-capacity glass containers in the light at 20°C. Avoid shaking the containers during the first few days. It is recommended to fix the grafted seedling by pressing into the medium.
3. All axillary shoots that arise from cotyledonary nodes of seedlings need to be removed during the following weeks.

3.5. Transfer to Soil and Hardening

1. Based on the appearance of good leaf development, rooted shoots and grafted seedlings are transferred to autoclaved soil. Prior to transfer to soil, remove all agar by rinsing the roots under running tap water.
2. Plantlets are transferred to pots (6 cm diam.) containing sterile soil and covered with a transparent plastic bag.
3. The two upper edges of the bag are cut off after 1-2 weeks respectively to allow hardening of plants. After another 4-6 weeks, plants are transferred to larger pots (20 cm diam.) and cultivated under glasshouse conditions.

4. NOTES

1. Faba bean contains one of the highest levels of phenolic compounds compared to many other legumes. It is also known that bacterial elicitors, wounding and UV light induce the production of protective phenolics in faba bean (*71, 72*). Although associated with antioxidant activity, phenolic compounds represent a serious constraint in *V. faba* tissue culture due to the deterioration of explant material and cultivated tissue. The following strategies limit phenolic production:
   - submerge explants when cutting for inoculation
   - use only fresh scalpels or razor blades
   - heat blades only briefly, since intense burning reduce sharpness
   - reduce light during early stages of culture
   - reduce temperature during later stages of culture
   - avoid any unnecessary mechanical manipulation or wounding of explants
   - select varieties low in phenolics since crossing within faba bean is not problematic
2. We found no positive effects on the use of acetosyringone or other elicitors in terms of improving T-DNA transfer. A longer duration of cocultivation may increase the transformation efficiency but this effect is outcompeted by the damage due to bacterial overgrowth.
3. Within 4 months of culture, 10–30% of explants (dependent from *A. tumefaciens*-strain/binary vector/selection marker) form callus on selection medium.
4. During the regeneration phase we omit selection agents, since they can disturb the regeneration process considerably.
5. Calli growing on MTN medium can reach a large size. In order to reduce the size of these calli, do not use blades or scalpels, but "break" the calli by using forceps.
6. Regeneration frequencies between 3-6% can be expected from kanamycin-resistant calli in transformation experiments, compared to ca. 10% for untransformed calli (percentage of callus lines producing shoots). However, almost 50% of regenerated shoots and plants showed morphological aberrations like dwarfism, no flowering, formation of abnormal flowers and pods, or failure to set seed. Determination of DNA indices by flow cytometry revealed a correlation between aberrant phenotypes and an index >1 (*68*). It is known that prolonged callus culture can result in genomic aberrations (*73*). In addition, transgenesis itself may contribute to an increase in cytogenetic aberrations in oat (*74*) and pea (*75*). However, in the case of faba bean, we attribute aberrant phenotypes mainly to the long culture period (up to 24 months).
7. In general, G418 (geneticin) is more efficient in selecting for transformed shoots compared to kanamycin. However, in the protocol presented here, G418 was found less efficient in selecting callus when cultured on TNZ2 medium. In contrast, during shoot development and propagation on MB1 medium, kanamycin is unsuitable as a selection agent and so G418 is used.
8. The use of the *bar* gene (*69*) as a plant selection marker reduces the number of callus lines capable of undergoing shoot formation. At present, we have failed to generate phosphinothricin-resistant stable lines.

9. DNA isolation from young leaf tissue is carried out either by CTAB extraction or by Qiagen DNeasy Maxiprep columns according to manufacturer's instructions. CTAB extraction followed the procedure of Doyle and Doyle (*76*), using a 50 mM Tris buffer containing 2% CTAB, 10 mM EDTA, 1% PVP and 4 M LiCl released sufficient DNA for PCR and non-radioactive Southern blot analysis, using random-primed DIG labelled (Roche) probes (*68*). Physical localization of transgenes was also carried out by fluorescence *in situ* hybridization (*77*).

CONCLUSIONS

Stably transformed lines of *V. faba* were repeatedly generated in independent transformation experiments, using different binary vectors in combination with *A. tumefaciens* strains EHA101 and EHA105. Although these results clearly demonstrate the reproducibility of the applied protocol, it is clear that the approach has certain disadvantages: a relatively low regeneration frequency, a long period to obtain seed producing primary transformants (16–24 months) and a high percentage of morphologically and cytogenetically aberrant regenerants.

Transformation protocols including a phase of unorganised callus growth and a *de novo* regeneration-step, need more time and are more prone to undesired somaclonal variation than systems based upon direct shoot organogenesis from 'complex' explants like embryo axes or cotyledonary nodes (*37, 38*). This suggests that the use of such an approach could be productive also for the transformation of *V. faba*. However, as already mentioned, a serious constraint in *V. faba* tissue culture is the deterioration of explant material and cultured tissues as a result of the action of phenolic compounds. Due to these compounds that are produced especially in young tissues, our attempts to apply direct shoot induction from 'complex' explants for transformation of *V. faba* have been unsuccessful. However, to pursue this approach, we have used genotypes that produce only minor amounts of phenolic compounds. This approach has resulted in some promising results in terms of producing transgenic plants of faba bean (Hanafy *et al.*, in prep.).

ACKNOWLEDGEMENTS

The authors wish to thank Otto Schieder, who passed away in May 1998, for giving us the opportunity to perform this work in his research group. In addition, we would like to thank Verena Schade for skilful technical assistance, Elizabeth Hood (now ProdiGene,USA), Gad Galili (Weizman Institute of Science/Rehovot, Israel) and Helmut Bäumlein (IPK-Gatersleben, Germany) for strains and vectors, as well as the "Norddeutsche Pflanzenzucht" for supplying *V. faba* seed material. The work was supported by the German Scientific Foundation (DFG), the German Ministry for Education and Science (BMBF), and a grant from the Free University of Berlin to P. Böttinger.

REFERENCES

1. FAO, FAOSTAT (2002). http://apps.fao.org/page/collections?subset=agriculture.
2. Hanounik SB, Jellis GJ and Hussein MM (1993). Screening for disease resistance in faba bean. In: Singh KB, Saxena MC (eds.), *Breeding for stress tolerance in cool-season food legumes*. John Wiley & Sons, Chichester, UK, pp. 97-106.
3. Robertson LD and Saxena MC (1993). Problems and prospects of stress resistance breeding in faba bean. In: Singh KB, Saxena MC (eds.), *Breeding for stress tolerance in cool-season food legumes*. John Wiley & Sons, Chichester, UK, pp. 37-50.
4. Vanraamsdonk V (1995). The effect of domestication on plant evolution. *Acta Botanica Neerlandica*, **44**: 421-438.
5. Maxted N (1993). A phenotypic investigation of *Vicia* L. subgenus *Vicia* (*Leguminosae, Vicieae*). *Botanical Journal of the Linnean Society*, **111**: 155-182.
6. Van de Ven WTG, Duncan N, Ramsay G, Phillips M, Powell W and Waugh R (1993). Taxonomic relationships between *V. faba* and its relatives based on nuclear and mitochondrial RFLPs and PCR analysis. *Theoretical and Applied Genetics*, **86**: 71-80.
7. McComb JA (1975). Is intergeneric hybridisation in the *Leguminosae* possible? *Euphytica*, **24**: 497-502.
8. Smartt J (1979). Interspecific hybridisation in the grain legumes – a review. *Economic Botany*, **33**: 329-337.
9. Newell CA and Hymowitz T (1982). Successful wide hybridisation between the soybean and a wild perennial relative, *G. tomentella* Hayata. *Crop Science*, **22**: 1062-1065.
10. Mejía-Jiménez A, Muñoz C, Jacobsen HJ, Roca WM and Singh SP (1994). Interspecific hybridisation between common and tepary beans: increased hybrid embryo growth, fertility and efficiency of hybridisation through recurrent and congruity backcrossing. *Theoretical and Applied Genetics*, **88**: 324-331.
11. Ramsay G and Pickersgill B (1986). Interspecific hybridisation between *Vicia faba* and other species of *Vicia*: Approaches to delaying embryo abortion. *Biologisches Zentralblatt*, **105**: 171-179.
12. Lazaridou TB, Roupakias DG (1993). Intraspecific variation in mean endosperm cell cycle time in *Vicia faba* L. and interspecific hybridisation with *Vicia narbonensis* L. *Plant Breeding* **110**: 9-15.
13. Zenketeler M, Tegeder M, Schieder O (1998). Embryological studies of reciprocal crosses between *Vicia faba* and *Vicia narbonensis*. *Acta Societatis Botanicorum Poloniae*, **67**: 37-43.
14. Pickardt T, Meixner M, Schade V and Schieder O (1991). Transformation of *Vicia narbonensis* via *Agrobacterium tumefaciens*-mediated gene transfer. *Plant Cell Reports*, **9**: 535-538.
15. Potokina E, Tomooka N, Vaughan DA, Alexandrova T and Xu RQ (1999). Phylogeny of *Vicia* subgenus *Vicia* (*Fabaceae*) based on analysis of RAPDFs and RFLP or PCR-amplified chloroplast genes. *Genetic Resources & Crop Evolution*, **46**: 149-161.
16. Jaaska V (1997). Isozyme diversity and phylogenetic affinities in *Vicia* subgenus *Vicia* (*Fabaceae*). *Genetic Resources & Crop Evolution*, **44**: 557-574.
17. Clough SJ and Bent AF (1998). Floral dip: a simplified method for *Agrobacterium*-mediated transformation of *Arabidopsis thaliana*. *The Plant Journal*, **16**: 735-743.
18. Trieu AT, Katagi H, Dewbre GR, Weigel D Harrison MJ (2000). Transformation of *Medicago truncatula* via infiltration of seedlings of flowering plants with *Agrobacterium*. *The Plant Journal*, **22**: 531-541.
19. Venketeswaran S (1962). Tissue culture studies on *Vicia faba* L. Establishment of culture. *Phytomorphology*, **12**: 300-306.
20. Grant M and Fuller KW (1968). Tissue culture of root cells of *Vicia faba*. *Journal of Experimental Botany*, **19**: 667-680.
21. Mitchell JP and Gildow FE (1975). The initiation and maintenance of *Vicia faba* tissue cultures. *Physiologia Plantarum*, **34**: 250-253.
22. Röper W (1979). Growth and cytology of callus and cell suspension cultures of *Vicia faba* L. *Zeitschrift für Pflanzenphysiologie*, **93**: 245-257.

23.   Martin C, Carré M and Duc G (1979). Note sur les cultures de tissus de féverole (*Vicia faba* L.). Bouturage, culture de cals, culture de méristèmes. *Annales de l'amélioration des Plantes*, **29**: 277-287.

24.   Cheyne V and Dale PJ (1980). Shoot tip culture in forage legumes. *Plant Science Letters*, **19**: 303-309.

25.   Galzy R and Hamoui M (1981). Induction de l'organogénèse sur des cals de *Vicia faba* minor provenant d'apex. *Canadian Journal of Botany*, **59**: 203-207.

26.   Schulze S, Grunewaldt J and Schmidt H (1985). Zur *in vitro* Regeneration von *Vicia faba* L. *Zeitschrift Pflanzenzüchtung*, **94**: 244-250.

27.   Busse G (1986). *In vitro* cultivation of *Vicia faba* and induction of morphogenesis. *Biologisches Zentralblatt*, **105**: 97-104.

28.   Fakhrai H, Fakhrai F and Evans PK (1989). *In-vitro* culture and plant regeneration in *Vicia faba* ssp. *equina* cultivar spring blaze. *Journal of Experimental Botany*, **40**: 813-818.

29.   Selva E, Stouffs M and Briquet M (1989). *In vitro* propagation of *Vicia faba* L. by micro-cutting and multiple shoot induction. *Plant Cell, Tissue and Organ Culture*, **18**: 167-179.

30.   Taha RM and Francis D (1990). The relationship between polyploidy and organogenetic potential in embryo and root-derived tissue cultures of *Vicia faba* L. *Plant Cell, Tissue and Organ Culture*, **22**: 229-236.

31.   Skoog F and Schmitz RY (1972). Cytokinins. In: Steward FC (ed.), *Plant Physiology: A treatise* Vol VIB. Academic Press/New York/London, pp. 181-212.

32.   Steeves TA and Sussex IM (1989). *Patterns in Plant Development*. Cambridge University Press, UK.

33.   Cheng T, Saka H and Voqui-Dinh TH (1980). Plant regeneration from soybean cotyledonary node segments in culture. *Plant Science Letters*, **19**: 91-99.

34.   Kartha K, Pahl K, Leung N and Mroginski LA (1981). Plant regeneration from meristems of grain legumes: soybean, cowpea, peanut, chickpea, and bean. *Canadian Journal of Botany*, **59**: 1671-1679.

35.   Martins I (1983). Multiple shoot formation from shoot apex cultures of *Phaseolus vulgaris* L. *Journal of Plant Physiology*, **115**: 205-208.

36.   Malik KA and Saxena PK (1992). Thidiazuron induces high-frequency shoot regeneration in intact seedlings of pea (*Pisum sativum*), chickpea (*Cicer arietinum*) and lentil (*Lens culinaris*). *Australian Journal of Plant Physiology*, **19**: 731-740.

37.   Schroeder HE, Schotz AH, Wardley-Richardson T, Spencer D and Higgins TJ (1993). Transformation and regeneration of 2 cultivars of pea (*Pisum sativum* L.). *Plant Physiology*, **101**: 751-757.

38.   Di R, Purcell V, Collins GB and Ghabrial SA (1996). Production of transgenic soybean lines expressing the bean pod mottle virus coat protein precursor gene. *Plant Cell Reports*, **15**: 746-750.

39.   McCabe D, Swain W, Martinell B and Christou P (1988). Stable transformation of soybean (*Glycine max*) by particle acceleration. *Bio/Technology*, **6**: 923-926.

40.   Brar GS, Cohen BA, Vick CL and Johnson GW (1994). Recovery of transgenic peanut (*Arachis hypogaea* L.) plants from elite cultivars utilizing ACCELL(R) technology. *The Plant Journal*, **5**: 745-753.

41.   Aragao FJL, Barros LMG, Brasileiro ACM, Ribeiro SG, Smith FD, Sanford JC, Faria JC and Rech EL (1996). Inheritance of foreign genes in transgenic bean (*Phaseolus vulgaris* L.) co-transformed via particle bombardment. *Theoretical and Applied Genetics*, **93**: 142-150.

42.   Christou P and McCabe DE (1992). Prediction of germ-line transformation events in chimeric R (0) transgenic soybean plantlets using tissue-specific expression patterns. *The Plant Journal*, **2**: 283-290.

43.   Griga M, Kubalakova M and Tejklova E (1987). Somatic embryogenesis in *Vicia faba* L. *Plant Cell, Tissue and Organ Culture*, **9**: 167-171.

44.   Albrecht C and Kohlenbach HW (1989). Induction of somatic embryogenesis in leaf-derived callus of *Vicia narbonensis* L. *Plant Cell Reports*, **8**: 267-269.

45.   Pickardt T, Huancaruna Perales E and Schieder O (1989). Plant regeneration via somatic embryogenesis in *Vicia narbonensis*. *Protoplasma*, **149**: 5-10.

46.   Ammirato PV (1983). Embryogenesis. In: Evans DA, Sharp WR, Ammirato PV, Yamada Y (eds.), *Handbook of Plant Cell Culture* Vol 1, Macmillan/New York, USA, pp. 82-123.

47.   Saalbach I, Pickardt T, Machemehl F, Saalbach G, Schieder O and Müntz K (1994). A chimeric gene encoding the methionine-rich 2S albumin of Brazil nut (*Bertholletia excelsa* H.B.K.) is stably expressed and inherited in transgenic grain legumes. *Molecular and General Genetics*, **242**: 226-236.

48.   Pickardt T, Saalbach I, Waddell D, Meixner M, Müntz K and Schieder O (1995). Seed specific expression of the 2S albumin gene from Brazil nut (*Bertholletia excelsa*) in transgenic *Vicia narbonensis*. *Molecular Breeding*, **1**: 295-301.

49.   Pickardt T, Ziervogel B, Schade V, Ohl L, Bäumlein H and Meixner M (1998). Developmental-regulation and tissue-specific expression of two different seed promoter GUS-fusions in transgenic lines of *Vicia narbonensis*. *Journal of Plant Physiology*, **152**: 621-629.

50.   Czihal A, Conrad B, Buchner P, Brevis R, Farouk AA, Manteuffel R, Adler K, Wobus U *et al.* (1999). Gene farming in plants: Expression of a heatstable *Bacillus* amylase in transgenic legume seeds. *Journal of Plant Physiology*, **155**: 183-189.

51.   Binding H and Nehls R (1978a). Regeneration of isolated protoplasts of *Vicia faba* L. *Zeitschrift für Pflanzenphysiologie*, **88**: 327-332.

52.   Binding H and Nehls R (1978b). Somatic cell hybridization of *Vicia faba* and *Petunia hybrida*. *Molecular and General Genetics*, **164**: 137-143.

53.   Röper W (1981). Callus formation from protoplasts derived from cell suspension cultures of *Vicia faba* L. *Zeitschrift für Pflanzenphysiologie*, **101**: 75-78.

54.   Tegeder M, Gebhardt D, Schieder O and Pickardt T (1995). Thidiazuron-induced plant regeneration from protoplasts of *Vicia faba* cv. Mythos. *Plant Cell Reports*, **15**: 164-169.

55.   Mok MC, Mok DWS, Armstrong DJ, Shudo K, Isogai Y and Okamoto T (1982). Cytokinin activity of N-phenyl-N'-1,2,3-thiadiazol-5-ylurea (thidiazuron). *Phytochemistry*, **21**: 1509-1511.

56.   Thynn M and Werner D (1987). Plantlet regeneration and somatic differentiation in faba bean (*Vicia faba* L.) from callus culture of various explants. *Angewandte Botanik*, **61**: 483-492.

57.   Tegeder M, Kohn H, Nibbe M, Schieder O and Pickardt T (1996). Plant regeneration from protoplasts of *Vicia narbonensis* via somatic embryogenesis and shoot organogenesis. *Plant Cell Reports*, **16**: 22-25.

58.   Lehminger-Mertens R and Jacobsen HJ (1989). Plant regeneration from pea protoplasts via somatic embryogenesis. *Plant Cell Reports*, **8**: 379-382.

59.   Böhmer P, Meyer B and Jacobsen HJ (1995). Thidiazuron-induced high frequency of shoot induction and plant regeneration in protoplast derived pea callus. *Plant Cell Reports*, **15**: 26-29.

60.   Wei Z and Xu Z (1988). Plant regeneration from protoplasts of soybean (*Glycine max* L.). *Plant Cell Reports*, **7**: 348-351.

61.   Schiemann J and Eisenreich G (1989). Transformation of field bean *Vicia*-faba l. cells expression of a chimeric gene in cultured hairy roots and root-derived callus. *Biochemie und Physiologie der Pflanzen*, **185**: 135-140.

62.   Ramsay G and Kumar A (1990). Transformation of *Vicia faba* cotyledon and stem tissues by *Agrobacterium rhizogenes* - Infectivity and cytological studies. *Journal of Experimental Botany*, **41**: 841-847.

63.   Bevan MW (1984). Binary *Agrobacterium tumefaciens* vectors for plant transformation. *Nucleic Acids Research*, **12**: 8711-8721.

64.   Jelenic S, Mitrikeski PT, Papes D and Jelaska S (2000). *Agrobacterium*-mediated transformation of broad bean *Vicia faba* L.. *Food Technology & Biotechnology*, **38**:167-172.

65.   Böttinger P, Gebhardt D, Steinmetz A, Schieder O and Pickardt T (1997). *Agrobacterium*-mediated transformation of *Vicia faba* cv. 'Mythos'. III. *International Food Legume Research Conference*, Adelaide/Australia, Abstracts p. 84.

66.   Hood EE, Helmer GL, Fraley RT, Chilton MD (1986). The hypervirulence of *Agrobacterium tumefaciens* A281 is encoded in a region of pTiBo542 outside of the T-DNA. *Journal of Bacteriology*, **168**: 1291-1301.

67.   Hood EE, Gelvin SB, Melchers LS and Hoekema A (1993). New *Agrobacterium* helper plasmids for gene transfer to plants. *Transgenic Research*, **2**: 208-218.

68.   Böttinger P, Steinmetz A, Schieder O, and Pickardt T (2001). *Agrobacterium*-mediated transformation of *Vicia faba*. *Molecular Breeding*, **8**: 243-254.

69.   White J, Chang SY, Bipp MJ, and Bipp M (1990). A cassette containing the *bar* gene of *Streptomyces hygroscopicus*: a selectable marker for plant transformation. *Nucleic Acids Research*, **18**: 1062.

70.   Murashige T and Skoog F (1962). A revised medium for rapid growth and bioassays with tobacco tissue cultures. *Physiologia Plantarum*, **15**: 473-497.
71.   Shetty P, Atallah MT and Shetty K (2001). Enhancement of total phenolic, L-DOPA and proline contents in germinating fava bean (*Vicia faba*) in response to bacterial elicitors. *Food Biotechnology*, **15**: 47-67.
72.   Shetty P, Atallah MT and Shetty K (2002). Effects of UV treatment on the proline-linked pentose phosphate pathway for phenolics and L-DOPA synthesis in dark germinated *Vicia faba*. *Process Biochemistry*, **37**:1285-1295.
73.   Jelaska S, Pevalek B, Papes D and Devide Z (1981). Developmental aspects of long-term callus culture of *Vicia faba* L. *Protoplasma*, **105**: 285-292.
74.   Choi HW, Lemaux PG and Cho M (2000). High frequency of cytogenetic aberration in transgenic oat (*Avena sativa* L.) plants. *Plant Science*, **156**: 85-94.
75.   De Kathen A, Wegelin T, Kiesecker H, Meyer B and Jacobsen HJ (1998). Transgenic grain legumes from protoplasts?! *3rd International Conference on Grain Legumes*, Valladolid, Spain: 370-371.
76.   Doyle JJ and Doyle JL (1990). Isolation of plant DNA from fresh tissue. *Focus*, **12**: 13–15.
77.   Snowdon RJ, Böttinger P, Pickardt T, Köhler W and Friedt W (2001). Physical localisation of transgenes on *Vicia faba* chromosomes. *Chromosome Research*, **9**: 607-610.

Chapter 26

GENE TECHNOLOGY IN PEA

K.E. MCPHEE, S. GOLLASCH[1], H.E. SCHROEDER[2] AND T.J.V. HIGGINS[1]

USDA-ARS, P.O. Box 646434, Pullman, WA, 99164-6434, USA; [1]CSIRO Plant Industry, GPO Box 1600, Canberra, ACT, 2601, Australia;[2]Retired, CSIRO Plant Industry, GPO Box 1600, Canberra, ACT, 2601, Australia. Corresponding author: K.E. McPhee. E-mail: kmcphee@mail.wsu.edu

1. INTRODUCTION

1.1. Worldwide Importance of Pea

Pea (*Pisum sativum* L.) is an important crop worldwide and is valued for its nutritional composition and vital role in crop production systems. It ranks second only to dry bean (*Phaseolus vulgaris* L.) among grain legumes in total world production and is consumed in human diets as a substitute for animal protein and included in livestock rations as an inexpensive source of energy and protein (*1*). In addition to its dietary and feed value, it has been the subject of some of the earliest genetic studies and was the subject of Gregor Mendel's experiments to establish the laws of genetics (*2*). The pea plant continues to serve as a model crop for genetic and physiological studies due to its ease of handling and manipulation.

Crop improvement efforts in pea have focused on biotic and abiotic stress which adversely affect crop growth. Conventional breeding has made great strides in overcoming many production constraints; however, limited natural variation within pea germplasm has forced geneticists to look beyond natural hybridization barriers for genes necessary to overcome the most serious and important production constraints. The advent of modern gene technology, also referred to as genetic modification or genetic engineering, will allow geneticists to access necessary traits from distantly related or unrelated species not possible through conventional breeding methods.

Gene technology also serves as a valuable tool in the investigation of fundamental genetics underlying enzymology, a metabolic pathway or phenotypic expression. Overexpression or knockout mutants of a single gene can be generated to determine the role of a given enzyme. These mutants can also be used to generate a desired and novel phenotype. An additional application of gene technology is to verify a putative function by introduction of a gene into a plant lacking that function. Gene technology has been applied in a variety of crops to address a wide range of issues (*3*). Future applications of gene technology will only be limited by individual creativity and promises to provide geneticists with a valuable research tool to answer

I.S. Curtis (ed.), Transgenic Crops of the World - Essential Protocols, 351-359.

fundamental questions as well as provide solutions to crop production and dietary limitations.

1.2. History of Gene Transfer in Pea

The first report of gene transfer in pea using *Agrobacterium* as a vector was by De Kathen and Jacobsen (*4*) where resistance genes *npt*II and *hpt* were successfully integrated and expressed. Several laboratories have modified and improved the transformation process subsequent to this early report. Despite the modifications and improvements in the process transformation efficiency remains at less than 5% and often between 1-3%. Similar to other species including other legumes, several factors are reported to affect gene transfer and selection efficiency including explant source, *Agrobacterium* strain, plant genotype, duration of co-cultivation and selection agent (*5*).

At least four antibiotic or herbicide resistance genes have been used to identify transformed pea plants, *hpt*, *npt*II, *bar* and acetohydroxy acid synthase (*ahas3r*) (*4, 6-11*). Use of the *npt*II gene has generally been considered ineffective as a selection agent due to the requirement for a long selection period and large number of shoots escaping selection (*12, 8*). However, Grant *et al.* (*10*) and Polowick *et al.* (*11*) were able to achieve satisfactory selection using the *npt*II gene for resistance to kanamycin. Polowick *et al.* (*11*) reported the first use of the *ahas3r* gene for resistance to the herbicide chlorsulfuron in pea and found it to be superior to the *bar* and *npt*II genes. Overall, the literature suggests that use of the herbicide resistance genes, *bar* and *ahas3r*, for resistance to Basta™ and chlorsulfuron, respectively, are superior to the antibiotic resistance genes, *hpt* and *npt*II.

The procedure described in this contribution yields transformation of existing meristematic tissue in the pea embryonic axis at a frequency of 1.5-2.5% of the starting explants and requires approx. 9 months from initial transformation to seed producing plants (*7*). Modifications have been made to this procedure including choice of explant, *Agrobacterium tumefaciens* strain and selectable marker system, but have not significantly increased the efficiency of gene transfer (*8, 9, 11, 12*). Each procedure has in common the use of relatively organized explant tissue containing meristematic regions and minimal disruption of cell structure and organization. The average time required to regenerate transformed plants is 6 months, but has been reported to be as short as 3 months and longer than 9 months.

1.3. Transformation in Other Legumes

All of the major grain legume species have been successfully transformed with the exception of cowpea (*Vigna unguiculata*) (*5*). *A. tumefaciens* has been the primary vector to transform legumes, however, other techniques such as electroporation, biolistics and pollen-mediated gene transfer have been attempted (*13*). With the exception of biolistics, these alternative techniques have not been widely adopted and appear to be difficult to duplicate.

2. MATERIALS

2.1. Agrobacterium tumefaciens *Strains*

Agrobacterium strains used in genetic transformation have been disarmed of their oncogenic ability and differ in the type of 'opine' production encoded by the genes contained on the Ti plasmid. Opines are a type of amino acid required by the bacterium for growth. Three main strains are currently available, octopine, nopaline and L,L-succinomopine. Several *A. tumefaciens* strains are available and have been used successfully to transform pea. Nodalska-Orczyk and Orczyk (*14*) reported that the hypervirulent L, L-succinomopine strain, EHA105, gave the highest frequency of regenerated pea transformants. The octopine strains, AGL0 and AGL1, have been used successfully with the procedure described below.

2.2. Agrobacterium tumefaciens *Culture Media*

1. MGL broth: 5.0 g/l mannitol, 1.0 g/l L-glutamic acid, 0.25 g/l potassium phosphate, 0.1 g/l sodium chloride, 0.1 g/l magnesium sulphate, 5 g/l tryptone (Difco, Becton Dickinson and Company, Sparks, MD, USA), 2.5 g/l yeast extract (Difco), 1 µg/l biotin (Sigma) (10 µl of 0.1 mg/ml stock), pH 7.0 (1 M NaOH)
2. Rifampicin: 20 mg/ml stock in methanol and filter-sterilized using 0.22 µm pore size filter (Advantec MFS Inc., Pleasanton, CA, USA).
3. Liquid broth medium: MGL medium containing 20 mg/l rifampicin (added after autoclaving).
4. Agar culture medium: MGL medium containing 20 mg/l rifampicin (added after autoclaving), 15 g/l agar (Difco).

2.3. *Explant Source*

Choice of explant and its treatment during the transformation process is fundamental to successful transformation. Pea explants which have been used to date include stem or hypocotyl segments, cotyledonary node explants, lateral cotyledonary meristems of germinating seed, thin slices of embryonic axes from immature seeds, and immature cotyledons. Overall, the most successful explants have been meristematic regions associated with the immature seed, either the embryonic axis or cotyledonary nodes.

Genetic background of the pea variety used for transformation affects the transformation frequency; however, more than 20 different genotypes have been successfully transformed. However, two preferred cultivars are Greenfeast (a garden pea) and Mukta (a field pea). Explants used for transformation are thin slices of the embryonic axis and can be acquired either from immature embryos from glasshouse-grown plants or from germinated mature seed.

2.4. Selection Agents

1. Glufosinate-ammonium, PESTANAL (Sigma-Aldrich GmbHD-30918 Seelze): 10 mg/ml stock in distilled water; filter-sterilized using a 0.22 µm filter.
2. Kanamycin sulphate (Sigma): 100 mg/ml stock in water; filter-sterilized using a 0.22 µm filter.

2.5. Tissue Culture

1. Surface sterilant: 70% ethanol; 1:4 dilution of commercial bleach solution, 50% 7.4 M orthophosphoric acid (BDH Merck).
2. B_5H Macro salts: 15 g/l potassium nitrate, 4.48 g/l calcium chloride, 2.5 g/l magnesium sulphate, 0.75 g/l sodium dihydrogen orthophosphate, 0.67 g/l ammonium sulphate.
3. B_5H Micro salts: 600 mg/l boric acid, 2 g/l manganese sulphate, 400 mg/l zinc sulphate, 150 mg/l potassium iodide, 50 ml/l sodium molybdate stock (1 mg/ml), 5 ml/l cupric sulphate stock (1 mg/ml), 5 ml/l cobalt chloride stock (1 mg/ml).
4. B_5H Sequestrene: 2.8 g/l sequestrene (ferrous-EDTA) (Phytotechnology Laboratories).
5. B_5H Vitamin stock #1: 100 g/l thiamine-HCl, 100 mg/l nicotinic acid, 100 mg/l pyridoxine-HCl.
6. B_5H Vitamin stock #2: 16 g/l L-glutamine, 2 g/l serine, 200 mg/l glutathione.
7. MS macro salts (20X): 33 g/l ammonium nitrate, 8.8 g/l calcium chloride, 38 g/l potassium nitrate, 7.4 g/l magnesium sulphate, 3.4 g/l potassium dihydrogen orthophosphate.
8. MS Micro salts: 1.25 g/l boric acid, 4.46 g/l manganese sulphate, 1.72 g/l zinc sulphate, 165 mg/l potassium iodide, 50 ml/l sodium molybdate stock (1 mg/ml), 5 ml/l cupric sulphate stock (1 mg/ml), 5 ml/l cobalt chloride stock (1 mg/ml).
9. MS Iron: 5.4 g/l ferrous chloride.
10. MS NaEDTA: 6.7 g/l sodium EDTA.
11. 2,4-D: 1 mg/ml stock solution.
12. Kinetin: 0.1 mg/ml stock solution.
13. NAA: 1mg/ml stock solution.
14. BAP: 1 mg/ml stock solution.
15. Co-cultivation medium: 200 ml/l B_5H Macro Salts, 5 ml/l B_5H Micro Salts, 10 ml/l B_5H Vitamin stock #1, 50 ml/l B_5H Vitamin stock #2, 10 ml/l B_5H Sequestrene, 30 g/l sucrose, 1 g/l *myo*-inositol, 0.5 g/l proline, 0.7 g/l MES, 1 ml/l 2,4-D (1 mg/ml stock), 3.8 g/l Agar, pH 5.8
16. Regeneration Medium (P245): 50 ml/l MS Macro salts, 5 ml/l MS Micro salts, 10 ml/l B_5H vitamin stock #1, 5 ml/l MS Iron, 5 ml/l MS sodium EDTA, 30 g/l sucrose, 0.1 g/l *myo*-inositol, 0.7 g/l MES, 0.02 ml/l NAA, 4.5 ml/l BAP, 3.8 g/l Agar, pH 5.8

17. Shooting Medium (P21): 50 ml/l MS Macro salts, 5 ml/l MS Micro salts, 10 ml/l B$_5$H vitamin stock #1, 5 ml/l MS Iron, 5 ml/l MS sodium EDTA, 20 g/l sucrose, 0.1 g/l *myo*-inositol, 0.7 g/l MES, 0. 2 ml/l NAA, 1 ml/l BAP, 3.8 g/l Agar, pH 5.8

3. METHODS

3.1. Growing and Maintaining Agrobacterium Strains

1. Initiate cultures in three sterile 15 ml Falcon tubes (USA Scientific, Ocala, FL, USA) for each experiment by adding 0.2 ml of *Agrobacterium* glycerol stock to 10 ml of MGL broth with 0.01 ml appropriate 1000X stock solution of an appropriate selection agent to select bacteria containing the plasmid of interest (*see* Note 1).
2. Secure the Falcon tube lid and place on a shaker at 28°C overnight for 18-24 h.

3.2. Explant Preparation from Immature Seed

1. Remove pods from glasshouse-grown plants when seeds are just beyond the maximum fresh weight stage. Pods should appear light in colour and have a leathery feel. The dorsal suture appears green and is easily distinguished from the pod wall and individual seeds are bright green in colour (Fig. 2A).
2. Surface-sterilize pods by rinsing in 70% ethanol for 60 sec followed by 30 min in a 1:5 (vol:vol) dilution of a commercial bleach solution.
3. Rinse 4-5 times with sterile water prior to dissection.
4. Remove sterilized pods from the container using flame-sterilized tweezers and place in a lid of a large sterile Petri dish.
5. Open the pod by making a single cut along the dorsal suture with a flame-sterilized scalpel blade (Fig. 2B). The seed is removed from the pod and placed in a second sterile container.
6. Place a small amount (2-5 ml) of *A. tumefaciens* suspension grown at 28°C for 18-24 h in a small (35 x 10 mm) sterile Petri dish prior to seed dissection.
7. Taking care not to cut through the embryonic axis, cut the seed coat along the divide between the cotyledons and separate the cotyledons. The embryonic axis should remain attached to one cotyledon and the other cotyledon is discarded.
8. Remove approximately one-third of the radicle and any remaining testa.
9. Dip the scalpel blade in the *Agrobacterium* suspension and make 3-5 longitudinal slices through the embryonic axis. The slices should be as thin as possible and may require the use of a dissecting microscope to easily view the embryo (Fig. 2C).
10. Immerse the slices in the *Agrobacterium* suspension (approx. 200 slices per 5 ml of suspension) for 1-2 h. It is usually necessary to change the scalpel blade after 40-60 slices to maintain efficiency of slicing.

3.3. Explant Preparation from Mature Seed

1. Sterilize mature dry seeds by rinsing in 70% ethanol for 60 sec and then soaking in 50% orthophosphoric acid for 30 min (*see* Note 2).
2. Rinse seeds 3-5 times with sterile water to remove all the acid and add approx. 3 volumes of sterile water and allow the seed to imbibe at room temperature overnight.
3. Place imbibed seeds into sterile containers one layer deep with sufficient sterile water to maintain seed moisture during germination, but do not fully immerse the seed. Seal the containers and place them at 24°C on a rotating shaker overnight (*see* Note 3).
4. Explants are prepared from the embryonic axis of seeds having radicles approx. 5 mm in length by making serial sections longitudinally through the embryo similar to the procedure described above for immature seed.

3.4. Culture Conditions and Selection

1. Following 1-2 h of immersion in the *Agrobacterium* suspension as much of the liquid suspension as possible is removed.
2. Plate the slices as densely as possible on co-cultivation medium for 72 h (Fig. 2D; *see* Notes 4 and 5).
3. Transfer explants to regeneration medium (P245) containing appropriate selective agent and space widely to facilitate maximum selection. The explants are subsequently transferred to fresh medium every 18-21 days for approx. 9 cycles (Fig. 2E). Dead shoots are discarded. All cultures are grown at 24°C under fluorescent light with a 16 h photoperiod at 40 $\mu mol/m^2/sec$ (*see* Note 6).
4. Divide callus with multiple shoots into clumps with no more than 2-4 shoots to ensure maximum selection on each shoot and to reduce the frequency of "escapes".
5. Individual shoots are cultured separately during the later stages with a minimum of basal callus to reduce the chance of a non-transformed plant being maintained in culture (*see* Note 7). It is important to keep detailed notes on the origin of each shoot so that duplicates of a transformation "event" are recorded.
6. Individual shoots surviving on P245 media are transferred to shoot elongation medium (P21) containing appropriate selection agent for up to an additional three cycles (approx. 60 days) or until the shoots are greater than 3 cm in length (Figs. 2F and G).

Figure 1. Stages during pea transformation; A) centre pod is collected at optimum stage of seed development, B) immature seeds 2-5 days beyond maximum fresh weight, C) embryonic slices, D) explants co-cultivated with A. tumefaciens *for 3 days, E) explants on regeneration media (P245) 17 days after transformation, F) shoot proliferation from individual explants, G) single shoots on shoot elongation media (P21) with selection, H) single shoot grafted onto root stock. Bars included in frames indicate scale of 1 cm.*

3.5. *Grafting and Transfer to Soil*

Individual surviving shoots can be transferred to the glasshouse either by grafting or by the initiation of roots in culture and then transplanting to soil. Initiating roots in culture is possible on a high auxin medium, but is not consistent among shoots and may take an additional 30 days before plants can be transferred to the glasshouse. Grafting is slightly more delicate, but allows immediate transfer of shoots and yields consistent results among shoots.

1. Establish rootstocks in the glasshouse approx. 7 days prior to the expected date for grafting (light intensity of 400 μmol/m^2/sec). Seed for rootstocks, should be sown 3-4 cm deep to produce a long epicotyl. Grafting should be done when the rootstocks are just emerging from the soil and prior to straightening of the plumule hook.
2. Excavate the soil to expose the epicotyl and remove the shoot above the lower scale nodes leaving approx. 2-3 cm of epicotyl.
3. Make a longitudinal incision approx. 1 cm down the central axis of the epicotyl.
4. Place a thin slice of silicon tubing with an inner diameter approximately equal to the epicotyl diam. (2 mm) around the epicotyl below the incision.
5. Prepare the transgenic shoot by making a v-shaped cut approx. 1 cm along the lower portion of the shoot.
6. Insert the shoot in the rootstock and slide the silicon tubing gently upward to firmly hold the graft in place (Fig. 2H).
7. Cover the grafts with a transparent pot and shade the plant using cloth for approx. 2 weeks to minimize moisture loss until the graft is firmly established. Any shoots emerging from the cotyledonary nodes should be removed since they will jeopardize the success of the graft. The transparent pots should be progressively removed from the grafts over 3-5 days to allow the graft to adjust to ambient conditions.

4. NOTES

1. Inclusion of rifampicin will adversely affect plant tissue health and reduce transformation efficiency.
2. Orthophosphoric acid breaks down the testa and aids uniform imbibition of all seeds.
3. If the radicles have begun to break through the testa within 8 h, the seed should be placed at 4°C overnight to slow their development.
4. The dense placement of the explants minimizes bacterial growth which may overgrow the explants and reduce their vigour and ability to regenerate.
5. The time period of co-cultivation affects transformation efficiency, but 3 days is considered optimum.
6. Maximum selection is maintained by direct contact of individual shoots with the medium. Excess callus growth at the base of individual shoots is removed during each transfer and large clumps of shoots are divided.
7. Individual shoots may produce healthy branches while the main shoot dies. This is an indication of a chimeric transformation event or survival of a non-transformed shoot that escaped earlier selection. Non-chimeric transformants may be recovered by subculturing the branches.

CONCLUSION

Gene technology in pea has the potential to positively impact crop production and nutritional value. Although the procedure described above is relatively time consuming (9-15 months), laborious and yielding efficiencies between 1-3%, it is a robust and reproducible method. Relatively few genes have been transferred to pea to date; however, the potential benefits of introducing novel genes improving seed quality and agronomic performance of pea offer a positive future for the crop as well as the continued use of this technology.

Gene technology is also being used to study the genetic basis of various biochemical pathways through overexpressing or knocking-out individual gene function and promises to expand our scientific knowledge of plant and gene function. Specific applications of the technology to improve crop production or nutritional components will benefit mankind in both developing and developed countries.

REFERENCES

1. FAO (2002). http://apps.fao.org/cgi-bin/nph-db.pl?subset=agriculture.
2. Mendel G (1866). Experiments in plant hybridization (1865). *Versuche uber plflanzen-hybriden.* Verhandlungen des naturforschenden Ver-eines in Brunn, Bd. IV fur das Jahr 1865. Abhandlungen. pp. 3-47.
3. Hansen G and Wright MS (1999). Recent advances in the transformation of plants. *Trends in Plant Science,* **4**: 226-231.
4. De Kathen A and Jacobsen H-J (1990). *Agrobacterium tumefaciens*-mediated transformation of *Pisum sativum* L. using binary and cointegrate vectors. *Plant Cell Reports,* **9**: 276-279.
5. Somers DA, Samac DA and Olhoft PM (2003). Recent advances in legume transformation. *Plant Physiology,* **131**: 892-899.
6. Puonti-Kaerlas J, Eriksson T and Engstrom P (1992). Inheritance of a bacterial hygromycin phosphotransferase gene in the progeny of primary transgenic pea plants. *Theoretical and Applied Genetics,* **84**: 443-450.
7. Schroeder HE, Schotz AH, Wardley-Richardson T, Spencer D and Higgins TJV (1993). Transformation and regeneration of two cultivars of pea (*Pisum sativum* L.). *Plant Physiology,* **101**: 751-757.
8. Davies DR, Hamilton J and Mullineaux P (1993). Transformation of peas. *Plant Cell Reports,* **12**: 180-183.
9. Grant JE, Cooper PA, McAra AE and Frew TJ (1995). Transformation of peas (*Pisum sativum* L.) using immature cotyledons. *Plant Cell Reports,* **15**: 254-258.
10. Grant JE, Cooper PA, Gilpin BJ, Hoglund SJ, Reader JK, Pither-Joyce MD and Timmerman-Vaughn GM (1998). Kanamycin is effective for selecting transformed peas. *Plant Science,* **139**: 159-164.
11. Polowick PL, Quandt J and Mahon JD (2000). The ability of pea transformation technology to transfer genes into peas adapted to western Canadian growing conditions. *Plant Science,* **153**: 161-170.
12. Bean SJ, Gooding PS, Mullineaux PM and Davies DR (1997). A simple system for pea transformation. *Plant Cell Reports,* **16**: 513-519.
13. Christou P (1994). The biotechnology of crop legumes. *Euphytica,* **74**: 165-185.
14. Nodalska-Orczyk A and Orczyk W (2000). Study of the factors influencing *Agrobacterium*-mediated transformation of pea (*Pisum sativum* L.). *Molecular Breeding,* **6**: 185-194.

Chapter 27

AGROBACTERIUM-MEDIATED TRANSFORMATION OF CABBAGE

D.C.W. BROWN AND H.Y. WANG

Southern Crop Protection and Food Research Centre, Agriculture and Agri-Food Canada, 1391 Sandford Street, London, Ontario, Canada, N0M2A0.
Corresponding author: D. Brown E-mail: browndc@agr.gc.ca

1. INTRODUCTION

1.1. Importance, Distribution and Biosystematics

Cabbages are a widely grown and economically important vegetable that were domesticated during the first millennium B.C. They belong to the highly diversified *Brassicaceae* family comprising about 3000 species. In North America and Europe, *B. oleracea* white cabbage (*B. oleracea* var. *capitata*) is the predominant agronomic type; whereas in Asia, Chinese cabbage (*Brassica rapa* subsp. *pekinensis*) is the most commonly cultivated cabbage species. The FAO listed the world area of cabbages harvested at 3,016,059 hectares (ha) in 2002 with the largest planted area in China (1,469,684 ha) followed by India (280,000 ha), the Russian Federation (180,000 ha), USA (104,000 ha), Indonesia (100,000 ha), Korea(s) (89,531 ha), Japan (61,000 ha) and Columbia (48,000 ha), Kenya (39,051 ha) and Belarus (27,300 ha). The remaining 30% of world planting is spread over 125 additional countries (*1*).

The origin, evolution, taxonomy and genomic relationships of the crop *Brassicas* have been studied extensively (*see* Note 1; Fig. 1) and the cabbages fall into two species groups *Brassica oleracea* and *Brassica rapa*. These plants have a great diversity of morphotypes which have been given various species, varietal and/or subspecific ranks. The wide variety of cabbage types referred to in the literature cited in this article have several synonymous names and are found in both the *B. oleracea* and *B. rapa* species (Table 1). *B. oleracea* is a member of the CC genome complex or *B. oleracea* cytodeme. It includes a number of interfertile Mediterranean species (*B. cretica, B. hilarionis, B. incana, B. insularis, B. macrocarpa, B. montana, B. rupestris,* and *B. villosa*), wild *B. oleracea* from coastal areas of western Europe and *B. bourgeaui* from the Canary Islands. The most important crops in *B. oleracea* are:
- Cabbages (var. *capitata* and var. *sabauda*), such as headed cabbages, savoy cabbage, etc., are characterized by the formation of heads formed from tightly-packed leaves;
- Kales (var. *viridis*, var. *costata*, var. *medullosa*, var. *sabellica*), including kitchen kale, green kale, dwarf Siberian kale, marrow-stem kale, collards, and

I.S. Curtis (ed.), Transgenic Crops of the World - Essential Protocols, 361-378.

tronchuda, which develop a strong main stem and are used for their edible foliage;

- Branching bush kales (var. *ramosa*), including thousand-headed kale and perpetual kale, were formerly much cultivated for their edible foliage;
- Brussel sprouts (var. *gemmifera*); axillary buds form edible heads of tightly packed leaves;
- Kohlrabi (var. *gongylodes*) is cultivated for its above-ground thickened stem;
- Inflorescence kales (var. *botrytis* and var. *italica*), such as cauliflower, broccoli, sprouting broccoli, are cultivated for their thickened edible inflorescences; and
- Chinese kale (var. *alboglabra*;), a cultivated white-flowering crop grown in China, is generally assumed to be an ancient import from the Mediterranean region and often treated as a separate species *B. alboglabra*.

The species names *B. rapa* and *B. campestris* are used interchangeably in the literature. *Brassica rapa* and *B. campestris* were described by Linnaeus in 1753 to indicate turnip and wild-weedy plants, respectively. The two taxa are interfertile and were first combined in 1833 by Metzger under the name *B. rapa*. *B. rapa* is a highly polymorphic species and contains many crops that have been domesticated over a long period in Europe as well as in Asia (Table 1). Little is known about its existence in the wild as plants found under natural conditions seem to be escapes from cultivation. *B. rapa* is most closely related to *B. oleracea* and both have arisen from ancestral members of the C genome cytodeme (*see* Note 1). The most important crops in *B. rapa* (subspecies in brackets) are:

- Chinese cabbage (subsp. *pekinensis*), Asiatic heading vegetable, with petioles winged;
- Pakchoi (subsp. *chinensis*), a Chinese non-heading leaf vegetable, with petioles fleshy but not winged;
- Vegetable turnip (subsp. *rapa*);
- Fodder turnip (subsp. *rapa*) which forms a leaf rosette and/or a turnip;
- Turnip rape (subsp. *oleifera*) and toria (subsp. *dichotoma*, black seeded with annual spring and biennial winter types, used for oil extraction;
- Yellow sarson (subsp. *trilocularis*), annual, yellow seeded, used for oil extraction;
- Mizuma, mibuna, komatsuna or leaf turnip (subsp. *nipposinica*), Asiatic non-heading leafy vegetable, with many tillers and either pinnate (mizuma) or entire leaves (mibuna);
- Broad-beak mustard or Chinese savoy (subsp. *narinosa*), an Asiatic non-heading leafy vegetable, flat rosette of many small, leaves;
- Broccoletto (formerly treated as *B. ruvo*, assigned to subsp. *oleifera*), European vegetable with an enlarged, compact inflorescence.

Most of the above mentioned crops have originally been described as separate species, but crossing studies have shown they readily intercross and belong to the *n* = 10, AA genome, *B. rapa* cytodeme (*see* Note 1). Members of the complex are now generally treated taxonomically as subspecies of *B. rapa*.

—

Table 1. Cabbage species and variety relationships

Species	Synonym	Common Name
Brassica oleracea var. *acephala*	*Brassica oleracea* var. *viridis*	Collards Cow cabbage Spring-heading cabbage Tall kale Tree kale
Brassica oleracea var. *capitata*	*Brassica oleracea* var. *sabauda*	Cabbage Head cabbage Red cabbage White cabbage Savoy cabbage
Brassica oleracea var. *chinensis*	*Brassica campestris* ssp. *chinensis* *Brassica rapa* ssp. *chinensis* *Brassica chinensis* *Brassica rapa* var. *chinensis*	Pakchoi Chinese mustard Bok-choy Celery mustard Chinese white cabbage
Brassica rapa subsp. *parachinesis*	*Brassica parachinensis* *Brassica chinensis* var. *parachinensis* *Brassica rapa* cv. *Gr. Caixin*	Flowering white cabbage False pakchoi Mock packchoi
Brassica rapa subsp. *pekinensis*	*Brassica campestris* ssp. *pekinensis* *Brassica chinensis* var. *pekinensis* *Brassica pe-tsai* *Brassica pekinesnsis* *Brassica rapa* subvar. *pe-tsai* *Brassica rapa* var. *amplexicaulis* *Sinapis pekinensis*	Chinese cabbage Pe-tsai cabbage Celery cabbage Hakusai Shantun cabbage Peking cabbage Long white cabbage Bejing cabbage Napa cabbage Tientsin cabbage Michihli cabbage

1.2. Need for Genetic Improvement of Cabbage

Breeding programmes in the vegetable *Brassicas* including cabbage usually address the following areas (in order of priority): uniformity, disease resistance, appearance, crop yield, pest resistance, nutritional quality and new crop types. Traits such as disease and pest resistance and new crop traits such as herbicide resistance are most amenable to direct genetic modification as they tend to be single gene traits. In particular, engineering of resistance in cabbage to insect pests by transfer of *cry* genes from the bacterium *Bacillus thuringiensis* has been shown to be an effective strategy   for   pest   control   under   controlled   conditions *cry1A(b)*, *cry1A(c)* and *cry1A(b3)* have been expressed in head cabbage *Brassica oleracea* var. *capitata* (*5, 7, 8*), *cry1C* in Chinese cabbage *Brassica rapa* ssp. *pekinesis* (*6, 9*), and *cry1A(b)* and *cry1A(c)* in Chinese flowering cabbage *Brassica campestris* ssp. *parachinensis* (*10*). All studies showed that *cry*-expressing plants could give effective control against that target insect pest; in this case, diamondback moth (*Plutella xylostella*). In the case of Chinese cabbage (*9*), and head cabbage (*8*) effective control at the glasshouse level was demonstrated against not only diamondback moth but also against cabbage looper (*Trichooplusa ni*) and imported cabbage worm (*Pieris rapae*). Typically, in detached leaf assays and whole plant tests, highly expressing plants showed minimal damage due to insect larvae followed rapidly by larvae death.   However, in some lines, protection was poor presumably due to low expression levels of the *cry* gene (*9*). Modification of *Brassica oleracea* var. *capitata* targeted toward resistance against Black rot caused by *Xanthomonas campestris* by expression of an *Aspergillus niger* glucose oxidase (GO) gene under the control of a constitutive promoter (*11*) did appear to show some promise with a correlation of expression of GO and the delay of onset of symptoms of Black rot disease. As glucose oxidase catalyses the oxidation of β-D-glucose releasing H_2O_2, it is thought that the potential increased H_2O_2 levels in transgenic cabbage may have acted as a signal to induce the plant defence system. A problem with the approach appeared to be the occurrence of many undesirable phenotypes and the lack of direct evidence that known defence related genes (e.g. PR-1a) were induced by the transgene.   Herbicide resistance, although not a high agromomic priority in cabbage, has also been shown to be feasible (*12, 13*).

1.3. Genetic Modification of Cabbage

There have been a number of studies in the last decade that have reported the successful   transformation   of   *Brassica*   cabbage   types   (*55;*   Table   2*).* *Agrobacterium*-mediated gene transfer has been the most widely reported approach and is described in detail here.   Other approaches such as whole plant transformation (*13*), particle bombardment (*18*), intergeneric fusion of isolated protoplasts (*61-63*), pollin tube gene transfer (*64*) and electroporation of protoplasts (*65*) have also been reported with some success.   With the *Agrobacterium* approach, the vector of choice has been *A. tumefaciens* but *A. rhizogenes* has also been used in some cases (Table 2). With respect to *Agrobacterium* strain, LBA4404 appears to be more common but a range of strains have been used; in addition, a study (*33*) found the LBA4404 strain with a TiAch5 chromosomal background (*34*) to be significantly better than three other strains with a C58 chromosomal background. The *npt*II selectable marker gene along with kanamycin selection has been the most widely used approach for tranformant recovery and embryonic explants; namely, hypocotyls and/or cotyledon explants which have been the target tissue of choice (Table 2). As with many species, the regeneration of cabbage *in vitro* is genotype dependant. We and others (*8, 66*), have found that medium components such as agar, silver nitrate and growth regulator concentrations, as well as the use of a 2,4-D pretreatment of explants can be determining factors in the efficiency of plant regeneration and subsequently in the efficiency of recovery of transformed plants. We have found that a wide range of genotypes can be efficiently regenerated *in vitro* from hypocotyl explants using variations in a protocol which uses a 3 day pretreatment on Murashige and Skoog (MS, *38*) medium modified to contain 2,4-D, 0-0.01 mg/l NAA, 2-4 mg/l BAP and 0-1 mg/l silver nitrate (*8*). Range of responses of 70-100% of introduced explants and shoot numbers of 6-58 per explant have been recorded. We have found that with efficiencies of 85% and shoot regeneration in the 8+ shoot per explant range in the target plant material, transformation efficiencies in the 4% range are possible with a range of genotypes. There are a number of reports (*8, 9, 19, 22, 29, 36, 39, 40, 54*) of this or higher regeneration and transformation efficiencies in cabbage germplasms with similar observations of genotype variability.

2. MATERIALS

2.1. A. tumefaciens *Strain*

A wide range of *Agrobacterium* strains representative of octopine, nopaline and succinamopine/agropine types have been reported to be useful for cabbage and Chinese cabbage (*see* Note 2). In our laboratory, *A. tumefaciens* strain GV3101/pMP90 (*34, 67*) containing the disarmed Ti plasmid was used to carry a binary vector for transformation. The T-DNA of the binary vector contained a *npt*II gene under the control of the *nos* promoter and terminator, which was used as a marker for the transgenic plant selection. Apart from the selectable marker, an expression cassette containing a reporter gene *uid*A or a target gene such as *cry1Ac* was used under the control of a tCUP promoter (*see* Note 3) and a NOS terminator (*30*).

2.2. Culture Medium for A. tumefaciens

The following stocks of medium can be stored for several months at the indicated temperature and condition. All medium should be prepared from the stored stocks just prior to use.
1. Kanamycin sulphate: 25 mg/ml stock in water. Sterilize by filtration through a 0.2 μm membrane. Dispense 1 ml liquots into Eppendorf tubes and store at -20 °C.

2. Gentamycin: 25 mg/ml stock in water. Sterilize by filtration through a 0.2 μm membrane. Dispense 1 ml aliquots into Eppendorf tubes and store at -20°C.
3. Rifampicin: 25 mg/ml stock in DMSO. Dispense 1 ml aliquots into Eppendorf tubes and store at -20°C.
4. LB medium: 10 g/l tryptone, 5 g/l yeast extract, 10g/l NaCl, pH 7.0-7.2
5. Liquid LB medium (for *Agrobacterium* culture): LB medium containing 50 mg/l kanamycin sulphate, 50 mg/l gentamycin, and 50 mg/l rifampicin.
6. Semi-solid LB medium: LB medium containing 150 mg/l kanamycin sulphate, 100 mg/l gentamycin, 150 mg/l rifampicin and 10 g/l agar.
7. Inoculation liquid medium (1/2MSi): half-strength MS salts and half-strength B5 vitamins, 0.095% *myo*-inositol, 1.5% sucrose, pH 5.8.

2.3. Plant Materials

1. Stored seeds of cabbage (*Brassica oleracea* var. *capitata*) or Chinese cabbage (*B. rapa* ssp. *pekinensis*) are used as the source of plant material. In our laboratory, we have used a wide variety of seed material. Much of the work described here was done using inbred lines provided by industry (*8*). All lines were responsive to the regeneration protocols but varied in their degree of response. One of the least regeneration-responsive lines (SSC-1) responded to the transformation protocol and one of the most regeneration-responsive lines (SSC-4) responded very well. We expect most cabbage lines will respond to this approach especially if the regeneration protocol is optimized so that 85% of explants form shoots in culture and the average shoot number per explant is 8 or more.
2. Cabbage hypocotyls were taken from 7-14 day-old seedlings. For Chinese cabbage, hypocotyls or cotyledons explants are harvested from 4-5 day-old seedlings (*9, 10, 35, 36*).

2.4. Tissue Culture

1. Vitamin stock: B_5 vitamin (*37*) 1000X stock, store at -20°C. The stock is prepared in the laboratory based on the formulation of Gamborg *et al.* (*37*). The prepared 1000x stock is dispensed into 15 ml tubes and stored at -20°C. It is thawed in warm water just prior to use and 1 ml/l of prepared medium is used.
2. Timentin stock: dissolve 3 g timentin (SmithKline Beecham Pharma) in 30ml water, and sterilize by filtration through a 0.2 μm membrane to produce a stock concentration of 100 mg/ml.
3. Surface sterilant: 20% (v/v) commercial sodium hypochlorite bleach. Prepare before use.
4. Seed germination medium (SGM): 2.15 g/l Murashige and Skoog (MS) salts (*8*) and 0.5 ml/l B_5 vitamins 15 g/l sucrose and 6 g/l agar (Bacto-Agar, DIFCO), pH 5.7.

5. Basic MS medium (BMS): 4.3 g/l MS salts, 1 ml/l B5 vitamin stock, 0.5 g/l MES, 30 g/l sucrose, 6 g/l agar, pH 5.7. No agar added in liquid BMS (*see* Note 4).
6. Pre-treatment medium (PTM): BMS with 1 mg/l 2,4-D and 0.1 mg/l kinetin.
7. Co-cultivation medium (CCM): BMS with 2 mg/l BAP and 0.01 mg/l NAA.
8. Selection shoot regeneration medium (SSM): BMS with 2 mg/l BAP, 0.01 mg/l AA, 1 mg/l AgNO$_3$, 300 mg/l timentin and 25 mg/l kanamycin (*see* Note 5).
9. Selection/Rooting medium (SRM): BMS with 0.1 mg/l NAA, 300 mg/l timentin and 25 mg/l kanamycin.
10. Kanamycin callusing medium (KCM): BMS with 1 mg/l BAP, 0.2 mg/l NAA, 300 mg/l timentin and 25 mg/l kanamycin.
11. Basic rooting medium (BRM): BMS with 0.1 mg/l NAA.

For Chinese cabbage, the protocol is similar with the exception of the use of a nurse culture stage (*see* Note 6).

3. METHODS

3.1. Explant Preparation and Pre-culture

1. Rinse cabbage seeds in a sterile Petri dish with 70% ethanol for 60 sec, then wash once with sterile distilled water, followed by surface-sterilization with 20% (v/v) bleach for 20 min constantly stirring and rinse thoroughly 3-4 times with sterile distilled water (*see* Note 7).
2. Sow sterile seeds onto a growth regulator-free SGM medium, usually 10 seeds in each 60 x 15 mm Petri dish. Typically, we use 15 duplicate Petri dishes per experiment (*see* Note 8).
3. Harvest hypocotyls from 7-14 day-old cabbage seedlings for transformation. Carefully remove the meristem, cotyledon and root tissues and place whole hypocotyls onto the pre-treatment medium PTM. Seal the plate with parafilm to keep a high humidity. Culture for 3 days at 24° C in the light (30 µE/m^2/sec, 16 h photoperiod; *see* Note 9).

3.2. Agrobacterium Preparation

1. Streak by scraping *Agrobacterium* cells from a glycerol stock onto semi-solid LB medium with appropriate antibiotics and incubate at 28°C for 2-3 days.
2. Pick single colonies and inoculate into a 125 ml flask containing 20 ml of LB medium with appropriate antibiotics. Incubate culture overnight in the dark at 28°C on shaker at 100 rpm.
3. Measure OD of overnight culture at a wavelength of 600 nm (OD$_{600}$ ≈ 1; *see* Note 10).
4. Centrifuge the *Agrobacterium* culture at 5000 rpm for 5 min, discard the supernatant, and re-suspend in liquid 1/2MSi medium.
5. Dilute *Agrobacterium* culture 1:10 with 1/2MSi medium (*see* Note 11).

3.3. Inoculation and Co-cultivation

1. After the 3 day pre-treatment on PTM medium, dissect the hypocotyls into 3-5 mm segments in the presence of the diluted *Agrobacterium* culture. At the same time, lightly score each explant 2-3 times on one side with a sharp scalpel blade (*see* Note 12).
2. Transfer explants, scored-side up, onto CCM (20 hypocotyl segment explants/ 9 cm Petri dish).
3. Add 5 μl of diluted *Agrobacterium* solution to each explant (this is also called the "droplet inoculation method").
4. Seal the plates (optional) with micropore^{TM} tape (3M 1530-0) and co-culture in dark for 2 days at 24°C followed by 2 days at 14°C until there is a halo of bacteria culture around each explant (*see* Note 13).

3.4. Selection and Recovery of Whole Transgenic Plants

1. After 4 days co-cultivation, blot explants on sterile filter paper to remove excess *Agrobacterium* and then transfer onto S0 medium (SSM without selection antibiotic). Culture the dishes at 25°C under a 16 h photoperiod (irradiance of 30 μE/m^2/sec) for 7 days and then transfer to SSM medium (*see* Note 14).
2. Maintain explants on selection medium SSM for 6-8 weeks with transfer to fresh medium every 2 weeks. All the cultures are maintained in the same culture condition as mention above. During this selection and regeneration period, all the plates are sealed with micropore^{TM} (*see* Note 15).
3. Putative green shoots are usually found at 6-8 weeks after co-cultivation. Green shoots (0.5-1.0 cm in length) are excised from the explant at this stage and transferred to a Magenta GA-7 container with SRM medium under the same conditions for further selection and rooting (*see* Note 16).
4. Transgenic cabbage shoots usually form roots in 10-15 days. Carefully remove the shoot tips from the rooted plantlets on SRM medium, without touching the bottom leaves and medium, and then transfer the shoot tips to BRM medium without antibiotics (*see* Note 17) and culture under the same conditions.
5   Once regenerated on SRM, plants that remain green should be subcultured, pending analysis and confirmation of transformation status.
6   Clean and rooted plantlets growing on BRM are then transferred to soil (*see* Note 18).
7   Transgenic plantlets with 5-6 leaves can be transplanted from *in vitro* to 8 cm seed trays and maintained in the glasshouse under conditions of 21-22°C, 16 h photoperiod. We use Pro-mix BX (Premier Horticulture Inc., *see* www.premierhort.com) transplanting medium. After 3 weeks, the established plants are transferred to 10 cm pots and finally transferred to 18 cm pots when the plants form heads. Cabbage is a biennial (taking 2 seasons to produce seed), and requires a cold treatment (vernalization) in order to flower. After 6-7 months growth in the glasshouse, plants are moved to a dark and cool storage room set at 3.5-4.5°C (late November) for 10-12 weeks. When flower stalks

appear (March), plants are moved to the glasshouse. Hand pollination is done in mid-March and seeds are collected 3 months later (early June).

8. Storage temperature, relative humidity and seed moisture are important factors in determining the longevity of seeds without loss of germination. Dry cabbage seed (transgenic and non-transgenic) should be placed in packages, labelled and stored in moisture-proof containers. Containers such as sealed cans or jars with air-tight caps work satisfactorily. Storage temperatures between 2.5°C-10°C are satisfactory when the moisture content of the seed is low. We use 10°C and 25% relative humidity in the dark. An alternate method of keeping seeds dry is to place them in a sealed jar with calcium chloride or silica gel. These substances should not touch the seed. Under cool, dry conditions, the seeds of cabbage can be stored for approx. 5 years (*68*).

9. PCR can be used to detect the presence of transgenes. Southern blotting should also be performed to confirm the stable integration of the transgenes. Northern blot and/or Western blot analysis can be used to confirm the expression of the transgenes and the protein product in the transgenic plants. We usually use material from glasshouse–established plants due to the greater availability of tissue and the reduced risk of plant loss during *ex vitro* sampling. We have used a Bt-cry1Ac ImmunoStrip test (Agdia) system for detecting Bt endotoxin protein for each plantlet before transplantation.

10. Collect seeds from T0 transgenic cabbage plants and store as described (*see* Notes 7 and 8). Surface-sterilize and sow seeds as outlined (*see* Sections 3.1.1. and 3.1.2.), but sow only 1 seed per tube (10 x 1.5 cm) and label each tube. After 10 days, most of the seeds germinate and the hypocotyl grows to approx. 2-3 cm in length. Transfer each shoot tip (including cotyledons and meristems) to another set of tubes containing 10 ml of rooting medium (MS medium with 0.1 mg/l NAA). These shoots should be kept for further growth and analysis. The remaining seedling, that is, the main part of each hypocotyl, should be transferred to a 1.5 ml sterile Eppendorf tube and labelled the same as the original seedling. These materials are used a source tissue for PCR analysis. To prevent DNA contamination, each seedling should be cut using a clean and sterile scalpel (*see* Note 19).

4. NOTES

1. Cultivated *Brassicas* are represented by 6 interrelated species; including, *Brassica rapa* ($n = 10$, genome AA), *B. nigra* ($n = 8$, genome BB), and *B. oleracea* ($n = 9$, genome CC), and amphidiploid derivatives *B. carinata* ($n = 17$, BBCC), *B. juncea* ($n = 18$, AABB) and *B. napus* ($n = 19$, AACC) (Fig. 1). The latter 3 are derived by hybridization and polyploidization of 2 of the diploid taxa. The cytogenetic relationship of the 3 amphidiploid species was first proposed in 1935, and is known as the triangle of U (*2*). The origin and genetic relationships of the amphidiploids relative to the diploid crop *Brassicas* has been confirmed by chromosome pairing and artificial synthesis of the amphidiploids, nuclear DNA content, DNA analysis and the use of genome specific markers (*3, 4*). With respect to the 3 diploid taxa, recent DNA analyses, including both nuclear and chloroplast restriction site data, suggest separate evolutionary pathways, with *B. rapa* and *B. oleracea* (including wild CC genome species) assigned to one group with *Diplotaxis erucoides* ($n = 7$) or a

close relative as the primary progenitor, and *B. nigra* assigned to a second group with *Sinapis arvensis* or a close relative as the primary progenitor species (*4*).

2.   *Agrobacterium*-mediated DNA delivery is commonly used for cabbage type vegetable transformation. Although the transformation rate is relatively low, a wide range of *Agrobacterium* strains representative of octopine, nopaline and succinamopine/agropine have been reported to be useful f or cabbage and Chinese cabbage. *A. tumefaciens* strains ASE-1, A281, ABI, EHA101, EHA105, GV3101, LBA4404, LGV3850, MOG 301 and *A. rhizogenes* strains A4 and LBA 9402 were used successfully to obtain transgenic cabbage or Chinese cabbage (*10, 14-16, 20, 22, 23, 28, 39, 40, 45, 46, 48-50, 56-58*).

3.   The tCUP promoter is a novel cryptic constitutive promoter isolated from the non-coding regions of repetitive DNA of *Nicotiana tabaccum*. It has the structure of a typical promoter and is comprised of a 60 bp core region, an upstream transcriptional *Cee* enhancer sequence and an enhancer sequence in the untranslated leader sequence tCUP shows patterns of expression similar to that of the CaMv 35S viral promoter in most dicotyledonous plants. For more detailed information consult references (*41-43*).

4.   MES is added to the medium as a buffer. Adjust pH within the range 5.0-8.5, buffering can be achieved by the addition of 0.5 g/l MES or HEPES.

5. $AgNO_3$ was found to be helpful for reducing ethylene in some cultivars but may not be effective in all cultivars. The combinations of plant regulators are also variable depending on genotype (*8, 26, 32, 66*).

6.   Materials for Chinese cabbage transformation are basically the same as that of cabbage (*47, 51-53, 59, 60*). It was reported that before transfer of the inoculated explants to the selection medium, co-cultivation with tobacco suspension culture for 2 days resulted a transformation rate of 5-9% in Chinese cabbage transformation. For this protocol, a 6-8 day-old tobacco (species not reported) cell suspension culture should be prepared (*9*). The use of a co-cultivation medium with a low pH of 5.2 and supplemented with 10 mg/l acetosyringone was reported to yield a high infection frequency in Chinese cabbage transformation (*35*). To prepare acetosyringone , dissolve 10 mg acetosyringone in 1 ml DMSO, store at -20°C.

7.   After a 70% ethanol rinse, place the cabbage seeds and a stir bar into a 125 ml flask, pour in 50-70 ml of 20% (v/v) bleach, and place the flask on a stir platform for 20 min, stir, and then rinse thoroughly 3-4 times with sterile water.

8.   Seedlings from stored seeds are commonly used as the source of plant material for cabbage and Chinese cabbage transformation. It is suggested to sow sterile seeds on SGM medium in a 60 x 15 mm Petri dish, place the seeded plates with lids removed, onto the upturned lids of sterile Magenta boxes and cover with the transparent polystyrene boxes. Typically, we use 150-180 seeds per experiment with approx. 10 seeds per dish. When the seedlings grow to a height of 4-5 cm, remove the polystyrene boxes and harvest the hypocotyls.

9.   Pre-treatment is required in most procedures of cabbage and Chinese cabbage transformation studies. Our histological studies on the timing and origin of shoot regeneration of cabbage hypocotyl explants, found that, active cell division of shoot meristem regeneration initiated between 4-8 days after pre-culture.The efficiency of *Agrobacterium* infection greatly increased during this period.

10.   Use LB medium with antibiotics as a background reference (blank) for determining OD.

11.   Dilute *Agrobacterium* culture with ½ MSi to 1:10: Take the *Agro*-culture ($1 \div OD_{reading}$) X 1 ml, add ½ MSi $10 - (1 \div OD_{reading})$ ml. Example: $OD_{reading}$ = 0.6, *Agro*-culture = ($1 \div 0.6$) x 1 ml = 1.67 ml, 1½ MSi = 10 ml - 1.67 ml = 8.33 ml. Total diluted *Agro*-solution = 1.67 ml + 8.33 ml = 10 ml (*21*).

12.   Our histological studies on the timing and origin of shoot regeneration of cabbage hypocotyls also found that the source of shoot meristems is relatively deep in the cells adjacent to the vascular bundles. The vacuum infiltration inoculation or the droplet inoculation methods aimed at deeper infection combined with a longer co-cultivation period is recommended for increasing the infection efficiency. The immersion-inoculation method can also be applied to the transformation of cabbage and Chinese cabbage. In this approach, explants are inoculated with *Agrobacterium* by immersion into the bacterial solution for 10-30 min. Explants are blotted on a sheet of sterile paper to remove excess bacteria, then returned to the co-cultivation medium (*35, 36*).

13.   The co-cultivation period also depends on the virulence of the *Agrobacterium* strain and the plant material (*21*). Because cabbage transformation needs a longer co-cultivation period, it is recommended to decrease the concentration of especially virulent strains and to decrease the

incubation temperature. The co-cultivation period in Chinese Cabbage transformation is usually 2-3 days (*35, 36*).

14. After the co-cultivation period, there should be a white "halo" of *Agrobacterium* surrounding each explant. Explants with a very dense halo indicate overgrowth by the *Agrobacterium*. Combine explants showing overgrowth of *Agrobacterium* and wash them thoroughly and several times in ½ MSi liquid medium, with 300 mg/l timentin in the last rinse. Blot the treated explants thoroughly on dry, sterile filter paper before transferring to S0 medium.

15. Sealing with micropore™ tape gives better results by reducing humidity and probably also by preventing build-up of ethylene (*54*).

16. Some early-forming green shoots are probably escapes. The putative transformants are usually formed later during the culture period.

17. When putative green shoots are transferred to SRM medium for rooting, several small pieces of the bottom leaves can be excised and placed on KCM medium (selective callus introducing medium) to confirm resistance to kanamycin, as exhibited by the ability of the explant to enlarge, produce callus and/or form roots. The non-kanamycin resistant tissue always comes from an escape (non-transformed or poorly expressing plant). In our experience, the inclusion of antibiotics does not effect the rooting of true transformants. The rooting medium for transgenic plant recovery has been optimized; that is, in the selection rooting medium (SRM - basic MS medium, 0.1 mg/l NAA and 25 mg/l kanamycin for transgene selection) with 300 mg/l timentin (for *Agrobacterium* elimination), a well developed root system can form within 2 weeks. Plantlets with rooting difficulty or rooting delay are suspected escapes.

18. Transplanting clean plantlets from the antibiotic-free medium BRM confirms that there is no risk of contamination of *Agrobacterium* being transferred to the soil.

19. A typical result of PCR analysis of the introduced gene is as follows: 60 seeds produced by a cross of a *Bt*-containing cabbage line and a GUS-containing cabbage line resulted in recovering 60 T1 seeds. These seeds germinated at 100%. PCR analysis of these 60 T1 plants by PCR showed 46 of the T1 plants contained both genes. Four plants contained the *Bt* gene only and 8 plants contained the GUS gene only. Two plants had neither gene.

CONCLUSIONS

A review on *Brassica oleracea* transformation in 1996 (*44*) found the majority of research was confined to the introduction of marker and reporter genes. The authors speculated that activity would be forthcoming in the area of expression of fungal and bacterial resistance genes, insect toxins especially *Bt* genes, and genes involved in the modification of morphological characteristics. It was speculated that the latter approach would be possible with advancements in the development of molecular markers as well as the fortuitous co-linearity of the *Brassica* and *Arabidopsis* genomes (*55*). Although there has been some demonstration of progress in disease and insect resistance, much of the potential for agronomic improvement of the species has yet to be realised. The present review of progress in culture regeneration and transformation indicates that the technologies and protocols are largely in place to support the genetic improvement of the major cabbage types although a recent review (*55*) pointed to genotype responses and *in vitro* selection as major limiting factors in this progress. Recent work indicates that the technologies and protocols are mostly reproducible within and between laboratories and widely applicable across cabbage types.

ACKNOWLEDGEMENTS

The authors wish to thank Agriculture and Agri-Food Canada, the Ontario Research Enhancement Program and the Ontario Fruit and Vegetable Growers Association for financial support and Sakata Seed America, Inc. for the gift of seed material used for research leading to the protocols described here. We thank Dr. Suzanne Warwick, Eastern Cereal and Oilseed Research Centre, Agriculture and Agri-Food Canada, Ottawa for insights and information on *Brassica* biosystematics used in the Introduction and Note 1.

REFERENCES

1. FAOSTAT Agriculture database @ http://www.fao.org/waicent/portal/statistics_en.asp.
2. U, N (1935). Genomic analysis in *Brassica* with special reference to the experimental formation of *B. napus* and peculiar mode of fertilization. *Japanese Journal of Botany*, 7: 389-452.
3. Song KM, Osborn TC and Williams PH (1988). *Brassica* taxonomy based on nuclear restriction fragment length polymorphisms (RFLPs). 1. Genome evolution of diploid and amphidiploid species. *Theoretical and Applied Genetics*, 75: 784-794.
4. Warwick SI and Black LD (1991). Molecular systematics of *Brassica* and allied genera (subtribe *Brassicinae*, tribe *Brassiceae*) - chloroplast genome and cytodeme congruence. *Theoretical and Applied Genetics*, 82: 81-92.
5. Earle ED, Metz TD, Roush RT, and Shelton AM (1996). Advances in transformation technology for vegetable *Brassica. Acta Horticulturae*, 407: 161-168.
6. Jun SI, Kwon SY, Paek KY, and Paek KH. (1995). *Agrobacterium*-mediated transformation and regeneration of fertile transgenic plants of Chinese cabbage (*Brassica campestris* ssp. *pekinensis* cv. 'Spring Flavor'). *Plant Cell Reports*, 14: 620-625.
7. Bhattacharya RC, Viswakarma N, Bhat SR, Kirti PB and Chopra VL (2002). Development of insect-resistant transgenic cabbage plants expressing a synthetic *cry1A(b)* gene from *Bacillus thuringiensis. Current Science*, 83:146-150.
8. Wang H, Tsang E, McNeil J, Hannam C, Brown D and Miki B (2003). Expression of *cry1Ac* and GUS in cabbage and caulilfower. *Acta Horticulturae*, 625: 475-464.
9. Cho HS, Cao J, Ren JP and Earle ED (2001). Control of Lepidopteran insect pests in transgenic Chinese cabbage (*Brassica rapa* ssp. *pekinensis*) transformed with a synthetic *Bacillus thuringiensis cry1C* gene. *Plant Cell Reports*, 20: 1-7.
10. Xiang Y, Wong WKR, Ma MC and Wong RSC (2000). *Agrobacterium*-mediated transformation of *Brassica campestris* ssp. *Parachinensis* with synthetic *Bacillus thuringiensis cry1Ab* and *cry1Ac* genes. *Plant Cell Reports*, 19: 251-256.
11. Lee YH, Yoon LS, Suh SC and Kim HI (2001). Enhanced disease resistance in transgenic cabbage and tobacco expressing a glucose oxidase gene from *Aspergillus niger. Plant Cell Reports*, 20: 857-863.
12. Lee YH, Lee SB, Suh SC, Byun MO and Kim HI (2000). Herbicide-resistant cabbage (*Brassica oleracea* ssp. *capita*) plants by *Agrobacterium*-mediated transformation. *Journal of Plant Biotechnology*, 2: 35-41.
13. Cao MQ, Fan L, Lei Y, Bouchez, D, Tourneur C, Yan L and Robaglia C (2000). Transformation of Pakchoi (*Brassica rapa* L. ssp. *chinensis*) by *Agrobacterium* infiltration. *Molecular Breeding*, 6: 67-72.
14. Bai YY, Mao HZ, Cao XL, Tang T, Wu D, Chen DD, Li WG and Fu WJ (1992). Transgenic cabbage plants with insect tolerance. In: You CB, Chen ZL (eds.), *Biotechnology in Agriculture. Proceeding of the First Asia-Pacific Conference on Agriculture Biotechnology*, Beijing, China, 20-24 Aug.1992, pp. 56-159.
15. Berthomieu P, Beclin C, Charlot F, Dore C and Jouanin L (1994). Routine transformation of rapid cycling cabbage (*Brassica oleracea*) - molecular evidence for regeneration of chimeras. *Plant Science*, 96: 223-235.

16. Liu F, Yao L, Li Y and Cao MQ (1998). Transgenic plants with herbicide resistance obtained by using microspore-derived embryos of Chinese cabbage. *Acta Agriculturae Boreali-Sinica*, **13**: 93-98.

17. Beclin C, Charlot F, Botton E, Jouanin L and Dore C (1993). Potential use of the *aux2* gene from *Agrobacterium rhizogenes* as a conditional negative marker in transgenic cabbage. *Transgenic Research,* **2**: 48-55.

18. Cho HS, Lee YH, Suh SC, Kim DH and Kim HI (1994). Transformation of beta-glucuronidase (GUS) gene into Chinese cabbage *(Brassica campestris* var. *pekinensis*) by particle bombardment. *RDA Journal of Agricultural Science, Biotechnology*, **36**: 181-186.

19. Christey MC, Sinclair BK, Braun RH and Wyke L (1997). Regeneration of transgenic vegetable brassicas (*Brassica oleracea* and *B. campestris*) via Ri-mediated transformation. *Plant Cell Reports*, **16**: 587-593.

20. He YK, Wang JY, Wei ZM, Xu ZH and Gong ZH (1995). Effects of whole Ri T-DNA and auxin genes alone on root induction and plant phenotype of Chinese cabbage. *Acta Horticulturae*, **402**: 418-422.

21. Gartland JS (1995). *Agrobacterium* Virulence. In: Gartland KMA, Davey MR (eds.), Methods in Molecular Biology, Vol. 44, *Agrobacterium* Protocols. Humana Press Inc., Totowa, NJ, USA, pp. 15-28.

22. Berthomieu P and Jouanin L (1992). Transformation of rapid cycling cabbage (*Brassica oleracea* var. *capitata*) with *Agrobacterium rhizogenes*. *Plant Cell Reports*, **11**: 334-338.

23. Cho YN, Park SY, Noh TK, Song MJ, Park YS and Min BW (2003). Transformation of Chinese cabbage with L-gulono-gamma-lactone oxidase (GLOase) - encoding gene using *Agrobacterium tumefaciens*. *Korean Journal of Horticultural Science & Technology*, **21**: 9-13.

24. Cai L, Cui HZ and Zhang YJ (1999). Transgenic cabbage with a *Bt* gene resistant to insects. *China Vegetables*, **4**: 31-32.

25. Cabrera JG, Padron RIV, Samsonov PD, Pardo AC, Menendez E, Lok MLC, Gonzalez PL, Arozarena NJ *et al.* (1998). Genetic transformation of cabbage (*Brassica oleracea* var. capitata) with *Bacillus thuringiennsis cry 1Ab* and *cry1B* genes. In: Crane JH (ed.), *Proceedings of the International Society for Tropical Horticulture,* Vol. 42, pp. 367- 373.

26. Eisner GI, Mar'yakhina IY and Shemyakin MF (1992). Optimization of conditions *for in vitro* plant regeneration for cabbage transformation. *Soviet Agricultural Sciences*, **11/12**: 15-19.

27. Fang HJ, Li DL, Wang GL and Li YH (1997). An insect-resistant transgenic cabbage plant with the cowpea trypsin inhibitor (*CpTi*) gene. *Acta Botanica Sinica*, **39**: 940-945.

28. Cai XN, She JM, Zhu Z, Zhu WM, Yuan XH and Su XJ (1997). Establishment of an *Agrobacterium*-mediated genetic transformation system for common Chinese cabbage (*Brassica chinensis*). *Jiangsu Journal of Agricultural Sciences*, **13**: 110-114.

29. Pius PK and Achar PN (2000). *Agrobacterium*-mediated transformation and plant regeneration of *Brassica oleracea* var. *capitata*. *Plant Cell Reports*, **19**: 888-892.

30. Sawant S, Singh PK, Madanala R and Tuli R (2001). Designing of an artificial expression cassette for the high-level expression of transgenes in plants. *Theoretical and Applied Genetics,* **102**: 635-644.

31. She JM, Cai XN, Zhu WM, Zhang CX, Ding WX, Li JB, Li B and Zhu Z (2001). Resistant characteristics of insect-resistant transgenic plants and progenies of cabbage (*Brassica oleracea* L. var. *capitata*). *Jiangsu Journal of Agricultural Sciences*, **17**: 73-76.

32. She JM, Cai XN, Zhu Zh, Xu HL, Zhu WM, Wu JY, Wang AM and Ding WX (1996). Studies on plant regeneration and conditions for gene transformation of cabbage (*Brassica oleracea* L. var. *capitata*). *Jiangsu Journal of Agricultural Sciences*, **12**: 6-9.

33. Lim HT, You YS, Park EJ, Song YN and Park HK (1998). High plant regeneration, genetic stability of regenerants, and genetic transformation of herbicide resistance gene (*bar*) in Chinese cabbage (*Brassica campestris* ssp. *pekinensis*). *Acta Horticulturae*, **459**: 199-208.

34. Hellens R, Mullineaux P and Klee H (2002). A guide to *Agrobacterium* binary vectors. *Trends in Plant Science*, **10**: 446-451.

35. Zhang FL, Takahata Y, Watanabe M and Xu JB (2000). *Agrobacterium*-mediated transformation of cotyledonary explants of Chinese cabbage (*Brassica campestris* L. ssp. *pekinensis*). *Plant Cell Reports*, **19**: 569-575.

36.   Kuginuki Y and Tsukazaki H (2001). Regeneration ability and *Agrobacterium*-mediated transformation of different cultivars in *Brassica oleracea* L. and *B. rapa* L. (syn. *B. campestris* L.). *Journal of the Japanese Society for Horticultural Science*, **70**: 682-690.

37.   Gamborg OL, Miller RA and Ojima K (1968). Nutrient requirements of suspension cultures of soybean root cells. *Experimental Cell Research*, **50**: 151-158.

38.   Murashige T and Skoog F (1962). A revised medium for rapid growth and bio-assays with tobacco tissue cultures. *Physiologia Plantarum*, **15**: 473-497.

39.   Jin RG, Liu YB, Tabashnik BE and Borthakur D (2000). Development of transgenic cabbage (*Brassica oleracea* var. *capitata*) for insect resistance *by Agrobacterium tumefaciens*-mediated transformation. *In Vitro Cellular and Developmental Biology–Plant*, **36**: 231-237.

40.   Cogan N, Harvey E, Robinson H, Lynn J, Pink D, Newbury HJ and Puddephat I (2001). The effects of anther culture and plant genetic background on *Agrobacterium rhizogenes*-mediated transformation of commercial cultivars and derived doubled-haploid *Brassica oleracea*. *Plant Cell Reports,* **20**: 755-762.

41.   Foster E, Hattori J, Labbe H, Ouellet T, Fobert P, James L, Iyer V and Miki B (1999). A tobacco cryptic constitutive promoter, tCUP, revealed by T-DNA tagging. *Plant Molecular Biology*, **41**: 45-55.

42.   Wu K, Malik K, Tian L, Hu M, Martin T, Foster E, Brown D and Miki B (2001). Enhancer and core promoter elements are essential for the activity of a cryptic gene activation sequence from tobacco, tCUP. *Molecular and General Genomics*, **265**: 763-770.

43.   Malik K, Wu K, Li X, Martin-Heller T, Hu M, Foster E, Tian L, Wang C *et al.* (2002). A constitutive gene expression system derived front the *tCUP* cryptic promoter elements. *Theoretical and Applied Genetics*, **105**: 505-514.

44.   Puddephat IJ, Riggs TJ and Fenning TM (1996). Transformation of *Brassica oleracea* L.: a critical review. *Molecular Breeding,* **2**: 185-210.

45.   Tsukazaki H, Kuginuki Y, Aida R and Suzuki T (2002). *Agrobacterium*-mediated transformation of a doubled haploid line of cabbage. *Plant Cell Reports*, **21**: 257-262.

46.   Wang F, Li HX, Ye ZB and Lu Y (2002). Genetic transformation of *Brassica campestris* L. ssp. *pekinensis via Agrobacterium* with a *Bt* gene. *Hunan Agricultural Science & Technology Newsletter,* **3**: 10-14.

47.   Cheng XH, Zhou XY, Liu F and Yao L (2001). Transformation of Chinese cabbage (*B. campestris* L. subsp. *pekinensis*) using *Agrobacterium*. *Journal of Hunan Agricultural University,* **27**: 463-466.

48.   Gonzalez Cabrera J, Vazquez Padron RI, Prieto Samsonov D, Ayra Pardo C, Menendez E, Lok MLC, Gonzalez PL, Arozarena NJ *et al.* (1998). Genetic transformation of cabbage (*Brassica oleraceae* var. *capitata*) with *Bacillus thuringiensis cryIAb* and *cryIB* genes. *Proceedings of the Interamerican Society for Tropical Horticulture*, **42**: 367-373.

49.   He YK, Gong ZH, Wang F and Wang M (1991). Transformation efficiency of *Brassica* crops with *Agrobacterium rhizogenes* harboring the binary vector Bin19. *Chinese Journal of Biotechnology*, 7: 382-385.

50.   Iqbal A, Qazi SHA, Hasnain S, Ahmad M and Shakoori AR (1996). Optimization of conditions for T-DNA transfer in callus culture of *Brassica oleracea*. *Proceedings of Pakistan Congress of Zoology*, **16**: 143-153.

51   Kang BK and Park YD (2001). Effect of antibiotics and herbicide on shoot regeneration from cotyledon and hypocotyl explants of Chinese cabbage. *Korean Journal of Horticultural Science & Technology*, **19**: 17-21.

52.   Kim BK, ChoYN, Noh TK, Park YS, Harn CH, Yang SG and Min BW (2003). *Agrobacterium*-mediated transformation of Chinese cabbage with a synthetic protein disulfide isomerase gene. *Journal of the Korean Society for Horticultural Science*, **44**: 5-9.

53.   Kim DH, Cho HS, Lee YH, Suh SC and Kim HI (1995). Plantlet regeneration of hygromycin resistant Chinese cabbage (*Brassica campestris* ssp. *pekinensis*) from callus induced from *Agrobacterium* infected hypocotyls. *RDA Journal of Agricultural Science, Biotechnology*, **37**: 161-166.

54.   Metz TD, Dixit R and Earle ED (1995). *Agrobacterium tumefaciens*-mediated transformation of broccoli (*Brassica oleracea* var. *italica*) and cabbage (*B. oleracea* var. *capitata*). *Plant Cell Reports*, **15**: 287-292.

55. Paul KA, Kumar PA and Saradhi PP (2002). Genetic transformation of vegetable *Brassicas*: a review. *Plant Cell Biotechnology and Molecular Biology*, **3**: 1-10.

56. Radchuk VV, Blume YAB, Ryschka U, Schumann G and Klocke E (2000). Regeneration and transformation of some cultivars of headed cabbage. *Russian Journal of Plant Physiology*, **47**: 400-406.

57. Sretenovic-Rajcic T, Mijatovic M, Stevanovic D and Vinterhalter D (2002). *In vitro* culture as a tool for improvement of cabbage cultivars in Yugoslavia. *Acta Horticulturae*, **579**: 209-213.

58. Wei ZM, Huang JQ, Xu SP and Xue HW (1998). High efficiency regeneration and *Agrobacterium*-mediated transformation of hypocotyls in *Brassica oleracea* var. *capitata* with a *B.t.* gene. *Acta Agriculturae Shanghai*, **14**: 11-8.

59. Yu PT, Wang W, He YK and Shen RJ (2000). Transformation of Barnase gene in Chinese cabbage. *Acta Agriculturae Shanghai*, **16**: 17-19.

60. Zhu CX, Song YZ, Zhang S, Guo XQ and Wen FJ (2001). Production of transgenic Chinese cabbage by transformation with the CP gene of turnip mosaic virus. *Acta Phytopathologica Sinica*, **31**: 257-264.

61. Ryschka U, Klocke E, Schumann G and Warwick S (2003). High frequency recovery of intergeneric fusion products of *Brassica oleracea* (+) *Lepidium meyenii* and their molecular characterization by RAPD and AFLP. *Acta Horticulturae*, **625**: 145-151.

62. Motegi T, Noi I, Zhou J, Kanno A, Kameya T and Hirata Y (2003). Obtaining an ogura-type CMS line from asymmetrical protoplast fusion between cabbage (fertile) and radish (fertile). *Euphytica*, **129**: 319-323.

63. Hou X, Cao S, She J and Lu W (2001). Synthesis of cytoplasm hybrid of non-heading Chinese cabbage through asymmetrical electric fusion of protoplast cell. *Acta Horticulturae Sinica*, **28**: 532-537.

64. Liao FS, Do YY, Lee GC and Huang PL (1998). Studies on the application of pollen tube gene transfer method to black rot-resistant cabbage breeding. *Journal of the Chinese Society for Horticultural Science*, **44**: 55-63.

65. Xu HQ, Cai GP, Hu Y, Zhao NM, Tang HX and Jia SR (1991). Transferring genes into protoplasts of Chinese cabbage and cucumber by electroporation. *Acta Botanica Sinica*, **33**: 7-13.

66. Zhang FL, Takahata Y and Xu JB (1998). Medium and genotype factors influencing shoot regeneration from cotyledonary explants of Chinese cabbage (*Brassica campestris* L. ssp. *pekinensis*). *Plant Cell Reports*, **17**: 780-786.

67. Koncz C and Schell J (1986). The promoter of TL-DNA gene 5 controls the tissue specific expression of chimeric genes carried by a novel type of *Agrobacterium* vector. *Molecular and General Genetics*, **204**: 383-396.

68. http://www.ianr.unl.edu/pubs/horticulture/g503.htm#tabi.

69. Yang GD, Zhu Z, Li Y, Zhu ZJ, Xiao GF and Wei XL (2002). Obtaining transgenic plants of Chinese cabbage resistant to *Pieris rapae* L. with modified *CpTI* gene (*sck*). *Acta Horticulturae Sinica*, **29**: 224-228

70. Zhang ZQ, Zhou Y, Zhong WJ, Zhang JJ, Yin LQ, Chen QQ, Gong ZZ, Xie WJ *et al.* (1999). Introduction of an arrowhead proteinase inhibitor gene in *Brassica campestris* ssp. *chinensis* L. and expression in transgenic plants. *Acta Agriculturae Shanghai*, **15**: 4-9.

71. Sato T (2000). Studies on breeding methodology using isolated microspores culture and genetic transformation in cruciferous vegetables. *Bulletin of the National Research Institute of Vegetables, Ornamental Plants and Tea*, **15**: 209-283.

Table 2. Transformation of cabbage vegetable types

Species / Variety	Common Name	Explant	Method (Agro type)	Selection Marker	Transgene	Ref.
B. oleracea var. *capitata*	cabbage	cotyledon	*A. tumefaceins* (ASE-1)	*npt*II	*cry1Ac*	*12*
		cotyledon	*A. tumefaciens* (LBA4404)		*gusA*, Bt.	*24*
		cotyledon	*A. tumefaciens* (A281)	*npt*II	*gusA*	*29*
		cotyledon	*A. rhizogenes* (*A4T*)	*npt*II	*cry1Ac*	*19*
		hypocotyl	*A. tumefaciens* (LBA4404, EHA101, EHA105, AGL0)	*npt*II	*gusA*	*45*
		hypocotyl petiole	*A. tumefaciens*	*npt*II	*cry1Ab3*	*39*
		hypocotyl	*A. tumefaciens* (GV3101)	*npt*II	*cry1Ac*	*8*
		hypocotyl	*A. tumefaciens*	*npt*II	*cry1B*	*25*
		hypocotyl	*A. tumefaciens* (ABI)	*npt*II	*cry1Ac*	*54*
		hypocotyl	*A. tumefaciens* (LBA4404)	*npt*II	*cpt*I	*27, 31*
		hypocotyl	*A. tumefaciens* (LBA4404)	*hpt*	GO	*11*
		unknown	*A. tumefaciens* (wildtype and disarmed)		*gusA*	*15*
		unknown	*A. tumefaciens*	*npt*II	PVY-cr	*56*
B. oleracea var. *chinensis*	pakchoi	cotyledon with petiole	*A. tumefaciens* (LBA4404)	*npt*II	API	*70*
		whole plant	*A. tumefaciens* (C58/pMP90)	*bar*	*bar*	*13*
B. campestris ssp. *pekinensis*	Chinese cabbage	unknown	*A. tumefaciens*		LMV-CP	*47*

(cont.)

Table 2.(cont.)

Species / Variety	Common Name	Explant	Method (Agro type)	Selection Marker	Transgene	Ref.
		cotyledonary petiole	A. tumefaciens (LBA4404)	hph	GLOase	23
		cotyledon	A. tumefaciens (LBA 4404)	nptII	PDI	52
B. rapa ssp. pekinensis	Chinese cabbage	cotyledon	A. tumefaciens (LBA4404)	nptII	PDI	52
		cotyledon, cotyledonary petiole	A. tumefaciens (EHA101)	nptII, hpt	gusA	35, 36
		hypocotyl	A. tumefaciens (ABI)	hph	Cry1c	9
		cotyledon with petiole	A. tumefaciens	nptII	CpTI(sck)	69
		cotyledon with petiole	A. tumefaciens		TuMV-CP	60
B. campestris ssp. parachinesis	Flowering Chinese cabbage	microspores			SLG-GUS	71
		hypocotyl	A. tumefaciens (MOG301)	nptII	cry1Ab, cry1Ac	10
Unknown	cv Hawke F1	hypocotyl	A. rhizogenes (LBA9402)	nptII	gfp	40

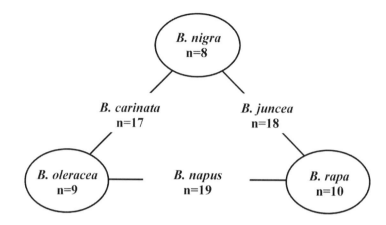

Figure 1. Triangle of U (2; see Note 1) Brassica *cytogenetic relationships.*

Chapter 28

AGROBACTERIUM-MEDIATED
TRANSFORMATION OF CANOLA

V. CARDOZA AND C. N. STEWART Jr

Department of Plant Sciences, University of Tennessee, Knoxville, TN 37996, USA.
Corresponding author: C.N. Stewart Jr., E-mail: Nealstewart@utk.edu

1. INTRODUCTION

1.1. Importance of Canola

Canola (*Brassica napus* L.) is an important oil crop, ranking third only to soybean and palm oil in global production. It is a member of the family *Brassicacea* (syn. *Cruciferae*). It is a winter or spring crop and is amenable to growth in cooler climates. Once considered a specialty crop for Canada, it is now a global crop. Many other countries including the USA, Australia and those in Europe also grow canola. However, Canada and the United States account for most of the canola crop. It is grown mostly in Western Canada and North Central United States. In the year 2002, in Canada alone, 9.6 million acres of canola was grown (*1*) and in the USA, 1.5 million acres was devoted to canola cultivation (*2*). The term 'canola' was adopted by Canada apparently as an acronym of the Canadian Oilseed Association in 1979. Although canola has been commonly also known as rapeseed or oilseed rape, in the strict sense canola oil is defined as an oil that must contain less than 2% erucic acid, and the solid component of the seed must contain less than 30 µM of any one or a mixture of 3-butenyl glucosinolate, 4-pentenyl glucosinolate, 2-hydroxy-3 butenyl glucosinolate, and 2-hydroxy-4-pentenyl glucosinolate per gram of air-dry, oil-free solid.

Canola oil is widely used as cooking oil, salad oil, and making margarine. Of all edible vegetable oils widely available today, it has the lowest saturated fat content, making it appealing to health-conscious consumers. Canola oil is also used in lubricants and hydraulic fluids especially when there is a significant risk of oil leaking to waterways or into ground water (*3*). It is also used in the manufacture of inks, biodegradable grease, in pharmaceuticals and cosmetics.

Canola meal, which is the leftover product of the seeds after extracting oil, is used as a protein supplement in dairy, beef, swine and poultry rations and is recognized for its consistent high quality and value. The increased processing activity within Canada has led to rising consumption and exports of meal and product mixtures. Feed mixtures include such items as pea-can meal (combining dry peas and canola meal) as well as incorporating canola meal with dehydrated alfalfa products. The bulk of Canada's canola meal exports supply the USA feed ingredient market, which imports over a million tonnes annually and represents 80% or more of

I.S. Curtis (ed.), Transgenic Crops of the World - Essential Protocols, 379-387.

total meal exports. Feed deficit Asian countries such as Japan, Taiwan and South Korea are also consistent markets for Canadian canola meal.

1.2. Need for Genetic Improvement of Canola

With the increasing demand for canola oil and the need to meet the demands of consumers, there is more research being pursued in improving canola breeding. Conventional breeding techniques are time consuming and laborious. It takes at least 8-10 generations to develop a new variety using conventional breeding. An alternative to trait improvement without conventional breeding techniques is by genetic engineering which reduces the time needed to develop a new variety. A considerable amount of research has already been undertaken in this direction and canola has been exploited for genetic engineering purposes. However of the total genetically modified (GM) crops grown in the world, GM canola represents only 7% of the total area and there is a need to increase the world cultivation of genetically modified canola in order to develop high yielding and pathogen-resistant canola plants. Genetic engineering can also be used to improve the oil quality in canola.

1.3. Genetic Improvement of Canola

Genetic engineering approaches in canola have mainly focused on improving oil quality (4-8) or making it herbicide tolerant (9, 10) and these plants are now commercially available. In Canada in the year 2002, 900,000 hectares of transgenic canola were grown which corresponds to 20% of the total canola cultivation area in the country. In addition to these traits found in the commercial varieties, canola has also been engineered for insect (11) and fungal (12) resistance. *Brassica juncea,* which is closely related to *Brassica napus* is known for its hyperaccumulator properties and is a well known phytoremediator. The prospects of using *Brassica napus* in phytoremediation remains unexploited. Although there have been a few reports of transformation of *Brassica napus* for heavy metal tolerance (13), there is still much research to be explored in this area. By genetically engineering canola with genes that make it tolerant to heavy metals, canola could be a potential candidate for phytoremediation. In addition to phytoremediation, there are numerous non-food traits that might prove useful. Canola has been used in the production of biopolymers (14). There is also significant interest in increasing fuel-oil capabilities or the oil and using the whole plant for biofumigation as a substitute for noxious gases. Canola might also be especially useful as a vehicle to overproduce pharmaceutically active proteins and edible vaccines (15). One especially attractive feature is the ability to target proteins to oil bodies for easy downstream processing. Since canola is very closely related to *Arabidopsis thaliana*, which is the principle genomic dicot model, the crop should benefit widely from bioinformatics advances that can be parlayed into transgenic improvements. Canola has an increasing market share in food-stuffs that should continue to drive agronomic and quality trait improvement.

1.4. Tissue Culture and Transformation Techniques Applied to Canola

Tissue culture techniques are quite advanced for canola, which has made it an attractive candidate for use in transformation studies. Various explants have been used for regeneration, however, hypocotyls remain the most attractive explants for transformation and have been used in most transformation experiments. Transformation has been performed using stem internodes (*16*), stem segments (*17*), cotyledonary petioles (*18*), hypocotyl segments (*11, 19-21*), microspores (*22-25*) and protoplasts (*26*).

Spring type canola varieties have proven to be most amenable for transformation. The model variety that has been transformed most often has been Westar, an old spring type cultivar from Canada. However, Westar is not black leg resistant and has other agronomic deficiencies. Technical solutions to address genotype dependency are still needed.

Various techniques such as microprojectile bombardment, electroporation, and *Agrobacterium tumefaciens*-mediated transformation have been used for canola transformation. A combined treatment of DNA-imbibition of desiccated embryos and microprojectile bombardment has been used to enhance transformation efficiency (*27*). However, most transformation procedures have been carried out using *Agrobacterium* because of its ease and cost effectiveness. The virulence of *Agrobacterium* can be increased by the use of acetosyringone, a phenolic compound and it is now being routinely used in canola transformation (*28, 21*). The efficiency of transformation using *Agrobacterium* is enhanced by preconditioning of the explant on callus inducing media before co-cultivation (*19, 29, 30, 21*). Co-cultivation period also plays a critical role in transformation with *Agrobacterium*. A Co-cultivation period of 2 days was found to be optimal (*21*).

The protocol we have described here *(21)* gives the highest transformation efficiency in our hands (25%) compared with other canola transformation protocols published. It is also consistent and reproducible. Increasing transformation efficieny is desirable to decrease the amount of resources to produce transgenic plants, and to also potentially provide a higher baseline for subsequent transformation of other canola varieties. This protocol describes an increase in efficiency by optimizing two important parameters; preconditioning and co-cultivation periods. *A. tumefaciens*-mediated transformation is described using hypocotyls as explant tissue. After obtaining the transgenic shoots, the two important prerequisites for efficient transgenic plant recovery are the promotion of healthy shoots by avoiding hyperhydration and the subsequent rooting of these shoots. This protocol describes methods to circumvent problems of hyperhydration of transgenic shoots and also a very efficient rooting system for transformed shoots, which gives 100% rooting.

2. MATERIALS

Seeds used: *B. napus* L. cv. Westar.

Surface-sterilization: 10 % commercial sodium hypochlorite bleach with 0.1% Tween 20 added as a surfactant followed by 95% ethanol with sterile distilled water rinses.

Germination: Seeds are germinated in Magenta GA-7 boxes containing Murashige and Skoog (MS) (*31*) basal medium comprising of MS macro and micro salts and vitamins (*see* Note 1) to which 20 g/l sucrose is added and semi-solidified with 2 g/l Gelrite (Sigma).

A. tumefaciens strain and plasmid: *A. tumefaciens* strain GV3850 strain harbouring pBinmGFP5-ER which contains the mGFP5-ER gene (reporter) and *npt*II gene (selectable marker) which confers resistance to kanamycin (*32*) were used. The mGFP5-ER gene is driven by the CaMV 35S promoter and *npt*II is driven by the NOS promoter (*see* Note 2).

Media: All media stock solutions were made with Type I water as solvent. Stock solutions have a refrigerated shelf-life of approx. 90 days, after which they are discarded.

Callus induction medium: MS medium with 1 mg/l 2,4-D (Sigma) and 30 g/l sucrose and semi-solidified with 2 g/l Gelrite.

Selection medium: Callus induction medium with 400 mg/l timentin (GlaxoSmithkline, USA) (to kill *Agrobacterium*) and 200 mg/l kanamycin (Sigma) to select for transformed cells (*see* Note 3). Stock solutions for the antibiotics were made by dissolving in water and filter-sterilization using a 0.2 μm membrane filter.

Organogenesis medium: MS medium with 4 mg/l BAP (Sigma), 2 mg/l zeatin and 5 mg/l silver nitrate (Sigma), antibiotics as mentioned above and 30 g/l sucrose and semi-solidified with 2 g/l Gelrite.

Shoot regeneration medium: MS medium containing 3 mg/l BAP, 2 mg/l zeatin, antibiotics and 30 g/l sucrose and 2 g/l Gelrite for shoot development.

Shoot elongation medium: MS medium with 0.05 mg/l BAP and 30 g/l sucrose and semi-solidified with 3 g/l Gelrite. Antibiotics are added as above.

Rooting medium: Half-strength MS salts, 10 mg/l sucrose, 3 g/l Gelrite and 0.5 mg/l IBA (Sigma). Antibiotics are added as above.

3. METHODS

3.1. Tissue Culture and Transformation

1. Seeds are surface-sterilized for 5 min with 10 % sodium hypochlorite with 0.1 % Tween 20 added as a surfactant. The sterilization is followed by a 60 sec rinse with 95% ethanol. The seeds are washed thoroughly 3 times with sterile distilled water. Seeds are surface-sterilized by shaking with the sterilants, manually in 1.5 ml Eppendorf tubes.

2. Seeds are germinated in Magenta GA-7 boxes (10 seeds per box) with 50 ml MS medium for 8-10 days. The cultures are grown in a light intensity of 40-60 μmol/m^2/sec.

3. Hypocotyls from the 8-10 day-old seedlings are cut into 1 cm pieces and preconditioned for 3 days on callus induction medium (*see* Note 4) in 100 x 15 mm polysterene disposable dishes (Petri dishes). Make sure to excise the apical meristem completely. Ten to fifteen explants were placed in each Petri dish containing 25 ml of the medium.

4. *Agrobacterium* cells are grown overnight in flasks to an OD$_{600}$ = 0.8 in 50 ml liquid LB medium, pelleted, and re-suspended in 15-20 ml liquid callus induction medium to which acetosyringone (Sigma) is added to a final concentration of 0.05 mM (*see* Note 5).

5. The preconditioned hypocotyls are inoculated with *Agrobacterium* for 30 min in a Petri dish, shaking the dish at intervals manually to make sure the explants are in constant contact with *Agrobacterium*. About 100-150 explants are used per 15-20 ml of the *Agrobacterium* solution.

6. The explants are picked carefully from the *Agrobacterium* solution (so that excess *Agrobacterium* is drained) and then co-cultivated with *Agrobacterium* for 48 h on callus induction medium in 100 x 15 mm Petri dishes (*see* Note 6). Twenty explants can be placed on each dish.

7. For selection of transformed cells, the explants are transferred to the callus induction medium with antibiotics (400 mg/l timentin and 200 mg/l kanamycin).

8. After 2 weeks, the transformed organogenetic calli are transferred to 25 ml of organogenesis medium in 100 x 15 mm Petri dishes.

9. After another 2 weeks, the calli are transferred to shoot induction medium in Petri dishes. Shoots are produced after 2-3 weeks on this medium.

10. Shoots (2-4 per Magenta box) are then transferred to the shoot elongation medium in Magenta boxes which contain 25 ml of the culture medium. The shoots elongate in approx. 3 weeks (*see* Note 7).

11. The elongated shoots are transferred to the root induction medium. Roots develop in 1-2 weeks (*see* Note 8). Only one shoot is rooted in each Magenta box which allows the formation of well developed roots and easy removal of the rooted shoot from the Magenta box.

12. The rooted plants are transferred to soil (from STA-GREEN, California, USA which is a mixture of composted forest products, sphagnum peat moss, perlite, ground dolomitic limestone, a wetting agent and water holding polymer) in 1 litre pots and grown in a plant growth chamber. They are covered with a plastic dome to retain humidity.

13. After 3 days, the dome is gradually removed and the plants are transferred to 3.5 litre pots and grown for seed set in the glasshouse.

3.2. Culture Conditions

All the cultures should be maintained at 25 ± 2 °C under a 16 h photoperiod using cool white daylight fluorescent lights at 40-60 μmol/m^2/sec. The rooted shoots are transferred to soil and grown at a photoperiod of 16 h at 20°C in a plant growth chamber at 300 μmol/m^2/sec.

3.3. Progeny Analysis

The progeny of transformed plants are analyzed using PCR for transgene presence. Western blots or RT-PCR may be used for expression analysis. Transgene segregation of the T1 seedlings can be performed using kanamycin selection (200 mg/l kanamycin in MS media—seeds germinated in Petri dishes). A 3:1 Mendelian ratio is expected for single-locus insertion inheritance. Southern analysis and more recently RT-PCR, which is a fluorescence based kinetic RT-PCR, is used to confirm gene copy number.

4. NOTES

1. MS micro and macro salts and vitamins are used in the medium wherever we state MS medium.
2. The *gfp* gene allows real-time monitoring of the transformed callus and shoots. GFP positive callus and plants glow bright green when excited by UV light.
3. Appropriate antibiotics need to be used according to the selectable marker genes on the plasmid. We use kanamycin, since the selectable marker gene in this work is *npt*II which confers resistance to kanamycin.
4. A preconditioning time of 72 h was found to be optimal for high transformation efficiency.
5. Acetosyringone increases the virulence of *Agrobacterium* thus increasing the transformation efficiency.
6. A co-cultivation time of 48 h is critical to obtain good transformation efficiency. With a co-cultivation time of less or more than 48 h the transformation efficiency decreases significantly.
7. A major problem encountered during transformation of canola is the hyperhydration of transformed shoots. This problem is overcome by increasing the Gelrite concentration from 2 g/l to 3 g/l in the rooting and elongation medium. By increasing the Gelrite concentration, the water availability for the shoot is reduced which would otherwise make the plant hyperhydrate.
8. Rooting is very efficient using half-strength MS medium vs. full-strength MS medium and reducing the sucrose concentration from 30 g/l to 10 g/l. The rooting medium described gives 100% rooting in a very short duration of time (1-2 weeks). When full-strength medium is used the plants grow tall instead of producing roots, hence we use a low strength, low sugar medium which facilitates rooting.

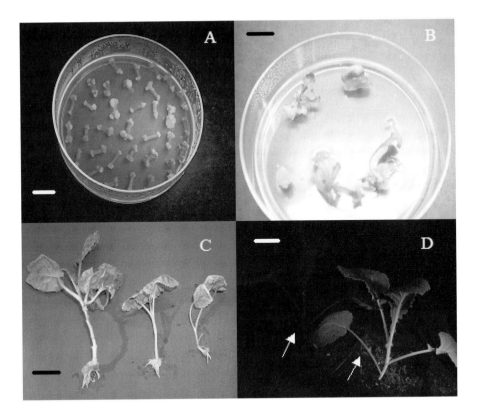

Figure 1. Progression of genetic transformation of canola. A. Hypocotyl segments are inoculated with A. tumefaciens *harbouring a GFP gene. B. Shoots regenerating from transgenic calli. C. Transgenic rooted shoots. D. Transgenic plant expressing GFP (right arrow). Panels A and D are under UV illumination. Horizontal scale bars in images A and B represent 2 mm; in C and D, 5 mm.*

CONCLUSION

Using the transformation method we have described, it is possible to obtain a transformation efficiency of 25%. All the factors which include, preconditioning and co-cultivation time, overcoming hyperhydricity and improving rooting together has brought about this improvement in transformation efficiency. We have used only one variety of canola. We see no reason why these techniques cannot be extended to other varieties of canola. With the increase in demand for canola oil, there is still a need for variety improvement. Various genes that improve oil quality can be genetically engineered in canola using the transformation technique we have described.

ACKNOWLEDGEMENTS

We thank Jim Haseloff for providing the mGFP5-ER construct. We thank Brenda Kivela for assisting in this project. We would also like to thank the NASA HEDS program for funding this research.

REFERENCES

1. Statistics Canada (2003). http://www.statcan.ca/daily/English/030424/do30424a.htm.
2. USDA (2003). http://usda.mannlib.cornell.edu/reports/nassr/field/pcp-bba/acrg0602.txt.
3. Sovero M (1993). Rapeseed, a new oilseed crop for the United States. In: Janick J, Simon JE (eds.), *New Crops*. Wiley, New York, USA, pp. 302-307.
4. Facciotti MT, Bertain PB and Yuan L (1999). Improved stearate phenotype in transgenic canola expressing a modified acyl-acyl carrier protein thioesterase. *Nature Biotechnology*, **17**: 593-597.
5. Knutzon DS, Hayes TR, Wyrick A, Xiong H, Davies HM and Voelker TA (1999). Lysophosphatidic acid acyltransferase from coconut endosperm mediates the insertion of laurate at the sn-2 position of triacylglycerols in lauric rapeseed oil and can increase total laurate levels. *Plant Physiology*, **120**: 739-746.
6. Stoutjesdijk PA, Hurlestone C, Singh SP and Green AG (2000). High-oleic acid Australian *Brassica napus* and *B. juncea* varieties produced by co-suppression of endogenous Delta 12-desaturases. *Biochemial Society Transactions*, **28**: 938-940.
7. Liu JW, DeMichele S, Bergana M, Bobik E, Hastilow C, Chuang LT, Mukerji P and Huang YS (2001). Characterization of oil exhibiting high gamma-linolenic acid from a genetically transformed canola strain. *Journal of the American Oil Chemists's Society*, **78**: 489-493.
8. Katavic V, Friesen W, Barton DL, Gossen KK, Giblin EM, Luciw T, An J, Zou JT, MacKenzie SL, Keller WA, Males D and Taylor DC (2001) Improving erucic acid content in rapeseed through biotechnology: What can the *Arabidopsis FAE1* and the yeast *SLC1-1* genes contribute? *Crop Science*, **41**: 739-747.
9. Oelck MM, Phan CV, Eckes P, Donn G, Rakow G and Keller WA (1991). Field resistance of canola transformants (*Brassica napus* L.) to Ignite (Phosphinothricin) In: McGregor DI (ed.), *Proceedings of the GCIRC Eighth International Rapeseed Congress*. July 9-11, 1991 Saskatoon, Saskatchewan, Canada, pp. 292-297.
10. Monsanto (2002). Safety assessment of Roundup Ready canola event GT73. http://www.monsanto.com/monsanto/content/our_pledge/roundupcanola_product.pdf.
11. Stewart CN Jr, Adang MJ, All JA, Raymer PL, Ramachandran S and Parrott WA (1996). Insect control and dosage effects in transgenic canola containing a synthetic *Bacillus thuringiensis cryIAC* gene. *Plant Physiology*, **112**: 115-120.
12. Grison R, Grezesbesset B, Schneider M, Lucante N, Olsen L, Leguay JJ and Toppan A (1996). Field tolerance to fungal pathogens of *Brassica napus* constitutively expressing a chimeric chitinase gene. *Nature Biotechnology*, **14**: 643-646.
13. Basu U, Good AG and Taylor GJ (2001). Transgenic *Brassica napus* plants overexpressing aluminum-induced mitochondrial manganese superoxide dismutase cDNA are resistant to aluminum. *Plant, Cell and Environment*, **24**: 1269-1278.
14. Houmiel KL, Slater S, Broyles D, Casagrande L, Colburn S, Gonzalez K, Mitsky TA, Reiser SE *et al.* (1999). Poly (β-hydroxybutyrate) production in oilseed leukoplasts of *Brassica napus*. *Planta*, **209**: 547-550.
15. Giddings G, Allison G, Brooks D, and Carter A (2000). Transgenic plants as factories for biopharmaceuticals. *Nature Biotechnology*, **18**: 1151-1155.
16. Fry J, Barnason A and Horsch RB (1987). Transformation of *Brassica napus* with *Agrobacterium* based vectors. *Plant Cell Reports*, **6**: 321-325.
17. Pua EC, Mehra Palta A, Nagy F and Chua NH (1987). Transgenic plants of *Brassica napus* L. *Bio/Technology*, **5**: 815-817.

18. Moloney MM, Walker JM and Sharma KK (1989). High efficiency transformation of *Brassica napus* using *Agrobacterium* vectors. *Plant Cell Reports,* **8**: 238-242.

19. Radke SE, Andrews BM, Moloney MM, Crouch ML, Krid JC and Knauf VC (1988). Transformation of *Brassica napus* L. using *Agrobacterium tumefaciens*: developmentally regulated expression of a reintroduced napin gene. *Theoretical and Applied Genetics,* **75**: 685-694.

20. De Block M, De Brower D and Tenning P (1989). Transformation of *Brassica napus* and *Brassica oleracea* using *Agrobacterium tumefaciens* and the expression of the *bar* and *neo* genes in the transgenic plants. *Plant Physiology,* **91**: 694-701.

21. Cardoza V and Stewart CN Jr (2003). Increased *Agrobacterium* mediated transformation and rooting efficiencies in canola (*Brassica napus* L.) from hypocotyls explants. *Plant Cell Reports,* **21**: 599-604.

22. Souvre A, Jardinaud MF and Alibert G (1996). Transformation of rape (*Brassica napus* L.) through the haploid embryogenesis pathway. *Acta Societatis Botanicorum Poloniae,* **65**: 194-195.

23. Jones-villeneuve E, Huang B, Prudhomme I, Bird S, Kemble R, Hattori J and Miki B (1995). Assessment of microinjection for introducing DNA into uninuclear microspores of rapeseed. *Plant Cell, Tissue and Organ Culture,* **40**: 97-100.

24. Fukuoka H, Ogawa T, Matsuoka M, Ohkawa Y and Yano H (1998). Direct gene delivery into isolated microspores of rapeseed (*Brassica napus* L.) and the production of fertile transgenic plants. *Plant Cell Reports,* **17**: 323-328.

25. Nehlin L, Moellers C, Bergman P and Glimelius K (2000). Transient beta-*gus* and *gfp* gene expression and viability analysis of microprojectile bombarded microspores of *Brassica napus* L. *Journal of Plant Physiology,* **156**: 175-183.

26. Hervé C, Rouan D, Guerche P, Montané MH and Yot P (1993). Molecular analysis of transgenic rapeseed plants obtained by direct transfer of 2 separate plasmids containing, respectively, the cauliflower mosaic virus coat protein gene and a selectable marker gene. *Plant Science,* **91**: 181-193.

27. Chen JL and Beversdorf WD (1994). A combined use of microprojectile bombardment and DNA imbibition enhances transformation frequency of canola (*Brassica napus* L.). *Theoretical and Applied Genetics,* **88**: 187-192.

28. Halfhill MD, Richards HA, Mabon SA and Stewart CN Jr (2001). Expression of GFP and *Bt* transgenes in *Brassica napus* and hybridization with *Brassica rapa*. *Theoretical and Applied Genetics,* **103**: 659-667.

29. Ovesná J, Ptacek L and Opartny Z (1993). Factors influencing the regeneration capacity of oilseed rape and cauliflower in transformation experiments. *Biologia Plantarum,* **35**: 107-112.

30. Schröder M, Dixelius C, Råhlèn L and Glimelius K (1994). Transformation of *Brassica napus* by using the *aadA* gene as selectable marker and inheritance studies of the marker genes. *Physiologia Plantrum,* **92**: 37-46.

31. Murashige T and Skoog F (1962). A revised medium for rapid growth and bioassay with tobacco tissue cultures. *Physiologia Plantarum,* **15**: 473-497.

32. Harper, BK, Mabon SA, Leffel SM, Halfhill MD, Richards HA, Moyer KA and Stewart CN Jr (1999). Green fluorescent protein in transgenic plants indicates the presence and expression of a second gene. *Nature Biotechnology,* **17**: 1125-1129.

Chapter 29

TRANSFORMATION OF CAULIFLOWER

D.C.W. BROWN AND H.Y. WANG

Southern Crop Protection and Food Research Centre, Agriculture and Agri-Food Canada, 1391 Sandford Street, London, Ontario, Canada, N0M2A0. Corresponding author: D. Brown E-mail: browndc@agr.gc.ca

1. INTRODUCTION

1.1. Importance of Cauliflower

Cauliflower is an important vegetable crop grown in about 80 countries worldwide (*1*). The world acreage has been increasing steadily over the last four decades; that is, it has increased from 3,595,442 metric tonnes (MT) in 1962 to 15,053,841 MT in 2002 with the World's biggest supplier, China, increasing not only its production but also is proportional contribution from 12% to 42% of world production over the same period. As of 2002, other important producing areas (comprising 89% of world production) are: India (4,800,000 MT), Italy (481,454 MT), France (441,611 MT), Spain (294,100 MT), USA (291,540 MT), Poland (201,154 MT) Pakistan (200,792 MT), Mexico (200,000 MT) and Germany (128,005 MT). The remaining 11% of world production is shared among 67 additional countries (*1*).

   Cauliflower *(Brassica oleracea* var. *botrytis)* belongs to the highly diversified *Brassicaceae* family comprising about 3000 species. The origin, evolution, taxonomy and genomic relationships of the crop *Brassicas* including *Brassica oleracea* is outlined (*see* Chapter 27, Note 1). Caulifower and its close relative broccoli (*Brassica oleracea* var. *italica*) are vegetable types cultivated for their thickened edible inflorescence also referred to as a "curd". Cauliflower is considered to be a "cool season" crop and grows well in daytime temperatures of 18–20°C. A profile of typical North American production practices can be found in information supplied by North Carolina State University (*2*).

1.2. Genetic Improvement of Cauliflower

The priorities for genetic improvement in cauliflower are similar to other Brassicas: uniformity, disease resistance, appearance, crop yield, pest resistance, nutritional quality and new crop types. Of high importance is the need for specific age and temperature requirements for proper development in the field in each location. This last requirement, in particular, has led to a wide array of cultivar availability.

   Genetic transformation in cauliflower (*3*), *Brassica olerecea* (*4*) and related *Brassicas* (*5- 7*) has been reviewed in some detail. In cauliflower, only a few traits have been introduced using genetic transformation (Table 1); notably, insect, virus and bacteria resistance (*8, 10, 12, 15, 16, 19, 20*). Genetic transformation of

389

I.S. Curtis (ed.), Transgenic Crops of the World - Essential Protocols, 389-403.
© 2004 Kluwer Academic Publishers. Printed in the Netherlands.

cauliflower can be achieved by various transformation methods including electroporation of protoplasts, PEG-mediated direct uptake into protoplasts and *Agrobacterium tumefaceins* and *A. rhzogenes*-mediated transformastion of cotyledon, hypocotyls and stem explants. The majority of the transgenic cauliflower plants produced to date have been created through the use of *Agrobacterium*-mediated gene transfer methods (*7, 8, 11, 13, 14, 19, 21, 27*). A wide range of *Agrobacterium* strains have been tested (*28*) and successfully used for the transformation of cauliflower (*10, 12*). In general, nopaline strains of *A. tumefaciens* and *A. rhizogenes* are most virulent (*4, 20, 30, 31*), although a range of strains have been reported to be useful (Table 1). Genotype dependant responses appear to be a minor consideration based on the wide range of cultivars and lines that have been reported to be amenable to gene transfer technology (Table 1). In addition, although most selection studies have been done using antibiotic resistance genes (*8-10, 12-24*), there has been reports using herbicide (*9, 21*) and morphological selection (*11, 14*). This suggests that there may be a number of options available for improving the efficiencies of existing systems beyond the 26% we have observed (*see* Note 8). In contrast, the explant of choice in most recent studies has been the hypocotyl, presumably due to its easy generation of large uniform numbers as well as its amenability to *in vitro* manipulation.   Here, we outline an *A. tumefaciens*-based protocol that has given consistently good results in our hands.

There are no reports of stable transformation of cauliflower using the biolistics approach, but is a very useful tool for evaluating gene expression and especially promoter activity analysis. Biolistic transformation, first reported in 1987 (*32*) is a method by which DNA may be introduced into intact plant cells and tissues via acceleration of microprojectile-coated DNA particles to a high velocity in order to penetrate the target tissue. This method is particularly useful for transient gene expression assays such as studies of promoter activity but has been effective in other systems (e.g., maize, soybean) for stable transformation. Successful delivery of genes into a target tissue by particle bombardment requires optimization of a number of the parameters involved (*33*). The following protocol was developed in our laboratory (*34, 35*) and used for extensive transient expression studies of the tCUP promoter in cauliflower tissue (*16*).

2. MATERIALS

2.1. Biolistic Transformation

2.1.1. Medium and Plant Tissue

1. Seed germination medium (SGM): Half-strength Murashige and Skoog (MS, *36*) salts (PhytoTechnology Laboratories), half-strength B_5 vitamins (*37*), 15 g/l sucrose and 6 g/l agar (Bacto-Agar, DIFCO), pH 5.7.
2. Basic MS medium (BMS): MS salts and B5 vitamin, 30 g/l sucrose, 6 g/l agar and pH 5.7.
3. Cauliflower leaves from approx. 4 week-old *in vitro*-grown plantlets.

2.1.2. Buffers, Solutions and Particles

1. TE buffer: 10 mM Tris-HCl, 1 mM EDTA, pH 8.0.
2. TAE (1X): 40 mM Tris, 0.5 mM EDTA.
3. $CaCl_2$ solution (2.5M): Dissolve 36.75g of $CaCl_2$ (Sigma) in 100 ml de-ionized H_2O and sterilize the solution by passing it through a 0.2 μm filter. Store the solution in 500 μl aliquots at 4°C.
4. Spermidine solution (0.1 M): Dissolve 145 mg of spermidine (Sigma) in 10 ml deionised water and sterilize the solution by passing it through a 0.2 μm filter. Store at -20°C. The spermidine solution should be made in small aliquots (such as 50 μl) which are enough for one bombardment experiment. Discard any remaining solution.
5. Ethanol (70% and 95%).
6. Tungsten particle (microcarriers): 100 mg of tungsten of 1.3 μm in diam. (Bio-rad, Inc).

2.1.3. Plasmid DNA

QIAGEN plasmid maxi kits (QIAGEN, Inc): For plasmid DNA isolation (*see* Note 1).

2.2. Agrobacterium-*mediated Transformation*

2.2.1. A. tumefaciens *Strain*

In our laboratory, nopaline strain GV3101 containing the disarmed Ti plasmid pGV3850/pMP90 (*45*) was used to carry a binary vector for transformation. The T-DNA of the binary vector contained an *npt*II gene under the control of the NOS promoter and terminator, which was used as a marker for the transgenic plant selection. Apart from the selectable marker, an expression cassette containing a reporter gene *uid*A or a target gene such as *cry1Ac* was under the control of a tCUP promoter (*16*) and a NOS terminator.

2.2.2. Culture Medium for A. tumefaciens

1. Kanamycin sulphate: 25 mg/ml stock in water. Sterilize by filtration through a 0.2 μm membrane. Dispense 1 ml aliquots into Eppendorf tubes and store at -20°C.
2. Gentamycin: 25 mg/ml stock in water. Sterilize by filtration through a 0.2 μm membrane. Dispense 1 ml aliquots into Eppendorf tubes and store at -20°C.
3. Rifampicin: 25 mg/ml stock in DMSO. Dispense 1 ml aliquots into Eppendorf tubes and store at -20°C.

4. Luria broth (LB) medium: 10 g/l tryptone, 5 g/l yeast extract, 10g/l NaCl, pH 7.0-7.2
5. Liquid LB medium: LB medium containing 50 mg/l kanamycin sulphate, 50 mg/l gentamycin, and 50 mg/l rifampicin.
6. Semi-solid LB medium: LB medium containing 150 mg/l kanamycin sulphate, 100 mg/l gentamycin, 150 mg/l rifampicin and 10 g/l agar.
7. Inoculation liquid medium (1/2MSi): half-strength MS salts and half-strength B5 vitamins, 0.095% *myo*-inositol, 1.5% sucrose, pH 5.8.

2.2.3. Plant Material

Hypocotyls taken from 4-10 day-old seedlings are used as explants for transformation of cauliflower (Fig. 1, Step 1).

2.2.4. Tissue Culture

1. Vitamin stock: B_5 vitamin 1000X stock, store in -20°C.
2. Timentin stock: dissolve 3 g timentin (SmithKline Beecham Pharma) in 30 ml water and sterilize by filtration through a 0.2 µm membrane. Stock concentration: 100 mg/ml.
3. Surface sterilant: 20% (v/v) commercial sodium hypochlorite bleach.
4. Seed germination medium (SGM): half-strength MS salts and half strength of B_5 vitamins, 15 g/l sucrose and 6 g/l agar (Bacto-Agar, DIFCO), pH 5.7.
5. Basic MS medium (BMS): Full-strength MS salts and full-strength B5 vitamin, 0.5 g/l MES (Sigma), 30 g/l sucrose, 6 g/l agar and pH 5.7.
6. Liquid BMS: Same composition as semi-solid BMS except no agar added.
7. Pre-treatment medium (PTM): BMS with 1 mg/l 2,4-D and 0.1 mg/l kinetin.
8. Co-cultivation medium (CCM): BMS with 2 mg/l BAP, 0.01 mg/l NAA.
9. Selection shoot regeneration medium (SSM): BMS with 2 mg/l BAP, 0.01 mg/l NAA, 2 mg/l $AgNO_3$, 300 mg/l timentin and 25 mg/l kanamycin (*9, 48*).
10. Selection /Rooting medium (SRM): BMS with 300 mg/l timentin and 20 mg/l kanamycin.
11. Kanamycin callusing medium (KCM): BMS with 1 mg/l BAP, 0.2 mg/l NAA, 300 mg/l timentin and 25 mg/l kanamycin.
12. Basic rooting medium (BRM): BMS without growth regulators.

All media should be sterilized by autoclaving at 121°C for 20 min. All growth regulator stock solutions should be sterilized by filtration through a 0.2 µm membrane. Growth regulators and antibiotics should be added to the medium after autoclaving and cooled to 50°C.

3. METHODS

3.1. Biolistic Transformation

3.1.1. Preparation of Plant Tissue

1. Sow surface-sterilized cauliflower seeds on SGM medium (Fig. 1).
2. After seed germination and the hypocotyls elongate (approx. 5-7 days), excise the apical meristems by cutting the hypocotyls with sterile scissors approx. 0.5 cm below the pair of cotyledonary leaves.
3. Place the apical meristems on BMS medium in Magenta GA-7 boxes and allow to grow for about 4 weeks until the leaves are of a suitable size (2.5-3 cm).
4. Excise the leaves and cut into 2.25 cm^2 pieces with the mid-rib in the centre of the square. Small horizontal cuts are made along the mid-rib and outer edges in order to allow the pieces to lay flat.
5. Establish new *in vitro* cultures from seed every 3 months. It has been observed that transient expression levels decrease as the age of the *in vitro* cultures increase.

3.1.2. Plasmid DNA Isolation and Quantification

1. Use QIAGEN plasmid maxi kits (QIAGEN, Inc) to isolate plasmid DNA.
2. After DNA precipitation, use two washes instead of one wash with 70% ethanol.
3. Quantify plasmid DNA using a spectrophotometer and adjust the concentration of DNA to 1 µg/µl using TE buffer (*see* Note 2).
4. Since RNA can be present in DNA samples, it is highly recommended that the plasmid DNA should be checked by running 1 µg sample on a 0.8% agarose gel in 1X TAE to ensure purity and accurate quantification.

3.1.3. Tungsten Particle Preparation

1. Weigh 100 mg tungsten particles (1.3 µm in diam.) (Bio-red, Inc) in a 1.5 ml Eppendorf tube.
2. Add 1 ml of 95% ethanol to the particles.
3. Mix the contents of the tube by vortex for 60 sec then centrifuge at 10,000 rpm for 60 sec in a table top centrifuge.
4. Remove the ethanol and add 1 ml sterile water to wash.
5. Vortex the tube again for 60 sec then centrifuge at 10,000 rpm for 60 sec. Gently remove the supernatant.
6. Repeat the water wash (*see* Section 2.2.3.4.) twice.
7. Re-suspend the tungsten in 1 ml sterile water.

8. While continuously mixing using a vortex mixer, pipette a 25 µl aliquot into a 0.5 ml tube (*see* Note 3).
9. Store the tungsten solution at -80°C (*see* Note 4).

3.1.4. DNA Coating onto Tungsten Particles

1. Thaw the tungsten on ice and mix using a vortex mixer for 2 min.
2. During mixing, add 5 µl of the 1 µg/µl plasmid DNA, 25 µl of the 2.5 M CaCl$_2$ and 5 µl of the 0.1 M spermidine solutions should be quickly added to the tungsten. When performing a co-bombardment using different plasmids, keep each plasmid in a separate tube then add the tungsten at this stage.
3. After all ingredients are added to each tube, continue mixing the tube by vortex for an additional 3 min. It is very important that the tube be continuously mixed by vortex during this procedure to ensure uniform coating of the particles with plasmid DNA.
4. Place the tubes on ice and allow to stand for 5 min to allow the tungsten particles to settle to the bottom of the tube.
5. Remove 54 µl of the supernatant. For bombardment, 2 µl of the remaining tungsten/DNA solution is removed by pipetting. Using this approach, the remaining tungsten/DNA solution in the tube should be enough to bombard 3 leaves. It is recommended that three tubes at a time be processed as described in this step (*see* Note 5).

3.1.5. Bombardment

1. After the DNA is coated onto the particles as outlined above, bombardment should be conducted as soon as possible.
2. Following the manufacturer's directions, pipette 2 µl of the tungsten/DNA onto a macrocarrier in the case of PDS 1000-type (*38*) equipment or onto a holding screen in the case of Particle In Flow-type (*34, 35, 39*) equipment. Load into the biolistic equipment and bombard a plate of cauliflower tissue. Bombardment pressure, bombardment distance, particle size, pre-treatment conditions and other factors must be experimentally determined for each type of equipment and each type of tissue (*34, 35, 38-40*).

3.1.6. Post-bombarment

Bombarded tissue cultures should be maintained on BMS culture medium at 24°C for at least 24 h before further testing. For transient expression, the tissue can be assessed using different assays; for GFP activity, examine the tissue with a microscope under 450-490 nm illumination. For GUS expression, assay for β-glucuronidase activity by histochemical staining or fluorometrical analysis (*14, 25, 41, 42*). For other gene products, analyse the newly synthesized protein by radioimmunoassay, by immunoblotting or by assays of appropriate enzymatic activity in cell extracts (*43, 44*). The optimum time after bombardment for target tissue analysis must be determined experimentally (*34*).

3.2. Agrobacterium-*mediated Transformation*

3.2.1. *Explant Preparation and Pre-culture*

1. Seeds are rinsed in a sterile Petri dish containing 70% ethanol for 60 sec, washed once with sterile distilled water, then surface-sterilized with 20% (v/v) bleach for 20 min while constantly stirring and, finally, rinsed thoroughly 3-4 times with sterile distilled water.
2. Sow sterile seeds onto growth regulator-free SGM medium.
3. Harvest hypocotyls from 4-10 day-old seedlings for transformation (Fig. 1, Step 2). Carefully remove the meristem, cotyledon and root tissues and place whole hypocotyls onto the pre-treatment medium PTM (Fig. 1, Step 3). Seal the plates with Parafilm to keep humidity. Culture for 3 days at 24°C in light (30 μE/m^2/sec, 16 h photoperiod).

3.2.2. Agrobacterium *Preparation*

1. Streak *Agrobacterium* cells from a stored glycerol stock onto a semi-solid LB medium containing appropriate antibiotics and incubate at 28°C for 2-3 days.
2. Pick single colonies and inoculate into a 125 ml flask containing 20 ml of LB medium with appropriate antibiotics. Incubate culture overnight in the dark at 28°C on a shaker (Fig. 1, Step 4).
3. Measure the OD of the overnight culture at 600 nm (OD$_{600}$ = 1.0). Use LB with antibiotics as a background reference (blank) for determining OD.
4. Centrifuge the *Agrobacterium* culture at 5000 rpm for 5 min, discard the supernatant, and re-suspend in liquid 1/2MSi medium.
5. Dilute *Agrobacterium* culture 1:10 with 1/2MSi medium (*see* Note 6).

3.2.3. *Inoculation and Co-cultivation*

1. After the 3 day pre-treatment, the hypocotyls are dissected into 3-5 mm segments in the presence of the diluted *Agrobacterium* culture. At the same time, lightly score each explant 2-3 times on one side with a sharp scalpel blade (Fig. 1, Step 5).
2. Transfer explants, scored-side up, onto CCM medium.
3. Add 5 μl of diluted *Agrobaterium* solution to each explant (Fig. 1, Step 6). This method of inoculation is called the "Droplet Method". Further approaches are also described (*see* Note 7).

4.   Seal the plates (optional) with micropore™ tape.
5.   Co-culture in dark for 4-6 days (*see* Note 8) at 24°C and/or until there is a halo of bacteria culture around each explant (Fig. 1, Step 7).

3.2.4. Selection and Recovery of Transgenic Plants

1.   After 4 days co-cultivation, blot explants on sterile filter paper to remove excess *Agrobacterium* and then transfer onto SSM medium without antibiotics (*18*). Culture the explants at 24°C under a 16 h photoperiod (irradiance 30 µE/m^2/sec) for 7 days, and then transfer to SSM medium with antibiotics (Fig. 1, Step 8).
2.   Explants are maintained on SSM selection medium for 6-8 weeks with transfer to fresh medium every 2 weeks. All cultures are maintained at the same culture conditions as mentioned above. During this selection and regeneration period, all the plates are sealed with micropore™ tape (*see* Note 9).
3.   Green shoots usually appear 2-3 weeks after selection. The early developing shoots are usually escapes. Green, putatively transgenic shoots usually form about 6-8 weeks after co-cultivation (Fig. 1, Step 9). Transfer green shoots to KRM medium under the same conditions for further selection and rooting. Several small pieces of the bottom leaves can be taken and placed on KCM medium (selective callus introducing medium) to confirm resistance to kanamycin, as evidenced by the ability to enlarge, produce callus or form roots. (Fig. 1, Step 10). Non-kanamycin resistant tissues are derived from escape shoots.
4   Transgenic cauliflower shoots usually form roots in 10-15 days in a growth regulator-free MS medium as shown (Fig. 1, Step 10). After well-rooted plants develop, carefully remove the shoot tips from the rooted plantlets onto KRM medium, without touching the bottom leaves and medium (Fig. 1 Steps 10 and 11), transfer to BRM medium without any antibiotics and culture under the same conditions. This step of transplanting the putative bacteria-free plantlets from the antibiotic-free medium (BRM) ensures that there is no transfer of *Agrobacterium* to the soil.
5.   Rooted plantlets growing on BRM medium are transferred to soil and grown under glasshouse conditions.

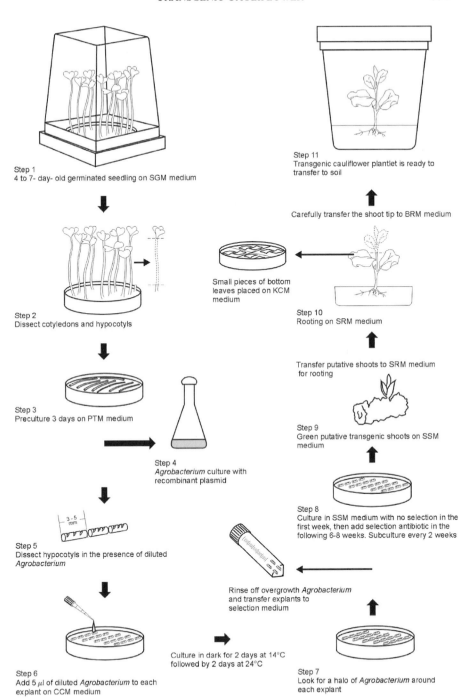

Step 1
4 to 7- day- old germinated seedling on SGM medium

Step 2
Dissect cotyledons and hypocotyls

Step 3
Preculture 3 days on PTM medium

Step 4
Agrobacterium culture with recombinant plasmid

Step 5
Dissect hypocotyls in the presence of diluted *Agrobacterium*

3 - 5 mm

Step 6
Add 5 μl of diluted *Agrobacterium* to each explant on CCM medium

Culture in dark for 2 days at 14°C followed by 2 days at 24°C

Step 7
Look for a halo of *Agrobacterium* around each explant

Rinse off overgrowth *Agrobacterium* and transfer explants to selection medium

Step 8
Culture in SSM medium with no selection in the first week, then add selection antibiotic in the following 6-8 weeks. Subculture every 2 weeks

Step 9
Green putative transgenic shoots on SSM medium

Transfer putative shoots to SRM medium for rooting

Step 10
Rooting on SRM medium

Small pieces of bottom leaves placed on KCM medium

Carefully transfer the shoot tip to BRM medium

Step 11
Transgenic cauliflower plantlet is ready to transfer to soil

Figure 1. Agrobacterium-*mediated transformation of cauliflower by the droplet method.*

6. Transgenic plantlets at the 4-5 leaf stage of development are transplanted from *in vitro* conditions to 8 cm diam. seed trays and maintained under glasshouse conditions (21-22°C, 16 h photoperiod). We use Pro-mix BX (Premier Horticulture Inc., *see* www.premierhort.com) as a transplanting medium. After 2 weeks, the established plants are transferred to 10 cm diam. plastic pots and then finally transferred to 18 cm diam pots. Cauliflower requires a cold treatment in order to bolt and flower; therefore, after the curd begins to form, the plants are transferred to a lower temperature for 4-6 weeks (18-24°C day, 7-13°C night, under a 16 h photoperiod). During this cold treatment, the soil is kept dry and fertilized twice (N:P:K; 1:1:1) at 5-10 g per pot. When flower stalks appear, the plants are returned to the glasshouse (21-22°C, 16 h photoperiod). Hand pollination is followed by T1 seed collection approx. 2 months later. The following is an outline of the transformation cycle for cauliflower. The whole process from sowing the seeds *in vitro* to collecting T1 seeds in the glasshouse takes 33-42 weeks: sowing seeds (1 week), preculture and transformation (1 week), regeneration on non-selection medium (1 week), regeneration on selection medium (4-6 weeks), rooting (1 week), transplantation and growth (8-10 weeks), curd development (5-6 weeks), cold treatment for bolting and flowering (4-6 weeks), pollination and seed set (8-10 weeks).

7. Seed storage and T1 seedling assessment is as described for cabbage (*see* Chapter 27, Sections 3.4.8.-3.4.10.).

4. NOTES

1. Even minor contamination of the plasmid DNA with endotoxin/lipopolysaccharide reduces the frequency of transfection. The quality of plasmid DNA is crucial. Different protocols and company kits can be used for plasmid DNA isolation, the CsCl method and Qiagen plasmid DNA isolation kits give satisfactory and consistent results for bombardment and can be employed. Currently, Qiagen kits are mostly used in our laboratory.

2. DNA is usually stable in TE buffer when kept frozen. However, after long term storage (e.g. > 6 months) DNA quantity and quality should be checked again by gel electrophoresis and using a spectrophotometer before use in bombardment. If a large quantity of DNA is isolated, a small portion (e.g. 20-30 µg) may be taken out for the next several bombardment experiments and the remaining is kept frozen. Avoid too many freeze-thaw steps.

3. Tungsten particles settle quickly to the bottom of the tube. Maintain a vigorous and constant vortex mixing while transferring by a pipette to ensure each tube receives the same amount of tungsten.

4. The tungsten solution may be stored at room temperature for 1-2 weeks. Longer storage can result in oxidation of the metal beads and a decline in transformation efficiency.

5. It is strongly recommended 3 tubes of tungsten/DNA be prepared at one time. Transient gene expression decreases if the DNA/metal particle preparation is left too long before bombardment. It appears that the particles aggregate into clumps and give a very poor distribution after bombardment and clumped particles may prevent DNA release (*46*). Typically, we use 3 tubes (three replications) of each plasmid. A skilled scientist can handle between 8-24 plasmids per day.

6. Dilute *Agrobacterium* culture with ½ MSi to a 1:10 (vol:vol). Take the *Agrobacterium* culture (1) OD $_{reading}$ X 1 ml and add ½ MSi 10) (1) OD $_{reading}$) ml. Example: OD $_{reading}$ = 0.6, *Agrobacterium* culture = (1) 0.6) x 1 ml = 1.67 ml, ½ MSi = 10 ml - 1.67 ml = 8.33 ml. Therefore, total diluted *Agrobacterium* solution = 1.67 ml + 8.33 ml = 10 ml.

7. We have tested different inoculation methods on cauliflower transformation. Our data showed that the transformation efficiency of droplet method or infiltration method was almost 10 times more than the immersion method. The following transformation protocols were tested:

a. Immersion method: After a 3 day pre-treatment, dissect the hypocotyls into 3-5 mm segments in the presence of the diluted *Agrobacterium* culture (Fig. 1, Step 5).

    1. Soak the explants in the *Agrobacterium* culture for 10 min.

    2. Blot-dry the explants on sterile filter paper to remove extra liquid and bacterial cells.

    3. Transfer the blotted explants to CCM medium (*see* Section 3.2.4.1.).

b. Droplet method: (*see* Section 3.2.3., Fig. 1)

c. Infiltration method: After a 3 day pre-treatment, dissect the hypocotyls into 3-5 mm segments in the presence of the diluted *Agrobacterium* culture (Fig. 1, Step 5).

    1. Transfer explants into a 20 ml syringe and mix with *Agrobacterium* culture. Then, cover the front of the syringe with your finger (wear sterile glove), and then pull the plunger to create vacuum inside the syringe.

    2. Release the plunger, hold it for 10 minutes and then transfer explants to CCM medium. Although this method seems to enhance infection efficiencies, explant damage sometimes occurs.

Under our laboratory conditions, the transformation efficiency (number of recovered transgenic plants/number of initial explants x 100) of the infiltration method (11.3%) is superior compared to the immersion (1.6%) and droplet (10.5%) methods.

8.   The co-cultivation period mostly depends on the virulence of the *Agrobacterium* strain used and the plant material. We have not recovered transformants from a 3 day (or less) co-cultivation period using the GV3101 strain. As the co-cultivation period increases, the frequency of transformation also increases. The highest transformation efficiency of explants of cauliflower in our laboratory was 26% (*15, 16*) using the droplet method. However, in some cases, damage may occur due to overgrowth of *Agrobacterium*. Under these circumstances, it is recommended to a) decrease the concentration of especially virulent strains, and/or b) decrease the incubation temperature (i.e. 2 days at 24°C followed by 2 days at 10 or 15°C).

9.   Sealing plates with Micropore™ tape gives better results by reducing humidity and probably by preventing build-up of ethylene (*5, 47*).

ACKNOWLEDGEMENTS

The authors wish to thank Agriculture and Agri-Food Canada, the Ontario Research Enhancement Program and the Ontario Fruit and Vegetable Growers Association for financial support for research leading to the protocols described here.

CONCLUSIONS

The methods described here have been used in our laboratory for several years and have given consistently good results. As we have outlined, a number of independent groups have used hypocotyls explants of cauliflower from a number of cultivars as target tissue for *Agrobacterium*-mediated gene transfer. In addition, a number of target genes have been expressed in cauliflower plants to date. In our opinion, the technical limitations for genetic improvement of cauliflower have been largely overcome and future progress will depend on gene availability and agronomic/business opportunities.

REFERENCES

1. FAOSTAT-Agriculture (2003). http://apps.fao.org/default.htm.
2. North Carolina State University (2003). http://pestdata.ncsu.edu/cropprofiles/docs/.
3. Dixit S and Srivastava DK (2000). Genetic transformation in cauliflower – a mini review. *Journal of Plant Biology*, **27**: 99-104.
4. Puddephat IJ, Riggs TJ and Fenning TM (1996). Transformation of *Brassica oleracea* L.: a critical view. *Molecular Breeding*, **2**: 185-210.
5. Earle ED, Metz TD, Roush RT and Shelton AM (1996). Advances in transformation technology for vegetable *Brassicas*. *Acta Horticulturae,* **407**: 161-168.
6. Poulsen GB (1996). Genetic transformation of *Brassica*. *Plant Breeding*, **115**: 209-225
7. Paul KA, Kumar PA and Saradhi PP (2002). Genetic transformation of vegetable *Brassicas*: a review. *Plant Cell Biotechnology and Molecular Biology*, **3**: 1-10.
8. Bhalla PL and Smith N (1998). *Agrobacterium tumefaciens*-mediated transformation of cauliflower, *Brassica oleracea* var. *botrytis*. *Molecular Breeding*, **4**: 531-541.
9. de Block M, de Brouwer D and Tenning P (1989). Transformation of *Brassica napus* and *Brassica oleracea* using *Agrobacterium tumefaciens* and the expression of the *bar* and *neo* genes in transgenic plants. *Plant Physiology*, **91**: 694-701.
10. Chakrabarty R, Viswakarma N, Bhat SR, Kirti PB, Singh BD and Chopra VL (2002). *Agrobacterium*-mediated transformation of cauliflower: optimization of protocol and development of *Bt*-transgenic cauliflower. *Journal of Biosciences*, **27**: 495-502.
11. David C and Tempe J (1988). Genetic transformation of cauliflower (*Brassica oleracea* L. var. *botrytis*) by *Agrobacterium rhizogenes*. *Plant Cell Reports*, **7**: 88-91.
12. Ding LC, Hu CY, Yeh KW and Wang PJ (1998). Development of insect-resistant transgenic cauliflower plants expressing the trypsin inhibitor gene isolated from local sweet potato. *Plant Cell Reports*, **17**: 854-860.
13. Hasegawa I, Terada E, Sunairi M, Wakita H, Shinmachi F, Noguchi A, Nakajima M and Yazaki J (1997). Genetic improvement of heavy metal tolerance in plants by transfer of the yeast metallothionein gene (CUP*1*). *Journal of Plant and Soil*, **196**: 277-281.
14. Puddephat IJ, Robinson HT, Fenning TM, Barbara DJ, Morton A and Pink DAC (2001). Recovery of phenotypically normal transgenic plants of *Brassica oleracea* upon *Agrobacterium rhizogenes*-mediated co-transformation and selection of transformed hairy roots by GUS assay. *Molecular Breeding*, **7**: 229-242.
15. Tian L, Wang H, Wu K, Latoszek-Green M, Hu M, Miki B and Brown DCW (2002). Efficient recovery of transgenic plants through organogenesis and embryogenesis using a cryptic promoter to drive marker gene expression. *Plant Cell Reports*, **20**: 1181-1187.
16. Tsang E (2001). Evaluation in cauliflower of genetic elements of a novel tobacco constitutive cryptic promoter. M.Sc. Thesis, Faculty of Graduate Studies, University of Guelph, Guelph, Canada. pp. 159.
17. He YK, Gong ZH, Wang F and Wang M (1991). Transformation efficiency of *Brassica* crops with *Agrobacterium rhizogenes* harbouring the binary vector Bin 19. *Chinese Journal of Biotechnology*, 7: 382-385.
18. Ovesna J, Ptacek L and Opatrny Z (1993). Factors influencing the regeneration capacity of oilseed rape and cauliflower in transformation experiments. *Biologia Plantarum*, **35**: 107-112
19. Braun RH, Reader JK and Christey MC (2000). Evaluation of cauliflower transgenic for resistance to *Xanthomonas campestris* pv. *campestris*. *Acta Horticulturae*, **539**: 137-143.
20. Passelegue E and Kerlan C (1996). Transformation of cauliflower (*Brassica oleracea* var. *botrytis*) by transfer of cauliflower mosaic virus genes through combined cocultivation with virulent and avirulent strains of *Agrobacterium*. *Plant Science*, **113**: 79-89.
21. Mukhopadhyay A, Topfer R, Pradhan AK, Sodhi YS, Steinbib HH, Schell J and Pental D (1991). Efficient regeneration of *Brassica oleracea* hypocotyls protoplasts and high frequency genetic transformation by direct DNA uptake. *Plant Cell Reports*, **10**: 375-379.
22. Radchuk VV, Ryschka U, Schumann G and Klocke E (2002). Genetic transformation of cauliflower (*Brassica oleracea* var. *botrytis*) by direct DNA uptake into mesophyll protoplasts. *Physiologia Plantarum*, **114**: 429-438.
23. Eimert K and Siegemund F (1992). Transformation of cauliflower (*Brassica oleracea* L. var.

botrytis): an experimental survey. *Plant Molecular Biology*, **19**: 485-490.

24.   Xue HW, Wei ZM and Xu ZH (1997). Regeneration of transgenic plants of *Brassica oleracea* L. via PEG- mediated genetic transformation. *Acta Botanica Sinica*, **39**: 28-33.

25.   Yang ZN, Xu ZH, Bai YY and Wei ZM (1994). The transient gene expression of exogenous GUS genes in cauliflower hypocotyl protoplasts. *Acta Phytophysiologica Sinica*, **20**: 272-276.

26.   Sonali D and Srivastava DK (2000). Genetic transformation in cauliflower - a mini review. *Journal of Plant Biology*, **27**: 99-104.

27.   Baranski R and Puddephat I (2003). Variation of beta-glucuronidase activity in cauliflower plants with *gus* gene introduced by *Agrobacterium rhizogenes* mediated transformation. *Acta Physiologiae Plantarum*, **25**: 63-68.

28.   Qazi SHA and Shahida H (1997). Impact of *Agrobacterium tumefaciens* co-cultivation time and temperature on T-DNA transfer and expression in plant cells. *Punjab University Journal of Zoology*, **12**: 59-72.

29.   Mora-Aviles MA and Earle ED (2000). Factors affecting *Agrobacterium tumefaciens*-mediated transformation of broccoli. *Serie Ingenieria Agricola*, **6**: 157-164.

30.   Henzi MX, Christey MC and McNeil DL (2000). Factors that influence *Agrobacterium rhizogenes*-mediated transformation of broccoli (*Brassica oleracea* L. var. *italica*). *Plant Cell Reports*, **19**: 994-999.

31.   Stipic M, Rotino GL and Piro F (2000). Regeneration and genetic transformation attempts in the cauliflower 'Tardivo di Fano'. (Translated from :Rigenerazione *in vitro* ed esperimenti di trasformazione genetica del cavolfiore 'Tardivo di Fano'). *Italus Hortus*, **7**: 20-26.

32.   Klein TM, Wolf ED, Wu R and Sanford JC (1987). High-velocity microprojectiles for delivering nucleic acids into living cells. *Nature*, **327**: 70-73.

33.   Fromm M, Klein TM, Goff SA, Roth B, Morrish F and Armstrong C (1991). Transient expression and stable transformation of maize using microprojectiles. In: Herrmann RG, Larkins BA (eds.), *Plant Molecular Biology 2, Proceedings of a NATO Advanced Study Institute*, May 14-23 1990, Elmau, Germany, pp. 219-224.

34.   Brown DCW, Tian LN, Buckley DJ, Lefebvre M, McGrath A and Webb J (1994). Development of a simple particle bombardment device for gene transfer into plant cells. *Plant Cell, Tissue and Organ Culture*, **37**: 47-53.

35.   Buckley DJ, Brown DCW, Lefebvre M, McGrath A and Tian L (1995). A particle accelerating device for delivering DNA material into plant cells. *Computers and Electronics in Agriculture*, **13**: 67-74.

36.   Murashige T and Skoog F (1962). A revised medium for rapid growth and bioassays with tobacco tissue cultures. *Physiologia Plantarum*, **15**: 473-497.

37.   Gamborg OL, Miller RA and Ojima K (1968). Nutrient requirements of suspension cultures of soybean root cells. *Experimental Cell Research*, **50**: 151-158.

38.   Kikkert JR (1993). The Biolistic PDS-1000/He device. *Plant Cell, Tissue and Organ Culture*, **33**: 221-226.

39.   Vain P, Keen N, Murillo J, Rathus C, Nemes C and Finer JJ (1993). Development of the Particle Inflow Gun. *Plant Cell, Tissue and Organ Culture*, **33**: 237-246.

40.   McCabe D and Christou P (1993). Direct DNA transfer using electric discharge particle acceleration (ACELL technology). *Plant Cell, Tissue and Organ Culture*, **33**: 227-236.

41.   Gallagher SR (1992). GUS protocols: using the GUS gene as a reporter of gene expression. San Diego, CA, USA, Academic Press, New York, USA, pp. 221.

42.   Puddephat IJ, Thompson N, Robinson HT, Sandhu P and Henderson J (1999). Biolistic transformation of broccoli (*Brassica oleracea* var. *italica*) for transient expression of the beta-glucuronidase gene. *Journal of Horticultural Science and Biotechnology*, **74**: 714-720.

43.   Chen LFO, Hwang JY, Charng YY, Sun CW and Yang SF (2001). Transformation of broccoli (*Brassica oleracea* var. *italica*) with isopentenyltransferase gene via *Agrobacterium tumefaciens* for post-harvest yellowing retardation. *Molecular Breeding*, **7**: 243-257.

44.   Cogan NOI, Lynn JR, King GJ, Kearsey MJ, Newbury HJ and Puddephat IJ (2002). Identification of genetic factors controlling the efficiency of *Agrobacterium rhizogenes*-

mediated transformation in *Brassica oleracea* by QTL analysis.   *Theoretical and Applied Genetics,* **105**: 568-576.

45.   Koncz C and Schell J (1986). The promoter of TL-DNA gene 5 controls the tissue specific expression of chimeric genes carried by a novel type of *Agrobacterium* vector.   *Molecular and General Genetics,* **204**: 383-396.

46.   Sautter C (1993). Development of a microtargeting device for particle bombardment of plant meristems. *Plant Cell, Tissue and Organ Culture,* **33**: 251-257.

47.   Metz TD, Dixit R and Earle ED (1995). Agrobac*terium tumefaciens*-mediated transformation of broccoli (*Brassica oleracea* var. *italica*) and cabbage (*B. oleracea* var. *capitata*). *Plant Cell Reports,* **15**: 287-292.

48.   Sonali D and Srivastava DK (1999). Kanamycin sensitivity in cultured tissues of cauliflower. *Journal of Applied Horticulture,* **1**: 94-96

.

Table 1. Genetic transformation of cauliflower

Cultivar	Method (*Agro* strain)	Explant	Selection Gene	Target Gene	Ref.
WG-11, B-4	*A. tumefaceins* (LBA 4404)	Cotyledon hypocotyl	*nptII*	*Bcp1*	8
	A. tumefaceins (C58CI)	hypocotyl	*nptII, bar*		9
Pusa Snowball K-1	*A. tumefaceins* (GV 2206)	hypocotyl	*nptII*	*gusA, cry1A(b)*	10
ES5-1	*A. rhizogenes* (pRi 8196)	hypocotyl	hairy roots	Mannopine, agropine	11
Known You Early, Snow Lady	*A. tumefaceins* (LBA 4404)	hypocotyl	*nptII*	Tripsin inhibitor T1, *gusA*	12
Beauty Lady Snow Crown	*A. tumefaceins* (LBA 4404)	hypocotyl	*nptII*	Metallothionein CUP1	13
Lateman White Rock	*A. tumefaceins* (C58CI-pRiA4)	hypocotyl	*NptII* hairy roots	*gusA*	14
Yukon	*A. tumefaceins* (GV3101)	hypocotyl	*nptII*	*cry1Ac, gusA*	15, 16
	A. rhizogenes (LBA 9402)	hypocotyl	*nptII*	*ags, mas*	17
	A. tumefaceins	stem, leaf	*nptII*		18
Hormade, Phenomenal Early Space Saver, All Year Round	*A. tumefaceins* (AGL1) *A. rhizogenes*	peduncle	*nptII*	Shiva 1, Magainin II	19
8.70	*A. tumefaceins* (C58pMP90) *A. tumefaceins* (82.139)	stem	*nptII, hpt*	CaMV VI, CaMV capsid	20
Early Kunwari	PEG-mediated	Hypocotyls protoplasts	*nptII, hpt bar*		21
Korso	PEG-mediated	leaf protoplasts	*nptII, hpt*	*gusA*	22
	PEG-mediated, Electroporation *A. tumefaceins* (6048)	protoplast	*nptII*	23
80 Days	PEG-mediated	hypocotyl protoplasts	*nptII, hpt*		24
	PEG-mediated	hypocotyl protoplasts		*gusA*	25

Chapter 30

TOMATO TRANSFORMATION – THE NUCLEAR AND CHLOROPLAST GENOMES

J. VAN ECK, A.M. WALMSLEY[1] AND H. DANIELL[2]

The Boyce Thompson Institute for Plant Research, Tower Rd., Ithaca, NY 14853, USA; [1]School of Life Sciences, Arizona State University, Tempe, AZ, 85287-1601, USA; [2]Department of Molecular Biology & Microbiology, University of Central Florida, Biomolecular Science Building, Orlando FL 32816-2364, USA. Corresponding author: A. Walmsley. E-mail: Amanda.Walmsley@asu.edu

1. INTRODUCTION

1.1. Tomato Biotechnology

Once thought to be poisonous, tomato has become the second most commonly grown vegetable crop in the world behind potato. Traditional plant breeding has resulted in great progress in increasing yield, disease and pest resistance, environmental stress resistance and quality and processing attributes. However, tomato plant breeding programmes still strive to generate a better product. To assist in this goal, some plant breeding programmes have been expanded to include biotechnological techniques. Tomato has long been recognized as an excellent genetic model for molecular biology studies (*1*). This has resulted in a flood of information including markers and genetic maps, identification of individual chromosomes, promoter isolation, chloroplast and nucleus genome sequences and identification of genes and their function. In turn, this information has made tomato biotechnology more precise and arguably more meaningful.

1.2. Plant Transformation Methods

Plant transformation is another molecular biology technique that may assist a plant breeding programme by increasing the scope as well as decreasing the duration of the programme. Plant transformation commonly employs either *Agrobacterium*-mediated transformation or particle bombardment. *Agrobacterium* is a plant pathogen that in the process of infection transfers a segment of its DNA (T-DNA) into the genome of its host. Molecular biologists have taken advantage of this process to transfer a gene of interest, in a plant expression cassette, into plant genomes. Upon incubation of the *Agrobacterium* with plant materials, transfer of the T-DNA from the bacterium into the host's genome occurs through a process similar to conjugation. During tissue culture, transformed plant cells are positively selected and regenerated into transgenic plants. Although *Agrobacterium*-mediated transformation does have limitations, it is generally preferred for transformation of the plant cell nucleus. This

405

I.S. Curtis (ed.), Transgenic Crops of the World - Essential Protocols, 405-424.
© *2004 Kluwer Academic Publishers. Printed in the Netherlands.*

is due to the relatively low copy number of transgenes inserted and the simple pattern of transgene insertion. Both of these properties have bearings on the probability of transgene silencing, a phenomenon responsible for decreased or no expression of the transgene and possibly a related homologue in the host genome.

Particle bombardment is a direct gene transfer method that does not require a biological vector such as *Agrobacterium*, but relies on physical force to transfer foreign DNA into the target cells. The DNA of interest is coated onto microparticles, generally made from gold or tungsten, that are then blasted into the target cells originally by devices using gun powder but which have since evolved to use a pressure wave of helium. As in *Agrobacterium*-mediated transformation, the target cells that have been bombarded are regenerated through tissue culture using positive selection for transformed cells.

1.3. Agrobacterium-*mediated Transformation of Tomato*

The process of inserting a defined number of genes into the genome of a plant can greatly advance a breeding programme particularly when the plant transformation system is routine. While the tomato nucleus may be transformed using the particle gun (*2*), the preferred method is *Agrobacterium*-mediated transformation. *Agrobacterium*-mediated transformation of the nucleus of tomato was first performed in 1985 by Horsch *et. al.* (*3*). However it was not until 1986 that it was shown that cultivated tomato could be transformed and regenerated into fertile plants (*4*). The tomato transformation system was still far from routine and hampered by poor regeneration characteristics. The study by Fillatti *et. al.* (*5*) was a distinct step forward in determining optimal conditions for an *Agrobacterium*-mediated transformation system of tomato. It is now common knowledge that optimal conditions for tomato transformation are dependent on the tomato variety and the bacterial strain used. Additional factors to be considered include bacterium concentration, explant source, age and size of explant pre-incubation, co-cultivation time and co-cultivation media (*6*), culture media (*7*), orientation of cotyledon explants on selective regeneration medium, gelling agent, plate sealant, frequency of transfer (*8*) and *Agrobacterium* selection agent (*9, 10*).

While tobacco is still the golden subject for plant transformation, gene transfer into the tomato nucleus is relatively quick and efficient. The use of *Agrobacterium* means that the gene of interest is inserted at a low copy rate (low number per nucleus) and in a simple manner thus decreasing the chance of transgene silencing. Transformation of the nucleus has the added benefit of using eukaryotic machinery to process a gene and the resulting protein; hence complex proteins that require post-translational modification can be expressed correctly. This has lead to a wide range of traits being stably transformed into the nucleus of tomato including resistance to environmental stress (*11-13*), disease resistance (*14-16*), genes improving agronomic traits (*17, 18*) and vaccine antigens (*19-22*).

1.4. Chloroplast Transformation

Transformation of the plant chloroplast genome is an attractive alternative to nuclear transformation. Advantages of chloroplast genetic engineering include high-level transgene expression (23), multi-gene engineering in a single transformation event (24-26), transgene containment via maternal inheritance (27), lack of gene silencing (24, 28), lack of position effect due to site-specific transgene integration (23) and lack of pleiotropic effects (28, 29). Chloroplast genetic engineering was first accomplished through the use of autonomously replicating chloroplast vectors in dicot plastids (30), transient expression in monocot plastids (31) and by stable integration of selectable marker genes into chloroplast genomes of *Chlamydomonas* and tobacco, using particle bombardment (32, 33). This field has progressed steadily and there are now biotechnology companies in the USA solely based on this technology (i.e., Chlorogen Inc.).

Chloroplast genetic engineering is most suitable for hyper-expression of vaccine antigens and production of valuable therapeutic proteins. Ever since the demonstration of expression of human-elastin derived polymers for various biomedical applications (34), this approach has been used to express vaccines antigens for cholera; anthrax (29, 35), monoclonal antibody (36) and human therapeutic proteins, including human serum albumin (37), magainin (38), interferon (35) and insulin like growth factor (35). Several other laboratories have expressed human somatotropin (39) and interferon-GUS fusion proteins to improve stability (40) in transgenic chloroplasts. The chloroplast genome has been engineered to confer herbicide resistance (41), insect resistance (24, 42, 43), disease resistance (38), drought tolerance (28) and phytoremediation of toxic metals (25).

1.5. Molecular Characterization of Transgenic Plants

No matter how they are produced, plantlets resulting from the genetic engineering process require characterization. The traits examined include the presence of the foreign gene(s) or transgene(s), the number of gene copies, and the size and concentration of the resulting protein. The presence of the transgene is usually tested by PCR. PCR is a powerful technique used for amplifying DNA *in vitro*. Two synthetic oligonucleotide primers, which are complementary to two regions of the target DNA (one for each strand) to be amplified, are added to the target DNA in the presence of excess dNTPs and the heat stable Taq polymerase. In a series of cycles (usually 30-40) the strands of the target DNA are denatured (typically at 94 or 95°C), the primers annealed to the target DNA (typically 50-60°C), and a daughter strand extended from the primer by the action of Taq (usually at 72°C). The daughter strands become template for the following series of reaction thus leading to the amplification of the region of interest. By using primers specific for the transgene, its presence can be detected from a DNA extraction of a transgenic plant.

The number of insertion sites and number of copies of the transgene present is typically determined through Southern analysis. In this process, DNA from the plants of interest is extracted, separated on a gel, and transferred to a membrane that is then probed for the region of interest. When determining the number of copies of a transgene, the probe is usually a PCR product of the transgene. PCR is usually first used as a screening technique to determine transgenic lines since it requires less time. Southern analysis is usually then performed on the PCR positive lines to determine how many insertion sites and copies of the transgene and to determine whether the transgene has been integrated into the genome.

1.6. Characterization of the Products of Transgene Expression

The size of the protein resulting from the processing of the transgene is determined through Western analysis. This technique is similar to Southern analysis except proteins are extracted from the plants of interest. Antibodies specific for the protein of interest are used as the probe. The concentration of the protein of interest is often determined through ELISA. ELISA is based upon the protein of interest being bound by an antibody that is linked to a plate. The protein-antibody complex is then detected through use of peroxidase coupled to the antibody. The peroxidase when supplied with an appropriate substrate is responsible for a detectable colour or luminescent reaction. The protein concentration is determined through comparison of the intensity of colour or light in the test reactions to that of a standard curve.

Within this chapter we describe a transformation protocol for stable integration of transgenes into the tomato nucleus using *Agrobacterium*-mediated transformation and into the chloroplast genome using the Biolistic PDS-1000/ He Particle Delivery System.

2. MATERIALS

2.1. Nuclear Transformation

2.1.1. A. tumefaciens *Strains*

We have used two different *A. tumefaciens* strains for transformation, LBA4404 and EHA105. Various expression cassettes have been used containing the genes of interest and several different selectable markers, which confer resistance to the antibiotics kanamycin and hygromycin, and the herbicide component bialaphos. The expression cassettes were contained in binary plasmids that were introduced into the *Agrobacterium* by electroporation.

2.1.2. Culture Media for A. tumefaciens

1. LB medium: 10 g/l Bacto-tryptone (Becton Dickinson and Company, Sparks, MD), 5 g/l yeast extract (Fisher Scientific, Suwanne, GA), 10 g/l NaCl, 15 g/l Bacto Agar (Becton Dickinson and Company). After autoclaving, cool to 55°C, and add appropriate antibiotics for selection depending upon the vector.

2. YM medium: 400 mg/l yeast extract, 10 g/l mannitol (Fisher Scientific), 100 mg/l NaCl, 200 mg/l MgSO$_4$•7H$_2$0. After autoclaving, cool to 55°C or lower, and add appropriate antibiotics for selection depending upon the vector.
3. Antibiotics for selection: prepare the appropriate antibiotics for selection, and filter-sterilize.

2.1.3. Tissue Culture

1. Seeds of *Lycopersicon esculentum*, cultivars and varieties of interest (*see* Note 1).
2. Surface sterilant: 20% (v/v) commercial bleach solution with 0.1% (v/v) Tween 20 (Fisher Scientific, Suwannee, GA).
3. 1/2 strength MSO medium: 2.15 g/l Murashige and Skoog (MS, *44*) salts (Caisson Laboratories, Rexburg, ID), 2 mg/l thiamine-HCl, 0.5 mg/l pyridoxine-HCl, 0.5 mg/l nicotinic acid, 10 g/l sucrose, 8 g/l Agar-Agar (Sigma), pH 5.8.
4. 2,4-D (Sigma): 1 mg/ml stock. Prepare by dissolving the powder in a few drops of 1M KOH and adding purified water (Milli-Q purification system, Millipore, Billerica, MA) to volume. Store at 4°C for 3 months.
5. Kinetin (Sigma): 1 mg/ml stock. Prepare by dissolving the powder in a few drops of 1M HCl and adding purified water to volume. Store at 4°C.
6. Zeatin (PhytoTechnology Laboratories): 1 mg/ml stock. Prepare by dissolving the powder in a few drops of 1M HCl and adding purified water to volume. Filter-sterilize and store in 1 ml aliquots in sterile 1.5 ml-capacity Eppendorf tubes at 0°C.
7. Nitsch vitamins: 1000x stock. For 50 ml: 100 mg glycine, 500 mg nicotinic acid, 25 mg pyridoxine-HCl, 25 mg thiamine-HCl, 25 mg folic acid, 2 mg d-biotin (Sigma). Adjust to pH 7.0 to dissolve. Store at 0°C for 1 year.
8. KCMS medium: 4.3 g/l MS salts, 100 mg/l *myo*-inositol, 1.3 mg/l thiamine-HCl, 0.2 mg/l 2,4-D, 200 mg KH$_2$PO$_4$ (Fisher Scientific), 0.1 mg/l kinetin, 30 g/l sucrose, 5.2 g/l Agargel (Sigma), pH 5.5. Dispense into 100 x 20 mm Petri dishes.
9. KCMS liquid medium: Same as KCMS medium, but without the Agargel.
10. NT1 suspension culture: To initiate a NT1 suspension culture, transfer approx. 1.8 g of friable callus to 50 ml of KCMS liquid medium in a 250 ml Erlenmeyer flask. Maintain on a gyratory shaker at 100 rpm in the dark. Subculture every 7 days by transferring 2 ml of suspension to 48 ml of KCMS liquid medium.
11. MS liquid medium: 4.3 g/l MS salts, 2 mg/l glycine, 0.5 mg/l nicotinic acid, 0.5 mg/l pyridoxine-HCl, 0.4 mg/l thiamine-HCl, 0.25 mg/l folic acid, 0.05 mg/l d-biotin, 30 g/l sucrose, pH 5.6. Store at room temperature for 1 month.
12. Selection agent: The appropriate selection agent will depend upon the selectable marker in the gene construct (*see* Note 2).
13. Timentin (PhytoTechnology Laboratories): 100 mg/ml stock. Prepare by dissolving the powder in purified water. Filter-sterilize, dispense into 3 ml aliquots, and store at 0°C (*see* Note 3).

.

14. 2Z selective medium: 4.3 g/l MS salts, 100 mg/l *myo*-inositol, 1 ml/l Nitsch vitamins stock (1000x), 5.2 g/l Agargel, pH 6.0. After autoclaving, cool to 55°C, and add: 2 mg/l zeatin, 300 mg/l timentin, and appropriate selective agent for selectable marker. Dispense into 100 x 20 mm Petri dishes.

15. 1Z selective medium: Same as 2Z selective medium, however, add only 1 mg/l zeatin. Dispense 25 ml into 100 x 20 mm Petri dishes or 50 ml into Magenta GA-7 boxes.

16. Selective rooting medium: 4.3 g/l MS salts, 1 ml/l Nitsch vitamins stock, 30 g/l sucrose, 8 g/l Bacto Agar, pH 6.0. After autoclaving, cool to 55°C. Add 300 mg/l timentin and appropriate selection agent for selectable marker.

17. Sterile paper towels.

18. Transfer to soil: 4 or 6 inch pots, soil-less potting mix (not commercially available) [1 lb 3 oz Unimix + III (Griffin Greenhouse and Nursery Supplies, Tewksbury, MA), 5 lbs lime (Hummert International, Earth City, MO, USA), 5.7 cu. ft. peat moss (Hummert International), 12 cu. ft. vermiculite (Hummert International), 5 lbs Osmocote 17-7-12 (Grower Supply Inc., Forest Hill, LA, USA)], clear plastic bags or containers.

2.2. Chloroplast Transformation

2.2.1. Isolation of Genomic DNA from Plants

Extract the genomic DNA from fresh green leaves using the DNeasy Plant Kit (Qiagen Inc. Valencia, CA) following the vender's instructions.

2.2.2. Amplification of Tomato Chloroplast Flanking Sequence

Species-specific flanking sequences from the chloroplast DNA of tomato plants are amplified through PCR using primers based on sequences known to be highly conserved within the tobacco chloroplast genome.

1. Extract genomic DNA from tomato.

2. Set up a PCR reaction mix containing 1 x PCR buffer (Quagen Inc. Valencia, CA), 1.5 mM $MgCl_2$, 1 mM dNTPs (Invitrogen Inc. Carlsbad, CA), 20 µM of each primer, a 1 in 100 dilution of Taq polymerase (Qiagen Inc. Valencia, CA) and 5-10 ng of genomic DNA.

3. Place reaction mix in a thermocylcer programmed to initially denature the template DNA at 94°C for 4 min followed by 30 cycles of 94°C for 15 sec, 55°C for 30 sec, and 72°C for 90 sec. A final extension step is performed at 72°C for 5 min before soaking at 4°C.

4. Purify the resulting PCR product using the PCR purification kit (Qiagen Inc., Valencia, CA) and use in vector construction.

2.2.3. Tissue Culture Media

Tomato shoot induction medium: 4.3 g/l MS salts (Gibco, Invitrogen Corporation, NY), 10 ml/l 10x B5 vitamins (1 mg/l *myo*-inositol, 0.01 mg/l nicotinic acid, 0.01 mg/l pyridoxine-HCl, 0.1 mg/l thiamine-HCl), 0.2 mg/l IAA (use 0.2 ml from 1 mg/ml stock), 3 mg/l of BAP (use 3 ml from 1 mg/ml stock). All these reagents are available from Sigma. Adjust to pH 5.8 with 1N KOH or 1N NaOH and add 6 g/l Phytagel (Sigma) before autoclaving at 121°C for 20 min. For preparation of 1 mg/ml stock of BAP or IAA, weigh 10 mg powder and dissolve first in 1 or 2 drops of 1N NaOH and make up the final volume to 10 ml. Stock solutions of plant growth regulators may be stored at 4°C for 2-3 months.

2.2.4. Construction of the Chloroplast Transformation Vector

The left and right flanks are the regions in the chloroplast genome that serve as homologous recombination sites for stable integration of transgenes. A strong promoter and the 5' and 3' untranslated regions (DNA sequence between the promoter and start codon or downstream of the stop codon) are necessary for efficient transcription and translation of the transgenes within chloroplasts. For multiple gene expression, a single promoter may regulate the transcription of the operon, and individual ribosome binding sites must be engineered upstream of each coding sequence (*23*). The following steps are used in vector construction.

1. Amplification of flanking sequences of plastid with primers that are designed on the basis of known sequence of the tobacco chloroplast genome. Chloroplast genome sequences are highly conserved, especially within the same family. Alternatively, tobacco chloroplast DNA flanking sequencing have been used to transform the tomato chloroplast genome (*45*).
2. Insert the PCR product containing the flanking sequence of the chloroplast genome into pUC19 plasmid digested with *Pvu*II restriction enzyme (to eliminate the multiple cloning site), dephoshorylated with the help of calf intestinal phoshatase (CIP) (New England Biolabs, Beverley, MA., USA) for 5 min at 50°C (to prevent recircularization of cloning vector). Inactivate the CIP at 68°C for 10 min.

Clone chloroplast transformation cassette (which is made blunt with the help of T4 DNA polymerase or Klenow filling, (New England Biolabs) into a cloning vector cut at the unique *Pvu*II site in the spacer region.

3. METHODS

3.1. Nuclear Transformation

3.1.1. Preparation of A. tumefaciens

1. Three days prior to transformation, streak the *Agrobacterium* onto plates of LB medium containing the appropriate selection agent. Incubate at 28°C for 48 h.
2. Select four, well-formed colonies, and transfer to 50 ml of YM selective medium. Culture in a shaking incubator at 250 rpm at 28°C overnight for 24 h until it reaches an OD_{600} of 0.4–0.6. If $OD_{600} > 1.0$, dilute the culture until the OD_{600} reading is below 0.5 and grow for approx. 1 h. Check the OD_{600} reading periodically during this time.
3. Centrifuge at 8000 rpm for 10 min at 20°C. Resuspend the pellet in 50 ml MS liquid medium.

3.1.2. Seedlings

1. Soak seeds in 20% (v/v) commercial bleach solution for 10 min, followed by 3 rinses in sterile distilled water (*see* Note 4).
2. Transfer 25 seeds to each Magenta GA-7 box containing 50 ml of 1/2 MSO medium. Maintain at 24 ± 1°C, in the light under a mixture of cool and warm fluorescent bulbs (F40CW and F40WW) (Philips Lighting Co. http://www.lighting.philips.com /index.htm) 16 h photoperiod at 45 μmol/m²/sec.

3.1.3. Feeder Layer

1. Prepare feeder layer plates 2 days prior to transformation. Using a 10 ml wide-bore pipette, pipette 2 ml of a 7 day-old (7 days after subculture to fresh KCMS liquid medium) NT1 suspension culture onto KCMS medium. Swirl gently to distribute evenly around the centre of each plate (*see* Note 5).
2. Cover plated suspension cultures with a sterile Whatman filter paper (7 cm circle) per plate (Fig. 1A). Seal plates with Micropore tape (McKesson, San Francisco, CA) and incubate in the dark at 24 ± 1°C.

3.1.4. Preculture of Cotyledon Explants

1. One day prior to transformation, excise cotyledons from 6-8 day-old seedlings (Fig. 1B) by cutting at the point where the petiole meets the cotyledon. It is important that the first true leaves have not emerged from the seedlings. If the cotyledons are 1 cm or longer, cut crosswise into 2 equal sections (*see* Notes 6 and 7).
2. Place the adaxial (top) sides down onto the feeder layer plate. As many as 80 explants can be cultured per dish (Fig. 1C).

3. Seal plates with Micropore tape and maintain at $24 \pm 1°C$, under a 16 h photoperiod at 45 µmol/m²/sec.

3.1.5. Infection and Cocultivation

1. After 24 h preculture, transfer cotyledon explants to 25 ml of prepared *Agrobacterium* in a sterile Magenta GA-7 box or other wide-mouthed container. Incubate for 5 min (*see* Note 8).
2. Remove *Agrobacterium* from container with a pipette.
3. Transfer explants onto sterile paper towels to remove excess *Agrobacterium*, and place with the adaxial (top) sides down onto the original feeder layer plates (Fig. 1D).
4. Maintain in the dark at 19-25°C for 48 h (*see* Note 9).

3.1.6. Selection for Transformants

1. Transfer 25 cotyledon explants with the adaxial sides up onto 2Z selective medium.
2. Seal plates with Micropore tape, and maintain at $24 \pm 1°C$, under a 16 h photoperiod at 45 µmol/m²/sec.
3. Every 3 weeks transfer to 1Z selective medium. From this point on, culture 10 explants per dish (*see* Note 10).

3.1.7. Shoot Regeneration

1. When shoots begin to regenerate from the callus (approx. 8–10 weeks after transformation, Fig. 1E), maintain cultures on 1Z selective medium in Magenta GA-7 boxes instead of Petri dishes (*see* Note 11).
2. Excise shoots (approx. 2 cm in length) from the callus and transfer to Magenta GA-7 boxes containing 50 ml of selective rooting medium (Fig. 1F). Maintain at $24 \pm 1°C$, under a 16 h photoperiod at 45 µmol/m²/sec.
3. Take shoot tip and nodal cuttings every 3–4 weeks, and transfer to fresh selective rooting medium depending upon the rate of growth. Shoots should be maintained on selective rooting medium for at least 2 additional transfers. For *in vitro* propagation for field or glasshouse, do not add timentin to the selective rooting medium to observe for possible internal contamination with *Agrobacterium* (*see* Note 12). The selective agent may be excluded if desired.

3.1.8. Transfer to Soil

1. Select lines which have well-formed root systems for transfer to soil (Fig. 1G).
2. Gently remove a plant from its culture vessel, and use tepid water to wash the medium from the roots (*see* Note 13).
3. Transfer each plant to a 4- or 6-inch pot containing a soil-less potting mix that has been thoroughly wetted.

4. Cover each plant with a plastic bag secured to the pot or use a clear, plastic container (Fig. 1H, *see* Note 14).
5. Transfer the plants to a growth chamber, in a shaded area of a glasshouse, or a laboratory setting with lighting (*see* Note 15).
6. After 7 days, if a plastic bag was used, cut a small hole each day during the course of that week for the gradual acclimation of plants, then remove the bag on the 8th day. If a plastic container was used, gradually lift the container each day during the week, and then remove the container on the 8th day.
7. Gradually acclimatize the plants to glasshouse conditions (Fig. 1I).

3.1.9. Analysis of the T1 Generation

1. Harvest the fruit from T0 plants when they are light red to red in colour (*see* Note 16). Since tomato fruit are perishable, it is advantageous to collect the seeds as soon as possible. However, this may not be feasible with a large harvest. We have found ripe fruit can last up to 2 weeks when stored at around 24°C with a low relative humidity (30-40%).
2. Collect the seeds by halving the fruit through its poles and scooping out the seeds and jelly-like parenchyma in the locular cavity onto a paper towel. Spread the seeds over the surface to reduce the amount of parenchyma on the seeds.
3. Using forceps, pick the seeds onto the surface of 6 x 6 inch weighing paper (VWR Scientific, West Chester, PA) and allow to dry for 2 days on a bench at room temperature.
4. Store the seeds at 4°C in appropriately labelled envelopes still attached to the weigh paper.
5. Germinate T1 seeds as described (*see* Section 3.1.2.), except add the appropriate selection agent to 1/2 strength MS media. For example, we germinate kanamycin resistant lines on MS media containing 300 mg/l kanamycin and bialaphos resistant lines on MS media containing 3 mg/l bialaphos.
6. Take note of how many seeds successfully germinate. Should the gene of interest follow Mendelian inheritance, 3 out of 4 seeds should germinate on selection medium.
7. Transfer the seedlings to soil (*see* Section 3.1.8.).
8. Molecular analysis of plantlets can begin once they reach a size that allows samples to be taken without jeopardizing the health of the plant.

3.2. Chloroplast Transformation

3.2.1. Particle Bombardment

3.2.1.1. Preparation of Gold Particle Suspension

1. Suspend 50-60 mg gold particles in 1 ml of 100% ethanol and vortex for 2 min.
2. Spin at maximum speed ~10,000 rpm (using tabletop microcentrifuge) for 3 min.
3. Discard the supernatant.

4. Add 1ml fresh 70% ethanol and vortex for 1 min.
5. Incubate at room temperature for 15 min and shake intermittently.
6. Spin at 10,000 rpm for 2 min.
7. Discard supernatant, add 1 ml sterile distilled H_2O, vortex for 1 min, leave at room temperature for 1 min, and spin at 10,000 rpm for 2 min.

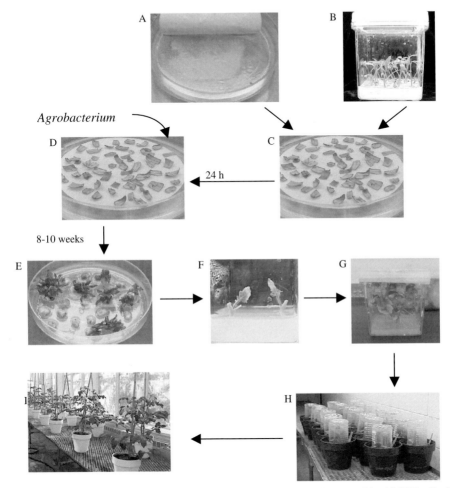

Figure 1. Nuclear transformation of tomato using A. tumefaciens *(not to scale). (A) NT1 cell feeder plates (9 cm diam.) are prepared 24 h in advance to excising the cotyledons of 8–10 day old seedlings (B). (C) The 1 cm long cotyledons are plated adaxial side down onto the plates and left to recover for 24 h. (D) The explants are co-cultivated with* Agrobacterium *for 48 h before transfer to selection media where regeneration of shoots occurs between 8–10 weeks after transformation (E). (F) Shoots that are about 2 cm in height are excised from the callus clumps and plated into rooting media. (G) Once a well established root system has developed, the plantlets (about 9 cm tall) are transferred to soil in 6 inch pots (H) and acclimatized to glasshouse conditions (I).*

8. Repeat above washing process 3 times with H_2O (step 7).
9. Resuspend the gold pellet in 1 ml of 50% glycerol, and store the microparticle stock at –20°C.

3.2.1.2. Precipitation of the Chloroplast Vector on Gold Particles

1. Vortex the microparticle stock for 1 min, then remove 50 µl of the gold particles to a 1.5 ml tube.
2. Add 10 µl of DNA (about 1 µg/µl plasmid DNA) and vortex the mixture for 30 sec.
3. Add 50 µl of 2.5 M $CaCl_2$ and vortex the mixture for 30 sec.
4. Add 20 µl of 0.1 M spermidine and vortex the mixture for 20 min at 4°C.
5. Add 200 µl of 100% ethanol and vortex for 30 sec.
6. Spin at 3000 rpm for 40 sec.
7. Pour off the ethanol supernatant.
8. Repeat the ethanol washings 5 times.
9. In the last step, pour off the ethanol carefully and add 35-40 µl ethanol (100%).

3.2.1.3. Preparation of Macrocarriers

1. Sterilize the macrocarriers by incubating in 100% ethanol for 15 min, allow to air dry, then insert into a sterile steel ring holder with the help of a plastic cap.
2. Vortex the gold-plasmid DNA suspension and pipette 8-10 µl in the centre of macrocarrier and allow to air dry.

3.2.1.4. Gene Gun Setup for Bombardment

1. Wipe the gun chamber and holders with 100% ethanol using fine tissue paper (do not wipe the door with alcohol).
2. Turn on the vacuum pump.
3. Turn on the valve (helium pressure regulator) of helium gas tank (anti-clockwise).
4. Adjust the gauge valve (adjustable valve) to approx. 200-250 psi above the desired rupture disk pressure (clockwise) using the adjustment handle.
5. Turn on the gene gun.
6. Place the rupture disc (sterilized by dipping in 100% ethanol for 5 min) in the rupture disc-retaining cap and tightly screw to the gas acceleration tube.
7. Place a stopping screen in the macrocarrier launch assembly and above that place the macrocarrier so that the gold particles covered with the chloroplast vector is facing down towards the screen. Screw assembly with a macrocarrier coverlid and insert in the gun chamber.
8. Place an intact leaf or explants to be bombarded on a sterile filter paper (Whatman No. 1) soaked in MS liquid medium containing no antibiotics. Place the sample plate over the target plate shelf, insert in the gun chamber and close the bombardment chamber door.

9. Press the "Vac" switch to build pressure (up to 28 inches of Hg in the vacuum gauge display). Turn the same switch down until the hold point and press the "Fire" switch until you hear the rupture disc burst.

10. Press the "Vent" switch to release the vacuum and open the chamber to remove the sample.

11. Incubate the bombarded sample plates in the culture room for 2 days in the dark (i.e. covered with aluminum foil) and on the third day cut explants in appropriate pieces and place on the selection medium.

3.2.2. Transformation

Using the tobacco chloroplast vector, tomato (*Lycopersicon esculentum* cv. IAC Santa Clara) plants with transgenic plastids are generated using very low intensity of light and the following protocol.

1. Bombard excised leaves from 4 week-old tomato seedlings and incubate in the dark for 2 days on selection-free medium.

2. Cut the bombarded leaves into 3 mm pieces using a sterile scalpel and place on shoot induction medium containing 0.2 mg/l IAA, 3 mg/l BAP, 3% sucrose and 300 mg/l spectinomycin, in deep Petri dishes.

3. Select spectinomycin-resistant primary calli after a 3-4 months duration.

4. Regenerate shoots in approx. 4 weeks by transferring the transgenic calli to shoot induction medium containing 0.2 mg/l IAA, 2 mg/l ZR, 2% sucrose and 300 mg/l spectinomycin and then root on hormone-free medium. Transfer regenerated transgenic plants to the glasshouse. Conditioning plants for glasshouse is as described (*see* Section 3.1.8.).

3.2.3. Molecular Analysis of Transgenic Plants

3.2.3.1. PCR Screening of Transgenic Shoots

1. Extract the genomic DNA from transgenic leaf tissue using DNeasy Plant Kit (QIAGEN Inc.) by following manufacturer's instructions. For lower amount of transgenic tissues, volume of buffers may be reduced appropriately.

2. Run PCR reaction with Taq DNA Polymerase using appropriate primers following the same conditions as described above for amplification of flanking sequences.

3. Run reaction on an agarose gel and look for a band of the expected size (*see* Note 17).

3.2.3.2. Southern Analysis of Transgenic Plants

1. Extract the genomic DNA from transgenic leaf tissue using DNeasy Plant Kit (QIAGEN Inc.) by following manufacturer's instructions.

2. Analyze samples containing 15 μg DNA as described by Sambrook *et al.* (*46*) (*see* Note 18).

3.2.4. Analysis of the Protein of Interest

3.2.4.1. Crude Protein Extracts

1. Suspend plant materials in appropriate volume of ice-cold extraction buffer (50 mM sodium phosphate, pH 6.6, 100 mM NaCl, 1 mM EDTA, 0.1% Triton X-100, 10 μg/ml leupeptin, 1 mM PMSF) (*see* Note 19).
2. Homogenize samples in a QBiogene (Carlsbad, CA, USA) Fast Prep machine.
3. Pellet insoluble materials by centrifugation at 14,000 rpm in an Eppendorf 5415C microcentrifuge at 4°C for 5 min.
4. Keep the supernatant on ice during analysis then store at –80°C.

3.2.4.2. Western Analysis

1. Add 30 μl of each sample to 6 μl of 6 X SDS gel loading buffer (300 mM Tris-HCl, pH 6.8, 600 mM DTT, 12% SDS, 0.6% Bromophenol Blue, 60% glycerol), boil for 10 min then place on ice.
2. Centrifuge the samples at 12,000 rpm for 5 min at 4°C in an Eppendorf 5415C microcentrifuge then load into a 12% SDS polyacrylamide gel electrophoresis (SDS-PAGE).
3. Run the SDS-PAGE in Tris-glycine buffer (25 mM Tris-HCl, 250 mM glycine, pH 8.3, 0.1% SDS) at 30 milliamps per gel until the dye front runs about 5 mm from the gel bottom.
4. Transfer the separated proteins from the gel to a PVDF membrane using a Bio Rad Trans Blot Cell (50 V for 2 h).
5. Block the membrane with 1% dry milk, PBS and 0.1% Tween 20 (PBSTM) for 1 h at room temperature using slow rotation in a Hybaid (Ashford, Middlesex, UK).
6. At room temperature, briefly wash twice in PBST before incubating with the primary antibody diluted appropriately in PBSTM.
7. Briefly rinse the membrane in PBST before a 15 min wash and 3 x 5 min washes in PBST. The membrane was then incubated in a 1:12,000 dilution of an anti-rabbit IgG horseradish peroxidase (HRP) conjugate (Sigma) for 1 h with slow rotation. The membrane was rinsed in PBST, and then subjected to a 15 min wash and 3 x 5 min washes. Detection was performed using the Amersham ECL + kit according to manufacturer's instructions (*see* Note 20).

3.2.4.3. ELISA Analysis

To determine the concentration of a specific protein in a crude plant protein extract, the ELISA protocol performed depends upon the reagents you have available to you, in particular the type of antibodies. Good sources to determine which protocol to use and how to optimize the protocol include Crowther (*47*) and Bruyns *et al.* (*48*).

4. NOTES

1. This *Agrobacterium*-mediated transformation method has been used successfully for the cultivars Moneymaker, Yellow Pear, Rio Grande, Micro-Tom and several breeding lines.

2. We have used the following selection agents: kanamycin (PhytoTechnology Laboratories), hygromycin (Phytotechnology Laboratories), and bialaphos (PhytoTechnology Laboratories). The effective concentration for each selection agent varies upon the tomato line being transformed. The concentrations we have used are as follows: kanamycin 75 and 100 mg/l, hygromycin 6 mg/l, bialaphos 3 mg/l.

3. If timentin cannot be acquired, 500 mg/l carbenicillin can be substituted. In addition to being more economical, timentin is preferable to carbenicillin to eliminate *Agrobacterium* since it is light stable and resistant to inactivation by β-lactamase (*9*). Timentin also has beneficial effects on tomato regeneration by increasing shoot and root formation (*10*).

4. If contamination occurs, the seeds can be treated with 70% ethanol for 2 min prior to the bleach treatment.

5. The use of a feeder layer is not essential, but may increase the transformation efficiency especially for lines difficult to transform. We have found feeder layers to be beneficial when expression of a protein from an introduced gene has a negative effect on the number of transgenics regenerated.

6. Hypocotyl explants have been tried; however, a high number of non-transformed shoots were recovered. Therefore, we use only cotyledons.

7. If seedlings with the first true leaves are used, the transformation efficiency will be very low, and for some lines, transformants may not be recovered.

8. A longer period of incubation should be avoided because of the possibility of water-soaked damage to the explant.

9. Lower cocultivation temperatures have been reported to increase the transformation efficiency for some crops. This has not been reported for tomato; however, we have observed an increase, although not significant, of transgenics recovered when expression of an introduced gene had a negative effect on transformation efficiency.

10. It is important to keep the number of explants to 10 per plate as a larger number per plate results in reduced transformation efficiency.

11. The transformation efficiency (the percent of cotyledon explants that give rise to transformants) varies with different cultivars, selectable markers, and introduced genes. Some genes when expressed can have a detrimental effect on the recovery of transformants. It is important to include an empty vector construct (no gene of interest, only the vector with a selectable marker) as a transformation control when using a new gene construct to determine if the introduced gene is having a negative effect on transformation efficiency. We observe an overall transformation efficiency of 10-14%.

12. An alternative to ensuring the lack of internal contamination by *Agrobacterium* is to use PCR primers specific to *Agrobacterium*. Plants having internal contamination should be discarded because it is not possible to remove the bacterial contaminant.

13. It is important to wash the medium from the roots because it contains sucrose. This could lead to contamination in the soil and death of the plant.

14. For plastic containers, we use either old Magenta boxes or clear, plastic juice containers that have the tops removed at a position that allows plants to be completely covered. It is important to cover each plant immediately after transfer to soil to prevent wilting.

15. Do not place in direct sunlight. Heat will build up under the plastic bag or container and kill the plant.

16. The stage at which the fruit should be harvested varies on the protein expressed. Of seven genes we have expressed in tomato, two of the transgene products have been found to decrease in concentration as the fruit ripened. In such a case, the fruit should be harvested when green.

17. This method has been used to distinguish between mutants, nuclear and chloroplast transgenic plants. By landing one primer on the native chloroplast genome adjacent to the point of integration and a second primer on the *aad*A gene (*24, 28, 29*). PCR product of an appropriate size should be generated in chloroplast transformants. Since this PCR product cannot be obtained in nuclear transgenic plants or mutants, the possibility of nuclear integration or mutants should be eliminated (*24, 28, 29*).

18. In Southern blot analysis, plastid genome digested with suitable restriction enzymes should produce a smaller fragment (flanking region only) in wildtype plants compared to transgenic chloroplast that include transgene cassette as well as the flanking region (*24, 28, 29*). In addition, homoplasmy in transgenic plants is achieved when only the transgenic fragment is observed. This establishes that all of the chloroplast genomes are transformed. However, in order to insure homoplasmy, seeds obtained from transgenic plants should be germinated in 500 mg/l of spectinomycin. Maternal inheritance is confirmed by 100% of offspring germinating on the selection medium as opposed to Mendelian segregation via nuclear genetic engineering.
19. Volume used depends upon the water content of the plant materials, for example in ELISA analysis of fresh tomato fruit we start with a dilution of 1 mg fruit material per 10 μl of extraction buffer while with fresh leaf material we start with a dilution of 1 mg per 14 μl of extraction buffer.
20. The temperature and incubation times given in this protocol are good guide, however, along with antibody dilutions, these parameters require optimization. The Stratagene ECL+ detection kit gives a rudimentary optimization protocol in its instruction booklet.

CONCLUSIONS

The *Agrobacterium*-mediated transformation protocol described in this chapter has been used successfully to generate transgenic lines from several different tomato cultivars and lines. These transgenic lines contained various genes of interest and selectable markers. The transformation efficiency using this protocol may vary depending upon: a) the introduced gene; b) selectable marker used; and 3) the chosen tomato variety, therefore, if the efficiency is very low, it will be necessary to make modifications. The highest transformation efficiency we achieved was 14%. The ability to generate nuclear transformants has proven to be an invaluable tool in the advancement of tomato biotechnology. In contrast, investigations to engineer the tomato chloroplast genome are at very early stages of development. It took 2 years to achieve the first stable chloroplast transgenic plant and only 6 transgenic calli among 540 plates were observed in 60 bombarded leaves. Therefore, more optimization of plastid transformation protocols in tomato is necessary before this approach can be used in research laboratories or to modify commercial cultivars with useful traits.

ACKNOWLEDGEMENTS

We would like to thank A. Snyder, B. Randall, C. Xu, S. Abend, and J. Lenz for their technical assistance in the development of the *Agrobacterium*-mediated transformation protocol. Tomato transformation and characterization in the Walmsley Laboratory is supported by Dow AgroSciences LLC. Chloroplast genetic engineering investigations in the Daniell laboratory are supported by grants from the National Institutes of Health (R 01GM 63879) and United States Department of Agriculture (3611-21000-017-00D).

REFERENCES

1. Fobes JF (1980). The tomato as a model system for the molecular biologist. In :*PMB Newsletter*. pp. 64-68.

2. Van Eck JM, Blowers AD and Earle ED (1995). Stable transformation of tomato cell cultures after bombardment with plasmid and YAC DNA. *Plant Cell Reports*, **14**: 299-304.

3. Horsch RB, Fry JE, Hoffmann N, Eichholtz D, Rogers SG, and Fraley RT (1985). A simple and general method for transferring genes into plants. *Science*, **227**: 1229-1231.

4. McCormick S, Niedermeyer J, Fry JE, Barnason A, Horsch RB and Fraley RT (1986). Leaf disc transformation of cultivated tomato (*L. esculentum*) using *Agrobacterium tumefaciens*. *Plant Cell Reports*, **5**: 81-84.

5. Fillatti JJ, Kiser J, Rose R and Comai L (1987). Efficient transfer of a glyphosate tolerance gene into tomato using a binary *Agrobacterium tumefaciens* vector. *Bio/Technology*, **5**: 726-730.

6. Adang M.J., Brody MS, Cardineau G, Eagan N, Roush RT, Shewmaker CK, Jones A, Oakes JV *et al.* (1993). The reconstruction and expression of a *Bacillus thuringiensis cryIIIA* gene in protoplasts and potato plants. *Plant Molecular Biology*, **21**: 1131-45.

7. Roekel JSC, Damm B, Melchers LS and Hoekema A (1993). Factors influencing transformation frequency of tomato (*Lycopersicon esculentum*). *Plant Cell Reports*, **12**: 644-647.

8. Frary A and Earle ED (1996). An examination of factors affecting the efficiency of *Agrobacterium*-mediated transformation of tomato. *Plant Cell Reports*, **16**: 235-240.

9. Ling H-Q, Kriseleit D and Ganal MW (1998). Effect of ticarcillin/potassium clavulanate on callus growth and shoot regeneration in *Agrobacterium*-mediated transformation of tomato (*Lycopersicon esculentum* Mill.). *Plant Cell Reports*, **17**: 843-847.

10. Costa MGC, Nogueira FTS, Figueira ML, Otoni WC, Brommonschenket SH and Cecon PR (2000). Influence of the antibiotic timentin on plant regeneration of tomato (*Lycopersicon esculentum* Mill.) cultivars. *Plant Cell Reports*, **19**: 327-332.

11. Hsieh TH, Lee JT, Yang PT, Chiu LH, Charng YY, Wang YC and Chan MT (2002). Heterology expression of the *Arabidopsis* C-repeat/dehydration response element binding factor 1 gene confers elevated tolerance to chilling and oxidative stresses in transgenic tomato. *Plant Physiology*, **129**: 1086-1094.

12. Jia GX, Zhu ZQ, Chang FQ and Li YX (2002). Transformation of tomato with the *BADH* gene from Atriplex improves salt tolerance. *Plant Cell Reports*, **21**: 141-146.

13. Mishra SK, Tripp J, Winkelhaus S, Tschiersch B, Theres K, Nover L and Scharf KD (2002). In the complex family of heat stress transcription factors, *HsfA1* has a unique role as master regulator of thermotolerance in tomato. *Genes & Development*, **16**: 1555-67.

14. Gubba A, Gonsalves C, Stevens MR, Tricoli DM and Gonsalves D (2002). Combining transgenic and natural resistance to obtain broad resistance to tospovirus infection in tomato (*Lycopersicon esculentum* Mill). *Molecular Breeding*, **9**: 13-23.

15. Li L and Steffens JC (2002). Overexpression of polyphenol oxidase in transgenic tomato plants results in enhanced bacterial disease resistance. *Planta*, **215**: 239-47.

16. Lincoln JE, Richael C, Overduin B, Smith K, Bostock R and Gilchrist DG (2002). Expression of the antiapoptotic baculovirus p35 gene in tomato blocks programmed cell death and provides broad-spectrum resistance to disease. *Proceedings of the National Academy of Sciences USA*, **99**: 15217-15221.

17. Carey AT, Smith DL, Harrison E, Bird CR, Gross KC, Seymour GB and Tucker GA (2001). Down-regulation of a ripening-related beta-galactosidase gene (*TBG1*) in transgenic tomato fruits. *Journal of Experimental Botany*, **52**: 663-668.

18. Mehta RA, Cassol T, Li N, Ali N, Handa AK and Mattoo AK (2002). Engineered polyamine accumulation in tomato enhances phytonutrient content, juice quality, and vine life. *Nature Biotechnology*, **20**: 613-618.

19. Kim CH, Kim KI, Hong SH, Lee YH and Chung IS (2001). Improved production of recombinant rotavirus VP6 in sodium butyrate-supplemented suspension cultures of transgenic tomato (*Lycopersicon esculentum* Mill.) cells. *Biotechnology Letters*, **23**: 1061-1066.

20.   Jani D, Meena LS, Mohammad Q, Haq R-U, Singh Y, Sharma AK and Tyagi AK (2002). Expression of cholera toxin B subunit in transgenic tomato plants. *Transgenic Research*, **11**: 447-454.

21.   Walmsley AM, Kirk DD and Mason HS (2003). Passive immunization of mice pups through oral immunization of dams with a plant-derived vaccine. *Immunology Letters*, **86**: 71-76.

22.   Walmsley AM, Alvarez ML, Jin Y, Kirk DD, Lee SM, Pinkhasov J, Rigano MM, Arntzen CJ *et al.* (2003). Expression of the B subunit of *Escherichia coli* heat-labile enterotoxin as a fusion protein in transgenic tomato. *Plant Cell Reports*, **21**: 1020-1026.

23.   Daniell H, Khan MS and Allison L (2002). Milestones in chloroplast genetic engineering: an environmentally friendly era in biotechnology. *Trends in Plant Science*, **7**: 84-91.

24.   De Cosa B, Moar W, Lee SB, Miller M and Daniell H (2001). Overexpression of the *Bt cry2Aa2* operon in chloroplasts leads to formation of insecticidal crystals. *Nature Biotechnology*, **19**: 71-74.

25.   Ruiz ON, Hussein HS, Terry N and Daniell H (2003). Phytoremediation of organomercurial compounds via chloroplast genetic engineering. *Plant Physiology*, **132**: 1344-1352.

26.   Daniell H and Dhingra A (2002). Multigene engineering: dawn of an exciting new era in biotechnology. *Current Opinion in Biotechnology*, **13**: 136-41.

27.   Daniell H (2002). Molecular strategies for gene containment in transgenic crops. *Nature Biotechnology*, **20**: 581-586.

28.   Lee SB (2003). Accumulation of trehalose within transgenic chloroplasts confers drought tolerance. *Molecular Breeding*, **11**: 1-13.

29.   Daniell H, Lee SB, Panchal T and Wiebe PO (2001). Expression of the native cholera toxin B subunit gene and assembly as functional oligomers in transgenic tobacco chloroplasts. *Journal of Molecular Biology*, **311**: 1001-1009.

30.   Daniell H, Vivekananda J, Nielsen BL, Ye GN, Tewari KK and Sanford JC (1990). Transient foreign gene expression in chloroplasts of cultured tobacco cells after biolistic delivery of chloroplast vectors. *Proceedings of the National Academy of Sciences USA*, **87**: 88-92.

31.   Daniell H, Krishnan M and McFadden BA (1991). Expression of ß-glucuronidase gene in different cellular compartments following biolistic delivery of foreign DNA into wheat leaves and calli. *Plant Cell Reports*, **9**: 615-619.

32.   Goldschmidt-Clermont M (1991). Transgenic expression of aminoglycoside adenine transferase in the chloroplast: a selectable marker of site-directed transformation of chlamydomonas. *Nucleic Acids Research*, **19**: 4083-4089.

33.   Svab Z and Maliga P (1993). High-frequency plastid transformation in tobacco by selection for a chimeric *aadA* gene. *Proceedings of the National Academy of Sciences USA*, **90**: 913-7.

34.   Guda C, Lee SB and Daniell H (2000). Stable expression of biodegradable protein based polymer in tobacco chloroplasts. *Plant Cell Reports*, **19**: 257-262.

35.   Daniell H (2003). *Medical molecular farming: expression of antibodies, biopharmaceuticals and edible vaccines via the chloroplast genome*. In: Vasil IK (ed.), *Plant Biotechnology 2002 and Beyond*. Proceedings of the 10th IAPTC&B Congress, June 23-28, 2002, Orlando, USA. Kluwer Academic Publishers: The Netherlands, pp. 371-376.

36.   Daniell H, Dhingra A and San-Milan AF (2001). Chloroplast transgenic approach for the production of biopharmaceuticals and resolution of basic questions on gene expression. In: *12th International Congress on Photosynthesis*. Brisbane, Australia: CSIRO.

37.   Fernandez-San Millan A, Mingo-Castel AM and Daniell H (2003). A chloroplast transgenic approach to hyper-express and purify human serum albumin, a protein highly susceptible to proteolytic degradation. *Plant Biotechnology Journal*, **1**: 71-79.

38.   DeGray G, Rajasekaran K, Smith F, Sanford J and Daniell H (2001). Expression of an antimicrobial peptide via the chloroplast genome to control phytopathogenic bacteria and fungi. *Plant Physiology*, **127**: 852-62.

39.   Staub JM, Garcia B, Graves J, Hajdukiewicz PT, Hunter P, Nehra N, Paradkar V, Schlittler M *et al.* (2000). High-yield production of a human therapeutic protein in tobacco chloroplasts. *Nature Biotechnology*, **18**: 333-338.

40.   Leelavathi S and Reddy VS (2003). Chloroplast expression of His-tagged GUS-fusion: a general strategy to overproduce and purify foreign protein using transplastomic plants as bioreactors. *Molecular Breeding*, **11**: 49-58.

41.   Daniell H, Datta R, Varma S, Gray S and Lee SB (1998). Containment of herbicide resistance through genetic engineering of the chloroplast genome. *Nature Biotechnology*, **16**: 345-348.

42.   McBride KE, Svab Z, Schaaf DJ, Hogan PS, Stalker DM and Maliga P (1995). Amplification of a chimeric *Bacillus* gene in chloroplasts leads to an extraordinary level of an insecticidal protein in tobacco. *Biotechnology (NY)*, **13**: 362-365.

43.   Kota M, Daniell H, Varma S, Garczynski SF, Gould F and Moar WJ (1999). Overexpression of the *Bacillus thuringiensis* (*Bt*) Cry2Aa2 protein in chloroplasts confers resistance to plants against susceptible and *Bt*-resistant insects. *Proceedings of the National Academy of Sciences USA*, **96**: 1840-1845.

44.   Murashige T and Skoog F (1962). A revised medium for rapid growth and bioassays with tobacco tissue cultures. *Physiologia Plantarum*, **15**: 473-497.

45.   Ruf S, Hermann M, Berger IJ, Carrer H and Bock R (2001). Stable genetic transformation of tomato plastids and expression of a foreign protein in fruit. *Nature Biotechnology*, **19**: 870-875.

46.   Sambrook J, Fritsch EF and Maniatis T (1989). Molecular Cloning - A Laboratory Manual. 2nd edn., Nolan C (ed.). Plainview, New York: Cold Spring Harbor Laboratory Press, USA.

47.   Crowther JR (1995). ELISA. Theory and practice. *Methods in Molecular Biology*, **42**: 1-218.

48.   Bruyns A-M, De Neve M, De Jaeger G, De Wilde C, Rouzé P and Depicker A (1998). Quantification of Heterologous Protein Levels in Transgenic Plants by ELISA. In: Cunningham C, Porter AJR (eds.), *Recombinant Proteins from Plants*. Humana Press: Totowa, New Jersey, USA, pp. 251-269.

Chapter 31

GENETIC TRANSFORMATION OF WATERMELON

M.E. COMPTON, D.J. GRAY[1] AND V.P. GABA[2]

School of Agriculture, University of Wisconsin-Platteville, 1 University Plaza, Platteville, WI 53818, USA; [1]Mid Florida Research and Education Center, University of Florida, Institute of Food and Agricultural Sciences, Apopka, FL, USA; [2]Department of Virology, Institute of Plant Protection, The Volcani Center, Bet Dagan, Israel. Corresponding author: M. Compton. E-mail: compton@uwplatt.edu

1. INTRODUCTION

1.1. Importance and Distribution of Watermelon

Watermelon, *Citrullus lanatus* (Thunb.) Matsum. and Nakai, is an important vegetable crop globally because of its high vitamin [25% and 20% of the USA recommended daily allowance (RDA) of vitamins C and A, respectively, per 0.280 g fwt.] and nutrient content (8% US-RDA of potassium, 4% US-RDA iron and 2% US-RDA of calcium per 280g fwt.). Watermelon flesh is also high in lycopene, a potent antioxidant that has been shown to reduce human risk to cancer of the prostrate, pancreas and stomach (*1, 2*).

The species *Citrullus* originated in tropical Africa but there is evidence of North American origin as well (*3*). China produces about 71% (57.65 million metric tonnes, MT) of the world watermelon crop with Turkey (3.9 million MT), Iran (1.9 million MT), the USA (1.78 million MT) and Egypt (1.45 million MT) rounding out the top five producing countries (*4*). In the USA, watermelon is cultivated mainly in the south eastern (Florida and Georgia), south western (California, Texas and Arizona) and central (Indiana) states. The 2002 watermelon crop in the USA was valued at over 326 million dollars (*4*).

Fruit of different watermelon cultivars vary in size, shape, rind colour and pattern, flesh and seed colour, and maturity date. Most have a striped rind with red flesh and small black seeds, and weigh 20-30 lbs at maturity. However, triploid seedless cultivars with high sugar content are becoming more prevalent (*5*). Tetraploid lines have been developed and are used primarily for breeding triploid seedless hybrids (*6*).

1.2. Need for Genetic Improvement

Watermelon plants are monoecious, producing separate male and female flowers on the same plant. Because of its reproductive biology, many early breeding schemes for the species employed several rounds of open pollination (*7*). More recently, breeders have concentrated on developing F_1 hybrid cultivars to take advantage of heterosis and improve uniformity among progeny. However, because of the absence of high quality male-sterile genotypes, considerable labour is required to perform hand pollinations. Because of reproductive biology and breeding objectives, cultivar development in watermelon may require more than 15 years (*8*).

I.S. Curtis (ed.), Transgenic Crops of the World - Essential Protocols, 425-433.

Another breeding strategy for watermelon has been the development of triploid seedless hybrids. Triploid watermelons produce parthenocarpic fruit (9), and are preferred by consumers because they are seedless and taste sweeter than diploid seeded cultivars (10). Polyploid genetics is involved in the production of seedless watermelon. Triploid plants are obtained by crossing tetraploid (♀, female) and diploid (♂, male) plants (9). Tetraploid plants are produced by treating young diploid seedlings with colchicine (11, 12) or by regenerating plants through tissue culture and selecting tetraploids (6). Once identified, tetraploid plants are repeatedly self-pollinated to generate a population of plants to be used for triploid seed production.

The development of disease-resistant cultivars has been another important objective in watermelon breeding. Watermelons are susceptible to a number of bacterial, fungal and viral diseases, requiring annual field rotation, frequent chemical sprays and development of disease-resistant cultivars. While traditional breeding has resulted in cultivars resistant to anthracnose, fusarium wilt and gummy stem blight (7, 8), the development of plants resistant to important bacterial diseases such as watermelon fruit blotch (13) and viruses such as papaya ringspot virus (PRSV) (formerly known as watermelon mosaic virus 1) (14), watermelon mosaic virus (WMV) (formerly known as watermelon mosaic virus 2) (15) and zucchini yellows mosaic virus (ZYMV) (16) has been elusive.

Susceptibility to diseases makes watermelon an excellent candidate for improvement through genetic engineering. Insertion of bacterial, fungal and viral resistance genes through recombinant DNA technology would facilitate the development of new disease resistant genotypes without significantly altering the genetic composition and desirable phenotypic qualities of accepted cultivars.

1.3. Tissue Culture and Plant Regeneration of Watermelon

The base work of developing an efficient plant regeneration system for the wide scope of watermelon cultivars has been achieved. Plant regeneration protocols for shoot tip micropropagation (17, 18), somatic embryogenesis (19) and adventitious shoot regeneration (20-22) have been long developed for watermelon. Of the three procedures, adventitious shoot regeneration from cotyledon pieces has been the most rapid, repeatable and efficient protocol for genetic improvement of American and Asian cultivars (20, 21, 23). For all *in vitro* regeneration procedures, the macro and micro salts plus vitamins outlined by Murashige and Skoog (MS, 24) are used with additions of (per litre) 0.1 g *myo*-inositol and 30 g sucrose. BAP (4.4-10 µM) is typically used for adventitious shoot regeneration (20, 22, 23, 25, 26). However, 2.85 µM IAA is sometimes added (21). Addition of kinetin (5-40 µM), zeatin, 2iP or TDZ (0.1-10 µM) failed to improve organogenesis (20, 21). Plant regeneration is optimum if the medium is semi-solidified with 7-10 g TC agar (PhytoTechnology Laboratories, Shawnee Mission, KS, USA) (20), 5 g Agar-Gel (26) or 4 g Gelrite (25) per litre. For adventitious shoot organogenesis, it is often beneficial to transfer explants with shoots to medium without growth regulators (20) or 0.92 µM kinetin (21) to stimulate shoot elongation. Shoots can be rooted with high success on medium with 1 µM IBA (20) or 0.54 µM NAA (21). Shoots root when they are at least 6 mm long and plantlets easily acclimatized once they are at least 2 cm tall (18). Successful plant regeneration has been obtained from diploid, triploid and tetraploid genotypes (20, 23, 27).

1.4. Genetic Transformation of Watermelon

Although stable genetic transformation of *Citrullis* has been reported, there are few published reports documenting stable integration of recombinant DNA into the watermelon genome (*25*). In this report, cotyledon explants were transformed using a disarmed strain of *Agrobacterium tumefaciens* LBA4404 containing the binary vector pBI121, which carried the CaMV 35S promoter-β-glucuronidase (GUS) gene fusion product as a reporter gene and NOS promoter-*npt*II marker gene. Approx. 16% of regenerated shoots examined by histological assay expressed GUS. However, examination of plants beyond *in vitro* culture was not reported. Transient gene expression following particle bombardment has been reported for watermelon (*28*) but stable integration of recombinant DNA using this method has yet to be published.

Adaptations of the method of Choi *et al.* (*25*) were used by Seminis Vegetable Seeds, Inc (Saticov, CA, USA) to produce virus-resistant transgenic watermelon plants (*29*). Researchers incorporated the coat protein genes of ZYMV and WMV under control of the cauliflower mosaic virus 35S promoter to produce transgenic inbred lines resistant to both viruses. Unfortunately, the protocols used by the inventors to produce transgenic watermelon plants were not thoroughly described in their patent.

Recovery of transgenic watermelon shoots and plantlets *in vitro* during transformation has generally been by use of the *npt*II gene for selection in media containing kanamycin (*21, 29*). Use of the phosphomannose isomerase positive selection method *in vitro* has been successfully employed with watermelon to produce transgenic plants (*30*).

Transgenic watermelon lines have been developed by several commercial companies and field-tested a number of times, predominantly in the USA, but also in Europe. Data on field tests of transgenic crops is available on a website (*31*), from which the information here was derived. However, much important data has been classified as confidential business information and not released into the public domain. The majority (80%) of the field tests were for virus resistance, as this is the most limiting factor in watermelon crop production. Most of the watermelon field tests for virus resistance were with plants bearing coat protein genes of both ZYMV and WMV, as these are the most important viruses infecting watermelon. Additionally, a field test has been conducted on watermelon plants bearing four coat protein genes (from ZYMV, WMV, PRSV and cucumber mosaic virus, although the last is not a problem in watermelon). Transgenic watermelon has also been engineered to produce parthenocarpic seedless fruit. No transgenic watermelon crop has been yet released for sale.

The purpose of this chapter is to outline existing shoot regeneration and transformation protocols designed for watermelon. It should be noted that the genetic transformation of watermelon is still in its infancy and the procedures outlined in this chapter may need refinement to be efficient for a wide range of watermelon genotypes.

2. MATERIALS

2.1. A. tumefaciens *Strain*

A. tumefaciens strain LBA4404 containing the binary vector pBI121 can be used (*25, 28, 32*). The binary vector contains a β–glucuronidase (GUS)-intron and *npt*II coding sequences driven by CaMV 35S and NOS promoters, respectively.

2.2. Culture Medium for **A. tumefaciens**

1. YEP medium: (per litre) 10 g peptone (Sigma), 5 g NaCl, 10 g yeast extract (Sigma), and 10 g Difco Bacto agar (Sigma), pH 7.0.
2. Kanamycin sulphate stock: weigh and dispense 0.5 g in 100 ml of distilled water. Filter-sterilize through a 0.2 μm membrane. Dispense into YEP medium at a rate of 10 ml/l.
3. Streptomycin sulphate stock: weigh and dispense 0.25 g in 100 ml distilled water. Filter-sterilize through a 0.2 μm membrane. Dispense into YEP medium at a rate of 10 ml/l.

2.3. Plant Materials

1. All watermelon cultivars tested to date have been competent for adventitious shoot organogenesis. However, the regeneration rate varies among cultivars. The most competent to date (in alphabetical order by common name and ploidy level) are the diploid cultivars Crimson Sweet, Minilee, Mickeylee, New Crown, Peace, Yellow Baby and Yellow Doll; triploid cultivars King of Hearts and Jack of Hearts; and tetraploid breeding lines F92U6 and SP90-2.
2. The freshest quiescent seeds should be obtained. Seeds can be stored at 4°C for about 5 years and remain regenerable.

2.4. Watermelon Embryo Germination and Plant Regeneration Media

1. Watermelon Embryo Germination Medium: MS salts and vitamins with (per litre) 0.1 g *myo*-inositol, 30 g sucrose, 5 g Agar-Gel (Phytotechnology Laboratories) at pH 5.8.
2. Watermelon Shoot Regeneration Medium: MS salts and vitamins with (per litre) 0.1 g *myo*-inositol, 30 g sucrose, 5 μM BAP and 5 g Agar-Gel at pH 5.8.
3. Watermelon Shoot Elongation Medium: MS salts and vitamins with (per litre) 0.1 g *myo*-inositol, 30 g sucrose, 5 g Agar-Gel at pH 5.8.
4. Watermelon Rooting Medium: MS salts and vitamins with (per litre) 0.1 g *myo*-inositol, 30 g sucrose, 1 μM IBA, 5 g Agar-Gel at pH 5.8.

2.5. Cocultivation Medium

1. Watermelon Shoot Regeneration Medium: MS salts and vitamins with (per litre) 0.1 g *myo*-inositol, 30 g sucrose, 5 μM BAP, 200 μM acetosyringone (filter-sterilize and add to cooled autoclaved medium) and 5 g Agar-Gel at pH 5.8.

3. METHODS

The methods described are a compilation of plant regeneration and *Agrobacterium*-mediated transformation protocols described previously (*20, 21, 25, 26, 28, 34*).

3.1. Preparation of Watermelon Cotyledon Explants and Culture Conditions

1. Imbibe seeds for 4-15 h in sterile distilled water. Incubate in darkness at 20-25EC (*see* Note 1). Approx. 75 seeds should be prepared for every 100 explants required.
2. Embryo Extraction: hold seeds at their pointed end using the toothed forceps (*see* Note 2). Position seeds on their side and make an incision in the seed coat using a scalpel (#10 blade) by applying a sharp downward motion at the rounded end of the seed. Several cuts are required to make an opening large enough so that the seed coat can easily be removed. Flick back about one-half of the seed coat and remove the embryo by gently working it side-to-side using the blunt side of the scalpel. Store embryos in sterile distilled water until all have been extracted.
3. Surface disinfest extracted embryos for 25 min in a commercial bleach solution (1.3 % NaOCl containing 1 ml/l of surfactant). Agitate manually as needed. However, containers may be placed on a shaker at 150 rpm if desired. Rinse embryos 3-6 times with sterile distilled water (*see* Note 3).
4. Transfer disinfested embryos to Magenta GA-7 vessels containing 50 ml of watermelon germination medium. Culture 6-9 embryos per vessel. Incubate in darkness for 5-7 days (*see* Note 4).
5. Remove seedlings from their culture vessel and transfer to sterile plastic Petri dishes for dissection (*see* Note 5). Prepare cotyledon explants by making a shallow cut across the width of the cotyledon about 1-2 mm above the point of attachment to the hypocotyls. Gently flick the cotyledons to detach them from the hypocotyl. Place cotyledons abaxial side (i.e., the "bottom" side) up and remove the margins (*see* Note 6).
6. Transfer cotyledon halves to 100 x 15 mm plastic Petri dishes containing 25 ml of regeneration medium. Place cotyledon pieces abaxial side down, culturing 5 explants per Petri dish. Seal dishes with parafilm before light incubation at 25°C. Explants should be incubated for 2-5 days, depending on cultivar, before inoculation with *Agrobacterium*.

3.2. Preparation of Agrobacterium

1. Culture *A. tumefaciens* in 100 x 15 mm Petri dishes containing 25 ml of semi-solidified YEP medium supplemented with selection antibiotics (*see* Note 7).
2. Once small colonies form, transfer a single colony to a 125 ml Erylenmeyer flask containing 50 ml liquid YEP medium supplemented with selection antibiotics.
3. Grow the *Agrobacterium* suspension for 24-48 h in darkness at 28°C to an O.D. of 1.0 at 600 nm (*see* Note 8).
4. On the day of explant inoculation, transfer *A. tumefaciens* bacterial cell suspension to a 15 ml centrifuge tube and centrifuge at 4,000 rpm for 10 min. Remove the supernatant and resuspend the bacteria in sufficient cocultivation medium to adjust the cell density to 10^6 cells per ml (*see* Note 9).

3.3. Inoculation of Watermelon Cotyledons with Agrobacterium and Regeneration of Transformed Plants

1. Gently wound precultured cotyledon pieces before inoculation with *A. tumefaciens* (*see* Note 10). Explants can be wounded by pricking their adaxial surface 10-15 times with the tip of a scalpel or by bombarding the regeneration site with tungsten particles accelerated using a particle inflow gun (*see* Note 11).

2. Transfer the wounded explants to the bacterial suspension and cocultivate for 48 h (*see* Note 12).
3. Remove cotyledon pieces from the bacterial suspension and lightly blot-dry with sterile filter paper before transfer to 15 x 100 mm Petri dishes containing 25 ml of watermelon regeneration medium supplemented with 250 mg/l carbenicillin and 100 mg/l kanamycin. Incubate cultures in the light (30-50 μmol/m^2/sec) at 25°C and transfer to fresh medium every 4 weeks (*see* Note 13).
4. Once shoots are visible, cultures can be transferred to watermelon elongation medium containing 250 mg/l carbenicillin and 100 mg/l kanamycin to facilitate shoot elongation and subsequent rooting (*see* Note 14).
5. After 3-4 weeks on elongation medium, shoots longer than 2 cm in length can be excised from the primary explant and all expanded leaves removed, leaving the shoot apex and first true leaf. Shoots are transferred to Magenta GA-7 vessels containing 50 ml of watermelon rooting medium supplemented with 250 mg/l carbenicillin and 100 mg/l kanamycin (*see* Note 15). Up to 9 shoots can be transferred to a single Magenta vessel.

3.4. Acclimatization Plantlets to the Glasshouse

1. Remove plantlets from the culture vessels and gently wash away any agar surrounding the roots (*see* Note 16).
2. Transplant plantlets to cell packs (3.3 x 5.1 cm; 72 cells per tray) containing soilless medium [1 Pro Mix BX (Premier Horticulture, CA, USA): 1 Coarse Vermiculite (Hummart Intl., Earth City, MO, USA)] (*see* Note 17).
3. Place transplanted plantlets into a plastic flat and cover with a clear plastic lid. Seal the lid to the flat with wide masking tape and place in a well-lit area (50 μmol/m^2/sec and 16 h photoperiod) for approx. 7 days before removing the tape and gradually removing the lid (*see* Note 18).
4. Plantlets should be fully acclimatized within 21 days (*see* Note 19).

3.5. Maintenance of Plants in the Glasshouse

1. Plants can be grown in 18.75 litre plastic pots containing soil-less growing medium (Fafard 3B, Agawam, MA, USA) and trained to a single-stemmed vine on a line suspended from the glasshouse trusses (*see* Note 20).
2. Plants should be fertilized weekly with 250 ppm (based on the N fraction) nitrogen from a 20-20-20 (N-P$_2$O$_5$-K$_2$O) or 20-10-20 fertilizer. Addition of a time-release fertilizer (Osmocote or Sierra) is also recommended (*see* Note 21).

3.6. Screening of Transformed Individuals

1. Leaves are collected from putatively transformed shoots *in vitro* or plants grown in the glasshouse. Genomic DNA is extracted according to procedures outlined by Choi *et al.* (*25*).
2. Digest DNA with *Eco*RI, *Hind*III, *Bam*HI or *Eco*RI/*Bam*HI for 4 h before electrophoresis on 0.8% agarose gel.
3. Transfer DNA bands to a positively charged nylon membrane and probe with 2.2 kb GUS NOS poly(A) labelled with digoxigenin (Boehringer Mannheim).
4. Transformed plants will exhibit a band at the 2.2 kb marker.

4. NOTES

1. It is helpful to soak seeds in a 50% (2.6% NaOCl) solution of commercial bleach for 30 min before soaking overnight in sterile distilled water. The bleach treatment removes a portion of the seed coat and facilitates embryo extraction. Cutting dry seeds is stressful on workers and is almost impossible
2. It is recommended that workers use special toothed forceps (# 6-134 Miltex Instrument Company, Inc., Lake Success, N.Y., USA) to facilitate handling of seeds during embryo extraction.
3. Disinfestation times can be reduced to 15 min for small seeded cultivars like Minilee.
4. Incubating embryos in darkness facilitates germination and improves shoot regeneration (26).
5. Excising the root and half of the hypocotyl before removing the cotyledons facilitates handling.
6. Care must be taken not to include the shoot apex. Cotyledons have been properly removed when the shoot tip can be seen remaining on the embryo axis after detaching the cotyledons. The proximal portion of the cotyledon has greater regeneration competence than the distal region (33). Likewise, larger explants (6 x 5 mm cotyledon bases) possess a greater regeneration potential than 3 x 5 mm cotyledon quarters.
7. The binary vector pBI121 carrying the CaMV 35S promoter-gusA gene-NOS terminator fusion and NOS promoter-nptII gene-NOS terminator fusion can be transformed into A. tumefaciens strain LBA4404 using triparental mating or electroporation (35).
8. It is important that A. tumefaciens cultures are not more than 48 h old when used for transformation.
9. A simpler inoculation medium consisting of 12.5 mM MES, 1 g/l NH$_4$Cl and 0.3 g/l MgSO$_4$ may be used.
10. Explants should be precultured for 1-5 days before inoculation with Agrobacterium. The number of days of preculture varies with experimental procedure.
11. Wounding explants with 1.1 μm tungsten particles delivered using a particle inflow gun (36) before inoculation with A. tumefaciens improves transformation frequency (32). To position explants for wounding via particle bombardment, insert watermelon cotyledon pieces vertically, so that the regeneration site is upward, into watermelon regeneration medium (12 ml in 15 x 60 mm Petri dishes) semi-solidified with 15 g/l TC agar. Only about 2 mm of the cotyledon base should extend above the medium surface. Explants are placed in a 4 x 4 grid arrangement in the centre of the plate with about 1-2 mm between explants. Preculture explants for 2 days and air dry in the laminar hood for 5-30 min before bombardment. Place 10 μl of autoclaved tungsten stock (50 mg/500 μl distilled water) onto a sterile particle suspension holder. Bombard under conditions of 120 psi helium and 27 cu in vacuum.
12. Instead of soaking entire explants in A. tumefaciens suspension, it may be desirable to place small droplets on the regeneration site only. To do this, add 2 drops (10-20 μl) of bacterial suspension onto the upper surface of a bombarded explant using a 5 ml syringe. Apply the suspension so that the droplet remains on the upper side of all explants. Cocultivate for 48-72 h.
13. The kanamycin concentration can be increased to 125-135 mg/l to reduce the number of escapes.
14. All brown leaves and non-responding portions of the cotyledon are trimmed before transferring shoot cultures to vessels containing elongation medium. The kanamycin concentration can be reduced to 75 mg/l, as 100 mg/l is often too high a dose at this stage.
15. The kanamycin concentration may be reduced to 50-75 mg/l during rooting.
16. Plants can be acclimatized once they have formed roots that are at least 1 cm in length (10-14 days).
17. An equal portion of coarse vermiculite should be added to soil-less medium to improve drainage. Cell packs filled with soil-less medium should be placed in clean 27.5 x 55 cm flats without drainage holes. Moisten mixture with 300 ml of freshly autoclaved distilled water the day before transplanting. Cover the flat with moistened medium with a clear plastic lid to raise humidity.
18. At least 75% of plantlets usually survive acclimatization.
19. It can be expected that between 5-25% of regenerants are polyploids (37). Plant ploidy can be estimated by counting the number of chloroplasts per guard cell pair of fully expanded leaves (6). Guard cell preparations can be obtained by performing epidermal peels from leaves of plants grown in the glasshouse (32) or by painting the lower epidermis of fully expanded leaves obtained from shoots on elongation medium with fluorescein diacetate and observing guard cells using a microscope equipped with an epifluorescent illuminator and a fluorescein isothiocyanate filter combination (38). Diploid regenerants contain an average of between 8-12 chloroplasts per guard cell pair while triploids and mixoploids possess 13-16 chloroplasts per guard cell pair and tetraploids between 17-21 chloroplasts per guard cell pair (6, 37, 38).
20. Plants can be grown hydroponically using perlite or rockwool bag culture.
21. Plants in bag culture should not be irrigated with unamended tap water.

CONCLUSION AND FUTURE PROSPECTS

Watermelon is a globally important crop ready for the application of genetic engineering technology for cultivar improvement. Efficient tissue culture methods for regenerating plants have been in place for over a decade and have paved the way for genetic transformation technology. Plant regeneration through adventitious shoot organogenesis from cotyledon pieces is the most efficient means of regenerating watermelon plantlets. Regeneration appears to occur directly from regenerative cells as indicated by a lack of evidence supporting significant somaclonal variation among regenerants. The most significant form of somatic variation appears to be the regeneration of polyploids, which is often an advantage when breeding seedless cultivars.

Agrobacterium-mediated transformation currently holds the most promise for watermelon biotechnology. Previous research has demonstrated that watermelon can be transformed using this vector. However, there is little information on the inheritance of recombinant DNA in transgenic watermelon plants. There is evidence that expression of viral coat protein genes in transgenic watermelon provides protection against economically devastating diseases that hamper grower production. The next step is to incorporate these genes into cultivars used by growers for food production.

REFERENCES

1. Anonymous (2003). Fields of green. Watermelon Promotion Board. Retrieved 22 May 2003 from http://www.watermelon.org.
2. Garster H (1997). The potential role of lycopene for human health. *Journal of American College of Nutrition*, **16**: 109-126.
3. Decoteau DD (2000). *Vegetable Crops*. Prentice Hall, Upper Saddle River, NJ, USA.
4. Anonymous (2003). Lycopene. Retrieved 22 May 2003 from http://www.lycopene.org.
5. Lucier G and Lin BH (2001). Factors affecting watermelon consumption in the United States. pp. 23–29. In: Anonymous (eds.), *Vegetables and Specialties: Situation and Outlook*, VGS-287. USDA-ERS.
6. Compton ME, Gray DJ and Elmstrom GW (1996). Identification of tetraploid regenerants from cotyledons of diploid watermelon cultured *in vitro*. *Euphytica*, **87**: 165-172.
7. Mohr HC (1986). Watermelon breeding. In: Bassett MJ (ed.), *Breeding Vegetable Crops*, AVI Publishing Co., Inc, Westport, Connecticut, USA.
8. Crall JM, Elmstrom GW and McCuistion Jr FT (1994). SSDL: A high-quality icebox watermelon breeding line resistant to fusarium wilt and anthracnose. *HortScience*, **29**: 707-708.
9. Kihara H (1951). Triploid watermelons. *Proceedings of American Society for Horticultural Science*, **58**: 217-230.
10. Marr CW and Gast KLB (1991). Reactions by consumers in a 'farmers' market to prices for seedless watermelon and ratings of eating quality. *HortTechnology*, **1**:105-106.
11. Andrus CF, Seshadri VS and Grimball PC (1971). Production of seedless watermelons. *Agricultural Research Service, United States Department of Agriculture Technical Bulletin No. 1425.*
12. McCuistion G and Elmstrom GW (1993). Identifying polyploids of various cucurbits. *Proceedings of the Florida State Horticulture Society*, **106**: 155-157.
13. Rane KK and Latin RX (1992). Bacterial fruit blotch of watermelon: association of the pathogen with seed. *Plant Disease*, **76**: 509-512.
14. Fehèr T (1993). Watermelon *Citrullus lanatus* (Thunb.) Matsum. & Nakai. In: Kalloo G, Bergh BO (eds.), *Improvement of Vegetable Crops*. Pergamon Press, Oxford, UK, pp. 295–311.
15. Gillaspie Jr AG and Wright JM (1993). Evaluation of *Citrullus* sp. germplasm for resistance to watermelon mosaic virus 2. *Plant Disease*, **77**: 352-354.
16. Boyhan GJ, Norton D, Jacobsen BJ and Abrahams BR (1992). Evaluation of watermelon and related germplasm for resistance to zucchini yellow mosaic virus. *Plant Disease*, **76**: 251-252.
17. Gray DJ and Elmstrom GW (1991). Process for the accelerated production of triploid seeds for seedless watermelon cultivars. *United States Patent No.* 5,007,198.

18.   Compton ME, Gray DJ and Elmstrom GW (1993). A simple protocol for micropropagating diploid and tetraploid watermelon using shoot-tip explants. *Plant Cell, Tissue and Organ Culture*, **33**: 211-217.

19.   Compton ME and Gray DJ (1993). Somatic embryogenesis and plant regeneration from immature cotyledons of watermelon. *Plant Cell Reports*, **12**: 61-65.

20.   Compton ME and Gray DJ (1993). Shoot organogenesis and plant regeneration from cotyledons of diploid, triploid and tetraploid watermelon. *Journal of the American Society for Horticultural Science*, **118**: 151-157.

21.   Dong JZ and Jia SR (1991). High efficiency plant regeneration from cotyledons of watermelon (*Citrullus vulgaris* Schrad.). *Plant Cell Reports*, **9**: 559-562.

22.   Srivastava DR, Andrianov VM and Piruzian ES (1989). Tissue culture and plant regeneration of watermelon (*Citrullus vulgaris* Schrad. cv. Melitopolski). *Plant Cell Reports*, **8**: 300-302.

23.   Jaworski JM and Compton ME (1997). Plant regeneration from cotyledons of five watermelon cultivars. *HortScience*, **32**: 469.

24.   Murashige T and Skoog F (1962). A revised medium for rapid growth and bioassays with tobacco tissue cultures. *Physiologia Plantarum*, **15**: 473-497.

25.   Choi PS, Soh WY, Kim YS, Yoo OJ and Liu JR (1994). Genetic transformation and plant regeneration of watermelon using *Agrobacterium tumefaciens*. *Plant Cell Reports*, **13**: 344-348.

26.   Compton ME (1999). Dark pretreatment improves adventitious shoot organogenesis from cotyledons of diploid watermelon. *Plant Cell, Tissue and Organ Culture*, **58**: 185-188.

27.   Compton ME and Gray DJ (1994). Adventitious shoot organogenesis and plant regeneration from cotyledons of tetraploid watermelon. *HortScience*, **29**: 211-213.

28.   Compton ME, Gray DJ, Hiebert E and Lin CM (1993). Expression of the β-glucuronidase gene in watermelon cotyledon explants following particle bombardment or infection with *Agrobacterium tumefaciens*. *HortScience*, **28**: 138.

29.   Tricoli DM, Carney KJ, Russell PF, Quemada HD, McMaster RJ, Reynolds JF and Deng RZ (2002). Transgenic plants expressing DNA constructs containing a plurality of genes to impart virus resistance. *US Patent 6,337,431*.

30.   Reed J, Privalle L, Powell ML, Meghji M, Dawson J, Dunder E, Suttie J, Wenck A, Launis K, Kramer C, Chang YF, Hansen G, Wright M and Chang YF (2001). Phosphomannose isomerase: an efficient selectable marker for plant transformation. *In Vitro Cellular and Developmental Biology-Plant*, **37**: 127-132.

31.   ISB (2002). Information Systems for Biotechnology website. Retrieved January 2002; from http://www.isb.vt.edu/cfdocs/fieldtests1.cfm.

32.   Compton ME, Gray DJ, Hiebert E and Lin CM (1994). Microprojectile bombardment prior to co-cultivation with *Agrobacterium* improves GUS expression in watermelon cotyledons. *In Vitro Cellular and Developmental Biology-Plant*, **30A**: 62.

33.   Compton ME (2000). Interaction between explant size and cultivar impacts shoot organogenic competence of watermelon cotyledons. *HortScience*, **35**: 749-750.

34.   Compton ME and Gray DJ (1999). Shoot organogenesis from watermelon cotyledon explants. In: Trigiano RN, Gray DJ (eds.), *Plant Tissue Culture Concepts and Laboratory Exercises*, 2nd edn., CRC Press, Boca Raton, Florida, USA, pp. 149-158.

35.   Maldonado-Mendoza IE, Lépez-Meyer M and Nessler CL (1999). Transformation of tobacco and carrot using *Agrobacterium tumefaciens* and expression of the β-glucuronidase (GUS) reporter gene. In: Trigiano RN, Gray DJ (eds.), *Plant Tissue Culture Concepts and Laboratory Exercises*, 2nd edn., CRC Press, Boca Raton, Florida, USA, pp. 305-319.

36.   Finer JJ, Vain P, Jones MW and McMullen M (1992). Development of the particle inflow gun for DNA delivery to plant cells. *Plant Cell Reports*, **11**: 323-328.

37.   Compton ME, Gray DJ and Elmstrom GW (1994). Regeneration of tetraploid plants from cotyledons of diploid watermelon. *Proceedings of the Florida State Horticulture Society*, **107**: 107-109.

38.   Compton ME, Barnett N and Gray DJ (1999). Use of fluorescein diacetate (FDA) to determine ploidy of *in vitro* watermelon shoots. *Plant Cell, Tissue and Organ Culture*, **58**: 199-203.

Chapter 32

GENETIC TRANSFORMATION OF SUNFLOWER (*HELIANTHUS ANNUUS* L.)

T. HEWEZI, G. ALIBERT AND J. KALLERHOFF

Laboratoire de Biotechnologies et Amélioration des Plantes (BAP), INP-ENSAT, Avenue de l'Agrobiopôle, BP 107, Auzeville Tolosane, F-31326 Castanet Tolosan Cedex, France. Corresponding author: J. Kallerhoff. E-mail: kallerho@ensat.fr

1. INTRODUCTION

1.1. Market of Oilseed Crops

The world production of oilseed crops has more than doubled in the last 30 years, estimated at 318 million metric tonnes (MT) against 128 million MT in 1973-74 (*1*). Soybean accounts for 57% of the world production, followed by rape and cotton (each approx. 12%), ground nut (8%) and sunflower (7%). United States are the leaders for global oilseed crop production, providing 90 million MT (30%) to the world market. The European Union (EU), with a production of 15 million MT in 2002, is the 6th largest producer after the USA, China, Brazil, Argentina and India. Although sunflower seed production seems to be lagging behind other crops, its culture has regained interest since 1973 in Europe, with the anticipation of dependency on America for provision of plant proteins. In 2001, the EU was third for sunflower seed production (3.6 million MT) after Ex-USSR (6.33 million MT) and Argentina (5.4 million MT). Sunflower oil world production now ranks 4th after soybean, palm and rape, the leaders being Russia and Ukraine, followed by the EU and Argentina. Within the EU, the production of oilseed crops has multiplied 5-fold between 1980 (3 million MT) and 2001 (13.6 million MT). Sunflower is the second most important cultivated crop after oilseed rape. Cultivated surfaces of EU oil-seed crops near 5.4 million acres, 37% being devoted to sunflower.

Sunflower oil is a major target for the food and feed industry. It is not only known for its richness in polyunsaturated fatty acids but also for its relative high content of minor constituents such as tocopherols (Vitamin E) and phytosterols, known to lower plasma cholesterol levels. It is also poor in squalene, a precursor of cholesterol, making it a heart-friendly oil to consumers. The recent European directives aiming at developing new sources of renewable energies as substitutes to fossil fuels is contributing to the creation of new research avenues for the enhancement of added value of by-products issued from the extraction and refining process of vegetable oils. Biodiesel production from vegetable oils is one of the priorities of the European Commission. According to the amendment of Directive

I.S. Curtis (ed.), Transgenic Crops of the World - Essential Protocols, 435-451.
© 2004 *Kluwer Academic Publishers. Printed in the Netherlands.*

92/81/EEC, the Member States would be under the obligation of using 2% of biofuels out of their global consumption as from 2005, to reach at least 5.75% in 2010. Methyl esters incorporated into fuel are derived from transesterification reactions of lipids from oilseed rape. Considering the objectives of the European Commision, oilseed rape culture exclusively, will not be able to meet the requested demands. Sunflower oil has been proposed to complement oilseed rape (2). Moreover, green chemistry has contributed to the development of the use of natural raw material for surfactants and new bio-lubricants, derived mainly from rape oils but also from sunflower (3). Use of plant resources in these industries is justified by their renewable nature and because they present original structural qualities, notably double bonds that petrochemicals cannot offer. Furthermore, these natural biomolecules have improved ecotoxicological and environmental profiles compared to petrochemicals. To this end, sunflower oil is a good candidate to complement rape oil, not only for its high content for polyunsaturated fatty acids, but also for the quality and quantity of its minor constituents. The application of products derived from sunflower edible-oil processing, other than those for food and feed are also extendable to the pharmaceutical, textile and cosmetical industries.

1.2. New Research Vistas for Sunflower

Conventional breeding has up to now, focused on introducing tolerance traits to diverse pathogens, increasing yields and modifying fatty acid spectra. Although successful, there remains hurdles to overcome, and hopes reside in the association of genetic engineering and molecular breeding to improve biotic (white rot, downy mildew, phoma and phomopsis) and particularly abiotic stresses (drought tolerance, water deficit and high temperature). The development of genetic transformation methodologies for sunflower could allow the introduction of specific traits of industrial interest into desired genomic backgrounds and would circumvent the difficulty of introgression by sexual crossing. Modifications targeted by industries are mainly the reduction of carbon chain length or modification of the degree of polyunsaturation levels to produce rare fatty acids. Recently, it has been proposed that sunflower could be manipulated for the endogenous production of a lipase in cellular compartments where lipids do not normally accumulate. The spatial compartment between enzyme and substrate should avoid all interactions during the normal life cycle of seeds. However, after grinding, oils and lipase would come in contact and would generate free fatty acids which can thereafter be extracted and tailored chemically to suit industrial demands. This concept has been patented (4) and validated on tobacco (5). Another area of research in sunflower genetic transformation, concerns the enrichment of phytosterol levels in sunflower, for use either in the food industry as a neutraceutic, in the medical sector for decreasing blood cholesterol content, or in the cosmetical industry for their rejuvenating properties. An original challenge will be to increase latex concentrations in sunflower by genetic transformation, thereby decreasing dependencies on imports from developing countries (6).

1.3. History of Sunflower Genetic Transformation

Genetic engineering of plants is now routinely achieved for most species except for sunflower, which still remains a challenge, mainly because of low transformation rates. However, several private companies have field-trialled transgenic sunflower carrying resistance to fungi and viruses, herbicides, insects, drought tolerance, modification of seed storage proteins and fatty acid spectra. As would be expected, these informations are tied up by patents (7).

The possibility of transferring genes into sunflower tissues using *Agrobacterium tumefaciens* as a vector has been recognised for many years (8). In the case of sunflower, the application of protocols, proven to be successful for other model species, has been limited, largely due to the inability of coinciding the transformation and the regeneration event within the same cell (9, 10). The broad genetic background of sunflower has prompted wide genotypic screening for aptitude towards regeneration (11, 12). Sunflower tissue culture systems and their uses in biotechnology have been reviewed (13). Fertile plants have been efficiently regenerated either through organogenesis or somatic embryogenesis of several highly tissue culture-adapted genotypes, using a variety of explants including hypocotyls, leaves, cotyledons and shoot apices (14-17). Immature zygotic embryos are currently used as starting material in sunflower regeneration systems (18-20). Described procedures for sunflower regeneration have mostly concerned public lines, and namely Ha300b, which shows high regeneration potential, whatever the system used. Few reports concern regeneration of economically important genotypes, in view of genetic transformation studies. Flores-Berrios *et al.* (21, 22) have described the organogenic and embryogenic potential of cotyledons and epidermal thin cell layers of RHA 266, a genotype resistant to *Plasmopara halstedii* (23). More recently, Hewezi *et al.* (24) described a new approach for efficient regeneration of this very recalcitrant genotype of sunflower for which neither direct nor indirect method of regeneration was efficient. The procedure is based on successive excision of the apical and axillary shoots originating from pre-existing meristems. This elimination process stimulates the formation of adventitious shoot buds, which develop into phenotypically and genotypically conformed plants respective to the parent.

The first transgenic sunflower was obtained by Everett in 1987, who succeeded in obtaining one transgenic shoot on hypocotyl-derived calli, co-cultured with *A. tumefaciens* (25). However, this protocol is highly genotype-dependent and could not be extended to other genotypes. Transforming callus, induced on organs other than hypocotyls, have not given rise to routine established protocols, due to the very low regeneration efficiencies under selective conditions. Using indirect organogenesis on hypocotyl induced calli, the transformation rate of a public line Ha300B, known for its good aptitude to tissue culture and high morphogenic potential, did not exceed 0.1% (26). Transformation of cotyledons from germinating seedlings of the same genotype did not lead to the recovery of stable transformants (10), in spite of exceptionally high regeneration efficiencies.

A major breakthrough in the history of sunflower genetic transformation has been achieved with the report of Schrammeijer *et al.* (27) based on earlier works of Ulian *et al.* (28) with petunia shoot tips. The strategy is to infect split embryonic axes of immature embryos, with an *A. tumefaciens* suspension and promote shoot elongation from the persisting meristemmatic tissues. There are two advantages to this method: a) shoots which emerge from developing meristems are vigorous and allow the recovery of fertile plants without an intervening callus phase; b) genotype dependency is circumvented as the method and is based on meristem development rather than on regeneration systems. Using this technique on sunflower, the authors obtained only one sterile plant from 1500 initial meristemmatic explants (*27*). Since then, the use of embryonic axes from mature (*27, 29-35*) or immature embryos (*26, 36*) as starting material proved to be generally applicable for the production of transgenic sunflower. Although, in most cases, the genotypes used were public lines showing high morphogenic response and adapted to *in vitro* culture conditions, the overall transformation efficiencies remain quite low and rarely exceed 0.1%. Organogenesis induced on immature embryos of R105, a highly morphogenic genotype, allowed the recovery of transgenic plants with an efficiency of 0.5% (*36*). The technique could not be reproduced for the poor tissue culture-responding RHA266 genotype, which is resistant to mildew. In view of transferring additional agronomic traits to this important genotype, we attempted to optimize a genetic transformation protocol. Instead of bombarding explants, dehydration of immature embryos of RHA266, followed by rehydration with *A. tumefaciens* allowed the recovery of transgenic plants, with a reported efficiency of 0.22% (*37*).

In further attempts to improve sunflower transformation protocols, Alibert *et al.* (*34*), treated meristematic explants from germinating seedlings with pectinases, before wounding using a glass bead beater (*32*) and immediately followed by infection by *A. tumefaciens*. Pectinase treatment enhanced penetration of bacteria within the inner cell layers of meristematic explants. This technique has since, allowed the recovery of transgenic plants of a recalcitrant genotype RHA266 with an efficiency of 0.45% (Hewezi *et al.*, in prep.).

From the experience we have gained during the last 15 years in the field of sunflower genetic transformation, we describe in this chapter, two detailed *Agrobacterium*-based sunflower transformation protocols which are currently used in our laboratory. The first concerns transformation of immature embryos using particle bombardment which is applied to highly responding genotypes such as R105. The second method relies on glass bead wounding of meristemmatic axes of mature embryos and is more appropriate to recalcitrant genotypes such as RHA266.

2. MATERIALS

2.1. Transformation of Immature Embryos by Particle Gun Bombardment

2.1.1. Plant Materials

Sunflower (*Helianthuus annuus* L.) R105, Ha300B and cultivar Albena were obtained from Rustica Prograin Génétique, Mondonville, France.

2.1.2. A. tumefaciens *Strain*

A. tumefaciens strain EHA105, harbouring the pTIBo542 plasmid (provided by S.B. Gelvin, Purdue University, USA) and the 14.7 kb binary vector P35S GUS INT (36) were used throughout this study. The P35S GUS INT vector is a pBin19 derivative, containing the *gus*A gene, truncated by the potato LS-1 intron, under the control of CaMV 35S promoter, and the *npt*II gene conferring kanamycin resistance under the control of the NOS promoter.

2.1.3. *Media for* A. tumefaciens

1. Liquid culture medium: 20 g/l LB (Sigma) dissolved in purified distilled water. Dispense into 100 ml aliquots, sterilize by autoclaving at 121°C for 20 min and store at room temperature.
2. Agar culture medium: 40 g/l LB Agar (Sigma). Suspend 40 g of the powder in 1 litre of purified distilled water. Autoclave at 121°C during 20 min. After cooling to 60°C, add sterile kanamycin, rifampicin and ampicillin (*see* Section 2.1.4.) to a final concentration of 50 mg/l.
3. Inoculation medium (IM): For 1 litre, dissolve 2.44 g MES, 1 g NH_4Cl and 0.3 g $MgSO_4$. Adjust to pH 5.6 and autoclave at 121°C for 20 min.

2.1.4. *Antibiotics*

1. Kanamycin sulphate (Sigma): 50 mg/ml stock solution dissolved in distilled water. Sterilize by filtration through a 0.2 μm Minisart membrane, (Sartorius, Göttingen, Germany) dispense in 1 ml aliquots and store at −20°C.
2. Ampicillin (Sigma): 50 mg/ml stock solution dissolved in purified water. Filter-sterilize as above and store as 1 ml aliquots at −20°C.
3. Rifampicin (Sigma): 50 mg/ml stock solution dissolved in methanol. Filter-sterilize and store as 1 ml aliquots at −20°C.
4. Augmentin (SmithKline Beecham, Heppignies, Belgium): 250 mg/ml stock solution dissolved in distilled water just before use. Filter-sterilize and use immediately.

2.1.5. *Growth Regulators*

1. BAP (Sigma): 1 mg/ml stock solution dissolved in 1N NaOH, then made to final volume with water. Store at 4°C for a maximum of 2 weeks.
2. IBA (Sigma): 1mg/l stock solution dissolved in water. IBA is heat labile and should be filter-sterilized and added to autoclaved media.
3. Acetosyringone (Fluka Chemie GmbH, Buchs, Switzerland): 200 mM stock solution dissolved in DMSO (Sigma). Filter-sterilize using Acrodisc 0.2 μm filter units (PALL-Gelman Sciences, France) and store at −20°C as 1 ml aliquots.

2.1.6. Tissue Culture

All media used for particle gun bombardment are based on Murashige and Skoog (MS, *38*) medium (Sigma).

1. Co-cultivation medium (M1): MS medium containing 9 g/l Bacto Agar, 20 g/l sucrose, 0.1 mg/l BAP and 200 μM acetosyringone (Sigma) (20 ml per 90 mm diam. Petri dish).
2. Regeneration medium (M2): MS medium containing 9 g/l Bacto Agar, 20 g/l sucrose, 0.1 mg/l BAP and 500 mg/l augmentin (20 ml per 90 mm diam. Petri dish).
3. Selection medium (M3): MS medium containing 9 g/l Bacto Agar, 20 g/l sucrose, 0.1 mg/l BAP, 500 mg/l augmentin and 50 mg/l kanamycin (20 ml per 100 mm Petri dish).
4. Root induction medium (M4): Half-strength liquid MS medium containing 20 g/l sucrose, 1 mg/ml IBA.
5. Rooting medium (M5): 60 ml MS medium, containing 9 g/l Bacto Agar, 20 g/l sucrose, without growth regulators and dispensed into Magenta GA-7 vessels (Sigma).

All media are sterilized by autoclaving at 121°C for 20 min. Antibiotics and IBA are added to the medium after autoclaving.

2.1.7. Particle Gun Bombardment

1. Biorad accessories for the PDS-1000/He particle gun are as follows: 900 psi rupture disks, macrocarriers, stopping screens, Tungsten 1.1 μm M17 microcarriers.
2. 2.5 $MCaCl_2$ stock solution dissolved in water, filter-sterilized through a 0.2 μm Minisart (Sartorius) membrane, stored at –20°C as 1 ml aliquots.
3. Spermidine (Sigma) 0.1 mM stock solution dissolved in water and filter-sterilized through a 0.2 μm Minisart (Sartorius) membrane, stored at –20°C as 1 ml aliquots.
4. 50% (w/v) glycerol diluted in water and sterilized by autoclaving.
5. Absolute ethanol.
6. Sterile 3MM Whatmann filter paper.

2.2. Transformation of Mature Embryos by Glass Bead Wounding

2.2.1. Plant Materials

Helianthuus annuus var. RHA 266 was provided by INRA, Clermont-Ferrand, France.

2.2.2. **A. tumefaciens** *Strain* (*see* Section 2.1.2.).

2.2.3. Culture Media for **A. tumefaciens** (*see* Section 2.1.3.)

2.2.4. Antibiotics (*see* Section 2.1.4.).

2.2.5. Growth Regulators (*see* Section 2.1.5.).

2.2.6. Enzymatic Digestion Medium (*ED*)

Macerozyme R10 (Onozuka, Yakult Honssha Co Ltd, Tokyo, Japan) 10 mg/ml stock solution prepared by dissolving the powder in IM medium. Centrifuge at 3000 rpm for 10 min and filter-sterilize the supernatant. Store 1 ml aliquots at –20°C.

2.2.7. Tissue Culture

Tissue culture media are based on Gamborg B5 medium (Sigma, *39*).
1. Germination medium (GT): 9 g/l bacto-agar in water. Autoclave at 121°C for 20 min and dispense 20 ml into 90 mm diam. Petri dishes.
2. Co-cultivation medium (B5-1): B5 medium, pH 5.5 containing 20 g/l sucrose, 9 g/l Bacto-agar. Autoclave at 121°C for 20 min and dispense 20 ml into 90 mm diam. Petri-dishes.
3. Selection medium (B5-2): B5 medium pH 5.5 containing 9 g/l Bacto-agar, 500 mg/l augmentin and 50 mg/l kanamycin. Autoclave at 121°C for 20 min and dispense 60 ml into Magenta GA-7 vessels (Sigma).

2.2.8. Preparation of Glass Beads

Soak glass beads (425-600 mm diam., Sigma) overnight in 12N HCl and rinse thoroughly approx. 10 times with distilled water during 3 h, until pH 7. Dry glass beads at 60°C. Weigh 0.5 g of beads in 2 ml screw cap tubes (Sarstedt, Nûmbrecht, Germany) and autoclave at 121°C for 20 min.

3. METHODS

3.1. Transformation of Immature Embryos by Particle Gun Bombardment

3.1.1. Preparation of Agrobacterium Inoculum

1. Prepare a preculture by inoculating 5 ml of LB medium supplemented with kanamycin, rifampicin and ampicillin at a concentration of 50 mg/l with 100 μl of a 50% glycerol stock of *Agrobacterium* or with a colony from *Agrobacterium* cultured on LB agar medium containing the same antibiotics.
2. Incubate overnight at 28°C on a rotary shaker at 220 rpm.

3. Inoculate 100 ml of LB medium containing the same antibiotics and 200 μM acetosyringone, with 1 ml of pre-culture on a rotary shaker at 220 rpm for 18 h.
4. Spin the culture at 1700 rpm for 10 min at 4°C.
5. Wash the pellet twice with IM medium and resuspend the final pellet in fresh IM medium. Adjust optical density at 600 nm to 1.

3.1.2. Sterilization of Akenes

1. Excise akenes from sunflower heads.
2. Remove the pericarp, leaving the integuments intact, and surface-sterilize the embryos in 50 ml 6% calcium hypochloride (Merck) solution under a laminar hood. After 20 min, discard the hypochloride solution and wash the embryos 5 times with 100 ml sterile distilled water.

3.1.3. Preparation of Micro-particles

This is performed in sterile conditions under a laminar flow hood and can be prepared 7 days before bombardment.

1. Into a 1.5 ml sterile Eppendorf tube, weigh 30 mg of micro-particles Tungsten M-17.
2. Add 1 ml of absolute ethanol. Vortex vigorously for 2 min, centrifuge at 6000 rpm for 2 min. Carefully remove the supernatant. Repeat this step 3 times.
3. Suspend the micro-particles in 1 ml of sterile 50% glycerol by vortexing for 2 min. Centrifuge at 13000 rpm for 5 min. Carefully remove supernatant. Repeat this step once.
4. Suspend the micro-particles in 0.5 ml of 50% glycerol to obtain a final concentration of 60 mg/ml.
5. Store the micro-particles at room temperature for a maximum of 2 weeks.

3.1.4. Preparation of Macro-carriers

This should be performed just prior the bombarding process. Use 50 μl of micro-particles (*see* Section 3.1.3.) for 6 bombardments.

1. Vortex vigorously the micro-particles suspended in 50% glycerol (60 mg/ml), for 5 min.
2. Transfer 50 μl (3 mg) of micro-particles to a sterile 1.5 ml Eppendorf tube. Add 50 μl of 2.5 M CaCl$_2$ and vortex for 2 min.
3. Add 20 μl of 0.1 M spermidine. Vortex for 3 min, centrifuge at 6000 rpm for 60 sec. Carefully remove the supernatant.
4. Wash the pellet with 250 μl of absolute ethanol. Centrifuge at 6000 rpm for 60 sec and carefully remove the supernatant.
5. Suspend the micro-particles in 60 μl of absolute ethanol by vortexing for 60 sec.
6. Sterilize 6 macrocarriers in a Petri dish, by immersion in absolute ethanol for 15 min. Blot-dry the macrocarriers on sterile Whatmann 3MM paper.

7. Quickly flick the tube containing micro-particles and transfer 10 µl (500 µg) to the centre of the macrocarriers. It is important that micro-particles be placed in the centre of the macro-carriers. Air-dry the macrocarriers for approx. 10 min and use within the next hour for bombardment.

3.1.5. Step by Step Methodology for Particle Gun Bombardment of Immature Embryos

1. Excise akenes from sunflower heads 21 days after self-pollination and surface-sterilize as described (*see* Section 3.1.2.).
2. Dissect immature embryos, as presented (Fig. 1A), under a Zoom 2000 stereo-microscope (Leica, France), placed in a laminar flow hood, maintaining sterile conditions. Using sharp forceps, eliminate carefully the papery thin translucent integuments beneath the pericarp. Remove one cotyledon. Excise the second cotyledon and discard the root tip. Cut the resulting embryonic axes, longitudinally, between the leaf primordia to produce two split embryonic axes (Fig. 1A; *see* Note 1).
3. Group 40 split embryonic axes (cut ends facing up, Fig. 1B) in the centre of 90 mm diam. Petri dishes containing 20 ml of M1 medium (*see* Note 2).
4. Prepare macro-carriers as described (*see* Section 3.1.4.).
5. Using the Biorad PDS 1000-He particle gun, bombard the explants to induce wounding by uncoated M-17 particles at a distance of 3 cm, an acceleration pressure of 900 psi, with a chamber vacuum of 28 inches Hg (*see* Note 3).
6. Immediately after bombardment, spread 200 µl of *Agrobacterium* inoculum adjusted to an optical density of 1.0, onto the surface of explants.
7. Incubate the explants at 20°C in the same dishes, in the dark for 3 days (*see* Note 4).
8. After co-cultivation, the leaf primordia will have started to develop and the hypocotyl will show elongation. Trim all elongated hypocotyls, leaving 5 mm of tissue below the newly formed leaves (*see* Note 5). Transfer the explants (40 explants per dish) to M2 medium and allow to develop in a growth chamber at 25°C with a 16 h photoperiod, at 55 µE/m^2/sec provided by daylight fluorescent Philips tubes.
9. Seven days after culture on M2 medium, remove the main shoot derived from the apical meristem and subculture the remaining explants on M2 medium under the same conditions for an additional 7 days (Fig. 1C, *see* Note 6). Remove thereafter, any newly formed callus growing around the base of hypocotyl (*see* Note 7).
10. Discard any fast developing axillary shoots (Fig. 1C, *see* Note 8) and transfer the remaining explants, devoid of shoots, to M3 selection medium (20 explants/dish).
11. After 2-3 weeks, adventitious shoots develop at the surface of explants (Fig. 1 D). Transfer those shoots which are more than 1 cm in height, to Magenta vessels containing 60 ml of M3 medium. Shoots are allowed to grow for a further 3 weeks with weekly subcultures onto fresh M3 medium before rooting (*see* Note 9).

12. Cut the base of completely dark green shoots (*see* Note 10) growing on medium containing kanamycin and make one vertical scratch (over 1 cm in length), with a dissecting needle on the side of the hypocotyl. Dip the shoots in M4 liquid root induction medium for 60 sec and then blot dry on sterile filter paper. Transfer the shoots to Magenta vessels containing 60 ml of M5 medium to allow development, rooting and elongation of plantlets (Fig. 1E; *see* Note 11). Growth conditions are as described (*see* Section 3.1.5.8.). Plantlets that do not root under these conditions should be grafted (Fig. 1G) as described (*see* Sections 3.1.5.16.-3.1.5.18.).

13. After 2 weeks, carefully remove the rooted shoots from the agar gel by thorough washing. Insert the rooted shoots individually into hydrated Jiffy–7 peat pellets (Jiffy Products International AS, Norway) and place them in a closed 50 cm high mini-glasshouse, themselves put in a growth chamber at 25°C, 16 h photoperiod, light intensity of 55 µE/m^2/sec provided by daylight fluorescent Philips tubes and under 75% relative humidity. High humidity is maintained by spraying the plantlets with tap water regularly.

14. After 2 weeks, reduce progressively the relative humidity by opening the cabinets until the plants can be nursed to glasshouse conditions (Fig. 1F).

15. Transfer the acclimatized putative transformants to 19 x 15 cm pots containing 3 litres of a special horticulture substrate of a mixture of granular clay, white peat, bark humus and perlite, commercialized by Hawita, Vechta, Germany. Allow the plants to grow to maturity. Self-pollinate by hand the flowers and screen immature embryos 14 days after fertilization for resistance to kanamycin (*see* Section 3.1.6.; Note 12).

16. Four week-old sunflower hybrid cultivar Albena plants, at the 4-leaf stage, are grown in the glasshouse and used as rootstocks. Cut the apical portion of the rootstock above the two basal leaves. Excise a conical slit about 1 cm long with a sharp scalpel.

17. Plantlets (*see* Section 3.1.5.13.) are cleared of any callus proliferating at the base of hypocotyls. The epidermal layer of approx. 1 cm high scions, is removed to reveal the conducting vessels.

18. Insert the scion immediately into the slit of the rootstock and secure both rootstock and scion with parafilm sterilized during 5 min with absolute ethanol (*see* Note 13). Cover grafts with a light plastic bag (such as food-freezer bags) and tighten with tape. In order to ensure maximum humidity of the grafts, make a hole at the top corner of the plastic bag and use this hole to spray the scion with tap water. Try to avoid contact of scion with the plastic bag. Incubate grafts in a growth chamber at a constant temperature of 25°C and 16 h photoperiod under a light intensity of 55 µE/m^2/sec. After 7 days, open the plastic bag progressively and transfer the grafts to the glasshouse to grow to maturity. Self-pollinate heads regularly. Fourteen days after fertilization, harvest immature embryos to screen for kanamycin-resistant progeny (*see* Section 3.1.6.).

3.1.6. Selection of Kanamycin-resistant Progeny

1. De-husk 2 week-old immature T1 embryos by removing the pericarp and the thin translucent layer beneath (*see* Note 14).
2. Transfer a maximum of 100 seeds into 50 ml-capacity Falcon conical tubes, containing 8 ml of screening germination medium (2 g/l Algospeed NPK 18:18:18 (v:v:v), Duclos International, France) dissolved in distilled water containing 50 mg/l kanamycin and 500 mg/l augmentin (AK medium).
3. Place the tube in a vacuum chamber and apply 10 kpa of vacuum pressure for 60 sec. Gently break the vacuum.
4. Repeat the vacuum infiltration 3 times.
5. Sow no more than 40 seeds on a 10 cm diam. Petri-dish containing 4 layers of filter paper, moistened with 8 ml of selective germination medium, (AK médium; *see* Section 3.1.6.2.). Close the dishes with parafilm to prevent evaporation of solution.
6. Allow the seeds to germinate in the culture room (16 h photoperiod, light intensity of 55 $\mu E/m^2/$ sec, 25°C) for 3 days.
7. Transfer the green, kanamycin-resistant seedlings into individual hydrated Jiffy–7 peat compost. Culture in a growth chamber until plants reach a 6-leaf stage of development. Test the kanamycin resistant plants by PCR and Southern blotting analysis (*see* Note 15).

3.2. Transformation by Glass Bead Wounding of Mature Embryos

3.2.1. Step by Step Methodology for Transformation using Glass Bead Wounding

1. Soak mature sunflower seeds, var. RHA266, in distilled water for 2 h. Remove the pericarp and surface-sterilize as described (*see* Section 3.1.2.).
2. Place seeds on Petri dishes (28 seeds/dish), containing GT medium. Incubate the seeds in the dark at 25°C for 16 h (*see* Note 16).
3. Excise both cotyledons from the germinating seedlings and transfer 14 of the resulting meristematic explants (Figs. 2A and 2B) to sterile 5 ml-capacity tubes containing 1.8 ml of IM médium and 0.2 ml of ED medium (final concentration of 1 mg/ml Macerozyme R10). Incubate for 4 h at 25°C.
4. Rinse the treated explants 4 times with 2 ml of IM. Introduce the explants into autoclaved 2 ml screw cap tubes (Sarstedt, Nümbrecht, Germany) containing 0.5 g of glass vedas and 100 µl of *Agrobacterium* suspension adjusted to an OD$_{600nm}$ of 2.
5. Shake the tubes in a Micro-Beedbeater (Biospec Products, Bartlesville, OK, USA) at 3800 strokes per min for 40 sec (Fig. 2C, *see* Note 17).
6. Immediately after wounding, place 56 explants into Petri dishes containing 10 ml of *Agrobacterium* suspension with an optical density of 1.0 for 1 h.
7. Blot-dry the explants on autoclaved filter paper and culture in 90 mm Petri dishes (28 per dish) containing B5-1 co-cultivation médium (Fig. 2D).
8. Incubate under a 16 h photoperiod, at 55 $\mu E/m^2/$sec at 25°C for 2 days.
9. Transfer explants to Magenta GA-7 vessels, containing B5-2 selection medium (*see* Note 18).

10. Select those plantlets that form roots (Fig. 2E) and exhibiting dark green leaves under selection conditions and transfer to hydrated Jiffy-7 peat pellets (1 plant/pellet, Fig. 2F).

11. Nurse the plants progressively to glasshouse conditions as described (*see* Section 3.1.5.14.-3.1.5.16.).

12. Transfer plants to the glasshouse and harvest seeds 2 weeks after self-pollination (Fig. 2G).

13. Select immature embryos on kanamycin-containing medium as described (*see* Section 3.1.6.).

14. Assay all green plants surviving on kanamycin by PCR and further characterize by Southern blotting (*see* Note 15).

4. NOTES

1. Dissection of 40 meristematic axes takes 30 min to an experienced worker.
2. It is advisable not to allow explants lose turgor pressure before being transferred to M1 medium.
3. Bombarding varieties other than Ha300B and R105 can result in browning of tissues. Addition of 1% PEG 6000 to M1 medium (Merck) should inhibit browning.
4. Co-cultivation in the dark results in higher transformation efficiencies.
5. In some cases, it is difficult to distinguish between the two ends of the explants. Great care should be taken to avoid elimination of the apical part of the explant, which appears to be round in shape.
6. Delay of excision of apical or axillary shoots decreases the regeneration efficiency.
7. Overgrowing calli around the hypocotyl is a barrier to kanamycin transport throughout the cells and thus reduces transformation efficiencies.
8. Elimination of axillary shoots stimulates adventitious bud formation in approx. 80% of treated explants for Ha300B and R105. This could be different for other genotypes.
9. Longer selection periods lead to precocious flowering.
10. Approx. 2% of the shoots entering selection remain green.
11. Rooting efficiency approximates to 80%.
12. In sunflower, it is easy to recognize fertilized flowers by the style retraction.
13. It is essential to work quickly to avoid desiccation of scions. Under the conditions described, 50% of R105 scions and 80% of HA300B scions survived grafting procedures.
14. It is important that immature embryos, of the stated genotypes, be sampled exactly 14 days after fertilization. Earlier than 2 weeks, embryos fail to germinate and exhibit high mortality rates, even under non selective conditions. If they are sampled later, germination efficiency is reduced due to dormancy and discrimination between green and bleached cotyledons is ambiguous. Therefore, it is recommended that the developmental stage at which immature embryos to be screened for transgene activity to be optimized for each genotype.
15. Escapes can be recovered amongst green seedlings. This is because sunflower heads contain embryos at different developmental stages. Older embryos always show greater resistance to kanamycin.
16. Seedlings exhibiting signs of contamination or damage during sterilization should be discarded.
17. Survival rates of wounded enzyme-treated explants by glass beads, is highly genotype-dependent. It is recommended to optimize speed and duration of wounding.
18. During the co-cultivation period, some explants can develop primary roots. For efficient selection of potentially transformed shoots, it is important to eliminate these roots before transferring to selection medium.

CONCLUSIONS

The advent of functional genomics has allowed the discovery of many agronomically important genes which have been utilized to improve different plant species in terms of pest resistance, drought and salinity tolerance, or nutritional enhancement, amongst other traits. Genetic improvement of sunflower, allying conventional and non-conventional methods, relies on the availability of routine established protocols for genetic transformation of genotypes which can readily enter breeding programmes. These genotypes are usually, not suited to *in vitro* culture manipulation and thus inaccessible to transformation technology. We have optimized two techniques, which are regularly reproduced in our laboratory. Wounding of immature embryos by particle gun bombardment, followed by co-cultivation with Agrobacteria resulted in 0.5% transformation rates on R105, a good-responding genotype in tissue culture. This method could not be adapted to RHA266, a recalcitrant genotype to *in vitro* culture. Glass-bead wounding of meristemmatic tissues of pectinase-digested mature embryos, followed by *Agrobacterium* co-cultivation, allowed the recovery of transgenic plants with an efficiency of about 0.5%. We anticipate that this methodology could give higher efficiencies with genotypes that respond well to *in vitro* culture. Absolute values of transformation efficiencies are still low, compared to those of other species. However, for sunflower, this is most acceptable, considering that the method is reproducible, transgenic plants recovered are phenotypically-normal and fertile and that introduced traits are stably inherited in a Mendelian manner throughout several generations.

REFERENCES

1. Amsol: Le marché du tournesol-Février 2002 http//www.amsol.asso.fr.
2. Antolin G, Tinaut FV, Briceno Y, Castano V, Perez C and Ramirez AI (2002). Optimisation of biodiesel production by sunflower oil transesterification. *Bioresource Technology*, **83**: 111-114.
3. Parant B (2001). Les huiles de colza et de tournesol: une source d'approvisionnement majeure pour les tensioactifs de demain. *Oléagineux, Corps gras, Lipides*, **8**:152-154.
4. Alibert G, Mouloungui Z and Boudet AM (1996). Method for producing fatty acids or derivatives thereof from oil plants. International Patent WO96/03511.
5. Alibert G, Mouloungui Z, Grison R and Romestan M (2001). Libération des acides gras par autolyse enzymatique des triglycérides des graines oléoprotéagineuses. *Oléagineux, Corps gras, Lipides*, **8**: 98-102.
6. Pearson CH, Brichta JL, Van Fleet JE and Cornish K (2002). The potential of sunflower as a rubber-producing crop for the United States. ASA-CSSA-SSSA Annual Meetings, Nov. 10-14, Indianapolis, Indiana, USA.
7. Hahne G (2002). Sunflower seed. In: Khachatourians GG, McHugen A, Scorza R, Nip WK, Hui YH (eds.), *Transgenic Plants and Crops*. Marcel Dekker, New York, USA.
8. De Ropp RS (1946). The isolation and behavior of bacteria-free crown-gall tissue from primary galls of *Helianthus annuus*. *Phytopathology*, **37**: 201-206.
9. Potrykus I (1990). Gene transfer to cereals: an assessment. *Bio/Technology*, **8**: 535-542.
10. Laparra H, Burrus M, Hunold R, Damm B, Bravo-Angel AM, Bronner R and Hahne G (1995). Expression of foreign genes in sunflower (*Helianthus annuus* L.) - evaluation of three gene transfer methods. *Euphytica*, **85**: 63-74.
11. Espinasse A and Lay C (1989). Shoot regeneration of callus derived from globular to torpedo embryos from 59 sunflower genotypes. *Crop Science*, **29**: 201-205.

12.   Punia MS and Bohorova NE (1992). Callus development and plant regeneration from different explants of six wild species of sunflower (*Helianthus* L.). *Plant Science*, **87**: 79-83.

13.   Alibert G, Aslane-Chanabé C, Burrus M (1994). Sunflower tissue and cell cultures and their use in biotechnology. *Plant Physiology & Biochemistry*, **32**: 31-44.

14.   Ceriani MF, Hopp HE, Hahne G and Escandon AS (1992). Cotyledons: an explant for routine regeneration of sunflower plants. *Plant Cell Physiology*, **33**: 157-164.

15.   Greco B, Tanzarella OA, Carrozzo G and Blanco A (1984). Callus induction and shoot regeneration in sunflower *(Helianthus annuus L.)*. *Plant Science Letters*, **36**: 73-77.

16.   Paterson KE (1984). Shoot tip culture of *Helianthus annuus* – flowering and development of adventitious and multiple shoots. *American Journal of Botany*, **71**: 925-931.

17.   Power CJ (1987). Organogenesis from *Helianthus annuus* inbreds and hybrids from the cotyledons of zygotic embryos. *American Journal of Botany*, **74**: 497-503.

18.   Bronner R, Jeannin G and Hahne G (1994). Early cellular events during organogenesis and somatic embryogenesis induced on immature zygotic embryos of sunflower *(Helianthus annuus)*. *Canadian Journal of Botany*, **72**: 239-248.

19.   Charrière F and Hahne G (1998). Induction of embryogenesis versus caulogenesis on *in vitro* cultured sunflower (*Helianthus annuus* L.) immature zygotic embryos: role of plant growth regulators. *Plant Science*, **137**: 63-71.

20.   Charrière F, Sotta B, Miginiac E and Hahne G (1999). Induction of adventitious shoots or somatic embryos on *in vitro* cultured zygotic embryos of *Helianthus annuus*: Variation of endogenous hormone levels. *Plant Physiology & Biochemistry*, **37**: 751-757.

21.   Flores Berrios E, Gentzbittel L, Kayyal H, Alibert G and Sarrafi A (2000a). AFLP mapping of QTLs for *in vitro* organogenesis traits using recombinant inbred lines in sunflower (*Helianthus annuus* L.). *Theoretical and Applied Genetics*, **101**: 1299-1306.

22.   Flores Berrios E, Sarrafi A, Fabre F, Alibert G and Getzbittel L (2000b). Genotypic variation and chromosomal location of QTLs for somatic embryogenesis revealed by epidermal layers culture of recombinant inbred lines in the sunflower (*Helianthus annuus* L.). *Theoretical and Applied Genetics*, **101**: 1307-1312.

23.   Vear F, Gentzbittel L, Philippon J, Mouzeyar S, Mestries E, Roeckel-Drevet P, Tourvieille de Labrouhe D and Nicolas P (1997). The genetics of resistance to five races of downy mildew (*Plasmopara halstedii*) in sunflower (*Helianthus annuus* L.). *Theoretical and Applied Genetics*, **95**: 584-589.

24.   Hewezi T, Jardinaud F, Alibert G and Kallerhoff J (2003). A new approach for efficient regeneration of a recalcitrant genotype of sunflower (*Helianthus annuus* L.) by organogenesis induction on split embryonic axes. *Plant Cell, Tissue and Organ Culture*, **73**: 81-86.

25.   Everett NP, Robinson KEP and Mascarenhas D (1987). Genetic engineering of sunflower *(Helianthus annuus* L.). *Bio/Technology*, **5**: 1201-1204.

26.   Müller A, Iser M and Hess D (2001). Stable transformation of sunflower *(Helianthus annuus L.)* using a non-meristematic regeneration protocol and green fluorescent protein as a vital marker. *Transgenic Research*, **10**: 435-444.

27.   Schrammeijer B, Sijmons PC, van den Elzen JM and Hoekema A (1990). Meristem transformation of sunflower via *Agrobacterium*. *Plant Cell Reports*, **9**: 55-60.

28.   Ulian EC, Smith RH, Gould JH and McKnight TD (1988). Transformation of plants via the shoot apex. *In Vitro Cellular and Developmental Biology-Plant*, **24**: 951-954.

29.   Bidney D, Scelonge C, Martich J, Burrus M, Sims L and Huffman G (1992). Microprojectile bombardment of plant tissues increases transformation frequency by *Agrobacterium tumefaciens*. *Plant Molecular Biology*, **18**: 301-313.

30.   Knittel N, Gruber V, Hahne G and Lénée P (1994). Transformation of sunflower (*Helianthuus annuus* L): a reliable protocol. *Plant Cell Reports*, **14**: 81-86.

31.   Malone-Schoneberg JB, Scelonge CJ, Burrus M and Bidney D (1994). Stable transformation of sunflower using *Agrobacterium* and split embryonic axis explants. *Plant Science*, **103**: 199-207.

32.   Grayburn WS and Vick BA (1995). Transformation of sunflower (*Helianthuus annuus* L.) following wounding with glass beads. *Plant Cell Reports*, **14**: 285-289.

33.   Burrus M, Molinier J, Himber C, Hunold R, Bronner R, Rousselin P and Hahne G (1996). *Agrobacterium*-mediated transformation of sunflower (*Helianthus annuus* L.) shoot apices: transformation patterns. *Molecular Breeding*, **2**: 329-328.

34.   Alibert B, Lucas O, Le Gall V, Kallerhoff J and Alibert G (1999). Pectolytic enzyme treatment of sunflower explants prior to wounding and cocultivation with *Agrobacterium tumefaciens*, enhances efficiency of transient β-glucuronidase expression. *Physiologia Plantarum*, **106**: 232-237.

35.   Rao KS and Rohini VK (1999). *Agrobacterium*-mediated transformation of sunflower (*Helianthus annuus* L): a simple protocol. *Annals of Botany*, **83**: 347-354.

36.   Lucas O, Kallerhoff J and Alibert G (2000). Production of stable transgenic sunflowers (*Helianthus annuus* L.) from wounded immature embryos by particle bombardment and co-cultivation with *Agrobacterium tumefaciens*. *Molecular Breeding*, **6**: 479-487.

37.   Hewezi T, Perrault A, Alibert G and Kallerhoff J (2002). Dehydrating immature embryo split apices and rehydrating with *Agrobacterium tumefaciens*: A new method for genetically transforming recalcitrant sunflower. *Plant Molecular Biology Reporter*, **20**: 335-345.

38.   Murashige T and Skoog F (1962). A revised medium for rapid growth and bioassays with tobacco tissue cultures. *Physiologia Plantarum*, **15**: 473-497.

39.   Gamborg OL, Miller RA, Ojima K (1968). Nutrient requirements of suspension culture of soybean root cells. *Experimental Cell Research*, **50**: 151-158.

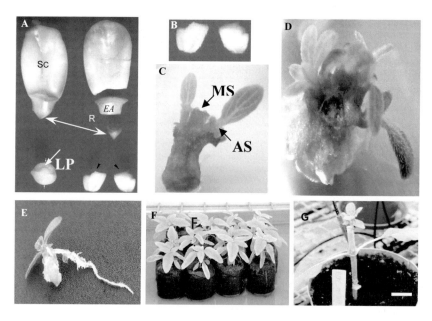

Figure 1. Procedure for the genetic transformation of sunflower immature embryos by particle gun bombardment. A. Dissection of immature embryos. After elimination of the thin papery translucent integuments, the first cotyledon is removed. The second cotyledon (Sc) is also discarded and the root tip (R) is cut. The resulting embryonic axes (EA), are sectioned longitudinally, between the leaf primordia (LP) to produce two split embryonic axes (B), used for bombardment. After co-cultivation, the main shoot (MS) and the axillary shoots (AS), indicated by arrows, arisen on regeneration medium are removed (C). This promotes development of adventitious shoots on the surface of explants (D). Shoots are either rooted (E) and acclimatized into Jiffy pots (F) or grafted (G). Bar = 2 mm (A and C), 1 mm (B), 3.2 mm (D), 4 mm (E), 36 mm (F) and 64 mm (G).

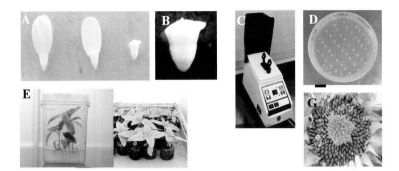

Figure 2. Procedure for the genetic transformation of sunflower mature embryos by glass bead wounding. A. Dissection of mature embryos. The two cotyledons are removed and the resulting explant (B) is digested by macerozyme for 4 h. After thorough washing with IM medium, explants are wounded in the presence of Agrobacteria, using the micro-bead beater (C). They are then incubated for 1 h with 10 ml of Agrobacterium *culture, before being co-cultivated for 2 days (D). After co-cultivation, the explants are cultured in Magenta GA-7 boxes (E) on kanamycin-containing medium. Rooted shoots are transferred to Jiffy peat pellets (F), and then further transferred to the glasshouse where they are allowed to flower. Two weeks after self-pollination, immature embryos are excised from the heads (G) and germinated on kanamycin-containing medium. Bar = 2 mm (A); 0.8 mm (B); 32 mm(C); 14 mm (D); 16 mm (E); 32 mm (F); 14 mm (G).*

KEY ABBREVIATIONS

ABA	Abscisic acid
AFLP	Amplified Fragment Length Polymorphism
als	acetolactate synthase
AVG	L-α-2-aminoethoxyvinylglycine
BAP	6-Benzylaminopurine
Bialaphos	Phosphinothricylanalyanaline, sodium
BR	Brassinosteroid
CaMV	Cauliflower Mosaic Virus
CAPS	Cleaved Amplified polymorphic Sequences
cat	chloramphenicol acetyltransferase
CIP	Calf Intestinal Phosphatase
CMS	Cytoplasmic Male Sterility
cpSSRs	chloroplast simple sequence repeats
CTAB	Cetyltrimethylammonium bromide
CTV	Citrus Tristeza Virus
dhfr	dihydrofolate reductase
2,4-D	2,4 Dichlorophenoxyacetic acid
DIG	Dioxygenin
DMSO	Dimethyl Sulphoxide
dNTPs	deoxynucleotides
DTT	Dithiothreitol
ECSs	Embryogenic Cell Suspensions
EDTA	Ethylenediamine-N,N,N',N'-tetraacetic acid
ELISA	Enzyme-Linked-Immunosorbant Assay
FISH	Fluorescent *In Situ* Hybridization
GA	Gibberellic Acid (GA$_3$)
GD	Gap Distance
GFP	Green Fluorescent Protein
GM	Genetically-Modified
G418	Geneticin Disulphate
gusA/uidA	β-glucuronidase
HPLC	High Pressure Liquid Chromatography
hpt	hygromycin phosphotransferase
IAA	Indole-3-Acetic Acid
IASP	Indole-3-Aspartic Acid
IBA	Indole Butyric Acid
2iP	6-(γ,γ-dimethylallylamino)purine
LB	Luria Broth
luc	luciferase
MES	2-(4-morpholino)-ethane sulphonic acid
MFD	Macro-Carrier Flight Distance
MOPS	3-[N-Morpholino]propanesulphonic acid

MS	Murashige and Skoog (1962)
MUG	4-Methylumbelliferyl β-D-glucuronide
NAA	α-Naphthaleneacetic Acid
NOS	Nopaline Synthase
nptII	neomycin phosphotransferase II
NOA	Naphthoxyacetic Acid
OD	Optical Density
PAGE	Polyacrylamide Gel Electrophoresis
PAR	Photosynthetic Active Radiation
PAT	Phosphinothricin Acetyltransferase
PBS	Phosphate Buffered Saline
PCR	Polymerase Chain Reaction
PEG	Polyethylene-Glycol
PIG	Particle Inflow Gun
PLRV	Potato Leaf Roll Virus
pmi	phosphomannose isomerase
PMSF	Phenylmethanesulphonyl Fluoride
PPT	Phosphinothricin
PR	Pathogenesis-Related
PVDF	Polyvinylidene Difluoride
RT-PCR	Reverse Transcriptase-PCR
QTL	Quantative Trait Loci
RAPD	Random Amplified Polymorphic DNA
RFLP	Restriction Fragment Length Polymorphism
SCARs	Sequence-Characterized Amplified Regions
SDS	Sodium Dodecyl Sulphate
SSRs	Single Sequence Repeats
STS	Silver Thiosulphate
TD	Target Distance
T-DNA	Transfer-DNA
TDZ	Thidiazuron
TE Buffer	Tris-EDTA Buffer
Tris	Tris[hydroxymethyl]aminomethane
X-gluc	5-bromo,4-chloro,3-indolyl-β-D-glucuronide
YMB	Yeast Mannitol Broth
ZR	Zeatin Riboside

Pneumatic Actuating Systems for Automatic Equipment
Structure and Design

Pneumatic Actuating Systems for Automatic Equipment

Structure and Design

Igor L. Krivts
German V. Krejnin

Taylor & Francis
Taylor & Francis Group
Boca Raton   London   New York

A CRC title, part of the Taylor & Francis imprint, a member of the
Taylor & Francis Group, the academic division of T&F Informa plc.

Published in 2006 by
CRC Press
Taylor & Francis Group
6000 Broken Sound Parkway NW, Suite 300
Boca Raton, FL 33487-2742

Library of Congress Cataloging-in-Publication Data

Krivts, Igor Lazar.
    Pneumatic actuating systems for automatic equipment : structure and design / Igor Lazar Krivts and German Vladimir Krejnin.
        p. cm.
    Includes bibliographical references and index.
    ISBN 0-8493-2964-7 (9780849329647)
    1. Pneumatic control. 2. Actuators. I. Krejnin, German Vladimir. II. Title.

TJ219.K75 2006
629.8'045--dc22   2005022839

Taylor & Francis Group
is the Academic Division of T&F Informa plc.

Visit the Taylor & Francis Web site at
http://www.taylorandfrancis.com

and the CRC Press Web site at
http://www.crcpress.com

Preface

When design engineers begin to develop some form of automatic equipment, they are confronted with two important problems: the first one is related to the mechanical and control design of a device that has functional property; the second problem is a commercial one that pertains to designing with reference to the cost of manufacture.

In order to solve the first problem, especially when an automatic control of complex motion is required, a wide knowledge of the principles underlying those mechanical movements, which have proved to be successful, is very helpful, even to the design engineer who has had extensive experience.

The second problem mentioned, that of cost, is directly related to the design itself, which should be reduced to the simplest form consistent with successful operating. Simplified designs usually are not only less costly, but more durable. Almost any action or result can be obtained mechanically if there are no restrictions as to the number of parts used and as to manufacturing cost, but it is evident that a design should pass the commercial as well as the purely mechanical test. In this connection it is advisable for the design engineer to study carefully the mechanical movement systems, which actually have been applied to commercial machines.

Currently, electromechanical, hydraulic, and pneumatic drives are most widely used as actuation systems in automation equipment. However, all these actuation systems have serious deficiencies, limiting their inherent performance characteristics.

Electrically driven actuators are normally used where movement is required for a number of intermediate positions, particularity when these positions need to be changed easily. They can also control speed and acceleration rate to a very high accuracy independently of the load. This allows very smooth motion to be performed in situations where this is a critical performance factor. In addition, electromechanical actuators can be used where more complex motion profiles are needed and for advanced motion control functions such as registration, contouring, following and electronic cam generation.

The use of electrical motors without torque-magnifying reducers is limited to direct-drive systems, which must employ large DC torque motors that are heavy and inefficient. To increase the torque output to useful levels, gear reducers are almost universally employed. However, there is an increase in torque-to-weight and power-to-weight ratio must be traded off against the large increase in reflected inertia, which increases with the square of the gear reduction value.

Using conventional rotating electrical motors to achieve linear motion requires transformational elements such as screw (ball screw or ACME screw and nut) or a timing belt. Where precision, thrust, and duty cycle are of paramount importance, ball screw models are frequently the best solution. The ACME screw models, which tend to be lower in cost, are an excellent solution unless the application calls for high thrusts. Screw-drive actuators are limited in speed and stroke by screw critical speed, and are offered in rod and guided-rod configurations for thrusting applications.

Belt drive actuators are available for longer stroke lengths or when higher speeds are required. However, because their belts are usually elastic, the screw drive models are typically more accurate and better suited for applications requiring rapid settling. Belt-drive actuators often require a gear reduction to get the mechanical advantage required by the motor to move the load.

The main advantage of electrical motors with transformational elements is that they allow using a low-cost motor that delivers high torque but runs at low speeds.

Electrical linear motors are used in applications requiring the highest speeds, acceleration, and accuracy. These direct-drive linear motors represent a departure from traditional electromechanical devices. Assemblies such as ball screws, gear trains, belts, and pulleys are all eliminated. As the name implies, the motor and load are directly and rigidly connected, improving simplicity, efficiency, and positioning accuracy. The acceleration available from direct-drive systems is remarkable compared with traditional motor drives that convert rotary motion to linear motion. The performance benefit is also substantial. There is no backlash, and because feedback resolution is high, direct-drive systems can be counted on to deliver superior repeatability and stiff, true positioning. However, direct-drive systems are more sensitive to the actuator's force/torque ripple, and they also suffer from lower continuous force/torque compared to geared actuators. Moreover, they are sensitive to load because of the lack of the attenuation effect of a gearbox.

The primary limitation of electromechanical drives is their relatively low power-to-weight, power-to-volume ratio, and payload-to-weight ratio. Table 0.1 represents these characteristics for electrical, hydraulic, and pneumatic motors. From this table it can be seen that the electrical motor has the poorest ratios, and this limits its application.

Generally, the linear motion systems with electrical motor and transformational elements have positioning accuracy of about 5–10 µm (best case) and velocity of up to 500–600 mm/s (second). For systems with electrical linear motors these parameters are the following: positioning accuracy up to 0.1 µm, and velocity up to 1.5 m/s.

Electrohydraulic servo systems provide positioning accuracy on a par with electromechanical systems, offering considerable force, excellent stiffness, and moderate speeds.

Hydraulic actuators (this actuator type is a direct-drive system), which have the highest torque and power density characteristics of any of the

TABLE 0.1

Characteristics of Motors

	Pneumatic Motor	Hydraulic Motor	Electrical Motor
Power-to-Weight Ratio (kW/kg)	0.3–0.4	0.5–1	0.03–0.1
Power-to-Volume Ratio (kW/m³)	$1 \cdot 10^{-3}$–$1.2 \cdot 10^{-3}$	$\sim 2 \cdot 10^{-3}$	$0.05 \cdot 10^{-3}$–$0.2 \cdot 10^{-3}$
Payload-to-Weight Ratio (N/kg)	11	20	3.5

actuation methods, are capable of performing tasks that involve the application of thousands of Newton-meters of torque and many kilowatts of power output. Other aspects that make a hydraulic actuator useful are the low compressibility of hydraulic fluids and high stiffness which leads to an associated high natural frequency and rapid response. This means that automatic equipment using hydraulic actuators can execute very quick movements with great force. Additionally, the actuators tend to be reliable and mechanically simple as well as having a low noise level and relative safety during operation. As for this method of actuation, design characteristics are well known, so the process of design is made easier.

One of the larger concerns with hydraulic systems is the containment of the fluid within the actuation system. Not only can this cause contamination of the surrounding environment, but leakage can contaminate the oil and possibly lead to damage of interior surfaces. In addition, the hydraulic fluid is flammable and pressurized, so leaks could pose an extreme hazard to equipment and personnel. This adds to undesirable additional maintenance to maintain a clean, sealed system. Other drawbacks include lags in the control of the system due to the transmission lines and oil viscosity changes from temperature changes. In fact, such temperature changes in the fluid can be drastic enough to form vapor bubbles when combined with the changes in fluid pressure in a phenomenon called *cavitation*. During operation, as temperature and pressure fluctuate, these bubbles alternately form and collapse. At times, when a vapor bubble is collapsing, the fluid will strike interior surfaces that have vapor-filled pores and high surge pressures and will be exhibited at the bottom of these pores. The cavitation can dislodge metal particles in the pore area and leave a metallic suspension within the fluid. The degradation of the interior surfaces and contamination of the fluid can result in a marked drop in the performance of the system.

Basically, the hydraulic actuation systems can develop controlled stroke speeds of up to 1 m/s, and positioning accuracy of about 1–5 µm.

Nearly 70% of today's positioning applications move loads of between 1 and 10 kg with accuracy between ±0.02 and ±0.2 mm. Electromechanical and electrohydraulic systems are overdesigned for these requirements. Electropneumatic motion systems have high application potential in this field.

Pneumatic actuators are still among the most widely used in automation equipment. As a rule, these actuators are direct-drive systems, too. Pneumatic actuators have been used in devices when lightweight, small-size systems with relatively high payload-to-weight ratio are needed. These actuators are selected for automation tasks as a preferred medium because they are relatively inexpensive (this technology costs approximately 15 to 20% of an electrical system), simple to install and maintain, offer robust design and operation, are available in a wide range of standard sizes and design alternatives, and offer high cycle rates. In addition, pneumatics is cleaner and nonflammable, making it more desirable in certain environments. Furthermore, pneumatic devices are less sensitive to temperature changes and contamination.

Pneumatic actuators are ideally suited to fixed travel applications and the control of force, where precise control of speed is not a prime requirement. In this case, hard mechanical stops are usually positioned along the length of the actuator. Though this adds a certain amount of adaptability, the stops are not truly programmable. They will need to be moved manually should an alternate position be desired.

New technologies today integrate the power of air with electronic closed-loop control. The combination of these technologies can provide much higher acceleration and deceleration capabilities than either one used alone. This position, velocity and force-control system technology is typically lower in cost compared with electrical motion systems. Such servo pneumatic systems retain the advantages of standard pneumatics and add the opportunity for closed-loop, controlled, programmable positioning to within fractions of a millimeter in systems in which positions can be approached rapidly and without overshoot, and provide stability under variable loads and conditions and adaptive control for optimized positioning. Generally, servo pneumatic actuators are similar to hydraulic servo actuators and use proportional or servo pneumatic valves, relying on the integration of electronic closed-loop controlled servo techniques. However, these actuators have the following major disadvantages: poor damping, high air compressibility, strong nonlinearities, and significant mechanical friction. Now, thanks to advances in pneumatic control theory, the combination of fast-acting valves, advanced electronics, and software, servo pneumatic systems are capable of positioning accuracy on the order of 0.05 mm. That level of precision is sufficient for an estimated 80% of typical industrial positioning requirements.

Generally, the linear motion systems with pneumatic actuators and hard mechanical stops have positioning accuracy about 10 μm (best case) and velocity of up to 2.5 m/s. For systems with servo or proportional valves these parameters are positioning accuracy up to 50 μm, and velocity up to 2.5 m/s.

The development of modern pneumatic actuation systems is to be seen as an evolution in mechatronic systems, when integrated with mechanical and electrical technologies, electronic control systems, and modern control algorithms. Trends in actuator pneumatic development can be broken down into the following areas:

- Development of new actuators (frictionless and flexible units in particular) and specialized actuators
- Optimization of component performance and reliability (with particular attention to miniaturizing valves and actuators, reducing frictional force, and standardization)
- Development of new control algorithms and control units with new interfaces between very low control signals and high power pneumatic signals
- Integration with sensors and control electronics to implement intelligent servo systems

All design problems are compromises. It is often practical to have a few parameters that make the compromise explicit. These parameters are called *design parameters*, and it is very important to define those that allow reaching the maximum efficiency in a fine-tuning process of the design on line of the automatic equipment.

This book describes many of the most-applied pneumatic actuating systems, which can be used in various classes of mechanisms, a study of such mechanical movements is particularly important to the designers and students of designing practice owing to the increasing use of automatic equipment in almost every branch of manufacture. The book discusses not only these actuator embodiment principles, their mathematical models, and methods of parameter calculation; but also included are many practical examples and exercises designed to enhance the reader's understanding of the concepts.

Practically, all the pneumatic actuating systems and their components shown in this treatise have been utilized on automatic machines of various classes.

This book is intended for engineers, system designers, and component manufacturers working in the field of pneumatics used in factory automation.

Feedback on the Book

We look forward to receiving readers' comments and corrections of any errors in this book. We encourage you to provide precise descriptions of any errors you may find. We are also open to suggestions on how to improve the textbook. For this, please e-mail the first author: igor_krayvitz@amat.com.

Acknowledgments

Most often, a book is based on previous experience that is complemented with modern advances, and this is especially true for this book. The majority of material described in this book was prepared during collaboration at the Mechanical Engineering Institute of the Russian Academy of Sciences.

We wish to thank our colleagues, both past and present, at the Mechanical Engineering Institute (IMASH), all of whom have given us help and encouragement in writing this book. We especially want to thank our colleagues in the Department of Actuating Systems; most of these people are mentioned by referencing their work in the bibliography at the end of the book. We would like to particularly acknowledge Doctors E.V. Hertz, V.I. Ivlev, K.S. Solnzewa, V.M. Bozrov, A.R. Taichinov, E.A. Tsuhanova, M.A. Jashina, V.V. Lunev, and S.J. Misurin for their collaboration, support, and help in our research.

For many years we have had effective cooperation with Professor V. Frank and Doctor A. Ulbricht from Dresden University of Technology. We are deeply grateful to them for their inspiring and fruitful work.

We would like to pay tribute to the support and help from staff of the Mechanical Department at Applied Materials (PDC Business Group), especially to Doctor E.Y. Vinnizky for his useful discussions about the configuration and dynamic behavior of pneumatic actuators.

Special thanks must go to Mrs. F. Swimmer for her expertise in the English language, as well as for her support in text preparation.

Authors

Igor Lazar Krivts, Ph.D., is a senior mechanical engineer in the Mechanical Department of the PDC Business Group of the Applied Materials Inc. Prior to his current affiliation, he was a senior researcher in the Mechanical Engineering Institute (IMASH), Russian Academy of Sciences from 1980 through 1991.

Dr. Krivts received his M.Sc. (1975) in mechanical engineering from the Penza Technical University (Russia) and a Ph.D. (1985) in mechanical engineering from the Mechanical Engineering Institute at the Academy of Sciences (Russia).

His current research and engineering interests include pneumatic and hydraulic servo systems and their components, precision mechanisms, robotics, motion systems, and vacuum devices for the semiconductor industry. Dr. Krivts' recent research and development activities involve mechanisms in the areas of instrumentation for semiconductor device manufacturing, especially metrology equipment.

As a result of his wide experience in the development and research field, Dr. Krivts holds 23 patents in collaboration with his colleagues. He has authored or coauthored one book and more than thirty scientific papers published in professional journals and proceedings of scientific conferences.

German Vladimir Krejnin, Ph.D., D.Tech.Sc., is a professor in the Mechanical Engineering Institute (IMASH), Russian Academy of Sciences. He is head of the Department of Actuating Systems and holds the academic secretary position in this institute.

Dr. Krejnin received his M.Sc. (1950) in mechanical engineering from the Bauman Moscow Technical University (Russia). He obtained his Ph.D. (1961) and D.Tech.Sc. (1970) degrees in mechanical engineering, both from the Mechanical Engineering Institute at the Academy of Sciences (Russia). Since 1980 he has been a full professor at the Department of Actuating Systems at Mechanical Engineering Institute (IMASH), Russian Academy of Sciences.

His main research interests lie in the areas of pneumatic and hydraulic servo systems and their components, the dynamics and control of mechanical systems with various types of drives, and the methods of optimal synthesis of such systems.

Dr. Krejnin deals with most branches of mechanical engineering, especially those that involve applications of the actuating systems in different kinds of machines and mechanisms. He has supervised 15 doctoral students, and written 7 books and more than 150 scientific journal articles on various topics in mechanical engineering.

List of Symbols

Subscript "i" indicates the working chamber index, superscript "+" indicates the upstream parameters, and superscript "–" indicates the downstream parameters.

Latin Symbols

A	Effective area of the actuator piston [m^2]
A_D	Effective area of the shock absorber piston [m^2]
A_V	Effective area of the control valve [m^2]
b_V	Viscous friction coefficient for linear motion $\left[\dfrac{N \cdot s}{m}\right]$
b_ω	Viscous friction coefficient for rotary motion [$N \cdot m \cdot s$]
C_P	Air heat capacity for constant pressure $\left[\dfrac{J}{kg \cdot K}\right]$
C_v	Flow coefficient [gal/min]
C_V	Air heat capacity for constant volume $\left[\dfrac{J}{kg \cdot K}\right]$
E	Energy [J]
E_M	Modulus of elasticity (Young's modulus) [Pa]
f_A	Actuator bandwidth [Hz]
F	Force [N]
F_D	Dynamic coulomb friction force [N]
F_{DE}	Desired value of the control force [N]
F_F	Friction force [N]
F_L	External force load [N]
F_M	Electromagnetic force [N]
F_S	Static coulomb friction force [N]
F_{SA}	Shock absorber force [N]
G	Air mass flow [kg/s]
h	Discrete time index
h_S	Specific enthalpy of the air flow [J/kg]
I	Current of the control signal [A]
J	Moment of inertia [$kg \cdot m^2$]
K_A	Acceleration gain $\left[\dfrac{V \cdot s^2}{m}\right]$
K_D	Derivative gain $\left[\dfrac{V \cdot s}{m}\right]$
K_P	Proportional gain [V/m]
K_{PA}	Gain of electrical power amplifier
K_{SL}	Slope coefficient of control valve [$1/V$]
K_v	Water flow rate [m^3/h]

K_*	Constant coefficient $\left(K_* = \sqrt{\dfrac{2 \cdot k \cdot R \cdot T_S}{k-1}} \approx 760 \dfrac{m}{s} \right)$
k	Adiabatic exponent (for air, $k = 1.4$)
k_S	Stiffness of spring [N/m]
k_B	Bellows spring rate [N/m]
k_D	Diaphragm spring rate [N/m]
L_S	Stroke of pneumatic cylinder [m]
L_A	Displacement of the actuator acceleration part (open-loop actuator) [m]
L_C	Actuator displacement where it moves with constant velocity (open-loop actuator) [m]
M	Torque [$N \cdot m$]
m	Mass of load (moving mass) [kg]
m_A	Mass of air [kg]
P	Absolute pressure in actuator working chamber [Pa]
P_A	Absolute atmospheric pressure ($P_A = 0.1 \cdot 10^6$ Pa)
P_D	Absolute pressure in working chamber of the shock absorber [Pa]
P_S	Absolute supply pressure [Pa]
Q_n	Standard nominal flow rate [m^3/s]
R	Gas constant $\left(\text{for air, } R = 287 \dfrac{J}{kg \cdot K} \right)$
s_D	Shock absorber working stroke [m]
t	Time [s]
t_{CR}	Carrier period [s]
t_D	Sampling period [s]
t_L	Delay time of the control signal [s]
t_M	Mechanical time constant [s]
t_P	Pneumatic time constant [s]
t_{PA}	Time constant of control valve with power amplifier [s]
t_{SM}	Time for actuator starting motion (open-loop actuator) [s]
t_V	Switching time of control valve [s]
t_*	Time scale factor coefficient [s]
T	Temperature of air [K]
U_C	Control signal [V]
U_{CRM}	Amplitude of carrier signal [V]
U_R	Regulating signal [V]
V	Volume of pneumatic chamber [m^3]
W^2	Dimensionless inertial load
x	Position of the cylinder piston [m]
x_D	Desired value of the control position [m]
$\dot{x}$	Velocity of cylinder piston [m/s]
$\dot{x}_C$	Constant velocity of steady-state motion [m/s]
$\dot{x}_{CS}$	Velocity set point [m/s]
$\ddot{x}$	Acceleration of the cylinder piston [m/s^2]
y	Displacement of the control valve plug [m]

Greek Symbols

α_A	Ratio of the effective areas of the piston actuator ($\alpha_A = A_2/A_1$)
α_R	Ratio of the effective areas of the rod and piston actuator ($\alpha_R = A_R/A_1$)
β	Opening coefficient of the control valve
δ_A	Steady-state positioning accuracy [m]

Δ_F	Admissible force error [N]
Δ_R	Admissible position error [m]
μ	Poisson's ratio
ν	Dimensionless viscous friction coefficient
ξ	Actuator dimensionless displacement
ρ	Density [kg/m^3]
ρ_{an}	Density of air under standard conditions ($\rho_{an} = 1.293\ kg/m^3$)
ρ_F	Shock absorber fluid density [kg/m^3]
σ	Pressure ratio
σ_A	Atmospheric and supply pressure ratio ($\sigma_A = P_A/P_S$)
τ	Dimensionless time
$\varphi(*)$	Flow function
φ_*	Value of flow function saturation ($\varphi_* = 0.259$)
Φ	Magnetic flux [Wb]
χ	Dimensionless force load
Ω	Effective area ratio of control valve

Contents

1

Structure of Pneumatic Actuating Systems

The pneumatic actuating servo systems used in automatic devices have two major parts: the power and control subsystems (Figure 1.1).

The main part of the power subsystem is the motor, which may be of the rotating or linear type. Basically, this device converts pneumatic power into useful mechanical work or motion. The linear motion system widely uses the pneumatic cylinder, which has two major configurations: single or double action. For the single-action configuration, the cylinder can exert controllable forces in only one direction and uses a spring to return the piston to the unenergized position. A double-action actuator can be actively controlled in two directions. In the case of rotary actuation, the power unit is a set of vanes attached to a drive shaft and encased in a chamber. Within the chamber, the actuator rotates by differential pressure across the vanes and the action transmits through the drive shaft.

Most often, the pneumatic actuator has the direct-drive structure; that is, the output motor shaft or rod is the actuator output link. However, sometimes the transmission mechanisms are installed after the motor; in this case, the output shaft is the actuator output link (e.g., in the rotating actuator where the pneumatic cylinder is used as the motor).

Actuator state variable sensors are the input elements of the control subsystems. In general, the displacement, velocity, acceleration, force, moment, and pressure can be measured in the pneumatic actuator. Different sensor designs can read incrementally or absolutely; they can contact a sensed object or operate without contact; and they span a broad range of performance and pricing levels. Linear-position sensors are widely used as feedback elements for motion control in pneumatic actuating systems; there are precision linear potentiometers, linear-variable-differential transformers (LVDTs), magnetostrictive sensors, and digital optical or magnetic encoders.

The important part of the control subsystem is the command module (or task controller), which stores the input information (such as desired positioning points, trajectory tracking, velocity, or force value) and selects them via input combinations. For example, in the positioning actuator, the positions can be stored in the command module (as position list records), and move commands can include additional parameters such as velocity and acceleration.

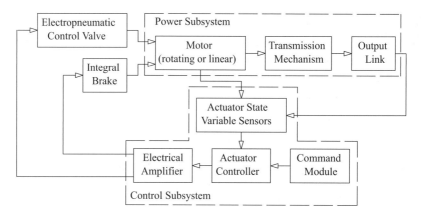

FIGURE 1.1
Block diagram of the pneumatic actuating system.

The central element of the control subsystem is the controller, which provides control, processing, comparing, and diagnostic functions. In general, the controller may be of both types: analog and digital. Currently, more than 90% of all controllers in industry are of the digital type. The main role of this device is to form the control signal according to the control algorithm. The most common form of process controller used industrially is the PID (proportional + integral + derivative) controller. PID control is an effective method in cases where the plant is expressed as a linear model, and the plant parameters do not change with wide or prolonged use. Owing to the compressibility characteristic of the air and high friction force, the pneumatic actuator system is very highly nonlinear, and the system parameters are time variant with changes in the environment. There are main causes, which are limited application of PID control in the pneumatic actuator systems.

For pneumatic actuators, the most common and successful controller is the so-called state controller or PVA (position, velocity, acceleration) controller.[163] In this case, the control signal is a function not only of the positioning signal, but also of the velocity and acceleration signal of the output link motion (for the positioning actuator).

As noted above, in pneumatic actuators, the dynamics of the plant change during performance. In this case, to improve the control performance, an adaptive control system with the controller adjusted bases on the identification results of the plant can be used.

Neural network control and a control algorithm using fuzzy inference are effective for a nonlinear plant. These techniques are applied in the pneumatic actuator controller.[22]

The controller output signals are sent to the electropneumatic control valve via the electrical amplifier. In the pneumatic actuator, the control valve is the interface between the power and control subsystems. This device is a key element in which a small-amplitude, low-power electrical signal is used to provide high response modulation in pneumatic power.

In general, there are three types of electropneumatic control valves — servo, proportional, and solenoid — used in the pneumatic actuator. These valves are available in one-, two-, or three-stage designs. A single stage is a directly operated valve. Two-stage valves consist of a pilot stage and a main stage. Three-stage valves are similar, except that the pilot stage itself is a two-stage valve. Three-stage valves are used in situations where one anticipates very high flow.

The distinction between servo valves and proportional valves is inconsistently defined, but in general, servo valves provide a higher degree of closed-loop control. Traditionally, the term "servo valve" describes valves that use a closed-loop stage spool position back to the pilot stage or drive, either mechanically or electronically.

Proportional valves displace the main-stage spool in proportion to a control signal but normally do not have any means of automatic error correction (feedback) within the valve. Many proportional valves are modified versions of four-way, on/off solenoid valves, in which proportional solenoids replace conventional solenoids. In operation, the solenoid force is balanced by a spring force to position the spool in proportion to the input signal. Removing the centering springs and adding a positioning sensor to the end of the spool can improve the positioning accuracy. The sensor signal then cancels the solenoid signal when the spool reaches the specified position.

Some manufacturers are producing proportional valves that are essentially servo valves made to mass-production specifications, with much greater tolerance allowances and looser fits than in their standard servo line. However, adding electronic feedback results in performance characteristics almost as good as those of a servo valve. In many cases, this results in performance that is perfectly suited to an application at a lower cost.

Solenoid valves are electromechanical devices that use a solenoid to control valve actuation. These devices are a fundamental element of the pneumatics and have high reliability and compact size. Standard models are available in both AC and DC voltages. The solenoid valve is low cost and universal in pneumatic systems operating with on/off control (e.g., it can be an effective solution for repeated stops in two positions). Using the on/off solenoid valve with a PWM (pulse width modulation) control method allows one to achieve the equivalent performance in proportional continually operation of the flow or pressure control. In this case, there is able to replace the solenoid valve instead of the expensive servo or proportional valve.

In some pneumatic positioning and speed control systems, the actuator consists of an integral brake. Usually, a proportional brake is linked to the actuator output link. A programmable controller provides a control signal to the brake and electropneumatic valves based on the stored program. In this case, for one of the possible configurations, the actuator has the on/off solenoid valve, which drives the pneumatic cylinder, and servo function can be achieved via the electric current that is sent to the brake. This type of combined technology system is low in cost, and provides moderate dynamic and accuracy performance.

Usually, pneumatic actuating systems are connected to compressed air lines with pressure from 0.3 to 1 MPa. Air compressors with pump technologies include positive displacement (piston, diaphragm, rotary vane, and screw styles) and nonpositive displacement (centrifugal, axial, and regenerative blowers) and provide air at the necessary pressure. As a rule, the compressor has an integral tank for compressed air storage, a coarse filter, an air dryer, and a pressure regulator.

The removal of moisture from compressed air is important for servo pneumatic systems. Moisture in an air line can create problems that can be potentially hazardous, such as the freezing of control valves. This can occur, for example, if very high-pressure air is throttled to very low pressure at a high flow rate. The Venturi effect of the throttled air produces very low temperatures, which will cause any moisture in the air to freeze into ice. This makes the valve (especially the servo or proportional valve) either very difficult or impossible to operate. Also, droplets of water can cause serious water hammer in an air system, which has high pressure and a high flow rate and can cause corrosion, rust, and dilution of lubricants within the system. For these reasons, air dryers (dehydrator, air purifier, or desiccator) are used to dry the compressed air. Major dryer groupings include refrigerant forced condensation (which removes the water by cooling the air) and desiccants (which adsorb the water in the air with granular material such as activated alumina, silica gel, or molecular sieves). The air can be dried in single or multiple stages.

An additional compressed air service unit is installed on every pneumatic line of the users. A service unit combination usually consists of the following individual units (Figure 1.2): on/off solenoid valve (2), filter (3), pressure regulator (4), and pressure gauge (5). In this case, the system has one input line (1) and two output lines (6) and (8). A lubricator (7) is installed on the output line (8) and supplies lubricant to the pneumatic components. In the

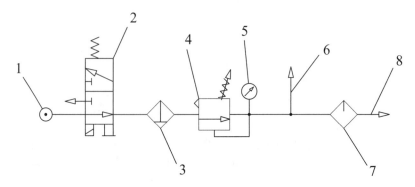

FIGURE 1.2
Block diagram of the compressed air line service unit.

output line (6) the compressed air is clear (without lubrication), which is very important, for example, in a clean-room application in the semiconductor industry. The input line (1) has an on/off control via the solenoid valve (2).

A compressed air filter (3) is used to remove water, oil, oil vapor, dirt, and other contaminants from the compressed-air supply. These contaminants can have a serious effect on the wear and operation of pneumatically operated machinery. In almost all applications, contamination of the air supply could lead to serious performance degradation and increased maintenance costs in terms of actual repairs and production time lost. The proper use and maintenance of compressed-air filters is one sure way to help cut down on these costs. Porous metal and ceramic elements are commonly used in filters that are installed in the compressed-air supply lines. Most pneumatic filters have a removable bowl in which liquids are separated. The condensate that collects in the filter bowl is drained from time to time, as otherwise the air would entrain it.

When selecting a compressed-air filter, it is important to note that the rate particle size of the device is the low end of the size range that is filtered or blocked by the filter. Other important specifications to consider when determining which compressed air filter is best for your system include the standard nominal flow rate or the maximum air volume that will be passed through the filter (generally measured in liters per minute), and the resistance to flow (pressure drop), which is measured in pascals (Pa).

Air-pressure regulators are devices that control the pressure in the air lines of pneumatic tools and machines. These regulators eliminate fluctuations in the air supply and are adjusted to provide consistent pressure. The inlet pressure must always be greater than the working pressure. Usually, the regulator has attached gauges. Just as for the filter, the regulator selection process is very important because its parameters, such as pressure drop, standard nominal flow rate, hysteresis, and transient response, have a significant influence on the dynamics and accuracy of the pneumatic device.

In particular, if positioning servo actuators are required to behave in a large piston stroke range as designed, the supply pressure should be as constant as possible. It is good if the supply pressure variations remain less than 5% of the designed value.

In addition, the extra volume between the pressure regulator and the electropneumatic control valve might improve the system's dynamic behavior. By increasing the value of the extra volume, the dropped pressure can be lowered.[197]

In some applications, where a few drives are operated, two separate supply lines are used: one with high pressure and the other with low pressure. In this case, the drive that moves on the idling mode may be connected to the low-pressure supply line, and the actuator works with high pressure only for the working stroke. This supply system allows for high efficiency.[102]

1.1   Pneumatic Positioning Systems

A pneumatic positioning system has been used widely in robots and manip-ulators, welding and riveting machines, pick-and-place devices, vehicles, and in many other types of equipment.

Pneumatic positioning actuators can generally be divided into two groups: (1) open-loop and (2) closed-loop position control. Usually, the open-loop pneumatic positioning actuator contains hard mechanical stops. In the sim-plest case, the system has a pneumatic cylinder, in which two covers play the role of the hard mechanical stops that define the stop positions. Figure 1.3 shows the block diagram of such an actuator. This construction has a piston (1, Figure 1.3) with two cylindrical parts (2 and 3), which are made with two cover cavities (4 and 5) and provide the air-cushioning mechanism. The solenoid control valve (6) connects the pneumatic cylinder to the supply pressure and exhaust port according to the control algorithm. Adjustable throttles (7 and 8) define the maximum value of the piston velocity. The piston (1) includes two permanent magnets (9 and 10) and two proximity sensors (11 and 12) attached to the outside of the cylinder tube. These provide a noncontact indication of cylinder piston position. As the piston approaches,

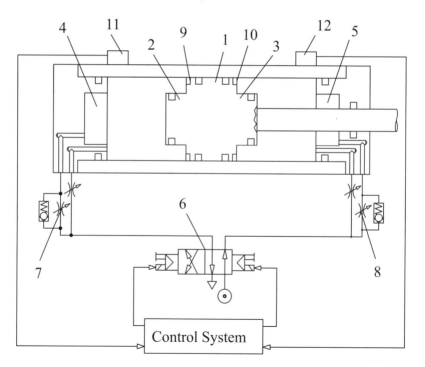

FIGURE 1.3
Block diagram of the pneumatic cylinder with two positioning stops on the ends.

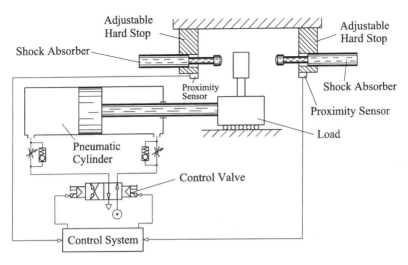

FIGURE 1.4
Schematic diagram of the pneumatic positioning actuator with two adjustable hard stops.

the magnetic field closes the switch, completing an electrical circuit and producing an electrical signal.

The basic function of the air cushioning is to absorb and dissipate the impact kinetic energy so that deceleration is reduced to a tolerable level. Linear and radial "float" of the cushion seals allows one to solve the problem associated with misalignment. Usually, adjustable air cushioning is used in the pneumatic cylinder if the piston velocity exceeds 0.2 m/s (second). Another major benefit of using air cushioning is that noise pollution, a hazard for workers and the environment, is greatly reduced. Because the contact surface at the stroke end is metal, stopping position repeatability is quite high (~0.01 mm).

Figure 1.4 shows the schematic diagram of the pneumatic actuator, which has the ability to stop the piston in the two adjustable positioning points within the whole piston stroke. The structure of this system is similar to the actuator illustrated in Figure 1.3, the main difference being the use of shock absorbers instead of the air cushions. This system can provide high speed (about 2 to 3 m/s) and positioning repeatability (up to 0.01 mm). The major weakness of this actuator is poor adaptability because the hard stops are not truly programmable. They must be moved manually to achieve a desired alternate position.

Shock absorber construction and parameters depend on the speed of the cylinder, the mass being moved, the external forces acting on the system, the system pressure, and piston diameter.

For implementation of the multiposition open-loop system, the so-called "multiposition pneumatic cylinder" is used, which typically consists of several connected cylinders (usually two or three). Figure 1.5 shows such an actuator with three pneumatic cylinders, which can reach four positioning

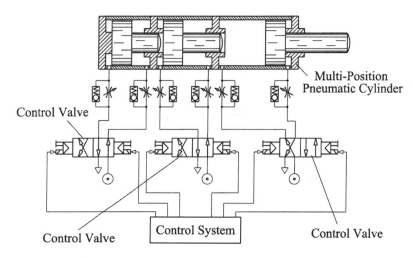

FIGURE 1.5
Schematic diagram of the actuator with multi-position cylinder.

points. The number of positioning points is defined by $N = n + 1$, where N is the number of positioning points and n is the number of connected pneumatic cylinders. Each rod that stands within the cylinder is the mechanical hard stop for the sequential piston, in essence; in this case, the two left pistons with their rods move the hard stop for the right piston. Although the construction is simple and affords high reliability, this actuator has impacts during the stop process that sometimes disturb the stability of the positioning. In addition, in a number of cases, such a system is bulky because of the numerous quantity of the control solenoid valves.

A similar positioning system with multiple stop points is shown in Figure 1.6. There is a positioning actuator with a so-called "digital" pneumatic cylinder, which consists of several pneumatic cylinders installed within the common sleeve. The stroke of the left-most cylinder is minimal, and each subsequent cylinder has double the stroke of the previous cylinder. Also, the rod of each cylinder is coupled to the body of the subsequent one that carries out the summation function for the cylinder's movement. Communication of the cylinder's pneumatic chamber with the supply pressure and exhaust line in variable combinations can achieve $N = 2^n$ positioning points with steps equal to the movement of the left cylinder. This allows using this construction not only as an actuator, but also as a digital converter. The use of diaphragm actuators provides more compact construction.

One can use the actuator depicted in Figure 1.7 for the implementation of multi-position open-loop systems with adjustable hard stops. Each sliding hard stop has its own actuator; in this case, there is a single-acting pneumatic cylinder, in which the rod is the hard stop. The sliding adjustable hard stops are assembled on a common base that can move along the main cylinder-moving axis. The base movement is limited by two mechanical hard stops

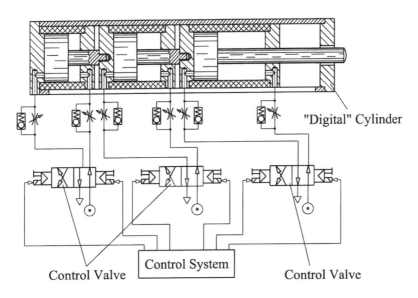

FIGURE 1.6
Positioning actuator with "digital" cylinder.

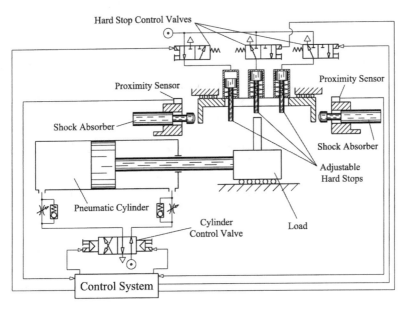

FIGURE 1.7
Multi-position actuator with hard stops.

with shock absorbers. Actuator stopping adjustment is achieved by mounting the pneumatic cylinders in the necessary positions. The pneumatic cylinder rods pass through the base slot. Usually, such a positioning actuator is used in cases where the number of the stop positions is not more than five or six; otherwise, actuator construction has a bulky build. The system positioning repeatability is approximately 0.03 to 0.04 mm.

The closed-loop pneumatic positioning actuator contains a transducer to measure and convert the actuator output signal to an electrical signal. This feedback signal is compared with the command signal, and the resulting error signal is applied to reach the necessary positioning or tracking of the movement. Two well-known technologies are widely used for point-to point closed-loop positioning systems: (1) airflow regulation using servo or proportional control valves and (2) a braking mechanism.

Usually, an actuator with a braking mechanism uses a pneumatically or electrically driven external mechanical brake, which consists of spring-loaded friction pads that act on the rod (or other moving component) of the pneumatic actuator. Typically, the application of air pressure causes the brake to release, providing hold actuation. Positioning is achieved in pneumatic braking systems by applying the pneumatic brakes at a predetermined point prior to reaching the target position. Braking is applied in an "on" or "off" manner, negating the possibility of programmable velocity control or a sophisticated deceleration profile.

The schematic diagram of such a positioning actuator is shown in Figure 1.8. It contains the pneumatic cylinder (1), a mechanical brake (2) drive by pneumatic cylinder (3), positioning sensor (4) that measures the load displacement, valves ($V_1 - V_5$), throttles ($R_1 - R_4$), and a control system. Four valves ($V_1 - V_4$) control the pneumatic cylinder (1), and they are arranged in pairs in series, which allows one to achieve independent adjustment of the high speed $\dot{x}_m$ and low speed (or creeping speed) $\dot{x}_c$ of the load. These

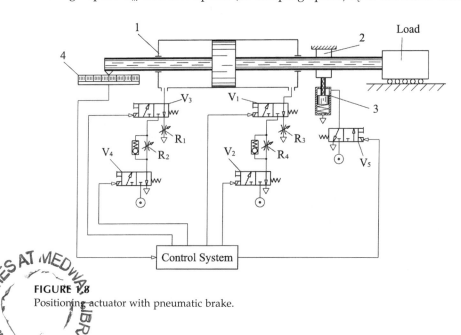

FIGURE 1.8
Positioning actuator with pneumatic brake.

adjustments are performed by four throttles (R_1 and R_3 for high speed, R_2 and R_4 for low speed). The valve (5) is used to control the brake pneumatic cylinder (3). To decrease the brake response time, a quick exhaust valve can be applied (not shown in Figure 1.8). In this case, the positioning (stop) process has two stages. In the first phase, the load speed is reduced from a high speed ($\dot{x}_m$) to a low (creeping) speed ($\dot{x}_c$) by pneumatic means. At the second stage, the mechanical brake is switched on and holds the load in the desired position. Figure 1.9 represents a typical velocity curve for this process and the control algorithm for this actuator, which is represented in Table 1.1. Here, x and $\dot{x}$ are the load position and velocity, respectively; V_i determines the valve's state (i = valve number: 1 = valve is energized, solenoid action; 0 = valve is deenergized, spring action); x_d is the coordinate of the positioning point (desired position); x_1 is the distance from the positioning point where the cylinder starts to change the velocity from $\dot{x}_m$ to $\dot{x}_c$; and x_2 is the distance from the positioning point where the brake is switched on.

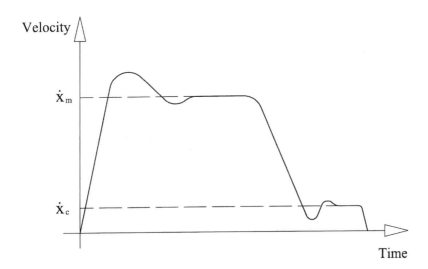

FIGURE 1.9
Velocity changing curve.

TABLE 1.1

Control Algorithm

x	V_i	$\dot{x}$	x	V_i	$\dot{x}$
$x \le x_d - x_1$	00111	$\dot{x}_m$	$x \ge x_d + x_1$	11001	$-\dot{x}_m$
$x_d - x_1 < x < x_d - x_2$	00111	$\dot{x}_c$	$x_d + x_2 < x < x_d + x_1$	11101	$-\dot{x}_c$
$x \le x_d - x_2$	11110	0	$x \le x_d + x_2$	11110	0

System tuning becomes sensitive to pressure variation, brake-pad wear, variation of the brake switching time, and the friction force in the pneumatic cylinder. Usually, the x_1 value is maintained constant, independent of any variation in the actuator operating conditions. The main criterion in its value selection is the nature of the transitional process — up to the moment of approaching the position $x = x_d - x_2$ or $x = x_d + x_2$, the oscillations of the creeping speed $\dot{x}_c$ must be negligible. Compensation for the influence of variation in operation conditions of the actuator is carried out by a change in the x_2 value. The simplest compensation algorithm has a correction for x_2, to be made in the next cycle, and is proportional to $(-\Delta_x)$, where Δ_x is the difference between the actual position of the stopping point in the preceding cycle and the preset position. Using this compensation algorithm, system positioning repeatability of ±0.15 mm can be achieved.[155,177]

Another positioning actuator involves placing magnetorheological braking devices functionally in parallel with a pneumatic cylinder or motor. Magnetorheological fluids are materials that respond to an applied field and the result is a dramatic change in rheological behavior. These fluids' essential characteristic is their ability to reversibly change from a free-flowing, linear, viscous liquid to a semisolid with controllable yield strength in milliseconds when exposed to a magnetic field. A typical magnetorheological fluid consists of 20 to 40%, by volume, relatively pure iron particles suspended in a carrier liquid such as mineral oil, synthetic oil, water, or glycol.

Magnetorheological brakes provide a braking force or torque that is proportional to the applied current. Through closed-loop feedback of the positioning sensor, accurate and robust motion control is achieved. The schematic diagram of such a positioning actuator is shown in Figure 1.10. The function of the three-position solenoid control valve is to ensure that the cylinder is

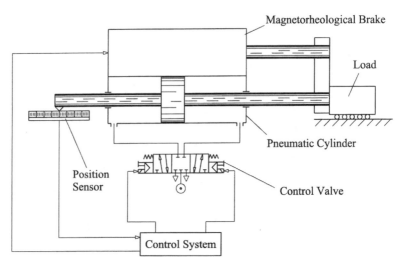

FIGURE 1.10
Positioning actuator with magnetorheological brake.

always directed toward the desired position and is commanded to a center position when the load is within some tolerance band around this point. Ideally, the magnetorheological brake function is substantial enough to "stall" the pneumatic cylinder. This will then provide the control authority with the ability to command a broad dynamic range of velocity control. Magnetorheological devices use a special liquid that undergoes property changes as a result of the action of the control current. In this sense, it is somewhat analogous to a conventional "air-over-oil" actuator. Because of significant hydraulic leveraging, this concept has a high force capacity.

The control algorithm can be of several types; one of them has the following description. Pneumatic logic is simple Boolean logic based on the sign of the position error. This logic also commands the valve to its neutral position when position x is within the tolerance band (Δ). The magnetorheological braking logic commands the application of braking (either step-wise or progressive) when position x is within Δ. The position is differentiated to provide an estimation of system velocity. System velocity is the basis for an error function that passes through a controller. This signal is summed with the magnetorheological braking signal to provide point-to-point velocity control.

This linear positioning system has a wide range of velocity control; for example, an actuator with pneumatic cylinder of 32-mm diameter bore and 160-mm stroke has the ability to move with constant velocity from 20 to 500 mm/s; the system positioning repeatability is ±0.15 mm.[89]

A positioning actuator is usually used for the servo or proportional control valve, which achieves the desired position by regulating the volume and flow rate of air into and out of pneumatic actuators. This linear positioning system (Figure 1.11) comprises the pneumatic cylinder, the positioning sensor (transducer), an electronic control system, and a continuously-acting valve as the control element.

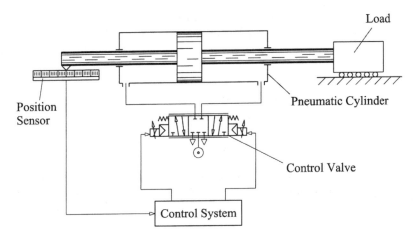

FIGURE 1.11
Positioning actuator with servo or proportional control valve.

Accurate, high-speed actuator movement requires quick and exacting valve response to commands from the control system. As the pneumatic cylinder approaches a set point, the valve shifts over-center to build up pressure that opposes piston motion. Internal algorithms control rapid shifting of the valve from one side to the other, giving smooth deceleration to the required position with the necessary dynamic characteristics.

Control algorithms provide the key to making positioning servo pneumatics work. As in most closed-loop systems, velocity and acceleration are controlled using three-loop position feedback (state controller or PVA controller). Three-loop algorithms measure position directly from the feedback transducer. Velocity and acceleration are derived from the position vector. The controller sums up these three different signals and generates a correction signal to the valve.

A practical consideration is that any servo system must be tuned. In this case, the control system calculates baseline loop parameters for stable operation based largely on the type of control valve and cylinder, as well as the payload, and motion parameters. The adaptive control algorithms also measure the quality of motion after every cycle to constantly optimize performance. For example, if overshoot is too high or low, it adjusts the filter parameters to improve response. Self-tuning also comes into play when the payload suddenly changes or seal and bearing characteristics change with use. Gain adjustments, critical damping, and overall system sensitivity can also be set manually.

A positioning actuator with a servo or proportional control valve can operate both the point-to-point modes and tracking motion.

In point-to-point mode, a velocity profile usually has a "trapezoid" form. In this case, the acceleration and motion with high constant velocity is realized by switching off the control valve to the saturation position. Only around the positioning point does the control valve move within the regulation range that provides the deceleration and stop process in the desired position.

In the tracking motion mode, the control valve permanently operates in regulation range, and valve effective areas change until the load stops.

For both tracking and point-to-point positioning, high performance control has nearly the same meaning: fast and accurate response to the reference. However, for tracking, the concern focuses on the response behavior along the entire reference trajectory, while for point-to-point positioning, the concern focuses on the response behavior around the reference point. Basically, high-quality, point-to-point implies high-quality tracking. It is well known that in the presence of uncertainly and disturbance; the point-to-point positioning quality is primarily decided by the feedback control quality. Well-designed feedback control will directly give high-performance, point-to-point positioning and facilitate the tracking control performance improvement.

In these working modes, the friction force in the pneumatic actuator plays a very important role. Friction will cause a steady-state error in point-to-point positioning and a tracking lag in a tracking motion. For precision tracking motion and positioning, the significant adverse friction effect must be compensated for.

Another important factor is the control valve nonlinearities (hysteresis, valve friction, dead zone, variation of the flow coefficient) that decrease the actuator positioning accuracy and dynamic characteristics. In part, this problem can be solved using a fast solenoid control valve (instead of the servo or proportional valve), which operates with PWM or Bang-Bang controller. In this case, if chattering exists, it may act as a dither, which is a classic friction compensation technique.

1.2 Pneumatic Systems for Velocity Control

The application fields of servo pneumatic actuators with speed motion control include arc welding machines; painting and printing equipment; scanning motion systems in inspection devices; cutting machines for plastic, wood, and fabric materials; gluing; and others.

In practice, open-loop pneumatic actuators are seldom used in these applications because of the poor ability to maintain constant velocity stabilization owing to low internal damping, high sensitivity to load and friction force changes, as well as the actuator's nonlinear characteristics.

The pneumatic actuator with magnetorheological braking devices (Figure 1.10) can also be used in velocity control systems. For example, the linear system with a pneumatic cylinder of 32-mm diameter bore and 160-mm stroke has the ability to move with constant velocity from 20 to 500 mm/s. In this case, the control accuracy is about 10% of the programmed value.

Figure 1.12 is a schematic diagram of the rotary pneumatic actuator, which has the ability to control the rotation velocity. The function of the three-position solenoid control valve is to ensure that the motor is rotated in the desired direction (clockwise or counterclockwise). The motor is stopped when the valve is at the center position. The actuator rotates the shaft, on which the magnetorheological brake, load, and velocity sensor are installed. Changing the brake impedance torque controls the shaft's angular velocity. For this system, the control accuracy is about 15% of the programmed value.

A pneumatic actuator for velocity control with a servo or proportional valve is based on the same principle of error-signal generation as the positioning servo actuator, except that the velocity of the output is sensed rather than the position of the load. When the velocity loop is at correspondence, an error signal is still present and the load moves at the desired velocity.

Most pneumatic servo applications require position control in addition to velocity control. The most common way to provide position control is to add a position loop "outside" the velocity loop, which is known as cascading

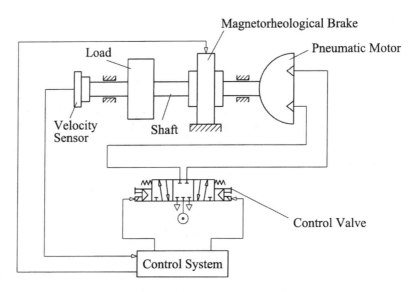

FIGURE 1.12
Rotary actuator with magnetorheological brake.

loops. In this case, the position error is scaled by the position loop gain to produce the velocity command.

Ripple, linearity, and low-speed performance are generally the most important characteristics of the velocity transducers or an algorithm that allows reaching the velocity signal because these characteristics determine the static errors in the velocity servo.

Figure 1.13 shows the schematic diagram of the linear pneumatic actuator with a solenoid valve for velocity control. Basic components include a standard rodless cylinder (1) with adjustable end-position cushioning at both ends; and the control valve (2), which is a standard double-solenoid valve with closed-neutral position and two flow positions (5/3-way valve). The position transducer (3) measures the load displacement, and the control system determines the velocity command and forms the control signal for the valve (2) according to the control algorithm. Using a nonreturn valve (4) in the supply port allows for energy recuperation in the motion reverse process; in this case, the kinetic energy of the moving mass is used for its acceleration in the opposite direction. Using two piloted nonreturn valves (5) with two single solenoid valves (6) results in an increase in the actuator efficiency because in this case the additional exhaust lines allow for an optimal ratio between the effective areas of the supply and exhaust port, which improves the steady-state velocity accuracy and the acceleration (deceleration) process.[102] For effective performance, the nonreturn valves (5) with solenoid valves (6) must be fitted directly into the ports of the pneumatic cylinder (1).

This system operates with PWM (pulse width modulation) or Bang-Bang controller with a four-loop feedback algorithm, which measures the load position directly from the feedback transducer; and velocity, acceleration,

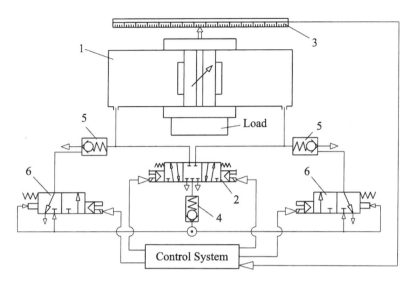

FIGURE 1.13
Velocity control actuator with solenoid valve.

and jerk signals are computed as the first, second, and third derivatives accordingly.

Application of the 5/3-way control valve (2) in the closed-neutral position allows one to obtain the positioning mode (the load stops in the desired position when the valve is in neutral). In this case, the positioning accuracy is not high; but for this actuator, the load stops are necessary only for the waiting period between technological actions, and accuracy in the range of 1 to 2 mm is quite sufficient.

The speed motion control with a pneumatic servo actuator for low velocity (<10 mm/s) is a difficult problem because the actuator internal friction force for the standard pneumatic cylinder is relatively high and inconsistent in the dynamic and static modes. Using a pneumatic cylinder with low friction force can solve this problem. In the past few years, great progress has been made in this area and now it is possible to find more pneumatic cylinders of the different types with better lubrication and friction features than ever before.

1.3   Pneumatic Systems for Force Control

Servo pneumatic actuators with force control are applied in the following fields: dynamic and static material test systems, spot-welding equipment, vehicle suspensions, manipulator grippers, physiotherapy and assembly robots, paint spraying systems, and others.

Force or torque is a function of the load on the actuator. A pressure regulator is commonly used to control the pressure in a circuit and a pneumatic cylinder or motor converts this pressure into a corresponding force or torque. Also, a directional control valve can be used in certain instances. The force or torque can be sensed by a force cell, which is installed between the load and pneumatic cylinder rod (pneumatic motor shaft), or by a pressure transducer that measures the pressure in the actuator working chambers. The feedback signal from the sensor is analyzed by the control system, which controls the pressure regulators or control valves. As with a position and velocity circuit, the regulator or valve is actuated to achieve the desired pressure.

The pressure feedback circuit provides the control of force without taking into account the friction characteristics of the cylinder. However, in many applications it is not viable to ignore the resulting errors. Particularly in cases where the point of application of force is not constant, there are noticeable deviations from the proportional ratio between pressure and force. Typical applications are, for example, the exertion of force on elastic materials or maintaining a constant force acting on a moving piston rod. In cases such as these, a high level of static accuracy can often only be achieved through a force servo loop involving directly measuring of force.

Figure 1.14 represents the schematic diagram of a linear pneumatic actuator for force control that contains an electronic closed-loop proportional pressure regulator. This system uses a pressure transducer to close the control loop, keeping the output pressure the same as the pressure command signal. The major weakness in this case is slow response time, as well as response time variation as a function of the load position. Therefore, such systems

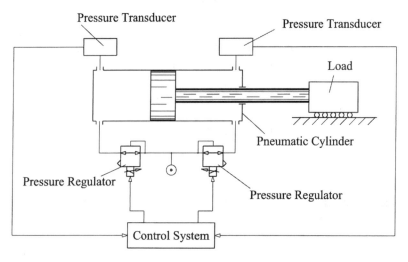

FIGURE 1.14
Force control actuator with pressure regulator.

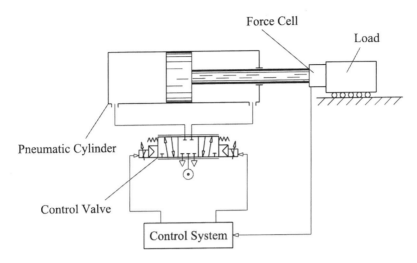

FIGURE 1.15
Force control actuator with servo control valve.

can be utilized in applications where such weaknesses can be tolerated (such as a damper control, exhaust control, and other applications that do not require fast response times).

The benefits offered by force control using a control valve include a highly dynamic response to changing force values because the two chamber pressures change inversely, simple compensation for force due to weight where units are mounted in a vertical position. The schematic diagram of such an actuator is shown in Figure 1.15. The function of the servo control valve (the three-position solenoid control valve can also be used) is to ensure the desired force between cylinder rod and load. A force cell senses this force.

Unlike the change of the positioning servo control, a transition from an analog controller to a digital controller would not afford any distinct benefits for force control. For one thing, the transducers used in industry work on analog principles and therefore usually also supply an analog output signal. This can be processed directly by an analog control system. The accuracy of a force control system is not determined primarily by the method of signal processing in the controller; rather, it is the quality of the transducer and the design aspects governing the application of the force that are decisive.

It is important to note that using the pressure or force control loop will greatly facilitate pneumatic motion control. The additional measurement (pressure or force) will give more flexibility in the control design. With the extra (inner) feedback control loop, the motion control loop's tolerance burden on plant uncertainty and nonlinearities will be considerably reduced. Application pressure or force inner control loop can cope with nonlinear friction to increase motion control performance.[25]

2

Pneumatic Actuators

A wide variety of pneumatic actuators have been developed for air services that convert the energy of pressured air into useful work. Pneumatic actuators can be divided into three groups: (1) linear actuators, (2) rotary actuators, and (3) pneumatic motors. Basically, rotary actuators provide a single rotation of up to 360°; however, in a number of cases, rotary actuators can be used with a rotating angle of up to 720°.

Pneumatic actuators are the components whose performance and cost can be the deciding factor in selecting pneumatic technology rather than another actuation technique. The main trends of pneumatic actuator development include increasing the efficiency, improving the power-to-weight ratio, and construction rationalization, as well as proposing new types of devices.

2.1   Linear Actuators

For linear motion control applications, there are three actuator constructions from which to choose: (1) a pneumatic cylinder, (2) a diaphragm actuator, or (3) an actuator with bellows. Many features and design factors influence the selection of a specific type of linear actuator. Before making a choice, it is necessary to examine each type based on operation, performance, environmental issues, and cost factors.

2.1.1   Pneumatic Cylinders

Basically, the conventional pneumatic cylinder has one moving member, which is a piston and rod assembly that converts the air-pressurized flow into linear motion. Figure 2.1 shows a regular double-acting pneumatic cylinder that has two ports through which the air supply is reversed to cause displacement in both directions. This construction contains the sleeve (1), piston (2), piston rod (3), back cover (4), and front cover (5). The piston rod (3) moves through the front cover (5) that houses the rod bearing (6) with a

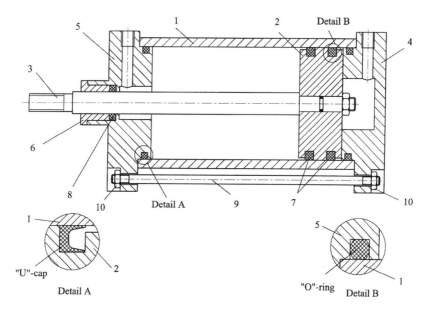

FIGURE 2.1
Double-acting pneumatic cylinder.

scraper ring and the rod seal (8). The piston (2) has piston seals (7). The covers are held on the sleeve by various methods, including screws, rods, threaded connections, and metal inserts. In this case (Figure 2.1), threaded rods (9) and fastening nuts (10) are used.

The sleeve of a cylinder can be made from cast iron, bronze, steel, aluminum, or other materials. Often, when the sleeve is of cast construction, one cover is cast integral with the sleeve.

Covers for the pneumatic cylinder can be made from different materials, but, in general, aluminum, steel, or bronze is used. Piston rods are made of polished stainless steel, nickel, or chrome-plated steel or bronze. Bronze, nylon, acetal, vesconite, pure polytetrafluoroethylene (PTFE), and PTFE-containing compounds have proven successful as slide ring materials for rod bearings. The function of a scraper ring is to prevent dirt particles from entering the components in pneumatic cylinders. In most cases, a scraper ring has one lip made of a wear-resistant elastomeric or thermoplastic material, which is sufficient, and some have metal cases for stiffness. It is important to protect the scraping edge from damage and to have adequate contact with the groove and piston rod diameter. The lip of the scraper ring is designed to have a preload with the piston rod, and this has an influence on the breakout friction.

In all cases where protection against corrosion and especially clean conditions are mandatory, stainless steel is used for cylinder components. Brass-based cylinder parts are ideally suited in arduous and relatively severe, high-temperature operating environments (e.g., at 200 to 220°C).

The sealing function of modern elements is guaranteed by compression of an expander made of synthetic or natural rubber within its elastic deformation range. For static seals, this type of material is used as a single element in most cases; for dynamic seals, it serves as an adjusting element of a slide ring, for example, and as a static sealing element in the groove depth.

The functional reliability and service life of a seal depends to a very great extent on the quality and surface finish of the mating surface for the seal. Scores, scratches, pores, concentric marks, or spiral-machining marks are not permitted. Ideally, the dynamic mating surface should be ground spiral-free. For normal application, the piston rod seals, which seal against moving surfaces, can be damaged by fine abrasive particles that might adhere to a rough surface. Piston rods therefore should have a low surface roughness value, and a surface similar to hard chrome or nickel. The ideal surface roughness for rods lies between 0.16 and 0.4 μm Ra. Piston seals seal against the inner surface of a cylinder. They are not affected to the same extent by abrasive particles entering from the atmosphere and can therefore have a rougher surface. The ideal surface roughness for bores lies between 0.25 and 0.63 μm Ra. The surface finish for seal housing, where the seal is in static operation, should be about 1.6 μm Ra.

The seal used for sealing a reciprocating member must meet static sealing requirements at its contact area with the stationary member, and also must seal effectively at its contact area on the reciprocating member. The ideal seal should:

- Prevent leakage over the pressure ranges encountered
- Have long life with minimum maintenance
- Be compatible with air at operating temperatures and pressures
- Have sufficient integrity to avoid air contamination
- Be easy to install, remove, and replace
- Exclude foreign material
- Be inexpensive

Curiously, the first two properties are at odds with each other in a dynamic sealing situation. A good compression force holds the seal against a reciprocating member, thus accelerating seal wear and shortening service life. Therefore, every dynamic seal design is a compromise to produce an acceptable balance between these two desirable properties. The great variety of seal shapes and materials allows the designer to select the degree of compromise for a particular application while, at the same time, satisfying the other requirements on the list.

In pneumatic actuators, "O"-ring and "U"-cup seals are widely used. The "O"-ring (Figure 2.1, detail A) is ideally suited for use as a static seal between nonmoving parts (e.g., between the covers and the sleeve). Use of an "O"-ring as the seal element between linear moving parts is sometimes not recommended because it creates a high friction force, which is the main cause of the "stick-slip" phenomenon.

The "U"-cup seal is installed with the cup facing in the direction of the pressure side; then air pressure "inflates" the "U"-cup and the seal edge is held against the rod or sleeve surface (Figure 2.1, detail B). The "U"-cup is ideal for use on linear moving rods and piston heads because, unlike an "O"-ring, the shape does not try to roll with the movement and does not create high friction.

All elastomeric sealing compounds are a combination of many chemical ingredients. These ingredients can be classified as:

- Polymers or elastomers
- Inert fillers (carbon or mineral fillers and reinforcing agents) to improve physical properties
- Accelerators, activators, retardants, and curing agents to assist in curing or vulcanization
- Inhibitors to inhibit undesirable chemical reactions during compounding
- Plasticizers to decrease stiffness and improve low-temperature properties

For standard pneumatic applications, the primary compound used currently is a thermoplastic elastomer based on polyurethane (PUR). This material can be processed in rapid injection molding procedures, shows good resistance toward mineral oils, and has excellent elasticity and wear values. In addition to having a good running behavior against steel (without stick slip), it also has a long service life in pneumatics. Its nonresistance to water, which would destroy the material by hydrolysis, limits the general use of PUR. In addition, polyurethane delivers the pneumatic requirements between −35 and +120°C.

Another compound that can be used with most of the above-mentioned parameters are acrylonitrile-butadiene rubbers (NBR or Perbunan), which have a particularly high level of quality and desirable properties. This material is used in all kinds of static and dynamic seals. Provided that the compounds are correctly formulated and processed, the vulcanized rubbers have very good resistance to liquid fuels, mineral oils, and greases; good resistance to aging; high resistance to wear and abrasion; low permeability to gases; and good physiological properties. The temperature range is −20 to +80°C.

Fluorocarbon elastomers have been compounded to meet a wide range of chemical and physical requirements. Under the tradenames Viton®, Fluorel, and Kel-F, fluorocarbon seals have been employed where other materials cannot survive the severe chemical conditions. The working temperature range of fluorocarbons is between −30 and +200° C. New compounds have greatly improved the compression set of fluorocarbon "O"-ring seals.

Also, PTFE is used as part of dynamic seals because this material has an extremely low coefficient of friction and high wear resistance. By adding additives (compounds) such as carbon-graphite, bronze, glass, or molybdenum disulfide(MoS_2), PTFE can be tailored to specific applications. The disadvantages of PTFE — namely, low creep resistance and only limited resilience — could be avoided by developing new compounds. Blending with thermoplastic portions is advantageous for the material when used in dynamic applications with the same good chemical and thermal resistance. Within certain limits, it is now possible to substitute elastic elements for these PTFE types.

The evolution of pneumatic cylinder construction is directed at increasing efficiency, improving the power-to-weight ratio, and rationalizing construction, as well as proposing and developing new types of devices. The result of this effort is instrumental in:

- Reducing friction and adhesion forces
- Ensuring effective sealing throughout the operating pressure range
- Eliminating the risk of extrusion
- Guaranteeing a long service life without the use of lubricated air

Using a composite material is an alternative to carbon steel, honed and chromed steel, stainless steel, aluminum, or brass cylinder sleeve. Constructed of fiber-reinforced thermoset epoxy, the composite material cylinder has an inner surface of evenly dispersed low-friction additives, for example, molybdenum disulfide (MoS_2), tungsten disulfide (WS_2), or other. In this manner, a self-lubricating cylinder wall with a low coefficient of friction can be achieved.

For added cylinder strength and stability, some kinds of fibers are available as reinforcements. In the composite industry, more than 90% of all fibers are glass. Electrical or E-glass is the most commonly used and most economical glass fiber, while structural or S-type glass has slightly higher strength and corrosion resistance. Advanced fibers such as carbon and Kevlar exhibit higher tensile strength and stiffness than glass fibers. Due to the higher costs of these fibers, they are typically reserved for applications demanding exceptional performance.

Another key to reducing the friction force in pneumatic cylinders is a pneumatic actuator with flexible chambers, which is shown in Figure 2.2. This construction consists of two covers (1 and 5), two flexible chambers (2 and 4), a sleeve (3), a piston (7), and a rod (6).

The problem of sealing a piston can be solved in the following manner. A chamber made from a flexible material is fitted into the working cavities of the cylinder and secured to the covers, the sleeve, and the piston. The space between the outer surface of the chambers and the inner surface of the cavities is then filled with grease. Under the action of air pressure, the chambers press against the walls of the cavities and fold randomly on their surfaces. As the piston moves, one of the chambers straightens out while the other crumples. The surface area of the chamber must exceed the internal area of the cylinder surface by 5 to 10% in the unstressed state to ensure that

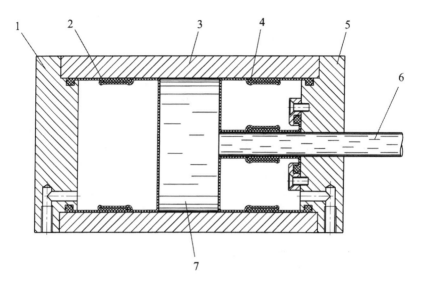

FIGURE 2.2
Double-acting pneumatic cylinder with flexible chamber.

the chamber will be relieved of tension. Chambers are connected to the covers.

This pneumatic cylinder design almost completely eliminates the leakage and blow-by of compressed air. It also reduces the dry-friction force by an order of magnitude. It does not require the vaporization of oil in the compressed air, and therefore workspace contamination is reduced.

Figure 2.3 is a schematic diagram of the single-acting pneumatic cylinder with low friction force. The inside surface of the piston (2) glides at the outside surface of the housing (1). The working chamber is formed by the inner surface of the flexible chamber (5) and clamping hollow bushing (6) via it the air pressure is supplied to the cylinder chamber. An opposing spring (4) is placed between the piston (2) and the sleeve (3). The profile of the flexible chamber has a special shape that makes the axial motion.

An equally important factor in many servo pneumatic applications is the rodless pneumatic cylinder. Rodless cylinders, as the name implies, differ from their more conventional counterparts in that no piston rod extends from the cylinder body. Rather, an internal piston is connected by some means to an external carriage. Thus, a major advantage of rodless cylinders is that they require considerably less mounting space.

Another area where rodless cylinders excel is in long-stroke applications. By design, rodless cylinders are unaffected by problems such as rod overhang and bending, piston cocking, and uneven seal wear.

The three main types of rodless cylinders are (1) the cable cylinder, (2) the piston-lug cylinder, and (3) the flexible-wall cylinder.

Cable cylinders were the first pneumatic rodless cylinders. In this design (Figure 2.4.), the cables (1 and 2) are attached to both sides of the piston (3).

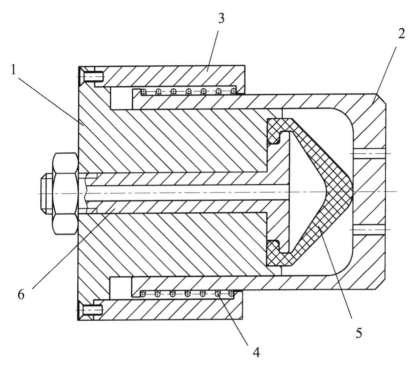

FIGURE 2.3
Single-acting pneumatic cylinder with flexible chamber.

The cables pass through the seals (4 and 5) at the cylinder covers (6 and 7), around pulleys (8 and 9), and are joined at the yoke (10). In addition, this construction may include cable tensioning, cable tracks for greater load stabilization and capacity, disk brakes on the cable pulley, and switches. As the double-acting piston moves in one direction, the yoke travels in the opposite direction because of the wrap around the pulleys. A variation of the cable cylinder relies on a metal band running over the pulleys instead of a cable.

Cable cylinders are typically the lowest-priced, rodless linear drives. However, they can handle only longitudinal loads. Direct radial loading or bending moment can be carried if a separate additional external load carriage is used.

The piston-lug rodless pneumatic cylinder has a slotted sleeve in which a slot runs the length of the barrel tube and is typically made from an aluminum extrusion. This allows the load to be driven directly by the piston through a lug that extends out the slot. The slot holds dynamic seals (usually steel bands) on both the inner and outer surfaces. Carriage travel moves the seals out of the slot and guides the repositioning of the seals after the carriage has passed. This system further stiffens the structure, while restricting unwanted pitch, yaw, and roll errors. And, of course, because the loads never travel beyond the end of the cylinder barrel, there are no overhanging loads.

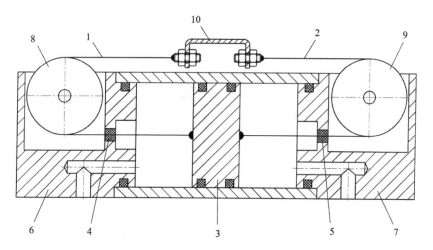

FIGURE 2.4
Double-acting cable pneumatic cylinder.

When the long, rigid aluminum housing is extruded with V-shaped slots running along the length, these slots can accommodate center-support fixtures that provide anchoring points anywhere along the axis.

Rodless cylinders are noted for their streamlined design; they are more compact than cable cylinders. However, the dynamic seal that runs the length of the cylinder sleeve poses a potential leakage source (particularly if dirt passes through the outer seal). From this point of view, the piston-lug rodless pneumatic cylinder shown in Figure 2.5 has some advantages: simple construction and low price. This cylinder contains (Figure 2.5a) a sleeve (1) and piston assembly that has two cylindrical parts (2 and 3) and a coupling rod (4). Two cylinder covers (5 and 6) have canals for the pressurized air supply. The slot (7) in the sleeve (1) is sealed by a sealing band (8), which is connected to the two cylinder covers (fastening is not shown). On the coupling rod (4), a bolt (9) is installed that connects this rod to the outside yoke (10). The inside hole of the sleeve (1) has a noncentral position relative to the outside sleeve diameter (Figures 2.5b and c). The maximum sleeve thickness is on the opposite side of the slot (7). It allows the optimal sleeve mechanical stiffness to be attained.[102]

The sealing band (8) has an "I"-form cross section (Figures 2.5b and c) and on one side it has two ledges (11 and 12) that are in constant contact with the sleeve (1). There are two other ledges (13 and 14) on the opposite side of the sealing band (8) and serve the cylinder sealing and passing of the bolt (9). The outside ledges (11 and 13) are designed for dirt protection, and the inside ones (12 and 14) provide the sealing function. A metallic strip (15) reinforces the sealing band (8); it can be used as a base for a positioning sensor.

The major disadvantage of the construction shown in Figure 2.5 is the complexity and expense of sealing band fabrication. The rodless pneumatic

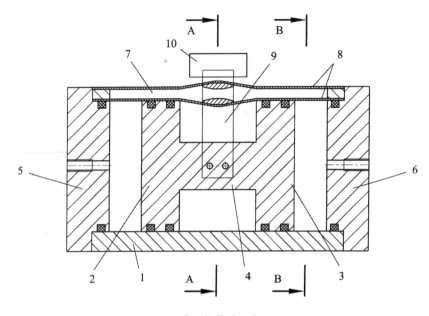

a. Longitudinal section

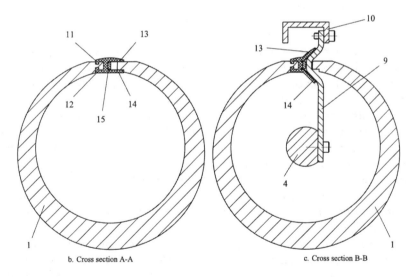

b. Cross section A-A   c. Cross section B-B

FIGURE 2.5
Piston-lug rodless slotted sleeve cylinder: (a) longitudinal section, (b) cross section A-A, and
(c) cross section B-B.

cylinder with flexible chambers (Figure 2.6) is devoid of these shortcomings.
This construction contains the cylinder sleeve (1) with slot (2); the drive lug
(3) passes via this slot and is connected to piston (4) by two pins (5 and 6).
The protection band (7) is attached to the cylinder covers (8 and 9) and passes
through the gap between the piston (4) and the drive lug (3). The protection

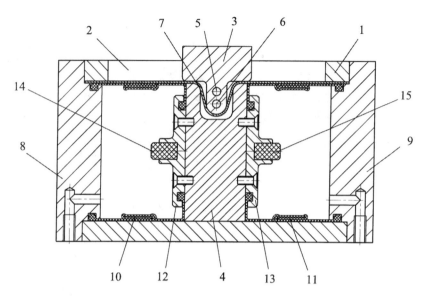

FIGURE 2.6
Double-acting piston-lug rodless cylinder with flexible chambers.

band (7) performs the function of a support element for the flexible chambers (10 and 11) (i.e., seals the cylinder working cavities). In addition, these flexible chambers are connected to cylinder covers (8 and 9) and to the piston (4) by two flanges (12 and 13); this cylinder has two elastic cushioning pins (14 and 15). In this construction, the problem of sealing a piston is solved in the same manner as that shown in Figure 2.2.

A variation of the piston-lug cylinder (Figure 2.7) uses permanent magnets in the piston to create a magnetic field that links the piston to the external yoke through the cylinder sleeve. The yoke moves on the sleeve's exterior. This construction consists of the cylinder sleeve (1) and two covers (2 and 3) with canals for the pressurized air supply. The piston (4) and yoke (5) have permanent magnets (6 and 7) that form the magnetic coupling between the yoke and the piston. The piston (4) has two seals (8 and 9), and the yoke (5) has two guiding rings (10 and 11) that sweep on the external surface of the sleeve (1). One advantage of this rodless cylinder is that it eliminates the need for a longitudinal slot in the sleeve and thus the need for any dynamic seals between the piston and the yoke. In addition, this construction is not prone to wear or be affected by harsh operating conditions. Thus, there is no air leakage, which offers potential energy savings, and it is recommended for use in a clean room, as well as in the food industry and similar applications.

Another type of rodless cylinder that uses magnetic coupling between the piston and the yoke for precise motion and position control is shown in Figure 2.8. This construction consists of the cylinder sleeve (1) and two covers (2 and 3) with canals for the pressurized air supply. Like the rodless cylinder shown in Figure 2.7, the piston (5) and yoke (4) have permanent magnets

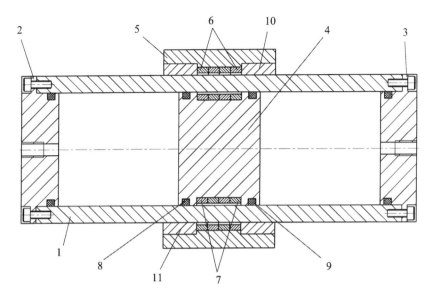

FIGURE 2.7
Rodless cylinder with magnetic coupling.

(6 and 7), which form the magnetic coupling between the yoke and the piston. The yoke (4) has a ball slide bushing (8), which consists of the outer cylinder and a ball retainer (not shown in Figure 2.8) that guide the circulation of the ball elements, resulting in smooth linear motion of the yoke (4). The flexible chambers (9 and 10) are sealed to the cylinder's working cavities. These flexible chambers are connected with cylinder covers (2 and 3) and with the piston (5) by two flanges (11 and 12). This cylinder has two elastic cushioning pins (13 and 14) installed on the flanges (11 and 12). In this construction, the problem of sealing a piston is solved in the same manner as that shown in Figure 2.2.

All mechanical parts of the rodless cylinder with magnetic coupling between the piston and yoke are made from nonmagnetic materials. The cylinder permanent magnets are usually produced from alloys containing rare-earth metals (e.g., samarium, yttrium, and others).

The rodless cylinder shown in Figure 2.8 has very low friction force, both in the piston and the yoke; therefore, it can be used in positioning pneumatic servo actuators with high positioning accuracy and repeatability.

Really, flexible-wall cylinders should not be designated as a "cylinder" because they do not contain a piston. However, in practice, these constructions refer to a pneumatic cylinder. Current designs for the flexible-wall cylinder can be broken down into two groups:

1. Pneumatic muscle
2. Linear peristaltic actuator

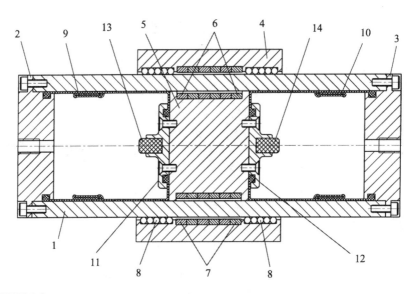

FIGURE 2.8
Rodless cylinder with magnetic coupling and flexible chamber.

The pneumatic muscle is known as a McKibben actuator and as a braided pneumatic muscle actuator. In practice, these actuators are built by wrapping a synthetic or natural rubber tube with no elastic fibers (such as Kevlar or nylon) in a rhomboidal fashion at a predetermined angle. The fiber wrapping is then given a protective rubber coating, and metal flanges are attached at each end (Figure 2.9). The resulting three-dimensional grid pattern deforms when actuated by compressed air. As the internal pressure increases, an axial pulling force develops that causes the tubular unit to contract, turning it into

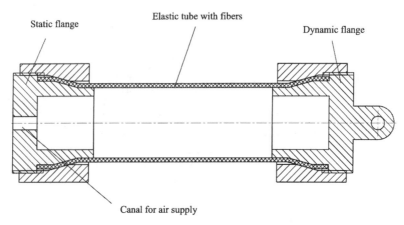

FIGURE 2.9
Schematic diagram of the pneumatic muscle.

a linear actuator. The contraction force depends on the applied pressure and on the muscle's length, ranging from an extremely high value at maximum length (i.e., zero contraction) to zero at minimum length or maximum contraction. The actual achievable displacement (contraction) depends on construction and loading but is typically 25 to 30% of the dilated length (this is comparable with the contraction achieved in natural muscle). This linear pneumatic actuator has the following advantages[80,151,192]:

- High power-to-weight (1 kilowatt/kg) and power-to-volume (1 watt/cm³) ratios, and a long lifetime (at least to 10 million switching cycles)
- Highly flexible, soft in contact, and having excellent safety potential (because there is no contact between the air inside the muscle and the surrounding air, the actuator exhibits high clean-room efficiency)
- Lower cost of manufacturing, installation, and maintenance
- No "sticktion" and self-damping characteristics allow using this actuator in closed-loop position and force control systems

The major drawback of pneumatic muscles is their nonlinear dynamic behavior, which makes it difficult to perform closed-loop, position-tracking tasks. In addition, these devices are single acting; therefore, some outside force must be used to retract them to their minimum, or starting height. Frequently, this outside force is gravity, or it can be another air spring or coil spring. A pneumatic muscle with return coil spring is shown in Figure 2.10. This construction comprises the static flange (1) and the dynamic flange (2), which are attached at each end of the tubular unit (3). The static flange (1) has a channel (4) to supply the pressurized air. In contrast to the construction shown in Figure 2.9, this pneumatic muscle has a plunger (5), a bushing (6), and a return coil spring (7). The plunger and bushing form the linear guide

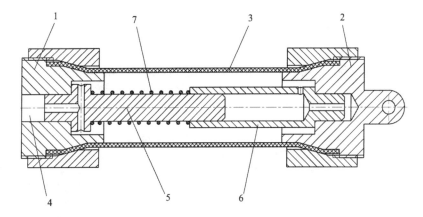

FIGURE 2.10
Schematic diagram of the pneumatic muscle with return coil spring.

that provides the accurate linear motion of the flange (2). In addition, this guide is very effective in cases where a lateral load is exerted.

A peristaltic linear pneumatic actuator, or the so-called hose pneumatic actuator, has a low and stable friction force because it has only dynamic friction forces. In addition, this type of pneumatic linear actuator is cheaper than a pneumatic cylinder. In this actuator, the roller that pinches the hose plays a key role (like the piston in the pneumatic cylinder) and it is the output link. Pressurized air is supplied to one end of the hose; the other end of the hose is connected to atmospheric pressure. Frequently, additional supporting rollers are used to achieve the stable motion of the output link. In practical applications, most often a peristaltic linear actuator is used in systems where long travel (up to 20 m) or curvilinear motion trajectory is needed.

The major problem associated with peristaltic linear actuators is their short lifetime. To increase this characteristic and decrease the leakage rate between the pressurized air part of the hose and the atmospheric air hose part, it is necessary to prevent formation of folds in the hose during actuator motion. These folds are the cause of hose wear and the increase in leakage.

The peristaltic linear pneumatic actuator with a semicircular elastic hose is shown in Figure 2.11. This construction consists of the hose (1), a moveable carriage (2) with supporting rollers (3), and a pinch roller (4), which pinches the hose (1). The base (5) has a guide surface (6), and its perimeter is equal to the inside perimeter of the semicircular hose (1). Therefore, the hose (1) adjoins tightly to the guide surface (6) under the force of the pinch roller (4). The working chambers are formed by the guide surface (6) and the inside surface of the hose (1), which is connected to the base (5) by plates (7) and screws (8). The springs (9) are pressed to the punch roller bearings and to the clasp of the roller to the guide surface (6) via the hose (1).

One of the advantages of this construction is that the guide surface has a wide range of change of curvature at which the conditions of pressing the hose remain constant.[102]

2.1.2 Diaphragm Actuators

In the diaphragm pneumatic actuators, a diaphragm is utilized as the sealing element between the piston (moving member) and the housing (stationary part), which is usually called the cylinder. These actuators are widely used in valves, regulators, vibration isolation systems, braking systems, precision polishing and grinding tools, and others mechanisms.

Diaphragm actuators provide a number of distinct advantages when used in place of pneumatic cylinders. Most notably, their cost is generally much lower than that of a pneumatic cylinder with comparable capability. In addition, they are more compact and can be installed more easily with side load flexibility. In terms of performance, diaphragm actuators offer a wide range of advantages. They do not require lubrication, which leads to greater

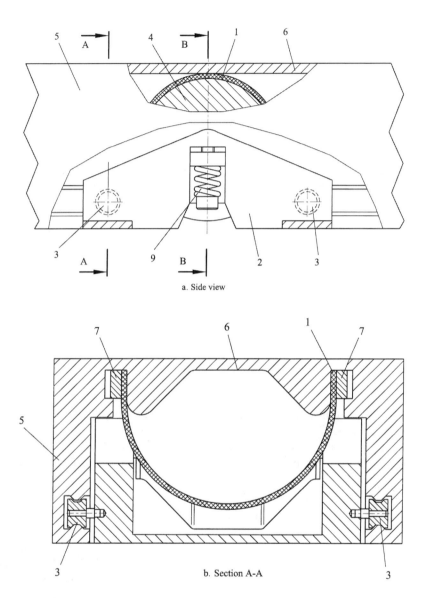

a. Side view

b. Section A-A

FIGURE 2.11
Peristaltic linear pneumatic actuator: (a) side view, (b) section A-A, and (c) section B-B.

system cleanliness and less overall maintenance, and they continue performing at a high level. Because they do not need lubrication, diaphragm actuators are designed without dynamic seals, which cause them to induce less
friction and make them better able to handle constant force. Because there
are no dynamic seals sliding against exposed surfaces, a diaphragm actuator
can often survive in abrasive and corrosive environments that require special
consideration when a conventional cylinder is used.

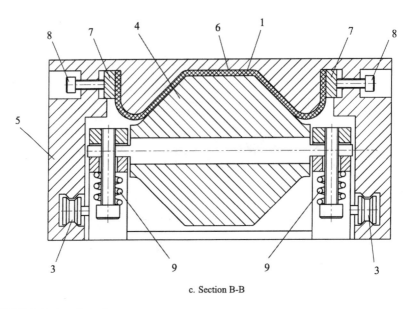

c. Section B-B

FIGURE 2.11 (continued)

Major disadvantages of the diaphragm actuator are the change of the diaphragm effective area as a function of its stroke, and its relatively short displacement.

Unlike pneumatic cylinders, diaphragm actuators are not designed around industrial standards. Diaphragms are produced in a wide range of shapes, different types of construction, and many different types of materials. Diaphragm actuators can be divided into two groups:

1. Actuators with flat diaphragm
2. Actuators with convoluted diaphragm

Figure 2.12 provides a schematic diagram of a single-acting, flat diaphragm actuator. The cylinder (housing) consists of (see Figure 2.12a) the rear (1) and front (2) covers, which are clamped to the diaphragm (3). The rear cover has a hole (4) to supply the pressurized air, and the front cover (2) contains vented holes (5). On the other side of the diaphragm, the piston contains the base (6) and the clamping disk (7), which are also clamped to the diaphragm (3). The actuator rod (8) is connected to the base (6) and the coil spring (9), which provides the return force. The rod (8) has a bearing bushing (10), which is usually made from sintered bronze.

In practice, two types of diaphragms are used in pneumatic actuators:

1. Homogeneous elastomer or metal diaphragm
2. Fabric-reinforced elastomer type

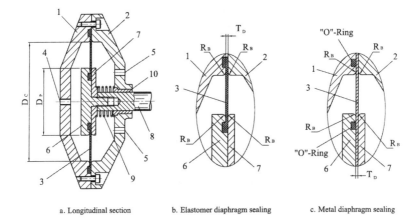

a. Longitudinal section   b. Elastomer diaphragm sealing   c. Metal diaphragm sealing

FIGURE 2.12
Single-acting flat diaphragm actuator: (a) longitudinal section, (b) elastomer diaphragm sealing, and (c) metal diaphragm sealing.

Most often, the metal diaphragm is made of stainless steel, a beryllium-copper alloy, a copper-nickel alloy, or a nickel-chrome alloy. The elastomer is selected on the basis of compatibility with the application gases and the operating temperatures. In pneumatic actuators, the following types of elastomers are used: Viton, nitrile, Buna N, neoprene, silicone, and others. With a differential pressure below 0.4 MPa, fabric reinforcements can be utilized that will optimize fabric-elastomer adhesion and cycle life. Fabric-supported diaphragms have a distinct advantage over homogeneous elastomeric diaphragms. Supported diaphragms have less tendency to stretch under pressure conditions. Because the rubber is continuously under stress conditions, its physical properties tend to more rapidly degrade the physical properties. Most importantly, a fabric substrate takes all the stress from the rubber, leaving the rubber to work in the least stressful conditions. Fabric-reinforced diaphragms are produced from a wide variety of materials and weaves. In the diaphragm pneumatic actuator, the fabric materials are polyester, nylon, silk, and cotton. Depending on the application, the elastomer is bonded either to one or both sides of the fabric base. It is always desirable to use a double-coated design because then the diaphragm is protected from abrasion and delamination.

The reduction in diaphragm internal stress is the key factor in a long diaphragm lifetime. From this point of view, sharp corners should be avoided in the vicinity of the diaphragm and mating parts. The blending radius (R_B) in the diaphragm clamping areas (Figure 2.12b) should be about double the thickness (T_D) of the diaphragm.

Clamped diaphragm seals are very important elements in pneumatic diaphragm actuators; Figure 2.12b shows one of the most popular flange designs for the elastomer diaphragm (homogeneous or fabric-reinforced), called the beaded type. This style of flange enables the designer to control the amount of "squeeze" applied to the diaphragm's flange without concern for the amount of force applied to the flange during assembly. Sealing of the homogeneous metal diaphragm is shown in Figure 2.12c; in such cases, the "O"-ring sealing technique is used.

For a pneumatic diaphragm actuator, the effective area is the part of the diaphragm area that is effective in producing a stem force. Usually, the effective area will change as the actuator piston is stroked. This most frequently affects flat diaphragms, while convoluted diaphragms will improve the actuator performance, and a rolling diaphragm will provide a constant stem force throughout the entire stroke of the actuator.

If one considers the flat diaphragm as a thin ring disk that is clamped in place around the outside and inside edge and displacement is less than the diaphragm thickness, then its effective area can be defined as:

$$A_{EF} = \frac{\pi \cdot D_C^2 \cdot (n_D^2 - 1) \cdot (n_D^4 - 1 - 4 \cdot n_D^2 \cdot \ln n_D)}{12 \cdot n_D^2 \cdot (n_D^4 - 2 \cdot n_D^2 + 1 - 4 \cdot n_D^2 \cdot \ln^2 n_D)} \tag{2.1}$$

For this case, the diaphragm spring rate is:

$$k_D = \frac{16 \cdot \pi \cdot E_M \cdot T_D^3 \cdot n_D^2 \cdot (n_D^2 - 1)}{3 \cdot D_C^2 \cdot (1 - \mu^2) \cdot [(n_D^2 - 1) - 4 \cdot n_D^2 \cdot \ln^2 n_D]} \tag{2.2}$$

where D_C is the cylinder diameter (outside clamp diameter), $n_D = D_C/D_P$, D_P is the piston diameter (inside clamp diameter), E_M is the modulus of elasticity (Young's modulus), and μ is Poisson's ratio.

From these formulas one can see that increasing the piston diameter (D_P) will increase the value of the effective area (A_{EF}), and this process also increases the value of the diaphragm spring rate (k_D). Usually, the piston diameter is $D_P = (0.4 \div 0.85) \cdot D_C$.

It is important to note that Equation 2.2 is relevant only for the homogeneous diaphragm. For a fabric-reinforced diaphragm, the spring rate depends on the material, structure, and size of the fabric; the material and thickness of the elastomer coating; the geometric parameters of the diaphragm; and the flange design. In practice, the spring rate for the fabric-reinforced diaphragm is defined by experimental testing.

The function of a diaphragm actuator is the most important factor in its design and manufacture. A simple actuator moves from one position to another, for example, in pressure switches or solenoid valves. In these cases, a linear actuator with a nonclamped flat diaphragm can be used.

Figure 2.13 provides a schematic diagram of the double-acting, nonclamped diaphragm actuator. This construction consists of (see Figure 2.13a) left (1) and right (2) covers that have holes to supply the pressurized air. The housing (3) has the two "V"-grooves for the flat diaphragms (4 and 5), which

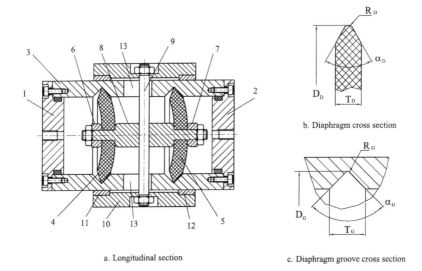

a. Longitudinal section

b. Diaphragm cross section

c. Diaphragm groove cross section

FIGURE 2.13
Double-acting actuator with nonclamped diaphragm: (a) longitudinal section, (b) diaphragm cross section, and (c) diaphragm groove cross section.

have a "V"-shaped outside surface. Such a configuration allows one to achieve effective and durable sealing between the housing and diaphragm (the sealing is realized by a circular contact "V"-groove surface with the surface of the diaphragm in a "V"-shape). The sealing between the actuator rod (8) and the two diaphragms is carried out by two clamping disks (6 and 7) and nuts. The nonclamped sealing technique can be used in this part of the actuator construction. The movement from the rod (8) to the actuator bushing (10) is transmitted by the transmitting rod (9), which is attached to both the rod (8) and the actuator bushing (10). Between the actuator bushing (10) and the housing (3) are the bearing bushings (11 and 12), which are usually made from sintered bronze. The housing (3) contains slots (13) as the passageway for the transmitting rod (9); simultaneously, these slots play the role of vented holes.

In this type of diaphragm actuator, the actuator rod usually takes only two stable positions (at the end of the movement point). In general, the diaphragm is made from homogeneous elastomer or fabric-reinforced elastomer. Basically, in the actuators with flat, nonclamped diaphragms, the angle of the diaphragm "V"-shape is $\alpha_D = 60°$ (Figure 2.13b) and the angle of the "V"-groove is $\alpha_G = 90°$ (Figure 2.13c). The fillet radius of the diaphragm is $R_D = (0.2 \div 0.3) \cdot T_D$, and in the groove this parameter should be $R_G = 0.5 \cdot R_D$. For other parameters, the correlations include the following: $T_G = (1.3 \div 1.5) \cdot T_D$ and $D_G = (0.95 \div 0.98) \cdot D_D$, where T_G is the width of the groove, D_G is the groove diameter, and D_D is the outside diameter of the diaphragm (Figures 2.13b and c).

Actuators with a convoluted diaphragm can be divided into two groups:

1. Actuators with a low convolution diaphragm
2. Actuators with a rolling (deep convolution) diaphragm

Construction of the actuator with a low convolution diaphragm has the same form as an actuator with a flat diaphragm (Figure 2.14a). Usually, in these constructions the convolution width is

$$W_C = (0.4 \div 0.7) \cdot \left(\frac{D_C - D_P}{2} \right),$$

and the convolution height is $H_C = (4 \div 18) \cdot T_D$. These actuators have a longer stroke (in comparison with the flat diaphragm actuators) and their effective area does not vary more than about $0.3 \cdot H_C$ in the stroke. The effective area in this range can be defined by[49]:

$$A_{EC} = \frac{\pi \cdot (D_C^2 + D_C \cdot D_P + D_P^2)}{12} \qquad (2.3)$$

In actuators with a low convolution diaphragm, both metals and the fabric-reinforced elastomer diaphragms are used. In practice, these actuator types are widely used in precision instruments such as pressure and flow sensors, altimeters, manometers, etc.

A double-acting actuator with a low convolution diaphragm is shown in Figure 2.14b. This construction comprises the left (1) and right (2) covers

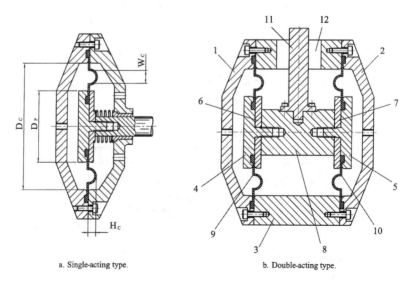

a. Single-acting type.

b. Double-acting type.

FIGURE 2.14
Linear actuator with low convolution: (a) single-acting type and (b) double-acting type.

with holes for air supply. These covers and housing (3) clamp two diaphragms (9 and 10) from the outside, and from the inside the diaphragms are clamped by piston heads (4 and 5) and clamping disks (6 and 7). In this design, the diaphragms are attached to both sides of the actuator rod (8), which has the yoke (11) that passes through a slot (12) in the housing (3). This construction allows one to avoid the back-pressure problem. In general, such diaphragms can withstand a high pressure differential only in one direction (in this case, from actuator outside to its inside direction). When the pressure on the low-pressure side of the diaphragm exceeds the pressure on the high-pressure side of the diaphragm, the convolution collapses and becomes wrinkled. This wrinkle will cause scrubbing and lead to premature failure. The problem with back-pressure usually occurs when the user is unaware that it even exists.

A single-acting actuator with a rolling (deep convolution) diaphragm is shown in Figure 2.15. Basically, this actuator is the same as that shown in Figure 2.14a. A rolling diaphragm is formed in a convoluted shape. It gets its name because as the stem moves, the diaphragm "rolls" at the convolution. It is frequently used in a manner similar to a "U"-cap seal, that is, to

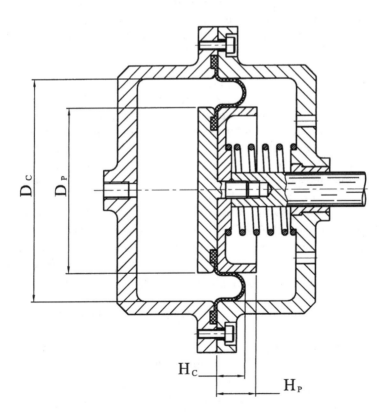

FIGURE 2.15
Linear actuator with rolling (deep convolution) diaphragm.

seal the gap between a linear moving piston and the actuator housing. But unlike the "U"-cup, the rolling diaphragm is permanently affixed to both the piston and the actuator housing. Because it is affixed to the piston and also "inflates" from air pressure, the force of the pressurized air directly impacts the movement of the piston. In an actuator with a rolling diaphragm, the convolution width is always equal to the width of the gap created by the difference in the piston diameter and the cylinder diameter.

Usually, the rolling diaphragm is made from a fabric-reinforced elastomer. In the hardware design of the actuator with rolling diaphragm, it is recommended that the supporting surfaces of the construction be no rougher than 0.8 micron (R_a), and, if necessary, should be finished to 0.4 micron in higher-cycle applications. Although diaphragms do not require lubrication, they can be coated with molybdenum disulfide (MoS_2) prior to installation to help reduce abrasion. The piston surface can also be coated with Teflon to reduce friction force when the diaphragm shifts against it, or with an elastomeric coating, which helps prevent abrasion by preventing the diaphragm from shifting.

The quickest failure occurs when the sidewall of the diaphragm comes into contact with itself. When this happens, the two elastomeric surfaces lock together while the piston continues to travel. This generally results in the sidewall of the diaphragm being jammed between the piston and cylinder wall, resulting in the elastomer and fabric being torn. Usually, there are two causes for this phenomenon:

1. *Alignment between the piston and cylinder.* In general, in high-pressure applications, this problem does not exist because the pressure force equalizes on the diaphragm and centralizes the piston. However, in low-pressure applications, the gravity can take over and pull the piston to one side, causing a problem. This can be avoided with a bushing for the piston or some other way of keeping the piston centered throughout its stroke.

2. *Back-pressure.* In general, the diaphragm can take a high-pressure differential in one direction only. If the pressure gets higher on the low-pressure side, the convolution collapses. This failure can be avoided if the vent holes are sized correctly to obtain sufficient flow rates.

In an actuator with a rolling diaphragm, the standard sintered bronze bearing can be replaced with a linear ball bearing, which reduces the friction force of the rod. This option gives a smooth-moving, accurate actuator.

The gap between the piston and the cylinder (the convolution width) is designed to minimize the rolling stress in the diaphragm. The stroke requirement will define the height of the convolution. The stroke capability of a diaphragm is usually calculated or specified using the "neutral position" (neutral plane) as a basis. It is defined as that point in the stroke where the piston head is in the same plane as the clamping cover. Many designs involve stroking the piston and diaphragm in only one direction from the neutral plane. "Half-stroke" is the term used to designate the stroke capability of the diaphragm under this condition. The total stroke capability is twice the half-stroke.

The convolution height (H_C) of the rolling diaphragm can be defined in the following form:

$$H_C = 0.5 \cdot (S_H + W_C + 2 \cdot R_B + 2 \cdot T_D) \qquad (2.4)$$

where S_H is the diaphragm half-stroke, W_C is the convolution width that is defined as $W_C = 0.5 \cdot (D_C - D_P)$, R_B is the blending radius, and T_D is the diaphragm thickness.

In the calculation of the rolling diaphragm, it is assumed that its effective area is constant on the half-stroke and can be defined as:

$$A_{ER} = \frac{\pi \cdot (D_C + D_P)^2}{16} \qquad (2.5)$$

Another important parameter of the actuator with rolling diaphragm is the length of the piston skirt H_P (Figure 2.15), which is:

$$H_P = \frac{H_C + S_H}{2} \qquad (2.6)$$

2.1.3 Actuators with Bellows

Bellows in pneumatic actuating systems are a compressible and extensible component providing an ultra-leak-tight connection while accommodating relative movement. They also shield the operating environment from particular contamination by encapsulating moving mechanical components. In these actuators, metal bellows are usually used. This is critical in applications demanding exceptionally clean conditions, such as vacuum and semiconductor fabrication equipment. Metal bellows are also used to eliminate elastomers because of temperature limits. The temperature range in which the metal bellows operate is from −190 to +540°C. Unlike rubber seals, metal bellows are virtually impervious to most liquids and gases, and do not degrade in areas of high radiation.

Deflection of a metal bellows results in stresses. These deflections are frequently repetitive; thus, the bellows must be designed to minimize stresses in order to enable the bellows to endure the high cycle deflection and avoid premature fatigue failure. To reduce the bending stresses, the bellows wall thickness, or convolution thickness, can be reduced. To withstand

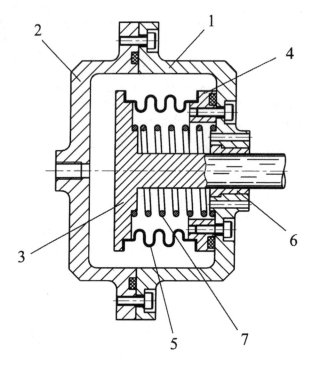

FIGURE 2.16
Single-acting linear actuator with bellows.

the forces of high pressure, the thickness can be increased. Selecting the right thickness to achieve the delicate balance of thickness for pressure, thinness for reduced stress, long life, and flexibility is the challenge faced by the bellows designer.

Figure 2.16 depicts a single-acting actuator with bellows. This actuator consists of the right cover (1) and left cover (2) with holes for supplying the pressured air. The bellows (5) are attached to the piston with rod (3) and to the sealing ring (4), which is installed on the right cover (1) that has the vented holes and the bearing bushing (6). The coil spring (7) provides the return force. In general, the ends of the bellows are welded to the piston and sealing ring.

Different processing can form the individual bellows convolutions:

- Convoluted, stamped plates can be welded. This is the most popular technique. "V"-type and "M"-type bellows start with disks that are precision stamped from a thin metal strip. Pairs of these disks are welded at their inside diameter. They are then stacked and welded together at their outside diameter. The entire bellows is made in a standard length called a *block*. Multiple blocks can be welded together to make longer bellows. Welded bellows can be fabricated from materials that do not form well (e.g., titanium). In welded bellows, the span (or distance from inside to outside) can be relatively large. This allows very low spring rates and a long stroke for a given package size. The ratio of free length to solid height can exceed 5:1.

- Rolling corrugations into thin metal tubes can form bellows; they can also be manufactured by being hydraulically formed in metal dies. In this case, a seamless tube is placed inside a movable set of plates in a hydraulic corrugating machine. Hydraulic pressure is applied internally to produce a complete, convoluted bellows in one operation. This fabrication method produces excellent-quality bellows at the lowest possible cost for high-volume production.

- The bellows configuration can be plated onto a wax mold that can later be melted away to leave the bellows configuration. The resultant thin bellows section eliminates this style for mechanical seals but they are frequently used in instrumentation.

Usually, metal bellows are used in pneumatic actuators. Most often, the bellows are made from brass (UNS C24000, ISO CuZn20). This material has excellent resistance to corrosion. It is relatively free from creep, drift, and hysteresis. All sizes and types of bellows detailed in brass are also offered in phosphor bronze (UNS C51000, UNS C52100, ISO CuSn5, and ISO CuSn8P). The phosphor bronze bellows exhibit characteristics similar to brass bellows. In the smaller sizes, these are identical; but in the larger sizes, the spring rate of the brass bellows may be as much as 7% lower than for the phosphor bronze equivalent. Also, in pneumatic actuators, the bellows are made from stainless steel. Alloys used for bellows manufacturing include AISI 321, AISI 316, AISI 316L, and AISI 347.

The effective area of the bellows can be defined as:

$$A_{EB} = \frac{\pi \cdot (D_{BO} + D_{BI})^2}{16} \tag{2.7}$$

where D_{BO} and D_{BI} are the bellows outside and inside diameters, respectively. This calculation is not theoretically accurate but gives results close to the actual value. The bellows effective area has a constant value on the full range of its stroke.

The spring range of the welded bellows is calculated as:

$$k_B = \frac{2 \cdot E_M \cdot t_B^3}{K_B \cdot N_B \cdot D_{BO}^2} \tag{2.8}$$

where t_B is the bellows wall thickness, N_B is the number of active convolutions in the bellows, and the coefficient K_B can be determined from:

$$K_B = \frac{3 \cdot (1 - \mu^2)}{4 \cdot \pi} \cdot (\frac{n_B^2 - 1}{n_B^2} - \frac{4 \cdot \ln^2 n_B}{n_B^2 - 1})$$

where $n_B = D_{BO}/D_{BL}$.

As opposed to welded bellows, in formed bellows, the wall thickness is not constant. In the point $r_B = 0.5 \cdot D_{BI}$ (r_B is the distance from the bellows center axis), the wall thickness is equal to the tube wall thickness (t_{BT}) before forming. In the point $r_B = 0.5 \cdot D_{BO}$, the wall thickness has the minimum value; in general, the following form can determine the wall thickness for the formed bellows

$$t_B = t_{BT} \cdot \frac{D_{BI}}{2 \cdot r_B}$$

The spring range of the formed bellows is calculated from Equation 2.8, in which the wall thickness is some averaged magnitude. In this case, the formula

$$t_B^3 = \frac{t_{BT}^3}{3} \cdot [1 + (\frac{2}{1 + n_B})^3 + \frac{1}{n_B^3}]$$

gives an adequate calculated result.[139]

Basically, pneumatic actuator systems with bellows have very low internal friction forces and, as result, their positioning accuracy is sufficiently high.

2.1.4   Combined Pneumatic Actuators

The two degree-of-freedom pneumatic cylinder allows one to gain improvement in the accuracy of automatic equipment-positioning mechanisms. In this case, the resulting motion of the output link can be arranged as the sum of two separate motions. The first motion is performed by a relatively rough but fast-acting actuator to transfer the object to the place near the position point. At the second stage, the object is moved by a precise drive to the given position with the accuracy required.

Usually, the second stage is fixed to the output link of the first stage, and a specially arranged control system puts the drives into operation in a definite sequence. However, in the case of pneumatic actuators, the change of operation mode can be realized automatically on the base of the same cylinder, provided the piston rod (output element) is connected to the piston by means of a flexible element (e.g., a diaphragm). When the piston stops

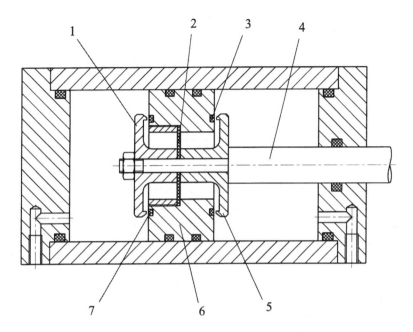

FIGURE 2.17
Two degree-of-freedom pneumatic cylinder with flat diaphragm.

under the action of friction forces, the piston rod continues its motion, providing the precise positioning of the output element.[101]

Figure 2.17 shows the two degree-of-freedom pneumatic cylinder design where the piston (6) and rod (4) are coupled through a diaphragm (2) (the flexible element). It is evident that the best accuracy is achieved when the flexible element spring rate is at a minimum. However, in this case, the high reliability of the drive is important. To that end, a pneumatic cylinder was developed with two disk valves: on the rod (4) are two disks (1 and 5), and on the piston (6) there are two "O"-rings (3 and 7). These valves have fixed stops to limit flexible element deflection. As in ordinary positioning devices with one degree-of-freedom, only one sensor is used in measuring the rod position.

Figure 2.18 depicts the two degree-of-freedom rodless pneumatic cylinder. In this construct, the motion transmission of the piston (8) to the outer slide (3) is obtained by magnetic coupling, which consists of permanent magnets (4 and 5) on the slide (3) and two piston bushings (1 and 2) with permanent magnets (6 and 7). The piston (8) is connected to the piston bushings (1 and 2) by a flat diaphragm (11), which is the flexible element. Two disk valves with two "O"-rings (9 and 10) form the fixed stops to limit the flexible element deflection.

As is well known, the pneumatic actuator has a poor damping characteristic; however, the pneumatic cylinder shown in Figure 2.19 has high damping because the second stage is within the chamber that is filled with oil.

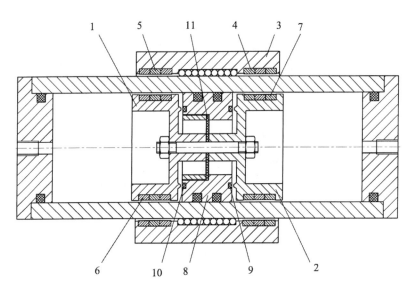

FIGURE 2.18
Two degree-of-freedom rodless pneumatic cylinder with flat diaphragm.

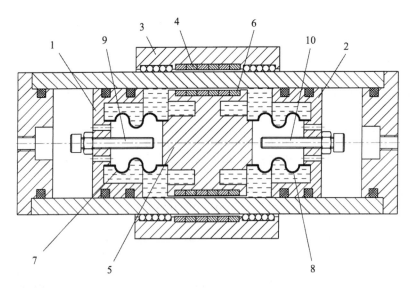

FIGURE 2.19
Two degree-of-freedom rodless pneumatic cylinder with bellow.

There is the two degree-of-freedom rodless pneumatic cylinder with bellows, in which the motion transmission of the piston (5) to the outer slide (3) is obtained by magnetic coupling via permanent magnets (4 and 6). The main piston (5) connects to the additional pistons (1 and 2) by bellows (7 and 8), which are the flexible elements. The stroke of the main piston (5) is limited

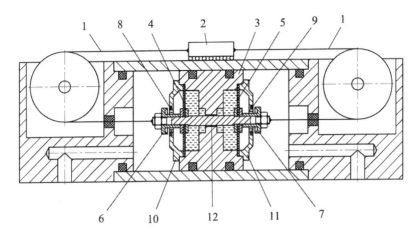

FIGURE 2.20
Two degree-of-freedom cable rodless pneumatic cylinder with flat diaphragm.

by the hard stops (9 and 10). The oil fills the internal cavity between additional pistons (1 and 2), which greatly increases actuator damping.

Figure 2.20 shows yet another example of the two degree-of-freedom rodless pneumatic cylinder with high damping force. In this design, rigid cables (1) attach to both ends of the piston (3) through the diaphragms (4 and 5), which are the flexible elements. The cables pass through seals at the covers of the cylinder, around the pulleys, and are joined at the carrier (2). To limit the flexible element's deflection, a pneumatic cylinder was developed with two disk valves, which are installed on the spool (12). These valves consist of two disks (6 and 7) and two covers (10 and 11) with two "O"-rings (8 and 9). The internal piston cavity between the diaphragms is filled with oil that flows through the orifice between the spool and inside surface of the piston central hole.

Two degree-of-freedom pneumatic actuators provide much better accuracy than ordinary systems (the repeatability is about 5 to 8 times better). In a pneumatic cylinder with flexible coupling between the piston and rod, the positioning error is a function of the flexible element spring rate, its maximum deflection, and the effective area ratio between the piston and the flexible element. However, although these actuators can be utilized generally in point-to-point positioning systems, their application in the tracking mode is very difficult.[101,105]

2.2   Rotary Actuators and Pneumatic Motors

Rotary actuators can be used to produce limited rotary motion. Most standard models only turn 180° or less, but some models can turn as much as

450°. With some models, the user can adjust the rotation within the maximum limit. Similarly, most standard pneumatic rotary actuators have only two stopping points, although some have as many as five.

Pneumatic rotary actuators can be divided into two major groups:

1. Vane actuators
2. Rack-and-pinion actuators

The use of any one type should be evaluated on the basis of four primary criteria: (1) working torque, (2) bearing load, (3) kinetic energy, and (4) the environment (e.g., clean room, hazardous, corrosive, etc.). Similar to pneumatic cylinders, rotary actuators have nonlinear characteristics because of the compressibility of air and friction forces.

The vane or blade-type rotary actuator is the direct-drive type that provides a rigid bearing system that allows for mounting the load directly on the actuator output shaft.

Figure 2.21 depicts the single-vane rotary actuator. This construct consists of the housing (1) with inlet and outlet ports (5 and 6). The shaft (2) has the vane (3), which is built as an integral part of the shaft. The internal barrier (4) is located inside the actuator and is affixed to the housing (1) with screws. Thus, the barrier (4), shaft (2), and vane (3) partition the inner actuator space into two working chambers. The shaft (2) with the vane (3) is supported by the two bearings (7 and 8). In this construct, the vane swings and rotates in a normal and reverse direction because of air pressure, wherein a torque is transmitted via the shaft.

Single-vane actuators are limited in rotation and most do not exceed 300°. These actuators are made with different rotating angles, the most common

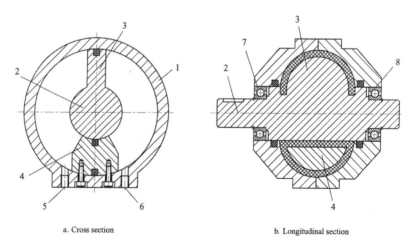

a. Cross section   b. Longitudinal section

FIGURE 2.21
Single-vane rotary actuator: (a) cross section and (b) longitudinal section.

being 90°, 180°, and 270°. Vane rotary actuators provide a high torque-to-size relationship. These actuators were designed to eliminate the "vane leakage" inherent in many vane actuators using a special "vane seal" with "rounded" corners rather than hard-to-seal square corners. This seal design permits low operating and start-up pressures; smooth, step-free operation; and long life in rugged service applications. The angle of rotation can be infinite, and the adjustable stop system is separated from the rotary actuator; usually, this system is installed on the backside of the actuator. The shock absorbers are installed together with the adjustable stop system.

Another construction of the vane actuator is the double-vane rotary actuator, as shown in Figure 2.22. This actuator type is the same as the single-vane actuator; it has the housing (1), shaft (2), and two pairs of the vanes (3 and 4) and internal barriers (5 and 6). The double-vane actuator has twice the torque of a similarly sized single-vane actuator. In addition, this design produces no side loading forces on the output shaft or actuator bearings; however, the rotating angle is not more than 150°.

The vane rotary actuator is less expensive than a rack-and-pinion rotary actuator; it has zero backlash and usually requires less maintenance. However, these compact rotary actuators are designed for manipulation where particular accuracy and high loads are not required.

The rack-and-pinion rotary actuator can provide multiple revolution motion. The actuator converts linear motion from a cylinder into rotary motion. The rack is a straight set of gear teeth attached to the cylinder's piston. The

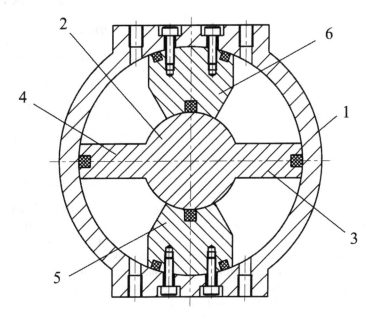

FIGURE 2.22
Double-vane rotary actuator.

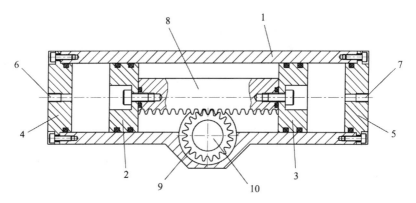

FIGURE 2.23
Rack-and-pinion rotary actuator with single rack.

rack is pushed linearly as the piston moves. The rack's teeth mesh with the circular gear teeth of the pinion, forcing it to rotate. The pinion rotates back to its original position when air pressure is supplied to the opposite side of the cylinder. The maximum rotation of a rack-and-pinion rotary actuator is limited by its size. In practice, the maximum angle of rotation angle is not more than 720°.

Figure 2.23 gives a schematic diagram of the rotary actuator with single rack, which consists of the pneumatic cylinder sleeve (1) with two pistons (2 and 3). Two covers (4 and 5) close the cylinder sleeve (1) on both sides and form two pneumatic working chambers, which are charged or discharged by pressurized air via the holes (6 and 7). The rack (8) is attached to the two pistons and forms the engagement with the pinion (9), which is installed on the output shaft (10). The action at the end of the rotation can be controlled using adjustable cushions and hard stops, which are available as an option.

In this construction, the pinion is keyed on the rotating part and is rotated by means of a rack. Two pneumatic pistons drive the rack — one on either side. As there is no play take-up system, it is not suitable for applications requiring accurate positioning.

Figure 2.24 depicts a rotary actuator with double rack; it has a shaft housing (1), two pneumatic cylinder sleeves (2 and 3), and two pistons (4 and 5) with racks (6 and 7). The pinion (8) with output shaft (9) is within an intermediate chamber, which has a hole (14) for the charging and discharging process. The two covers (10 and 11) have two holes (12 and 13) that are used for supplying pressurized air to the pneumatic chambers or their connection to the exhaust line. The output shaft rotates clockwise if two outer pneumatic chambers are connected to the supply line (holes 12 and 13), and the intermediate chamber (hole 14) — with the exhaust port. The counterclockwise rotation implements if the pneumatic chambers are connected in the reverse direction.

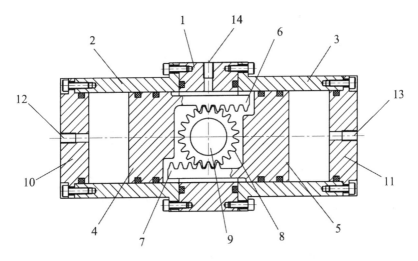

FIGURE 2.24
Rack-and-pinion rotary actuator with double rack.

Two opposing racks, each driven by pneumatic pistons, turn the pinion keyed onto the rotating element. The system has an automatic play take-up, which means that it is suitable for applications requiring accurate positioning. The parallel-piped shape of the design makes it ideal for use on mechanical arms.

Pneumatic motors are continuously rotating rotary actuators and can be divided into three groups:

1. Vane motors
2. Piston motors
3. Turbine motors

The most popular type of air motor is the vane motor. Small vane motors usually operate at speeds exceeding 20,000 rpm (revolutions per minute) and gear down to usable speeds with planetary gearing. This maintains their very small diameter and high power-to-weight ratio, making them excellent for use in hand tools.

A typical vane-type pneumatic motor is shown in Figure 2.25. This motor provides rotation in either direction. The rotating element is a slotted rotor (1) mounted on a drive shaft. Each slot of the rotor is fitted with a freely sliding rectangular vane (2). The rotor (1) and vanes (2) are enclosed in the housing (3), the inner surface of which is offset from the drive shaft axis. When the rotor is in motion, the vanes (2) tend to slide outward because of centrifugal force. The shape of the rotor housing limits the distance the vanes can slide. This motor operates on the principle of differential areas of the vanes (2). The two ports (4 and 5) can be alternately used as inlet and outlet, thus providing

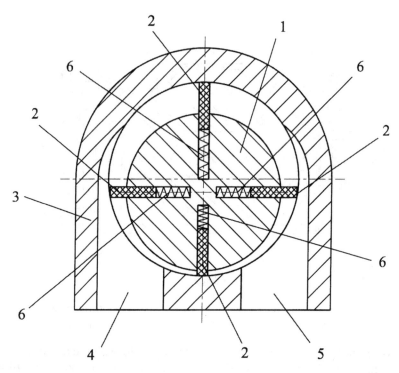

FIGURE 2.25
Bidirectional vane motor.

rotation in either direction. The springs (6) inserted in the slots of the rotor hold the vanes against the housing (3) during the initial starting of the motor because centrifugal force does not exist until the rotor begins to rotate. Practically, the number of vanes ranges from three to ten. An increase in the number of vanes increases the rotation smoothness but also increases the friction force, which decreases the motor's efficiency. The torque of a vane motor consists of three parts: (1) the torque from air displacement, (2) the torque from its expansion, and (3) the torque required to compress the air in the other chamber.

The final operating speed of vane and piston pneumatic motors is very unpredictable. It usually results in a balance between the resisting torque (the load), the working pressure realized inside the motor, and, of course, the available unrestricted compressed-air flow. Pneumatic motors should be sized to operate at maximum peak power, which is usually about 60% of the free running speed.

Piston-type motors are not the most commonly used in pneumatic systems. Principally, radial-piston motors are applied where high torques at a low speed are required (e.g., for a winch drive). In this case, a gearbox is not necessary because of the low output speed. The schematic diagram in Figure 2.26 depicts the radial-piston motor. This construct has a housing (1)

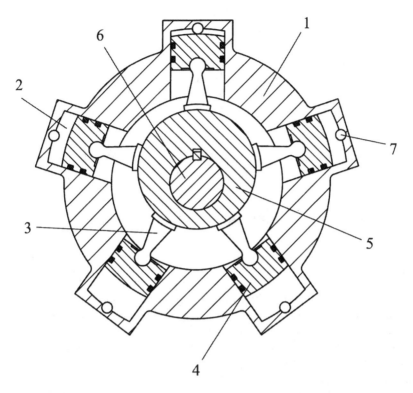

FIGURE 2.26
Radial-piston motor.

with radial cylinders (2). The connecting rods (3) of the pistons (4) are pushed on the eccentric part (5) of the output shaft (6). A rotating sleeve valve (not shown in this figure) is driven by the output shaft and takes care of the cylinder charging or discharging process via the holes (7). By changing the direction of the compressed air supply to the motor, the direction of rotation can be changed. Normally, the radial-piston pneumatic motor contains seven or nine pistons, and it has high starting torque that is ideal for heavy starting loads; in addition, this motor has high smoothness. For the radial-piston pneumatic motor, it is not recommended for operation at speeds greater than 75% of the free rotation speed (without load). In industrial robots, radial-piston motors with bellows actuators (instead of pneumatic cylinders) are used. In this case, the actuating system has a simple and inexpensive construct.[205]

Turbines are used in pneumatic systems to convert the kinetic energy of air to mechanical energy. The basic parts of a turbine are the rotor, which has blades projecting radially from its periphery, and nozzles, through which the gas is expanded and directed. The conversion of kinetic energy to mechanical energy occurs on the blades. The basic distinction between types of turbines is the manner in which the gas causes the turbine rotor to move. When the rotor is moved by a direct push or impulse from the gas impinging upon the blades, the turbine is called an "impulse turbine." When the rotor moves by the force of reaction, the turbine is called a "reaction turbine." Although the distinction between impulse turbines and reaction turbines is a useful one, it should not be considered as an absolute distinction in real turbines. An impulse turbine uses both the impulse of the gas jet and, to a lesser extent, the reactive force that results when the curved blades cause the gas to change direction. A reaction turbine is moved primarily by reactive force, but some motion of the rotor is caused by the impact of the gas against the blades.

Air turbine motors are only efficient at high speeds. These devices can operate at speeds up to 500,000 rpm. Turbines are used to drive electric generators, to convert mechanical energy into electrical energy, and to drive pumps to supply fluid flow in hydraulic systems. In practice, this actuator type is not used in position, velocity, or force servo pneumatic control systems.

Air motors can be regulated easily for speed and torque, and can stop and reverse very quickly. They are commonly used in many industrial applications and are noted for their economical power delivery, straightforward maintenance, and safety in spark-prohibited applications. The most important performance specifications to consider when searching for air motors include required torque, maximum air pressure, air consumption, rated free speed (output), and operating noise level.

Output styles for air motors include output shaft of hollow or collet. With a shaft, the output is a solid shaft, typically cylindrical. A hollow shaft or collet is an output shaft with a center hole for tool mounting, or a collet style for adjustable tool clamping. The diameter of the shaft or collet is an important parameter. Dimensions of length, side, or outside diameter, and weight are also important considerations. Common materials for air motor construction include aluminum, cast iron, steel, stainless steel, and plastic.

2.3   Mathematical Model of Pneumatic Actuators

The mathematical model that describes the dynamics of the pneumatic actuator comprises the equation of motion for the output element (rod, shaft, or other) and equations of changing for the pressure in the actuator working chambers.

The charging and discharging process, and also the thermodynamic process within the chambers, determine the pressure change in the working chambers. These processes assume that:

- The compressed air is the perfect gas.
- The pressure and temperature within the chamber are homogeneous.

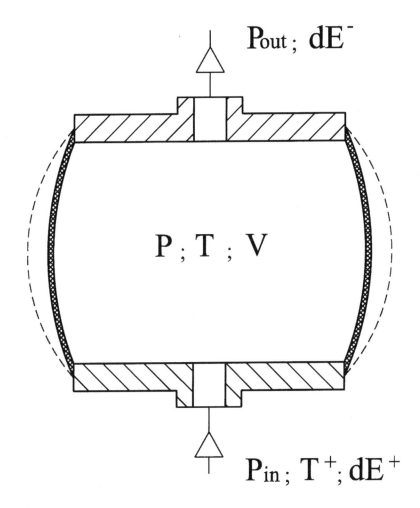

FIGURE 2.27
Schematic diagram of the pneumatic working chamber.

The pressure-changing process can be considered for a chamber with uncertain form and variable volume that is connected to the supply and exhaust port (Figure 2.27). This model is suitable for all types of pneumatic actuators considered hereinafter.

The first law of thermodynamics describes the pressure-changing processes in the cylinder working chambers. The energy balance in the variable chamber volume can be written as (heat exchange is negligible):

$$dE^+ = dE_{CH} + dE_{AE} + dE^-$$

where

      E^+ = air energy of input line
      E_{CH} = chamber internal air energy
      E_{AE} = work of the air expansion
      E^- = air energy of output line

Considering that:

$$dE^+ = h_S^+ \cdot G^+, \quad dE^- = h_S^- \cdot G^-, \quad dE_{CH} = C_V \cdot m_A \cdot dT + C_V \cdot T \cdot dm_A,$$

$$dE_{AE} = P \cdot dV, \quad h_S^+ = C_P \cdot T^+, \quad h_S^- = C_P \cdot T$$

where

      T = air temperature in the chamber
      m_A = air mass in the chamber
      h_S = specific enthalpy of the air flow
      C_P, C_V = air heat capacity for constant pressure and volume, respectively,

we obtain:

$$C_P \cdot T^+ \cdot G^+ - C_P \cdot T \cdot G^- = CV \cdot m_A \cdot dT + C_V \cdot T \cdot dm_A + P \cdot dV \quad (2.9)$$

The air condition equation $P \cdot V = m_A \cdot R \cdot T$ can be written as:

$$m_A \cdot dT = \frac{P \cdot dV + V \cdot dP - R \cdot T \cdot dm_A}{R} \quad (2.10)$$

Substituting Equation 2.10 into Equation 2.9, and considering that $C_P/C_V = k$ and $C_P - C_V = R$, we obtain:

$$k \cdot R \cdot T^+ \cdot G^+ - k \cdot R \cdot T \cdot G^- = k \cdot P \cdot dV + V \cdot dP \quad (2.11)$$

The mass flow rate G is determined by the following formula:

$$G = A_V \cdot P_{in} \cdot \sqrt{\frac{2 \cdot k}{R \cdot T_{in} \cdot (k-1)}} \cdot \varphi(\sigma) \quad (2.12)$$

where V is the working chamber volume, G is the air mass flow, k is the adiabatic exponent (for air, $k = 1.4$), R is the gas constant (for air, $R = 287 \, J/kg \cdot K$), $\sigma = P_{out}/P_{in}$, P_{in} is the absolute input pressure, P_{out} is the absolute output pressure, and A_V is the valve nozzle effective area.

    The flow function $\varphi(\sigma)$ can be determined as:

$$\begin{cases} \varphi(\sigma) = \sqrt{\sigma^{\frac{2}{k}} - \sigma^{\frac{(k+1)}{k}}}, & \text{for } \sigma_* \leq \sigma \leq 1 \\ \varphi(\sigma) = \varphi(\sigma_*) = \varphi_* = 0.259, & \text{for } 0 < \sigma < \sigma_* \end{cases} \tag{2.13}$$

where $\sigma_* = 0.528$.

In practical calculations, the flow function $\varphi(\sigma)$ can be determined from the following formulas:

$$\begin{cases} \varphi(\sigma) = 2 \cdot \varphi_* \cdot \sqrt{\sigma \cdot (1 - \sigma)}, & \text{for } 0.5 \leq \sigma \leq 1 \\ \varphi(\sigma) = \varphi_* = 0.259, & \text{for } 0 < \sigma < 0.5 \end{cases} \tag{2.14}$$

Equation 2.11 can be rewritten as:

$$\dot{P} = \frac{k}{V} \cdot (G^+ \cdot R \cdot T^+ - G^- \cdot R \cdot T - P \cdot \dot{V}) \tag{2.15}$$

Using Equation 2.10 and $G = G^+ - G^-$, $m = \dfrac{P \cdot V}{R \cdot T}$ then becomes:

$$\dot{T} = \frac{T \cdot \dot{V}}{V} + \frac{T \cdot \dot{P}}{P} - \frac{R \cdot T^2 \cdot (G^+ - G^-)}{P \cdot V} \tag{2.16}$$

Equation 2.15 and Equation 2.16 determine the change in the pressure P and temperature T in the working chamber for the adiabatic law.

Actually, the thermodynamic process in the pneumatic working chamber has polytropic behavior, and for this process the pressure changing equation can be written as[4, 153, 154]:

$$\dot{P} = \frac{1}{V} \cdot (\alpha^+ \cdot G^+ \cdot R \cdot T^+ - \alpha^- \cdot G^- \cdot R \cdot T - \alpha \cdot P \cdot \dot{V}) \tag{2.17}$$

where the coefficients α^+, α^-, and α take on values between 1 and k, depending on the actual heat transfer during the process. In Equation 2.17, one does not have to know the exact heat exchange characteristics, but can merely estimate the coefficients α^+, α^-, and α. For the charging process, a value of α^+ close to k is recommended; while for the discharging of the chamber, α^- should be chosen close to 1. The thermal characteristic of the compression/expansion process is better described using $\alpha = 1.2$.

In practice, the influence of heat exchange in pneumatic actuators becomes apparent if the temperature difference between the source of the compressed air and the ambient is large. In the industrial condition, where this difference is small, the energy exchange due to heat exchange is negligible.[102]

In industrial environments, supply regulators limit the pneumatic supply pressure. Most often, the supply pressure maximum value is 1.2 MPa and air temperature is about 290K. Under such conditions, changing the air temperature in the pneumatic working chambers is not sufficiently intensive and can be neglected.

Under these circumstances (i.e., the heat exchange and temperature changing in the actuator working chambers are negligible), the isothermal low (T = constant) can be applied, and then in the working chamber only the pressure changes. In this case, the pressure can be determined[78, 102]:

$$\dot{P} = \frac{1}{V} \cdot (G^+ \cdot R \cdot T_S - G^- \cdot R \cdot T_S - P \cdot \dot{V}) \tag{2.18}$$

where T_S is the ambient temperature (T_S = 290K). This hypothesis is the most reasonable for the major types of the pneumatic actuators.[152] Hereafter, Equation 2.18 is applied for describing the pressure changing process in the actuator's working chambers.

Pneumatic cylinder. There are two sources for the flow entering a cylinder chamber:

1. The pressure vessel, through the control valve and connecting tube
2. The neighboring chamber if it has a higher pressure and piston seals leak

The compressed air can flow out to the atmosphere through the control valve or piston rod seal or to the second chamber if it has a lower pressure. The leakage between pneumatic chambers and through the piston rod seals can be neglected for regular pneumatic cylinders with rubber-type seals, but it can be significant for cylinders that have graphite or Teflon seals.[153]

For a standard double-acting pneumatic cylinder (Figure 2.28), the volume of each chamber can be expressed as (the origin of piston displacement at the middle of the stroke):

$$V_i = V_{0i} + A_i \cdot (0.5 \cdot L_S \pm x) \tag{2.19}$$

where $i = 1,2$ is the cylinder chamber index, V_{0i} is the inactive volume at the end of stroke and admission ports, A_i is the piston effective area, L_S is the piston stroke, and x is the piston position. The difference between the piston effective areas for each chamber A_1 and A_2 is due to the piston rod (effective area is A_R). Then, substituting Equation 2.19 into Equation 2.18, the time derivative for the pressure in the pneumatic cylinder chambers becomes:

$$\dot{P}_i = \frac{1}{V_{0i} + A_i \cdot (0.5 \cdot L_S \pm x)} \cdot (G_i^+ \cdot R \cdot T_S - G_i^- \cdot R \cdot T_S \mp P_i \cdot A_i \cdot \dot{x}) \tag{2.20}$$

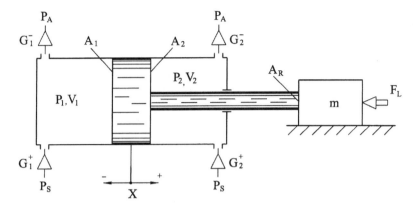

FIGURE 2.28
Schematic diagram of the standard double-acting pneumatic cylinder.

In this equation, the first term in the parenthesis represents the charging process in the chamber where

$$G_i^+ = A_{Vi}^+ \cdot P_S \cdot \sqrt{\frac{2 \cdot k}{R \cdot T_S \cdot (k-1)}} \cdot \varphi(\sigma_i) \tag{2.21}$$

The second term represents the discharging process, for which the term G_i^- can be written as:

$$G_i^- = A_{Vi}^- \cdot P_i \cdot \sqrt{\frac{2 \cdot k}{R \cdot T_S \cdot (k-1)}} \cdot \varphi\left(\frac{\sigma_A}{\sigma_i}\right) \tag{2.22}$$

And the third term represents the air compression or expansion due to piston movement. In Equation 2.21 and Equation 2.22 for the determination of the flow function, there are

$$\sigma_i = \frac{P_i}{P_S} \text{ and } \sigma_A = \frac{P_A}{P_S},$$

where P_S is the absolute supply pressure and P_A is the absolute atmospheric pressure.

The mechanical subsystem can be modeled as a second-order system with friction. The equation of motion for the piston-load assembly can be expressed as:

$$m \cdot \ddot{x} + b_V \cdot \dot{x} + F_F + F_L = P_1 \cdot A_1 - P_2 \cdot A_2 - P_A \cdot A_R \tag{2.23}$$

where m is the external load, the piston, and the rod assembly moving mass; b_V is the viscous friction coefficient; and F_L is the external force. In Equation 2.23, F_F represents the coulomb friction force.

In general, the friction properties are unclear because they depend on a number of factors, such as lubrication conditions, operating conditions, and interface temperature. In practice, the coulomb friction force is usually defined as[16, 92, 101, 180, 214]:

$$F_F = F_D \cdot \lambda(\dot{x}) + F_S \cdot [1 - \lambda(\dot{x})] \qquad (2.24)$$

where the function $\lambda(\dot{x})$ is:

$$\begin{cases} \lambda(\dot{x}) = 1, & \text{for } |\dot{x}| > 0 \\ \lambda(\dot{x}) = 0, & \text{for } |\dot{x}| = 0 \end{cases} \qquad (2.25)$$

Here, F_S is the static coulomb friction force, F_D is the dynamic coulomb friction force, which is represented by $F_D = F_D \cdot sign(\dot{x})$; here,

$$sign(\dot{x}) = \begin{cases} -1, & \text{if } \dot{x} < 0 \\ 0, & \text{if } \dot{x} = 0 \\ 1, & \text{if } \dot{x} > 0 \end{cases} \qquad (2.26)$$

F_S represents the friction force when $|\dot{x}| = 0$. In this case, when the magnitude of the applied force is less than the limits of F_S, the friction is equal to the applied force. Generally speaking, the magnitude of F_D is not greater than F_S.

As mentioned above, the sealing of the piston and rod of standard pneumatic cylinders can be made of some kind of elastomer. Acrylonitrile-butadiene rubbers (NBR or Perbunan) have primarily been used as the static and dynamic seals material. In addition, anodized aluminum as the material of the cylinder barrel and stainless steel as the piston rod material have been most often used in pneumatic cylinders in industrial equipment. The values of the static (F_S) and the dynamic (F_D) friction forces, and the viscous friction coefficient (b_V) of the double-acting standard cylinder, are given in Table 2.1.

Determination of the pneumatic cylinder friction forces (both static and dynamic) and their viscous friction coefficients was carried out with two experimental tests. In the first experiment, the piston was at rest and the air ports were connected to atmospheric pressure. Increasing force was applied at the rod end; this force was measured using a strain gage force cell. Simultaneously, the piston displacement, its velocity, and the pressure in the cylinder chambers were recorded. In the second experiment, the displacement, velocity, and pressure in the cylinder chambers were measured during the

TABLE 2.1

Values of Static (F_S) and Dynamic (F_D) Friction Forces,
and the Viscous Friction Coefficient (b_V) of the Double-Acting
Standard Cylinder

Piston Diameter (mm)	Static Friction Force (N)	Dynamic Friction Force (N)	Viscous Friction Coefficient $\left(\dfrac{N \cdot s}{m}\right)$
10	3–4	1–2	8–25
12	4.5–6	1.5–2.5	10–30
16	6–8	2.5–4	12–35
20	12–19	8–12	15–40
25	20–28	10–16	20–50
32	25–35	12–20	25–60
40	30–50	15–25	30–70
50	40–60	18–30	45–90
63	45–75	22–40	50–100
80	55–95	28–48	60–110
100	65–110	32–55	70–140

motion for the open-loop responses with the different valve control signals. Using Equation 2.23 through Equation 2.26 and the results of these experiments, we obtained the friction forces and the viscous friction coefficients for the different conditions (for new cylinders, for the pneumatic cylinder that made about 100,000 cycles, at the beginning of the cylinder operation, and for the pneumatic cylinder after 15 to 20 minutes of continuous work).

Usually, in the simulation process of the pneumatic actuator dynamic, the velocity value of ε is set to a small positive number only to ensure the stability of the numerical integration algorithm; in this case, the function $\lambda(\dot{x})$ is described by:

$$\begin{cases} \lambda(\dot{x}) = 1, & \text{for } |\dot{x}| > \varepsilon \\ \lambda(\dot{x}) = 0, & \text{for } |\dot{x}| \le \varepsilon \end{cases} \tag{2.27}$$

Diaphragm actuators and actuators with bellows. For the double-acting flat diaphragm actuator (see Figure 2.29), the time derivative for the pressure in the pneumatic chambers can be obtained using Equation 2.20, in which the effective area $A_1 = A_{EF}$ and $A_2 = A_{EF} - A_R$, where the diaphragm effective area A_{EF} is defined by Equation 2.1. The equation of motion for the diaphragm-load assembly can be expressed as:

$$m \cdot \ddot{x} + b_V \cdot \dot{x} + k_D \cdot x + F_F + F_L = P_1 \cdot A_1 - P_2 \cdot A_2 - P_A \cdot A_R \tag{2.28}$$

where k_D is the diaphragm spring rate as determined by Equation 2.2.

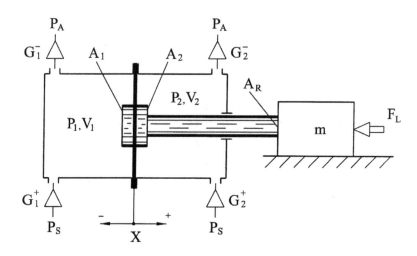

FIGURE 2.29
Schematic diagram of the flat diaphragm double-acting actuator.

For the diaphragm actuator with rolling diaphragm (Figure 2.14a) in Equation 2.20, the effective areas are $A_1 = A_{ER}$ and $A_2 = A_{ER} - A_R$, where the diaphragm effective area A_{ER} is defined by Equation 2.5. In this case, the equation of motion for the diaphragm-load assembly is Equation 2.23.

For the actuator with bellows in Equation 2.20, the effective areas are $A_1 = A_{EB}$ and $A_2 = A_{EB} - A_R$, where the bellows effective area A_{EB} is defined by Equation 2.7. In this case, the equation of motion for the mass m can be expressed as:

$$m \cdot \ddot{x} + b_V \cdot \dot{x} + k_B \cdot x + F_F + F_L = P_1 \cdot A_1 - P_2 \cdot A_2 - P_A \cdot A_R \qquad (2.29)$$

where k_B is the bellows spring rate as determined by Equation 2.8.

Two degree-of-freedom pneumatic cylinder. The schematic diagram of this cylinder is shown in Figure 2.30; this construct includes a flat diaphragm that couples the piston and the rod. The equations of piston and rod-load assembly motion are written in different forms, depending on the flat diaphragm (flexible element) deflection. If the deflection is $\Delta L_D = x_P - x_R < \Delta L_m$, there are two separate equations of motion:

$$\begin{cases} m_P \cdot \ddot{x}_P + b_{VP} \cdot \dot{x}_P + k_D \cdot E_R + F_{FP} = A_P \cdot (P_1 - P_2) \\ m \cdot \ddot{x}_R + b_{VR} \cdot \dot{x}_R - k_D \cdot E_R + F_{FR} + F_L = A_D \cdot (P_1 - P_2) + A_R \cdot (P_2 - P_A) \end{cases} \qquad (2.30)$$

where F_{FP} and F_{FR} are the coulomb friction forces acting on the piston and rod, respectively, and are determined by Equation 2.24; m_P is the piston mass;

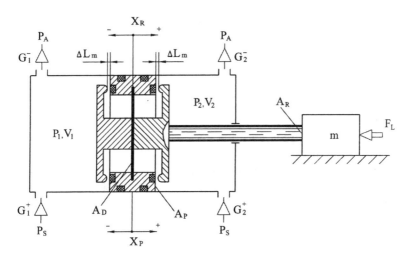

FIGURE 2.30
Schematic diagram of the two degree-of-freedom double-acting cylinder.

b_{VP} and b_{VR} are the viscous friction coefficients for the piston and rod, respectively; k_D is the diaphragm spring rate defined by Equation 2.2; and A_D is the effective area of the diaphragm, which is determined from Equation 2.1.

When the flexible element (diaphragm) deflection reaches ΔL_m, it contacts the fixed stop and the piston and rod begin to move together. In this situation, their motion is described by:

$$m_\Sigma \cdot \ddot{x}_R + b_{V\Sigma} \cdot \dot{x}_R + F_{F\Sigma} + F_L = A_\Sigma \cdot (P_1 - P_2) + A_R \cdot (P_2 - P_A) \quad (2.31)$$

where $m_\Sigma = m_p + m$, $b_{V\Sigma} = b_{VP} + b_{VR}$, $F_{F\Sigma} = F_{FP} + F_{FR}$, $A_\Sigma = A_P + A_D$.

After each impact between the piston and the rod, the initial velocity $\dot{x}_R^+$ of both elements at contact is determined by the well-known relationship describing inelastic impact:

$$\dot{x}_R^+ = \frac{m_p \cdot \dot{x}_p^- + m \cdot \dot{x}_R^-}{m_\Sigma} \quad (2.32)$$

where $\dot{x}_p^-$ and $\dot{x}_R^-$ are the piston and the rod velocities, respectively, before the impact.

Before the impact, when the piston and the rod move separately, the time derivative for the pressure in the cylinder working chambers is determined by:

$$\dot{P}_i = B_i \cdot \left[G_i^+ \cdot R \cdot T_S - G_i^- \cdot R \cdot T_S \mp P_i \cdot (A_p \cdot \dot{x}_p + A_{di} \cdot \dot{x}_R) \right] \quad (2.33)$$

where the mass flow rates G_i^+ and G_i^- are determined by Equation 2.21 and Equation 2.22, respectively; effective areas A_{di} can be defined as $A_{d1} = A_D$ and $A_{d2} = A_D - A_R$; and the coefficient B_i is:

$$B_i = \frac{1}{V_{0i} + 0.5 \cdot L_S \cdot (A_P + A_{di}) \pm (A_P \cdot x_P + A_{di} \cdot x_R)} \tag{2.34}$$

After each impact, the time derivative for the pressure in chambers is described by Equation 2.20.

Vane rotary actuator. For this type of pneumatic actuator (similar to the pneumatic cylinder), the time derivative for the pressure in the working chambers is obtained from:

$$\dot{P}_i = \frac{1}{V_i} \cdot (G_i^+ \cdot R \cdot T_S - G_i^- \cdot R \cdot T_S \mp P_i \cdot \dot{V}_i) \tag{2.35}$$

where (Figure 2.31) $V_i = V_{0i} + 0.5 \cdot L_V \cdot (R_V^2 - r_V^2) \cdot (0.5 \cdot \varphi_S \pm \varphi)$, φ_S is the full vane turning angle, φ is the vane angular position, L_V is the vine width, R_V is the vane outside radius, and r_V is the vane inside radius.

The equation of motion for the vane-load assembly can be expressed as:

$$J \cdot \ddot{\varphi} + b_\omega \cdot \dot{\varphi} + M_F + M_L = 0.5 \cdot L_V \cdot (R_V^2 - r_V^2) \cdot (P_1 - P_2) \tag{2.36}$$

where J is the reduced moment of inertia of the system actuator-load (not depending on the vane turning angle), b_ω is the viscous friction coefficient for the vane's rotary motion, M_F is the resistance torque that is defined similar to the coulomb friction force (see Equation 2.24), and M_L is the load torque.

Vane motor. In this actuator type (see Figure 2.32), the active (driving) torque depends on the pressure differential on the rotor vanes ($\Delta P_{AM} = P_i - P_{i-1}$, where P_i and P_{i-1} are the pressures in the adjoining working chambers). In general, the pressure and temperature-changing process can be described by Equation 2.15 and Equation 2.16, where it is necessary to place the $V(\varphi)$ and $\dot{V}(\varphi)$ instead of V and $\dot{V}$, respectively. For the approximate calculation, it is possible to use the following equation:

$$G_{M1,2}^+ \cdot R \cdot T_S - G_{M1,2}^- \cdot R \cdot T_S \mp P_{1,2} \cdot \dot{V}_M = 0 \tag{2.37}$$

Here, the volumetric flow $\dot{V}_M$ per revolution can be obtained as:

$$\dot{V}_M = \frac{V_{ME} \cdot \dot{\varphi}_M}{2\pi},$$

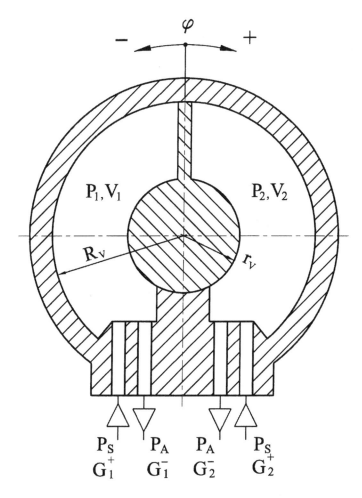

FIGURE 2.31
Schematic diagram of the vane rotary actuator.

where V_{ME} is the effective motor capacity per revolution, and $\dot{\varphi}_M$ is the speed of rotation. The mass flow rates $G_{M1,2}^+$ and $G_{M1,2}^-$ are determined by Equation 2.21 and Equation 2.22. The pressures P_1 and P_2 are determined from Equation 2.37 for given values of the supply pressure (P_S), the effective areas of the input and output nozzles (A_{M1}^+, A_{M1}^-, A_{M2}^+, and A_{M2}^-), the effective motor capacity per revolution (V_{ME}), and the speed of rotation ($\dot{\varphi}$). For the vane motor that provides rotation in either direction (Figure 2.25), the average active torque is

$$M_M = \frac{Z_V \cdot E_M}{2 \cdot \pi},$$

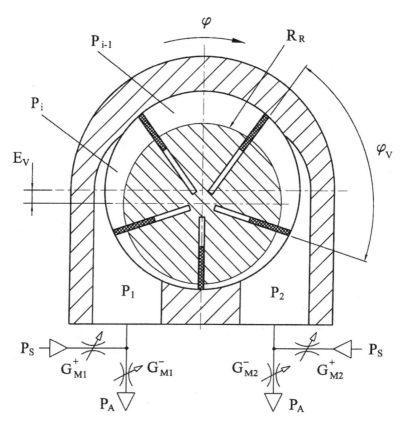

FIGURE 2.32
Schematic diagram of the vane motor.

where Z_V is the vane number, E_M is the work done per revolution by one vane that is determined as:

$$E_M = C_M \cdot L_V \cdot R_R \cdot (P_1 - P_2),$$

where the parameter C_M is:

$$C_M = \frac{1}{a_m}\left[\varphi_V\left(a_m + \frac{1}{2}\right) - 2(a_m + 1)Cos\left(\varphi_1 + \frac{\varphi_V}{2}\right)Sin\frac{\varphi_V}{2}\right.$$
$$\left. + \frac{1}{2}Cos\left(2\varphi_1 + \frac{\varphi_V}{2}\right)Sin\beta_m\right],$$

where L_V is the vane width, R_R is the outside rotor radius, $a_m = R_R/H_V$, H_V is the motor eccentricity, φ_V is the angle between two neighboring vanes (φ_V

$= 2\pi/Z_V$), and φ_1 is the angle that determines the end of the charging process. Finally, the volumetric flow per revolution can be determined from:

$$\dot{V}_M = \frac{C_M \cdot R_R^2 \cdot L_V \cdot Z_V \cdot \dot{\varphi}_M}{2\pi} \qquad (2.38)$$

and the average active torque is:

$$M_M = \frac{C_M \cdot R_R^2 \cdot L_V \cdot Z_V}{2\pi} \cdot (P_1 - P_2) \qquad (2.39)$$

The equation of motion for the vane motor can be expressed as:

$$J_M \cdot \ddot{\varphi} = M_M - M_F - M_L \qquad (2.40)$$

where J_M is the reduced moment of inertia of the system motor-load (not depending on the vane turning angle), M_M is the active torque determined from Equation 2.39, M_F is the resistance torque, and M_L is the load torque.

The vane motors have relatively high resistance torque owing to the friction force between the vane's outside surface and inside surface of the motor housing. Usually, this torque is defined by the following formula:

$$M_F = C_{\omega 1} \cdot \dot{\varphi}^2 \qquad (2.41)$$

However, the estimate calculation is often used in the simple form:

$$M_F = C_{\omega 2} \cdot \dot{\varphi} \qquad (2.42)$$

If the vane motor operates at low rotating speed, the air leakage between the neighboring working chambers is relatively high, which is the cause of the active torque decrease. This must be taken into consideration for the motor starting process and for its operation in low-speed rotating mode in positioning or speed motion control systems.

3

Electropneumatic Control Valves

In pneumatic actuating systems, the electropneumatic control valve is the key element. This device controls the air flow to the actuator according to the drive current or voltage from the control system. Thus, an electropneumatic servo valve is the power element (together with actuator), in which a small-amplitude, low-power electrical signal is used to provide a high response modulation in pneumatic power. The dynamic response, null shift, threshold, and hysteresis are the most critical valve parameters that strongly influence the dynamic and static characteristics of the pneumatic actuator.

There are many available designs for the electropneumatic control valves; the designs differ with respect to the geometry of the active orifice, the type of flow sealing (regulating) element (plug), the number of paths and ports, etc. Current construction of electropneumatic control valves can be categorized into the following three groups: (1) on/off solenoid valves, (2) proportional valves, and (3) servo valves.

On/off solenoid valves are normally used in the fully opened or fully closed position and not for throttling purposes. One method of classifying on/off solenoid valves is by the number of flow paths, or ways, within the valve in its various operating conditions. Others factors to consider are the number of individual ports, the number of flow paths for which the valve is designed, and the internal connection of the ports by the movable element. For naming the valve, the following strategy is usually used: if a valve has "n" ports and "m" positions, it is called an n/m valve.

Usually, on/off solenoid valves have two-position or three-position construction. In three-position valves, the two extreme positions are directly related to the actuator's direction of motion, first in one direction and then in the other, just as a two-position valve does (Figure 3.1a).

The center position of a three-position valve satisfies other system requirements (the center position also is commonly referred to as the neutral condition). The most common center conditions are blocked center (with all ports blocked), exhaust center (with both cylinder ports open to exhaust and pressure blocked), and pressure center (with both cylinder ports open to pressure).

The blocked-center condition blocks all working ports and is often called closed center (Figure 3.1b). Depending on the circuit design and cylinder

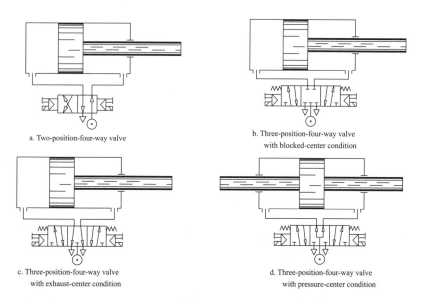

a. Two-position-four-way valve

b. Three-position-four-way valve
with blocked-center condition

c. Three-position-four-way valve
with exhaust-center condition

d. Three-position-four-way valve
with pressure-center condition

FIGURE 3.1

On/off solenoid valves: (a) two-position, four-way valve; (b) three-position, four-way valve with blocked-center condition; (c) three-position, four-way valve with exhaust-center condition; and (d) three-position, four-way valve with pressure-center condition.

loading conditions, this center condition can provide a holding action on the device to which the valve is connected. With suitable controls (taking into account the compressibility of air), this type of valve may stop a cylinder at intermediate points along its stroke as it travels in either direction.

With the exhaust-center condition, the double-acting cylinder is free to float with the valve in its neutral state because both ports of the cylinder are open to exhaust (Figure 3.1c). The cylinder rod can be moved manually (subject to internal friction force and external loading).

The pressure-center type of valve might be used both with the double-acting cylinder (Figure 3.1d) and with alternating the motion control of two or three groups of single-acting cylinders. When the valve is in its center (or neutral) position, pressure is directed to both cylinders, causing both to extend. If the valve is shifted to either extreme position, one cylinder group retracts while the other remains extended. When the valve is shifted in the opposite direction, the motion of the cylinder groups is reversed.

In pneumatic actuating systems, the electropneumatic control valves are usually used in one- or two-stage design. A single stage valve is a directly operated (direct-drive) valve. A two-stage valve consists of a pilot stage and a main stage. If in a one-stage on/off solenoid valve the actuator is a solenoid, force/torque motor, or piezoelectric actuator, then in a two-stage type valve, the second stage is moving by pilot pressure.

Traditionally, the term "servo valve" describes valves that use closed-loop control. These valves monitor and feed the main stage spool position back to the pilot stage or driver.

Proportional valves were developed as a less expensive option to servo valves when remote electronic control is required, but fast, accurate control is not critical. Most proportional valve applications are open-loop control, which provides less control over the actuator; however, the design of some proportional valves provides a performance level close to that of a servo valve. The term "proportional" means that the output is directly related to the input. The basic principle behind the operation of a proportional valve is the use of a specially wound solenoid designed to give a varying output force directly proportional to the input electrical signal. The desired output is usually obtained by balancing the proportional solenoid force against an opposing spring. The spring characteristics are calibrated to the solenoid characteristics to give accurate control. In this way, for example, the spool position in a directional control valve can be accurately controlled for metering purposes.

In practice, the proportional pressure control valves, proportional flow control valves, and proportional direction and flow control valves have wide-ranging applications. It is also possible to have different settings for different functions in one cycle with one proportional valve.

The most important decision (from the point of view of economy of operation) is matching the valve's flow capacity to the application requirements. In the past, the common, unscientific practice used in selecting or sizing a valve was simply to match the port size of an actuator to the valve. This method is acceptable if energy conservation is ignored. Today, however, it is no longer even suggested as a way to select an air-valve size. This is because technological advancements now allow physically smaller valves to pass greatly increased flow, and port size has become even less significant.

Smaller valves tend to have several generic advantages over larger valves. In general, these include:

- Faster shifting
- Less leakage
- Lower power consumption (because smaller solenoids can move lighter internal parts)
- Mounting flexibility (because valve footprints are smaller)
- Reduced cost

With these advantages, smaller valves will save money initially and then continue to save money because of decreased operating expenses.

3.1 Electromechanical Valve Transducers

The electromechanical valve transducer provides a means of converting an electrical input into a mechanical output. In general, the electromechanical transducers can be categorized into the following groups (depending on the principle of operation): electromagnetic, electrodynamic, piezoelectric, magnetostrictive, magnetorheologic, electrostatic, and others. However, according to the technical resources and performance attributes, the electromagnetic transducers with mobile armatures are the most acceptable. Recently, in the one-stage valves and in the pilot stage of the two-stage valves, piezoelectric mechanical transducers are also used.

Electromagnetic transducers are divided into two groups:

1. Solenoid actuators (linear type)
2. Torque motor (rotary type)

Solenoid actuators can be classified into two main groups with respect to the desired actuation task:

1. On/off (switching) solenoids that are typically used to realize an on/off behavior. The force and displacement characteristics are strongly nonlinear.
2. Proportional solenoids that are used, if mechanical characteristics (force or displacement) are required, which are proportional to an electrical signal (current or voltage). To obtain good controllability, the force or displacement characteristic is almost linear.

While on/off solenoids are relatively cheap and simple, proportional solenoids normally require integrated electronics for position sensing and controlling in order to compensate for mechanical and magnetic hysteresis and nonlinearities.

Figure 3.2 presents a schematic diagram of the on/off solenoid actuator. It consists of housing (1), armature (2), stem (3), sleeve bearing (4), coil (5), and spring (6). The coil (5) is connected to an external power supply (not shown in Figure 3.2). The spring (6) rests on the armature (2) to force it. The armature (3) moves inside the coil (5) and transmits its motion through the stem (3) to the valve. When current flows through the coil, a magnetic field forms around the coil. This magnetic field attracts the armature (2) toward the center of the coil (5). As the armature (2) moves upward, the spring (6) collapses and the valve opens.

The stem (3) is used for external attachment to the solenoid. It is normally made of nonmagnetic stainless steel (e.g., AISI 303). The sleeve bearing (4) provides a guide for the stem (3) and is usually made of phosphor bronze (e.g., UNS C54400). It can be made of other materials for different applications requiring longer life. The magnetic static and moving components of a solenoid are typically made from low carbon steel (e.g., AISI 1006).

A major objective in the design of a solenoid is to provide an iron path capable of transmitting maximum magnetic flux density with a minimum energy input. Another object is to obtain the best relationship between the

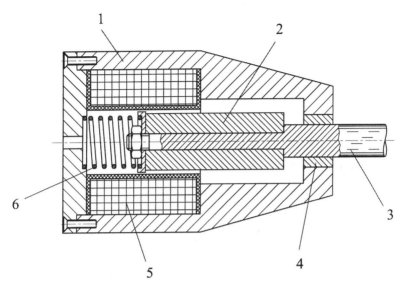

FIGURE 3.2
On/off solenoid actuator.

variable ampere turns and the working flux density in the air gap between the coil shell and the armature. When applying a solenoid, it is extremely important to consider the effects of heat because in a constant voltage application, an increase in coil temperature will reduce the work output. Ambient temperature rises all affect the net output force.

Single on/off solenoid valves are termed "normally opened" or "normally closed," depending on the position of the valve with the solenoid deenergized. A normally opened solenoid valve is opened by spring force and closed by energizing the solenoid. A normally closed solenoid valve is closed by spring force and opened by energizing the solenoid. A double solenoid valve is typically termed "as is; that is, the valve position does not change when both solenoids are de-energized.

The key to the operation of a proportional valve is establishing a balance between the forces in action on the armature. These balanced forces include a mechanical force provided by a spring specially developed for proportional valves and a magnetic force created by the current level passing through the coil. The spring force is proportionally opposed by the magnetic force.

The major differences between an on/off solenoid actuator and a proportional solenoid actuator include:

- The centering springs in proportional solenoids are stronger than the centering springs used in on/off constructions.
- Proportional actuators are engineered to produce more force than on/off solenoids.
- Proportional designs always use DC solenoids.

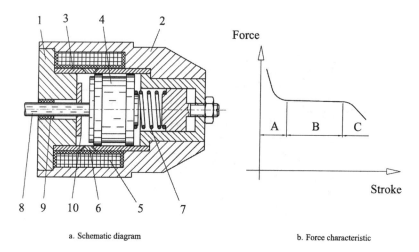

a. Schematic diagram   b. Force characteristic

FIGURE 3.3
Proportional solenoid actuator: (a) schematic diagram and (b) force characteristic.

Figure 3.3a is a schematic diagram of the proportional solenoid actuator..
It consists of the pole piece (1), housing (2), core tube (3), and armature (4),
which are made of magnetic materials. The core ring (6) is made from
nonmagnetic material in order to concentrate the magnetic field into the gap
between the armature (4) and pole piece (1). Usually, the coil (5) is encapsu-
lated in glass-filled polyester. A nonferrous washer (10) prevents the armature (4) from sticking to the pole piece (1) when the solenoid is deenergized.
The stem (8) is attached to the armature (4). The spring (7) ensures that the
stem (8) is in contact with the valve spool (not shown). The stem bearing
(9) provides a guide for the stem (8).

Regarding the working functions of proportional solenoids, they can be
divided into two groups:

1. Stroke proportional solenoid actuators
2. Force proportional solenoid actuators

Actuators in the first group give displacement of the stem (8) proportional
to the current through the solenoid coil (5). In such a construction, the stem
(8) has a long stroke (about 5 mm), the spring rate of the spring (7) is low,
and the active magnetic force for the armature motion is also low. In general,
this solenoid actuator type is used in one-stage directional and flow control
valves.

The second group of actuators gives variable force to hold the stem (8).
Force will be proportional to the current passing through solenoid. In this case,
the armature has a short motion stroke (less than 2 mm) and the spring (7)
has a high spring rate. These solenoid actuators are used in two-stage valves
(as the pilot stage) and in proportional relief valves.[76]

There are three specific regions in the typical force-stroke characteristic (force as a function of displacement of the armature) of the proportional solenoid actuator (Figure 3.3b). As depicted in the figure, the operating characteristic of the proportional actuators provides a linear relationship between force and length of stroke over a wide range "B" of displacement. Usually, there is a working area of the proportional actuator. For the displacement range "A," the solenoid force is increased if the distance between the armature and the pole piece decreases. To eliminate this nonlinear region from the solenoid working range, the movement of the armature (4) to the left relative to the pole piece (1) is limited by the engagement of the surface of the armature with the nonferrous washer (10). In general, the area "C" of the force-stroke curves to meet the solenoid idling.

The response time of the proportional actuator as well as the turn-on threshold are a function of the amount of force produced by the device. The amount of force that can be generated by proportional actuators of this type is related to the coaxial diameters of the magnetic pole and the armature, the number of turns of solenoid coil, and the current that is applied to the solenoid coil. The size of the solenoid coil generally determines the dimensions of the device because the solenoid coil is wound on the magnetic pole. Thus, methods of maximizing the force generated by such devices are usually directed toward optimizing the magnetic circuit of the device.[39]

Accuracy and repeatability of the proportional solenoid actuator are a function of the spring's symmetry and the ability of the design to minimize nonlinear effects of spring and machining-tolerance variations.

One problem associated with known proportional solenoid actuators is their hysteresis phenomenon (hysteresis is the difference in actuator input currents required to produce the same actuator output as the actuator slowly cycles between plus and minus rated current). The total hysteresis value has two components; the first one is the magnetic reversal of the magnetic solenoid parts and the second is connected with friction force in the mobile parts.

A standard practice for reducing hysteresis has been to physically support the solenoid's movable armature within the bore of its surrounding drive coil by means of low-friction bearings, such as Teflon rings. In addition, the housing, the armature, and sometimes the bearings arranged between the armature and the housing all must be machined accurately for the sliding axial fit of the armature within the housing to have minimal friction. Diameters and the concentricity of bearing surfaces must be highly accurate to minimize friction, and the location and shape of bearing surfaces must be considered to minimize the effects of side or off-axis loading. These needs have necessitated that all contacting surfaces be accurately machined, and errors that inevitably occur in attempting to accomplish this adversely affect solenoid performance. However, even with the use of such a low-friction material and a high machining accuracy, there is still significant "dead band" current, which limits the operational precision of the valve.

Another solution for dealing with this physical contact-created hysteresis problem is to remove the armature support mechanism from within the bore of the solenoid coil (where the unwanted friction of the armature support bearings is encountered) to an end portion of the coil, and to support the armature for movement within the bore by means of a spring mechanism located outside the solenoid coil.

Sometimes, the improvement of the hysteresis characteristic can be achieved using the dither, that is, the armature oscillations at high frequency exist to overcome static friction. In this case, low amplitude (about 10 to 15% from the maximum value of the control signal), a relatively high-frequency (about 10 to 15 times more than the solenoid's natural frequency) periodic signal is superimposed on the input current signal.[48,63,194]

Figure 3.4 depicts another design of the proportional linear actuator. The "linear force motor" is a permanent magnet differential motor that usually operates together with a position transducer and acts in PWM (pulse width modulation) mode. This construction consists of two flanges (1 and 2), armature (3), stem (4), two bearings (5 and 6), coil (7), two centering springs (8 and 9), and two permanent magnets (10 and 11). Usually, the coil (7) is connected to the integrated electronics of the proportional valve (not shown).[48] The permanent magnets provide part of the required magnetic force. For the linear force motor, the current needed is considerably lower than what would be required for a comparable proportional solenoid. This motor has a neutral midposition from which it generates force and stroke in both directions. Force and stroke are proportional to current of the control signal.

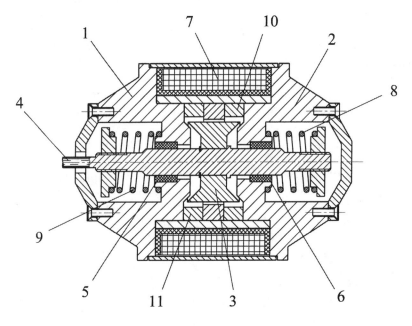

FIGURE 3.4
Linear force motor.

High spring stiffness and the resulting centering force plus external forces (air-flow forces, friction forces) must be overcome during out-stroking. During back-stroking to the center position, the spring force adds to the motor force and provides additional spool driving force, which makes the valve much less sensitive to contamination. The linear force motor requires very low current in the spring-centered position.

Torque motors serve as electromechanical transducers for either proportional displacement, force output, or a combination of the two. They can also be operated as on/off actuators. These virtually frictionless devices provide smooth, stepless variations of output in response to low-power level (less than two watts) electrical current input signals. The term "torque" refers to the armature rotational motion around its pivot point, resulting from electrical and magnetic forces. This torque is instrumental in pneumatic valve electrical-to-mechanical power transfer.

Depending on the amount of torque and stroke needed, there are two basic designs: single-coil or double-coil arrangement. Both the single-coil and double-coil units can operate in a bidirectional manner by simply reversing the current.

Double-coil torque motors have wide practical application in two-stage control valves. Typically, such a motor (Figure 3.5) is constructed from the casing (1), two permanent magnets (2 and 3), an armature (4), two coils (5 and 6), pole pieces (7 and 8), and base (9). A movable armature (4) is mounted on the frictionless spring (10) and rotates around the pivot point (12). The tip (11) is connected to the armature (4). The double-coil motor can be independently operated or connected for series aiding or parallel aiding.

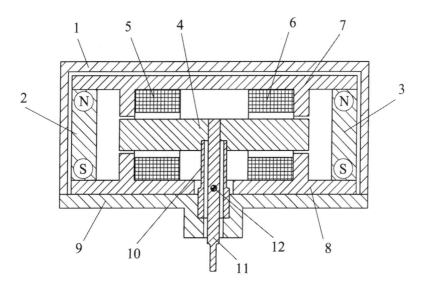

FIGURE 3.5
Dual-coil torque motor.

The torque motor has an armature mounted on a torsion pivot spring and suspended in the air gaps of a magnetic field. The two pole pieces, one polarized north and the other south by the permanent magnets, form the framework around the armature and provide pathways for magnetic flux flow. When current flows through the coils, the armature becomes polarized and each end is attracted to one pole piece and repelled by the other. The torque exerted on the armature is restrained by the torsion spring upon which the armature is mounted. This torsion spring makes armature output motion proportional to input current.

The permanent magnets are horizontally magnetized and consist of a material of high magnetic remanence, preferably neodymium-iron or samarium-cobalt. The first-mentioned alloy, which has a high iron content, can be produced in an inexpensive way and has an especially high magnetic remanence. As a consequence, a permanent magnet of small dimension can be used in the improved torque motor, resulting in low cost and saving of space.

The rotational torque is created directly proportional to the amount of polarization or magnetic charge of the armature — increased armature polarization creates a higher force attraction to the pole pieces. Because the amount of polarization of the armature is proportional to the magnetic flux created by the current through the coils, the torque output of the torque motor is proportional to the coil input current. The magnetic flux created by the coils depends on two factors: (1) the number of coil wire turns and (2) the strength of current that is applied. That is, the torque of the motor depends on the ampere-turns applied.

Single-coil torque motors are basically used as electromechanical transducers in pilot valves, which include a nozzle-flapper construction. This torque motor is shown in Figure 3.6 and consists of the casing (1), a permanent magnet (2), an armature (3), a coil (4), the pole piece (5), and the base (6). Just as in the double-coil motor, the armature (3) is mounted onto the spring (7), and the tip (8) is connected to the armature (3) and rotated around pivot point (9). The stiffness of the suspension spring in conjunction with the magnetic spring characteristic establishes the net spring rate of the torque motor.

The major advantage of the suspension design of constructions illustrated above is its incorporation of a semistatic "O"-ring at the pivot point. This allows dry motor operation and facilitates the additional damping force.

As the solenoid actuates, the torque motors can operate in three modes:

1. Force control, in which the device has a low net spring rate and the force output is essentially proportional to input current
2. Displacement control, in which the torque motor has a high net spring rate and the displacement output is in close proportion to the input current
3. On/off control, in which the unit has a very low or no suspension spring rate so the negative magnetic spring rate results in the desired bi-stable operation

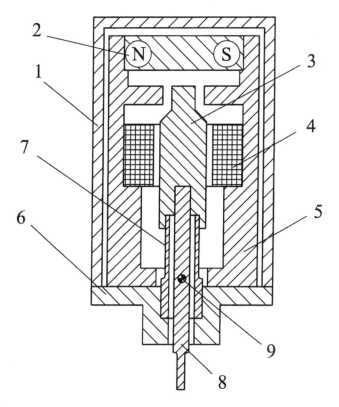

FIGURE 3.6
Single-coil torque motor.

One of the disadvantages of electromagnetic actuators (solenoid actuators or torque motors) is the limit of the force and bandwidth for a given device size. High energy density piezoelectric materials present a possible alternative to electromagnetic actuation to further improve the response time of control valves.[50]

Two main types of piezo actuators are available:

1. Low-voltage (multilayer) devices requiring 100 volts or less for full displacement
2. High-voltage devices requiring about 1000 volts for full extension

Lead/zirconium/titanium-based ceramic materials are most often used today. The maximum electrical field that the piezo ceramics can withstand is on the order of 1 to 2 kV/mm. To keep the operating voltage within practical limits, piezo actuators consist of thin layers of electroactive ceramic material electrically connected in parallel. The total displacement is the sum of the displacements of the individual layers. The thickness of the individual layer determines the maximum operating voltage for the actuator.

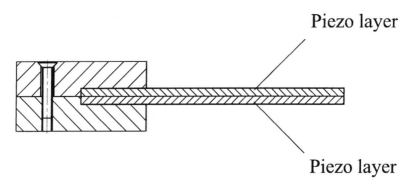

FIGURE 3.7
Bender type actuator.

High-voltage piezo actuators consist of bulk ceramic disks, which are 0.4 to 1 mm thick and glued together to form a stack. Low-voltage piezo actuators are manufactured in a lamination process, in which thick-film electrodes are printed on ceramic foils. The layers of ceramics and electrodes are then pressed together and co-fired to form a monolithic block. Typical layer thicknesses are in the range of 20 to 100 μm. After cutting the individual stacks to size, wire leads are applied.

In servo valve constructions, two types of piezo actuators are basically used: (1) bender type actuators (bimorph and multimorph design) and (2) stack type actuators.

Piezo bimorph actuators (Figure 3.7) bend up and down with applied voltage. One piezo layer of the actuators is driven with a positive voltage (with the "poling" direction), causing it to contract along the linear axis, while the other piezo layer is driven with a negative voltage (against the poling direction), causing it to extend along the linear axis. This extension and contraction work together to cause the device to bend in one direction. As the polarity of the voltage is reversed, the device bends in the other direction. Monolithic multilayer-type benders are also available. Bender type actuators provide large motion (up to 1 mm) in a small package at the expense of stiffness, force, and speed. In general, the bender type of piezo actuators is used in nozzle-flapper valve construction.

An electromechanical transducer with a stack type piezo actuator is shown in Figure 3.7. The active part of the positioning element consists of a stack of ceramic disks separated by thin metallic electrodes. However, like other stiff smart materials, piezo electrics do not produce large strokes. They are typically capable of only 0.1% strain, perhaps 0.15% for certain compositions. To increase piezo actuator displacement, a lever motion amplifier is usually used. Such a device (Figure 3.8) has the stack of the piezo ceramic disks (1), the top piece (2), preloading springs (3), the case (4), fastening screw (5), the base part (6), the flexure part (7), the lever part (8), an adjustable screw (9) with nut (10), a return spring (11), and the tip (12).

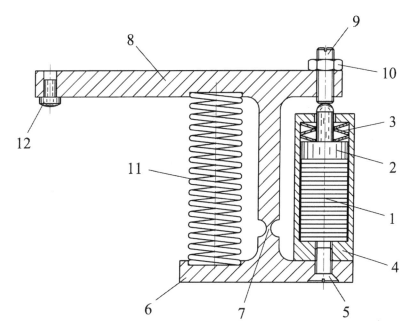

FIGURE 3.8
Electromechanical transducer with stack piezo-actuator and lever motion amplifier.

Piezo stack actuators are extremely sensitive to tensile loads. Therefore, for dynamic applications, these actuators must be preloaded to ensure that the actuators are never placed in tension. Usually, to protect the piezo ceramic against external influences, a metal case (4) is often placed around it. This case may also contain preloading springs (3) to compress the ceramic to allow both pull and push operation. Applying a preload in this manner slightly decreases the actuator available stroke because the stack has to work against the preload mechanism. To minimize the reduction in stroke, the stiffness of the preload mechanism should be much less than the stiffness of the stack.

The electromechanical transducer (Figure 3.8) has a level motion amplifier, which is integrated into the system. To maintain high resolution with the increased displacement range, the amplifier must be stiff, and backlash- and friction-free, which means ball or roller bearings cannot be used. In this case, a flexure guiding system is typically used. Piezo stack actuators with lever amplifiers have compact size compared with stack actuators with equal displacement; however, the increase in travel gained with a mechanical amplifier reduces the system stiffness and maximum operating frequency.

Usually, the lever amplifier transmission ratio changes from 2 to 20. If the amplifier transmission ratio is more than factor of 20, a hydraulic amplifier can be utilized. Figure 3.9 provides a schematic diagram of such an electro-mechanical transducer. It consists of the piezo stack actuator (1) with

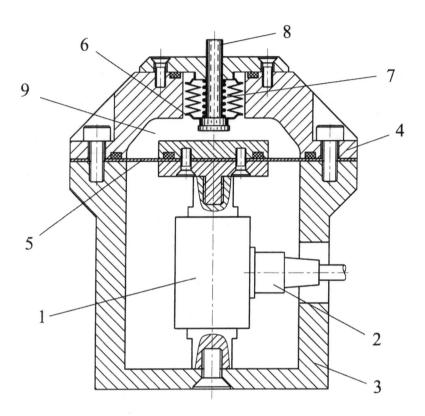

FIGURE 3.9
Electromechanical transducer with stack piezo-actuator and hydraulic amplifier.

connector (2), housing (3), cover (4), flat diaphragm (5), bellows (6), return spring (7), and tip (8). The cavity (9) is filled with mineral oil. In this case, the amplifier transmission ratio is $n_A = A_{ED}/A_{EB}$, where A_{ED} is the effective area of the flat diaphragm (5) and A_{EB} is the effective area of the bellows (6).

Integrating the piezo actuators into a standard pilot valve is an alternative to solenoid operation for most standard valves. Such systems provide significant application advantages, including:

- Fast operation: the reaction time is typically less than 2 ms
- No requirement for the circuit protection normally associated with solenoid valves
- With a low power rating (usually less than 0.1 W), piezo valves are inherently, intrinsically safe
- Switching, proportional, and pulse width modulation control modes are all available

Another advantage related to the use of piezo technology is that it eliminates the problem of air leakage in the steady-state condition. The pneumatic positioning system is efficient in its air consumption, with the control system flowing measured volumes of air during the positioning operation only. Once in position, the air flow is totally closed off, which eliminates the air waste associated with designs that require a constant air leak.

3.2   One-Stage Valves

One-stage control valves are classified according to flow-restricting elements and can be divided into three main groups[29]:

1. Seating types
2. Sliding types
3. Flow-dividing elements

In seating type valves, the air stream is pinched off by a pair of opposing blunt edges. Three main types of the seating control valves are shown in Figure 3.10. There are valves with a ball control (sealing) element (Figure 3.10a), with conical flow-controlling members (Figure 3.10b), and poppet types (Figure 3.10c). Usually, these valves are suited for applications that require high flow because they open relatively large ports with short strokes; in addition, they have an inherent characteristic of fast response with minimum wear. On the other hand, the seating type valve has several disadvantages, which have prevented its general use in servo applications:

- It is difficult to balance completely against static pressure, which usually requires large and somewhat unpredictable operating forces.
- Its flow-versus-displacement curve tends to be nonlinear at small openings.
- It has a small but sometimes significant time lag that might handicap it for high-speed applications.

In these valves, resilient seals shut off flow paths tightly and help absorb the kinetic energy of the moving members (for example, see Figure 3.10c). The design resists damage from foreign matter carried through the air line because the seats are self-cleaning. These valves are readily maintained with inexpensive parts. Their performance is not sensitive to air-line lubricants or to other materials carried in the air stream.

In the general way, an area of the seat base surface should provide the unit pressure not more than the maximum safe pressure. Usually, the plugs or sealing elements (ball, conical flow-controlling member, or poppet) are made from a material that has a surface hardness greater than the seat material hardness. In practice, the plugs are made from stainless steel (400

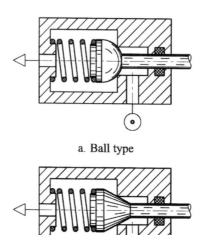

a. Ball type

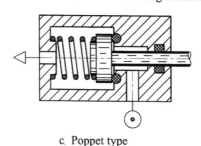

b. With conical flow-controlling member

c. Poppet type

FIGURE 3.10
Seating type one-stage valves: (a) ball type, (b) with conical flow-controlling member, and (c) poppet type.

series stainless steel according to AISI). Brass, aluminum, or 300 series stainless steel (AISI classification) is generally used for seat parts. For stainless steel, the pressure on the sealing surface between the sealing element and the seat should not be more than 80 MPa. If the seat is made from brass, this pressure parameter should not be more than 30 MPa, and for aluminum it should not be more than 12 MPa. In the case where resilient seals are used (e.g., "O"-ring construction, see Figure 3.10c), the unit pressure on the sealing surface should not be more than 2 MPa.

In many cases, the plug of the servo valves has the force-balancing mechanism. In its "balanced" form (the balance is usually far from complete), although the valve may not leak, balance pistons are usually used. Its flow-vs.-displacement and its force-vs.-displacement curves are nonlinear at small openings. It is subject to a peculiar kind of unsymmetrical "stiction," which

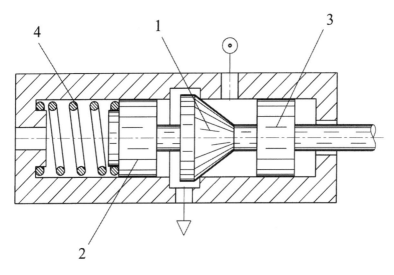

FIGURE 3.11
Seating type one-stage valve with force balancing mechanism.

in effect introduces a highly variable lag in operation. This last effect is negligibly small in most applications but may be serious if very fast operation is required and if large operating forces are not available. Attempts have been made to use ganged seating type valves in high-performance servo applications, but these have seldom if ever been successful. For shutoff applications, however, it is the most generally useful type.

Figure 3.11 provides a schematic diagram of a seating type valve with balancing mechanism. In this construct, the plug (1) has two pistons (2 and 3) that carry out the force balancing of the pressurized air influence. The diameter of the piston should be equal to the diameter of the seat sealing surface; thus, the plug is pressed to the seat by the force of the spring (4) only.

To minimize circuit problems with seating-type valves, consider the following points:

- Seating valve designs with direct-operated actuators may be bulky in large flow-port sizes.
- It is difficult to obtain some flow-path configurations with these valves. Also, their design inherently allows flow from the pressure port to escape to atmosphere as the valve sealing element shifts.
- Crossover flow may drop in pressure at the valve's inlet port below the minimum operating level, causing the valve to malfunction.

In sliding-type valves, the air flow is controlled by an element that slides across the flow paths. The sliding flow-restricting element can be defined as one in which the geometry of the flow-controlling members is essentially that of a pair of shear blades. The air flows between a pair of sharp edges and the width of the opening can be varied from a maximum down to zero. Ideally, the geometry of the opening is independent of the position of the sealing (moving) element, but in practice the necessity of providing a finite working clearance prevents the attainment of this ideal at very small openings.

There are four basic types of sliding valves: (1) spool valves, (2) plate valves, (3) suspension valves, and (4) rotary-plug valves. The sliding spool, plate, or rotary disk uses pressure unbalance to force the sealing mechanism against a mating surface. The effect is to control the flow of air to and from desired ports and seal the flow from others. This design can provide two-, three-, or four-way action.

In practice, spool-type valves are generally used not only in one-stage constructions, but also as the main stage in two-stage designs. There are three main options of the spool valve constructions: (1) metal-to-metal lapped spool designs, (2) packed-spool types, and (3) packed-bore valves.

The schematic diagram of the lapped spool valve is shown in Figure 3.12. Basically, this design contains the body, sleeve, and spool. Sealing between the sleeve and body is usually carried out by tension fitting or by the "O"-ring technique, as shown in Figure 3.12. Operating characteristics of these valves reveal that this design can provide almost any flow-path pattern desired in most porting and actuating configurations. External forces required to shift a balanced spool are low, and that is important when direct actuation of the spool is required. The force needed to position the spool tends to remain constant during a shifting stroke. This helps the spool complete its stroke once stiction (stick-slip operation) has been overcome.

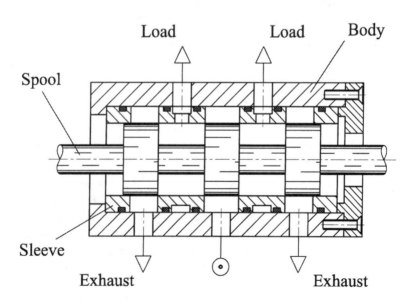

FIGURE 3.12
Metal-to-metal lapped spool valve.

The lapped spool design can be configured to prevent interconnection of pressure, outlet, and exhaust ports while the valve is shifted. The elimination of crossover flow while the spool is in transit reduces the chance of shift failure when operating at low pressure as a pilot-operated valve, and can eliminate spurious signal pulses. Because the spool is balanced, sudden pressure surges resulting from external forces on cylinders cannot cause the valve to lose its sealing capabilities.

Other factors must be considered to minimize potential problems when designing with metal-to-metal lapped spool valves. Long-stroke requirements for spools may call for excessive travel of electromechanical transducer to complete the shifting of the valve. Closely fitted parts are vulnerable to the ingress of foreign matter between mating parts, which can produce rapid wear and leakage, or cause the parts to stick.

Lapped spool valves require good filtration and consistent lubrication (it is often best to run on filtered, dry, unlubricated air). An oxidized airborne lubricant from a compressor or other material carried down an air line may cause the closely fitted parts to varnish in place. This is more likely to occur when valves must remain idle for long periods of time. The initial cost is higher and maintenance costs may be as well. Improper torque during installation may cause the valve spool to fail to shift due to sleeve distortion. If the valve is subjected to vibration, detent mechanisms may be necessary.

In the valve assembly process, the clearances between the sleeve and spool are usually controlled by individually matching each spool and sleeve to achieve a proper floating air bearing. The clearance has a narrow range between the minimum and maximum that every valve must achieve. The thermal expansion coefficients of the sleeve and spool are the same, so the clearance stays the same over a wide range of temperatures.

The packed-spool and packed-bore type valves are presented in Figure 3.13 and Figure 3.14, respectively.

Packed-spool valves offer the circuit designer several characteristics:

- The design maintains the seal shape and size for a long time. Packed-spool valves are less affected by improper torque during installation.
- The valve's spool is balanced and this construction is relatively insensitive to contamination.
- Maintenance is less costly from the standpoint of parts replacement.

Packed-bore valves offer the circuit designer the following characteristics:

- The valve is available in a variety of flow-path patterns in most porting and actuating configurations.
- The design results in a balanced spool, so shifting forces are only slightly greater than in metal-to-metal designs.
- Sudden pressure surges should not cause the valve to lose its sealing capabilities.

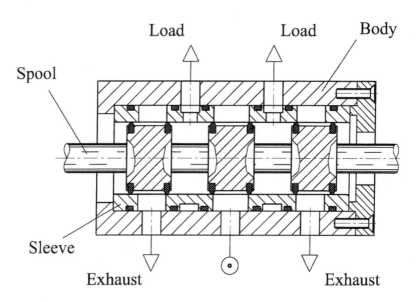

FIGURE 3.13
Resilient sealing packed-spool valve.

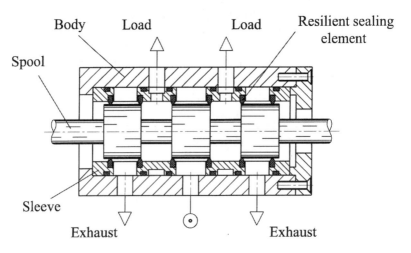

FIGURE 3.14
Resilient sealing packed-bore valve.

- The resilient seal makes the valve less vulnerable to abrasion by foreign material than in metal-to-metal designs.
- Limited contact areas reduce any tendency for the spool to varnish in place.
- The spool will not bind under the ambient temperatures found in most industrial applications. The seal can be changed without changing mating parts.

Other considerations are important to minimize circuit problems with packed-spool and packed-bore valves. The compatibility of seals and air-borne liquid contamination must be checked. Swollen seals require excessive shifting force on the spool; shrinking seals will increase leakage. Also, extreme temperatures may cause the seals to change size or harden. Lubrication of the air supply is suggested for longer valve life.

It is necessary to note that the spool valve at the present time is widely used for the continuous control of servo pneumatic actuators. For precise applications, the spool valve must be very accurately made because it depends on closeness of fit between the spool and sleeve to hold the leakage at an acceptable level.

Spool valves are classified into three types on the basis of the relationship between the land width and the port width:

1. Over-lap design (called closed-center), in which the spool land width is greater than the sleeve port width with the spool at null position where the air flow is blocked (Figure 3.15a)

2. Zero-lap type, where the spool land width is equal to the sleeve port width (Figure 3.15b)

3. Under-lap option (also called open-center), in which the spool land width is less than the sleeve port width (Figure 3.15c)

The spool valves often have an over-lap or "dead band" configuration. The presence of the dead band causes nonlinearity in the response of the system, for which the motion controller must compensate. This dead band is not critical for velocity control systems but can cause errors and instability in positioning systems. Although the size of the over-lap is small (about 10 to 15% of maximum spool movement for the proportional valve and 6 to 12% of maximum spool stroke for the servo valve), it does take time for the spool to move across it to open valve ports. After a solenoid-operated spool valve receives a signal to start an actuator, there is no flow until the spool shifts

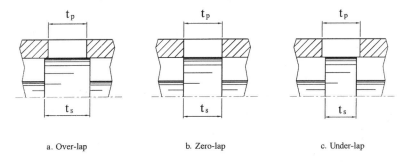

a. Over-lap b. Zero-lap c. Under-lap

FIGURE 3.15
Sliding spool valve lap:(a) over-lap, (b) zero-lap, and (c) under-lap.

through its over-lap. This time is short but adds to the overall cycle each valve shift that is very important for fast and accurate actuators.

Spool valves with zero-laps are usually used in servo valves as tight tolerances are required, and they are often necessary for optimum performance.

An under-lap configuration of the spool valves has high leakage at the central position of the spool, and hence reduced stiffness in closed-loop position actuating systems. In this case, the positioning error is more sensitive to load changes but can be compensated for with electronics. Spool valves with under-lap have higher damping in the central region, which is beneficial in terms of stability, particularly for positioning systems.

The material for making the spools and sleeves should exhibit high abrasion hardness. To double the lifetime, the working surfaces of the spool and sleeve are plated with chromium. Usually, the edges of the spool lands and port holes of the sleeve have a sharp form that allows for valve reliability to increase (large dirt particles are disrupted).

A reduction in spool valve friction force can be achieved using dither, but in this case the relative motion of the spool and the sleeve leads to intensive wear and heat generation on the contact surfaces, which may be the cause of the change (in these points) of the material structure with the loss of hardness.

The main disadvantage of the sliding spool valve is the complexity of the manufacturing process; it is difficult to achieve a high accuracy for the spool and fitted sleeve, and it is also very difficult to measure. From this point of view, the sliding plate valve is preferable to the spool valve. One type of plate valve is shown in Figure 3.16. In this construct, the flat sealing element (1) slides on the surface of the plate (2). The sealing element (1) is pressed against this surface by a spring (3) and by supplied air pressure that is connected to one of the load ports according to the position of the moving part (4) installed within the housing (5). Sliding plate valves can operate for many millions of cycles, even under adverse conditions, because they "wear–in" during use. Wear tends to keep mating parts in contact, thus controlling leakage over long periods of use. Temperature extremes do not seriously affect the valve. In addition to ease of manufacture, the plate valve has several other important advantages, such as the possibility of repair and renewal after wear or damage in service and significant freedom for design modifications. In addition, the plate valve can be more easily and completely force-compensated than the spool valve. The main disadvantage of the plate valve is a high friction force between the sealing element (1) and plate (2); therefore, these devices are mostly used in on/off sliding valve designs.

The suspension valve (see the schematic diagram in Figure 3.17) is one modification of the plate valve. In this construct, the moving plate (1) with two precise plugs (2 and 3) is installed on the two flexure levers (4). The static plate (5) has two holes for load connection, and pressurized air is supplied via the channel (6). Here, the proper clearance between the plugs and static plate is maintained by mechanical suspension, and there is never

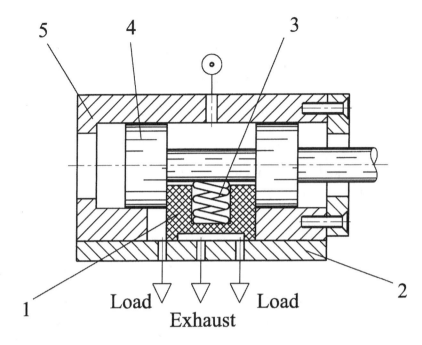

FIGURE 3.16
Sliding plate valve.

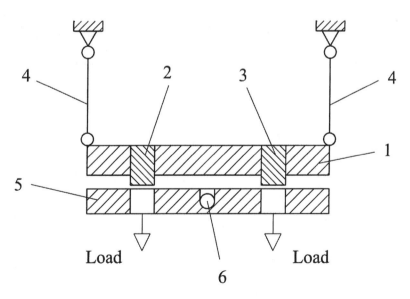

FIGURE 3.17
Suspension valve.

any metal-to-metal contact. This feature ensures that friction will be negligibly small (in the absence of large dirt particles, which would cause trouble in any closely fitted mechanism). The use of the hole-and-plug technique provides close dimensional tolerances at minimum expense, and the valve can be economically made in very small sizes where spool construction is prohibitively expensive and the spools work poorly. The main disadvantage of suspension valves is changing the clearance between the plugs and static plate as a function of the moving plate stroke. In addition, this gap can be changed as a result of the supply pressure changing and also the changing of the pressure in the actuator working chambers (load pressure). Sometimes, the clearance value in the end position of the moving plate can be two or three times greater than one in its null (middle) position. This is the cause of increasing the pressurized air leakage and nonlinear flow characteristics of the valve.

Another basic construct of the sliding valve type is the rotary-plug valve. When properly constructed and lubricated, and particularly when the mating surfaces are conical, it gives a very tight shut-off but requires relatively large operating forces. In pneumatic actuating systems, rotary valves with cylindrical plugs are commonly used. Utilization of needle bearings (1) allows one to achieve ultralow friction forces and small radial gaps (about 3 to 5 microns) between the rotary plug (2) and static sleeve (3) (Figure 3.18). Usually, both the rotary plug and valve sleeve are made of stainless steel, heat treated to 58–62 HRC hardness (e.g., AISI 440). This design exhibits minimal leakage of pressurized air and can be successfully used in pneumatic positioning systems.

Composite materials with high antifriction and wear-resistant performance can be utilized in rotary-plug valve construction. The schematic diagram of such a valve with inserts from composite material (e.g., graphite-filled Teflon) is shown in Figure 3.19. Here, two cross sections of one rotary-plug

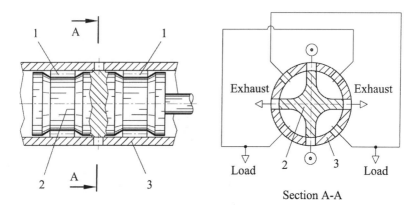

Section A-A

FIGURE 3.18
Rotary-plug valve with needle bearing.

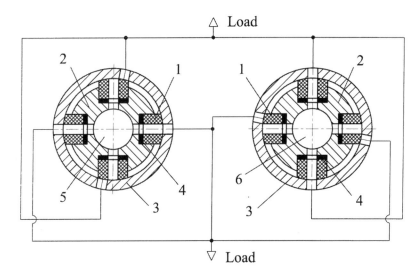

FIGURE 3.19
Rotary-plug valve with inserts.

valve are shown in which the inserts (1) are the sealing members. These inserts are installed within grooves of the plug (2) and are pressed to the inside surface of the sleeve (3) by resilient shims (4). The plug (2) comprises two separate chambers: one of them is connected to the supply line (for example, chamber (5), and the second is connected to the exhaust port, in this case the chamber (6). This design has low levels of pressurized air leakage and small friction force, which allow its use not only in on/off configurations, but also in servo and proportional constructs.

In the flow-dividing type valve, the air stream is divided between the receiver ports, each of which is connected to the load working chamber. This valve type includes the jet-pipe constructs and deflector-jet designs. Both are available commercially in pneumatic actuating systems.

A typical jet-pipe valve is shown schematically in Figure 3.20a. The working principle of this type of amplifier is based on a movable jet-pipe (projector jet). In a jet-pipe servo valve, air flow is directed through a jet-pipe, which is basically a tube (3) with a nozzle (2) on the end. The jet-pipe assembly is attached to the torque motor armature (not shown in Figure 3.20). The air flow in the jet-pipe's nozzle is directed toward a receiver (1) that has two holes into which air from the nozzle is directed, and each hole is connected to the load working chamber. At null position (no signal to the torque motor), the jet stream impinges equally on each receiver hole; therefore, equal pressure is applied at each load working chamber. When an electrical input signal is applied to the coils of the torque motor, an electromagnetic force is created. This force causes the armature and jet-pipe assembly to rotate around the pivot axis, that is, the shaft bearing (4) and hollow shaft (5), resulting in more pressurized air impinging on one receiver hole than the other. This is

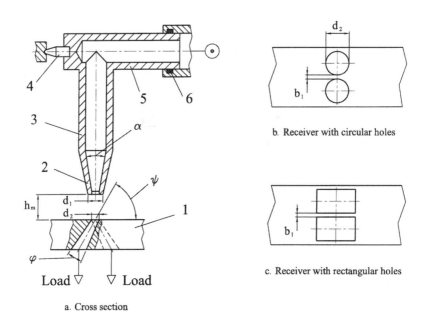

b. Receiver with circular holes

c. Receiver with rectangular holes

a. Cross section

FIGURE 3.20
Jet-pipe valve: (a) cross section, (b) receiver with circular holes, and (c) receiver with rectangular holes.

the cause of differential pressure between the receiver's hole, and this differential pressure is created across the actuator piston, for example, and moves it in the direction opposite to the jet displacement. The angle of the jet-pipe is proportional to the input current applied to the torque motor. The differential pressure rises in proportion to the angle of the jet-pipe, and the resulting load displacement is proportional to the rise in differential pressure. The sealing of the hollow shaft (5) is carried out by an "O"-ring (6).

Usually, the receiver holes have a circular form (Figure 3.20b). However, in a number of cases, rectangular holes (Figure 3.20c) allow for increasing valve efficiency.

The main difference with jet-pipe valves is the double transformation of energy. During flow out of the tube nozzle, potential energy is transformed into the kinetic energy of the air stream. The second energy transformation (from kinetic to potential) is carried out when the air stream comes into the receiver hole. These processes have high performance if both the pipe nozzle and receiver holes have a conical shape (Figure 3.20a). Basically, the conical angle of the receiver holes is $\varphi = 10°$; on the other hand, the angle of the pipe nozzle is $\alpha \approx 13°$.

The receiver holes are oriented with angle ψ for the perpendicular plane of the jet-pipe moving plane. In this case, the discharged stream that exits the receiver hole does not react to the face surface of the jet-pipe assembly (otherwise, the additional torque on the jet-pipe would decrease the valve dynamic and accuracy requirements). Hence, the minimum distance h_m

between the face of the jet-pipe assembly and the upper surface of the receiver can be determined as:

$$h_m = \frac{d_2 + d_1 \cdot Sin\psi}{2 \cdot Cos\psi} \tag{3.1}$$

where d_1 is the diameter of the jet-pipe face, d_2 is the diameter of the receiver holes, and ψ is the angle between center axis of the receiver holes and the upper surface of the receiver (Figure 3.20a).

Distance "b" (Figures 3.20b and c) between the receiver holes is usually 0.1 to 0.5 mm; decreasing the "b" size increases the accuracy of the jet-pipe valve.

In the jet-pipe valve, the jet flow force may be considerable and is of such a nature that it causes serious instability. Therefore, the vector of the air stream reactive force should cross the jet-pipe pivot axis, and then the torque from this reactive force is equal to zero. However, the reactive force creates additional friction force in the shaft bearings and for decreasing this force the jet pipe has a vertical orientation as shown in Figure 3.20a. In this case, the weight of the pipe reduces the influence of the reactive force. Another solution to compensate for the jet flow force is the application of bi- or quadricjet construction, in which the jet-pipes are installed opposite each other.

The important part of the jet-pipe valve is the air supply construction element. The main requirement for this unit is a minimum amount of friction force. From this point of view, for example, the scheme shown in Figure 3.20a has a drawback because the "O"-ring is the sealing element. The supply construction shown in Figure 3.21 has an apparent advantage because its friction force is ultralow. It is achieved using flexible tubes (2) as the sealing elements between the rotating shaft (1) and the static terminal (3).

In general, the jet-pipe valves have low effective area (inside diameter of the nozzle pipe is about 2 to 4 mm). The main cause of this is the large air leakage in the null position of the pipe. The jet-pipe valve shown in Figure 3.22 does not have this drawback; therefore, the inside diameter of the nozzle pipe can be increased up to 10 to 12 mm. This construction consists of the rotating tube (1) with two nozzles (2 and 3) on the ends. Receiver (8) has four inlet holes (3, 4, 5, and 6) where holes (3) and (4), and also (5) and (6), are connected together. The special valve apparatus (15) consists of two non-clamped diaphragms (11 and 12) with holes for the check valves (13 and 14) and two relief valves (9 and 10). A special feature of this design that distinguishes it from regular valves of the jet-pipe type is complete separation of the charge and discharge air flows. The charge stream comes from the supply line through the pressure channels of the jet-pipe, the inlet channels of the receiver, the chambers (A) or (D) of the valve apparatus (15), through the check valves, and then via chamber (B) or (C) into the working chamber of the load. The discharge flow from the load working chamber passes through

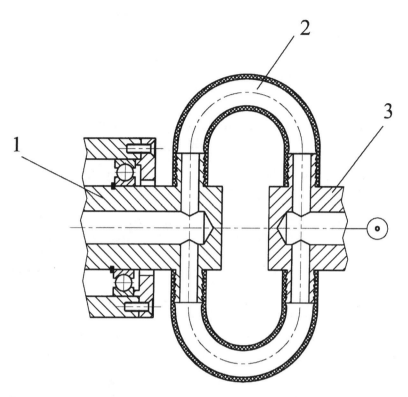

FIGURE 3.21
Supply of pressurized air to the jet-pipe valve.

the chamber (B) or (C) of the valve apparatus (15), the relief valve (9) or (10), and is connected to the exhaust port, which bypasses the jet-pipe. This allows for a considerable decrease in the gap between the end of the jet-pipe and the receiver surface. Loss of compressed air decreases correspondingly, and pressure is also lost in the working chamber of the load. As a result of separation of air streams, it becomes possible to select the flow rates of channels at the inlet and outlet independently of each other. The flow rate of the outlet channel in the present valve construction is considerably greater than the channel at the inlet; as a result, the efficiency of the actuator increases.

The main advantage of the jet-pipe valve is that it offers high reliability. One of the most important features of this type of design is its superior contamination resistance. The product features a design to pass particles up to 300 microns in size. It also has the highest chip shear capability, which adds to the valve's efficiency. The wear-resistant pipe nozzles and receiver design add greater dependability to the product.

Another type of the flow-dividing valve is the deflector-jet design. This construction is based on the same jet momentum–recovery principle as the jet-pipe valve. Instead of turning the jet-pipe itself, a deflector is used to do it. Normally, the jet of air from the pressure nozzle impinges directly between

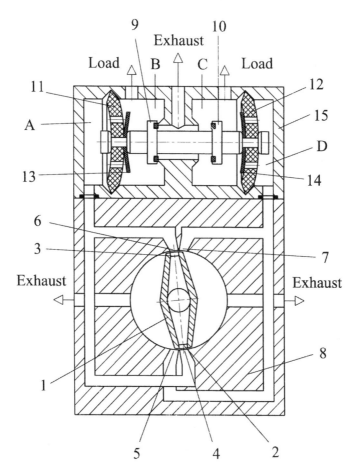

FIGURE 3.22
Double-jet-pipe valve with stream divider.

the two receiver holes, giving equal pressures in the two receiver ports (Figure 3.23). Motion of the deflector to one side or the other causes a jet reaction from the deflector sidewall (Figure 3.23a). This reaction changes the momentum vector of the jet and thus produces a difference of jet impingement at the two receiver holes. Most often, for such a construction both the supply jet and receiver holes are rectangular in shape. The hole in the deflector is sufficiently wide that zero pressure drop occurs. Thus, the efficiency of jet momentum recovery is unaffected by the deflector. The jet reaction force on the deflector produces only a small centering force at the deflector so that large flows can be controlled with relatively small forces.

In flow-dividing valve construction, a combination flow-dividing and flow-restricting technique is sometimes used. A schematic diagram of such a design is shown in Figure 3.23b.

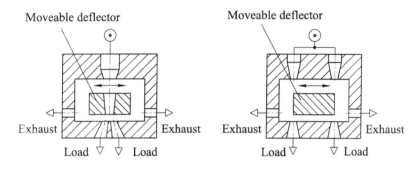

a. Deflector with hole

b. Blank deflector

FIGURE 3.23
Deflector-jet valves: (a) deflector with hole and (b) blank deflector.

Unlike the jet-pipe valve, the absence of a flexing supply pressure connection to the deflector-jet valve eliminates potential vibration problems. In addition, this valve type has very high pressure gain.

The nozzle-flapper pneumatic valves can be placed among the seating class valves; however, sometimes they are related to the so-called restricted valve type. The nozzle-flapper pneumatic valves are the most common among two-stage servo valves, where they are used as a pilot (first-stage) valve. Two types of nozzle-flapper valves must be distinguished, as their properties are different. The first one has only a single nozzle (Figure 3.24), and the second type has a symmetrical double-nozzle structure (Figure 3.25).

In operating the single nozzle-flapper valve, the flapper is positioned against the nozzle opening. This device consists of the fixed orifice (1), a nozzle (2), and a flapper (3) that is moved by a torque motor (not shown in Figure 3.24). The nozzle pressure P_1 is controlled by the distance (y) between the face of the flapper and the air nozzle. As the flapper approaches the nozzle, the opposition to the flow of air through the nozzle increases, with the result that the nozzle pressure P_1 increases. If the nozzle is completely closed by the flapper, the nozzle pressure becomes equal to the supply pressure. When the flapper is moved away from the nozzle so that the nozzle-flapper distance is wide, then there is practically no restriction to flow and the nozzle pressure takes on a minimum value. In this construction with a constant supply pressure, it is possible to provide a variable output pressure that is proportional to the input signal over the range of 20 to 98% of the supply pressure. Leakage is at a maximum when the control pressure is at a minimum and depends on the size of the constant orifice (1), the distance (y) between the face of the flapper and the air nozzle, and the supply pressure. Minimum leakage is about 2% of maximum load flow.

Single nozzle-flapper valves are seldom used, owing to their disadvantages. The chief weakness is a null shift of the output signal with varying supply pressure. In addition, this type of valve can control only one working

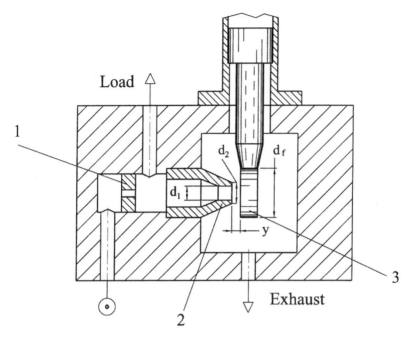

FIGURE 3.24
Single nozzle-flapper valve.

chamber of the load and it has static force unbalance of the flapper element. These problems were overcome by the use of a symmetrical double-nozzle structure, which is shown schematically in Figure 3.25.

The double nozzle-flapper valve consists of two nozzles (1 and 2), two fixed orifices (3 and 4), and the flapper (5). In this valve as in the single nozzle-flapper construction, the torque motor (not shown in Figure 3.25) controls the pressure differential between the two working chambers of the load. In this case, the flapper changes the air flow from the two nozzles in an inversely proportional manner. As the flow from one nozzle increases, the flow from the other decreases. The resultant change in the flow of pressurized air creates a differential pressure between two output ports. Because the double nozzle-flapper structure consists of four orifices arranged in a bridge configuration, the differential flow or pressure created across the bridge by flapper motion remains independent of changes in supply pressure. This valve has flapper force unbalance, which is almost exactly equal to the pressure differential multiplied by the area of the nozzle face. This force is often useful because it provides a built-in pressure feedback of good accuracy.

Double nozzle-flapper valves are normally used in the force balance mode of operation to provide a differential pressure output proportional to input current. Output is limited to a maximum of ±80% of the supply pressure. This valve serves as an excellent pneumatic amplifier in which the pressure

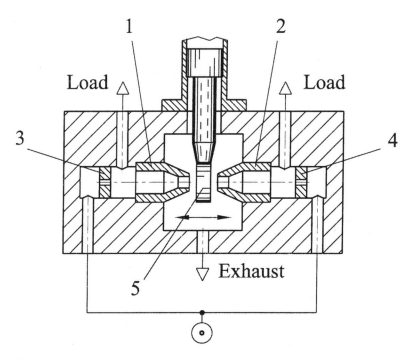

FIGURE 3.25
Double nozzle-flapper valve.

gain is high. Such construction can be designed to have low quiescent leakage, but the maximum of flow recovery is always less than 50%. The design process usually keeps the following relationships between the structural parameters (Figure 3.24):

$$s_n < y_{\min}; \quad y_{\max} < d_1/5$$

$$d_2 < (1.2 \div 1.5) \cdot d_1; \quad d_f = (3 \div 5) \cdot d_1$$

where $s_n = 0.5 \cdot (d_2 - d_1)$, d_1 is the inside diameter of the nozzle, d_2 is the outside diameter of the nozzle, d_f is the face diameter of the flapper, and y is the distance between the nozzle and the flapper face.

The double-nozzle valve with revolving eccentric flopper (cam-flopper) has simple construction and high reliability (Figure 3.26). The cam flopper (5) is rotated by the torque motor, which has relatively large rotation angles and a small torque. The radius of the cam-flopper (5) and orientation of the nozzle (1 and 2) faces (which have a cylindrical shape) are chosen so that the rotation of the cam-flopper changes the distance between the flopper and nozzle faces in an inversely proportional manner. This valve also consists of two fixed

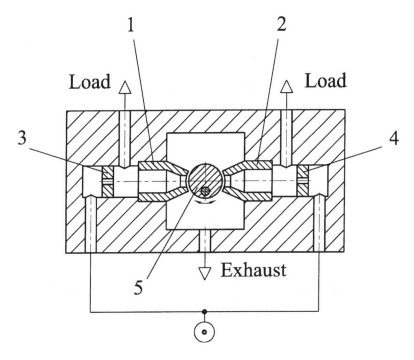

FIGURE 3.26
Double nozzle valve with revolving eccentric flapper.

orifices (2 and 3) and two variable orifices (nozzle-flopper), which form the bridge configuration, as in the construction shown in Figure 3.25.

Nozzle-flapper valves are primarily used for positioning the main stage of a two-stage servo valve. In this case, the flow capacity of the pilot stage establishes the main-stage velocity, which achieves a low flow recovery and a small quiescent flow limit frequency response.

3.3 Two-Stage Valves

Two-stage electropneumatic control valves are commonly used in actuator systems that require high power gain. In general, the power gain in these valves is about $10^4 \div 10^5$.

In a two-stage servo valve, the amplifier is typically used as a first stage (pilot stage) to selectively displace a second-stage (main-stage) valve spool relative to a sleeve. In this case, the first-stage amplifier is used to create a pressure differential, which is applied to the spool end areas. Displacement of the second-stage spool is used to vary the effective area of the orifices through which pressurized air can flow, respectively, from the source to the load and from the load to the exhaust.

Traditionally, in pneumatic two-stage valves, the nozzle-flapper, jet-pipe, or deflector-jet valve can be used as the pilot stage. The spool design is most often used as the main stage, but the two-seat construction is seldom utilized. Sometimes, the main-stage is referred to as the power stage.

As stated above, the double nozzle-flapper and jet-pipe valves are most often used as the pilot stage. The electromagnetic circuit of a nozzle-flapper and jet-pipe torque of force motor is essentially the same. The differences between the two lie in the pneumatic bridge design; nevertheless, two-stage control valves with the nozzle-flapper and jet-pipe pilot stage have competed for similar applications that require high dynamics. Typically, the nozzle-flapper pilot valve gives better overall response, whereas improved pressure recovery of the jet/receiver bridge design gives the jet-pipe constructions higher spool driving forces. Both valves require low command currents and therefore offer a large mechanical advantage.

In practice, high-performance two-stage control valves are usually classified as either servo or proportional, a distinction that gives an indication of the expected performance. Unfortunately, this classification tends to generalize and blur the true differences between various valve styles.

Traditionally, the term "servo valve" describes valves that use closed-loop control. They monitor and feed back the main-stage spool position to the pilot-stage, either mechanically or electronically. Proportional valves, on the other hand, displace the main-stage spool proportional to command, but they usually do not have any means of automatic error correction (feedback) within the valve (open-loop control). Typically, proportional valves displace the spool by driving it against a set of balanced springs (positioning springs), which makes the position proportional to the driving current. These springs also center the main-stage spool.[46]

Figure 3.27 shows a two-stage valve with jet-pipe pilot valve that has an open-loop configuration. The pilot stage consists of a torque motor having a pivotable armature (1) and two coils (2 and 3); a jet-pipe/receiver assembly (first stage) has the jet-pipe (4) with receiver (5). A jet-pipe is connected to the armature and a pair of receivers, and a single sliding spool valve (second stage) having a spool (6) that controls two output load ports. Air at system pressure is fed to a jet-pipe nozzle that directs a fine jet stream of air at the two receivers. Each receiver is connected to a corresponding end of the main-stage spool. At null (no signal to the torque motor), the jet stream impinges equally on each receiver and equal recovery pressure is generated for each one. Thus, the forces at each end of the second-stage spool are equal and it remains in the null position. When an electrical input signal is applied to the torque motor, it causes the armature and the jet-pipe to rotate around the armature pivot point, causing more air to impinge on one receiver than the other. The resulting differential pressure at the ends of the spool causes the spool to move, which opens the second-stage ports, causing air to flow to one load port and out the other load port. The two positioning springs

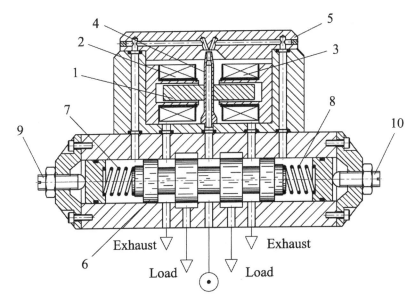

FIGURE 3.27
Open-loop, two-stage valve with jet-pipe pilot stage.

(7 and 8) with adjusting screws (9 and 10) at each end of the second-stage spool are used to counteract the force of the differential pressure and stabilize the spool in its new position.

Figure 3.28 depicts a two-stage control valve with a symmetrical double nozzle-flapper pilot valve. The force motor of the pilot stage consists of the armature (1) and two coils (2 and 3). The armature (1) has a flapper (4) that is attached to the diaphragm (5). The pilot-stage force motor receives an electrical signal applied as voltage or current to the coils, and converts it into a mechanical force on the armature and flapper assembly. The force output is directly proportional to the input signal. Deflection of the diaphragm (5) with flapper (4) changes the air flow through two nozzles (6 and 7), and the resulting differential pressure at the ends of the spool (8) causes the spool to move. The second stage has similar construction and operation mode. The major difference from the previous construction is the two fixed orifices (9 and 10) installed on the spool (8). Two positioning springs (11 and 12) with two adjusting screws (13 and 14) provide the center position of the main-stage spool (8).

Figure 3.29 presents yet another open-loop, two-stage valve. This device was designed for pneumatic positioning systems. The valve consists of the torque motor (1) that moves the flapper (2) between two nozzles (3 and 4). Each of the two main stages has a double-seated valve: on the left side there is a pair (9 and 7), and on the right side another pair (8 and 10). If, for example, the flapper (2) is moved in the left direction, then the gap between the nozzle (3) and the face of the flapper (2) decreases and the pressure in

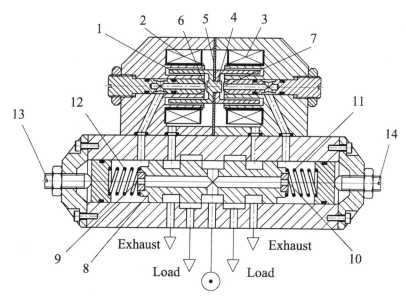

FIGURE 3.28
Open-loop, two-stage valve with nozzle-flapper pilot stage.

the working chamber (5) increases. As a result of this process, the left double-seated valve moves in the left direction, and the valve (9) connects the supply pressure with the left output port. At the same time, the gap between the nozzle (4) and the face of the flapper (2) increases and the pressure in the working chamber (6) decreases. In this case, the right double-seated valve also moves in the left direction, and the valve (8) connects the exhaust port with the right output port. This construction has built-in load pressure feedback that allows one to obtain a stable valve operation.

The two-stage control valves with open-loop structure have a simple adjustment, manufacture, and assembly process; consequently, they tend to be cheaper. However, these valves have several common shortcomings. Their main-stage position accuracy and repeatability depends on the positioning spring's symmetry and the ability of the design to minimize the nonlinear effects of spring hysteresis, friction force, and machining-tolerance variations. In addition, the overall valve power gain depends on supply pressure, which causes these constructs to be sensitive to supply pressure changes. Maximum spool driving forces are only available when the valve is commanded to maximum opening. For a low signal, fine positioning the forces are proportionally law, thus making the valve more sensitive to contamination than a valve with closed-loop control.

Feedback from the main-stage spool to the pilot stage decreases the valve sensitivity to the supply pressure change and changes the force loads that are activated over the spool (friction, positioning spring's force, and others). Therefore, it improves the control valve response speed.

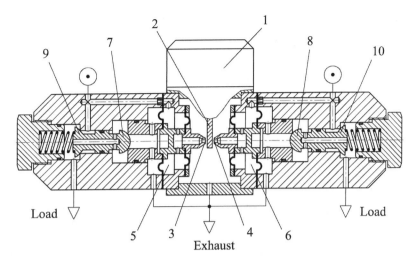

FIGURE 3.29
Open-loop, two-stage valve with nozzle-flapper pilot stage and double-seat main stage.

In two-stage servo valves, internal mechanical feedback of the main-stage spool position is most often used. In the double nozzle-flapper pilot stage, the feedback achieved with mechanical connection between the flapper and the main spool is very common. This construct is simple and reliable in the nozzle-flapper pilot configuration, but difficult to implement in the other types of pilot stages. A cross-sectional view of the two-stage control valve with mechanical closed-loop control is shown in Figure 3.30. As with previ-

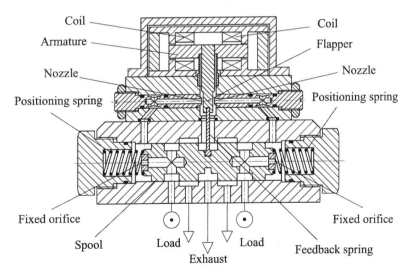

FIGURE 3.30
Closed-loop, two-stage valve with nozzle-flapper pilot stage.

ous designs, a torque motor receives an electrical input and the motor arma-ture moves in response to the electrical flux created by the current flowing through the coils. A flapper is connected to the armature by a thin-walled flexure sleeve. The flapper is positioned between two opposing nozzles. Pressurized air flows through these nozzles. Branching off from the inlet passage of each of the nozzles is a connection to each end of the main-stage spool. When no input current is applied to the torque motor, the flapper remains centered between the two nozzles, the pressure between each of the nozzles and the flapper remains balanced, and the pressure at both ends of the spool also remains balanced. Also connected to the flapper is a thin length of stainless steel wire, termed a "feedback spring." The free end of the feed-back spring rests in a groove in the main-stage spool. In this way, the spool position is mechanically fed back into the nozzle-flapper pilot valve assembly.

If, for example, the torque motor armature is rotated counterclockwise, this moves the right side of the flapper closer to the right nozzle, creating higher pressure in the right passage and lower pressure in the left passage. The higher pressure in the right passage acts on the right end of the spool, applying force to displace the spool to the left. At the same time, lower pressure is acting on the left side of the spool, creating a force imbalance and facilitating the spool's movement to the left. As the spool moves to the left, the small jewel on the end of the feedback spring pulls the feedback spring to the left, thereby putting a feedback force on the flapper, thus counteracting the force generated by the torque motor and recentering the flapper between the nozzles. Once the flapper is centered between the noz-zles, the pressure between each of the nozzles becomes equalized, and so does the pressure acting on the ends of the spool. Once the pressures at the ends of the spool are equalized, the spool stops moving yet remains dis-placed and controls the flow to and from the load ports.

The junction between the feedback spring and the spool is often a precision steel ball on the end of a cantilever feedback spring. Wear on this ball creates play in the feedback path, affects stability, and reduces valve life. The two-stage control valve shown in Figure 3.31 does not have this drawback. The two nozzles of the pilot stage are installed on the main-stage spool and enables realization of position feedback between these valve parts. The prin-ciple of operation of this device is similar to the two-stage control valve shown in Figure 3.30.

Sometimes, the jet-pipe pilot valve is used in a two-stage configuration where the mechanical feedback from the main-stage spool to the pilot stage is replaced with electronic feedback. Electrical feedback valves are becoming the preferred solution for both servo valves and proportional valves. Depending on the required control, many two-stage valves close a position loop on the main stage using a short stroke transducer, for example, the linear variable differential transducer (LVDT), which monitors the spool position. In such constructs, closed-loop electronics provide valve control either externally to the valve or increasingly on board. This configuration of the two-stage control valve is shown in Figure 3.32. In case of pneumatic

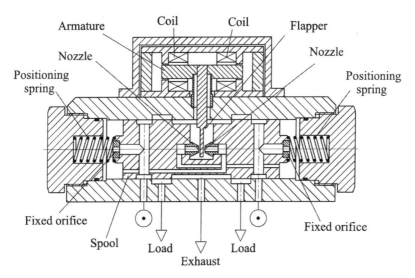

FIGURE 3.31
Closed-loop, two-stage valve with nozzle-flapper pilot stage.

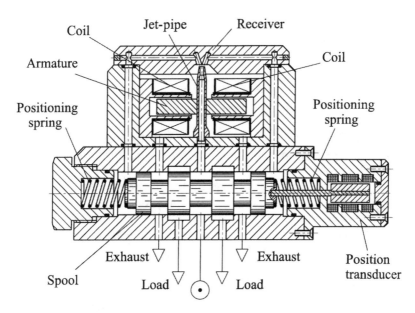

FIGURE 3.32
Closed-loop, two-stage valve with jet-pipe pilot stage and electronic feedback.

power loss, the springs on the opposite sides of the main-stage spool return it to a neutral position.

By providing closed-loop control of the spool position via a mechanical feedback spring, the valve characteristics and contamination sensitivity can be improved significantly. A tension spring was used to generate force (torque) on the torque motor proportional to the spool position, thus closing a mechanical loop around the spool position. By combining the frictionless first stage with mechanical feedback from the second stage, it is possible to improve the threshold, improve the dynamic response because of lower first-stage displacement, and reduce the temperature and pressure caused by spool position anomalies because of increased first-stage gain allowed by feedback.

Various spool actuating devices can be used with electrical closed-loop positioning to achieve the best control, and contamination resistance. A short transducer measures spool position and closed-loop electronics control may be either external to the valve or increasingly onboard. Electrical feedback has improved this one or two steps further than mechanical feedback by allowing higher-gain loops that are more effective in suppressing any disturbance, whether it be a contaminant or flow force. If the first stage is linear, a tight feedback loop around a good feedback transducer is the next step in maximizing dynamic linearity. Electrical feedback with electrical loop closure provides the most potent solution.

The advantages of closed-loop spool control are many, and the primary advantage is very high performance. Nevertheless, the disadvantages of such two-stage control valves include the tendency to be expensive because of the stringent manufacturing tolerances and the complicated assembly process.

3.4 Operating Mode of Electropneumatic Control Valves

As discussed previously, electropneumatic control valves can be divided into two main groups: (1) on/off solenoid valves, which can only take two or three possible positions; and (2) "continuous valves," which can take any position within some interval defined by the electromechanical valve transducers. As a rule, continuous valves are the proportional or servo valves. Depending on the valve operating mode, which can be analog (continuous) or digital (discrete), an appropriate for valve type is usually used.

3.4.1 Analog (Continuous) Operating Mode

Figure 3.33 presents a block diagram of a control valve in analog (continuous) operating mode. The output of the actuator (actuator state variable) is measured with a transducer device (sometimes calculated by some kinds of the algorithms and techniques) to convert it to an electric signal, which is fed back to the controller and compared with the command signal. The resulting regulating signal, which is denoted U_R, is then amplified by the electric servo amplifier and used as an input control signal (U_C) to the servo valve. The

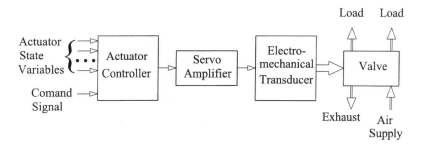

FIGURE 3.33
Block diagram of control valve with analog operating mode.

servo valve controls the air flows to/from the actuator in proportion to the drive current from the amplifier. In that case, all elements operate in analog (continuous) mode, and an electromechanical transducer is usually a proportional solenoid actuator.

Pneumatic analog valves have a significant influence on the behavior of open-loop and closed-loop systems. The nonlinearities of valves influence the steady-state accuracy of a system, especially in position servo systems. The most serious nonlinearities, from an accuracy point of view, are hysteresis, dead band, individual over/under laps of control edges, zero point drift, and the pressure gain of servo valves. Load pressure causes varying flow modes and influences the valve flow gain, and, as a result, both open- and closed-loop achieve steady-state and dynamic behavior. The pressure gain of a valve influences system stiffness, especially in position servo applications under force disturbances. The pressure gain of a valve depends on the structure of the spool of a valve and on leakage.[198]

The main advantage of analog control valves is their long lifetime; however, this parameter strongly depends on the cleanliness and chemical composition of the pressurized air. Cleanliness is a function of selecting the appropriate filters and application environment. Chemical properties relate to factors such as water content and various forms of breakdown that can occur due to chemical contamination.

In control valves with analog operation mode, servo valves and proportional valves are usually the most important area for filtration design. Inherently, they perform critical control functions for the machine, requiring consistent operation and high reliability. From practical experience, the focus has been on the design issues for servo valves and proportional valves, with secondary consideration for compressors, motors, and other devices.

In control valves with analog operation mode, the dither technique is often used. In this technique, a periodic signal is added to the actuator control signal. The frequency of the dither should be higher than the bandwidth of the actuator system dynamics, and the dither amplitude is adjustable from 10 to 15% of the rated maximum control signal. The optimum dither amplitude is attained when a small input signal change registers similar changes in current output (pressure or flow through the control valve). Dither amplitude and frequency depend on the specific valve. The purpose of the dither is to eliminate stiction and to increase the dirt tolerance within the control valve. In addition, owing to the dither, the valve dead zone can be significantly decreased.

3.4.2 Digital (Discrete) Operating Mode

In actuating systems with control valves that operate in the digital (discrete) mode, the control signal is pulse-modulated, in which some parameters are changed as a function of time. The pulse modulation mode can be divided into three different forms: (1) pulse amplitude modulation, (2) pulse frequency modulation, and (3) pulse width modulation.

In pulse amplitude modulation, pulses of the control signal are generated with amplitudes corresponding to the regulation signal value. This operating mode is very sensitive to noise, and, in addition, this technique is difficult to use in the pneumatic actuator system. There are two main reasons why this operating mode is not used in pneumatic actuators. In pulse-frequency modulation, a frequency of the pulses of the control signal is a function of the regulation signal amplitude. This method of modulation is also not implemented in the pneumatic actuator because of complicated pulse frequency modulation generation methods.

Pulse width modulation (PWM) is more popular in pneumatic actuating systems at present. In this operating mode, pulses of the control signal are generated with widths corresponding to the regulation signal value. This technique involves the direct control of pressure or flow rates using a train of pulse flows that are transformed by switching valves. It can obtain control accuracy using cheap on/off solenoid valves via modulation technology. In this technique, an on/off transistor located on the amplifier card turns the current to the electromechanical transducer coils on and off very rapidly. Because the pulse frequency remains constant, the duration of the pulse is varied; that is, for example, if the width of the pulse is 50% of its maximum duration, theoretically the valve should shift enough to deliver a 50% output of air flow.

From the standpoint of pneumatic circuit design with PWM, a number of different configurations have been used. In practice, the following combinations of solenoid valves can be used:

- One 3/2 (three ports and two positions) slow-speed valve (for reversing motion) driven by a conventional on/off relay in combination with a second 3/2 high-speed valve that operates with PWM
- Two 3/2 high-speed solenoid valves operated in the PWM regime
- One 5/3 high-speed solenoid valve with blocked-center condition

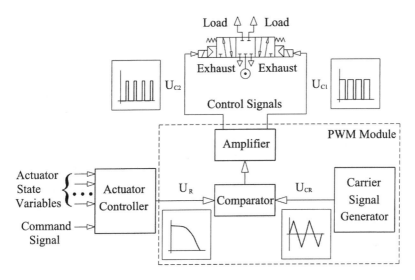

FIGURE 3.34
Block diagram of control valve with pulse width modulation.

Most often, the last combination is used in positioning and speed motion control systems. The rationale for this circuit is the reduction in the number of active valves and cost, and in addition it improves the reliability of the system. A block diagram of the control valve with such control mode is shown in Figure 3.34; it illustrates the basic principles of PWM. Typically, a periodic saw-tooth signal is used as the carrier signal. The regulating signal (U_R) and the carrier signals (U_{CR}) are compared at a given interval. And as result, pulse-width modulated signals, after electrical amplification, are used as input control signals $(U_{C1}$ and $U_{C2})$ to the solenoid valve. The most important parameter in PWM is the carrier period (t_{CR}). In theory, t_{CR} should be as small as possible. In practice, the choice of t_{CR} is limited by the valve response time. Thus, these characteristics must be identified in order for the user to be able to choose the appropriate PWM parameter values.

If the regulating signal (U_R) has a time-varying form, the control signals $(U_{C1}$ and $U_{C2})$ generated by the PWM module and supplied for two electro-mechanical valve transducers have constant-amplitude pulses whose width is proportional to the values of U_R (Figure 3.35). It is significant to note that if the absolute value of the regulating signal is greater than the absolute value of the carrier signal $(|U_R| > |U_{CR}|)$, the system is in the saturation condition and the valve-sealing element is in one of its end positions.

In PWM systems, the on/off solenoid valve has a simple structure and is insensitive to air contamination. This system procedure simplifies the design of electropneumatic elements and improves the overall reliability of the control system. An additional advantage of PWM is that the signal remains

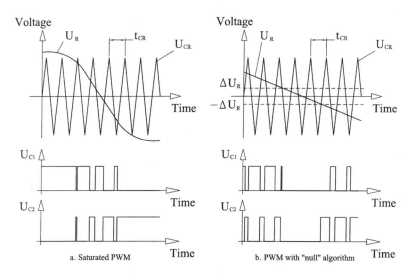

FIGURE 3.35

Basic operating principle of PWM: (a) saturated PWM and (b) PWM with "null" algorithm.

digital all the way from the processor to the power element; no digital-to-analog conversion is necessary. By keeping the signal digital, noise effects are minimized. The trade-off is that a disturbance is introduced into the system at the PWM frequency. In this case, the demodulation element is the actuator (cylinder or motor) itself or the solenoid valve with their low-pass filter characteristics. For example, if the low-pass filter is a solenoid valve, then its plunger does not reach at both stroke ends and it keeps floating from valve sealing surfaces.

In the past, PWM was implemented using analog electronics but suffers the imprecision and drift of all analog computations, as well as having difficulty in generating multiple edges when the signal has even a little additional noise. Now, many PWM modulators are implemented digitally. This can be realized by a special control algorithm in a microcomputer or by using digital signal processors.

The main disadvantage of the control valves with PWM is their short lifetimes due to the relatively high frequency of the carrier signals. This parameter can be improved using a special algorithm ("null" algorithm) that allows holding the valve in the blocked-center position if the absolute value of the resulting error signal is less than or equal to the tolerated error ($|U_R| \leq \Delta U_R$, where ΔU_R is the tolerated error). This can reduce valve cycling significantly because these null signals will not move the valve-sealing element by the solenoid, and a spool will be returned to its central position by the spring.

More recently, in servo pneumatic closed-loop systems, the "Bang-Bang" control mode has become widely used. In such a system, the valve closes or opens in response to the sign of the resulting error signal (like sign control flip-flop). The structure of the Bang-Bang control module is the same as the

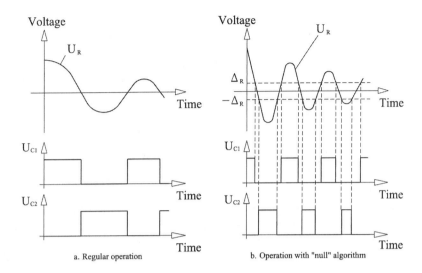

FIGURE 3.36
Basic principle of "Bang-Bang" operation mode: (a) regular operation and (b) operation with "null" algorithm.

module of PWM; a distinction is the absence of a carrier signal generator, and then its input in the comparator is connected to the ground (Figure 3.34).

Figure 3.36 represents the basic operating principle of the Bang-Bang valve operation mode. Each time the regulating signal (U_R) changes sign, the energizing of the electromechanical transducer also changes. This style of valve operation promotes the chatter phenomenon of the cylinder or motor about the positioning set point, for example. Sometimes, this is not acceptable from the accuracy and dynamic behavior points of view. To improve this situation, two techniques are usually used: (1) high-order control law (for servo pneumatic actuator; state controller) and (2) the so-called null algorithm. In the latter case (Figure 3.36b), if the resulting error signal is less than or equal to the tolerated error, the valve-sealing element is in the blocked-center position (as in PWM algorithm). Thus, the amplitude of the chatter can be significantly decreased.

3.5 Mathematical Model of Electropneumatic Control Valves

The electropneumatic control valve assembly must be developed as a unit that consists of an electromechanical subassembly (electromechanical transducer), mechanical construction of the sealing mechanism, and the pneumatic structure, which defines the mass flow rate of the valve. The particular design features of the control valve are very important in achieving good process control under dynamic conditions. A mathematical model relating the various physical parameters to performance can also be used to predict and improve the performance when designing the valves. The mathematical model for the electropneumatic control valve can be separated into three major, distinct parts: (1) the magnetic circuit with its electromagnetic equations; (2) the mechanical subsystem, which consists of the spring-mass system with friction force; and (3) the air flow part that describes valve flow ability.

Magnetic circuit. The electromechanical transducer in the control valve represents an electromechanical interface, where the electric current is transformed into an electromagnetic force acting on the armature that is connected to the valve plug.

In general, the mathematical model of an on/off solenoid is nonlinear because of the existence of hysteresis and saturation of the ferromagnetic materials.[138] However, in practice, calculations usually take into account that these characteristics are negligible. In the design process of the solenoid actuator, the following characteristics are used:

- Tractive (force or torque) characteristic: change in electromagnetic force as a function of the control signal and armature movement
- Steady-state characteristic: relation between the movement of the armature and the value of the control signal
- Maximum value of the electromagnetic force acting on the armature
- Maximum value of the armature movement for the specified load

The electromagnetic force required to develop magnetic flux is broken up into components for the ferromagnetic material and the air gap between the armature and solenoid core. Obviously, the major circuit reluctance is concentrated at the air gap.

An electromagnetic force on the armature can be defined in the following way:

$$F_M = \frac{B^2 \cdot A_\delta}{2 \cdot \mu_A}$$

where B is the magnetic flux density, A_δ is the cross-sectional area in the air gap, and μ_A is the magnetic permeability of air. Taking into account that the magnetic flux density is

$$B = \frac{\Phi}{A_\delta} \text{ and } \Phi = \frac{\mu_A \cdot A_\delta \cdot I \cdot n_C}{L_M}$$

(here, Φ is magnetic flux), the formula for an electromagnetic force can be rewritten as:

$$F_M = \frac{\mu_A \cdot A_\delta}{2 \cdot \phi_M} \cdot \left(\frac{I \cdot n_C}{L_M} \right)^2 \tag{3.1}$$

where ϕ_M is the coefficient of the magnetic flux dispersal, I is the current of the control signal, n_C is the number of turns in the solenoid coil, and L_M is the magnetic circuit length.

A major drawback of on/off solenoid valves is the nonlinear characteristic of both the movement of armature and the electromagnetic force as a function of the control signal current, particularly for the long stroke of the valve armature.

As noted above (Section 3.1), the proportional solenoid has three different sections in the typical force — stroke characteristic (Figure 3.3b). On the working area "B," the force — current of the control signal characteristic — can be written as:

$$F_M = \frac{\partial F}{\partial I} \cdot (I - I_I) + \frac{\partial F}{\partial y} y$$

where $\partial F / \partial I$ is the electromagnetic stiffness by a control signal current, $\partial F / \partial y$ is the electromagnetic stiffness by an armature movement, I_I is the value of the control signal current when the valve armature begins to move, and y is the stroke of the armature.

Usually, for the proportional solenoid, the electromagnetic stiffness by a control signal current can be obtained as:

$$\frac{\partial F}{\partial I} = \frac{\mu_A \cdot n_C \cdot A_\delta}{2 \cdot \phi_E \cdot \phi_R}$$

where ϕ_E is the equivalent coefficient of the magnetic flux dispersal (ϕ_E is about 250 to 320), and ϕ_R is the relation of the magnetic flux dispersal to the full magnetic flux.

The electromagnetic stiffness by an armature movement $\partial F / \partial y$ depends on the stiffness of the valve springs.

Mechanical subsystem. The mathematical model of both the one-stage control valves and two-stage valves can be modeled as a second-order system with friction. Figure 3.37 depicts the schematic diagram of the one-stage seating type control valve structure. In this case, the solenoid actuator strokes the valve plug directly. For the non-balanced valve (Figure 3.37a) a differential pressure $(P_1 - P_2)$ across the plug acts on the valve effective area A_{VE}. Other forces that act on the sealing part include spring force, coulomb friction, viscous friction, and the force produced by the solenoid actuator. Usually, in electropneumatic control valves, the flow forces acting on the

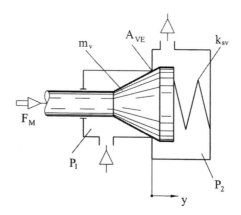

a. Regular structure

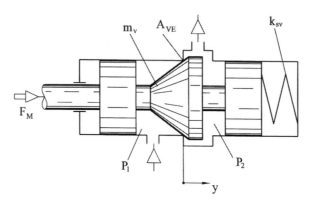

b. Force balancing structure

FIGURE 3.37
Schematic diagram of the seating type control valve structure: (a) regular structure and (b) force balancing structure.

plug are assumed as negligible. For negative displacement y ($y \leq 0$), the plug is not moved and remains on the sealing surface, that is, if $y \leq 0$, then $y = 0$ and $\dot{y} = 0$. Under these conditions, the equation for the motion of the valve plug together with solenoid armature can be written as:

$$m_V \cdot \ddot{y} + b_{VV} \cdot \dot{y} + k_{SV} \cdot (y_0 + y) + F_{VF} = (P_1 - P_2) \cdot A_{VE} + F_M \qquad (3.2)$$

where y is the valve plug displacement; m_V is the armature and plug assembly mass; b_{VV} is the viscous friction coefficient; k_{SV} is the valve spring constant; y_0 is the valve spring compression at the valve closed position; F_{VF} is the coulomb friction force, which can be defined as F_F (Equation 2.24); A_{VE} is the valve plug effective area; and F_M is the solenoid electromagnetic force (Equation 3.1).

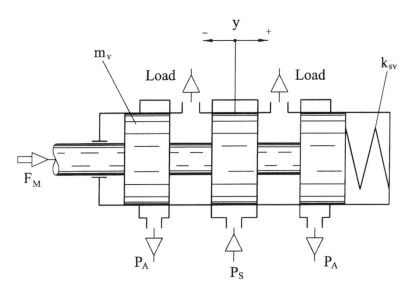

FIGURE 3.38
Schematic diagram of the sliding-type control valve structure.

The mathematical model of the mechanical subsystem for the seating type one-stage control valve with force balancing mechanism (Figure 3.37b) can be written as:

$$m_V \cdot \ddot{y} + b_{VV} \cdot \dot{y} + k_{SV} \cdot (y_0 + y) + F_{VF} = F_M \tag{3.3}$$

According to Equation 3.3, the force of the differential pressure $(P_1 - P_2)$ does not influence the motion of the plug of the control valve.

Figure 3.38 provides a schematic diagram of the one-stage sliding type (with spool) control valve structure. In this case, the dynamic behavior of the spool can be described by Equation 3.3. The origin of the spool displacement is at the middle of its full stroke. Most often in these types of control valves, the dither technique is applied, in which a low-amplitude (about 10 to 15% of the maximum value of the control signal), relatively high-frequency (about 10 to 15 times more than the solenoid natural frequency) periodic signal is superimposed on the input current signal. The spool will slightly vibrate around the equilibrium position, and the coulomb friction force will significantly decrease ($F_{VF} \approx 0$).

A schematic diagram of the double nozzle-flapper one-stage control valve is shown in Figure 3.39. The flapper with armature is rotated around the pivot point by the torque motor (the result of input current). The armature-flapper system is subjected to the electromagnetic force, the force of the balancing spring, the damping moment, and the nozzle flow forces (in this case, the moment due to pivot stiffness is negligible).

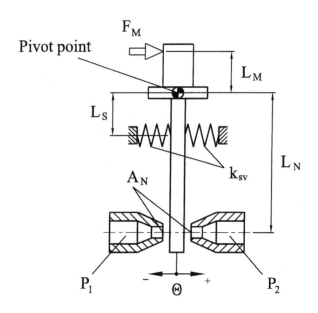

FIGURE 3.39
Schematic diagram of the double nozzle-flapper control valve structure.

Summing moments about the pivot point, and applying Newton's second law, the equation for the dynamic behavior of the armature-flapper system has the following form[6]:

$$J_{VF} \cdot \ddot{\Theta} + b_{VF} \cdot \dot{\Theta} + 2 \cdot k_{SV} \cdot L_S^2 \cdot \Theta = F_{NF} \cdot L_N - F_M \cdot L_M \qquad (3.4)$$

where Θ is the angular displacement of the armature-flapper system; J is the moment of inertia of the armature-flapper system; b_{VF} is the angular damping coefficient; L_S, L_N, and L_M are the arms of the balancing spring, the nozzle flow, and the electromagnetic forces, respectively; and F_N is the nozzle flow force on the flapper, which can be defined as $F_N = (P_1 - P_2) \cdot A_N$ (here, A_N is the effective area where the flow force acts on the flapper).

In Equation 3.4, only the static pressure force is taken into account because the dynamic component is negligible. The flapper's original position is at the middle point between two nozzles. As can been seen in Equation 3.4, the pressure difference $(P_1 - P_2)$ has the effect of restoring the flapper to its neutral position. The relationship between the angular armature-flapper system displacement and its linear displacement on the nozzle central line has the following form: $y = L_N \cdot \Theta$.

The dynamic behavior of the one-stage jet-pipe control valve can be described by Equation 3.4; however, in this case, the nozzle flow force is absent ($F_N = 0$) because the vector of the air stream reactive force crosses the jet-pipe pilot axis, and the moment from this reactive force is equal to zero.

Both the nozzle-flapper and the jet-pipe valves are referred as a frictionless type devices in which only viscous friction force is enacted, but in pneumatic systems this force has a low value.

As indicated above in two-stage control valves, the pilot pressures (P_1 and P_2) act on the two ends of the main-stage spool. Because the spool is spring centered, the displacement of the spool is roughly proportional to the differential pilot pressure and inversely proportional to the stiffness of the springs. In that case, the equation of the dynamic behavior of the second-stage spool has the following form:

$$m_V \cdot \ddot{y} + b_{VV} \cdot \dot{y} + 2 \cdot k_{SV} \cdot y + F_{VF} = (P_1 - P_2) \cdot A_{SP} \qquad (3.5)$$

In Equation 3.5, m_V is the spool mass and A_{SP} is the effective area of the spool ends (see Figure 3.40). Notice that a feedback wire between the main stage and pilot stage is not used in this construction.

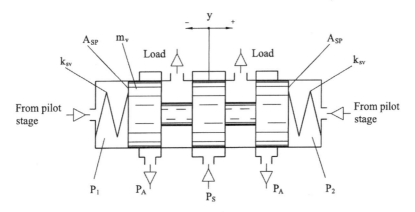

FIGURE 3.40
Schematic diagram of the second-stage spool valve structure.

The moment of inertia of the armature-flapper system is quite easy to calculate. The effective stiffness of the armature-flapper is a composite of several effects, the most important of which is the centering effect of the permanent magnet flux. This is set by the charge level of the torque motor, and is individually adjusted in each valve to meet prescribed dynamic response limits. The damping force on the armature-flapper is likewise a composite effect. Here, it is known from experience that the equivalent damping coefficient is about 0.4.

Valve flow ability. One of the most important parameters of control valves is their flow ability, or the mass flow rate. In general, this characteristic can be determined by Equation 2.12; however, for electropneumatic control valves, the mass flow rate is determined by:

$$G = A_V \cdot \beta \cdot P_{in} \cdot \sqrt{\frac{2 \cdot k}{R \cdot T_{in} \cdot (k-1)}} \cdot \varphi(\sigma) \qquad (3.6)$$

where β is the degree to which the control valve is open, and remaining parameters correspond to parameters in Equation 2.12.

The pressure drop across the valve orifice is usually large, and the flow must be treated as compressible and turbulent. If the downstream-to-upstream pressure ratio is small, then a critical value ($\sigma = P_{out}/P_{in} < 0.5$) for the flow will attain sonic velocity (choked flow) and will depend linearly on the upstream pressure (P_{in}). If this pressure ratio is larger than the critical value, the mass flow depends nonlinearly on both pressures (Equation 2.14). Determining the upstream and downstream pressures in Equation 3.6 is different for the charging and discharging processes of the actuator working chambers. For the charging process, the pressure in the supply line should be considered as the upstream pressure, and the pressure in the working chamber is the downstream pressure. For discharging, the pressure in the working chamber is the upstream pressure, and the pressure in the exhaust line is the downstream pressure.

Various companies define the flow rate of pneumatic elements in various ways. Usually, the standard nominal flow rate Q_n, water flow rate K_v, and flow coefficient C_v are used. These parameters provide a method to compare the flow capabilities of different valves. In addition, they allow one to determine valve size, which helps in selecting the appropriate valve for a given application.

In European and Japanese companies, the standard nominal flow rate Q_n is most often used for the flow characteristic of pneumatic components, including the control valves. Figure 3.41 is a diagram of the circuit that is used to measure the standard nominal flow rate. Such a measurement is carried out under typical nominal conditions:

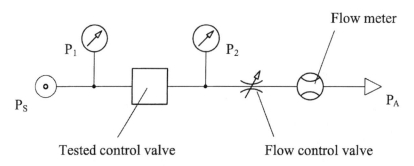

FIGURE 3.41
Schematic diagram of flow rate measurement.

- Test medium air temperature is 20 ± 3°C.
- Test control valve is at room temperature.
- Upstream pressure (P_1) is 0.6 MPa (absolute).
- Downstream pressure (P_2) is 0.5 MPa (absolute).

In the United States, it is often convenient to express the capacities and flow characteristics of control valves in terms of a flow coefficient C_v, which is defined as the flow of water at 60°F in U.S. gallons per minute (gal/min) at a pressure drop of 1 lb/in² across the valve.

In theoretical calculations and computer simulations, it is very important to know the control valve effective area A_V. This parameter can be obtained using Equation 3.6:

$$A_V = \frac{G}{\beta \cdot P_{in} \cdot \varphi(\sigma)} \cdot \sqrt{\frac{R \cdot T_{in} \cdot (k-1)}{2 \cdot k}}.$$

Taking into account that $G = Q_n \cdot \rho_{an}$, the effective area becomes:

$$A_V = \frac{Q_n \cdot \rho_{an}}{\beta \cdot P_{in} \cdot \varphi(\sigma)} \cdot \sqrt{\frac{R \cdot T_{in} \cdot (k-1)}{2 \cdot k}} \qquad (3.7)$$

where ρ_{an} is the density of air under standard conditions ($\rho_{an} = 1.293$ kg/m³). With regard to $\sigma = P_2/P_1 = 0.833$, $\varphi(\sigma) = 0.193$ (according to Equation 2.14), $\beta = 1$ (control valve fully open), $P_{in} = 0.6$ MPa, $R = 287$ Joul/kg·K, $T_{in} = 290$ K, and $k = 1.4$, the effective area is $A_V = 1.2 \cdot 10^{-3} \cdot Q_n$.

The water flow rate K_v and flow coefficient C_v are related to the standard nominal flow rate Q_n by the following formulae: $K_v = 54.4 \cdot Q_n$ and $C_v = 60.9 \cdot Q_n$.

Finally, Table 3.1 represents formulas for the valve effective area definition.

Sometimes, for the theoretical and experimental investigation, the non-dimensional charge coefficient is also used. In general, this parameter is the ratio of the valve effective area to its actual geometric cross-sectional area; that is, $\mu_V = A_V/A_{VG}$. The nondimensional charge coefficient is a function of the valve geometric parameters, the surface finish of the valve passages, and the Reynolds number. This coefficient is commonly determined by experimental testing.[49]

TABLE 3.1

Valve Effective Area Definition

	Effective Area A_V [m²]
For standard nominal flow rate, Q_n (m³/s)	$A_V = 1.2 \cdot 10^{-3} \cdot Q_n$
For water flow rate, K_v (m³/h)	$A_V = 2.2 \cdot 10^{-5} \cdot K_v$
For flow coefficient, C_v (gal/min)	$A_V = 2.0 \cdot 10^{-5} \cdot C_v$

The valve effective area (A_V), just as the standard nominal flow rate (Q_n) or flow coefficient (C_v), is a valve property that allows for calculating the flow and pressure at all operating conditions.

The effective area of proportional or servo valves is a function of the control signal (U_C). It is convenient for simulation of the valve behavior to describe the changing of effective area by the degree to which the valve is open (β). This parameter is a nondimensional quantity that changes between 0 and 1. Equal increments of the control signal provide equal increments of the plug movement; that is, equal increments of the opening coefficient β. For example, for 25% of the control signal, the plug will pass 25% of its movement (β = 0.25). For β = 0, the control valve is closed; and for β = 1, the valve is fully open. In this approach, the valve steady-state characteristic is the opening coefficient change for the applied control signal. In actual conditions, the valve movement occurs with some response lag, dead band, hysteresis, etc.. In general, these characteristics should be taken into consideration. However, for an approximate analysis, the steady-state valve characteristic can be applied. From this point of view, the control valve itself is only part of the energy-throttling mechanism because it is driven by an electromagnetic actuator, itself driven by an electronic amplifier. The last stage in the energy-throttling mechanism is based on forcing the pressed air through orifices whose areas can be controlled, and the last stage of energy conversion is accomplished by simply letting the air pressure act differently on the sides of actuator pistons or vanes.

The type of control valve steady state characteristic depends on the valve type and its geometry. Figure 3.42 shows the steady-state characteristic of a seating type valve, where the control signal is U_C and U_{CS} is the control signal saturation value. This diagram has a line section with an inclination α_V where

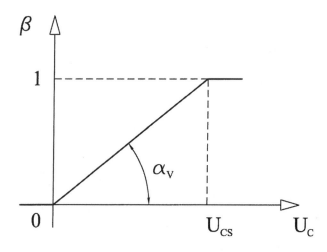

FIGURE 3.42
Steady-state characteristics of seating-type valves.

the opening coefficient β and control signal U_C have direct proportionality to the coefficient $K_{SL} = tg\alpha = \beta/U_C$. The valve has a maximum flow rate value if the opening coefficient $\beta = 1$. For this situation, the control signal $U_C = U_{CS}$ and the control valve achieve the saturation condition, for which subsequently increasing the control signal does not change the value of the valve effective area. For this case, the mathematical model has the following form:

$$\beta = \begin{cases} 1, & \text{if } U_C > U_{CS} \\ K_{SL} \cdot U_C, & \text{if } 0 \leq U_C \leq U_{CS} \\ 0, & \text{if } U_C < 0 \end{cases} \qquad (3.8)$$

The steady-state characteristics of spool (sliding) type control valves are shown in Figure 3.43. Typically, these types of valves use two output lines (load ports) (see Figure 3.12).

For the zero-lap design (Figure 3.43a) in the neutral spool position ($U_C = 0$), the supply, exhaust, and two load ports are closed. If the spool has any shift from this position, then one of the load ports is connected to the supply line and another to the exhaust port; in this case the opening coefficients for these load ports are equal, that is, $\beta_1^+ = \beta_2^-$ and $\beta_2^+ = \beta_1^-$ (where 1 and 2 are the load ports' index). For this design, the mathematical model is:

$$\beta_1^+ = \beta_2^- = \begin{cases} 1, & \text{if } U_C > U_{CS} \\ K_{SL} \cdot U_C, & \text{if } 0 \leq U_C \leq U_{CS} \\ 0, & \text{if } U_C < 0 \end{cases} \qquad (3.9)$$

$$\beta_2^+ = \beta_1^- = \begin{cases} 1, & \text{if } U_C < -U_{CS} \\ -K_{SL} \cdot U_C, & \text{if } -U_{CS} \leq U_C \leq 0 \\ 0, & \text{if } U_C > 0 \end{cases}$$

For an under-lap design (Figure 3.43b) in the neutral spool position ($U_C = 0$), the supply and exhaust ports are simultaneously connected to the load ports. In this case, the mathematical model has the following form:

$$\beta_1^+ = \beta_2^- = \begin{cases} 1, & \text{if } U_C > U_{CS} \\ K_{SL} \cdot U_C + \beta_0, & \text{if } -\dfrac{\beta_0}{K_{SL}} \leq U_C \leq U_{CS} \\ 0, & \text{if } U_C < -\dfrac{\beta_0}{K_{SL}} \end{cases} \qquad (3.10)$$

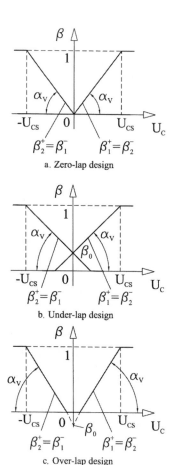

FIGURE 3.43
Steady-state characteristics of spool (sliding) type valves: (a) zero-lap design, (b) under-lap design, and (c) over-lap design.

$$\beta_2^+ = \beta_1^- = \begin{cases} 1, & \text{if } U_C < -U_{CS} \\ -K_{SL} \cdot U_C + \beta_0, & \text{if } -U_{CS} \leq U_C \leq \dfrac{\beta_0}{K_{SL}} \\ 0, & \text{if } U_C > \dfrac{\beta_0}{K_{SL}} \end{cases}$$

For the over-lap spool control valve (Figure 3.43c), the steady-state characteristic has a dead zone. From one point of view it increases the actuator dynamic stability and improves its efficiency; but on the other hand, such valve construction decreases actuator accuracy. In this case, the mathematical model has the following form:

$$\beta_1^+ = \beta_2^- = \begin{cases} 1, & \text{if } U_C > U_{CS} \\ K_{SL} \cdot U_C - \beta_0, & \text{if } \dfrac{\beta_0}{K_{SL}} \leq U_C \leq U_{CS} \\ 0, & \text{if } U_C < \dfrac{\beta_0}{K_{SL}} \end{cases} \tag{3.11}$$

$$\beta_2^+ = \beta_1^- = \begin{cases} 1, & \text{if } U_C < -U_{CS} \\ -K_{SL} \cdot U_C - \beta_0, & \text{if } -U_{CS} \leq U_C \leq -\dfrac{\beta_0}{K_{SL}} \\ 0, & \text{if } U_C > -\dfrac{\beta_0}{K_{SL}} \end{cases}$$

In this case, the coefficient K_{SL} is

$$K_{SL} = \frac{1 - \beta_0}{U_{CS}}.$$

The steady-state performance of double nozzle-flapper valves (Figure 3.44) is similar to that for the under-lap spool valve design (see Figure 3.43b). The major difference is that the charging flows in the nozzle-flapper construction are constant all the time ($\beta_1^+ = \beta_2^+ = 1$); curves in Figure 3.44 describe the changing effective areas of the discharge valve lines. In this case, the mathematical model of the steady-state performance has the following form:

$$\beta_1^+ = 1$$

$$\beta_1^- = \begin{cases} 1, & \text{if } U_C < -U_{CS} \\ -K_{SL} \cdot U_C + 0.5, & \text{if } -U_{CS} \leq U_C \leq U_{CS} \\ 0, & \text{if } U_C > U_{CS} \end{cases} \tag{3.12}$$

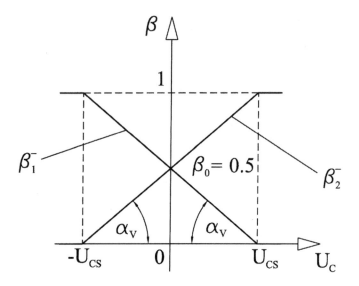

FIGURE 3.44
Steady-state characteristics of double nozzle-flapper valves.

$$\beta_2^+ = 1$$

$$\beta_2^- = \begin{cases} 1, & \text{if } U_C > U_{CS} \\ K_{SL} \cdot U_C + 0.5, & \text{if } -U_{CS} \le U_C \le U_{CS} \\ 0, & \text{if } U_C < -U_{CS} \end{cases}$$

Typically, the steady-state performance of jet-pipe control valves reveals the following correlation between charge and discharge opening coefficients: $\beta_{1,2}^- = 1 - \beta_{1,2}^+$. This correlation is correct if the distance "b" (Figure 3.20) between the receiver holes is negligible and these holes have the rectangular form (Figure 3.20c). The resultant expression for the mathematical model of the steady-state characteristics is:

$$\beta_1^+ = \begin{cases} 1, & \text{if } U_C > U_{CS} \\ K_{SL} \cdot U_C + 0.5, & \text{if } -U_{CS} \le U_C \le U_{CS} \\ 0, & \text{if } U_C < -U_{CS} \end{cases}$$

$$\beta_1^- = 1 - \beta_1^+ \qquad\qquad (3.13)$$

$$\beta_2^+ = \begin{cases} 1, & \text{if } U_C < -U_{CS} \\ -K_{SL} \cdot U_C + 0.5, & \text{if } -U_{CS} \leq U_C \leq U_{CS} \\ 0, & \text{if } U_C > U_{CS} \end{cases}$$

$$\beta_2^- = 1 - \beta_2^+$$

The variation in the opening coefficients for the solenoid valve operating in PWM with the null algorithm (Figure 3.35b) can be described by:

$$\beta_1^+ = \beta_2^- = \begin{cases} 1, & \text{if } U_R \geq U_{CR} \\ 0, & \text{if } U_R < U_{CR} \end{cases}$$

$$\beta_2^+ = \beta_1^- = \begin{cases} 1, & \text{if } U_R < U_{CR} \\ 0, & \text{if } U_R \geq U_{CR} \end{cases} \tag{3.14}$$

$$\beta_1^+ = \beta_2^+ = \beta_1^- = \beta_2^- = 0, \quad \text{if } -\Delta U_R \leq U_R \leq \Delta U_R$$

For the regular PWM operating mode (Figure 3.35a), the value of the tolerated error (ΔU_R) in Equation 3.14 is zero. The carrier signal (U_{CR}) usually has a sawtooth periodic form. However, the sinusoidal signal is sometimes used; in that case

$$U_{CR} = U_{CRM} \cdot Sin\left(\frac{2 \cdot \pi \cdot t}{t_{CR}}\right),$$

where U_{CRM} is the carrier signal amplitude and t_{CR} is its period. Using the sinusoidal carrier signal simplifies the PWM control system and control algorithm, while at the same time this technique adds an additional nonlinear element to the control system.

The variation in the opening coefficients for the solenoid valve operating in Bang-Bang control mode (Figure 3.36) is described by Equation 3.14, in which the carrier signal is zero ($U_{CR} = 0$).

Equation 3.14 describes the ideal valve switching process, which implies that the control valve has a higher bandwidth (at least 30 to 50 times more than the overall actuator bandwidth).

Control valves are complex devices and have many nonlinear characteristics that are significant in their operation. These nonlinearities include electrical hysteresis of the torque motor, change in torque motor output with displacement, change in valve charge coefficient with pressure ratio, friction force of the flow sealing elements, and others.

Many control valve parts are small and thus have a shape, which is analytically nonideal. Therefore, the practical design from a performance standpoint is not necessarily the ideal design from an analytical standpoint. Experience has shown that these nonlinear and nonideal characteristics limit the usefulness of theoretical analysis of control valve dynamics in system design. The analytic representation of control valve dynamics is useful during preliminary design of a new valve configuration, or when attempting to alter the response of a given design by parameter variation. Analysis also contributes to a clearer understanding of control valve operation.

3.6 Performance Characteristics of Control Valves

The static and dynamic performance characteristics of control valves are extremely important in achieving the necessary control process of pneumatic actuators under given dynamic conditions. The dynamic characteristics of valves can be described by examining their response to either a step function input or a sinusoidal input. One of the most critical parameters of a control valve is the time or dynamic response. It is (in simple terms) the lag between the input and the output when the valve is exposed to a dynamic input.

For the step function input, the time parameters related to valve response are stated as follows (according to control technology terminology):

- Time constant is the time to reach about 63% of the demand output level.

- Settling time is the time required for the demand output to reach and stay in a defined tolerance band.

- Delay time is the time required to reach 50% of the demand output signal.

- Rise time is the time required to rise from 10 to 90% of the demand output level.

Valve dynamic response for a sinusoidal function input is normally called a frequency response analysis. In this approach, two principal parameters are usually used: (1) the overall amplitude ratio and (2) the phase angle shift. The frequency response information is normally given as a graph of the attenuation (amplitude ratio) and the phase lag vs. frequency. In this case, the Bode diagram is often used. Typically, the Bode diagram is a log-log graph of the magnitude and phase of impedance as a function of the frequency of a sinusoidal excitation. The magnitude is usually expressed in decibels (dB), where each 20-dB increment represents a factor of 10 in the amplitude ratio

$$G_D = 20 \cdot \log_{10}\left(\frac{a_{out}}{a_{nor}}\right),$$

where a_{out} is the output amplitude and a_{nor} is amplitude of the output signal for very low frequency when the transfer function magnitude tends to 1. In this analysis, the bandwidth frequency is defined as the frequency at which the magnitude response is equal to

−3dB, in this case

$$\frac{a_{out}}{a_{nor}} = \frac{1}{\sqrt{2}} = 0.707.$$

For this condition, the position output lag is in phase by 90° from the input signal.

In addition to the dynamic performance characteristics as discussed above, one must take into account the static performance characteristics in the valve specification. Static performance indicators include flow capacity, internal leakage, hysteresis, symmetry, linearity, threshold, etc.

The threshold is one of the most critical parameters of control valves. This parameter is a measure of internal friction within the servo valve (for the frictionless first stage; there is friction force in the second stage) and internal force (spool driving force). The threshold represents the amount of valve current change necessary to cause a corresponding change in servo valve output.

For a two-stage control valves with jet-pipe or nozzle-flapper pilot stage (Figure 3.27 and Figure 3.28), the mathematical model is obtained by assuming that:

- The hypothesis of an adiabatic process is reasonable.
- The gas is perfect.
- The pressure and temperature within the valve chambers are homogeneous.

The mathematical model for these valves consists of five differential equations: an equation for main-stage spool dynamics, two equations for the rate of change in pressure in main-stage chambers, and two equations for the rate of the change in temperature in these chambers. Finally, the equations that describe the dynamic process in the valve can be written as:

$$\begin{cases} m_V \cdot \ddot{y} = A_{SP} \cdot (P_1 - P_2) - b_{VV} \cdot \dot{y} - 2 \cdot k_{SV} \cdot y - F_{VS} \\[2ex] \dot{P}_1 = \dfrac{k}{V_1} \cdot (G_1^{\,+} \cdot R \cdot T_S - G_1^{\,-} \cdot R \cdot T_1 - P_1 \cdot \dot{y} \cdot A_{SP}) \\[2ex] \dot{P}_2 = \dfrac{k}{V_2} \cdot (G_2^{\,+} \cdot R \cdot T_S - G_2^{\,-} \cdot R \cdot T_2 + P_2 \cdot \dot{y} \cdot A_{SP}) \\[2ex] \dot{T}_1 = \dfrac{T_1 \cdot \dot{y} \cdot A_{SP}}{V_1} + \dfrac{T_1 \cdot \dot{P}_1}{P_1} - \dfrac{R \cdot T_1^2 \cdot (G_1^{\,+} - G_1^{\,-})}{P_1 \cdot V_1} \\[2ex] \dot{T}_2 = \dfrac{T_2 \cdot \dot{y} \cdot A_{SP}}{V_2} + \dfrac{T_2 \cdot \dot{P}_2}{P_2} - \dfrac{R \cdot T_2^2 \cdot (G_2^{\,+} - G_2^{\,-})}{P_2 \cdot V_2} \end{cases} \qquad (3.15)$$

In Equation 3.15, T is the air temperature in the spool working chambers, G is the mass flow rate, T_S is the air temperature in the supply channel, V is the control volume of the main-stage working chambers, which can be expressed as $V_1 = A_{SP} \cdot (y_0 + y)$ and $V_2 = A_{SP} \cdot (y_0 - y)$, where y_0 is the length of the main-stage spool chamber's initial volume.

As indicated in the schematic diagrams shown in Figure 3.45 and Figure 3.46, the main difference between these configurations is that the charging

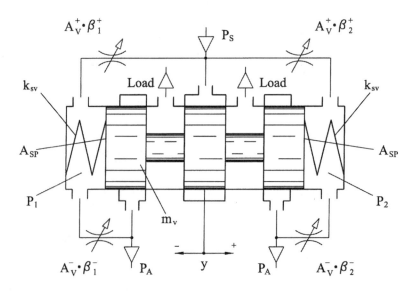

FIGURE 3.45
Schematic diagram of the two-stage control valve with jet pipe pilot stage.

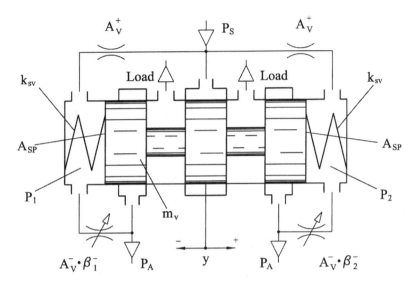

FIGURE 3.46
Schematic diagram of the two-stage control valve with nozzle-flapper pilot stage.

flow in the nozzle-flapper structure is constant, while in the jet-pipe construction this flow is variable. Hence, the mass flow rates for the two-stage control valve with a jet-pipe pilot stage can be described by:

$$G_{1,2}^{+} = A_{V}^{+} \cdot \beta_{1,2}^{+} \cdot P_{S} \cdot \sqrt{\frac{2 \cdot k}{R \cdot T_{S} \cdot (k-1)}} \cdot \varphi(\sigma_{1,2})$$

and

$$G_{1,2}^{-} = A_{V}^{-} \cdot \beta_{1,2}^{-} \cdot P_{1,2} \cdot \sqrt{\frac{2 \cdot k}{R \cdot T_{1,2} \cdot (k-1)}} \cdot \varphi\left(\frac{\sigma_{A}}{\sigma_{1,2}}\right)$$

where A_{V}^{+} is the maximum effective area of the jet-pipe valve charge line, A_{V}^{-} is the maximum effective area in its discharge line, opening coefficients β are defined by Equation 3.13, the flow function $\varphi(\sigma)$ can be determined by Equation 2.14, and the pressure ratios are $\sigma_{1,2} = P_{1,2}/P_{s}$ and $\sigma_{A} = P_{A}/P_{S}$.

For the two-stage control valve with nozzle-flapper pilot stage, the mass flow rates can be determined as:

$$G_{1,2}^{+} = A_{V}^{+} \cdot \beta_{1,2}^{+} \cdot P_{S} \cdot \sqrt{\frac{2 \cdot k}{R \cdot T_{S} \cdot (k-1)}} \cdot \varphi(\sigma_{1,2})$$

and

$$G_{1,2}^- = A_V^- \cdot \beta_{1,2}^- \cdot P_{1,2} \cdot \sqrt{\frac{2 \cdot k}{R \cdot T_{1,2} \cdot (k-1)}} \cdot \varphi\left(\frac{\sigma_A}{\sigma_{1,2}}\right)$$

Here, A_V^+ is the fixed orifice effective area, A_V^- is the maximum effective area of the nozzle-flapper discharge line, and opening coefficients $\beta_{1,2}^-$ are defined by Equation 3.12.

Determination of the higher pressure gain of the pilot stage can be obtained for the steady-state condition of the null spool position. In this case, $y = 0$, $\dot{y} = 0$, $\ddot{y} = 0$, and $T_1 = T_2 = T_S$ = constant. Then, it follows from Equation 3.15 for jet-pipe pilot stage that:

$$\begin{cases} \dfrac{\beta_0 + \Delta\beta}{\Omega \cdot (\beta_0 - \Delta\beta)} = \dfrac{\sigma_1 \cdot \varphi\left(\dfrac{\sigma_A}{\sigma_1}\right)}{\varphi(\sigma_1)} \\[4mm] \dfrac{\beta_0 - \Delta\beta}{\Omega \cdot (\beta_0 + \Delta\beta)} = \dfrac{\sigma_2 \cdot \varphi\left(\dfrac{\sigma_A}{\sigma_2}\right)}{\varphi(\sigma_2)} \end{cases} \tag{3.16}$$

and for the symmetrical double nozzle-flapper pilot stage:

$$\begin{cases} \dfrac{1}{\Omega \cdot (\beta_0 - \Delta\beta)} = \dfrac{\sigma_1 \cdot \varphi\left(\dfrac{\sigma_A}{\sigma_1}\right)}{\varphi(\sigma_1)} \\[4mm] \dfrac{1}{\Omega \cdot (\beta_0 + \Delta\beta)} = \dfrac{\sigma_2 \cdot \varphi\left(\dfrac{\sigma_A}{\sigma_2}\right)}{\varphi(\sigma_2)} \end{cases} \tag{3.17}$$

In Equation 3.16 and Equation 3.17, the $\Delta\beta$ values characterize the control signal, which is a function of the feedback gains, and the effective area ratio Ω is determined by:

$$\Omega = \frac{A_V^-}{A_V^+}$$

The optimal value of this parameter can be defined by assuming that in most industrial environments, the absolute supply pressure maximum value is 1.1 MPa and its minimum value is 0.5 MPa; that is, $0.09 \le \sigma_A = P_A/P_S \le 0.2$. In general, in these pneumatic systems, there may be three combinations of charge and discharge flow conditions (Figure 3.47).

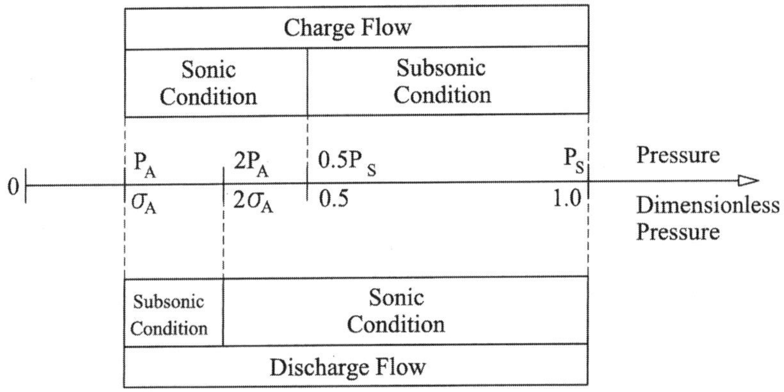

Charge Flow				
Sonic Condition		Subsonic Condition		

FIGURE 3.47
Flow conditions in working spool chambers.

If the charge flows in the working spool chambers are moving at the sonic condition and discharge flows are mowing at the subsonic condition (first combination), the flow functions should be described by the followings equations:

$$\varphi(\sigma_1) = \varphi_*$$

$$\varphi\left(\frac{\sigma_A}{\sigma_1}\right) = 2\varphi_* \cdot \sqrt{\frac{\sigma_A}{\sigma_1} \cdot \left(1 - \frac{\sigma_A}{\sigma_1}\right)}$$

$$\varphi(\sigma_2) = \varphi_*$$

$$\varphi\left(\frac{\sigma_A}{\sigma_2}\right) = 2\varphi_* \cdot \sqrt{\frac{\sigma_A}{\sigma_2} \cdot \left(1 - \frac{\sigma_A}{\sigma_2}\right)}$$

From Equation 3.16, for the jet-pipe pilot stage we obtain:

$$\frac{\beta_0 + \Delta\beta}{\Omega \cdot (\beta_0 - \Delta\beta)} = \frac{2 \cdot \varphi_* \cdot \sqrt{\frac{\sigma_A}{\sigma_1} \cdot \left(1 - \frac{\sigma_A}{\sigma_1}\right)}}{\varphi_*} \cdot \sigma_1$$

and

$$\frac{\beta_0 - \Delta\beta}{\Omega \cdot (\beta_0 + \Delta\beta)} = \frac{2 \cdot \varphi_* \cdot \sqrt{\frac{\sigma_A}{\sigma_2} \cdot \left(1 - \frac{\sigma_A}{\sigma_2}\right)}}{\varphi_*} \cdot \sigma_2$$

Then, the dimensionless pressure in the spool chambers is

$$\sigma_1 = \frac{a^2}{4 \cdot \sigma_A \cdot b^2 \cdot \Omega^2} + \sigma_A$$

$$\sigma_2 = \frac{b^2}{4 \cdot \sigma_A \cdot a^2 \cdot \Omega^2} + \sigma_A$$

and the dimensionless pressure differential is:

$$\Delta\sigma = \sigma_1 - \sigma_2 = \frac{a^4 - b^4}{4 \cdot \sigma_A \cdot a^2 \cdot b^2 \cdot \Omega^2} \tag{3.18}$$

For the symmetrical double nozzle-flapper pilot stage, such formulae can be obtained using Equation 3.17, and the dimensionless pressure in the spool chambers is:

$$\sigma_1 = \frac{1}{4 \cdot \sigma_A \cdot b^2 \cdot \Omega^2} + \sigma_A$$

$$\sigma_2^+ = \frac{1}{4 \cdot \sigma_A \cdot a^2 \cdot \Omega^2} + \sigma_A$$

and dimensionless pressure differential is:

$$\Delta\sigma = \sigma_1 - \sigma_2 = \frac{a^2 - b^2}{4 \cdot \sigma_A \cdot b^2 \cdot a^2 \cdot \Omega^2} \tag{3.19}$$

In the second combination, the charge and discharge flows move at the sonic condition; in this case, the flow functions are described by the followings equations:

$$\varphi(\sigma_1) = \varphi_*,$$

$$\varphi(\sigma_2) = \varphi_*,$$

$$\varphi\left(\frac{\sigma_A}{\sigma_1}\right) = \varphi_*,$$

$$\varphi\left(\frac{\sigma_A}{\sigma_2}\right) = \varphi_*.$$

For the jet-pipe pilot construction, the dimensionless pressure in the spool chambers is

$$\sigma_1 = \frac{a}{b \cdot \Omega}, \quad \sigma_2 = \frac{b}{a \cdot \Omega},$$

and dimensionless pressure differential is:

$$\Delta\sigma = \sigma_1 - \sigma_2 = \frac{a^2 - b^2}{a \cdot b \cdot \Omega} \tag{3.20}$$

For symmetrical double nozzle-flapper pilot construction, the dimensionless pressure in the spool chambers is:

$$\sigma_1 = \frac{1}{b \cdot \Omega}$$

$$\sigma_2 = \frac{1}{a \cdot \Omega}$$

and the dimensionless pressure differential is:

$$\Delta\sigma = \sigma_1 - \sigma_2 = \frac{a - b}{a \cdot b \cdot \Omega} \tag{3.21}$$

In the final (third) combination, the charge flow moves at the subsonic condition and the discharge flow moves at the sonic condition. In this case, the pressures P_1 and P_2 are in the range between $0.5\,P_S$ and P_S. Flow functions can be described by:

$$\varphi(\sigma_1) = 2\varphi_* \cdot \sqrt{\sigma_1 \cdot (1 - \sigma_1)}$$

$$\varphi(\sigma_2) = 2\varphi_* \cdot \sqrt{\sigma_2 \cdot (1 - \sigma_2)}$$

$$\varphi\left(\frac{\sigma_A}{\sigma_1}\right) = \varphi_*$$

and

$$\varphi\left(\frac{\sigma_A}{\sigma_2}\right) = \varphi_*$$

Using Equation 3.16 for the jet-pipe pilot stage obtains:

$$\frac{\beta_0 + \Delta\beta}{\Omega \cdot (\beta_0 - \Delta\beta)} = \frac{\sigma_1 \cdot \varphi_*}{2\varphi_* \cdot \sqrt{\sigma_1 \cdot (1 - \sigma_1)}}$$

and

$$\frac{\beta_0 - \Delta\beta}{\Omega \cdot (\beta_0 + \Delta\beta)} = \frac{\sigma_2 \cdot \varphi_*}{2\varphi_* \cdot \sqrt{\sigma_2 \cdot (1 - \sigma_2)}}$$

Then the dimensionless pressure in the spool chambers is:

$$\sigma_1 = \frac{4 \cdot a^2}{4 \cdot a^2 + \Omega^2 \cdot b^2}$$

$$\sigma_2 = \frac{4 \cdot b^2}{4 \cdot b^2 + \Omega^2 \cdot a^2}$$

and the dimensionless pressure differential is:

$$\Delta\sigma = \sigma_1 - \sigma_2 = \frac{4 \cdot a^2}{4 \cdot a^2 + \Omega^2 \cdot b^2} - \frac{4 \cdot b^2}{4 \cdot b^2 + \Omega^2 \cdot a^2} \qquad (3.22)$$

For the symmetrical double nozzle-flapper pilot design, such correlations are obtained using Equation 3.17, and the dimensionless pressure in the spool chambers is:

$$\sigma_1 = \frac{4}{4 + \Omega^2 \cdot b^2}$$

$$\sigma_2 = \frac{4}{4 + \Omega^2 \cdot a^2}$$

and dimensionless pressure differential is:

$$\Delta\sigma = \sigma_1 - \sigma_2 = \frac{4}{4 + \Omega^2 \cdot b^2} - \frac{4}{4 + \Omega^2 \cdot a^2} \tag{3.23}$$

In Equation 3.18 through Equation 3.23, the parameters "a" and "b" are $a = \beta_0 + \Delta\beta$ and $b = \beta_0 - \Delta\beta$, respectively.

Analysis of Equation 3.18, Equation 3.20, and Equation 3.22 for the jet-pipe pilot valve construct, and also of Equation 3.19, Equation 3.21, and Equation 3.23 for the nozzle-flapper pilot design, reveals that the dimensionless pressure differential reaches its maximum value in the third part, where the charge flow moves at the subsonic condition and the discharge flow moves at the sonic condition. Then, for the jet-pipe construction, the derivation of the dimensionless pressure differential $\Delta\sigma$ (Equation 3.22) for Ω becomes:

$$\frac{d(\Delta\sigma)}{d(\Omega)} = 8 \cdot a^2 \cdot b^2 \cdot \Omega \cdot \left[\frac{1}{(4 \cdot b^2 + a^2 \cdot \Omega^2)^2} - \frac{1}{(4 \cdot a^2 + b^2 \cdot \Omega^2)^2} \right]$$

The maximum value of $\Delta\sigma$ can be determined by

$$\frac{d(\Delta\sigma)}{d(\Omega)} = 0, \text{ or } (\Omega^2 - 4) = 0,$$

from which $\Omega = 0.5$.

By carrying out the same calculations for the double nozzle-flapper option (Equation 3.23), it can be shown that the maximum value of the dimensionless pressure differential $\Delta\sigma$ is if $\Omega = 4$.

These results are very important for the optimization of the geometric parameters of the pilot stage. In addition, Figure 3.48 shows these conclusions, in an illustration how the curves have been obtained by using Equation 3.18 through Equation 3.23. For all cases, $\Delta\beta$ is 0.01 (1% of full range), $\Delta\sigma_A \approx 0.143$ ($P_S = 0.7$ MPa), and $\Omega = A_V^-/A_V^+$ is changed in the range from 0.5 to 5.

For the very same performance, the threshold and hysteresis characteristics of the jet-pipe pilot stage are nearly two times higher than the static characteristics of the double nozzle-flapper pilot stage.[103]

In addition, the computer simulation of the dynamic response of these valves (developed by integration of Equation 3.15 using the fourth rank Runge-Kutte stability criteria) shows that for the jet-pipe pilot stage, the point $\Omega = 2$ is the "critical" point, for which the settling time value (the time required for the spool to settle within 3% error band of its defined position) is minimum. That is, $\Omega = 2$ is the optimum value for the dynamic characteristics of this servo valve.

Similar processes can be obtained in the servo valve with a double nozzle-flapper pilot stage. In this case, the critical point for which the settling time has minimum volume is $\Omega = 4$.

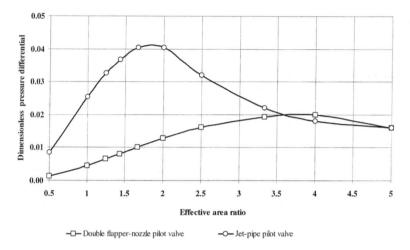

FIGURE 3.48
Dimensionless pressure differential in working spool chambers.

However, for the optimal geometry parameters of the pilot stage, the dynamic behavior of the two-stage control valve with jet-pipe and flopper-nozzle pilot design looks the same. For example, the frequency response of the two-stage control valves, which consists of the pilot stages with optimal parameters, is illustrated in Figure 3.49. In this case, the bandwidth is about 40 Hz for both constructions.

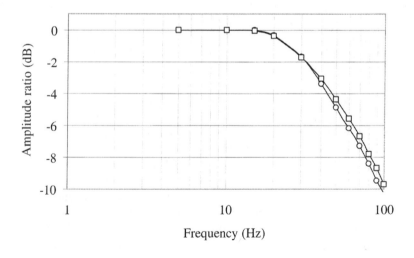

FIGURE 3.49
Frequency response of two-stage control valve (Bode diagram).

4

Determination of Pneumatic Actuator State Variables

High dynamic and static performances of the pneumatic actuators with closed-loop control systems can be achieved using "multilevel" (displacement, velocity, acceleration, and others) feedback modern control algorithms. An effort to achieve accurate state variables of the pneumatic actuator is the major aim of a design process in such systems. This issue can be solved by two approaches:

1. Measurement of these parameters using appropriate sensors
2. Application of parameter estimation techniques

In general, the displacement, velocity, acceleration, force, moment, and pressure can be measured in the pneumatic actuator. Different sensor designs may read incrementally or absolutely; they may contact a sensed object or operate without contact; and they probably will range through various levels of performance and price. Direct measurement of all state parameters of the pneumatic actuator during operation may be sometimes exceedingly difficult and expensive, if possible at all. Hence, the application of parameter estimation techniques can provide this key information indirectly. A variety of such techniques is available; however, one of the most powerful is the observer method. In state regulators, the differential method is often used; in this case, the velocity and acceleration of the actuator load are calculated by differentiating a measured displacement signal.

4.1 Position and Displacement Sensors

In the open-loop pneumatic positioning actuators, using hard mechanical stops positioned along the length of the cylinder can be an effective solution for repeated stops in the same location. These actuators consist of positioning sensors, which are usually attached to the outside of the actuator (on the cylinder tube or on the hard stop housing). The basic function of the positioning sensors indicates the desired position that the actuator reaches. Sometimes, the positioning sensors are used in the solenoid valves to indicate the valve's position. According to the operating mode of the positioning sensors, these sensors are discrete devices. In pneumatic actuators, basically two general categories of discrete positioning sensors are utilized:

1. Mechanical contact type, which consists of limit switches and magnetic reed switches
2. Noncontact type, which are inductive, Hall effect, and magneto-resistive sensors

Noncontact position-measurement devices offer several advantages over contact types. They provide higher dynamic response with higher measurement resolution and lower hysteresis. In addition, they can work in a highly dynamic process.

The discrete positioning sensors have the three most important characteristics: (1) response time, (2) hysteresis, and (3) repeatability.

Response time is simply the amount of time it takes a sensor or switch to turn "on" or "off." A fast response time becomes critical as end users operate cylinders at faster speeds, to lower process cycle times.

For discrete positioning sensors, hysteresis is normally specified as the maximum difference that occurs in a complete cycle between the "on" and "off" conditions.

Repeatability is the range that the switch will turn "on" or "off," given the same physical switching point. Repeatability is the absolute accuracy of the switch when subjected to any combination of normal operational environments. For example, a change in temperature or voltage to power a switch will cause a change in the position where the switch operates.

The limit switches that are used in pneumatic actuators with open-loop control (which consists of shock absorbers and hard stops) are usually inexpensive devices and their response time, hysteresis, and switch position repeatability are adequate for most pneumatic actuator applications. However, these switches have a number of disadvantages; they require mechanical contact with a mobile part, they have moving parts, and they are subject to wear, being jammed, or broken. Because of the movement of the actuator and components of the limit switch, they must often be readjusted to remain accurate. The lifetime for this type of switch is significantly less than that of solid-state devices. Basically, these are primary reasons why limit switches are not extensively used in pneumatic actuating systems.

The reed switches do not have mechanical contact with a mobile part of the actuator but they do contain hermetically sealed reeds or contacts. When a magnet moves close to the switch, the reeds become magnetized and the switch contacts will close or open, depending on switch configurations (normally open or normally closed). Although the reed switch is a mechanical device, if used properly, the lifetime for this type of switch is long. The advantages of a reed switch include high sensitivity to magnetic fields, no leakage current or voltage drop, the ability of switching a high voltage and current, low cost, and high repeatability. The disadvantages of this device are that it is electronically noisy, and it has a slow response time and a relatively large amount of hysteresis. The amount of hysteresis is due to the mechanical response of the reed contacts and the current flow through the device.

Inductive switches are a noncontact type of sensor; they detect the proximity of a metallic object by gauging the effect the metal target has upon a magnetic field emanating from the sensor. The inductive sensor generates an oscillating magnetic field that, in turn, induces surface currents on the metal target. These surface currents are known as eddy currents. When the metal target is outside the sensor's range, the magnitude of the oscillations is not affected. However, when the target is inside the range of the sensor, the oscillating field is attenuated. As the sensor moves closer to the target, the oscillations become smaller. Inductive sensors are very accurate, highly repeatable devices that employ solid-state technology and exhibit very low hysteresis. They typically have fast response times and exhibit long lifetimes. They do not suffer false triggering from nonmetallic objects; however, metal objects can deceive them. Nevertheless, inductive sensors are one of the most expensive types of discrete positioning sensors.

The Hall effect switch is a solid-state electronic device with no moving mechanical parts and therefore it is more reliable than a reed switch. The Hall effect switches are compact devices with fast response times, high repeatabilities, and long lifetimes. Their major disadvantage is that they are highly subject to thermally induced errors. As with other noncontact types of sensors, the Hall effect switch must be aligned with the magnetic field for false triggering to occur.

The magnetoresistive sensor combines the best features of the Hall effect device and the reed switch. It is a very small, medium-cost, high-speed, noncontact solid-state device with excellent repeatability. It has very low hysteresis and can be fabricated repeatedly at the specified gauss level. A magnetoresistor is a variable resistor that changes as a function of the applied magnetic field. Conventional magnetoresistance materials have very low saturation. These devices have a number of inherent advantages over reed switch technology and, over time, the market for this product has grown. The solid-state nature of the magnetoresistive sensor, its small size, and the elimination of mechanical contact closure offset the relatively higher costs associated with this technology.

The closed-loop pneumatic positioning actuator consists of a position transducer (displacement sensor) to measure and convert the actuator output signal to an electrical signal. This feedback signal is compared with the command signal, and the resulting error signal is applied to reach the necessary positioning or tracking of the movement.

Displacement sensors can be divided into two types:

1. Analog (absolute type), in which the output signal has continuous form. There are potentiometric sensors, inductive transducers, magnetostrictive devices, and capacitive sensors.

2. Digital (incremental type), in which the output signal has a discrete form. This type of sensor includes encoders and magnetostrictive sensors, which also operate in this mode.

An important difference between incremental and absolute sensors is that incremental sensors typically must be reinitialized after power-down by moving the monitored actuator to a home position at power-up. This limitation is unacceptable in some applications.

The accuracy of the closed-loop actuation system can be no more accurate than the displacement sensor itself. The displacement sensors have the following very important characteristics: permissible motion speed, measuring range, reliability, and cost. The most important types of displacement sensor inaccuracies include:

- *Repeatability.* When the load returns to a given position, will the sensor output always return to the same value, regardless of the direction of approach? Errors of this type can be caused by lost motion in the actuator as well as by the sensor itself.

- *Resolution.* The outputs of some sensors are not perfectly smooth. Instead, they look like a staircase. Wire-wound potentiometers are a classic example of this phenomenon.

- *Linearity.* Sometimes it is necessary that the positioning system output be a linear function of the command input. This might be important in position tracking actuators where both the command and feedback signals are generated by potentiometers, whose outputs must be matched with each other. Linearity on the order of ±0.5% of full scale is common, while ±0.1% or better is feasible. Sometimes the sensor mounting can create nonlinearity. Another source of sensor inaccuracy, which is often overlooked, is ripple. This is generally a characteristic of sensors excited by AC voltage, and is caused by imperfect filtering of the carrier signal. If the carrier frequency is selected properly, the response of the actuation system to the ripple can be minimized.

Potentiometric displacement sensors, also referred to as linear resistive transducers, are one of the simplest devices and are a common continuous displacement sensing device. The potentiometric sensor operates like a variable resistor, in which a wiper moves in correspondence with the object being measured. The wiper completes a circuit for a current flowing through a resistance track. The output resistance fluctuates, depending on its location on the track and thus measures the actuator position. A potentiometric sensor has three integral connections: (1) excitation voltage input, (2) common return, and (3) signal output connections. The input excitation (usually 10 V DC) is attached to the resistive side of the potentiometric sensor. The input signal flows through the resistive element until it contacts the precious metal wiper. The wiper couples the input excitation to the conductive side of the sensor, allowing the output signal to exit the sensor. The common return is that it connects the negative side of the transducer to the return path of the excitation source, thus completing the circuit. The feedback voltage measured from the output signal changes proportionally to the position of the electric wiper, which is typically attached to the actuator output link. This enables continuous tracking of the actuator position.

Potentiometric displacement sensors typically require a 10-V DC input excitation, making them easy to interface with either PLCs or other data acquisition devices. These devices are generally used because of their small size, low cost, ease of integration, and output, which can be either AC or DC. Their primary disadvantages are limited motion, limited life span due to wear, and high torque required to rotate the wiper contact.

In practice, two options of linear potentiometric displacement sensors for pneumatic actuators are available: (1) the internally mounted sensor and (2) the externally mounted sensor. Figure 4.1 shows the internally mounted linear potentiometric displacement sensor for pneumatic cylinders. In such constructions, a conductive plastic potentiometer is most often used. These potentiometers (internally or externally mounted) have the following requirements: a measuring range of up to 4 m, a maximum velocity of about 1.5 to 2 m/s, a resolution of up to 0.01 mm, a lifetime of up to $13 \cdot 10^6$ cycles with stroke of 4 mm (this parameter depends on the speed of motion), a temperature range from $-30°C$ to $+150°C$, and a maximum acceleration of

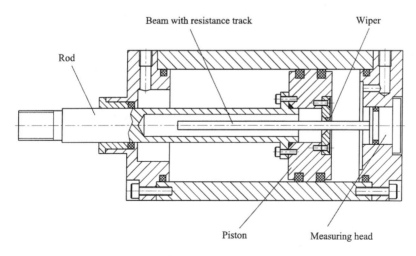

FIGURE 4.1
Internally mounted linear potentiometric displacement sensor for pneumatic cylinder.

about 200 m/s^2. In industrial practice, the potentiometric displacement sensors, where the wire-wound potentiometer is used, have the following requirements: the measuring range is up to 3 m, the maximum velocity is about 1 to 1.5 m/s, the resolution is up to 0.1 mm, the lifetime is up to $2.5 \cdot 10^6$ cycles with stroke of 4 mm (this parameter depends on the speed of motion), the temperature range is from $-30°C$ to $+120°C$, and maximum acceleration is about 200 m/s^2.

Inductive transducers used for continuous displacement measurements are typically linear variable differential transducers (LVDTs) for linear motion, and rotary variable differential transducers (RVDTs) for rotary motion. The LVDT, as well as the RVDT, is a transducer that converts mechanical motion into an electrical signal through mutual induction. A typical LVDT consists of a primary coil and two secondary coils symmetrically spaced on a cylindrical form. A free-moving, rod-shaped magnetic, usually nickel-iron, core inside the coil assembly provides a path for the magnetic flux leaking of coils. When the primary coil is energized by an external AC source, voltages are induced in the two secondary coils. These are connected series opposing so the two voltages are of opposite polarity. Therefore, the net output of the sensor is the difference between these voltages, which is zero when the core is at the center or null position. If the core is moved from the null position, the induced voltage in the coil toward which the core is moved increases, while the induced voltage in the opposite coil decreases. This action produces a differential voltage output that varies linearly with changes in core position. The phase of this output voltage changes abruptly by 180° as the core is moved from one side of null to the other. The core must always be fully within the coil assembly during operation of the LVDT; otherwise, gross nonlinearity will occur.

In pneumatic actuating systems, the LVDT is often used to measure the displacement of a spool in the proportional control valves and as the displacement feedback device in closed-loop positioning actuators with short stroke. For example, Figure 4.2 shows the LVDT with single-acting linear diaphragm actuator, where the core of the LVDT is attached to the actuator rod via the diaphragm piston, while the coil assembly is fastened to a stationary rear cover. Displacement of the core precisely represents the movement of the actuator rod. In practice, it is not cost effective to use the LVDT to perform over 150-mm displacement in pneumatic actuators.

The fact that the movable core is not mechanically connected to the frame of the LVDT makes it essentially a noncontact transducer. The noncontact measurement assures an almost infinite lifetime as well as input/output isolation. The inherent symmetry of the LVDT construction produces high null repeatability; its null position is extremely stable. Thus, this device can be used as an excellent null position indicator in high-gain, closed-loop control systems, such as proportional control valves. An LVDT is predominantly sensitive to the effects of axial core displacement motion and relatively insensitive to radial core motion. This means the LVDT can be used in

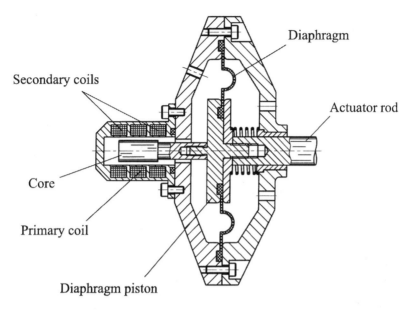

FIGURE 4.2
Single-acting linear diaphragm actuator with LVDT.

applications where the core does not move in an exact straight line, as, for example, in diaphragm and bellows actuators.

In practice, the AC- and DC-operated LVDT constructions have wide application. The DC LVDT is provided with onboard oscillator, carrier amplifier, and demodulator circuitry. The AC LVDT requires these components externally. Because of the presence of internal circuitry, the temperature limits of a DC LVDT typically range from −25 to +70°C. The AC LVDT is able to tolerate extreme variations in operating temperature (from −40 to +120°C) that the internal circuitry of the DC LVDT cannot tolerate. The major advantages of DC devices are the ease of installation, the ability to operate from dry cell batteries in remote locations, and lower system cost. The AC LVDT advantages include greater accuracy and a smaller body size.

In principle, the LVDT position resolution is only limited by electronic noise, which is kept low by using a simple phase-locked signal detection technique. Practically, the modern type of LVDT has a submicron-level resolution.

Another type of noncontact displacement sensor is the magnetostrictive linear position sensor, which also has a competitive position in pneumatic actuation systems. Magnetostriction is a property of ferromagnetic materials (iron, nickel, and cobalt) that expands or contracts when placed in a magnetic field. These changes are due to the behavior of the magnetic domains (tiny permanent magnets) within the material. When a material is not magnetized, the domains are randomly arranged; if a magnetic field is applied, those domains will align, causing a change in shape or length. A magnetostrictive position sensor takes advantage of this effect to induce a mechanical wave or strain pulse in a specially designed sensing element called a waveguide, which is a long, thin ferromagnetic wire or tube. The time-of-flight of this pulse is measured and can be equated to distance because the speed of traverse is both constant and repeatable. The pulse is by shortly causing interaction between two magnetic fields. The first field originates from a permanent magnet that passes along the outside of the sensor tube. The second one, encompassing the entire waveguide, is created when a current interrogation pulse is applied to the waveguide. The interaction point between these two magnetic fields, which is the current magnet position, produces a strain pulse. The strain pulse, or wave, travels at the speed of sound in the waveguide alloy (approximately 3000m/s) along the waveguide until the pulse is detected at the head or coil end of the sensor. The position of the moving magnet is determined precisely by measuring the elapsed time between the launching of the electronic pulse and the arrival of the strain pulse. Noncontact position sensing is thus achieved with absolutely no wear on any of the sensing elements.

Currently, magnetostrictive position sensors have wide application in pneumatic actuating systems, particularly in cylinder applications. This is because the position magnet can send its magnetic field through many solid nonmagnetic materials (e.g., aluminum, stainless steel, polymers, and others), from which many cylinder barrels are manufactured. This makes the sensor capable of reading between sealed areas. For example, the long, thin waveguide section of the sensor can extend within the length of the rod and piston part of the cylinder assembly. The position magnet would then be mounted externally on an adjacent area of the cylinder wall. However, the internally mounted structure allows one to achieve a more compact design. Figure 4.3 is a schematic diagram of such a construct; it looks similar to

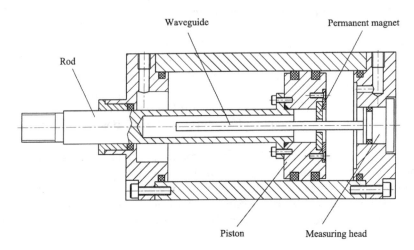

FIGURE 4.3
Internally mounted magnetostrictive position sensor for pneumatic cylinder.

cylinder construction with the internally mounted linear potentiometric displacement sensor.

This arrangement will provide a linear indication of the amount of extension of the piston rod. The magnetostrictive position sensor is unaffected by a wide range of temperatures as well as high shock and vibration environments often found in cylinder applications. For installation convenience, versions of magnetostrictive position sensors are available with sensor-head housings small enough to mount right on the cover of a cylinder.

In industrial practice, magnetostrictive position sensors have the following requirements: measuring range up to 7 m, displacement speeds up to 10 m/s, resolution of about 0.002 mm, linearity ranges from 0.1 to 0.025% of full-scale measurement, acceleration up to 100 m/s^2, and temperature range ranges from −30 to +75°C.

Capacitive displacement transducers are analog, noncontact devices. Usually, these sensors consist of two plates (electrodes), one of them moving relative to the other. A fixed electrode (called the probe) is installed on the transducer base, and the moving electrode (named the target) is connected to the moving part of the actuator. Because the electrode size and dielectric medium (usually air) remain unchanged, capacitance is directly related to the distance between the electrodes. Precise electronics convert the capacitance change information into a signal proportional to distance.

There are two types of these sensors: (1) two electrodes and (2) single-electrode transducers. The theoretical measurement resolution is limited only by quantum noise. In practical applications, stray radiation, electronics-induced noise, and geometric effects of the electrodes are the limiting factors. Resolution on the order of picometers is achievable with short-range, two-electrode devices. Basically, these sensors are used as a displacement feedback device in piezo actuators for control valves. In that case, the measuring range is up to 1 mm and a resolution of approximately 0.01 microns is achievable. Such a capacitive sensor has good dynamic requirements, for example, its typical bandwidth is up to 1 kHz.

The single-electrode capacitive displacement sensor provides significantly less resolution, linearity, and accuracy than two-electrode types. In general, this type of sensor is used for measuring displacement up to 15 mm, with an accuracy of 0.2% and repeatability of 0.1% of full-scale measurement. The temperature range is from 0 to +200°C. In pneumatic actuating systems, capacitive displacement sensors are seldom used as the feedback devices because they are not cost effective.

Encoders are also noncontact displacement sensors and they can be divided into two main groups: magnetic and optical. Traditionally, two very different types of encoders exist in automation: incremental and absolute. The two types vary greatly in their design and in the type of interface electronics typically used to read the encoder. Applications determine which type of encoder is required to satisfy a particular system requirement. Of the two types of encoders, incremental encoders are most commonly used because of their low cost and simple application. Absolute encoders, although more costly than incremental types, provide position information in a very different manner. In practice, absolute encoders are seldom used in pneumatic actuators.

Incremental magnetic encoders consist of three main components: magnetic tape, sensor head, and translator. The magnetic tape has flexible rubber tape bonded to the steel strip and provides the scale for the measuring system. The absolute information is magnetized onto the tape in a sequential code. This position information is enhanced by interpolation of sine/cosine signals provided by an additional incremental track that is magnetized on the tape. The magnetic tape is laminated onto a ferromagnetic steel strip, which is used as both a magnetic return path and a dimensionally stable mounting aid. The magnetic tape is supplied with an adhesive backing for mounting. A noncontact magnetic sensor head with integrated electronics is mounted on the apparatus whose displacement is to be measured. As the read head moves over the measuring tape, its position output signal has a resolution up to 1 micron. The measuring range is up to 40 m, the operating temperature ranges from −30 to +75°C, and the maximum displacement speed for high resolution (up to 1 micron) is about 1 m/s (this parameter usually depends on the resolution requirement).

There are two basic types of optical encoders: rotary and linear. While the technical principles behind them are similar, their specific applications most often are not.

Linear optical encoders operate in the reflective mode. The scale of these devices can be constructed from glass, metal, or tape (metal, plastic, etc.). The markings on the scale are read with a moving head assembly that contains the light source and photodetectors. The resolution of a linear encoder is specified in units of distance and is dictated by the distance between markings (an encoder system is typically based on a common 20-micron pitch scale). Reflective type encoders bounce collimated beam off a patterned reflective code scale. Fitting all the electronics of a reflective encoder onto one side of the code scale makes it a more compact design than transmissive types. In practice, the light source used for most popular types of encoders is a point source light-emitting diode (LED) rather than a conventional LED or filament. Overall, there are two physical versions of an encoder linear scale: exposed or enclosed. The versions themselves dictate, fairly accurately, the type of application.

With an *enclosed* or "sealed" scale, the reading head assembly mounts on a small carriage guided by ball bearings along the glass scale; the carriage connects to the actuator slide via a backlash-free coupling that compensates for alignment errors between the scale and the actuator slide. A set of sealing lips protects the scale from contamination. Typical applications for enclosed linear encoders primarily include machine tools and cutting-type machines, or any type of machine located in a harsh environment.

Exposed linear encoders also consist of a glass scale and a reading head assembly but the two components are physically separated. The typical advantages of a noncontact system include easier mounting and higher traversing speeds because no contact or friction between the reading head assembly and scale exists. Exposed linear scales can be found in measuring devices, translation stages, and materials handling equipment.

Another version of the scale and reading head assembly arrangement is one that uses a metal base rather than glass for the scale. With a metal scale, the line grating is a deposit of highly reflective material such as polished steel or gold coating steel that reflects light back to the scanning unit and onto photovoltaic cells. The advantage of this type of scale is that it can be manufactured in extremely long lengths, up to 30 m, for larger machines. Glass scales are limited in length to typically 3 m. In these cases, the maximum displacement speed for 0.5-micron resolution is about 0.5 m/s. Operating temperatures range from 0 to +50°C.

Most optical rotary encoders operate in transmissive mode. These devices consist of a glass disk with equally spaced markings, a light source mounted on one side of the disk, and a photodetector mounted on the other side. The components of rotary optical encoders are typically packaged in a rugged enclosed housing that protects the light path and the electronics from dust and other materials frequently present in hostile industrial environments. When the disk rotates, the markings on the disk temporarily obscure the passage of light, causing the encoder to output a pulse. Typical glass rotary encoders have from 100 to 6000 markings. To detect the direction of motion and increase the effective resolution of the encoder, a second photodetector is added and a mask is inserted between the glass disk and the photodetectors. The two photodetectors and the mask are arranged so that two sine waves (which are out of phase by 90°) are generated as the encoder shaft is rotated. These quadrature signals are either sent out of the encoder directly as analog sine wave signals or squared using comparators to produce digital outputs. To increase the resolution of the encoder, a method called interpolation is applied to either or both the sine wave or square wave outputs. Direction is derived by simply looking at the timing of the quadrature signals from the encoder. In general, this type of sensor is used for measuring angular displacement with a maximum rotating speed of 12,000 rpm. The operating temperature ranges from 0 to +60°C.

In most cases, the dynamic response of the position and displacement sensors is negligible compared to the servo-valve and load resonance; and, most often, in the mathematical model of the actuator, these sensors are described as ideal dynamic elements (without lag).

4.2 Measurement of Velocity, Acceleration, Pressure, Force, and Torque Signals

A digital or analog speed transducer can carry out the measurement of the rotary motion speed of a pneumatic actuator. Tachogenerators are analog speed transducers, and they are commonly used in providing velocity measurement and feedback for a rotary motion. Such a device operates on the same principle as a DC generator but is designed for use as a sensor. The voltage produced by windings rotation in a magnetic field is proportional to the rotary speed. A tachogenerator can also indicate the direction of rotation by the polarity of the output voltage. When a permanent-magnet style DC generator's rotational direction is reversed, the polarity of its output voltage will switch. In measurement and control systems where directional indication is needed, the tachogenerator provides an easy way to determine that. The main disadvantage of the tachogenerator is a ripple in the output signal due to the existence of a commutator with brush noises. Using a low-pass filter and increasing the number of the commutator poles can improve a poor signal-to-noise ratio at high speeds, but may not be effective at low speeds. One of the more common voltage signal ranges used with tachogenerators is 0- to 10-V. Obviously, because a tachogenerator cannot produce voltage when it is not turning, the zero cannot be "live" in this signal standard. Tachogenerators can be purchased with different "full-scale" (10-V) speeds for different applications. Although a voltage divider could theoretically be used with a tachogenerator to extend the measurable speed range in the 0 to 10 V scale, it is not advisable to significantly over-speed a precision instrument like this, or its life will be shortened. Typical tachogenerators can operate with maximum speeds up to 12,000 rpm and have linearity about 0.1% of full-scale measurement. Their operating temperature range is from −20 to +100°C. Special designs are available with a high output for very low speeds.

Tachometers are digital speed transducers that have several constructs, all of which are based on the use of a timer in conjunction with regularly spaced pulses produced by rotary motion. Essentially, they measure speed by dividing a known displacement by the time taken for it to occur. Because there are several alternatives available, these methods will be described both in terms of the methods of production of suitable pulses and their subsequent conditioning and timing. All tachometers derive from discrete, noncontact positioning sensors, which include inductive, Hall effect, magnetoresistive, capacitive, and optical sensors.

The magnetic tachometer is an active device that produces a voltage in the winding when a magnetically soft material (ferromagnetic) target passes close to the probe. The output voltage is sensitive to the speed of movement of the target, the air gap (typically 0.5 mm), and the shape of the target itself. Higher speeds and larger mass targets will produce larger outputs, which can be on the order of 50 V at peak. The probe is generally cheap and reasonably compact. Higher sensitivities are available with a larger probe suitable for lower speeds. The eccentricity of the mounting of rotating targets can cause problems, including false triggering or missed pulses. The minimum output level will define a minimum speed at which the probe can work. If a multi-toothed target is used, then 10 rpm should be possible. The maximum rotating velocity rate is about 4000 rpm.

A tachometer based on the Hall effect is a passive device that requires an energizing supply to provide the current as well as the magnetic field. In practice, this means the use of a permanent magnet either mounted on the target (with a spacing of about 2 mm) or incorporated into the sensor as a vane. It is sometimes possible to mount a powerful magnet behind the target. These transducers are rugged and very cheap. Maximum operating rotary speeds on the order of 15,000 rpm are possible.

Tachometers with inductive proximity detectors are usually based on the eddy current principle and contain two coils: one energizing and the other balancing. The change in inductance due to eddy currents on the target surface is detected, causing a change in the output state. Inductive devices will detect the presence of any metallic target. They will work from the slowest speeds up to maximum rotary speeds on the order of 150 rpm with air gaps of 1 to 2 mm.

Tachometers with capacitive proximity sensors are also available that will work with a wide range of nonmetallic materials, including plastics. However, practically these devices are not used in pneumatic actuators because of their high cost and low range of measurement speed (only a few tens of rpm).

Tachometers based on the use of optical switches operate with a triggering signal from a photodiode. These tachometers will operate at the slowest speeds; however, they are obviously not good in dusty or dirty environments. The reflective types suffer problems from extraneous light sources.

Optical encoders are generally used as position sensors but they give an output that is appropriate for use in speed measurements. Inherently, encoders produce waveforms that are sinusoidal but they normally have internal circuits that produce digital pulse train outputs in quadrature. Incremental encoders (or a special tachometer encoder) might have from 100 to 5000 pulses per revolution output, and edge-triggered conditioning could allow a multiple of four using both output waveforms. They are thus appropriate where lower rotary speeds will be detected with high accuracy. These devices are essentially shaft-mounted devices that should be connected co-axially through a coupling. The cost varies — from the cheaper end with open collector outputs and a "kit of parts" to the upper end, which is entirely enclosed with fully integrated conditioning circuits. The high pulse rate output is the principal advantage and typical rotating speeds of 2000 rpm are possible for measurement.

For linear applications, similar measurement techniques can usually be used. Linear velocity transducers work on the same principle, with a magnet attached to the moving component, producing a moving field inside a coil. These devices consist of high coercive force, permanent magnet cores, which induce a sizable DC voltage while moving concentrically within shielded coils. The basic design of the linear velocity transducer permits operation without external excitation, while the generated output voltage varies linearly with core (magnet) velocity. Commercial transducers usually have two coils connected to the sum of their induced voltage, with the magnet mounted inside a sleeve on a rod similar to a position sensor. A linear velocity transducer will only work through a limited stroke of movement while the magnet is inside the coils. This can be called the linear range of the transducer and refers to its physical operating stroke rather than the range of its output value. Linear velocity transducers are available with linear ranges up to 0.5-m stroke, with a nonlinearity of 1%. These devices tend to have high coil resistance to give a good sensitivity; they require impedance matching if the output is not to be affected by the load input resistance. The operating temperature range is −40 to +80°C, and maximum operating speeds on the order of 0.5 m/s are possible.

An optical or magnetic encoder can be used as a linear velocity transducer. A strip or bar with optical or magnetic marks is attached to the stationary base and the read head is mounted on the moving object whose velocity one wants to measure. The frequency of the pulses or their width is proportional to the measuring velocity. This technique allows reaching the maximum velocity measurement up to 1 m/s with a resolution of about 0.5% of full-scale measurement. Operating temperatures range from 0 to +50°C.

Rotary actuator feedback (rotary encoder or resolver) is relatively inexpensive and is not "length" dependent. Linear encoders are many times more expensive than their rotary counterparts. Therefore, sometimes the rack-and-pinion linear measurement system is used. Such a system consists of lengths of precision measuring racks, an enclosed channel, and an incremental rotary encoder or resolver with an integral pinion. Usually, racks are ground with a crown, which reduces the need for precision pinion alignment. As motion occurs, the pinion rotates along the racks, causing the encoder/resolver contained within the sealed housing to rotate. This rotation reflects the change in linear position, and the velocity is determined from the frequency of pulses coming from the encoder/resolver. In practice, the pinion is lightly spring loaded against the rack to keep the total backlash at less than 1 or 2 microns. The pinion is constructed of a metal softer than the ground rack to allow the pinion to be honed to the rack pattern. Basically, rack-and-pinion linear measurement systems are cost-effective devices with high reliability.

Accelerometers (acceleration transducers) are inertial measurement devices that convert mechanical motion to an electrical signal. A wide variety of accelerometers can be categorized as force sensing or displacement sensing devices. Force sensing accelerometers use various techniques for measuring forces such as piezoelectric crystal, silicon capacitive, strain gage, force balance, and micro-machined resonators. Accelerometers based on force sensing operate by directly detecting the force applied on a proof mass as a result of measuring. Based on Newton's second law, (force = mass × acceleration), as the accelerated force is applied to the sensing device (usually, a strain gage or piezoelectric material), the resistance changes in the sensing element, which will generate an output proportional to acceleration. Such accelerometers usually consist of a piezoelectric crystal and mass normally enclosed in a protective metal case. As the mass applies force to the crystal, the crystal creates a charge proportional to acceleration. Some sensors have an internal charge amplifier, while others have an external charge amplifier. The charge amplifier converts the charge output of the crystal to a proportional voltage output.

In a capacitive-type accelerometer, the sensing element consists of a small mass and flexure element chemically etched from a single piece of silicon. The flexure element supports the seismic mass between two plates that act as electrodes. Under acceleration, the mass moves from its center position, and the capacitive half-bridge is unbalanced as a function of the applied acceleration. Advances in microelectromechanical system technologies have made it possible to build silicon inertial sensors of very small size and with low power consumption. Using these technologies, "smart" acceleration capacitive transducers of the pick-off type are now available. The pick-off method has the advantages of high output levels, low sensitivity to temperature drift, and, most importantly, can be readily used in force-balancing configurations (closed-loop operation).[64]

For acceleration measurements, the sensor mounting technique is very important. In practice, three mounting methods are typically used: (1) bolt mounting, (2) adhesive mounting, and (3) magnetic mounting.

The magnetic mounting method is typically used for temporary measurements with a portable data collector or analyzer. This method is not recommended for permanent monitoring in pneumatic actuators. The transducer may be inadvertently moved and the multiple surfaces and materials of the magnet may interfere with or increase high-frequency signals.

The adhesive or glue mounting method provides a secure attachment without extensive machining. However, this mounting method will typically reduce the operational frequency response range. This reduction is due to the damping qualities of the adhesive. In addition, replacement or removal of the accelerometer is more difficult than with any other mounting method. Surface cleanliness is of prime importance for proper adhesive bonding. This mounting method is not recommended for pneumatic actuators.

The bolt-mounting method is the best method available for permanent mounting applications, and it is accomplished via a stud or machined block. The mounting location for the accelerometer should be clean and paint-free. The mounting surface should be spot-faced to a surface smoothness of Ra = 0.8. The spot-faced diameter should be 10% larger than the accelerometer-attached surface diameter. Any irregularities in the mounting surface preparation will translate into improper measurements or damage to the accelerometer.

Accelerometers are designed to measure vibration over a given frequency range. For pneumatic linear actuators, an accelerometer for measuring their acceleration will have an operating frequency range from 0.2 to 300 Hz. By design, accelerometers have a natural resonance, which is three to five times greater than the advertised high-end frequency response. The low-impedance accelerometer should be selected so that the expected peak value of acceleration lies within the measuring range. For pneumatic actuator applications, the maximum magnitude of the measuring range is about 50 g (where $g = 9.8$ m/s^2). High-impedance acceleration sensors can be used up to 240°C without problems. Because of internal electronics, low-impedance and capacitive sensors are typically operable up to 120°C as a maximum.

A pressure transducer is a device that converts pressure into an analog electrical signal. Depending on the reference pressure used, they could indicate absolute, gage, or differential pressure. Differential pressure transducers are often used for flow measurement where they can measure the pressure differential across orifices or other types of primary elements. "Gage" pressure is defined relative to atmospheric conditions. A pressure transducer might combine the sensor element of a gage with a mechanical-to-electrical converter and a power supply.

There are various types of pressure transducers, including strain-gage based transducer, piezoelectric type, capacitance sensor, potentiometric transducer, resonant-wire type, inductive sensor, and optical transducer. In pneumatic power systems, the strain gage, piezoelectric, and capacitance pressure transducers are most often used.

In the strain gage transducer, the conversion of pressure into an electrical signal is achieved by the physical deformation of a metal or silicon semiconductor strain gage. The gage material is sputtered onto a diaphragm or diffused into a silicon diaphragm structure, and then wired into a Whetstone bridge configuration. Essentially, the strain gage is used to measure the displacement of an elastic diaphragm due to a difference in pressure across the diaphragm. Strain gages are made of materials that exhibit significant resistance change when strained. This change is the sum of three effects. First, when the length of a conductor is changed, it undergoes a resistance change approximately proportional to the change in length. Second, in accordance with the Poisson effect, a change in the length of a conductor causes a change in its cross-sectional area and a resistance change that is approximately proportional to the change in area. Third, the piezoresistive effect, a characteristic of the material, is a change in the bulk resistivity of a material when it is strained. Usually, strain gage transducers are used for narrow-span pressures and for differential pressure measurements. This measuring technology is moderately accurate but has limited ability to achieve high accuracy. The adhesive used to bond the strain gage limits the operating temperature; also, long-term stability is an issue. The relatively high mass of the sensing diaphragm limits the response time, so these sensors are used mostly for static measurements. Strain gage transducers are available for inaccuracy ranges from 0.1 to 0.25% of full-scale measurement.

Piezoelectric devices can further be classified according to whether the crystal electrostatic charge, its resistivity, or its resonant frequency electrostatic charge is measured. Depending on which phenomenon is used, the crystal sensor can be called electrostatic, piezoresistive, or resonant. When pressure is applied to a crystal, it is elastically deformed. This deformation results in a flow of electric charge (which lasts for a period of a few seconds). The resulting electric signal can be measured as an indication of the pressure applied to the crystal. Quartz, tourmaline, and several other naturally occurring crystals generate an electrical charge when strained. Specially formulated ceramics can be artificially polarized to be piezoelectric, and they have higher sensitivities than natural crystals. Unlike strain gage transducers, piezoelectric devices require no external excitation. Because their output is very high impedance and their signal levels low, they require special signal conditioning (such as charge amplifiers and noise-treated coaxial cable).

Electrostatic pressure transducers are small and rugged. Force to the crystal can be applied longitudinally or in the transverse direction, and in either case will cause a high voltage output proportional to the force applied. The crystal's self-generated voltage signal is useful when providing power to the sensor is impractical or impossible. These sensors also provide high-speed responses (about 30 kHz), which makes them ideal for measuring transient phenomena.

Piezoresistive pressure sensors operate based on the resistivity dependence of silicon under stress. Similar to a strain gage, a piezoresistive sensor consists of a diaphragm onto which four pairs of silicon resistors are bonded. Unlike the construction of a strain gage sensor, here the diaphragm itself is made of silicon and the resistors are diffused into the silicon during the manufacturing process. The diaphragm is completed by bonding the diaphragm to an unprocessed wafer of silicon. Piezoresistive pressure sensors are sensitive to changes in temperature and must be temperature compensated. Piezoresistive silicon pressure sensors have the advantage of inherent accuracy when properly designed and fabricated, and offer excellent long-term stability. In mass production, they are one of the most cost-effective technologies. The low mass of the silicon diaphragm reduces shock and vibration sensitivity. Piezoresistive pressure sensors can be used from about 0.021 to 100 MPa.

Resonant piezoelectric pressure sensors measure the variation in resonant frequency of quartz crystals under an applied force. The sensor can consist of a suspended beam that oscillates while being isolated from all other forces. The beam is maintained in oscillation at its resonant frequency. Changes in the applied force result in resonant frequency changes. Because quartz is a common and naturally occurring mineral, these transducers are generally inexpensive. They can be used for absolute pressure measurements with spans from 0–100 kPa to 0–6 MPa or for differential pressure measurements with spans from 0–40 kPa to 0–275 kPa.

Although piezoelectric transducers are not capable of measuring static pressures, they are widely used to evaluate dynamic pressure phenomena associated with explosions, pulsations, or dynamic pressure conditions. They can detect pressures of 0.7 kPa to 70 MPa. Typical accuracy is 1% of full-scale measurement. Usually, semiconductor pressure sensors are sensitive, inexpensive, accurate, and repeatable.

Capacitance pressure transducers were originally developed for use in low-pressure environments. When one plate of a capacitor is displaced relative to the other, the capacitance between the two plates changes. One of the plates is the diaphragm of a pressure sensor, and the capacitance can be correlated to the pressure applied to it. This change in capacitance is used either to vary the frequency of an oscillator or to be detected by a bridge circuit. The diaphragm is usually metal or metal-coated quartz and is exposed to the process pressure on one side and to the reference pressure on the other. Depending on the type of pressure, the capacitive transducer can be either an absolute, gage, or differential pressure transducer. In a capacitance-type pressure sensor, a high-frequency, high-voltage oscillator is used to charge the sensing electrode elements. In a two-plate capacitor sensor design, the movement of the diaphragm between the plates is detected as an indication of the changes in process pressure. Variable capacitance sensors are normally stable and have very good performance characteristics, but require media isolation to isolate the capacitive cell from contamination and moisture. The primary advantages of these devices are low hysteresis and good linearity, stability, and repeatability. However, complicated electronics are required. Capacitance pressure transducers are widespread in industry because of their ability to operate over a wide measuring range, from high vacuums in the micron range to 70 MPa. Differential pressures as low as 1 mm of water can readily be measured. In addition, compared with strain gage transducers, they do not drift much. Better designs are available that are accurate to within 0.1% of reading or 0.01% of full-scale measurement.

Modern pressure transducers combine proven silicon sensor technology with microprocessor-based signal conditioning to provide an extremely powerful, accurate, and stable pressure transducer. The ability to provide high accuracy over a wide temperature range (from −40 to 85°C), coupled with many software features and a compact rugged design, makes these transducers the most versatile and cost-effective devices available on the market today.

Force feedback sensors are useful for planning pneumatic actuating systems with force control. In practice, when combined with force-control algorithms, it becomes possible to sense contact with external load and to control the active force of the actuator (this is important for robotic systems). Sometimes the force sensor system makes use of pneumatic cylinder pressure measurements and thereby measures machine force indirectly. Successful implementation of such an approach will eliminate the need for expensive, direct-force sensors. However, errors in an indirect method of the determination of pneumatic cylinder force depend on the viscous and friction forces. Therefore, in many cases, this technique is unacceptable, and force and torque transducers are usually used.

The fundamental operating principles of force and torque instrumentation are closely allied to the piezoelectric and strain gage devices used to measure static and dynamic pressure discussed above. It is often the specifics of configuration and signal processing that determine the measurement output. Many force transducers employ an elastic load-bearing element or combination of elements. Application of force to the elastic element causes it to deflect and a secondary transducer, which converts it into a measurable output, then senses this deflection. Such transducers are known generically as elastic devices, and form the bulk of all commonly used force transducers.

There are a number of different elastic transducer elements but, in general, they consist of solid and hollow cylinders, round and flat proving rings, cantilevers, simply supported and restrained beams, and washers. In these designs, the gages are attached to the sensing elements at places of concentrated stress. The most common arrangement of a cell is the combination of two gages sensing a change in length and two others responding to contraction. These gages, almost inevitably connected in the Whetstone bridge, form the output-voltage signal.

All elastic devices share this common basis but the method of measuring the distortion of the elastic element varies considerably. The most frequently used method is to make measurements of the longitudinal and lateral strain, and when electrical resistance undertakes these strain gages, such a transducer is known as a strain gage load cell. These constitute the most common commercially available type of force transducer. The rated capacities of strain gage load cells range from 5 N to more than $5 \cdot 10^7$ N. They have become the most widespread of all force measurement systems and can be used with high-resolution digital indicators as force transfer standards. In these devices, a resonant frequency up to 5 kHz and a linearity performance of between 0.01 and 0.02% of full-scale measurement are possible.

4.3 Computation of State Coordinates

Computation velocity and acceleration from a digital displacement sensor is a cost-effective strategy in the control of pneumatic actuators. In this approach, the requirement of the sampling period t_D and a resolution Δ_D of the digital displacement sensor are very important. Usually, the resolution of the displacement sensor should be about 0.1 to 0.2 of the desired accuracy of the actuator. In this case, the sensor counter bit-width should be of the following size:

$$N_{CS} = 3.33 \cdot \log(\frac{L_S}{\Delta_D}) \tag{4.1}$$

where
$\quad N_{CS}$ = size of the displacement sensor counter
$\quad L_S$ = stroke of the pneumatic actuator
$\quad \Delta_D$ = digital sensor resolution

Example 4.1. *Define the size of the displacement sensor counter for the pneumatic position linear actuator, in which the stroke is $L_S = 1$ m and the sensor resolution is $\Delta_D = 0.02$ mm. According to Equation 4.1, the size of the displacement sensor counter is $N_{CS} = 3.33 \cdot \log(1000/0.02) \approx 16.6$, and therefore the 16-bit sensor counter should be used.*

There are many approaches to define the sampling period t_D. One of them is based on the requirement that the sampling frequency should be at least two times greater than the maximum frequency of the processes in the pneumatic actuator. In that case, the displacement information has acceptable accuracy in the quantization process. In general, this frequency is about 50 to 70 Hz, and the sampling period is $t_D \leq 0.0067$ s. The current industry standard for sample rates is between 2.5 and 5 kHz; as a result, the sampling period is $t_D = 0.004 - 0.002$ s.

The simplest method to estimate velocity from discrete and quantized position measurements is the first-order approximation. The conventional approach to velocity estimation can be written as:

$$\dot{x}(h) = \frac{x(h) - x(h-1)}{t_D} \tag{4.2}$$

where h is the discrete time index. The displacement is read at the beginning of each velocity loop calculation, the difference $x(h) - x(h-1)$ formed, and a new velocity estimate is computed by multiplying the known constant $1/t_D$.

Estimation based on Equation 4.2 has an inherent accuracy limit directly related to the resolution of the displacement sensor Δ_D and the sampling period t_D. For example, consider a linear encoder with a resolution of 0.02 mm and a velocity loop sample of 1000 Hz ($t_D = 0.001$ s); this gives a velocity resolution of 20 mm/s. While this resolution may be satisfactory at moderate or high velocity (e.g., 2% error at 1 m/s), it would clearly prove inadequate at low speeds. In fact, at speeds less than 20 mm/s, the speed estimate would erroneously be zero much of the time.

At low velocity, the following equation provides a more accurate approach:

$$\dot{x}(h) = \frac{X}{t(h) - t(h-1)} \tag{4.3}$$

where t is time and X is a fixed displacement interval. In this case, the width of each pulse is defined by the sensor resolution and by measuring the elapsed time between successive pulse edges. Note that the velocity estimate is no longer updated at regular time intervals, but rather the update rate is proportional to actuator speed. The accuracy of this method is directly related to both the counter bit-width and the speed of the actuator. For example, consider a linear encoder with a resolution of 0.02 mm and a minimum actuator speed of 1 mm/s. At this minimum speed, the pulse width will be $(0.02 \text{ mm})/(1 \text{ mm/s}) = 0.02$ s. Usually, for such applications, minimum 16-bit timers are used. With a $50 \cdot 10^{-9}$-s digital signal processor clock, $4 \cdot 10^5$ clock counts will occur for each pulse. This exceeds the $(2^{16} - 1 = 65,535)$ count limit, and therefore a suitable prescaler should be used to clock the desired general-purpose timer. Most often in digital signal processors, a prescaler has a size of 32, which covers most practical applications.

For example, in this case, a minimum prescale value of 6.1 is needed. For this data, the value of $[t(h) - t(h-1)]$ is about 0.1 s, and the minimum actuator speed that can be estimated from Equation 4.3 is about 0.2 mm/s. However, this method suffers from the opposite limitation, as does Equation 4.2. A combination of relatively large actuator speed and high displacement sensor resolution makes the time interval $[t(h) - t(h-1)]$ small, and thus is more greatly influenced by timer resolution. This can introduce considerable error into high-speed estimates.

The actuator speeds for which the estimation calculation will produce valid results are bounded by both upper and lower limits. These limits are determined by the resolution of the displacement sensor Δ_D, sampling period t_D, pulse width X, the general-purpose timer prescale value, and the numerical scaling employed in the software. One approach to designing a system with a wide speed range is to keep the upper estimation limit relatively small so that good accuracy is obtained at low speed using Equation 4.3. When speeds reach higher values, the software switches the system to a new structure that has good accuracy for high speed, which can be reached using Equation 4.2.

In practice, digital differentiation of the displacement signal can be carried out with variable sampling period t_D. Sometimes, the value of the sampling period t_D is a function of the actuator velocity, which allows reaching some kind of optimization for decreasing the velocity ripple.

Another way to improve the resolution of the velocity and acceleration estimation is to use several samples of the digital displacement signal. For such estimation, least squares minimization can be used; then the following formula can determinate the actuator velocity[35]:

$$\dot{x}(h) = \frac{1}{t_D} \cdot \frac{N_S \cdot \displaystyle\sum_{d=1}^{N_S} x(h-d+1) - \sum_{d=1}^{N_S} d \cdot \sum_{d=1}^{N_S} x(h-d+1)}{N_S \cdot \displaystyle\sum_{d=1}^{N_S} d^2 - \left(\sum_{d=1}^{N_S} d\right)^2} \qquad (4.4)$$

where N_S is the number of digital displacement signal samples.

It is obvious that the minimum number of the digital displacement signal samples for the velocity estimation is $N_S = 2$. Then Equation 4.2 is used. For the three samples ($N_S = 3$), the following form (according to Equation 4.4) determines the actuator velocity:

$$\dot{x}(h) = \frac{x(h+1) - x(h-1)}{2 \cdot t_D} \qquad (4.5)$$

Estimation of the acceleration signal from a digital displacement sensor can be carried out by the same techniques as the velocity estimation; in that case, the double differential process is used. For the simplest calculation of the acceleration signal, a minimum of three digital displacement signal samples are needed. Equation 4.6 represents the acceleration estimation in this case:

$$\ddot{x}(h) = \frac{x(h+1) - 2 \cdot x(h) + x(h-1)}{t_D^2} \tag{4.6}$$

Another important feature in the digital differentiation process is the synchronization between the sensor update time and the sampling time of the controller. When the encoder sends a position value cyclically, for example, at a time interval of 0.5 ms and the sampling time of the controller is 1 ms, the occasional error in the velocity can be 50%. Furthermore, if the encoder is read in a time interval of 0.25 ms, the error can be 25%. This can be avoided using a deterministic system, which does not cause any problems even in a situation where the encoder position value is received twice in the control period. The other solution is to synchronize the bus traffic and set the encoder update time equal to the sampling time of the system. In the real system, the encoder could synchronize the entire control system and act like a real time clock. In other words, the control will be started when the feedback signal has been received. Thus, the encoder update time must be chosen so that the message can be transferred reliably. The encoder update time must be less than or equal to the sampling time of the controller.

In practice, estimation of the velocity and acceleration information by differentiation of a position signal is widely used. This can be carried out either in analog or digital form. It is, however, much more difficult in cases where the position signal has transducer and quantization noise present. Such noise will be magnified by the differentiation process and can cause excessive errors in the derived velocity and acceleration signal, thus requiring additional noise filters. A digital differentiating filter combined with a second-order, low-pass filter allows obtaining both the velocity and the acceleration of the pneumatic actuator with low noise and small phase delay.

4.4 Observer Technique for State Variables

There are many methods of estimating pneumatic actuator state variables. They include state observer methods, neural networks, statistical techniques, and others. For nonlinear systems, one of the most powerful estimation methods is the observer method, which can be basically divided into the following groups: Kalman-Buce filter and Luenberger observer, some kinds of the linearization methods, adaptive scheme observer, and sliding mode observer. In pneumatic actuating systems, Luenberger observers based on the Kalman-Buce filter have wide application.[135, 185]

The basic requirement of the system with observer is that both "observer" and "object" are dynamic elements with the same dynamic properties and approximately the same parameters. The observer technique is built on mathematical equations that provide an efficient computational (recursive) mean to estimate the state of a process, in a way that minimizes the mean of the squared error. The observer is very powerful in several aspects: it supports estimations of past, present, and even future states, and it can do so even when the precise nature of the modeled system is unknown. In general, the observer is the electronic unit or computer (mathematic) algorithm that has dynamic behavior identical to that of the pneumatic actuating system. The actuator and observer have the same input signal and they operate in the synchronous and coherent mode, which allows estimation of the actuator state variable. If, for example, the actuator displacement is a measuring variable, then the velocity, acceleration, and pressure in the working chamber, or pressure differential in the working chamber, can be estimated.

To compensate for the variation between actuator and observer dynamic behavior, complementary feedback between measuring variables in the actuator and adequate variables in the observer is used. This difference is called the "measurement innovation," or "residual." The residual reflects the discrepancy between the predicted measurement and the actual measurement. A residual of zero means that the two are in complete agreement.

These variations between the dynamic behavior of the actuator and the observer might be due to the lack of consideration in the initial conditions, the inaccuracy of the mathematical model, measurement noises, and others.

The observer algorithm, as is the case in most of system identification techniques commonly used in practice, operates under the assumption that the system being considered is linear and time invariant. Methods for selecting the observer structure and determination of its parameters are built on the assumption that the actuator dynamic is described by linear differential equations. However, practical applications show that these systems can be used even if the dynamic system is described by nonlinear differential equations. In the process of parameter determination, the actuator dynamic is described by n first-order differential equations. These equations establish the relations between input signals $u_1, u_2, \ldots, u_p$ and output signals $y_1, y_2, \ldots, y_q$ using the state variables (parameters) $x_1, x_2, \ldots, x_n$.

The state variables can be determined from these n first-order differential equations if input signals $u_1, u_2, \ldots, u_p$ are known in the time $t > t_0$, and the initial conditions $x_1(t_0), x_2(t_0), \ldots, x_n(t_0)$ are known too. In this case, the output signal can be defined by solving the algebraic equations.

The state variables can be considered components of the state vector X if the differential equations that describe the actuator dynamic behavior are linear or linearized, which allows using the matrix notation.

In a number of cases, the state variables are considered as n-measured Cartesian coordinates, which describe the state space. In this case, the variation of the state is some curve called the state path.

As is well known, the dynamic behavior of the pneumatic actuating systems can be approximately described by linear differential equations. In the vector notation, these equations have the following form:

$$\dot{X}(t)=A \cdot X(t)+B \cdot U(t)$$

$$Y(t)=C \cdot X(t)$$

where X is the $(n \times 1)$ vector of the state variables, Y is the $(q \times 1)$ vector of the output variables, U is the $(p \times 1)$ vector of the input variables, A is the $(n \times n)$ dynamic matrix of the pneumatic actuator, B is the $(n \times p)$ control matrix, and C is the $(q \times n)$ observation matrix that indicates the state variables, which are measured.

The mathematical model of the observer has the following form:

$$\dot{\hat{X}}(t)=A \cdot \hat{X}(t)+B \cdot U(t)$$
$$\hat{Y}(t)=C \cdot \hat{X}(t)$$
(4.7)

In these equations, the vectors with a "^" sign are referred for state and output variables of the observer. The schematic diagram of the actuator with observer is shown in Figure 4.4, and the mathematical model of the dynamic behavior of the observer can be written in the following form:

$$\dot{\hat{X}}(t)=[A-G \cdot C] \cdot \hat{X}(t)+G \cdot C \cdot X(t)+B \cdot U(t)$$
(4.8)

where G is the gain matrix of the error signal between vectors $Y(t)$ and $\hat{Y}(t)$, and $[A-G \cdot C]$ is the dynamic matrix of the observer.

The gain matrix G defines the dynamic and accuracy characteristics of the observer. Usually, the high dynamic and accuracy requirement can be achieved by increasing the components of the matrix G; however, in that case, the observer has a high sensitivity to the noise, that is, it becomes similar to the differentiation process.

Determination of the observer parameters is carried out using linear differential equations, which can be obtained by linearization of the following nonlinear equations (see Chapter 2):

$$\begin{cases} m \cdot \ddot{x}+b_V \cdot \dot{x}+F_F+F_L = P_1 \cdot A_1 - P_2 \cdot A_2 - P_A \cdot A_R \\[2mm] \dot{P}_1 = \dfrac{1}{V_{01}+A_1 \cdot (0.5 \cdot L_S + x)} \cdot (G_1^+ \cdot R \cdot T_S - G_1^- \cdot R \cdot T_S - P_1 \cdot A_1 \cdot \dot{x}) \quad (4.9) \\[2mm] \dot{P}_2 = \dfrac{1}{V_{02}+A_2 \cdot (0.5 \cdot L_S - x)} \cdot (G_2^+ \cdot R \cdot T_S - G_2^- \cdot R \cdot T_S + P_2 \cdot A_2 \cdot \dot{x}) \end{cases}$$

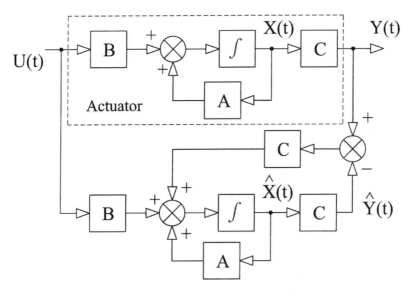

FIGURE 4.4
Schematic diagram of the actuator with observer.

The mathematical model (Equation 4.9) was obtained by assuming that:

- The standard double-acting pneumatic cylinder is used as the actuator.
- The actuating system operates with a servo valve in the trajectory tracking mode.
- The hypothesis of an isothermal process is reasonable.
- The gas is perfect.
- The pressure and temperature within the actuator chambers are homogeneous.
- The influence of the servo valve dynamic on the actuator dynamic behavior is negligible.

Linearization was carried out for the following assumptions:

1. The equilibrium state of the actuating system has the following parameters: $U_C = 0$, $P_1 = P_{10}$, $P_2 = P_{20}$, $\dot{P}_1 = \dot{P}_2 = 0$, $x = 0$ (the middle position of the piston), $\dot{x} = 0$, and $\ddot{x} = 0$.
2. The servo valve with the under-lap plug design is used (see Equation 3.10 for steady-state characteristics).
3. Pressure for the equilibrium state can be determined by assuming that the charge flows in the actuator working chambers are moving at the subsonic condition, and the discharge flows are moving at the sonic condition. In this case, the pressure is

TABLE 4.1

Differential of the Flow Function

σ	0.6	0.7	0.8	0.9	0.95	0.98
$\dot{\phi}(\sigma)$	−0.105	−0.226	−0.384	−0.69	−1.14	−1.76

$$P_{10} \approx P_{20} = P_0 = \frac{4 \cdot \Psi_A^2 \cdot P_S}{1 + 4 \cdot E^2},$$

where

$$\Psi_A = \frac{A_V^+ \cdot \beta_0}{A_V^- \cdot (1 - \beta_0)}.$$

4. The flow function can be represented in the linearization form as: $\phi(\sigma) = \phi(\sigma_0) + \dot{\phi}(\sigma_0) \cdot \dot{\sigma}_0 \cdot \Delta P$, where $\phi(\sigma_0) = \phi_0$, $\dot{\phi}(\sigma_0) = \dot{\phi}_0$, and $\dot{\sigma}_0$ are the magnitudes of the flow function, its derivative and derivate of the argument (accordingly) in the equilibrium point. For the charging process, the argument is $\sigma = P_i/P_S$; and for the discharging process, it is $\sigma = P_A/P$; then the flow function can be represented as: $\phi(P_i/P_S) = \phi_0 - K_\phi^+ \cdot \Delta P$, $\phi(P_A/P_i) = \phi_0 + K_\phi^- \cdot \Delta P_i$, where

$$K_\phi^+ = \frac{|\dot{\phi}_0|}{P_S} \quad \text{and} \quad K_\phi^- = \frac{P_A \cdot |\dot{\phi}_0|}{P_0^2}.$$

Table 4.1 represents the values of the flow function differential $\dot{\phi}(\sigma)$ for various σ.

Substituting into the second equation of the differential equations (Equations 4.7) results in the following correlations: $\beta_1^+ = \beta_0 + \Delta\beta$, $\beta_1^- = \beta_0 - \Delta\beta$, $\Delta\beta = K_{SL} \cdot \Delta U_C$, $P_1 = P_{10} + \Delta P_1$, $\phi_1^+ = \phi_0 - K_\phi^+ \cdot \Delta P_1$, $\phi_1^- = \phi_*$, $x = \Delta x$, $\dot{x} = \Delta \dot{x}$, $\ddot{x} = \Delta \ddot{x}$; after several transformations and rejection of the negligible members, the equation for the pressure derivative can be written as follows (variation sign Δ is not written):

$$\dot{P}_1 = K_* \cdot \frac{A_V^+ \cdot \phi_0 \cdot P_S}{V_{01} + 0.5 \cdot A_1 \cdot L_S} \cdot \left(2 \cdot K_{SL} \cdot U_C - \frac{\beta_0 \cdot K_\phi^+ \cdot P_S}{P_{10}^2} \cdot P_1 \right) - \frac{P_{10}}{\frac{V_{01}}{A_1} + 0.5 \cdot L_S} \cdot \dot{x},$$

where

$$K_* = \sqrt{\frac{2 \cdot k \cdot R \cdot T_S}{k-1}} \approx 760 \frac{m}{s}$$

for $k = 1.4$, $R = 287 \dfrac{J}{kg \cdot K}$, and $T_S = 290$ K.

Taking a new variable, which is the force of the pressure in the first working chamber $F_{P1} = P_1 \cdot A_1$, and carrying out some transformation, the following differential equation obtains:

$$\dot{F}_{P1} = C_{PR1} \cdot (K_{SL} \cdot K_{\beta1} \cdot U_C - K_{P1} \cdot F_1 - \dot{x}) \qquad (4.10)$$

Making the same transformation, the differential equation that describes the dynamic behavior of the pressure force in the second working chamber can be obtained in the following form:

$$\dot{F}_{P2} = C_{PR2} \cdot (-K_{SL} \cdot K_{\beta2} \cdot U_C - K_{P2} \cdot F_2 + \dot{x}) \qquad (4.11)$$

The constant members in Equation 4.10 and Equation 4.11 are defined by the following formulae:

$$C_{PRi} = \frac{P_{i0} \cdot A_i}{\dfrac{V_{0i}}{A_i} + 0.5 \cdot L_S},$$

$$\dot{x}_{iC} = \frac{K_* \cdot A_V^+ \cdot \varphi_0 \cdot P_S}{A_i \cdot P_{i0}},$$

$$K_{\beta i} = 2 \cdot \dot{x}_{iC},$$

$$K_{\varphi i} = 1 + \frac{K_\varphi^+ \cdot P_{i0}^2}{\varphi_0 \cdot P_S},$$

$$K_{Pi} = \frac{\beta_0 \cdot \dot{x}_{iC} \cdot K_{\varphi i}}{P_{i0} \cdot A_i},$$

$$K_{SL} = \frac{1 - \beta_0}{U_{CS}}$$

Thus, the linearized differential equations that describe the dynamic behavior of the linear pneumatic actuator with standard double-acting cylinder and trajectory tracking working mode have the following form (it can be assumed that the friction force has only viscous form):

$$\begin{cases} m \cdot \ddot{x} + b_V \cdot \dot{x} + F_L = F_{P1} - F_{P2} - P_A \cdot A_R \\ \dot{F}_{P1} = C_{PR1} \cdot (K_{SL} \cdot K_{\beta 1} \cdot U_C - K_{P1} \cdot F_{P1} - \dot{x}) \\ \dot{F}_{P2} = C_{PR2} \cdot (-K_{SL} \cdot K_{\beta 2} \cdot U_C - K_{P2} \cdot F_{P2} + \dot{x}) \end{cases} \quad (4.12)$$

For a symmetrical, double-acting pneumatic cylinder — for example, a rodless cylinder loaded with only the inertial load ($F_L = 0$) — it can be assumed that $A_1 = A_2 = A_A$, $P_{10} = P_{20} = P_0$, $V_{01} = V_{02} = V_0$, $C_{PR1} = C_{PR2} = C_{PR}$, $K_{\beta 1} = K_{\beta 2} = K_\beta$, and $K_{P1} = K_{P2} = K_P$. Then Equation 4.12 can be rewritten as:

$$\begin{cases} m \cdot \ddot{x} = F_P - b_V \cdot \dot{x} \\ \dot{F}_P = C_{PR} \cdot (2 \cdot K_{SL} \cdot K_\beta \cdot U_C - K_P \cdot F_P - 2 \cdot \dot{x}) \end{cases} \quad (4.13)$$

where the force of the pressure differential (active force) is $F_P = (P_1 - P_2) \cdot A_A$.

Taking new variables $x_1 = x$, $x_2 = \dot{x}$, and $x_3 = F_P$, the differential Equations 4.13 have the following form:

$$\begin{cases} \dot{x}_1 = x_2 \\ \dot{x}_2 = -\dfrac{b_V}{m} \cdot x_2 + \dfrac{1}{m} \cdot x_3 \\ \dot{x}_3 = -2 \cdot C_{PR} \cdot x_2 - K_P \cdot C_{PR} \cdot x_3 + 2 \cdot C_{PR} \cdot K_{SL} \cdot K_\beta \cdot U_{CS} \cdot e \end{cases} \quad (4.14)$$

where U_{CS} is the control signal saturation value and e is the control parameter, which changes from 0 to 1 and defines the degree of the control valve effective areas changing (in some cases, it is equivalent to the opening coefficient β).

The system of differential Equations 4.14 has the following matrix form:

$$\dot{X} = A \cdot X + B \cdot U$$

where

$$X = \begin{bmatrix} x_1 \\ x_2 \\ x_3 \end{bmatrix},$$

$$A = \begin{bmatrix} 0 & 1 & 0 \\ 0 & -\dfrac{b_V}{m} & \dfrac{1}{m} \\ 0 & -2 \cdot C_{PR} & -K_P \cdot C_{PR} \end{bmatrix},$$

$$B = \begin{bmatrix} 0 \\ 0 \\ 2 \cdot C_{PR} \cdot K_{SL} \cdot K_\beta \cdot U_M \end{bmatrix},$$

$$U = \begin{bmatrix} 0 \\ 0 \\ e \end{bmatrix}.$$

The matrix form of the observer differential equations has the following description:

$$\hat{X} = [A - G \cdot C] \cdot [\hat{X} - X] + B \cdot U$$

The observable parameter is the displacement x_1, and then the observation matrix is $C = [1 \ 0 \ 0]$. In that case, the gain matrix is $G = [g_1 \ g_2 \ g_3]^T$ and the dynamic matrix of the observer is:

$$A - G \cdot C = \begin{bmatrix} 0 & 1 & 0 \\ 0 & -\dfrac{b_V}{m} & \dfrac{1}{m} \\ 0 & -2 \cdot C_{PR} & -K_P \cdot C_{PR} \end{bmatrix} - \begin{bmatrix} g_1 \\ g_2 \\ g_3 \end{bmatrix} \cdot [1 \ 0 \ 0]$$

$$= \begin{bmatrix} -g_1 & 1 & 0 \\ -g_2 & -\dfrac{b_V}{m} & \dfrac{1}{m} \\ -g_3 & -2 \cdot C_{PR} & -K_P \cdot C_{PR} \end{bmatrix}$$

In pneumatic servo actuating systems, the control system is usually built as an integral part of the dynamic system actuator-observer. Then the observer estimates the actuator state variables with the control signal, which is made in the outside module, and, in this case, the parameters of the control

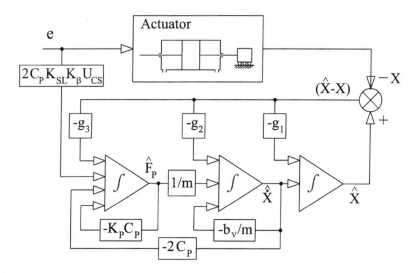

FIGURE 4.5
Observer for pneumatic actuator with trajectory tracking control mode.

system do not influence the behavior of the observer. A block diagram of such system is shown in Figure 4.5, and the system of differential equations has the following form:

$$
\begin{cases}
\hat{x}_1 = -g_1 \cdot (\hat{x}_1 - x_1) + \hat{x}_2 \\[2mm]
\hat{x}_2 = -g_2 \cdot (\hat{x}_1 - x_1) - \dfrac{b_V}{m} \cdot \hat{x}_2 + \dfrac{1}{m} \cdot \hat{x}_3 \\[2mm]
\hat{x}_3 = -g_3 \cdot (\hat{x}_1 - x_1) - 2 \cdot C_{PR} \cdot \hat{x}_2 - K_P \cdot C_{PR} \cdot \hat{x}_3 + 2 \cdot C_{PR} \cdot K_{SL} \cdot K_\beta \cdot U_{CS} \cdot e
\end{cases}
$$

The members $g_1, g_2,$ and g_3 of the gain matrix G are the gains in the feedback lines of the deviation between the signal from the last observer integrator ($\hat{x}$) and the actuator displacement (x). Increasing the values of these coefficients improves the observer speed of response and the estimation of the actuator state parameters. However, at the same time, it enhances the inclination for unstable oscillations of the observer.

Members of the gain matrix G ($g_1, g_2,$ and g_3) can be defined by the dynamic requirement of the observer (acceptable auto track for the actuator state variables). At the same turn, it is obvious that the members of the observer dynamic matrix define the property of the estimation of the state variables. One of the several methods of the observer dynamic matrix member's choice process is building on the selection of the secular equation members. The observer provides an acceptable estimation of the actuator state variables if the relation between succedent and previous members of the secular equation is defined by the series: $0.5 \cdot R_{OB} \cdot S, 0.25 \cdot R_{OB} \cdot S, 0.125 \cdot R_{OB} \cdot S....$ That is,

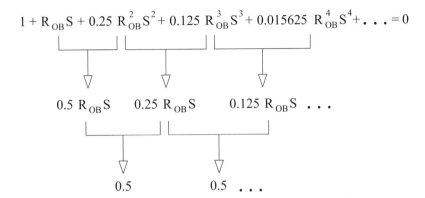

FIGURE 4.6
Choice of observer dynamic matrix members.

in this series, each previous member is two times more than the succeedent member (Figure 4.6). In this procedure, the value R_{OB} is usually defined as

$$R_{OB} = \frac{1}{N_{OB} \cdot \omega_A},$$

where ω_A is the actuator undamped natural frequency ($\omega_A = k_P \cdot C_P$) and N_{OB} is the coefficient of the observer speed of response, which is usually taken in the range of 40 to 60.[102]

In our case, the equation of the observer dynamic matrix can be written as:

$$\begin{vmatrix} -g_1 - S & 1 & 0 \\ -g_2 & -\dfrac{b_V}{m} - S & \dfrac{1}{m} \\ -g_3 & -2 \cdot C_{PR} & -K_P \cdot C_{PR} - S \end{vmatrix} = 0 \qquad (4.15)$$

or in another form:

$$S^3 + \left(\frac{b_V}{m} + K_P \cdot C_{PR} + g_1\right) \cdot S^2 + \left[\left(\frac{b_V}{m} + K_P \cdot C_{PR}\right) \cdot g_1 + g_2 + \frac{C_{PR} \cdot (b_V \cdot K_P + 2)}{m}\right] \cdot S$$

$$+ \left[\frac{C_{PR}}{m} \cdot (b_V \cdot K_P + 2) \cdot g_1 + K_P \cdot C_{PR} \cdot g_2 + \frac{g_3}{m}\right] = 0$$

The members g_1, g_2, and g_3 of the gain matrix can be defined using Equation 4.15 and the correlation in Figure 4.6:

$$\frac{\left(\dfrac{b_V}{m}+K_P \cdot C_{PR}\right) \cdot g_1 + g_2 + \dfrac{C_{PR} \cdot (b_V \cdot K_P + 2)}{m}}{\dfrac{C_{PR}}{m} \cdot (b_V \cdot K_P + 2) \cdot g_1 + K_P \cdot C_{PR} \cdot g_2 + \dfrac{g_3}{m}} = \frac{1}{N_{OB} \cdot \omega_A} = \frac{1}{N_{OB} \cdot K_P \cdot C_{PR}}$$

$$\frac{\dfrac{b_V}{m}+K_P \cdot C_{PR} + g_1}{\left(\dfrac{b_V}{m}+K_P \cdot C_{PR}\right) \cdot g_1 + g_2 + \dfrac{C_{PR} \cdot (b_V \cdot K_P + 2)}{m}} = \frac{0.5}{N_{OB} \cdot \omega_A} = \frac{0.5}{N_{OB} \cdot K_P \cdot C_{PR}}$$

$$\frac{1}{\dfrac{b_V}{m}+K_P \cdot C_{PR} + g_1} = \frac{0.25}{N_{OB} \cdot \omega_A} = \frac{0.25}{N_{OB} \cdot K_P \cdot C_{PR}}$$

These equations become:

$$g_1 = (4 \cdot N_{OB} - 1) \cdot K_P \cdot C_{PR} - \frac{b_V}{m}$$

$$g_2 = 2 \cdot N_{OB} \cdot K_P \cdot C_{PR} \cdot \left(g_1 + K_P \cdot C_{PR} + \frac{b_V}{m}\right) - g_1 \cdot \left(K_P \cdot C_{PR} + \frac{b_V}{m}\right)$$
$$- \frac{C_{PR} \cdot (b_V \cdot K_P + 2)}{m}$$

$$g_3 = C_{PR} \cdot [N_{OB} \cdot K_P \cdot g_1 \cdot (m \cdot K_P \cdot C_{PR} + b_V) + (N_{OB} - 1) \cdot m \cdot K_P \cdot g_2 +$$
$$+ (b_V \cdot K_P + 2) \cdot (N_{OB} \cdot K_P \cdot C_{PR} - g_1)]$$

Example 4.2. *Estimate the observer parameters for the pneumatic servo actuator with the following data:*

- *Moving mass is m = 40 kg.*
- *Piston effective area is $A_A = 78.5 \cdot 10^{-4} \ m^2$ (piston outside diameter is 0.1 m).*
- *Valve effective area is $A_V = 12 \cdot 10^{-6} \ m^2$.*
- *Piston stroke is $L_S = 0.5 \ m$.*
- *Inactive volume at the end of the stroke is $V_0 = 2 \cdot 10^{-4} \ m^3$.*
- *Coefficient β_0 is 0.5.*
- *Control signal saturation value is $U_{CS} = 10 \ V$.*

- *Viscous friction coefficient is $b_V = 100 \; N \cdot s/m$.*
- *Supply pressure is $P_S = 0.6 \; MPa$.*

For the given data, the following parameters can be determined: $P_0 = 0.48$ MPa, $C_{PR} = 1.37 \cdot 10^4 \; N/m$, $\varphi_0 = 0.163$, $\dot{\varphi}_0 = -0.384$, $K_\varphi^+ = 0.64 \cdot 10^{-6} \; m^2/N$, $K_\varphi = 2.5$, $K_{SL} = 0.05 \; 1/V$, $\dot{x}_C = 0.238 \; m/s$, $K_\beta = 0.48 \; m/s$, $K_P = 8 \cdot 10^{-5} \; m/N \cdot s$. Then, for $N_{OB} = 50$, the members of the gain matrix are: $g_1 = 212.8 \; 1/s$, $g_2 \approx 2.2 \cdot 10^4$ $1/s^2$, and $g_3 \approx 4.4 \cdot 10^7 \; kg/s^3$.

The velocity response of a linear pneumatic actuator with a closed-loop control system that operates in trajectory tracking control mode is shown in Figure 4.7. In this figure, three curves represent the following results:

1. Measurement by analog speed transducer
2. Dynamic simulation by integration of Equation 4.9 using the fourth-rank Runge-Kutte stability criteria
3. Estimation by analog observer carried out with integrated circuits according to the block diagram in Figure 4.5

The analog observer was realized using operational amplifiers. An operational amplifier is a high-gain electronic amplifier with feedback. Without feedback, the amplifier would be very unstable because of the high gain.

The difference between measurement results and the results of observer estimation and computer simulation is not more than 5% in velocity amplitude and 10% in its phase. First, there is acceptable estimation accuracy for the observer application; second, it indicates a high degree of accuracy for the mathematical model that describes the actuator dynamic behavior. The active force response of this actuator is shown in Figure 4.8. In this case, the

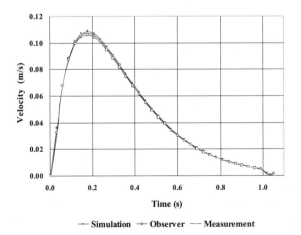

FIGURE 4.7
Velocity response of the actuator with trajectory tracking control mode.

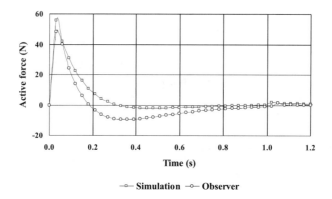

FIGURE 4.8
Active force response of the linear actuator with trajectory tracking control mode.

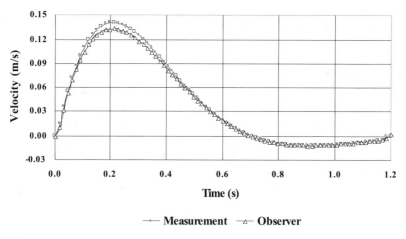

FIGURE 4.9
Velocity response of the linear actuator with smaller piston.

difference between computer simulation and observer estimation results is much more than in the velocity response, but it also allows the use of the estimation signal as a state coordinate. Analysis of the observer dynamic indicates that the largest influence has the value of the coefficient g_3. Figure 4.9 illustrates the velocity response of the linear actuator in which the piston effective area is half the actuator piston examined in the previous actuator (Figure 4.8). In both actuators, the values of the members g_i of the observer gain matrix were the same. It should be observed (Figure 4.9) that the difference between measurement results and results of the observer estimation stands on the same level (not more than 5%). It indicates that the dynamic behavior of the observer is not sensitive to changing the actuator inertia,[102] which can be estimated by the dimensionless coefficient

$$W = \frac{K_* \cdot A_V^+}{A_1} \cdot \sqrt{\frac{m}{A_1 \cdot L_S \cdot P_S}}$$

In the point-to-point control mode, the actuator velocity profile has a "trapeze" form (e.g., see Figure 1.9). In this case, the acceleration and motion with high constant velocity is realized by switching off the control valve to the saturation position. Only around position point the control valve moves within the regulation range that provide the deceleration and stop is the process in the desired position point. In this control mode of the positioning actuator, the observer should have the modulus of the control signal limitation. Then the dimensionless control signal e can be obtained by the following formulae:

$$e = \begin{cases} 1, & \text{if } U_C > U_{CS} \\ \dfrac{U_C}{U_{CS}}, & \text{if} -U_{CS} \le U_C \le U_{CS} \\ -1, & \text{if } U_C < -U_{CS} \end{cases}$$

where U_C is the control signal and U_{CS} is the control signal saturation value.

The schematic diagram of the observer for the point-to-point positioning actuator is shown in Figure 4.10. The proportional module with saturation is the common element for both the actuator and the observer, and it limits

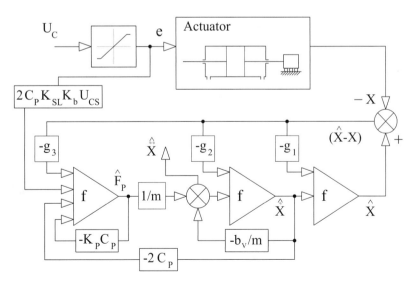

FIGURE 4.10
Observer for pneumatic actuator with point-to-point positioning mode.

the input signal for these parts. This schematic diagram (Figure 4.10) illustrates the ability to estimate the acceleration/deceleration signal ($\hat{\ddot{x}}$), which can be determined from the equation:

$$\hat{\ddot{x}} = \frac{\hat{F}_p - b_V \cdot \hat{\dot{x}}}{m} \qquad (4.16)$$

The velocity response of the linear actuator with point-to-point control mode is shown in Figure 4.11. Here, the observer estimation curve is very close to the measurement result (the difference is less than 5%). Figure 4.12 illustrates the acceleration response of this actuator; the estimation obtains from Equation 4.16. A comparison of these two curves shows the relatively good estimation result. The difference between computer simulation results and results of observer estimation is not more than 15% in acceleration amplitude and 20% in its phase. This is very important for the state controller application.

The sensitivity in observer behavior from the errors in the mathematical model of the actuator is an essential aspect of observer reliability. Usually, two types of model errors are considered:

1. Structure discrepancy
2. Variance between parameters of the real system and its mathematical model

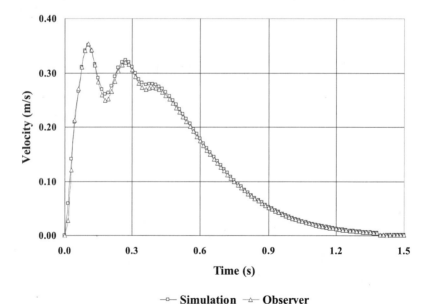

FIGURE 4.11
Velocity response of the linear actuator with point-to-point control mode.

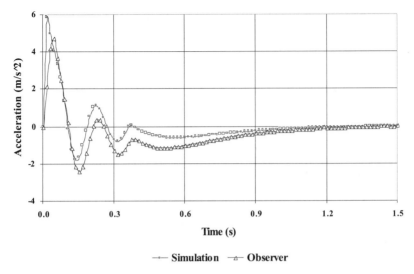

FIGURE 4.12
Acceleration response of the linear actuator with point-to-point control mode.

For the first type of error, it is important to note that the structure of the system consists of the one moving mass oscillator with viscous damping, which is serially connected to the integrator. This scheme can be used if the following conditions are satisfied:

- Natural frequency of the system pneumatic cylinder piston — displacement sensor should be at least two times greater than the natural frequency of the pneumatic cylinder.
- Natural frequency of the electropneumatic control valve should be a minimum of five times greater than the natural frequency of the pneumatic cylinder.

If these conditions do not exist, then the mathematical model should be based on a multi-moving mass structure.

The variations in moving mass, friction force, supply pressure, temperature, and others refer to the second type of model errors. Analysis of the influences of these parameters on the observer dynamic behavior shows that the values of the members g_i of the observer gain matrix are not sensitive to changes in these parameters. Variations in viscous friction and the presence of coulomb friction force in the real system do not significantly affect overall observer dynamics. However, variations in the moving mass and piston effective areas do influence observer dynamics. However, in practice, during actuator operation, these parameters remain constant. Variations in the supply pressure and air temperature have a weak influence on observer behavior.

In a discrete-time-controlled process, the observer estimation technique has an additional problem that is connected with the quantization process of the measurable variables. In that case, the discrete counterpart of the continuous state-space model of the observer (see Equation 4.7 and Equation 4.8) can be expressed as:

$$\hat{X}(h+1)=[A-G\cdot C]\cdot\hat{X}(h)+G\cdot C\cdot X(h)+B\cdot U(h)$$

$$\hat{Y}(h)=C\cdot\hat{X}(h)$$

where h is the discrete time index.

Dynamic simulation by integration of the differential equations of the discrete observer shows that its behavior is stable and the estimation of the state variable is acceptable if the sampling period t_D is less than 0.001 s.

5

Open-Loop Pneumatic Actuating Systems

Pneumatic actuating systems, similar to actuating systems of other types, can usually be divided into two groups: namely, open- and closed-loop control. Open-loop pneumatic systems remain widely used in manufacturing and the processing industry. This is due to the sturdiness, versatility, and ease of use of pneumatic systems, which are regarded chiefly as a means of achieving low-cost automation. Consequently, this means low initial investments, low setup costs, and low operating and maintenance costs.

However, the performance obtainable using open-loop control has several limits. First of all, open-loop systems are sensitive to initial conditions. For the actuators with repeated stops in the same locations this drawback is not critical, but for multi-location applications, where the initial conditions significantly vary, this factor is very important. Pneumatic actuators with open-loop control do not attenuate disturbances and do not mitigate sensitivity to plant parameter variations. The interpretation is that the output of such a system is sensitive to plant parameter variations and disturbances for inputs at all frequencies. A given change in a plant parameter and disturbances will cause a proportional change in system output. In a position system, these limitations change the system response time, and in many cases these variations are not important. However, for actuators with velocity and acting force controls, these disadvantages are very critical because they significantly influence the controlled output parameters. In such cases, closed-loop control systems are usually used.

5.1 Position Actuators

Open-loop pneumatic actuators are most often used in positioning applications. In these systems, the final positioning of the actuator is provided by a manually adjusted hard stop. Position status is indicated by position sensors, which are usually attached to the outside of the actuator (on the cylinder tube or on the hard stop). These actuators contain air-cushioning units or shock absorbers, which provide the absorption and dissipation of actuator kinetic energy. Because of these devices, deceleration of the actuator is reduced to a tolerable level and the positioning process takes place without impact and has high repeatability.

Shock absorber construction and parameters depend on the speed of the cylinder, the mass being moved, the external forces acting on the system, the system pressure, and the piston diameter. In pneumatic actuators, hydraulic and pneumatic shock absorbers are usually used.

The power consumption of pneumatic shock absorbers is less than for hydraulic devices; therefore, these constructions are utilized in pneumatic actuators with low carrying capacity. Pneumatic shock absorbers have the following advantages: low price, simplicity in installation and maintenance, robust design and operation, and low sensitivity to temperature change.

Hydraulic shock absorbers are usually used when the high energy of the actuator should be absorbed, and in the cases where the specific deceleration curves should be provided under dynamic environment conditions.

5.1.1 Shock Absorbers in Pneumatic Positioning Actuators

The basic function of the shock absorber is to absorb and dissipate the impact kinetic energy to the surrounding atmosphere, where decelerations of a moving actuator part are reduced to a tolerable level. For example, in a hydraulic shock absorber, when the pneumatic cylinder is stopped, the shock absorber rod is struck, the piston of the absorber is moved, and this increases the fluid pressure in the shock absorber working chamber. The fluid flows from one working chamber of the absorber to another, increasing in temperature. Thus, the kinetic energy of the actuator is converted to heat and the load is stopped. It does this smoothly and it takes only a few milliseconds to stop the actuator; thus, shock load and vibration are avoided.[17]

Pneumatic shock absorbers operate on the same basic principle of movement of the observer piston that increases the air pressure in the shock absorber working chamber, wherein air or another gas is forced via a small orifice. This avoids the disadvantage of hydraulic shock absorbers because there is significantly less heat dissipation from air than from oil. However, this does not provide a very powerful type of shock absorber because air is compressible and hence the force maintained through the stroke decreases more and more as the stroke progresses.

The efficiency and effectiveness of the absorber depend almost entirely on the flow path between the two working chambers. However, the energy-absorbing capacity depends on the size of the shock absorber and the method of returning the piston to its rest position. Spring-return constructions are more compact and convenient than accumulator models, but do not have as much energy capacity. Accumulator shock absorbers have more hydraulic fluid and more surface area from which to radiate heat. Therefore, they can be cycled more frequently at maximum capacity than spring-return models. The heat dissipation of the shock absorber can be improved by cooling it

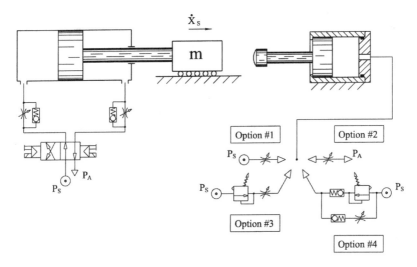

FIGURE 5.1
Schematic diagram of the pneumatic shock absorber.

with air, which is exhaust air from the pneumatic cylinder powering the retarded mass.

A single-acting pneumatic cylinder can be utilized as the simplest shock absorber. Figure 5.1 provides a schematic diagram of such construction, in which the shock absorber slows down the moving load with mass m_L and sink velocity $\dot{x}_S$ by applying a reaction force. This figure illustrates four possible options for connecting such a shock absorber to the supply or atmospheric line. The parameters of the absorber have been selected correctly if the working function of its reaction force equals the sum of the load kinetic energy and the working function of the pneumatic cylinder propelling force. The energy-absorbing capacity of these shock absorbers depends on the geometric parameters of their construction and the amount of pressurized air in its working chamber. For example, the energy-absorbing capacity of the absorber in which the working chamber is connected to the atmospheric line (option 2 in Figure 5.1) is less than in a shock absorber with a working chamber that is connected to the supply line (option 1 in Figure 5.1). However, for the design in option 1, it is necessary to comply with a rule that at the end of an absorber displacement, the braking force should be less than the actuator propelling force. The energy-absorbing capacity can be changed by adjustable flow control valves, as shown in Figure 5.1.

In a shock absorber with a working chamber connected to the supply line (option 1), the piston comes back to the initial position after the load with mass m begins moving to the opposite side. For option 2, the return of the absorber piston can be performed using an additional element (e.g., a return spring). Pneumatic absorber devices that are connected to the supply line store energy, rather than dissipate it, which causes the load to bounce back after stopping. From an efficiency and effectiveness point of view, this is an important advantage; however, the ability of the bounce back motion requires the best calculation of the absorber parameters for the elimination of this phenomenon.

In the shock absorber shown in Figure 5.1, option 3, the working chamber is connected to the supply line via a pressure regulator. This design allows reaching the best flexibility in the absorber adjustment process; therefore, the reduction in the initial pressure in the pneumatic working chamber allows for a decrease in the impact value in the initial contact or interact of the load and the absorber rod.

The shock absorber illustrated in Figure 5.1, option 4 has a greater ability of adjustment because the adjustable flow control valve is installed parallel to the pressure regulator.

The pneumatic shock absorber (Figure 5.2) consists of two nonclamped diaphragms. It can provide a smooth actuator stop process in the direct and inverse load motions; such constructs are usually referred to as double-action shock absorbers. This construction produces a pneumatic positioning actuator having a compact structure; however, there are limitations with regard to absorber adjustment because the same adjustment parts are used for both motions.

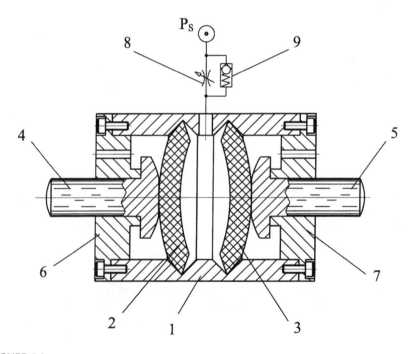

FIGURE 5.2

Schematic diagram of the double-acting pneumatic shock absorber with nonclamped diaphragm.

The shock absorber housing (1) has two inside grooves (Figure 5.2), in which the two nonclamped diaphragms (2 and 3) are installed. In the initial state, both diaphragms are bent in an outside direction owing to air pressure in the inside chamber (between diaphragms). In this position, the central part of the diaphragm is set against the head surface of the absorber rods (4 and 5). In addition, the heads of the rods rest against the inner surface of the covers (6 and 7). Under these conditions, only a small part of the air pressure force passes to the absorber rods; therefore, at the beginning of actuator braking, the impact between the actuator rod and shock absorber rod is practically absent. The supply pressure is connected to the shock absorber through the adjustable flow control valve (8), which is used for absorber adjustments and the check valve (9). The main disadvantage of pneumatic shock absorbers with diaphragms is the short displacement of its rods; for example, in the described construction (Figure 5.2), this value is not more than $0.2–0.4 \cdot D_D$ (it depends on the thickness and outside diameter (D_D) of diaphragm). From this point of view, the piston construction of the shock absorbers has a significant advantage; theoretically, its working stroke does not have limitations.

Figure 5.3 presents a schematic diagram of double-action piston construction of a pneumatic shock absorber. This device operates in the same manner as the absorber with diaphragms. Using two pistons instead of two diaphragms allows for a significant increase in the absorber working stroke. The inside pneumatic chamber between the two pistons (1 and 2) and sleeve (3) is connected with supply pressure (P_S) via a hole in the cover (4), the chamber between the cover (4) and the piston (1), and the fixed orifice (6). The chamber between the piston (2) and the cover (5) is also connected to the supply pressure. This construction of the pneumatic shock absorber has

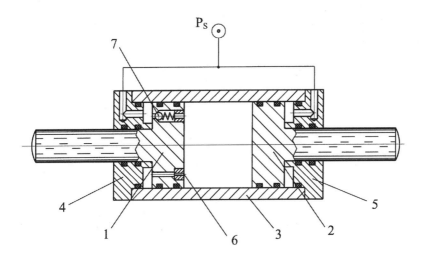

FIGURE 5.3
Schematic diagram of the double-acting pneumatic shock absorber with pistons.

high efficiency and effectiveness because the pressurized air enters the supply line during the braking process. Resetting the pistons to the initial position takes place automatically because there is a difference between the effective areas on the two sides of the pistons. The check valve (7) is used to provide the quick return process.

Pneumatic shock absorbers have a high sensitivity to variations in pneumatic actuator parameters (e.g., moving mass, sink velocity, supply pressure, etc.). Absorber constructions that have the ability to change the braking displacement and the discharge air mass flow can be used in applications under dynamic environmental conditions. Figure 5.4 shows the schematic diagram of such a modular pneumatic shock absorber that consists of several

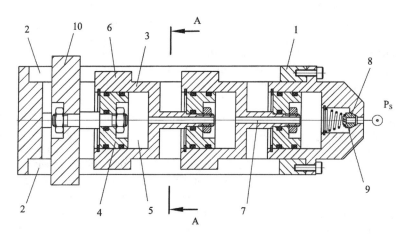

a. Longitudinal section

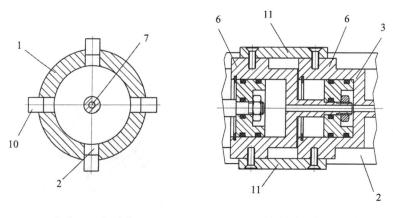

b. Cross section A-A   c. Module latching mechanism

FIGURE 5.4

Modular pneumatic shock absorber: (a) longitudinal section, (b) cross section A-A, and (c) module latching mechanism.

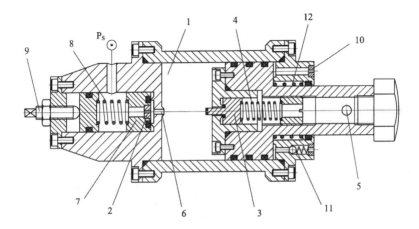

FIGURE 5.5
Single-acting pneumatic shock absorber.

single-type modules. The number of modules depends on the energy-absorbing capacity of one of them and the necessary energy-absorbing capacity of the shock absorber as a whole.

The single-type modules (Figure 5.4a) are in the sleeve (1), which has four slots (2). Each module has a housing (3) and a piston (4), which form the pneumatic working chamber (5). The module housing (3) has four lugs (6) that are in the slots (2) and serve as sliding elements for the modules. The supply pressure is connected to the shock absorber through the check valve (8) and the fixed orifice (9), which are used for absorber adjustments. Figure 5.4c depicts the latching mechanism of the shock absorber module; at that rate, stopping the separate module is carried out by crampons (11). During the braking process, the actuator interacts with the shock absorber via the stems (10).

The single-action pneumatic shock absorber, which is shown in Figure 5.5, has wide potential in the adjustment of the energy-absorbing capacity. In this construct, the rod pneumatic chamber is used as the additional working chamber where the vacuum environment is formed. There is an increase in the energy-absorbing capacity.

The main working chamber (1) can be connected with the supply pressure (via the constant orifice [2] and channel [6] in the back cover) or with the atmospheric pressure (through the cutoff valve [3] and hole [5] in the shock absorber rod). The cutoff valve (3) opens in two cases:

1. When the pressure in the chamber (1) exceeds the level, which is defined by adjustment of the spring (4)
2. In the end area of the piston displacement, where the stem of the cutoff valve (3) closes the channel that is connected to the chamber (1) with the supply pressure

The relief valve (7) also restricts the pressure in the main chamber (1); its open level is adjusted by a screw (9) that compresses the spring (8). In this case (when the valve is opened) the chamber (1) is connected to the supply pressure. Restriction of the air pressure allows reaching gradual braking because the deceleration process does not have rebound.

In the end of an absorber displacement, the main chamber (1) is connected to atmospheric pressure via the cutoff valve (3), which closes the supply channel (6) at the same time. This design prevents air lock formation, which might induce reciprocating motion.

The energy-absorbing capacity increases when the absorber rod chamber is used as the additional working chamber. In that case, this chamber is connected to atmospheric air via the constant orifice (10) and the check valve (11). The size of the orifice (10) defines the vacuum level in the chamber during the braking process, and the check valve (11) is necessary for the quick return of the absorber piston to the initial position.

Application of the compression spring (12) allows obtaining "soft" brake starting because the initial contact between the actuator and the shock absorber does not have a "hard" impact nature (spike of the deceleration).

Computation of the pneumatic shock absorber parameters is a complicated task because the nonlinear differential equations describe its dynamic behavior. In practice, the estimation methods of the parameters calculation are usually used; and finally, tuning of the braking behavior is carried out in the service adjustment. In addition, after parameters estimation, computer simulation of the dynamic system actuator-absorber allows for refining and tuning the shock absorber parameters.

For the simplest shock absorber shown in Figure 5.1a and Figure 5.1b, parameter estimation can be performed with the assumption that the absorber power-consumption is equal to the sum of the load kinetic energy and the working function of the pneumatic cylinder force. In this case, the schematic diagram of the system pneumatic cylinder — shock absorber is shown in Figure 5.6. Actually, before the braking process, the actuator moves with constant velocity $\dot{x}_S$ (sink velocity); then the load kinetic energy is E_{LK} = $0.5 \cdot m_L \cdot \dot{x}_S^2$. The maximum value of the working function of the pneumatic cylinder propelling force can be estimated as $E_{PC} = 0.5 \cdot A_1 (P_S - P_A) \cdot s_D$, where s_D is the absorber working stroke.

Taking into consideration that the discharging process in the shock absorber working chamber has isothermal behavior, the approximating computation of the pressure behavior in the absorber can be obtained from the following differential equation:

$$\dot{P}_D = \frac{P_D \cdot A_D \cdot \dot{x}_D - P_D \cdot A_{VD} \cdot \varphi_* \cdot K_*}{A_D \cdot (x_{D0} + s_D - x_D)} \tag{5.1}$$

where P_D is the absolute pressure in the shock absorber working chamber, A_D is the absorber piston effective area, x_D is the absorber piston displacement,

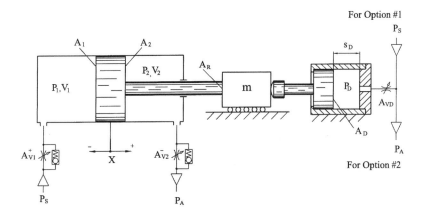

FIGURE 5.6
Schematic diagram of the pneumatic cylinder with pneumatic shock absorber.

x_{D0} is the reduced length of the absorber working chamber dead volume, and A_{VD} is the effective area of the flow control valve.

Most often, for the braking process, limitation of the deceleration is required. If we take into account that the deceleration has a constant value ($\ddot{x}_D$ = constant), then $\dot{x}_D = \dot{x}_S - \ddot{x}_D \cdot t$ and $x_D = 0.5 \cdot \ddot{x}_D \cdot t^2$. If we substitute these equations into Equation 5.1 and solve this differential equation, the pressure in the absorber working chamber can be approximated from:

$$P_D = P_{D0} \cdot \exp\left(\frac{B_D \cdot t - 0.5 \cdot C_D \cdot t^2}{N_D} \right) \tag{5.2}$$

where P_{D0} is the initial pressure, $B_D = A_D \cdot \dot{x}_S - A_{VD} \cdot \varphi_* \cdot K_*$, $C_D = A_D \cdot \ddot{x}_D$, and $N_D = A_D \cdot (0.5 \cdot s_D + x_{D0})$. Equation 5.2 exists under the assumption that the absorber chamber value is constant and equal to $A_D \cdot (0.5 \cdot s_D + x_{D0})$.

Absorber power consuming can be approximately determined from:

$$E_D = (0.5 \cdot P_{D0} + 0.5 \cdot P_{DM} - P_A) \cdot A_D \cdot s_D$$

where P_{DM} is the maximum pressure in the absorber chamber. From the equation $E_D = E_{LK} + E_{PC}$, the equation for the P_{DM} would be:

$$P_{DM} = \frac{E_{LK} + E_{PC}}{0.5 \cdot A_D \cdot s_D} - P_{D0} + 2 \cdot P_A \tag{5.3}$$

On the other hand, the maximum pressure P_{DM} in the absorber chamber can be found from Equation 5.2 for time t_M, which is determined from:

$$\frac{dP_D}{dt} = \frac{P_{D0} \cdot (B_D - C_D \cdot t)}{N_D} \cdot \exp\left(\frac{B_D \cdot t - 0.5 \cdot C_D \cdot t^2}{N_D}\right) = 0$$

From this it follows that

$$t_M = \frac{B_D}{C_D}$$

and

$$P_{DM} = P_{D0} \cdot \exp\left(\frac{B_D^2}{2 \cdot N_D \cdot C_D}\right) \tag{5.4}$$

Substituting this result into Equation 5.3, the equation that connects the absorber piston effective area A_D and the effective area A_{VD} of the flow control valve can be obtained:

$$\frac{E_{LK} + E_{PC}}{0.5 \cdot P_{D0} \cdot A_D \cdot s_D} - 1 + 2 \cdot \frac{P_A}{P_{D0}} = \exp\left(\frac{B_D^2}{2 \cdot N_D \cdot C_D}\right) \tag{5.5}$$

In this equation, the shock absorber working stroke can be estimated as:

$$s_D = \frac{\dot{x}_S^2}{2 \cdot \ddot{x}_D} \tag{5.6}$$

Given the size of the absorber piston effective area A_D, the value of the effective area of the flow control valve A_{VD} can be defined using Equation 5.5; in that case,

$$A_{VD} = \frac{A_D \cdot \dot{x}_S - \sqrt{2 \cdot N_D \cdot C_D \cdot \ln G_D}}{\varphi_* \cdot K_*} \tag{5.7}$$

where

$$G_D = \frac{E_{LK} + E_{PC}}{0.5 \cdot P_{D0} \cdot A_D \cdot s_D} - 1 + 2 \cdot \frac{P_A}{P_{D0}}$$

The restriction of the size of absorber piston effective area is defined by the relationship $P_{DM} > P_{D0}$, and A_D can then be found from Equation 5.3:

$$A_D < \frac{E_{LK} + E_{PC}}{s_D \cdot (P_{D0} - P_A)} \tag{5.8}$$

Example 5.1. *As an example, define the parameters of the pneumatic shock absorber for a pneumatic cylinder in which the piston diameter is 40 mm ($A_1 = 1.256 \cdot 10^{-3}$ m²), the rod diameter is 16 mm ($A_2 = 1.055 \cdot 10^{-3}$ m²), the absolute supply pressure is $P_S = 0.6$ MPa, the sink velocity is $\dot{x}_s = 1$ m/s, and the mass of the load is $m_L = 20$ kg. The necessary deceleration is about $\ddot{x}_D = 5$ m/s², and thus the shock absorber working stroke is $s_D = 0.1$ m. Take into consideration that the absolute initial pressure in the shock absorber working chamber is $P_{D0} = P_S = 0.6$ MPa and the reduced length of the absorber working chamber dead volume is $x_{D0} = 10$ mm. According to the relationship in Equation 5.8, the absorber piston effective area should be $A_D < 8.359 \cdot 10^{-4}$ m²; that is, the absorber piston diameter is less than 32 mm. Consider the shock absorber that has a 30-mm piston diameter ($A_D = 7.064 \cdot 10^{-4}$ m²). According to Equation 5.7, the effective area of the flow control valve is $A_{VD} = 2.184 \cdot 10^{-6}$ m².*

The procedure above is the technique for the approximate calculation of the shock absorber parameters. Computer simulation of the dynamic behavior of the actuator–shock absorber system allows the choice of accurate parameters for the absorber. A mathematical model of the dynamic behavior of this system has been developed in order to predict the parameters of the braking process (deceleration, braking force, time of the braking process, etc.) in terms of the physical parameters of the device.

The mechanical system actuator–absorber is described by Newton's second law, and the equation of motion for the piston-load assembly can be expressed by Equation 2.23, in which the shock absorber force (F_{SA}) is added:

$$m \cdot \ddot{x} + b_V \cdot \dot{x} + F_F + F_L + F_{SA} = P_1 \cdot A_1 - P_2 \cdot A_2 - P_A \cdot A_R \tag{5.9}$$

Here, the absorber force is $F_{SA} = A_D \cdot P_D$, and the absolute pressure in the shock absorber working chamber (P_D) can be obtained from the following differential equation:

$$\dot{P}_D = \frac{P_D \cdot A_D \cdot \dot{x} - A_{VD} \cdot P_D \cdot K_* \cdot \varphi(\frac{P_{Din}}{P_D})}{V_{0D} + A_D \cdot [s_D - (x - x_D)]} \tag{5.10}$$

where V_{0D} is the inactive volume at the end of the shock absorber stroke and admission port, x_D is the coordinate of the actuator piston where it has contact with the shock absorber rod (absorber begins to operate, together with the pneumatic cylinder), and $\varphi(P_{Din}/P_D)$ is the flow function as defined by Equation 2.14. If $x < x_D$, the value of the absorber force (F_D) in Equation 5.9

is equal to zero. For the option 1 (Figure 5.1), the initial pressure in the absorber working chamber is $P_{D0} = P_S$ and the pressure in the inlet port is $P_{Din} = P_S$. For option 2, $P_{D0} = P_{Din} = P_A$.

The differential equations that describe the change in pressure in the working chambers of a pneumatic cylinder can be obtained from Equation 2.18. However, it is necessary to take into account that in open-loop pneumatic actuators, the working chambers are connected to the supply port and exhaust line in rotation and inversely; that is, if first the chamber is connected to the supply port, and simultaneously, the second chamber is connected to the exhaust line and vice versa. Then the two differential equations are:

$$\dot{P}_1 = \frac{A_{V1}^+ \cdot P_S \cdot K_* \cdot \varphi\left(\dfrac{P_1}{P_S}\right) - P_1 \cdot A_1 \cdot \dot{x}}{V_{01} + A_1 \cdot (0.5 \cdot L_S + x)} \tag{5.11}$$

and

$$\dot{P}_2 = \frac{P_2 \cdot A_2 \cdot \dot{x} - A_{V2}^- \cdot P_2 \cdot K_* \cdot \varphi\left(\dfrac{P_A}{P_2}\right)}{V_{02} + A_2 \cdot (0.5 \cdot L_S - x)} \tag{5.12}$$

Figure 5.7 shows the dynamic curves for the braking process of the actuator–absorber system after parameter adjustment by computer simulation. In this case, the absorber piston diameter is 35 mm ($A_D = 961.6$ mm²), the effective area of the flow control valve is $A_{VD} = 1.5 \cdot 10^{-6}$ m², and the absorber working stroke is $s_D = 80$ mm. This construction has the compression spring on the rod side, similar to the spring (12) in the shock absorber illustrated in Figure 5.5. This design allows for obtaining a "soft" start of the braking process.

Analysis of the obtained results demonstrates the high accuracy of the approximate calculation of the shock absorber parameters. In addition, pneumatic shock absorbers have an acceptable performance for application in pneumatic positioning actuators.

It is important that pneumatic shock absorbers also be able to function as an acceleration device. In general, industrial equipment not only uses shock absorbers to absorb the energy during the retraction stroke, but also employs a separate accelerator or actuator to move the machine part in the reverse direction. It would be less costly if a single device could be employed as both a shock absorber and as an accelerator. Tremendous energy savings could be achieved by recycling energy used during the shock absorption and then reused for acceleration; a much lower propelling force would be needed, a low return force could be achieved, the heaving weight could be moved at high cycling frequency, and a high cycling frequency could be achieved. The pneumatic shock absorbers described above achieve these advantages.

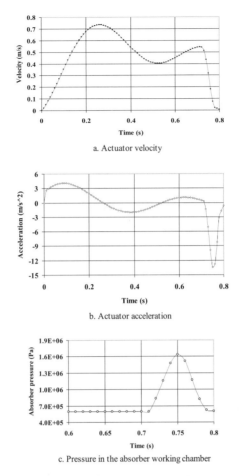

a. Actuator velocity

b. Actuator acceleration

c. Pressure in the absorber working chamber

FIGURE 5.7
Dynamics of the actuator–pneumatic shock absorber system: (a) actuator velocity, (b) actuator acceleration, and (c) pressure in the absorber working chamber.

Basically, in hydraulic shock absorbers, the load kinetic energy is transformed into heat energy. As a result, the system loses the ability to reuse the energy that was transformed into heat. In addition, the possibility of the system overheating greatly reduces the applicability of these shock absorbers at high frequency. The heat dissipation of the shock absorber can be improved by cooling it with air, for example, exhaust air from the pneumatic cylinder powering the decelerated mass. Cooling can be made particularly efficient by channeling the cooling air through a tube within the shock absorber. On the other hand, the advantage of using hydraulic shock absorbers is that they are very powerful because oil is virtually noncompressible. With the use of oil-based shock absorbers, a uniform braking force can be maintained throughout the actuator stroke. Because work is proportional to force and distance, the power of the shock absorber is maximized.

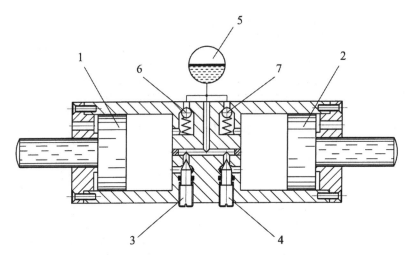

FIGURE 5.8
Double-acting hydraulic shock absorber.

Two common types of hydraulic shock absorbers, each having a cylinder and a piston, are mono-tube and twin-tube designs. The simplest mono-tube hydraulic shock absorbers, sometimes called dashpots, have one orifice for fluid flow. A schematic diagram of such a double-acting absorber is shown in Figure 5.8. This construct consists of two working chambers with two pistons (1 and 2), which move in turn. The two adjustable flow control valves (3 and 4) connect the working chambers with the hydraulic accumulator (5). When one of the pistons moves from the outside cover to the inside wall, the hydraulic fluid is displaced from the working chamber into the hydraulic accumulator via a flow control valve. In the process, a gas cushion is compressed in the accumulator (5). As a result, a counterforce builds up, which pushes the oil back out of the hydraulic accumulator and into the working chamber space through a flow control valve and a check valve. The two check valves (6 and 7) are used to provide the quick return process. In addition, the hydraulic accumulator (5) functions as a compensation chamber for the hydraulic fluid if the system has a hydraulic leak. In particular, each flow control valve is individually adjustable, and in particular can be set to a position equal to necessary value of the hydraulic resistance for the specified braking process. However, such a mono-tube hydraulic shock absorber with flow control valve has limitations re respect to the adjustment of the dynamic behavior of the braking process; thus, in some cases, using these devices increases the braking time.

The mono-tube hydraulic shock absorber shown in Figure 5.9 has a continuous variable orifice that provides the necessary dynamic conditions for the braking process. Most often, the compression orifice is merely a hole (3) in the back cover (4); the main piston (1) has the metering pin (2) extending through the hole (3); and by varying the pin (2) diameter, the orifice area is

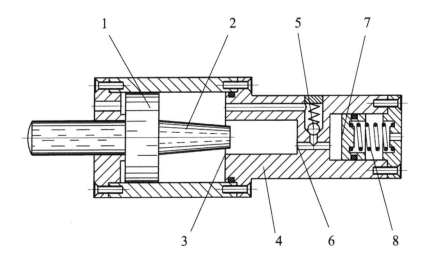

FIGURE 5.9
Single-acting hydraulic shock absorber with continuously variable orifice.

varied. This variation is adjusted so that the absorber braking force is necessary for the dynamic loading. The check valve (5) is used to enable the quick return process. Sometimes, in the line (6) that connects the absorber working chamber with the compensation chamber, an additional constant orifice is installed. In the compensation chamber, the piston (7) with a compression spring (8) is installed, and these elements provide the return of the main piston (1) to the initial position and compensate for the hydraulic fluid. The main advantage of this absorber is that it has nearly perfect deceleration while only working for one moving mass, velocity, and propelling force. If the shock is not properly sized for the application, the result is high collision and set-down forces. In addition, the dynamic behavior of such a design has high sensitivity that derives from the dimensional accuracy of the metering pin (2) and the hole (3).

A manually adjustable hydraulic shock absorber is shown in Figure 5.10. In its most general form, such a design consists of a twin-tube cylinder with an inner working chamber (1), known as the pressure chamber, and the second chamber (2), known as the reserve chamber, between the concentric inner and outer walls. It also has a piston (3) that moves within the inner cylinder. A series of orifices (4) are drilled through the inner cylinder wall; most often, the holes space along the length of the cylinder at exponential intervals. When a moving load contacts the piston rod (5), it drives the piston (3) inward, forcing oil through the orifices and into the second chamber (2). As the piston (3) moves, it progressively blocks the orifices (4) behind it. This reduces the effective metering area and maintains a uniform deceleration force, while the energy of the actuator load is converted to heat in the fluid. Orifice size and spacing are critical and best accomplished by sophisticated computer modeling. The inner tube with holes (4), which fits within

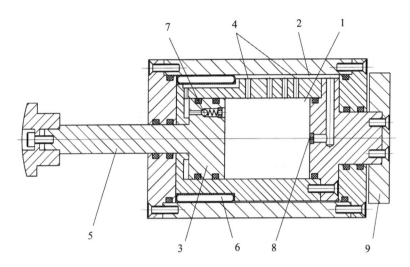

FIGURE 5.10
Single-acting, twin-tube hydraulic shock absorber.

the stationary outer tube, can be rotated via a handle (9) to adjust the total effective area and desired deceleration rate. When the inner tube is rotated toward the open position, the shock absorber provides maximum orifice area and minimum resistance. Conversely, movement toward the closed position reduces the total orifice area and increases the resistance. This adjustment method provides the capability to handle high propelling forces at low viscosities. In this construction, the function of the accumulator has the toroidal chamber (6) with pressured gas. The check valve (7) is needed to provide the quick return process. The constant orifice (8) is necessary for smooth end motion of the shock absorber when all orifices (4) are blocked by piston (3).

Sometimes, instead of a constant orifice (8), a pressure-relief valve can be used (Figure 5.11). In this case, the variable orifice (1) connects the inner working chamber (2) with the reserve chamber (3), and the adjustable spring (4) defines the desired pressure differential.

The adjustable construction overcomes the main disadvantage of mono-tube devices by adjusting the orifices to custom fit any input conditions. Therefore, a properly adjusted shock absorber can produce nearly perfect deceleration. In addition, these devices have the ability to handle a wide range of input conditions. Their major disadvantage is that they must be manually adjusted each time the input conditions change.

Figure 5.12 depicts a hydraulic shock absorber with cam mechanism. This design consists of the housing (1) and the main piston (2), which has a double-ended rod (3). These elements form two main working chambers (4 and 5) that are connected to each other through the constant orifice (7) and check valve (6), which are needed for the quick return of the main piston to the initial position. The additional piston (8) has a double-ended rod (9)

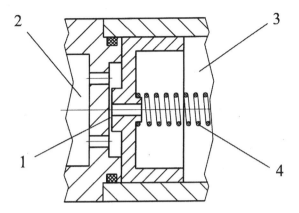

FIGURE 5.11
Schematic diagram of the pressure limiter.

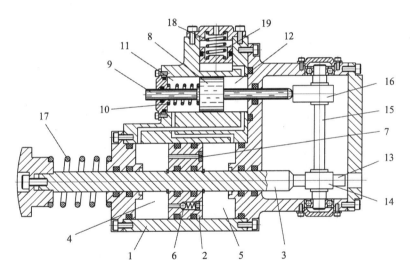

FIGURE 5.12
Hydraulic shock absorber with cam mechanism.

and forms with the housing (1) two additional working chambers (11 and 12), which are connected to the main working chamber (chamber [11], which is connected to chamber [5], and chamber [12], which is connected to chamber [4]). The additional piston (8) has a kinematic connection with the main piston (2) through the cam mechanism, which consists of the rack (13), the pinion (14), the shaft (15), and the cam (16). The rack (13) is the integral part of the rod (3), and the pinion (14) — just as the cam (16) — is mounted on the shaft (15). The helical spring (10) provides good contact between the end face of the rod (9) and the working surface of the cam (16). The shock absorber returns to the initial position by a spring (17). In addition, this device has a compensation chamber, which is formed by the housing (1) and the piston (19), and it has a spring (18). As the piston (2) moves, the rack (13) and pinion (14) mechanism rotates the shaft (15) with cam (16); as consequence of this, the additional piston (8) moves to the left or to the right side, depending on the cam (16) profile. As a result, the fluid flow displaced from the right working chamber (5) to the left chamber (4) via the orifice (7) can be changed according to the profile of the cam (16) and positioning of the main piston (2). If the additional piston (8) moves to the left side, the fluid flow that goes through the constant orifice (7) is increased; if the additional piston moves to the right, this flow decreases. Such a design has a wide range of possibilities for controlling the deceleration process. In addition, this shock absorber's "adjustability" derives from the use of different cams to custom fit various braking conditions.

For the simplest mono-tube hydraulic shock absorbers (Figure 5.8), the calculation of their effective area of the piston (A_D), the shock absorber working stroke (s_D), and the effective area of the flow control valve (A_{VD}) is carried out depending on the moving load mass (m_L), the sink velocity ($\dot{x}_S$), and the pneumatic actuator parameters. In addition, it is necessary to obtain the braking time (t_D) or the maximum deceleration value ($\ddot{x}_D$) for the assumption that the braking process has constant deceleration. Figure 5.13 is a schematic diagram of the system pneumatic cylinder–hydraulic shock absorber.

The motion of the dynamic system pneumatic actuator–hydraulic shock absorber can be described, as a first approximation, by the following equation:

$$m_L \cdot \ddot{x} = F_A - F_{SA} \tag{5.13}$$

where F_A is the pneumatic actuator force and F_{SA} is the force of the hydraulic shock absorber.

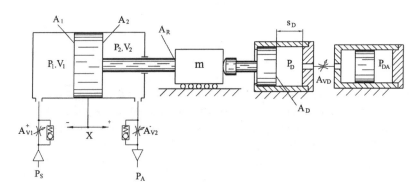

FIGURE 5.13
Schematic diagram of the pneumatic cylinder with hydraulic shock absorber.

Before the braking process, the pneumatic actuator has uniform motion, and it is assumed as a first approximation that the pneumatic actuator force in the deceleration process is changed by the linear magnification law. Then the sum of the load kinetic energy and the working function of the pneumatic actuator would be defined by:

$$E_\Sigma = \frac{A_1 \cdot (P_S - P_A) \cdot s_D + m_L \cdot \dot{x}_S^2}{2}$$

where the shock absorber working stroke (s_D) can be estimated by Equation 5.6. Hydraulic shock absorber power consumption is defined as:

$$E_D = P_D \cdot A_D \cdot s_D$$

where P_D is the fluid pressure in the absorber working chamber. In general, this power consumption is equal to the energy E_Σ; that is:

$$P_D \cdot A_D \cdot s_D = \frac{A_1 \cdot (P_S - P_A) \cdot s_D + m_L \cdot \dot{x}_S^2}{2} \tag{5.14}$$

The volumetric fluid flow of the constant orifice is described by (fluid compressibility in the shock absorber is negligible):

$$Q_D = \dot{x} \cdot A_D = A_{VD} \cdot \sqrt{\frac{2 \cdot (P_D - P_{DA})}{\rho_F}} \tag{5.15}$$

where P_{DA} is the gas pressure in the hydraulic accumulator and ρ_F is the shock absorber fluid density.

Bounding the pressure variable P_D by a value P_{DM}, which is determined by design limitation or working condition, and using Equation 5.14 and Equation 5.15, one can determine the effective area of the shock absorber piston and the effective area of the flow control valve as:

$$A_D = \frac{A_1 \cdot (P_S - P_A) \cdot s_D + m_L \cdot \dot{x}_S^2}{P_{DM} \cdot s_D} \tag{5.16}$$

and

$$A_{VD} = \frac{A_D \cdot \dot{x}_S}{\sqrt{\dfrac{2 \cdot (P_{DM} - P_{DA})}{\rho_F}}} = \frac{A_1 \cdot (P_S - P_A) \cdot s_D \cdot \dot{x}_S + m_L \cdot \dot{x}_S^3}{P_{DM} \cdot s_D \cdot \sqrt{\dfrac{2 \cdot (P_{DM} - P_{DA})}{\rho_F}}} \tag{5.17}$$

Equation 5.16 and Equation 5.17 are obtained by assuming that the fluid compressibility is negligible and the active force of the pneumatic actuator changes by linear law. It is necessary to refine the estimation parameters (the piston effective area and the effective area of the flow control valve) by computer simulation of the dynamic behavior of the pneumatic actuator–hydraulic shock absorber system.

The mathematical model of the dynamic behavior of the pneumatic cylinder–hydraulic shock absorber system can be described by Equation 5.9, Equation 5.11, and Equation 5.12, but in this case the differential equation that describes the change in pressure in the shock absorber working chamber (P_D) has the following form:

$$\dot{P}_D = \frac{E_F}{V_{0D} + A_D \cdot [s_D - (x - x_D)]} \cdot \left(A_D \cdot \dot{x} - A_{VD} \cdot \sqrt{\frac{2 \cdot (P_D - P_{DA})}{\rho_F}} \right) \tag{5.18}$$

where E_F is the fluid bulk modulus elasticity.

Example 5.2. *Define the parameters for the hydraulic shock absorber for a pneumatic cylinder in which the piston diameter is 50 mm ($A_1 = 1.963 \cdot 10^{-3}$ m^2) and the rod diameter is 20 mm ($A_2 = 1.649 \cdot 10^{-3}$ m^2), the absolute supply pressure is $P_S = 0.6$ MPa, the sink velocity is $\dot{x}_S = 1.5$ m/s, and mass of the load is $m_L = 30$ kg. The necessary deceleration is about $\ddot{x}_D = 15$ m/s^2, and then the shock absorber working stroke is about $s_D = 0.075$ m. The fluid density is $\rho_F = 900$ kg/m^3 and its bulk modulus elasticity is $E_F = 1.8 \cdot 10^9$ N/m^2. The maximum fluid pressure value is about $P_{DM} = 3$ MPa and the gas pressure in the hydraulic accumulator is about $P_{DA} = 0.3$ MPa. Then the shock absorber piston effective area, as defined by Equation 5.16, is about $A_D = 0.627 \cdot 10^{-3}$ m^2; then the shock absorber piston diameter is about $D_D = 28$ mm. The effective area of the flow control valve, as estimated from Equation 5.1), is about $A_{VD} = 1.214 \cdot 10^{-5}$ m^2.*

Figure 5.14 shows the dynamic curves of the braking process for the pneumatic cylinder–hydraulic shock absorber system after parameter adjustment by computer simulation. In this case, the absorber piston diameter is $D_D = 40$ mm ($A_D = 1.256 \cdot 10^{-3}$ m^2), the effective area of the flow control valve is $A_{VD} = 4 \cdot 10^{-6}$ m^2, and the absorber working stroke is $s_D = 55$ mm. As is obvious from this simulation, the analytical estimation of the absorber parameters is very rough and should only be used as an initial evaluation. In this case, the computer simulation is the sine qua non stage in the accurate calculation of the absorber parameters.

Figure 5.14 reveals the major drawback of the simplest mono-tube hydraulic shock absorber; namely, the spikes in actuator deceleration and in the fluid pressure of the absorber working chamber. From this point of view, the use of a shock absorber with a continuously variable orifice or with the several constant orifices has a significant advantage.

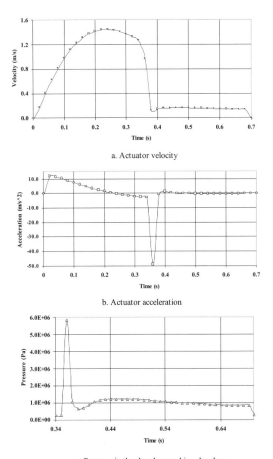

a. Actuator velocity

b. Actuator acceleration

c. Pressure in the absorber working chamber

FIGURE 5.14
Dynamics of the pneumatic cylinder–hydraulic shock absorber system: (a) actuator velocity, (b) actuator acceleration, and (c) pressure in the absorber working chamber.

Calculation of the parameters of the hydraulic shock absorbers with continuously variable orifice (Figure 5.9) can be carried out by bunch curves, which are obtained for constant deceleration.[189] In this case, the orifice effective area changes as a linear function of the shock absorber piston displacement. For such an estimation, the following assumptions are usually taken into account:

- The working fluid is an incompressible liquid.
- The actuator active force is constant.
- Maximum fluid pressure and deceleration values are limited by working conditions.

Shock absorbers with several constant orifices (Figure 5.10) have some peculiarity in their parameter estimation. The ultimate approximation of the continuously variable orifice by the step function reveals the appearance of spikes in the deceleration and fluid pressure in the working chamber. In particular, it is conspicuous because of the small number of orifice rows (less than three). For a given limitation in the deceleration value, the shock absorber parameters are estimated for minimum braking displacement or minimum time of the braking process. For acceptable dynamic behavior of a pneumatic actuator–hydraulic shock absorber system, six rows of constant orifices are admissible.

As in the simplest mono-tube shock absorbers, computer simulation provides a choice of accurate parameters for absorbers with a multi-orifice inner tube. Computer simulation is also used to maximize the performance of a shock absorber by predicting the ideal adjustment setting for a particular group of conditions.

Although some applications require custom-made shocks, most off-the-shelf models can be used. In these cases, a hydraulic shock absorber catalog and a calculator are sufficient to select the right part. Engineers can then use these specifications early in the design stage. To simplify the process, manufacturers provide numerous application examples with drawings and equations to help pick shock absorbers in advance. Some manufacturers have gone a step further by offering free software for selecting shock absorbers.

5.1.2 Parameters of Pneumatic Positioning Actuators

The open-loop pneumatic position actuators, both linear and rotary types, most often contain hard mechanical stops and shock absorber devices, which provide the soft stopping process. Usually, a graph of velocity vs. time is the familiar trapezoidal motion curve, which is the most commonly used. In this case, velocity ramps up to some level and holds steady for a specified time, then ramps down to zero at the end of the motion (Figure 5.15a). Seldom does the motion graph have a triangular profile (Figure 5.15b), which is typical for actuators with high inertial load.

Theoretically, in a trapezoidal profile, the acceleration and deceleration rates of a motion are constant. The velocity accelerates linearly until the profile reaches the required velocity value. During deceleration, the velocity decreases linearly until the motion reaches its stop position, where the value of the velocity should be negligible.

The trapezoid motion curve has three specific parts:

1. Acceleration part with duration t_A
2. Movement with steady-state velocity; running time is t_S
3. Deceleration motion with time t_B

Characteristics of the last part of the motion are defined by the actuator requirements specification and provided by using the shock absorber (compu-

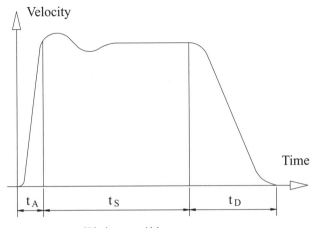

a. Velocity trapezoidal move curve

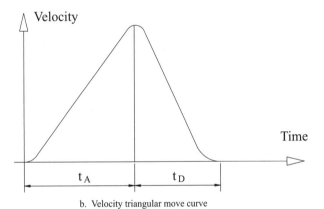

b. Velocity triangular move curve

FIGURE 5.15
Velocity changing for the open-loop position actuator: (a) velocity "trapezoidal" move curve and (b) velocity "triangular" move curve.

tation of parameters is shown in the Chapter 5.1.1). For this type of position actuator, the character of the transition between the first and second part of the movement is not determined because, in the point-to-point motion, the major requirement is the movement time while the kind of motion is not important.

For triangular actuator motion, the motion curve has only two parts: (1) acceleration motion (t_A) and (2) movement with deceleration (running time t_B). In open-loop pneumatic position actuators, this type of motion mode very rarely because the actuator efficiency is very low.

In the initial position, before the start of motion, the actuator output link is pressed to the hard stop surface by the force of the pressure differential in the actuator working chambers (Figure 5.16a). Acceleration and motion with steady-state velocity is realized when the working chambers are connected in the inverse form relative to the initial condition (Figure 5.16b). The

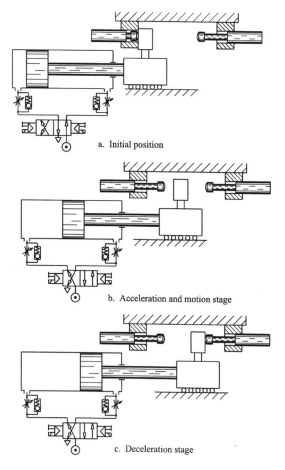

FIGURE 5.16
Operation steps of the open-loop position actuator: (a) initial position, (b) acceleration and motion stage, and (c) deceleration stage.

deceleration motion exists when the actuator moves together with the shock absorber device (Figure 5.16c).

Usually, the estimation of actuator parameters begins with an analysis of the requirement conditions of the point-to-point movement. Most often, in the position pneumatic actuator, this motion is determined by the following parameters: the moving mass (m); the maximum actuator stroke, which corresponds to the maximum piston stroke (L_S) in the linear actuator, for instance; the external force (F_L); the time of the motion (t_m) for maximum actuator stroke; the maximum value of the acceleration/deceleration process ($\ddot{x}_D$); and the supply pressure (P_S). Sometimes, the maximum value of the steady-state velocity ($\dot{x}_S$) is also required.

In general, the complete positioning time consists of various stages, and it can be determined by the following formula:

$$t_{MC} = t_V + t_{SM} + t_A + t_S + t_B \tag{5.19}$$

where t_V is the switching time of the electropneumatic control valve and t_{SM} is the time of the pressure change (in the actuator working chambers) until the start of motion of the actuator.

Standard types of the electropneumatic control valves have short switching times, which range from 0.01 to 0.05 s (that is, about 1 to 5% from the complete positioning time). In practice, the estimation of actuator parameters is not taken into account in this stage of actuator operation.

The second period of positioning time depends on the parameters of the actuator, and for the first estimation, this time is a negligible quantity. In the adjustment stage of the parameter estimation, this time is refined, and is shown below.

In the first step of parameter estimation, the deceleration time can be defined by the required values of the maximum deceleration ($\ddot{x}_D$) and the steady-state velocity ($\dot{x}_S$) or the shock absorber working stroke (s_D). A rough calculation of the deceleration time can be carried out using the following formula:

$$t_B = \frac{\dot{x}_S}{\ddot{x}_D} \quad \text{or} \quad t_B = \sqrt{\frac{2 \cdot s_D}{\ddot{x}_D}}$$

Calculation of actuator parameters is carried out for the requirement of acceleration and velocity steady-state motion. For this step, the mechanical subsystem can be modeled as a second-order system with friction, and the equation of motion for the piston-load assembly is described by Equation 2.23. The differential equations that describe the pressure changes in the actuator working chambers can be obtained from Equation 2.18. In open-loop position actuators that operate in point-to-point mode in this motion stage, the first chamber is connected with the supply pressure and the second is connected with the atmospheric line (Figure 5.16b). In this case, the dynamic behavior of the actuator is determined by the following differential equations (Figure 5.17):

$$\begin{cases} m \cdot \ddot{x} + b_V \cdot \dot{x} + F_F + F_L = P_1 \cdot A_1 - P_2 \cdot A_2 - P_A \cdot A_R \\[2mm] \dot{P}_1 = \dfrac{1}{V_{01} + A_1 \cdot (0.5 \cdot L_S + x)} \cdot \left[A_V^+ \cdot P_S \cdot K_* \cdot \varphi\left(\dfrac{P_1}{P_S}\right) - P_1 \cdot A_1 \cdot \dot{x} \right] \\[4mm] \dot{P}_2 = \dfrac{1}{V_{02} + A_2 \cdot (0.5 \cdot L_S - x)} \cdot \left[P_2 \cdot A_2 \cdot \dot{x} - A_V^- \cdot P_2 \cdot K_* \cdot \varphi\left(\dfrac{P_A}{P_2}\right) \right] \end{cases} \tag{5.20}$$

To simplify dynamic analysis and parameter estimation of the position actuator, and to achieve some general criteria for the actuator's dynamic

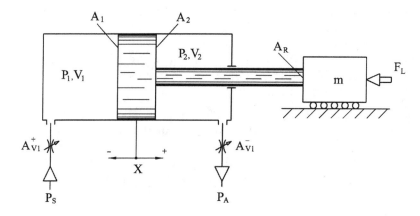

FIGURE 5.17
Schematic diagram of the open-loop pneumatic cylinder (movement with acceleration and steady-state velocity).

behavior, the differential equations that describe the actuator dynamic are rewritten in dimensionless form. For that, the following dimensionless similarity parameters are usually used:

- Actuator dimensionless displacement $\xi = x/L_S$
- Dimensionless pressure in the working chambers $\sigma_i = P_i/P_S$, where i is the actuator chamber index
- Dimensionless time $\tau = t/t_*$, where t_* is the time scale factor coefficient, which is determined as $t_* = \dfrac{A_1 \cdot L_S}{A_V^+ \cdot K_*}$

Then the dimensionless differential equations that describe the actuator's dynamic behavior have the following form:

$$
\left\{
\begin{aligned}
& W^2 \cdot \ddot{\xi} + v \cdot \dot{\xi} = \sigma_1 - \alpha_A \cdot \sigma_2 - \alpha_R \cdot \sigma_A - \chi \\[2mm]
& \dot{\sigma}_1 = \frac{1}{\xi_{01} + 0.5 + \xi} \cdot [\varphi(\sigma_1) - \sigma_1 \cdot \dot{\xi}] \\[2mm]
& \dot{\sigma}_2 = \frac{1}{\xi_{02} + 0.5 - \xi} \cdot \left[\sigma_2 \cdot \dot{\xi} - \frac{\Omega \cdot \sigma_2}{\alpha_A} \cdot \varphi\left(\frac{\sigma_A}{\sigma_2}\right) \right]
\end{aligned}
\right.
\tag{5.21}
$$

In these equations, W^2 determines the dimensionless inertial load, and this parameter is

$$
W = \frac{A_V^+ \cdot K_*}{A_1} \cdot \sqrt{\frac{m}{P_S \cdot A_1 \cdot L_S}} \ ;
$$

the dimensionless viscous friction coefficient is

$$v = b_V \cdot \sqrt{\frac{L_S}{2 \cdot P_S \cdot A_1 \cdot m}} \; ;$$

the dimensionless load is

$$\chi = \frac{F_F + F_L}{P_S \cdot A_1}.$$

Remainder dimensionless parameters are:

$$\alpha_A = \frac{A_2}{A_1}, \; \alpha_R = \frac{A_R}{A_1}, \; \xi_{01} = \frac{x_{01}}{L_S}, \; \xi_{02} = \frac{x_{02}}{L_S}, \; \sigma_A = \frac{P_A}{P_S}, \; \text{and} \; \Omega = \frac{A_V^-}{A_V^+}.$$

Here, x_{0i} is the length of the inactive volume at the end of the stroke and admission ports.

The dimensionless inertial parameter W, which defines the integrated dimensionless inertia load of the actuator, can be represented as $W = t_M/t_P$, where

$$t_M = \sqrt{\frac{m \cdot L_S}{P_S \cdot A_1}} \; \text{and} \; t_P = \frac{A_1 \cdot L_S}{A_V^- \cdot K_*}.$$

Here, t_M is the mechanical time constant and t_P is the pneumatic time constant. It is obvious that the mechanical time constant is equal to the time it takes the actuator to pass distance L_S under the maximum driving force, which is $P_S \cdot A_1$ (motion has constant acceleration). The pneumatic time constant is equal to the time it takes the actuator to pass this distance with constant velocity, which depends on the mass flow rate of the orifice with effective area A_V^- and pressure differential $(P_S - P_A)$.

When the value of W is large ($t_M \gg t_P$), the actuator has high mechanical inertia (for example, when the moving mass m is large), and this circumstance defines the behavior of the actuator motion. Usually, for large W ($W > 1$), the actuator has a triangular velocity curve. In the range $0 < W \leq 1$, the major factor that determines actuator motion behavior is the mass flow rate of the flow control valve. In that case, the graph of velocity vs. time has a trapezoidal form. A reduction in the value of W decreases the time of the acceleration part of the trapezoidal curve. For the critical value of the dimensionless parameter W ($W = 0$), the piston with load is considered a weightless partition, and actuator dynamic behavior is determined by the charging and discharging process in the working chambers.

In general, analysis of the dynamic behavior of an open-loop position actuator is carried out by computer integration of differential equations (Equations 5.21). Usually, the results of these calculations are represented by curves of the dimensionless positioning time vs. some dimensionless parameters (for example, W, Ω, σ_A, or χ). Data processing of the computer calculation results yields approximate formulae for the dimensionless positioning time (dimensionless stroke is $\xi_S = 1$)[79,102]:

$$
\begin{cases}
\tau_S = \dfrac{1.16 \cdot (\Omega + 3.05)}{\Omega \cdot (1 - 0.9 \cdot \chi)}, & \text{if } 0 \le W \le 1 \\[4mm]
\tau_S = \dfrac{0.35 \cdot (\Omega + 3.05) \cdot [(1.6 \cdot \Omega + \sqrt{\Omega} - 0.85) \cdot W + 5]}{\Omega \cdot (1 + \sqrt{\Omega}) \cdot (1 - 0.9 \cdot \chi)}, & \text{if } 1 < W < 5
\end{cases}
\tag{5.22}
$$

The use of Equation 5.22 is acceptable for the following range of dimensionless parameters: $0.15 \le \sigma_A \le 0.3$, $0.25 \le \Omega \le 2$, and $0 < \xi_{0i} \le 0.3$.

It is well known[78,79] that the change dimensionless motion time τ_S vs. the dimensionless inertial parameter W (the actuator moves at the maximum stroke without using the shock absorbers) exhibits nonlinear behavior. This graph can be approximated by a hyperbolic function. Consider that the pressure differential during motion time is not constant, and the behavior of the charge and discharge flow in the working chambers is nonlinear. In practice, the following modified formula can be used for a quick estimation of τ_S:

$$
\tau_S = \frac{(\Omega \cdot \sigma_A \cdot \alpha_A + \varphi_*)}{\Omega \cdot \sigma_A \cdot \alpha_A \cdot (1 - 0.9 \cdot \chi)} + \frac{(0.5 \cdot W)^2}{\sqrt{1 - (\alpha_A + \alpha_R) \cdot \sigma_A - \chi}}
\tag{5.23}
$$

An estimation of the dimensionless positioning time from Equation 5.23 allows one to obtain results with an error of about 10% from the results obtained from computer integration of Equation 5.21.

After estimating τ_S, the motion time t_M can be determined from:

$$
t_m = \frac{A_1 \cdot L_S \cdot \tau_S}{A_V^+ \cdot K_*}
\tag{5.24}
$$

Example 5.3. *Determine the motion time of the open-loop linear pneumatic actuator (pneumatic cylinder) for the following requirements: the moving mass is $m = 200$ kg , the maximum piston stroke is $L_S = 0.5$ m, and the external force is $F_L = 100$ N. The piston of the pneumatic cylinder has an outside diameter of $D_P = 0.08$ m and its rod diameter is $D_R = 0.025$ m (for example, the pneumatic cylinder DNU type of "FESTO"). The 5/2-way double solenoid valve type MVH ("FESTO") is used as a control valve; its effective areas are $A_V^- = 12.6 \cdot 10^{-6}$ m² and $A_V^+ = 12.6 \cdot 10^{-6}$ m². The supply pressure is $P_S = 0.6$ MPa.*

Using the above data, the effective piston areas are $A_1 = 5.024 \cdot 10^{-3}$ m² and $A_2 = 4.533 \cdot 10^{-3}$ m². The dimensionless parameters are $\alpha_A = 0.902$, $\alpha_R = 0.098$, $\sigma_A = 0.167$, $\Omega = 1$, $\chi = 0.033$, and $W = 0.7$. Then, using Equation 5.23, the dimensionless motion time is $\tau_S = 2.942$; and according to Equation 5.24, the motion time is $t_m \approx 0.77$ s. The motion time obtained by computer numerical integration of Equation 5.23 is $t_m \approx 0.86$ s, and exhibits an acceptable estimation error (about 10.5%). By the way, estimation of the motion time using Equation 5.22 and Equation 5.24 gives a value of $t_m \approx 1.2$ s, which shows the worst estimation accuracy.

5.1.2.1 Parameter Estimation of an Actuator with Trapezoidal Velocity Curves

According to the computer integration results of the differential Equations 5.21, the open-loop pneumatic actuator has trapezoidal velocity curves if $W \leq W_*$. Because the dimensionless parameter W is the dimensionless mass in the motion equation, this condition defines the range of immediate-action actuators. The value of W_* depends primarily on two dimensionless parameters: Ω and χ. For this, the uniformity of the actuator velocity contributes to decreasing Ω and increasing χ. Decreasing Ω is achieved by reducing A_V^- relative to the value of A_V^+; as a result, the pressure in the discharging actuator working chamber increases. Selecting the value of W_* can be carried out using Equation 5.22; however, in the initial stage, when the value of the parameters A_1, A_V^-, and A_V^+ are unknown, estimation of the dimensionless parameters W, and χ is impossible. In this case, using the dimensionless parameter

$$v = \dot{x}_A \cdot \sqrt{\frac{m}{L_S \cdot (F_F + F_L)}}$$

is recommended, which can be considered as the dimensionless average actuator speed on the displacement L_S. Calculating the value of v can be derived from the required data: maximum piston stroke (L_S), moving mass (m), external force (F_L), maximum value of the acceleration/deceleration process ($\ddot{x}_D$), and positioning time (t_{MC}); from that the average speed is:

$$\dot{x}_A = \frac{\ddot{x}_D \cdot t_{MC}}{2} - \sqrt{\frac{(\ddot{x}_D \cdot t_{MC})^2}{4} - L_S \cdot \ddot{x}_D}$$

The actuator friction force can be approximately defined by $F_F = 0.1 \cdot m \cdot \ddot{x}_D$.

In this approach, the trapezoidal curve of the actuator velocity is reached when $v \leq v_*$, where $v_* = 0.25$. Meeting this condition provides the specified type of motion and allows one to carry out actuator parameter estimation. For cases where $v > v_*$, a trapezoidal curve of actuator velocity is not achieved. Some causes for this include a very high average speed ($\dot{x}_A$), a big

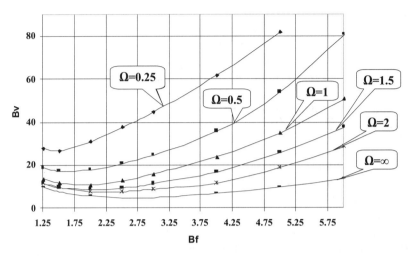

FIGURE 5.18
Parameter estimation of the point-to-point open-loop pneumatic actuator.

moving mass (m), or a short piston stroke (L_S). Exclusion is the case where a large value of υ is achieved by decreasing the external force; that is, the actuator has only an inertial load. In this case, actuator parameter estimation is carried out using a special procedure as described below.

For a pneumatic actuator with the condition $\upsilon \leq 0.25$, parameter estimation is performed using the curves shown in Figure 5.18. During the estimation process, it is necessary to find the design point on the curves in Figure 5.18 that has the coordinate $B_V = A_V^+ \cdot b_1$ and $B_f = 1/\chi = A_1 \cdot b_2$. The coefficients b_1 and b_2 are determined by the required data:

$$b_1 = \frac{P_S \cdot K_*}{(F_F + F_L) \cdot \dot{x}_A} \quad \text{and} \quad b_2 = \frac{P_S}{F_F + F_L}.$$

First of all, choose the value of Ω; then it is necessary to consider the following circumstance: decreasing the value of Ω increases the pressure in the discharging working chamber. It enhances the stability of the actuator motion with the steady-state velocity ($\dot{x}_S$), even if the external force has some variations. At the same time, by increasing the pressure in the discharging working chamber, the actuator internal resistance force is significantly increased, which decreases the actuator efficiency. Under these circumstances, the estimated piston effective area will be greater than the calculated diameter for other conditions. The increasing Ω value increases the actuator efficiency; however, the stability of the steady-state velocity degrades. Therefore, if the requirement for the stability of the steady-state velocity is the determining factor, then the value of Ω should be in range from 0.25 to 0.5. In the case where fluctuation in the steady-state velocity is acceptable, the value of Ω can be increased up to 1. A subsequent rise in Ω will render the actuator motion closer to uniform acceleration. The curve in Figure 5.18, which corresponds to $\Omega = \infty$, refers to a single-acting pneumatic cylinder.

After choosing the value of Ω, define the design point on the proper curve. Basically, the dimensionless parameter B_V can be considered the dimensionless effective area of the control valve and B_f the dimensionless area of the cylinder piston. From a quantity of design points for special values of Ω, only the minimum value of B_V provides the best correlation between the effective areas of the control valve and the actuator piston. In this case, the control valve effective area has a minimum volume. For each curve in Figure 5.18, the minimum value of B_V conforms to specific values of $B_f = B_{fC}$ in the range from 1.35 to 2.5. In practice, the position of the design point on the curve is chosen on the right side of the point that belongs to the minimum value of B_V. This performance provides better stability in the steady-state velocity under the external force variations.

Example 5.4. *Determine the parameters of the point-to point open-loop linear pneumatic actuator (pneumatic cylinder) with the following requirements: the moving mass is m = 10 kg , the maximum piston stroke is L_S = 0.5 m, the external force is F_L = 100 N, the maximum acceleration/deceleration is $\ddot{x}_D$ = 15 m/s2, the time of motion for the maximum piston stroke is about t_{MC} = 1 s, and the supply pressure is P_S = 0.6 MPa.*

For the above data, the average velocity is:

$$\dot{x}_A = \frac{\ddot{x}_D \cdot t_{MC}}{2} - \sqrt{\frac{(\ddot{x}_D \cdot t_{MC})^2}{4} - L_S \cdot \ddot{x}_D} \approx 0.5 \frac{m}{s}$$

The friction force is about $F_F = 0.1 \cdot m \cdot \ddot{x}_d = 15N$. Then the dimensionless parameter υ has the value

$$\upsilon = \dot{x}_A \cdot \sqrt{\frac{m}{L_S \cdot (F_F + F_L)}} \approx 0.21 .$$

That is, the value of υ is less than 0.25 and a trapezoidal curve of the actuator velocity can be obtained. Given $\Omega = 1$, the design point can be found from the proper curve in Figure 5.18. According to this data, B_V = 13 and B_f = 2. After that, parameters b_1 and b_2 should be found:

$$b_1 = \frac{P_S \cdot K_*}{(F_F + F_L) \cdot \dot{x}_A} \approx 8 \cdot 10^6 \frac{1}{m^2} \quad \text{and} \quad b_2 = \frac{P_S}{F_F + F_L} \approx 5.2 \cdot 10^3 \frac{1}{m^2} .$$

The effective areas of the control valve then become: $A_V^+ = B_V/b_1 \approx 1.6 \cdot 10^{-6}$ m^2 and $A_V = B_V/b_1 \approx 1.6 \cdot 10^{-6}$ m^2. The effective area of actuator piston is $A_1 = B_f/b_2 \approx 0.4 \cdot 10^{-3}$ m^2. In this case, the piston diameter is about 23 mm. The nearest standard piston diameter is 32 mm (the piston stroke is 500 mm), which is the one to select.

For this case, a control valve with a standard nominal flow rate of $Q_{NV} = 1.5 \cdot 10^{-3}$ to $2.5 \cdot 10^{-3}$ m³/s (according to Table 3.1, $A_V = 1.8 \cdot 10^{-6}$ to $3 \cdot 10^{-6}$ m²) is suitable; for instance, the double solenoid valve type JMYH-5/2 ("Festo"), which has an effective area of about $A_V^+ = A_V^- = 3 \cdot 10^{-6}$ m², can be used.

Figure 5.19 depicts the dynamic behavior of this pneumatic actuator. The slow-down process is performed by the pneumatic shock absorber, in which the piston diameter is 0.028 m, the working stroke is $s_D = 0.028$ m, and the effective area of the flow control valve is $A_{VD} = 3 \cdot 10^{-6}$ m² (see Figure 5.1, option 1). For this actuator, the effective area of the orifice in the discharging chamber is $A_V^- = 2 \cdot 10^{-6}$ m²; thus, the actuator has the structure shown in Figure 1.4, where in addition to the main control valve, two additional flow control valves are used.

Analysis of the obtained results demonstrates the high accuracy of the parameter calculation for the point-to point open-loop linear pneumatic actuator.

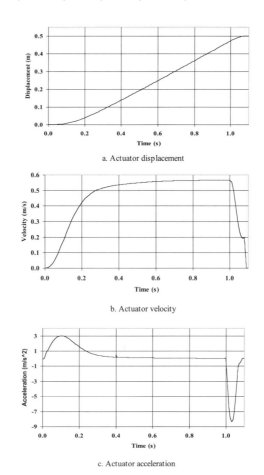

a. Actuator displacement

b. Actuator velocity

c. Actuator acceleration

FIGURE 5.19
Dynamics of the open-loop pneumatic cylinder with trapezoidal velocity profile: (a) actuator displacement, (b) actuator velocity, and (c) actuator acceleration.

In cases where the actuator external load is only inertial load ($F_L = 0$), parameter estimation is performed by another procedure. Primarily, the relation $\alpha_p = A_V^+/A_1$ is estimated from the following empirical equation:

$$\alpha_p = \frac{A_V^+}{A_1} = 5.25 \cdot 10^{-3} \cdot \frac{\dot{x}_A}{\Omega} \tag{5.25}$$

where $\dot{x}_A$ is an average actuator velocity in meters per second (m/s). As a first approximation, $\Omega = 1$ is an acceptable value. After that, the minimum value of the actuator piston area should be calculated. In this estimation step, the formula for the dimensionless inertial load is used:

$$A_{1MIN} = \frac{2 \cdot k \cdot m \cdot R \cdot T_S \cdot \alpha_p^2}{P_S \cdot L_S \cdot (k-1) \cdot W^2} \tag{5.26}$$

In this case, $W = 1$ is recommended as a first approximation. In the final step, the estimated parameters of the actuator should be refined by computer integration of the differential Equations 5.20.

Example 5.5. *Determine the parameters of a point-to-point open-loop linear pneumatic actuator (pneumatic cylinder) that has only an inertial load. The requirements of this actuator are: the moving mass is m = 20 kg , the maximum piston stroke is L_S = 0.3 m, the maximum acceleration/deceleration is $\ddot{x}_D$ = 20 m/s², the time of motion for the maximum piston stroke is about t_{MC} = 1 s, and the supply pressure is P_S = 0.6 MPa.*
The average velocity is:

$$\dot{x}_A = \frac{\ddot{x}_D \cdot t_{MC}}{2} - \sqrt{\frac{(\ddot{x}_D \cdot t_{MC})^2}{4} - L_S \cdot \ddot{x}_D} \approx 0.3 \frac{m}{s}.$$

According to Equation 5.25, the dimensionless parameter is α_p = 1.575 · 10⁻³. Using Equation 5.26, the minimum value of the actuator piston area is:

$$A_{1MIN} = \frac{2 \cdot k \cdot m \cdot R \cdot T_S \cdot \alpha_p^2}{P_S \cdot L_S \cdot (k-1) \cdot W^2} \approx 162 \cdot 10^{-6} \ m^2$$

Then the minimum piston diameter is 15 mm. Select the standard pneumatic cylinder that has a piston diameter of 25 mm (the piston stroke is 300 mm), and the effective piston area is A_1 ≈ 491 · 10⁻⁶ m². According to Equation 5.25, the effective areas of the control valve are: $A_V^+ = A_V^- = \alpha_p \cdot A_1$ ≈ 0.8 · 10⁻⁶ m².
As above in Example 5.3, this pneumatic actuator consists of one double solenoid-control valve and two flow-control valves (see Figure 1.4). The double solenoid valve of type JMZH-5/2 ("Festo") with an effective area of about $A_V^+ = A_V^-$ ≈ 1.3 · 10⁻⁶ m²

can be used as the main control valve. A one-way flow-control valve of type GRLZ-M5 ("Festo") can be used as additional flow-control valves (its effective area is adjusted in the range from 0 to $1.7 \cdot 10^{-6}$ m^2).

Figure 5.20 depicts the dynamic behavior of this pneumatic actuator. The braking process is performed by the hydraulic shock absorber, for which the piston diameter is 0.025 m, the working stroke is $s_D = 0.015$ m, the gas pressure in the hydraulic accumulator is about $P_{DA} = 0.3$ MPa, and the effective area of the flow control valve is $A_{VD} = 1 \cdot 10^{-6}$ m^2 (see Figure 5.13).

Analysis of the obtained results demonstrates the high accuracy of the parameter calculation for a point-to-point open-loop linear pneumatic actuator that has only inertial load. The velocity curve has a trapezoidal shape, and the time of motion meets the requirement.

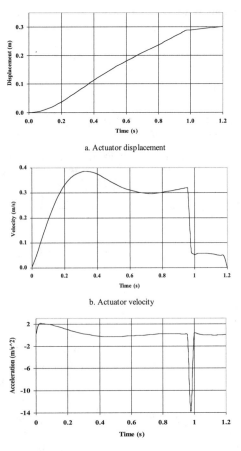

a. Actuator displacement

b. Actuator velocity

c. Actuator acceleration

FIGURE 5.20
Dynamics of the open-loop pneumatic cylinder with trapezoidal velocity profile and inertial load: (a) actuator displacement, (b) actuator velocity, and (c) actuator acceleration.

5.1.2.2 Parameter Estimation of an Actuator with Triangular Velocity Curves

As discussed above, a triangular velocity profile is seldom encountered in open-loop pneumatic position actuators. An exception would be an actuator with a high inertial load and a short working displacement. In such systems, estimation of the actuator parameters can be carried out by diagrams. For this situation, Equation 5.22 for the range $1 < W < 5$ can be used because this particular range defines the actuator with a triangular velocity curve. Most often, the estimation of parameters is performed for conditions that provide minimum actuator motion time. For that, after substituting the relations for the dimensionless parameters into the second equation (Equation 5.22), the equation for the moving time then becomes:

$$t_{MC} = b_3 \cdot \Psi(\chi) \tag{5.27}$$

where

$$\Psi(\chi) = \frac{b_4 \cdot \chi^{3/2} + 1}{\chi \cdot (1 - 0.9 \cdot \chi)} ,$$

$$b_3 = \frac{1.75 \cdot (\Omega + 3.05) \cdot (F_F + F_L) \cdot L_S}{\Omega \cdot (1 + \sqrt{\Omega}) \cdot A_V^+ \cdot P_S \cdot K_*}$$

$$b_4 = \frac{(1.6 \cdot \Omega + \sqrt{\Omega} - 0.85) \cdot A_V^+ \cdot P_S \cdot K_*}{5 \cdot (F_F + F_L) \cdot \sqrt{\dfrac{(F_F + F_L) \cdot L_S}{m}}} .$$

It follows from Equation 5.27 that for certain parameters Ω, F_L, F_F, L_S, m, P_S, and A_V^+, the minimum value $t_{MC} = t_{MCM}$ can be achieved for $\chi = \chi_M$, which conforms to the minimum value of the function $\Psi(\chi) = \Psi_M$.

To determine the value of χ_M from the condition that $t_{MC} = t_{MCM}$, the diagrams in Figure 5.21 are conveniently used. In this case, the coefficients b_3 and b_4 are estimated from the given data. Then, according to the value of b_4, one can define the dimensionless function Ψ_M and the parameter χ_M. Subsequently, the minimum value of $t_{MC} = t_{MCM}$ is calculated from Equation 5.27 and, in the end, the effective actuator piston area is estimated using:

$$A_1 = \frac{F_L + F_F}{P_S \cdot \chi_M} \tag{5.28}$$

For this type of actuator, the estimation of shock absorber parameters is performed assuming that the actuator motion is in the uniformly accelerated

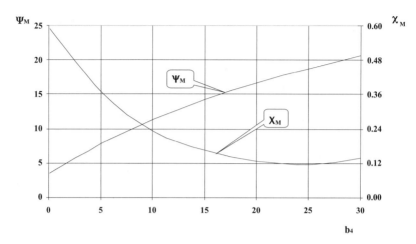

FIGURE 5.21
Curves for the estimation of actuator parameters.

mode. The value of the acceleration is a function of the effective area ratio Ω; for $\Omega \geq 1.5$, the actuator motion approaches uniform acceleration with a value

$$\ddot{x}_S \approx \frac{0.35 \cdot P_S \cdot A_1}{m}.$$

Example 5.6. *For the position pneumatic actuator with external force $F_L = 1000$ N, moving mass $m = 120$ kg, piston stroke $L_S = 0.3$ m, supply pressure $P_S = 0.6$ MPa, and control valve effective area $A_V^+ = 50 \cdot 10^{-6}$ m^2, define the piston effective area and minimum motion time for the actuator movement with a triangular velocity profile.*

Substituting this data into the formulae for the coefficients b_3 and b_4, and taking into account that $\Omega = 1.5$, the values of these coefficients are: $b_3 \approx 0.03$ s and $b_4 \approx 8$. For this value of b_4 (see Figure 5.21), the values $\Psi_M \approx 10.5$ and $\chi_M \approx 0.25$. According to Equation 5.27, the minimum moving time is $t_{MCM} = b_3 \cdot \Psi_M \approx 0.32$ s. The effective area of the piston is

$$A_1 = \frac{F_L + F_F}{P_S \cdot \chi_M} \approx 67 \cdot 10^{-4} \, m^2 \; ;$$

then the selected piston outside diameter is 0.1 m (the nearest standard size) with an effective area $A_1 = 78.5 \cdot 10^{-4}$ m^2 (rod diameter is 0.025 m).

The dynamic behavior of this pneumatic actuator is shown in Figure 5.22. The braking process is carried out by a pneumatic shock absorber, in which the piston diameter is 0.6 m, the working stroke is $s_D = 0.07$ m, and the effective area of the flow control valve is $A_{VD} = 1.7 \cdot 10{-5}$ m^2. The effective area of the orifice in the

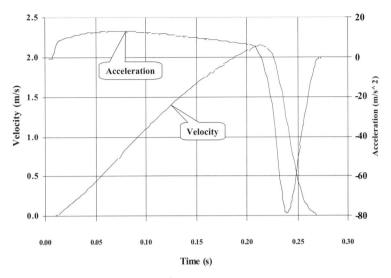

FIGURE 5.22
Dynamics of the open-loop pneumatic cylinder with triangular velocity profile.

discharging chamber of the pneumatic cylinder is $A_V^- = 75 \cdot 10^{-6} \ m^2$ (because $\Omega = A_V^-/A_V^+ = 1.5$); thereby, the actuator has the structure shown in Figure 1.4 (as in Example 5.4).

According to the results of a computer simulation, the parameter estimation has an acceptable accuracy; for example, the calculated minimum moving time is about 0.32 s, and using computer simulation this time is about 0.28 s.

It is obvious that actuator motion is in an approximate uniform acceleration mode with a triangular velocity curve (see Figure 5.22). In addition, the values of the dimensionless parameter W (W ≈ 1.4) and the parameter υ (υ = 0.63) also show that the actuator parameters are in the zone of triangular velocity motion.

In this case where the actuator parameters should be estimated for given values of the moving time (t_{MC}) and the specific control valve (A_V^+ and A_V^- are known), the curves in Figure 5.23 are usually used. In the first step of the estimation process, the values of the coefficients b_3 and b_4 are determined; after that, the value of the function $\Psi(\chi)$ is calculated from Equation 5.27. Figure 5.23 plots the horizontal line that conforms to the defined value of function $\Psi(\chi)$. Coordinates for the intersection points of this line with the curve that meets the determined value of b_4 provide the necessary values of the dimensionless parameter χ. In this case, two solutions can obtain, and both of them meet the requisite conditions. Choosing one of the two solutions depends on design issues and working conditions; for example, a pneumatic cylinder with a large piston diameter has poor sensitivity to external load fluctuation, which is an advantage; however, it has large overall dimension. The piston effective area is defined by Equation 5.28.

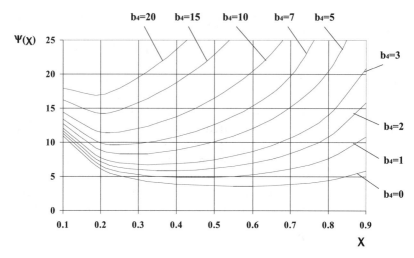

FIGURE 5.23
Curves for estimation of open-loop actuator parameters with triangular velocity profile.

Example 5.7. *A position pneumatic actuator with external force $F_L = 500$ N, moving mass $m = 80$ kg, piston stroke $L_S = 0.25$ m, supply pressure $P_S = 0.6$ MPa, motion time $t_{MC} = 0.4$ s, and a control valve effective area $A_V^+ = 12.6 \cdot 10^{-6}$ m² defines the piston effective area for the actuator that moves with a triangular velocity profile.*

For $\Omega = 1.5$, the coefficient b_3 is about 0.052 s and the coefficient b_4 is about 5.1. Then the value of function $\Psi(\chi)$ is $\Psi(\chi) = t_{MC}/b_3 \approx 7.69$. For this value of the function $\Psi(\chi)$ and for $b_4 \approx 5.1$, two values for the dimensionless parameter χ satisfy the requisite conditions (see Figure 5.23); they are $\chi_1 \approx 0.2$ and $\chi_2 \approx 0.4$. For these two solutions, two effective areas of the actuator piston $A_{11} \approx 41.7 \cdot 10^{-4}$ m² and $A_{12} \approx 20.8 \cdot 10^{-4}$ m² conform to specification. The standard pneumatic cylinder with piston diameter of 63 mm is perfectly suited to the technical requirements, and its dynamic behavior is similar to the actuator described in Example 5.5. In this case, the maximum velocity is about 1.5 m/s, the motion time is about 0.34 s, and a pneumatic shock absorber with a piston diameter of 0.04 m, a working stroke of $s_D = 0.041$ m, and an effective area of the flow control valve of $A_{VD} = 5 \cdot 10^{-6}$ m² should be used.

It can be assumed that for actuators that operate with only inertial load, the dimensionless parameter χ does not depend on the piston effective area because the external load is the friction force of the actuator, which, as a first approximation, is pro rata to the value of the effective area of the piston. Taking into account this assumption, substituting the relation for the dimensionless parameter W into Equation 5.22, and considering the formula for dimensionless time, the actuator moving time becomes:

$$t_{MC} = b_5 \cdot \left(\frac{b_6}{\sqrt{A_1}} + A_1 \right) \tag{5.29}$$

where

$$b_5 = \frac{1.75 \cdot (\Omega + 3.05) \cdot L_S}{\Omega \cdot \left(1 + \sqrt{\Omega}\right) \cdot (1 - 0.9 \cdot \chi) \cdot A_V^+ \cdot K_*}$$

and

$$b_6 = 0.2 \cdot \left(1.6 \cdot \Omega + \sqrt{\Omega} - 0.85\right) \cdot A_V^+ \cdot K_* \sqrt{\frac{m}{P_S \cdot L_S}}$$

In this case, the minimum actuator motion time is:

$$t_{MCM} = 3 \cdot b_5 \cdot \sqrt[3]{0.25 \cdot b_6^2} \tag{5.30}$$

and for this condition, the effective area of the actuator piston is:

$$A_1 = \sqrt[3]{0.25 \cdot b_6^2} \tag{5.31}$$

Example 5.8. *Define the minimum motion time and effective area of the pneumatic cylinder piston for the actuator in which the moving mass is m = 80 kg, the piston stroke is L_S = 0.25 m, the supply pressure is P_S = 0.6 MPa, and the control valve effective area is $A_V^+ = 12.6 \cdot 10^{-6} \, m^2$. From this it follows that the requirements are the same as those in Example 5.6 except for the external load, which here is inertial load only. Then, it may be assumed that $F_L = 0.1 \cdot P_S \cdot A_1$ and $\chi = 0.1$.*

For Ω = 1.5, the coefficient b_5 is about 68.62 s/m² and the coefficient b_6 is about $0.122 \cdot 10^{-3} \, m^3$. According to Equation 5.30, the minimum motion time is $t_{MCM} \approx 0.31$ s; and using Equation 5.3), the piston effective area is $A_1 \approx 1.55 \cdot 10^{-3} \, m^2$, which conforms to the diameter of 44 mm. Finally, the standard pneumatic cylinder with a piston diameter of 50 mm can be used. In this case, a pneumatic shock absorber with a piston diameter of 0.03 m, a working stroke of s_D = 0.0475 m, and an effective area for the flow control valve of $A_{VD} = 3 \cdot 10^{-6} \, m^2$ should be used.

As shown in Figure 5.24 for this actuator, the minimum working time is about 0.28 s, the maximum actuator velocity is about 1.8 m/s, and the value of uniform acceleration is about 8 m/s². This data indicates that the parameter estimation method exhibits acceptable accuracy.

5.1.2.3   Estimation of the Time for Actuator Starting Motion

After switching the electropneumatic control valve, there is a change in pressure in the actuator working chambers. This process continues until there is a pressure differential that is able to overcome the actuator's resistance force (after this, the actuator starts its motion). Estimation of this process time (t_{SM}) can be carried out after parameter estimation of the pneumatic

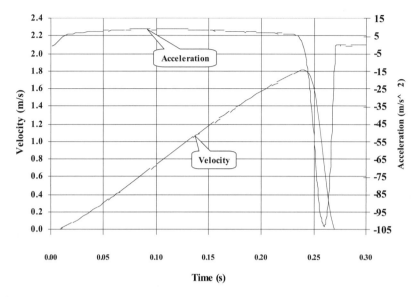

FIGURE 5.24
Dynamics of the open-loop pneumatic cylinder with inertial load and triangular velocity profile.

actuator. Usually, the initial conditions are as follows (for example, see Figure 5.16a): in the charge-working chamber, the pressure is at atmospheric levels or $P_1 = P_A$ (Figure 5.17); in the discharge-working chamber, the pressure is the supply pressure ($P_2 = P_S$) and the actuator piston is in a stationary position ($\dot{x} = 0$ and $\ddot{x} = 0$). The charging process in the first working chamber and the discharging process in the second working chamber can be described by the differential Equations 5.20, in which the initial conditions should be considered. Then the change in pressure in the first actuator chamber is described by the following equation:

$$\dot{P}_1 = \frac{A_V^+ \cdot P_S \cdot K_* \cdot \varphi\left(\dfrac{P_1}{P_S}\right)}{V_{01}} \tag{5.32}$$

and the change in pressure in the second chamber is:

$$\dot{P}_2 = -\frac{A_V^- \cdot P_2 \cdot K_* \cdot \varphi\left(\dfrac{P_A}{P_2}\right)}{V_{02} + A_2 \cdot L_S} \tag{5.33}$$

Integration of the Equation 5.32 and Equation 5.33 is carried out with the assumption that until the start of piston motion, the charging process in the first chamber and discharging process in the second chamber take place under sonic condition. In this case, $\varphi(P_1/P_S) = \varphi_*$ and $\varphi(P_A/P_2) = \varphi_*$; then, solving these differential equations obtains:

$$P_1 = P_A + B_{CH} \cdot t \tag{5.34}$$

and

$$P_2 = P_S \cdot \exp(-B_{DC} \cdot t) \tag{5.35}$$

where

$$B_{CH} = \frac{A_V^+ \cdot P_S \cdot K_* \cdot \varphi_*}{V_{01}}$$

and

$$B_{DC} = \frac{A_V^- \cdot K_* \cdot \varphi_*}{V_{02} + A_2 \cdot L_S}$$

The estimation of the time for the actuator starting motion can be confirmed by solving Equation 5.34 and Equation 5.35, together with the first equation of the system (Equation 5.20) that considered the static condition ($\dot{x} = 0$ and $\ddot{x} = 0$). After a few transformations, the equation for this time is:

$$B_{CH} \cdot t - \alpha_A \cdot P_S \cdot \exp(-B_{DC} \cdot t) = \frac{F_L + F_F + A_R \cdot P_A}{A_1} - P_A \tag{5.36}$$

Because the interval of this process is quite short, the solution of Equation 5.36 as a first approximation, can be obtained as follows:

$$t_{SM} = \frac{F_L + F_F + A_R \cdot P_A - A_1 \cdot P_A + \alpha_A \cdot P_S \cdot A_1}{A_1 \cdot (B_{CH} + \alpha_A \cdot P_S \cdot B_{DC})} \tag{5.37}$$

In practice, if this time is less than 5% of the total motion time, its value is not taken into account.

Example 5.9. *Define the starting motion time for the actuator described in Example 5.4. In this actuator, the moving mass is m = 10 kg, the maximum piston stroke is L_S = 0.5 m, the external force is F_L = 100 N, the supply pressure is P_S = 0.6 MPa, the piston effective area is A_1 = 803.8 · 10^{-6} m² (piston diameter is 32 mm), A_2 = 690.8 · 10^{-6} m², A_R = 113 · 10^{-6} m² (rod diameter is 12 mm), control valve effective areas are $A_V^+ = A_V^-$ = 12.6 · 10^{-6} m², and the inactive volumes of the cylinder are $V_{01} = V_{02} \approx 2 \cdot 10^{-5}$ m³. For Equations 5.34 and 5.35, the coefficients B_{CH} and B_{DC} are $B_{CH} \approx 1.8 \cdot 10^7$ kg/m·s³ and $B_{DC} \approx 1.6$ l/s. The coefficient α_A is about 0.86.*

Finally, the time for the actuator starting motion is $t_{SM} \approx 0.03$ s. It can be seen from the result of this estimation and from the curves in Figure 5.19 that the actuator starting motion time is less than 5% of the total motion time. Comparison of the estimation results and results of the computer simulation demonstrates an insignificant difference between them.

In Equation 5.19, the passage time of the pressure wave that goes from the control valve to the pneumatic actuator is not included because, usually, the pipe length between the valve and actuator is short in such systems. As is well known, the passage time of the pressure wave in the pneumatic pipe can be estimated as $t_p = L_p/v_A$, where L_p is the length of the pipe between the control valve and the actuator, and v_A is the sound speed under typical nominal conditions ($v_A \approx 340$ m/s)(). In practice, in the open-loop pneumatic actuators, the time t_p is taken into account if the pipe length is $L_p \geq 10$ m.

The development of a decentralized fieldbuses for the valve manifolds lets designers reduce the distance between valves and actuators. This reduces the pipe volume. In fact, from an efficiency standpoint, the ideal place to mount a valve is directly on the actuator, and this eliminates the piping.

Reducing the pipe volume between the valve and the actuator saves energy even if the pipe volume increases between the service unit and valve manifold. That is because the volume between the valve and actuator pressurizes and empties every cycle, whereas the volume between the service unit and the valve manifold rarely empties.

In open-loop positioning systems with vane rotary actuator, the positioning time can be determined from Equation 5.19. In this case, as above for linear systems, the actuator's dynamic behavior is described by the dimensionless differential Equations 5.21, where:

- Actuator dimensionless displacement $\xi = \dfrac{\varphi}{\varphi_S}$
- Time scale factor coefficient $t_* = \dfrac{V_V \cdot \varphi_S}{A_V^+ \cdot K_*}$
- Dimensionless length of the inactive volume at the end of stroke and admission ports $\xi_{0i} = \dfrac{\varphi_{0i}}{\varphi_S}$
- Dimensionless actuator area ratio $\alpha_A = 1$ ($A_1 = A_2$)
- Dimensionless load $\chi = \dfrac{M_F + M_L}{P_S \cdot V_V}$,
- Dimensionless inertial load coefficient $W = \dfrac{A_V^+ \cdot K_*}{V_V} \cdot \sqrt{\dfrac{J}{P_S \cdot V_V \cdot \varphi_S}}$

The remaining dimensionless parameters are the same as the parameters described in Equation 5.21. In addition, here, as also referred to in Chapter 1, φ_S is the full vane turning angle, φ_{0i} is the angle equivalent of the inactive volume, J is the reduced moment of inertia of the system actuator-load, M_F is the resistant torque, and M_L is the load torque. Here, the parameter $V_V =$

$0.5 \cdot L_V \cdot (R_V^2 - r_V^2)$, where L_V is the vane width, R_V is the vane outside radius, and r_V is the vane inside radius.

5.2 Pneumatic Actuators with Constant Velocity Motion

In manufacturing and process industry equipment, open-loop pneumatic actuators with velocity control are rarely used. This is due to the limitations in their performance; for example, the output of such systems is sensitive to plant parameter variation and, in a few cases, to the change in the force disturbance (both external and internal). Therefore, these systems usually are used where the output stability requirement is not critical. Such systems can be utilized, for example, in painting equipment, rough inspection devices, and wood and fabric material industries.

It is obvious that in an open-loop actuator with constant velocity motion, the graph of velocity vs. time has the trapezoidal form. The dynamic behavior of this actuator is described by the differential Equations 5.20 and the corresponding dimensionless differential Equations 5.21. For steady-state conditions, when the actuator moves with constant speed and the acceleration is zero ($\dot{\sigma}_1 = 0$, $\dot{\sigma}_2 = 0$, $\ddot{\xi} = 0$, and $\dot{\xi} = \dot{\xi}_C$, where $\dot{\xi}_C$ is the dimensionless steady-state constant velocity), Equations 5.21 can then be rewritten in the following form:

$$
\left\{
\begin{aligned}
&v \cdot \dot{\xi}_C = \sigma_{1C} - \alpha_A \cdot \sigma_{2C} - \alpha_R \cdot \sigma_A - \chi \\[2mm]
&\dot{\xi}_C = \frac{\varphi(\sigma_{1C})}{\sigma_{1C}} \\[2mm]
&\dot{\xi}_C = \frac{\Omega}{\alpha_A} \cdot \varphi\left(\frac{\sigma_A}{\sigma_{2C}}\right)
\end{aligned}
\right.
\tag{5.38}
$$

where σ_{iC} is the dimensionless pressure in the actuator working chambers for steady-state conditions.

Most often in such actuation systems, rodless pneumatic cylinders are used (see Figure 5.25) because in this design, the adjustment of the speed in both motion directions is identical and simple (this is very important for these actuators). In such a construct, both shock absorbers, as shown in Figure 5.25, and air cushioning mechanisms (see, for example, Figure 1.3) can be used. Then the dimensionless equations that describe the actuator dynamic behavior are:

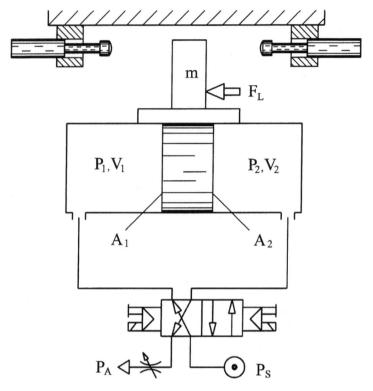

FIGURE 5.25
Schematic diagram of the open-loop pneumatic cylinder with constant velocity motion.

$$
\begin{cases}
v \cdot \dot{\xi}_C = \sigma_{1C} - \sigma_{2C} - \chi \\[2ex]
\dot{\xi}_C = \dfrac{\varphi(\sigma_{1C})}{\sigma_{1C}} \\[2ex]
\dot{\xi}_C = \Omega \cdot \varphi(\dfrac{\sigma_A}{\sigma_{2C}})
\end{cases}
\tag{5.39}
$$

In general, the graph of dimensionless output velocity ($\dot{\xi}_C$) vs. dimensionless load (χ) can be obtained by solving nonlinear algebraic equations (Equations 5.39). Real actuators typically have instantaneous limits on force and velocity output capabilities. The maximum load force and the maximum actuator velocity define the envelope in which the actuator can operate. In general, there may be four combinations of charge and discharge flow moving conditions in these actuators. However, it is important to point out that open-loop pneumatic actuators with constant velocity motion are typically designed to be highly stabilized toward variations in external load. Analysis of Equations 5.39 reveals that there is no influence from external load if the discharge flows are moving at the sonic condition; then the flow function is $\varphi(\sigma_A/\sigma_{2C}) = \varphi_*$ and the dimensionless velocity is:

$$\dot{\xi}_C = \Omega \cdot \varphi_* \tag{5.40}$$

That is, its value depends only on the dimensionless valve effective areas ratio Ω.

In this case, the actuator has high robustness and, in addition, it is easier to adjust its steady-state velocity. In particular, for this condition, the open-loop actuator with velocity control will be considered.

If the charge flow in the first working chamber and the discharge flow in the second chamber are moving at the sonic condition, the flow functions are described by $\varphi(\sigma_{1C}) = \varphi_*$ and $\varphi(\sigma_A/\sigma_{2C}) = \varphi_*$. In this case, the dimensionless pressure σ_{1C} is in the range $0 \le \sigma_{1C} \le 0.5$, and σ_{2C} is in the range $2 \cdot \sigma_A < \sigma_{2C} \le 1$. Then from Equations 5.39 after several transformations, the following can be obtained:

$$\sigma_{1C} = \frac{1}{\Omega}$$

$$\sigma_{2C} = \frac{1}{\Omega} - v \cdot \Omega \cdot \varphi_* - \chi \tag{5.41}$$

The dimensionless velocity is defined by Equation 5.40. Going over to the dimension parameter, the steady-state process has the following characteristics:

$$P_{1C} = \frac{A_V^+}{A_V^-} \cdot P_S$$

$$P_{2C} = \left(\frac{A_V^+}{A_V^-} - \frac{A_V^- \cdot b_V \cdot \varphi_*}{A_V^+} \cdot \sqrt{\frac{L_S}{2 \cdot P_S \cdot A_1 \cdot m}} - \frac{F_F + F_L}{P_S \cdot A_1} \right) \cdot P_S \tag{5.42}$$

$$\dot{x}_C = \frac{A_V^- \cdot \varphi_* \cdot K_*}{A_1}$$

where P_{1C} and P_{2C} are the absolute pressures in the first and second actuator working chamber, respectively, and $\dot{x}_C$ is the value of the steady-state velocity.

For a definition of the limitation of the dimensionless actuator parameters, Equations 5.41 and inequality equations for σ_{1C} and σ_{2C} should be considered together. For the first equation in Equation 5.41, a limitation for σ_{1C} allows reaching the dimensionless valve effective areas ratio, which is $\Omega \geq 2$. Considering the second equation in Equation 5.41 and limitation for σ_{2C} together gives the dimensionless load parameter range, which is

$$0 \leq \chi \leq \frac{1}{\Omega} - v \cdot \Omega \cdot \varphi_* - 2 \cdot \sigma_A .$$

Taking into account that the minimum Ω is 2, and

$$\frac{1}{\Omega} - v \cdot \Omega \cdot \varphi_* - 2 \cdot \sigma_A \geq 0 ,$$

then the parameter σ_A should be $\sigma_A \leq 0.25$ or $P_S \geq 0.4$ MPa. Because the minimum value of σ_A is about 0.1, the dimensionless load parameter should be changed in the range $0 \leq \chi \leq 0.3$. In the final stage, this analysis can be stated that the first combination of the charging and discharging actuator flows takes place if $\Omega \geq 2$, $P_S \geq 0.4$ MPa, and $0 \leq \chi \leq 0.3$.

In the second combination, the charge flow in the first working chamber is moving at the subsonic condition and the discharge flow in the second chamber is moving at the sonic condition; thus, the flow functions are described as $\varphi(\sigma_{1C}) = 2\varphi_* \cdot \sqrt{\sigma_{1C} \cdot (1 - \sigma_{1C})}$ and $\varphi(\sigma_A / \sigma_{2C}) = \varphi_*$. For this condition, the dimensionless pressure σ_{1C} is in the range of $0.5 < \sigma_{1C} \leq 1$, and σ_{2C} is in the range $2 \cdot \sigma_A < \sigma_{2C} \leq 1$. In this case, the dimensionless pressures in the actuator working chambers can be defined as:

$$\sigma_{1C} = \frac{1}{\Omega^2 + 1}$$

$$\sigma_{2C} = \frac{1}{\Omega^2 + 1} - v \cdot \Omega \cdot \varphi_* - \chi \tag{5.43}$$

As in the first combination, the dimensionless velocity is defined by Equation 5.40. The dimension parameters of the steady-state process have the following forms:

$$P_{1C} = \frac{P_S}{\left(\dfrac{A_V^-}{A_V^+}\right)^2 + 1}$$

$$P_{2C} = \left[\frac{1}{\left(\dfrac{A_V^-}{A_V^+}\right)^2 + 1} - \frac{A_V^- \cdot b_V \cdot \varphi_*}{A_V^+} \cdot \sqrt{\frac{L_S}{2 \cdot P_S \cdot A_1 \cdot m}} - \frac{F_F + F_L}{P_S \cdot A_1} \right] \cdot P_S \tag{5.44}$$

$$\dot{x}_C = \frac{A_V^- \cdot \varphi_* \cdot K_*}{A_1}$$

In this case, the definition of the limitation of the dimensionless actuator parameters is performed in the same manner as described for the first combination. Consider the first equation in Equations 5.43 and that the limitation for σ_{1C} allows reaching the range for the dimensionless valve effective areas, which is $0 < \Omega < 1$. Then, the second equation in Equations 5.43 and the limitation for σ_{2C} yield:

$$0 \le \chi \le \frac{1}{\Omega^2 + 1} - v \cdot \Omega \cdot \varphi_* - 2 \cdot \sigma_A;$$

and as in the previous case, $P_S \ge 0.4$ MPa and $0 \le \chi \le 0.3$. Finally, this condition of the charging and discharging actuator flows are: $0 < \Omega < 1$, $P_S \ge 0.4$ MPa and $0 \le \chi \le 0.3$.

In the process of actuator parameter estimation, the nature of the transient response and its duration are very significant. These requirements, together with the requirements referred to above, provide a basis for actuator parameter estimation.

It is important to note that pneumatic actuators with constant velocity motion, in which the parameter Ω is greater than 2, are not used in practice. In such actuators, the counterpressure is almost absent, their approach to the single-action devices that have the long transient response time.

Usually, actuators operating in the field meet the following conditions: $0 < \Omega < 1$, $P_S \ge 0.4$ MPa and $0 \le \chi \le 0.3$. In these devices, the dimensionless inertial load W has a strong effect on the nature of the transient response. The graphs in Figure 5.26 illustrate the influence of the parameter W on the nature of the transient response. From this point of view, the range from 0.1 to 0.5 for W is recommended. As seen from this figure, for $W \ge 1$, the transient response has a long duration and oscillation type, which is unacceptable for proper operation.

The dimensionless valve effective areas ratio Ω has a weak influence on the nature of the transient response. This parameter has a strong effect on the response time and the value of the steady-state velocity (Figure 5.27). The range $0.6 < \Omega < 1$ is recommended for the practical applications.

The magnitude of the dimensionless load χ by no means influences the nature of the transient response (see Figure 5.28). Increasing this parameter increases only the transient response time, which is visible. According to Figure 5.28, the recommended values of χ ($0 \le \chi \le 0.3$; see above) are acceptable.

Another important matter in the design of an open-loop actuator with constant velocity is the estimation of its stroke for the acceleration and deceleration parts of motion. In practice, the most essential is the stroke where the actuator moves with constant velocity; however, the total actuator

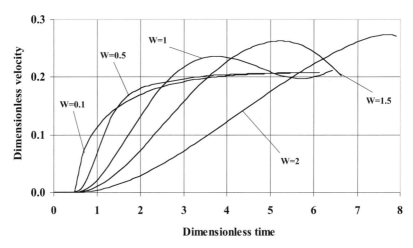

FIGURE 5.26
Influence of the dimensionless inertial load on the transient response of an actuator with constant velocity motion.

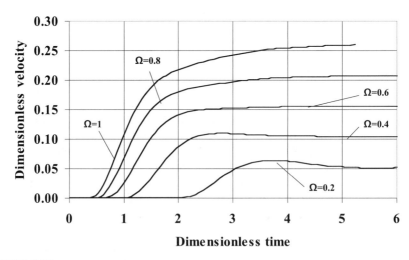

FIGURE 5.27
Influence of the dimensionless valve effective areas ratio on the transient response of an actuator with constant velocity motion.

stroke should also be included in the acceleration and deceleration displacements. For the acceleration part, when $0.6 < \Omega < 1$ and $0.1 \leq W \leq 0.5$, it can be assumed that the equivalent uniform acceleration is

$$\ddot{x}_S = \frac{B_{AC} \cdot P_S \cdot A_1}{m} ,$$

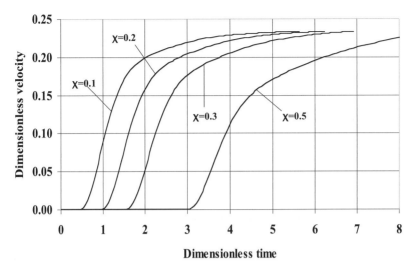

FIGURE 5.28
Influence of the dimensionless load on the transient response of an actuator with constant velocity motion.

where the coefficient B_{AC} can be defined as $B_{AC} \approx 0.01\text{–}0.02$ (this estimation was obtained by analysis of the experimental and computer simulation data). Then the displacement on the acceleration part is:

$$L_A = \frac{m \cdot \dot{x}_C^2}{2 \cdot B_{AC} \cdot P_S \cdot A_1} \tag{5.45}$$

The value of the deceleration ($\ddot{x}_D$) is usually given, and the displacement of this part (s_D) can be estimated from Equation 5.6. Then, the total stroke of the actuator is:

$$L_S = L_A + L_C + s_D \tag{5.46}$$

where L_C is the actuator displacement where it moves with constant velocity (usually, is given).

The key parameters of the actuator, which are the piston effective area (A_1) and the effective areas of the control valve (A_V^+ and A_V^-), should be estimated by the following sequence:

1. Using the given value of the actuator constant velocity and the third equation in Equations 5.44, the ratio

$$B_A^- = \frac{A_V^-}{A_1} = \frac{\dot{x}_C}{\varphi_* \cdot K_*}$$

is defined.

2. For the required parameters Ω $(0.6 < \Omega < 1)$ and W $(0.1 \leq W \leq 0.7)$, the ratio

$$B_A^+ = \frac{A_V^+}{A_1} = \frac{B_A^-}{\Omega}$$

and the effective area of the piston

$$A_1 = \frac{(B_A^+)^2 \cdot K_*^2 \cdot m}{2 \cdot W^2 \cdot P_S \cdot L_S} \qquad (5.47)$$

are defined, where the total actuator stroke (L_S) should be taken into consideration in the first approximation as $L_S = L_C + s_D$ (value of s_D is estimated as

$$s_D = \frac{\dot{x}_S^2}{2 \cdot \ddot{x}_D}).$$

3. Using Equation 5.45, the displacement of the acceleration part of the actuator motion (L_A) and according to Equation 5.46, the actuator total stroke (L_S) is also defined.

4. Using Equation 5.47, the adjusted value of the piston effective area (A_1) is determined and, after that, the effective areas of the control valve $(A_V^- = B_A^- \cdot A_1$ and $A_V^+ = A_V/\Omega)$ are defined.

Example 5.10. *Define the parameters for an open-loop actuator with constant velocity motion that has moving mass $m = 10$ kg, supply pressure $P_S = 0.6$ MPa, external force $F_L = 100$ N, maximum deceleration $\ddot{x}_D = 10$ m/s2, constant velocity $\dot{x}_C = 0.5$ m/s, and actuator displacement where it moves with constant velocity $L_C = 0.5$ m.*

From Equation 5.44, the ratio B_A^- is

$$B_A^- = \frac{\dot{x}_C}{\varphi_* \cdot K_*} \approx 2.5 \cdot 10^{-3}.$$

Assuming that the value of $\Omega = 0.8$ and $W = 0.25$, then the ratio B_A^+ is

$$B_A^+ = \frac{B_A^-}{\Omega} \approx 3.2 \cdot 10^{-3}.$$

For the maximum deceleration value, the displacement of the deceleration part of the actuator motion is

$$s_D = \frac{\dot{x}_C^2}{2 \cdot \ddot{x}_D} \approx 0.013 m,$$

and the first approximation of the total actuator stroke is $L_S = L_C + s_D \approx 0.52 \ m$. *Then the piston effective area is:*

$$A_1 = \frac{(B_A^+)^2 \cdot K_*^2 \cdot m}{2 \cdot W^2 \cdot P_S \cdot L_S} \approx 3 \cdot 10^{-3} m^2$$

Using Equation 5.45, the displacement of the acceleration part of the actuator motion is

$$L_A = \frac{m \cdot \dot{x}_C^2}{0.02 \cdot P_S \cdot A_1} \approx 0.11 m,$$

and the actuator total stroke is $L_S = L_A + L_C + s_D \approx 0.65 \ m$. *In this case, the adjusted value of the piston effective area is* $A_1 \approx 2.4 \cdot 10^{-3} \ m^2$, *and effective areas of the control valve are* $A_V^- = B_A^- \cdot A_1 \approx 6 \cdot 10^{-6} \ m^2$ *and* $A_V^+ = A_V^-/\Omega \approx 7.5 \cdot 10^{-6} \ m^2$.

The nearest standard pneumatic cylinder has a piston diameter of 0.063 m (its effective area is $A_1 \approx 3.1 \cdot 10^{-3} \ m^2$). The double solenoid valve with standard nominal flow rate of 600 l/min ($A_V^+ = A_V^- \approx 10 \cdot 10^{-6} \ m^2$) can be used as the main control valve (Figure 5.25); and as the adjustment unit for the discharging flow, the flow control valve with a standard nominal flow rate from 0 to 1000 l/min ($A_V^- = 0 \div 17 \cdot 10^{-6} \ m^2$) can also be used. For this actuator, the dimensionless load is $\chi \approx 0.1$.

The dynamic behavior of this pneumatic actuator is shown in Figure 5.29. Analysis of the obtained results demonstrates the high accuracy of the parameter calculation of such an actuator.

5.3 Adjustment of Acting Force in Pneumatic Actuators

For an actuating system, it is very important to be able to control the interaction forces between the actuator and the environment. Usually, this is referred to as force control and denotes a controlled force/torque vector.

Force control is a technology that has been developed for many manufacturing processes, for example, in machining, injection molding, casting, and forging. Force control systems are also widely used in materials testing equipment, robots, and physiotherapy devices.

Force control systems can be divided into two major groups: (1) open-loop and closed-loop systems. Open-loop controlled force pneumatic actuators are most widely used, wherein actuator output link movement is prevented

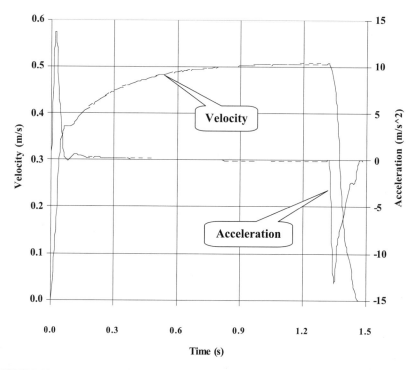

FIGURE 5.29
Dynamics of the open-loop pneumatic cylinder with constant velocity motion.

(fixed-position force control). Sometimes, these systems are called passive force control systems; they have no mechanism to adjust for force errors.

With the piston position fixed, the control valves supply the pneumatic cylinder chambers (constant volume for fixed position) with air. The control valves have within them analog circuitry that regulates the output pressure to be proportional to the input voltage. When the two valves are combined with a pneumatic cylinder, the voltage input to the pair can be considered proportional to the cylinder piston force output, thus enabling force control without feedback. This technique is simple and inexpensive, but not very accurate or flexible.

The most common applications of open-loop controlled force pneumatic actuators are the clamping devices, holding mechanisms, and counterbalanced designs. However, these open-loop actuators have some disadvantages. The open-loop control scheme does not directly monitor the force being applied to the surface; rather, the control valves try to maintain a constant pressure. Maintaining constant pressure is not the same as maintaining constant force because the actuator friction force affects the resultant force vector. From this point of view, using a pneumatic actuator with low friction force, for example, a pneumatic cylinder with flexible chambers (Figure 2.2), diaphragm actuators, or actuators with bellows, is preferable.

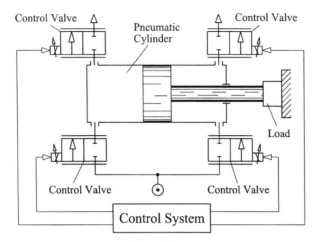

a. Schematic diagram

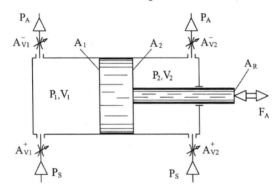

b. Schematic diagram for mathematical model

FIGURE 5.30

Open-loop force control actuator with proportional flow control valves: (a) schematic diagram and (b) schematic diagram for mathematical model.

As discussed in Chapter 1, both proportional pressure regulators (Figure 1.14) and electropneumatic control valves (Figure 1.15) can be used in pneumatic actuators for force control. Systems with pressure regulators are rarely used in practice because of slow response time. Force control systems with control valves are generally used not only owing to their good dynamic response, but also because these systems can be utilized in another control mode (e.g., as a positioning actuator). The most flexible control can be achieved using four proportional flow control valves, as shown in Figure 5.30a. In this actuator, each working chamber has a pair of control valves; one of them controls the supply flow and the other controls the flow that goes through the exhaust port.

The steady-state condition is described by an equation that can be obtained from Equation 2.18 and Equation 2.20 for the following assumptions: $\dot{P}_1 = 0$,

$\dot{P}_2 = 0$, $\dot{x} = 0$, and $\ddot{x} = 0$. Then, taking into consideration that the actuator friction force is negligible, it becomes:

$$
\begin{cases}
F_A = P_1 \cdot A_1 - P_2 \cdot A_2 - P_A \cdot A_R \\[4mm]
P_1 = \dfrac{\varphi(\dfrac{P_1}{P_S})}{\Omega_1 \cdot \varphi(\dfrac{P_A}{P_1})} \cdot P_S \\[6mm]
P_2 = \dfrac{\varphi(\dfrac{P_2}{P_S})}{\Omega_2 \cdot \varphi(\dfrac{P_A}{P_2})} \cdot P_S
\end{cases}
\tag{5.48}
$$

where F_A is the force that the actuator develops.

In this type of actuator, the working chambers are connected simultaneously to both a supply line and to an exhaust port. Usually, such pneumatic chambers are known as two-port or flow-type chambers.

In general, there are four combinations of charge and discharge flow conditions in the actuator working chamber. In the first combination, the charge and discharge flows move at the sonic condition and then the flow function is described by the following equations: $\varphi(\sigma_i) = \varphi_*$ and $\varphi(\sigma_A/\sigma_i) = \varphi_*$ (here, $\sigma_i = P_i/P_S$, $\sigma_A = P_A/P_S$, and i is the actuator working chamber index). This condition takes place for $\sigma_i = P_i/P_S \leq 0.5$ and $\sigma_A/\sigma_i = P_A/P_i \leq 0.5$; or after rearranging these inequalities: $2 \cdot \sigma_A \leq \sigma_i \leq 0.5$ and $0.2\ \text{MPa} \leq P_i \leq 0.5 \cdot P_S$, or $P_S > 4 \cdot P_A$ ($P_S > 0.4\ \text{MPa}$). Then, from Equation 5.48:

$$
\sigma_i = \frac{1}{\Omega_i}
$$

and

$$
P_i = \frac{A_{Vi}^+}{A_{Vi}^-} \cdot P_S \ .
$$

In this case, the dimensionless parameter Ω_i changes in the range

$$
2 \leq \Omega_i \leq \frac{1}{2 \cdot \sigma_A} \ .
$$

In the second combination, the charge flow moves at the sonic condition and the discharge flow moves at the subsonic condition. In this case, the flow functions are described by the followings equations: $\varphi(\sigma_i) = \varphi_*$ and

$$\varphi\left(\frac{\sigma_A}{\sigma_i}\right) = 2\varphi_* \cdot \sqrt{\frac{\sigma_A}{\sigma_i} \cdot \left(1 - \frac{\sigma_A}{\sigma_i}\right)} \ .$$

Also, $\sigma_i \leq 0.5$ and $0.5 < \sigma_A/\sigma_i \leq 1$, and after rearrangement becomes: $\sigma_i \leq 2 \cdot \sigma_A$ or $P_i \leq 0.2$ MPa. Then, from Equation 5.48:

$$\sigma_i = \frac{1}{4 \cdot \sigma_A \cdot \Omega_i^2} + \sigma_A \quad \text{and} \quad P_i = \frac{P_S^2 \cdot (A_{Vi}^+)^2}{4 \cdot P_A \cdot (A_{Vi}^-)^2} + P_A$$

For this combination, the parameter Ω_i changes in the range

$$\Omega_i > \frac{1}{2 \cdot \sigma_A}.$$

In the third combination, the charge flow moves at the subsonic condition and the discharge flow moves at the sonic condition; then the flow functions are: $\varphi(\sigma_i) = 2\varphi_* \cdot \sqrt{\sigma_i \cdot (1 - \sigma_i)}$ and $\varphi(\sigma_A/\sigma_i) = \varphi_*$. This combination exists for $0.5 < \sigma_i \leq 1$ and $\sigma_A/\sigma_i \leq 0.5$, which gives the following condition: $\sigma_i \geq 2 \cdot \sigma_A$ or $P_i \geq 0.2$ MPa, and $0.5 \cdot P_S < P_i \leq P_S$. Using Equation 5.48, the dimensionless charging pressure is:

$$\sigma_i = \frac{1}{1 + 0.25 \cdot \Omega_i^2}$$

and

$$P_i = \frac{P_S}{1 + 0.25 \cdot \left(\dfrac{A_V^-}{A_V^+}\right)^2}$$

In this case, the dimensionless parameter Ω_i changes in the range $0 \leq \Omega_i < 2$.

In the fourth combination, the charge and discharge flows move at the subsonic condition, and then the flow function is described by the following equations:

$$\varphi(\sigma_i) = 2\varphi_* \cdot \sqrt{\sigma_i \cdot (1 - \sigma_i)} \quad \text{and} \quad \varphi\left(\frac{\sigma_A}{\sigma_i}\right) = 2\varphi_* \cdot \sqrt{\frac{\sigma_A}{\sigma_i} \left(1 - \frac{\sigma_A}{\sigma_i}\right)}.$$

TABLE 5.1

Pressure Computation

Changing Range of Ω_i	Dimensionless Pressure σ_i	Pressure P_i
$0 \le \Omega_i < 2$	$\sigma_i = \dfrac{1}{1+0.25 \cdot \Omega_i^2}$	$P_i = \dfrac{P_S}{1+0.25 \cdot \left(\dfrac{A_{Vi}^-}{A_{Vi}^+}\right)^2}$
$2 \le \Omega_i \le \dfrac{1}{2 \cdot \sigma_A}$	$\sigma_i = \dfrac{1}{\Omega_i}$	$P_i = \dfrac{A_{Vi}^+}{A_{Vi}^-} \cdot P_S$
$\Omega_i > \dfrac{1}{2 \cdot \sigma_A}$	$\sigma_i = \dfrac{1}{4 \cdot \sigma_A \cdot \Omega_i^2} + \sigma_A$	$P_i = \dfrac{P_S^2 \cdot (A_{Vi}^+)^2}{4 \cdot P_A \cdot (A_{Vi}^-)^2} + P_A$

In this case, $0.5 < \sigma_i \le 1$ and $0.5 < \sigma_A/\sigma_i \le 1$. That is to say, the value of the pressure P_i satisfies two conditions: $0.5 \cdot P_s \le P_i \le P_S$ and $P_A \le P_i < 2 \cdot P_A$, from which one can obtain: $P_A \le 2 \cdot P_A = 0.2$ MPa. As referred to above, in most industrial environments, the absolute supply pressure maximum value is 1.1 MPa and its minimum value is 0.5 MPa. For these conditions, this combination of the charge and discharge flow conditions does not exist.

It is seen that in the actuator working chambers, the pressure is a function of the valve effective areas ratio $\Omega_i = A_{Vi}^-/A_{Vi}^+$; in addition, the parameter $\sigma_A = P_A/P_S$ or the value of P_S also influences the pressure change process. Table 5.1 presents the equations for the determination of the pressure (both dimensionless and dimension values) as a function of the parameter Ω_i.

The curves of the changing of dimensionless pressure in the actuator working chambers are shown in Figure 5.31. In industrial environments, the dimensionless parameter σ_A changes in range from 0.09 to 0.2, since the supply pressure is 0.5 MPa $\le P_S \le$ 1.1 MPa. In Figure 5.31 two curves are shown, which conform to the two marginal cases ($\sigma_A = 0.09$ and $\sigma_A = 0.2$). It is seen that the difference between these graphs exist if the parameter σ_i is in the range $\Omega_i \ge 3$; therefore, it may be assumed that the major influence over the dimensionless pressure in the actuator working chambers has the valves effective areas ratio $\Omega_i = A_{Vi}^-/A_{Vi}^+$.

In this actuator (Figure 5.30), the steady state characteristic of the proportional control valves is defined by the graph shown in Figure 3.42. There are many kinds of control algorithms that may be used in this type of actuator; however, the following law of control is most often utilized:

$$\begin{cases} A_{V1}^+ = A_{V2}^- = A_{VM} \cdot \beta \\ A_{V1}^- = A_{V2}^+ = A_{VM} \cdot (1-\beta) \end{cases} \qquad (5.49)$$

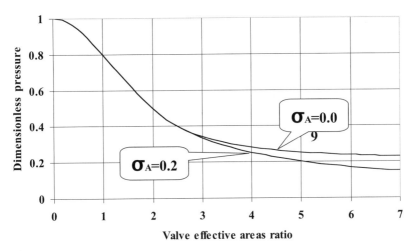

FIGURE 5.31
Dimensionless pressure in the actuator working chambers as a function of the valve effective areas ratio.

where A_{VM} is the maximum value of the valves effective area, β is the valve opening coefficient, which is defined by Equation 3.8.

The maximum actuator contracting force F_{ACM} is in the condition when and $A_{V1}^+ = A_{V2}^- = A_{VM}$ and $A_{V1}^- = A_{V2}^+ = 0$ that is $F_{ACM} = (P_s - P_A) \cdot A_1$. The maximum actuator expanding force F_{AEM} exists in the inverse condition for which $A_{V1}^+ = A_{V2}^- = 0$ and $A_{V1}^- = A_{V2}^+ = A_{VM}$, then $F_{AEM} = (P_s - P_A) \cdot A_2$. For the equilibrium point, where the actuator force is $F_A = 0$, the valve opening coefficient β is at a point $\beta = 0.5$, and the parameter Ω_i changes in the area around 1 (the first range in Table 5.1). Then, for the double-acting rodless cylinder, this condition may be achieved by $\beta = 0.5$, and for the cylinder with rod this state of the actuator force exists when the valve opening coefficient is

$$\beta \approx \frac{1}{1 + \sqrt{2 \cdot \frac{A_R}{A_2} + \sqrt{4 \cdot \frac{A_R^2}{A_2^2} + \frac{A_1}{A_2}}}} \tag{5.50}$$

Example 5.11. *Define the condition of the equilibrium point for the pneumatic actuator with force control that consists of the double-acting cylinder, which has a piston diameter of 63 mm and ($A_1 \approx 3.12 \cdot 10^{-3}$ m^2) rod diameter of 20 mm ($A_R = 0.314 \cdot 10^{-3}$ m^2 and $A_2 = 2.806 \cdot 10^{-3}$ m^2). The supply pressure is $P_S = 0.6$ MPa.*

Using Equation 5.48 the opening coefficient is $\beta \approx 0.467$, then the valves effective areas ratio is

$$\Omega_1 = \frac{A_{V1}^-}{A_{V1}^+} = \frac{1-\beta}{\beta} \approx 1.141 \text{ and } \Omega_2 = \frac{A_{V2}^-}{A_{V2}^+} = \frac{\beta}{1-\beta} \approx 0.876$$

According to Table 5.1 (first row) the pressure in the piston side chamber is $P_1 \approx$ 0.453 MPa, and the pressure in the rod side chamber is $P_2 \approx 0.503$ MPa.

Estimation of the actuator force F_A value ($F_A = P_1 \cdot A_1 - P_2 \cdot A_2 - P_A \cdot A_R \approx 0$) shows that the calculating values of the opening coefficient and pressure in the actuator working chambers relate to the condition of the equilibrium point. In this case, the control signal, which applies to the supply valve of the first chamber and exhaust valve of the second one (see Figure 5.30a), is $0.467 \cdot U_{CS}$ (U_{CS} is the maximum or saturation value (see Equation 3.8 and Figure 3.43) of the control signal). For the exhaust valve of the first chamber and the supply valve of the second, one such control signal is $0.533 \cdot U_{CS}$.

The dynamic behavior of the actuator is determined by equations that describe the changes in the actuator force and the pressure in the working chambers (see Figure 5.30b):

$$
\left|
\begin{aligned}
& F_A = P_1 \cdot A_1 - P_2 \cdot A_2 - P_A \cdot A_R \\
& \dot{P}_1 = \frac{K_*}{V_1} \cdot \left[A_{V1}^+ \cdot P_S \cdot \varphi\left(\frac{P_1}{P_S} \right) - A_{V1}^- \cdot P_1 \cdot \varphi\left(\frac{P_A}{P_1} \right) \right] \\
& \dot{P}_2 = \frac{K_*}{V_2} \cdot \left[A_{V2}^+ \cdot P_S \cdot \varphi\left(\frac{P_2}{P_S} \right) - A_{V2}^- \cdot P_2 \cdot \varphi\left(\frac{P_A}{P_2} \right) \right]
\end{aligned}
\right.
\tag{5.51}
$$

In the case where the effective area of the actuator and the capacity of the working chambers are equal (i.e., $A_1 = A_2$ and $V_1 = V_2$), the changes in the valves effective areas are described by Equation 5.49, and the charge flow in the actuator moves at the subsonic condition and the discharge flow moves at the sonic condition (this combination is the most frequent); then Equation 5.51 can be rewritten in the dimensionless form as:

$$
\left|
\begin{aligned}
& \chi_A = \sigma_1 - \sigma_2 \\
& \dot{\sigma}_1 = 2 \cdot \beta \cdot \sqrt{\sigma_1 \cdot (1 - \sigma_1)} - (1 - \beta) \cdot \sigma_1 \\
& \dot{\sigma}_2 = 2 \cdot (1 - \beta) \cdot \sqrt{\sigma_2 \cdot (1 - \sigma_2)} - \beta \cdot \sigma_2
\end{aligned}
\right.
\tag{5.52}
$$

Here, the dimensionless actuator force is

$$
\chi_A = \frac{F_A}{A_1 \cdot P_S},
$$

and the dimensionless time is

$$\tau = \frac{t}{t_{SF}},$$

where t_{SF} is the time-scale factor coefficient, which is determined as:

$$t_{SF} = \frac{V_1}{A_{VM} \cdot \varphi_* \cdot K_*} \tag{5.53}$$

The nonlinear function $\sqrt{\sigma_i \cdot (1 - \sigma_i)}$ can be represented in the linearization form as:

$$\sqrt{\sigma_i \cdot (1 - \sigma_i)} \approx \psi_{*i} - N_{*i} \cdot \sigma_i,$$

where

$\psi_{*i} = \sqrt{\sigma_{Ei} \cdot (1 - \sigma_{Ei})}$ is the magnitude of the function in the equilibrium point,

$N_{*i} = \dfrac{1 - 2 \cdot \sigma_{Ei}}{\sqrt{\sigma_{Ei} \cdot (1 - \sigma_{Ei})}}$ is the magnitude of derivative of the function in this point,

σ_{Ei} = the magnitude of the dimensionless pressure in the equilibrium point.

In Equations 5.52, another nonlinear function is $\beta \cdot \sigma_i$, which can be represented in the linearization form as: $\beta \cdot \sigma_i \approx \beta_E \cdot \sigma_{Ei} + \beta_E \cdot \sigma_i + \beta \cdot \sigma_{Ei}$, where β_E is the magnitude of the opening coefficient in the equilibrium point.

Taking into consideration the results of this linearization, Equations 5.52 can be rewritten in the following form:

$$\begin{cases} \chi_A = \sigma_1 - \sigma_2 \\ \dot{\sigma}_1 = A_{11} + A_{12} \cdot \sigma_1 + A_{13} \cdot \beta \\ \dot{\sigma}_2 = A_{21} - A_{22} \cdot \sigma_2 - A_{23} \cdot \beta \end{cases} \tag{5.54}$$

where:

$A_{11} = (1 - 2 \cdot N_{*1}) \cdot \beta_E \cdot \sigma_{E1}$

$A_{12} = (1 - 2 \cdot N_{*1}) \cdot \beta_E - 1$

$A_{13} = (1 - 2 \cdot N_{*1}) \cdot \sigma_{E1} + 2 \cdot \psi_{*1}$

$A_{21} = 2 \cdot \psi_{*2} - (1 - 2 \cdot N_{*2}) \cdot \beta_E \cdot \sigma_{E2}$

$A_{22} = (1 - 2 \cdot N_{*2}) \cdot \beta_E + 2 \cdot N_{*2}$

$A_{23} = (1 - 2 \cdot N_{*2}) \cdot \sigma_{E2} + 2 \cdot \beta_E$

Taking into account that for this case the dimensionless parameters are $\sigma_{E1} = \sigma_{E2} = 0.8$, $\psi_{*1} = \psi_{*2} = 0.4$, $N_{*1} = N_{*2} = -1.5$, and $\beta_E = 0.5$; and in addition, going over to the new argument, that is, $\Delta\sigma = \sigma_1 - \sigma_2$, then from Equations 5.54, one obtains:

$$\dot{\chi}_A = \chi_A + 2.4 + 8.2 \cdot \beta \tag{5.55}$$

The solution to the secular equation of the differential Equation 5.55 is (initial condition: $\tau = 0$, $\chi_A = 0$):

$$\chi_{A*} = 2.4 \cdot [\exp(\tau) - 1] \tag{5.56}$$

The dimension solution of the secular equation has the following form:

$$F_{A*} = 2.4 \cdot A_1 \cdot P_S \cdot \left[\exp\left(\frac{t}{t_{SF}} \right) - 1 \right] \tag{5.57}$$

It is seen that the actuator speed of response is the inverse of the value of the time-scale factor coefficient t_{SF}, which depends on the capacity of the working chambers and effective areas of the control valves. It is clear (see Equation 5.53) that a decrease in the working chamber capacity increases the actuator speed of response, and an increase in the effective area of the control valve also increases this parameter.

6

Closed-Loop Pneumatic Actuating Systems

Pneumatic drives are widely used in the manufacturing and process industry. The capability of conventional pneumatic drives to perform high-speed actuation is broadly recognized. The challenge is to produce a high-performance system (employing pneumatic actuation) in terms of accuracy (positioning, timing, and contouring) and flexibility (modularity in mechanical configuration, capability of integrating with other machine elements, versatility in the changeover of application tasks, and responsiveness to system failure). The basic reasons for using closed-loop systems in contrast to open-loop systems include the need to improve transient response times, reduce the steady-state errors, and reduce the sensitivity to load parameters. Servo control and computer/digital technology are the two major factors that enable realization of the performance potential of pneumatic drives to their best advantage.

In a closed-loop pneumatic actuating system, the variable to be controlled (i.e., controlled variable or controlled signal) is continuously measured and then compared with a predetermined value (i.e., reference variable or desired signal). If there is a difference between these two variables, adjustments are made until the measured difference is eliminated and the controlled signal equals the desired one. Hence, the characteristic feature of a closed-loop control is a closed-action flow.

The design of closed-loop pneumatic actuators is a very complicated problem area because there are many factors that must be considered; for example, attenuation of the nonlinearity influence on the actuator dynamic, variation and uncertainties in process behavior, reduction of the effect of inside and outside disturbances, and others.

When system performance was not very high, design engineers were able to successfully rely on steady-state evaluation to ensure that the system they created would accomplish the intended mission. However, the successful design of a high-performance actuating system requires much more concern re the dynamics of the system. In fact, most of the problems encountered in the performance of engineered actuators involve their dynamics. Programs for computer simulation have been developed that can be used to perform dynamic pneumatic actuator design analysis. These software programs usually require component characteristics to complete the actuating system description.

6.1 Control Systems and Control Algorithms

The basic elements of a pneumatic closed-loop actuator are shown in Figure 6.1. The output of the pneumatic closed-loop actuator is measured with a transducing device to convert it into an electrical signal. This feedback signal is compared with the command signal, and the resulting error signal is used to obtain the control signal, which is then amplified and used to drive the control valve. The proportional or servo valve controls the air flow to the pneumatic actuator in proportion to the drive voltage or current from the power amplifier. The pneumatic actuator forces the load to move. Thus, a change in the command signal generates an error signal, which causes the load to move in an attempt to zero the error signal. If the amplifier gain is high, the output will very rapidly and accurately follow the command signal. Ideally, the amplifier gain would be set high enough that the accuracy of the actuating system will depend only on the accuracy of the transducing device. Even if infinite loop gain were achievable, the accuracy of the actuator cannot be more accurate than that of the transducer. In practice, however, the power amplifier gain is limited by stability considerations.

Each element of the closed-loop actuator has the input-output structure with specific gain. The controller (which consists of the command signal module, central processing unit and power amplifier) accepts an input voltage from the transducer and delivers an output current or voltage (mA or V). The control valve accepts an input current or voltage from the controller and delivers an output flow (m^3/s). The cylinder accepts an input flow from the control valve and delivers an output movement (m). Moreover, the feedback device accepts an input movement from the actuator and delivers an output voltage (V), which is sent to the controller.

The motion controller is a significant component of the motion control system that the designer must choose. The controller with its power amplifier is the "brain" of a closed-loop system. The motion controller, in addition to

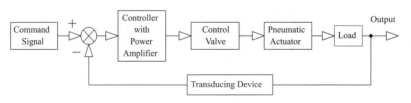

FIGURE 6.1
Schematic diagram of the closed-loop pneumatic actuator.

performing control functions, generates the command signal to the power amplifier. The error signal (the difference between the desired position and the actual position of the actuator) that is sent to a current-driving amplifier will saturate either a (+) or (−) transistor, which delivers the power to the control valve torque motor. Based on the gain setting, the power amplifier will saturate the control valve coil with the maximum current available from the supply. The output from the amplifier will drive the control valve torque motor to its full position, thus allowing the actuator to travel at its designed maximum velocity. When the error falls below the gain setting, the valve will balance the pressures across the area of the piston and the cylinder will stop. Any over-travel or under-travel will generate an error signal, thus making the amplifier respond accordingly. If any external disturbances alter the position of the actuator, the feedback device will measure the true position of the actuator. The summing amplifier will measure the difference between the desired position and the actual position, and the process will repeat.

Typically, the controller with power amplifier has extremely high frequency response. This allows the controller to send an alternating (+) and (−) current to the control valve at a very high frequency.

Motion controllers have evolved considerably in the past few years, following the trend of improving price/performance ratios for microprocessors, digital signal processors, and programmable logic devices. Equipment designers faced with a build-or-buy decision usually realize quickly that the level of development time and technical expertise contained in the hardware and software of these specialized products often rules out a competitive in-house design. The wide variety of motion controllers available gives a designer considerable flexibility; however, selecting a vendor focused on motion control is typically the best choice.

Space requirements and cost are usually about the same, whether a designer chooses a stand-alone or card controller. A bus-based card typically requires an external breakout board to allow transmission of the many signals from its single high-density edge connector to the outside environment. In many cases, a powerful stand-alone motion controller with significant analog and digital input/output (I/O) can function as the entire machine controller and eliminate the need for a computer.

Additional considerations for choosing motion controllers include the ease of use and the power of the programming language and setup software tools; multitasking capabilities; number of I/O points; coordinated motion requirements, such as linear and circular interpolation and electronic gearing; synchronization to internal and external events; and error-handling capabilities.

6.1.1 Control Algorithms

A proportional-integral-derivative (PID) algorithm is the most popular feedback controller used across the industry. It is a robust, easily understood

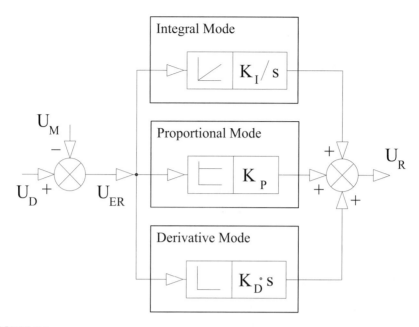

FIGURE 6.2
Block diagram of the PID controller.

algorithm that can provide excellent control performance despite the varied dynamic characteristics of a process plant. It is easy to implement and relatively easy to tuning. Figure 6.2 represents a block diagram of the PID algorithm (or controller). In general, a PID controller takes as its input $(U_{ER} = U_D - U_M)$ the error signal (or the difference) between the desired set point (U_D) and the output (or measurable) signal (U_M). It then acts on the input such that a regulating output (or regulating signal) (U_R) is generated. Gains K_P, K_I, and K_D are the proportional, integral, and derivative gains by the system to act on the input, integral of the input, and derivative of the input, respectively. The PID regulating signal can be expressed as:

$$U_R = K_P \cdot U_{ER} + K_I \cdot \int U_{ER} \cdot dt + K_D \cdot \dot{U}_{ER} \tag{6.1}$$

The control signal, which is the input of the control valve, is determined as:

$$U_C = K_{PA} \cdot U_R \tag{6.2}$$

where K_{PA} is the gain of the electrical power amplifier.

The proportional mode adjusts the output signal in direct proportion to the input (U_{ER}). A proportional controller reduces error but does not eliminate it (unless the process has naturally integrating properties); that is, an offset between the actual and desired values will normally exist. As the gain K_p increases, the actuator responds more rapidly to changes in the set point, and the final (steady-state) error is smaller, but the system becomes less stable because it is increasingly under-damped. Further increases in gain will result in overshoots and ultimately undamped oscillation.

The additional integral mode (often referred to as reset) corrects any offset (error) that might occur between the desired value (set point) and the process output automatically over time. A controller with integral action, combined with an actuator that becomes saturated, can give some undesirable effects. If the error signal (U_{ER}) is so large that the integrator saturates the actuator, the feedback path will be broken because the actuator will remain saturated even if the process output changes. The integrator, being an unstable system, can then integrate up to a very large value. When the error signal is finally reduced, the integral may be so large that it takes considerable time until the integral assumes a normal value again. This effect is called *integrator windup*.

The integral action improves static accuracy (eliminates the steady-state error); however, integral action alone generally leads to an unstable system and therefore in this case, there is a trade-off between oscillatory response and sluggish behavior.

Derivative action anticipates where the process is heading by looking at the time rate of the change in the controlled variable. In theory, derivative action should always improve dynamic response, and it does in many loops. This control mode can improve the stability, reduce the overshoot that arises when proportional or integral terms are used at high gain, and improve response speed by anticipating changes in the error. The derivative gain, or the "damping constant" (K_D), can usually be adjusted to achieve a critically damped response to changes in the set point or the regulated variable. Sometimes, the derivative control is viewed as electronic damping. However, derivative action depends on the slope of the error, unlike the proportional and integral mode; and if the error is constant, the derivative action has no effect. Too little damping and the overshoot from the proportional control main remains. Too much damping may cause an unnecessarily slow response. The designer should also note that differentiators amplify high-frequency noise, which appears in the error signal.

It is important to note that the fixed-gain PID controller is a very effective method in cases where the plant expresses as a linear model and the plant parameter does not change during operation. On the other hand, the simplicity of the controller puts limitations on its capabilities in dealing with complex control problems, such as the plant nonlinear model and changing its parameters while in operating mode. It is difficult to achieve satisfactory position, velocity, and force control of a pneumatic actuator via fixed-gain PID control because of the inherent actuator nonlinearities. The most serious nonlinearities are those of control valves, load, and friction forces. In addition, the nonlinear elements (e.g., hysteresis) tend to vary with temperature, the friction of internal components, and manufacturing tolerance stack-up.

Therefore, it is not surprising to see a significant difference in characteristics from actuator to actuator.

Recently, a number of developments have been dedicated to modified PID algorithms that can be utilized in closed-loop pneumatic actuators. The main trend in this approach is a design mechanism that allows tuning or pre-scheduling the control gains.

The basic idea in gain scheduling is to select a number of operating points that cover the range of the system operation. Then, at each of these points, the designer makes a linear time-invariant approximation of the plant dynamics and designs a linear controller for this linearized plant.[142, 144]

Using a fixed-gain PID controller with a feed-forward term, an antiwinding mechanism, and bang-bang mode demonstrates better command following and disturbance rejection qualities than with a conventional PID scheme. The pneumatic proportional valve that operates with this controller provides better step response and greater bandwidth than conventional methods. Furthermore, it is demonstrated that robust control is achieved in the presence of significant dynamic variations in the valve.[73]

The most common and successful controller in pneumatic positioning systems is the "state controller," in which three feedback signals are used: position, velocity, and acceleration. The state controller is a useful solution when fast response and high accuracy are required. These controllers are generally called PVA controllers. A block diagram of the PVA algorithm is shown in Figure 6.3. In the PVA controller, the main input is the error between the desired signal and the output signal ($U_{ER} = U_D - U_M$). The additional feedback signals are the velocity ($\dot{U}_M$) and acceleration ($\ddot{U}_M$) of

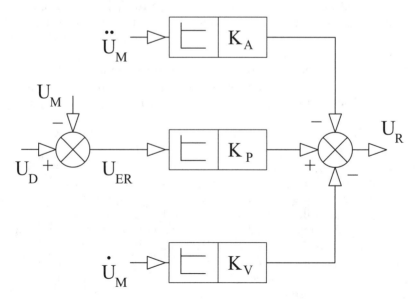

FIGURE 6.3
Block diagram of the PVA controller.

the output signal (U_M). Gains K_P, K_V, and K_A are the proportional, velocity and acceleration gains by which the system acts on the input, the velocity of the output signal, and its acceleration, respectively. PVA control action can be expressed as:

$$U_R = K_P \cdot (U_D - U_M) - K_V \cdot \dot{U}_M - K_A \cdot \ddot{U}_M \qquad (6.3)$$

From a general standpoint, the proportional gain controls the "stiffness" of the system, and the velocity and acceleration feedback is useful in improving the damping of lightly damped systems; however, the application of the acceleration signal has several problems, including:

- High acceleration could lead to amplifier saturation.
- Static data is needed.
- Amplification of noise occurs if acceleration is obtained from the derivative of velocity or the second derivative of position.

A practical difficulty with PVA controllers is the necessity of the acceleration signal. Its measurement is difficult and its numerical calculation is noise contaminated. For this reason, several systems make use of a modified state controller, wherein acceleration is substituted by the pressure difference in the actuator working chambers. The PVA and modified state controllers have similar performance, with good dynamic and precision characteristics in the position actuators, but they are not sufficiently robust to payload changes and to friction force effects.[137, 184]

The state loop control algorithm achieves good results; however, in the case where actuator parameters change, it is necessary to readjust the state variable gains.

An adaptive control system with a controller parameter adjustment is usually used to improve the control performance of the actuating system in the case where the dynamics of the plant (in this case, the term "plant" refers to the pneumatic actuator) change during the operation. Adaptive control is the capability of the system to modify its own operation to achieve the best possible mode of operation. A general definition of adaptive control implies that an adaptive system must be capable of performing the following functions:

1. Providing continuous information about the present state of the system or identifying the process
2. Comparing present system performance to the desired or optimum performance and making a decision to change the system to achieve the defined optimum performance
3. Initiating a proper modification to drive the control system to the optimum

These three principles (identification, decision, and modification) are inherent in any adaptive system.

In practice, two main approaches to adaptive control design are used: (1) model reference adaptive control and (2) self-tuning regulators.

A self-tuning regulator assumes a linear model for the process being controlled (which is usually nonlinear). It uses a feedback-control law that contains adjustable coefficients and self-tuning algorithms that change the coefficients.

These controllers typically contain an inner and an outer loop. The inner loop consists of an ordinary feedback loop and the plant. This inner loop acts on the plant output in conventional ways. The outer loop adjusts the controller parameters in the inner feedback loop. The outer loop consists of a recursive parameter estimator combined with a control design algorithm. The recursive estimator monitors plant output and estimates plant dynamics by providing parameter values in a model of the plant. These parameter estimates go to a control-law design algorithm that sends new coefficients to the conventional feedback controller in the inner loop.

In model reference adaptive control, a reference model describes system performance. The adaptive controller is then designed to force the system or plant to behave like the reference model. Model output is compared to the actual output, and the difference is used to adjust feedback controller parameters. Most of the work of these controllers has focused on the adaptation mechanism. This mechanism must account for the output error and determine how to adjust the controller coefficients. It must also remain stable under all conditions. One problem with this approach is that there is no general theoretical method for designing an adapter. Thus, most adapter functions are specially keyed to some kind of end application.

An advantage of model reference adaptive control is that it provides quick adaptations for defined inputs. A disadvantage is that it has trouble adapting to unknown processes or arbitrary disturbances.

Neural network controllers are also used in pneumatic positioning systems. These systems are based on the learning algorithm of the human brain and are excellent in identifying an arbitrary input-output relation and in compensating for nonlinearity. However, the disadvantage of a neural network is that it requires a lot of learning time and it cannot express the learning process clearly.

It is well known that fuzzy control algorithms are effective for nonlinear plants. The characteristics of conventional fuzzy control are as follows: the control rule is expressed using "If" and "Then" type fuzzy functions, and the qualitative input-output relation of the plant, in which the control algorithm contains "fuzziness" and robustness, determines the control rule. However, if the plant is complicated, it is difficult for the fuzzy control to obtain satisfactory control performance because it lacks adaptability and learning ability. In this case, it would be ideal to utilize the known information of the plant to improve its control performance.[169,171]

The sliding mode control has been recognized for many years as one of the key approaches to the systematic design of robust controllers for complex nonlinear dynamic systems operating under uncertain conditions. For pneumatic positioning systems, the sliding mode control ensures high accuracy and sufficient robustness. Sliding control enables separating of the overall system motion into independent partial components of lower dimensions and, as a result, reducing the complexity of the control design. It is a kind of variable structure control where the gains are constant but discontinuous and are switched around a sliding surface in the state space. The basic features of sliding control regimes are[30, 133, 191]:

- Theoretical invariance to the external load and internal perturbations (parameters uncertainty) if actuator conditions are satisfied and practical robustness.
- The character of the system's movement is known in advance.
- Unlike adaptive control algorithms, the sliding mode controller is easy to implement, does not require parameter estimation, and only requires knowledge of the bounds of parameters and disturbances.

Apart from the above listed features, actuating systems with sliding mode control have several shortcomings, which include:

- The necessity of measurability of the full state of the actuator
- The occurrence of vibration in the control signal, which may cause excitation of nonmodeled dynamics of the actuator and undesired movement (chattering) in the area of the predicted trajectory or around the desired positioning set point

The first shortcoming can be solved successfully by the application of the observer, which partly alleviates the second shortcoming. However, algorithms that can solve one or both of these problems have also been developed.

6.1.2 Types of Control Systems

In pneumatic actuating systems, the analog, digital, and hybrid (where both analog and digital components are utilized) control systems are generally used.

Simple control systems can be carried out using analog (continuous) integrated circuits, which usually consist of operational amplifiers. Analog-type control systems are available that can consider several variables at once for more complex control functions. These are very specific in their applications, however, and thus are not commonly used.

An analog closed loop will provide all the intercommunications in analog format. Analog signals typically have two categories: voltage and current. Voltage loops typically fall into the following categories: from 0 to +10 V DC, from −10 to +10 V DC, from 0 to 5 V DC, and from −5 to +5 V DC.

Typically, the proportional or servo valves accept a bipolar continuous voltage signal (usually ±10 V DC) for control. Because the current is more immune to electrical noise, current loops are becoming increasingly popular. Commonly used current signals are from 4 to 20 mA and from 0 to 20 mA.

Analog differs from digital in the ability to measure between two points. With a digital system, only the end points are defined. With an analog system, there are an infinite number of possibilities between these two points; that is, analog control systems have a continuously varying value, with infinite resolution (theoretically) in both time and magnitude. Based upon this, one would naturally think that analog circuitry is the way to go, but one must look at linearity and repeatability as well as resolution in determining which type of control to use.

As intuitive and simple as analog control might seem, it is not always economically attractive or otherwise practical. For one thing, analog circuits tend to drift over time and can be very difficult to tune. Precision analog circuits, which solve that problem, can be very large, heavy, and expensive. Analog circuits can also get very hot; the power dissipated is proportional to the voltage across the active elements multiplied by the current through them. Analog circuitry can also be sensitive to noise. Because of its infinite resolution, any perturbation or noise on an analog signal can change the current value.

With the development of very reliable models in the late 1960s, digital (discrete) control systems (based on computer applications) quickly became popular elements of industrial-plant-control systems. Computers are applied to industrial control problems in three ways: (1) for supervisory or optimizing control, (2) direct digital control, and (3) hierarchy control. The advantage offered by the digital control system over the analog control system is that the computer can be readily programmed to carry out a wide variety of separate tasks. In addition, it is fairly easy to change the program so as to carry out a new or revised set of tasks should the nature of the process change or the previously proposed system prove inadequate for the proposed task. With computers, this can usually be done with no change in the physical equipment of the control system.

A fully digital system eliminates potentiometer tuning and personality modules. Configuration settings are not stored in the drive, so it can be replaced with a minimum amount of setup time. In addition, drive and machine parameters can be quickly and easily coupled with other production machines without extensive readjusting.

The use of digital controllers has advantages, on one hand, regarding the repeatability, the utilizing of complex nonlinear structures, and the comfortable changing of control parameters and structures. On the other hand, the digital controller has some disadvantages, which are based on the time discrete operation of the processor and signal quantization. The use of extended nonlinear controller structures could compensate for the disadvantages of a digital controller.

In practice, a hybrid controller has wide application in the motion control technique. Such a control system consists of both analog and digital components. The control function in a process environment is implemented by programmable logic controllers, a process control system, or an industrial personal computer. Because these devices are digital systems operating with process-specific software, all analog signals must be converted to digital numbers before a computer can read them.

The wide variety of hybrid motion controllers available gives the designer considerable flexibility; however, of all the component choices made in designing a motion control system, the choice of motion controller can have the most serious ramifications. The motion controller usually involves a software component, which also includes a learning curve and test-and-debug process. In addition, controllers can differ in their feature sets, communication protocols, or hardware interfaces.

In industrial practice, the programmable logic controller (PLC) is most often utilized. The PLC is a microprocessor-based device with either modular or integral I/O circuitry that monitors the status of field-connected "sensor" input and controls the attached output "actuator" according to a user-created logic program stored in the microprocessor's memory. Most PLCs today have the capability of accepting and delivering analog signals. They are available in both voltage (usually ±10 V DC) and current (from 4 to 20 mA).

Where accuracy requirements are looser and processor time is available, the PLC can be used to close the loop using analog input and output. However, it is not always practical in pneumatic actuating systems. In this case, where the servo amplifier is removed and the control valve is driven by PLC, it is necessary to monitor the analog signal, convert it to digital form, compare the digital signals, and then convert the signal back to an analog signal to drive the control valve. This process usually requires too much time to be effectively used in closed-loop control.

With the increased demands for tighter controls, PLC manufacturers are offering the addition of stand-alone motion controllers. These controllers are typically installed in an expansion slot on the PLC backplane. The controllers are generally minicomputers configured to close a loop digitally. These controllers typically require a digital feedback device so that the analog-to-digital process is eliminated. These controllers offer either a current output (±100 mA) or a voltage output (±10 V DC). This allows easy interface to most existing proportional or servo control valves.

Motion controllers offer many features that are not obtainable from standard analog loops. Because a motion controller accepts a digital feedback device (counts), the output can be configured to give both velocity and position feedback simultaneously. With the controller delivering the analog output to the servo control valve through a digital-to-analog converter, the programmer can select a desired velocity that causes the actuator to travel to its commanded position. This feature also allows control of acceleration and deceleration, which may be significant to the application. In this case, high-speed processors within the controller provide loop updates, thus eliminating the processing load from the servo amplifier and keeping computing in the controller.

Noise immunity for digital PWM (pulse width modulation) command signals is inherently better for digital systems than for DC analog counterparts, and only digital systems allow digital serial feedback. Given a sufficient bandwidth, any analog value can be encoded with PWM; and briefly, PWM is a way of digitally encoding analog signal levels. Many microcontrollers include on-chip PWM units.

Figure 6.4 is a schematic diagram of a pneumatic actuator with a digital control system. The system consists of the actuator with a sensor, which is an analog device. The power amplifier is connected to the proportional control valve, which is also an analog apparatus.

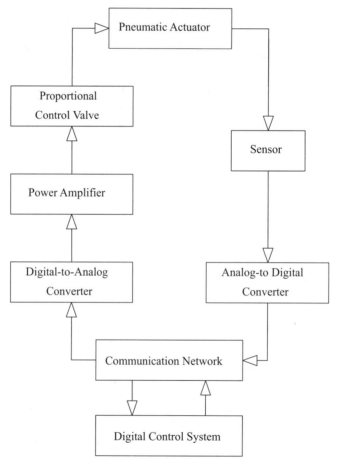

FIGURE 6.4
Schematic diagram of a pneumatic actuator with digital control system.

The analog-to-digital converter converts the analog output signal of the actuator into a finite precision digital number, depending on how many bits or levels are used in the conversion. The analog-to-digital conversion comprises three main processes, the first of which is sampling (at discrete time intervals, samples are taken from the analog signal). Each sampled value is maintained for a certain time interval, during which the next processes can take place. The second step is quantization. This is the rounding of the sampled value to the nearest limited number of digital values. This quantization is a digitization of the signal amplitude. Finally, the quantized value is converted into binary code.

Digital control systems consist of a computer with a clock, software for real-time applications, and control algorithms. The control algorithm receives quantized data both in time and in level, and consists of a computer program that transforms the measurements into a desired control signal. The control signal is transferred to the digital-to-analog converter, which with finite precision converts the number in the computer into a continuous-time signal. This implies that the digital-to-analog converter contains both a conversion unit and a hold unit that translates a number into a voltage or current signal that is applied as the actuator control signal. Communication in such an actuating system is done through a communication link or network. The clock, with the requirement that all computations be performed within a given time, controls all activities in the computer-controlled system.

Using digital systems for motion control implies more than replacing a DC analog signal with bits and bytes. For an effective analysis and design digital control system, an engineer must understand the effects of all sample rates (fast and slow), as well as the effects of quantization of large and small word sizes.

In general, the use of digital control systems reduces the actuating systems to dynamic systems with time delay. The design of controllers for such systems depends critically on knowledge of the delays. It is well known that control system behavior is more sensitive to time delay than to other linear system parameters. In fact, a closed-loop control system may be unstable or may exhibit unacceptable transient response characteristics if the time delay used in the system model for controller design does not coincide with the actual process time delay. Ignorance of the computation delay during analysis and design of digital control systems may lead to unpredictable and unsatisfactory system performance.

Basically, the delay time in digital control system consists of two major parts:

1. Communication network time delay (communication between the sensor and the controller, and between the controller and the actuator)
2. Sampling rate of the controller (computational delay)

In most actuating systems, the time delay of the communication network is small compared with the time scale of the processes, but poor choices of sampling rates in multi-rate systems, for example, can create long delays in the transfer of information between different parts of the system.

Most often in pneumatic actuators with digital control, the communication network time delay is negligible because, usually, the sensor update time must be less than the sampling time of the controller. However, synchronization of the signals to and from I/O devices with the sampling time of the controller is very critical for the acceptable performance of actuating systems.

The sampling process in the control of the actuator typically consists of the following steps:

Step 1: Read the new sample from the sensor and make the analog-to-digital conversion of this signal (in the case where the sensor is an analog device).

Step 2: Compute (recalculate and adjust) the control signal.

Step 3: Digital-to-analog conversion of the control signal (if actuator consists of a proportional control valve).

Step 4: Update the state of the controller.

Notice that the implementation of the control algorithm is done so that the control signal is sent out to the process in Step 3 before the state to be used at the next sampling interval is updated in Step 4. This is to minimize the computational delay in the algorithm.[89,90]

The time between two sampling instants is the sampling period and is denoted t_D. Periodic sampling is normally used, implying that the sampling period is constant; that is, the output is measured and the control signal is applied to each (t_D)th time unit. The sampling frequency is $f_D = 1/t_D$.

Choice of sampling period (rate) is an important issue in the design of digital control. The choice of sampling period depends on many factors. Acceptable sample rates depend on actuator inertia and control system frequency bandwidth, where bandwidth determines how fast and how far a load can move in response to a motion command signal.

It seems quite obvious that the behavior of a computer-controlled actuating system will be close to the continuous-time system if the sampling period is sufficiently small. In general, to preserve full information in the analog signal, it is necessary to sample at twice the maximum frequency of this signal. This is known as the Nyquist rate. The sampling theorem states that an analog signal can be exactly reproduced if it is sampled at a frequency f_D, where f_D is greater than twice the maximum frequency in the continuous-time signal. The problem with this theory is that sampling at twice the signal frequency is not enough in most practical situations. Usually, for practical applications, the sampling period can be defined by the following relation:

$$\omega_C \cdot t_D \approx 0.05 \div 0.14 \tag{6.4}$$

where ω_C is the crossover frequency (in radians per second) of the continuous-time actuating system.

In pneumatic actuating systems, position loop bandwidths are typically less than 40 Hz. Then, according to Equation 6.4, the minimum sampling period value should be about 0.5 ms.

In practical applications, a very short sampling period will lead to an increase in the computation power and will create a serious dimensionality problem. The period of all tasks must be chosen so that the central processing unit is not overloaded.

It is important to understand how jitter in the sampling interval and computational delay affect control performance. Compensation for sampling interval jitter can be performed in several ways. A simplistic approach is to keep the sampling interval (t_D) as a parameter in the design and to update this in every sample. For a sampled control system, this corresponds to resampling the continuous-time system. For a discretized continuous-time design, the changes in sampling interval only influence the approximation of the derivatives.

In digital control systems, the quantization is digitization of the signal amplitude. In this process, the analog-to-digital converter converts a continuous-time signal into a sequence of numbers. This transformation could be an introduction to the path process of the continuous-time signal through the device with multistage output characteristics. In the digital domain, amplitude is measured against a grid of discrete stair steps. The binary word length determines the resolution of the stair steps. An N_D-bit word offers 2^{N_D} discrete levels with which to define the momentary amplitude. Typically, the converter selects the digital value that is closest to the actual sampled value. The more bits in the word, the finer the resolution. A word with more bits can more accurately define the amplitude at each sample point. In reality, each additional bit adds a significant amount of resolution because the increase is calculated exponentially, not through simple addition. Most importantly, the increased amplitude resolution (decreased step sizes between quantization levels) yields less quantization noise.

The bits resolution of the analog-to-digital converter depends on the dynamic range of the output signal and the desired accuracy. The minimum value of the bits resolution can be obtained from the following equation:

$$\delta_A = \frac{x_R}{2^{N_D}} \tag{6.5}$$

where N_D is the minimum bits resolution of the analog-to-digital converter, x_R is the dynamic range of the output signal, and δ_A is the desired accuracy.

Example 6.1. *Define the minimum bits resolution of the analog-to-digital converter for a pneumatic position linear actuator that has a displacement of 0.5 m $(x_R = 500$ mm$)$ and whose accuracy should be $\delta_A = 0.01$ mm. According to Equation 6.5, the value of 2^{N_D} is $x_R/\delta_A = 500/0.01 = 50,000$, and then $N_D \approx 15.6$; that is, the analog-to-digital converter with 16-th bits resolution is suitable.*

Typically, the analog-to-digital converter has 12- to 16-bit resolution, giving 2^{12} to 2^{16} levels of quantization. This is normally a much higher resolution than the precision of linear displacement sensors used in a pneumatic actuator. The resolution of digital-to-analog converters is 8 to 12 bits. It is better to have good resolution of the measured signal (i.e., at the analog-to-digital converter) because a control system is less critical for a quantized input signal. In some cases where high actuator accuracy is required, 32-bit-resolution devices are utilized.

Computer control systems represent quite a mature field. Nonlinearities, such as quantization and sampling, will normally introduce oscillations or limit cycles in the actuating system. However, it is quite difficult to estimate the influence of the sampling period and converter resolution on the dynamic behavior of nonlinear systems, which are pneumatic actuators. This is very serious because it requires many ad hoc approaches when applying the powerful theory of nonlinear systems. The standard approach is to sample fast and to use simple difference approximations. In addition, the final choice of resolution and sampling interval must be based on computer simulation of the dynamic behavior of such systems.

The dynamic analysis of the pneumatic systems has always been a great challenge for the design engineer. To conduct a worthwhile dynamic analysis, it is necessary to develop a mathematical model that adequately describes each component and its interaction within the system. These models are nonlinear and require a great amount of time to produce. In addition, the engineer who is charged with the task of performing a dynamic analysis is normally not the engineer who designed any of the components of the system. When a component is used in a control system, the dynamic performance of that component is extremely important in the analysis of the system; therefore, the system design engineer must rely on the component manufacturer to provide the necessary data for the computer modeling process. One of the most critical parameters of a pneumatic actuating system component (for the most part, these are the control valves) used for control purposes is the time or dynamic response. In simple terms, the time response of a control component refers to the lag between the input and output when the component is exposed to a dynamic input.

In recent years, more useful computer-aided design analysis software has become available for pneumatic systems. These programs permit the system designer to use component manufacturer-supplied information to define the components of a system and topographical information to define the system configuration.

6.2 Pneumatic Positioning Actuators

The most common "loop" in pneumatic actuating systems is position. As stated in Chapter 1, the closed-loop position actuator consists of a controller, a control valve, an actuator, and a device to measure the actuator's displacement (see Figure 1.11).

In general, closed-loop pneumatic positioning actuators can be broken into two fundamental groups. The first group deals with continuous position tracking systems. It addresses the question of how well the actuator motion follows the command signal, which is the mechanical trajectory. It can be thought of as what internal commands are needed so that the user's motion commands are followed without any error, assuming of course that a sufficiently accurate model of the actuator is known. In this case, a basic servo controller generally contains both a trajectory generator and a controller. Sometimes, the tracking motion is carried out using a buffer of positions and then creates a smooth path or spline through those positions.

The second general group addresses point-to-point positioning systems. Such a system is one of the most basic types of moves. For the point-to-point motion control system, the actuator should be required to move a given distance within a specified time and with a desired accuracy (sometimes positioning repeatability is required). In this case, the graph of velocity vs. time typically has a trapezoidal form and, sometimes in addition to the position parameter, the motion controller requires velocity and acceleration parameters.

In the design process of pneumatic positioning actuators, it is important to achieve the desired positioning accuracy and dynamic performance. One peculiarity of pneumatic actuators is the limitation of the bandwidth restricting the high gains that can be applied. In addition, combined with their poor damping and low stiffness arising from the compressibility of air, as well as significant coulomb friction from moving parts, both accuracy and repeatability are limited by variations in payload and supply pressure. Therefore, the analyses of actuator dynamic behavior and the estimation of actuator parameters are complicated.

6.2.1 Continuous Position Tracking Systems

In contrast with point-to-point actuator systems, where the fixed set point control is used, in continuous position tracking systems the desired signal does not remain constant but changes over time. Usually, the desired signal is predetermined by the plant operator or by external equipment (trajectory generator). A desired signal that changes fast requires a control loop with good reference action. If, in addition, considerable disturbances must be eliminated, the disturbance reaction must also be taken into account when designing the controller. Such a position system can take any position within some interval defined by the actuator.

Sometimes, in continuous position tracking actuators, the motion controller provides gearing and complex motion control with splines. With the spline function, implementing smoothly curving motion profiles is as easy as providing the motion controller with endpoint coordinates and instructing it to connect the dots.

The motion curve defined by the spline represents the position of the axis, which is a function of time. Velocities and accelerations are determined by differentiating the spline equation at each axis position. With splines, complex motion profiles can be easily specified graphically. The designer defines only the positions; the spline algorithm computes the acceleration and velocity necessary to get smoothly from one point to another. In an ideal situation, these points can be defined graphically with a CAD-type tool, and the designer is relieved of the tedious calculations for each segment between the defining points.

The main attribute of pneumatic actuators that operate in continuous position tracking mode is that the proportional control valve does not reach a saturation condition. In this case, the graph of velocity vs. time has a triangular shape. The parameter estimation for a linear actuator can be carried out using a mathematical model that is obtained for the following assumptions (in addition to the assumption obtained in Chapter 2):

- The control proportional valve has under-lap or zero-lap (particular case of the under-lap) design.
- The dynamic behavior of the proportional control valve with an electrical power amplifier is described as a first-order dynamic system.
- The supply pressure and temperature to the control valve are constant.
- The external and internal leakage of the pneumatic actuator are negligible.
- The piping connecting the control valve and actuator is quite short and offers negligible resistance to flow.

Then, using Equation 2.20 and Equation 2.23, the nonlinear differential equations of the mathematical model can be expressed as:

$$
\begin{cases}
m \cdot \ddot{x} + b_V \cdot \dot{x} + F_F + F_L = P_1 \cdot A_1 - P_2 \cdot A_2 - P_A \cdot A_R \\[2mm]
\dot{P}_1 = \dfrac{1}{V_{01} + A_1 \cdot (0.5 \cdot L_S + x)} \cdot (G_1^+ \cdot R \cdot T_S - G_1^- \cdot R \cdot T_S - A_1 \cdot P_1 \cdot \dot{x}) \\[2mm]
\dot{P}_2 = \dfrac{1}{V_{02} + A_2 \cdot (0.5 \cdot L_S - x)} \cdot (G_2^+ \cdot R \cdot T_S - G_2^- \cdot R \cdot T_S + A_2 \cdot P_2 \cdot \dot{x}) \\[2mm]
t_{PA} \cdot \dot{U}_C + U_C = K_{PA} \cdot U_R
\end{cases}
\tag{6.6}
$$

where the mass flow rate G_i^+ can be determined by Equation 2.21 and the mass flow rate G_i^- by Equation 2.22, t_{PA} is the time constant of the control valve with power amplifier, U_C is the control signal, and U_R is the regulating signal.

As discussed above, in pneumatic actuators, the state regulator is usually used; then in the displacement term, the regulating signal (U_R) can be expressed as:

$$U_R = K_P \cdot (x_D - x) - K_V \cdot \dot{x} - K_A \cdot \ddot{x} \tag{6.7}$$

where x_D is the desired moving track or position.

In the equations for the mass flow rates (G_i^+ and G_i^-), the effective areas of the control valve are described in the following forms: $A_{V1}^+ = A_{VM}^+ \cdot \beta_1^+$, $A_{V1}^- = A_{VM}^- \cdot \beta_1^-$, $A_{V2}^+ = A_{VM}^+ \cdot \beta_2^+$, and $A_{V2}^- = A_{VM}^- \cdot \beta_2^-$; where A_{VM}^+ is the valve effective area for the completely open supply line and A_{VM}^- is the valve effective area for the completely open exhaust line. Because the control valve operates in the mode in which it does not reach the saturation condition, and the control valve has under-lap or zero-lap construction (see Equation 3.10 and Figure 3.43), the valve effective areas can be expressed as:

$$A_{V1}^+ = A_{VM}^+ \cdot (K_{SL}^+ \cdot U_C + \beta_0^+)$$
$$A_{V1}^- = A_{VM}^- \cdot (-K_{SL}^- \cdot U_C + \beta_0^-)$$
$$A_{V2}^+ = A_{VM}^+ \cdot (-K_{SL}^+ \cdot U_C + \beta_0^+)$$
$$A_{V2}^- = A_{VM}^- \cdot (K_{SL}^- \cdot U_C + \beta_0^-) \tag{6.8}$$

where β_0^+ and β_0^- are the opening coefficients of the supply and exhaust lines, respectively (for the zero-lap valve design $\beta_0^+ = \beta_0^- = 0$), and K_{SL}^+ and K_{SL}^- constitute the slope of the valve steady-state characteristic of the supply and exhaust lines, respectively (for example, see Figure 3.42). In practice, most often $\beta_0^+ = \beta_0^- = \beta_0$ and

$$K_{SL}^+ = K_{SL}^- = K_{SL} = \frac{1 - \beta_0}{U_{CS}}$$

(see Chapter 3.5), and these assumptions are used in the following considerations. Then, using Equation 6.8, the mass flow rates are described by the following equations:

$$G_1^+ = A_{VM}^+ \cdot (K_{SL} \cdot U_C + \beta_0) \cdot P_S \cdot \sqrt{\frac{2 \cdot k}{R \cdot T_S \cdot (k-1)}} \cdot \varphi(\sigma_1)$$

$$G_1^- = A_{VM}^- \cdot (-K_{SL} \cdot U_C + \beta_0) \cdot P_1 \cdot \sqrt{\frac{2 \cdot k}{R \cdot T_S \cdot (k-1)}} \cdot \varphi\left(\frac{\sigma_A}{\sigma_1}\right)$$

$$G_2^+ = A_{VM}^+ \cdot (-K_{SL} \cdot U_C + \beta_0) \cdot P_S \cdot \sqrt{\frac{2 \cdot k}{R \cdot T_S \cdot (k-1)}} \cdot \varphi(\sigma_2) \tag{6.9}$$

$$G_2^- = A_{VM}^- \cdot (K_{SL} \cdot U_C + \beta_0) \cdot P_2 \cdot \sqrt{\frac{2 \cdot k}{R \cdot T_S \cdot (k-1)}} \cdot \varphi\left(\frac{\sigma_A}{\sigma_2}\right)$$

In the closed-loop pneumatic actuator with continuous position tracking operation mode, design criteria are a very important issue in the design and parameter estimation process. Classical design procedures such as response to a step function, characterized by a desired rise time, settling time, peak overshoot or positioning accuracy to a steady-state value; or frequency criteria such as phase margin, gain margin, peak amplitude, or bandwidth are most suitable for systems described by linear differential equations. For nonlinear systems (to which the pneumatic actuators refer), parameter estimation is usually implemented in a two-step process. In the first step, the crude estimate is carried out by using linearized differential equations. In the second step, these parameters are adjusted by computer simulation and mathematical analysis using the nonlinear mathematical model.

A crude estimation of actuator parameters can be carried out using linear differential equations, which can be obtained by linearizing the fifth-order mathematical model given by Equation 6.6 through a Taylor series expansion around an equilibrium point. According to the linearization principle, the first-order approximations are sufficient to characterize the local behavior of the nonlinear model. The term "local" refers to the fact that satisfactory behavior only can be expected for those initial conditions that are close to the point about which the linearization was made.[176]

The linearization is made for the following assumptions:

- At the equilibrium point (initialization point of the linearization process), the pressure in the working chambers and their capacity undergo small changes. At this point the actuator has the following initial conditions: $U_R = 0$, $U_C = 0$, $x = 0$, $\dot{x} = 0$, $\ddot{x} = 0$, and $P_1 = P_2 = P_0$. These conditions designate that the piston moves a small distance closer to its center position, the pressure in the working chamber only differs slightly from the initial value P_0, and the spool of the control valve is centered.

- In an industrial application, the supply pressure (P_S) is about 0.5 to 0.7 MPa and then the value of the initial pressure in the working chamber (P_0) is usually greater than 0.2 MPa. In this condition, the flow function for the discharging process is:

$$\varphi\left(\frac{\sigma_A}{\sigma_i}\right) = \varphi_*$$

(6.10)

and the linearized equation that defines the flow function for the charging process has the following form:

$$\varphi(\sigma_i) = \varphi_0 + N_0 \cdot \sigma_i$$

(6.11)

where

$$\varphi_0 = 2 \cdot \varphi_* \cdot \sqrt{\sigma_0 \cdot (1 - \sigma_0)},$$

$$N_0 = \left| \frac{\varphi_* \cdot (1 - 2 \cdot \sigma_0)}{\sqrt{\sigma_0 \cdot (1 - \sigma_0)}} \right|,$$

and

$$\sigma_0 = \frac{P_0}{P_S}.$$

In this case, when the charge flow moves at the subsonic condition and the discharge flow moves at the sonic condition, the initial value of the pressure in the working chambers is

$$P_0 = \frac{P_S}{1 + 0.25 \cdot \Omega_M^2},$$

where

$$\Omega_M = \frac{A_{VM}^-}{A_{VM}^+}.$$

- The friction force on the pneumatic cylinder is viscous, while the coulomb friction force (F_F) and the external load force (F_L) are negligible.

In this case, where all second-order terms in the Taylor series expansion are assumed negligible, the second differential equation of the system (Equation 6.6) can be rewritten in the following linear form:

$$\dot{P}_1 = B_{x1} \cdot (B_{U1} \cdot U_C - B_{P1} \cdot P_1 - \dot{x} + B_1) \tag{6.12}$$

where

$$B_{x1} = \frac{P_0}{x_{01} + 0.5 \cdot L_S}$$

$$B_{U1} = K_{SV} \cdot \left(\frac{A_{VM}^+ \cdot P_S}{A_1 \cdot P_0} \cdot \varphi_0 + \frac{A_{VM}^+}{A_1} \cdot N_0 + \frac{A_{VM}^-}{A_1} \cdot \varphi_* \right) \cdot K_*$$

$$B_{P1} = \beta_0 \cdot \left(\frac{A_{VM}^+}{A_1 \cdot P_0} \cdot N_0 + \frac{A_{VM}^-}{A_1 \cdot P_0} \cdot \varphi_* \right) \cdot K_*$$

$$B_1 = \frac{A_{VM}^+ \cdot P_S}{A_1 \cdot P_0} \cdot \varphi_0 \cdot \beta_0 \cdot K_*$$

In a similar way, for the third differential equation of the system (Equation 6.6), the linear form can be rewritten as:

$$\dot{P}_2 = B_{x2} \cdot (-B_{U2} \cdot U_C - B_{P2} \cdot P_2 + \dot{x} + B_2) \tag{6.13}$$

where

$$B_{x2} = \frac{P_0}{x_{02} + 0.5 \cdot L_S}$$

$$B_{U2} = K_{SL} \cdot \left(\frac{A_{VM}^+ \cdot P_S}{A_2 \cdot P_0} \cdot \varphi_0 + \frac{A_{VM}^+}{A_2} \cdot N_0 + \frac{A_{VM}^-}{A_2} \cdot \varphi_* \right) \cdot K_*$$

$$B_{P2} = \beta_0 \cdot \left(\frac{A_{VM}^-}{A_2 \cdot P_0} \cdot \varphi_* - \frac{A_{VM}^+}{A_2 \cdot P_0} \cdot N_0 \right) \cdot K_*$$

$$B_2 = \frac{A_{VM}^+ \cdot P_S}{A_2 \cdot P_0} \cdot \varphi_0 \cdot \beta_0 \cdot K_*$$

The linearized mathematical model can be modified into a third-order form if the following assumptions are taken into account: the pneumatic cylinder has a symmetrical design ($A_1 = A_2 = A_C$), the length of the inactive volume at each end of stroke and admission port is equal ($x_{01} = x_{02} = x_{0C}$), and the time constant of the control valve with power amplifier is negligible ($t_{PA} = 0$). Then, the linear differential equations of the mathematical model can be expressed as:

$$\begin{cases} m \cdot \ddot{x} + b_V \cdot \dot{x} = \Delta P \cdot A_C \\ \Delta \dot{P} = B_x \cdot (B_U \cdot U_C - B_P \cdot \Delta P - \dot{x}) \end{cases} \tag{6.14}$$

where

$$B_x = \frac{2 \cdot P_0}{x_{0C} + 0.5 \cdot L_S}$$

$$B_U = \frac{A_{VM}^+}{A_C} \cdot Q_1$$

$$\Omega_M = \frac{A_{VM}^-}{A_{VM}^+}$$

$$B_P = \frac{A_{VM}^+}{A_C} \cdot Q_2$$

$$\Delta P = P_1 - P_2$$

$$U_C = K_{PA} \cdot [K_P \cdot (x_D - x) - K_V \cdot \dot{x} - K_A \cdot \ddot{x}]$$

$$Q_1 = K_{SL} \cdot \left(\frac{P_S}{P_0} \cdot \varphi_0 + N_0 + \Omega_M \cdot \varphi_* \right) \cdot K_*$$

$$Q_2 = \frac{\beta_0 \cdot N_0 \cdot K_*}{P_0}$$

The system of differential Equations 6.14 can be rewritten as a third-order differential equation:

$$m \cdot \dddot{x} + B_1 \cdot \ddot{x} + B_2 \cdot \dot{x} + B_3 \cdot x = B_3 \cdot x_D \qquad (6.15)$$

where

$$B_1 = b_V + B_x \cdot (B_U \cdot K_{PA} \cdot K_A \cdot A_C + B_P \cdot m)$$

$$B_2 = B_x \cdot (B_U \cdot K_{PA} \cdot K_V \cdot A_C + B_P \cdot b_V + A_C)$$

$$B_3 = B_x \cdot B_U \cdot K_P \cdot A_C$$

The main cause of dissymmetry in the pneumatic cylinder is the change in the working chamber's volume due to the piston motion. In linear pneumatic actuators, the stability margin is a function of piston position; and in the middle point of the cylinder stroke, this parameter has the lower value. Therefore, the estimation of the parameters, which are carried out to the middle position and which provide the stable dynamic behavior, and it also guarantees the stable dynamic to any other position of the actuator. Even if Equation 6.15 is valid only for small changes in x, P_1, and P_2, it gives qualitative information about the relationship between the cylinder pressure, the movement of the piston, and the mass flow from the control valve.

To simplify the crude estimation of actuator parameters and to achieve the general criteria of the actuator's dynamic behavior, the differential Equations 6.14 are rewritten in the dimensionless form. For that, the following dimensionless parameters are used:

- Actuator dimensionless displacement: $\xi = x/x_D$
- Dimensionless time $\tau = t/t_*$, where t_* is the time scale factor coefficient, which is determined as

$$t_* = \sqrt[3]{\frac{m}{B_3}} = \sqrt[3]{\frac{m}{B_x \cdot B_U \cdot K_P \cdot A_C}}$$

Then the linearized dimensionless differential equation that describes the actuator dynamic behavior has the following form:

$$\dddot{\xi} + B_{1D} \cdot \ddot{\xi} + B_{2D} \cdot \dot{\xi} + B_{3D} \cdot \xi = B_{3D} \tag{6.16}$$

where

$$B_{1D} = B_1 \cdot \frac{t_*}{m}, \quad B_{2D} = B_2 \cdot \frac{t_*^2}{m}, \quad B_{3D} = B_3 \cdot \frac{t_*^3}{m} = 1$$

The dimensionless velocity and acceleration are connected with the dimension parameters by the following relations:

$$\dot{\xi} = \dot{x} \cdot \frac{t_*}{x_D} \quad \text{and} \quad \ddot{\xi} = \ddot{x} \cdot \frac{t_*^2}{x_D}$$

Typical performance specifications for a pneumatic actuator in continuous position tracking operation mode are:

- Moving mass m
- Actuator stroke L_S
- Steady-state positioning accuracy δ_A
- Actuator bandwidth f_A
- Supply pressure P_S
- Opening coefficient of the control valve (for the zero control signal) β_0
- Slope of the control valve steady-state characteristic K_{SL}
- Gain of the power servo amplifier K_{PA}

The actuator parameters that should be determined include:

- Pneumatic cylinder piston effective area A_C
- Maximum effective areas of the control valve A_{VM}^+ and A_{VM}^-
- Parameters of the state regulator, which are the gains K_P, K_V, and K_A

As stated previously, in the first step a crude estimation of the actuator parameters should be performed using linearized differential Equation 6.15 and Equation 6.16.

The frequency response analysis assists in understanding the effects of each actuator component on the overall system stability and bandwidth. Bandwidth is the measure of the actuator's ability to follow a command signal. The traditional approach for measuring bandwidth is to issue a sinusoidal command to the system and compare it with the system's response. Typically, an actuator bandwidth is defined as the frequency that yields a 70.7% (−3 dB) response of the command value. The bandwidth may be taken from either a magnitude or a phase vs. frequency plot (called a Bode diagram). In a magnitude plot, bandwidth is the frequency at which the amplitude falls to 0.707 of the input. In a phase plot, it is the frequency at which a 45° phase shift occurs. It should be noted that the Bode diagram of a pneumatic actuator has characteristics that depend both on the amplitude of the sinusoidal command signal and on the position of its equilibrium point. In practice, the Bode diagrams for the closed-loop position actuators are built for the middle point of the piston stroke and the amplitude of the command signal is up to 0.05 from its maximum displacement.

The equation that states the correlation between the actuator natural frequency (f_{AN}) and the bandwidth (f_A) is useful for estimating actuator parameters. For example, if a second-order model describes an actuator, the bandwidth of this system is given by[110]:

$$f_A = f_{AN} \cdot \sqrt{1 - 2 \cdot \varsigma^2 + \sqrt{2 - 4 \cdot \varsigma^2 + 4 \cdot \varsigma^4}}$$

where ς is the damping factor.

For actuators described by a differential equation of greater than second order, the value of the bandwidth and natural frequency ratio is a function of the step response profile. For a well-damped system, when the step response is monotonic, it can be assumed that $f_A \approx f_{AN}$. If the system has a bad damping characteristic (e.g., the step response has an oscillating profile), the bandwidth can be defined as $f_A \approx 1.5 \cdot f_{AN}$.

Because most closed-loop pneumatic actuating systems of practical interest should have monotonic step response, the correlation between the bandwidth and natural frequency has the following form:

$$f_A \approx f_{AN} \tag{6.17}$$

From Equation 6.15 the natural frequency of the closed-loop system can be estimated as:

$$f_{AN} = \frac{1}{2 \cdot \pi} \cdot \sqrt{\frac{B_x \cdot Q_1 \cdot A_{VM}^+ \cdot A_C \cdot K_P}{A_C \cdot b_V + B_x \cdot Q_2 \cdot A_{VN}^+ \cdot m}} \tag{6.18}$$

and from this equation the value of the proportional gain (K_P) is:

$$K_P = \frac{4 \cdot \pi^2 \cdot f_A^2 \cdot (A_C \cdot b_V + B_x \cdot Q_2 \cdot A_{VN}^+ \cdot m)}{B_x \cdot Q_1 \cdot A_{VM}^+ \cdot A_C} \tag{6.19}$$

One of the important requirements of an actuating system is the position error (δ_A) in the steady-state condition. From Equations 6.14 the maximum value of the position error for the steady-state condition can be described by the following equations (here, the friction force [F_F] is considered):

$$\begin{cases} F_F = \Delta P \cdot A_C \\ B_U \cdot K_{PA} \cdot K_P \cdot \delta_A = B_P \cdot \Delta P \end{cases}$$

From these equations the relationship between the piston effective area (A_C), the proportional gain (K_P), and the position error (δ_A) can be obtained:

$$K_P = \frac{F_F \cdot Q_2}{Q_1 \cdot K_{PA} \cdot \delta_A \cdot A_C} \tag{6.20}$$

As discussed previously, in the continuous position tracking mode, the piston velocity graph has a triangular form. In this case, the dimensionless inertial load W should be in the range between 1 and 3 (see Chapter 5). For practical applications, it is reasonable to take into account that this parameter is $W = 2$; that is,

$$W = \frac{A_{VM}^+ \cdot K_*}{A_C} \cdot \sqrt{\frac{m}{A_C \cdot P_S \cdot L_S}} = 2,$$

and then

$$A_{VM}^+ = \frac{2 \cdot A_C}{K_*} \cdot \sqrt{\frac{A_C \cdot P_S \cdot L_S}{m}} \qquad (6.21)$$

Solving Equation 6.19 through Equation 6.21 for A_C yields:

$$A_C = \frac{K_*^2 \cdot m \cdot b_V^2}{4 \cdot P_S \cdot L_S \cdot \left(\dfrac{F_F \cdot Q_2 \cdot B_x}{4 \cdot \pi^2 \cdot f_A^2 \cdot K_{PA} \cdot \delta_A} - B_x \cdot Q_2 \cdot m \right)^2} \qquad (6.22)$$

then the estimation of the proportional gain (K_P) and effective area (A_{VM}^+) of the control valve can be performed using Equation 6.20 and Equation 6.21, respectively.

The characteristic equation of the dimensionless differential Equations 6.16 is:

$$s^3 + B_{1D} \cdot s^2 + B_{2D} \cdot s + 1 = 0 \qquad (6.23)$$

For a monotonic step transient response that has dynamic behavior similar to the step response of a critically damped second-order process, the ratio between coefficients B_{1D} and B_{2D} may have some sizes.[213] However, to achieve the minimum mean square deviation, the ratio between these dimensionless coefficients should be $B_{2D}/B_{1D} = 2$, and the coefficient B_{1D} should be in the range between 8 and 10. Assuming that the first dimensionless coefficient is $B_{2D} = 8$ and the second one is $B_{2D} = 16$, the velocity and acceleration gains of the state regulator can be estimated in the following way.

Taking into account that the dimensionless coefficient $B_{1D} = B_1 \cdot t_* / m = 8$, and substituting the relationship for B_1 and t_* into this equation, the value of the acceleration gain is:

$$K_A \approx \frac{1}{B_U \cdot K_{PA} \cdot A_C} \left(\frac{8 \cdot \sqrt[3]{m^2 \cdot B_x \cdot B_U \cdot K_P \cdot A_C} - b_V}{B_x} - B_P \cdot m \right) \qquad (6.24)$$

From the condition of $B_{2D} = B_2 \cdot t_c^2/m = 20$ and after some substituting and rearranging, the equation for estimating the velocity gain has the following form:

$$K_V \approx \frac{16 \cdot m}{B_x \cdot B_U \cdot K_{PA} \cdot A_C} \cdot \sqrt[3]{(\frac{B_x \cdot B_U \cdot K_P \cdot A_C}{m})^2} - \frac{B_P \cdot b_V + A_C}{B_U \cdot K_{PA} \cdot A_C} \qquad (6.25)$$

Thus, in the first step of parameter estimation using Equation 6.22, the effective area (A_C) of the pneumatic cylinder is determined. After that, the control valve effective area (A_{VM}^+) and the proportional gain (K_P) are defined using Equation 6.21 and Equation 6.20, respectively. In the last stage of the crude estimation, the values of the acceleration gain (K_A) and the velocity gain (K_V) are obtained using Equation 6.24 and Equation 6.25.

In the second step of the estimation process, these values are adjusted by computer simulation of the actuator's dynamic behavior. This adjustment (servo tuning) process sets the K_P, K_V, and K_A parameters of the PVA algorithm to achieve better motion performance. Such a tuning procedure can be divided into three major steps:

1. Always start tuning with proportional gain K_P, which gives adequate response speed and the desired steady-state accuracy. In this step, set the gains K_V and K_A to zero. Excite the actuator with a step command, in which the positioning point should be in the middle of the actuator stroke and the step magnitude is up to 5% from the actuator's maximum displacement. Set the gain K_P to a value that is determined using Equation 6.20. Usually in this case, the actuator begins to oscillate. In this condition, record the value of the actuator oscillation amplitude (Δ_{OS}).

2. Then set the gain K_V to a value that is obtained from Equation 6.25. In general, the motion profile is the same as the under-damped, second-order system response to a step function input. By changing the K_V value, a maximum overshoot amplitude of about $(0.2-0.4) \cdot \Delta_{OS}$ should be obtained.

3. In the final step, set the gain K_A to a value that is calculated from Equation 6.24. By changing its magnitude, find the minimum value of K_A that allows achieving the monotonic (aperiodic) step transient response (without overshoot).

Example 6.2. *Define the parameters of a linear pneumatic actuator with continuous position tracking operation mode that has the following performance specifications:*

- *Moving mass is m = 5 kg.*
- *Actuator stroke is L_S = 0.3 m.*
- *Steady state accuracy is δ_A = 0.1 mm.*

- *Actuator bandwidth is $f_A = 10$ Hz.*
- *Supply pressure is $P_S = 0.7$ MPa.*
- *Opening coefficient of the proportional control valve (for the zero control signal) is $\beta_0 = 0.2$.*
- *Maximum value of the control signal is $U_{CS} = 10$ V.*
- *Gain of the power servo amplifier is $K_{PA} = 10$.*

Assuming that the dimensionless parameter Ω_M is $\Omega_M = 1$, then the control valve effective areas can be written as $A_{VM}^+ = A_{VM}^-$.

The slope of the control valve steady-state characteristic can be defined as (see Chapter 3.5)

$$K_{SL} = \frac{1-\beta_0}{U_{CS}} = 0.08 \frac{1}{V} .$$

For given characteristics, the following parameters and coefficients can be determined:

$$P_0 = \frac{P_S}{1+0.25 \cdot \Omega_M^2} = 0.56 \, MPa$$

$$\sigma_0 = \frac{P_0}{P_S} = 0.8$$

$$\varphi_0 = 2 \cdot \varphi_* \cdot \sqrt{\sigma_0 \cdot (1-\sigma_0)} \approx 0.207 \quad (\varphi_* = 0.259)$$

$$N_0 = \left| \frac{\varphi_* \cdot (1-2 \cdot \sigma_0)}{\sqrt{\sigma_0 \cdot (1-\sigma_0)}} \right| \approx 0.39$$

$$Q_1 = K_{SL} \cdot \left(\frac{P_S}{P_0} \cdot \varphi_0 + N_0 + \Omega_M \cdot \varphi_* \right) \cdot K_* \approx 55.4 \frac{m}{V \cdot s}$$

$$Q_2 = \frac{\beta_0 \cdot N_0 \cdot K_*}{P_0} \approx 1 \cdot 10^{-4} \frac{m^2 \cdot s}{kg}$$

In addition, assuming that the length of the actuator inactive volume is $x_{OC} = 0.02$ m, then the parameter B_x is

$$B_x = \frac{2 \cdot P_0}{x_{OC} + 0.5 \cdot L_S} \approx 6.6 \cdot 10^6 \frac{kg}{m^2 \cdot s^2} .$$

For the approximating computation of the actuator parameters, the actuator friction force (F_f) and the viscous friction coefficient (b_V) should also be taken into account. For a standard pneumatic cylinder with a piston diameter from 25 to 63 mm, the static coulomb friction force changes from 20 to 75 N, and the viscous friction coefficient is in the range from 20 to 100 N·s/m (see Table 2.1). For the first step in the estimation, these parameters have the following values: $F_F \approx 30$ N and $b_v \approx 35$ N·s/m. Then effective piston area is:

$$A_C = \frac{K_*^2 \cdot m \cdot b_V^2}{4 \cdot P_S \cdot L_S \cdot \left(\dfrac{F_F \cdot Q_2 \cdot B_x}{4 \cdot \pi^2 \cdot f_A^2 \cdot K_{PA} \cdot \delta_A} - B_x \cdot Q_2 \cdot m \right)^2} \approx 1.28 \cdot 10^{-3} m^2$$

The closest standard pneumatic cylinder has a piston diameter of 0.04 m. (its effective area is $A_C \approx 1.26 \cdot 10^{-3}$ m^2).

For this standard pneumatic cylinder, the maximum effective area of the control valve is:

$$A_{VM}^+ = \frac{2 \cdot A_C}{K_*} \cdot \sqrt{\frac{A_C \cdot P_S \cdot L_S}{m}} \approx 2.5 \cdot 10^{-5} m^2$$

According to Equation 6.20, the proportional gain in the state regulator is:

$$K_P = \frac{F_F \cdot Q_2}{Q_1 \cdot K_{PA} \cdot \delta_A \cdot A_C} \approx 45 \frac{V}{m}.$$

Using Equation 6.25, the velocity gain is:

$$K_V \approx \frac{16 \cdot m}{B_x \cdot B_U \cdot K_{PA} \cdot A_C} \cdot \sqrt[3]{\left(\frac{B_x \cdot B_U \cdot K_P \cdot A_C}{m} \right)^2} - \frac{B_P \cdot b_V + A_C}{B_U \cdot K_{PA} \cdot A_C} \approx 1.563 \frac{V \cdot s}{m}.$$

And using Equation 6.24, the value of the acceleration gain is:

$$K_A \approx \frac{1}{B_U \cdot K_{PA} \cdot A_C} \left(\frac{8 \cdot \sqrt[3]{m^2 \cdot B_x \cdot B_U \cdot K_P \cdot A_C} - b_V}{B_x} - B_P \cdot m \right) \approx 0.018 \frac{V \cdot s^2}{m}$$

The computer simulation of pneumatic actuator dynamic behavior is carried out for the rodless pneumatic cylinder with a piston diameter of 0.04 m ($A_1 = A_2 = A_C = 1.26 \cdot 10^{-3}$ m^2), and a stroke length is $L_S = 0.3$ m. In this case, for instance, the pneumatic linear drive unit type DGP ("Festo") can be used. The actuator

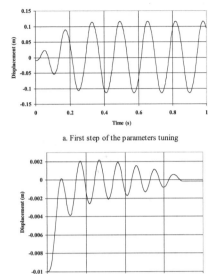

a. First step of the parameters tuning

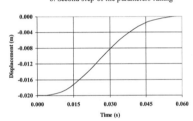

b. Second step of the parameters tuning

c. Third step of the parameters tuning

FIGURE 6.5
Three steps of the parameter tuning process for the PVA regulator: (a) first step of the parameters tuning, (b) second step of the parameters tuning, and (c) third step of the parameters tuning.

consists of the proportional control valve, in which the effective areas are $A_{VM}^{+} = A_{VM}^{-} = 2.5 \cdot 10^{-5} \, m^2$ (maximal standard nominal flow rate is about 1500 l/min). The resolution of the displacement sensor is 0.01 mm (for example, the linear encoder type RGH by "Renishow" can be used).

Figure 6.5 presents the results of the three steps of the servo tuning process for the K_P, K_V, and K_A parameters of the PVA algorithm. In the first step (Figure 6.5a), the proportional gain is $K_P = 45 \, V/m$ ($K_V = 0$ and $K_A = 0$), and the actuator oscillation amplitude is $\Delta_{OS} \approx 0.11m$. In the second step of the tuning process, the value of the velocity gain is

$$K_V = 1.1 \frac{V \cdot s}{m},$$

which gives a maximum overshoot value (Figure 6.5b) of about 0.0022 m (that is, approximately 2% of Δ_{OS}). In the third step, the value of the acceleration gain is chosen; the value of

$$K_A = 0.01 \frac{V \cdot s^2}{m}$$

allows one to achieve the monotonic step transient response (Figure 6.5c).

Figure 6.6 shows the actuator response to the step position command, where the position point is in the middle of the pneumatic cylinder stroke and the value of the command step is 0.01 m (it is about 3.3% from maximum actuator displacement). In this case, the state regulator has the parameters that were estimated above. It can

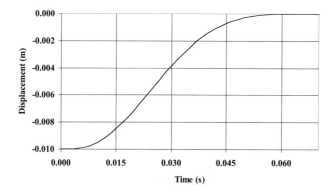

a. Position response

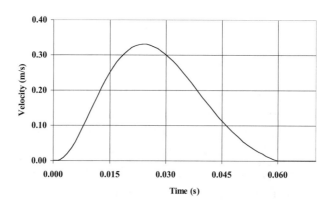

b. Velocity response

FIGURE 6.6
Actuator response to the step position command (step magnitude is 10 mm): (a) position response and (b) velocity response.

be seen that the position response (Figure 6.6a) has the monotonic form (without overshoot) and the velocity (see Figure 6.6b) is changing by the triangular law.

Figure 6.7 shows the frequency response of a linear pneumatic actuator with continuous position tracking operation mode. The Bode diagrams (Figure 6.7a), which obtained for command amplitudes of 0.01 m (about 3.3% from maximum actuator displacement) and of 0.05 m (about 16.7% from maximum actuator displacement) are significantly different. This indicates that the actuator is characterized by several dynamic nonlinearities, as described above. The actuator response to a sinusoidal command (command frequency is 5 Hz and amplitude is 0.01 m) is illustrated in Figure 6.7b.

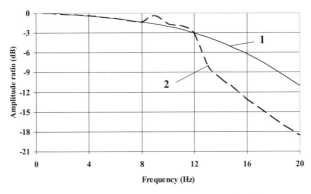

1 – Command amplitude is 10mm   2 – Command amplitude is 50 mm

a. Magnitude Bode diagram

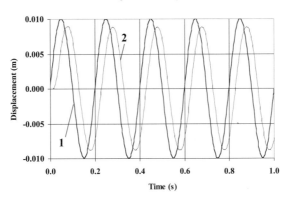

1 – Command signal   2 – Actuator output

b. Actuator response to a sinusoidal command

FIGURE 6.7
Actuator frequency response: (a) magnitude Bode diagram (1, command amplitude is 10 mm; 2, command amplitude is 50 mm); and (b) actuator response to a sinusoidal command (1, command signal; 2, actuator output).

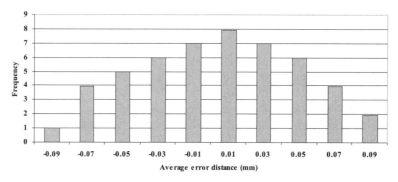

FIGURE 6.8
Histogram of steady-state positioning errors.

To estimate the steady-state positioning accuracy, the histogram of positioning errors is obtained (Figure 6.8). This histogram is generated from 50 samples. It can be seen that its form is not fit for normal distribution; therefore, an estimation of the positioning accuracy is carried out by the root-mean-squared (RMS) error and the value of the sampling range. In this case, the RMS is 0.04565 mm, and the sampling range is 0.02 mm (±0.1 mm). This plot shows the acceptable positioning accuracy because the desired value is ±0.1 mm.

Analysis of the obtained results demonstrates the high accuracy of the parameter calculation of the linear pneumatic actuator with continuous position tracking operation mode.

In practice, there are two primary ways to go about selecting PVA gains. The operator uses either a trial-and-error or analytical approach. Using a trial-and-error approach relies significantly on the operator's own prior experience with other servo systems. The one significant downside to this is that there is no physical insight into what the gains mean and there is no way to know if the gains are optimal by any definition.

To address the need for an analytical approach, the method of tuning the K_P, K_V, and K_A parameters of the PVA regulator, as described above, can be successfully used. This approach allows one to obtain not only a short adjustment time, but also high dynamic performance and steady-state accuracy.

6.2.2 Point-to-Point Systems

In point-to-point pneumatic positioning systems, a fixed set point control is generally used. In such an actuator, the desired signal is set to a fixed value, and fixed set point controllers are used to eliminate disturbances and are therefore designed to show good disturbance reaction.

In a point-to-point position application, the position loop gain is very high and the control valve saturates most of the stroke time. The error signal that is sent to a current-driving amplifier will saturate either a (+) or (–) transistor, which delivers the power to the control valve torque motor. Based on the gain setting, the power amplifier will saturate the control valve coil with the maximum current available from the supply. The output (current or voltage) from the amplifier will drive the torque motor of the control valve to its full position, thus allowing the actuator to travel at its designed maximum velocity.

When the error falls below the gain setting, the valve will balance pressures across the area of the piston and the cylinder will stop. Any over-travel or under-travel will generate an error signal, thus making the amplifier respond accordingly. If any external disturbances alter the position of the actuator, the feedback device will measure the true position of the actuator. The summing amplifier will measure the difference between the desired position and the actual position, and the process will repeat.

Typically, servo amplifiers have extremely high frequency responses. This allows the amplifier to send an alternating (+) and (−) current/voltage to the servo valve at a very high frequency. The servo valve also has a relatively high response characteristic. This allows the valve to oscillate at a higher frequency than the actuator can respond to. Because a servo valve is nothing more than an electronic flow control, it will be metering the air in and out of the actuator to obtain the desired position.

Basically, the motion profile of the pneumatic actuator with point-to-point positioning mode is similar to the open-loop position pneumatic actuator with a trapezoidal velocity curve. In this case, the graph of velocity vs. time has three specific parts (see Figure 5.15a): acceleration part, movement with steady-state velocity, and deceleration motion with a successive stop in the desired position. In contrast to open-loop systems, in closed-loop, point-to-point actuators, the gentle deceleration and stop process is performed using the control valve. As well as continuous position tracking systems, point-to-point actuators most often have the PVA control algorithm. For this type of position actuator, the character of the transition between the first and second parts of the movement is not determined because in the point-to-point motion, the major requirement is the movement time and the kind of motion is not important. Naturally, each point-to-point move should be as fast as possible to increase overall throughput.

In general, it may be considered that the first and second parts of the actuator motion take place in the open-loop operation mode. In this motion stage, one of the working chambers is connected to the supply line and another chamber is connected to the exhaust port. Only in the deceleration motion part does the actuator operate as a closed-loop system.

Most often, the parameters of a point-to-point, closed-loop pneumatic actuator are estimated for the following given characteristics:

- Moving mass (m)
- Maximum actuator stroke (L_S)
- Time (t_m) of the motion for the maximum actuator stroke
- Supply pressure (P_S)
- Steady-state positioning accuracy (δ_A)

As well as for position tracking actuators in point-to-point systems, the following parameters should be determined:

- Pneumatic cylinder piston effective area A_C
- Maximum effective areas of the control valve A_{VM}^+ and A_{VM}^-
- Parameters of the state regulator, which are the gains K_P, K_V, and K_A

Most of the time, this type of position system has only an inertial load, and then in the nonlinear differential Equations 6.6, which describe its dynamic behavior, the parameter $F_L = 0$. In this case, in the first step of parameter computation, the dimensionless parameter W is $W = 0.5$ and the relation $\alpha_p = A_V^+ / A_C$ can be determined by the empirical Equation 5.25, where the crude estimation is carried out. In this equation, as a first approximation, if it can be assumed that the ratio of the control valve effective areas is $\Omega = 1$ and the average actuator velocity is $\dot{x}_A = L_S / t_m$, then the minimum value of the piston effective area is:

$$A_C = \frac{4 \cdot K_*^2 \cdot m \cdot \alpha_p^2}{P_S \cdot L_S} \tag{6.26}$$

For closed-loop position pneumatic actuators with a "trapezoidal" velocity curve (in first approximation), the steady-state velocity can be estimated as $\dot{x}_C = 1.3 \cdot \dot{x}_A$. Then, using Equation 5.40, the maximum effective area for the actuator discharge flow (i.e., the maximum effective area of the control valve exhaust line) can be defined as:

$$A_{VM}^- \approx \frac{1.3 \cdot \dot{x}_A \cdot A_C}{\varphi_* \cdot K_*} \tag{6.27}$$

The maximum effective area of the actuator charge flow (maximum effective area of the control valve supply line) can be estimated under the condition that the dimensionless parameter W is equal to 0.5, and then:

$$A_{VM}^+ \approx \frac{0.5 \cdot A_C}{K_*} \cdot \sqrt{\frac{P_S \cdot A_C \cdot L_S}{m}} \tag{6.28}$$

After computation of the control valve and pneumatic cylinder parameters (A_C, A_{VM}^+, and A_{VM}^-), the parameters of the state regulator should be crudely estimated. In this stage of computation, the proportional gain (K_P) can be estimated from Equation 6.20.

The velocity (K_V) and acceleration (K_A) gains are estimated on the condition that the deceleration part of the actuator motion have a monotonic profile (without overshoot). Then, in the dimensionless differential Equation 6.16, which describes the actuator's dynamic behavior, the dimensionless coefficients should be $B_{1D} = 8 \div 10$ and $B_{2D} = 2 \cdot B_{1D}$. As for continuous position tracking systems (see Chapter 6.2.1), the value of the velocity gain is obtained from Equation 6.25, and the acceleration gain is estimated from Equation 6.24.

In the second step of the estimation process, these values are adjusted by computer simulation of the actuator's dynamic behavior. This tuning process can be carried out in the following sequence:

1. Similar to continuous position tracking actuators, the tuning process starts with proportional gain K_P. In this step, set the gains K_V and K_A to zero. Excite the actuator with a step command, in which the positioning point should be in the middle of the actuator stroke and the step magnitude 50% from the actuator maximum displacement. Set the gain K_P to a value that is determined using Equation 6.20. Usually, the actuator begins to oscillate around the positioning point. In this case, record the value of the actuator oscillation amplitude (Δ_{OS}).

2. In the second step, set the gain K_V to a value obtained from Equation 6.25. In general, in this case, the motion profile is the same type as an under-damped, second-order system response to a step function input. By changing the K_V value, a maximum overshoot amplitude of about $(0.02–0.04) \cdot \Delta_{OS}$ should be obtained.

3. In the third step, set the gain K_A to a value calculated from Equation 6.24. By changing its magnitude, find the minimum value of K_A that allows achieving the monotonic step transient response (without overshoot).

4. In this final step, the time (t_m) of the motion for the maximum actuator stroke is verified. If this value is more than demand size, then increase the size of A_{VM}^+ and A_{VM}^- directly proportional to the ratio of t_{ma}/t, where t_{ma} is the measured motion time. After that, it is necessary to repeat the tuning process of K_V and K_A again.

Example 6.3. *Define the parameters of a linear pneumatic actuator with point-to-point operation mode that has the following performance specifications:*

- *Moving mass is $m = 15$ kg.*
- *Maximum actuator stroke is $L_S = 0.5$ m.*
- *Time of motion for the maximum actuator stroke is $t_m = 1$ s.*
- *Steady-state accuracy is $\delta_A = 0.1$ mm.*
- *Supply pressure is $P_S = 0.6$ MPa.*
- *Opening coefficient of the proportional control valve (for the zero control signal) is $\beta_0 = 0.2$.*
- *Maximum value of the control signal is $U_{CS} = 10$ V.*
- *Gain of the power servo amplifier is $K_{PA} = 10$.*

The slope of the control valve steady-state characteristic is

$$K_{SL} = \frac{1 - \beta_0}{u_{CS}} = 0.08 \frac{1}{V} \; ,$$

and the average actuator velocity is $\dot{x}_A = L_S/t_m = 0.5$ m/s. According to Equation 5.25, the dimensionless relation is $\alpha_P = A_V^+/A_C \approx 2.6 \cdot 10^{-3}$; and then using Equation 6.26, the minimum value of the piston effective area is:

$$A_C = \frac{4 \cdot K_*^2 \cdot m \cdot \alpha_P^2}{P_S \cdot L_S} \approx 8 \cdot 10^{-4} \, m^2$$

That is to say, the actuator piston diameter should be more than 32 mm. The nearest standard pneumatic cylinder has a piston diameter of 0.04 m (its effective area is $A_C \approx 1.26 \cdot 10^{-3} \, m^2$).

According to Equation 6.27, the maximum effective area of the valve exhaust line is:

$$A_{VM}^- \approx \frac{1.3 \cdot \dot{x}_A \cdot A_C}{\varphi_* \cdot K_*} \approx 4.21 \cdot 10^{-6} \, m^2$$

and using Equation 6.28, the maximum effective area of the valve supply line is:

$$A_{VM}^+ \approx \frac{0.5 \cdot A_C}{K_*} \cdot \sqrt{\frac{P_S \cdot A_C \cdot L_S}{m}} \approx 4.24 \cdot 10^{-6} \, m^2$$

That is, the ratio of the effective areas of the control valve is

$$\Omega_M = \frac{A_V^-}{A_V^+} \approx 1,$$

which correlates well with the man-made assumption.

To estimate the proportional gain (K_p), one should determine the following parameters and coefficients:

$$P_0 = \frac{P_S}{1 + 0.25 \cdot \Omega_M^2} = 0.48 \, MPa$$

$$\sigma_0 = \frac{P_0}{P_S} = 0.8$$

$$\varphi_0 = 2 \cdot \varphi_* \cdot \sqrt{\sigma_0 \cdot (1 - \sigma_0)} \approx 0.207 \quad (\varphi_* = 0.259)$$

$$N_0 = \left| \frac{\varphi_* \cdot (1 - 2 \cdot \sigma_0)}{\sqrt{\sigma_0 \cdot (1 - \sigma_0)}} \right| \approx 0.39$$

$$Q_1 = K_V \cdot \left(\frac{P_S}{P_0} \cdot \varphi_0 + N_0 + \Omega_M \cdot \varphi_* \right) \cdot K_* \approx 55.4 \frac{m}{V \cdot s}.$$

$$Q_2 = \frac{\beta_0 \cdot N_0 \cdot K_*}{P_0} = 1.24 \cdot 10^{-4} \frac{m^2 \cdot s}{kg}$$

In addition, assume that the length of the actuator inactive volume is $x_{0C} = 0.02$ m, and then the parameter B_x is

$$B_x = \frac{2 \cdot P_0}{x_{0C} + 0.5 \cdot L_S} \approx 3.56 \cdot 10^6 \frac{kg}{m^2 \cdot s^2}.$$

For the pneumatic cylinder with a piston diameter of 0.04 m, the static friction force is about $F_F \approx 30$ N and the viscous friction coefficient is $b_V \approx 35$ N·s/m. Then, according to Equation 6.20, the proportional gain is:

$$K_P = \frac{F_F \cdot Q_2}{Q_1 \cdot K_{PA} \cdot \delta_A \cdot A_C} \approx 52.3 \frac{V}{m}$$

Using Equation 6.25, the velocity gain is:

$$K_V \approx \frac{16 \cdot m}{B_x \cdot B_U \cdot K_{PA} \cdot A_C} \cdot \sqrt[3]{\left(\frac{B_x \cdot B_U \cdot K_P \cdot A_C}{m} \right)^2} - \frac{B_P \cdot b_V + A_C}{B_U \cdot K_{PA} \cdot A_C} \approx 5.3 \frac{V \cdot s}{m}$$

And using Equation 6.24, the value of the acceleration gain is:

$$K_A \approx \frac{1}{B_U \cdot K_{PA} \cdot A_C} \left(\frac{8 \cdot \sqrt[3]{m^2 \cdot B_x \cdot B_U \cdot K_P \cdot A_C} - b_V}{B_x} - B_P \cdot m \right) \approx 0.2 \frac{V \cdot s^2}{m}$$

Similar to Example 6.2, a rodless pneumatic cylinder (for example, type DGP "Festo") with piston diameter 0.04 m can be used ($A_1 = A_2 = A_C = 1.26 \cdot 10^{-3}$ m²).

The maximum stroke length of this cylinder should be L_{SC} = 0.55 m (usually, this parameter is chosen by formula: L_{SC} = 1.1 · L_S).

The proportional control valve with effective areas A_{VM}^+ = A_{VM}^- = 4.21 · 10^{-6} m^2 (maximal standard nominal flow rate is about 250 l/min) should be used as the control valve.

The resolution of the displacement sensor is 0.01 mm (for example, the linear encoder type RGH by "Renishow" can be used).

Figure 6.9 shows the actuator response to the step position command, where the position point is in the middle of the pneumatic cylinder stroke and the value of the command step is 0.25 m (it is 50% from maximum actuator displacement). In this case, after the adjustment of the regulator parameters as stated in the following values: K_P = 53 V/m, K_V = 5.1 V · s/m, and K_A = 0.21 V · s^2/m. It can be seen that

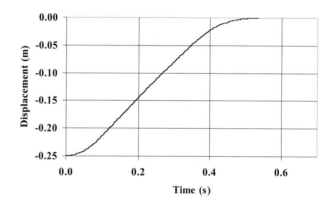

a. Position response

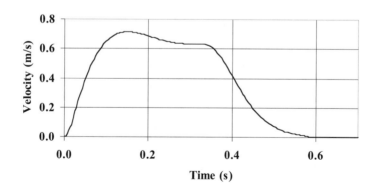

b. Velocity response

FIGURE 6.9

Step response of the point-to-point position actuator (positioning point is in the middle displacement): (a) position response and (b) velocity response.

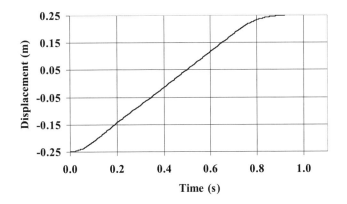

a. Position response

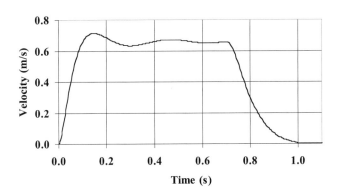

b. Velocity response

FIGURE 6.10
Step response of the point-to-point position actuator (for maximum displacement): (a) position response and (b) velocity response.

the position response (Figure 6.9a) has the monotonic form (without overshoot) and the velocity (Figure 6.9b) changes according to the trapezoidal law.

Figure 6.10 shows the step response for the maximum actuator stroke ($L_S = 0.5$ m). This plot shows that the time of motion is about $t_m = 1$ s. Both the positioning curve and the actuator velocity curve meet the requirements.

6.2.3 Actuators with Pulse Width Modulation (PWM)

As discussed in Chapter 3, PWM is currently very popular in pneumatic actuating systems. The on/off control solenoid valve used in such systems has a simple structure and is insensitive to air contamination. This system procedure simplifies the design of electropneumatic elements and improves overall actuator reliability.

It is important to note that the on/off type of control always results in excessive pressure oscillation when the regulating signal (U_R) is close to zero. For this reason, the PWM term is set to zero when the regulating signal is $|U_R| \leq \Delta U_R$, where ΔU_R is the tolerated error. Such a null algorithm allows one to hold the control valve in the blocked-center position (if in the system the 5/3-way control solenoid valve is used) and block the pneumatic actuator in the desired position. In addition, such a control mode can reduce valve cycling significantly and improve actuator reliability.

As discussed in Chapter 3, the most popular is an actuator that consists of one 5/3-way solenoid valve (Figure 6.11a). However, the actuator that has four 2/2-way solenoid valves (Figure 6.11b) is also widely used. In this case, the valves are usually mounted on the cylinder covers; this allows for decreasing the energy loss in the pneumatic line between the control valves

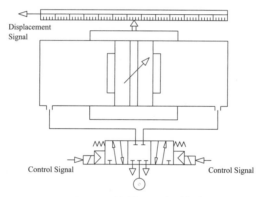

a. Actuator with 5/3-way solenoid valve

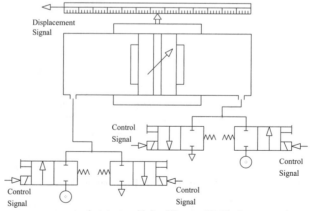

b. Actuator with four 2/2-way solenoid valves

FIGURE 6.11

Block diagram of control valve with pulse-width modulation: (a) actuator with 5/3-way solenoid valve and (b) actuator with four 2/2-way solenoid valves.

and the actuator. In addition, this scheme allows one to obtain the various values of $\Omega = A_V/A_V^+$ that enable reaching the required dynamic behavior of the actuator.

Pneumatic actuators with PWM can be used both in position tracking systems and in systems that operate in point-to-point position mode. In general, the actuator (pneumatic cylinder or motor) in these systems is the demodulation element with its low-pass filter characteristics. Such actuators are related to quasi-proportional control actuators, in which the control valve operates with PWM behavior without defacement of the square-topped pulses.

The dimensionless charging duration $\gamma_i^+ = t_i^+/t_{CR}$ and the dimensionless discharging duration $\gamma_i^- = t_i^-/t_{CR}$ define the pressure behavior in the working chamber. Here, t_i^+ is the time that the actuator working chamber is connected to the supply line, t_i^- is the time that the actuator working chamber is connected to the exhaust line, and t_{CR} is the carrier period of the PWM (Figure 3.34 and Figure 3.35). Usually, when the regulating signal is $U_R = 0$, the dimensionless charging and discharging duration is $\gamma_i^+ = \gamma_i^- = 0.5$; thus, the working chamber is connected to the supply pressure and to the exhaust port at equal time slots (half of the carrier period). In general cases, such a working chamber is designated as the quasi flow-type chamber, in which the supply and exhaust pressures are connected in turn. In such a chamber, there is a steady-state pressure oscillation. The amplitude of the pressure in the actuator working chambers and the carrier period (t_{CR}) of the PWM have an influence on the behavior of the actuator output link.

In general, in pneumatic systems where the absolute supply pressure maximum value is 1.1 MPa and the minimum value is 0.5 MPa ($0.09 \leq \sigma_A = P_A/P_S \leq 0.2$), three combinations of charge and discharge flow conditions can exist (see Figure 3.47). Determination of the maximum value (P_H) of the pressure oscillation (Figure 6.12) and its minimum level (P_L) is carried out with the assumption that the pressure change takes place without the change-over of the combination of charge and discharge flow condition. In this connection, the definition of the combination of charge and discharge flow condition can be determined by the steady-state pressure value (P_{SS}) in the equivalent flow-type chamber. In such a chamber, the opening coefficient for the charge mass flow rate is $\beta^+ = \gamma^+$; and for the discharge mass flow rate, the opening coefficient is $\beta^- = \gamma^-$. Then, the dimensionless steady-state pressure ($\sigma_{SS} = P_{SS}/P_S$) is defined by:

$$\sigma_{SS} = \frac{\gamma^+ \cdot \varphi(\sigma_{SS})}{\Omega \cdot \gamma^- \cdot \varphi\left(\dfrac{\sigma_A}{\sigma_{SS}}\right)} \tag{6.29}$$

The pressure of the charging process in the quasi flow-type chamber is described by the following differential equation:

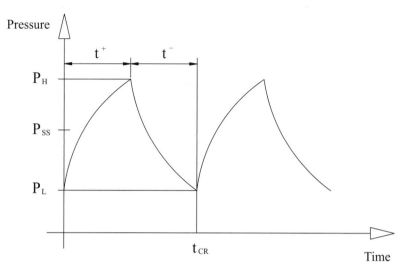

FIGURE 6.12
Pressure changing curve in the quasi flow-type chamber.

$$\dot{P} = \frac{A_V^+ \cdot P_S \cdot K_*}{V} \cdot \varphi\left(\frac{P}{P_S}\right) \qquad (6.30)$$

and for the discharging process, the pressure should be obtained from the differential equation:

$$\dot{P} = -\frac{A_V^- \cdot P \cdot K_*}{V} \cdot \varphi\left(\frac{P_A}{P}\right) \qquad (6.31)$$

where P is the absolute pressure in the quasi flow-type chamber, and V is the chamber volume.

Solving Equation 6.29 through Equation 6.31 for the different combinations of charge and discharge flow moving conditions in the quasi flow-type chamber yields the formulas for the determination of σ_{SS}, P_H, P_L, and pressure behavior.

- First combination: the charge flow is moving at the sonic condition, which in this case is $\varphi(P/P_s) = \varphi_*$; and the discharge flow is moving at the subsonic condition, which is

$$\varphi\left(\frac{P_A}{P}\right) = 2 \cdot \varphi_* \cdot \sqrt{\frac{P_A}{P}\left(1 - \frac{P_A}{P}\right)}.$$

Thus:

$$\sigma_{SS} = \frac{1}{\sigma_A} \cdot \left(\frac{\gamma^+}{2 \cdot \Omega \cdot \gamma^-} \right)^2 + \sigma_A \tag{6.32}$$

$$P_H = 0.25 \cdot P_A \cdot \left[B_W^- \cdot \varphi_* \cdot t_{CR} \cdot (1 - \gamma^+) + \frac{B_W^+ \cdot P_S \cdot \gamma^+}{B_W^- \cdot P_A \cdot (1 - \gamma^+)} \right]^2 + P_A \tag{6.33}$$

$$P_L = P_H - B_W^+ \cdot P_S \cdot \varphi_* \cdot \gamma^+ \cdot t_{CR} \tag{6.34}$$

The pressure difference in the working chamber is:

$$\Delta P = P_H - P_L = B_W^+ \cdot P_S \cdot \varphi_* \cdot \gamma^+ \cdot t_{CR} \tag{6.35}$$

The pressure in the charging process $(0 \le t < t^+)$ is described by:

$$P = B_W^+ \cdot P_S \cdot \varphi_* \cdot t + P_L \tag{6.36}$$

The pressure in the discharging process $(t^+ \le t < t_{CR})$ is described by:

$$P = \frac{\left[P_A \cdot B_W^- \cdot \varphi_* \cdot (\gamma^+ \cdot t_{CR} - t) + \sqrt{P_A \cdot (P_H - P_A)} \right]^2}{P_A} + P_A \tag{6.37}$$

- Second combination: the charge and the discharge flows are moving at the sonic condition; that is $\varphi(P/P_s) = \varphi_*$ and $\varphi(P_A/P) = \varphi_*$. Then, one can obtain:

$$\sigma_{SS} = \frac{\gamma^+}{\Omega \cdot \gamma^-} \tag{6.38}$$

$$P_H = \frac{B_W^+ \cdot P_S \cdot \varphi_* \cdot \gamma^+ \cdot t_{CR}}{1 - \exp[-B_W^- \cdot \varphi_* \cdot t_{CR} \cdot (1 - \gamma^+)]} \tag{6.39}$$

The minimum pressure value (P_L) should be determined by Equation 6.34, the pressure difference is estimated from Equation 6.35, and the pressure in the charging process $(0 \le t < t^+)$ is described by Equation 6.36.

The pressure in the discharging process ($t^+ \leq t < t_{CR}$) is described by:

$$P = P_H \cdot \exp[-B_W^- \cdot \varphi_* \cdot (t - \gamma^+ \cdot t_{CR})] \qquad (6.40)$$

- Third combination: the charge flow is moving at the subsonic condition, and the discharge flow is moving at the sonic condition; that is,

$$\varphi\left(\frac{P}{P_S}\right) = 2 \cdot \varphi_* \cdot \sqrt{\frac{P}{P_S} \cdot \left(1 - \frac{P}{P_S}\right)} \quad \text{and} \quad \varphi\left(\frac{P_A}{P}\right) = \varphi_*.$$

$$\sigma_{SS} = \frac{1}{1 + \left(\dfrac{\Omega \cdot \gamma^-}{2 \cdot \gamma^+}\right)^2} \qquad (6.41)$$

$$P_H = P_S \cdot \left[\sigma_{SS} + \frac{Sin(2 \cdot B_W^+ \cdot \varphi_* \cdot \gamma^+ \cdot t_{CR})}{8 \cdot \sqrt{\sigma_{SS} \cdot (1 - \sigma_{SS})}}\right] \qquad (6.42)$$

$$P_L = P_H \cdot \exp[-B_W^- \cdot \varphi_* \cdot t_{CR} \cdot (1 - \gamma^+)] \qquad (6.43)$$

The pressure difference in the working chamber is:

$$\Delta P = P_H - P_L =$$

$$P_S \cdot \left[\sigma_{SS} + \frac{Sin(2 \cdot B_W^+ \cdot \varphi_* \cdot \gamma^+ \cdot t_{CR})}{8 \cdot \sqrt{\sigma_{SS} \cdot (1 - \sigma_{SS})}}\right] \cdot \left\{1 - \exp\left[-B_W^- \cdot \varphi_* \cdot t_{CR} \cdot (1 - \gamma^+)\right]\right\} \quad (6.44)$$

The pressure in the charging process ($0 \leq t < t^+$) is described by:

$$P = 0.5 \cdot P_S \cdot \{1 - Sin[arcSin(\frac{P_S - 2P_L}{P_S}) - 2 \cdot B_W^+ \cdot \varphi_* \cdot t]\} \qquad (6.45)$$

and the pressure in the discharging process ($t^+ \leq t < t_{CR}$) is described by Equation 6.40.

In Equation 6.32 through Equation 6.45, the coefficients B_W^+ and B_W^- are:

$$B_W^+ = \frac{A_V^+ \cdot K_*}{V} \quad \text{and} \quad B_W^- = \frac{A_V^- \cdot K_*}{V}$$

Equation 6.32 through Equation 6.45 permit an estimation of the behavior of the pressure in the quasi flow-type chamber with constant volume. In addition, computation of pneumatic actuators with PWM may be useful in the design of pneumatic actuators for test facilities. In any case, the estimation process begins with the calculation of the dimensionless steady-state pressure (σ_{SS}), which defines the combination of the charge and discharge moving condition. After that, the maximum and minimum values (P_H and P_L) of the pressure oscillation can be defined for the given effective areas A_V^- and A_V^+, the chamber volume V, the period of the carrier signal t_{CR}, and the supply pressure P_S. In addition, the carrier signal period t_{CR} can be defined if the value of the pressure difference (ΔP) is given, which is very important for the estimation of PWM parameters.

Example 6.4. *Define the maximum and the minimum values (P_H and P_L) of the pressure oscillation in a constant value chamber ($V = 2l = 2 \cdot 10^{-3}$ m³) that is controlled by a solenoid valve with $A_V^- = A_V^+ = 3.5 \cdot 10^{-6}$ m² (standard nominal flow rate is about 200 l/min) and $t_{CR} = 0.05$ s (carrier frequency is $f_{CR} = 20$ Hz). The supply pressure is $P_S = 0.6$ MPa.*

The combination of charge and discharge flow moving is defined by the dimensionless steady-state pressure σ_{SS}, and the third combination should be considered first because this combination is most often encountered. Then, using Equation 6.41, where $\Omega = 1$ and $\gamma^- = \gamma^+ = 0.5$, becomes $\sigma_{SS} = 0.8$; that is, the third combination actually exists for the given parameters. In this case, the coefficients B_W^+ and B_W^- are

$$B_W^+ = B_W^- = 1.336\frac{1}{s}.$$

Using Equation 6.42 and Equation 6.43, the maximum and the minimum values of the pressure oscillation are $P_H = 0.4833$ MPa and $P_L = 0.4791$ MPa. The pressure difference in the working chamber is estimated by Equation 6.44: $\Delta P = P_H - P_L = 4.178 \cdot 10^3$ Pa.

Figure 6.13 shows the pressure behavior in this quasi flow-type pneumatic chamber. The first curve belongs to the analytical estimation and the second one is obtained by experimental testing. The difference between the oscillation magnitudes is about 0.2%. In the experimental test, the pressure difference is $\Delta P \approx 4.226 \cdot 10^3$ Pa; in this case, the difference is about 1%. These results have shown the good convergence between the analytical estimation and experimental results.

The dynamic behavior of pneumatic actuators with PWM is similar to that of actuators with continuous operating mode; therefore, these actuators are sometimes called "quasi continuous actuators." This fact allows one to use

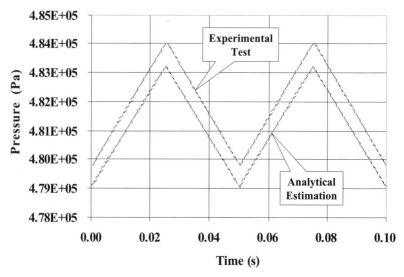

FIGURE 6.13
Pressure behavior in the quasi flow-type pneumatic chamber.

the method of parameter estimation of actuators with continuous operating mode (see Chapter 6.2.1 and Chapter 6.2.2). In this approach, the differential equations (Equations 6.6) describe the actuator motion and pressure behavior in the working chambers. However, in the system with PWM, the carrier signal amplitude U_{CRM} should be taken into account instead of the maximum value of the control signal U_{CS}. In addition, in the parameter estimation process the parameter β_0 equal 0.5 should be considered. Equations 3.14 are used for the definition of the opening coefficients for the solenoid control valve in the computer simulation of the actuator's dynamic behavior.

For actuators with PWM, the important task is to estimate the carrier signal period t_{CR}. Regarding this issue, an unknown quantity can be estimated for the condition where the oscillation amplitude of the actuator piston does not exceed the predetermined positioning accuracy. In this case, the following assumptions were taken into account: the changing of the pressure difference is the sine function and friction force is not considered. Then, the piston motion is described by the following differential equation:

$$m \cdot \ddot{x} = 0.5 \cdot \left[A_1 \cdot (P_{H1} - P_{L1}) + A_2 \cdot (P_{H2} - P_{L2}) \right] \cdot Sin\left(\frac{2 \cdot \pi \cdot t}{t_{CR}} \right) \qquad (6.46)$$

Considering the third combination of the charge and discharge flow moving (this combination exists most often), using Equation 6.42 through Equation 6.44 and taking into account the relationship $\Delta x_W \leq \delta_A$, where Δx_W is the amplitude of the piston oscillation in the steady-state condition, the following relation obtains:

$$\delta_A \geq t_{CR}^3 \cdot (B_{F1} + B_{F2} \cdot t_{CR}) \tag{6.47}$$

where

$$B_{F1} = \frac{P_S \cdot \sigma_{SS} \cdot \varphi_* \cdot (1 - \gamma^+)}{8 \cdot \pi^2 \cdot m} \cdot (A_1 \cdot B_{W1}^- + A_2 \cdot B_{W2}^-)$$

$$B_{F2} = \frac{P_S \cdot \varphi_*^2 \cdot \gamma^+ \cdot (1 - \gamma^+)}{32 \cdot \pi^2 \cdot m \cdot \sqrt{\sigma_{SS} \cdot (1 - \sigma_{SS})}} \cdot (A_1 \cdot B_{W1}^+ \cdot B_{W1}^- + A_2 \cdot B_{W2}^+ \cdot B_{W2}^-)$$

and

$$B_{W1}^+ = \frac{A_V^+ \cdot K_*}{V_1} \ , \ B_{W1}^- = \frac{A_V^- \cdot K_*}{V_1} \ , \ B_{W2}^+ = \frac{A_V^+ \cdot K_*}{V_2} \ , \ B_{W2}^- = \frac{A_V^- \cdot K_*}{V_2}$$

Estimation of the carrier signal period from Equation 6.47 gives slightly overstated results; however, for the first approximation, it is good enough. In general, for the above-mentioned condition, the value of t_{CR} is a function of the control valve effective areas (A_V^+ and A_V^-), the effective areas of the pneumatic cylinder (A_1 and A_2), the supply pressure (P_S), the value of the moving mass (m), and the coordinate of the position point (V_1 and V_2 is the volume of the first and second working chamber accordingly, which is a function of the coordinate of the position point). In particular, for each position point, there is a specific carrier signal period when the piston oscillation amplitude does not exceed the specified positioning accuracy. From this standpoint, the maximum carrier signal period (minimum modulation frequency) is in the middle stroke position, where the actuator has minimum stiffness. For this modulation in the two-end position, the amplitude of the piston oscillation exceeds the value of δ_A. However, in practice, it is quite difficult to design an actuator with PWM that has variable modulation frequency, which is a function of the position point. In this case, the application of the null algorithm is very efficient, which is illustrated in Figure 3.35b. When the actuator reaches the desired position with the specified accuracy, then both chambers can be closed to stop air flow. An additional advantage of this approach is that air consumption decreases and actuator reliability increases.

Example 6.5. *Define the parameters of the linear pneumatic actuator with PWM that work in point-to-point operation mode. The actuator has the following performance specifications:*

- *Moving mass is $m = 40$ kg.*
- *Maximum actuator stroke is $L_S = 0.5$ m.*
- *Time of motion for the maximum actuator stroke is $t_m = 1$ s.*
- *Steady-state position accuracy is $\delta_A = 0.05$ mm.*
- *Supply pressure is $P_S = 0.6$ MPa.*
- *Amplitude of the carrier signal is $U_{CRM} = 10$ V.*
- *Gain of the power servo amplifier is $K_{PA} = 10$.*

The slope of the control valve steady-state characteristic is

$$K_{SL} = \frac{1-\beta_o}{U_{CRM}} = 0.05 \, \frac{1}{V}$$

($\beta_0 = 0.5$), and the average actuator velocity is $\dot{x}_A = L_S/t_m = 0.5$ m/s. According to Equation 5.25, the dimensionless relation is $\alpha_p = A_V^+/A_C \approx 2.6 \cdot 10^{-3}$; and then using Equation 6.26, the minimum value of the piston effective area is:

$$A_C = \frac{4 \cdot K_*^2 \cdot m \cdot \alpha_P^2}{P_S \cdot L_S} \approx 2.14 \cdot 10^{-3} \, m^2$$

That is to say, the actuator piston diameter should be greater than 0.052 m. The nearest standard pneumatic cylinder has a piston diameter of 0.063 m (its effective area is $A_C \approx 3.12 \cdot 10^{-3}$ m²).

According to Equation 6.27, the maximum effective area of the valve exhaust line is:

$$A_{VM}^- \approx \frac{1.3 \cdot \dot{x}_A \cdot A_C}{\varphi_* \cdot K_*} \approx 1.02 \cdot 10^{-5} \, m^2$$

and using Equation 6.28, the maximum effective area of the valve supply line is:

$$A_{VM}^+ \approx \frac{0.5 \cdot A_C}{K_*} \cdot \sqrt{\frac{P_S \cdot A_C \cdot L_S}{m}} \approx 0.99 \cdot 10^{-5} \, m^2$$

That is, the ratio of the effective areas of the control valve is $\Omega_M = A_V^-/A_V^+ \approx 1$, which correlates well with the man-made assumption.

To estimate the proportional gain (K_P), the following parameters and coefficients should be determined:

$$P_0 = \frac{P_S}{1 + 0.25 \cdot \Omega_M^2} = 0.48 \, MPa$$

$$\sigma_0 = \frac{P_0}{P_S} = 0.8$$

$$\varphi_0 = 2 \cdot \varphi_* \cdot \sqrt{\sigma_0 \cdot (1-\sigma_0)} \approx 0.207 \quad (\varphi_* = 0.259)$$

$$N_0 = \left| \frac{\varphi_* \cdot (1-2\cdot\sigma_0)}{\sqrt{\sigma_0 \cdot (1-\sigma_0)}} \right| \approx 0.39$$

$$Q_1 = K_{SL} \cdot \left(\frac{P_S}{P_0} \cdot \varphi_0 + N_0 + \Omega_M \cdot \varphi_* \right) \cdot K_* \approx 34.6 \frac{m}{V \cdot s}.$$

$$Q_2 = \frac{\beta_0 \cdot N_0 \cdot K_*}{P_0} = 3.1 \cdot 10^{-4} \frac{m^2 \cdot s}{kg}$$

In addition, assuming that the length of the inactive volume of the actuator is $x_{0C} = 0.05$ m, then the parameter B_x is

$$B_x = \frac{2 \cdot P_0}{x_{0C} + 0.5 \cdot L_S} \approx 3.2 \cdot 10^6 \frac{kg}{m^2 \cdot s^2}.$$

For a pneumatic cylinder with a piston diameter of 0.063 m, the static friction force is about $F_F \approx 45$ N and the viscous friction coefficient is $b_V \approx 50$ N·s/m. Then, according to Equation 6.20, the proportional gain is:

$$K_P = \frac{F_F \cdot Q_2}{Q_1 \cdot K_{PA} \cdot \delta_A \cdot A_C} \approx 250 \frac{V}{m}$$

Using Equation 6.25, the velocity gain is:

$$K_V \approx \frac{16 \cdot m}{B_x \cdot B_U \cdot K_{PA} \cdot A_C} \cdot \sqrt[3]{\left(\frac{B_x \cdot B_U \cdot K_P \cdot A_C}{m} \right)^2} - \frac{B_P \cdot b_V + A_C}{B_U \cdot K_{PA} \cdot A_C} \approx 20.5 \frac{V \cdot s}{m}$$

and using Equation 6.24, the value of the acceleration gain is:

$$K_A \approx \frac{1}{B_U \cdot K_{PA} \cdot A_C} \left(\frac{8 \cdot \sqrt[3]{m^2 \cdot B_x \cdot B_U \cdot K_P \cdot A_C} - b_V}{B_x} - B_P \cdot m \right) \approx 0.5 \frac{V \cdot s^2}{m}$$

The dimensionless steady-state pressure is $\sigma_{SS} = 0.8$, which is actually the third combination and exists for the given parameters. To estimate the carrier signal period (t_{CR}), the following parameters and coefficients should be determined:

$$B^{+}_{W1} = B^{-}_{W1} \approx 9.8 \frac{1}{s}, \ B^{+}_{W} = B^{-}_{W2} \approx 10.9 \frac{1}{s}, \ B_{F1} \approx 1.2 \frac{m}{s^3}, \ B_{F1} \approx 1.26 \frac{m}{s^4}$$

These parameters have been calculated for the middle position of the piston and for a cylinder with a piston diameter of 0.063 m and a rod diameter of 0.02 m. Then, according to the Equation 6.47, the carrier signal period is $t_{CR} \approx 0.037$ s; that is, the modulation frequency is $f_{CR} \approx 27.2$ Hz. The curve of the modulation frequency (f_{CR}) vs. the dimensionless position of the actuator piston (x_D/L_S) is shown in Figure 6.14. It can be seen that the minimum frequency ($f_{CR} \approx 27.2$ Hz) is in the middle stroke position and its maximum value ($f_{CR} \approx 40.3$ Hz) is in the two end positions.

Figure 6.15 shows the step response for maximum actuator stroke ($L_S = 0.5$ m). This plot shows that the time of the motion is about $t_m = 0.9$ s. In this case, after the adjustment of regulator parameters, which are stated in the following values:

$$K_P = 250 \frac{V}{m}, \ K_V = 19 \frac{V \cdot s}{m}, \text{and } K_A = 4 \frac{V \cdot s^2}{m},$$

the modulation frequency is $f_{CR} = 30$ Hz. Both the positioning curve and the curve of the actuator velocity meet the requirements.

Actuator behavior in the middle stroke position is shown in Figure 6.16. It can be seen that the piston amplitude oscillation does not exceed the position accuracy $\delta_A = 0.05$ mm.

Analysis of the obtained results demonstrates the acceptable accuracy of parameter estimation for a linear pneumatic actuator with PWM.

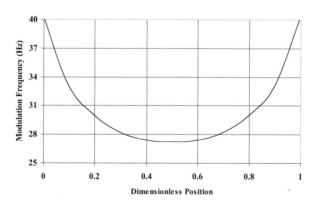

FIGURE 6.14
Modulation frequency vs. dimensionless position for the positioning actuator with PWM.

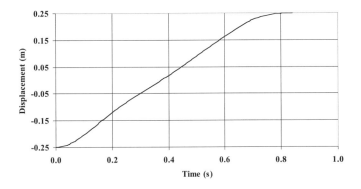

a. Position response

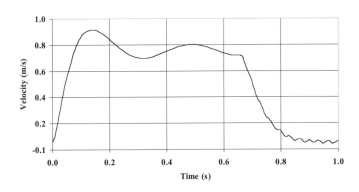

b. Velocity response

FIGURE 6.15
Step response of a point-to-point position actuator with PWM (maximum displacement): (a) position response and (b) velocity response.

6.2.4 Actuators with Bang-Bang Control Mode

Similar to PWM application, Bang-Bang control can be performed using either one 5/3-way solenoid valve or four 2/2-way valves (Figure 6.11). However, the second option is preferable because in this case, in addition to decreasing the energy loss, one can also achieve minimum lag in the control signal (the response time of a 2/2-way valve is shorter than that of a 5/3-way valve), which is very important for actuator dynamic behavior.

Ideally, the control valve should have a minimum rise time for any position application so that the actuator system will have a good step response. To this end, it is shown that an On-Off or Bang-Bang controller is the best

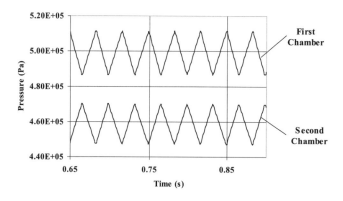

a. Pressure behavior in the working chambers

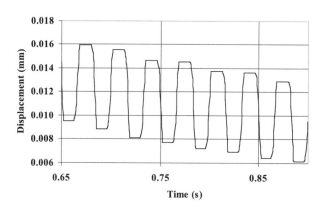

b. Position response

FIGURE 6.16
Actuator dynamic in the middle stroke position (modulation frequency is 30 Hz): (a) pressure behavior in the working chambers and (b) position response.

mechanism for achieving minimum rise times (the time required to rise from 10 to 90% of the demand output).[62,73]

Based on the Bang-Bang control principle, the control law has the property that each control variable is always at either its upper or lower bound. That is, if one wants to bring the state of the process back to its original state as fast as possible, then the largest available effort in the proper direction must be used. Therefore, in Bang-Bang control mode, the controller output is no longer a smooth signal proportional to the regulation signal. It is always saturated. Such systems are called variable structure controls.

In variable structure control, the control structure is changed to a certain rule. The slide mode control is one kind of variable structure control based on control laws that are defined with the objective of driving the system trajectory in the state space toward hypersurfaces known as sliding surfaces.

These surfaces, once reached, must confine the system trajectory in such a way that it slides over the surface toward the equilibrium point. These behavior characteristics are the so-called sliding mode. In such actuators, the system's phase trajectories all go into the given sliding surface. Because it is selected so that it passes through the outcome of the space of state, which represents the equilibrium state, the asymptotic stability of the system is also ensured. In this way, the system is brought into equilibrium according to a predetermined trajectory, which may also have attributes of optimality. In position actuating systems with Bang-Bang control, the sliding surface (sometimes this surface is called the switching surface) is most often defined in the following form:

$$S = (x_D - x) - K_{VB} \cdot \dot{x} - K_{AB} \cdot \ddot{x} \qquad (6.48)$$

where K_{VB} and K_{AB} are the velocity and acceleration gains, respectively.

In such systems, the control signal is determined as $U_{CB} = U_{CS} \cdot sign(S)$, where U_{CS} is the saturation value of the control signal and sign(*) is the sign function (see Equation 2.26, for example). Then the opening coefficients β_i^+ and β_i^-, which are a function of the regulation signal, should be described in the following form:

$$\beta_i^+ = 1 - \beta_i^-$$

$$\beta_1^+ = \beta_2^- = 1 \text{ , if } S \geq 0 \qquad (6.49)$$

$$\beta_1^+ = \beta_2^- = 0 \text{ , if } S < 0$$

For these nonlinear systems, it is difficult to obtain an analytical solution; therefore, most often, the phase trajectories (phase portraits) are used for qualitative analysis. The theoretical phase trajectories of the actuators with Bang-Bang control are shown in Figure 6.17. In general, the movement of these systems has three phases: (1) reaching the sliding surface, (2) operating in the sliding mode, and (3) the phase of the steady-state condition.

In these systems, which consist of only the position feedback signal ($K_{VB} = 0$ and $K_{AB} = 0$), the sliding surface has a vertical orientation (Figure 6.17a); and in the steady-state condition, the actuator has self-excited oscillations, called the encounter limit cycle. For actuators with negligible dead band values and hysteresis, the limit cycle amplitude is quite large; therefore, such systems are not used in industrial applications.

In actuators with position and velocity feedback ($K_{AB} = 0$), the sliding (switching) surface is inclined and, depending on this incline, two operation modes are available. In the first case, where the coefficient K_{VB} is less than the critical value, the actuator has the encounter limit cycle (see Figure 6.17a); in the second case, when the coefficient K_{VB} is greater than the critical value, the actuator moves close to the sliding surface with small oscillations. However,

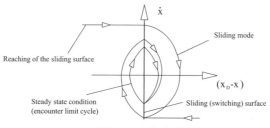

a. With only position feedback

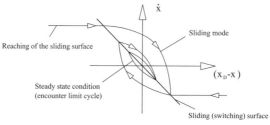

b. With position and velocity feedback

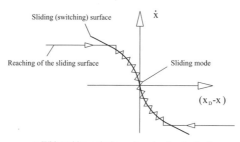

c. With position, velocity, and acceleration feedback

FIGURE 6.17
The theoretical phase trajectories of an actuator with Bang-Bang control: (a) with only position feedback; (b) with position and velocity feedback; and (c) with position, velocity, and acceleration feedback.

this option is not used in industrial applications because the position transient response is quite long.

Position actuators with position, velocity, and acceleration feedback have good dynamic and accuracy performance, which allows for their use in several industrial applications. In this case, the sliding surface has the nonlinear form, and the phase-plane trajectory is shown in Figure 6.17c. It can be seen that the actuator moves close to the sliding surface with small oscillations, and the transient response time has an acceptable value (in this option, the actuator has a negligible value of dead band and hysteresis).

It is important to note that actuators with Bang-Bang control have essential nonlinearity; therefore, the feedback control system has self-excited oscillations. These oscillations tend to cause poor position accuracy and operational stability. Using the null algorithm, which consists of the tolerated value of the dead zone, one can avoid the problem of the limit cycle and improve the reliability of the actuator. The dead zone inevitably brings about a steady-state position error, so the dead band is set according to position accuracy. When the position error is less than ΔR (ΔR is the admissible error), the Bang-Bang controller exports zero signal and control valves block the actuator. In this case, the coefficients β_i^+ and β_i^- are defined as:

$$\left.\begin{array}{l} \beta_1^+ = \beta_2^- = 1 \\ \beta_1^- = \beta_2^+ = 0 \end{array}\right\} \quad \text{if } S \geq 0 \text{ and } (x_D - x) \geq \Delta_R$$

$$\left.\begin{array}{l} \beta_1^+ = \beta_2^- = 0 \\ \beta_1^- = \beta_2^+ = 1 \end{array}\right\} \quad \text{if } S < 0 \text{ and } (x_D - x) < -\Delta_R \tag{6.50}$$

$$\beta_i^+ = \beta_i^- = 0 \quad \text{if } |x_D - x| \leq \Delta_R$$

The estimation process for actuator parameters can be divided into two stages:

1. Estimation of the mechanical parameters (effective areas of the piston and control valve)
2. Estimation of the control system parameters (feedback gains)

Calculation of the mechanical parameters of actuators depends on the type of the operation mode. The triangular velocity curve can be taken into consideration for parameter estimation of continuous position tracking systems (Chapter 6.2.2). In the point-to-point operation mode, the mechanical parameters are computed with the assumption that the velocity curve has a trapezoidal profile (Chapter 6.2.3).

In the second step, the control parameters (feedback gains) are approximately estimated. This stage deals with a nonlinear dynamic model that has variable structure, and obtaining the analytical solution is a very difficult task. In practice, using an empirical estimation of the feedback gains K_{VB} and K_{AB} is admissible. In this case, the values of

$$K_{VB} = \frac{2 \cdot A_1 \cdot L_S \cdot \varphi_*}{A_V^+ \cdot (1 + \Omega) \cdot K_*} \tag{6.51}$$

and

$$K_{AB} = \frac{3 \cdot (\Omega + 1) \cdot L_S}{K_*} \cdot \sqrt{\frac{m \cdot L_S}{P_S \cdot A_1}} \tag{6.52}$$

allow one to achieve acceptable accuracy for the first approximation. As noted above, the final adjustment for both the mechanical and control parameters is carried out by computer simulation.

In an actuator with Bang-Bang control mode, the delay time of the control signal is very important, both for the dynamic behavior and for the position accuracy. The switching will lag the output signal because of valve response delay and the sampling period; that is, the delay time (t_L) includes two major components: sampling period (t_D) and switching time of the control valve (t_V):

$$t_L = t_D + t_V \tag{6.53}$$

In the first approximation, the maximum permissible value of the delay time can be estimated as:

$$t_L \approx \frac{\delta_A \cdot t_m}{(0.01 \div 0.02) \cdot L_S} \tag{6.54}$$

In this case, actuator movement around the set point does not exceed the predetermined position accuracy.

In the final step of the estimation process, the actuator parameters are adjusted by computer simulation of the actuator dynamic behavior, which is performed in the following sequence:

1. Always start the tuning process with the velocity feedback gain K_{VB}. In this step, set the acceleration gain K_{AB} to zero; the gain K_{VB} should be equal to the value determined by using Equation 6.51. Excite the actuator with a step command, in which the positioning point should be in the middle of the actuator stroke. Usually, in this condition, the actuator has the encounter limit cycle. By increasing the value of the gain K_{VB}, find its minimum value that allows achieving the damped oscillations.

2. In this step, set the acceleration gain K_{AB} to some value, the value obtained from Equation 6.52. By changing its magnitude, find the minimum value of K_{AB} that allows for achieving the monotonic step transient response (without overshoot).

Example 6.6. *Define the parameters of the linear pneumatic closed-loop actuator with Bang-Bang control that work in point-to-point operation mode. The actuator has the following performance specifications:*

- *Moving mass is m = 5 kg.*
- *Maximum actuator stroke is L_S = 0.3 m.*
- *Time of motion for the maximum actuator stroke is t_m = 0.5 s.*
- *Steady-state position accuracy is δ_A = 0.1 mm.*

- *Supply pressure is $P_S = 0.6$ MPa.*

The average velocity is $\dot{x}_A = L_S/t_m = 0.6$ m/s, and according to Equation 5.25, the dimensionless relation is $\alpha_P = A_V^+/A_C \approx 3.15 \cdot 10^{-3}$. Then, using Equation 6.26, the minimum value of the piston effective area is $A_C \approx 6.423 \cdot 10^{-4}$ m², and the nearest standard pneumatic cylinder has the piston diameter of 0.032 m (its effective areas are $A_1 \approx 8.038 \cdot 10^{-4}$ m² and $A_2 \approx 6.91 \cdot 10^{-4}$ m²). Using Equation 6.27 and Equation 6.28, the effective areas of the control valve are $A_V^- \approx 3.172 \cdot 10^{-6}$ m² and $A_V^+ \approx 2.833 \cdot 10^{-6}$ m². In this case, the solenoid valve with $\Omega_M = 1$ and effective areas of $A_V^+ = A_V^- = 3.4 \cdot 10^{-6}$ m² (maximal standard nominal flow rate is about 200 l/min) can be used. According to Equation 6.51 and Equation 6.52, the feedback gains K_{VB} and K_{AB} are $K_{VB} \approx 0.024$ s and $K_{AB} \approx 1.32 \cdot 10^{-4}$ s².

Figure 6.18 shows the position and velocity response of the actuator with the estimated parameters (after the final step of estimation, the feedback gains are $K_{VB} \approx 0.03$ s and $K_{AB} \approx 2.4 \cdot 10^{-4}$ s²). The input signal is the step function for the maximum actuator stroke ($L_S = 0.3$ m). Here, the delay time is negligible ($t_L = 0$). It can be

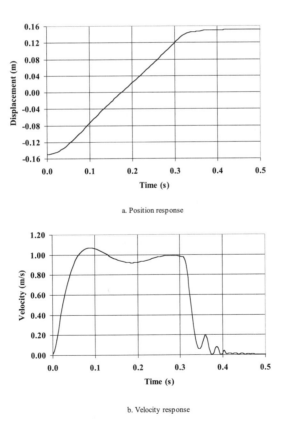

a. Position response

b. Velocity response

FIGURE 6.18
Step response of a point-to-point position actuator with Bang-Bang control (maximum displacement): (a) position response and (b) velocity response.

clearly seen that both the positioning curve and the curve of the velocity meet the requirements.

The phase-plane trajectories shown in Figure 6.19 illustrate the actuator dynamic behavior in the different stages of the final step of parameter estimation. In this actuator, the null algorithm is used, where the value of the dead zone is $\Delta_R = \delta_A = 0.1$ mm, and the displacement of the input step signal corresponds to 50 mm.

According to Equation 6.54, the maximum permissible delay time is $t_L = 0.0083–0.017$ s. Results of actuator dynamic simulation for the different values of the delay time t_L are shown in Figure 6.20. It is seen that the permissible value is about 0.011 s; that is, the switching time of the control valve should not be more than 0.01 s.

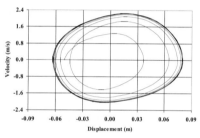

a. Only position feedback signal

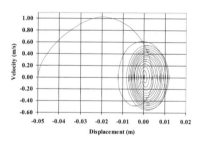

b. Position and velocity feedback signals

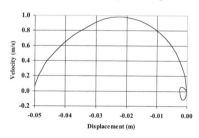

c. Position, velocity, and acceleration feedback signals

FIGURE 6.19

Phase-plane trajectories for different feedback structures: (a) only position feedback signal, (b) position and velocity feedback signals, and (c) position, velocity, and acceleration feedback signals.

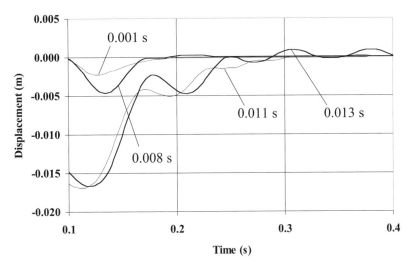

FIGURE 6.20
Position response for the different values of delay time.

The major advantage of position pneumatic actuators with Bang-Bang control mode is the invariance of their dynamic behavior to external load and internal perturbations (parameter uncertainty). In addition, these actuators have very high robustness. The shortcoming of such systems is the necessity of measurability for the full state of the actuator. This disadvantage can be successfully overcome by the application of the observer (see Chapter 4).

6.3 Actuators for the Velocity Control

Actuating systems used for metrology, inspection, printing, DNA assaying, and laser machining are typified by a need for smooth motion and constant velocity. Constant velocity, in this context, is the difference between the actual velocity deviation and a theoretical, desired value at a particular sampling rate as quantified by a power spectrum.

A velocity loop is very similar to a positional loop, with the exception that the device measuring actuator movement is different. Velocity is determined by displacement over time. In analog systems, the feedback device is scaled to deliver an analog voltage at a predetermined velocity (speed). More sophisticated controls use a digital feedback device and a time constant to determine velocity. This feature is beneficial because position feedback can be derived as well. The schematic diagram of such an actuator is shown in Figure 1.13.

In general, pneumatic actuating systems for velocity control can be divided into two major groups: (1) actuators with continuous control, where the proportional or servo valves operate in analog mode; and (2) discrete systems,

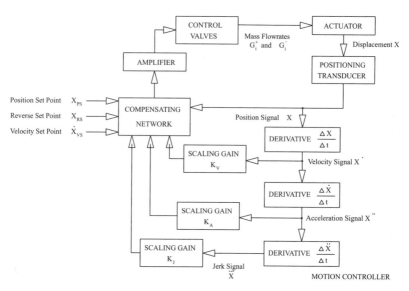

FIGURE 6.21
Block diagram of a closed-loop actuator for the velocity control.

which consist of solenoid control valves that operate in PWM or Bang-Bang control mode. As discussed in Chapter 1, the most common way to provide position control is the additional loop, which is necessary only for actuator stops in the period between technological actions. In this case, the actuator positioning accuracy is not high — 1 to 2 mm is sufficient.

Figure 6.21 shows the block diagram of a closed-loop speed motion control with pneumatic actuator. As in position systems, a state regulator is used for the outer control loop. For position control, the state variables are position x, velocity $\dot{x}$, and acceleration $\ddot{x}$. For speed motion control, the state variables are velocity $\dot{x}$, acceleration $\ddot{x}$, and jerk $\dddot{x}$.

An actuator with speed motion control can operate in two working modes:

1. Motion mode, in which the actuator has a reciprocating motion with the necessary constant velocity (predetermined velocity).

2. Positioning mode, when the actuator piston, after the last movement with constant velocity, should be held in the stop position. Usually, this position point is close to one of the two end actuator positions.

The motion mode consists of the following stages: start-up period (acceleration range), movement at a predetermined velocity, and shutdown time (deceleration range). Because the movement is of the reciprocating motion type, the deceleration and acceleration stages are connected in the reverse motion step.

Estimation of the actuator mechanical parameters is performed for the following given parameters: the predetermined constant velocity $(\dot{x}_C)$, the moving mass (m), the value of the movement (L_C) with constant velocity, and the supply pressure (P_S). Sometimes, the maximum acceleration/deceleration value $(\ddot{x}_R)$, or the time of the reverse motion step (t_R), is specified.

A better velocity profile brings the load to constant acceleration more gradually than an exponential curve or a sinusoidal ramp. However, the best curve is called a jerk-limited profile or an S-curve profile. It starts the load acceleration with a constant jerk $(\dddot{x})$, accelerates at a linearly increasing rate, and moves at a constant acceleration for a time. However, the S-curves take longer to reach command velocity; and the more the jerk is limited, the less efficient is the profile.

In the pneumatic cylinder, quick acceleration is performed by connecting the supply line to the first working chamber and the exhaust line to the second working chamber. For effective deceleration or reversing move, these chambers are connected in the opposite direction; that is, the first chamber is connected to the exhaust line and the second chamber is connected to the supply pressure. In general, the nonlinear differential equations (Equations 6.6) describe the dynamic behavior of this actuator, and the mechanical parameters of the actuator (piston effective area and effective areas of the control valve) are usually estimated with the assumption that the dimensionless inertial load W is in the range between 1 and 3, and the saturated value of the velocity actuator is $\dot{x}_{CM} = (1.3-1.5)\cdot\dot{x}_C$. Taking into account that for this application the rodless actuator construction is the preferred option (that is, $A_1 = A_2 = A_C$), the above-mentioned parameters can be estimated by the following formulas:

$$A_C \approx \frac{1.4 \cdot \dot{x}_C^2 \cdot m}{P_S \cdot L_C \cdot \Omega^2 \cdot \varphi_*^2} \tag{6.55}$$

$$A_{VM}^+ = \frac{2 \cdot A_C}{K_*} \cdot \sqrt{\frac{A_C \cdot P_S \cdot L_S}{m}} \tag{6.56}$$

In such actuators, the stroke of the shutdown stage (L_R) is a very important parameter because the minimum value of the actuator movement is $L_S = L_C + 2\cdot L_R$. The estimation of this parameter is performed by considering the differential equations that describe the dynamic behavior of the deceleration process. These equations have the following form:

$$\begin{cases} m\cdot\ddot{x} + b_V\cdot\dot{x} + F_F + F_L = (P_1 - P_2)\cdot A_C \\[2mm] \dot{P}_1 = \frac{1}{V_{01} + A_C\cdot(0.5\cdot L_S + x)}\cdot\left[-A_{VM}^-\cdot P_1\cdot K_*\cdot\varphi\left(\frac{P_A}{P_1}\right) - P_1\cdot A_C\cdot\dot{x}\right] \\[2mm] \dot{P}_2 = \frac{1}{V_{02} + A_C\cdot(0.5\cdot L_S - x)}\cdot\left[P_2\cdot A_C\cdot\dot{x} + A_{VM}^+\cdot P_S\cdot K_*\cdot\varphi\left(\frac{P_2}{P_S}\right)\right] \end{cases} \tag{6.57}$$

In this case, before the beginning of the slowdown process, the first actuator chamber is connected to the supply pressure, the second chamber is connected to the exhaust line, and the actuator moves with a predetermined constant velocity. The initial conditions for the deceleration stage are $x = 0.5 \cdot L_C$, $\dot{x} = \dot{x}_C$, $P_1 = P_{1C}$, and $P_2 = P_{2C}$, where P_{1C} and P_{2C} are the pressures in the first and second working chamber, respectively, which can be estimated as:

$$P_{1C} = \frac{(2 \cdot A_{VM}^+ \cdot \varphi_* \cdot K_*)^2 \cdot P_S}{\dot{x}_C^2 \cdot A_C^2 + (2 \cdot A_{VM}^+ \cdot \varphi_* \cdot K_*)^2} \tag{6.58}$$

$$P_{2C} = P_{1C} - \frac{b_V \cdot \dot{x}_C + F_F}{A_C} \tag{6.59}$$

The shutdown stroke (L_R) can be approximately estimated from:

$$L_R = \frac{2 \cdot m \cdot x_C^2}{(P_S - P_A) \cdot A_C} \tag{6.60}$$

and the shutdown time (t_{SD}) can be found from the following equation:

$$t_{SD} = \frac{2 \cdot L_R}{\dot{x}_C} \tag{6.61}$$

As mentioned above for the pneumatic actuator with velocity control mode, the control valve can be operated in different modes. However, the Bang-Bang control mode is preferred because, in this case, the actuator's dynamic behavior does not depend on the changing of the external load and the internal perturbations.

In these actuating systems, the sliding surface of the velocity control mode does not include the jerk signal; its value is defined according to:

$$S_V = (\dot{x}_{CS} - \dot{x}) - K_{ABV} \cdot \ddot{x} \tag{6.62}$$

Here, the velocity set point ($\dot{x}_{CS}$) is determined as follows:

$$\begin{cases} \dot{x}_{CS} = \dot{x}_C, & if \begin{cases} x \le x_{RS1} \ and \ \dot{x} \ge 0 \\ x \le x_{RS2} \ and \ \dot{x} < 0 \end{cases} \\ \\ \dot{x}_{CS} = -\dot{x}_C, & if \begin{cases} x > x_{RS1} \ and \ \dot{x} \ge 0 \\ x > x_{RS2} \ and \ \dot{x} < 0 \end{cases} \end{cases} \tag{6.63}$$

where x_{RS1} and x_{RS2} are the coordinates of the two reverse set points ($x_{RS1} > x_{RS2}$) and K_{ABV} is the acceleration gain.

In velocity control mode, the regular Bang-Bang algorithm is usually used. Then the control signal (U_{CBV}) and opening coefficients (β_i^+ and β_i^-) are described in the following form:

$$U_{CBV} = U_{CS} \cdot sign(S_V)$$

$$\beta_i^+ = 1 - \beta_i^-$$

$$\beta_1^+ = \beta_2^- = 1 \text{ , if } S_V \geq 0 \tag{6.64}$$

$$\beta_1^+ = \beta_2^- = 0 \text{ , if } S_V < 0$$

In positioning mode, the Bang-Bang null algorithm should be used; and in this case, the sliding surface (S_P), control signal (U_{CBP}), and opening coefficients (β_i^+ and β_i^-) are defined as:

$$S_P = (x_{PS} - x) - K_{VBP} \cdot \dot{x} - K_{ABP} \cdot \ddot{x}$$

$$U_{CBP} = U_{CS} \cdot sign(S_P)$$

$$\left. \begin{matrix} \beta_1^+ = \beta_2^- = 1 \\ \beta_1^- = \beta_2^+ = 0 \end{matrix} \right\} \text{ if } S_P \geq 0 \text{ and } (x_{PS} - x) \geq \Delta_{RP} \tag{6.65}$$

$$\left. \begin{matrix} \beta_1^+ = \beta_2^- = 0 \\ \beta_1^- = \beta_2^+ = 1 \end{matrix} \right\} \text{ if } S_P < 0 \text{ and } (x_{PS} - x) < -\Delta_{RP} \text{ ,}$$

$$\beta_i^+ = \beta_i^- = 0 \text{ if } \left| x_{PS} - x \right| \leq \Delta_{RP} \text{ ,}$$

where x_{PS} is the coordinate of the stop point, K_{VBP} is the velocity gain, K_{ABP} is the acceleration gain, and Δ_{RP} is the admissible position error.

In positioning mode, the control parameters (K_{VBP} and K_{ABP}) can be approximately estimated using Equation 6.51 and Equation 6.52. The final adjustment is carried out via computer simulation (Chapter 6.2.4).

In velocity control mode, the parameter of the control system (the acceleration gain K_{ABV}) can be estimated for the desired value of the delay time

(t_L), which includes two major components: (1) sampling period and (2) switching time of the control valve. In this case, substituting Equations 6.65 into the differential Equations 6.6 and carrying out linearization by the describing function method of the nonlinear members, the formulas for the frequency (f_{SE}) and amplitude ($\Delta \dot{x}_{SE}$) of the self-excited oscillation can be obtained:

$$f_{SE} = \frac{1}{2 \cdot \pi} \cdot \sqrt{\frac{B_{V1} \cdot t_L + B_{V2} + B_{V1} \cdot K_{ABV}}{m \cdot t_L - (B_{V2} \cdot t_L + m) \cdot K_{ABV}}} \tag{6.66}$$

$$\Delta \dot{x}_{SE} = \frac{B_{V3}}{4 \cdot f_{SE}^2 \cdot \pi^3 \cdot (B_{V2} \cdot t_L + m) + B_{V1} \cdot \pi} \tag{6.67}$$

Using Equation 6.66 and Equation 6.67, the acceleration gain can be approximated if the value of the velocity ripple ($\Delta \dot{x}_A$) is specified. Then:

$$K_{ABV} \approx$$

$$\frac{m \cdot t_L \cdot (B_{V3} - \pi \cdot B_{V1} \cdot \Delta \dot{x}_A) - \pi \cdot \Delta \dot{x}_A \cdot (B_{V1} \cdot t_L + B_{V2}) \cdot (B_{V2} \cdot t_L + m)}{B_{V3} \cdot (B_{V2} \cdot t_L + m)} \tag{6.68}$$

In Equation 6.66 through Equation 6.68, the coefficients B_{V1}, B_{V2}, and B_{V3} are determined as:

$$B_{V1} = \frac{4 \cdot K_* \cdot A_V^+ \cdot \varphi_* \cdot b_V + 2 \cdot P_S \cdot A_C^2}{A_C \cdot (x_0 + 0.5 \cdot L_S)}$$

$$B_{V2} = \frac{K_* \cdot A_V^+ \cdot \varphi_* \cdot m}{A_C \cdot (x_0 + 0.5 \cdot L_S)} + b_V \tag{6.69}$$

$$B_{V3} = \frac{8 \cdot K_* \cdot A_V^+ \cdot \varphi_* \cdot P_S}{x_0 + 0.5 \cdot L_S}$$

In the final step of the estimation process, the actuator parameters are adjusted by computer simulation of the actuator dynamic behavior.

Example 6.7. *Define the parameters of the linear pneumatic closed-loop actuator that should be moved with constant velocity. The actuator has the following performance specifications:*

- *Moving mass is m = 30 kg.*
- *Value of movement with constant velocity is L_C = 0.5 m.*
- *Value of the constant velocity is $\dot{x}_C$ = 0.5 m/s.*
- *Supply pressure is P_S = 0.6 MPa.*

Using Equation 6.55, the minimum value of the piston effective area is $A_C \approx 7.305 \cdot 10^{-4}$ m^2, and the nearest standard rodless pneumatic cylinder has a piston diameter of 0.04 m (its effective area is $A_C = A_1 = A_2 \approx 1.256 \cdot 10^{-3}$ m^2). According to Equation 6.60, the shutdown stroke is $L_R \approx 0.024$ m; then the minimum value of the cylinder stroke is $L_S = L_C + 2 \cdot L_R = 0.548$ m. Finally, the rodless pneumatic cylinder with a piston diameter of 0.04 m, a working stroke of 0.56 m, and length of the inactive volume of $x_0 = 0.02$ m is used (in this case, the viscous friction coefficient is

$$b_V \approx 35 \frac{N \cdot s}{m}$$

and the dynamic friction force is $F_F \approx 40$ N).

Using Equations 6.56, the effective area of the control valve is $A_V^+ \approx 1.166 \cdot 10^{-5}$ m^2. In this case, the solenoid valve with $\Omega_M = 1$ and effective areas of $A_{VM}^+ = A_{VM}^- = 1.126 \cdot 10^{-5}$ m^2 (maximal standard nominal flow rate is about 750 l/min) can be used.

For steady-state motion with constant velocity $\dot{x}_C = 0.5$ m/s, the pressure in the working chambers is estimated from Equation 6.58 and Equation 6.59; then $P_{1C} = 0.5968$ MPa and $P_{2C} = 0.559$ MPa.

To estimate the acceleration gain using Equations 6.69, the following coefficients can be determined:

$$B_{V1} \approx 5.76 \cdot 10^3 \ kg/s^2$$

$$B_{V2} \approx 226.9 \ kg/s$$

$$B_{V3} \approx 3.86 \cdot 10^4 \ kg \cdot m/s^3$$

Then, for the velocity ripple $\Delta \dot{x}_A = 0.01 \cdot \dot{x}_C = 0.005$ m/s and for the delay time $t_L = 0.025$ s, the acceleration gain is $K_{ABV} \approx 0.021$ s.

Figure 6.22 shows the velocity curve of the reciprocal motion of this actuator, which only has velocity feedback ($K_{ABV} = 0$). Even for a very short response time of the control valve (in this case, the delay time is $t_L = 0.001$ s), the velocity ripple is quite large (ripple amplitude is about $\Delta \dot{x}_C = 0.025$ m/s).

As stated previously, using the acceleration feedback allows reaching the acceptable dynamic behavior of the pneumatic actuator with velocity control. The velocity curve of such an actuator is shown in Figure 6.23, where the acceleration gain is $K_{ABV} = 0.037$ s. This value was obtained after an adjustment by computer simulation. In this case, the ripple amplitude is less than 0.004 m/s, and the frequency of the self-excited oscillation is about 28 Hz.

The phase-plane trajectory shown in Figure 6.24 illustrates the actuator dynamic behavior. From this illustration it can be seen that the shutdown stroke is $L_R \approx 0.02$ m.

Analysis of the obtained results demonstrates the acceptable accuracy of the parameter estimation of the linear pneumatic actuator with velocity control.

One can see from Equations 6.66 through 6.68 that actuator behavior and the value of the velocity ripple depend on the value of the delay time (t_L).

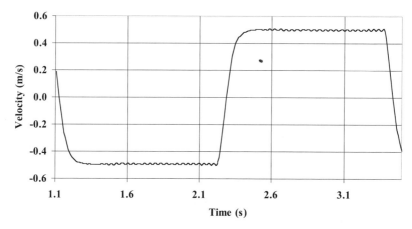

FIGURE 6.22

Velocity curve of an actuator with velocity and acceleration feedback.

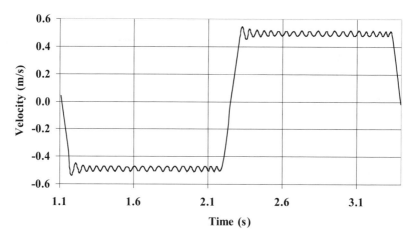

FIGURE 6.23

Velocity curve of an actuator with velocity feedback only.

From this point of view, using 2/2-way and 3/2-way control valves is the preferable solution because, in this case, the minimum delay time can be achieved. A schematic diagram of such actuator is shown in Figure 6.25. Here, the two control valves (1) and (2) are used only for the positioning mode when the actuator piston should be held in the stop position (these valves close the inlet ports of the cylinder). Two control valves (3) and (4) are used for the motion mode, where the actuator moves with the desired velocity (valves [1] and [2] are opened).

A very important issue in pneumatic actuating systems with velocity control is reaching low and extremely low speeds, which is usually less than 0.02 m/s. In a low-speed pneumatic cylinder, the trade-off between the driving force is generated by air pressure, and the seal friction must be considered

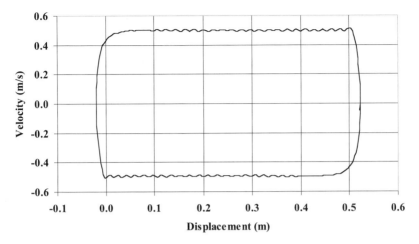

FIGURE 6.24
Phase-plane trajectory of the actuator with velocity and acceleration feedback.

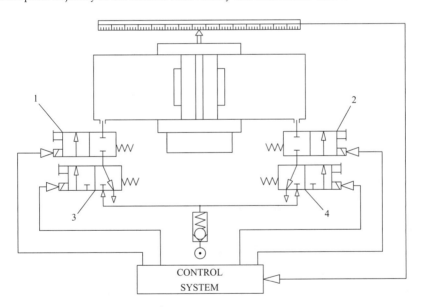

FIGURE 6.25
Velocity control actuator with 2/2-way and 3/2-way solenoid valves.

to avoid stick-slip motion. Friction forces act at contacting surfaces between two mechanical elements. There is a sudden change between static and dynamic friction, a phenomenon called stiction (Stiction is a combination of stick and friction. Combining these two words gives "stiction." In general, stiction is represented as the force necessary to start a body in motion.) Basically, friction properties are unclear because they depend on a number of factors, such as lubricating condition, operating condition, interface temperature, manufacturing irregularities, and others. If nonlinear friction acts in a servo actuator, the so-called stick-slip motion, in which the actuator moves intermittently, occurs at low speed. This problem can be overcome using a pneumatic cylinder with a relatively large piston diameter and low friction force (see Chapter 2). In practice, for such an application, the pneumatic cylinders with piston diameters from 40 to 120 mm are used.

For low-speed applications (constant velocity is less than 0.02 m/s), the mechanical parameters are estimated using the assumption that the dimensionless parameter W is 0.05 to 0.1. In this case, the effective area of the control valve can be estimated as:

$$A_{VM}^+ = \frac{0.05 \cdot A_C}{K_*} \cdot \sqrt{\frac{A_C \cdot P_S \cdot L_S}{m}} \qquad (6.70)$$

where the minimum piston effective area can be determined from:

$$A_C = \frac{4 \cdot K_*^2 \cdot m \cdot \alpha_{PV}}{P_S \cdot L_S} \qquad (6.71)$$

Here, the dimensionless coefficient α_{PV} is defined by the empirical formula:

$$\alpha_{PV} = 0.00525 \cdot \frac{\dot{x}_C}{\Omega} \qquad (6.72)$$

Example 6.8. *Define the parameters of the linear pneumatic closed-loop actuator that should be moved with low constant speed. The actuator has the following performance specifications:*

- *Moving mass is $m = 40$ kg.*
- *Value of the movement with constant velocity is $L_C = 0.6$ m.*
- *Value of the constant velocity is $\dot{x}_C = 0.005$ m/s.*
- *Supply pressure is $P_S = 0.6$ MPa.*

Assuming that the dimensionless parameter $\Omega = 1$, and using Equation 6.72 and Equation 6.71, the piston effective area is $A_C \approx 6.797 \cdot 10^{-3} m^2$ (piston diameter is 93 mm), the nearest standard rodless pneumatic cylinder has the piston diameter of 100 mm (its effective area is $A_C = 7.85 \cdot 10^{-3} m^2$). In this application, the pneumatic cylinder with flexible chambers (see Figure 2.2) has been utilized. The inside cylinder sleeve diameter is 100 mm, the rod diameter is 20 mm, and the stroke is 650 mm. The chambers are made of polyethylene 0.3 mm thick. The space between the outer surface of the chambers and the inner surfaces of the cylinder cavities are filled with lithium soap grease. This cylinder has the following friction characteristics: the static coulomb friction force is $F_S \approx 0.8$ N, the dynamic coulomb friction force is $F_D \approx 0.6$ N, and the viscous friction coefficient is $b_V \approx 450$ N·s/m.

Using Equations 6.70, the effective area of the control valve is $A_V^+ \approx 4.3 \cdot 10^{-6} m^2$. In practice, the one-stage jet-pipe control valve with $\Omega_M = 1$, effective areas of $A_{VM}^+ = A_{VM}^- = 4 \cdot 10^{-6} m^2$, and a response time of 0.004 s has been used.

To estimate the acceleration gain, the coefficients B_{V1}, B_{V2}, and B_{V3} are estimated using Equations 6.69; then,

$$B_{V1} \approx 2.7 \cdot 10^4 \frac{kg}{s^2}$$

$$B_{V2} \approx 411 \frac{kg}{s}$$

and

$$B_{V3} \approx 1.1 \cdot 10^4 \frac{kg \cdot m}{s^3}$$

Then, for the velocity ripple $\Delta \dot{x}_A = 0.02 \cdot \dot{x}_C = 1 \cdot 10^{-4}$ m/s and for the delay time $t_L = 0.005$ s (the sampling period is 0.001 s), the acceleration gain is $K_{ABV} \approx 0.005$ s.

The velocity curves shown in Figure 6.26 were obtained by computer simulation and experimental examination. In these cases, the acceleration gain was $K_{ABV} \approx 0.013$ s, which is obtained after adjustment. This figure illustrates that the actuator motion has a stable nature with an acceptable value of the velocity ripple.

It is important to note that the delay time (t_L) plays a major role in actuator dynamic behavior. Therefore, in this application the use of control valves with short response times is extremely important.

6.4 Pneumatic Systems for Acting Force Control

Open-loop systems have limits when trying to increase the capability of force control equipment. For actuators to have the capability to perform more general tasks of force control, there is a need for active control. This leads to the need for active feedback of measurements to modulate the control of force on the actuator output link.

In automatic equipment, various active force control methods are known, including explicit force control, stiffness control, virtual model control, impedance control, and hybrid position/force control.

It is very difficult to control force and position simultaneously, and in pneumatic actuating systems, the hybrid position/force control is used most often. In this case, the actuator has two control loops: position and force. The environment dictates natural constraints (such as being in contact with

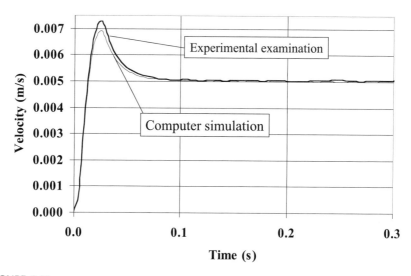

FIGURE 6.26
Velocity response of an actuator with low-speed control mode.

a surface) where only force control can be used. Similarly, position control is used in directions where there are no constraints and the actuator can move freely. Here, the accuracy and dynamic behavior in the position mode are sacrificed due to force demands.

Using pneumatic actuators to implement position/force control has major benefits. The inherent low stiffness of the pneumatic system and direct drive capabilities enable smooth compliant motion, which is difficult to obtain from the conventional geared electric motor systems.

Friction force, intermittent environment contact (impacts), transmission dynamics, and control valve saturation are limiting factors that should be overcome in order to achieve the high-level desired force.

Every effort should be made to minimize friction, both in the load and by selecting a pneumatic actuator with low breakout pressure. From this standpoint, the use of ultralow friction pneumatic cylinders with flexible chambers (Figure 2.2, Figure 2.6, and Figure 2.8) or pneumatic cylinders with anti-friction materials — for example, a cylinder that consists of a graphitized carbon piston and borosilicate glass cylinder sleeve with a precision, fire-polished bore (Airpot Pneumatic Actuators) — allows for achieving acceptable performance.

In force control pneumatic systems, either proportional or servo valve (Figure 1.15) or solenoid control valves (Figure 6.27) can be used. Control valves are the most important and influential components in this application. Their small response times and the short length of the pneumatic lines between the actuator and valves play a prime role in the dynamic behavior of the actuator. From this point of view, the 2/2-way control valves have apparent advantages, because they have a very short response time and also

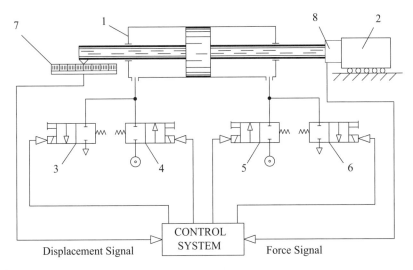

FIGURE 6.27
Acting force control actuator with 2/2-way solenoid valves.

the ability to mount on the inlet/outlet ports of the actuator. In this case, a PWM or Bang-Bang controller can be used; a recent investigation has shown that the Bang-Bang controller with dead zone (null algorithm) has excellent performance, is easy to program, and requires little computer power.

The value of the actuator acting force can be obtained using two major principles: (1) measurement by force transducer and (2) using the indirect method of measurement. Because force is defined as a restriction to movement, position can be used to determine the force output but the potential for mathematical error would be too high. Strain gages are the preferred measurement device. However, in this case, when a strain gage is used in a pressure transducer, the acting force can be defined by some mathematical calculations with actuator effective areas. Here again, the potential for mathematical and scaling errors is high because the friction force in a cylinder remain outside the control loop. For this reason, the most common means of measuring force output is with a load cell mounted between the actuator and the load.

A force loop is similar to a positional or velocity loop, with the exception that a feedback device measuring the force generated by the cylinder is used. Typically, force loops are applied in the tensile testing area and engineers select a high-end controller to interface with the system. Such controllers use a positional feedback device and a force feedback device for the closed-loop control. Typically, the cylinder selected has equal piston area (double rod or rodless) and the output link of the cylinder is connected to the positional feedback device.

Figure 6.27 provides a schematic diagram of the pneumatic actuator with position/force control. This system consists of the cylinder (1), load (2), four

2/2-way solenoid valves (3, 4, 5, and 6), position transducer (7), force transducer (8), and control system. The force transducer is used to force feedback, and the position transducer is utilized only for motion mode, when the load is moved to the desired position, where the force control mode begins to operate. Most often in the motion control mode, the actuator moves with predetermined constant velocity; after achieving a specific position, the actuator begins to move with low velocity, in which the contact with a processing surface is performed. After that, the force control mode begins to operate. Such operating conditions, for example, are inherent in the grinding, polishing, and lapping processes, and also in testing equipment.

The dynamic behavior of the actuator in motion control mode is described by the differential Equations 6.6, where the Bang-Bang controller is used. In this case, the sliding surfaces and control signal can be defined as:

$$S_V = (\dot{x}_{C1} - \dot{x}) - K_{A1} \cdot \ddot{x} \quad \text{if} \ \ x \le x_1$$

$$S_V = (\dot{x}_{C2} - \dot{x}) - K_{A1} \cdot \ddot{x} \quad \text{if} \ \ x_1 < x \le x_2 \qquad (6.73)$$

$$U_C = U_{CS} \cdot sign(S_V)$$

where $\dot{x}_{C1}$ is the predetermined constant velocity, $\dot{x}_{C2}$ is the low constant velocity, x_1 is the coordinate of the switching to the low velocity motion, x_2 is the coordinate of the contact point with a processing surface, and K_{A1} is the acceleration gain.

In force control applications, the motion mode is the secondary operation mode; therefore, the values of $\dot{x}_{C1}$ and $\dot{x}_{C2}$ are defined after determining the actuator mechanical parameters for the force control condition, which is the primary working regime. The acceleration gain K_{A1} can be estimated from Equation 6.68.

The low-friction cylinder minimizes the stiction effect and enables accurate position, velocity, and force control. The paired control valves in each working pneumatic chamber regulate its pressure. This enables the pressure difference across the cylinder to be specified by software changes alone, and also allows the individual chamber pressure to be regulated by the valves themselves.

In the position/force control mode, the control system also commands the beginning and the end of the force feedback loop, based on actuator collision detection during operation. Sometimes in these actuators, the accelerometer mounted on the output link is also used. In this case, this device provides a continuous output signal that does not depend on the update speed of the actuator controller. This means that the accelerometer signal can be read every millisecond or more often, which results in the direct reading of inertial effects induced by moving over the part. However, such application is used only in specific cases where the inertial load plays a significant role.

In general, the force control mode operates in the stationary condition where the piston is in a contact point position; that is, $x = x_2$, $\dot{x} = 0$ and $\ddot{x} = 0$. Then the pressure dynamic behavior in the actuator working chambers is described by the following equations (the cylinder has the equal piston areas, $A_C = A_1 = A_2$):

$$
\begin{cases}
F_A = A_C \cdot (P_1 - P_2) - F_F - F_L \\[2ex]
\dot{P}_1 = \dfrac{K_*}{V_{01} + A_C \cdot (0.5 \cdot L_S + x_2)} \cdot \left[A_{V1}^+ \cdot P_S \cdot \varphi\left(\dfrac{P_1}{P_S}\right) - A_{V1}^- \cdot P_1 \cdot \varphi\left(\dfrac{P_A}{P_1}\right) \right] \\[3ex]
\dot{P}_2 = \dfrac{K_*}{V_{02} + A_C \cdot (0.5 \cdot L_S - x_2)} \cdot \left[A_{V2}^+ \cdot P_S \cdot \varphi\left(\dfrac{P_2}{P_S}\right) - A_{V2}^- \cdot P_2 \cdot \varphi\left(\dfrac{P_A}{P_2}\right) \right]
\end{cases}
\qquad (6.74)
$$

where F_A is the active actuator force, and the effective areas of the control valves are $A_{Vi}^+ = A_{VM}^+ \cdot \beta_i^+$ and $A_{Vi}^- = A_{VM}^- \cdot \beta_i^-$.

The values of the opening coefficients (β_i^+ and β_i^-) depend on the type of control valve and on the valve operating mode. For example, if the control valve is a proportional or servo valve with zero-lap design, and operates in continuous mode, the opening coefficients are described by Equations 3.9. For the PWM operating regime, the opening coefficients are usually obtained from Equations 3.14. As stated previously, the Bang-Bang controller has excellent performance in the force control application; in this case, the opening coefficients for the regular algorithm are:

$$
\beta_i^+ = 1 - \beta_i^- ,
$$

$$
\beta_1^+ = \beta_2^- = 1 , \text{ if } S_F \geq 0 \qquad (6.75)
$$

$$
\beta_1^+ = \beta_2^- = 0 , \text{ if } S_F < 0
$$

and for the null algorithm, these coefficients are:

$$
\left.
\begin{aligned}
\beta_1^+ = \beta_2^- &= 1 \\
\beta_1^- = \beta_2^+ &= 0
\end{aligned}
\right\}
\text{ if } S_F \geq 0 \text{ and } (F_{DE} - F) \geq \Delta_F ,
$$

$$
\left.
\begin{aligned}
\beta_1^+ = \beta_2^- &= 0 \\
\beta_1^- = \beta_2^+ &= 1
\end{aligned}
\right\}
\text{ if } S_F < 0 \text{ and } (F_{DE} - F) < -\Delta_F , \qquad (6.76)
$$

$$
\beta_i^+ = \beta_i^- = 0 \text{ if } \left| F_{DE} - F \right| \leq \Delta_F ,
$$

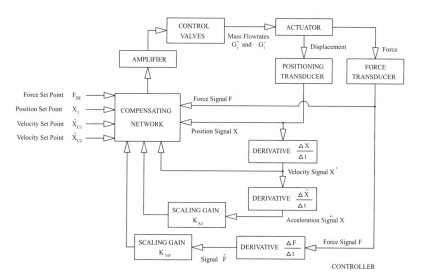

FIGURE 6.28
Block diagram of an actuator with position/force control.

where F_{DE} is the desired force value, F is the measured force (by force transducer), Δ_F is the admissible force error, and S_F is the sliding surface, which is usually determined as:

$$S_F = (F_{DE} - F) - K_{VF} \cdot \dot{F} \qquad (6.77)$$

The null algorithm allows reaching the force control process without oscillations about the desired value.

Figure 6.28 presents a block diagram of the actuator with position/force control.

As stated in Chapter 6.2.4, in the actuator with Bang-Bang control, the delay time of the control signal is a very important parameter. In general, this time includes three components: (1) sampling period, (2) switching time of the control valves, and (3) the time delay due to the connecting tubes between the control valves and the cylinder. However, because the solenoid control valves can be mounted on the input/output cylinder ports, the third component has negligible value. Quite often, the delay time is a compromise between cost and performance. The response with the short delay time is quite well behaved and not very sensitive to disturbances. In the first approximation, the maximum permissible value of the delay time can be estimated as:

$$t_L \approx \frac{\Delta_F \cdot t_{FR}}{(0.01 \div 0.02) \cdot F_M} \qquad (6.78)$$

where F_M is the maximum value of the control force, and t_{FR} is the rise time of the force changes (the time required to rise from 10 to 90% of the maximum force change value).

Analysis of the pressure behavior in the working chambers for the different types of combination of charging and discharging flows allows one to obtain the same results as in Chapter 3.6. That is the maximum control accuracy and actuator efficiency that could be obtained if the ratio between the maximum effective areas of the control valve exhaust line and supply channel is equal to 2 ($\Omega = A_{VM}^-/A_{VM}^+ = 2$).[103, 192] This result is very important for the estimation of the actuator mechanical parameters.

In force control actuating systems, the estimation of the effective area of the piston is the most important issue because this parameter defines the power ability of the actuator. In this design step, the maximum value of the control force (F_M) is taken into account. Then the actuator effective area can be estimated as:

$$A_C \approx \frac{1.5 \cdot F_M}{P_S - P_A} \tag{6.79}$$

It is clear from Equation 6.79 that the maximum pressure differential develops the acting force, which is 1.5 times greater than the required maximum force value (F_M).

A rough estimation of the control valve effective areas can be performed with the assumption that the pneumatic time constant of the actuator in the force control mode is less than or equal to the required value of the rise time (t_{FR}). Because the volumes of the actuator working chambers are constant, then

$$t_{FR} \geq \frac{V_M}{A_{VM}^+ \cdot K_*};$$

and from this inequality, the following can be obtained:

$$A_{VM}^+ \approx \frac{V_M}{t_{FR} \cdot K_*} \tag{6.80}$$

where V_M is the volume of the largest working chamber.

As stated previously, the ratio between the maximum effective area of the control valve exhaust line and the supply channel is $\Omega = A_{VM}^-/A_{VM}^+ = 2$; and then $A_{VM}^- = 2 \cdot A_{VM}^+$.

Pneumatic actuators with Bang-Bang force control mode are nonlinear systems with a variable structure, and they belong to a closed-loop self-excited system. In this system, periodic steady-state oscillations are called limit cycles, and their amplitude and frequency may be independent of the initial conditions on the dynamics. The control input may not be coherently persistent while the system is operating in its limit cycle. That is, the energy from the input can be disguised by the limit cycle energy because modifying the free response of a self-excited system is more complex than perturbing a stable linear system from its origin. In such cases, the results of the linear analysis are unavailable to estimate the global nature of these actuators. As a result, in practice, the empirical estimation of the feedback gain K_{VF} is usually used. In this case, its value can be obtained from the following equation:

$$K_{VF} = \frac{\varphi_* \cdot (V_1 + V_2) \cdot F_F \cdot \left(1 + \dfrac{F_F}{\Delta_F}\right)}{A_{VM}^+ \cdot (1 + \Omega) \cdot K_* \cdot A_C \cdot P_S} \tag{6.81}$$

where $V_1 = V_{01} + A_C \cdot (0.5 \cdot L_S + x_2)$ is the volume of the first working chamber and $V_2 = V_{02} + A_C \cdot (0.5 \cdot L_S - x_2)$ is the volume of the second working chamber.

Equation 6.81 is obtained by assuming that the delay time (t_L) and external force (F_L) are negligible. In this case, the Bang-Bang control algorithm is the regular type.

In practice, the use of computer simulation is the only way to study the influence of control valve characteristics and system delay time on the performance of a force control actuator. Therefore, in the final step of the estimation process, the actuator parameters are adjusted by computer simulation of the actuator dynamic behavior.

In general, the estimation process of the actuator parameters can be performed in the following sequence:

1. The mechanical parameters should be estimated using Equation 6.79 and Equation 6.80. In this stage, the maximum permissible value of the delay time is calculated using Equation 6.78.

2. In the second step, the force controller parameter (feedback gain K_{VF}) is estimated using Equation 6.81.

3. In the third step, for given motion time and working displacement, the velocity values of $\dot{x}_{C1}$ and $\dot{x}_{C2}$ are determined.

4. In the final step, the controller parameters of the motion working mode are estimated (see Chapter 6.3).

Example 6.9. *Define the parameters of the linear pneumatic closed-loop actuator that operates in position/force control mode. This actuator is the power part of the polishing tool for semiconductor specimens. The maximum value of the control force between the polishing head and pad should be $F_M = 1000\,N$. The external force (F_L) is negligible because the counterbalance mechanism is utilized. The actuator has the following performance specifications:*

- *Moving mass m = 35 kg.*
- *Distance between polishing head and pad in the load/unload position is $L_G = 0.25$ m.*
- *Maximum time of the position operation mode is $t_P = 2$ s.*
- *Maximum value of the control force is $F_M = 1000$ N.*
- *Admissible force error is $\Delta_F = 2$ N.*
- *Maximum rise time of the force change in force operation mode is $t_{FR} = 0.3$ s.*
- *Supply pressure is $P_S = 0.6$ MPa.*

Using Equation 6.79, the actuator effective area is $A_C \approx 4 \cdot 10^{-3} m^2$; then the rodless pneumatic cylinder with an outside piston diameter of 0.08 m (effective area is $A_C \approx 5.024 \cdot 10^{-3} m^2$) can be used. In such a low friction pneumatic cylinder, the static friction force is in the range between 3 and 6 N. In the first approximation, the value of $F_F = 5$ N can be considered.

Because the desired distance between polishing head and pad in the load/unload position is $L_G = 0.25$ m, the cylinder with stroke $L_S = 0.3$ m is considered. Then the coordinate of the load/unload position is $x_L = -0.15$ m and the coordinate of the contact point, where the force control mode begins to operate is $x_2 = 0.1$ m $x_2 = 0.1$m. In this case, the volumes of the working chambers are $V_1 = 1.356 \cdot 10^{-3} m^3$ and $V_2 = 0.352 \cdot 10^{-3} m^3$.

The desired maximum rise time in force operating mode is $t_{FR} = 0.3$ s and then, using Equation 6.80, the effective area of the control valve is $A_{VM}^{+} \approx 0.6 \cdot 10^{-5} m^2$. In this case, the standard 2/2-way solenoid valve with an effective area of $A_{VM}^{+} \approx 0.8 \cdot 10^{-5} m^2$ (maximum standard nominal flow rate is about 400 l/min) can be used. Because the parameter Ω of the control valve is $\Omega = 2$, the effective area of the exhaust line should be $A_{VM}^{-} \approx 1.6 \cdot 10^{-5} m^2$. For this implementation, two solenoid valves with $A_{VM}^{-} \approx 0.8 \cdot 10^{-5} m^2$ can be used in parallel.

According to Equation 6.78, the permissible value of the delay time (for the admissible force error of $\Delta_F = 2$ N) is $t_L \approx 0.03$ s. A standard 2/2-way solenoid valve with a standard nominal flow rate of 400 l/min has a 0.015- to 0.018-s switching time; that is, a maximum value for the actuator delay time of 0.02 s can be achieved. The value of the gain K_{VF} is estimated using Equation 6.81 and, in this case, its value is $K_{VF} \approx 1.4 \cdot 10^{-4}$s.

Because the maximum time of the position operating mode is $t_P = 2$ s and the distance between the load/unload position and contact point is $L_G = 0.25$ m, the average velocity is about 0.125 m/s. Taking into account that the contact process between the polishing head and pad should be performed with low velocity (which is $\dot{x}_{C1} \approx 0.02$ m/s) and the distance from the contact point where the actuator goes into low speed motion is about $0.015 \div 0.02$m ($x_1 \approx 0.08 \div 0.085$m), one can then consider that the low velocity value is $\dot{x}_{C1} \approx 0.25$ m/s. In this case, using Equation 6.68, the acceleration gain in the motion controller is $K_{A1} \approx 0.02$ s.

The velocity curve of the position operating mode is shown in Figure 6.29. Here, the acceleration gain is $K_{A1} = 0.03$ s, which is obtained after adjustment. This figure illustrates that the velocity ripple is about 10 to 15%, both in the motion with high

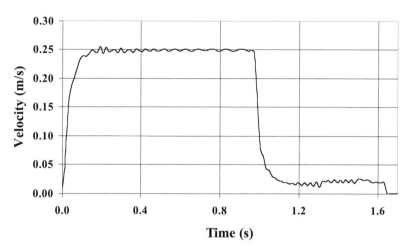

FIGURE 6.29
Velocity curve of the position mode for a Bang-Bang position/force control actuator.

and low velocity values. It is the result of a large delay time. For stable motion, this parameter should be $t_L \leq 0.01$ s. However, for this application, the positioning mode is the secondary operation mode, and provides only auxiliary motion of the polishing pad. Therefore, the motion with such a ripple is acceptable, and allows for the use of inexpensive standard solenoid control valves.

Figure 6.30 shows the active force response of an actuator that has the estimated parameters. The input signal is the step function for the desired force of 1000 N. Here, the delay time is $t_L = 0.02$ s. It is clearly seen that without force derivative feedback ($K_{VF} = 0$), the system has the limited cycle with an amplitude of ~25 N and a frequency of ~45 Hz (see Figure 6.30a). Using force derivative feedback ($K_{VF} = 0.0014$ s) allows for reaching an acceptable transient process a rise time of ~0.08 s and a settling time of ~0.22 s (see Figure 6.30b), which conforms to the specification.

The frequency response of a pneumatic actuator with force control mode, which has the parameters as estimated in Example 6.9, is illustrated in Figure 6.31 (in this case, the amplitude of the desired force signal is 10% of maximum value). It is clearly seen that the force control dynamic behavior of the actuator with $\Omega = 2$ has, in addition to maximum control accuracy and actuator efficiency, a better frequency response.

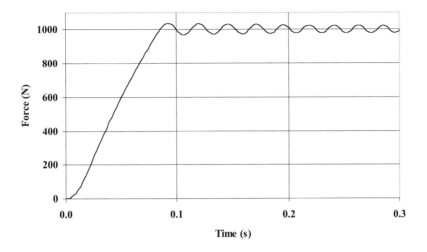

a. Without force derivative feedback

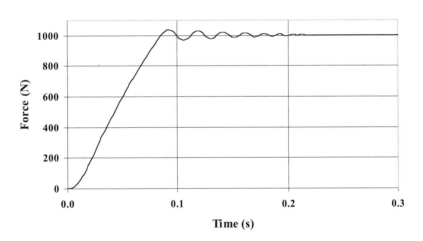

b. With force derivative feedback

FIGURE 6.30
Active force response of the actuator with force control mode.

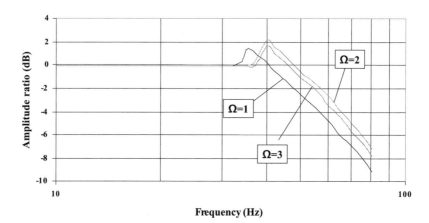

FIGURE 6.31

Frequency response of the actuator with force control mode.

Bibliography

1. Acarman, T., Hatipoglu, C., and Ozguner, U. (2001), "A Robust Nonlinear Controller Design for a Pneumatic Actuator," *Proceedings of the American Control Conference*, Vol. 6, pp. 4490—4495.
2. Ahn, K.-K., Pyo, S.-M., Yang, S.-Y., and Lee, B.-R. (2003), "Intelligent Control of Pneumatic Actuator Using LVQNN," *Science and Technology, Proceedings KORUS 2003, The 7th Korea-Russia International Symposium*, Vol. 1, pp. 260–266.
3. Al-Dakkan, K.A., Goldfarb, M., and Barth, E.J. (2003), "Energy Saving Control for Pneumatic Servo Systems," *Proceedings 2003 IEEE/ASME International Conference on Advanced Intelligent Mechatronics (AIM 2003)*, Vol. 1, pp. 284–289.
4. Al-Ibrahim, A.M. and Otis, D.R. (1992), "Transient Air Temperature and Pressure Measurements During the Charging and Discharging Processes of an Actuating Pneumatic Cylinder," *Proceedings of the 45th National Conference on Fluid Power*.
5. Andersen, B. (2001), "The Analysis and Design of Pneumatic Systems." Krieger Publishing, 302 p.
6. Anderson, R.T. and Li, P.Y. (2000), "Mathematical Modeling of a Two Spool Flow Control Servovalve Using a Pressure Control Pilot," *ASME Symposium on Modeling and Control Electrohydraulic Systems*, Orlando, FL, pp. 134–141.
7. Anglani, A., Gnoni, D., Grieco, A., and Pacella, M. (2002), "A CAD Environment for the Numerical Simulation of Servo Pneumatic Actuator Systems," *Proceedings of 7th International Workshop on Advanced Motion Control — AMC'02*, pp. 593–598.
8. Araki, K. (1986), "Frequency Response of a Pneumatic Valve Controlled Cylinder with an Overlap Four-Way Valve. I. Theoretical Analysis," *Journal of Fluid Control*, Vol. 17, No. 1, pp. 7–43.
9. Araki, K. (1987), "Frequency Response of a Pneumatic Valve Controlled Cylinder with an Overlap Four-Way Valve. II. Experimental Analysis," *Journal of Fluid Control*, Vol. 17, No. 2, pp. 33–50.
10. Araki, K., Chen, N., and Ishino, Y. (1995), "Characteristics of a Force-Balance Nozzle-Flapper Type Pneumatic Pressure Control Proportional Valve," *Journal Japanese Hydraulic, Pneumatic Society*, Vol. 26, No. 2, pp. 184–190.
11. Araki, K., Tanahashi, T., and Ehana, M. (1986), "Frequency Response of Pressure Control with an Electro-Pneumatic Proportional Valve," *Fluid Control and Measurement*, Vol. 1, pp. 39–44.
12. Armstrong-Helouvry, B. (1993), "Stick Slip and Control in Low Speed Motion," *IEEE Transactions on Automatic Control*, Vol. 38, No. 10, pp. 1483–1496.
13. Astrom, K.J. and Wittenmark, B. (1997), *Computer Controlled Systems, 3rd edition*, Prentice Hall, Englewood Cliffs, NJ.
14. Aziz, S. and Bone, G.M. (2000), "Automatic Turning of Pneumatic Servo Actuators," *Advanced Robotics*, Vol. 13, No. 6, pp. 563–576.

15. Backe, W. (1986), "Application of Servo-Pneumatic Drives for Flexible Mechanical Handling Techniques," *Robotics*, Vol. 2, No. 1, pp. 45–56.

16. Backe, W. and Ohligschlager, O. (1989), "A Model of Heat Transfer in Pneumatic Chambers," *Journal of Fluid Control*, Vol. 20, pp. 61–78.

17. Balandin, D. (2001), Optimal Protection from Impact, Shock and Vibration, T&F STM, 472 p.

18. Barth, E.J., Zhang, J., and Goldfarb, M. (2003), "Control Design for Relative Stability in a PWM-Controlled Pneumatic System," *Journal of Dynamic Systems, Measurement, and Control*, Vol. 124, No. 3, pp. 504–508.

19. Basso, R. (1993), "A Simplified Model Simulating the Dynamic Behavior of an Electro-Pneumatic Proportional Valve for Pressure Control," *International Journal of Modeling and Simulation*, Vol. 13, No. 2, pp. 77–81.

20. Belforte, G., Romiti, A., Ferraresi, C., and Raperelli, T. (1984), "Pneumatic Cylinder with Dynamic Seals," *Journal of Fluid Control*, Vol. 15, No. 4, pp. 42–50.

21. Belforte, G., D'Alfio, N., and Raparelli, T. (1989), "Experimental Analysis of Friction Force in Pneumatic Cylinders," *Journal of Fluid Control*, Vol. 20, No. 1, pp. 42–60.

22. Belforte, G. (2000), "New Developments and New Trends in Pneumatics," *Proceedings of the Sixth Triennial International Symposium on Fluid Control, Measurement and Visualization*, Sherbrooke, Quebec, Canada.

23. Belforte, G., Raparelli, T., Viktorov, V., Eula G., and Ivanov, A. (2000), "Theoretical and Experimental Investigations of an Opto-Pneumatic Detector," *Journal of Dynamic Systems, Measurement, and Control*, Vol. 122, No. 1, pp. 168–173.

24. Belgharbi, M., Thomasset, D., Scavarda, S., and Sesmat, S. (1999), "Analytical Model of the Flow Stage of a Pneumatic Servo-Distributor for Simulation and Nonlinear Control," *Sixth Scandinavian International Conference on Fluid Power*, SICFP'99, Tampere, Finland, pp. 847–860.

25. Ben-Dov, D. and Salcudean, S.E. (1995), "A Force-Controlled Pneumatic Actuator," *IEEE Transaction on Robotics and Automation*, Vol. 11, No. 6, pp. 906–911.

26. Bequette, B.W. (2002), *Process Control: Modeling, Design, and Simulation*, Prentice Hall PTR, Englewood Cliffs, NJ, 800 p.

27. Bigras, P., Wong, T., and Botez, R. (2001), "Pressure Tracking Control of a Double Restriction Pneumatic System," *IASTED International Conference on Control and Application*, pp. 273–278.

28. Bigras, P. and Khayati, K. (2002), "Nonlinear Observer for Pneumatic System with Non Negligible Connection Port Restriction," *American Control Conference ACC 2002*, pp. 3191–3195.

29. Blackburn, J.F., Reethof, G., and Shearer, J.L. (1960), *Fluid Power Control*, The Technology Press of MIT, Cambridge, MA, 710 p.

30. Blagojevic, V., Milosavljevic, C., Radovanovic, M., and Stojiljkovic, M. (2003), *Improvement of the Work of the Pneumatic Machine for Bending by Using the Digital Sliding Mode*, FACTA UNIVERSITATIS, Series: Mechanical Engineering, Vol. 1, No. 10, pp. 1347–1354.

31. Bobrow, J.E. and Jabbari, F. (1991), "Adaptive Pneumatic Force Actuation and Position Control," *Journal of Dynamic Systems, Measurement, and Control*, Vol. 113, pp. 267–272.

32. Bobrow, J.E. and McDonell, B.W. (1998), "Modeling, Identification, and Control of a Pneumatically Actuated, Force Controllable Robot," *IEEE Transactions on Robotics and Automation*, 14(10), pp. 732–741.

33. Bourdat, S., Richard, E., and Scavarda, S. (1991), "A Modified Linear Reduced Observer for a Pneumatic Servodrive," *Proceeding of the Third Bath International Fluid Power Workshop*, pp. 359–371.

34. Bouri, M., Thomasset, D., Richard, E., and Scavarda, S. (1994), "Non Linear Sliding Mode Control of an Electropneumatic Servodrive," *Proceeding of the Seventh Bath International Fluid Power Workshop*, pp. 201–220.

35. Bronstein, I.N. and Semendjajew, K.A., (2004), *Handbook of Mathematics, 15th edition*, Springer, Berlin, 973 p.

36. Brun, X., Belgharbi, M., Sesmat, S., Thomasset, D., and Scavarda, S. (1999), "Control of an Electropneumatic Actuator, Comparison between Some Linear and Nonlinear Control Law," *Journal of Systems and Control Engineering*, Vol. 213, No. 15, pp. 387–406.

37. Brun, X., Thomasset, D., and Sesmat, S. (1999), "Limited Energy Consumption in Positioning Control of an Electropneumatic Actuator," *Bath Workshop on Power Transmission & Motion Control*, pp. 199–211.

38. Brun, X., Thomasset, D., Bideaux, E., and Scavarda, S. (2000), "An Accurate Tracking Control of an Electropneumatic Actuator," *Proc. of 1st FPNI-PhD Symposium, Hamburg 2000*, pp. 215–226.

39. Bu, F. and Yao, B. (2000), "Performance Improvement of Proportional Directional Control Valves: Methods and Experiments," Orlando, FL, *Proceedings of the ASME Dynamic Systems and Control Division*, Vol. 1, pp. 297–304.

40. Caracciolo, R., Gallina, P., and Gasparetto, A. (1998), "Fuzzy Control of a Pneumatic Actuator," *Proceedings of Symposium on Robot Control*, France, Vol. 1, pp. 275–281.

41. Chiavervini, S. and Sciavicco, L. (1993), "The Parallel Approach to Force/Position Control of Robotic Manipulators," *IEEE Transactions on Robotics and Automation*, Vol. 9, No. 4, pp. 361–373.

42. Chillari, S., Guccione, S., and Muscato, G. (2001), " An Experimental Comparison between Several Pneumatic Position Control Methods," *Proceedings of the 40th IEEE Conference on Decision and Control, USA*, pp. 1168–1173.

43. Colombi, S. (1995), "Comparison of Different Control Strategies and Friction Compensation Algorithm in Position and Speed Controls," *Workshop on Motion Control*, Munich, pp. 173–180.

44. Daerden, F. and Lefeber, D. (2001), "The Concept and Design of Plated Pneumatic Artificial Muscles," *International Journal of Fluid Power*, Vol. 2, No. 3, pp. 41–50.

45. Daerden, F., Lefeber, D., Verrelst, B., and Van Ham, R. (2001), "Pleated Pneumatic Artificial Muscles: Actuators for Automatisation and Robotics," *IEEE/ASME International Conference on Advanced Intelligent Mechatronics*, Italy, pp. 738–743.

46. D'Amore, M. and Pellegrinetti, G. (2001), "Dissecting High-Performance Electrohydraulic Valves," *Machine Design*, April 5, pp. 84–88.

47. Davis, S.T. and Caldwell, D.G. (2001), "The Bio-Mimetic Design of a Robot Primate Using Pneumatic Muscle Actuators," *Proceedings of the 4th International Conference on Climbing and Walking Robots, CLAWAR 2001*, Germany, pp. 197–204.

48. DeRose, D. (2003), "Proportional and Servo Valve Technology," *Fluid Power Journal*, March/April, pp. 8–12.

49. Dmitriev, V.N. and Gradezky, V.G. (1973), *Fundamentals of the Pneumatic Automation*, Moscow, Machinostroenie, 360 p. (in Russian).

50. Dong, S., Du, X.H., Bouchilloux, P., and Uchino, K. (2002), "Piezoelectric Ring-Morph Actuators for Valve Application," *Journal of Electroceramics*, Vol. 8, No. 2, pp. 155–161.

51. Drakunov, S., Hanchin, D., Su, W.C., and Ozguner, U. (1997), "Nonlinear Control of a Rodless Pneumatic Servoactuator, or Sliding Modes versus Coulomb Friction," *Automatica*, Vol. 33, No. 7, pp. 1401–1408.

52. Dunbar, W.B., De Callafon, R.A., and Kosmatka, J.B. (1999), "Coulomb and Viscous Fault Detection with Application to a Pneumatic Actuator," *Proc. IEEE/ASME International Conference on Advanced Mechatronics, AIM '99*, Atlanta, GA, pp. 1239–1245.

53. Dunbar, W.B., De Callafon, R.A., and Kosmatka, J.B. (2001), "Coulomb and Viscous Friction Fault Detection with Application to a Pneumatic Actuator" *Proceedings of the 2001 International Conference on Advanced Intelligent Mechatronics*, Como, Italy, pp. 22–31.

54. Dupont, P.E. and Dunlap, E.P. (1995), "Friction Modeling and Proportional-Derivative Compensation at Very Low Velocities," *Journal of Dynamic Systems, Measurement, and Control*, Vol. 117, No. 1, pp. 8–14.

55. Eker, J., Hagander, P., and Arzen, K.-E. (2000), "A Feedback Scheduler for Real-Time Control Tasks," *Control Engineering Practice*, Vol. 8, No. 12, pp. 1369–1378.

56. Erlbacher, E.A. (1993), "A Discussion of Passive and Active Pneumatic Constant Force Devices," *Session Twenty of the International Robots & Vision Automation Conference*, Detroit, MI.

57. Esposito, A. (2003), *Fluid Power with Applications, 6th edition*, Prentice Hall, Englewood Cliffs, NJ, 656 p.

58. Ferretti, G., Magnani, G., and Rocco, P. (1997), "Towards the Implementation of Hybrid Force/Position Control in Industrial Robots," *IEEE Transactions on Robotics and Automation*, Vol. 13, No. 6, pp. 838–845.

59. Figliolini, G. and Sorli, M. (2000), "Open-Loop Force Control of a Three-Finger Gripper through PWM Modulated Pneumatic Digital Valves," *Journal of Robotics and Mechatronics*, Vol. 12, No. 4, pp. 480–493.

60. Fok, S.C. and Ong, E.K. (1999), "Position Control and Repeatability of a Pneumatic Rodless Cylinder System for Continuous Positioning," *Robotics and Computer Integrated Manufacturing*, Vol. 15, pp. 365–371.

61. Fraden, J. (1996), *AIP Handbook of Modern Sensors*, American Institute of Physics, New York, 264 p.

62. Friedland, B. (1996), *Advanced Control System Design*, Prentice Hall, Englewood Cliffs, NJ, 368 p.

63. Gamble, J.B. and Vaughan, N.D. (2000), "Comparison of Sliding Mode Control with State Feedback and PID Control Applied to a Proportional Solenoid Valve," Chicago, IL, *Proceedings of the ASME Dynamic Systems and Control Division*, Vol. 1, pp. 51–58.

64. Gaura, E., Steele, N., and Rider, R.J. (1999), "A Neural Network Approach for the Identification of Micromachined Accelerometers," *Proceedings of MSM'99*, San Juan, Puerto Rico, pp. 245–248.

65. Goodman, R.B. (1997), *A Primer on Pneumatic Valves and Control*, Krieger Publishing, Melbourne, FL, 95 p.

66. Gorinevsky, D.M., Formalsky, A.M., and Schneider, A.Y. (1997), *Force Control of Robotic Systems*, CRC Press, Boca Raton, FL.

67. Gorce, P. and Guihard, M. (1999), "Joint Impedance Pneumatic Control for Multilink Systems," *Journal of Dynamic Systems, Measurement, and Control*, Vol. 121, pp. 293–297.

68. Granosik, G. and Borenstein, J. (2004), "Minimizing Air Consumption of Pneumatic Actuators in Mobile Robots," *IEEE International Conference on Robotics and Automation*, New Orleans, LA, pp. 3634–3639.

69. Gross, D.C. and Rattan, K.S. (1998), "An Adaptive Multilayer Neural Network for Trajectory Control of a Pneumatic Cylinder," *IEEE International Conference on Systems, Man, and Cybernetics*, Vol. 2, pp. 1662–1667.

70. Guihard, M. and Gorce, P. (2001), "Tracking of High Acceleration Movements with a Pneumatic Impedance Controller," *Journal of Dynamic Systems, Measurement, and Control*, Vol. 123, No. 3, pp. 549–551.

71. Guihard, M. and Gorce, P. (2004), "Dynamic Control of a Large Scale of Pneumatic Multichain Systems," *Journal of Robotic Systems*, Vol. 21, No. 4, pp. 183–192.

72. Haessig, D.A. and Friedland, B. (1991), "On the Modeling and Simulation of Friction," *Journal of Dynamic Systems, Measurement, and Control*, Vol. 113, pp. 354–362.

73. Hamdan, M. and Gao, Z. (2000), "A Novel PID Controller for Pneumatic Proportional Valves with Hysteresis," *35th IEEE Industrial Application Society 2000. Annual Meeting and World Conference on Industrial Application of Electrical Energy, LAS'2000*, Rome, Italy, Oct. 8–12, 2000, Vol. 2, pp. 1198–1201.

74. Hamiti, K., Voda-Besancon, A., and Roux-Buisson, H. (1996), "Position Control of a Pneumatic Actuator under the Influence of Stiction," *Control Engineering Practice*, Vol. 4, No. 8, pp. 1079–1088.

75. Han, B.J., Kawashima, K., Fujuta, T., and Kagawa, T. (2001), "Flow Rate Characteristics Measurement of Pneumatic Valves by Pressure Response," *Fifth International Conference on Fluid Power Transmission and Control (ICEP 2001)*, April 2001, Hangzhou, China.

76. Hardwick, D.R. (1984), "Understanding Proportional Solenoids," *Hydraulics and Pneumatics*, Vol. 37, No. 8, pp. 58–60.

77. Henri, P.D., Hollerbach, J.M., and Nahvi, A. (1998), "An Analytical and Experimental Investigation of a Jet Pipe Controlled Electropneumatic Actuator," *IEEE Transactions on Robotics and Automation*, Vol. 14, No. 4, pp. 601–611.

78. Hertz, E.V. (1969), *Pneumatic Drives*, Machinostroenie, Moscow, 356 p. (in Russian).

79. Hertz, E.V. and Krejnin, G.V. (1975), *Analysis of the Pneumatic Drives*, Handbook, Machinostroenie, Moscow, 272 p. (in Russian).

80. Hildebrandt, A., Sawodny, O., Neumann, R., and Hartmann, A. (2002), "A Flatness Based Design for Tracking Control of Pneumatic Muscle Actuators," *Seventh International Conference on Control, Automation, Robotics and Vision (ICARCV'02)*, December, Singapore, pp. 1156–1161.

81. Hitchcox, A.L. (2002), "Sorting through Air Valve," *Hydraulics & Pneumatics*, April, pp. 20–32.

82. Hitchcox, A.L. (2003), "Proportional Valve Overcomes Limitations," *Hydraulics & Pneumatics*, July, pp. 22, 24, 26.

83. Imaizumi, T., Oyama, O., and Yoshimitsu, T. (2000), "Study of Pneumatic Servo System Employing Solenoid Valve Instead of Proportional Valve by Keeping the Solenoid Valve Plunger To Be Floating," *Proceedings of the Sixth Triennial International Symposium on Fluid Control, Measurement and Visualization*, Sherbrooke, Quebec, Canada, pp. 261–267.

84. Isidori, A. (1989), *Nonlinear Control Systems, 2nd edition*, Springer-Verlag, New York, 479 p.

85. Ivlev, V.I., Krejnin, G.V., and Krivts, I.L. (1985), "On Stabilizing Low Speed in a Pneumatic Motor," Soviet Machine Science (Academy of Sciences of the USSR), *Machinovedenie*, No. 4, 1985, Allerton Press, New York, pp. 34–39.

86. Ivlev, V.I., Krejnin, G.V., Krivts, I.L., and Lyuttsau, V.G. (1985), "Certain Promising Air-Cylinder Components Design," Soviet Machine Science (Academy of Sciences of the USSR), *Machinovedenie*, No. 1, 1985, Allerton Press, New York, pp. 47–50.

87. Jacobsen, S.C., Wood, J.E., Knutti, D.F., and Biggers, K.B. (1984), "The Utah/MIT Dextrous Hand: Work in Progress," *International Journal of Robotics Research*, Vol. 3, No. 4, pp. 21–50.

88. Jeong, Y., Lee, Y., Kim, K., Hong, Y.-S., and Park J.-O. (2001), "A 7 DOF Wearable Robotic Arm Using Pneumatic Actuators," *Proceedings of the 32nd International Symposium on Robotics (ISR)*, Seoul, South Korea.

89. Jolly, M.R. (2001), "Pneumatic Motion Control Using Magnetorheological Technology," *8th International Symposium on Smart Structures and Materials*, March 4–8, Newport Beach, CA, pp. 35–42.

90. Kagawa, T., Tokashiki, L.R., and Fujita, T. (2002), "Influence of Air Temperature Change on Equilibrium Velocity of Pneumatic Cylinder," *Journal of Dynamic Systems, Measurement, and Control*, Vol. 124, No. 2, pp. 336–341.

91. Kaitwanidvilai, S. and Parnichkun, M. (2005), "Force Control in a Pneumatic System Using Hybrid Adaptive Neuro-Fuzzy Model Reference Control," *Mechatronics*, Vol. 15, No. 1, pp. 23–41.

92. Karnopp, D. (1985), "Computer Simulation of Stick-Slip Friction in Mechanical Dynamic Systems," *Journal of Dynamic Systems, Measurement, and Control*, Vol. 107, pp. 100–103.

93. Karpenko, M. and Sepehri, N. (2004), "Design and Experimental Evaluation of a Nonlinear Position Controller for a Pneumatic Actuator with Friction," *Proceedings of the 2004 American Control Conference (AAC)*, USA, Vol. 6, pp. 5078–5083.

94. Kawakami, Y., Akao, J., Kawai, S., and Machiyama, T. (1988), "Some Considerations on the Dynamic Characteristics of Pneumatic Cylinder," *Journal of Fluid Control*, Vol. 19, No. 2, pp. 22–36.

95. Keller, H. and Isermann, R. (1993), "Model-Based Nonlinear Adaptive Control of a Pneumatic Actuator," *Control Engineering Practice*, Vol. 1, No. 3, pp. 505–511.

96. Khayati, K., Bigras, P., and Dessaint, L.-A. (2004), "A Robust Feedback Linearization Force Control of a Pneumatic Actuator," *International Conference on Systems, Man and Cybernetic*, IEEE SMC 2004, Netherlands, pp. 6113–6119.

97. Kimura, T., Hara, S., Fujita, T., and Kagawa, T. (1997), "Feedback Linearization for Pneumatic System with Static Friction," *Control Engineering Practice*, Vol. 5, No. 10, pp. 1385–1394.

98. Kosaki, T. and Sano, H. (2000), "An Analytical and Experimental Study of Chaotic Oscillation in a Pneumatic Cylinder," *Proc. of 1st FPNI-PhD Symposium*, Hamburg, 2000, pp. 303–310.

99. Krejnin, G.V., Krivts, I.L., Solnzewa, K.S., Frank, V., and Ulbricht, A. (1986), "Positional Translational Pneumatic Drive," Soviet Machine Science (Academy of Sciences of the USSR), *Machinovedenie*, No. 2, 1986, Allerton Press, New York, pp. 41–47.

100. Krejnin, G.V., Krivts, I.L., and Smelov, L.A. (1991), "Analog and Digital Observers for Position Pneumatic Actuators," *Devices and Control Systems*, No. 8, pp. 19–20. (in Russian).

101. Krejnin, G.V., Krivts, I.L., and Smelov, L.A. (1992), "Improved Positioning of Pneumatic Cylinder by Using Flexible Coupling between the Piston and Rod," *Journal of Mechanical Engineering Science*, Vol. 206, pp. 431–435.

102. Krejnin, G.V., Krivts, I.L., Vinnizky, E.Y., and Ivlev, V.I. (1993), *Hydraulic and Pneumatic Drives of The Manufacturing Robots and Automatic Manipulators*, Moscow, Machinostroenie, 301 p. (in Russian).

103. Krivts, I.L. (2004), "Optimization of Performance Characteristics of Electro-pneumatic (Two-Stage) Servo Valve," *Journal of Dynamic Systems, Measurement, and Control*, Vol. 126, No. 2, pp. 416–420.

104. Krivts, I.L. (2004), "New Pneumatic Cylinders for Improving Servo Actuator Positioning Accuracy," *Journal of Mechanical Design*, Vol. 126, No. 4, pp. 744–747.

105. Krivts, I.L. and Eshel, R. (1994), "Improving the Positioning Accuracy of Pneumatic Servo Actuators," *The 25th Israel Conference on Mechanical Engineering*, Technion City, Haifa, Israel, pp. 204–206.

106. Kunt, C. and Singh, R. (1990), "A Linear Time Varying Model for On-Off Valve Controlled Pneumatic Actuators," *Journal of Dynamic Systems, Measurement, and Control*, Vol. 112, pp. 740–747.

107. Lai, J.Y., Meng, C.H., and Singh, R. (1990), "Accurate Position Control of a Pneumatic Actuator," *Journal of Dynamic Systems, Measurement, and Control*, Vol. 112, pp. 734–739.

108. Latino, F. and Dandoval, D. (1996), " Quit Overspending for Servomotion Systems," *Machine Design*, April 18, pp. 93–96.

109. Lee, H.K., Choi, G.S., and Choi, G.H. (2002), "A Study on Tracking Position Control of Pneumatic Actuators," *Mechatronics*, Vol. 12, No. 6, pp. 813–831.

110. Lee, T.H. (2004), *The Design of CMOS Radio-Frequency Integrated Circuits, 2nd edition*, Cambridge University Press, Cambridge, UK, 816 p.

111. Lin, F., Yao, X., Liang, Z., and Zhang, M. (2003), "The Testing System of Friction Forces in the Pneumatic Servo Actuators at High Pressure: Analysis and Design," *5th International Symposium on Test and Measurement, ISTM/2003*, Vol. 4, pp. 2874–2976.

112. Lin, S. and Aker, A. (1991), "Dynamic Analysis of a Flapper-Nozzle Valve," *Journal of Dynamic Systems, Measurement, and Control*, Vol. 113, pp. 163–167.

113. Lin, X., Spettel, F., and Scavarda, S. (1996), "Modeling and Test of an Electro-pneumatic Servovalve Controlled Long Rodless Actuator," *Journal of Dynamic Systems, Measurement, and Control*, Vol. 118, pp. 457–462.

114. Lin-Chen, Y.Y., Wang, J., and Wu, Q.H. (2003), "A Software Tool Development for Pneumatic Actuator System Simulation and Design," *Computers in Industry*, Vol. 51, No. 1, pp. 73–88.

115. Liu, S. and Bobrow, J.E. (1988), "An Analysis of a Pneumatic Servo System and its Application to a Computer-Controlled Robot," *Journal of Dynamic Systems, Measurement, and Control*, Vol. 110, pp. 228–235.

116. Mahgoub, H.M. and Craighead, I.A. (1995), "Development of a Microprocessor Based Control System for a Pneumatic Rotary Actuator," *Mechatronics*, Vol. 5, No. 5, pp. 541–560.

117. Mattiazzo, G., Mauro, S., Raparelli, T., and Velardocchia, M. (2002), "Control of Six-Axis Pneumatic Robot," *Journal of Robotic Systems*, Vol. 19, No. 8, pp. 363–378.

118. McDonell, B.W. and Bobrow, J.E. (1993), "Adaptive Tracking Control of an Air Powered Robot Actuator," *Journal of Dynamic Systems, Measurement, and Control*, Vol. 115, pp. 427–433.

119. McDonell, B.M. and Bobrow, J.E. (1997), "Modeling, Identification and Control of a Pneumatically Actuated Robot," *Proceedings of the IEEE International Conference on Robotics and Automation*, pp. 124–129.

120. Miyajima, T., Sakaki, K., Shibukawa, T., Fujita, T., Kawashima, K., and Kagawa, T. (2004), "Development of Pneumatic High Precise Position Controllable Servo Valve," *Proceedings of the 2004 IEEE International Conference on Control Applications*, Vol. 2, Part No. 2, pp. 1159–1164.

121. Moore, P.R., Ssenkungo, F.W., Weston, R.H., Thatcher, T.W., and Harrison, R. (1986), "Control Strategies for Pneumatic Servo Drives," *International Journal of Production Research*, Vol. 24, No. 6, pp. 1363–1382.

122. Morioka, H., Nishiuchi, A., Kurahara, K., Tanaka, K., and Oka, M. (2000), "Practical Robust Control Design of Pneumatic Servo Systems," *26th Annual Conference of the IEEE Electronics Society, IECON 2000*, Vol. 3, pp. 1755–1760.

123. Ning, S. and Bone, G.M. (2002), "High Steady-State Accuracy Pneumatic Servo Positioning System with PVA/PV Control and Friction Compensation," *Proceedings of the 2002 IEEE International Conference on Robotics & Automation*, Washington, D.C., pp. 2824–2829.

124. Noritsugu, T. (1987), "Electro-Pneumatic Feedback Speed Control of a Pneumatic Motor. I. With an Electro-Proportional Valve," *Journal of Fluid Control*, Vol. 17, No. 3, pp. 17–37.

125. Noritsugu, T. (1988), "Electro-Pneumatic Feedback Speed Control of a Pneumatic Motor. II. With a PWM Operated On-Off Valve," *Journal of Fluid Control*, Vol. 18, No. 2, pp. 7–21.

126. Noritsugu, T. and Takaiwa, M. (1995), "Robust Positioning Control of a Pneumatic Servo System with Pressure Control Loop," *IEEE International Conference on Robotics and Automation*, pp. 2613–2618.

127. Noritsugu, T. and Takaiwa, M. (1997), "Positioning Control of Pneumatic Parallel Manipulator," *Journal of the Robotics Society of Japan*, Vol. 15, No. 7, pp. 14–20.

128. Nouri, B.M.Y., Al-Bender, F., Swevers, J., Vanherek, P., and Van Brussel, H. (2000), "Modelling a Pneumatic Servo Positioning System with Friction," *Proceedings of the American Control Conference*, Chicago, IL, pp. 1067–1071.

129. Nyce, D.S. (2003), *Linear Position Sensors: Theory and Application*, Wiley, New York, 184 p.

130. Ottaviano, E., Toti, M., and Ceccarelli, M. (2000), "Grasp Force Control in Two-Finger Grippers with Pneumatic Actuation," *Proceedings — IEEE International Conference on Robotics and Automation*, Vol. 2, pp. 1976–1981.

131. Outbib, R. and Richard, E. (2000), "State Feedback Stabilization of an Electropneumatic System," *Journal of Dynamic Systems, Measurement, and Control*, Vol. 122, No. 3, pp. 410–415.

132. Pandian, S.R. (2002), "Force Control of a Pneumatic Cylinder by Learning Control," *6th World Multiconference on Systemics, Cybernetics and Informatics. Proceedings 2002*, Pt. 6, Vol. 6, pp. 284–289.

133. Pandian, S.R., Hayakawa, Y., Kanazawa, Y., Kamoyama, Y., and Kawamura, S. (1997), "Practical Design of a Sliding Mode Controller for Pneumatic Actuators," *Journal of Dynamic Systems, Measurement, and Control*, Vol. 119, pp. 666–674.

134. Park, N.-C., Park, H.-W., Yang, H.-S., and Park, Y.-P. (2002), "Robust Vibration/Force Control of a 2 D.O.F. Arm Having One Flexible Link with Artificial Pneumatic Actuators," *Journal of Vibration and Control*, Vol. 8, No. 3, pp. 405–423.

135. Park, S.H. and Chang, P.H. (2000), "An Enhanced Time Delay Observer for Nonlinear Systems," *Transactions on Control, Automation and Systems Engineering*, Vol. 2, No. 3, pp. 149–156.

136. Paul, A.K., Mishra, J.K., and Radke, M.G. (1994), "Reduced Order Sliding Mode Control for Pneumatic Actuator," *IEEE Transactions on Control Systems Technology*, Vol. 2, No. 3, pp. 271–276.

137. Perondi, E.A. and Guenther, R. (2000), "Control of Servopneumatic Drive with Friction Compensation," *First FPNI — PhD Symposium*, Hamburg, pp. 117–127.

138. Pohl, J., Sethson, M., Krus, P., and Palmberg, J.O. (2001), "Modeling and Simulation of a Fast 2/2 Switching Valve," *Fifth International Conference on Fluid Power Transmission and Control (ICEP 2001)*, April, Hangzhou, China, pp. 123–128.

139. Ponomarev, S.D. and Andreeva, L.A. (1980), *Calculation of the Spring Elements for the Machines and Instruments*, Moscow, Machinostroenie, 326 p. (in Russian).

140. Prudnikov, A.P., Brichkov, U.A., and Marichev, O.I. (1981), *Integrals and Series*, Moscow, Nauka, 800 p. (in Russian).

141. Pu, J. and Weston, R.H. (1989), "A New Generation of Pneumatic Servo for Industrial Robot," *Robotica*, Vol. 7, pp. 17–23.

142. Pu, J., Moore, P.R., and Weston, R.H. (1991), "High-Gain Control and Tuning Strategy for Vane-Type Reciprocating Pneumatic Servo Drive," *International Journal of Production Research*, Taylor-Francis, Vol. 29, No. 8, pp. 1587–1601.

143. Pu, J., Moore, P.R., and Weston, R.H. (1991), "Digital Servo Motion Control of Air Motors," *International Journal of Production Research*, Taylor-Francis, Vol. 29, No. 3, pp. 599–618.

144. Pu, J., Moore, P.R., Harrison, R., and Weston, R.H. (1993), "A Study of Gain-Scheduling Method for Controlling the Motion of Pneumatic Servos," *Sixth Bath International Fluid Power Workshop – Modeling and Simulation*, England, 23–24 September.

145. Qi, Y. and Surgenor, B.W. (2003), "Pulse-Width Modulation Control of a Pneumatic Positioning System," *2003 ASME International Mechanical Engineering Congress & Exposition*, Washington, D.C., November, pp. 1–10.

146. Radcliffe, C.J. and Southward, S.C. (1991), "Robust Nonlinear Stick-Slip Friction Compensation," *Journal of Dynamic Systems, Measurement, and Control*, Vol. 113, pp. 639–645.

147. Raibert, M.H. and Craig, J.J. (1981), "Hybrid Position/Force Control of Manipulators," *Journal of Dynamic Systems, Measurement, and Control*, Vol. 103, pp. 126–133.

148. Raparelli, T., Mattiazzo, G., Mauro, S., and Velardocchia, M. (1999), "Design and Development of a Pneumatic Anthropomorphic Hand," *Journal of Robotic Systems*, Vol. 17, No. 1, pp. 1–15.

149. Rathbun, D.B., Berg, M.C., and Buffinton, K.W. (2004), "Stiction and Coulomb Friction," *Journal of Dynamic Systems, Measurement, and Control*, Vol. 126, No. 1, pp. 131–138.

150. Renn, J.-C. and Liao, C.-M. (2004), "A Study on the Speed Control Performance of a Servo-Pneumatic Motor and the Application to Pneumatic Tools," *International Journal of Advanced Manufacturing Technology*, Vol. 23, No. 7-8, pp. 572–576.

151. Repperger, D.W., Johnson, K.R., and Phillips, C.A. (1999), "Nonlinear Feedback Controller Design of a Pneumatic Muscle Actuator System," *Proceedings of the American Control Conference*, San Diego, CA, pp. 1525–1529.

152. Richard, E. and Scavarda, S. (1996), "Comparison between Linear and Nonlinear Control of an Electropneumatic Servodrive," *Journal of Dynamic Systems, Measurement, and Control*, Vol. 118, pp. 245–252.

153. Richard, E. and Hurmuzlu, Y. (2000), " A High Performance Pneumatic Force Actuator System. Part 1. Nonlinear Mathematical Model," *Journal of Dynamic Systems, Measurement, and Control*, Vol. 122, pp. 416–425.

154. Richard, E. and Hurmuzlu, Y. (2000), " A High Performance Pneumatic Force Actuator System. Part 2. Nonlinear Controller Design," *Journal of Dynamic Systems, Measurement, and Control*, Vol. 122, pp. 426–434.

155. Richardson, R., Brown, M., and Plummer, A.R. (2000), "Pneumatic Impedance Control for Physiotherapy," *European Advanced Robotics Systems. – Robotics 2000*, University of Salford, 12–14 April.

156. Richardson, R., Plummer, A.R., and Brown, M.D. (2001), "Self-Tuning Control of a Low Friction Pneumatic Actuator under the Influence of Gravity," *IEEE Transactions on Control Systems Technology*, Vol. 9, No. 2, pp. 330–334.

157. Robinson, D.W. (2000), "Design and Analysis of Series Elasticity in Closed-Loop Actuator Force Control," Ph.D. thesis, Massachusetts Institute of Technology, Mechanical Engineering, Cambridge, MA, 123 p.

158. Royston, T. and Singh, R. (1993), "Development of a Pulse-Width Modulated Pneumatic Rotary Valve for Actuator Position Control," *Journal of Dynamic Systems, Measurement, and Control*, Vol. 115, No. 3, pp. 495–505.

159. Ruan, J., Li, S., Li, M., and Yang, J. (1999), "One-Stage Pneumatic Digital Servo Valve," *Proceedings of the 3rd International Symposium on Fluid Power Transmission and Control (ISFP'99)*, pp. 488–493.

160. Ruan, J., Burton, R., and Ukrainetz, P. (2002), "An Investigation into the Characteristics of a Two Dimensional '2D' Flow Control Valve," *Journal of Dynamic Systems, Measurement, and Control*, Vol. 124, No. 1, pp. 214–220.

161. Sanville, F.E. (1971), "A New Method of Specifying the Flow Capacity of Pneumatic Fluid Power Valve," *Hydraulic Pneumatic Power*, Vol. 17, No. 195, pp. 37–47.

162. Sawodny, O. and Hildebrandt, A. (2002), "Aspects of the Control of Differential Pneumatic Cylinders," Proceedings of 10th German Japanese Seminar on Nonlinear Problems in Dynamic Systems," Kanazawa, Japan.

163. Scavarda, S. (1993), "Some Theoretical Aspects and Recent Developments in Pneumatic Positioning Systems," *Proceedings of the Second JHPS International Symposium on Fluid Power*, Tokyo, Japan, pp. 29–48.

164. Scavarda, S., Betemps, M., and Jutard, A. (1992), "Models of a Pneumatic PWM Solenoid Valve for Engineering Applications," *Journal of Dynamic Systems, Measurement, and Control*, Vol. 114, pp. 680–688.

165. Schroeder, L.E. and Singh, R. (1993), "Experimental Study of Friction in a Pneumatic Actuator at Constant Velocity," *Journal of Dynamic Systems, Measurement, and Control*, Vol. 115, pp. 575–577.

166. Schulte, H. and Hahn, H. (2004), "Fuzzy State Feedback Gain Scheduling Control of Servo-Pneumatic Actuators," *Control Engineering Practice*, Vol. 12, No. 5, pp. 639–650.

167. Shearer, J.E. (1956), "Study of Pneumatic Process in the Continuous Control of Motion with Compressed Air" I, II, *Transactions of ASME*, February, pp. 233–249.

168. Shen, T., Tamura, K., Kaminaga, H., Henmi, N., and Nakazawa, T. (2000), "Robust Nonlinear Control of Parametric Uncertain Systems with Unknown Friction and its Application to a Pneumatic Control Valve," *Journal of Dynamic Systems, Measurement, and Control*, Vol. 122, No. 2, pp. 257–262.

169. Shibata, S., Jindai, M., and Shimizu, A. (2000), "Neuro-Fuzzy Control for Pneumatic Servo System," *Industrial Electronics Society, 26th Annual Conference of the IEEE, IECON 2000*, Japan, Vol. 3, pp. 1761–1766.

170. Shih, M.C. and Ma, M.A. (1998), "Position Control of a Pneumatic Rodless Cylinder Using Sliding Mode M-D-PWM to Control the High Speed Solenoid Valves," *JSME International Journal, Series C*, Vol. 41, No. 2, pp. 236–241.

171. Shimizu, A., Shibata, S., and Jindai, M. (2000), "Pneumatic Servo Systems Controlled by Self-Tuning Fuzzy Rules," *Proceedings of the Sixth Triennial International Symposium on Fluid, Measurement and Visualization, FLUCOM 2000*, Canada, pp. 528–534.

172. Situm, Z. and Crnekovic, M. (2002), "Control of a Pneumatic Actuator Using Proportional Pressure Regulators," *CIM 2002 Computer Integrated Manufacturing and High Speed Machining — 8th International Scientific Conference on Production Engineering*, Croatia, pp. 11035–11046.

173. Situm, Z., Pavkovic, D., and Novakovic, B. (2004), "Servo Pneumatic Position Control Using Fuzzy PID Gain Scheduling," *Journal of Dynamic Systems, Measurement, and Control*, Vol. 126, No. 2, pp. 376–387.

174. Song, G., Cai, L., and Wang, Y. (1999), "Robust Friction Compensation for Precise and Smooth Position Regulation," *Journal of Systems and Control Engineering*, Vol. 213, No. 2, pp. 157–161.

175. Song, J. and Ishida, Y. (1997), "A Robust Sliding Mode Control for Pneumatic Servo Systems," *International Journal of Engineering Science*, Vol. 35, No. 8, pp. 711–723.

176. Sontag, E.D. (1998), *Mathematical Control Theory: Deterministic Finite Dimensional Systems, 2nd edition*, Springer-Verlag, New York, 531 p.

177. Sorli, M., Ferraresi, C., and Pastorelli, S. (1995), "Force Controlling Pneumatic Servoactuator via Digital PWM Modulated Valve," *Second International Symposium on Fluid Power Transmission and Control*, Shanghai, pp. 480–485.

178. Sorli, M., Figliolini, G., and Pastorelli, S. (2001), "Dynamic Model of a Pneumatic Proportional Pressure Valve," *IEEE/ASME International Conference on Advanced Intelligent Mechatronics, AIM*, Vol. 1, pp. 630–635.

179. Sorli, M. and Pastorelli, S. (2000), "Performance of a Pneumatic Force Controlling Servosystem: Influence of Valves Conductance," *Robotics and Autonomous Systems*, Vol. 30, No. 3, pp. 283–300.

180. Southward, S.C., Radcliffe, C.J., and MacCluer, C.R. (1991), "Robust Nonlinear Stick-Slip Friction Compensation," *Journal of Dynamic Systems, Measurement and Control*, Vol. 113, pp. 639–644.

181. Spiegel, M.R. and Liu, J. (1999), *Mathematical Handbook of Formulas and Tables, 2nd edition*, McGraw-Hill, New York, 278 p.

182. Surgenor, B.W., Vaughan, M.D., and Ueblin, M. (1995), "Continuous Sliding Mode Control of a Pneumatic Positioning System," *Proceedings of the Eighth Bath International Fluid Power Workshop*, pp. 270–283.

183. Surgenor, B.W. and Vaughan, N.D. (1997), "Continuous Sliding Mode Control of a Pneumatic Actuator," *Journal of Dynamic Systems, Measurement, and Control*, Vol. 119, No. 3, pp. 578–581.

184. Surgenor, B.W. and Iordanou, H.N. (1994), "Experience with Advanced Position Control of Pneumatic Systems," *IFAC Workshop: Trends in Hydraulics and Pneumatics*, Chicago, IL.

185. Takemura, F., Pandian, S.R., Kawamura, S., and Hayakawa, Y. (1999), "Observer Design for Control of Pneumatic Cylinder Actuator," *PTMC 99, Bath Workshop on Power Transmission and Motion Control*, Bath, September 8–10, 1999, pp. 223–236.

186. Tang, J. and Walker, G. (1995), "Variable Structure Control of a Pneumatic Actuator," *Journal of Dynamic Systems, Measurement, and Control,* Vol. 117, pp. 88–92.

187. Tao, G. and Lewis, F.L. (2001*), Adaptive Control of Nonsmooth Dynamic Systems,* Springer, 407 p.

188. Thomas, M.B. (2003), "Advanced Servo Control of a Pneumatic Actuator," doctoral thesis, The Ohio State University, Columbus, 222 p.

189. Tsuhanova, E.A. (1978), *Dynamic Synthesis of the Hydraulic Control Devices with the Restricted Flow Elements,* Moscow, Nauka, 255 p. (in Russian).

190. Uebling, M., Vaughan, N.D., and Surgenor, B.W. (1997), "On Linear Dynamic Modeling of a Pneumatic Servo System," *The Fifth Scandinavian International Conference on Fluid Power, SICFP'97,* Linkoping, Sweden, pp. 363–378.

191. Utkin, V.I. (1992), *Sliding Modes in Control and Optimization,* Spring-Verlag, Berlin.

192. Van Ham, R., Verrelst, B., Daerden, F., Vanderborght, B., and Lefeber, D. (2005), "Fast and Accurate Pressure Control Using On-Off Valves," *International Journal of Fluid Power,* Vol. 6, No. 1, pp. 53–58.

193. Van Varseveld, R.B. and Bone, G.M. (1997), "Accurate Position Control of a Pneumatic Actuator Using On/Off Solenoid Valves," *IEEE/ASME Transactions on Mechatronics,* Vol. 2, No. 3, pp. 195–204.

194. Vaughan, N.D. and Gamble, J.B. (1996), "The Modeling and Simulation of a Proportional Solenoid Valve," *Journal of Dynamic Systems, Measurement and Control,* Vol. 118, No. 1, pp. 120–131.

195. Virvalo, T. (1995), "Modeling and Design of a Pneumatic Position Servo System Realized with Commercial Components," Ph.D. thesis, Tampere University of Technology, Finland, 197 p.

196. Virvalo, T. (1997), "Nonlinear Model of Analog Valve," *The Fifth Scandinavian International Conference on Fluid Power, SICFP 97,* pp. 199–214.

197. Virvalo, T. and Makinen, E. (2000), "Dimensioning and Selecting Pressure Supply Line Equipment for a Pneumatic Position Servo," *Proceedings of the Sixth Triennial International Symposium on Fluid Control, Measurement and Visualization,* Sherbrooke, Quebec, Canada, pp. 32–38.

198. Virvalo, T. (2001), "The Influence of Servo Valve Size on the Performance of a Pneumatic Position Servo," *Fifth International Conference on Fluid Power Transmission and Control (ICEP 2001),* April, Hangzhou, China, pp. 278–283.

199. Vuorisalo, M. and Virvalo, T. (2002), "Use of Shape Memory Alloy for a Pilot Stage Actuator of a Water Hydraulic Control Valve," *Proceedings of 5th International Symposium on Fluid Power,* Nara, Japan, pp. 153–158.

200. Wang, X.G., and Kim, C. (2000), "Self-Tuning Predictive Control with Forecasting Factor for Control of Pneumatic Lumber Handling Systems," *International Journal of Adaptive Control and Signal Processing,* Vol. 14, No. 5, pp. 533–557.

201. Wang, J., Pu, J., and Moore, P.R. (1999), "A Practical Control Strategy for Servo-Pneumatic Actuator Systems," *Control Engineering, Practice,* Vol. 7, pp. 1483–1488.

202. Wang, J., Pu, J., Moore, P.R., and Zhang, Z.M. (1998), "Modeling Study and Servo-Control of Air Motor Systems," *International Journal of Control,* Vol. 71, No. 3, pp. 459–476.

203. Wang, Y.T., Singh, R., Yu, H.C., and Guenther, D.A. (1984), "Computer Simulation of a Shock-Absorbing Pneumatic Cylinder,," *Journal of Sound and Vibration,* Vol. 93, pp. 353–364.

204. Whitmore, S.A., Lindsey, W.T., Curry, R.E., and Gilyard, G.B. (1990), "Experimental Characterization of the Effects of Pneumatic Tubing on Unsteady Pressure Measurements," NASA Technical Memorandum 4171, pp. 1–26.

205. Will, D., Frank, W., et al. (1987), "Freipositionierbare, pneumatische angetriebene Achsen für drehende und lineare Bewegung," *Jlhydraul. und Pneum.*, Vol. 31, No. 2, pp. 127–136.

206. Wu, J., Goldfarb, M., and Barth, E. (2003), "The Role of Pressure Sensors in the Servo Control of Pneumatic Actuators," *Proceedings of the American Control Conference*, Vol. 2, pp. 1710–1714.

207. Wu, J., Goldfarb, M., and Barth, E. (2004), "On the Observability of Pressure in a Pneumatic Servo Actuator," *Journal of Dynamic Systems, Measurement, and Control*, Vol. 126, No. 4, pp. 921–924.

208. Wu, R-H. and Tung, P-C. (2002), "Studies of Stick-Slip Friction, Presliding Displacement, and Hunting," *Journal of Dynamic Systems, Measurement, and Control*, Vol. 124, No. 1, pp. 111–117.

209. Xiang, F. (2001), "Block-Oriented Nonlinear Control of Pneumatic Actuator Systems," Doctoral thesis, Mechatronics Lab, Department of Machine Design, Royal Institute of Technology, KTH, Sweden, 62 p.

210. Xiang, F. and Wikander, J. (2003), "QFT Control Design for an Approximately Linearized Pneumatic Positioning System," *International Journal of Robust and Nonlinear Control*, Vol. 13, No. 7, pp. 675–688.

211. Xiang, F. and Wikander, J. (2004), "Block-Oriented Approximate Feedback Linearization for Control of Pneumatic Actuator System," *Control Engineering Practice*, Vol. 12, No. 4, pp. 387–399.

212. Xu, Y., Hollerbach, J.M., and Ma, D. (1995), "A Nonlinear PD Controller for Force and Contact Transient Control," *IEEE Control Systems*, Vol. 15, No. 1, pp. 15–21.

213. Yavorsky, V.N., Makshanov, V.I., and Ermolin, V.P. (1978), "Design of the Nonlinear Servo Systems," Leningrad, Energia, 208 p. (in Russian).

214. Yen, J.Y., Huang, C.J., and Lu, S.S. (1997), "Stability of PDF Controller with Stick-Slip Friction Device," *Journal of Dynamic Systems, Measurement, and Control*, Vol. 119, pp. 486–490.

Index

Waste to energy conversion technology

Related titles:

Biomass combustion science, technology and engineering
(ISBN 978-0-85709-131-4)

Handbook of biofuels production: Processes and technologies
(ISBN 978-1-84569-679-5)

Biodiesel science and technology: from soil to oil
(ISBN 978-1-84569-591-0)

Details of these books and a complete list of titles from Woodhead Publishing can be obtained by:

- visiting our web site at www.woodheadpublishing.com
- contacting Customer Services (e-mail: sales@woodheadpublishing.com; fax: +44 (0) 1223 832819; tel.: +44 (0) 1223 499140 ext. 130; address: Woodhead Publishing Limited, 80, High Street, Sawston, Cambridge CB22 3HJ, UK)
- in North America, contacting our US office (e-mail: usmarketing@ woodheadpublishing.com; tel.: (215) 928 9112; address: Woodhead Publishing, 1518 Walnut Street, Suite 1100, Philadelphia, PA 19102-3406, USA)

If you would like e-versions of our content, please visit our online platform: www. woodheadpublishingonline.com. Please recommend it to your librarian so that everyone in your institution can benefit from the wealth of content on the site.

We are always happy to receive suggestions for new books from potential editors. To enquire about contributing to our energy series, please send your name, contact address and details of the topic/s you are interested in to sarah.hughes@ woodheadpublishing.com. We look forward to hearing from you.

The team responsible for publishing this book:

Commissioning Editor: Sarah Hughes
Publications Coordinator: Steven Mathews
Project Editor: Cathryn Freear
Editorial and Production Manager: Mary Campbell
Production Editor: Mandy Kingsmill
Project Manager: Annette Wiseman, RCL
Copy Editor: Jonathan Webley
Proof Reader: Clare Dobson
Cover Designer: Terry Callanan

Woodhead Publishing Series in Energy: Number 29

Waste to energy conversion technology

Edited by
Naomi B. Klinghoffer and
Marco J. Castaldi

WP

WOODHEAD
PUBLISHING

Oxford   Cambridge   Philadelphia   New Delhi

Published by Woodhead Publishing Limited,
80 High Street, Sawston, Cambridge CB22 3HJ, UK
www.woodheadpublishing.com
www.woodheadpublishingonline.com

Woodhead Publishing, 1518 Walnut Street, Suite 1100, Philadelphia, PA 19102-3406, USA

Woodhead Publishing India Private Limited, 303, Vardaan House, 7/28 Ansari Road,
Daryaganj, New Delhi – 110002, India
www.woodheadpublishingindia.com

First published 2013, Woodhead Publishing Limited
© Woodhead Publishing Limited, 2013. Note: the publisher has made every effort to
ensure that permission for copyright material has been obtained by authors wishing to use
such material. The authors and the publisher will be glad to hear from any copyright
holder it has not been possible to contact.
The authors have asserted their moral rights.

British Library Cataloguing in Publication Data
A catalogue record for this book is available from the British Library.

Library of Congress Control Number: 2013936017

ISBN 978-0-85709-011-9 (print)
ISBN 978-0-85709-636-4 (online)
ISSN 2044-9364 Woodhead Publishing Series in Energy (print)
ISSN 2044-9372 Woodhead Publishing Series in Energy (online)

The publisher's policy is to use permanent paper from mills that operate a sustainable
forestry policy, and which has been manufactured from pulp which is processed using
acid-free and elemental chlorine-free practices. Furthermore, the publisher ensures that
the text paper and cover board used have met acceptable environmental accreditation
standards.

Typeset by RefineCatch Limited, Bungay, Suffolk
Printed by MPG Printgroup, UK

Contents

7   Waste firing in large combustion plants   98

P. VAINIKKA, M. NIEMINEN and K. SIPILÄ, VTT
Technical Research Centre of Finland, Finland

8   Waste to energy (WTE) systems for district heating   120

L. TOBIASEN and B. KAMUK, Ramboll, Denmark

9   Gasification and pyrolysis of municipal solid waste (MSW)   146

N. B. KLINGHOFFER, Columbia University, USA and
M. J. CASTALDI, City University of New York, USA

Contributor contact details

(* = main contact)

Editors and Chapter 9

Naomi B. Klinghoffer
Earth Engineering Center
Columbia University
New York
NY 10027
USA

E-mail: naomi.klinghoffer@gmail.com;
nbk2107@columbia.edu

Marco J. Castaldi*
Associate Professor
Chemical Engineering Department
The City College of New York
City University of New York
140th Street at Convent Avenue
Steinman Hall, Room 307
New York
NY 10031
USA

E-mail: mcastaldi@che.ccny.cuny.edu

Chapter 1

Naomi B. Klinghoffer and Nickolas
J. Themelis
Earth Engineering Center
Columbia University
New York
NY 10027
USA

E-mail: naomi.klinghoffer@gmail.com;
nbk2107@columbia.edu;
njt1@columbia.edu

Marco J. Castaldi*
Associate Professor
Chemical Engineering Department
The City College of New York
City University of New York
140th Street at Convent Avenue
Steinman Hall, Room 307
New York
NY 10031
USA

E-mail: mcastaldi@che.ccny.cuny.edu

Chapter 2

Ted Michaels
President
Energy Recovery Council
1730 Rhode Island Avenue, NW
Suite 700
Washington
District of Columbia
USA

E-mail: tmichaels@
    energyrecoverycouncil.org

Chapter 3

Scott Kaufman
Research Associate
Department of Earth &
    Environmental Engineering
Columbia University
New York
USA

E-mail: smk2108@columbia.edu

Chapter 4

Thomas F. McGowan
TMTS Associates, Inc.
399 Pavillion Street
Atlanta
GA 30315
USA

E-mail: tfmcgowan@mindspring.com

Chapter 5

Garrett C. Fitzgerald
Department of Earth &
    Environmental Engineering
Columbia University
918 S. W. Mudd Building
520 W 120th St
New York
NY 10027
USA

E-mail: Gcfitzgerald@gmail.com

Chapter 6

Leonard M. Grillo
Principal, Grillo Engineering
    Company
9 Ash Street
Hollis
NH 03049
USA

E-mail: len@grilloengineering.com

Chapter 7

Pasi Vainikka, Matti Nieminen and
    Kai Sipilä*
VTT Technical Research Centre of
    Finland
PO Box 1000
Biologinkuja 5
02044 Espoo
Finland

E-mail: kai.sipila@vtt.fi

Chapter 8

Lasse Tobiasen*
Chief Consultant, Waste to Energy
Ramboll Energy
Hannemanns Alle 53
DK-2300 Copenhagen
Denmark
E-mail: LST@ramboll.com

Bettina Kamuk
Head of Department, Waste to
   Energy
Ramboll Energy
Hannemanns Alle 53
DK-2300 Copenhagen
Denmark

E-mail: bkc@ramboll.com

Chapter 10

Jürgen Vehlow
Karlsruhe Institute of Technology
Institute for Technical Chemistry
Hermann-von-Helmholtz-Platz 1
76344 Eggenstein-Leopoldshafen
Germany

E-mail: juergen.vehlow@kit.edu

Chapter 11

John S. Austin
Facilities Management Department
University of Maryland
Service Building
College Park
MD 20742
USA

E-mail: jsaustin@umd.edu

Woodhead Publishing Series in Energy

Foreword
by Floyd Hasselriis

Albert Einstein observed that 'scientists investigate that which already is; engineers create that which has never been.' The history of waste to energy (WTE) is a long story of learning, experimenting and evolution. The chapters in this book tell this story from many different points of view. Engineers have been working on the recovery of the energy in wastes from the time that steam engines for power and electric generation were first in operation in the late 1800s. The first 'destructor' in England, built in Manchester in 1876, was reported to be operating 30 years later.[1]

Joseph G. Branch, the chairman of the American Society of Mechanical Engineers (ASME), self-published a comprehensive book on the subject of 'waste to energy' in 1906.[2] In this he tabulated the waste generated at the time, and over 100 British and foreign municipal incinerators using typically over 30 tons per day waste for lighting and power uses. He described the power plant built under the new Williamsburg Bridge, to burn New York City waste. At that time waste was carried by horse-carts mostly to be dumped in the rivers. Some carts burned the trash on the way!

After 1906, only a few public WTE plants were built. As internal combustion engines replaced the horses, trucks that were used to haul the waste to landfill or to dump it in the river; much of land areas of Manhattan and other boroughs were expanded by filling with the high-ash-containing trash. After about 1950, as land became more expensive, and population blossomed, local governments began to hire engineers to build refractory chambers to burn the waste, and even wash down the smoke somewhat.

A group of ASME members, as the *Incinerator Division*, published standards for refractory incinerators to destroy infectious medical waste in 1961.[3] In 1966 the ASME *Incinerator Committee* was formed 'to bring together the foremost authorities in the field of refuse incineration for an exchange of knowledge, experiences and expectations'. This group of engineers was subsequently renamed the *Solid Waste Processing Division* (SWPD) and is now the *Material and Energy Recovery Division* of the ASME. The *ASME Research Committee on Industrial and Municipal Waste*, most recently renamed the *ASME Research Committee on Energy Environment and Waste*, was organized specifically for research, and to reflect the extension of their activities.

In 1977 the issue of health and environmental effects from human exposure to dioxins suddenly arose. Loosely named here, dioxins are forms of a basic building block in nature, two benzene rings linked together, exhibiting various degrees of toxicity depending upon how many of the six outside points are chlorinated rather than oxygenated. Those with the two outermost joints, number 2,3 and 7,8 chlorinated (2,3,7,8 TCDD) being most toxic to human cells, causing hormonal damage, whereas with all six joints oxygenated, not being toxic.

Formation of dioxins was found in laboratory studies of fly ash from three Dutch plants at temperatures typical of fly ash from combustion of municipal waste, i.e. 450 °F (Olie, 1977). At the same time laboratories were measuring the effects of the dioxins on human health.

By 1977, the US Environmental Protection Agency (USEPA) laboratory in Cincinnati, Ohio, had investigated the temperature and oxygen conditions under which dioxins and other 'difficult' chlorinated compounds could be destroyed, i.e. reduced to essentially zero, under laboratory conditions, as gases, not solids.[4] Dioxins had also become the subject of research throughout Europe. Sweden declared a moratorium on new WTE facilities until a comprehensive research program could be completed.[5]

Particulate (dust) emissions were sampled from the stack of a new trash burning plant in Hempstead, Long Island, New York, by Midwest Research under contract for the newly established USEPA in 1980. Hearing of the dioxin laboratory tests in Europe, David Sussman of the USEPA decided to have the Hempstead samples analyzed for dioxins. Indeed, dioxins were found. The chief engineer of the Hempstead plant immediately brought this report to the attention of a meeting of the ASME Solid Waste Processing Division, stating that 'we will not be allowed to burn municipal waste any more until we solve the problem of dioxin emissions'.

Testing of existing WTE plants worldwide in fact showed an incredibly wide range of dioxins emissions, ranging from 10 000 nanograms per dry standard cubic meter (ng/dscm) (toxic equivalent) dioxin down to 300 ng/dscm, initially with no explanation as to why the range was so wide. A plant in Montreal tested in Canada in 1984 showed results as low as 0.1 ng/dscm, and that is now the international standard.

The ASME Adhoc Committee on Dioxins was immediately organized by Anthony Licata, chair of the SWPD. Arthur D. Little was asked to write a report on all that was known about dioxins, specifically dioxins from combustion of waste, and this report was published in 1980. At that point, no solution to the dioxin problem had been offered.

The first plant to burn shredded municipal waste in a conventional stoker-fired Babcock and Wilcox (B&W) steam boiler was started up in Hamilton, Ontario in 1984. Diagnostic tests were performed to observe the effects of different settings of combustion air dampers, and for the first time dioxins were measured along with CO, oxygen and furnace temperature. Hasselriis analyzed these data and plotted graphs showing the trends of these measurements under the various operating conditions. He presented these graphs at the 1984 Conference of Waste Management in Hofstra, Long Island, and at the Air and Waste Management Conference in Chicago that year. Meanwhile, the Swedish Government had been testing WTE units in Sweden and Denmark, and offered the explanation of non-ideal provision of combustion air in the furnace for the presence of dioxins.

Hasselriis' findings needed to be confirmed by full-scale testing of an operating 'mass-burn' WTE plant (the Hamilton boiler was burning shredded RDF (refuse-derived fuel)). A plant in Pittsfield, Massachusetts was ideal for comprehensive tests because it had the ability to be operated with flue-gas recirculation, along a full range of furnace temperatures, and excess oxygen levels could be used as part of the broad test program.

The Adhoc Committee on Dioxins obtained support from the New York State Energy Research and Development Authority (NYSERDA) and the USEPA. It was decided to perform comprehensive tests, burning normal garbage and dry paper, and also running tests with added vinyl plastics (as vinyl had been accused of being a major source of dioxins) at the recently started Pittsfield, plant. The most important overall finding was that carbon dioxide and dioxins followed the same trends with furnace temperature and oxygen, regardless of waste composition, moisture or the presence of vinyl in the waste. Hence CO was a good surrogate for dioxins: and continuous monitoring of CO became mandatory for operation of waste combustion. The results of the Pittsfield tests were published at a World Health Organization Conference in Copenhagen in 1987.[6]

Actually, it took another ten years to develop an understanding of why there was such an enormous range in dioxin emissions measured in stacks of different plants. The practice of injecting activated carbon into the flue gases prior to capturing the particulate in a fabric filter became a principal means of controlling the dioxins, in addition to optimizing combustion conditions. Environment Canada[7] had carried out extensive testing of emission controls on a pilot spray-dry/bag-house on an existing WTE plant that showed that such a relatively simple system could reduce emissions to acceptable levels. The scrubber/bag-house system became the standard for plants in the US and Canada.

After the Clean Air Act was passed in 1976, the USEPA first tested all measurable stack emissions, and the health and environmental risks were evaluated. The first tests had been performed in an incinerator in Boston, and included testing for all organic pollutants that it was possible to detect; this was later weeded down to those pollutants that are of significance to health and the environment, especially sulfur dioxide and particulate matter.

The EPA continued to test each plant as a permit condition, collecting a database that could be used to make estimates of emissions from future plants as they were built. Standard deviations of the data were calculated, from which it could be stated, for instance, that 99% of future test readings would be lower than that number.

Credit is due to the US Congress, which became an active supporter of WTE. Congress passed laws that required the utilities to pay 6 cents per kilowatt-hour for electricity for the period of time covered by the bond issue. This launched the WTE industry in the US. By the time this regulation expired in 1990, well over 50 new WTE plants had been built and brought under operation in the US.

The major boiler manufacturing companies of today, such as Babcock and Wilcox (B&W), were founded in the coal power and sugar industry well over 100 years ago. B&W is today building a new plant to burn 3000 tons per day of municipal garbage, and other wastes, to produce 80 megawatts of power to serve the community of West Palm Beach, Florida; enough to serve the modern needs of 56 000 homes in collaboration with perhaps the oldest manufacturers of WTE plants in the world, Vølund of Denmark (founded in 1898). Vølund built its first waste-to-power plant in 1978–1980, supplying thermal energy to the central district heating system of Aalborg. Today, this new plant is currently being rebuilt with the most advanced emissions control technology, to meet or exceed emissions conditions that will be the lowest of any renewable energy facility burning municipal waste in the US or elsewhere.

The ultimate control of all pollutants is described by Jurgen Vehlow in the final chapter of this history, describing the perfect controls that would take all consequences of waste reduction into consideration and find a safe resting place for all residues.

The drawback to the 'perfect system' is the cost: obviously from an economic point of view, only by compromises can WTE plants be kept at an economical level compared with alternatives, and relative risk to the community throughout the world, for all communities and countries.

The success story of WTE is the product of the community of engineers that kept developing the plants. Working within small and large organizations, writing papers, being involved in conferences, as parts of academic organizations, visiting plants, reading reports and papers, attending meetings of the Waste to Energy Research and Technology Council (WTERT), SWPD, etc.

This book should be of great interest to, and serve as an exceptional resource for, libraries, students, teachers, plant operators, engineers, managers, environmentalists and the general public. It chronicles the process by which engineers have systematically solved the problems of managing society's wastes and safely converting the wastes into energy, while at the same time recovering materials where practical.

References

1. Encyclopedia Britannica (1910) Destructors, *Encyclopedia Britannica, Vol. 8, Slice II*, 105–108, available from: http://www.gutenberg.org/files/30685/30685-h/30685-h.htm [Accessed April 24, 2013].
2. Branch, J. G. 'Heat and Light from Municipal and Other Waste', Wm. H. O'Brien Printing and Publishing Co., St. Louis, MO, 1906.
3. Incinerator Institute of America., Incinerator Standards', 1972.

4. Duvall, D.S. and Rubey, W.S. 'Laboratory Evaluation of High Temperature Destruction of Polychlorinated Biphenyls and Related Compounds', University of Dayton Research Institute, USEPA, MERL, Cincinnati, OH, 1977.
5. Bergvall, G. and Hult, J. 'Technology, Economics and Environmental Effects of Solid Waste Treatment', Final Report from the DRAV-Project, Naturvardsverket Rapport 3033, Svenska Renhallningsverks Foreningen, Publ. 85:11, 1985.
6. Hasselriis, F. 'Optimization of Combustion Conditions to minimize Dioxin Emissions', ISWA-WHO-DAKOTA Specialized Seminar, Emissions of Trace Organics from Municipal Solid Waste Incinerators, Copenhagen, 20–22 January 1987 (published in *Waste Management & Research* (1987) **5**, 311–326).
7. Environment Canada. 'The National Incinerator Testing and Evaluation Program: Air Pollution Control', 1986.

Part I
Introduction to waste to energy conversion

1

Waste to energy (WTE): an introduction

N. B. KLINGHOFFER and N. J. THEMELIS,
Columbia University, USA and M. J. CASTALDI,
City University of New York, USA

DOI: 10.1533/9780857096364.1.3

Abstract: Energy supply and waste management are great challenges that humans have faced for millennia. In order to meet these challenges in the future, it is necessary to progress to an atom economy where every atom is utilized in the best possible manner. This chapter provides an introduction to the subject, outlining the most important issues that must be considered.

Key words: waste hierarchy, waste to energy, sustainability ladder.

1.1 Energy supply and waste management

Energy supply and waste management are great challenges that humans have faced for millennia. Tremendous progress has occurred, yet these issues are still important today. To meet these challenges we must move to an atom economy where every atom is utilized in the best possible manner. To achieve this goal, a fundamental understanding of the underlying mechanisms and processes of energy and waste generation is necessary.

Demand for energy has increased rapidly in the last century. This will continue as the rate of consumption increases as people strive to improve their living standards. The requirement that the rate of energy consumption matches the rate of energy accumulation and conversion has led to a transition from wood (biomass) to coal to oil to natural gas: biomass needs to be grown, has to be harvested, dried and processed before use, coal needs to be mined, transported and processed before use and oil needs to be transported and refined before use, whereas gaseous fuels, which are ready to use immediately, only need to be transported. This trend in energy use – an increased emphasis on energy sources that are distributed and ready for immediate use with preferably no pre-processing – is expected to continue as populations spread out.

For thousands of years humans have had to take care of waste effectively and we will have to continue to do so if the human species is to survive. In fact, nearly all of societies' health problems can be traced to insufficient treatment or management of waste. Initially, waste management was only concerned with human excrement and sewage. It has now transitioned to what we term municipal solid waste (MSW), which includes other non-hazardous solid waste and is also

3

increasingly concerned with gaseous wastes, primarily CO_2 but also CH_4 and others (NO_x, SO_x, etc). In terms of the impact of waste on society, concern about the former as an emitted by-product or waste is increasing due to the threat of climate change associated with CO_2 emissions into the atmosphere.

Energy and waste are intimately connected in many ways. One of the more challenging is the gaseous and thermal waste associated with energy use. That is, useful work can be extracted from heat energy but is limited by the Carnot cycle and thus must yield thermal waste. Currently 80% of the world's energy is produced via combustion or gasification, thus there is gaseous waste (emissions) such as NO_x, SO_x, particulate matter and CO_2.

A convergence of energy and waste is occurring as evidenced by the growth in conversion of landfill gas to energy (LFGTE) and landfill gas (LFG) to fuels and the growth of WTE (nearly 100 new plants in the past few years). New technology development is focused on the extraction of energy in the most efficient manner and from all possible sources while reducing waste from those processes. Nations that do not have entrenched energy generation infrastructures are aggressively pursuing generation from all indigenous sources, particularly waste and biomass. Production of energy at optimal conditions from these 'alternative' sources comes from understanding the underlying reaction sequences (mechanisms) that occur. This enables the technologies to be designed in the most efficient way, extracting the maximum energy possible simultaneously producing the minimum amount of waste.

After reuse and recycling there are only two methods that can manage waste by matching the rate at which it is generated. The two proven means for disposal are burying municipal solid waste in landfill or thermally converting it in specially designed chambers at high temperatures, thereby reducing it to one tenth of its original volume and simultaneously recovering materials and energy. The heat generated by combustion or gasification is transferred to steam, which flows through a turbine to generate electricity. This process is called waste to energy (WTE). It recycles the energy and the metals contained in the MSW while most of the remaining ash by-product can be beneficially used for the maintenance of landfill sites (in the US) or for building roads and other construction purposes (in the EU and Japan). Waste to energy reduces the volume of MSW by 90%; if the remaining ash is reused, this is a nearly 'zero waste' solution.

1.1.1 Global municipal solid waste (MSW) generation and disposal

The economic development of the last century was accompanied by a greatly increased generation of solid wastes, which, when not managed properly, can adversely affect the environment and human health. In the generally accepted waste hierarchy adopted by USEPA and the EU, the first priority is waste reduction, followed by recycling of reusable materials, mostly paper fiber and metals, plus composting of fairly clean biodegradable yard and food wastes. In 2007, the Earth Engineering Center collaborated with NASA's Goddard Institute for Space Studies to produce a study on global landfill.[1] More recent information[2] on the amount of urban MSW landfill required in developing nations resulted in a downward revision of the global landfill to about one billion tonnes annually. This estimate is based on a combination of known tonnages landfilled in the US and other countries and interpolation to the rest of the world, as shown in Table 1.1. In comparison, the amount of MSW processed in thermal treatment facilities globally is estimated currently to be less than 200 million tons annually.[3]

The study developed four scenarios for WTE growth, ranging from very conservative, where the 2000–2007 growth in capacity was assumed to remain at a constant rate of 2.5% through to 2030, to assumed increases in the rate of growth of WTE capacity of 5%, 7.5% and 10% per year in the period 2010–2030. The overall conclusion was that although global WTE capacity had increased by about 4 million tons per year in the period 2000–2007, to about 170 million tons, this rate of growth will not be enough to curb landfill methane emissions by the year 2030; population growth and economic development will result in a much greater rate of MSW generation and landfilling. The only way to reduce landfill greenhouse gases (GHGs) between now and 2030 is by achieving a 7.5% growth in thermal treatment capacity on a global scale, or by increasing the amount of methane captured at landfills (Fig. 1.1 and Fig. 1.2). To appreciate the greenhouse gas impact of global landfill, one needs to consider that according to the IPCC, one unit volume of methane emitted into the atmosphere has the same effect as 21 unit volumes of carbon dioxide.[4]

1.1.2 The hierarchy of sustainable waste management

The BioCycle/Columbia bi-annual surveys in the US show that nearly 20% of US municipal solid wastes are recycled and about 10% are composted. As mentioned above, there are only two possible routes for the disposal of post-recycling wastes: thermal treatment with energy recovery, generally called waste to energy, and

Table 1.1 Estimate of MSW landfilled globally

	Population (millions)	Landfilled (million tonnes)	Landfilled (tonnes/capita)
US	304	225	0.7
EU15	380	104	0.3
Japan	127	Nil (putrescibles)	Nil
Rest of OECD nations	290	116	0.4
China	1322	180	0.14
Rest of the world	4280	385	0.09
Global	6703	1010	0.15

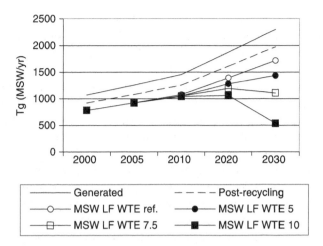

1.1 Generated and post-recycling MSW (constant for all scenarios) and landfilled MSW under four WTE scenarios.

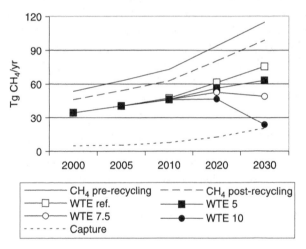

1.2 Impact of WTE growth on net methane emissions from MSW. The solid line, the potential maximum, is included for reference only.

landfill. In the US, approximately 7% of the US municipal solid waste is used for WTE while the remainder (64% or 240 million tons) is landfilled. Figure 1.3 shows the expanded hierarchy of waste management, which differentiates between the best landfills, i.e. those that collect and utilize landfill gas, and the worst, which add to the greenhouse gas problem of the planet by emitting methane to the atmosphere.

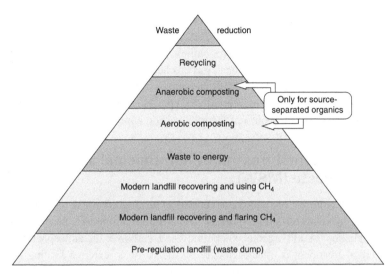

1.3 The expanded hierarchy of waste management.*

WTE is an integral part of MSW treatment meeting global standards. In Europe, the preferred solution for the portion of MSW that is not recycled or reused is to combust the remainder in waste to energy facilities that generate electricity and recover steam, metals and other resources, with only about one-tenth of the original volume going to ash landfills or used for road construction and other uses.

1.2 Biogenic fraction of carbon and calorific value of municipal solid waste (MSW)

On the basis of chemical analysis, the average composition of combustible materials in MSW (i.e., not including water, metals, or glass) can be expressed by the formula $C_6H_{10}O_4$. When this hypothetical compound is combusted with air in a WTE combustion chamber the ensuing reaction is:

$$C_6H_{10}O_4 + 6.5O_2 = 6CO_2 + 5H_2O + 23.5\,MJ/kg$$

However, the MSW does not consist entirely of combustible materials. On average, US MSW contains about 25% moisture and nearly an equal amount of metals and other inorganic materials, as indicated by the fact that the ash generated in US WTE plants amounts to an average of 25% of the MSW processed. Therefore, the chemical heat contained in MSW is about one half of the heat given by the

___

* Figure 1.3 has intentionally been placed in more than one chapter in this book due to its broad relevance.

above chemical equation, that is, about 11.75 MJ/kg. In fact, the average EU MSW, after recycling, contains 10 MJ/kg, equivalent to about 2800 kWh per tonne.

US MSW contains about 30% carbon. The ratio of carbon-14 to carbon-12 in the carbon dioxide of WTE stack gas has been shown to represent the biogenic to geogenic carbon ratio in the MSW accurately. Extensive analysis of the carbon-14 fraction in the stack gas of over 30 WTE plants in the US has shown that biogenic carbon constitutes 64–66% of the total carbon content in MSW.[5]

1.3   Thermal treatment of municipal solid waste (MSW)

Thermal treatment of solid waste consists of heating the waste to a temperature that results in volatilization of moisture and organic compounds usually followed by combustion of all organic compounds and carbon with atmospheric air. When thermal treatment is effected totally by means of an external heating source, e.g. electrical heaters, infra-red energy, or thermal plasmas, the process is called pyrolysis. Because of the cost of electricity, pyrolysis can be economic only with high calorific value wastes, e.g. source-separated plastics and textiles. These chemical compounds contain little oxygen and therefore can be converted into a liquid fuel.

Thermal treatment under conditions where the amount of oxygen provided for combustion is well below that required for complete combustion is called gasification. The combustible gases generated in the gasification process consist principally of hydrogen and carbon monoxide ('syngas'). This synthetic gas can then be combusted in a separate vessel to produce steam and then electricity, or used as fuel in a gas engine, or converted into a fuel oil.

For MSW-to-energy plants to increase their contribution to the reduction of landfill burden, while providing energy to local residents at competitive costs, improvements need to be made to the combustion efficiency of the current plants. The groundwork for improvements in combustion efficiency has already been prepared, setting the stage for marked improvements in WTE plants. The composition of MSW has been extensively characterized for various regions, times of the year, etc. Thus, there is a very good understanding of this man-made 'fuel,' especially with regard to its biomass fraction. However, most of the past work has been on the development of empirical expressions to approximate the performance and emission characteristics of WTE plants. While those efforts have contributed to considerable improvements in the evolution of WTE plants in the last 40 years, especially with regard to advanced emissions control, there is still a lot of room for further advances.

1.3.1   The dominant WTE technology: grate combustion

By far the dominant WTE technology is combustion on a moving grate, either inclined or horizontal. An estimated 80% of the global WTE industry uses this method because of its simplicity and reliability. The solid wastes are unloaded from the collection trucks into a concrete bunker, which can hold a week's feedstock or more. In cases where the WTE plant is some distance away from the collection points, waste transfer stations are used where the collection trucks (2–5 tons capacity) unload and the MSW is then loaded onto larger trucks (20–25 tons) for transport to the WTE facility.

Overhead claw cranes scoop the material from the bunker and load it into the furnace hopper. A piston at the bottom of the hopper pushes the MSW onto the feed end of the moving grate (Fig. 1.4). The width of the grate depends on the design capacity of the WTE furnace (anywhere from 2 tons to 40 tons per hour) while the length ranges from 6 to 10 m. The depth of the bed on the grate ranges from about 1 m at the feed end to less than 30 cm at the discharge end, from which the WTE 'bottom' ash falls into an ash pit below the grate. Typically the solids residence time in the furnace is one hour and the maximum temperature reached is less than 1150°C.

The steam generated in the water-cooled walls of the furnace is heated to temperatures up to 500°C in superheater tubes and is used to drive the steam turbine, which generates electricity. Depending on its size and the calorific value of the MSW combusted, a WTE plant can generate from 0.3 to 0.7 MWh of electricity per ton combusted. Co-generating WTE facilities can also recover 0.5 to 1 MWh of thermal energy.

1.3.2   WTE facilities in the US

WTE is currently classified as a renewable source of electricity by the Energy Policy Act of 2005, the US Department of Energy and 24 state governments. It has been proven through carbon-14 methods (ASTM D6866 protocol) that typical

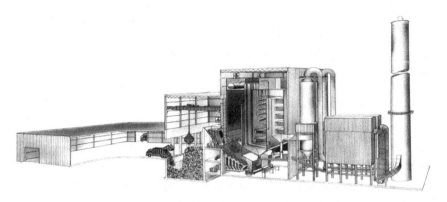

1.4 Typical arrangement of a WTE plant, with tipping floor on the left, bunker, hopper, furnace, boiler and air pollution control system on the right.

WTE stack emissions contain ~65% biogenic CO_2, i.e. bio-carbon. Waste to energy has the potential to provide disposal for more than 140 million tons of US municipal solid waste, approximately 35% of projected 2030 generation. This would prevent the emission of over 140 million tons of greenhouse gases annually. An expanded WTE industry at this scale would produce over 2% of US electricity in 2030, up from 0.5% in 2007. In addition, use of WTE is compatible with increasing recycling rates, both of which reduce landfill rates. There are 87 WTE facilities in the US, which collectively process more than 90 000 tons of MSW each day and supply electricity to more than 2 million US homes generating 2.3 GW of electricity. The US WTE industry has been in existence for more than 25 years and has developed state-of-the-art technology, making it one of the cleanest forms of energy generation, meeting or exceeding all standards set by the US Environmental Protection Agency.

WTE has two big advantages over other renewable electricity sources. It operates 24/7 to reduce baseload fossil-fuel generation and it is located in populated areas where the power is needed the most. As the US begins to focus on conservation and renewables, WTE has already proved to be a reliable technology. Excluding hydroelectric power, only 2% of the US electricity is generated from renewable energy sources. A third of this renewable energy is due to WTE. The World Economic Forum's Davos Report identifies eight emerging clean energy sectors including wind, solar and waste to energy. The National Research Council of the National Academy of Science has recognized that the ideal vision for sustainability in the year 2100 is to transition to an 'atom economy–zero waste.' Until we reach that highly integrated system, the waste that is generated today should be processed to extract energy.

1.4   Recycling and WTE

Sustainable management of municipal solid wastes requires that every possible effort be made to separate recyclable or compostable materials from the MSW stream. These materials should be separated at the source, i.e. at households, businesses and institutions, so that the cost of separation is shared by the generators (in terms of time and effort to separate recyclable materials) and by the municipalities (in terms of separate collection vehicles and processing systems). However, unless the source-separated materials are usable, they will end up in landfill. An example of the lack of markets for certain materials is the fact that over 80% of the plastic waste generated in the US is landfilled, despite considerable effort by the petrochemical industry and by many communities to increase the recycling of plastics.[6]

Opponents claim that WTE naturally deters people from recycling. Actually, the opposite is true: communities who are willing to spend money and effort on recycling come to realize that there are material properties and economic limits to how much of the MSW fraction can be recycled. Then, they look at the next available means for reducing their dependence on landfill: energy recovery by WTE. This effect is made obvious in the Sustainable Waste Management Ladder produced by the Earth Engineering Center (Fig. 1.5). It clearly shows that nations who have reduced or even eliminated landfilling have done so by a combination of recycling and waste to energy.

There was a time when uncontrolled combustion produced harmful emissions. Technology has improved and waste to energy today meets the maximum achievable control technology (MACT) standards by a large margin of safety. Today, a WTE facility operating for one year produces the amount of dioxins that are generated by 15 minutes of a fireworks display.[7] Before MACT regulation, WTE facilities emitted about 10 000 grams TEQ (toxic equivalent), today the total

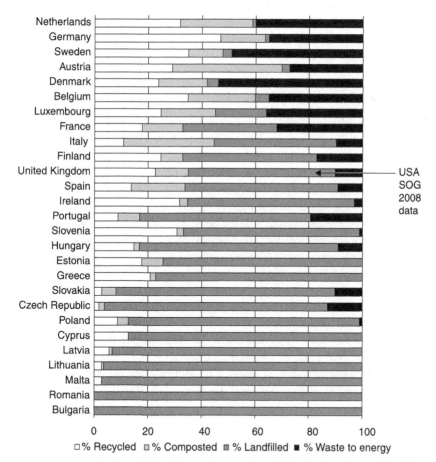

1.5 The sustainable waste management ladder produced by the Earth Engineering Center, Columbia University (based on Eurostat 2008 data). Countries can move up the ladder by recovering both materials and energy from municipal solid wastes.

dioxin emissions from all US WTE plants have been estimated by USEPA at 12 grams TEQ. In fact, right now a major source of dioxins is backyard barrel burning (580 grams TEQ). Technology has evolved and been implemented.

Finally, WTE processing of MSW has the additional benefit of reducing the transport of MSW to distant landfill sites and the attendant emissions and fuel consumption. It also reduces interstate truck traffic. According to US Department of Transportation traffic statistics, an average of seven deaths and over 40 serious injuries occur per year, based on the number of trucks required to transport New Jersey's 2 million tons of excess MSW to landfill sites in Pennsylvania, Virginia and Ohio. An article in the *New York Times* by Elizabeth Rosenthal described how Europe, in particular Denmark,[8] utilizes WTE extensively and how the US lags behind in implementation. A follow-up running commentary on the story in the 'Room for Debate' section, where the public could submit comments, demonstrated 57% were in favor and only 24% were against. So perhaps WTE is beginning to be recognized as an important solution to the waste management and energy supply issues our world faces.

1.5   Contents of this book

The contributions to this book were made by worldwide experts. The contributors are recognized for their research contributions, practical engineering solutions, committed interaction with government officials to keep them apprised of developments in WTE and tireless education efforts, both to the public and students. Each chapter is the sole responsibility of the authors. The editors ensured the chapters were reviewed by outside experts and only edited chapters for continuity and coherence throughout the book. The editors anticipate that this text will provide a foundation for accurate, as of its publication date, unbiased information on thermal conversion technologies and waste to energy in general. This book should be considered a starting point for engineers, students and professionals interested in waste to energy technologies. However, it is expected that developments will continue and it is incumbent on the reader to continue to seek out unbiased information to make informed decisions.

The book is assembled into self-contained chapters. Starting with Chapter 1, this chapter, it is clear that a case has been made for waste to energy technologies. That case is based on unbiased research through studies, site visits, conference attendances and engineering technical due diligences. It strives to put WTE into perspective and hopefully dismisses some myths that have surrounded the WTE and thermal conversion industry in general. Chapter 2 outlines some of the benefits and issues associated with implementation of waste to energy systems in communities. It covers some of the key policies that are relevant to WTE technologies, on topics such as air pollution control, ash handling and health and safety. In Chapter 3, life cycle assessment (LCA) is introduced as a tool for evaluating and comparing technologies. A quantitative comparison of different waste treatment technologies is presented, and resource conservation efficiency (RCE) is used as a metric for addressing goal-oriented needs. Chapter 4 addresses the variability of waste through a discussion on the many different types of waste and their compositions and properties. It includes an overview of the methods that are used to test different types of waste. One way to address the variability of MSW is through pre-processing of the waste, which is discussed in Chapter 5. It covers basic techniques such as visual screening and more complex technologies such as advanced separation, and biological or thermal treatment of the waste. Chapter 6 covers mass burn, refuse-derived fuel (RDF) and modular combustion systems. For each of these technologies, the process steps are discussed including waste storage and handling, feeding, grates, boilers and air pollution control. Current processes in operation as well as new developments are presented. Chapter 7 outlines options for waste firing in large combustion plants, including direct solid waste firing in pulverized-coal (PC) boilers or large fluidized-bed boilers, firing gasification gas produced from waste in PC boilers, and converting PC boilers to waste-firing fluidized-bed (FB) boilers. The advantages and disadvantages of each process are discussed. Energy recovery from WTE can be enhanced through district heating, which is discussed in Chapter 8. This chapter covers the energetic benefits of this process, and gives an overview of different types of boiler arrangements and the use of steam turbines for energy recovery. It discusses how systems can be modified to deliver electricity only, electricity and heat, or large amounts of heat (with slightly less electricity output) from waste depending on the needs of the community. Gasification is introduced in Chapter 9 as an alternative to combustion for energy recovery from MSW. The principles of gasification are discussed, as well as different process designs for gasification systems. An overview of current gasification facilities is presented. Chapter 10 outlines the progress of the WTE industry from before air pollution was controlled to the current reductions in pollutant emissions. It discusses the development of regulations and how these regulations influenced the development of WTE technologies. This chapter also presents a detailed discussion of the issue of dioxin emissions and the environmental impacts of emissions from current WTE plants. Chapter 11 discusses each component of air pollution control systems, such as acid gas scrubbing, particulate control devices, control of nitrous oxide emissions (NO_x), and control of hazardous air pollutants (HAP). Overall, this book covers a broad range of topics, from social and regulatory issues involved in implementing WTE in communities, to technology developments in order to improve waste conversion processes.

1.6   References

1   E. Matthews and N. J. Themelis, Potential for reducing global methane emissions from landfills, 2000–2030 in *11th International Waste Management and Landfill Symposium*, Sardinia, 2007.

2 N. J. Themelis, Reducing landfill methane emissions and expansion of the hierarchy of waste management in *Proceedings of Global Waste Management Symposium*, Rocky Mountains, CO, 7–10 September 2008.

3 C. Ellyin, Small scale waste to energy technologies, MS thesis, Earth Engineering Center, Columbia University 2012.

4 Bogner *et al.*, Mitigation of global greenhouse gas emissions from waste: conclusions and strategies from the Intergovernmental Panel on Climate Change, 4th Assessment, *Waste Management and Research*, 2008, 26: 11–32, wmr.sagepub.com/cgi/reprint /26/1/11.

5 E. Matthews and N.J. Themelis, Potential for reducing global methane emissions from landfills in *11th International Waste Management and Landfill Symposium*, Cagliari, Italy, 1–5 October 2007, pp. 2000–2030.

6 ASTM D6866 Testing Method.

7 N. J. Themelis, Thermal treatment review: Global growth of traditional and novel thermal treatment technologies, *Waste Management World*, July–August 2007, pp. 37–45.

8 The Public Acceptability of Incineration: Research undertaken by National Society for Clean Air and Environmental Protection, ISBN 0903 474 51 4, www.nsca.org.uk.

9 E. Rosenthal, Europe finds clean energy in trash, but US lags, *New York Times*, Environment Section, 12 April 2010.

Environmental and social impacts of waste to energy (WTE) conversion plants

T. MICHAELS, Energy Recovery Council, USA

DOI: 10.1533/9780857096364.1.15

Abstract: When considering waste to energy as a tool, communities must evaluate the benefits and challenges associated with waste to energy and compare it to the available alternatives. This chapter will examine the significant attributes of waste to energy and describe some of the challenges that exist in the development of such technology.

Key words: waste to energy, renewable energy, waste management, recycling compatibility.

2.1 Introduction

In any society, as the population grows and usable land decreases, the burden of waste management and energy supply gain significantly more attention. Since an abundance of land and natural resources is a luxury many communities do not enjoy, policymakers have long looked for sustainable methods to manage their detritus and to streamline their energy supply, whether through conservation or generation. Waste to energy (WTE) has existed longer than 'sustainability' has been in vogue, but the ability of facilities to manage waste and generate electricity must be weaved into the fabric of a community's approach to solving these critical issues.

As waste to energy is discussed in this chapter, how is it defined? There are many types of waste and many types of energy. Many terms have been utilized to refer to the same technology: waste to energy, energy from waste, incineration, resource recovery, energy recovery and more. In this chapter, waste to energy means the direct conversion of municipal solid waste (MSW) to steam or electricity. This chapter will not refer to facilities that process medical waste, hazardous waste, tire-derived fuel, biomass, or any other types of waste not characterized as MSW. This term also does not apply to the conversion of landfill gas to energy. The primary means of converting waste to energy in the world today is combustion. However, many of the issues that will be discussed in this chapter will apply to alternative technologies (e.g. gasification, pyrolysis and plasma arc) if and when they become more prevalent in the marketplace as a means to manage mixed municipal solid waste on a commercial scale.

Waste to energy plants are, in essence, sophisticated power plants that utilize MSW instead of a conventional fuel. Waste to energy facilities today have been primarily developed in Europe, North America and Asia. The vast majority of facilities are found in western Europe, which has high population density and an acute lack of free land to bury waste. In the United States, these facilities are concentrated most heavily on the east coast, which shares many of the same demographic challenges as Europe. In Asia, island nations such as Japan and Singapore have utilized waste to energy as a means of conserving their very limited land. However, China has very aggressively pursued waste to energy, not because of a lack of available land, but because of an acute need for energy.

When considering waste to energy as a tool, communities must evaluate the benefits and challenges associated with waste to energy and compare it to the available alternatives. This chapter will examine the significant attributes of waste to energy and describe some of the challenges that exist in the development of such technology.

2.2   Contributions of WTE conversion to waste reduction and energy generation

2.2.1   Waste management contributions of WTE

Waste is a by-product of consumption and is generated at every level of existence, from the cellular level to the macroeconomic level. Waste management is an issue faced by every society in the world, and has been for millennia. In fact, apart from law enforcement and education, waste management is perhaps the most important function of local government. Many politicians will tell you that while they have received little to no credit for successful waste management practices, political peril awaits on days when the trash is not collected.

Land, population density, economic strength and the importance placed on sustainability are the leading factors that impact the development of a waste management strategy. How a community prioritizes those factors is what leads different communities, regions and nations to manage their waste in such disparate ways. Many governments utilize a solid waste hierarchy to provide general guidance on how to prioritize the use of various waste management alternatives. The most common hierarchies share the following characteristics in descending order of preference: waste reduction, reuse, recycling and/or composting, waste to energy and landfilling. This hierarchy is shown in Fig. 1.3 in Chapter 1. The most successful waste management strategies in the world recognize that a successful system will include all options in varying degrees. The most unsuccessful strategies in the world rely entirely on the least preferable alternatives, or set unrealistic goals of utilizing only the most preferable. The lack of planning associated with choosing only the most preferable option usually leads to an eventual overuse of the least preferable option. A balanced approach with appropriate emphasis on the hierarchy has always been the most successful.

Communities should place a high priority on reducing, reusing and recycling waste. However, in every society in the world, there are limits to these options (both economic and technological) and there will be waste left over. The question then becomes, what do you do with that waste? It is at this point that the hierarchies prefer waste to energy to landfilling. Whereas waste to energy facilities are buildings in which waste is processed, landfills are traps that hold waste. Once a landfill is full, it must be closed and land for another landfill must be found. This is a pressing issue in many areas around the world, and will become an issue in all areas over time. Sustainable waste management is good practice in all communities; however, some will not feel the pinch for quite some time, which allows them to delay the hard decisions.

Waste to energy is also preferable to landfilling in the sense of energy recovery. Even at a landfill that has landfill-gas capture and energy generation, the amount of electricity that can be drawn from each ton of trash is significantly lower than what can be achieved at a waste to energy plant. Waste to energy plants extract an average of 600 kWh of electricity from one ton of MSW, whereas landfills extract an average of 65 kWh per ton (Kaplan *et al.*, 2009). In addition, the electricity generated from waste in a landfill setting is extracted slowly over time, whereas electricity is generated immediately in a waste to energy facility.

2.2.2 Energy generation attributes of WTE

While trash is a commodity that must be managed and mitigated, energy is a commodity that must be generated. All modern societies run on energy and the search for energy has consumed the human population ever since the Industrial Revolution. Ironically, energy companies have been digging holes in the Earth in the hope of finding energy at the very same time communities and companies have been digging holes in the ground to bury waste material containing energy. It is much more efficient to extract energy from waste and reduce the number of holes that are dug in the Earth.

Economies develop based on access to affordable energy. While access to energy has allowed developed nations to reach high levels, the limited nature of energy will stress those same civilizations, especially as developing nations seek to catch up with the world's leading economies. It is increasingly important that societies do not ignore substantial available energy sources – and waste is a significant source.

Europeans have long understood the intrinsic energy value of waste to energy and have used it to generate electricity, steam, combined heat and power, and district heating. Energy generation in the United States has focused primarily on electricity generation for the grid. While the electric capacity of waste to energy plants is small considering the electric capacity of a large coal-fired power plant, MSW will otherwise be buried in the ground if the energy is not recovered. Given that the energy demand will not be quenched in the near future and waste will continue to be generated, waste to energy is a technology that helps solve two issues and still has incredible potential. In the United States, 85 waste to energy facilities process only 7% of the national waste stream (van Haaren et al., 2010). With the proper incentive, the potential for waste to energy to grow and to provide much needed energy is high. In the United States, if all discarded wastes were utilized in waste to energy plants for electricity generation, there would be a generation capacity of approximately 14 000 MW (Kaplan et al., 2009).

2.3   Air quality and residue management considerations of WTE conversion

2.3.1   Air quality considerations of WTE

The air quality associated with waste to energy facilities is one of the most fiercely contested issues between people who support waste to energy and those who do not. Those who do not support waste to energy generally cite the opinion that the air quality issues cannot be overcome, as the reason for their position. However, it is not the case that air quality issues cannot be overcome. The combustion of municipal solid waste over time has taken many forms, but technology and the associated emission profile have continuously and dramatically improved. The emissions produced by waste to energy today should not be confused with those of back-yard burn barrels. Allusions to 'smokestacks' and 'incinerators' harken back to the days when units were used solely for volume reduction, without the benefits of emission control or energy recovery.

Of course, much has changed over the years. Today's waste to energy facilities are sophisticated power plants that use MSW to generate energy. In the United States, the Clean Air Act Amendments of 1990 required industries, including municipal waste combustors, to establish and achieve maximum achievable control technology (MACT) standards. Municipal waste combustor units processing more than 250 tons per day were required to meet the new standards by 2000 and units processing less than 250 tons per day were required to meet MACT standards by 2005. These standards are among the most stringent in the United States (US EPA, 1996). Companies and local governments have invested more than a billion dollars over the years to retrofit waste to energy plants and install the most modern emissions control equipment.

As a result of these investments, and after review of compliance data, the United States Environmental Protection Agency (EPA) in 2003 said that waste-to-energy facilities produce 'electricity with less environmental impact than almost any other source of electricity' (Holmstead and Horinko, 2003). In 2007, a memo from the EPA's Office of Air and Radiation showed that waste to energy achieved significant and remarkable reductions in criteria pollutants (Stevenson, 2007).

Emissions control equipment plays a major factor in the environmental performance of waste to energy facilities. However, there are several other factors that contribute to the superior emissions of waste to energy facilities today. Improvements in combustion engineering have helped limit the formation of pollutants in the boiler and facilities are always striving to achieve better performance than exists today.

Waste to energy will also always have detractors because there are those who would prefer that waste did not exist in the first place. There is a misconception that if waste to energy facilities and landfill sites were closed, people would have no choice but to stop generating trash. As there are those who prefer zero waste, some also desire zero emissions from waste to energy. Zero emissions policies, like zero waste policies, are laudable goals, but they are not reality. That is why the EPA and other regulatory agencies around the world set standards that are protective of human health and the environment with a margin of safety. In reality, waste to energy facilities meet those standards and usually operate at a fraction of the allowable limits. The truth of the matter is that waste to energy facilities are some of the most regulated facilities in the world and that the largest opportunities to reduce emissions can be achieved by focusing on other sources.

2.3.2 Ash management associated with WTE

Waste to energy facilities burn trash to generate electricity. After combustion, an ash residue remains that is approximately 10% of the original volume and 25% of the original weight of the MSW. Ash comprises both bottom ash, which is the residue of the waste combustion, and fly ash, which is the residue of the emissions control equipment and processes. This ash has historically been disposed in landfills. Disposal is not the only option. A significant amount of ash is reused as landfill roadbed material, daily and final landfill cover, road aggregate, asphalt-mixture and even in the construction of artificial reefs and cement blocks (Roethel and Breslin, 1995).

In the United States, in accordance with the federal law, waste to energy ash is tested to ensure it is non-hazardous. The US EPA has developed a test called the toxicity characteristic leaching procedure (TCLP), which subjects ash to an acidic liquid, causing metals to leach from the material. If metals leach in amounts greater than a fraction of a percent, the ash is considered hazardous. Years of testing ash from every waste to energy facility in the country has proven that ash is safe for disposal and reuse. Waste to energy ash consistently passes TCLP, despite the fact that the TCLP test greatly exaggerates the potential for metals to leach from ash into the environment (US EPA and Coalition on Resource Recovery and the Environment, 1990).

Test results and measurements taken in the field show that the levels of metals present in waste to energy ash leachate are close to the significantly more restrictive drinking water standards and far lower than the TCLP toxicity criteria (US EPA and Coalition on Resource Recovery and the Environment, 1990; Andrews, 1991).

Typical practice in the United States has been to combine the bottom ash and the fly ash on site. This combined ash stream makes the beneficial reuse of ash less desirable for reuse in construction activities. In the US, most ash is sent to landfill, either for disposal or as daily cover, avoiding the need for clean dirt. In Europe, there is a higher rate of reuse of bottom ash due to separated ash management; however, the fly ash is usually treated as hazardous waste.

2.4   Greenhouse gas profile of WTE

Waste to energy achieves a reduction of greenhouse gas emissions through three separate mechanisms: (1) by generating electrical power or steam, waste to energy avoids carbon dioxide (CO_2) emissions from fossil fuel-based electrical generation; (2) the waste to energy combustion process effectively eliminates all potential methane emissions from landfill, thereby avoiding any potential release of methane in the future; and (3) the recovery of ferrous and nonferrous metals from municipal solid waste by waste to energy is more energy efficient than production from raw materials.

These three mechanisms provide a true accounting of the greenhouse gas emission reduction potential of waste to energy. A life cycle analysis, such as the US Environmental Protection Agency's municipal solid waste decision support tool, is the most accurate method for understanding and quantifying the complete accounting of any waste management option. A life cycle approach should be used by decision makers to weigh and compare all of the effects of greenhouse gas emissions associated with various activities and management options. Life cycle analysis is discussed in Chapter 3.

The decision support tool is a peer-reviewed tool (Harrison *et al.*, 2001), which lets the user directly compare the energy and environmental consequences of various management options for a specific or general situation. Technical papers authored by the EPA (Thorneloe *et al.*, 2005, 2007) report on the use of the decision support tool to study municipal solid waste management options.

These studies used a life cycle analysis to determine the environmental and energy impacts for various combinations of recycling, landfilling and waste to energy. The results of the studies show that waste to energy yielded the best results – maximum energy with the least environmental impact (emissions of greenhouse gases, nitrogen oxide, fine particulate precursors, etc). In brief, waste to energy has been demonstrated to be the best waste management option for both energy and environmental parameters and specifically for greenhouse gas emissions.

When the decision support tool is applied to the US inventory of waste to energy facilities, which process 28 million tons of trash, it has been shown that the waste to energy industry prevents the release of approximately 28 million tons of carbon dioxide equivalents that would have been released into the atmosphere if waste to energy were not employed (Thorneloe *et al.*, 2002).

The ability of waste to energy to prevent greenhouse gas emissions on a life cycle basis and to mitigate climate change has been recognized in the actions taken by foreign nations trying to comply with Kyoto targets. The Intergovernmental Panel on Climate Change (IPCC), the Nobel Prize winning independent panel of scientific and technical experts, has recognized waste to energy as a key greenhouse gas emission mitigation technology. The World Economic Forum in its 2009 report, 'Green Investing: Towards a Clean Energy Infrastructure,' identifies waste to energy as one of eight technologies likely to make a meaningful contribution to a future low-carbon energy system.

In the European Union, waste to energy facilities are not required to have a permit or credits for emissions of CO_2, because of their greenhouse gas mitigation potential. In the 2005 report, 'Waste Sector's Contribution to Climate Protection,' the German Ministry of the Environment stated that 'waste incineration plants and co-incineration display the greatest potential for reducing emissions of greenhouse gases' (Dehoust *et al.*, 2005). The report concluded that the use of waste combustion with energy recovery coupled with the reduction in landfilling of biodegradable waste will assist the European Union 15 in meeting its obligations under the Kyoto Protocol. In a 2008 briefing, the European Environment Agency attributes reductions in waste management greenhouse gas emissions to waste to energy (EEA, 2008).

Under the Kyoto Protocol, by displacing fossil-fuel-fired electricity generation and eliminating methane production from landfill sites, waste to energy plants can generate tradable credits (certified emission reductions or CERs) through approved clean development mechanism protocols (UNFCCC, 2008). These CERs are accepted as a compliance tool in the European Union's emissions trading scheme.

The ability of waste to energy to reduce greenhouse gas emissions has been embraced in the United States as well. The US Conference of Mayors adopted a resolution in 2004 recognizing the greenhouse gas reduction benefits of waste to energy. In addition, the US Mayors' Climate Protection Agreement supports a 7% reduction in greenhouse gases from 1990 levels by 2012. The agreement recognizes waste to energy technology as a means to achieve that goal. As of 31 October 2010, 1044 mayors have signed the agreement (US Conference of Mayors, 2010).

The Global Roundtable on Climate Change (GROCC), convened by Columbia University's Earth Institute, issued a statement on 20 February 2007 identifying waste to energy as a means to reduce carbon dioxide emissions by the electricity generating sector and methane emissions from landfill. The GROCC, which brought together high-level, critical stakeholders from all regions of the world, recognized the importance of waste to energy's role in reducing greenhouse gas emissions.

2.5   Compatibility of WTE with recycling

Statistics compiled for nearly two decades have proven that waste to energy and recycling are compatible, despite many attempts by detractors to conclude otherwise. Since research on the subject began in 1992 (Kiser, 2003; Berenyi, 2009), communities that rely upon waste to energy maintain, on average, a higher recycling rate than the national EPA average.

Communities that employ integrated waste management systems usually have higher recycling rates and the use of waste to energy in these integrated systems plays a key role. There are several factors that explain why the recycling rates of communities with waste to energy facilities are higher than those without. First, communities with waste to energy plants tend to be more knowledgeable and forward thinking about recycling and MSW management in general. Second, communities with waste to energy plants have more opportunities to recycle since they handle the MSW stream more. Third, a municipal recycling program can be combined with on-site materials recovery at the waste to energy plant (e.g. metals recovered at a waste to energy plant post-combustion usually cannot be recycled curbside and would otherwise have been buried had that trash been landfilled).

In a paper entitled 'Recycling and waste-to-energy: Are they compatible? (2009 update),' Eileen Berenyi with Governmental Advisory Associates, Inc. researched the recycling characteristics surrounding 82 waste to energy facilities in 22 states. Recycling data from 567 local governments as well as statewide data from the 22 states were covered in the report. The report shows that in 2009 communities with waste to energy facilities have an average recycling rate of 33.2%. The national average for recycling as estimated by the EPA is estimated at 32.5%, while BioCycle/Columbia University estimate it to be 28.6%. However, Berenyi has calculated an 'adjusted' recycling rate for the EPA that more closely tracks the recycling rates calculated by others (Berenyi, 2009).

The unadjusted, US EPA computed, national recycling rate (32.5%) was computed using a waste stream model and includes commercial and industrial components and yard waste. These materials are often excluded in individual state and local recycling tonnages. In order to juxtapose comparable statistics, it is appropriate to use Berenyi's adjusted EPA rate of 27.8%. Regardless of what factor you use, communities with waste to energy facilities outperform communities without when it comes to recycling. If you compare the US EPA rate to the BioCycle or the Berenyi adjusted EPA rate, the difference is almost five percentage points. This is borne out by the recycling rates of European countries, which depend on their reliance upon waste to energy or landfilling. The most progressive countries recycle extensively, recover energy as much as possible, and landfill little. Less advanced countries landfill as much as possible, and recycle and combust almost nothing. Figure 1.5 in Chapter 1 shows the amount of landfilling, recycling and WTE in different countries.

Opponents of waste to energy often suggest that they oppose waste to energy because they would prefer to recycle instead. This line of reasoning creates a false choice when both options must work in conjunction.

2.6   Health and safety aspects of WTE

WTE facilities focus on health and safety for the simple reason that companies want all of their employees to go home at the end of each day alive and healthy. The sophistication of the facilities requires incredible attention to detail to ensure that health and safety are paramount.

Waste to energy facilities, like all other workplaces, must meet the tough standards put in place by the US Occupational Safety and Health Administration (OSHA). The waste to energy industry takes tremendous pride in its health and safety programs and often goes beyond what is required by law. Great importance is placed on developing and implementing successful programs that protect the people working in the plants.

OSHA has recognized the stellar accomplishments of 51 waste to energy facilities in the United States with the designation of Star status under the voluntary protection program (VPP). VPP Star status is the highest honor given to worksites with comprehensive, successful safety and health management systems. Star sites are committed to effective employee protection beyond the requirements of federal standards and participants develop and implement systems to effectively identify, evaluate, prevent and control occupational hazards to prevent injuries and illnesses. The keys to health and safety success under VPP are employee engagement and ongoing involvement in on-site health and safety program development combined with long-term commitment and support from management. Facilities that achieve Star status routinely have injury and illness rates that are at or below the state average for their specific industry.

Impressively, 51 of the 85 waste to energy facilities in the United States have earned VPP Star status. While only a small fraction of 1% of all worksites in the United States are enrolled in VPP, more than 59% of US waste to energy facilities have achieved Star status. This illustrates the commitment of this sector to health and safety.

In addition, through an alliance program with OSHA, waste to energy facilities in the United States created a program to promote the safety of haulers on the tipping floor of waste to energy facilities. The 'Safety: Do it for life' campaign was launched regionally in 2006, launched nationally in 2007, and takes place each year. The campaign has designated that June is 'hauler safety month' and waste to energy facilities make a special effort to educate the haulers that deliver trash to each facility of the tipping floor rules. The campaign has an underlying and unifying personal message to keep safety at the forefront of work practices in order to stay safe for family and loved ones.

2.7   Integrated planning for WTE plants

Waste to energy is an important part of an integrated solid waste management system. The most successful systems in the world are those that rely on a variety of tools to manage a complicated problem. A community must make a host of decisions with respect to managing municipal solid waste, which will evolve into a path forward on some or preferably all of the alternatives. Issues that are often evaluated and addressed by communities include cost, environmental impact, sustainability, public perception, job creation, greenhouse gas impacts, energy demand and siting.

Different communities place different values on factors associated with waste management, which is why an integrated system is the most successful because the various options can be tailored to best serve the community in need. Tools, such as EPA's municipal solid waste decision support tool, help local governments evaluate the characteristics associated with the different waste management systems and allow them to decide for themselves which factors have the most significance. Cost used to be the most important factor, and the environmental benefits would have been a perk in a cost-effective solution. However, more and more local governments are making decisions based on environmental aspects, or sustainability aspects, with cost a secondary factor. Many communities understand that landfill sites have finite capacity and that long-haul transport of waste is not often a financially or environmentally sustainable activity. Some waste to energy facilities are built to ensure that local landfill capacity is not depleted too soon. In densely populated areas, existing landfill sites may be a less costly alternative, but establishing the next landfill site may be an extraordinarily expensive prospect.

The ability of waste to energy to reduce greenhouse gases on a life cycle basis may also be a factor that enhances the attractiveness of waste to energy to a community. Many local governments have made political commitments to reduce greenhouse gas emissions. Europeans have used waste to energy to reduce baseline emissions (Dehoust et al., 2005). Waste to energy facilities in Asia are being developed and may receive CERs under the clean development mechanism (CDM) of the Kyoto Protocol. More and more, as public policies drive carbon constraints, waste to energy will be an active consideration.

The cost of waste disposal at a waste to energy facility is directly related to the revenue that is generated from energy sales. When considering a facility, the need for energy in a community is a vital consideration, especially if the product is steam or district heating. Electric generation is by far the most common energy product from waste to energy facilities in the United States due in large part to the fact that it can be put on the grid from almost any location. In addition, it removes the risk associated with having a symbiotic partner nearby. Several otherwise successful waste to energy facilities have been forced to close due to the fact that their steam customers either moved or went out of business. Locating in a place where there are long-term energy customers will greatly enhance the potential success of a facility. Europe has had much success with using waste to energy to feed district heating loops. While the infrastructure for district heating is lacking in the United States, such a consideration would make waste to energy even more attractive to a community.

The siting of a waste to energy facility is often the subject of intense debate. There will generally be opposition by at least someone any time you site a piece of energy infrastructure, whether it is a waste to energy plant, a landfill, a biomass facility, a natural gas power plant, a liquefied natural gas terminal, or even a wind turbine. Often the opposition comes from a small, but vocal minority, and it must be dealt with. Knowing that there will be opposition requires the developer to ensure that each aspect of the development plan is carefully laid out. Is the cost of waste to energy comparable to the cost of alternative disposal? How will the recycling program be bolstered by the presence of waste to energy? What air pollution control technologies will be employed to ensure superior performance? What are the benefits associated with the site location that make it preferable to other alternatives?

When a community can answer these questions, much of the uncertainty felt by residents who are unfamiliar with the technology will be alleviated. Of course, there will always be part of the population who fear change and will not support a facility. However, facilities in the United States that expand their existing operations face almost no opposition, which is not true of proposed new facilities. This is because communities that have facilities understand that the facilities are good neighbors and that many of the initial concerns were unfounded. In a community that is considering its first facility, it is generally hard to disprove a negative until the people have first-hand experience.

Siting of waste to energy facilities take leadership, knowledge and capital. Communities that are willing to put all three to use are those that end up with successful and sustainable solutions to their waste management challenges.

2.8   Future trends

Waste to energy facilities are sophisticated power plants with state-of-the art emissions control technology. These facilities require significant capital investment, more so than the alternatives for solid waste management. In Europe, waste to energy has benefitted from policies that make it more difficult or more expensive to landfill. High taxes on landfill tipping fees, EU regulations that require members to reduce the amount of landfilling (EU, 1999), and national policies banning outright landfilling of combustible material have all led to a culture in which waste to energy can thrive. The United States does not employ such policies, mainly because policymakers are reluctant to raise the price of landfilling when there is still so much open space available in parts of the country. However, there are policies that can be put in place that will enhance the opportunity for waste to energy to be developed. Those policies usually provide income on the non-waste side of the ledger so that tipping fees may be lowered to a level much more in line with landfill rates.

A popular mechanism for providing incentives to renewable energy sources in the United States is a renewable portfolio standard (RPS). Thirty states have enacted statewide policies that require electric utilities to provide a minimum percentage of their electricity from renewable sources. The utility may meet this requirement by owning a renewable generating technology, purchasing power from a renewable generating technology, or purchasing renewable energy credits from a renewable source. Renewable energy credits represent the environmental benefits of the electricity generated by a renewable source and can be sold independently from the power as a commodity, subject to market pricing. The price will be determined by the size of the renewable mandate and the availability of renewable generating capacity. Each state has the flexibility to determine which generating sources are deemed renewable for purposes of the RPS. Waste to energy has been defined as renewable in 21 states, the District of Columbia and Puerto Rico. Many federal proposals have been introduced and more are pending in Congress, which together would establish a national mandate. In 2010, all of the leading proposals included waste to energy as renewable; however, the likelihood of passage of this legislation is complicated by the regional and partisan politics.

In the United States, renewable energy production tax credits have been available for waste to energy since 2004 (26 USC § 45). Electricity generated by new waste to energy capacity placed in service while the tax credit is in place is eligible to receive a 1.1 cent/kWh (adjusted for inflation) tax credit for a period of ten years. Because of legislation enacted in 2009 (Pub. L. 111-5), eligible taxpayers can convert a production tax credit to an investment tax credit allowing the recipient to take a tax credit for up to 30% of the capital costs of a project in the year in which it was incurred. This investment tax credit can also be converted to a cash grant, which was preferred by companies that did not have sufficient profits against which to write off a tax credit. The cash grant program was further restricted to facilities that started construction early. The issue with tax policies is that they benefit taxpayers, which local governments are not. Therefore, a waste to energy facility owned by a governmental entity and operated by a private vendor (a common scenario) would not benefit from tax credits. Privately owned facilities do benefit from these tax policies as long as they meet the other restrictions.

The United States federal government also provides incentives to renewable energy sources by requiring federal agencies to procure electricity from renewable sources. The Energy Policy Act of 2005 (Pub. L. 109-58) requires federal agencies to purchase 7.5% of their electricity from renewable resources by 2013. Waste to energy is an eligible renewable resource under this law. Since the federal government is an incredibly large consumer of electricity, this policy helps ensure a marketplace for renewable energy. It can be expected that the percentage of electricity generated by renewable sources required to be purchased by federal agencies will increase over time.

Climate change legislation can have dramatic positive or negative impacts on waste to energy depending on how it is drafted. Already policies worldwide have had a positive effect on waste to energy. The Kyoto Protocol allows waste to energy facilities in developing countries to earn CERs, which has led to increased development. In Europe, waste to energy is an attractive alternative since it reduces baseline greenhouse gas emissions. In the United States, there is no national climate legislation. Federal proposals have worked their way through Congress, but prospects for sweeping new policies seem limited in the near future. Current proposals would exempt waste to energy from the cap, which allows facilities to operate without purchasing compliance credits to cover greenhouse gas emissions. This type of policy recognizes that waste to energy reduces greenhouse gas emissions on a life cycle basis. A more myopic view would lead to a policy that penalizes sources that emit any carbon, despite the life cycle benefits. Such a facility would be required to purchase compliance credits at significant cost.

The marketing of carbon offsets provides a ripe opportunity for waste to energy to benefit financially from a carbon-constrained marketplace. Many proposals allow sources of greenhouse gas reductions from non-capped sectors to sell offsets to those sources covered by the cap. Policies that recognize the net greenhouse gas reductions of waste to energy and permit those reductions to be quantified and sold as offsets are a significant opportunity to promote waste to energy. In the United States, the Lee County (FL) waste to energy facility is the first waste to energy facility in the nation to qualify for carbon offsets. Greenhouse gas reductions associated with the capacity of a new boiler added to the existing facility in 2007 are quantified under the voluntary carbon standard (VCS) and have been sold to entities that purchase offsets voluntarily on the market. While the future revenue potential associated with voluntary markets is limited, a mandatory carbon program could provide tangible and significant benefits to waste to energy, which will make these facilities much more attractive in the future.

2.9   References

Andrews C (1991), 'Analysis of laboratory and field leachate test data for ash from twelve municipal solid waste combustors,' in *Municipal Waste Combustion: Conference Papers and Abstracts* for the 2nd Annual International Specialty Conference.

Berenyi E (2009), 'Recycling and waste to energy: Are they compatible? (2009 update),' www.energyrecoverycouncil.org/userfiles/file/2009%20Berenyi%20recycling%20update.pdf.

Dehoust G, Wiegmann K, Fritsche U, Stahl H, Jenseit W, *et al.* (2005), 'Waste sector's contribution to climate protection,' www.energyrecoverycouncil.org/userfiles/file/2005AugGermanyclimate.pdf.

EEA (European Environment Agency) (2008), 'Better management of municipal waste will reduce greenhouse gas emissions,' www.eea.europa.eu/publications/briefing_2008_1/EN_Briefing_01-2008.pdf.

EU (European Union) (1999), Council Directive 1999/31/EC of 26 April 1999 on the landfill of waste. *Official Journal of the European Communities*, L182, 42, 1–19.

Harrison K, Dumas R, Solano E, Barlaz M, Brill E, *et al.* (2001), 'Decision support tool for life-cycle-based solid waste management,' *Journal of Computing in Civil Engineering*, 15, 44–58.

Holmstead J and Horinko M (2003), Letter to Zannes M, Integrated Waste Services Association, 14 February, www.energyrecoverycouncil.org/userfiles/file/epaletter.pdf.

Kaplan P, DeCarolis J and Thorneloe S (2009), 'Is it better to bury or burn waste for clean electricity generation?' *Environ. Sci. Technol.*, 43, 1711–1717.

Kiser, J (2003), 'Recycling and waste to energy: The ongoing compatibility success story,' *MSW Management*, reventurepark.com/uploads/1_CWR_ART_2.pdf.

Roethel F J and Breslin V T (1995), 'Municipal solid waste (MSW) combustor ash demonstration program The Boathouse,' US Environmental Protection Agency Report EPA/600/SR-95/129, August.

Stevenson W (2007), Memorandum re 'Emissions from large and small MWC units at MACT Compliance,' 10 August, www.energyrecoverycouncil.org/userfiles/file/2007_EPAemissions_memo.pdf.

Thorneloe S, Weitz K and Jambeck J (2005), 'Moving from solid waste disposal to management in the United States,' www.reventurepark.com/uploads/1_WM_ART_3.pdf.

Thorneloe S, Weitz K and Jambeck J (2007), 'Application of the US decision support tool for materials and waste management,' *Waste Management*, 27, 1006–20.

Thorneloe S A, Weitz K A, Nishtala S R, Yarkosky S and Zannes M (2002), 'The impact of municipal solid waste management on greenhouse gas emissions in the United States', *J. Air & Waste Manage. Assoc.*, 52, 1000–1011.

UNFCCC (2008), 'Avoided emissions from organic waste through alternative waste treatment processes,' AM0025 v11 and associated memorandum, cdm.unfccc.int/UserManagement/FileStorage/CDMWF_AM_PJSD36RRF6X16OA7CSTR7H38OXVJTG.

US Conference of Mayors (2010), www.usmayors.org/climateprotection/revised, (accessed 8 November 2010).

US EPA (US Environmental Protection Agency) (1996), 'Standards of performance for new stationary sources and guidelines for control of existing sources: municipal solid waste landfills,' Federal Register, Vol. 61, No. 49.

US EPA (US Environmental Protection Agency) and Coalition on Resource Recovery and the Environment (1990), 'Characterization of municipal combustion ash, ash extracts, and leachates,' nepis.epa.gov.

van Haaren R, Themelis N and Goldstein N (2010), 'The state of garbage in America', *BioCycle*, October, www.seas.columbia.edu/earth/wtert/sofos/SOG2010.pdf.

Lifecycle assessment (LCA) and its application to sustainable waste management

S. KAUFMAN, Columbia University, USA

DOI: 10.1533/9780857096364.1.29

Abstract: This chapter deals with quantitative ways to compare waste treatment technologies. It introduces lifecycle analysis and details a lifecycle-based method to evaluate the sustainability of waste treatment systems. It concludes that cities need a mix of waste treatment technologies to manage municipal solid waste flows sustainably.

Key words: lifecycle assessment (LCA), waste to energy (WTE) systems, sustainable waste management, recycling, landfill, waste treatment technology.

3.1 Introduction

Lifecycle assessment (LCA), otherwise known as lifecycle analysis – though still a relatively new field – is one of the leading quantitative methodologies for assessing the sustainability of human activities. Though there is some debate about who exactly performed the first LCA (it occurred sometime in the 1960s (Baumann, 2004)), it is clear that beginning in the 1990s, it went from an activity performed by a handful of companies and academics to a more widespread and broadly recognized tool. Today, most people in the world of sustainability have if not a working knowledge then at least a rudimentary understanding of what it is and what it involves. It is probably pretty safe to say that anyone with at least a cursory relationship to sustainability has read at least a few LCA reports.

So what is LCA? Broadly speaking, it is an accounting tool that takes all of the resource inputs (usually in the form of materials and energy) across the entire lifecycle of a product or service; sums up all of the emissions outputs across the same lifecycle stages; and wraps that all up in terms of environmental impacts that can be understood and reacted to by decision-makers.

All LCAs contain four fundamental building blocks (USEPA, 2006):

1. *The goal and scope definition.* This is the part of the LCA where the practitioner lays out the basic parameters of the study. What is the point of this particular LCA? Is it an internal, corporate study for product managers to better understand the environmental impacts of their product? Or is it a similar study resulting in a public report of lifecycle impacts? And which impacts will you be considering – a full range, or perhaps a focus simply on lifecycle energy or global warming potentials associated with your product? In addition, you will need to decide what are the boundaries of your study – which processes will

be included and excluded? For most full LCAs, all five main lifecycle stages (shown in Fig. 3.1) are considered but in some cases – and often in waste treatment scenario evaluation – a few stages are left out. We will explore this later in the chapter.

2. *The Lifecycle Inventory.* The LCI phase represents the real meat of the LCA, and is therefore very important. This is the part of the LCA where the user accounts for all of the inflows (material and energy requirements) and outflows (emissions to the environment in the form of atmospheric, land and aquatic discharges).

3. *The Lifecycle Impact Assessment.* The LCIA phase takes the quantitative data generated during the LCI step (step 2 above) and assesses and evaluates those numbers in terms of their impact on broader environmental categories. These categories often include, but are not necessarily limited to, the greenhouse effect (global warming potential); ozone depletion; acidification; eutrophication; and natural resource depletion. (There are also non-environmental impacts included in many LCIA methods, such as human health.) LCIAs include classification and weighting steps, so the results – while often presented as final 'numbers,' usually include some additional uncertainty.

4. *Interpretation.* This is the 'human' step – the results of the first three phases of the LCA are systematically reviewed by the person or team in charge of the study. The results are then interpreted based on the intended purpose and audience, and are usually formatted into a final report, which is made available to the identified audience.

Perhaps the easiest way to understand how an LCA works is by looking at the most commonly used type of LCA – the product carbon footprint (PCF). A product carbon footprint is an LCA in which the impact assessment stage is confined to the global warming potential (BSI, 2008). So, to calculate a product carbon footprint for a can of cola, for example, one would account for all of the inputs across the five main lifecycle stages, as illustrated in a simplified form in Fig. 3.1.

The next step would be to account for all of the greenhouse gas emissions resulting from inputs at each stage. All of these steps account for the LCI phase of the LCA/carbon footprint. The results of this LCI would then be fed into the LCIA – in this case, the greenhouse gas emissions would be expressed as an overall global warming potential (GWP), usually following the methodology of

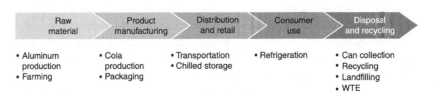

3.1 Lifecycle stages for a can of cola.

the Intergovernmental Panel on Climate Change (IPCC), which keeps an up-to-date method that is publicly available (IPCC, 2007).

A full-scale LCA differs from a product carbon footprint type of LCA only in the sense that the scope of environmental burdens is broader and accounts for a wider range of impacts. Usually, the material and energy inputs that need to be collected for a product carbon footprint LCI are almost exactly the same as those required for a full-scale LCA.

3.2 Energetic comparison of waste to energy (WTE) systems and alternative waste options

In this section, we explore in some detail the ways LCA helps quantify the energetic tradeoffs between the major waste treatment options (recycling, composting, WTE and landfilling). We begin by exploring the difference between energy as it is produced and utilized by facilities (traditional power plants, WTE plants, landfill gas to energy plants, etc) and the 'lifecycle' energy that LCA seeks to incorporate. We then illustrate the quantitative differences between MSW management technologies from a lifecycle energy perspective, using actual cumulative energy demand values from the literature.

The energetic implications of the management of municipal solid waste (MSW) vary considerably depending on the treatment methods chosen. All three of the main contemporary methods of managing waste – recycling (including composting and, increasingly, anaerobic digestion), waste to energy and landfilling – have the potential to either save or recover energy. This differs from the situation only a few decades ago, when the majority of waste generated in the United States was simply landfilled or combusted in crude facilities without the capacity to recover the potential energy (or, indeed, emissions) generated by the processes (USEPA, 2009).

Lifecycle analysis has often been used to measure the energetic efficiency of different systems (Huijbregts *et al.*, 2006). This is often accomplished through a lifecycle impact assessment method known as cumulative energy demand (CED), also known as lifecycle embodied energy (LEE). CED is a cradle-to-grave account of the energy inputs and consumption necessary to manufacture, use and dispose of a product (Blok, 2006). As one would expect from an LCIA, each stage in the lifecycle is accounted for but instead of material inputs and emissions outputs, you simply sum the total energy demand across all the stages.

As its name would imply, CED is frequently employed to determine the energy payback periods for alternative energy generation technologies such as solar (Knapp and Jester, 2001), wind (Wagner and Pick, 2004) and biomass (Kim and Dale, 2004). It has also been used to evaluate the lifecycle energy impact of buildings and building materials (Thormark, 2002). A recent study utilized CED as the basis of a metric to evaluate the effectiveness of waste management systems (Kaufman *et al.*, 2010). This study will be referenced throughout the remainder of this chapter.

When people talk about the benefits of recycling, they often refer to the lifecycle energy savings of recycling a given material versus disposing of it and producing a new, similar product from virgin materials. This is indeed the case for many materials commonly found in the waste stream. Most metals require a great deal more energy to manufacture from virgin resources than to recover from the waste stream and recycle into a new product. This is particularly the case for aluminum – some studies show that producing a tonne of aluminum from recovered scrap can save as much as 175 GJ over virgin production, a dramatic difference (McDougall and White, 2001). Other metals such as steel cans also yield savings, though not nearly as dramatic as those realized by aluminum recycling.

No matter the material, the methodology for calculating the energy savings from recycling is the same: tabulate the lifecycle energy demand (CED) of the material produced from virgin resources; calculate the energy required to collect, sort, clean and re-manufacture the material into a new product (which is the energy needed to recycle the scrap into a new product); and finally subtract the energy required to recycle the scrap into a new product from the energy required to manufacture the product from virgin resources:

Energy savings from recycling = (energy required for virgin production) – (energy required for a recycled production)   [3.1]

Table 3.1 Energy savings from recycling, energy recovered by WTE and energy recovered in landfill gas for different waste materials

Material	E_{REC}, rec. energy (GJ/ton)	E_{WTE}, WTE energy (GJ/ton)	E_{LF}, LF energy (GJ/ton)
Newspaper	12.7	5.5	1.4
Kraft paper	13.3	5.5	1.4
Waxed old corrugated containers		5.5	1.4
High grade paper	21.8	5.5	1.4
Mixed low grade paper	13.3	5.5	1.4
Polycoated paper		5.5	1.4
Soiled paper	2.0	1.8	1.4
Other paper		5.2	1.4
Glass	4.0		
PET	33.1	8.9	
HDPE	38.7	16.4	
PP	34.2	12.6	
Aluminum	188.1	170.1	
Steel	16.4	14.8	
Yard waste	2.0	1.8	1.4
Food waste	2.0	1.8	1.4

Notes:

E_{REC} = energy savings from recycling (rec.); E_{WTE} = energy recovered from WTE; E_{LF} = energy recovered from capturing landfill gas. PP = polypropylene. HDPE = high-density polyethylene. PET = polyethylene terephthalate.

There is a meaningful amount of energy saved by recycling most materials commonly found in a MSW stream. So, from an energetic point of view, it nearly always pays to recycle. But where it is not possible to do so and the material enters the disposal stream, it becomes necessary to decide how best to handle it. This turns out to be relatively simple (at least from the point of view of energy) – in all cases where the material in question has a heating value, the energy extracted from that material will be greater if it is combusted with energy recovery than if it is landfilled (Table 3.1).

Later in this chapter, we will explore a lifecycle energy-based method of assessing the best ways to manage MSW. But first we'll take a look at the emissions profiles and the other major waste treatment methodologies.

3.3   Emissions comparison of WTE systems and alternative waste options

Building on the energetic analysis from the previous section, in this section we expand the toolkit to include analysis of emissions from WTE and other MSW management technologies. We highlight the key raw emissions of concern, both from a public perception and scientific perspective, and how WTE fares against other technologies. We then employ lifecycle impact assessment tools to examine the broader effects of these raw emissions from common lifecycle perspectives (such as human health, environmental impacts like global warming and resource depletion).

In terms of human health impacts and as analyzed by LCI and LCIA studies, the most significant emissions from WTE are carbon dioxide to the air. A significant amount of fossil CO_2 is released (roughly a half tonne per tonne of MSW combusted) and eco-indicator 99 (ecoinvent Centre, 2007) – a commonly used lifecycle impact assessment methodology – categorizes some of the impacts of global warming on human health (mostly due to increased incidence of disease in subtropical areas as well as displacement of populations from rising sea levels, etc).

The impacts from landfilling are most significantly realized in two areas – the release of potentially carcinogenic metals to waterways (mostly through leachate produced as water percolates through landfills) and the interaction of materials (such as disposed batteries). Modern landfills have elaborate systems for collecting and treating this leachate, but these are not 100% effective and some of this hazardous liquid reaches groundwater, streams or other nearby waterways.

Of course the best known environmental impact from landfill (aside, perhaps, from the significant use and concomitant degradation of the land itself) is methane emissions to the air; methane is a greenhouse gas that is 21 times more potent than carbon dioxide. A tonne of MSW deposited in landfill can produce anywhere from 200 to 400 m^3 of landfill gas, roughly half of which is methane (McDougall and White, 2001). Modern landfill sites have sophisticated systems for collecting this landfill gas, treating it and using the methane for fuel for heating or electricity (though in some cases the gas is simply flared to convert the methane to CO_2, which is released to the atmosphere). However, as with leachate collection systems, landfill gas collection systems are not 100% effective, so some methane necessarily escapes to the atmosphere anyway (Themelis and Ulloa, 2005).

The results of our rough comparison of the overall environmental impact of WTE vs. landfilling of typical MSW are shown in Fig. 3.2. A few things are notable in this comparison. First, though the impacts from landfilling are clearly higher than those from WTE, there is some uncertainty in the results (with several variations of both epistemic and aleatory certainty possible), which narrows the gap somewhat. In addition, these comparisons are for 1 tonne of generic MSW produced and treated in Switzerland. While the profile of MSW is relatively similar across most industrialized nations, there are differences that affect the outcome on a localized basis. Also, because these analyses are for Swiss technologies, other areas of the world might experience higher or lower emissions based on the levels of environmental controls, etc.

A final point worth noting is that 1 point on this graph represents one thousandth of the yearly environmental load of one average European inhabitant. So you can see that 1 tonne of waste that is landfilled or combusted does not have a very significant impact on the environment. However, with millions of tonnes of materials being landfilled and combusted around the world, waste management is definitely an important topic to consider. (Landfill accounts for around 2% of global greenhouse gas emissions annually.)

3.4   Advantages and limitations of using an LCA approach to evaluate waste management systems

In this section, we discuss some of the advantages and disadvantages of using an LCA approach to evaluate waste management systems. In particular, we will emphasize the robustness of LCA metrics as opposed to cruder metrics like 'recycling rate.' We will also touch on some of the drawbacks, including a lack of publicly available LCI data. We will also demonstrate how the lack of local data (most available LCI data is of a regional or national nature) can affect the outcome of LCA analyses.

MSW management is not a field that has used particularly rigorous quantitative methodologies to measure the effectiveness of decision-making processes. For most of modern MSW history, the closest we have come to any kind of metric has been the so-called recycling rate – that is, the percentage of materials diverted from the waste stream for recycling. This is reflected in the waste hierarchy pyramid (Fig. 3.3) popularized (in MSW circles) by the US EPA and recently updated by scientists at Columbia University in New York City (SEAS, 2011).*

---

* Figure 3.3 has intentionally been placed in more than one chapter in this book due to its broad relevance.

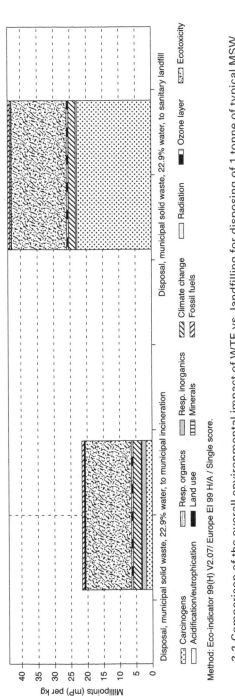

3.2 Comparison of the overall environmental impact of WTE vs. landfilling for disposing of 1 tonne of typical MSW.

Method: Eco-indicator 99(H) V2.07/ Europe EI 99 H/A / Single score.

Disposal, municipal solid waste, 22.9% water, to municipal incineration

Disposal, municipal solid waste, 22.9% water, to sanitary landfill

- Carcinogens
- Acidification/eutrophication
- Resp. organics
- Land use
- Resp. inorganics
- Minerals
- Climate change
- Fossil fuels
- Radiation
- Ozone layer
- Ecotoxicity

Millipoints (mP) per kg

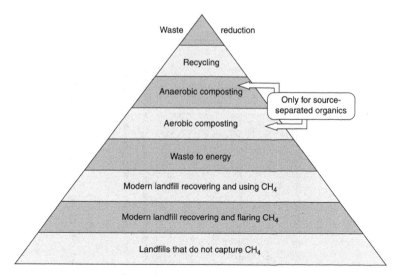

3.3 Waste hierarchy pyramid.

Even with the improvements to the pyramid introduced by Columbia University (namely the breaking down of the previously broad 'disposal' category into more granular variants – i.e. WTE, and the various forms of landfilling), it still does not really serve as a tool for goal-oriented waste management.

The recycling rate is not suitable as a measure of overall waste management sustainability. There are several reasons for this, but a few stand out as being the most significant. First, it does not account for the differences between landfilling and waste to energy for non-recycled wastes. Perhaps more importantly, it omits materials that are not recyclable at all (e.g., many plastics, contaminated wood, etc) with contemporary technologies. In other words, the maximum possible recycling rate is well below 100%. Without a theoretical maximum, it becomes nearly impossible to benchmark different systems against one another – it is not a clear, goal-oriented approach to waste management.

Instead of defining arbitrary goals with non-quantifiable targets in the hierarchal pyramid, the goal-oriented waste management approach seeks to set measurable targets that are based upon a foundation of values and practicality. The Austrian Waste Management Act, for example, outlines the following broad goals for waste management systems (Brunner and Rechberger, 2004, p. 305):

- protect human health and the environment
- resource conservation.

Fortunately, these are areas of assessment that LCA has evolved to deal with quite comprehensively. In fact, it could be argued that human health, environmental protection and resource conservation are the fundamental pillars of lifecycle impact assessment. What is needed then is a metric that more effectively incorporates these important goals and allows stakeholders to measure progress towards them. This will be the subject of the next section.

3.5   An alternative metric to evaluate waste management systems that addresses goal-oriented needs

In the last couple of years the author has set out – along with a team of researchers at Columbia University – to come up with a more suitable screening metric to help planners better assess their approach to municipal solid waste management. It was our desire to create a metric that is relatively quick and easy to use, does not require specialists to complete and allows users to benchmark different systems against one another (Kaufman *et al.*, 2010).

The metric is known as resource conservation efficiency (RCE). To understand how this metric works, it is necessary to quickly revisit the subset of LCA known as cumulative energy demand (CED), otherwise known as lifecycle embodied energy (LEE). As mentioned earlier, CED is really a lifecycle impact assessment methodology that accounts for all the energy inputs and consumption necessary across the full lifecycle of the examined product or service. It is well suited to screening analyses since energy requirements often correlate quite highly with environmental impacts (Huijbregts *et al.*, 2006).

The development work for RCE started with the fundamental assumption, often repeated by anyone even marginally familiar with waste management, that recycling 'saves energy' over other forms of waste treatment. This is in fact the case for nearly all materials currently recoverable under the typical 'blue bin' curbside recycling collection scheme. But one of the primary problems with recycling rates as a metric is the fact that not all materials are recyclable. So, while a pound of newspaper that is recycled instead of, say, landfilled certainly saves energy from a lifecycle perspective, this is not necessarily the case for many other materials. In fact, up to 60% of MSW is not recyclable with current infrastructure and logistics. This action needs to be accounted for when assessing the efficacy of any city's program.

As we also discussed earlier, the assumption that recycling 'saves energy' is because if the product is combusted or landfilled, a new product will have to be manufactured to serve the same function. Therefore, the energy saving from recycling is really the difference between the energy required to manufacture a new product minus all of the energy required to transport and reprocess the product in the recycling phase. If that difference is greater than the energy that would be realized from combusting that product and recovering the energy (or, for that matter, landfilling the material and capturing the landfill gas) then recycling is indeed the best option.

Because any kind of material being considered for waste treatment has an optimal amount of energy savings or realization under current waste management technologies, it is possible to design an 'efficiency metric' that considers waste treatment like any other process in which energy is consumed or saved (such as an internal combustion engine). We simply need to find a maximum level of energy savings or realization (or our optimal treatment regime) and compare it with the actual treatment regime and present it as an efficiency or RCE equation:

$$RCE_i = \left[\frac{(E_{LF,i} \times m_{LF,i}) + (E_{WTE,i} \times m_{WTE,i}) + (E_{REC,i} \times m_{REC,i})}{\max_j \{E_i m_i\}} \right] \qquad [3.2]$$

where RCE_i is the RCE for a city or region, E_{ij} is the energy savings for material i in process j (GJ/tonne), m_{ij} is the mass of material i in process j (tonnes), i is the index for materials and j is the index for treatments.

The most basic way to use the RCE equation is for a specific material. For example, if a user wanted to figure out how optimally his or her city was handling newspaper waste, they would simply need to reference the energy savings or realization for each process, and then simply input the fraction of newspaper going to each process in their current treatment regime.

As an example, recycling a tonne of newspaper saves 10 GJ over disposal plus the remanufacturing of the new product; combusting the material generates 5 GJ of usable energy; and landfilling (LF) the material and capturing the associated landfill gas from the biodegradation of the newsprint yields 1 GJ of energy. In this case, the maximum value to be placed in the denominator of the RCE equation is 10 GJ, or the recycling value. Using these hypothetical values, we can then calculate different treatment scenarios (Fig. 3.4).

As shown in Fig. 3.4, a 100% recycling rate under this scenario would yield a 100% resource conservation efficiency value. Sending all of the material to a combustion facility would yield around a 50% RCE, while landfilling all of the material would drop the RCE value to around 15%. A combination of treatment methods using 50% recycling, 25% landfilling and 25% WTE yields the second best result in terms of RCE.

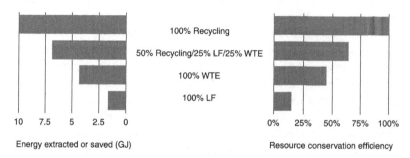

3.4 Energy saved and RCE for 1 tonne of newspaper under different scenarios (LF: landfill).

As part of our research we also compiled a table of reference energy values for different materials and their treatment options (Table 3.2, shown earlier as Table 3.1 but updated below to illustrate E_{max} values). This table needs to be constantly updated and improved upon, but facilitates screening analyses of different cities' treatment options (Kaufman *et al.*, 2010).

As the table shows, waste to energy performs favorably for a variety of different materials, and is often the best option for a given type of product. There is clearly no one-size-fits-all answer to the waste management problem, but WTE is an important contributor in any rational treatment regime.

As is evident in Eq. 3.2, the RCE metric is intended to be used for the aggregate total of all materials and a given city, not just a single type of waste. Equation 3.2 is the extension of the RCE from individual products to a 'systemwide' treatment.

To get a sense of the utility of the RCE metric for waste management planning, a test case comparing two American cities – Honolulu and San Francisco – was run. This test case modeled the base case scenario along with a few hypothetical alternatives (Fig. 3.5). Honolulu recycles about one quarter of its waste, with much of the remainder going to waste to energy. San Francisco, on the other hand, recycles roughly 35% of its waste, with all of the remaining MSW going to

Table 3.2 Energy savings from recycling, energy recovered by WTE, and energy recovered in landfill gas for different waste materials, as shown in Table 3.1 but with an indication of how the maximum energy saved can be achieved

Material	E_{REC}, rec. energy (GJ/ton)	E_{WTE}, WTE energy (GJ/ton)	E_{LF}, LF energy (GJ/ton)	$E_{MAX,sav}$ process
Newspaper	12.7	5.5	1.4	Recycling
Kraft paper	13.3	5.5	1.4	Recycling
Waxed OCC		5.5	1.4	WTE
High grade paper	21.8	5.5	1.4	Recycling
Mixed low grade paper	13.3	5.5	1.4	Recycling
Polycoated paper		5.5	1.4	WTE
Soiled paper	2.0	1.8	1.4	Recycling
Other paper		5.2	1.4	WTE
Glass	4.0			Recycling
PET	33.1	8.9		Recycling
HDPE	38.7	16.4		Recycling
PP	34.2	12.6		Recycling
Aluminum	188.1	170.1		Recycling
Steel	16.4	14.8		Recycling
Yard waste	2.0	1.8	1.4	Recycling
Food waste	2.0	1.8	1.4	Recycling

Notes:

E_{REC} = energy savings from recycling; E_{WTE} = energy recovered from WTE; E_{LF} = energy recovered from capturing landfill gas; $E_{MAX,sav}$ = maximum energy saved.

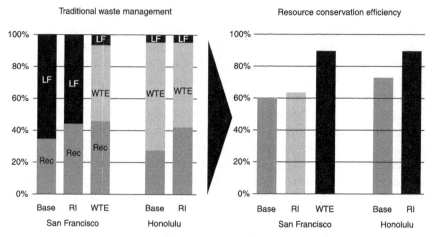

3.5 Traditional waste management and RCE compared for Honolulu and San Francisco. For explanation of test case scenarios see text.

landfill. As the figure shows, in the base case scenario San Francisco's RCE score is 60% while Honolulu's approaches 75%. This is due to the fact that non-recyclable, high-heating-value materials such as plastic wraps and packaging are converted into useful energy rather than 'stored' in a landfill.

The alternative scenarios represent best case hypotheticals, where recyclable material is captured and reprocessed at the highest possible levels, while most of the remaining MSW is sent to WTE facilities and converted to useful energy. The first alternative scenario (RI) assumes that San Francisco successfully captures a large fraction of its food waste for diversion to composting facilities but still landfills the remainder of its non-recycled waste. The second scenario (WTE) still assumes higher food waste composting rates, but this time the non-recycled waste is sent to a standard-sized US WTE facility that would easily handle this waste stream. The alternative scenario for Honolulu (RI) assumes increased recycling for several materials that are efficiently captured in San Francisco and could presumably be handled similarly in Hawaii. As the figure shows, both San Francisco's and Honolulu's RCE values increase under the alternate scenarios, and San Francisco's improvement is more dramatic with the addition of WTE.

3.6   Sources of further information

There has been much work done on using LCA to evaluate waste management systems. The EPA WARM model (www.epa.gov/warm) is a lifecycle-based planning tool that allows users to enter waste by type and treatment method and see instant results highlighting greenhouse gas emission impacts for the baseline scenarios and planned alternatives (USEPA, 2011). There is a variety of educational and non-profit organizations conducting research and work related to MSW and

LCA. A prominent example is RTI (www.RTI.org), which has done a lot of work developing LCIs to help with waste management models. Universities such as Berkeley, the University of Michigan, Carnegie Mellon, the Vienna University of Technology and many others specialize in LCA, waste management or both and are valuable sources of state-of-the-art knowledge and research. Also, many national and international governments and organizations are using LCA to develop toolsets (some more useful than others).

3.7   References

Baumann, H., Tillman, A.-M. (2004), *The Hitch Hiker's Guide to LCA: An Orientation in Life Cycle Assessment Methodology and Application*, Studentlitteratur: Lund.

Blok, K. (2006), *Introduction to Energy Analysis*, Techne Press: Amsterdam, p. 256.

British Standards Institute (2008), 'Specification for the caps assessment of the lifecycle greenhouse gas emissions of goods and services,' PAS2050:2008, London.

Brunner, P. H., Rechberger, H. (2004), *Practical Handbook of Material Flow Analysis*, CRC/Lewis: Boca Raton, FL.

ecoinvent Centre (2007), ecoinvent data v2.0., ecoinvent reports No.1-25, Swiss Centre for Life Cycle Inventories, Dübendorf, retrieved from: www.ecoinvent.org, accessed 1 November 2011.

Huijbregts, M. A. J., Rombouts, L. J. A., Hellweg, S., Frischknecht, R., Hendriks, A. *et al.* (2006), 'Is cumulative fossil energy demand a useful indicator for the environmental performance of products?' *Environ. Sci. Technol.* 40(3), 641–648.

IPCC (2007), IPCC Fourth Assessment Report (AR4) by Working Group 1, Geneva.

Kaufman, S., Krishnan, N., Themelis, N. (2010), 'A screening life cycle metric to benchmark the environmental sustainability of waste management systems,' *Environ. Sci. Technol.* 44(15), 5949–5955.

Kim, S., Dale, B. E. (2004), 'Cumulative energy and global warming impact from the production of biomass for biobased products,' *J. Ind. Ecol.* 7(3), 147–162.

Knapp, K., Jester, T. (2001), 'Empirical investigation of the energy payback time for photovoltaic modules,' *Solar Energy* 71(3), 165–172.

McDougall, F. R., White, P. (2001), *Integrated Solid Waste Management: A Life Cycle Inventory*, 2nd ed., Blackwell Science: Oxford, UK; Malden, MA.

SEAS (2011), Columbia University, www.seas.columbia.edu/earth/wtert/faq.html, accessed 7 November 2011.

Themelis, N. J., Ulloa P. (2005), 'Capture and utilization of landfill gas,' *Renewable Energy*, pp. 77–81, www.sovereign-publications.com/renewable-energy-art.htm, accessed 20 November 2011.

Thormark, C. (2002), 'A low energy building in a life cycle – its embodied energy, energy need for operation and recycling potential,' *Building and Environment* 37, 429–435.

USEPA (2006), 'Life Cycle Assessment: Principles and Practice,' May.

USEPA (2009), 'Municipal solid waste generation, recycling, and disposal in the United States,' Office of Resource Conservation and Recovery, November.

USEPA (2011), 'Waste reduction model (WARM),' www.epa.gov/climatechange/waste/calculators/Warm_home.html, accessed November 2011.

Wagner, H. J., Pick, E. (2004), 'Energy yield ratio and cumulative energy demand for wind energy converters,' *Energy* 29(12–15), 2289–2295.

4

Feedstocks for waste to energy (WTE) systems: types, properties and analysis

T. F. McGOWAN, TMTS Associates, Inc., USA

DOI: 10.1533/9780857096364.1.42

Abstract: This chapter discusses the types of waste used to fuel WTE plants, waste properties, why the properties are important in the design and operation of WTE plants, and the tests used to define the properties. Test methods are also listed.

Key words: WTE, waste testing, ASTM, waste properties, calorimeter, ash fusion temperature, moisture content, heating value.

4.1 Introduction

This chapter discusses waste types, properties and the testing of feedstock for waste to energy (WTE) systems. Waste categorization and testing have been in place for as many years as there have been incinerators. What is new is the composition of the waste, for example, waste has a higher plastic, chlorine and sulfur content than might have been the case five decades ago, and there is more interest in burning biomass for power production. Some additional tests have been introduced, such as measuring the bio-based carbon content of feedstocks, which is used for calculating their renewable fuel content. This is related to interest in reducing CO_2 emissions and governmental initiatives that promote use of sustainable fuels.

Why test the waste? If the properties of waste are not well known (and most waste properties for WTE can be found in the literature) because it is a new type of waste, or perhaps will become a larger part of the waste stream than usual, testing will yield data required for combustor design, the air pollution control system and for permitting. The waste properties will affect the air-to-fuel ratio, the need for lime or caustic soda in the air pollution control system, and the need for predrying if the moisture content is too high and the heating content is too low.

Useful standard references in the field (detailed at the end of this chapter) include ASTM test methods, and texts including Babcock & Wilcox's *Steam* book, the *North American Combustion Handbook* and *Perry's Chemical Engineers' Handbook*.

42

4.2   Types of feedstock for WTE systems and their characteristics

While many fuels are used in WTE plants, a mixture of trash, rubbish, refuse and garbage is the primary WTE fuel around the world. Not that others, like refuse-derived fuel (RDF) and tire-derived fuel (TDF), have not been tried, but in terms of the volume incinerated, the above fuel is most important. A quick list of major waste fuel sources for WTE plants follows. Only one deserves explanation now: Landfill relocation and remediation waste. This feedstock is from digging up landfill sites, perhaps due to groundwater issues, or more often, when a better and higher use of the land is planned. The waste most closely resembles municipal solid waste mixed with substantial amounts of the soil used for daily cover.

Major waste streams for municipal solid waste (MSW) incinerators follow:

- MSW
- RDF
- Biomass and organic waste
- Plastics (from production waste or mixed with other waste)
- Biosolids (wastewater sludge, animal waste)
- Liquid wastes
- Commercial and industrial waste
- Landfill relocation and remediation waste
- TDF
- Hazardous waste (coal tars from manufactured gas plants).

Table 4.1 lists heating values, moisture content and ash content for a wide range of potential WTE fuels. Much of it is based on the classification system produced by the Incineration Institute of America (IIA, a now defunct group that helped classify and codify waste for incineration). There is more detail on each waste stream in the following paragraphs.

Type 0, *trash*, is a mixture of highly combustible waste, such as paper, cardboard cartons, wood boxes and combustible floor sweepings, from commercial and industrial activities. The mixture contains up to 10% by weight of plastic bags, coated paper, laminated paper, treated corrugated cardboard, oily rags and plastic or rubber scraps.

Type 1, *rubbish*, is a mixture of combustible waste, such as paper, cardboard cartons, wood scrap, foliage and combustible floor sweepings, from domestic, commercial and industrial activities. The mixture contains up to 20% by weight of restaurant or cafeteria waste but contains little or no treated paper, plastic or rubber waste.

Type 2, *refuse*, is an approximately even mixture of rubbish and garbage by weight.

Type 3, *garbage*, is animal and vegetable waste from restaurants, cafeterias, hotels, hospitals, markets and similar installations.

Table 4.1 Waste type, heating content, moisture and ash

Waste	Btu/lb (cal/g)	Moisture%	Ash%	Comment
Type 0: trash	8500 (4720)	10%	5%	IIA waste standard
Type 1: rubbish	6500 (3610)	25%	10%	IIA waste standard
Type 2: refuse	4300 (2390)	Up to 50%	7%	IIA waste standard
Type 3: garbage	2500 (1400)	Up to 70%	5%	IIA waste standard
Type 4: human and animal remains	1000 (560)	Up to 85%	5%	IIA waste standard
Type 5: by-product waste (liquid, solid, sludge)	By test	Varies	Varies	IIA waste standard
Type 6: solid by-product industrial waste	By test	Varies	Varies	IIA waste standard
Wood fuel, 50% moisture content	4250	~ 50%	0.5–2%	Hog fuel, wet sawdust, chipped windfalls, yard waste
Wood fuel, 10% moisture content	7650	~10%	0.5–2%	Dry planer shavings, sander dust, C&D wood
Used or recycled oil	~19 000			Reclaimed engine lube, cutting oil, etc.
Tires and TDF	12 000–16 000			Sulfur content ~1.2%, ash from metal and glass/polyaramid cords
Used cooking oil	16 900–18 500			Need to decant water, filter grit if atomized
No. 2 fuel oil	19 600		Trace	Used for start-up and load control
No. 6 fuel oil	18 300			Used for start-up and load control, sulfur content
Natural gas	24 000	N/A	N/A	Used for start-up and load control

C&D, construction and demolition.

Type 4, *human and animal remains*, includes carcasses, organs and solid organic waste from hospitals, laboratories, abattoirs, animal pounds and similar sources.

Type 5, *by-product waste*, consists of gaseous, liquid or semi-liquid materials, such as tar, paints, solvents, sludge and fumes, from industrial operations. Heat values need to be determined for the individual materials to be destroyed. This category includes sewage sludge from municipal wastewater treatment plants.

Type 6, *solid by-product waste*, includes rubber, plastic and wood waste, from industrial operations. Heat values need to be determined for the individual materials to be destroyed.

Other wastes not specifically covered in the IIA standard types are:

Type C&D (*construction & demolition*) includes wastes from new construction and razed buildings, remodeling and repair of commercial and industrial buildings. The combustible fraction includes lumber and shingles.

Type *agricultural and silviculture* waste includes waste and residues from diverse agricultural activities. Wastes from the planting and harvesting of row, field, tree and vine crops, the production of milk, the production of animals for slaughter, and the operation of feedlots are collectively called agricultural wastes. Silviculture or forestry wastes come from land clearing, thinning, harvesting waste, as well as processing wastes such as sawdust from sawmills, sander dust and planer shavings from production of wood products, land clearing, windfalls and yard waste. Wood fuel also enters WTE plants inadvertently in trash, or by design as supplemental fuel. The heating value of wet wood is more than that of all but Type 0 and Type 1 waste, with a good starting point being 8500 Btu/lb dry, 4250 Btu/lb wet (50% moisture content as cut), with 0.5 to 2% ash, and 80% volatiles and 20% fixed carbon. While it is true that pine has somewhat more heating value than hardwood, most wood waste is a mixture of the two types, and moisture content, on a day-to-day basis, is more important than wood type in practice.

Type *other*: these include street sweepings, roadside litter pickup, catch-basin debris, dead animals, municipal wastewater and industrial waste-treatment sludge, which can be very high in water content. While this sludge could be burned in a WTE plant, it has low heating value, and is more often incinerated in multiple hearth furnaces and fluid-bed incinerators.

Type hazardous, *hazardous wastes*, designated as such by the US EPA for US applications, are regulated under the Resource Conservation and Recovery Act of 1976 (RCRA) and are excluded from WTE plants that are permitted solely for solid waste incineration. However, at least one WTE incinerator was permitted to utilize manufactured gas plant (MGP) coal tars as feedstock, as this waste was created by coal gasifier plants that were shut down before the RCRA went into effect. This waste has about 17 000 Btu/lb if it is pure tar and contains no emulsified water or debris. Coal tars from ongoing operations, such as coke ovens at steel mills, are considered a RCRA hazardous waste.

Table 4.1 notes several instances of moisture being 'up to' a certain percent, and the wood fuel items show approximate percentages. The bottom line is that except for auxiliary fossil fuels, waste properties can and will vary. The reader is advised to accept this as fact, and realize that while waste can be subjected to tests, the variation in properties must be addressed on a day-to-day basis by the worker running the front-end loader or overhead clam shell bucket, mixing the wet with the dry, the fine with the coarse, the dense with the lightweight material. This blend master or brew master is a key link in the chain of keeping an incinerator running smoothly, working mainly by visual examination. A simple example is that trash brought in after three days of hard rain should be expected to contain more moisture than that brought in a dry spell. It should be mixed with dryer waste from before or after the wet spell, or more auxiliary fuel should be fired to achieve the correct load and keep the furnace temperature in an appropriate range.

If fuel is purchased (e.g., RDF, TDF or wood), then periodic fuel value and moisture testing may be required to address payment and contract stipulations.

4.3   Testing of feedstocks for WTE systems

While there is little point in extensive testing of waste that has well-known properties available from the producer or literature, sampling of waste fuels and treatability studies (lab, bench scale or pilot scale burns to simulate WTE operation) are key factors impacting the success or failure of WTE systems, *which may be able to handle non-run-of-the-mill MSW from home and commercial trash pickups*. In short, a little knowledge and low-cost lab testing at the front end can save the owner a significant amount of money and grief by isolating design issues and addressing them prior to purchase of the WTE plant, thus preventing such fire-side problems as slagging, premature refractory failure and failure of stack tests.

There are three levels of tests that should be considered when designing a WTE system:

1. Standard lab tests that use small samples (typically 1–100 grams).
2. Larger (bench scale) treatability tests, which are usually carried out in a muffle furnace (a small, high-temperature laboratory oven; Fig. 4.1).
3. Pilot-scale firing.

The more exotic the fuel or waste, the more likely that tests will be run and if there are problematic features (e.g., slag formation), then tests will proceed upwards from standard to bench scale to pilot scale. A specific example of this is the use of chicken and turkey litter as fuel, which contains significant salts that can reduce the temperature at which slag forms.

The primary tests are for heating value and moisture content. Secondary tests are for ash content, elements in the ash, ash slagging temperature and acid-forming elements, such as sulfur and chlorine.

A calorimeter — a key component for heating value tests — is a simple device (Fig. 4.2). A small sample is placed in a cup, along with a firing wire. The cup is placed in a sealed, heavy-walled vessel, pressurized with pure oxygen, immersed in a water bath, and then the firing wire is ignited electrically. The resulting combustion causes a small temperature rise in the water bath, which is translated into a heating value in units of Btu/lb or kJ/kg. The test result is the higher heating value of the fuel, since it is carried out at ~60°F, and any water that vaporizes or

4.1 Lab oven or muffle furnace for testing samples.

4.2 Calorimeter vessel used to fire samples for measuring their heating value.

is created in the combustion process condenses and is in liquid form. For fuels with high moisture and low heating value (<2000 Btu/lb), the calorimeter test must be performed using a 'combustion aid', usually mineral oil or benzoic acid. The author favors mineral oil because it gives more reliable firing. It is best to use labs that are well established in fuel testing, such as those doing routine tests of coal, coke and fuel oil.

Preparation for tests always starts with the search for the elusive 'representative sample'. Various approaches are used, such as grinding up a considerable amount of waste and using statistical sampling techniques, then splitting and resplitting it prior to testing, or taking a grab sample each hour off a belt and compositing it. However, it is smart to do all tests in triplicate to see if the numbers have some repeatability.

Standard lab procedures and equipment are used prior to testing, such as grinding, sieving, drying, use of muffle furnaces, and weighing on scales.

The standard tests described in Table 4.2 are relatively inexpensive to conduct, and yield reliable results. These tests can be used quantitatively. For example, the heating value can be used to judge the feed rate of the waste and the amount of auxiliary fuel needed to produce the required amount of steam. The results of these tests can also be used to help an engineer make qualitative judgments, such as those related to selecting the appropriate material handling, incinerator and air pollution-control equipment. Note that the tests described are not the typical EPA SW-846 8000 series using gas chromatographs (GCs) to assess hazardous constituents for regulatory purposes. For those interested in such tests, a listing of EPA-846 methods can be found at the following URL, and at the end of that document there are hot links to the actual methods (www.epa.gov/region1/info/testmethods/pdfs/testmeth.pdf).

Table 4.2 ASTM combustion and thermal property tests

Test method and comments	Effect on design and operation
ASTM D5142 Proximate, for mc, vm, fc, ash, at 950°C	Ash content and vm can be used to infer the upper limit of organic content. Ash content used for ash handling.
ASTM D3176 Ultimate analysis, modified for C, H, N, Cl, S, O (by difference), ash, mc	Carbon and hydrogen can be used to infer organic content in the absence of carbonates; S and Cl, if organic, will impact scrubbing costs.
ASTM D1989 Higher heating value via automatic calorimeter	Required for combustion calculations, sizing the amount of combustion air, furnace, air pollution control system, and fuel feed rate.

(Continued)

Table 4.2 Continued

Test method and comments	Effect on design and operation
ASTM D1857 Ash fusion temperature (oxidizing and reducing)	Important for choice of type of combustor (e.g., grates vs. fluid bed) and potential for slag generation.
Moisture content, gravimetric at 105°C (not required if D5142 is done)	May be needed for payment and contractual issues.
ASTM D2795 Elemental analysis*: C, N, S, Cl	Adds to information from ultimate analysis.
D3862 Major and minor elements in ash*: Cl, P, Na, Mg, Al, Si, P, S, K, Ca, Ti, Mn, Fe	Required if slagging is a concern.
ASTM D3863 Trace metals analysis*	May be required if metal emissions are a concern.
ASTM D410 Screen fractionation (particle size)	May be required for feed and metering system.
ASTM D292 Bulk density	Required for storage and handling design.
ASTM D5142 Ash content (or as part of proximate)	For ash handling; also sets upper limit of organic content.
TGA/DSC Thermo-gravimetric analysis, differential scanning calorimetry*	Provides feedback on temperatures required to drive off volatiles.
ASTM D3174 Standard test method for ash in the analysis of coal and coke from coal, run at 550°C rather than 750°C	If it correlates well with organic content, can be used as cheap and fast method to estimate organic content or fuel value in the field based on TVS/LOI test at 550°C.

Notes:

*Optional; run tests based on waste properties and equipment design needs.
vm = volatile matter; mc = moisture content; fc = fixed carbon; TVS = total volatile solids; LOI = loss on ignition.

Source:
Table excerpted by special permission from *Chemical Engineering*, April 2004, page 59. Copyright © (2004), by Access Intelligence, New York, NY 10038, Remediating organic-laden soils: Do your homework before breaking ground, by Thomas F. McGowan.

Table 4.3 contains RDF test methods. These are derived from other, earlier test standards, refined for this particular waste fuel. Table 4.4 contains methods to detect acid-forming elements in the waste feed, and a test method to define the biomass content of fuels. That latter test may be required for grants and other government assistance.

Table 4.3 ASTM RDF test methods

Test method	Testing for
ASTM E887 – 88(1996)	Standard test method for silica in RDF and RDF ash.
ASTM D5115 – 90(1996)	Standard test method for collecting gross samples and determining the fuel quality of RDF.
ASTM E953 – 88(1998)	Standard test method for fusibility of RDF ash.
ASTM E897 – 88(1998)	Standard test method for volatile matter in an analysis sample of RDF.
ASTM E777 – 87(2004)	Standard test method for carbon and hydrogen in an analysis sample of RDF.

Table 4.4 Test methods for chlorine, sulfur, biocontent and metal leaching

Test method and comments	Effect on design and operation
Total organic chlorine (optional if chlorine is known to not exist in samples). Done by total chlorine via ASTM D808, oxygen bomb for sample prep, followed by D4327 (ion chromatography). D4327 is repeated on raw samples for inorganic chlorine. Organic chlorine is determined by difference.	Dictates the amount of caustic soda or lime required to neutralize HCl; dictates construction material for quench.
Total organic sulfur (requires inorganic and total to find organic fraction).	Dictates the amount of caustic soda or lime required to neutralize SO_2; dictates construction material for quench, and acid gas dewpoint design temperatures for corrosion.
ASTM WK30163 standard test methods for determining the biobased content of solid, liquid, and gaseous samples using radiocarbon analysis.	The intent of ASTM-D6866 is to utilize radiocarbon dating for the determination of biobased content within manufactured products, needed for USDA's Federal Biobased Product Preferred Procurement Program (7 CFR 2902, originally termed FB4P, now designated as Biopreferred (www.biopreferred.gov)).
EPA 1311 TCLP (toxicity characteristic leaching procedure)	Used to assess metal leaching potential of ash when landfilled.

4.4   References

ASTM Test Methods, American Society of Testing and Materials, 100 Barr Harbor Drive, West Conshohocken, PA, www.astm.org/DIGITAL_LIBRARY/index.shtml. 2009.

North American Combustion Handbook, Volume I third ed, 1986, Volume II third ed, 1997, North American Manufacturing Company, Division of Fives, Cleveland, Ohio, www.namfg.com/comb-handbook/gra49.pdf.

Perry's Chemical Engineers' Handbook (8th Edition), 2008, McGraw-Hill, NY.

Steam, Its Generation and Use, Babcock & Wilcox, 41st ed., Barberton, Ohio, shop.fullpond.com/bwco/pdf/STEAM41orderform.pdf.

Part II
Waste to energy systems, engineering and technology

5

Pre-processing and treatment of municipal solid waste (MSW) prior to incineration

G. C. FITZGERALD, Columbia University, USA

DOI: 10.1533/9780857096364.2.55

Abstract: This chapter outlines the primary municipal solid waste (MSW) treatment methods commonly practiced in the integrated solid waste management industry as precursors to MSW incineration. Beginning with the most basic processing techniques, used in mass burn plants, and expanding to advanced separation, and biological and thermal treatment of MSW, this chapter gives the reader a basic understanding of the machinery, techniques and fundamental theories employed in the various methods of processing MSW for energy and material recovery.

Key words: MSW shredding, WTE pre-processing, waste densification, mechanical–biological treatment, MSW treatment and handling.

5.1   Introduction

Processing municipal solid waste (MSW) prior to energy recovery has become an increasingly common and successful method for minimizing the environmental impact of waste disposal while taking advantage of the waste stream's inherent energy. During the initial development of waste to energy (WTE) facilities, the notion of waste pre-processing was non-existent. The majority of first-generation incineration facilities were designed with the singular goal of the reduction of garbage volume, with little or no concern for energy recovery and minimal concern for pollution abatement and material recovery. In recent years environmental regulations have become stricter and progressively more enforced, which has subsequently required WTE facilities to greatly improve pollution reduction and abatement systems. Creative engineering by the industry has developed air pollution control systems that can effectively reduce particulate and gaseous emissions to a level complying with federal and state requirements. Material processing can be used to improve the control of emissions by enhancing the combustion properties resulting in a more complete waste combustion process. Pre-sorting of MSW also gives WTE operators the opportunity to remove non-combustibles, bulky, and hazardous and pollutant-rich materials before they enter the combustion chamber, where they would be volatilized and released in the emitted effluent combustion gasses. Additionally, the pre-processing of MSW makes possible the recovery and reuse of valuable resources such as

55

plastics, paper fibers and metals that may have otherwise been deposited in landfill.

The need to minimize the entry of hazardous materials into the combustion chamber is motivation enough for the most primitive waste-processing method currently in use, that is, the visual screening or picking of the raw MSW delivered to the tipping floor of a WTE facility. This screening process is meant to prevent bulky, hazardous, flammable, explosive and toxic materials from entering the rest of the material-processing system and most importantly from entering the combustion chamber where these materials may damage the system. More advanced pre-screening and processing practices, such as the preparation of refuse-derived fuel (RDF) and the use of mechanical–biological treatment (MBT), involve several sorting, sieving and fuel upgrading stages, which result in a more complete and efficient use of the MSW stream, albeit at an added economic cost. The extent to which MSW is pre-treated and processed before thermal processing or WTE incineration depends on factors such as the inherent requirements of the post-processing technology, local regulations, cost of landfill dumping and MSW composition, and in general pre-treatment may not be globally advantageous.

5.2   Basic screening processes: mass burn

WTE facilities receive waste from a wide range of sources including construction and demolition (C&D) debris and commercial, industrial and residential refuse. The waste that is produced by each of these sectors is highly dissimilar in terms of material composition, heating value, homogeneity and toughness. In addition to these complexities all sources may contain bulky and hazardous materials. Gasification technologies often involve MSW pre-processing, particularly to reduce the size of the processed fuel to meet specifications. Combustion chambers for these technologies operate most efficiently when fed a consistent fuel stream of homogeneous particle size and constant heating value. Mass burn technology is capable of thermal processing of MSW in the as-delivered state and does not require pre-processing. All WTE plants practice some form of pre-screening to minimize the entry of bulky and hazardous materials onto the combustion grates. The most basic process typically occurs on the tipping floor of the facility or via the overhead crane, where undesired materials, such as white goods, are picked from the storage pit. The tipping floor is a flat concrete surface where waste trucks empty their contents either onto the floor or directly into a refuse storage pit, depending on the size and layout of the facility and the level of pre-screening required. Facility personnel then proceed to visually sift through the incoming refuse for large or bulky objects such as stoves, refrigerators and other appliances often referred to as white goods. White goods are removed for material recovery while dangerous items such as propane tanks and other flammable liquids and gasses are removed to avoid equipment damage and personnel injury. An additional processing step, referred to as fluffing, is performed in the storage pit to

5.1 Fluffing.

crudely address the heterogeneous nature of MSW. The crane operator uses the loading claw to pick up several tons of MSW and redistribute it across the pit as shown in Fig. 5.1. This process has the twofold effect of breaking bags and mixing the waste, resulting in a more homogeneous fuel, which can be fed more fluidly into the hopper.

In mass burn plants, screening and pit fluffing are the full extent of waste processing prior to combustion, aside from any basic source separation that may be implemented in participating communities. In situations where source separation is not common practice and MSW is disposed of in a mass burn facility, all glass, metal and non-combustibles are passed through the moving grate furnace. The result of this is a bottom ash that is typically disposed of in a sanitary landfill and may or may not contain unrecovered metals and glass. Many facilities implement post-combustion material recovery, which includes the separation of ferrous and non-ferrous metals from bottom ash; however, the separation of glass is seldom preformed in mass burn plants.

5.3   Fuel upgrading and enhancement processes

Municipal solid waste, although combustible in an unprocessed state, is occasionally treated mechanically, biologically, thermally or by some combination of the three to produce a higher-value fuel. The processing of MSW to enhance the fuel provides many positive benefits, such as increased heating value, additional material recovery, pollution reduction, ease of handling and a wider range of applicability. The extent of MSW processing depends on the desired end use of the material and the particular economics of the community providing the MSW management infrastructure. The various methods of waste treatment in practice result in quite different end products, yet they all share the same goal of improving the efficiency of the material or energy recovery process and in effect minimize landfilling.

Refuse-derived fuel (RDF) is made by trommelling, shredding, sorting and dehydrating MSW to generate a more homogeneous fuel with an increased heating value and improved combustion characteristics. RDF plants focus on generating revenue from both material recovery and fuel production and rely on profits from the material recovery side to remain a successful business operation that is competitive with mass burn facilities. RDF is primarily composed of organic matter such as plastics and biodegradable wastes (yard waste, wood and peat), which have been separated from the MSW stream by a series of shredding, magnetic separation and air knifing operations.

One of the more promising products of MSW processing is biomass, which may be produced as a sized homogeneous fuel in pelletized or briquette forms. Once the non-combustibles have been removed the biomass is compressed into pellets, logs or bricks and either combusted on site or sold to local combustion facilities. Biomass can be used in cement plants, WTE plants or co-combusted in a coal-fired power plant.

RDF fuels and biomass can be an attractive option for incineration plants because of the increased uniformity in chemical and physical composition. Additionally many governments recognize that biomass is a carbon-neutral fuel and regulations allow municipalities to report lower CO_2 emissions based on the bio-based composition of the RDF. The combustion of RDF compared to unprocessed MSW typically results in lower rates of ash and residual carbon production, higher energy recovery efficiency and lower heavy-metal effluent pollution. It is important to note that the composition of RDF and MSW in general varies widely so these trends may not always be the case.

The increased value of RDF as a fuel comes at an economic cost and is not feasible in all situations. The economics of RDF production in relation to mass burn facilities are complex and must be investigated on a case-by-case basis. Primary influencing factors in this decision include environmental regulations, degree of source separation, the cost of incineration, the population dispersion, the scale of processing as well as the type of WTE facility. RDF plants require greater personnel support, and their operation and maintenance costs are noticeably elevated compared to conventional mass burn facilities due to the additional systems required for MSW processing. This expense is attributed to the complex waste-handling system necessary for the various stages of material recovery, sorting and preparation of the RDF prior to combustion.

Some WTE plants are called shred-and-burn plants. At these plants, ferrous and non-ferrous materials are removed prior to combustion and in some cases further material separation and recovery are practiced. The concept of shred and burn is simple: prepared fuel burns better and more evenly if its size is reduced and it is partially homogenized.

5.4   Advanced screening, separation and processing

In municipalities where recycling and waste diversion have high priorities in the integrated waste management program, specific facilities are implemented to separate, process and prepare recyclable materials from the waste stream. These material recovery facilities (MRFs) are dedicated plants that receive raw or pre-sorted MSW and separate the recyclable materials common to MSW such as plastic bottles, aluminum cans, cardboard and paper fibers from the reject fraction that is to be disposed of in a landfill or WTE plant. MRFs that accept only curbside-collected comingled recyclables are called clean MRFs, while those facilities designed to process raw MSW are referred to as dirty MRFs. A typical MRF will have several stages of material sorting and separation based on density, magnetic properties, particle size and optical properties of the feed material. The most common automated separation techniques include trommel screening, air classification, magnetic and eddy current separation, optical sorting, shredding and bailing.

5.4.1   Trommel screening

Trommel screening is a separation method used in MSW sorting and separation for resource recovery. Rotary screens, generally called trommels, have been used primarily in mineral processing and other separation operations. The high throughput and effective size separation have supported the increased use of trommel screening in the MSW sector. A trommel screen or drum, as shown in Fig. 5.2, is simply a rotating tubular sieve, which operates based on a balance between gravitational forces and angular momentum in order to separate MSW by size in an automated fashion. The efficiency and range of separation achieved by

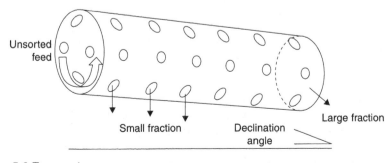

5.2 Trommel screen.

a trommel is governed by the declination angle of the rotating drum, the size of the screen opening and the rotational speed of the drum. Typical drum speeds are in the range 10–20 rpm and may be adjusted to govern the passage rate and efficiency. The rotational motion of the cylinder combined with its declination angle determines the rate of particle collision with the trommel surface, and hence can be used to tune the rate of particle separation. Rotational velocities common in MSW separation tend to be higher than in mineral processing because of the need to open bags, break up the material and increase separation. Increased trommel speeds allow the material to travel a greater distance up inside the trommel before centrifugal forces are overtaken by gravity, which causes the particles to drop back to the bottom of the trommel at greater velocities. The interaction of MSW particle size, trommel aperture size, declination angle, drum length and rotational speed are all necessary parameters that need to be addressed when designing a MSW screening tunit. The interested reader is directed to Wheeler *et al.* (1989) and Alter *et al.* (1981) for further details of trommel design.

5.4.2   Air classification

Air classification, a technique often used in material recovery facilities, is a separation process, which takes advantage of the density differences between the light fraction and heavy fraction of MSW. The light fraction consists of plastics, paper fibers and aluminum cans, while the heavy fraction is typically composed mostly of organic material or larger metallic items. In this process the raw material is fed into an upward moving air stream where the heavy fraction falls downward and the light fraction is lifted vertically upward and collected via a cyclone system. A similar technique to air classification is air knifing, which again takes advantage of the density variation of the material, where a horizontal air stream is blown across a vertically falling material stream and the lighter materials are forced into a collection area some horizontal distance away from the falling stream. Using this method the material can be separated into more than two components based on how far the material is carried by the air stream. More advanced separation systems can couple air separation techniques to new high-tech optical sorting devices that are capable of separating out each specific material based on the optical properties of the plastic.

5.4.3   Magnetic separation

The recovery of ferrous materials, such as saleable iron and steel from an MSW stream, is one of the oldest and simplest separation processes. Magnetic separation can be used to extract any ferromagnetic material from a mixed waste stream using magnetic fields. Two separation techniques are mainly used for material recovery: rotating drum magnets in a suspended drum or top feed configuration and overhead belt magnets. A rotating drum magnet system uses a permanent or electromagnet to attract ferrous materials as a conveyor belt travels around the drum. Figure 5.3(a) illustrates the suspended drum configuration where the feed material is picked up from the conveyor using a suspended drum magnet and deposited in a collection conveyor after traveling around the drum. The top feed configuration, as shown in Fig. 5.3(b), operates on a similar principle: the material

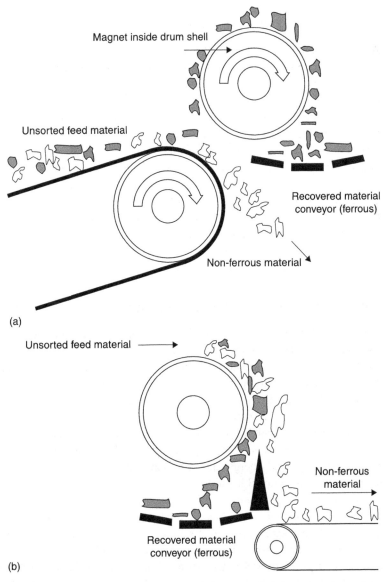

Magnet inside drum shell

Unsorted feed material

Recovered material conveyor (ferrous)

Non-ferrous material

(a)

Unsorted feed material

Non-ferrous material

Recovered material conveyor (ferrous)

(b)

5.3 Rotating drum magnet systems: (a) suspended drum configuration; (b) top feed configuration.

passes over the magnetic drum and all non-ferrous material falls off the drum while ferrous materials are retained on the drum for a longer arch resulting in a different trajectory into a different collection bin or chute. Overhead belt separators are customarily positioned above a conveyor carrying shredded MSW and similarly use a permanent or electromagnet to lift the ferrous material off the conveyor belt and deliver it along a different trajectory onto a separate conveyor.

5.4.4   Eddy current separation

Aluminum, a non-ferrous but valuable material, is a very common component of MSW and is separated from the bulk stream by eddy current separators. Eddy current separation is an advanced separation technique used to separate electrically conductive but non-ferrous materials such as aluminum, brass and copper from the other inert materials in the MSW stream. Eddy current separators use a Foucault or induced current in electrically conductive materials. When a conductive material is exposed to a changing magnetic field, a current is induced in that conductive material; this induced circular flow of electrons in the material is called an eddy current. The eddy currents themselves create their own inductance in the material being recovered and thus produce a magnetic field. This induced magnetic field is used to separate the non-ferrous material from the rest of the waste stream. Eddy current separators create a repulsive magnetic force and consequently project the non-ferrous material away from the eddy current generator, typically from a rotating drum and into a collection bin.

5.4.5   Mechanical–biological treatment

Mechanical–biological treatment (MBT) is defined as the processing and conversion of MSW containing biologically degradable components using a combination of mechanical and biological treatment processes. MBT facilities are essentially a combination of a material recovery facility and an aerobic or anaerobic digestion treatment process. These facilities are utilized to recover recyclable material and stabilize the organic fraction of the waste stream in order to reduce leachate and landfill gas production when the rejected material is eventually disposed of in a sanitary landfill. MBT is an alternative method for recovering embedded energy and minimizing pollution during waste disposal compared to conventional landfilling.

The biological component of MBT is performed via either anaerobic digestion or aerobic composting, depending on the product and process conditions desired. Biological decomposition is used to stabilize the biodegradable organic matter so as to minimize the production and release of landfill gas, odors and leachate during the disposal of MSW. Uncontrolled production of landfill gas, a mixture composed primarily of CO_2 and CH_4, is undesirable due to its greenhouse gas potential and explosive properties.

Anaerobic digestion

Anaerobic digestion (AD) is a microbial process in which microorganisms break down biodegradable material in an oxygen-free environment to produce a solid digestate along with biogas. Anaerobic digestion converts organic matter into carbon dioxide (~40%) and methane (~60%) via four defined stages: hydrolysis, fermentation, acetogenesis and finally methanogenesis. The digestion process can occur in a range of temperatures, typically between 30–60°C and at various moisture levels, which determine both the conversion rate and composition of the effluent. A typical AD process generally achieves a reduction of 50–55% of the organic content (Fricke *et al.*, 2005).

Aerobic degradation

Aerobic degradation, also known as composting, is an alternative biological treatment, which converts the organic fraction of MSW using aerobic microorganisms in the presence of oxygen. The aerobic process produces CO_2, H_2O, nitrates, sulfates and heat in addition to a stable solid material commonly used as fertilizer for agriculture crops. The aerobic bacteria that thrive in oxygen-rich environments break down and digest the waste. These processes are controlled or accelerated by manipulating the availability of oxygen in the system. Several different techniques are commonly used to facilitate the aerobic decomposition of organic matter including windrow composting, aerated static pile composing and in-vessel composting. In all of these processes, the aeration process acts to improve porosity and oxygen availability to enhance the decomposition rate. Windrow composting uses long piles or rows of organic matter, which are periodically turned mechanically or manually to aerate them. The size, geometry and turning frequency of the windrows are used to regulate the heat generation and oxygen flow through the core of a windrow. It is important that the exothermic nature of aerobic decomposition is utilized to keep the compost pile at the necessary elevated temperatures for the bacteria to thrive. In aerated static pile composting, as the name suggests, the material is arranged in piles rather than rows and does not undergo the same turning procedure as windrow composting. In static pile composting, bulking agents such as large woodchips or other bulk material are mixed into the compost to allow for greater air flow and keep the decomposition process in the aerobic zone. Additionally, air blowers may be added to the system to further assist in the aeration of the compost pile. Finally, in-vessel composting gives the best control of conditions and is performed in a closed drum or silo where conditions such as moisture, temperature and aeration can be more precisely controlled.

5.4.6 Torrefaction

The thermal pre-treatment of waste prior to combustion or gasification is sometimes implemented to improve transportation and storage properties and additionally often has beneficial processing effects. Torrefaction is a biomass process used to increase the fuel value of organic materials such as yard waste and woody materials commonly found in MSW streams. This method of waste upgrading results in three primary benefits: fuel densification, decreased moisture content and increased calorific value. Figure 5.4 shows a before and after picture of torrefied beech wood, where it is clear that the macro-physical changes are minimal. Chemically, torrefaction is simply the removal of oxygen and water from the biomass, producing a final solid product with a lower oxygen/carbon ratio than the original unprocessed feed. Torrefaction is often described as a high-temperature drying process, or in some cases, a mild form of pyrolysis in which biomass undergoes partial devolatilization. This low-temperature thermal treatment is performed at atmospheric pressure in the range 200–300°C and yields an improved fuel with enhanced combustion characteristics. Torrefied biomass contains roughly 70% of the original mass and yet retains 90% of the initial energy content in the solid product while 10% of the heating value is lost in torrefaction off-gassing. Torrefied material is versatile and can be co-fired with coal in a thermal conversion plant, fed to a gasifier to produce syngas, or used as a feed for pyrolysis. Due to the hydrophobic properties of torrefied biomass, it can be stored for longer and transported greater distances in the open air without absorbing water and lowering its fuel value. Additionally, the drying and roasting process of torrefaction produces a more brittle and harder material, which may result in less energy-intensive processing, such as crushing and grinding, that will need to be performed later in the processing line. Five main stages have been defined to quantify the torrefaction process (van der Stelt *et al.*, 2011):

- Heating.
- Pre-drying: Free water is removed via evaporation at 100°C.
- Post-drying: The temperature is increased to 200°C, light fractions evaporate, physically bound water is released.

5.4 Beech wood, before (left) and after (right) torrefaction.

- Torrefaction: The temperature is maintained at 200°C as the majority of mass loss occurs during this stage.
- Solids cooling: Torrefied biomass is cooled to room temperature.

5.5   Shredding and size reduction processes

Size reduction and homogenization of bulk MSW can be an integral part of both material recovery and material disposal. By processing MSW to give smaller particles and a more homogeneous particle size distribution, various subsequent sorting and separation processes become more effective and efficient. The shredding or comminution of MSW has several other beneficial effects on MSW processing, including volume reduction, enhanced combustion characteristics and in some cases improved biodegradability in sanitary landfills equipped with gas collection systems. Size-reduction technology has been widely used in mining and mineral processing for centuries to enable quicker and more effective recovery of valuable minerals and metals. The composition of MSW is so widely varied that machines designed for MSW must be robust enough to handle both soft, ductile materials as well as tough and resilient materials such as metal and dense plastics.

MSW size reduction is carried out by large machinery, which utilizes force to smash, break, shear or cut the material into smaller and potentially more uniform particles. There are two prominent categories of waste comminution devices used in the management of MSW: high-speed, low-torque (HSLT) hammermill shredders and low-speed, high-torque (LSHT) shear cutters. The method of size reduction used in HSLT and LSHT are dissimilar in nature and this often leads to some inherent advantages and disadvantages regarding the acceptable MSW feed, product size distribution and overall process capacity and efficiency. With a HSLT device, breakages are primarily accomplished via impact forces, while LSHT devices rely much more heavily on shearing mechanisms, requiring tight tolerances between the cutting surfaces. These two different shredding techniques will produce a distinctively different size distribution and will have optimum performance under different material feed, capacity, loading and size reduction.

5.5.1   Shredding

The hammermill shredder

Low-torque shredders, such as the top-fed horizontal hammermill, utilize high-speed rotating shafts (700–1200 rpm), which are equipped with fixed or pinned hammers, as shown in Fig. 5.5, to crush incoming material. These hammers are responsible for the primary size reduction of the MSW and subsequently require maintenance and repair to be kept in optimal condition. The principal difference between these machines and LSHT devices is that hammermills rely almost

5.5 Top-fed horizontal hammermill.

entirely on impact and abrasive forces to smash the refuse into smaller particles while LSHT devices predominantly use shear forces to cut the material. The degree of size reduction and retention time of the MSW in the grinder is controlled via sizing bars or sieves located below the hammers. The operator can install or remove bars to achieve the desired particle size. As a result of their impact-breaking mechanisms, hammermills are generally more effective in processing brittle materials, which shatter under the impact forces while softer material tends to absorb energy though deformation and softening mechanisms. Rags and other such stringy materials occasionally wrap around the shaft leading to over-torqueing of the drive system, which requires stoppage time to remove the problem material. This is directly related to the low torque of the system and can be avoided by keeping a close watch on the rotor during operation. Hammermills produce a fairly inhomogeneous product, with brittle materials making up a higher portion of the fines than ductile materials. In many cases this is of little concern, and in the case of material separation this is actually a beneficial quality. However, when shredding MSW prior to incineration, this material bias may lead to unnecessary size reduction of non-combustible brittle materials such as glasses and ceramics.

HSLT shredders have specific energy consumptions ranging from 5–25 kWh/ton depending on the characteristic size and composition of the shredded refuse. The energy consumption and process efficiency of high-speed rotational devices are sensitive to operation speed, i.e. running at higher rpm can lead to unwanted power losses. In practice, up to 20% of a hammermill's power is consumed in overcoming forces created by bearing friction and surface drag, which develops between the hammer's surface and entrained air.

Decreasing the particle size of combustible materials increases the surface to volume ratio, thus allowing for quicker heat and mass transfer and enhanced combustion rates. Due to the inherent heterogeneity of MSW streams, the heating value of unprocessed MSW entering a combustion chamber is quite variable. The effectiveness of combustion and pollution control systems are improved by the consumption of a more homogeneous and well-mixed fuel. The grate and boiler operator can better control operation when it can utilize a uniform and more well-known fuel source. Due primarily to bag breaking and mixing processes, the variability of shredded MSW is notably lower than that of the raw MSW. Additionally, smaller particles have better mixing properties leading to enhanced combustion. The extent of these benefits from an economic point of view is not uniformly advantageous and is site dependent.

It is important to note that over-processing MSW can lead to particles that are excessively fine, which can cause entrainment problems during combustion. Particles entering the freeboard increase pollution rates and place a heavier load on the baghouse or electrostatic particulate collection systems.

The moisture content of MSW also has an effect on the operation and efficiency of high-speed devices. The amount of moisture present in MSW is highly variable and can range from as low as 5% up to nearly 60% depending on the origin of the waste and the time of year. Some of the more common materials found in MSW, such as paper, lose their tensile strength when wet; thus, the energy required in tearing paper decreases with increased moisture content. However, Trezek *et al.* (1997) have shown that the specific energy consumed (the energy per unit of material) decreases with the moisture content of MSW up to about 35%; at higher moisture contents the required specific energy increases due to the energy consumed in moving the more dense wet material. The issue of water content is unique to HSLT devices since wet materials absorb the impact energy of the hammer and tend to deform rather than break. Wet material can also cause flow issues by disrupting the smooth rotation of the hammers, leading to an uneven mass distribution in the grinder, which causes excessive wear and vibration, similar to an unevenly loaded washing machine.

Safety of operation is a very serious concern in waste management and material processing. One of the more alarming safety issues encountered in MSW processing is that of unexpected explosions and ejected objects. Typically, explosions are caused by the accidental processing of combustible gas, such as in a propane tank, or other compressed flammable materials. The potential for explosions is high in both high-speed and low-speed devices. However, it is of particular concern for high-speed devices due to the rigorous mixing of air and potentially flammable gasses that may enter the device, combined with the potential ease of ignition due to sparking. It is necessary to pay particular attention to these issues.

For a more in-depth look at hammermill operation the reader is directed to Trezek *et al.* (1997).

Low-speed, high-torque shredders

Low-speed, high-torque shredders, such as rotary shear shredders, operate on a different fundamental principle than hammermills. Rotary shear devices rely on shear cutting and tearing forces with little to no impact force involved. Counter-rotating shafts are fitted with cutting knives that intermesh and create large shear forces on the material trapped between them. These cutting knives or hooks, as shown in Fig. 5.6, are designed to pull the material into the cutting region of the shafts where the major processing takes place. LSHT shredders can be configured in single-, double- or quad-shaft designs depending on the material and amount of size reduction desired. Shear shredders used in MSW processing are usually the two-shaft design in order to avoid excess size reduction. LSHT shredders generally have a speed of between 10 and 50 rpm so they are often only available in lower capacities than HSLT. The capacity of the shredder depends on the rotor speed and the volume available between the cutting knives. LSHT machines tend to have a lower specific power consumption, in the range 3–11 kWh/ton, compared to HSLT machines with 5–25 kWh/ton. This energy consumption is highly dependent on composition and feed rate.

The torque produced in this type of shredder is characteristically in the range 50–350 kN.m, significantly greater than the 1–4 kN.m typical of a hammermill. The high torque results in a more even particle distribution, because shear forces are the major breakage mechanism and are less sensitive to material properties.

Both HSLT and LSHT size-reduction devices are continually exposed to harsh operating conditions and experience severe wear and tear. When operating

5.6 Cutting hooks of a low-speed, high-torque shredder.

hammermills and shear shredders, it is essential to the productivity of the machine that the hammers and cutting surfaces are well maintained. Hammer tip replacement is the most common maintenance procedure and results in the most downtime for HSLT devices. For the same reason, LSHT devices require cutting surface maintenance or replacement to ensure proper performance.

Capacity

For both HSLT and LSHT shredding, the capacity of each machine is one of the many important criteria to be addressed when designing an MSW-processing system. Waste-fired power plants can have capacities in excess of 3000 tons per day; in these high-capacity operations it is common to have two or three separate boiler lines. Each separate line has its own hopper, grate and boiler, meaning that it may be desirable for each line to have its own shredding and processing system. To ensure optimum operation, the installed shredding capacity should exceed nominal demand by more than 25% to allow for downtime and processing of non-ideal materials of higher durability. Capacities of nearly 300 tons per hour for HSLT shredders and roughly 150 tons per hour for LSHT are typical of the larger devices used in MSW processing.

5.5.2   Size reduction

The particle size distribution of MSW ranges widely depending on its initial source and geographic location; for example C&D waste will generally have a larger mean particle size and contain tougher material than household garbage. The main goal of shredding is the reduction of average particle size; however, because material properties play a role in the degree of size reduction, different materials undergo different levels of size reduction. Separation facilities such as MRFs take advantage of this characteristic of shredders because different materials tend to break into distinctive size ranges allowing for easier sorting and recovery. The overall effect of shredding tends to reduce particles to between 1/3 and a 1/4 of their original size, depending on feed composition, rotor speed and sizing bars. These parameters can be adjusted by varying shaft speed, tightening or widening cutting bars, adjusting tolerances and selecting the right device for the material stream.

5.6   Conclusion

Integrated solid waste management is a wide-ranging practice that encompasses many technologies both in the mature and cutting-edge stages of development. The pre-processing and treatment of MSW prior to conversion in an incineration plant is only a small aspect of the entire waste-management infrastructure, only a fraction of which is discussed here. This sub-section of waste-management systems is in itself still developing and adapting to modern conversion and recovery processes. In terms of energy and material recovery, it is likely that the industry will continue to see improvements in efficiency and environmental impact. This chapter is meant to be an overview of the more common MSW processes in use today and has covered the basics of separation, size reduction and thermal and biological treatment. Material separation is addressed to the extent of common processes found in the majority of MRFs and introduces the fundamental techniques of separation based on size, density and magnetic properties. Finally we discussed the two primary technologies for MSW size reduction: high-speed hammermill shredders and low-speed shear cutters. Thermal processes such as pyrolysis and gasification are left for other chapters of this book; however, torrefaction is addressed in this section as it is considered a pre-treatment method prior to incineration or gasification.

The complexities associated with MSW pre-treatment, in terms of economics and finance, are not addressed in this chapter. In the field there is still debate on the absolute value of pre-treatment of MSW prior to incineration: it is clearly advantageous for environmental reasons in terms of enhanced material and energy recovery; however, it has not been broadly proven as a more profitable process over mass burn facilities.

5.7   Further reading

Each sub-topic of this chapter is a field of MSW management and processing in its own right and they have only been briefly discussed here. The following references represent a good starting point for the reader interested in a more in-depth understanding of these processes and techniques.

Harvey, A., Gavis, J., Renard, L.M. (1981) Design models of trommels for resource recovery processing, *Resources and Conservation* 6, 223–240.

Hasselriis, F. (1984) *Refuse-derived Fuel Processing*, Boston: Butterworths (available from Amazon).

Lorange, R., Redon, E., Lagier, T., Hebe, I., Carre, J. (2007) Performance of a low cost MBT prior to landfilling: Study of biological treatment of size reduced MSW without mechanical sorting, *Waste Management* 27, 1755–1764.

Mata-Alvarez, J., Mace, S., Llabres, P. (2000) Anaerobic digestion of organic solid waste. An overview of research achievements and perspectives, *Bioresource Technology* 74, 3–16.

Savage, G.M., Trezek, G.J. (1974) On grinder wear in refuse comminution, *Compost Science*, 15(4), 51–53.

Shiflett, G.R., Trezek, G.J. (1979) Parameters governing refuse comminution, *Resource Recovery and Conservation* 4, 31–42.

Soyec, K., Plickert, S. (2002) Mechanical-biological pretreatment of waste-state of the art and potentials of biotechnology, *Acta Biotechnologa* 22(3–4), 271–284.

Trezek, G.J. (1972) Mechanical properties of some refuse components, *Compost Science* 13, 10–15.

Trezek, G.J., Savage, G. (1975) Report on a comprehensive refuse comminution study, *WasteAge* 3, 49.

5.8   References

Alter H., Gavis, J., Renard, M.L. (1981) Design models of trommel for resource recovery processing, *Resource Conservation* 6, 223.

Fricke, K., Santen, H., Wallmann, R. (2005) Comparison of selected aerobic and anaerobic procedures for MSW treatment, *Waste Management* 25, 799–810.

Trezek, G.J. *et al.* (1997) Significance of Size Reduction in Solid Waste Management, EPA-600/2-77-131.

van der Stelt, M.J.C., Gerhauser, H., Kiel, J.H.A., Ptasinski, K.J. (2011) Biomass upgrading by torrefaction for the production of biofuels: A review, *Biomass and Bioenergy* 35(9), 3748–3762.

Wheeler, P.A., Barton, J.R., New, R. (1989) An empirical approach to the design of trommel screens for fine screening of domestic refuse, *Resources, Conservation and Recycling* 2, 261–273.

6

Municipal solid waste (MSW) combustion plants

L. M. GRILLO, Grillo Engineering Company, USA

DOI: 10.1533/9780857096364.2.72

Abstract: Energy was recovered from municipal solid waste (MSW) in the United States as early as the 1890s. During the early 1970s, MSW was considered as a fuel and the first 'modern' waste to energy facilities were built. Large mass burn facilities were constructed as field-erected grate-fired waterwall furnaces. Smaller facilities were shop-assembled modular systems with combustion taking place on hearths and the energy recovered in waste heat boilers. Refuse-derived fuel (RDF) facilities were developed to prepare a fuel suitable for a boiler designed for that fuel specification. This chapter discusses the mass burn, RDF and modular combustion systems that have been proven in over 30 years of successful operation in the United States.

Key words: waste to energy, mass burn, refuse-derived fuel, modular, waste combustion, grate.

6.1 Introduction

Energy has been recovered from municipal solid waste (MSW) in the United States since the 1890s. Those facilities generally provided electricity for in-plant or local use and did not sell excess power to the utilities (Stoller and Niessen, 2009).

Most early waste combustors had refractory-lined furnaces and spray chambers to reduce the flue-gas temperature and control particulate emissions. In the late 1960s and 1970s, many combustors were shut down due to public dissent over stack emissions and the passage of the Clean Air Act in 1970. During the early 1970s, with the first of several energy crises, MSW was looked upon as a fuel. The first 'modern' waste to energy facilities were built at that time. These were designed as furnaces with integral waterwall or waste heat boilers to recover the energy in the form of steam. Most facilities used the steam to generate electricity.

In the late 1970s, several attempts were made to recover recyclable materials from MSW. After recovery, the remaining material was refuse-derived fuel (RDF). Early attempts to use this form of RDF were unsuccessful. Many facilities attempted to co-combust RDF with conventional fossil fuels in power plants. The combustion characteristics of the fuel, the material handling problems and the resulting residue resulted in the utilities refusing to accept the RDF. The recovered materials were also too contaminated to be sold. As a result, these early RDF facilities were unsuccessful.

72

Later RDF facilities started with the concept of preparing a fuel suitable for firing in a boiler designed specifically for that fuel specification. The materials that were removed, with the exception of ferrous metals, were discarded. These RDF facilities had a great deal of success, and most are still operating after 25 years.

Mass burn facilities continued to be constructed through the 1980s and up to the mid-1990s. Larger facilities, generally greater than 400 tons per day, were built as field-erected waterwall furnaces. Smaller facilities, processing less than 400 tons per day, were generally shop-assembled modular systems with combustion taking place on hearths and the energy recovered in waste heat boilers. These units had the advantage of low capital cost, but had lower efficiency than the waterwall boilers. They also did not achieve the high degree of burnout that was typical of waterwall applications.

This chapter discusses the mass burn, RDF and modular combustion systems that have been proven through over 30 years of successful operation in the US.

6.2   Principles of combustion

Table 6.1 shows the theoretical amount of air needed to combust the elements contained in waste. The amount of oxygen required for combustion can be determined from the ratio of molecular weights and the following reactions:

Carbon:

$$C + O_2 = CO_2 \qquad\qquad [6.1]$$

Atomic weight: $12 + 32 = 44$

Ratio: $1 + 2.667 = 3.667$

Hydrogen:

$$2H_2 + O_2 = 2H_2O \qquad\qquad [6.2]$$

Atomic weight: $4 + 32 = 36$

Ratio: $1 + 8 = 9$

Sulfur:

$$S + O_2 = SO_2 \qquad\qquad [6.3]$$

Atomic weight: $32 + 32 = 64$

Ratio: $1 + 1 = 2$

The amount of air required for the above reactions is 4.32 times the weight of oxygen, based on air containing 23.15% oxygen by weight. The remainder of the weight of air is considered to consist of nitrogen only (Velzy and Grillo, 2007a).

Table 6.1 Stoichiometric air calculations

Combustible	Pounds per pound of combustible						
	Required for combustion			Products of combustion			
	O_2	N_{atm}	Air	CO_2	H_2O	SO_2	N_{atm}
Carbon	2.67	8.87	11.54	3.67			8.87
Hydrogen	8.00	26.57	34.57		9.00		26.57
Sulfur	1.00	3.32	4.32			2.00	3.32

An ultimate analysis of a solid fuel consists of the weight fractions of carbon, hydrogen, oxygen, nitrogen, sulfur, chlorine, moisture and ash. The weight fractions are determined using American Society of Testing and Materials (ASTM) standards for all components except oxygen, which is determined by difference (Velzy and Grillo, 2007b). An assumed ultimate analysis of waste is shown in Table 6.2, which will be used for sample calculations in the following sections.

The higher heating value (HHV) is the lower heating value (LHV) plus the latent heat contained in water vapor resulting from combustion (Hammerschlag, *et al.*, 2007). The HHV of waste can be estimated using Dulong's formula (Schlesinger, 2007):

$$\text{HHV (BTU/lb)} = 14\,544C + 62\,028(H - O/8) + 4050S \qquad [6.4]$$

where C is the weight fraction of carbon, H is the weight fraction of hydrogen, O is the weight fraction of oxygen and S is the weight fraction of sulfur.

For the example calculation, the HHV is calculated to be 5306 BTU/lb using Dulong's formula. The LHV of the waste is the total quantity of sensible heat released during combustion.

Table 6.2 Ultimate analysis of waste

Component	Percentage by weight
Carbon	30.0
Hydrogen	4.0
Oxygen	20.0
Nitrogen	0.5
Sulfur	0.3
Chlorine	0.5
Ash	21.0
Moisture	23.7
Total	100.0

The LHV is calculated as follows (Velzy and Grillo, 2007b):

$$LHV = HHV - [WH_2O + (9 \times WH)] \times 1050 \qquad [6.5]$$

where WH_2O is the weight fraction of moisture in the fuel and WH is the weight fraction of hydrogen in the fuel. For the example calculation, the LHV is calculated to be 4679 BTU/lb.

The stoichiometric quantity of air is that which is needed to completely oxidize all of the combustible matter in the waste. Excess air is the amount of additional air that is injected to ensure complete burnout. The stoichiometric quantity of air can be calculated from the ultimate analysis of the waste. Table 6.3 shows the sample theoretical air calculation for the ultimate analysis contained in Table 6.2. The theoretical stoichiometric air for the sample calculation is 3.991 pounds of air per pound of waste. This does not account for any excess air. A facility designed for 100% excess air would require 7.982 pounds of air per pound of waste.

Table 6.4 shows the products of combustion for the sample calculation. A total of 4.828 pounds of flue gas would be generated per pound of waste combusted, with no excess air. At 100% excess air, the flue gas is 8.871 pounds of flue gas per pound of waste. To check the calculation, the sum of the products of combustion should equal the theoretical air plus moisture in the combustion air plus the weight fraction of carbon, hydrogen, oxygen, nitrogen, sulfur and moisture from the ultimate analysis. In this case for stoichiometric conditions,

$$4.828 = 3.991 + 0.052 + 0.30 + 0.040 + 0.20 + 0.005 + 0.003 + 0.237$$

and for 100% excess air,

$$8.871 = 7.982 + 0.104 + 0.30 + 0.040 + 0.20 + 0.005 + 0.003 + 0.237.$$

Table 6.3 Sample theoretical air calculation

Substance	Weight fraction	Oxygen required for combustion, lb/lb of element	Theoretical oxygen, lb/lb of element[a]	Theoretical dry air, lb/lb of element[b]
Carbon	0.300	2.67	0.801	3.460
Hydrogen	0.040	8.00	0.320	1.382
Sulfur	0.003	1.00	0.003	0.013
Total			1.124	4.855
Less oxygen in the fuel			(0.200)	(0.864)
Theoretical air required			0.924	3.991

Notes:
[a] Weight fraction of the substance times the oxygen required for combustion.
[b] Theoretical oxygen times 4.32.

Table 6.4 Sample products of combustion calculation

Substance	lb/lb of element	Pounds of product at 0% excess air	Pounds of product at 100% excess air
Carbon dioxide	3.67	1.101[a]	1.101
Moisture from hydrogen	9.00	0.360[a]	0.360
Oxygen		0.000	0.924
Nitrogen		3.072[b]	6.139
Sulfur dioxide	2.00	0.006[a]	0.006
Moisture from fuel	1.00	0.237[a]	0.237
Moisture from air		0.052[c]	0.104[c]
Total moisture		0.649	0.701
Total		4.828	8.871

Notes:
[a] Weight fraction of substance times lb/lb of element.
[b] Theoretical dry air times 0.7686 plus weight fraction nitrogen in fuel.
[c] Specific humidity of air at 80°F and 60% relative humidity gives 0.013 pounds of moisture per pound of dry air.

The maximum temperature occurs at the point when stoichiometric air is completely combusted and there is no excess air. The temperature developed during combustion can be calculated using the following equation (Velzy and Grillo, 2007b):

$$T_{comb} = T_a + LHV / [(W_W \times C_pW) + (W_{FG} - W_W) \times C_pFG] \qquad [6.6]$$

where T_{comb} is the combustion temperature, T_a is the ambient temperature, LHV is the lower heating value of the waste, W_W is the weight of water in the flue gas, C_pW is the heat capacity of moisture in the flue gas (= 0.55), W_{FG} is the total weight of the flue gas and C_pFG is the heat capacity of dry flue gas (= 0.28).

The maximum temperature (at stoichiometric conditions) is calculated to be 3124°F. The combustion temperature at 100% excess air is calculated to be 2105°F. Excess air contributes to boiler efficiency loss, and should be kept to a minimum to maximize boiler efficiency.

6.3   Mass burn waterwall combustion systems

Mass burn waterwall facilities combust unprocessed waste that is fed directly into the furnace. These facilities generally process 400 tons per day of MSW or more. There is no presorting or materials recovery prior to combustion. The material is introduced into the furnace where it passes through drying, combustion and burnout stages before the residue is discharged.

Combustion occurs in the furnace and the flue gas passes through an integral waterwall boiler, superheater, generation bank and economizer, producing steam.

Some technologies also incorporate an air heater to preheat combustion air. Flue gases then pass through the air pollution control devices.

6.3.1   Waste storage and handling

Tipping areas are enclosed to prevent odors and litter from escaping the building. Combustion air is drawn from the tipping and storage area to induce a negative draft in the building to control the release of these materials further. Using this air for combustion also destroys any odors that are present when the air is exposed to the high temperature. The waste should be handled on a first in/first out basis to reduce the potential for odor and decomposition.

Approximately four days of storage is necessary to ensure the availability of fuel over long weekends and during periods when deliveries are low. Waste can be stored in a pit or on a floor. Most large facilities use a pit to store the waste. The pit is long, narrow and typically 30 to 40 feet deep, and runs in front of all of the units in the facility. The waste is dumped into the pit and mixed and stacked using an overhead crane that spans the width of the pit. The crane also retrieves the waste and feeds it to the charging hoppers.

Depending on the depth of the pit, the waste can have an average density of 15 to 25 lb/cf. The size of the pit is determined by the throughput of the facility and the ability to store about four days of waste. The available storage is calculated using the pit depth from the bottom of the pit to the tipping floor plus waste that is stacked above the tipping floor level. The available volume above the tipping floor is calculated from the edge of the tipping floor to the back wall of the pit, assuming the waste is stacked at a 45° angle.

During operation, a trench is dug by the crane along the front of the pit along the tipping floor. This trench allows trucks to dump into the pit when the pit is relatively full. The waste that is dug from the trench is stacked against the rear wall. Although the angle of repose of MSW is approximately 45° when dropped onto a pile, waste in a pit can be handled and stacked so it forms a vertical wall.

As time passes, the waste compacts in the pit and must be 'fluffed' by the crane before being fed to the charging hoppers. The fluffing breaks up clumps in the waste and allows for better combustion in the furnace. The waste is dug up by the grapple and released slowly above the waste in the pit, allowing it to loosen as it drops. It is then picked up again by the crane and delivered to the charging hoppers.

When excavation for construction of a pit is not possible due to high groundwater or other geologic conditions, a tipping floor may be used instead of a pit for storage. Operation of a tipping floor is more labor intensive, since the waste must be moved, stacked, retrieved and fed using front-end loaders. When a tipping floor is used, the density of the waste should be assumed to be 12 to 15 lb/cf when calculating the storage area. The density is lower because the loaders can only stack the waste to about 20 feet high. The sides of the piles should be assumed to have a 45° angle when calculating the storage area.

Some facilities use a bulldozer or similar equipment to densify the waste periodically to increase storage. Densifying the waste increases the energy and labor required to operate the facility. This procedure is usually employed when additional storage is needed because a unit is out of service and waste quantities are high. It is not the recommended normal operating mode when using a tipping floor for storage.

When a tipping floor is used, the trucks dump on the floor and a front-end loader pushes the waste into the storage pile. The loader retrieves waste from a different area to feed to the furnaces. Again, the oldest waste is recovered first to reduce odor and decomposition of the waste. The loaders feed the waste onto a conveyor, which elevates it to the charging hopper. The conveyors are usually steel apron pan conveyors because of the potential damage from impact of the waste onto the conveyor.

6.3.2   Feeding

All of the systems use a charging hopper and chute that accepts the waste and feeds it to the furnace. The charging hopper is designed with a slope of at least 45° and has steel plates on which the waste is dumped. The waste is deposited on the slope of the hopper, where it slides into the charging chute toward the boiler. Feeding waste directly over the charging chute often causes jams in the chute. The waste is held back on the slope by the waste in the chute. As the waste is pushed into the furnace from the bottom of the chute, the waste in the chute falls by gravity, allowing the waste in the hopper to slide slowly into the top of the chute. When the hopper is nearly empty, the crane deposits another load onto the hopper slope.

The waste in the charging chute forms an air seal to the boiler. The charging chute usually has a cut-off gate to prevent back fires and maintain an air seal during start-up and shutdown. The lower part of the charging chute is frequently water cooled to protect it from the high temperature in the furnace.

At the bottom of the charging chute is a hydraulic ram feeder that pushes the waste into the furnace and onto the grates. There is usually one ram feeder for each grate section. The rams may operate in parallel (all stroking together) or in alternating mode.

6.3.3   Grates

The basic difference between the various technologies is the means by which the waste is transported through the furnace during the combustion process. The primary purpose of the grate is to convey the waste automatically from the ram feeder to the residue discharge point. A second purpose is to tumble the waste to ensure that all of the waste is exposed to the combustion air and that complete combustion occurs. Grates can use a reciprocating motion, rotary drum, rollers or other means to move the waste through the furnace. Rocking grates were used in older combustors. However, they have not been used in facilities in the past 20 years.

Reciprocating grate

The reciprocating grate is the most common grate used in mass burn waterwall facilities. This type of grate uses a step action with alternating stationary and moving grates to push the waste through the furnace (see Fig. 6.1). As the waste is pushed over a stationary grate, it tumbles, exposing unburned particles to the combustion air and allowing good burnout.

Grate manufacturers generally have a standard length of grate with the same number of steps, air zones, etc. The width of the grate determines the throughput of the unit. The width of the grate can be increased to a maximum size, then additional grate sections are added. Thus, facilities with high throughputs have wide grates with several independent grate runs. Each grate run would be fed by a dedicated ram feeder, and contain separate underfire air compartments and grate drives. This allows for independent control of the various grate runs for better combustion control. The size of the grate is determined by the grate heat release rate, which should be between 250 000 and 300 000 BTU/sf/hr (Velzy and Grillo, 2007b).

Underfire air is injected into the hoppers below the grate. Air penetrates through the grate to keep the grate cool and promote combustion. The grates have between three and five zones from the inlet point to the discharge. At each zone change,

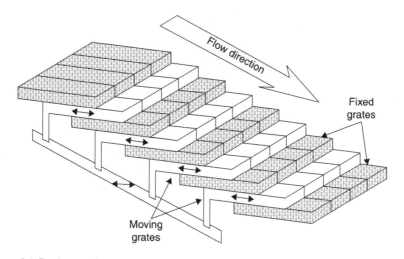

6.1 Reciprocating grate.

there is a drop of one to two feet to promote additional tumbling. The waste passes through drying, combustion and burnout zones. Most of the air is injected into the zones where active combustion is taking place, which begins about one-quarter of the way down the grate. Little air is fed into the final burnout zone.

At the end of the grate, the residue falls off the last step into the residue disposal system. Most systems use either a water-filled ash discharger or water-filled trough to quench the residue and maintain an air seal with the boiler. Residue is then transported by conveyor to the disposal system.

System suppliers that provide traditional reciprocating grate technology include Von Roll (Wheelabrator) (air-cooled or water-cooled grate) (Owens and Sczcepkowski, 2010), Foster Wheeler, Steinmuller, Detroit Stoker and Keppel Seghers (air-cooled or water-cooled grate).

Reverse reciprocating grate

The Martin grate (see Fig. 6.2) uses a reverse acting reciprocating grate. Alternating stationary and moving grates are on a 26° downward angle and push the waste

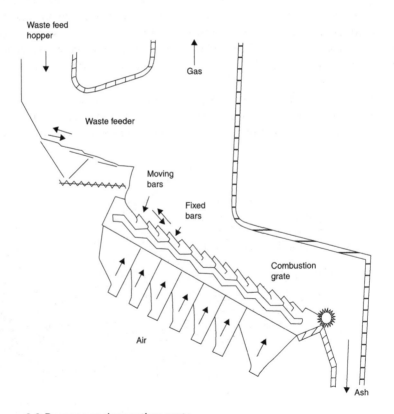

6.2 Reverse reciprocating grate.

upward, causing it to flip over the grate and tumble downward through the furnace. This motion provides a good deal of tumbling and mixing of the waste, exposing unburned material for good burnout. Three grate drive zones move the waste through the drying, combustion and burnout zones of the furnace. Each grate drive can be controlled individually so the combustion may be controlled within the burning zones. Underfire air is supplied through five individually-controlled air zones, providing the proper amount of air for the drying, combustion and burnout stages.

Roller grate

Figure 6.3 shows a cross section of the Deutsche Babcock roller grate. It consists of five or six cylinders, depending on the throughput, on a 30° downward angle that transport the waste through the furnace. The drums are made of cast iron grate sections that rotate at three to six revolutions per hour, causing a tumbling action on the grate. As the burning waste moves down the slope, the rotating speed of each successive roller is reduced to keep the fuel bed thickness approximately uniform. Combustion air, which can be controlled in each combustion zone, enters the interior of each drum from both ends and flows through the many small gaps in between the interlocking grate bars (Hickman, 2003).

Rotary combustor

Figure 6.4 shows a cross section of an O'Connor rotary combustor. The combustor is a cylinder approximately 40 feet long, approximately 14 feet in diameter and on

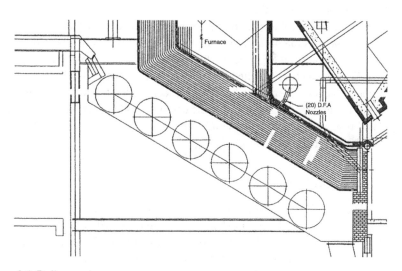

6.3 Roller grate.

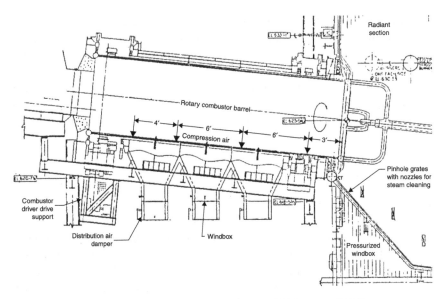

6.4 Cross section of a rotary combustor (O'Connor, 1984).

a 6° downward angle. The actual length and diameter vary depending on the throughput. The cylinder has a series of tubes and membranes running the length of the combustor. The membranes have holes to allow combustion air to enter the cylinder. The tubes and membranes are formed into a cylinder in which the MSW combusts, and are an integral part of the boiler, containing hot water and steam. The tubes of the combustor are connected at each end by a ring header, which is larger in diameter than the tubes.

MSW is pushed into the combustor by ram feeders. Heated combustion air is injected through the holes in the membrane. The combustor has four combustion air zones, and air entering below the MSW (underfire air) and above the MSW (overfire air) can be controlled to ensure good combustion. The combustor rotates at about six revolutions per hour. As the combustor rotates, MSW tumbles and moves toward the discharge end of the combustor. Combustion continues until the waste reaches the end of the combustor and falls onto an afterburning grate to complete the burnout.

6.3.4   Combustion air

Combustion air is injected into a furnace as underfire air (below the grate) and overfire air (into the flame above the grate). Generally, 50% to 70% of the total air is underfire air and the remaining portion is overfire air. Most mass burn furnaces operate with between 50% and 100% excess air. Combustion air and flue-gas handling equipment should be sized for 100% excess air.

Combustion air is drawn from the waste storage area for two reasons. First, it maintains a negative draft in the tipping area, preventing litter and odor from escaping the building. It also destroys the odor by exposing it to the high temperatures in the furnace.

Underfire air is frequently preheated using steam coil air heaters. Heated air assists with drying the waste and allows for more stable operation. Using heated air slightly reduces the throughput of the unit because of the heat added to the furnace in the air. Underfire air is directed by a forced draft fan that blows the air into a plenum below the grates. Most furnace designs have multiple air plenums along the length of the grate. This allows the operator to control the amount of combustion air as the waste passes through the drying, combustion and burnout stages. The underfire air enters the furnace through slots or spaces between the grate components. Underfire air passes through the fuel bed, providing air for drying and combustion.

Overfire air is generally not heated. It may be drawn from either a common air duct from the receiving and storage area or from a separate source, such as the upper level of the boiler house. Overfire air is injected through a series of nozzles, usually located on the back and/or front wall of the furnace above the grate. The air is injected at a higher pressure than underfire air, and often requires a booster fan if the source of the air is the same as underfire air. Overfire air is injected into the active flame zone to provide additional air to complete burnout of the volatile gases, which are produced by heating the waste. It provides the turbulence needed to completely mix the flue gas to ensure good combustion.

6.3.5 Boilers

Boiler designs for mass burn facilities generally include a radiant pass above the grate with integral waterwall tubes. The superheater generally is the next component with two to four sections. Most facilities have a desuperheater between the final two stages of the superheater to control steam temperature. The superheater is followed by an evaporator section and bare tube economizer. Some facilities also have a bare tube air heater to preheat the combustion air following the economizer.

The radiant section consists of vertical tubes, nominally two inches in diameter, with a continuously welded membrane between the tubes. This creates an airtight enclosure within which combustion takes place. The tubes are connected by headers along the grate and the roof that connect to the steam drum. The lower section of the tubes, along the grate line, frequently have a refractory coating to protect against abrasion from the fuel as it moves along the grate. The tubes and membranes above the refractory, generally to the top of the radiant section, are overlaid with Inconel. The Inconel protects the tubes from corrosion and erosion by the flue gases. Early mass burn installations did not have this protective overlay and deteriorated rapidly. It is imperative that some coating be used in the radiant section of the boiler to protect the tubes.

At the top of the radiant section, the flue gases turn and pass through a superheater. Here, saturated steam enters the superheater from the steam drum and rises in temperature. Most boilers have two to four superheater sections. At the inlet of the final stage, a desuperheater is used to maintain the desired outlet temperature by spraying boiler feedwater into the steam. Superheaters are generally either shielded or overlaid in part with Inconel to protect against corrosion and erosion.

After passing through the superheater, the flue gas passes through an evaporator section or generation bank. This may consist of an upper steam drum and lower mud drum connected with a series of tubes. Alternatively, additional waterwall passes provide for water evaporation without the drums. In this case, a series of headers connects the various boiler sections. The evaporator section adds sufficient energy to convert the saturated water to saturated steam.

The economizer follows the evaporator section of the boiler. Boiler feedwater is pumped at high pressure into the economizer. The economizer raises the temperature of the water up to the saturation point before it goes into the evaporator section. Economizer tubes in waste combustion facilities have bare tubes (no fins) to reduce fouling due to the presence of particulate matter.

Some facilities incorporate an air heater as the final heat trap. This raises the combustion air temperature to as much as 500°F. The air heater tubes usually do not have fins due to the presence of particulate matter. Combustion air flows through the tubes and flue gas flows outside the tubes.

6.3.6 Residue

After combustion is complete, the burned out residue is discharged from the grate into a water-filled discharger. The water cools the residue and maintains a seal to prevent tramp air from entering the boiler. Two types of residue dischargers are commonly used. The first is an extractor, which is a water-filled container below the end of the grate. The extractor has a ram that extends periodically to force the residue up an inclined chute where it is partially dewatered. When the chute is full of residue, each stroke of the ram forces some residue out of the extractor onto a conveyor that removes it for storage.

The second type of discharger is a water-filled trough that contains a drag conveyor. The residue is quenched by the water and removed by the drag conveyor. In small units with narrow grates, the trough may be oriented in the same direction as the grate. Each unit would have a dedicated conveyor for removing the residue. In larger units, with wide grates, the trough usually runs perpendicular to the grates and collects residue from several units. This orientation would have a redundant conveyor system, where the residue may be discharged into either of two conveyor systems through a bifurcated chute. The drag conveyors move slowly and dewater the residue as the conveyor inclines upward out of the water for discharge to another conveyor to take the residue to storage. The extractor type of residue discharger results in better dewatering, resulting in lower moisture in the residue.

In addition to residue that falls off the grate, some small particles of residue fall through the grate. These siftings, or riddlings, are collected and conveyed to the extractor or trough and removed with the bottom ash.

6.3.7 Air pollution control

After the flue gas leaves the boiler, it passes through a series of air pollution control devices to remove acidic gases, particulates, oxides of nitrogen, mercury, heavy metals and other pollutants. This equipment is discussed in detail in Chapter 11.

6.4   Refuse-derived fuel (RDF) combustion systems

Unlike mass burn combustion systems, which process waste directly with no presorting, RDF systems produce a more uniform material prior to combustion. The fuel processing may be done at a remote facility or at a facility adjacent to the combustion plant. Co-locating the processing and combustion facilities eliminates the additional expense of transporting the RDF.

6.4.1 Waste storage and handling

RDF facilities must have provision for storing both raw waste and the prepared fuel. Roughly two days of raw waste storage and four days of RDF storage are usually provided. RDF facilities generally use a tipping floor instead of a pit to store both unprocessed and processed waste. Raw waste and RDF are stored in separate buildings and pushed into piles using front-end loaders. The loaders also recover the waste and feed it to the processing lines and feed the RDF to the boiler feed lines.

As with mass burn facilities, the tipping and RDF storage areas are enclosed to prevent odor and litter from escaping the building. In most cases, combustion air is drawn from the areas to induce a negative draft in the building to control the release of these materials further.

6.4.2 RDF preparation

All RDF facilities include shredding and ferrous metal removal, and some incorporate additional screening steps to remove grit or use eddy current separation for non-ferrous metal recovery. Primary shredding may be accomplished with a flail mill, or bag breaker, to coarse shred the waste to a nominal 12 inch particle size and expose it to the downstream equipment. Ferrous metal separation can be accomplished with a drum or belt magnet. Further classification can be achieved using a trommel, disc or other type of screening equipment to remove grit and non-combustible material and sort the waste by size. Small particles of waste that are suitably sized for RDF bypasses further equipment, while material with a larger particle size goes on to a second stage of shredding. The desired final particle size is generally two to six inches. Details of the types of equipment used for RDF production are included in Chapter 5.

Some RDF facilities, called 'shred and burn', are of a simpler design. They include a single stage of primary shredding followed by ferrous metal removal. There is no further classification of the waste, and the resulting RDF is appropriately sized for the boiler.

6.4.3 Feeding

RDF is recovered from storage by front-end loaders, which push the fuel onto apron conveyors. Apron conveyors are used because they can withstand the impact of the weight of the fuel falling onto them and they can elevate the fuel at a relatively steep angle. A series of belt conveyors and diverters are used after the apron conveyors to deliver the fuel to the front of the boiler. One boiler feed chute per approximately five feet of boiler width are needed.

Two types of boiler feed systems are used: storage bin and direct feed by conveyor. The storage bin method consists of a bin that is the width of the boiler, which holds about a 15 minute supply of fuel. Screw augers on the floor of the bin remove the fuel and meter it into the individual feed chutes that lead to the boiler.

The direct feed method provides an oversupply of RDF to the boiler face. Each boiler feed chute has a small bin that holds about a 15 minute supply of fuel. The bins are constantly overfed and the excess fuel is returned to RDF storage. Each bin has a dedicated ram feeder and drag conveyor to remove the RDF and meter it to the boiler feed chute (Gittinger and Arvan, 1998).

6.4.4 Combustion systems

RDF is most commonly combusted on a spreader-stoker traveling grate (see Fig. 6.5), similar to the type used for coal combustion. The fuel flows by gravity down the feed chute at the front of the boiler and is blown to the back of the grate through an air swept spout. Much of the combustion of light materials occurs above the grate. Heavier materials that require more time for combustion fall onto the grate. The grate moves slowly from the back to the front, discharging the burned out residue into a water quench trough at the front of the boiler.

The boiler design is similar to that of mass burn systems. Underfire combustion air enters through a plenum beneath the grate, and passes through the fuel bed. Overfire air is injected above the grate to complete combustion of the volatile gases. Since the fuel is more homogeneous than unprocessed waste, the combustion process can be accomplished with a lower amount of excess air. RDF facilities

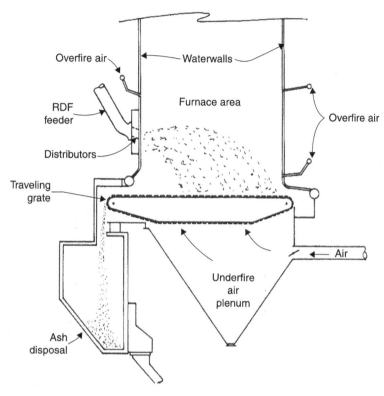

6.5 Spreader-stoker traveling grate (Russell and Roberts, 1984).

usually operate with between 30% and 50% excess air. This improves the efficiency of the process, and a greater amount of energy can be extracted from the fuel.

Another type of combustor that has been used with RDF with limited success is the fluidized bed. Fluidized beds are reactors that have a sand or similar medium and jets of air that are injected vertically from the bottom of the bed. The air jets cause the sand to become fluidized. Waste is injected into the hot sand bed, where combustion occurs. The turbulent mixing of solids provides a high degree of heat transfer. Once heated, the medium retains its heat, causing a stable combustion process. Flue gases rise upward and out of the reactor, then pass through a boiler for heat recovery.

Fluidized-bed combustion evolved from efforts to find a combustion process able to control pollutant emissions without external emission controls (such as scrubbers). A sulfur-absorbing chemical, such as limestone or dolomite, may be added to the bed. The mixing action of the fluidized bed brings the flue gases into contact with the sorbent (REI, 2005). Even with the ability to reduce sulfur emissions, additional scrubbing equipment is necessary in waste to energy facilities.

6.5   Modular combustion systems

Modular combustion systems are generally constructed in a shop and shipped to the field for erection. The capacity of each unit is generally about 150 tons per day or less. Modular units are similar to mass burn facilities in that they combust unprocessed waste that is fed directly into the furnace without presorting or materials recovery.

Primary combustion occurs in a refractory-lined furnace, and the flue gas passes into a secondary combustion chamber where combustion of the flue gas is completed. Flue gas is then directed through a waste heat boiler. The waste heat boiler includes a superheater, generation bank and economizer sections. Flue gases then pass through the air pollution control devices.

6.5.1   Waste storage and handling

With few exceptions, modular facilities use a tipping floor and front-end loaders to store and move the waste. As with larger facilities, tipping areas are enclosed to prevent odor and litter from escaping the building and combustion air is drawn from the tipping and storage area. Approximately four days of storage is also provided.

6.5.2   Feeding

Modular systems are batch fed by front-end loaders, which push the waste into a hopper in front of the combustion chamber. After the hopper is loaded, a door is closed over it to prevent flue gas from escaping as the waste is fed in. An isolation door separating the charging hopper from the furnace is raised and a hydraulic ram pushes the waste into the first stage of the combustion chamber. The ram retracts, the isolation door closes and the hopper door opens in preparation for the next load. Loads are charged about every ten minutes.

6.5.3   Combustion chamber

Most modular facilities burn the waste on refractory-lined hearths. A series of three to seven step hearths usually make up the primary combustion chamber. The hearths have about a one foot drop off the end to allow the waste to tumble as it proceeds through the furnace. A hydraulic ram under each hearth pushes the waste through the furnace. The last ram strokes first, providing space for the material falling from the hearth above. After the ram extends and retracts, the next hearth ram strokes. This continues until all of the rams have stroked. After the last ram has stroked at the inlet to the furnace, another hopper full of waste is charged, pushing the waste ahead of it onto the second hearth.

When the last ram strokes, it pushes the residue off the hearth into a water-filled trough. The water serves to quench the ash and maintain an air seal for the combustion chamber. The residue is dragged up an incline where it is partially dewatered prior to disposal.

6.5.4   Combustion air

Underfire combustion air is injected into the furnace through nozzles in the hearth or in the front face of the hearth rams. Overfire air is injected through nozzles in the roof. Because the waste is combusted on hearths and does not receive the high degree of agitation as with grate systems, a higher amount of excess air is needed to ensure good combustion. Excess air is generally in the 100% to 150% range. Also, burnout is not as good as with modular systems due to the limited agitation. Air is also injected into the secondary combustion chamber, where the products of primary combustion are completely burned out.

Some modular designs operate in a 'starved air' mode, where sub stoichiometric air conditions are maintained in the primary combustion chamber. This causes gasification of the waste, producing combustible flue gases. The flue gas is burned with excess air in a secondary combustion chamber. Some supplementary fuel such as natural gas may be added to enhance the combustion of the gases.

6.5.5   Boiler

The boiler design for modular facilities is different from that of mass burn facilities in that the boilers are not integral waterwall boilers, but are waste heat boilers that are prefabricated in sections in a shop and delivered to the site for assembly. Boiler sections consisting of evaporators, superheaters and economizers are shop assembled and shipped as components to the field for erection. This reduces construction costs substantially.

After leaving the boiler, flue gases enter air pollution control devices similar to those used in mass burn facilities.

6.5.6   Manufacturers

Enercon Systems

Combustors produced by Enercon consist of a two-stage excess air combustion process (see Fig. 6.6). Underfire air is injected through nozzles in the hearth rams. A system of 'push rods' driven by the ash transfer rams provides cleaning of these holes with every ram stroke (Clark, 1996). Dual-fuel burners, accepting gas or oil, are located in the primary chambers and are used for initial ignition of the refuse.

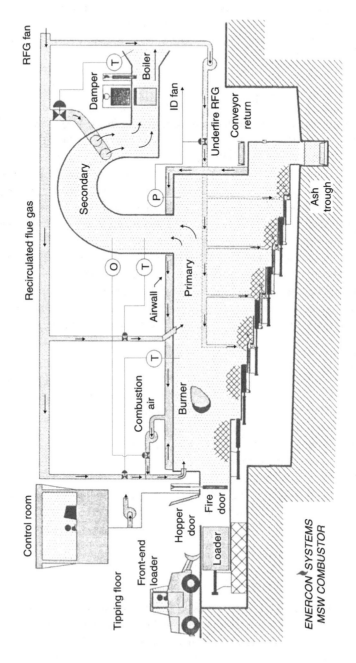

6.6 Enercon Systems MSW combustor (RFG: recirculated flue gas).

These are turned off after the fire is established (Clark, 1982). Overfire air is provided by nozzles in the roof of the primary chamber. Flue gas from the primary combustion chamber enters a secondary combustion chamber where secondary air is added to complete combustion. Fossil fuel burners are installed at the end of the secondary combustion chamber to maintain constant temperature of the flue gas entering the waste heat boiler. However, these burners are rarely used (Clark, 1982).

Recirculated flue gas is used for underfire air and for furnace cooling and combustion gas cooling prior to entering the waste heat boiler. Using recirculated flue gas for furnace cooling reduces the quantity of fresh air needed, thereby increasing thermal efficiency and minimizing thermal NO_x formation (Clark, 1996).

ConsuTech (Consumat)

Consumat combustors use a starved air process with two chambers: the primary (or lower) chamber and the secondary (or upper) chamber. Combustion in each chamber is controlled independently to ensure efficient waste processing in the lower chamber and complete combustion of flue gases in the upper chamber. The units comprise a refractory-lined combustion chamber (no grate is required) into which waste is loaded. Air is supplied to this chamber at a rate less than that required for complete combustion. The waste is initially ignited by an auxiliary burner and undergoes essentially a gasification/pyrolysis process. For normal wastes, the reaction will proceed without the need for additional fuel. Waste decomposes under quiescent conditions, therefore carry-over of particulate matter is minimized (Consumat, 2004).

The partial combustion products pass into an afterburning secondary chamber, which is mounted immediately above the main combustion chamber. The gases are mixed with additional air at an increased temperature to ensure successful burnout of particulate matter to eliminate smoke. This chamber is designed to retain the combustion gases at 1800°F with a retention time based on waste composition and regulatory guidelines.

Combustion air is supplied to the lower and upper chambers independently. The rate of air supply to both chambers is automatically controlled to provide the correct combustion conditions.

The lower chamber operates at low interior gas velocities under controlled temperature conditions. The amount of heat released from the burning waste is controlled by limiting the air added into the lower chamber to less than what is required for complete combustion of the waste. Underfire air is introduced into the lower chamber through air holes located under the waste. Sufficient heat is released to keep the waste burning for partial combustion.

The combustion gases then pass into the upper chamber through a turbulent mixing zone where ignition takes place and additional air is provided to complete the oxidation reactions. Oxidation of the combustible products is completed in the upper chamber (Consumat, 2004).

Laurent Bouillet

A Laurent Bouillet combustor is different from other modular systems. The combustors have a refractory-lined drum that rotates for 210° then reverses. The waste tumbles through the drum where combustion occurs. Flue gases exit the combustor and enter a tranquilization chamber where combustion is completed. Flue gases then pass through the boiler and air pollution control devices (MMWAC, 2010).

6.6   Advantages and limitations

The types of combustors discussed above have all been proven in commercial operation for over 25 years. While many systems are similar within each classification of combustor, each type has advantages and limitations, as discussed below.

6.6.1   Efficiency

Boiler efficiency can be calculated using ASME Performance Test Codes (PTC). PTC 34 is used for waste combustors with energy recovery using the boiler as a calorimeter. Using this method, the heat outputs, losses and credits must be calculated. Heat credits correct the efficiency to standard ambient conditions, such as air temperature. When comparing the efficiency of different systems, the heat credits do not matter, since they would be substantially the same for any system. Heat losses have the most significant impact on efficiency. Many parameters that contribute to heat loss, such as moisture and hydrogen in the fuel, are independent of the type of technology. Others, such as loss due to carbon monoxide in the flue gas, are small and can be ignored when comparing systems.

The most significant losses that can be attributed to the type of technology are the dry gas loss, water-from-fuel loss and loss due to unburned carbon in the residue. The dry gas loss is dependent on the flue-gas exit temperature from the boiler and the amount of excess air. The water-from-fuel loss is also dependent on the flue-gas exit temperature. The unburned carbon in the residue is related to how complete combustion is. The three main factors in determining efficiency are therefore flue-gas exit temperature, excess air and burnout, all of which are dependent on the technology.

The flue-gas exit temperature is related to boiler design rather than the combustor. A well-designed boiler can achieve low exit gas temperatures for any of the technologies.

Excess air and burnout are the two main contributors to heat loss that are strictly combustor related. They are also related to each other. Insufficient excess air will increase the amount of unburned carbon. Less excess air will improve efficiency, while the resulting higher amount of unburned carbon will reduce efficiency. A careful balance of excess air and burnout must be made to optimize efficiency in any system.

Of the three types of technologies discussed, modular combustors use the greatest amount of excess air and have the worst burnout, resulting in the lowest efficiencies, generally around 55% to 65%. Mass burn uses less excess air and has better burnout, and results in higher efficiencies, around 68% to 72%. RDF burned on a spreader-stoker traveling grate uses the lowest amount of excess air and also results in the best burnout, with efficiencies around 72% to 75%. The reason for the lower excess air and better burnout is that the fuel has been shredded and blown into the furnace, resulting in good mixing of the fuel and combustion air. Mass burn and modular facilities fire unprocessed waste. Items that tend to go through the furnace in clumps, such as telephone books, are not completely exposed to the combustion air, resulting in poorer burnout.

6.6.2 Cost

From 1995 to 2007, no new waste to energy facilities were constructed in the US. There were, however, additions to existing facilities and the complete overhaul of systems. Any cost data presented would be dated. In light of the changes in air pollution control based on current regulations, cost data from older facilities is not considered to be reflective of today's costs.

In general, modular combustion facilities are the least costly to construct on a dollars per installed ton of capacity. This is because many of the components are fabricated in the shop, which is a lower cost method of construction. RDF facilities are the most expensive plants to construct on a per ton basis. This is due to the additional need to process the waste prior to combustion.

The important factor when evaluating costs is the net cost. This is the amortized cost of construction, plus annual operating and maintenance costs, minus the revenues. The differences in the cost of construction discussed above are offset in both cases by the lower thermal efficiency of the modular units and the higher efficiency of the RDF units. This affects the revenue stream and brings the three types of facilities close in terms of bottom line costs.

6.7   New developments

6.7.1 Seghers PRISM

The Keppel Seghers PRISM technology is a means of increasing combustion and boiler performance, based on the installation of a prism-shaped body in the first

6.7 Keppel Seghers PRISM combustor.

empty boiler pass (see Fig. 6.7). The prism optimizes secondary air injection, leading to more homogeneous flue-gas conditions. This leads to increased performance and solves some of the corrosion problems that occur in the radiant and superheater sections. The flue-gas temperature measurements in the passages next to the prism regulate the secondary air injection and control the air distribution under the grate (Keppel Seghers, 2007).

The flue-gas flow is divided into two partial flows, A and B, prior to entering the radiant pass of the boiler. This division is achieved by means of a membrane-wall construction, in the shape of a prism, which is water cooled and integrated with the natural circulation system of the boiler and protected with a refractory lining (Perilleux and Eeraerts, 2002).

Secondary air is injected into the divided flue-gas streams, A and B, through multiple secondary air nozzles. The secondary air injection nozzles are installed in the boiler front and rear walls, and on both sides of the prism. This results in complete burnout of the flue gases just above the prism. Since the prism is located in a turbulent and high-temperature zone, its membrane walls are water cooled and protected with a ceramic coating. The water cooling is integrated into the natural circulation system of the boiler (Perilleux and Eeraerts, 2002).

The secondary air nozzles of the prism contain an on-line cleaning system to prevent slag build-up on the nozzles. This is achieved by periodically blowing low-pressure steam into the air nozzles. The fast expansion of the water in the steam removes any slag deposits on the secondary air nozzles during operation (Perilleux and Eeraerts, 2002).

6.7.2   Recirculated flue gas

Some of the underfire air that is used in furnaces is needed to penetrate the fuel bed. Using more air than is necessary for combustion raises the excess air and results in lower boiler efficiency. Flue-gas recirculation is the process where a slipstream of flue gas is returned to the furnace as underfire air. This provides the air needed to penetrate the fuel bed while improving the efficiency of the boiler since part of the combustion air is replaced by flue gas that has a lower oxygen content. This process has been incorporated into several operating facilities with good results.

6.7.3   Tube metallurgy and coatings

Corrosive conditions in waste to energy boilers produce rapid wastage rates of traditional boiler tube materials. It is not unusual to see corrosion rates in the range of 1 to 3 mm/y on carbon steel boiler tubes and occasionally corrosion occurs at even higher rates (Paul *et al.*, 2004).

Municipal waste contains various constituents and impurities that induce corrosion attacks on boiler tubing. Among the leading contributors are chlorine, sulfur, zinc, lead, sodium and potassium. During combustion, various metallic chlorides and sulfates as well as HCl and SO_2 are formed and then deposited on the cooler surfaces, such as the waterwall and superheater tubes. Based on the low melting points and high vapor pressures of many metallic chlorides, severe corrosion will occur when carbon and low alloy steels are in contact with these chloride salts. Additionally, flue-gas streams can contain HCl gas, which also can cause a chloride attack on steel (Joiner and Lai, 1999).

There are four principles for selecting the right metals for boilers:

* Reducing the iron content in the boiler tube material will reduce iron chloride formation.
* Replacing iron with nickel. Nickel chlorides have much lower vapor pressures hence better resistance to chloride attack.
* Relying on an alloy that will create a corrosion protective scale. Chromium is excellent in the formation of a chromium oxide scale even in lower temperature ranges.
* Utilizing molybdenum because of its resistance to high-temperature chloride attacks at temperatures up to 1100°F (Joiner and Lai, 1999).

Many materials have been evaluated in refuse-to-energy boilers. The most successful of these materials include Alloys 625, 50, 59, 825 and 45TM. The

Table 6.5 Chemistry of common alloys used in waste to energy boilers

Alloy	No.	Ni %	Cr %	Mo %	Fe %	Other %
FM625	N06625	63	22	9	<1	3.4 Nb
FM59	N06059	59	23	16	1	
FM50	N06650	53	19.5	11	14	0.25 Al, 0.25 Nb, 1.5 W
825	N08825	42	21.5	3	30.4	2.2 Cu, 0.9 Ti
45TM	N06045	46	27	–	23	2.75 Si, 0.1 Re

chemistry of these alloys is given in Table 6.5. Alloys 625, 50 and 59 are applied as a weld overlay onto carbon steel boiler tubes and Alloys 825 and 45TM are used as solid tubing only for superheater applications (Paul *et al.*, 2004).

6.8   Sources of further information

Materials and Energy Recovery Division, American Society of Mechanical Engineers, divisions.asme.org/MER.

North American Waste to energy Conference (NAWTEC, annual), www. nawtec.org.

Solid Waste Association of North America (SWANA), www.swana.org.

Waste to energy Research and Technology Council (WTERT), www.seas. columbia.edu/earth/wtert.

6.9   References

Clark, L (1982), Case history of a 240 ton/day refuse to energy project: Vicon, Crane & Co., Pittsfield, Massachusetts, in *Proceedings of the 10th National Waste Processing Conference*, New York, ASME, pp. 1–9.

Clark, L (1996), Case history of a 240 ton/day resource recovery project: Part II, in *Proceedings of the 17th Biennial Waste Processing Conference*, New York, ASME, pp. 235–248.

Consumat (2004), Consumat controlled-air incinerator, available from www.consutech. com/inciner.htm, accessed 19 January 2011.

Gittinger, J and Arvan, W (1998), Considerations for the design of RDF-fired refuse boilers, presented to *Power-gen Europe 1998*, Milan.

Hammerschlag, R, *et al.* (2007), Energy storage, transmission, and distribution, in Kreith, F and Goswami, D (eds), *Handbook of Energy Efficiency and Renewable Energy*, Boca Raton, CRC Press, pp. 18-1–18-33.

Hickman, L (2003), *American Alchemy: the History of Solid Waste Management in the United States*, Santa Barbara, CA, Forester Press.

Joiner, D and Lai, G (1999), Economic impacts and solutions for waste to energy boiler corrosion management, in *North American Waste to energy Conference (NAWTEC 7)*, New York, ASME, pp. 221–223.

Keppel Seghers (2007), Keppel Seghers PRISM – for waste to energy installations, available from www.kepcorp.com/en, accessed 21 January 2011.

MMWAC (2010), Mid-Maine Waste Action Corporation, Auburn, ME, available from www.eskerridge.com/MMWAC%20Brochure%20Package.pdf, accessed 19 January 2011.

O'Connor, C (1984), The Sumner County mass burning experience, in *Proceedings of the 11th Biennial National Waste Processing Conference*, New York, ASME, pp. 301–319.

Owens, E and Sczcepkowski, J (2010), Advancements in grate cooling technology, in *18th North American Waste to Energy Conference*, New York, ASME, pp. 1–3.

Paul, L, Clark, G, Eckhardt, M and Hoberg, B (2004), Experience with weld overlay and solid alloy tubing material in waste to energy plants, in *12th North American Waste to Energy Conference*, New York, ASME, pp. 111–119.

Perilleux, M and Eeraerts D (2002), Retrofit of WTE boiler: Case study on Bonn plant, in *10th North American Waste to Energy Conference*, New York, ASME, pp. 25–33.

REI (2005), Fluidized bed combustion, Renewable Energy Institute, available from www. fluidizedbedcombustion.com, accessed 24 January 2011.

Russell, S and Roberts, J (1984), Oxides of nitrogen: Formation and control in resource recovery facilities, in *Proceedings of the 11th Biennial National Waste Processing Conference*, New York, ASME, pp. 417–423.

Schlesinger, MD (2007), Fuels and Furnaces, in Avallone, E, Baumeister, T, and Sadegh, A, *Marks' Standard Handbook for Mechanical Engineers*, 11th edition, New York, McGraw Hill, p. 7–6.

Stoller, P and Niessen, W (2009), Lessons learned from the 1970s experiments in solid waste conversion technologies, *17th Annual North American Waste to energy Conference*, NAWTEC17-2348, 1.

Velzy, C and Grillo, L (2007a), Fuels and furnaces, in Avallone, E, Baumeister, T, and Sadegh, A, *Marks' Standard Handbook for Mechanical Engineers*, 11th edition, New York, McGraw Hill, pp. 7-48–7-53.

Velzy, C and Grillo, L (2007b), Waste to energy combustion, in Kreith, F and Goswami, D, *Handbook of Energy Efficiency and Renewable Energy*, Boca Raton, CRC Press, pp. 24-1–24-42.

7

Waste firing in large combustion plants

P. VAINIKKA, M. NIEMINEN and K. SIPILÄ,
VTT Technical Research Centre of Finland, Finland

DOI: 10.1533/9780857096364.2.98

Abstract: Large combustion plants generate >100 MW power as their primary function. They were not designed to utilise waste as fuel, so they do not utilise it as their primary fuel and operate at higher steam parameters than purely waste-fired units. The main options for firing waste are: direct solid waste firing in pulverised-coal (PC) boilers, direct firing of solid waste in large fluidised-bed boilers, firing waste gasification gas in PC boilers and converting PC boilers to waste-firing fluidised-bed boilers. Each option has been demonstrated at a large scale, some plants having been in commercial operation for many years. The plants have a high power-to-heat ratio and they require pre-treatment of the waste.

Key words: waste firing, combustion plants, pulverised coal (PC), high-temperature corrosion, fluidised-bed (FB) boilers.

7.1 Introduction

EU regulations differentiate between waste 'incineration' and 'co-incineration' depending on whether the primary purpose of the plant is treatment of waste or heat and/or power generation. In waste combustion in large plants these objectives are mixed and the roles of the waste management sector and the utilities meet. This results in tailor-made solutions for each plant and situation, which may be characterised by special factors compared to those typically driving the economics of power stations:

- *Fuel price.* While the price of fossil fuels is consistently increasing not least due to the emission trading scheme in place in the European Union, waste fuel prices are essentially negative. There is the potential for a huge fuel cost reduction for a power plant.
- *Tailor-made fuel qualities and mixtures.* The co-firing percentage, regardless of the actual technical solution at the power station, is always defined through combustion trials and requires close cooperation between the fuel supplier and boiler operator. The fuels used can be in a broad sense categorised as refuse-derived fuel (RDF), the quality of which varies as a function of the RDF processing feedstock. This, combined with the chemical characteristics of the base fuel, requires careful selection of material streams in RDF production.
- *Technical retrofit.* A technical retrofit is almost inevitable. At least it means adding a pneumatic injection line to a pulverised-coal (PC) furnace, or it may

mean a retrofit of the technology, for example from a PC unit to a bubbling fluidised bed (BFB).

The main drivers for utilising solid waste in large combustion plants are:

- Fuel costs.
- Wastes are partially renewable. Different EU member states have different practices and processes on how the fossil energy fraction is determined. Roughly speaking, municipal solid waste is 40% to 50% renewable on an energy basis.

Figure 7.1 illustrates the electric efficiency of steam power plants for different fuels. It can be seen that the waste to energy (WtE) sector has a huge potential for improvement in electric efficiency. The latest developments in biomass combustion (wood and agrobiomass) have improved the electric efficiency of units fired with these fuels so that it is close to coal-fired units. In wood-fired plants, steam values

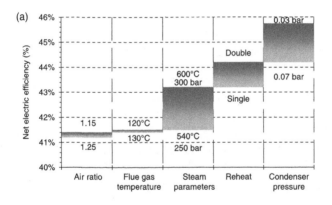

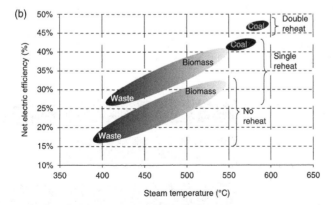

7.1 Electric efficiency of state-of-the-art solid fuel power plants as a function of design/operation-based improvements (a) (courtesy of Dr Klaus R. G. Hein) and fuel (b).

up to 540 °C/140 bar are commercially applied. State-of-the-art waste incinerators operate at 480 °C/70 bar. Wood-fired units reach a net electric efficiency of 41%, while waste to energy plants are close to 25% (based on the lower heating value). Combustion of gasification gas does not change the picture unless the gas is cleaned of corrosive gaseous salts and metals. Steam-side enhancements could achieve a 4% improvement in electric efficiency.

The technical options presented in this chapter are categorised based on whether the existing power plant is a fluidised-bed or pulverised-coal plant. This is because these two technical options require very different fuel treatment prior to combustion and result in completely different overall solutions for waste firing.

Figure 7.2 schematically presents the options. The left side indicates that an existing fluidised-bed unit may require only a small investment in additional fuel handling in order to utilise waste fuels. This is discussed using the example of one of the largest fluidised beds operating with fuels other than coal – the Alholmens Kraft circulating fluidised-bed (CFB), which is a 240 MWe biomass-fired CFB co-firing RDF (Oy Alholmens Kraft Ab, 2013). An existing PC boiler may require a dedicated system to receive waste fuel, and also a fine grinding system in order to prepare waste fuel with (the required) <3 mm particle size. This poses a technical challenge and requires more precise fuel processing in RDF production than for fluidised beds. This is illustrated by ENEL's Fusina 320 MWe power plant (Lattanzi, 2011; Gasparetti, 2011; Lattanzi *et al.*, 2010, Lattanzi 2010; Grosso *et al.*, 2010; Rossi *et al.*, 2009). The third option from the left in Fig. 7.2 is an investment in a gasifier and gas burners alongside the PC boiler. Gasification gas can then be fired in the PC boiler where it replaces the main fuel. The fourth option is to convert an existing PC boiler to a bubbling fluidised bed. This is also commercially available technology, and there are tens of references particularly in biomass combustion. There is the option of converting the PC unit so that there is a complete switch of the fuel to biomass, or co-firing of biomass and waste, or co-firing waste with coal. There are several examples of

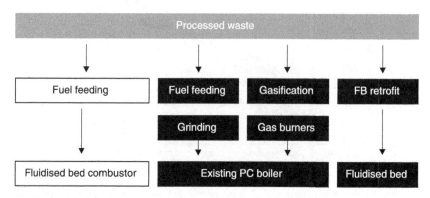

7.2 The technical options for waste combustion in large combustion plants.

this approach; the example unit presented here is the BFB operating at the Stora Enso paper mill in Anjalankoski, Finland. All these technologies have been demonstrated at a large scale.

7.2   Pulverised-coal (PC) units with direct co-firing

Direct waste co-firing in a PC unit requires only moderate additional investment and results in a moderate amount of waste in the fuel mixture. An example of this option is ENEL's Fusina Power Station close to Venice, Italy, where RDF is introduced into the furnace of an existing PC boiler through coal burners.

The Fusina plant has four units. Units #3 and #4 co-fire RDF with coal. Since 1974, these units have each produced 320 MWe. They are tangentially fired dry-bottom boilers. The steam values are 538°C/178 bar. The boilers are equipped with a low NO_x concentric firing system, selective catalytic reduction, electrostatic precipitator (ESP) and a flue-gas desulphurisation (FGD) plant. Both of the units have an environmental permit to fire RDF up to 5% on an energy basis (5 enb.%). Co-firing started in 2004 and since 2009 the permissions have allowed firing 5 enb.% in both units.

The RDF is produced from the residual waste from source-segregated municipal waste. It undergoes mechanical–biological treatment (MBT) to produce RDF. The waste is the residual waste after segregation of food waste, plastics and packaging, glass, aluminium cans and paper and cardboard. After MBT, RDF is pelletised to 16 mm diameter. The RDF is characterised by:

- moisture content: 8–18 wt.%
- ash content: 15–20 wt.%
- lower heating value (LHV): 17–21 MJ/kg
- chlorine content: 0.7–0.9 wt.%.

Aerobic bio-stabilisation is carried out for approximately seven days at about 40–45°C. This reduces the moisture content and stabilises the waste biodegradable organic fraction. Some 30–35% mass reduction results from the release of CO_2 and moisture, which increases the LHV by 35%. The moisture content is reduced to <15 wt.%. Microbial control measurements (hygienic control) are taken regularly at the power station in order to monitor any possible contamination in the working environment, and the storage system at the power plant is sanitised monthly.

The pelletised RDF is transported to a receiving station at the Fusina plant. From the receiving pit, the RDF is transported by means of walking floors, discharge screws and scraper conveyors to two storage tanks and finally to fine grinding. The grinding is accomplished by means of two high-speed blade mills with a power of 220 kW. After grinding, the RDF is transported pneumatically and injected into the pulverised-coal–air suspension after the coal mills. One mill serves a single level of the burners and the RDF is utilised in the second-level and third-level coal burners. This has been chosen to give sufficient residence time in the furnace and to avoid loss of unburned fuel at the bottom. After the RDF is injected into the line for each burner level, the stream is divided into four to direct it to the burner in each of the four corners of the boiler.

The main challenges in the system have been the wear of the mill blades and the transport system. The mill knives suffer wear due to impurities such as glass and metals. Wear-resistant materials have been shown to improve the operating hours significantly.

Milling performance is the most critical parameter for PC boilers. At Fusina this has been accomplished using high-performance blade mills. In the RDF, 90% of the particles are <4 mm, 80% <2.4 mm and 50% <1.2 mm, but the milling results get worse over time due to wear of the blades, which have to be replaced after processing some 500–600 tonnes.

The operation of the pneumatic system is relatively sensitive to fuel moisture. It has been found that <15% moisture content must be achieved during production, storage and transport in order to avoid clogging of the rotary valves and pneumatic lines. The transport lines also require wear-resistant material selection in some places. Although the same transport ducts are used for the air–coal–RDF mix (after the coal mills), this has not been found to cause major problems.

The unit operating with this proportion of RDF has not experienced any major corrosion or excessive formation of slag in the furnace. The proportion of the substitute fuel (whether biomass or RDF) is typically only up to a few per cent on an energy basis in this type of installation. This hardly affects plant performance. Furnace deposition (slagging) may occur as almost all of the ash-forming elements will vaporise or form melt at the PC furnace temperatures. Therefore slagging can be avoided by altering the particle flux on the waterwalls. At Fusina, RDF

Table 7.1 Emissions as measured at the Fusina PC plant with 5 enb.% RDF. Dry gases at 6 vol.% O_2

	Measured
Dust	2 mg/m³(n)
Total organic carbon (TOC)	0.003 mg/m³(n)
HCl	4 mg/m³(n)
HF	Not detected
SO_2	140 mg/m³(n)
NO_x	150 mg/m³(n)
CO	14 mg/m³(n)
Cd + Tl	<0.002 mg/m³(n)
Hg	0.002 mg/m³(n)
Sb + As + Pb + Cr + Co + Cu + Mn + Ni + V	0.03 mg/m³(n)
Dioxins and furans	0.0003 ng/m³(n)

n = normal cubic metre

co-combustion has not affected NO_x, CO or unburned carbon emissions from the boiler. In contrast, the higher concentration of HCl has been found to promote mercury oxidation, thereby resulting in a net decrease in mercury emissions.

The flue-gas treatment train comprises high-dust catalytic reduction of NO_x, an electrostatic precipitator and a flue-gas desulphurisation unit fed with limestone. With this gas cleaning train, the unit has achieved the emission levels in Table 7.1.

7.3   Direct fluidised-bed combustion

There is increasing global interest in fluidised beds at the utility scale. This is due to the fact that coals available for power generation are of decreasing quality (with higher ash and sulphur content, varying maceral composition and lower calorific value) and fluidised-bed technology has become commercially available in utility scale up to 800 MWe.

In fluidised beds the requirements for the physical properties (particle shape and size, moisture and density) are less stringent in comparison to those required for firing in pulverised-fuel combustion plants. Therefore introducing alternative fuels to a fluidised bed requires significantly less fuel pre-treatment and homogenisation than in PC units. However, fuel quality also depends on chemical properties (ash chemistry, reactivity and nitrogen content), which is the same for all types of units.

Fluidised beds are multifuel units capable of firing inhomogeneous fuel mixtures, with a high ash and moisture content. In a fluidised bed, the fuel is dried and volatilised by the intense heat transfer between the fuel particles and sand. In a bubbling bed there may be entrainment of some plastic fractions by combustion gases higher in the furnace, whereas in a CFB the fuel may comprise only some 5% of the solid inventory, which recirculates in the combustor until complete burnout.

One of the largest fluidised beds operating with fuels other than coal is the Alholmens Kraft CFB in Pietarsaari, Finland. The CFB boiler was designed as a multifuel boiler for wood (bark, logging residue and stumps) and peat, with coal as a support and back-up fuel. This unit can operate with any one or any mix of the design fuels. Commercial operation of the plant started in autumn 2001. The unit is a 240 MWe CFB generating 194 kg/s steam at 545°C and 165 bar in the high-pressure section and 177 kg/s at 545°C and 38 bar after the re-heater. The plant produces process heat for the adjacent paper mill and district heating for the town of Pietarsaari when operating in back-pressure mode. There is also a low-pressure section in the turbine; when the electricity price is high enough it is possible to shift to condensing mode.

This unit was equipped with a separate receiving and crushing station for RDF in 2008. The treatment system incorporates a receiving hall for unloading and crushing, a sieving station, and an intermediate storage silo and scraper and belt conveyors. The RDF arrives at the receiving station in a truck, which tips the waste into a receiving pocket. From there, the RDF is sieved and crushed and held in an intermediate storage silo. From the silo, the RDF is discharged onto a belt conveyor with peat, which feeds the mixture of peat/RDF to the boiler house silos.

The RDF treatment and feeding capacity is 80 000 tonnes/year, which corresponds to 6–8% of fuel usage (on an energy basis) in the CFB unit. The investment cost of the waste supply line at the plant was approximately €30/tonne/year. With the fuel prices in the operating environment the payback time for the investment is less than two years.

Typically, an RDF particle size of 60 mm is sufficiently small for a fluidised bed. With a large furnace, as at Alholmens Kraft, which is 8.5 × 24 × 40 m, this particle size requires an even distribution in the fuel feed. The CFB boiler house is equipped with four silos for biomass/peat/waste mixtures. Each silo has its own fuel feed line serving three fuel chutes. The feeding points are staggered so that stopping one feed line causes minimal disturbance to combustion. Three feed lines can achieve full boiler capacity. Coal is fed via one coal silo with two hoppers and feeders. After the feeders, the coal is distributed into the same feed lines as the other fuels. A schematic picture of the fuel feed system is shown in Fig. 7.3.

The waste is source-segregated household waste as well as commercial and industrial waste. The household waste is segregated at source: biodegradable waste and combustible waste are separated into different bags and commingled. Householders are instructed to include the following items in the energy waste bags:

- paper, cardboard, plastic and Styrofoam (polystyrene)
- empty milk cartons, egg cartons, etc.
- empty plastic bottles (shampoo, etc.)
- small plastic items (i.e. toothbrushes, hairbrushes)
- plastic materials except PVC products
- dust and vacuum cleaner bags
- textiles (except shoes, raincoats and leather garments).

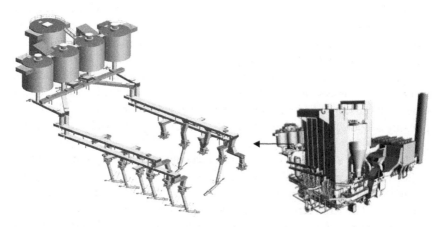

7.3 Fuel feed system of a multifuel CFB. Courtesy of Metso Power Oy.

Table 7.2 Emission limits for the Alholmens Kraft CFB in co-incineration. Dry gases at 6 vol.% O_2

	Emission limit	Measured
Dust	28	1 mg/m³(n)
Total organic carbon (TOC)	15	0 mg/m³(n)
HCl	90	10 mg/m³(n)
HF	1.5	0.5 mg/m³(n)
SO₂	185	170 mg/m³(n)
NOₓ	155	100 mg/m³(n)
CO	70	40 mg/m³(n)
Cd + Tl	0.05	Pass mg/m³(n)
Hg	0.05	Pass mg/m³(n)
Sb + As + Pb + Cr + Co + Cu + Mn + Ni + V	0.5	Pass mg/m³(n)
Dioxins and furans	0.1	Pass ng/m³(n)

There are drop-off collection points for the rest of the waste including landfill waste and recyclables such as paper, cardboard, glass, metals, batteries, hazardous waste, electrical and electronic scrap, etc.

This type of unit can comply with the European Waste Incineration Directive emission limits with ESP. The emission limits for the unit are given in Table 7.2. Additional measures include the injection of limestone into the furnace, which is typical for CFBs. However, CFB characteristics include the so-called SO_2 auto-reduction, which is due to the high alkaline content of biomass and waste (Fig. 7.4). At the CFB combustion temperatures and in oxidising conditions, the formation of alkali sulphates is favourable over alkali halides, which promotes the binding of sulphur in particles, and also reduces the concentration of corrosive alkali metal halides in the furnace.

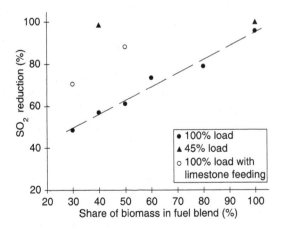

7.4 SO_2 reduction in the Alholmens CFB as a function of biomass share (enb.%) in the fuel mix. Filled symbols: auto-reduction.

If such units are not equipped with acid gas reduction measures, the emission of HCl may be the most important limiting factor on the proportion of waste in the fuel mix. This type of unit is typically equipped with a selective non-catalytic reduction system, which does not need to be operated continuously. A solution of 25% ammonia is injected into the cyclone.

From the data, it can be seen that SO_2 emissions are kept close to the emission limit value. This has been found to be a practical means to control chlorine-induced corrosion. The objective is to maintain a sufficiently high partial pressure of SO_2 in the furnace so that the alkali metal chlorides will form sulphates and thereby reduce high-temperature corrosion by alkali halides. It is well known that at fluidised-bed combustion temperatures, alkali metals form stable alkali sulphates if sulphur is present, and sulphates do not induce high-temperature corrosion in the steel at the operating temperatures.

The ashes from the boiler are utilised in land construction or are landfilled in landfill sites for conventional waste.

7.4   Co-combustion of gasification gas in a pulverised-coal boiler

Typical waste-derived fuels are fluffy materials and they may be difficult to pre-treat to give sufficiently fine particles and a homogeneous mix. A flexible way to utilise waste-derived fuels in large PC boilers is gasification of wastes and co-firing of the gasification gas in the PC boiler. This concept has been verified successfully on an industrial scale in Lahti, Finland, for more than ten years. At the Kymijärvi I power plant, a CFB gasifier is used for gasification of different types of waste fractions as well as biomass (Fig. 7.5). The gas produced is co-fired in a PC boiler in dedicated gas burners.

The Kymijärvi PC boiler was originally built as an oil-fired Benson-type, once-through boiler in 1976 and it was modified for pulverised-coal firing in 1982. The steam values are 125 kg/s 540 °C/170 bar and 540 °C/40 bar at re-heat. The maximum capacity is 167 MWe and the maximum production of district heat is 240 MWth.

The nominal capacity of the CFB gasifier is 60 MW but the actual capacity varies, depending on the energy content of the feedstock, from 40 MW to 70 MW. The gasification gas substitutes for about 15% of the coal fired in the PC boiler. The CFB gasifier uses a cyclone to separate the circulating bed material from the gas. The material is then sent to the bottom again. The hot gas flowing through the cyclone is cooled in an air pre-heater before it is fed into the PC boiler. Simultaneously, the gasification air will be heated in the pre-heater before being fed into the gasifier. The gas is then transported through a duct to two burners located below the coal burners in the main boiler.

An unusual feature of the system is that the fuel is not dried, although the moisture content can be up to 60% when biomass is co-gasified. This means that the heating

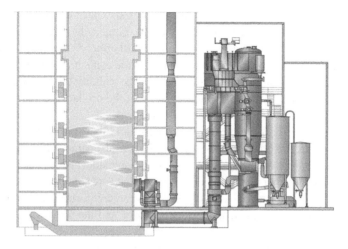

7.5 Lahti I gasifier. Courtesy of Foster Wheeler Energia Oy.

value of the gas is low. The burners are of a unique design developed through pilot-scale combustion tests and computational fluid dynamics (CFD) modelling. The gasifier design fuels were bark, woodchips, sawdust and uncontaminated wood waste. Later the unit began to utilise both source-segregated household and industrial waste. Shredded tyres and railway sleepers have also been used. The use of these fuels has forced some modifications to deal with such unusual impurities as nails, metal wire and concrete. The RDF used is mainly processed from source-separated household waste. By weight, it is composed approximately of 5–15% plastics, 20–40% paper, 10–30% cardboard and 30–60% wood.

The RDF is discharged from rear-unloading trucks, falls onto the floor of a receiving hall from where it is pushed via a bucket loader onto an apron conveyor and forwarded to the primary shredder. A conveyor below the primary shredder takes the RDF through magnetic separation, screening and secondary shredding. After this stage the RDF is less than 50 mm in size. After secondary shredding, the RDF is forwarded to the intermediate storage building. A travelling screw reclaimer on the floor discharges RDF onto a belt conveyor, and further onto chain conveyors to the gasifier feed bins.

The CFB gasifier started commercial operation in March 1998. The operation of the gasifier over the years has been excellent. On an annual basis, the gasifier's availability has been typically 96–99%. With regard to the gasification plant itself, the few problems have been related mostly to the use of shredded tyres. On several occasions the wire content of the tyres (there is no further separation of metal wires with magnets after the secondary shredding) was so high that a mass of wires blocked the bottom ash extraction system and the gasifier had to be shut down. Otherwise, with all other fuel fractions, the operation of the gasification process has been very satisfactory.

The stability of the main boiler steam cycle has also been excellent. The large openings that were made for the low btu gas burners have not caused any disturbances to the water/steam circulation. Furthermore, the combustion of product gas has been stable even though the moisture content of the solid fuel has been high and the heating value of the gas low. The stability of the main boiler's coal burners has been normal despite the fact that the product gas burners were integrated very close to the lowest level coal burners.

Gasifier fly ash makes up only a small proportion (3–5%) of total main boiler ash and, therefore, has little effect on the main boiler ash quality. Unburned carbon and alkali levels were unchanged, but some heavy metal levels increased slightly, depending on the type of feedstock. For example, the zinc content increased when shredded tyres were gasified. No changes in trace organics, such as dioxins, were detected. Leachability test results were satisfactory, and the plant is permitted to use boiler ash as before.

The only residue from the gasifier is bottom ash, which consists mainly of sand and limestone. Furthermore, small amounts of solid impurities such as metal pieces, pieces of concrete, glass, etc., have been found in bottom ash. The carbon content is typically less than 0.5% and chlorine levels are negligible. Ash also contains trace metals, however, with low leachability. The following elements have been analysed: arsenic, cadmium, chromium, copper, nickel, lead, zinc and mercury. Elements such as chromium, copper and zinc have been found at the level of hundreds of ppms. When shredded tyres were used as a fuel in the gasifier, the zinc content in the gasifier bottom ash increased to 3000 ppm. All other analysed elements were measured to be in the range of a few ppms or tens of ppms. Most elements exit the gasifier in the gas phase or as fine fly ash particles. As a result of the low content of trace metals and the low leachability of trace metals, the gasifier bottom ash is currently disposed of in a landfill for conventional waste.

The combustion of the gasification gas caused small changes in the emissions from the PC boiler. The boiler is not equipped with $DeNO_x$ or $DeSO_x$ plants and the emission limit values for the emissions were as follows: NO_x 240 mg/MJ (as NO_2) and SO_x 240 mg/MJ. Table 7.3 summarises the effects of co-combustion of the gasifier product gas on the main boiler's emissions.

The dust content in the flue gas after the ESP decreased approximately by 10–20 mg/m^3(n). The most probable reason was the increase in the flue-gas moisture content, which has enhanced the operation of the ESP. Perhaps the most positive phenomenon has been the decrease in NO_x emissions. According to the measurements, NO_x emissions from the main boiler decreased typically by approximately 10 mg/MJ, equivalent to a decrease of 5 to 10% from the base level. This was evidently due to the cooling effect of the low btu, high-moisture gasification gas in the lower part of the boiler. Obviously, due to the cooling effect, the formation of thermal NO_x was lower in the coal burners located at the lower part of the boiler. Furthermore, because of the low sulphur content of waste and

Table 7.3 The effects of co-combustion of the gasifier product gas on the main boiler's emissions

Emission	Change caused by gasifier
NO_x	Decrease by 10 mg/MJ (= 5 to 10%)
SO_x	Decrease by 20–25 mg/MJ
HCl	Increase by 5 mg/MJ
CO	No change
Particulates	Decrease by 15 mg/m^3(n)
Heavy metals	Slight increase in some elements, base level low
Dioxins	No change
Furans	No change
Polycyclic aromatic hydrocarbons (PAHs)	No change
Benzenes	No change
Phenols	No change

biomass, SO_x emissions from the main boiler decreased by approximately 20–25 mg/MJ. In contrast, because of the very low chlorine content (0.01%) of the coal used in the main boiler, the HCl content of the flue gas increased by approximately 5 mg/MJ when the gasifier was in operation. The reason for this was the use of SRF fuel and shredded tyres in the gasifier. Both of these fuels contain chlorine. No changes were observed in CO emissions or the carbon content of the fly ash from the main boiler.

With regard to heavy metal stack emissions, increases in some elements could be seen, but because of the low base levels in coal combustion, the changes that were measured were in practice very small. No changes could be seen compared to the results from 100% coal combustion for dioxins, furans, polyaromatic hydrocarbons, chlorinated phenols and chlorinated benzenes.

7.4.1   Incorporation of gas cleaning

The Kymijärvi I gasifier is not equipped with gas cleaning equipment and thus the selection of fuels is limited to relatively 'clean' fuels only. Most recovered fuels contain alkali and heavy metals as well as halogens, which can cause severe corrosion or deposition problems in a combustion boiler and they can lead to increased emission levels of regulated compounds. These problems can be avoided by cleaning the gasification gas prior to combustion.

The temperature of the gas leaving a fluidised-bed gasifier is typically 750–850°C depending on the air pre-heater and the product gas has to be cooled prior to gas cleaning. The gas can be cleaned by wet scrubbing; however, dry scrubbing gives significantly better economy. In dry scrubbing, the gasification gas is cooled to 350–500°C and then it is passed through high-temperature filters (Dia-Schumalith® by Pall Corporation or Cerafil® by Clear Edge). If necessary, a chlorine-removing sorbent such as calcium hydroxide can be injected into the gasification gas prior to the gas filters and the chlorine is then removed efficiently simultaneously with particulates. Alkali metals as well as heavy metals, excluding mercury, are removed by filtration and all these impurities are concentrated in filter ash. A schematic figure of the process is shown in Fig. 7.6.

Typically filter ash contains 20–50% of unburnt carbon depending on the fuel type. Filter ash is classified as a hazardous waste because it also contains heavy metals and often high concentrations of water-soluble chlorine. Dry scrubbing of waste-derived gasification gas has been verified at a pilot scale but it has not yet been demonstrated in full scale. The most challenging technical questions are related to gas cooling and filtration of the tar-loaded gas. Cooling of tar-loaded gas to 300–400°C may lead to deposit formation and thus specific attention has to be paid to the design and cleaning of the heat transfer surfaces. During filtration, the high tar loading may also lead to severe problems due to the stickiness of the dust. Sticky dust may be difficult to clean off the filter elements, which may lead further to an increased pressure drop and eventually filter clogging.

Another type of advanced stand-alone gasification WtE process is where the cleaned waste gasification gas is combusted in a gas-fired boiler as a single fuel. It is easy to combust this clean gas, which allows the use of high steam parameters without risk of corrosion or deposit formation.

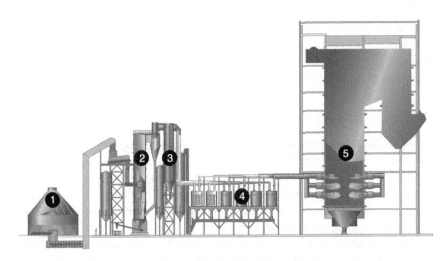

7.6 Gasifier with gas cleaning train connected to a PC boiler. 1: fuel handling, 2: gasification, 3: gas cooling, 4: gas cleaning, 5: gas combustion. Courtesy of Metso Power Oy.

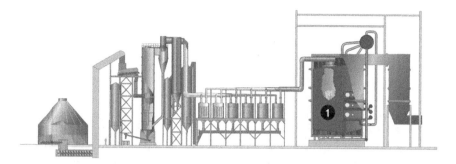

7.7 Kymijärvi II. 1: gas boiler. Courtesy of Metso Power Oy.

This technology is under demonstration in the Kymijärvi II unit. The new plant produces $160\,MW_{fuel}$ and the plant is based on circulating fluidised-bed gasification of RDF, gas cooling followed by gas combustion as a single fuel in a gas boiler. The unit started operating in 2012.

The plant contains two parallel gasification and gas cleaning trains, each producing $80\,MW_{fuel}$, and one gas boiler with four gas burners. The plant is a combined heat and power plant: the electric output is about 50 MWe and the district heat capacity 90 MWth. The electric efficiency is approximately 31%, which is significantly higher than in conventional incinerators. The steam values of the boiler are 540 °C/140 bar, which could allow >35% electric efficiency in condensing mode. As the boiler steels are not exposed to the severe high-temperature chemistry typically found in waste incinerators it has been possible to use less expensive alloys resulting in reduced investment cost for the boiler. A schematic figure of Kymijärvi II is given in Fig. 7.7.

Gasification-based WtE plants have a high power-to-heat ratio for waste fuel firing. Fluidised-bed gasification is additionally highly flexible on the fuel used but requires a level of fuel quality control. For example, untreated household waste cannot be directly used: the waste has to be processed at least mechanically to a form of RDF before gasification. Processing is mostly the homogenisation of particle size and the removal of large particles of impurities. This waste processing has been commercially demonstrated, but typically creates residue streams that are not allowed in landfills due to the high organic carbon content and so they are incinerated in traditional waste incinerators. The key questions are how tolerant the process actually is for quality fluctuations of the waste and which are the most optimal combinations in an energy system if different technologies are applied. The processes involved in gasification gas cleaning are in the demonstration phase.

7.4.2   Precipitation of corrosive salts

Figure 7.8(a) is a plot of the saturation pressures and melting temperatures of selected halides. The cations of these salts are in waste fractions typically found

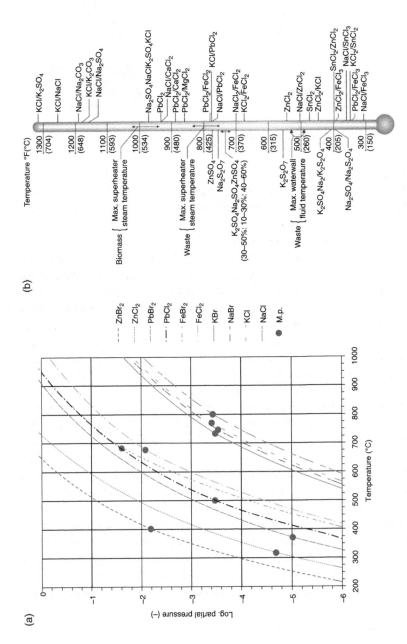

7.8 Calculated saturation pressures and melting points (M.p.) for selected chlorides and bromides (a) and melting points of selected salt mixtures (b).

in SRF, i.e. in packaging material such as cardboard and paper (potassium, sodium and calcium) and plastics and plastic additives (stabilisers, acid scavengers, catalyst residues or colorants: zinc, lead, copper, barium and cadmium), rubber and various metallic/alloy impurities. Iron halides are potential corrosion products.

In gas cleaning, the gas is cooled and the corrosive salts precipitate out from the gas phase as solid particles according to the saturation curves. At a sufficiently low gas temperature, elemental heavy metals and their halides are precipitated. This procedure results in a salt-free gas, which can be fired in a unit with high steam values without risk of high-temperature corrosion and ash melt (as in Kymijärvi II).

It is essential to remove all the salt and metals from the gases as boiler deposits are typically mixtures of different salts. The first melting temperature of a salt mixture is in most cases lower than the melting point of the pure salts that the mixture consists of. In particular, chlorides (and bromides) lower the first melting point of boiler deposits, see Fig. 7.8(b). From the figure it can be seen that when a boiler ash system changes from being a purely biomass-fired system with potassium, sodium and calcium dominance to a typical waste-originated ash system with lead and zinc salts, the melting points are drastically reduced. For example, in a $NaCl–Na_2SO_4$ salt system, the first melting point is approximately 630°C whereas for $NaCl–ZnCl_2$ it is approximately 250°C.

7.5   Retrofitting a pulverised-coal plant with fluidised-bed units

Another form of waste combustion for large plants is to convert an existing PC boiler into a BFB. This is a routine practice for fluidised-bed boiler technology suppliers. A schematic picture of such conversion is shown in Fig. 7.9. Several such conversions have been carried out, especially in those European countries where progress in biomass utilisation for combined heat and power (CHP) has been promoted. One example of such a conversion where waste is utilised in large quantities is the Stora Enso 80 MW BFB unit in Anjalankoski, Finland.

The main features modified include the grate, the fuel feed and the air distribution system. Flue-gas cleaning and heat transfer surfaces may be updated but this may not be necessary. Combustion air fans and air staging are modified to correspond to those in a BFB process, enabling fluidisation and air staging of primary, secondary and tertiary air. It is essential that the fuel feeding system from receiving the fuel all the way to the furnace wall is carefully designed so that the unit can achieve high availability and combustion stability. The advantage of this approach is that the investment is approximately one fourth to one third of that of a new unit depending upon the extent of flue-gas cleaning.

The Anjalankoski unit started as a 170 MWth pulverised-coal boiler in 1971 with a small fixed grate for bark combustion. The first commercial BFB boiler in Finland was connected to this coal boiler in 1983 to combust wet sludge from the

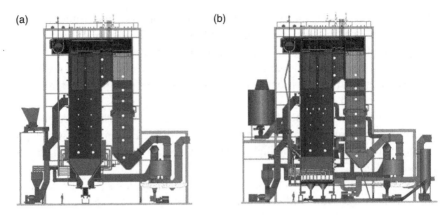

7.9 Example of the conversion of an existing PC boiler (a) to BFB (b). Courtesy of Metso Power Oy.

adjacent paper mill's wastewater treatment plant. The grate of the coal boiler was converted to an 80MWth BFB in 1995. A wet scrubber was also installed after the ESP to improve flue-gas cleaning and recover heat from the flue gases. With this flue-gas cleaning system the plant can comply with the EU Waste Incineration Directive emission limits.

After the installation of the BFB, co-combustion of RDF was initiated to manage the paper mill's waste treatment, and was later widened to use packaging wastes that could not be recycled. Simultaneously, the boiler's live steam values were reduced to 500°C/80 bar from 525°C/87 bar. The lower superheat temperature gave a safety margin for the more corrosive combustion gases in comparison to coal firing. The hottest superheaters were upgraded to a high alloy stainless steel.

An indirect thermal sludge dryer was commissioned in the year 2000 to combust all of the paper mill's wastewater treatment sludge with a dry matter content of about 90%. This also made it possible to increase the amount of RDF in the fuel mix, due to the (high sulphur) sludge's favourable reduction of the chlorine-induced corrosion risk due to RDF firing. In 2006, the proportion of SRF from all fuels was increased to 60% (on an energy basis) after long-term tests and investigation.

The boiler utilises on average 30% Scandinavian spruce bark and 10% dried paper mill sludge alongside RDF on an energy basis. All the fuels are introduced into the boiler house in separate streams. RDF arrives in a receiving hall where trucks unload fluff or baled RDF. After crushing, magnetic metal separation and storage in an intermediate silo, the RDF is supplied to the boiler house by a series of scraper conveyors (see Fig. 7.10). The RDF is crushed to approximately 60 mm particle size. In the boiler house, the RDF is discharged from a scraper conveyor by screws, which forward the fuel to two parallel fuel chutes. The RDF proportion in the fuel mix is adjusted using these discharge screw feeders. There are four fuel

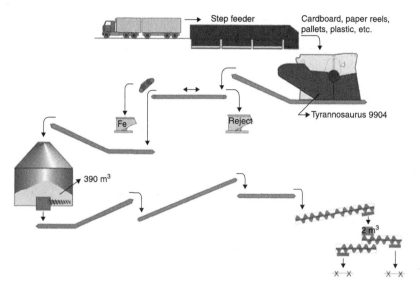

7.10 RDF reception, treatment and feeding system for a fluidised-bed. Courtesy of BMH Energy.

feed chutes in the boiler walls, two on each side of the furnace. Both the bark and RDF discharge screws have been calibrated to supply a volume per revolution that enables accurate fuel proportioning in the two parallel fuel chutes, and eventually in the furnace.

The flue-gas is cleaned by the ESP, which consists of two parallel lines in the flue-gas flow direction, both with two sequential fields, and a wet scrubber. The emission limits for the unit are given in Table 7.4.

It has been found technically feasible to fire approximately 50–60 enb.% RDF alongside bark and sludge in the unit. The RDF utilised originates from offices, wholesale businesses and small and medium-sized industrial facilities. Suitable waste is source-separated packaging and other solid industrial non-recyclable wastes, which are collected separately by the RDF supplier. In offices and commercial premises, the waste is collected in bins labelled 'energy fraction'. The labelling indicates that the following materials should be placed into the bins:

- packaging plastics (not PVC)
- contaminated paper and cardboard
- wooden boxes
- plastics (not PVC)
- expanded polystyrene
- paper towels
- clothes and textiles

Table 7.4 Emissions limits for co-incineration in the
Anjalankoski BFB. Dry gases at 6 vol.% O_2

	Emission limits
Dust	25 mg/m³(n)
Total organic carbon (TOC)	15 mg/m³(n)
HCl	15 mg/m³(n)
HF	2 mg/m³(n)
SO_2	155 mg/m³(n)
NO_x	440 mg/m³(n)
CO	150 mg/m³(n)
Cd + Tl	0.05 mg/m³(n)
Hg	0.05 mg/m³(n)
Sb + As + Pb + Cr + Co + Cu + Mn + Ni + V	0.5 mg/m³(n)
Dioxins and furans	0.1 ng/m³(n)

The waste collected from commercial premises (such as shopping centres and supermarkets) is mostly packaging materials, i.e. paper, cardboard and plastics in various forms. Industry-sourced energy fractions include different types of demolition and packaging discards. Production waste from furniture manufacturing contains mainly textiles, wood and plastics. In the SRF preparation plant, the material is crushed to 50–100 mm particle size and magnetic materials are separated out. The RDF is transported to the BFB plant as wrapped bales or fluff.

7.6   Controlling high-temperature corrosion in co-fired units

One method for controlling high-temperature corrosion in co-fired units is to add sulphur to the combustion gases prior to the hottest superheaters. In oxidising conditions, alkali metals form sulphates rather than chlorides. This results in non-condensable (at in-furnace temperatures) HCl forming in the combustion gases and reduced deposition of chlorine on heat transfer surfaces. Figure 7.11 shows a schematic view of a spraying system in a BFB. The main patented liquid additives are ammonium sulphate, ferric sulphate and aluminium sulphate. These have been commercialised by Vattenfall and Metso Power under the trade names ChlorOut and CorroStop, respectively.

When a water solution of these chemicals is sprayed into the furnace, the salts crystallise from the solution after the water evaporates and finally the salts heat up and decompose to form SO_3:

$$(NH_4)_2SO_4 \rightarrow 2NH_3 + SO_3 + H_2O$$

$$2KCl + SO_3 + H_2O \rightarrow K_2SO_4 + 2HCl$$

The Vattenfall system incorporates an *in situ* alkali chloride monitor (IACM), which can detect the corrosive alkali halide concentration in the combustion gases downstream from additive spraying, see Fig. 7.12. Thus, a sufficient level

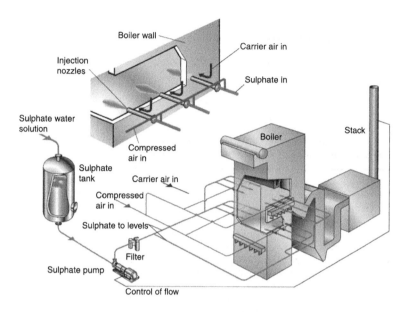

7.11 Liquid-sulphate spraying system. Courtesy of Metso Power Oy and Kemira.

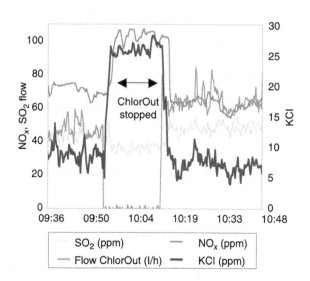

7.12 Effect of ammonium sulphate on KCl and NO_x concentration in a boiler furnace (Vattenfall, 2005).

of chemical dosing may be achieved. Since the additive contains ammonium ions it also reduces NO_x:

$$4NH_3 + 4NO + O_2 \rightarrow 4N_2 + 6H_2O$$

7.7   Conclusion

A summary of the advantages and disadvantages of different waste combustion options is presented in Table 7.5.

Table 7.5 Summary of advantages and disadvantages of different waste combustion options in large combustion plants

	Advantages	Disadvantages
• Direct firing in PC boiler	• Low additional investment cost • High steam cycle efficiency • 'Infinite' treatment capacity available • With a small proportion of waste it is easy to keep below emission limits • Simplicity of installation	• High purity requirement for the RDF (metals, glass) • Fine grinding needed, risk of significant decrease in burnout due to mill wear • Practical maximum of 5–10 enb.% RDF in the fuel mix • Stringent particle size requirement, <2–3 mm • Waste and coal ashes are mixed, which may compromise ash utilisation
• Direct firing in fluidised bed	• Low additional investment cost • High fuel flexibility • Less stringent particle size requirements, <60 mm	• High RDF proportion may restrict ash utilisation
• PC retrofit to fluidised bed	• High fuel flexibility • Large particle size allowed • Moderate investment cost, 1/3 for new biomass unit, 1/5 for new stand-alone RDF-fired unit	• Retrofit downgrades capacity from original coal firing (by ~25%) • Still not capable of 100% waste firing without significant upgrade in heat transfer surface alloys
• Gasification, gas combustion	• Moderate investment cost, comparable to PC-BFB retrofit • High power-to-heat ratio. Utility scale steam cycle efficiency • High power-to-heat ratio may be achieved in stand-alone unit if gas cleaning is applied	• High proportion may affect ash utilisation • Without gas cleaning, there are the same ash chemistry restrictions as direct waste co-firing units • With gas cleaning, residual ash management has to be solved

(Continued)

Table 7.5 Continued

Advantages	Disadvantages
• Low operational cost, no extra personnel needed • High fuel flexibility • Minor modifications to existing main boiler (minor technical risk) • Low risk to availability of main boiler	

7.8   References

Gasparetti, S., 2011. 'Assessment on air pollution control device performance at fusina power plant.' Wrocklaw, *IFRF*, 22–23 September.

Grosso, M., Rigamonti, L., Paoli, S. and Teardo, G., 2010. 'Co-combustion of RDF in a coal fired power plant: An evaluation using the life cycle approach.' Venice, Italy, *CISA*, 8–11 November.

Lattanzi, S., 2010. 'Activities at Fusina Power Plant (ENEL).' Pisa, *IFRF*, 23–24 September.

Lattanzi, S., 2011. 'CFD analysis of Fusina 4 boiler in co-firing.' Wrocklaw, *IFRF*, 22–23 September.

Lattanzi, S., Rossi, N., La Marca, C., Gasperetti, S., Dalle Mura, D., Molina, G. *et al.*, 2010. 'Advancements in RDF co-firing demonstration project at Enel Fusina Power Plant.' Lyon, France, ETA-Florence, Italy and WIP–Munich, Germany, 3–7 May.

Oy Alholmens Kraft Ab, 2013. www.alholmenskraft.com/en/production/fuel, accessed 21 January 2013.

Rossi, N., La Marca, C. and Lattanzi, S., 2009. 'RDF co-firing at Enel Fusina power plant.' Hamburg, Germany, ETA-Florence, Italy and WIP–Munich, Germany, 29 June–3 July.

Vattenfall, 2005. Vattenfall Business Services Nordic AB, ChlorOut, product brochure.

Waste to energy (WTE) systems for district heating

L. TOBIASEN and B. KAMUK, Ramboll, Denmark

DOI: 10.1533/9780857096364.2.120

Abstract: This chapter deals with energy recovery in waste-fired boilers, starting with a review of types of boilers suitable for waste incineration. The chapter highlights how combined heat and power (CHP) production can substantially increase overall energy recovery from a waste-fired facility. Technical aspects of district heating and district heating networks are described, as well as general methods of energy recovery optimization.

Key words: waste to energy (WTE), district heating, large district heating networks, steam turbines, grate-fired waste boilers, energy recovery optimization, decentralized power production, waste CHP, combined heat and power, energy from waste (EfW).

8.1 Introduction

This chapter deals with energy recovery in waste-fired boilers using the Rankine steam cycle for electricity production. This thermal cycle is used to produce 90% of the world's electricity.

Figure 8.1 illustrates a simple Rankine process where water is pressurized in a pump, superheated in a boiler and expanded in a steam turbine. It can be shown thermodynamically that the electricity yield at the turbine, $\dot{W}_{turbine}$, is much higher than the power consumed by the pump $\dot{W}_{pump}$.

The cycle efficiency, from fuel to turbine output, increases for higher steam temperatures and pressures (point 3) and a higher efficiency is achieved by lowering the steam-condensing temperature (point 4). Steam is condensed to water (point 1) and then pumped to the boiler (point 2) for evaporation. And so the cycle continues. The condensing temperature depends ultimately on the cooling temperature level available at the condenser. The condenser may be either water cooled (seawater or river or lake water) or air cooled. This is why power stations located in cold regions (with access to cold seawater for instance) have a higher electrical efficiency than power stations without access to such cooling.

Electricity-only waste to energy (WTE) facilities in Europe are often equipped with an air-cooled condenser, which can be either wet (water consuming) or dry, and configured with either a forced air flow or a natural draft (a cooling tower). Air-cooled condensers are the most common choice for WTE facilities in the EU, if there is no access to a district heating network.

120

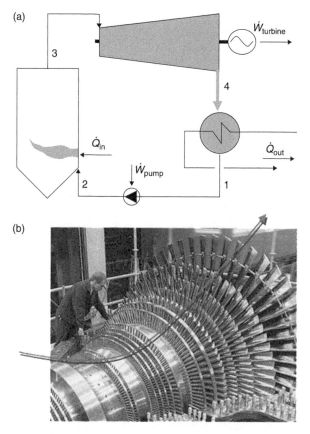

8.1 (a) The Rankine water-steam cycle for electricity generation. For explanation, see text. (b) Steam turbine rotor, used with permission from MAN Diesel & Turbo SE (MAN, 2009).

Figure 8.2 shows the principle of such an air-cooled condenser system. In the top image, low-pressure, high-volume steam from a steam turbine exhaust is routed to a four-row air-cooled condenser unit, where the steam is condensed in finned tube bundles; large fans (often 10 m in diameter) pass air across the condenser's heat surfaces, thus removing the heat. Air-cooled condenser installations are necessary where no other means of cooling is available.

Waste to energy facilities with no access to a market for waste heat thus lose large amounts of energy through condensers, such as those described above. Typically the net electrical efficiency of WTE facilities is in the region of 25% (Rambøll, 2012a); see Table 8.1. Hence, for a plant operating at a fuel throughput of 100 MW (corresponding to 32 t/h of waste at about 11 GJ/tonne), roughly 25 MW of electricity can be sold to the grid. Of the 100 MW energy input, about 55 MW is lost in the air-cooled condenser (if this is the chosen cooling device). As such, a relatively large amount of energy is not recovered.

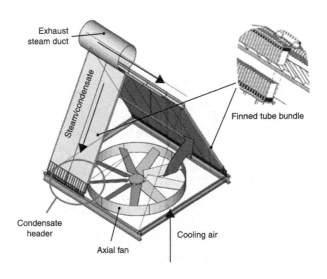

8.2 Air-cooled condensers. Courtesy of GEA Energietechnik (GEA, 2012).

Table 8.1 Where the energy goes

Fuel input (waste)	100%
Electricity from waste	~25%
Heat lost in air-cooled condenser	~55%
Other losses	~20%
Useful heat potential (hot water or steam)	~55–65%[a]

Notes:

[a] Flue-gas condensation, a method of extracting 10–15% of the fuel input as useful heat, is not included in this figure.

If there is a demand for waste heat, such as space heating, in the vicinity of the WTE facility, it may very well be feasible to provide this heat from the WTE facility – thereby utilizing part of the 55% or more heat that is not used. The heat from the WTE facility can be transferred either as hot water or steam. Other options include the supply of process steam to a nearby factory or industry, which would otherwise be produced by oil or gas.

There are various methods of producing heat from a WTE plant, which will be further discussed in the sections below. In addition, WTE boilers and industrial steam turbines are described.

8.2   Waste boilers

This chapter is mainly based on *Solid Waste Technology Management* (Hulgaard and Vehlow, 2010), as well as on knowledge and experience within Rambøll (Rambøll, 2012b). WTE plants are designed to treat waste where there is a great variation in the composition of the incoming waste. This is the primary difference between waste incineration and other combustion systems, and it has significant implications for the design of the incineration plant. Waste is not an ideal fuel and it is important to link the fuel characteristics to the combustion process. The main technologies described in this section are mass burn technologies by moving grate and rotary kiln.

Common to waste boilers is the concept of the capacity diagram, which determines the limits of thermal capacity (energy input) for a particular boiler, given the variation in heating value of the waste. In the example shown in Fig. 8.3, the boiler can accept waste with a (lower) heating value from 9 MJ/kg to 15 MJ/kg, and a throughput from 7.2–14.4 t/h of waste.

Typical heating values for some waste fractions are shown below (ECN, 2012; Patel *et al.*, 2001; Vølund, 2009). Note the very large variations, and

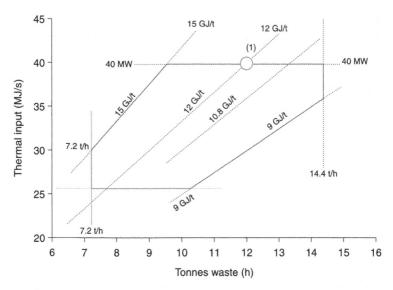

8.3 Typical waste boiler capacity diagram. The nominal load point in this example is indicated by point (1), where the waste has a heating value of 12 MJ/kg.

hence the need for a waste boiler to handle fuel input with a variety of lower heating values:

- Fresh wood (50% moisture):   10 GJ/tonne
- oven-dried wood: 17–19 GJ/tonne
- paper and cardboard: 14–15 GJ/tonne
- plastic: 28–35 GJ/tonne
- metal and glass: 0 GJ/tonne
- organic waste: 4 GJ/tonne
- municipal waste (Copenhagen):   9–12 GJ/tonne
- municipal waste (Bangkok):   2.5–11 GJ/tonne.

The minimum thermal load of a modern grate-fired waste boiler is 65–70%, limited by the need to ensure proper burnout and uphold legislative requirements with regard to emission limits (like CO, dioxins, etc) and flue-gas temperature limits (850°C for a minimum of 2 s).

8.2.1   Rotary kiln technology

In a mass-burning incinerator based on a rotary kiln, the waste burns in layers in a rotating cylinder. This technology is used when thermal treatment at high temperature with a long residence time is necessary, for instance in order to thermally destroy high levels of toxic materials.

The material is transported through the furnace by the rotation of the inclined cylinder (Fig. 8.4). A rotary kiln is usually refractory lined but can also be equipped with waterwalls. The diameter of the cylinder may be 1 to 5 m and the length 8 to 20 m. The capacity may be as low as 2.4 t/day (0.1 t/h) and is limited to a maximum of approximately 480 t/day (20 t/h). The kiln rotates with a speed of typically 3–5 rotations per hour.

8.2.2   Grate-fired boilers

A combustion grate is a transport device that moves the burning fuel from the inlet through the furnace to the bottom ash outlet (Fig. 8.5). The fuel is mixed on the grate, and combustion air is added. Volatile material is released into the furnace and fixed carbon is burned on the grate. The grate is an integral part of the furnace, where the fuel is converted into energy. The grate ensures drying, ignition, combustion and energy release, and complete burnout before the bottom ash outlet (Vølund, 2009).

Flue gases leaving the grate enter the boiler heat exchange passes, where steam generation takes place. A modern European WTE boiler often consists of three empty flue-gas passes, where the flue gas is cooled to approximately 600–700°C while providing heat to generate steam. The primary means of heat transfer is

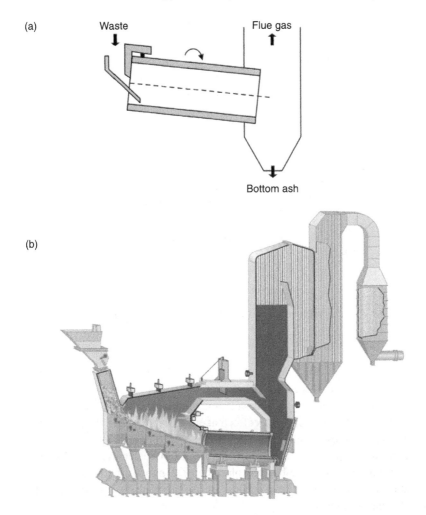

(a)

Waste

Flue gas

Bottom ash

(b)

8.4 (a) Rotary kiln (Hulgaard and Vehlow, 2010, Fig. 8.1.4). (b) Rotary kiln furnace connected to a steam-generator with a vertical boiler. Figure courtesy of boiler supplier B&W Vølund (Vølund, 2009).

radiation in this part of the boiler, due to the high temperature of the flue gases, and the first three passes are often called radiation passes.

Boilers for waste facilities are drum-type boilers, which are characterized by a drum (and by being below the critical steam pressure of 221 bar). The drum is a large high-pressure cylinder containing water situated at the top of the boiler, which physically separates water evaporation from where the steam is superheated. Water circulates naturally via down-comers and up through the evaporator tubes, since evaporated water (steam) has a lower density than the water in the

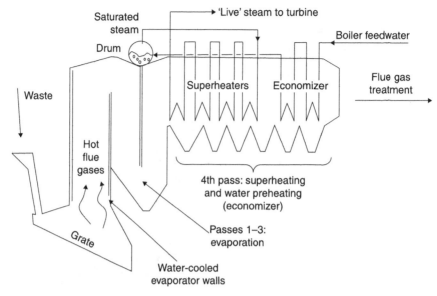

8.5 Modern European four-pass WTE boiler (Rambøll, 2012a).

down-comers. Typically the circulation ratio is 15–20, i.e. the circulating mass flow in the evaporator section is 15–20 times the dry steam flow exiting the drum. A very simple analogy to the circulation principle in a drum-type boiler is a household electric drip-brewing coffeemaker, where hot water also moves without the need for pumps because of the thermosyphon method.

In the fourth pass, the primary method of heat transfer from the flue gases is convection: superheater tubes are placed in the part of the flue gases in tube bundles, thus providing a large surface area to facilitate the transfer of heat from the flue gases to the steam.

The economizer is the last set of tube bundles in the boiler, seen from the flue-gas point of view. Here the feedwater is heated to somewhere below the boiling point for the particular water pressure (for instance, feedwater may be heated to 200°C) while the flue gas is cooled to a desired set point temperature, say 180°C.

Typically the pressure in the drum is 50–70 bar, and the final steam temperature 380–440°C. Figure 8.6 shows a cross section of such a boiler.

8.2.3   Different types of boiler

Depending on the needs and requirements of the purchaser of a WTE facility, the boiler heat transfer surfaces can be arranged in various ways, thus influencing up-front costs, operating and maintenance costs, availability and efficiency. The most common boiler configurations are described in Table 8.2.

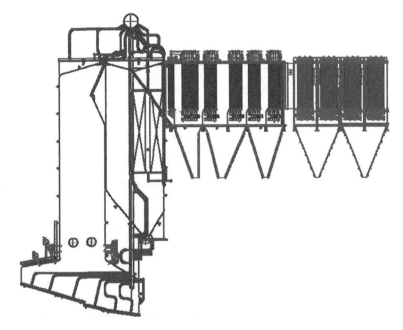

8.6 Cross section of a horizontal arrangement boiler. Figure courtesy of boiler supplier B&W Vølund (Vølund, 2009).

Table 8.2 Typical types of waste incineration boiler

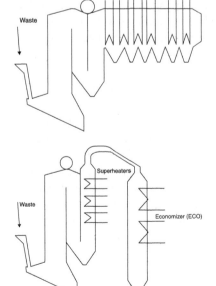	**Horizontal waste boiler** Often used in Europe for larger WTE facilities. Advantages: – high efficiency, high availability – good access to cleaning tubes in the fourth pass, hence lower maintenance costs Disadvantages: – higher initial investment **Vertical waste boiler** Often used in smaller WTE facilities. Advantages: – low initial investment (smaller boiler) – smaller plant footprint Disadvantages: – replacement of superheaters, and in general maintenance, more difficult due to poorer access conditions – economizer tubes cleaned with a soot blower (lower efficiency)

(*Continued*)

Table 8.2 Continued

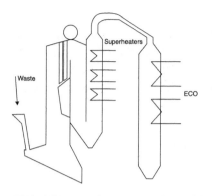

Vertical waste boiler with superheater tubes in the radiation pass
This boiler configuration is often seen in the USA.

Advantages:
– low initial investment (smaller boiler)
– smaller plant footprint

Disadvantages:
– the superheater tubes in the radiation pass may have an increased risk of corrosion and hence higher maintenance costs

8.3   Electricity production in waste to energy (WTE) facilities

8.3.1   Steam turbines for electricity production

The steam turbine is the component that transfers energy in the steam (temperature and pressure) to rotating energy, via which a generator produces electricity. The thermodynamic process is expansion of the high-pressure steam, which thus produces kinetic energy (high velocity), which in turn makes the blades of the steam turbine turn. The process can, in very simple terms, be compared to a wind turbine with the medium passing the blades being replaced by hot steam under pressure (Fig. 8.7).

It is often useful to plot steam turbine expansion in a so-called Mollier or *hs* diagram (Fig. 8.8). *h* is the steam enthalpy and *s* is the entropy, both of which are thermodynamic variables of state, and can be thought of as variables that can be looked up in tables given that the temperature and pressure are known (in the wet steam region this is not enough – the steam dryness must also be known).

In such a diagram the enthalpy change during the steam expansion is shown visually. This is of particular interest due to the first law of thermodynamics, which gives the relationship between mechanical work, heat loss and enthalpy drop (see the equation below). The validity is subject to various assumptions for steam turbines, like steady state operation and negligible heat loss.

$$W = m \times (h_{in} - h_{out})$$

where W is the work (kW), m is the mass flow of the steam (kg/s), and h_{in} and h_{out} are the inlet and outlet enthalpies (kJ/kg), respectively.

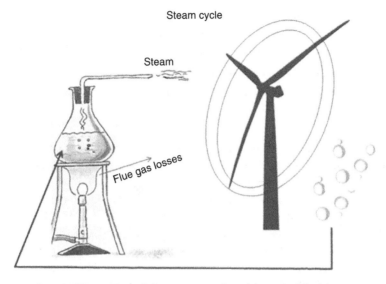

8.7 Simple illustration of the steam cycle, with a wind-turbine analogy (Rambøll, 2012a).

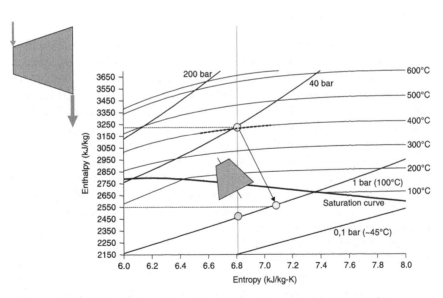

8.8 The *hs* diagram for a hypothetical steam turbine expansion (Rambøll, 2012a).

Since the heat loss from a steam turbine is negligible, and given that approximately 97–99% of the mechanical shaft work can be translated into electrical power, this equation shows that the steam turbine power output mainly depends on:

- the mass flow of the steam (given by the size of the boiler)
- the inlet and outlet enthalpies (given by the steam parameters and available cooling, which sets the final expansion pressure).

A steam expansion from 40 bar to 1 bar is shown in Fig. 8.8, see the line between the two dots. The enthalpy drop on the y-axis can be read to be (3240 – 2560) kJ/kg = 680 kJ/kg. So, for every kg/s of steam subject to such an expansion, roughly 680 kW of electricity can be produced.

Figure 8.8 also shows that each kg/s of steam would produce additional electricity if the inlet pressure and temperature were higher (see for instance the 200 bar line) and if the outlet pressure were lower (see the 0.1 bar line).

Thermal cycles are more efficient at higher steam temperatures and pressures. However, due to high-temperature corrosion of the boiler tubes because of chloride and other substances in the waste, WTE boilers are for practical purposes limited to approximately 440–450°C and 60–90 bar, depending on the levels of maintenance a plant owner will accept.

8.3.2 Process concepts for a complete electricity-only waste to energy facility

When the boiler and steam turbine technologies are combined, along with appropriate flue-gas treatment, a waste-fired power plant concept is formed. This is illustrated in Fig. 8.9, which shows a horizontal boiler with a vertical economizer, a 400°C/40 bar steam turbine connected to an air-cooled condenser, and a dry flue-gas cleaning system. The thermal cycle also shows other necessary components like a de-aerator, which removes non-condensable gases from the boiler water/condensate.

Experience with waste boilers has shown that high-temperature corrosion of the superheater tubes is greatly dependent on the flue-gas temperature at the entrance to the fourth pass. It is therefore often a boiler design criterion that the flue gas does not exceed 625–650 °C when passing across the hottest superheater tubes. Selection of the appropriate material (a steel alloy) for the tubes is also important, although developments of anti-corrosive steel seem not to have solved the corrosion problems in waste boilers.

Superheater tubes producing 400°C steam in a municipal waste facility are estimated by Rambøll to have a lifetime of up to 10–15 years, under the best conditions and with a modern boiler design. Even moderate increases to this temperature, say up to 450°C, may cause severe corrosion and reduce the lifetime of superheater tubes to perhaps 3–5 years. Rambøll estimates that further increases, say to 500°C, may entail boiler tube maintenance several times a year due to severe high-temperature corrosion.

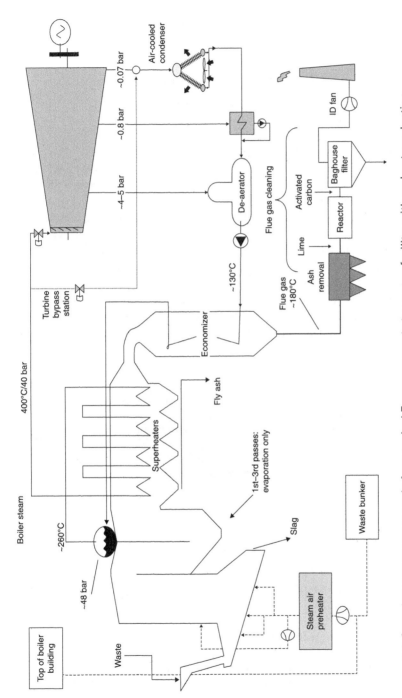

8.9 Complete water-steam cycle for a typical European waste to energy facility, with no heat production (source: Rambøll).

Boiler manufacturers generally do not recommend higher steam temperatures than 440–450°C for WTE unless special circumstances dictate otherwise. Methods of reducing corrosion include choosing high alloy steels or specially resistant coatings.

8.4   WTE facilities as sources of heat

8.4.1   WTE facilities delivering small to medium amounts of heat

An electricity-only waste to energy plant can easily be designed to produce small to medium amounts of heat, for instance if heat or process steam can be sold for a number of months of the year (winter) while for the remaining months the plant operates to produce electricity only.

A technical solution to this is to have an extraction turbine, with a cross-over valve placed somewhere along the turbine expansion line. When actuated towards the closed position, the valve forces steam out to a hot water condenser. The plant shown in Fig. 8.10 produces 80–110°C water. The cross-over valve can thus divert steam to where it is most needed: either to produce heat or to produce electricity, as required. As such, this type of turbine enables both heat and power operation as well as electricity-only operation. This could be the choice if the plant was connected to a small district heating (DH) network, or indeed if the plant was to be prepared for future district heating operation.

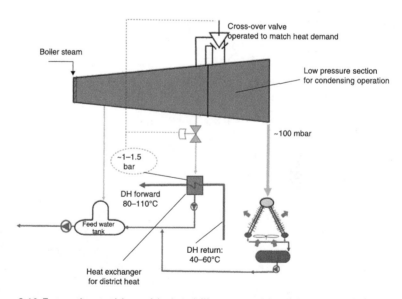

8.10 Extraction turbine with the ability to provide variable quantities of hot water, depending on the heating requirement. The depicted solution is applicable for a plant that can be operated either in electricity-only mode or in combined heat and power mode (DH=district heat) (source: Rambøll).

Figure 8.11 shows a cross section of an extraction steam turbine with the ability to operate freely between electricity-only and combined heat and power modes. Figure 8.12 is an extension of Fig. 8.10 to show the major control mechanisms relevant for an extraction turbine connected to a small district heating network.

The water-side district heating pumps (a set of three pumps for redundancy reasons) are used to achieve the desired differential pressure in the district heating network, dP, for instance 2 bar. This allows enough pressure to provide hot water to all heating installations in the network. If the heating consumers in the network turn on their radiators (call for more heat), the pressure in the network will drop and the district heating pumps at the plant will respond by increasing the flow to again uphold the desired differential pressure.

The steam turbine cross-over valve, as well as the steam valve in the line to the DH condenser, is regulated to keep a constant district heating forward temperature. So, if the consumer calls for more heat, the district heating pumps will increase the flow and the initial consequence will be a lower DH temperature. This is picked up by the turbine controls, which respond by providing more steam to the condenser until the set-point district heating temperature is obtained. These controls in combination will thus match the consumer demand for heat (in megawatts) at the desired forward temperature, which is a set point determined by the heating network operators. For large transmission heat networks the forward temperature is 100–120°C or more, whereas smaller heat networks may have forward temperatures as low as 75–85°C.

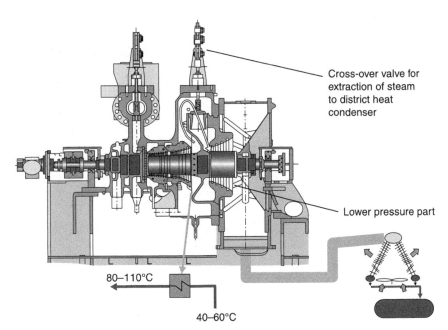

8.11 Cross section of an extraction steam turbine. The cross-over valve is used to drive steam to the district heating condenser depending on the required heating demand. Figure courtesy of MAN Diesel and Turbo, modified by Rambøll (MAN, 2008).

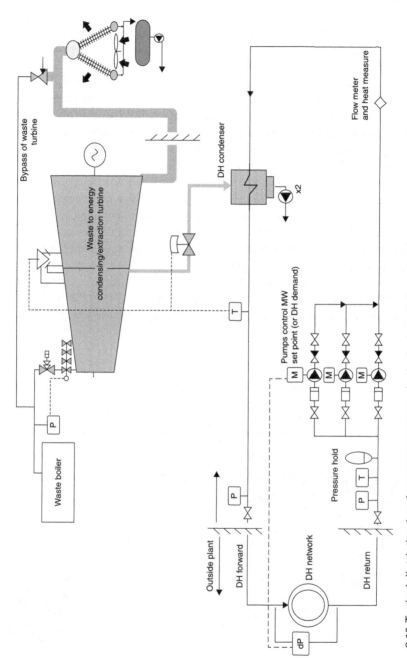

Waste boiler

Bypass of waste turbine

Waste to energy condensing/extraction turbine

DH condenser

Flow meter and heat measure

x2

Pumps control MW set point (or DH demand)

Pressure hold

Outside plant

DH forward

DH network

DH return

dP

8.12 Typical districting heating components and principal control mechanisms.

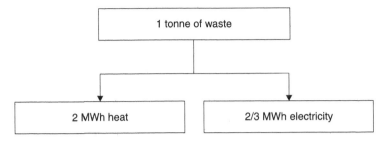

8.13 1-2-3 rule of thumb for waste to energy facilities configured
for district heating. 1 tonne of waste yields 2 MWh of heat, and
2/3 MWh of electricity.

A simple rule of thumb can be used to calculate the overall energy recovery in
a waste-fired district heating plant: the rule of thumb is that 1 tonne of waste
yields 2 MWh of heat and 2/3 MWh of electricity. See also Fig. 8.13.

Generally the forward temperature is a function of the ambient temperature,
since a higher forward temperature will allow the transfer of more heat in the
district heating pipework, which is also bound by hydraulic limits. Table 8.3
shows an example of set points for district heating network forward temperatures
as a function of ambient temperature.

8.4.2 WTE facilities for district heating networks

When a WTE facility has access to a large district heating network, such as in
some cold weather countries like Denmark, Sweden and Finland, it is common to
establish so-called back-pressure turbines, where all the steam exiting the turbine
is used to produce district heat.

This configuration is illustrated in Fig. 8.14. Notice that the air-cooled condenser,
which is physically a very large component, has been replaced by two (much
smaller) water cooled surface steam condensers, where all the condensing heat is
transferred to the district heating water. To exemplify the options for selling heat, the
extraction of 14 bar process steam to a (hypothetical) nearby factory is also shown.

Table 8.3 Example of DH forward temperatures versus ambient temperature for a
large Swedish DH network

	T ambient (°C)	T forward (°C)	T return (°C)
Very cold winter	< –15	115	52
Cold winter	–10 to –15	105	50
Winter	0 to –10	95	45
Spring and autumn	+10 to 0	85	42
Summer	> +10	75	45

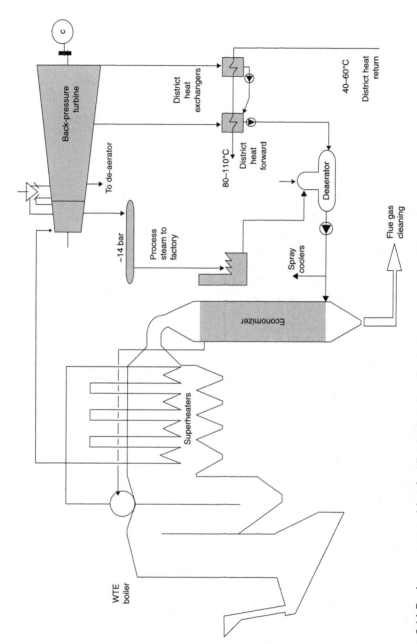

8.14 Back-pressure turbine for district heating. Such a set-up does not require a large air-cooled condenser, since all the heat is transferred to the water-based district heating network.

Table 8.4 Estimates of electrical and total efficiencies for district heating versus electricity-only energy for waste facilities.

Size: 100 MW thermal, 36 t/h of waste (~300 000 tonnes/yr) CV = 10 GJ/tonne	Net electrical efficiency (example)	Heat efficiency (example)	Total efficiency (example)
Back-pressure (district heat) plant (without air-cooled condenser)	22.5% (22.5 MW sold)	63% (63 MJ/s heat sold)	85.5%
Electricity-only plant (with air-cooled condenser)	26% (26 MW sold)	0%	26%

Assumptions: electricity-only back pressure = 100 mbar; district heating network temperatures = 50–90°C; live steam parameters: 425°C/52 bar

Table 8.4 compares the electrical and total efficiencies of an electricity-only plant and a CHP plant, configured as discussed in this chapter. Since the district heating water induces a higher back pressure (~1 bar) than an air-cooled condenser (~0.1 bar), the electrical efficiency of the electricity-only plant is higher. As a result, the electricity-only plant will have an electrical efficiency approximately 3–6 percentage points higher than the corresponding district heating plant (based on various assumptions). The table shows typical figures based on a 100 MW thermal plant as an example, showing that the district heating plant produces 3.5 MW less electricity. However, a total of 63 MW of heat is produced instead – hence instead of 1 MW of electricity, 18 MW of heat is produced (63/3.5 = 18). This is 18 times more efficient than electric space heating, and also much more efficient than high efficiency heat pump installations, which can produce up to approximately 4–6 MW of heat for every megawatt of electricity.

The conclusion is that a waste to energy district heating turbine is an extremely efficient means of producing heat, assuming that the waste in any case needs to be combusted. The 'value' of the produced heat is very much dependent on the size of the connected district heating network. Figure 8.15 shows the district heating network in the Copenhagen region, stretching an impressive distance of about 40 km from east to west. The high total efficiency of WTE plants connected to (medium-large) district heating networks results in lower waste gate fees, since there is generally a higher energy sales income compared to electricity-only plants.

The configuration of pumps and main control strategies for a district heating WTE facility can be seen in Fig. 8.16. The mechanisms in Fig. 8.16 are slightly different from those shown in Fig. 8.12. In the plant depicted in Fig. 8.12, the production of district heat can be turned up or down by the extraction valve (independently of waste firing), thus passing more or less steam to the final stage condensing turbine. The plant in Fig. 8.16 does not have the final stage condensing turbine, and thus district heat production is determined by the amount of waste combusted. The plant

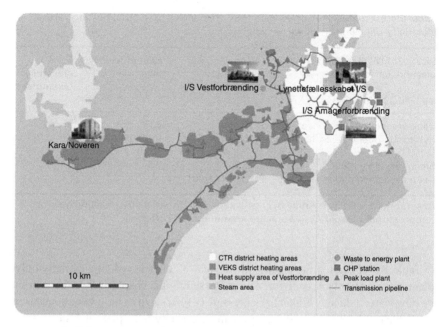

8.15 Copenhagen district heating network spanning about 40 km from east to west (Rambøll, 2012b).

configuration shown in Fig. 8.16 is called a heat-only or back-pressure plant, and typically used if the plant is connected to a large district heating network. In this plant the district heating pumps are controlled to maintain a given district heating forward temperature. The result is that the district heating pumps ensure that variations in the steam flow are taken up by the district heating water flow.

If, for instance, the thermal load on the waste boiler was to go down (say due to a lower heating value of the waste) the steam flow to the turbine would be lower. The immediate consequence would be a lower district heating forward temperature. The district heating pumps would respond by turning down the flow, to again obtain the desired district heating forward temperature. In such a system, the heating network relies on other production units to balance the heat demand and heat production at any given point in time.

It is common during summer that a waste back-pressure turbine produces more heat than is necessary in the heating network. Since waste incineration usually is upheld throughout the year, it is not always possible to reduce the thermal boiler load. Instead, plants often install water-to-air summer coolers to remove the excess district heat.

Ultimately the size of the district heating network compared to the size of the waste to energy facility will determine whether it is more cost effective to install an extraction turbine with an air-cooled condenser (Fig. 8.12) or a district heating back-pressure turbine (Fig. 8.16).

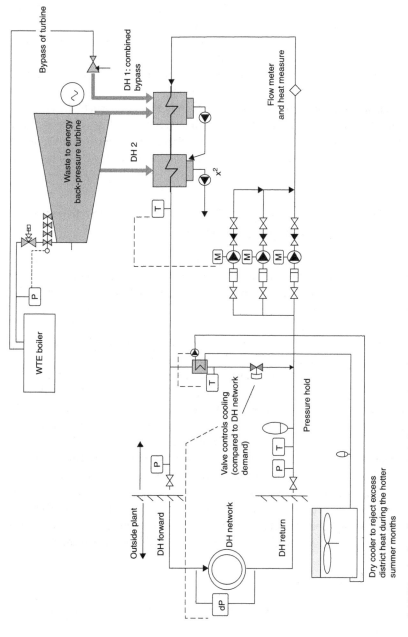

Bypass of turbine

DH 1: combined bypass

Waste to energy back-pressure turbine

DH 2

Flow meter and heat measure

WTE boiler

x^2

M

M

M

T

T

Valve controls cooling (compared to DH network demand)

Pressure hold

P

T

P

T

Outside plant

P

DH forward

DH network

dP

DH return

Dry cooler to reject excess district heat during the hotter summer months

8.16 District heating conceptual diagram for a heat-only back-pressure plant (Rambøll, 2012a).

8.4.3 Advanced solutions: heat pumps and district cooling

WTE plants (and other thermal installations) can be equipped to supply both (district) heating and cooling. This could be a preferred solution for a geographical placement with cold winters but hot summers.

For this application an electrical or steam-driven chiller could be used. The electricity or driving steam could be supplied by the waste to energy steam turbine, thus allowing for the complete installation to supply power, district heating and district cooling, depending on the current demands.

A schematic of such tri-generation can be seen in Fig. 8.17, where the chiller is driven by steam from a waste-fired steam turbine. District cooling is less common than district heating, and in most cases only makes sense in hot countries, or for the industrial cooling of large office buildings, airports or the like. District cooling on the household level is not common.

8.5   Optimizing energy efficiency in WTE combined heat and power (CHP) facilities

A WTE facility typically has efficiencies on the orders of magnitude (based on Rambøll's calculations and experience from Rambøll projects; some plants may deviate from these values) shown in Table 8.5. Here, MWt is thermal input. The lower efficiency of the combined heat and power facilities is due to a higher turbine back-pressure, as previously discussed. See also Fig. 8.8. However, the heat and power facilities have total efficiencies ranging from 82–86%, hence a large amount of the energy in the waste is recovered for useful purposes.

WTE facilities can also be equipped with so-called flue-gas condensation equipment, where the boiler flue gas is cooled to 35–45°C to produce additional district heat. This flue-gas temperature is below the flue-gas dew point of normally 52–60°C, and hence moisture in the flue gas is condensed with a subsequent large energy release as a result. The flue-gas dew point depends on the excess air ratio and absolute pressure, as well as the moisture content in the flue gas. The latter is closely related to the waste's lower heating value (LHV), since a lower heating value usually arises from a wetter fuel. A dew point of 51°C corresponds roughly to a LHV of 12 GJ/tonne, and a dew point of 60°C corresponds roughly to a LHV of 8 GJ/tonne. Facilities using flue-gas condensation can have total energy recovery efficiencies of more than 100% measured on a lower heating value basis, and are thus extremely efficient.

8.5.1 Optimizing for increased electrical output

Generally, the electrical efficiency of WTE facilities depends on three factors:

- size of the plant
- high- and low-temperature corrosion of the boiler tubes
- connection to district heating network or 'electricity only'.

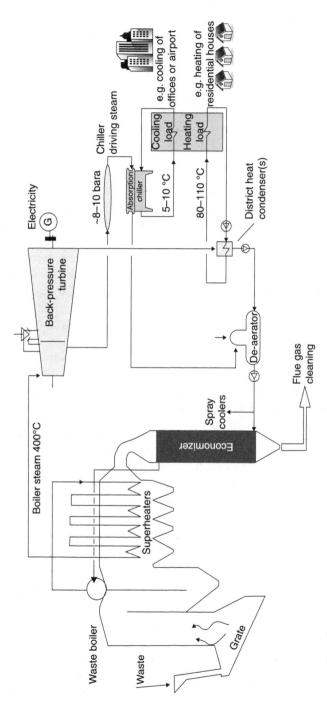

8.17 Generation of power, heat and cooling: tri-generation (Rambøll, 2012a).

Table 8.5 Net electrical efficiencies for WTE plants

WTE facility	Size	Net electrical efficiency
Electricity only	Up to 100 MWt	23–26%
Electricity only	>100 MWt	up to 30%
Heat and power	Up to 100 MWt	17–23%
Heat and power	>100 MWt	up to 25%

The maximum size of a single boiler WTE facility is around 100–115 MW fired, and generally limited by the maximum grate size for the boiler, which is currently about 35–40 tonnes of waste an hour. For larger waste flows, two or more boilers can provide steam to a common steam turbine. However, also due to the relatively large catchment area of the waste, it is uncommon even for the largest WTE facilities to combust more than 500 000 tonnes of waste a year, corresponding to about 200–250 MW thermal input (depending on the waste heating value). As such, even very large waste to energy facilities have a much lower thermal input than utility boilers, and as a result the electrical efficiencies will probably never reach utility level since optimization measures become less cost effective for smaller energy inputs.

Waste to energy facilities are also prone to boiler tube corrosion, which in practice has limited the live steam temperature to approximately 440–450°C. For comparison, steam temperatures at utility boilers can be up to 600°C.

A WTE facility can generally be optimized by focusing on the following areas:

- Boiler efficiency (generating more steam): Lowering the excess air ratio and lowering the flue-gas temperature will lower the resulting thermal loss to the chimney, and hence produce more steam.
- Water–steam cycle optimizations: Using feedwater and condensate preheaters, as well as preheating the combustion air, can increase the thermal cycle efficiency.
- Turbine back-pressure: Besides the live steam temperature, as already discussed, the turbine back-pressure can be decreased, for instance by having larger condensers.

Figure 8.18 shows the change in net electrical efficiency for a WTE facility (in percentage points) when changing the live steam temperature (range 350–500°C), and Fig. 8.19 shows the same change for different temperatures of the district heating water (for a plant producing district heat).

Even for relatively big changes in steam temperature and district heating temperature, both of which would result in large or expensive impacts on the mechanical design, the effect on electrical efficiency is only 1–5 percentage points. These relatively slight changes in electrical efficiency are in stark contrast

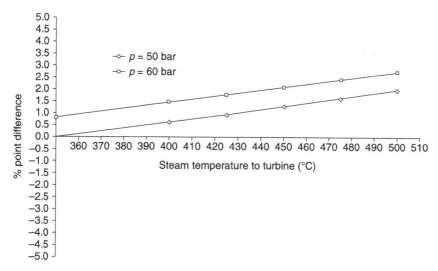

8.18 Change in net electrical efficiency for a 100 MW European WTE facility producing district heating as a function of final boiler live steam parameters (temperature and pressure) (Rambøll, 2012a).

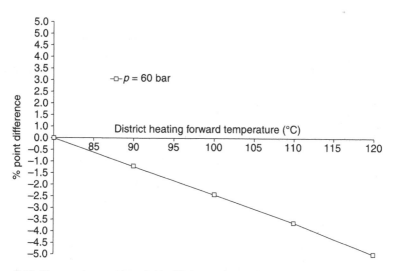

8.19 Change in net electrical efficiency for a 100 MW European WTE facility producing district heating as a function of the forward temperature in the district heating network (Rambøll, 2012a). $\Delta T_{water} = 30°C$.

to the benefits of combined heat and power, where total plant efficiency can be increased by up to 60 percentage points (from 25% to 85%) by using the waste heat for useful purposes.

Obviously heat and power requires a heat sales opportunity, which in practice means an energy infrastructure of district heating pipework and pump installations. The district heating network is not cost effective in the short run, and certainly not for the individual producers of heat. However, municipalities, large energy companies and governments interested in the long-term benefits of highly efficient energy facilities (for all types of fuel), can drive forward such infrastructure investments to the benefit of the whole community.

When using flue-gas condensation in combination with a standard heat and power configured plant, the total increase in efficiency can be up to 80 percentage points or more, resulting in total energy recovery rates above 100% (measured on LHV). Figure 8.20 shows the potential for flue-gas condensation: up to approximately 20 MW of additional heat output for a 90 MWt plant. As can be seen, the output is very sensitive towards the district heating return water, since the water is used to cool the flue gas below the dew point of 55–60°C. It is possible, and relatively common in Scandinavia, to incorporate absorption chillers to thus boost the output of a flue-gas condensation installation, if the district heating return temperature is high.

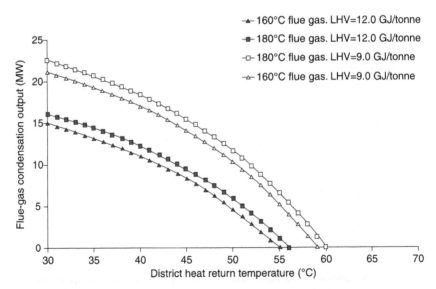

8.20 Additional heat output as a function of return temperature, LHV, and flue-gas exit temperature from the boiler for flue-gas condensation for a 90 MW plant. The additional increase in total efficiency is in the range 0–20% (Rambøll, 2012a).

8.6   Conclusion

If a heat market exists or can be made available, waste to energy facilities can be a cheap source of heat – for example, the total energy recovery in electricity-only facilities of about 25% can be increased to 85–105% for combined heat and power installations. The loss in electrical efficiency by doing so is moderate: the electrical efficiency is reduced by only 3–6 percentage points. Configuring a steam turbine for heat and power instead of power only results in heat production for only a slight loss of electrical power: up to 18 MW of heat can be generated instead of 1 MW of electricity.

Even for smaller amounts of heat extraction from a waste plant, it is not uncommon to achieve an electricity-to-heat ratio of up to 1:10. Using such heat is particularly interesting given the fact that waste incineration is used mainly to reduce the mass and volume of (non-recyclable) waste, and hence the alternative is to reject large amounts of heat in cooling installations with no real benefit.

8.7   References

ECN, 2012, online fuel database 'Phyllis' hosted by ECN: www.ecn.nl/phyllis. Accessed November 2012.

GEA, 2012. Available from www.gea-energytechnology.com/opencms/opencms/gas/en/products/Direct_Air-Cooled_Condensers.html. Accessed November 2012.

Hulgaard T, Vehlow J, 2010, *Solid Waste Technology and Management*, Ed. T H Christensen, Volume 1, Chapter 8.1 (Incineration: Process and technology), Wiley.

MAN, 2008, MARC Steam Turbines – The modular Steam Turbine Generation. Sales Brochure by MAN Turbo AF, December 2008.

MAN, 2009, MAN Turbo AG Brochure Industrial Steam Turbine.

Patel N, Gordon G, Howlett L, 2001, Accomplishments from IEA Bioenergy Task 23: Energy from Thermal Conversion of MSW and RDF.

Rambøll, 2012a, Own calculations and figures by Rambøll (not published).

Rambøll, 2012b, General knowledge in-house in Rambøll, as well as the following specific references available for download:

Kleis H (Babcock & Wilcox Vølund) and Dalager S (Ramboll), 2004, 100 years of waste incineration in Denmark. Available from www.ramboll.com/wte.

Rand T, Haukohl J, Marxen U, 2000, Municipal Solid Waste Incineration – A Decision Maker's Guide, World Bank Publications, Washington, DC. Available from www.ramboll.com/wte.

Renosam and Ramboll, 2006, Waste to energy in Denmark – The most efficient waste management system in Europe. Available from www.ramboll.com/wte.

Vølund, 2009, 21st century advanced concept for waste-fired power plants, Babcock & Wilcox. Available from www.volund.dk/en/Waste_to_Energy/Self_study. [accessed January 2013] Direct download: www.volund.dk/en/Waste_to_Energy/~/media/Downloads/Brochures%20-%20WTE/Advanced%20concept%20for%20waste-fired%20power%20plants.ashx.

9

Gasification and pyrolysis of municipal solid waste (MSW)

N. B. KLINGHOFFER, Columbia University, USA
and M. J. CASTALDI, City University of New York, USA

DOI: 10.1533/9780857096364.2.146

Abstract: This chapter discusses gasification and pyrolysis as alternative methods to grate combustion. These processes are important alternatives to combustion since they produce fuels, chemicals, electricity or heat from waste. First, the theory behind gasification and pyrolysis is discussed, along with key differences between these processes and full combustion. This chapter also provides an overview of the products that can be made from gasification, environmental issues and different types of gasification processes. Specific examples of facilities that are in operation are discussed later in this chapter.

Key words: gasification, pyrolysis, synthetic gas, syngas, partial oxidation.

9.1   Introduction

This chapter introduces gasification and pyrolysis as alternative methods to grate combustion for energy recovery from waste. Gasification is the partial oxidation of a solid fuel in order to produce synthesis gas (also called synthetic gas or syngas), which can be used for electricity generation or in the synthesis of fuels or chemicals. It is an important alternative to combustion since it can be used to produce different forms of energy. Gasification technologies have been around for many years and have been used to convert coal to gaseous or liquid fuels. Sasol, a gasification company based in South Africa, has been in operation since 1950. In 1955 it produced automotive fuels from coal. However, since the 1950s, gasification has been recognized as a promising technology in the conversion of unconventional fuels to common products. More recently, gasification technologies have been used to treat biological waste such as agricultural waste, sewage sludge or wood waste.[1-5] However, there is now a growing interest in gasification of municipal solid waste,[6,7] commercial and industrial wastes,[8] and plastics.[9,10] This chapter introduces the theory behind gasification and pyrolysis and outlines the key differences between it and full combustion. An overview of the types of products that can be made from gasification and the applications of these products are presented. In addition, environmental issues relating to gasification are addressed. Different types of gasification processes and specific examples of facilities that are in operation are discussed later in this chapter.

146

9.2   Gasification and pyrolysis

Gasification is a process for converting solid carbonaceous materials to a combustible gas (e.g., a mixture of H_2, CO, CH_4 and CO_2). The overall objective is to convert these gases into fuels that can be easily integrated into current energy technologies. In general, gasification involves the reaction of a solid fuel with a co-reactant at temperatures in the range of 550–1000°C. Co-reactants are introduced in sub-stoichiometric quantities in order to partially oxidize the fuel to CO and H_2 rather than completely oxidize it to CO_2 and H_2O. Gasification in the absence of oxygen, i.e. using an external source of energy, is called pyrolysis. A conceptual schematic of the reaction path is shown in Fig. 9.1. Here, the relative enthalpy change is shown for different gasification environments. Gasification with air or O_2 is an exothermic reaction, so heat is released during the reaction and the products have a lower enthalpy than the reactants. Further combustion of the syngas will form combustion products (CO_2 and H_2O) that have a lower enthalpy than the gasification products. The two steps (waste to syngas and syngas to combustion products) have approximately the same enthalpy change. Reaction in an inert environment such as N_2 will require heat input to initiate the reaction. Hence, the products will have a slightly higher enthalpy than the reactants. In this reaction environment, hydrocarbons will form and there will be reduced conversion to H_2 and CO. Reaction with steam is highly endothermic and heat must be added to the process. The presence of steam results in high concentrations of H_2, which are formed from the water, so the products have a higher enthalpy than the reactants. The reaction of CO_2 with waste is also endothermic and the addition of CO (via conversion of CO_2 to CO) creates a higher enthalpy product stream. CO_2 is a more stable molecule than H_2O (the heat of formation is -394 kJ. mol^{-1} for CO_2 and -229 kJ.mol^{-1} for steam) so the increase in enthalpy is slightly higher for CO_2 gasification than for steam gasification (in other words, combustion of one mole of CO will release more heat than combustion of one mole of H_2). The conversion of synthesis gas to chemicals or fuels is an exothermic process, and is therefore accompanied by a decrease in enthalpy.

As Fig. 9.1 illustrates, gasification with air is an exothermic process, whereas gasification with stable co-reactants, such as steam or CO_2, is endothermic. Both endothermic and exothermic gasification processes operate in temperature regimes that are much lower than combustion temperatures. In these temperature regimes, heat losses through reactor walls, and the heating of product and diluent gases (such as N_2 in air gasification) can result in quenching of the reactions. In other words, if a combustion system is operating at 950°C, then even if there are heat losses the temperatures will still be high enough to auto-thermally sustain the reactions. If a gasification system is operating at a lower temperature, for example 750°C, then the heat losses can reduce the temperature to a point where reactions are quenched. Therefore, gasification systems often require combustion of some of the product gases to heat the gasification reactor or the diversion of some fuel (waste) in order to maintain gasification temperatures. In addition, these lower

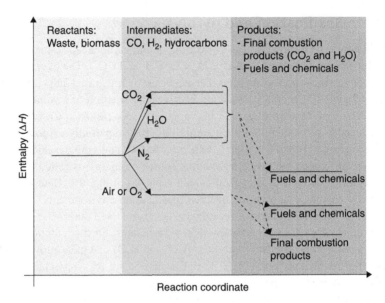

9.1 Conceptual pathway for conversion of solid fuel to different products. The enthalpy change is shown for different co-reactants. An increase in enthalpy is observed for endothermic reactions, such as reactions with CO_2, H_2O, or in an inert N_2 environment. Exothermic reactions with air or O_2 show a decrease in overall enthalpy. Final products can be either fuels and chemicals, or combustion products, which are very stable and therefore have low enthalpies.

temperatures lead to the formation of intermediates, commonly referred to as tars, which need to be processed further in order to convert them to products. So why gasify waste, when it can be combusted? The benefit of gasification is its ability to create many different types of fuel or chemicals. Therefore, gasification is a method for producing common fuels or chemicals from unconventional feedstock. If electricity generation is the desired outcome, then the synthesis gas can be combusted in a gas turbine with heat recovery steam generation (HRSG) for further electricity production from steam. Alternatively, the synthesis gas can be converted into gaseous or liquid fuels, or chemicals, which is discussed later in this chapter. Conventional combustion systems produce steam, which drive a steam turbine for electricity generation, but do not provide the option of synthesis of other fuels or chemicals. Another benefit of gasification is the scalability of the process; small-scale gasifiers can be used to recover energy from waste in smaller communities.

Gasification can be described as a three-step process where the first step is dehydration, followed by pyrolysis of volatiles and finally gasification of char.[11] This can be understood by observing the mass loss of a solid material during gasification. Figure 9.2 shows the results of gasification of a wood chip where a

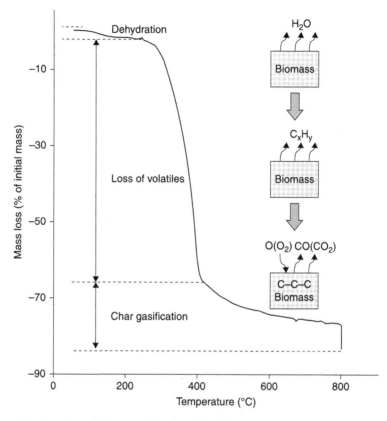

9.2 Mass loss during gasification of poplar wood where three regimes are visible: dehydration, loss of volatiles, and gasification of char. Below 200°C water loss takes place. In the intermediate temperature regimes, volatiles are removed and may react with co-reactants in the system. The char reacts in the high-temperature regime and generally requires a co-reactant such as O_2, CO_2, or H_2O in order to decompose to CO or CO_2. Temperature regimes shown here are approximate and will vary based on the heating rate, residence time, and reactants in a system. (These data were produced by the Combustion and Catalysis Lab (CCL) group at Columbia University.)

small mass loss from dehydration is observed at temperatures below 200°C, followed by significant and rapid mass loss at intermediate temperatures and a more gradual mass loss at high temperatures, where the char decomposes. The reactive gas will affect both the rate of reaction as well as the total mass loss. Figure 9.3 compares gasification under air and N_2 where the extent of the reactions, which is represented by mass loss, is much more significant with air compared to N_2. Steam is commonly used as a co-reactant in gasification processes since it can increase the overall H_2 yield. CO_2 can also be used as a co-reactant during

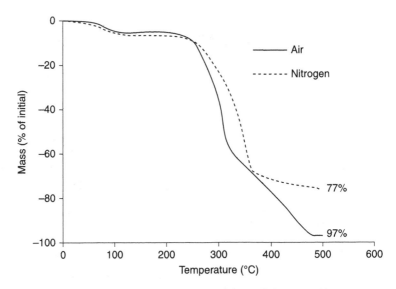

9.3 Influence of co-reactant on gasification of biomass. Air
gasification results in higher mass loss compared to N_2. (These data
were produced by CCL at Columbia University.)

gasification and has been shown to reduce energy requirements to the gasifier as
well as increase overall conversion of the solid fuel.[12,13]

Gasification of waste forms a gaseous mixture composed of CO, H_2, CO_2, CH_4,
as well as some C_2 and C_3 hydrocarbons. While the main desired product from
gasification is synthesis gas, by-products are formed. One of these by-products is
tar, which is a mix of liquid organic hydrocarbons. Since many biologically based
materials contain cellulose and lignin, which are made of multi-carbon rings, tars
are primarily composed of aromatic hydrocarbons, which may be oxygenated.
While the composition of the tars is dependent on the fuel that is gasified and the
reaction conditions, an example of the composition of tars is shown in Fig. 9.4.
Another by-product from gasification is char or ash, which is a solid residue that
includes metals, minerals and carbon that has not been converted to gas or liquid
products. Ash is defined as the solid residue that consists primarily of minerals
and metals. Char can be primarily ash with a high carbon fraction or a nearly pure
carbon residue that maintains a similar skeletal structure to the original feedstock.
An example of the distribution of carbon in char, tar and synthesis gas is shown
in Fig. 9.5. In that work, biomass feedstocks (corn stover, switchgrass, wheat
straw and wood) were gasified with steam at reaction temperatures in the range
600–710°C. Higher temperatures gave higher gas yields and lower char and tar
formation (in that work, condensate refers to tar and water recovered in the
condenser). However, even at high temperatures, tar was still formed. During
gasification of wood in a fluidized-bed reactor at temperatures in the range

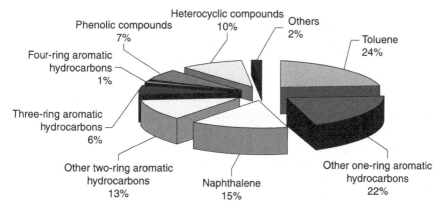

9.4 Typical composition of tar from gasification of biomass.[14]

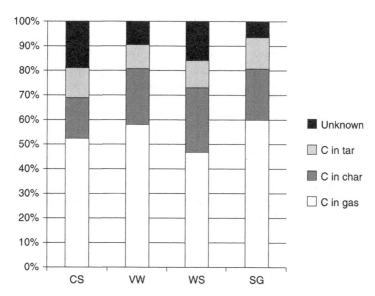

9.5 Distribution of carbon between three phases after the gasification of four types of biomass with steam at 650°C and a thermal tar cracker at 875°C.[1] (CS: corn stover, VW: Vermont mixed wood, WS: wheat straw, SG: switchgrass)

750–900°C, the gas yield increased with temperature but even at 900°C the gas yield did not exceed 75%.[15]

9.2.1 Tar removal in gasification processes

Tars must be removed from synthesis gas since they can deposit on the walls of a reactor and downstream equipment, and can cause clogging. They can also decompose further, which can result in the formation of solid carbon or hazardous by-products. Tar formation represents a decrease in efficiency of the process since any carbon and hydrogen in the tars has been diverted from the gas-phase product stream.

Tar handling is one of the major barriers in the commercialization of gasification technologies, and there is a significant amount of work on methods for tar decomposition. Tar formation can be reduced by operating at higher reactor temperatures (temperatures above 750°C will result in lower overall tar formation)[16] or by increasing the gas residence time.[17] Operating with higher concentrations of air will result in lower tar yields due to the increase in oxygen, which reacts with the hydrocarbons.[18] The addition of steam can also decompose tars via steam reforming, according to:[19]

$$C_nH_m + nH_2O \leftrightarrow nCO + (n + m/2)H_2 \qquad [9.1]$$

One method that is gaining increasing attention because it can be selective and does not require high temperatures is catalysis. Catalytic decomposition of tars can take place at lower temperatures than thermal decomposition of tars. Conventional catalysts, such as precious or base metal catalysts, and unconventional catalysts such as ash or char are being considered for these processes.[14,20–27]

9.2.2 Comparison of gasification and combustion systems

Combustion and gasification processes are similar in that both involve reactions of an oxidant with a fuel at elevated temperatures. The parameter that defines whether a process is considered combustion or gasification is the air to fuel ratio, which is often described by the equivalence ratio. The equivalence ratio is the stoichiometric air to fuel ratio divided by the actual air to fuel ratio:

$$\phi = \frac{(air/fuel)_{stoichiometric}}{(air/fuel)_{actual}} \qquad [9.2]$$

where ϕ is the equivalence ratio.

The reactor temperature will depend on the equivalence ratio of the system. Figure 9.6 shows the adiabatic temperature for different air fractions in a waste combustion system. As shown, the maximum temperatures are achieved at stoichiometric air (equivalence ratio of 1), where complete combustion of the fuel is theoretically possible. When air is used as the oxidizer, nitrogen is introduced according to the ratio of nitrogen to oxygen in air (79:21). Nitrogen does not participate in the reactions but absorbs heat, so in order to achieve high reaction temperatures the amount of nitrogen should be low. Therefore, operating at an equivalence ratio of 1 minimizes the amount of excess nitrogen introduced while still providing enough oxygen for complete combustion to occur. If a system operates at an equivalence ratio of 1 then the overall reaction rate will likely be

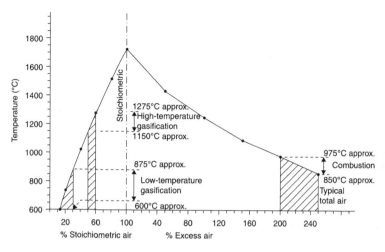

9.6 Adiabatic temperature vs air supply per 1 tonne of 20 MJ.kg^{-1} waste or biomass (image modified from Entech[28]).

limited by transport, rather than chemical kinetics. For example, if the residence time is very low (which is desirable in order to achieve high throughput or small system volume) then the transport of the oxidant to the fuel may not be fast enough to completely oxidize all of the fuel before it exits the reactor. One can show that due to transport the equivalence ratio at the point of reaction is always 1.

Combustion systems typically operate with excess air in order to ensure complete oxidation of the fuel. Excess air also helps to keep temperatures within the material limitations and affects the emissions that are produced. Because municipal solid waste (MSW) is variable in composition, the equivalence ratio is typically calculated based on an assumed MSW composition. Gasification processes operate at sub-stoichiometric conditions with the oxygen supply controlled such that both heat and a refined gaseous fuel are produced as the feed material is consumed. Typically, a gasifier operates with less than 40% of the amount of oxygen theoretically required for complete combustion.

In combustion processes where MSW is the fuel, the primary emissions that must be controlled are dioxins, particulates, chlorine, sulfur and the products of incomplete combustion. NO$_x$ is also a concern since combustion takes place at temperatures where thermal NO$_x$ can form. In pyrolysis, hydrocarbons are the primary product that must be removed from the flue gases. NO$_x$ formation is much less of a concern, since gasification and pyrolysis take place in a reducing environment, where fuel-bound NO$_x$ is more likely to be converted to N$_2$ or NH$_3$ rather than NO$_x$, which is formed in an oxidizing environment. In addition, gasification and pyrolysis take place at temperatures that are below the temperature where thermal NO$_x$ will form.

Some benefits of combustion over gasification are that combustion achieves very high conversion of the fuel and does not produce significant amounts of tar.

However, while gasification produces more by-products, it offers flexibility due to the ability to produce electricity, fuels or chemicals. Producing fuels and chemicals is particularly beneficial because they store energy: a liquid fuel can be stored for a long time whereas electricity is best used as it is produced. Gasification plants can be smaller than incinerators since much less oxidant (for example, air) is introduced. This presents an opportunity for smaller gas cleanup systems because the gas stream is smaller than with combustion systems. Reducing the volume of gas at the reactor outlet means that downstream equipment can be designed for a lower throughput. In addition, this implies that the concentrations of pollutants being removed will be higher (since there is less diluent), which is beneficial for gas cleanup.

9.3   Products and their applications

9.3.1   Uses for synthetic gas (syngas)

Three types of synthetic gas can be produced: low heating value, medium heating value and high heating value. Synthetic natural gas (SNG) can be produced directly from synthesis gas with minimal processing and is typically considered as a fourth type of synthetic gas. SNG can be produced via the methanation reaction:

$$CO + 3H_2 \rightarrow CH_4 + H_2O, \qquad \Delta H = 210 \, kJ.mol^{-1} \qquad [9.3]$$

If the syngas is to be converted to SNG and supplied to a pipeline, the nitrogen (and other diluents) must be removed because the presence of N_2 will lower the heating value of the gas. A typical N_2 concentration for pipeline SNG is about 2%. For these reasons it is best to remove N_2 from air (and gasify with pure O_2), which is a very mature technology, while removing N_2 from a product stream with multiple constituents is not well developed. An outline of the process steps for conversion of biomass into SNG is shown in Fig. 9.7.

Low heating value gas (3.5 to $10 \, MJ.m^{-3}$ or 100 to $270 \, Btu.ft^{-3}$) typically consists of syngas mixed with relatively high amounts of N_2 or CO_2, which result in its low heating value. This fuel is typically used for combustion applications where high purity or specific CO/H_2 ratios are not needed. For example, it can be used as gas turbine fuel in an integrated gasification combined cycle (IGCC) system, as boiler fuel for steam production and as fuel for smelting and iron ore

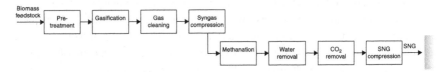

9.7 Process steps for conversion of biomass waste to synthetic natural gas.[29]

reduction applications. An IGCC system uses synthesis gas to drive a gas turbine and then the hot exhaust from the gas turbine is used to heat steam, which drives a steam turbine. Hence, both steam turbines and gas turbines are used to generate electricity from the combustion of synthesis gas. Low heating value gas is not well suited as a natural gas replacement or for chemical synthesis because of its high nitrogen content and low heating value. The primary reason is that nitrogen can participate in the synthesis reactions and become incorporated into the products, producing hydrocarbons with fuel-bound nitrogen. When fuels with bound nitrogen are subsequently combusted, NO_x will form. However, as discussed above, one advantage of gasification is the ability to avoid NO_x emissions. In the gasification chamber the atmosphere is primarily a reducing environment therefore any nitrogen in the fuel will either form N_2 or NH_3 with N_2 being dominant. If the produced gas is to be combusted, NO_x can form from N_2. Therefore if the produced gas is to be combusted, it is better to have low N_2 content.

Medium heating value gas (10 to 20 MJ.m^{-3} or 270 to 540 Btu.ft^{-3}) can be used as fuel gas for gas turbines in IGCC applications, for SNG in combination with methanation (Eq. 9.3), for hydrogen production, as a fuel cell feed, and for chemical and fuel synthesis. High heating value gas (20 to 35 MJ.m^{-3} or 540 to 940 Btu.ft^{-3}) can also be used as fuel gas for gas turbines in IGCC applications, for SNG and hydrogen production, as a fuel cell feed and for chemical and fuel synthesis. However, it does not require as much upgrading. Syngas (over 35 MJ.m^{-3} or 940 Btu.ft^{-3}) can be easily substituted for natural gas and is suitable for hydrogen and chemical production and as a fuel cell feed.

As mentioned before, synthesis gas can be used to produce many different fuels and chemicals from methane and methanol to gasoline and diesel fuel. This is the primary engineering driver in designing a gasification system rather than a combustion system. Chemical synthesis is done via the Fischer–Tropsch (F-T) process. F-T synthesis is an exothermic process that is done catalytically at elevated pressures. While the overall F-T reactions are very complex and numerous, F-T synthesis is generally described by:

$$CO + (1 + m/2n)H_2 \rightarrow (1/n)C_nH_m + H_2O \qquad [9.4]$$

Process parameters, such as reactor pressure, catalyst selection and H_2/CO ratio of the synthetic gas can be tuned in order to create specific products. For example, operating at high reactor pressures will lead to the production of hydrocarbons with a higher chain length, such as gasoline, diesel and waxes. The H_2/CO ratio will also affect the product distribution. In F-T synthesis, higher hydrocarbons are formed from single carbon molecules (CO) and H_2, where H_2 blocks the formation of long hydrocarbon chains. Therefore, higher H_2 concentrations will lead to the formation of smaller hydrocarbons whereas operating a F-T reactor with a feed containing a low H_2/CO ratio will result in the production of longer hydrocarbon chains. Operating with different catalysts will also vary the product distribution. The two types of catalysts that are most commonly used for F-T synthesis are iron and cobalt.[30] Iron promotes the water gas shift (WGS) reaction, which produces H_2 and CO_2 from CO and H_2O:

$$CO + H_2O \leftrightarrow CO_2 + H_2 \qquad [9.5]$$

This reaction is therefore appropriate when the synthesis gas has a low H_2/CO fraction, since H_2 can be produced *in situ*. Cobalt catalysts produce mostly alkanes, with much lower concentrations of olefins and these catalysts produce high concentrations of methane (which is undesirable) at high temperatures.[30–32] Detailed information about catalysts and reaction mechanisms for F-T synthesis can be found in Davis and Occelli.[33]

Since F-T synthesis is used to make many different products, it is desirable to be able to adjust the H_2/CO ratio of the synthesis gas. One way this can be done is by altering the co-reactants that are used during the gasification process. This is illustrated in Fig. 9.8, which shows how changing the amount of CO_2 introduced into a system can alter the H_2/CO ratio of the products. Here we see that with different types of biomass, a range of H_2/CO ratios are achieved by changing the concentration of CO_2. So, for example, if a H_2/CO ratio around 1 is desired, then the gasifier should be at 680°C with 0% CO_2 injection if the feedstock is poplar or at 640°C with 15% CO_2 injection if the feedstock is beachgrass. In F-T synthesis N_2 is inert because it does not react in the temperature regimes of F-T reactions.

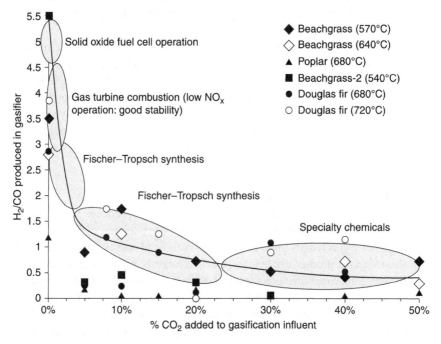

9.8 Influence of CO_2 concentration in gasification on the H_2/CO ratio of the syngas produced.

While some processes can accommodate up to about 15% N_2, it will increase the heat duty for the process plant. The same is true for CO_2, yet CO_2 can participate in some reforming reactions. N_2 is inert, thus it would only serve to absorb heat that would normally be used for the reaction. F-T synthesis is an exothermic reaction, but even with exothermic reactions, a high temperature needs to be maintained in order for the kinetics of the reaction to be sufficiently fast for high production yields. Therefore, it is common for heat to be added to the process to achieve fast reaction kinetics.

Synthetic gas used in fuel cells for electricity production needs the highest amount of H_2, as shown in Fig. 9.8. The synthetic gas can be upgraded by removing the CO to create a pure H_2 stream, which can be used in a proton exchange membrane (PEM) fuel cell. Alternatively, synthesis gas containing CO and H_2 can be used directly in solid oxide fuel cells (SOFCs), which operate at higher temperatures than PEM fuel cells. Fuel cell efficiencies can be as high as 80%.[34] Another application for syngas is in combustion systems. It can be combusted directly to produce heat or combusted in a gas turbine to generate electricity and heat. Integrated gasification combined cycle (IGCC) systems allow for higher energy recovery from the synthesis gas. The type of gasification process selected will depend on product specifications and on the fuel choice (e.g., coal, pet coke, oil, biomass, municipal waste, etc.) and whether the system objectives are electrical power only, electrical power with co-production of hydrogen, power with co-production of syngas, co-production of power with liquid fuels or co-production of power and chemicals.

9.3.2 Alternative products

As discussed above, gasification products are sometimes eventually converted to liquid fuels. Therefore, solid and liquid (for example, tar) components are broken down into gases and then converted back into liquids. Attempts have been made to reduce the number of process steps by converting the waste directly into liquid or solid fuels. Liquid or solid fuel production from MSW has the advantage that the product is storable and transportable. However, the products are very heterogeneous, which can result in variable heat quality, and it can be more difficult to predict or control the final distribution of components in the product. The advantages of syngas are that separation of ash, metals and aromatic compounds is possible and fast reaction times are more easily achieved in the gas phase. Therefore, the applicability of products from direct liquefaction or densification (to create solid fuel) of waste can be limited.[35] However, it is still being considered for treatment of some types of waste.

Refuse-derived fuel (RDF)

MSW can be minimally processed in order to produce a fuel that is suitable for specific applications, and this product is referred to as refuse-derived fuel (RDF).

Some examples of RDF are MSW that has been compacted into pellets, which are used as fuel, MSW with organic fractions removed or MSW with metals and glass removed. RDF has been practically implemented in co-firing applications such as electricity produced by steam from boilers and cement kilns. Combustion of only pelletized RDF in fixed-bed gasifiers may cause problems because of reduced oxygen diffusivity in the material resulting in a tendency for tars to form.[36] In other words, a solid pellet in an oxidative environment will have excess oxygen at the surface but the inside of the pellet will have little or no oxygen available. In the absence of oxygen, solid organic materials will decompose to tars (liquid hydrocarbons) at high temperatures. The difficulty of producing stand-alone solid and liquid products suggests that the direct production of syngas may be the most economical solution. Caputo and Pelagagge reviewed the production of refuse-derived fuel, and analyzed various different process configurations from a technical and economic perspective.[36] They found that mixing tires with MSW-derived RDF was often necessary in order to achieve the required RDF quality.

Bio-oil

Fast pyrolysis involves very rapid heating of the fuel and high flow rates, resulting in reactor residence times of <1 s.[37] Exposing the waste to rapid heating at temperatures around 500°C can produce a pumpable liquid, which can be manufactured ash free or with some solid char particles in the liquid. If the final application for the fuel is combustion, for example for heating applications, then this is primarily a densification procedure, and hence solid particulates in the system are permitted. Pilot and small commercial-scale plants have been built (10–100 ton/day). Flash pyrolysis can also be used to produce bio-oil, which requires a cleaner fuel where the char fraction must be removed. A diagram of this process is shown in Fig. 9.9. Some key design features of fast pyrolysis are summarized in Table 9.1. A summary of pyrolysis production processes as of 2000 has been published by Bridgwater and Peacocke.[38]

Torrefied biomass

An alternative to direct conversion of solid fuel to syngas is torrefaction. In general, torrefaction is used with biomass rather than MSW. Torrefaction involves exposing the biomass to low temperatures (200–300°C) in an inert environment to produce a product that is similar to low-grade char. During this process most of the water will be removed. The resulting product has a higher energy content than raw biomass, is more easily ground and has less moisture uptake.[39] Therefore, torrefaction is appropriate when biomass is to be stored for long periods of time, transported or ground to be fed into a continuous process. The resulting product (torrefied biomass) can be used in similar ways to conventional biomass – either for direct heat, production of syngas or production of liquid fuels.

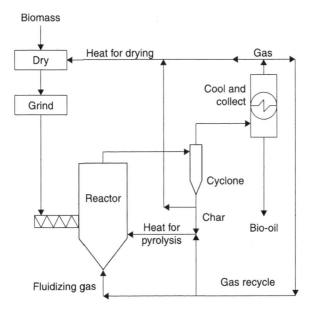

9.9 Overview of flash pyrolysis system.[38]

Table 9.1 Key design features of fast pyrolysis systems[38]

Key fast pyrolysis design features	
Pretreatment	
Feed drying	Essential to ~ 10%
Particle size	Small particles needed. Costly
Washing and additives	For chemicals production
Reactor	
Reactor configuration	Many configurations have been developed, but there is no best one
Heat supply	High heat transfer rate needed
Heat transfer	Gas-solid and/or solid-solid
Heating rates	Wood conductivity limits heating rate
Reaction temperature	500°C maximizes liquids from wood
Product conditioning and collection	
Vapor residence time	Critical for chemicals, less for fuels
Secondary cracking	Reduces yields
Char separation	Difficult from vapor or liquid
Ash separation	More difficult than char separation
Liquids collection	Difficult. Quench and EP seem best

9.4   Process analysis and reactor design

9.4.1   Exergy analysis

The energy distribution of the conversion process can be obtained from the energy balance:

$$\sum_{In} H_j = \sum_{Out} H_k \qquad [9.6]$$

The exergy (available energy) balance takes a similar form, where I is the irreversibility, according to:

$$\sum_{In} E_j = \sum_{Out} E_k + I \qquad [9.7]$$

The difference between an exergy and energy balance is that exergy is not conserved but subject to dissipation. This means that the exergy leaving any process step will always be less than the exergy in. The difference in exergy between all entering and exiting streams is called irreversibility I. Irreversibility represents the internal exergy loss in a process through the loss of quality of materials and energy due to dissipation. The losses of exergy are due to entropy production generated during fluid flow, heat and mass transfer, and chemical reactions. Figure 9.10 shows that MSW and biomass have the lowest available energy that can be extracted either as heat or chemicals. This analysis helps in

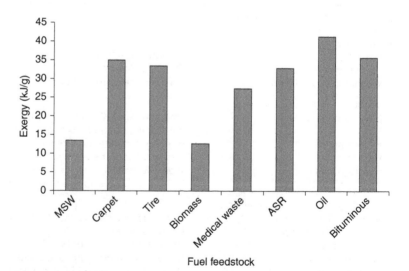

9.10 Comparison of the available energy (exergy) of different feedstocks used for candidate fuels (calculated using the methods described in a paper by Prins et al.[40]). ASR, auto-shredder residue.

understanding the overall efficiency of a MSW gasification system compared to conventional coal gasification systems, which are widely used at much larger scales.

9.4.2 Reactor design for gasification systems

Gasification systems operate under fuel-rich conditions, where ideally just enough oxygen is introduced to oxidize the carbon to CO but not to CO_2; similarly hydrogen should be converted to H_2 but not H_2O. This mixture can lead to a condition where the transport of oxygen to the fuel limits the conversion as discussed before. Therefore, it is very important to have a well-mixed system where the oxygen can quickly mix with the solid fuel while reactions are occurring. In the absence of proper mixing, some of the solid fuel will be completely oxidized, since there will be regions of high oxidizer concentration relative to the solid fuel. The remaining fuel will be in regions where there is insufficient oxygen to convert the carbon to CO, yet the temperature will still be high enough to thermally break down the fuel. This will lead to the formation of intermediates (tars, as discussed above), such as aromatic hydrocarbons, which is not desired. The resulting combination will have too much CO_2 and significant tar and will need further processing.

Gasification reactors are designed to achieve good mixing to avoid these problems. Two prominent gasifier types are fixed beds and fluidized beds, classified by the mobility of the reacting material. In a fixed-bed reactor, the solid material is introduced from the top and flows downward towards a fixed bed of material where it reacts. This is called a downdraft reactor. The co-reactant gas can be introduced either from the top (co-current) or from the bottom (countercurrent). Circulating fluidized-bed reactors, which employ a recycle loop, are the most popular models but they operate at marginal profits when MSW is the feed due to the reduction in throughput from circulation of the gases. The different types of reactors are shown in Fig. 9.11. Reviews by Bridgwater discuss specific reactors used for gasifying wastes.[41,42] Downdraft fixed beds have occupied a large fraction of research expenditures and are the most cost-effective option for smaller operations. Circulating fluidized beds have been studied to a lesser extent and present some advantages at large scales. Fluidized-bed combustors are more appropriate for MSW because they offer higher gas–solid reaction rates, which are necessary to sustain the combustion of low-quality fuel. However, fluidized-bed reactors require a small particle size for efficient operation so fluidized beds are not appropriate for materials that are expensive to shred, such as tires.

Both fixed-bed gasifiers and fluidized-bed gasifiers have lower outlet temperatures (below the ash melting or slagging temperatures) than single-stage entrained gasifiers. The syngas from these gasifiers contains significant methane content, representing typically 10–15% of the carbon content at 400–500 psig. Because of their lower operating temperature, fluidized-bed gasifiers have lower

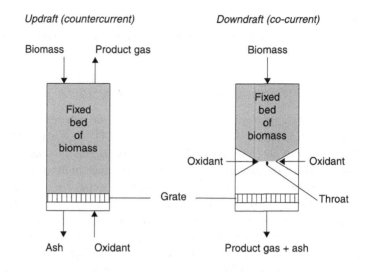

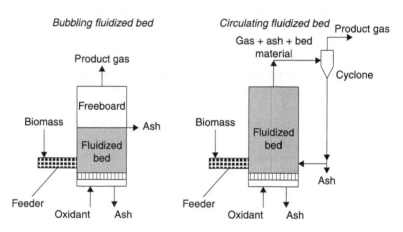

9.11 Different types of reactors used for gasification.[41]

carbon conversion and their ash contains more carbon than the slag from other slagging gasifiers.

Other process parameters, such as space velocity, are important in determining the types of products that will be produced. For example, low space velocities can lead to high amounts of tar formed in the reactor whereas higher space velocities can lead to charcoal dusting. These issues, as well as considerations for scale-up of gasifiers, are discussed in detail later in this chapter when looking at the InEnTec process.

Since gasification involves the introduction of an oxidant in sub-stoichiometric quantities, introduction of all of the oxidant at once can result in complete oxidation of some of the fuel while the rest of the fuel has no access to oxygen and tars will form. In order to minimize this effect, some gasifier designs use two stages of oxidant introduction (in other words, part of the oxidant is introduced at the inlet and more oxidant is introduced further downstream while maintaining an overall oxygen-lean environment) to improve gasifier cold gas efficiency. In addition, this reduces the sensible heat in the product gas and lowers the oxygen requirements. Similarly, fuel can be introduced in a staged fashion. In a two-stage entrained gasifier, the material fed to the second stage reduces the outlet temperature and there is some methane survival in the syngas. The methane content will increase if a higher proportion of the fuel is fed to the second stage. It should be noted that entrained-flow gasifiers have been selected for the majority of commercial-sized IGCC applications that operate on coal. A major advantage of the high-temperature entrained-flow gasifiers is that they avoid tar formation. An entrained-flow gasifier produces little methane and is very suitable for hydrogen and synthesis gas production. The high reaction rate also allows single gasifiers to be built with large gas outputs, sufficient to fuel large commercial gas turbines.

9.4.3 Considerations for reactor scale-up

During scale-up, many issues need to be considered, for example physical parameters such as mass and heat transport and mixing of feed species. Dependencies on inlet and reactor temperatures and varying feed component concentrations have to be understood. These relationships allow for accurate prediction of performance at any size, provided the reactor operates within the residence time, temperature and concentration ranges previously studied. When scaling up the reactor, the transport characteristics change, albeit in a predictable manner. For example, in a smaller-scale reactor the surface area to volume ratio is large, therefore in an exothermic reaction, a reactor can lose substantial amounts of heat and measures must be taken to ensure heat loss is minimized. Conversely, in a large, full-scale reactor the surface area to volume ratio is much smaller, therefore it would be expected that the heat loss will be much less compared to the smaller-scale reactor. Precautions may need to be taken then to ensure heat can be removed at an adequate rate to prevent the reactor temperature from exceeding a predetermined value.

Another concern when scaling up a reactor is the ease of mixing the reactants prior to entering the reactor. In a small-scale experiment, the reactants are usually considered to be fully mixed prior to entering the reactor and usually that is the case. In a large-scale reactor, the reactants must mix over a bigger volume compared to their introduction points. If proper measures are not taken to ensure the reactants are sufficiently mixed prior to entering the reactor, hot spots can occur within the reactor or parts of the reactor may not be efficiently used (dead zones). Reactant mixing and reaction zones are also influenced by the final full-scale geometry, therefore the closer the final design is to the pilot-scale the better.

An important reactor parameter to consider is the superficial velocity (SV), which is defined by:

$$SV = \frac{\text{Gas production rate}}{\text{Cross-sectional area}} \qquad [9.8]$$

where the SV is measured in $m.s^{-1}$, the gas production rate in $m^3.s^{-1}$ and the cross-sectional area in m^2. The SV is independent of gasifier size, and so permits comparison of gasifiers of very different dimensions. It is easily estimated or measured from the air/oxygen and fuel throughput and gasifier dimensions. Superficial velocity is very important as it is the primary driver, all other parameters remain constant, to achieve the desired performance of the system. Further, it controls the rate at which air, then gas, passes down through a gasifier. This in turn exercises a primary effect on heat transfer around each particle during gasification of the feedstock and the pressure drop through the system. Table 9.2 gives some representative SV values for recent gasifiers.

Table 9.2 Space velocity values for different gasifier systems

Gasifier type	Pyrolysis zone SV $(m.s^{-1})$
Imbert	0.63
Biomass Corp.	0.24
SERI, air	0.28
SERI, oxygen	0.24
Syngas, air	1.71
Syngas, oxygen	1.07
Buck Rogers (Chern)	0.23
Buck Rogers (Wallawender)	0.13

9.5   Process modifications for gasification systems

9.5.1   High-pressure operation for CO_2 capture

In some cases, it is desirable to remove the CO_2 after gasification since, as mentioned before, the presence of inert compounds in the syngas stream will lead to increased compression and heating costs. CO_2 removal can be done by physical or chemical absorption methods. If physical absorption is used, the amount of CO_2 absorbed into the liquid phase depends on the concentration in the gas phase and the thermodynamic properties of that specific gas–liquid mixture. This relationship can be described using Henry's law:

$$y_i P = x_i H \qquad [9.9]$$

where y_i is the gas-phase mole fraction, P is the pressure, x_i is the liquid-phase mole fraction and H is Henry's constant.

For these processes, operating at higher pressures will give increased CO_2 dissolution into the liquid phase, hence operating the gasifier at higher pressures can be advantageous. Two common methods for this are the Rectisol process and the Selexol process. Another method of CO_2 capture is by chemical absorption, which is more appropriate for gas streams where the CO_2 has a low partial pressure. In this process, the CO_2 chemically reacts with a compound, which must then be regenerated in a stripper at high temperatures, which comes with an energy penalty. The most common chemical absorbent is monoethanolamine (MEA). The regeneration energy is highly dependent on process conditions such as CO_2 concentration and relative flow rates of solvent, but to give an idea of the scale, regeneration energies have been reported to be in the range of 4–5 GJ/tCO$_2$$^{-1}$ with MEA.[43,44]

9.5.2 Hydrothermal treatment

Some types of waste, for example sewage sludge or waste with high concentrations of biomass, have high moisture content. Because of the high cost of evaporating so much water, it is not economical to process this type of waste in combustion, gasification or pyrolysis systems. One way to treat waste with high moisture content is through hydrothermal treatment, which can produce liquid fuels directly from waste. Hydrothermal treatment exposes the high moisture material to elevated temperatures and pressures, which remove trapped bulk water from within the structure of the material. This process typically takes place at temperatures between 280–370°C and pressures of 10–25 MPa.[45] Catalysts can be used to enhance conversion.[46] Hydrothermal treatment is relatively simple and can be envisioned as producing pressurized dilute slurry, which is heated to temperatures of 200°C or higher for ten minutes or less. When heated under pressure to the final temperature, structural groups attached to the chemical structure decompose to form carbon dioxide, which helps to force liquid water out of pores and into the carrier medium. The exposure to the extreme conditions also causes both chemical and physical changes that significantly reduce the propensity for the substance to reabsorb the removed moisture, a vast improvement over other drying techniques.

Work on sewage sludge has shown that a pumpable 35–45% solids mixture can be produced by hydrothermal treatment without deleterious side effects (odor, bacteria, etc). The ash, nitrogen, sulfur and energy content were similar on a dry basis to the untreated feed slurry. Hammerschmidt et al. are working on a process for catalytic hydrothermal conversion of waste sludge into biocrude oil.[47] The process is compared to conventional processes for production of biodiesel and

Table 9.3 Characteristics of biomass treatment processes. The present process is catalytic hydrothermal conversion[47]

	Present process	Biodiesel	Bioethanol
Product	Biocrude oil	Biodiesel	Ethanol
Feed	Waste biomass	Rapeseed	Sugar or starch-rich plants
Competition with food production	No	Yes	Yes
Art of the process	Thermo-chemical	Chemical	Fermentation
Mean residence time	10 min	45 min	72–96 h, can be as long as 18 d
Process temperature	<350°C	<45°C	<90°C
Process pressure	25 MPa	0.1 MPa	0.1 MPa
Complexity of the process	Low	Middle	High
Stage of development	R&D work	Widely used	Widely used

bioethanol in Table 9.3. Yoshikawa and colleagues are developing a process for hydrothermal treatment of waste where mixed MSW is converted into a uniform powder, making it easier to process.[48] Texaco and Ameritech are working on hydrothermal processing technologies.

9.5.3 Fuel types

Table 9.4 shows the equilibrium gas phase composition for different feedstocks when they are gasified in an air/steam environment. The calculations for different fuel feeds were performed on the basis of maintaining an operating temperature of 1200°C and the concentrations of steam and oxygen were adjusted for each feedstock in order to achieve this temperature at atmospheric pressure. It is clear from Table 9.4 that the composition of the feedstock has a large impact on the product composition and importantly on the ratio of hydrogen to carbon monoxide.

Table 9.4 Product gas mix for gasification of various types of feedstock in vol% (ASR: auto-shredder residue)

	CO	H_2	CO_2	CH_4	H_2O	HCL	H_2S	H_2/CO
MSW (typical)	41.0	33.7	13.8	4.1	6.3	0.13	0.13	0.82
Carpet	33.2	43.1	6.8	8.8	4.9	0.02	0.03	1.30
Tire	56.9	18.9	1.5	22.2	0.3	0.04	0.00	0.33
Medical waste	27.9	37.8	18.2	1.8	13.7	0.03	0.65	1.36
ASR	29.8	37.4	17.3	2.1	12.0	0.00	0.64	1.26
Oil	48.8	25.6	2.2	21.1	0.6	1.61	0.00	0.52
Bituminous	55.9	23.9	4.1	12.8	1.0	1.71	0.00	0.43

Potential advantages of co-gasification

Combustion systems convert essentially all of the fuel into CO_2, H_2O and heat whereas gasification systems only partially oxidize the fuel, creating intermediates, such as CO, H_2 and hydrocarbons. Therefore, in combustion systems, understanding the reaction pathways is less important than it is for gasification systems, since intermediates rarely exit the combustor. As a result, gasification systems are designed to achieve specific desired products while achieving high fuel conversion and minimal energy input to the gasifier. One way to modify the gasification reaction is to introduce co-reactants. There are systems that attempt to co-gasify waste with coal. The combined gasification of waste (or biomass) and coal provides several potential advantages over the separate gasification of each feedstock, such as:

- Reduction in specific investment costs, which are very high for the small size, stand-alone waste gasifiers presently operating or under development (typically in the 3000–5000 $.kWe^{-1} range).
- There are fewer problems related to seasonality and availability of the fuel.
- Possibility of exploiting synergies due to the properties of the two fuels. A study of the comparative thermodynamic efficiency of coal and biomass, for example, concluded that biomass was less efficient than coal but co-processing could increase the overall efficiency compared to using separate processing.[40]

Trippe *et al.* report that co-gasification of biomass and coal reduces syngas production costs by 50%.[49] Seo *et al.* observed synergistic effects when co-pyrolyzing RDF with coal.[50] An example of the increase in syngas production by co-gasifying coal and biomass is shown in Fig. 9.12.

9.6   Environmental effect of gasification

As mentioned, one benefit of gasification is the production of a lower volume of gas, which allows for smaller gas cleanup equipment, and the ability to retain some of the impurities (metals and minerals) in the solid phase. Some compounds, such as sulfur and chlorine, will still be converted to gas-phase products and will need to be removed, but other metals and minerals will remain as solid ash and are therefore easy to separate. Some waste products, for example plastics, have high chlorine content, whereas others, such as scrap tires, have low sulfur and chlorine content. Therefore, the pollution control systems need to be designed based on the feedstock and operating conditions for a given process. A report published in 2009 by the University of California, Riverside, investigated the emissions produced by many different gasification facilities, including fluidized-bed, plasma arc, pyrolysis and high-temperature gasification systems. For all of the systems, the following emissions were reported: particulate matter (PM), hydrogen chloride (HCl), nitrous oxides (NO_x), oxides of sulfur (SO_x), mercury (Hg), dioxins and

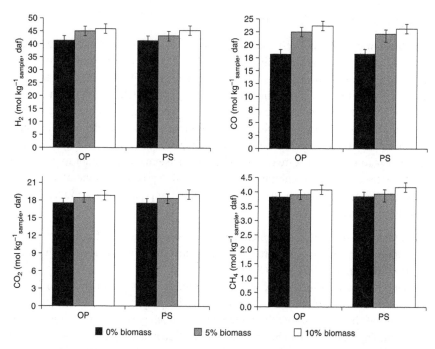

9.12 Effect of co-gasification of biomass with coal. 0% indicates gasification of coal only. Addition of biomass increases the amount of gas-phase products produced per kilogram of fuel, indicating that synergistic effects are present.[51] (OP: olive pulp, PS: pine sawdust, daf: dry ash-free.)

furans.[52] Dioxins and furans are hazardous compounds, which are composed of benzene rings with oxygen and chlorine compounds. They are formed at a temperature of 250–700°C with a maximum formation rate around 315°C. These compounds are typically formed downstream of the reactor when oxygen reacts with HCl, which is formed in the gasifier, to produce chlorine gas, which can react with the aromatic radical compounds in the flue gas. Since there is generally little or no oxygen in the flue gas in gasification or pyrolysis systems, the production of dioxins and furans is significantly reduced compared to combustion systems.[52] Chapter 10 discusses in detail the regulatory issues surrounding pollutant emissions from WTE plants and some existing technologies for pollutant removal. The range of emissions reported from various gasification systems are shown in Table 9.5. The major economic barrier to gasification is the cost of cleaning the syngas to remove acid and small particles so the gas can be used in turbine-powered electric generators or as a chemical feedstock.[53]

In the first part of this chapter we introduced gasification as a method of creating synthesis gas from solid fuel. Synthesis gas can be converted into gaseous or liquid fuels or chemicals, burned to produce heat or electricity, or upgraded to

Table 9.5 Emissions from gasification systems (data taken from University of California, Riverside[52])

Compound	Range (approximate)	Units
PM	1–18	mg.Nm^{-3} at 7% O$_2$
HCl	<5–55	mg.Nm^{-3} at 7% O$_2$
NO$_x$	<10–255	mg.Nm^{-3} at 7% O$_2$
SO$_x$	<1–50	mg.Nm^{-3} at 7% O$_2$
Hg	<0.008	mg.Nm^{-3} at 7% O$_2$
Dioxins and furans	~0–0.1	ng TEQ.Nm^{-3} at 7% O$_2$

produce pure hydrogen. This offers much more flexibility than a conventional WTE process, which produces electricity and heat. Unlike conventional combustion, where nearly all of the fuel is converted to CO$_2$ and H$_2$O, by-products such as tars are produced. The decomposition of tars is one of the main barriers to the commercialization of gasification technologies. We introduced some alternatives to gasification, such as the production of RDF, torrefied biomass or liquid fuels via fast pyrolysis. Exergy was discussed as a concept for evaluating the available energy in a fuel, accounting for losses during conversion processes. Reactor design is a key part in gasification systems, and gasifiers can be fixed or fluidized beds, updraft or downdraft, and may include staged introduction of either fuel or oxidant. Process modifications such as hydrothermal treatment, high-pressure gasification coupled with CO$_2$ capture and co-gasification of coal with waste or biomass were discussed. Finally, typical emissions profiles for gasification systems were presented.

In the next section we look at technologies that are in operation for the conversion of waste or biomass to gaseous or liquid fuels. We will introduce the theory behind the technologies and the statuses of various operations. This section is intended to provide an overview of the state of the industry at the time of publication of the book.

9.7 Technologies in operation

9.7.1 Thermoselect gasification

The Thermoselect process has a standard design developed for a typical two-line system that has a nominal capacity of 528 tons per day (tpd). There are many units operating in steel-producing facilities and municipalities in Japan. However, most systems in operation use about 300 tpd. The Thermoselect system gasifies MSW as received and thus does not require pre-processing. There is a pre-gasification chamber, which also serves as a lock-hopper type system that introduces the fuel into the main gasifier. The main gasification section uses oxygen lances to fully convert the entire organic fraction to syngas, leaving a molten slag to exit the bottom in a homogenizer. The syngas is quenched via a water spray prior to entering the gas cleanup system. Because of the gasification reactions and the quenching of the high-temperature syngas, the Thermoselect process has an inherent loss of about $400\,kWh.ton^{-1}$ in the conversion of MSW to syngas. Two other energy consumers in the Thermoselect system are natural gas for the homogenizer, and in the production of the oxygen needed for the process (about $100\,kWh.ton^{-1}$). By comparison, calculations based on the data provided by JFE show that the net electricity available for sale from a 125 tpd unit would be about 650 kWh using MSW with a lower heating value (LHV) of $12.7\,MJ.kg^{-1}$, and 450 kWh for MSW with LHV $10.6\,MJ.kg^{-1}$. These values are in the range of net electricity generation from large WTE facilities combusting a similar quality and quantity of MSW. A diagram of the process is shown in Fig. 9.13.

9.7.2   PROLER Syngas system

Another gasification plant operating is the PROLER Syngas system. This system was originally developed to gasify auto-shredder residue but has had limited operation with MSW. The feedstock must be pre-shredded to about 2 inches prior to entering the gasifier. The products are syngas and vitrified by-products suitable for landfill. The only demonstration unit has been a 48 tpd plant in Houston, TX, USA. The feedstock is introduced to a dual-ram feeder that

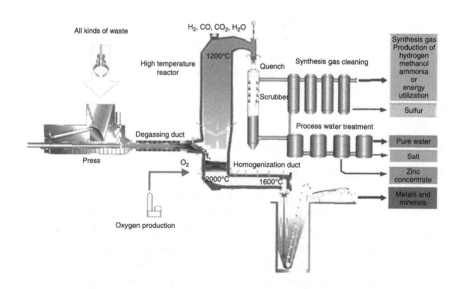

9.13 Thermoselect process (used with permission from Thermoselect[54]).

pushes the waste into a refractory-lined horizontal vessel rotating at between 1 and 2 rpm. Natural gas and oxygen are introduced through burners to maintain a temperature of about 850°C. The produced gas and solids exit into a hot pneumatic separator (HPS), which is basically a series of baffles that reverse the flow several times. At the points of flow reversal, solids drop out by impaction and momentum. The plant has demonstrated a gas product with a heating value of about 278 Btu.ft^{-3} (10.35 MJ.m^{-3}). This is similar to the value of 239 Btu.ft^{-3} targeted for IET's 125 tpd plant. The capital cost for a two-line facility (a facility that has two units installed) has been reported by Niessen *et al.* and is shown in Table 9.6.

The size of a reactor (surface area/volume and formation of the hot zone relative to the volume and proximity to the sides) governs the generation, maintenance and control of the hot zone. That is the hot zone needs to occupy a certain portion of the reactor volume, while low temperatures (<600°C) need to be maintained at the reactor walls. There is no theoretical limitation on the gasifier or the thermal residence chamber. There are practical limitations on how the shredded fuel can

Table 9.6 Economic overview of a 1247 tpd two-line RDF gasification facility[55]

Facility capital investment			Source
Fuel preparation		$52 000 000	CDM
Process/heat recovery/ APC train:			
Feeding system	$ 1 500 000		Developer
Reactor equipment	28 000 000		Developer
Product gas treatment	12 600 000		Developer
Water treatment	225 000		Developer
CEM system	2 000 000		CDM
Process core cost	$44 325 000		
Engineering and contingency (30% of process core)	13 300 000		CDM
Subtotal		57 625 000	
Electrical generation (gas turbine)		44 000 000	CDM
Total		$153 625 000	
		per Mg/d MSW:	$123 200
		per t/d MSW:	$ 112 100

System: 1247 Mg/d (1370 t/d) MSW, 872 Mg/d (960 t/d) RDF, two-line furnace/boiler systems. Air pollution control (APC): hot, pneumatic separator; dry cyclone; scrubber; fabric filter.

be introduced evenly and how the oxygen addition can be done uniformly over very large volumes.

9.7.3   Plasma-assisted gasification

As discussed above, while gasification can be an exothermic process overall, it requires heat input to initiate and sustain the reactions, since they typically take place at low temperatures. Conventional gasification uses combustion or direct electricity for heat. Heat can also be supplied using a plasma torch, which generates heat via the passage of an electric current through a gas flow. Some advantages of plasma gasification are that very high temperatures can be achieved, allowing for high heat transfer rates. Also, installations of plasma systems can be smaller compared to combustion systems, and startup and shutdown procedures are fast. While electricity can give good control of a gasification unit's temperature, it is an expensive form of energy, which is currently one of the limiting factors in making plasma gasification competitive with completely combustion-based processes. Some companies that are currently using or developing plasma gasification technologies are Westinghouse Plasma (owned by AlterNRG), Plasco Energy Group, Europlasma and InEnTec (owned by Waste Management Inc.). A demonstration plant was reported by Byun *et al.* in the Republic of Korea, which can process 10 tpd of MSW. This facility produces $287 \, \mathrm{Nm^3.ton_{MSW}^{-1}}$ of H_2 and $395 \, \mathrm{Nm^3.ton_{MSW}^{-1}}$ of CO. It uses $1.14 \, \mathrm{MWh.ton_{MSW}^{-1}}$ and $7.37 \, \mathrm{Nm^3.ton_{MSW}^{-1}}$ of liquefied petroleum gas.[56]

Plasma can be used in the primary gasification chamber or can be used in combination with traditional gasification. Europlasma and Plasco use plasma gasification in conjunction with traditional gasification. In these systems, recycled process heat is used to drive the gasification of the raw MSW and then the gases from this process (synthesis gas with many impurities such as organic hydrocarbons) pass through a plasma-heating zone, which decomposes the organics into synthesis gas. The plasma torch is also used to melt the slag in order to produce an inert material, which can then be disposed of safely, or used in construction applications. Both Europlasma and Plasco use the synthesis gas to produce electricity in a gas turbine. Heat recovery from the turbines via steam generation can also be integrated. AlterNRG uses plasma to directly heat the MSW. The plasma torches are located at the bottom of the gasifier and the MSW is introduced at the top and is heated as it falls down.

9.7.4   Ethanol and methanol production from MSW via gasification

Enerkem is a Canadian-based company, which produces ethanol and methanol from gasification of solid waste. Ethanol can be readily integrated into transportation fuels and is the primary component of E85, which is widely used as automotive fuel. Methanol is considered a building block for many different fuels and chemicals. The Enerkem process uses a bubbling fluidized-bed gasifier, with air as the oxidizer to create syngas. At the time of publication of this book, Enerkem has one commercial facility in operation and is building two more. The Enerkem facility in Quebec, Canada, which has been in operation since 2009, uses treated wood from electricity poles as a feedstock to produce biomethanol. This facility has a capacity of 1.3 million gallons per year. The facilities that are under development are in Alberta, Canada, and Mississippi, USA, and will accept MSW for conversion into ethanol and methanol. These facilities are projected to produce 10 million gallons of fuel per year.

9.7.5 Production of liquids from waste sludge – SCF's CatLiq® Process

SCF Technologies, which is based in Denmark, has developed a catalytic process for converting waste with high moisture content into liquid fuels. The process is designed to accept sludge from wastewater treatment plants, manure from livestock production, algae, as well as residues from food processing or production of ethanol or biodiesel. The CatLiq® process takes place in the supercritical regime where high pressures assist the breakdown of organic matter into liquid oils. The product has a heating value of approximately $35\,MJ.L^{-1}$. At the time of the publication of this book the technology has not yet been commercialized.

9.7.6 Other technologies

There are a number of other companies who currently have waste gasification technologies in operation or under development. Ebara's TwinRec technology is a circulating fluidized-bed gasifier with a bed temperature around 550°C. The synthesis gas is combusted to generate electricity, the metal is recycled, and the ash is vitrified to produce a stable glass granulate. Between 1995 and 2007, 13 facilities were commissioned in Japan, with capacities ranging from 7 tpd to 550 tpd and thermal capacities ranging from 1.5 to 82.8 MW. Most of these facilities are still in operation. A diagram of the process is shown in Fig. 9.14.

Entech has developed a process to gasify waste to produce syn-oil by pyrolysis or syngas by low- or high-temperature gasification. The syngas is fired into a boiler to produce steam for electricity generation. Entech has facilities in Poland (waste capacity of 15 tpd, thermal capacity of 5.6 MWt), Malaysia (waste capacity 60 tpd, thermal capacity of 6.9 MWt), Taiwan (waste capacity 30 tpd, thermal capacity of 3.5 MWt), Singapore (waste capacity 72 tpd, thermal capacity of 5.8 MWt) and Republic of Korea (waste capacity 20 tpd, thermal capacity of 5.7 MWt). OE Gasification is a company based in Waterloo, Canada, which provides small-scale gasification systems. Each module can accept between 3500–7500 tonnes of waste per year with a thermal output of 1.5–2.5 MWt. These

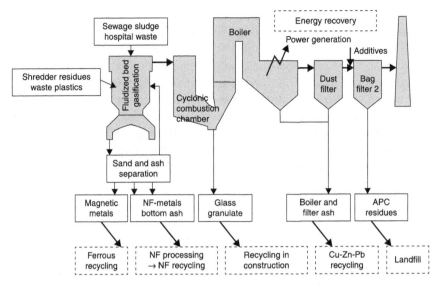

9.14 Process diagram of the Ebara TwinRec gasification process.[57]
(APC: air pollution control, NF: non-ferrous)

systems can be stacked in order to accommodate varying amounts of throughput. The system is designed to accept primarily organic waste and can accommodate waste with a moisture content of up to 60% by weight. OE Gasification has facilities in operation in Norway and Republic of Korea (both facilities have a waste capacity 20 tpd and thermal capacity of 2 MWt). Other gasification facilities that are either in operation or expected to be in operation soon include Mitsui R21, INEOS Bio, Nippon Steel DMS, Taylor Biomass Energy, Energos (small-scale gasification facilities in the UK), Chinook Energy (Europe and around the world) and Ze-Gen, which uses liquid metals to transport the heat in the gasification processes (one facility in New Bedford another one to be built in Attleboro).

9.8   Conclusion

This chapter has presented gasification as an alternative to combustion for waste conversion. Unlike combustion, which completely converts the fuel to CO_2 and H_2O, gasification produces partially-oxidized products (CO and H_2). This enables the production of many different fuels or chemicals whereas combustion produces heat and/or electricity. Some challenges associated with gasification are the production of by-products such as tar and ash, and the need for advances in F-T synthesis processes for high conversion of synthesis gas to the desired fuels. Current gasification technologies include CO_2 or steam-assisted gasification, plasma-assisted gasification, supercritical gasification, co-gasification of MSW with coal, hydrothermal treatment, and fast pyrolysis.

9.9   References

1   Carpenter D.L., Bain R.L., Davis R.E., Dutta A., Feik C.J., Gaston K.R., *et al.*, *Industrial & Engineering Chemistry Research*, 49 (2010) 1859–1871.

2   Aksoy B., Cullinan H., Webster D., Gue K., Sukumaran S., Eden M., *et al.*, *Environ. Prog. Sustain. Energy*, 30 (2011) 720–732.

3   Erlich C., Fransson T.H., *Appl. Energy*, 88 (2011) 899–908.

4   Ghani W., Moghadam R.A., Salleh M.A.M., Alias A.B., *Energies*, 2 (2009) 258–268.

5   Nipattummakul N., Ahmed I., Kerdsuwan S., Gupta A.K., *Appl. Energy*, 87 (2010) 3729–3734.

6   Chen C., Jin Y.Q., Yan J.H., Chi Y., *Journal of Zhejiang University-Science A*, 11 (2010) 619–628.

7   Velghe I., Carleer R., Yperman J., Schreurs S., *J. Anal. Appl. Pyrolysis*, 92 (2011) 366–375.

8   Lupa C.J., Ricketts L.J., Sweetman A., Herbert B.M.J., *Waste Management*, 31 (2011) 1759–1764.

9   Toledo J.M., Aznar M.P., Sancho J.A., *Industrial & Engineering Chemistry Research*, 50 (2011) 11815–11821.

10   Kim J.W., Mun T.Y., Kim J.O., Kim J.S., *Fuel*, 90 (2011) 2266–2272.

11   Bridgwater A.V., *Chemical Engineering Journal*, 91 (2003) 87–102.

12   Castaldi M.J., Dooher J.P., *Int. J. Hydrog. Energy*, 32 (2007) 4170–4179.

13   Butterman H.C., Castaldi M.J., *Environmental Science & Technology*, 43 (2009) 9030–9037.

14   Abu El-Rub Z., Bramer E.A., Brem G., *Industrial & Engineering Chemistry Research*, 43 (2004) 6911–6919.

15   Gomez-Barea A., Nilsson S., Barrero F.V., Campoy M., *Fuel Processing Technology*, 91 (2010) 1624–1633.

16   Narvaez I., Orio A., Aznar M.P., Corella J., *Industrial & Engineering Chemistry Research*, 35 (1996) 2110–2120.

17   Arena U., Zaccariello L., Mastellone M.L., *Waste Management*, 29 (2009) 783–791.

18   Kinoshita C.M., Wang Y., Zhou J., *J. Anal. Appl. Pyrolysis*, 29 (1994) 169–181.

19   Devi L., Ptasinski K.J., Janssen F., *Biomass & Bioenergy*, 24 (2003) 125–140.

20   Klinghoffer N., Castaldi M.J., Nzihou A., in: *19th Annual North American Waste to energy Conference (NAWTEC19) ASME*, Lancaster, PA, 2011.

21   El-Rub Z.A., Bramer E.A., Brem G., *Fuel*, 87 (2008) 2243–2252.

22   Chaiwat W., Hasegawa I., Mae K., *Industrial & Engineering Chemistry Research*, 49 (2010) 3577–3584.

23   Furusawa T., Tsutsumi A., *Applied Catalysis A: General*, 278 (2005) 195–205.

24   Li D.L., Wang L., Koike M., Nakagawa Y., Tomishige K., *Applied Catalysis B: Environmental*, 102 (2011) 528–538.

25   Constantinou D.A., Efstathiou A.M., *Applied Catalysis B: Environmental*, 96 (2010) 276–289.

26   Noichi H., Uddin A., Sasaoka E., *Fuel Processing Technology*, 91 (2010) 1609–1616.

27   Wang D., Yuan W.Q., Ji W., *Appl. Energy*, 88 (2011) 1656–1663.

28   ENTECH, www.entech-res.com/wtgas, accessed 13 January 2012.

29   Vitasari C.R., Jurascik M., Ptasinski K.J., *Energy*, 36 (2011) 3825–3837.

30   Eliseev O.L., *Russ. J. Gen. Chem.*, 79 (2009) 2509–2519.

31   Raje A.P., Davis B.H., *Catalysis Today*, 36 (1997) 335–345.

32   Fernandes F.A.N., *Industrial & Engineering Chemistry Research*, 45 (2006) 1047–1057.

33   Davis B.H., Occelli M.L., *Advances in Fischer-Tropsch Synthesis, Catalysts, and Catalysis*, C. Press (Ed.), CRC Press, Taylor & Francis Group, Boca Raton, 2009.

34   Heck R.M., Farrauto R.J., *Catalytic Air Pollution Control*, Wiley-Interscience, New York, 2002.

35   Cunliffe A.M., Williams P.T., *Energy and Fuels*, 13 (1999) 166.

36   Caputo A.C., Pelagagge P.M., *Applied Thermal Engineering*, 22 (2002) 423–437.

37   Vamvuka D., *International Journal of Energy Research*, 35 (2011) 835–862.

38   Bridgwater A.V., Peacocke G.V.C., *Renewable & Sustainable Energy Reviews*, 4 (2000) 1–73.

39   Ciolkosz D., Wallace R., *Biofuels Bioprod. Biorefining*, 5 (2011) 317–329.

40   Prins M.J., Ptasinski K.J., Janssen F., *Energy*, 32 (2007) 1248–1259.

41   Bridgwater A.V., *Fuel*, 74 (1995) 631.

42   Bridgwater A.V., *Chemical Engineering Journal*, 91 (2003) 87.

43   Mangalapally H.P., Hasse H., *Chem. Eng. Res. Des.*, 89 (2011) 1216–1228.

44   Rackley S.A., Absorption capture systems in *Carbon Capture and Storage*, Butterworth-Heinemann, 2010. Online version available at http://www.knovel. com/web/portal/browse/display?_EXT_KNOVEL_DISPLAY_bookid= 2739&VerticalID=0.

45   Toor S.S., Modelling and Optimization of Catliq® Liquid Biofuel Process, PhD thesis Department of Energy Technology, Aalborg University, Aalborg, 2010.

46   Azadi P., Farnood R., *Int. J. Hydrog. Energy*, 36 (2011) 9529–9541.

47   Hammerschmidt A., Boukis N., Hauer E., Galla U., Dinjus E., Hitzmann B., *et al.*, *Fuel*, 90 (2011) 555–562.

48   Lu L., Namioka T., Yoshikawa K., *Appl. Energy*, 88 (2011) 3659–3664.

49   Trippe F., Frohling M., Schultmann F., Stahl R., Henrich E., *Fuel Processing Technology*, 92 (2011) 2169–2184.

50   Seo M.W., Kim S.D., Lee S.H., Lee J.G., *J. Anal. Appl. Pyrolysis*, 88 (2010) 160–167.

51   Fermoso J., Arias B., Gil M.V., Plaza M.G., Pevida C., Pis J.J., *et al.*, *Bioresource Technology*, 101 (2010) 3230–3235.

52   Evaluation of Emissions from Thermal Conversion Technologies Processing Municipal Solid Waste and Biomass, report, University of California, Riverside, 2009.

53   Belgiorno V., Feo G.D., Rocca C.D., Napoli R.M.A., *Waste Management*, 23 (2003) 1–15.

54   Thermoselect (2003). Available from: www.thermoselect.com/index.cfm?fuseaction =Verfahrensuebersicht&m=2, accessed 16 December 2012.

55   Niessen W.R., Marks C.H., Sommerlad R.E., Evaluation of Gasification and Novel Thermal Processes for the Treatment of Municipal Solid Waste, report, National Renewable Energy Laboratory, Golden, 1996.

56   Byun Y., Namkung W., Cho M., Chung J.W., Kim Y.S., Lee J.H., *et al.*, *Environmental Science & Technology*, 44 (2010) 6680–6684.

57   Selinger A., Steiner C., Shin K., in: *Waste 2003: 4th Int. Symposium on Waste Treatment Technologies*, Sheffield, 2003.

Part III
Pollution control systems for waste to energy technologies

10

Transformation of waste combustion facilities from major polluters to pollution sinks

J. VEHLOW, Karlsruhe Institute of Technology, Germany

DOI: 10.1533/9780857096364.3.179

Abstract: This chapter describes the development of waste combustion in terms of its environmental compatibility. Starting in the 1980s, public opinion, legislation and regulations, and technological advances have resulted in significantly reduced air emissions, which have had an almost negligible impact on health and the environment. The improvement stages are discussed as examples for the United States of America, Japan, Germany and the Netherlands. Due to their importance in terms of public perception, dioxins are a special focus and their formation as well as technologies for their abatement are discussed in more detail.

Key words: legislative emission regulations, air pollution control, de-dusting techniques, wet and dry scrubbing, dioxin formation, dioxin abatement.

10.1 Introduction

Although landfill is worldwide still the prevailing waste disposal strategy, and scavenging on landfills is common practice in many poor countries even today, in the early 16th century, when the Black Death devastated Europe, the medical doctor Johannes Bökel suggested that the epidemic might have originated from a lack of hygiene (Frilling and Mischer, 1994). His assumption caused some European cities to reorganize their waste disposal strategies. However, his statement was only taken seriously in the second half of the 19th century when waste disposal became a big problem in the densely populated and highly industrialized regions in the United Kingdom (UK).

At that time a campaign for the total destruction of germs in waste by combustion was started in London and Manchester. The first attempts in 1870 to burn waste in closed furnaces in Paddington, a London ward, were not successful. In 1874 Alfred Fryers erected the first 'destructor' in Manchester and this type of furnace was in use for more than 30 years (Encyclopedia Britannica, 1910). The emissions from these plants, especially the odor caused by the poor burnout of the flue gas, were a nuisance for neighbors. Nevertheless, medical doctors tried to convince complaining citizens that the smell from waste combustors was far less hazardous to their health than the handling and disposal of raw waste (de Fodor, 1911). So combustion of waste – in Europe generally called waste incineration – became common in the UK and within 30 years, 250 of these destructors were built.

179

The second country to use waste combustion was the United States of America. The US Army built the first facility on Governors Island in New York Harbor in 1885; in the same year the first public combustor started operation in Allegheny, PA, and other cities followed quickly (Hickmann, 2003). By 1910, approximately 300 combustors had been built in the USA.

In 1893 the first Japanese waste combustor was built in Tsuruga and here too, this strategy of waste disposal expanded with time. Surprisingly late, considering the geographical situation, waste combustion was implemented on the European continent as a consequence of a cholera epidemic in Hamburg, Germany, in 1892. The first combustor started regular operation in 1896 in that city (Zwahr, 1996). Very soon many other Central European cities followed this example and shifted to waste combustion; in 1900 there were more than 60 plants in operation across continental Europe.

Almost all of these first waste combustors generated power, some, like the one in Frederiksberg, a district of Copenhagen, Denmark, produced combined heat and power (CHP) by also delivering heat to a nursery (Kleis and Dalager, 2004). In contrast to this rather modern-looking practice, environmental measures were not even thought about. The destructors were batch operated with manual feeding, their combustion performance was poor and no measures for air emission control existed. The odor nuisance from unburned matter and high levels of dust were the reasons that the neighbors started complaining, and after 1910 many of these first-generation waste combustors were shut down in Europe as well as in the USA. The aftermath of World War I, when the widespread poverty reduced the calorific value of the household waste almost everywhere to extremely low values, prompted almost all European countries to turn away from waste combustion.

This tendency was slowly reversed by the development of new continuously operated furnaces with mechanical feeding in Europe at the end of the 1920s and the 1930s. For example, in 1926 Josef Martin received a German patent for his reverse-acting grate, an improved version of which is still marketed today (Martin, 2004). A combination of a short reciprocating grate combined with a rotary kiln for final burnout of the ashes, the Vølund furnace was first implemented in Gentofte, near to Copenhagen, in 1931 (Kleis and Dalager, 2004). Other types of combustion systems based on traveling and reciprocating grates were developed during the following years, especially after World War II. One more important development in Germany was the Düsseldorf roller grate, first introduced in the Düsseldorf Flingern 250 000 Mg/a plant in 1965 (Reimann, 1991). This grate type is still found today in installations with high waste throughput.

Waste combustion developed differently in different parts of the world. In the USA more than 700 plants were in use by the end of the 1930s (Hickmann, 2003). The years after World War II saw a further expansion of waste combustion capacity in various parts of the world, which was to a great extent fuelled by the triumphant proliferation of plastics in all areas of life.

In a number of countries, especially in Japan and the USA, waste management policy followed the principle of on-site disposal: the disposal of waste in close vicinity to its occurrence. For instance, New York issued a law in 1961, which made it compulsory that each newly built apartment house would have its own waste burning facility (Fenton, 1975). Soon some 17 000 small combustors were burning waste inside the city, all of them without energy recovery and, of course, without any air pollution control (APC) measures. The 1961 law was withdrawn in 1966 and the many tiny systems were slowly shut down during the next decade.

A similar development took place in Japan resulting in there being approximately 200 000 waste combustors in 1970. The phasing out of the many small batch-operated plants did not take place till the 1990s. This was also a time when many bigger plants were taken out of operation due to poor environmental performance.

For several decades, considerable research and development (R&D) effort was spent to improve the design and operation of grate and boiler systems. However, attempts to improve environmental compatibility did not start before the 1950s. In the 1960s and 1970s the first air emission standards were issued in the USA, Europe and Japan.

The Seveso accident in 1976 (Sambeth, 1983) in particular, and also the detection of dibenzo-p-dioxins (PCDDs) and dibenzofurans (PCDFs) in filter ashes from three Dutch municipal solid waste combustors by Olie, Vermeulen and Hutzinger one year later (Olie *et al.*, 1977) initiated violent public discussions concerning waste combustion, especially in Europe. This public opposition initiated extensive R&D activities around the world to understand the formation of these compounds and to develop countermeasures. This growing opposition against waste combustion triggered the waste authorities to issue tighter legislative regulation of air emissions, which were a strong driver for the development of improved technology. The environmental impact of waste combustion changed dramatically and within less than 30 years this process was transformed from a source of pollutants into a real sink.

The following will consider waste combustion before air emissions were controlled. It will then discuss the development of air emission standards and their effect on the development of technology. Activities to reduce the emission of particulate matter and acid gases will be presented. Specific attention is paid to the dioxin problem, its scientific solution and the resulting abatement technologies.

10.2   Status of waste combustion before 1970

Waste combustion plants were initially designed to burn waste and, partly, to make use of the released heat for power production; APC measures were not built in. Even after the closure of the first generation of batch-operated furnaces in the early 20th century, the new plants replacing them were characterized by improved burnout, thus reducing the smell from unburned organic matter, but again they had no specific APC systems.

Among the first reported special gas-cleaning devices implemented in waste combustors were the soot pockets in the boiler and economizer of the Aarhus plant in Denmark, a plant equipped with a Vølund furnace. The plant was also equipped with a 'special flue gas cleaner,' a cyclone, which was 'able to perform a complete cleaning of the flue gas,' as was claimed in a description of the plant (Kleis and Dalager, 2004). A rough estimate based on the primary fly-ash release of a Vølund furnace of the order of 20 kg/Mg of burnt waste and a fly-ash reduction of approximately 80% by the cyclone indicates dust stack emissions of approximately 500–800 mg/m³. A cross section of the Aarhus plant is shown in Fig. 10.1.

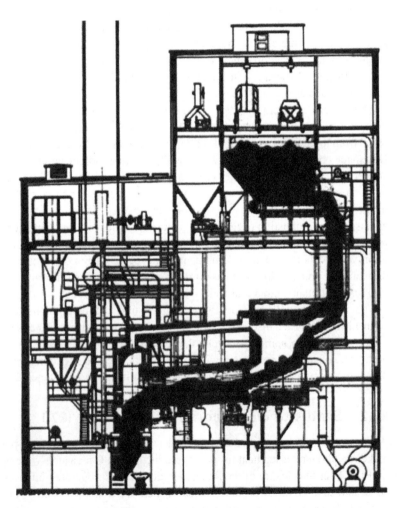

10.1 Cross section of the Aarhus waste combustor, which started operation in 1934 (adapted from Kleis and Dalager, 2004).

Considering the continuous operation of this plant and the generally limited release of dust from this furnace type, it must have been one of the cleanest of its time. Dust emissions from the batch-operated combustors prevailing at that time have been estimated as being much higher, in particular directly after feeding.

Fly ash was – and still is – to a great extent made up of alkaline constituents like CaO, which largely neutralizes the acidic gases HCl and SO_2 in the flue gas. Hence the emission of HCl was of minor environmental concern, the more so since waste before the plastic era contained less chlorine than it does today.

Regular control of stack emissions, as is common practice today, was not required and was widely not performed. An early method to characterize the quality of dust emissions from combustion furnaces including waste combustion plants was the Ringelmann Smoke Chart, which was introduced by the French Professor Maximilian Ringelmann at the end of the 19th century (US Bureau of Mines, 1967). The chart allows the classification of smoke plumes exiting from stacks by comparison with four cards, which have graduated shades of gray as shown in Fig. 10.2. The chart was soon used in Europe and in the USA, first in St. Louis in 1904, and, by 1910, it had been recognized officially in the smoke ordinance for Boston. Later many regulatory bodies developed standards and permits for the operation of combustion plants. The chart is even used today in various countries, e.g. in Germany for measuring the soot content in the off-gas of small stoves.

To comply with local plume darkness standards (i.e. dust emission standards), it became common practice to install cyclones to remove fly ash. In a cyclone, the gas enters a cylindrical chamber tangentially at high velocity and is forced into a cylindrical path. The centripetal force causes the particles to impinge on the wall and settle down into the discharge hopper; the gas is extracted through a central tube. A sketch of a cyclone is shown in Fig. 10.3.

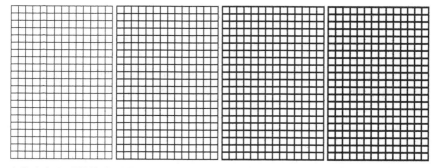

1. Equivalent to 20% black   2. Equivalent to 40% black   3. Equivalent to 60% black   4. Equivalent to 80% black

10.2 Ringelmann Smoke Charts as distributed by the US Bureau of Mines in 1967 (size reduced).

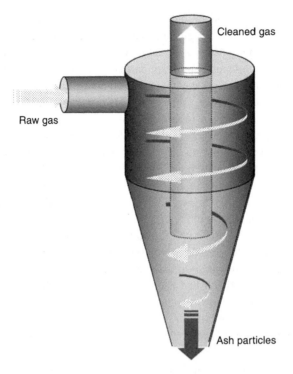

Cleaned gas

Raw gas

Ash particles

10.3 Cyclone.

The removal efficiency of cyclones is of the order of 50–90% for particles >10 μm, though for fine particulate matter it is rather low. Considering a raw gas dust load of several g/m³ for a cyclone-equipped waste combustor, then dust emissions of 500–800 mg/m³ have been estimated for the old Aarhus plant described above. In some US plants built in the 1960s, water spray systems, wet-baffle waterwalls and even wet scrubbers have been installed. Their only purpose, however, was to reduce dust emissions (Hickmann, 2003).

Few data on the dust emissions of early US full-scale waste combustion plants have been published; these are compiled in Table 10.1 (Achinger and Daniels, 1970; Cross *et al.*, 1970; Ellison, 1970; Jackson *et al.*, 1970). The values are presented in mg/m³ for consistency and they have been rounded: the reported gas conditions did not always allow standardization. The numbers vary between about 100 mg/m³ and more than 3000 mg/m³; there is no correlation with the type of combustor or APC system. The Manhattan combustor is an exception with its new Venturi scrubber, which had obviously higher dust removal efficiency than all the other installations.

At the time of the tests, 1968 or 1969, few dust emission limits existed in the USA. The federal guideline was 0.2 gr/ft³ (approximately 460 mg/m³) and that of New Jersey was 0.1 gr/ft³ (230 mg/m³). New York City had issued a dust emission

Table 10.1 Particulate matter (PM) emissions of selected US waste combustors in the late 1960s

Name	Furnace type	Capacity (Mg/d)	APC system	Year built	PM emission (mg/m³)	Reference
A	Reciprocating grate	300	Wetted column scrubber	1966	1260	Achinger and Daniels (1970)
B	Rocking grate		Flooded baffle walls	1967	2560	Achinger and Daniels (1970)
D	Traveling grate		Flooded baffle walls	1965	1050	Achinger and Daniels (1970)
E	Vølund system	500	Water sprays and baffle wall	1963	1670	Achinger and Daniels (1970)
F	Vølund system	600	Water sprays and baffle wall	1963	1650	Achinger and Daniels (1970)
G	Reciprocating grate	400	Multitube dry cyclones and wet-baffle wall	1967	3100	Achinger and Daniels (1970)
Manhattan 73rd Street		220	Venturi scrubber		90–470	Ellison (1970)
Milwaukee South Side					335–450	Jackson *et al.* (1970)
Waterbury	Shaft furnace batch fed	150	Water sprays and baffle wall	1957 (APC 1967)	610–730	Cross *et al.* (1970)

limit of 2 lb/ton of burnt waste, which can be recalculated to approximately 200 mg/m³. It is obvious that most existing plants could not meet these limits. Even the Venturi scrubber had problems with the New York standard. This can be globalized since combustors in other parts of the world, which used more or less the same technology, had emissions of the same order of magnitude.

As mentioned above, no emission standards for acidic gases existed before 1970 and consequently no measures for acidic gas removal were in place. Hence it has to be assumed that the emission concentration of acidic gases and heavy metals was of the order of what we find in the raw gas of modern waste combustion plants today. A cautious estimate of such emission concentrations is compiled in the first column of Table 10.2. The data resemble those published by Vogg in 1988 (Vogg, 1988) and document the poor environmental compatibility of waste combustion plants in those days. The situation was very similar wherever waste combustion was practiced in the world.

10.3   Air emission regulations and their influence upon technology

10.3.1 Changes in the 1970s and early 1980s

The first serious attempts to change this situation started in a number of countries, e.g. in the USA, Japan and Germany, in the 1960s. The development over time will be described using as examples the USA, Germany, the European Union and somewhat less so Japan.

The US federal government enacted the Air Pollution Control Act in 1955. Real changes in improving air quality were enforced later by the Clean Air Acts (CAA), the first of which was issued in 1967. In 1970, the CAA Extension was released, which was followed by amendments in 1977 and 1990. The CAA prescribes responsibilities and relationships for federal, state and local agencies. The Environmental Protection Agency (EPA) was founded in 1970 in the USA to coordinate and harmonize the situation. It determined whether state regulations were in line with federal regulations, which were guidelines only. Its role is complicated because all EPA rules promulgated under the CAA can be subject to court review and any citizen can file a civil action against the EPA.

The Clean Air Act quickly focused regulators' concerns on the quality of emissions from waste combustors. The first standards applied to dust emission only were issued by local and state agencies in the late 1960s. A few examples were discussed above. In 1975, the first air emission limits for five pollutants were released.

Current federal emission regulations for large municipal combustion plants date back to 1998 (EPA, 1998); the limits are shown in Table 10.3, which has the latest air emission limits for various parts of the world. New regulations are on the way for some federal states: these address small plants as well as commercial and industrial waste combustors, but not large combustion plants.

Table 10.2 Air emission data from various waste combustors between the 1960s and the 2000s (standardization partly unknown)

	Pre-1970 typical data	United Kingdom (8 plants)	Germany	USA	Austria (2 plants)	Germany (typical data)
Years	1960s	1980	1985	1989	1998	2004
PM (mg/m³)	300–3000	200–700	20–100	4.5–34	0.8–2.9	1
CO (mg/m³)	500–1000	16–140	50–600		13–22	10
TOC (mg/m³)		20–1360	<20		0.4–0.6	1
HCl (mg/m³)	700–1000	50–305	50–900	15–76	0.5–1.9	1
HF (mg/m³)		4	0.5–9	0.8–1.6	0.03–0.04	0.1
SO₂ (mg/m³)	300–800	115–545	70–300	13–130	3.9–8.4	1.5
NOₓ (mg/m³)	300–500	190–390	150–300	50–150	16–21	
Hg (mg/m³)	0.5		<0.01–0.1	<0.1–0.7	0.003–0.025	
Cd + Tl (mg/m³)	0.5			<0.01–0.1	0.002–0.015	
PCDD/F (ng(I-TE)/m³)					0.03–0.04	0.005
Reference	Author's estimate	Wallin and Clayton (1982)	Barniske (1987)	Donnelly (1991)	Schuster (1999)	BMU (2005)

Notes:

TOC = total organic carbon
PCDD/F = PCDD + PCDF
I-TE = international toxicity equivalents

Table 10.3 Comparison of actual air emission limits for large-scale waste combustion plants in various countries

	EU[1]	USA[2]	China[1]	Japan[3]
PM (mg/m^3)	10	24	80	40
CO (mg/m^3)	50	5.8	150	116
TOC (mg/m^3)	10			
HCl (mg/m^3)	10	38	75	700
SO_2 (mg/m^3)	50	80	260	
NO_x (mg/m^3)	200	190	400	330
Hg (mg/m^3)	0.05	0.08	0.2	
Cd (mg/m^3)	0.05[4]	0.02	0.1	
PCDD/F $(ng(I\text{-}TE)/m^3)$	0.1	0.3	1.0	0.1

Notes:

[1] 273 K, 101.3 kPa, 11 vol% O_2.
[2] 273 K, 101.3 kPa, 7 vol% O_2.
[3] 273 K, 101.3 kPa, 14 vol% CO_2.
[4] Cd + Tl.

The Japanese government began to consider a comprehensive pollution-control policy in the early 1960s and from the very beginning waste combustion was considered. Due to Japan's high population density and its high number of waste combustion plants, most of which were located directly in urban centers, the environmental impact was evident at that time.

In 1967 Japan enacted the world's first overarching environmental law, the Basic Law for Environmental Pollution Control. The law constituted a framework for environmental protection and on its basis a number of specific measures like air emission standards were released in the following years. As in the USA, national regulations were guidelines; under this umbrella, the actual permits and standards for combustion plants took site-specific conditions into account (Hershkowitz and Salerni, 1989).

This situation has not changed today. Current national emission regulations, based on the Waste Management Law and the Waste Management and Public Cleansing Law, both released in 2001, set a nationwide air emission limit for dioxins of $0.1 \, ng(I\text{-}TE)/m^3$; all other limits are, compared to those valid in other parts of the world, rather high. Local authorities are responsible for the permits of specific plants and the limits laid down in these permits vary from plant to plant, but fall far below those in the national regulations. The Japanese limits are also shown in Table 10.3.

The development of air emission standards in Europe started in Germany, Switzerland and the Netherlands. The German situation will briefly be presented as an example of the development in Europe. The German government released its first air emission control regulation, the Technical Instructions on Air Quality Control, in 1964 (TA Luft 64), which set a dust or particulate matter (PM) emission

Table 10.4 Waste combustion air emission limits in Germany and in the Netherlands (daily average values, 273 K, 101.3 kPa, 11 vol% O_2)

	TA Luft 64	TA Luft 74	Netherlands 1977	TA Luft 86	17. BlmSchV 1990	Netherlands 1990
PM (mg/m³)	540	100	14.3×vol% CO_2	30	10	5
CO (mg/m³)		1000		100	50	50
TOC (mg/m³)				20	10	10
HCl (mg/m³)		100	1000	50	10	10
HF (mg/m³)		5	20	2	1	1
SO_2 (mg/m³)			600	100	50	40
NO_x (mg/m³)			300	500	200	70
Hg (mg/m³)		20		0.2	0.03	0.05
Cd + Tl (mg/m³)		20		0.2	0.05	0.05
PCDD/F (ng(I-TE)/m³)					0.1	0.1

limit of 540 mg/m³ for combustion plants emitting >20 000 m³/h (see second column of Table 10.4). This was, as can be seen from Table 10.1, hard to comply with for the vast majority of existing plants and caused the need for upgrades to give better dust removal.

The most common improvement was the installation of electrostatic precipitators (ESP). In an electrostatic precipitator, the flue gases pass between grounded metal plates and central high-voltage wires (approximately 100 kV) at low velocity (few cm/s). The fly-ash particles are charged and move along the lines of the electric field to impinge on the grounded collector plates. The plates are periodically cleaned by rapping and the accumulated dust particles fall down into a discharge hopper. A modern ESP comprises at least two and often three sectors or fields and guarantees dust removal efficiencies of >99% at particle sizes >2.5 µm. An ESP is shown in Fig. 10.4.

An alternative to an ESP is a fabric filter in which the raw gas passes through the material of a fabric bag, which is supported by a metal cage. The gas flows from the outside of a bag to the interior and the fly ash collects on the outer side of the filter bag where it is periodically removed by air pulses in a direction opposite to the filtration direction. Fabric filters have higher removal efficiencies than ESPs but are not often used directly downstream of a boiler due to the risk of fabric damage by sparks in the flue gas. A fabric filter is shown in Fig. 10.5.

The installation of ESPs in waste combustion plants reduced dust emissions down to values in the region of 100 mg/m³. Consequently the amended German regulations, TA Luft 74, reduced the dust emission limit to 100 mg/m³. In parallel the emission of some heavy metals, which are mainly transported by the fly ash, was reduced as well. Since fly ash contains high amounts of alkaline ingredients like CaO, it neutralizes a significant fraction of the acidic gases in the flue gas. As

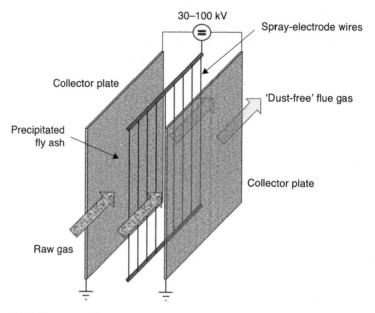

10.4 Electrostatic precipitator.

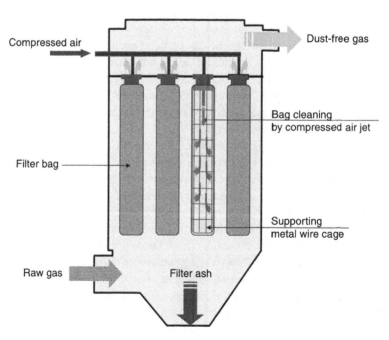

10.5 Fabric filter.

a consequence of the reduction of fly-ash emissions, the acidic components of the flue gas came to the fore. This was taken care of by emission limits for HCl and HF. Furthermore the emission of CO was limited to $1000\,mg/m^3$, which required more careful combustion control. The limits in the TA Luft 74 are shown in the third column of Table 10.4.

Similar regulations came into force in other countries around the same time, e.g. in the Netherlands in 1977, which are given in the fourth column of Table 10.4. The Dutch standards were much higher for HCl and HF, but unlike the German TA Luft 74, limits for the acidic gases SO_2 and NO_x were also included.

The reduction of acidic components requires more extensive gas treatment through the addition of chemical cleaning stages. This can be performed using wet scrubbers or alternatively dry systems with the injection of dry reagents or slurries into the flue gas. HCl and HF abatement can be accomplished by a single scrubber operated with water at a low pH value. The removal of SO_2, however, requires a second scrubber operating close to the neutral point to prevent the separation of CO_2, which would result in excessive consumption of the neutralizing agent, typically either NaOH or $Ca(OH)_2$. Two-stage wet scrubbing became the typical configuration of the first European APC systems. In most cases the first scrubber was of the Venturi type and the second neutral one was a packed tower. The effluents of both scrubbers were combined, neutralized and cleaned of heavy metals and finally discharged into a sewage treatment plant. The typical configuration of such two-stage wet scrubbing systems is shown in Fig. 10.6.

In many countries, such as Germany, the water authorities started to ban the discharge of scrubber effluent into public sewers. Hence the operators of wet scrubbing systems were forced to evaporate the effluent. This was sometimes done in externally heated driers. In most cases, however, evaporation was achieved by injecting the neutralized effluent into a spray drier downstream of the ESP,

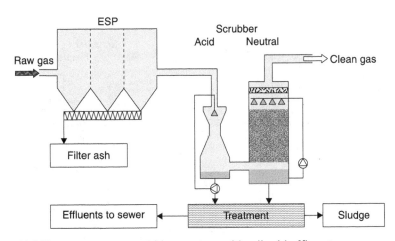

10.6 Two-stage wet scrubbing system with a liquid effluent.

where the hot flue gas evaporates the water. The dried scrubbing residues, sodium or calcium salts depending on the neutralizing agent, were precipitated in a second filter installed between the spray drier and the wet scrubber. Fabric filters were mainly used for this. This configuration, which is shown in Fig. 10.7, is still prevalent where wet scrubbing is applied today.

In the late 1970s, engineers in Germany as well as in Japan tested new dry scrubbing systems to avoid the high requirement for water cleaning, which was necessary if wet scrubbing was applied (Miyanohara and Kitami, 1979; Schuch, 1979). Dry scrubbing can be performed with different reagents; the most common ones are limestone ($CaCO_3$), calcium oxide (CaO), lime ($Ca(OH)_2$), and sodium hydrogen carbonate ($NaHCO_3$), which are injected into the flue gas to neutralize the acidic gas components. Often the agent is injected as slurry for better reactivity; such systems are called semi-dry. In most cases activated carbon is added to the reagent to adsorb low-volatile organic compounds and mercury. The scrubbing residues are typically removed from the flue gas by a fabric filter.

Figure 10.8 depicts a dry scrubbing system. As can be seen, this configuration is very simple and requires lower investment costs than wet systems. In many cases even the ESP can be omitted and the fly ash precipitates with the dry scrubbing residues.

Dry gas cleaning is able, especially if $NaHCO_3$ is used as reagent (which was introduced in the 1990s as the Neutrec® Process), to meet all current acidic gas emission limits. The disadvantage of using calcium-based reagents is the low reactivity of the two-phase system, which necessitates higher stoichiometry and hence generates higher amounts of residues than wet scrubbing. After becoming prominent in the 1980s, dry scrubbing systems are still often preferred in new installations due to their low investment and operational costs.

The use of chemical gas cleaning in waste combustors produced a significant reduction in the environmental impact of these systems. The third column in

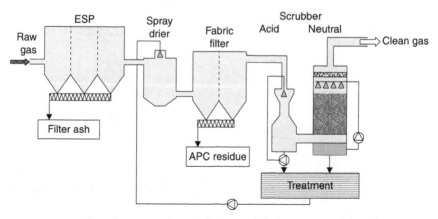

10.7 A two-stage wet scrubbing system without liquid effluent.

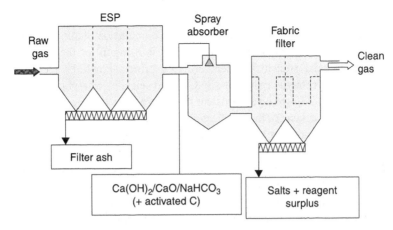

10.8 Dry scrubbing system.

Table 10.2 shows the ranges of stack emissions for eight plants in the UK in 1980, which have low values for dust and acidic gas emissions compared to the situation before 1970 (Wallin and Clayton, 1982). Slightly lower results were measured in a number of German waste combustion plants, which took part in an extended test campaign organized by the German environmental protection agency, the Umweltbundesamt (UBA), in 1985 (Barniske, 1987).

10.3.2 Changes after 1985

At the end of the 1970s, the detection of dioxins in waste combustor filter ashes caused violent public opposition against waste combustion, especially in some European countries. This was one reason to investigate the dioxin issue, which will be described below. Another reason was to further tighten up emission limits. Germany produced an amended regulation in 1986, TA Luft 86. The strongly reduced air emission limits are found in the fifth column of Table 10.4. However, it soon became obvious that technical progress in the mid-1980s allowed for further tightening of the emission limits, which was realized in 1990 by the release of the 17. BImSchV, the 17th ordinance for the implementation of the German inmission protection law. At about the same time, very similar regulations were issued in the Netherlands.

The respective standards of both decrees are shown in the last two columns of Table 10.4. Both were used as the basis of the EU Waste Incineration Directive (WID) issued in 2000 (European Parliament and Council, 2000), which had to be adopted into national law by all EU member states and is today the European air emission guideline. The Dutch emission regulation, which has lower limits than the WID for particulate matter, SO_2 and NO_x, is still valid in the Netherlands. So is the German 17. BImSchV, and its mercury limit was reduced from 50 to $30 \mu g/m^3$ in 2003.

Whereas the NO_x limit of 500 mg/m³ in TA Luft 74 can typically be met without any special abatement measure by all waste combustion plants, the 200 mg/m³ of 17. BImSchV and the WID and more so the 70 mg/m³ of the Dutch guideline, requires special strategies. There are two principal options for NO_x control: selective non-catalytic reduction (SNCR) and selective catalytic reduction (SCR). SNCR, which is based on a US patent from 1975, uses the injection of ammonia or urea into the backend of the combustion chamber to reduce NO to N_2 (Lyon, 1975).

The SCR technology was developed for power plants in Japan in the 1970s and is widely used worldwide today. The reduction of NO by NH_3 is achieved at the surface of a catalyst, e.g. on V_2O_5 stabilized in TiO_2. The first catalysts operated at temperatures between 300°C and 350°C. There are now catalysts available, which show satisfactory efficiency at temperatures between 220°C and 250°C. The typical design of catalysts is the honeycomb type. Layers, each with many modules of catalyst elements, form the full-scale catalyst system as shown in Fig. 10.9. Both of these NO_x abatement technologies were implemented starting in the late 1980s.

The effects of the new regulations and the newly developed technologies became visible in the tests of numerous waste combustion plants, e.g. in the results of tests in a US waste combustor in 1989 (Donnelly, 1991). The data are presented in the fifth column of Table 10.2.

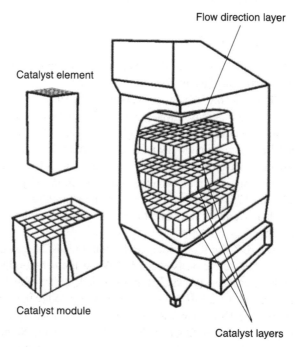

Flow direction layer

Catalyst element

Catalyst module

Catalyst layers

10.9 SCR system (adapted from European Commission, 2003).

Stack emission tests of two waste combustors in Vienna, both equipped between 1986 and 1992 with highly efficient wet scrubbing and SCR NO$_x$ removal systems, showed that a modern waste combustion plant is easily able to meet the most stringent air emission limits. The results of tests performed in 1998 are shown in the sixth column of Table 10.2 (Schuster, 1999).

The high quality of emissions from modern combustion plants is underlined by typical stack emission data published by the German UBA in 2005 and shown in the final column of Table 10.2 (BMU, 2005).

10.4   Dioxin emissions

Dioxins, more precisely the group of 75 polychlorinated dibenzo-p-dioxins and 135 dibenzofurans, a group of some (but not all) highly toxic compounds, became a controversial issue in the public perception of waste combustion in the 1980s. As mentioned above, one year after the Seveso accident, Olie, Vermeulen and Hutzinger were the first to report on the dioxins found in filter ashes from three Dutch municipal solid waste combustors (Olie *et al.*, 1977). This finding initiated, besides a fierce controversy, research activities around the world to understand their formation in waste combustion and to develop countermeasures.

This research was urgently needed since the mass flow of dioxins in a combustion plant in the early 1980s, as calculated from numerous published dioxin measurements and shown in Fig. 10.10, shows emissions up to roughly 70 ng(I-TE)/m^3. Furthermore, APC residues, especially filter ashes, carried extremely high loads of dioxins.

Within only ten years the fundamentals of the synthesis of PCDD/F in waste combustion were determined. The key in identifying the main formation route

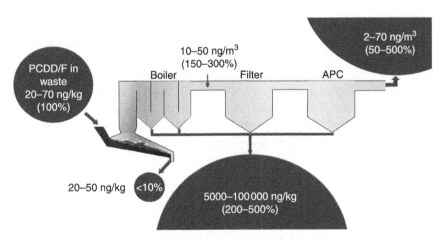

10.10 Mass flow of dioxins in a waste combustor in the early 1980s (concentrations are calculated as I-TE values).

was the accidental detection that during annealing experiments with waste combustion filter ashes their dioxin inventory increased approximately ten times at a temperature of 300°C (Vogg and Stieglitz, 1986).

Detailed investigations of the reaction mechanism quickly revealed that the basic reaction is a heterogeneous oxy-chlorination process (Stieglitz and Vogg, 1987; Hagenmaier *et al.*, 1987; Hiraoka *et al.*, 1987; Vogg *et al.*, 1987; Gullet *et al.*, 1990). The ingredients and conditions necessary to form dioxins are:

- products of incomplete combustion (PICs), e.g. soot
- halides, mainly chlorides, but also bromides
- an oxidizing atmosphere
- a catalyst – copper salts being most effective.

This slow reaction is called *de novo* synthesis and mainly takes place in the dust deposits in the backend of the boiler and, if the dust removal system is operated at temperatures above 200°C, also in this stage (Vogg *et al.*, 1989). Careful investigations in test plants (Hunsinger *et al.*, 2002) confirmed that dioxins fed into the combustor along with the waste are totally destroyed inside the combustion chamber and that the dioxins found in the flue gas were newly synthesized in the boiler.

On the basis of this knowledge, strategies were soon developed to minimize their formation by suitable primary or headend measures (Vogg *et al.*, 1991):

- optimization of combustion control to achieve a better burnout and hence to reduce the content of PICs in the raw gas, in the fly ashes and in the deposits inside the boiler
- reduction of the gas velocity in the fuel bed to minimize the release of fly ash
- adequate cleaning of the boiler to minimize the deposition of fly ash
- operation of the dust removal system at low temperature.

Following these principles, it was possible in most existing plants, without major upgrades, to reduce dioxin emissions easily, quickly and without additional costs. Measurements in a German plant in 1990, in which these principles were applied, documented surprisingly low dioxin concentrations in the raw gas, of the order of 1–5 ng(I-TE)/m^3 (Reeck *et al.*, 1991). This is the level that can be achieved in state-of-the-art waste combustion plants. However, to comply with common international dioxin emission limits of the order of 0.1–0.3 ng(I-TE)/m^3 (see Table 10.3), secondary abatement measures are needed.

In the 1980s, the carbon content of fly ash of most combustors was high and so was the formation of dioxins in the boiler. These dioxins, however, were mainly adsorbed on the soot particles in the fly ash. Hence good de-dusting was a rather efficient removal step. However, there was still sufficient gaseous PCDD/F present that it needed to be removed by additional process stages like adsorption on charcoal fixed-bed filters (Dannecker and Hemschemeier, 1990) or by injection of pulverized activated carbon into the flue gas at the backend of the APC system

(Nielsen and Møller, 1989). This process is frequently found as a polishing stage for dioxins, but also for mercury, and is therefore called a police filter.

Also catalysts, if operated in oxidative mode, can have a high destruction potential for dioxins (Hiraoka *et al.*, 1989). This technology is often used if SCR for NO_x abatement is practiced: one or two catalyst layers are added at the backend of the system and operated under oxidizing conditions. These post-combustion abatement technologies were implemented in full-scale plants starting in the early 1990s. In the following years, further technologies were developed, which are characterized by the fact that they combine different tasks in one stage.

The REMEDIA® process uses an existing device in dry scrubbing systems, the fabric filter, which was originally used only for dust separation. By incorporating a low-temperature catalyst in the cloth of the filter, it can be used to eliminate gaseous organic compounds (Bonte *et al.*, 2002). The operation temperature can vary between 140°C and 260°C with a recommended range of 200–230°C. The process is used in several medical and municipal solid waste combustion plants in Europe and Japan.

For wet scrubbers, a carbon-filled plastic material called Adiox® has been developed, which is used to produce tower packings and demisters (Andersson *et al.*, 2003). The plastic absorbs dioxins and other low volatile organic pollutants, which are fixed in the material by the carbon filler. The dioxin-loaded packings can be burnt in the plant's furnace to destroy the dioxins totally. The process has been installed in a number of thermal treatment plants.

A compilation of published stack emission data measured in modern waste combustion plants proves the huge progress achieved during recent decades. This was possible by the fast transfer of the results of many successful research projects into full-scale facilities. Figure 10.11 shows the dramatically changed dioxin mass flows in a modern waste combustor compared to the situation in the 1980s (Vehlow, 2005). Today the emission of dioxins from state-of-the-art plants is in most cases well below $0.01 \, ng(I\text{-}TE)/m^3$.

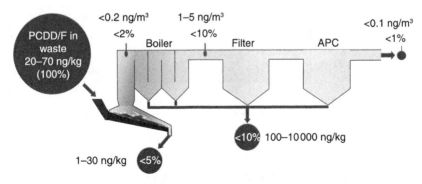

10.11 Mass flow of dioxins in a state-of-the-art waste combustor (concentrations are calculated as I-TE values).

This is reflected in the contribution of waste combustion to the total German dioxin emissions to air. As Table 10.5 shows, waste combustion at 400 g(I-TE)/a was the second largest source of emissions in 1990, whereas its contribution at 0.5 g(I-TE)/a was almost marginal in 2000 – although the throughput of German waste combustors increased in that period by more than 50%.

Not only can emissions into the air be reduced significantly, the same holds for the dioxin inventory in solid residues. A respective review of data published between 1985 and 2005 (Vehlow *et al.*, 2006) reveals that for bottom as well as for filter ashes this value dropped down by one to two orders of magnitude (Fig. 10.12).

Table 10.5 Annual dioxin emissions in Germany between 1990 and 2000 (adapted from BMU, 2005)

	1990	1994	2000
Total emissions to air (g(I-TE))	1200	330	<70
Emissions from waste combustors (g(I-TE))	400	32	0.5
Throughput of waste combustors (10^6 Mg)	9.2	10.9	14

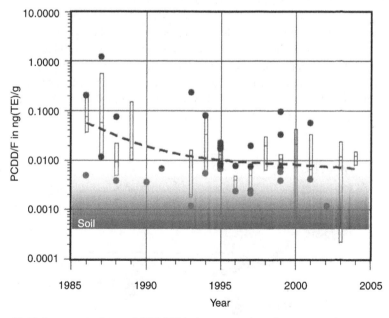

10.12 Concentrations of PCDD/F in bottom ashes from grate furnaces; the shaded area indicates typical concentration ranges in natural soil in Central Europe (adapted from Vehlow *et al.*, 2006).

The decline is definitely the result of optimized combustion control, which entails more stable combustion conditions, less dust released and optimized burnout of the gas phase. However, this combustion mode reduced not only the PIC concentration in fly ash, but consequently the formation of dioxins in the boiler. A side effect was an improved burnout of bottom ash and consequently a much lower dioxin load in this residue stream, too. In the mid-1990s dioxins in bottom ash from modern plants fell to a level that is comparable to or not much higher than that found in uncontaminated soils in Central Europe (Fiedler, 1996).

Filter ash followed the same pattern as the bottom ash; however, its dioxin inventory is higher by a factor of 1000. This means that the disposal of filter ash is still a problem. However, in real life this problem is more due to the high load of soluble salts and heavy metals than to dioxins since these can easily be destroyed by simple thermal treatment.

10.5   Environmental impact of emissions from modern waste combustion plants

When evaluating the environmental compatibility of modern waste combustion and speculating about future R&D needs, it might be useful to validate the environmental impact of a state-of-the-art municipal solid waste combustor. For this reason, a distribution calculation of stack emissions of a plant with a flue-gas flow of $100\,000\,m^3/h$ and a stack height of 65 m has been performed (Vehlow, 1996). Table 10.6 shows the results for a number of selected components.

The second column in the table shows typical current emission levels. The third column gives the mean concentrations of the components in ambient air in Germany around 1990 (Vogg, 1992), and the fourth column contains the calculated emissions for the component's concentration at the point where the plume hits the ground. This is the location were the ambient air should receive the highest input from the stack emissions.

The final column is a sum of the ambient air concentrations and the additional emissions. The numbers furnish evidence that the stack emissions of a modern

Table 10.6 Environmental relevance of stack emissions from waste combustion (Vehlow, 1996)

	Stack emissions	Background concentration	Additional emissions	Contaminated air
HCl	$5\,mg/m^3$	$30\,mg/m^3$	$0.15\,mg/m^3$	$30.15\,mg/m^3$
SO_2	$20\,mg/m^3$	$20\,mg/m^3$	$0.6\,mg/m^3$	$20.6\,mg/m^3$
Dust	$1\,mg/m^3$	$40\,mg/m^3$	$0.03\,mg/m^3$	$40.03\,mg/m^3$
Cd	$5\,mg/m^3$	$3\,ng/m^3$	$0.15\,ng/m^3$	$3.15\,ng/m^3$
Hg	$10\,mg/m^3$	$5\,ng/m^3$	$0.3\,ng/m^3$	$5.3\,ng/m^3$
PCDD/F	$0.05\,ng(I\text{-}TE)/m^3$	$50\,fg(I\text{-}TE)/m^3$	$1.5\,fg(I\text{-}TE)/m^3$	$51.5\,fg(I\text{-}TE)/m^3$

waste combustion plant give rise to only minimal increases in the low level of existing background concentrations. These additional emissions do not cause any environmental risk, even for the critical dioxins and mercury.

From this perspective, a major reduction of emission limits as well as further efforts to improve the efficiency of flue-gas cleaning systems will not result in a significant difference concerning the protection of human health and the environment. However, further reductions in air emission limits will entail higher costs in upgrading, and eventually in the operation, of waste combustion plants.

Nevertheless, it can be envisaged that in the EU in the near future, the emission levels typically achieved in modern plants, as listed in the *Reference Document on the Best Available Techniques for Waste Incineration* (European Commission, 2006), will become guidelines for the emission limits. This, however, implies a significant reduction of the operational clean gas levels, which need to be kept well below the emission limits to guarantee permanent compliance.

Further attempts have been made to limit the national emission of specific gases as is already done for a few pollutants as specified in the 'EU directive 2001/81/ EC on national emission ceilings for certain atmospheric pollutants', e.g. for NO_x. In particular, this directive caused a change to be made to the German 17. BImSchV: starting in 2013 the mean annual NO_x emission of new waste combustors will be limited to $100\,mg/m^3$. All such changes, however, should be carefully evaluated to determine whether the effort and the resulting higher costs can be justified regarding the eco-efficiency of the measures, before they are enacted.

Another aspect concerning the environmental compatibility of waste combustion was added to the agenda only recently. Current discussions on energy supply and the efforts to replace fossil fuels by renewable energy sources have drawn attention to the energy efficiency of waste combustors, the more so since in most countries at least 50% of the energy inventory of municipal solid waste is of biogenic origin. Hence approximately 50% of the CO_2 emitted by waste combustors can be accounted for as carbon neutral. This together with the low impact of harmful pollutants is another driver for using waste combustion not only for inertization of residual waste but also as a valuable energy source.

10.6   Conclusion

Waste combustion is a key component in waste management strategies since it is an efficient and reliable process for inertizing that fraction of residential waste that is left over after material recycling. The technology is well developed and different processes can be applied according to local conditions. Emissions from waste combustion are very low and have no discernible impact on human health and the environment. In particular, the control of dioxins has been developed to a degree that the air emissions as well as the concentrations in bottom ash cause no harm. Bottom ash reaches a leaching stability, which allows their utilization in the building sector. Concerning dioxins, but also other contaminants, waste combustion has since the 1980s successfully been transformed from a source of pollutants into an efficient sink.

10.7 References

Achinger W C and Daniels L E (1970) An evaluation of seven incinerators, *Proceedings, 1970 National Incinerator Conference*, Cincinnati, American Society of Mechanical Engineers, 17–20 May, 32–64.

Andersson S, Kreisz S and Hunsinger H (2003) Innovative material technology removes dioxins from flue gases, *Filtration and Separation*, 40, 22–25.

Barniske L (1987) Emission of trace organics from municipal solid waste incinerators – Rationale of national guidelines in the Federal Republic of Germany, *Waste Management & Research*, 5, 347–354.

BMU (2005) *Waste incineration – A potential danger*, Berlin, Federal Ministry for the Environment, Nature Conservation and Nuclear Safety, available from: http://www.bmu.de/english/waste_management/downloads/doc/35950.php [accessed 26 March 2012].

Bonte J L, Fritsky K J, Plinke M A and Wilken M (2002) Catalytic destruction of PCDD/F in a fabric filter: Experience at a municipal waste incinerator in Belgium, *Waste Management*, 22, 421–426.

Cross F L, Drago R J and Francis H E (1970) Metal and particulate emissions from incinerators burning, *Proceedings, 1970 National Incinerator Conference*, Cincinnati, American Society of Mechanical Engineers, 17–20 May, 189–195.

Dannecker W and Hemschemeier H (1990) Level of activated-coke technology for flue gas dust collection behind refuse destruction plants looking at the problem from the special aspects of dioxin separation, *Organohalogen Compounds*, 4, 267–272.

de Fodor E (1911) *Elektrizität aus Kehrricht*, Budapest, K.U.K. Hofbuchhandlung von Julius Benkö.

Donnelly J R (1991) Overview of air pollution controls for municipal waste combustors, in *Municipal Waste Combustion*, Pittsburgh, Air & Waste Management Association, 125–144.

Ellison W (1970) Control of air and water pollution from municipal incinerators with the wet-approach venture scrubber, *Proceedings, 1970 National Incinerator Conference*, Cincinnati, American Society of Mechanical Engineers, 17–20 May, 157–166.

Encyclopedia Britannica (1910) Destructors, *Encyclopedia Britannica, Vol. 8, Slice II*, 105–108, available from: www.gutenberg.org/files/30685/30685-h/30685-h.htm [accessed 26 March 2012].

EPA (1998) Federal plan requirements for large municipal waste combustors constructed on or before September 20, 1994, *Federal Register, Vol. 63*, No. 218, Thursday, November 12, 1998 Rules and Regulations, 63191.

European Commission (2003) *Integrated Pollution Prevention and Control (IPPC) Draft Reference Document on Best Available Techniques for Large Combustion Plants*, Institute for Prospective Technological Studies, Seville, Spain, Draft March 2003.

European Commission (2006) *Integrated Pollution Prevention and Control (IPPC) Reference Document on the Best Available Techniques for Waste Incineration*, Seville, Institute for Prospective Technological Studies, European IPPC Bureau, August, available from: eippcb.jrc.es/reference/BREF/wi_bref_0806.pdf [accessed 26 March 2012].

European Parliament and Council (2000) Directive 2000/76/EC of the European Parliament and of the Council of 4 December 2000 on the incineration of waste, *Official Journal of the European Communities*, 28 December, L332/91.

Fenton R (1975) Current trends in municipal solid waste disposal in New York City, *Resource Recovery and Conservation*, 1, 167–176.

Fiedler H (1996) Sources of PCDD/PCDF and impact on the environment, *Chemosphere*, 32, 55–64.

Frilling H and Mischer O (1994) *Pütt und Pann'n – Geschichte der Hamburger Hausmüllbeseitigung*, Hamburg, Ergebnisse Verlag GmbH.

Gullett B K, Bruce K R and Beach L (1990) The effect of metal catalyst on the formation of polychlorinated dibenzo-p-dioxins and polychlorinated dibenzofurans precursors, *Chemosphere*, 20, 1945–1952.

Hagenmaier H, Kraft M, Brunner H and Haag R (1987) Catalytic effects of fly ash from waste incineration facilities on the formation and decomposition of polychlorinated dibenzo-p-dioxins and polychlorinated dibenzofuranes, *Environmental Science & Technology*, 21, 1080–1084.

Hershkowitz A and Salerni E (1989) Municipal solid waste incineration in Japan, *Environmental Impact Assessment Review*, 1989, 257–278.

Hickmann H L Jr. (2003) *American Alchemy: The History of Solid Waste Management in the United States*, Santa Barbara, CA, Forester Press.

Hiraoka M, Takeda N, Tsumura K, Fujiwara T and Okajima S (1989) Control of dioxins from municipal solid waste incinerator, *Chemosphere*, 19, 323–330.

Hiraoka M, Takizawa Y, Masuda Y, Takeshita R, Yagome K, Tanaka M, *et al.* (1987) Investigation on generation of dioxins and related compounds from municipal incinerators in Japan, *Chemosphere*, 16, 1901–1906.

Hunsinger H, Jay K and Vehlow J (2002) Formation and destruction of PCDD/F inside a grate furnace, *Chemosphere*, 46, 1263–1272.

Jackson M R, Lieberman A, Townsend L B and Romanek W (1970) Prototype fly ash monitor for municipal incinerator stacks, *Proceedings, 1970 National Incinerator Conference*, Cincinnati, American Society of Mechanical Engineers, 17–20 May, 182–188.

Kleis H and Dalager S (2004) *100 Years of Waste Incineration in Denmark*, Esbjerg, DK, Babcock & Wilcox Vølund ApS, Esbjerg and Rambøll, Virum, DK.

Lyon R K (1975) *Method for the reduction of the concentration of NO in combustion effluents using ammonia*, United States Patent 3900554.

Martin J (2004) Global use and future prospects of waste-to-energy technologies. *WTERT Fall Meeting*, New York, Oct. 7–8, 2004.

Miyanohara T and Kitami S (1979) Gegenwärtiger Stand der Technologie zur Beseitigung von HCl bei Müllverbrennungsanlagen in Japan, in *Recycling '79* (Thomé-Kozmiensky K J, ed.), Berlin, EF-Verlag für Energie und Umwelttechnik GmbH, 540–547.

Nielsen K K and Møller J T (1989) Reduction of chlorinated dioxins and furans in the flue gas from incinerators with spray absorbers and electrostatic precipitators, *Chemosphere*, 19, 367–372.

Olie K, Vermeulen, P L and Hutzinger O (1977) Chlorodibenzo-p-dioxins and chlorodibenzofurans are trace components of fly ash and flue gas of some municipal incinerators in the Netherlands, *Chemosphere*, 6, 455–459.

Reeck G, Schröder W and Schetter G (1991) Zukunftsorientierte Abfallverbrennung in der MVA Ludwigshafen, *Müll und Abfall*, 23, 661–673.

Reimann D O (1991) *Rostfeuerungen zur Abfallverbrennung*, Berlin, EF-Verlag für Energie und Umwelttechnik GmbH.

Sambeth J (1983) The Seveso accident, *Chemosphere*, 12, 681–686.

Schuch P-G (1979) Trockensorption von Chlorwasserstoff, Fluorwasserstoff und Schwefeldioxid aus Rauchgasen in einer halbtechnischen Versuchsanlage, in *Recycling '79* (Thomé-Kozmiensky K J, ed.), Berlin, EF-Verlag für Energie und Umwelttechnik GmbH, 515–520.

Schuster H (1999) *Waste Incineration Plants in Austria*, Vienna, Greenpeace Austria.

Stieglitz L and Vogg H (1987) On formation conditions of PCDD/PCDF in fly ash from municipal waste incinerators, *Chemosphere*, 16, 1917–1922.

US Bureau of Mines (1967) *Ringelmann Smoke Chart*, Washington, DC, US Dep. of the Interior, Bureau of Mines, Information circular 8333.

Vehlow J (1996) Simple, reliable and yet efficient – Modern strategies in waste incineration, *UTA Technology and Environment*, 2/96, 144–160.

Vehlow J (2005) Dioxins in waste combustion – Conclusions from 20 years of research, *Bioenergy Australia 2005*, Melbourne, 12–13 December, Proceedings on CD-ROM.

Vehlow J, Bergfeldt B and Hunsinger H (2006) PCDD/F and related compounds in solid residues from municipal solid waste incineration – A literature review, *Waste Management & Research*, 24, 404–420.

Vogg H (1988) Von der Schadstoffquelle zur Schadstoffsenke – neue Konzepte der Müllverbrennung, *Chem.-Ing.-Tech.*, 60, 247–255.

Vogg H (1992) Abfallverbrennung – Fakten und Perspektiven, *BWK/TÜ/UT-Special*, October, M14–M21.

Vogg H and Stieglitz L (1986) Thermal behaviour of PCDD/PCDF in fly ash from municipal incinerators, *Chemosphere*, 15, 1373–1378.

Vogg H, Metzger M and Stieglitz L (1987) Recent findings on the formation and decomposition of PCDD/PCDF in municipal solid waste incineration, *Waste Management & Research*, 5, 285–294.

Vogg H, Merz A, Stieglitz L and Vehlow J (1989) Chemical-process engineering aspects of dioxin reduction in waste incineration processes, *VGB Kraftwerkstechnik*, 69, 693–699.

Vogg H, Hunsinger H, Merz A, Stieglitz L and Vehlow J (1991) Head-end-Techniken zur Dioxinminderung, *VDI Berichte*, 895, 193–210.

Wallin S C and Clayton P (1982) Emissionen aus der Verbrennung von Hausmüll und BRAM, in *Recycling International* (Thomé-Kozmiensky K J, ed.), Berlin, EF-Verlag für Energie und Umwelttechnik GmbH, 271–277.

Zwahr H (1996) 100 Jahre thermische Müllverwertung in Deutschland, *VGB Kraftwerkstechnik*, 76, 126–133.

11

Air quality equipment and systems for waste to energy (WTE) conversion plants

J. S. AUSTIN, University of Maryland, USA

DOI: 10.1533/9780857096364.3.204

Abstract: Air quality regulations for municipal waste combustion facilities have a significant cost impact for developers of modern waste to energy (WTE) plants. Modern facilities must use good combustion practices and design, acid gas scrubbing, particulate collection, nitrogen oxide control and technology to minimize the release of toxic air pollutants. Many technology and equipment options are available, but most new WTE facilities use spray dryer adsorbers, fabric filters, selective non-catalytic reactions and activated carbon injection to reduce air pollution emissions below acceptable levels. Economic analysis determines the most cost-effective solution for air pollution control. Advances in air pollution control can reduce emissions and provide economic benefits to facility owners and operators.

Key words: municipal waste combustor, incinerator, air pollution, acid gas, nitrogen oxides, dioxin, mercury.

11.1 Air quality considerations and regulations for municipal waste combustors

Regulatory requirements are a significant cost faced by developers of modern waste to energy (WTE) plants. They are the most important factor that determines the wide cost variations of specific facilities. Selection of air quality control equipment is driven by air, water and waste regulations and permit requirements. While these are specific to a particular locality, they are surprisingly similar throughout the world because the underlying public health concerns are the same. Prevalent emission standards set the minimum requirements for equipment and stack emissions, but other factors are important when selecting specific technologies and equipment. Local ambient air concerns often drive developers to choose technology that provides the most stringent control of air pollutants emitted from the facility. Requirements for the beneficial use of residues may drive a developer to select a particular technology. Water use and discharge requirements may drive selection of equipment. Many factors, including local public acceptance of the project, will ultimately result in the specific requirements for air pollution control. Development companies often have requirements for their facilities that go beyond those imposed upon them. Economic analysis will play a role in determining the most cost-effective solution to meet or exceed

204

permit conditions. When a developer can meet all of these requirements at a cost that the customer can afford, only then will a specific project have a chance of being built.

The passage of the Clean Air Act in the United States in 1963 marked the first enactment of comprehensive air pollution control legislation by a major industrial nation.[1] The Clean Air Act of 1963 is rightfully credited as such in many textbooks because of its comprehensive mechanisms to reduce air pollution and improve ambient air quality. However, the United Kingdom enacted its first Clean Air Act in 1956, and many nations soon followed due to the international public, who had become affected by and concerned about air pollution.[2] Localities such as London had air quality legislation as early as 1273. Regulations and requirements vary between jurisdictions, but air pollution control approaches are very similar with four basic elements:

- *Emissions standards*: These limit the mass of pollutants emitted. Typically for waste combustion facilities these are based on the normalized volume of flue gas or mass of waste processed.
- *Percentage removal requirements*: These are based on a percentage reduction of uncontrolled emissions.
- *Fuel requirements*: These are less used for waste to energy than other sources. However, pre-processing and recycling requirements have been included for many facilities.
- *Technology requirements*: Specific technologies may be required, and methods such as best available control technology (BACT) analysis may be mandated for determining specific permit requirements.

Regulations tighten as air pollution control (APC) technology improves. This is specifically codified in the United States as the most achievable control technology (MACT) emissions standards.[3] European and other international regulations have similar means of continuing to decrease the allowable stack emissions from WTE facilities. The MACT rule requires the US Environmental Protection Agency to review and set new standards for new and existing facilities based on the most stringent 12% of permit requirements for specific classes of WTE facilities. While the technical ability to control pollution emissions has always been a factor in determining new plant APC requirements, the MACT rule extends this to continuous improvements for existing facilities as well. Advances in APC technologies can often provide economic benefits to facility owners and operators.

Emission standards for specific pollutants that drive APC equipment selection are presented in Table 11.1 for the United States and the European Union.[4] These emissions standards set the minimum performance standards only, but many facilities have stricter requirements. Emissions measurements and calculations have traditionally been done in System International (SI) units worldwide. In the United States older regulations and permits had limits in imperial units, but since the early 1990s all US regulations and permit requirements have been in SI or

Table 11.1 Emission standards for municipal waste incineration facilities

Pollutant	Europe (EU)	United States*
Dioxins, furans (ng/dscm)	0.1 ITEQ	13 Total
Total particulate dust (mg/dscm)	10	20
Cadmium	0.1	0.01
Lead	1	0.14
Mercury	0.1	0.05
Hydrogen chloride	10 mg/dscm	25 ppmv
Sulfur dioxide	50 mg/dscm	30 ppmv
Nitrogen oxides	200 mg/dscm	150 ppmv
Carbon monoxide	50 mg/dscm	100 ppmv

Note:

*EU limits are standardized to 11% oxygen and the US limits are standardized to 7%; for direct comparison the US limits should be multiplied by a factor of 0.71 to compensate for this difference.

dual units. Emissions are normalized to standard conditions of temperature, pressure and percentage of oxygen or carbon dioxide. This is to ensure that pollution dilution cannot be a solution.

Facility owners have an inherent need to go beyond what is required by regulations or permits: the primary reason is the need to meet permit requirements during upsets and off days. Going beyond also helps to avoid costly retrofits due to tightening of requirements during a facility's lifespan. Often owners have to go even further to build the public support needed to proceed with a project.

11.1.1 Optimizing the combustion process

Good combustion practices and design are critical factors for any air pollution control strategy, and these are often incorporated in the legislation and regulations. Combustor operator training has always placed a focus on proper fuel/air ratios and the three Ts: time, temperature and turbulence. Combustor design incorporates the latest advances and technology to minimize pollution before any post-combustion air pollution controls. These factors have a large effect on the formation and emission of particulates, carbon monoxide, nitrogen oxides and organic pollutants.

Good combustion practices and design are critical performance and economic factors. All air pollution controls have pollution reduction efficiencies. The less pollution that goes into an APC system, the less emissions come out. With less pollution going in, the operation and maintenance costs of the equipment are less. Obviously lower capital costs can be incurred when good combustion practices and designs are utilized. Combustion practices and design cannot be understated. This section is brief because this topic will be covered later in this chapter.

11.2   Acid gas scrubbing in municipal waste combustors

The author has chosen to cover acid gas scrubbing first because the choice of acid gas scrubbing technology factors into the choice of the equipment used for particulate control and technologies used throughout the air pollution control process. Emissions from waste combustors have high concentrations of acid gasses such as hydrogen chloride and sulfur dioxide. The selection of technology and APC system design may also be effected by requirements such as byproduct utilization, water discharge regulations and physical space available, especially for retrofits.

Acid gasses are removed from flue gasses by absorption into a liquid or adsorption into porous solid particles. Both mechanisms can occur with or without chemical reactions, but typically an alkaline sorbent is used to improve removal of the acid gas.[5] Calcium and sodium sorbents are typically used in waste combustion facilities, and Table 11.2 shows the primary chemical reactions that occur.

Stoichiometry and other kinetic factors affect these reactions. The concept of stoichiometry is easier to understand than to say. Stoichiometry applies the laws of definite proportion and conservation of mass to chemical reactions, or in simpler terms, it gives the recipes for chemical reactions.[6] A stoichiometric ratio (SR) of 1:1 represents the exact theoretical amount of sorbent chemical required to completely react all of the acid gases in the flue gas. Several factors limit the reaction so that more sorbent is actually needed to get emission reduction near 100%. These factors are more predominant in the dry and semi-dry processes, so a higher amount of sorbent is required compared to the wet processes. Active sorbent chemical is always entrapped within the particles, so not all of the sorbent is available for reaction. Uneven mixing in the process creates different concentrations of pollutants and sorbent within the process flow. Variations of waste fuel cause an ever-changing pollution concentration, so a stoichiometric ratio of 1:1 cannot achieve 100% removal of the acid gasses.

How much sorbent chemical will be needed is an important question for operations and economics. Acid gas scrubbers by design have varying removal efficiencies. The example calculations in Fig. 11.1 apply to equipment that will provide 75% removal efficiency with an SR of 1.6 for sulfur dioxide and an SR of 1.1 for hydrogen chloride.

Table 11.2 Acid gas scrubbing reactions

Calcium sorbents	Sodium sorbents
$Ca(OH)_2 + 2HCl \rightarrow CaCl_2 + 2H_2O$	$NaHCO_3 + HCl \rightarrow NaCl + CO_2 + H_2O$
$Ca(OH)_2 + 2HF \rightarrow CaF_2 + 2H_2O$	$NaHCO_3 + HFl \rightarrow NaFl + CO_2 + H_2O$
$Ca(OH)_2 + SO_2 \rightarrow CaSO_3 + H_2O$	$2NaHCO_3 + SO_2 + \frac{1}{2}O_2 \rightarrow Na_2SO_4 + 2CO_2 + H_2O$
$Ca(OH)_2 + SO_2 + \frac{1}{2}O_2 \rightarrow CaSO_4 + H_2O$	

$$\text{Acid gas (kg)} \times \frac{\text{(moles sorbent)}}{\text{(moles acid gas)}} \times \frac{\text{(MW sorbent)}}{\text{(MW acid gas)}} \times \frac{SR}{\% \text{ Sorbent}} = \text{Sorbent (kg)}$$

Sulfur dioxide, 100 kg: 75% removal with SR = 16, 96% pure $Ca(OH)_2$
Formula: $SO_2 + Ca(OH)_2 + \frac{1}{2}O_2 \longrightarrow CaSO_4 + H_2O$

$$100 \text{ kg } (SO_2) \times \frac{(1)}{(1)} \times \frac{(74)}{(64)} \times \frac{(1.6)}{(0.96)} = 193 \frac{\text{kg}}{\text{hr}} Ca(OH)_2$$

Hydrogen chloride, 50 kg: 75% removal with an SR = 1.1, 96% pure $Ca(OH)_2$
Formula: $2HCl + Ca(OH)_2 \longrightarrow CaCl_2 + 2H_2O$

$$50 \text{ kg } (HCl) \times \frac{(1)}{(2)} \times \frac{(74)}{(36.5)} \times \frac{(1.1)}{(0.96)} = 58 \frac{\text{kg}}{\text{hr}} Ca(OH)_2$$

Total $Ca(OH)_2$ sorbent required = 251 kg/hr

11.1 Example calculation for calcium sorbent usage.

Three types of equipment for scrubbing acid gases are predominant for municipal waste combustors: wet scrubbers, dry sorbent injection and spray dryer adsorbers (SDAs). Plant-specific operational and economic considerations are the primary reasons for choosing one method over another. Rotary or dual spray dryer adsorbers have become the technology of choice for new waste combustors.

Dry injection of sorbent has the least capital cost for equipment, but has relatively poor stoichiometric efficiency, requiring more sorbent for a given reduction of acid gasses. This method has seen most use in small combustors, since the low capital costs provide a lower life-cycle cost. This is especially true in the United States where low-cost natural sodium sorbents are available, and small combustors have less stringent emission reduction requirements. Utilizing limestone products for dry injection is more difficult because reactions for hydrogen chloride and sulfur dioxide must happen at two different temperature ranges. Lime sorbents must be injected near the high temperatures of the furnace and further down the flue-gas train in the range 450°C to 600°C. Trona, natural sodium sesquicarbonate or sodium bicarbonate can be injected in one place at much lower mass rates than the limestone products. These are injected at temperatures above 550°C and a calcination process produces a very porous carbonate. The acid gasses react well with the high surface area of the carbonate. The effectiveness of the process requires stable temperatures at the injection point. Capital and the specific sorbent costs are primary selection criteria.

Wet scrubbers have been used to a large extent in Europe for waste combustion facilities, although they have seen little use elsewhere. Wet scrubbers do offer higher stoichiometric efficiency and allow better separation of APC byproducts for potential reuse. Typically particulates are removed prior to the scrubber to limit solids in the caustic solution. Wet scrubbers spray a caustic solution of sodium hydroxide or calcium carbonate to cause a wet chemical acid base reaction. The spray can be in an open spray tower or over a packed bed. Sulfur can be recovered as calcium sulfate in a simple process when calcium sorbents are used. Sulfuric acid can be simply formed by mixing with water, but sulfuric acid is not very soluble in pure water. Sodium hydroxide is often used to control the acidity of the scrubber solution. The sodium is then extracted from the solution with further processing. Recovering sulfur does provide additional revenue, but increased material recovery is the typical reason why the higher capital costs are invested in sulfur recovery.

There are many patented and proven acid gas control technologies, but the unique needs of waste combustors and tightening requirements have significantly reduced these options to a few. Most prominent in the US and around the world are the semi-dry processes, especially spray dry adsorbers. These processes control exhaust gas temperature and scrub all prominent acid gasses in a single unit (see Fig. 11.2), thus reducing capital costs while simplifying operations. Spray dry adsorbers also can be operated with a wide variety of sorbents such as limestone, hydrated lime, sodium bicarbonate and natural trona.

Spray dry adsorbers are semi-dry processes that introduce water and sorbent chemicals into the flue-gas stream through dual fluid nozzles or rotary atomizers.

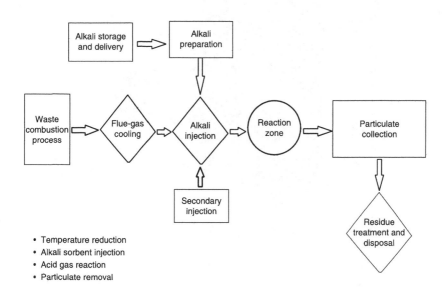

11.2 Acid gas scrubbing process.

The latter is more effective with larger diameter units, because drier and more homogeneous mixing is possible. These units can be of up-flow design, but the down-flow design is more widely used for a variety of factors including the predominant use of fabric filters as the downstream particulate collectors. Fabric filters generally introduce the exhaust flue gasses into the bottom of the units. A primary design factor for an SDA is the flue-gas residence time after the gas is fully dried. Prior to the gas drying wet alkaline reactions occur, but it is not until the gas has evaporated completely that the true chemical adsorption processes are optimized. Reactions continue with sorbent that remains in the ashcake deposited on the bags of a fabric filter, increasing the stoichiometric efficiency of the process.

With all acid gas sorbents, the transportation costs to the facility are a large part of the cost per ton of the sorbent. Production of sorbents also requires

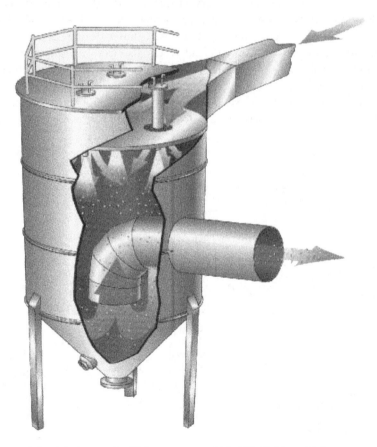

11.3 Dual fluid nozzle SDA, courtesy of McGill AirClean LLC.

significant energy. Therefore the entire analysis and cost projections for acid gas scrubbing are very sensitive to energy costs. When choosing an acid gas scrubbing system, careful consideration of the long-term sorbent and energy costs are important.

11.3   Particulate control devices utilized at waste combustion facilities

Particulate emissions from waste to energy plants consist primarily of fly ash and air pollution control byproducts. Combustion fly ash is traditionally thought of as unburnt materials or soot, and unburnable materials or ash. Soot is minimal in modern combustion facilities where little unburnt carbon and sulfur leaves the furnaces except during start-up, shutdown and periods of malfunction. Municipal waste fuel has a high proportion of unburnable ash. Municipal waste fuel has unburnable material in the form of silica from glass and dirt, clay from cat litter, mineral salts and metals. These items are found pure or in products. Municipal waste combustion (MWC) fly ash has higher concentrations of particulate of concern such as lead, cadmium, arsenic and zinc. These often are found as metal salts in the fly ash. MWC fly ash is also contaminated by semi-volatiles such as mercury and aromatic compounds like dioxins and furans. Particulate control devices effectively remove fly ash from the flue gases emitted from the combustors.

Mechanical separating devices and cyclones have low collection efficiencies, especially for fine particles. These devices are now used in one of two ways. Mechanical devices can be used to precondition the flue gases prior to the primary scrubber or particulate collection devices. This can be useful with small units to reduce the size and cost of air pollution control equipment. Mechanical devices may also be found with dry scrubber systems that recycle flue dust. Heavy particles are separated and returned directly to the furnace for sintering, and the finer particles are recycled back into the dry scrubber section.

Electrostatic precipitators (ESPs) work by placing a charge on the particles.[5] Particles are then attracted to oppositely charged collection plates. Recent advances with electrostatic precipitators have generally involved the control and transforming circuitry. Modern ESP control systems have increased the average voltage while reducing the amount of electric power going to ground from arcing and the DC rectifying process. These devices have much better collection efficiencies while using about a third of the power than in the past.

Typically banks of electrode wires are charged to over 18 000 volts DC. At some point determined by the unit's geometry the electrode reaches and exceeds the corona onset voltage (E_O), and ionized free electrons form a cloud around the wire. This corona cloud imparts electrons to the flue-gas particles giving them a negative charge. The electric power will continue to rise to the spark voltage (E_S); at that point an electric spark leaves the wire and goes to the opposite charged plate or to the grounded casing. Typically the braided wires are free hanging with heavy weights restrained within a hoop (Fig. 11.4). Some European precipitators use rigid rods instead of wires in waste combustion applications. Some ESP configurations include rods within tubes or charged plates, but they are not used for waste to energy applications due to fouling of the tubes and plates.

Polyphase alternating current is converted to high voltage direct current by transformer rectifier (TR) sets, which supply the power to the charging wires. ESP control systems can detect corona onset, power and the onset of sparking by monitoring the power characteristics of silicon core reactor transformers that are now widely in use. When the onset of sparking occurs the voltage is reduced slightly to prevent sustained arcing, which increases electric power use. Older control systems reduced the voltage once arcing actually occurred. Newer control systems also rectify the alternating current to a square wave instead of a half sign wave. This increases the average power, which increases the collection efficiency of the electrostatic field.

Various mechanisms are used to dislodge the particles from the wires and plates allowing them to fall to the ash collection hoppers. The wire electrodes are typically cleaned online with vibrator mechanisms. The collection plates are cleaned with weighted rappers. When the plates are properly rapped, the ash moves in waves down the collection plates falling off at the bottom. When the wires and plates are cleaned, a portion of the collected fly ash becomes re-entrained in the flue gas and is not collected. Online cleaning limits the collection efficiency to less than 95% for each ESP collection bank. Multiple collection banks are installed in series for increased collection efficiency, but re-entrainment will allow fly ash to exit from the final ESP bank.

Several considerations specific to municipal waste combustors limit the use of electrostatic precipitators as the sole particulate collection devices for MWC air pollution control. Flue gasses emitted from waste combustors have high concentrations of acid gasses such as hydrogen chloride and sulfur dioxide. ESPs

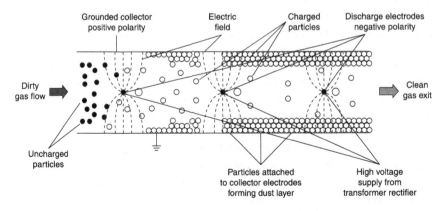

11.4 Single field view of an electrostatic precipitator, courtesy of the US EPA.

remove unreacted chemical sorbents rapidly from the process with the fly ash. This requires using more chemicals and may require further treatment of the ash. This factor is the predominant reason the industry is turning to fabric filters for particulate collection. APC residues can be recycled back to the inlet of the ESP to get more use from the sorbent. This, however, increases the dust load going into the ESP requiring the equipment to be sized larger and have more collection banks to achieve the same removal efficiency.

Electrostatic precipitators also provide a surface where unreacted hydrocarbons can reform to benzene and aromatic compounds. These devices have been shown to contribute to *de novo* reformation of dioxins and furans in MWC air pollution control systems with hot-side ESPs.[7] Special temperature considerations must be made to avoid reformation of these aromatic compounds. Typically this requires the inlet temperature not to exceed 230°C. In general electrostatic precipitators are also less effective than fabric filters for collection of semi-volatile compounds such as dioxins and furans, and may actually increase these emissions.

Electrostatic precipitators do have two advantages over fabric filters: they do not need to be by-passed during start-up, shutdown and failures. Generally emission limits are exempted for these periods, so this advantage has not been factored in during the development of regulations and when determining the best available air pollution control technologies. ESPs also use much less electric power than fabric filters, so they have a smaller carbon footprint.

Fabric filter baghouse technologies have become the best available control technology for controlling particulates from municipal waste combustors. A cake of fly ash and sorbent forms on the fabric bags, which provide additional acid gas scrubbing and collect semi-volatile emissions. Fabric filters comprise multiple filter compartments, each having up to a hundred filter bags. Filter bags are generally about 30 ft long and stretched over a metal cage. Fabric bags need to be cleaned, or eventually no flue gasses will pass through them. Filter bags come in a variety of fabrics with different service lives and performance enhancements. In general an operator should choose one of the bag options that the fabric filter manufacturer recommends. Interconnecting ductwork and control dampers are used to direct the dirty and clean gasses to and from the filter compartments.

Fabric filters are classified by the method by which the filters are cleaned. Reverse air fabric filters are cleaned offline by passing the clean gas exiting from an online compartment through the bags in a reverse flow to dislodge the ash. Most fabric filters used for waste combustors are the pulse jet type (Fig. 11.5). Pulses of compressed air from nozzles send an acoustic wave down the bag to dislodge the ash. Pulse jet units can be cleaned online, so some of the ashcake reforms on the bag surface. Since the ashcake itself acts as a filter this results in higher average collection efficiencies. During proper online cleaning no fly ash is entrained in the exiting flue gas, so overall collection efficiencies exceed 99%. When a clean reverse air bag goes online it has low collection efficiency, and this steadily increases as the bag becomes dirty.

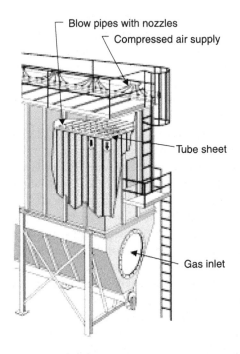

11.5 Pulse jet fabric filter, courtesy of McGill AirClean LLC.

Primary design factors for fabric filters are the air-to-filter (A/F) ratio and the can velocity.[5] Can velocity is the velocity of the gases entering through the cross section in the bottom of the bags or cans. High can velocity will reduce collection on the lower portions of the bags, and will hinder the falling of the ash to the hopper during online cleaning. High can velocity will also cause premature failure of filter bags. Design can velocity is lowered by spacing the filter bags further apart. Longer bags require greater spacing to get the proper amount of gasses into the filters while keeping can velocity at a reasonable rate.

The air-to-filter ratio is the ratio of the volume of gasses being filtered in cubic meters per second to the filter surface area in square meters, so the measurement unit for the A/F ratio is meters per second or feet per second. Standard practice for waste combustors is for an A/F ratio of 1.2 m/s online and 1.6 m/s total including the offline spare compartments. This may vary slightly based on waste composition and collection efficiencies desired. A higher A/F ratio means the gasses pass more slowly through the filter, and this reduces the filter differential pressure. Lower differential pressure reduces the amount of energy needed for fans and online cleaning. Fabric filter systems have advantages and disadvantages when compared to electrostatic precipitators (Table 11.3).

Capturing and handling particulate emissions and acid gas scrubber byproducts are also important considerations for air pollution control designs. Generally dry

Table 11.3 Comparison of fabric filters and electrostatic precipitators

Property	Electrostatic precipitator	Fabric filter
Collection efficiency	98%	99.5%
Fine particle collection	95%	99%
Pressure loss	8 kPa	24 kPa
Fan power	1/3 of fabric filter	300% plus
Overall power	40% of fabric filter	250% plus
Maintenance cost	50% of fabric filter	100% plus
Capital costs	75% of fabric filter	150% plus
Unit start-up and shutdown	Online	By-passed under 250°C
Combustor malfunction	Online except during major tube ruptures	By-passed for many malfunction events
Acid gas collection	Much sorbent unreacted	Greatly increases the amount of sorbent reacted

and semi-dry systems capture a dry powder residue, which is then conveyed by drag or screw conveyers. Drag conveyors consist of single or double strands of chain with attached paddles known as flytes. Friction and the abrasiveness of the ash require hard materials to be used for these conveyers. Screw conveyers are large wide-pitched screws, which channel the ash through enclosed tubes. Air locking devices are required to keep the ash from being sucked back into the negative pressure APC devices. These devices can be rotary plug valves or double dump paddle valves.

Ash handling devices must also condition the APC residue with water and often chemicals. This wetting is typically done in pug mills or shaftless screw mixers. Water wets the residue, which reduces the dust hazard. Chemicals such as phosphoric acid and dolomite lime are used to reduce the leaching potential of the APC residue, facilitating safer handling, transport and disposal.

11.4   Control of nitrogen oxide emissions and hazardous air pollutants from waste combustors

11.4.1   Control of nitrogen oxide emissions from waste combustors

Many waste combustion plants began implementing controls for nitrogen oxides (NO_x) for the first time over the last decade. Nitrogen oxides facilitate one of the most severe air pollution problems we face today: atmospheric photochemical smog.[5] Recycling of nitrogen dioxide (NO_2) into nitric oxide (NO) generates an oxygen molecule known as ozone (O_3). Tropospheric ozone, unlike diatomic oxygen (O_2), has widespread effects on human health and the environment. Ozone in the presence of volatile organic compounds (VOCs) sets off a chain of chemical reactions that often makes the air we breathe polluted (Fig. 11.6). Sunlight is the source of energy that fuels this chemical factory in the sky, so smog is most prominent during the summer months.

Vehicles and combustion sources are the primary emitters of NO, which then reacts with water vapor to form nitrogen dioxide (NO_2) and hydroxyl radicals (OH). Nitrogen dioxide has a light brown color, which causes the brown haze of urban smog. The NO_2 absorbs ultraviolet radiation from sunlight to decompose back to NO and molecular oxygen. Cycling from nitrogen oxide to nitric oxide is naturally reversible, but in the atmosphere when volatile organic compounds are present the reaction is not fully reversible. The ozone produced reacts with hydrocarbons to form alkylperoxyl radicals (RO_2), which react to form peroxyacetyl nitrates (PANs) that form other organic compounds. These

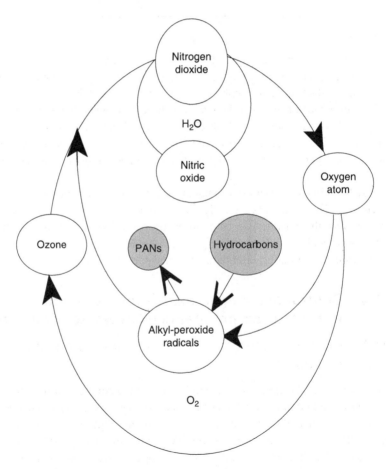

11.6 Ozone and smog propagation in the atmosphere.

out-of-balance reactions create a toxic mix of ozone, PANs, aldehydes, ketones, akyl nitrates and carbon monoxide, and this mix is referred to as photochemical smog. Since ozone is most prominent in smog, it is most often measured, reported and talked about.

Unfortunately all natural and anthropic combustion creates NO_x, as well as VOCs to a certain extent. Municipal waste combustors emit a small amount of VOCs during normal operations, but during start-up, shutdown and malfunction elevated levels of VOCs and NO_x occur. Regulators are now placing greater emphasis on minimizing off-normal conditions at operating facilities to reduce the impacts. The transportation sector is the largest contributor to NO_x and VOCs, so smog problems are predominant in large metropolitan areas. Current technologies reduce nitrogen oxides much less than the other major pollutants, so particular interest and activity are now focused on this sector of air pollution control technology.

Control technologies for NO_x fall into three categories: catalytic reactors, non-catalytic reactors and combustion improvements that reduce the amount of NO_x produced by the combustion process. Optimization is always implemented no matter what post-combustion technology is used. Combustion optimization can greatly reduce the amount of NO_x emissions, known as thermal NO_x, which is formed at high temperatures through burning nitrogen in air. A first step in optimization is keeping the amount of combustion air to an absolute minimum. Secondary air to give complete combustion is the second optimization step, which is used with liquid and solid fuel combustion. This is referred to as overfire air in waste combustors. A third optimization step is the re-injection of flue gas into the combustion process to use more of the oxygen and hence reduce the overall atmospheric nitrogen/oxygen ratio in the supply of the needed combustion oxygen. Flue-gas recirculation also provides heat recovery and stabilizes the combustion process. At the cutting edge is the use of pure oxygen instead of air for combustion. This has been done in specialized processes for almost a century. It has been tried in practice by injecting supplemental oxygen into the combustion. Using pure oxygen is a prominent feature in many of the new solid waste thermal technologies now being developed.

Solid waste contains a second major source of NO_x, fuel NO_x. This is released when the nitrogen bound in the waste fuel combusts. Solid waste has a relatively high nitrogen content, and this will ultimately lead to the need for post-combustion technologies to meet tightening emissions requirements placed on municipal waste combustors. A third minor source of NO_x is known as prompt NO_x, and results from the nitrogen in the fuel reacting with hydrocarbon radicals in the combustion process. This author believes, from many years of process observations, that this formation mechanism is much more prominent in municipal waste combustors during periods of start-up, shutdown and malfunction.

Reactions of nitrogen oxides, with catalysts or not, use ammonia (NH_3) to react with NO_x to form nitrogen and water vapor. The ammonia is available in three ways. Anhydrous ammonia is 99% pure mixed with a small amount of water, but in this form it is very toxic and hazardous. When accidentally released it reacts with atmospheric moisture to form a highly toxic and explosive plume. Current regulations will make this less likely to be used for air pollution control except at facilities that use a lot of ammonia or have limited storage space. Aqueous ammonia is mixed with 60% or more water making it a more stable product to store and handle. Regulations in the United States now consider aqueous ammonia at concentrations of 20% or more to be a hazardous substance, so many facilities use 19% aqueous ammonia.[1] When the percentage of water increases so do the transportation and storage costs. Using aqueous ammonia increases the fan power required, and more moisture leaves the stack causing a visible plume more often.

The third form of ammonia used is urea $(2(NH_2)CO)$. Urea is available as a liquid, solid or powder. Typically it is delivered as a powder then mixed with de-ionized water on site. When urea is used with a catalytic reactor it must first be transformed to ammonia with a thermal reaction that also produces water and carbon dioxide. Additional equipment and energy is required, so urea is used less often for catalytic reactors.

Selective non-catalytic reaction (SNCR) is now the most common technology for waste combustors. These systems have lower capital costs, but have lower efficiencies and higher operating costs per ton of NO_x removed. SNCR systems typically are limited to removing only about 50% of the NO_x emissions. The non-catalytic systems inject ammonia or more often urea directly into the upper furnaces of the combustor. The reactions are highly dependent on having a stable temperature of the order of 1000°C. Since furnace temperatures vary, injection is done at several different heights of the furnace, and often supplementary gas firing is used to achieve the optimum reaction temperature. Additional residence time is also needed for the urea to react into ammonia, and then the ammonia to react with the NO_x. Overall the lower capital costs of SNCR more than make up for the added operational costs.

Selective catalytic reactors (SCRs) can obtain much higher emission reduction, up to 90%. Catalysts are materials that speed up or are involved in a chemical reaction without being used up in the reaction. Catalytic reactions occur on the active surface of a SCR, which increases the efficiency and effectiveness of NO_x removal. Higher efficiency means less ammonia, and reduces the amount of ammonia slip. Ammonia slip refers to the percentage of ammonia that is unreacted and leaves the stack. Ammonia itself is a regulated hazardous air pollutant. As NO_x limits are lowered, facilities have yet another challenge: complying with ammonia emissions limits. This is another reason why more facilities may use SCR technology in the future.

Catalytic reactors come in several physical forms such as plates, corrugated plates, honeycombs (Fig. 11.7) and corrugated fibers.[1] Catalysts are packaged in box-like modules that are then loaded into racks. The pitch and sizing of the elements are important design considerations that affect the reduction efficiency

11.7 Honeycomb-type catalyst module.

and pressure drop across the SCR unit. Maximizing the active catalyst surface area, while minimizing fouling, is an important design consideration.

Catalyst materials come in three different types: base metals, zeolites or precious metals. Zeolites and precious metals are seldom used for waste combustion applications. The catalyst material is homogeneous, layered or coated over an active metal base, which provides extra strength. Most catalyst base-metal formulations are proprietary, but generally use titanium and vanadium oxides with other minor constituents. Catalyst base metals can be poisoned by alkali metal oxides, especially lead oxide, and this can quickly make a catalyst ineffective. That is why SCRs are generally installed downstream of the fabric filter of a municipal waste combustor. Catalysts are more effective at higher temperature, so this requires reheating of the flue gasses, which is another added cost of SCR.

11.4.2 Control of hazardous air pollutants (HAP) from waste combustors

Any amount of designated hazardous air pollutants emitted by a municipal waste combustor will alarm the public. In fact, huge reductions of these emissions, up to 90%, were made by operating facilities during the 1980s before any regulations were issued. During the 1990s, industry innovations and new regulations again dramatically reduced the levels of HAP emissions.[7] Hazardous air pollutants of concern emitted by municipal waste combustors fall into three categories: organic emissions such as dioxins and furans, heavy metals including lead, cadmium and mercury, and the acid gasses, hydrogen chloride, hydrogen fluoride and hydrogen bromide.

Acid gas control technologies are very effective in controlling acid gasses. Waste combustors generally are allowed to emit fairly high concentrations of hydrogen chloride, but it is the least toxic of the three. High concentrations of all three of these gasses are very toxic to humans, but at lower concentrations the effect on human health has not been quantified. In fact humans consume the liquid form of hydrogen chloride, hydrochloric acid, when we consume vinegar and other foods. It is very likely these have an effect on ecosystems, which may accumulate halogen compounds as a result of these emissions, so our diligence and expense are justified.

Heavy metals, with the exception of mercury, are well controlled by modern particulate control devices used at municipal waste combustors. With proper treatment, disposal and re-use of ash, the lead and cadmium problems are well controlled. Mercury is semi-volatile, and is in the flue gas as an aerosol and vapor, which particulate control systems have a limited ability to collect. Activated carbon injection has now improved the capture of mercury in many waste combustors. For many years the amount of mercury in municipal waste has steadily dropped as less is being used in batteries and lights. With the new requirements for compact fluorescent lighting, combustors are likely to see an increased concentration of mercury and fluorine in waste.

The neurological effects of mercury on humans have been well documented. The ecological effects are now being recognized as producing a major pathway for human exposure. For many years elemental mercury has been used for filling teeth with little documented health effects. It is now understood that mercury compounds have the more adverse effects, and elemental mercury is converted to these compounds in the ecosystem. Mercury compounds bio-accumulate in fish and other seafood that humans then consume. Coal-fired utility boilers are by far the largest source of anthropic mercury emissions, and the natural sources are also significant. The baseline mercury level in the environment is already of concern, so any additional emissions by humans must be kept to very low rates. Lower emission limits for waste combustors and coal-fired boilers are certain in the future.

Municipal waste combustor organics, as defined by the US Environmental Protection Agency, are dioxin-like compounds. While the limits are specifically for dioxins and furans, it is recognized that similar aromatic organic compounds are also controlled when these are controlled. Dioxins, or polychlorinated dibenzo-dioxins (PCDDs), are formed when two six-carbon benzene rings are joined by two oxygen atoms. The only commercial use of dioxins was in herbicides produced prior to 1980, such as Agent Orange, which was used extensively by the US Army during the Vietnam War. Significant exposure to dioxin-like compounds can result in the acute disfiguring condition known as chloracne. Medical follow-ups on people exposed have not resulted in definitive links to an increased incidence of cancer. Whatever the evidence, though, the molecular structure of these chlorinated compounds has the geometric configuration to damage DNA. These compounds are currently listed by the World Health Organization as probable carcinogens, and rightfully so.

Stack testing at municipal waste combustors includes a panel of twenty-eight dioxin and furan species. Furan molecules are almost identical to dioxins except the benzene rings are bound by a single oxygen molecule. It is recognized that dioxin molecules with a chlorine atom in the eighth position, such as 2,3,7,8 PCDD, are the most toxic of the compounds (Fig. 11.8). European facilities are regulated in the units of international toxic equivalents (ITEQ). Dioxin 2,3,7,8 has an equivalency of 1, and the other dioxins and furans have much lower equivalency factors. In the United States, the EPA has decided to regulate total dioxins since the ratio of ITEQ to total is reasonably constant.

11.8 2,3,7,8 PCDD molecule.

When the industry became aware in the 1980s of high dioxin emissions, mitigating action began well before any regulations were written. The first waste combustor stack measurements were made at the Hampton/NASA Steam Plant, where this author worked for over 20 years. Immediate research and corrective actions reduced the dioxin emission rates by 99% before any regulatory requirements were even proposed. There, and at other waste combustion facilities, it was determined that dioxins were reforming in the electrostatic precipitators, a process referred to as *de novo* formation. Products of incomplete combustion (PICs), especially carbon monoxide, formed benzene rings, which combined and picked up chlorine from the flue gas. Improving combustion and reducing the ESP operating temperature resulted in significant emission reductions. Over the past ten years further huge reductions have been recorded at facilities using fabric filters and injecting activated carbon. Reducing the products of incomplete combustion, fabric filters and activated carbon injection are the methods used to control these emissions.

11.5   Air pollution control cost–benefit analysis

Best available control technology (BACT) analysis compares the costs and benefits of several alternative air pollution control options.[5] BACT analysis is a requirement for permitting new facilities in the United States, and similar requirements exist in most other countries. With the advent of technology-based standards, it plays less of a role in determining actual equipment selection, but it remains an important step in developing a project's economic and financial plan.

Prior to starting the BACT analysis, project decisions must be made, and the parameters for the analysis must be set by the decision makers. The waste processing capacity of the facility must be determined based on present and future solid waste availability. The location must be determined to evaluate utility availability and costs. How much space (its footprint) will be available for the APC system? Economic analysis parameters must be set: expected useful life of the equipment, the value of money (expected rate of return) used in the analysis, tax considerations, inflation rate or index, energy cost index and any contingency factors. Expected emission targets need to be understood, as well as the ash disposal or re-use requirements. Once these are determined, the engineers can proceed with the economic analysis of the competing options.

Three or more competing APC options should be developed with a preliminary design study. When the project involves a facility retrofit the existing APC should be Option 1. New developments should use the minimum technology standard as Option 1. Options could involve using different configurations, sorbent chemicals or vendor technologies.

Estimate the total capital investment (TCI) for each option: depreciable investment plus non-depreciable investment such as land and working capital costs. Depreciable investment includes direct and indirect costs. Indirect installation costs are engineering, supervision, permitting costs, construction fees, start-up, performance testing and contingencies. Direct costs, including site preparation and buildings, may be calculated from a reliable construction estimating program using current construction cost indices. These items should vary only slightly between the competing options.

Direct equipment purchase costs and installation should be estimated from methods provided in a reliable process engineering or air pollution control engineering handbook.[5] Factors used for these calculations are regularly published in process engineering and air pollution control journals. Process producer cost indices are also published annually in *Chemical Engineering* magazine including the chemical engineering plant cost index and the Marshall and Swift equipment cost index.

It is tempting to simply extrapolate equipment costs from recent similar projects based on cost per ton or per megawatt. However, direct extrapolation does not always work well. Costs of flue-gas scrubbers escalate according to a power law with different exponents for different items. Scrubber system costs escalate with plant size to the 0.8 power and to acid gas concentrations to the 0.3 power. Fabric filter costs have two components, which can be calculated using regression constants derived from recent project experience. Constant 'a' represents the basic unit cost with add-ons such as advanced metals and insulation options. Constant 'b' is multiplied by the gross cloth area for the specific unit. Calculations made with a reliable handbook and cost indices should be used with knowledge of other recent projects. Certainly there is commercial and proprietary software to help with the analysis, but the engineer must be knowledgeable and experienced enough to select the methodology, factors and cost indices and to validate the model and results.

Total annual costs (TACs) include variable and semi-variable direct annual costs and indirect annual costs and these should be offset by recovery credits from tax credits, energy sales, tipping fees and material recovery revenue. Variable annual costs are those dependent on the process throughput such as utilities, sorbents, waste treatment and residue disposal. Semi-variable costs include labor, maintenance materials and replacement parts. Indirect costs include capital recovery, interest, plant overheads, property taxes, insurance and administrative charges. Experience from similar projects in conjunction with reliable cost-estimating methods and indices are essential for properly calculating the TAC.

Once the TCI and TAC are determined for each option over the life of the project, a standard economic analysis is done for each of the options with two minor additional steps: the author recommends the calculation of the present worth (PW) and the equivalent uniform annual cost (EUAC).

To complete the BACT analysis the incremental benefits of the options must be calculated. Calculate the actual cost of emission control (ACE): divide the total number of kilograms of emissions reduced for regulated and non-regulated pollutants that are not emitted by the capitalized cost for that period. Divide PW by the life-cycle emissions reduction and EUAC by the annual emissions reduction. The incremental cost of emission control (ICE) is the cost difference between the competing options for reducing one kilogram of emissions. The ACE and ICE calculations should also be done separately for each of the regulated pollutants as well, but those results would not be cumulative for total ACE and ICE. Health benefit factors may be used to distribute the costs among the different pollutants, but that analysis is beyond the realm of the practicing engineer; other expertise would be needed.

The BACT analysis will provide valuable information for the project's decision makers as well as to regulators and the public. The analysis may show surprising results, such as the least expensive option having the highest cost per ton of pollution reduction. It may demonstrate that the most expensive option has a prohibitively higher incremental cost of control, giving only an additional small reduction in the amount of pollution emitted. Maybe it would show to everyone that more investment is the best economic choice. In any case it will show that the developers and their engineers did due diligence for public health and the environment while putting together the project plan.

11.6   Air quality technology innovations for municipal waste combustors

These are exciting times for technology improvements for municipal waste combustion. As always the industry is using the most modern techniques to lead the way with research and development efforts for combustion optimization, materials and air pollution control equipment. Improved computational fluid dynamic (CFD) modeling has produced great advances in combustion optimization, combustor design and air pollution control equipment design. Advanced materials allow improvements to combustors, boilers and selective catalytic reactors. Materials and technologies are now available to develop new processes for gasification and thermal treatment of wastes. We could present an entire book detailing technology developments, but instead we will briefly present some significant advances and proprietary technologies recently demonstrated at full scale and presented at industry conferences.

A major success in air pollution control has been the deployment of activated carbon injection in waste combustors. Significant reductions in the amount of mercury emitted from municipal waste combustors are well beyond expectations, and the technology has worked well without major equipment problems. Activated carbon has the secondary benefit of reducing emissions of dioxins and furans to very low levels. These results have led to the US Environmental Protection Agency and several environmental groups praising the industry.[7] Many now see waste to energy as the best acceptable technology for the immediate reduction of greenhouse gas emissions.

Several proprietary technologies have recently been implemented with great success for optimizing combustion and reducing NO_x emissions. A Keppel Seghers prism is an innovative prism-shaped bridge placed in the furnace just above the stoker grates (Fig. 11.9).[8] Overfire air is injected through air ports in the prism, releasing more overfire air where it is needed most. The prism provides

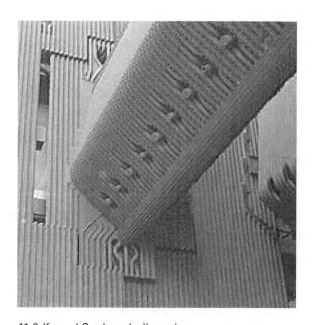

11.9 Keppel Seghers boiler prism.

better combustion gas flow control, a more stable temperature profile and stable boundaries between reducing and oxidizing zones.

Janson Combustion and Boiler Technologies has deployed its deNOx technology in municipal waste combustors and biomass boilers with great success.[9] The technology uses high-pressure metered overfire air, SNCR injection and advance temperature monitoring equipment to control the combustion process in a more exact and effective way.

Most prominent in research and development is Covanta Energy and its European partner Martin GbM. Covanta has now deployed a very low NO_x (VLN) system.[10] This combines combustion optimization using integrated CFD modeling, recirculating combustion gases from above the stoker to the upper furnace and SNCR injection. Covanta has been able to achieve sustained NO_x emissions below 70 ppm, a control level previously only obtainable with the use of catalysts. Further Covanta research and development has included plant-scale testing of supplementing combustion air with oxygen.

Babcock Power Environmental has recently developed and deployed scalable air pollution control technologies at coal-fired and waste-fired plants. Most notably is the regenerative selective catalytic reactor, which combines high-efficiency SCR with ceramic regenerative heaters.[11] This technology significantly reduces the reheat energy costs associated with tail-end SCR.

In 2007, the Amsterdam Waste and Energy Company started operation of their completely rebuilt waste to energy plant (Fig. 11.10).[12] This has become known as the world's first fourth-generation waste to energy plant. The facility is the most efficient waste-fired power plant ever built. It uses integrated advanced material recovery and air pollution control technologies. The APC system has several types of modern air pollution control equipment in series, and it achieves the lowest overall stack emissions of any facility built to date.

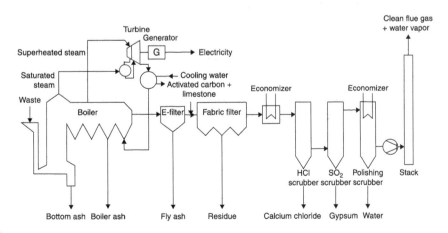

11.10 Amsterdam's waste-fired power plant.

Robust research and development efforts, renewed industry enthusiasm, and a strong record of industry achievements are now setting the stage for a new generation of waste to energy plant development.

11.7   References

1. Kitto J.B. and Stulz S.C., 2005, *Steam 41st Edition*, Barberton, Ohio, USA, Babcock & Wilcox.
2. UK Environment Agency, History of Air Pollution in the UK, 2011, London, available at: www.environment-agency.gov.uk [accessed 6 August 2011].
3. US Environmental Protection Agency, *40 CFR Part 60,* 2011, Washington, available at: www.epa.gov [accessed 6 August 2011].
4. The European Parliament and Council, *Directive 2000/76/EC on the Incineration of Waste*, 2000, Brussels, available at: www.eur-lex.europa.eu [accessed 17 October 2011].
5. Benitz J., 1993, *Process Engineering and Design for Air Pollution Control*, Englewood Cliffs, NJ, PTR Prentice Hall.
6. Brady J. and Humiston G., 1986, *General Chemistry: Principles and Structure 4th Edition*, New York, John Wiley & Sons.
7. US Environmental Protection Agency, *Dioxin Reassessment – A Science Advisory Board Review*, 2001, Washington, available from National Technical Information Service, Springfield, Virginia 22161.
8. Keppel Seghers, *Keppel Seghers Boiler Prism*, available at: www.keppelseghers.com [accessed 8 August 2011].
9. Pethe S. *et al.*, *Elements of a Successful Waste To energy Boiler Upgrade*, Proceedings of NAWTEC 17, 2009, New York, ASME Publishing.
10. White M. *et al.*, *New Process for Achieving Very Low NO$_x$*, Proceedings of NAWTEC 17, 2009, New York, ASME Publishing.
11. Abrams F. and Faia R., *RSCR System to Reduce NOx Emissions from Boilers*, Proceedings of NAWTEC 17, 2009, New York, ASME Publishing.
12. de Waart H., *Amsterdam Waste Fired Power Plant, First Year Operating Experience*, Proceedings of NAWTEC 17, 2009, New York, ASME Publishing.

Index

227